D1084310

KIRK-OTHMER

ENCYCLOPEDIA OF
CHEMICAL
TECHNOLOGY

FOURTH EDITION

VOLUME **13**

HELIUM GROUP
TO
HYPNOTICS

EXECUTIVE EDITOR
Jacqueline I. Kroschwitz

EDITOR
Mary Howe-Grant

KIRK-OTHMER

ENCYCLOPEDIA OF CHEMICAL TECHNOLOGY

FOURTH EDITION

VOLUME 13

HELIUM GROUP
TO
HYPNOTICS

A Wiley-Interscience Publication
JOHN WILEY & SONS

New York · Chichester · Brisbane · Toronto · Singapore

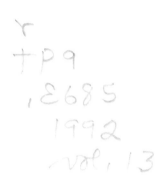

This text is printed on acid-free paper.

Copyright © 1995 by John Wiley & Sons, Inc.

All rights reserved. Published simultaneously in Canada.

Reproduction or translation of any part of this work
beyond that permitted by Sections 107 or 108 of the
1976 United States Copyright Act without the permission
of the copyright owner is unlawful. Requests for
permission or further information should be addressed to
the Permissions Department, John Wiley & Sons, Inc.,
605 Third Avenue, New York, NY 10158-0012.

Library of Congress Cataloging-in-Publication Data

Encyclopedia of chemical technology/executive editor, Jacqueline
  I. Kroschwitz; editor, Mary Howe-Grant.—4th ed.
    p. cm.
  At head of title: Kirk-Othmer.
  "A Wiley-Interscience publication."
  Includes index.
  Contents: v. 13, Helium group to hypnotics
  ISBN 0-471-52682-7 (v. 13)
  1. Chemistry, Technical—Encyclopedias. I. Kirk, Raymond E.
(Raymond Eller), 1890–1957. II. Othmer, Donald F. (Donald
Frederick), 1904– . III. Kroschwitz, Jacqueline I., 1942–
IV. Howe-Grant, Mary, 1943– . V. Title: Kirk-Othmer encyclopedia
of chemical technology.
TP9.E685 1992   91-16789
660'.03—dc20

Printed in the United States of America

10 9 8 7 6 5 4 3 2

CONTENTS

EDITORIAL STAFF FOR VOLUME 13

Executive Editor: **Jacqueline I. Kroschwitz**
Editor: **Mary Howe-Grant**
Assistant Editor: **Cathleen A. Treacy**
Associate Managing Editor: **Lindy J. Humphreys**
Copy Editors: **Christine Punzo**
          Lawrence Altieri

CONTRIBUTORS TO VOLUME 13

Carmen Avendaño, *Universidad Complutense, Madrid, Spain*, Hydantoin and its derivatives

Roger G. Bates, *University of Florida, Gainesville*, Hydrogen-ion activity

Gerald W. Becker, *Lilly Research Laboratories, Indianapolis, Indiana*, Human growth hormone (under Hormones)

Kenneth E. Bett, *University of London*, High pressure technology

Roger E. Billings, *International Academy of Science, Independence, Missouri,* Hydrogen energy

Gary T. Blair, *Haarmann & Reimer Corporation, Springfield, New Jersey,* Hydroxy dicarboxylic acids

Tilak V. Bommaraju, *Occidental Chemical Corporation, Grand Island, New York,* Hydrogen chloride

Judith M. Bradow, *United States Department of Agriculture, New Orleans, Louisiana,* Herbicides

Steven R. Childers, *Wake Forest University, Winston-Salem, North Carolina,* Brain oligopeptides (under Hormones)

Michel Costantini, *Rhône-Poulenc Recherches, France,* Hydroquinone, resorcinol, and catechol

T. A. Czuppon, *M. W. Kellogg Company, Houston, Texas,* Hydrogen

Rathin Datta, *Consultant, Chicago, Illinois,* Hydroxycarboxylic acids

William J. Dawson, *Chemical Materials International, Dublin, Ohio,* Hydrothermal processing

Jeffrey J. DeFraties, *Haarmann & Reimer Corporation, Springfield, New Jersey,* Hydroxy dicarboxylic acids

Alan D. Denniston, *Unocal Corporation, San Ramon, California,* Hydraulic fluids

Lisa A. Dixon, *R. W. Johnson Pharmaceutical Research Institute, Raritan, New Jersey,* Sex hormones (under Hormones)

Christopher P. Dionigi, *United States Department of Agriculture, New Orleans, Louisiana,* Herbicides

Richard A. Durst, *Cornell University, Geneva, New York,* Hydrogen-ion activity

Alan D. Eastman, *Phillips Petroleum, Bartlesville, Oklahoma,* C_1-C_6; Survey (both under Hydrocarbons)

William C. Engeland, *University of Minnesota, Minneapolis,* Survey (under Hormones)

Richard E. Gannon, *Textron Defense Systems, Everett, Massachusetts,* Acetylene (under Hydrocarbons)

Joseph W. Gunnet, *R. W. Johnson Pharmaceutical Research Institute, Raritan, New Jersey,* Sex hormones (under Hormones)

J. Z. Guo, *R. W. Johnson Pharmaceutical Research Institute, Raritan, New Jersey,* Estrogens and antiestrogens (under Hormones)

D. W. Hahn, *R. W. Johnson Pharmaceutical Research Institute, Raritan, New Jersey,* Estrogens and antiestrogens (under Hormones)

Wayne T. Hess, *E. I. du Pont de Nemours & Co., Inc., Memphis, Tennessee,* Hydrogen peroxide

Mohamed W. M. Hisham, *Occidental Chemical Corporation, Grand Island, New York,* Hydrogen chloride

Charles C. Hobbs, *Consultant, Corpus Christi, Texas,* Hydrocarbon oxidation

Mark W. Holladay, *Abbott Laboratories, Abbott Park, Illinois,* Hypnotics, sedatives, anticonvulsants, and anxiolytics

John H. Hong, *Rockwell International Science Center, Thousand Oaks, California,* Holography

Shuen-Cheng Hwang, *The BOC Group, Inc., Murray Hill, New Jersey,* Gases (under Helium group)

Monique M.-L. Janssens, *Janssen Research Foundation, Beerse, Belgium,* Histamine and histamine antagonists

Richard M. Johnson, *United States Department of Agriculture, New Orleans, Louisiana,* Herbicides

S. A. Knez, *M. W. Kellogg Company, Houston, Texas,* Hydrogen

Pieter Krijgsman, *CEC Company, Switzerland,* Hydrothermal processing

Léon Krumenacker, *Rhône-Poulenc Recherches, France,* Hydroquinone, resorcinol, and catechol

Robert Leurs, *Leiden/Amsterdam Center for Drug Research,* Histamine and histamine antagonists

R. Derric Lowery, *Exxon Chemical Company, Baton Rouge, Louisiana,* Hydrocarbon resins

Warren C. MacKellar, *Lilly Research Laboratories, Indianapolis, Indiana,* Human growth hormone (under Hormones)

Christian Maliverney, *Rhône-Poulenc Recherches, France,* Hydroxybenzaldehydes

Robert M. Manyik, *Union Carbide Corporation, South Charleston, West Virginia,* Acetylene (under Hydrocarbons)

Robert L. Matteri, *United States Department of Agriculture, Columbia, Missouri,* Anterior pituitary hormones; Anterior pituitary-like hormones (both under Hormones)

David E. Mears, *Unocal Corporation, Brea, California,* C_1-C_6; Survey (both under Hydrocarbons)

J. Carlos Menéndez, *Universidad Complutense, Madrid, Spain,* Hydantoin and its derivatives

Michel Mulhauser, *Rhône-Poulenc Recherches, France,* Hydroxybenzaldehydes

Irving Moch, Jr., *E. I. du Pont de Nemours & Co., Inc., Wilmington, Delaware,* Hollow-fiber membranes

D. S. Newsome, *M. W. Kellogg Company, Houston, Texas,* Hydrogen

Malcolm Polk, *Georgia Institute of Technology, Atlanta,* High performance fibers

P. Pontal, *Rhône-Poulenc Recherches, France,* Hydroquinone, resorcinol, and catechol

Ralph M. Riggin, *Lilly Research Laboratories, Indianapolis, Indiana,* Human growth hormone (under Hormones)

Henry W. Schiessl, *Olin Corporation, Cheshire, Connecticut,* Hydrazine and its derivatives

Gary J. Schrobilgen, *McMaster University, Ontario, Canada,* Compounds (under Helium group)

J. Sentenac, *Rhône-Poulenc Recherches, France,* Hydroquinone, resorcinol, and catechol

Arno F. Spatola, *University of Louisville, Kentucky,* Posterior pituitary hormones (under Hormones)

Norman S. Stoloff, *Rensselaer Polytechnic Institute*, High temperature alloys

Edward A. Sullivan, *Morton International, Danvers, Massachusetts*, Hydrides

Norman S. Thompson, *Consultant, Appleton, Wisconsin*, Hemicellulose

Hendrik Timmerman, *Leiden/Amsterdam Center for Drug Research*, Histamine and histamine antagonists

Albin F. Turbak, *Consultant, Sandy Springs, Georgia*, High performance fibers

Tyrone L. Vigo, *United States Department of Agriculture, New Orleans, Louisiana*, High performance fibers

M. P. Wachter, *R. W. Johnson Pharmaceutical Research Institute, Raritan, New Jersey*, Estrogens and antiestrogens (under Hormones)

William R. Weltmer, Jr., *The BOC Group, Inc., Murray Hill, New Jersey*, Gases (under Helium group)

J. Marc Whalen, *McMaster University, Ontario, Canada*, Compounds (under Helium group)

Michael Williams, *Abbott Laboratories, Abbott Park, Illinois*, Hypnotics, sedatives, anticonvulsants, and anxiolytics

Suhad Wojkowski, *United States Department of Agriculture, New Orleans, Louisiana*, Herbicides

Victor J. Wroblewski, *Lilly Research Laboratories, Indianapolis, Indiana*, Human growth hormone (under Hormones)

Marek Zaidlewicz, *Nicolaus Copernicus University, Poland*, Hydroboration

NOTE ON CHEMICAL ABSTRACTS SERVICE REGISTRY NUMBERS AND NOMENCLATURE

Chemical Abstracts Service (CAS) Registry Numbers are unique numerical identifiers assigned to substances recorded in the CAS Registry System. They appear in brackets in the *Chemical Abstracts* (CA) substance and formula indexes following the names of compounds. A single compound may have synonyms in the chemical literature. A simple compound like phenethylamine can be named β-phenylethylamine or, as in *Chemical Abstracts*, benzeneethanamine. The usefulness of the *Encyclopedia* depends on accessibility through the most common correct name of a substance. Because of this diversity in nomenclature careful attention has been given to the problem in order to assist the reader as much as possible, especially in locating the systematic CA index name by means of the Registry Number. For this purpose, the reader may refer to the CAS Registry Handbook—Number Section which lists in numerical order the Registry Number with the *Chemical Abstracts* index name and the molecular formula; eg, **458-88-8**, Piperidine, 2-propyl-, (*S*)-, $C_8H_{17}N$; in the *Encyclopedia* this compound would be found under its common name, coniine [*458-88-8*]. Alternatively, this information can be retrieved electronically from CAS Online. In many cases molecular formulas have also been provided in the *Encyclopedia* text to facilitate electronic searching. The Registry Number is a valuable link for the reader in retrieving additional published information on substances and also as a point of access for on-line data bases.

In all cases, the CAS Registry Numbers have been given for title compounds in articles and for all compounds in the index. All specific substances indexed in *Chemical Abstracts* since 1965 are included in the CAS Registry System as are a large number of substances derived from a variety of reference works. The CAS Registry System identifies a substance on the basis of an unambiguous computer-language description of its molecular structure including stereochemical detail. The Registry Number is a machine-checkable number (like a Social Security number) assigned in sequential order to each substance as it enters the registry system. The value of the number lies in the fact that it is a concise and unique means of substance identification, which is independent of, and therefore bridges, many systems of chemical nomenclature. For polymers, one Registry Number may be used for the entire family; eg, polyoxyethylene (20) sorbitan monolaurate has the same number as all of its polyoxyethylene homologues.

Cross-references are inserted in the index for many common names and for some systematic names. Trademark names appear in the index. Names that are incorrect, misleading, or ambiguous are avoided. Formulas are given very frequently in the text to help in identifying compounds. The spelling and form used, even for industrial names, follow American chemical usage, but not always the usage of *Chemical Abstracts* (eg, *coniine* is used instead of *(S)-2-propylpiperidine*, *aniline* instead of *benzenamine*, and *acrylic acid* instead of *2-propenoic acid*).

There are variations in representation of rings in different disciplines. The dye industry does not designate aromaticity or double bonds in rings. All double bonds and aromaticity are shown in the *Encyclopedia* as a matter of course. For example, tetralin has an aromatic ring and a saturated ring and its structure

appears in the *Encyclopedia* with its common name, Registry Number enclosed in brackets, and parenthetical CA index name, ie, tetralin [*119-64-2*] (1,2,3,4-tetrahydronaphthalene). With names and structural formulas, and especially with CAS Registry Numbers, the aim is to help the reader have a concise means of substance identification.

CONVERSION FACTORS, ABBREVIATIONS, AND UNIT SYMBOLS

SI Units (Adopted 1960)

The International System of Units (abbreviated SI), is being implemented throughout the world. This measurement system is a modernized version of the MKSA (meter, kilogram, second, ampere) system, and its details are published and controlled by an international treaty organization (The International Bureau of Weights and Measures) (1).

SI units are divided into three classes:

BASE UNITS

length	meter[†] (m)
mass	kilogram (kg)
time	second (s)
electric current	ampere (A)
thermodynamic temperature[‡]	kelvin (K)
amount of substance	mole (mol)
luminous intensity	candela (cd)

SUPPLEMENTARY UNITS

plane angle	radian (rad)
solid angle	steradian (sr)

[†]The spellings "metre" and "litre" are preferred by ASTM; however, "-er" is used in the *Encyclopedia*.

[‡]Wide use is made of Celsius temperature (*t*) defined by

$$t = T - T_0$$

where T is the thermodynamic temperature, expressed in kelvin, and $T_0 = 273.15$ K by definition. A temperature interval may be expressed in degrees Celsius as well as in kelvin.

DERIVED UNITS AND OTHER ACCEPTABLE UNITS

These units are formed by combining base units, supplementary units, and other derived units (2–4). Those derived units having special names and symbols are marked with an asterisk in the list below.

Quantity	Unit	Symbol	Acceptable equivalent
*absorbed dose	gray	Gy	J/kg
acceleration	meter per second squared	m/s^2	
*activity (of a radionuclide)	becquerel	Bq	1/s
area	square kilometer	km^2	
	square hectometer	hm^2	ha (hectare)
	square meter	m^2	
concentration (of amount of substance)	mole per cubic meter	mol/m^3	
current density	ampere per square meter	$A//m^2$	
density, mass density	kilogram per cubic meter	kg/m^3	g/L; mg/cm^3
dipole moment (quantity)	coulomb meter	C·m	
*dose equivalent	sievert	Sv	J/kg
*electric capacitance	farad	F	C/V
*electric charge, quantity of electricity	coulomb	C	A·s
electric charge density	coulomb per cubic meter	C/m^3	
*electric conductance	siemens	S	A/V
electric field strength	volt per meter	V/m	
electric flux density	coulomb per square meter	C/m^2	
*electric potential, potential difference, electromotive force	volt	V	W/A
*electric resistance	ohm	Ω	V/A
*energy, work, quantity of heat	megajoule	MJ	
	kilojoule	kJ	
	joule	J	N·m
	electronvolt[†]	eV[†]	
	kilowatt-hour[†]	kW·h[†]	
energy density	joule per cubic meter	J/m^3	
*force	kilonewton	kN	
	newton	N	$kg·m/s^2$

[†]This non-SI unit is recognized by the CIPM as having to be retained because of practical importance or use in specialized fields (1).

Quantity	Unit	Symbol	Acceptable equivalent
*frequency	megahertz	MHz	
	hertz	Hz	1/s
heat capacity, entropy	joule per kelvin	J/K	
heat capacity (specific), specific entropy	joule per kilogram kelvin	J/(kg·K)	
heat-transfer coefficient	watt per square meter kelvin	W/(m²·K)	
*illuminance	lux	lx	lm/m²
*inductance	henry	H	Wb/A
linear density	kilogram per meter	kg/m	
luminance	candela per square meter	cd/m²	
*luminous flux	lumen	lm	cd·sr
magnetic field strength	ampere per meter	A/m	
*magnetic flux	weber	Wb	V·s
*magnetic flux density	tesla	T	Wb/m²
molar energy	joule per mole	J/mol	
molar entropy, molar heat capacity	joule per mole kelvin	J/(mol·K)	
moment of force, torque	newton meter	N·m	
momentum	kilogram meter per second	kg·m/s	
permeability	henry per meter	H/m	
permittivity	farad per meter	F/m	
*power, heat flow rate, radiant flux	kilowatt	kW	
	watt	W	J/s
power density, heat flux density, irradiance	watt per square meter	W/m²	
*pressure, stress	megapascal	MPa	
	kilopascal	kPa	
	pascal	Pa	N/m²
sound level	decibel	dB	
specific energy	joule per kilogram	J/kg	
specific volume	cubic meter per kilogram	m³/kg	
surface tension	newton per meter	N/m	
thermal conductivity	watt per meter kelvin	W/(m·K)	
velocity	meter per second	m/s	
	kilometer per hour	km/h	
viscosity, dynamic	pascal second	Pa·s	
	millipascal second	mPa·s	
viscosity, kinematic	square meter per second	m²/s	
	square millimeter per second	mm²/s	

Quantity	Unit	Symbol	Acceptable equivalent
volume	cubic meter	m^3	
	cubic diameter	dm^3	L (liter) (5)
	cubic centimeter	cm^3	mL
wave number	1 per meter	m^{-1}	
	1 per centimeter	cm^{-1}	

In addition, there are 16 prefixes used to indicate order of magnitude, as follows:

Multiplication factor	Prefix	Symbol	Note
10^{18}	exa	E	
10^{15}	peta	P	
10^{12}	tera	T	
10^{9}	giga	G	
10^{6}	mega	M	
10^{3}	kilo	k	
10^{2}	hecto	h[a]	[a]Although hecto, deka, deci, and centi
10	deka	da[a]	are SI prefixes, their use should be
10^{-1}	deci	d[a]	avoided except for SI unit-multiples
10^{-2}	centi	c[a]	for area and volume and nontech-
10^{-3}	milli	m	nical use of centimeter, as for body
10^{-6}	micro	μ	and clothing measurement.
10^{-9}	nano	n	
10^{-12}	pico	p	
10^{-15}	femto	f	
10^{-18}	atto	a	

For a complete description of SI and its use the reader is referred to ASTM E380 (4) and the article UNITS AND CONVERSION FACTORS which appears in Vol. 24.

A representative list of conversion factors from non-SI to SI units is presented herewith. Factors are given to four significant figures. Exact relationships are followed by a dagger. A more complete list is given in the latest editions of ASTM E380 (4) and ANSI Z210.1 (6).

Conversion Factors to SI Units

To convert from	To	Multiply by
acre	square meter (m^2)	4.047×10^3
angstrom	meter (m)	1.0×10^{-10}[†]
are	square meter (m^2)	1.0×10^{2}[†]

[†]Exact.

To convert from	To	Multiply by
astronomical unit	meter (m)	1.496×10^{11}
atmosphere, standard	pascal (Pa)	1.013×10^{5}
bar	pascal (Pa)	$1.0 \times 10^{5\dagger}$
barn	square meter (m²)	$1.0 \times 10^{-28\dagger}$
barrel (42 U.S. liquid gallons)	cubic meter (m³)	0.1590
Bohr magneton (μ_B)	J/T	9.274×10^{-24}
Btu (International Table)	joule (J)	1.055×10^{3}
Btu (mean)	joule (J)	1.056×10^{3}
Btu (thermochemical)	joule (J)	1.054×10^{3}
bushel	cubic meter (m³)	3.524×10^{-2}
calorie (International Table)	joule (J)	4.187
calorie (mean)	joule (J)	4.190
calorie (thermochemical)	joule (J)	$4.184^{\dagger}$
centipoise	pascal second (Pa·s)	$1.0 \times 10^{-3\dagger}$
centistokes	square millimeter per second (mm²/s)	$1.0^{\dagger}$
cfm (cubic foot per minute)	cubic meter per second (m³/s)	4.72×10^{-4}
cubic inch	cubic meter (m³)	1.639×10^{-5}
cubic foot	cubic meter (m³)	2.832×10^{-2}
cubic yard	cubic meter (m³)	0.7646
curie	becquerel (Bq)	$3.70 \times 10^{10\dagger}$
debye	coulomb meter (C·m)	3.336×10^{-30}
degree (angle)	radian (rad)	1.745×10^{-2}
denier (international)	kilogram per meter (kg/m)	1.111×10^{-7}
	tex‡	0.1111
dram (apothecaries')	kilogram (kg)	3.888×10^{-3}
dram (avoirdupois)	kilogram (kg)	1.772×10^{-3}
dram (U.S. fluid)	cubic meter (m³)	3.697×10^{-6}
dyne	newton (N)	$1.0 \times 10^{-5\dagger}$
dyne/cm	newton per meter (N/m)	$1.0 \times 10^{-3\dagger}$
electronvolt	joule (J)	1.602×10^{-19}
erg	joule (J)	$1.0 \times 10^{-7\dagger}$
fathom	meter (m)	1.829
fluid ounce (U.S.)	cubic meter (m³)	2.957×10^{-5}
foot	meter (m)	$0.3048^{\dagger}$
footcandle	lux (lx)	10.76
furlong	meter (m)	2.012×10^{-2}
gal	meter per second squared (m/s²)	$1.0 \times 10^{-2\dagger}$
gallon (U.S. dry)	cubic meter (m³)	4.405×10^{-3}
gallon (U.S. liquid)	cubic meter (m³)	3.785×10^{-3}
gallon per minute (gpm)	cubic meter per second (m³/s)	6.309×10^{-5}
	cubic meter per hour (m³/h)	0.2271

†Exact.

‡See footnote on p. xiii.

To convert from	To	Multiply by
gauss	tesla (T)	1.0×10^{-4}
gilbert	ampere (A)	0.7958
gill (U.S.)	cubic meter (m^3)	1.183×10^{-4}
grade	radian	1.571×10^{-2}
grain	kilogram (kg)	6.480×10^{-5}
gram force per denier	newton per tex (N/tex)	8.826×10^{-2}
hectare	square meter (m^2)	$1.0 \times 10^{4\dagger}$
horsepower (550 ft·lbf/s)	watt (W)	7.457×10^{2}
horsepower (boiler)	watt (W)	9.810×10^{3}
horsepower (electric)	watt (W)	$7.46 \times 10^{2\dagger}$
hundredweight (long)	kilogram (kg)	50.80
hundredweight (short)	kilogram (kg)	45.36
inch	meter (m)	$2.54 \times 10^{-2\dagger}$
inch of mercury (32°F)	pascal (Pa)	3.386×10^{3}
inch of water (39.2°F)	pascal (Pa)	2.491×10^{2}
kilogram-force	newton (N)	9.807
kilowatt hour	megajoule (MJ)	$3.6^{\dagger}$
kip	newton (N)	4.448×10^{3}
knot (international)	meter per second (m/S)	0.5144
lambert	candela per square meter (cd/m^3)	3.183×10^{3}
league (British nautical)	meter (m)	5.559×10^{3}
league (statute)	meter (m)	4.828×10^{3}
light year	meter (m)	9.461×10^{15}
liter (for fluids only)	cubic meter (m^3)	$1.0 \times 10^{-3\dagger}$
maxwell	weber (Wb)	$1.0 \times 10^{-8\dagger}$
micron	meter (m)	$1.0 \times 10^{-6\dagger}$
mil	meter (m)	$2.54 \times 10^{-5\dagger}$
mile (statute)	meter (m)	1.609×10^{3}
mile (U.S. nautical)	meter (m)	$1.852 \times 10^{3\dagger}$
mile per hour	meter per second (m/s)	0.4470
millibar	pascal (Pa)	1.0×10^{2}
millimeter of mercury (0°C)	pascal (Pa)	$1.333 \times 10^{2\dagger}$
minute (angular)	radian	2.909×10^{-4}
myriagram	kilogram (kg)	10
myriameter	kilometer (km)	10
oersted	ampere per meter (A/m)	79.58
ounce (avoirdupois)	kilogram (kg)	2.835×10^{-2}
ounce (troy)	kilogram (kg)	3.110×10^{-2}
ounce (U.S. fluid)	cubic meter (m^3)	2.957×10^{-5}
ounce-force	newton (N)	0.2780
peck (U.S.)	cubic meter (m^3)	8.810×10^{-3}
pennyweight	kilogram (kg)	1.555×10^{-3}
pint (U.S. dry)	cubic meter (m^3)	5.506×10^{-4}
pint (U.S. liquid)	cubic meter (m^3)	4.732×10^{-4}

†Exact.

To convert from	To	Multiply by
poise (absolute viscosity)	pascal second (Pa·s)	$0.10^{\dagger}$
pound (avoirdupois)	kilogram (kg)	0.4536
pound (troy)	kilogram (kg)	0.3732
poundal	newton (N)	0.1383
pound-force	newton (N)	4.448
pound force per square inch (psi)	pascal (Pa)	6.895×10^3
quart (U.S. dry)	cubic meter (m³)	1.101×10^{-3}
quart (U.S. liquid)	cubic meter (m³)	9.464×10^{-4}
quintal	kilogram (kg)	$1.0 \times 10^{2\dagger}$
rad	gray (Gy)	$1.0 \times 10^{-2\dagger}$
rod	meter (m)	5.029
roentgen	coulomb per kilogram (C/kg)	2.58×10^{-4}
second (angle)	radian (rad)	$4.848 \times 10^{-6\dagger}$
section	square meter (m²)	2.590×10^6
slug	kilogram (kg)	14.59
spherical candle power	lumen (lm)	12.57
square inch	square meter (m²)	6.452×10^{-4}
square foot	square meter (m²)	9.290×10^{-2}
square mile	square meter (m²)	2.590×10^6
square yard	square meter (m²)	0.8361
stere	cubic meter (m³)	$1.0^{\dagger}$
stokes (kinematic viscosity)	square meter per second (m²/s)	$1.0 \times 10^{-4\dagger}$
tex	kilogram per meter (kg/m)	$1.0 \times 10^{-6\dagger}$
ton (long, 2240 pounds)	kilogram (kg)	1.016×10^3
ton (metric) (tonne)	kilogram (kg)	$1.0 \times 10^{3\dagger}$
ton (short, 2000 pounds)	kilogram (kg)	9.072×10^2
torr	pascal (Pa)	1.333×10^2
unit pole	weber (Wb)	1.257×10^{-7}
yard	meter (m)	$0.9144^{\dagger}$

†Exact.

Abbreviations and Unit Symbols

Following is a list of common abbreviations and unit symbols used in the *Encyclopedia*. In general they agree with those listed in *American National Standard Abbreviations for Use on Drawings and in Text (ANSI Y1.1)* (6) and *American National Standard Letter Symbols for Units in Science and Technology (ANSI Y10)* (6). Also included is a list of acronyms for a number of private and government organizations as well as common industrial solvents, polymers, and other chemicals.

Rules for Writing Unit Symbols (4):

1. Unit symbols are printed in upright letters (roman) regardless of the type style used in the surrounding text.
2. Unit symbols are unaltered in the plural.
3. Unit symbols are not followed by a period except when used at the end of a sentence.
4. Letter unit symbols are generally printed lower-case (for example, cd for candela) unless the unit name has been derived from a proper name, in which case the first letter of the symbol is capitalized (W, Pa). Prefixes and unit symbols retain their prescribed form regardless of the surrounding typography.
5. In the complete expression for a quantity, a space should be left between the numerical value and the unit symbol. For example, write 2.37 lm, *not* 2.37lm, and 35 mm, *not* 35mm. When the quantity is used in an adjectival sense, a hyphen is often used, for example, 35-mm film. *Exception:* No space is left between the numerical value and the symbols of degree, minute, and second of plane angle, degree Celsius, and the percent sign.
6. No space is used between the prefix and unit symbol (for example, kg).
7. Symbols, not abbreviations, should be used for units. For example, use "A," not "amp," for ampere.
8. When multiplying unit symbols, use a raised dot:

$$N \cdot m \quad \text{for} \quad \text{newton meter}$$

In the case of W·h, the dot may be omitted, thus:

$$Wh$$

An exception to this practice is made for computer printouts, automatic typewriter work, etc, where the raised dot is not possible, and a dot on the line may be used.

9. When dividing unit symbols, use one of the following forms:

$$m/s \quad or \quad m \cdot s^{-1} \quad or \quad \frac{m}{s}$$

In no case should more than one slash be used in the same expression unless parentheses are inserted to avoid ambiguity. For example, write:

$$J/(mol \cdot K) \quad or \quad J \cdot mol^{-1} \cdot K^{-1} \quad or \quad (J/mol)/K$$

but *not*

$$J/mol/K$$

10. Do not mix symbols and unit names in the same expression. Write:

$$\text{joules per kilogram} \quad or \quad \text{J/kg} \quad or \quad \text{J·kg}^{-1}$$

but *not*

$$\text{joules/kilogram} \quad nor \quad \text{joules/kg} \quad nor \quad \text{joules·kg}^{-1}$$

ABBREVIATIONS AND UNITS

A	ampere	AOAC	Association of Official Analytical Chemists
A	anion (eg, HA)		
A	mass number	AOCS	American Oil Chemists' Society
a	atto (prefix for 10^{-18})		
AATCC	American Association of Textile Chemists and Colorists	APHA	American Public Health Association
		API	American Petroleum Institute
ABS	acrylonitrile–butadiene– styrene	aq	aqueous
abs	absolute	Ar	aryl
ac	alternating current, *n.*	*ar-*	aromatic
a-c	alternating current, *adj.*	*as-*	asymmetric(al)
ac-	alicyclic	ASHRAE	American Society of Heating, Refrigerating, and Air Conditioning Engineers
acac	acetylacetonate		
ACGIH	American Conference of Governmental Industrial Hygienists		
		ASM	American Society for Metals
ACS	American Chemical Society		
		ASME	American Society of Mechanical Engineers
AGA	American Gas Association		
Ah	ampere hour	ASTM	American Society for Testing and Materials
AIChE	American Institute of Chemical Engineers	at no.	atomic number
AIME	American Institute of Mining, Metallurgical, and Petroleum Engineers	at wt	atomic weight
		av(g)	average
		AWS	American Welding Society
		b	bonding orbital
AIP	American Institute of Physics	bbl	barrel
		bcc	body-centered cubic
AISI	American Iron and Steel Institute	BCT	body-centered tetragonal
		Bé	Baumé
alc	alcohol(ic)	BET	Brunauer-Emmett-Teller (adsorption equation)
Alk	alkyl		
alk	alkaline (not alkali)	bid	twice daily
amt	amount	Boc	*t*-butyloxycarbonyl
amu	atomic mass unit	BOD	biochemical (biological) oxygen demand
ANSI	American National Standards Institute		
		bp	boiling point
AO	atomic orbital	Bq	becquerel

C	coulomb	DIN	Deutsche Industrie Normen
°C	degree Celsius		
C-	denoting attachment to carbon	*dl*-; DL-	racemic
		DMA	dimethylacetamide
c	centi (prefix for 10^{-2})	DMF	dimethylformamide
c	critical	DMG	dimethyl glyoxime
ca	circa (approximately)	DMSO	dimethyl sulfoxide
cd	candela; current density; circular dichroism	DOD	Department of Defense
		DOE	Department of Energy
CFR	Code of Federal Regulations	DOT	Department of Transportation
cgs	centimeter-gram-second	DP	degree of polymerization
CI	Color Index	dp	dew point
cis-	isomer in which substituted groups are on same side of double bond between C atoms	DPH	diamond pyramid hardness
		dstl(d)	distill(ed)
		dta	differential thermal analysis
cl	carload		
cm	centimeter	(*E*)-	entgegen; opposed
cmil	circular mil	ϵ	dielectric constant (unitless number)
cmpd	compound		
CNS	central nervous system	*e*	electron
CoA	coenzyme A	ECU	electrochemical unit
COD	chemical oxygen demand	ed.	edited, edition, editor
coml	commercial(ly)	ED	effective dose
cp	chemically pure	EDTA	ethylenediaminetetra-acetic acid
cph	close-packed hexagonal		
CPSC	Consumer Product Safety Commission	emf	electromotive force
		emu	electromagnetic unit
cryst	crystalline	en	ethylene diamine
cub	cubic	eng	engineering
D	debye	EPA	Environmental Protection Agency
D-	denoting configurational relationship		
		epr	electron paramagnetic resonance
d	differential operator		
d	day; deci (prefix for 10^{-1})	eq.	equation
d	density	esca	electron spectroscopy for chemical analysis
d-	*dextro*-, dextrorotatory		
da	deka (prefix for 10^1)	esp	especially
dB	decibel	esr	electron-spin resonance
dc	direct current, *n.*	est(d)	estimate(d)
d-c	direct current, *adj.*	estn	estimation
dec	decompose	esu	electrostatic unit
detd	determined	exp	experiment, experimental
detn	determination	ext(d)	extract(ed)
Di	didymium, a mixture of all lanthanons	F	farad (capacitance)
		F	faraday (96,487 C)
dia	diameter	f	femto (prefix for 10^{-15})
dil	dilute		

FAO	Food and Agriculture Organization (United Nations)	hyd	hydrated, hydrous
		hyg	hygroscopic
fcc	face-centered cubic	Hz	hertz
FDA	Food and Drug Administration	i (eg, Pr^i)	iso (eg, isopropyl)
		i-	inactive (eg, i-methionine)
FEA	Federal Energy Administration	IACS	International Annealed Copper Standard
FHSA	Federal Hazardous Substances Act	ibp	initial boiling point
		IC	integrated circuit
fob	free on board	ICC	Interstate Commerce Commission
fp	freezing point		
FPC	Federal Power Commission	ICT	International Critical Table
		ID	inside diameter; infective dose
FRB	Federal Reserve Board		
frz	freezing	ip	intraperitoneal
G	giga (prefix for 10^9)	IPS	iron pipe size
G	gravitational constant = 6.67×10^{11} N·m^2/kg^2	ir	infrared
		IRLG	Interagency Regulatory Liaison Group
g	gram		
(g)	gas, only as in H_2O(g)	ISO	International Organization Standardization
g	gravitational acceleration		
gc	gas chromatography	ITS-90	International Temperature Scale (NIST)
gem-	geminal		
glc	gas–liquid chromatography	IU	International Unit
		IUPAC	International Union of Pure and Applied Chemistry
g-mol wt; gmw	gram-molecular weight		
GNP	gross national product	IV	iodine value
gpc	gel-permeation chromatography	iv	intravenous
		J	joule
GRAS	Generally Recognized as Safe	K	kelvin
		k	kilo (prefix for 10^3)
grd	ground	kg	kilogram
Gy	gray	L	denoting configurational relationship
H	henry		
h	hour; hecto (prefix for 10^2)	L	liter (for fluids only) (5)
ha	hectare	l-	levo-, levorotatory
HB	Brinell hardness number	(l)	liquid, only as in NH_3(l)
Hb	hemoglobin	LC$_{50}$	conc lethal to 50% of the animals tested
hcp	hexagonal close-packed		
hex	hexagonal	LCAO	linear combination of atomic orbitals
HK	Knoop hardness number		
hplc	high performance liquid chromatography	lc	liquid chromatography
		LCD	liquid crystal display
HRC	Rockwell hardness (C scale)	lcl	less than carload lots
		LD$_{50}$	dose lethal to 50% of the animals tested
HV	Vickers hardness number		

LED	light-emitting diode	N-	denoting attachment to nitrogen
liq	liquid		
lm	lumen	n (as n_{D}^{20})	index of refraction (for 20°C and sodium light)
ln	logarithm (natural)		
LNG	liquefied natural gas	^{n} (as Bu^{n}),	
log	logarithm (common)	n-	normal (straight-chain structure)
LOI	limiting oxygen index		
LPG	liquefied petroleum gas	n	neutron
ltl	less than truckload lots	n	nano (prefix for 10^9)
lx	lux	na	not available
M	mega (prefix for 10^6); metal (as in MA)	NAS	National Academy of Sciences
M	molar; actual mass	NASA	National Aeronautics and Space Administration
$\overline{M}_w$	weight-average mol wt		
$\overline{M}_n$	number-average mol wt	nat	natural
m	meter; milli (prefix for 10^{-3})	ndt	nondestructive testing
		neg	negative
m	molal	NF	*National Formulary*
m-	meta	NIH	National Institutes of Health
max	maximum		
MCA	Chemical Manufacturers' Association (was Manufacturing Chemists Association)	NIOSH	National Institute of Occupational Safety and Health
		NIST	National Institute of Standards and Technology (formerly National Bureau of Standards)
MEK	methyl ethyl ketone		
meq	milliequivalent		
mfd	manufactured		
mfg	manufacturing		
mfr	manufacturer	nmr	nuclear magnetic resonance
MIBC	methyl isobutyl carbinol		
MIBK	methyl isobutyl ketone	NND	New and Nonofficial Drugs (AMA)
MIC	minimum inhibiting concentration	no.	number
min	minute; minimum	NOI-(BN)	not otherwise indexed (by name)
mL	milliliter		
MLD	minimum lethal dose	NOS	not otherwise specified
MO	molecular orbital	nqr	nuclear quadruple resonance
mo	month		
mol	mole	NRC	Nuclear Regulatory Commission; National Research Council
mol wt	molecular weight		
mp	melting point		
MR	molar refraction	NRI	New Ring Index
ms	mass spectrometry	NSF	National Science Foundation
MSDS	material safety data sheet		
mxt	mixture	NTA	nitrilotriacetic acid
μ	micro (prefix for 10^{-6})	NTP	normal temperature and pressure (25°C and 101.3 kPa or 1 atm)
N	newton (force)		
N	normal (concentration); neutron number		

NTSB	National Transportation Safety Board	qv	quod vide (which see)
O-	denoting attachment to oxygen	R	univalent hydrocarbon radical
o-	ortho	(*R*)-	rectus (clockwise configuration)
OD	outside diameter	*r*	precision of data
OPEC	Organization of Petroleum Exporting Countries	rad	radian; radius
o-phen	*o*-phenanthridine	RCRA	Resource Conservation and Recovery Act
OSHA	Occupational Safety and Health Administration	rds	rate-determining step
		ref.	reference
owf	on weight of fiber	rf	radio frequency, *n.*
Ω	ohm	r-f	radio frequency, *adj.*
P	peta (prefix for 10^{15})	rh	relative humidity
p	pico (prefix for 10^{-12})	RI	Ring Index
p-	para	rms	root-mean square
p	proton	rpm	rotations per minute
p.	page	rps	revolutions per second
Pa	pascal (pressure)	RT	room temperature
PEL	personal exposure limit based on an 8-h exposure	RTECS	Registry of Toxic Effects of Chemical Substances
		ˢ (eg, Buˢ);	
pd	potential difference	*sec-*	secondary (eg, secondary butyl)
pH	negative logarithm of the effective hydrogen ion concentration	S	siemens
		(*S*)-	sinister (counterclockwise configuration)
phr	parts per hundred of resin (rubber)	S-	denoting attachment to sulfur
p-i-n	positive-intrinsic-negative		
pmr	proton magnetic resonance	*s-*	symmetric(al)
p-n	positive-negative	s	second
po	per os (oral)	(s)	solid, only as in $H_2O(s)$
POP	polyoxypropylene	SAE	Society of Automotive Engineers
pos	positive		
pp.	pages	SAN	styrene-acrylonitrile
ppb	parts per billion (10^9)	sat(d)	saturate(d)
ppm	parts per million (10^6)	satn	saturation
ppmv	parts per million by volume	SBS	styrene–butadiene–styrene
ppmwt	parts per million by weight	sc	subcutaneous
PPO	poly(phenyl oxide)	SCF	self-consistent field; standard cubic feet
ppt(d)	precipitate(d)		
pptn	precipitation	Sch	Schultz number
Pr (no.)	foreign prototype (number)	sem	scanning electron microscope(y)
pt	point; part		
PVC	poly(vinyl chloride)	SFs	Saybolt Furol seconds
pwd	powder	sl sol	slightly soluble
py	pyridine	sol	soluble

soln	solution	*trans-*	isomer in which substituted groups are on opposite sides of double bond between C atoms
soly	solubility		
sp	specific; species		
sp gr	specific gravity		
sr	steradian		
std	standard	TSCA	Toxic Substances Control Act
STP	standard temperature and pressure (0°C and 101.3 kPa)	TWA	time-weighted average
		Twad	Twaddell
sub	sublime(s)	UL	Underwriters' Laboratory
SUs	Saybolt Universal seconds	USDA	United States Department of Agriculture
syn	synthetic		
ᵗ (eg, Buᵗ),		USP	*United States Pharmacopeia*
t-, tert-	tertiary (eg, tertiary butyl)	uv	ultraviolet
T	tera (prefix for 10^{12}); tesla (magnetic flux density)	V	volt (emf)
		var	variable
t	metric ton (tonne)	*vic-*	vicinal
t	temperature	vol	volume (not volatile)
TAPPI	Technical Association of the Pulp and Paper Industry	vs	versus
		v sol	very soluble
		W	watt
TCC	Tagliabue closed cup	Wb	weber
tex	tex (linear density)	Wh	watt hour
T_g	glass-transition temperature	WHO	World Health Organization (United Nations)
tga	thermogravimetric analysis		
		wk	week
THF	tetrahydrofuran	yr	year
tlc	thin layer chromatography	(*Z*)-	zusammen; together; atomic number
TLV	threshold limit value		

Non-SI (Unacceptable and Obsolete) Units		Use
Å	angstrom	nm
at	atmosphere, technical	Pa
atm	atmosphere, standard	Pa
b	barn	cm^2
bar†	bar	Pa
bbl	barrel	m^3
bhp	brake horsepower	W
Btu	British thermal unit	J
bu	bushel	m^3; L
cal	calorie	J
cfm	cubic foot per minute	m^3/s
Ci	curie	Bq
cSt	centistokes	mm^2/s
c/s	cycle per second	Hz

†Do not use bar (10^5 Pa) or millibar (10^2 Pa) because they are not SI units, and are accepted internationally only for a limited time in special fields because of existing usage.

Non-SI (Unacceptable and Obsolete) Units		Use
cu	cubic	exponential form
D	debye	C·m
den	denier	tex
dr	dram	kg
dyn	dyne	N
dyn/cm	dyne per centimeter	mN/m
erg	erg	J
eu	entropy unit	J/K
°F	degree Fahrenheit	°C; K
fc	footcandle	lx
fl	footlambert	lx
fl oz	fluid ounce	m^3; L
ft	foot	m
ft·lbf	foot pound-force	J
gf den	gram-force per denier	N/tex
G	gauss	T
Gal	gal	m/s^2
gal	gallon	m^3; L
Gb	gilbert	A
gpm	gallon per minute	(m^3/s); (m^3/h)
gr	grain	kg
hp	horsepower	W
ihp	indicated horsepower	W
in.	inch	m
in. Hg	inch of mercury	Pa
in. H_2O	inch of water	Pa
in.-lbf	inch pound-force	J
kcal	kilo-calorie	J
kgf	kilogram-force	N
kilo	for kilogram	kg
L	lambert	lx
lb	pound	kg
lbf	pound-force	N
mho	mho	S
mi	mile	m
MM	million	M
mm Hg	millimeter of mercury	Pa
mμ	millimicron	nm
mph	miles per hour	km/h
μ	micron	μm
Oe	oersted	A/m
oz	ounce	kg
ozf	ounce-force	N
η	poise	Pa·s
P	poise	Pa·s
ph	phot	lx
psi	pounds-force per square inch	Pa
psia	pounds-force per square inch absolute	Pa
psig	pounds-force per square inch gage	Pa
qt	quart	m^3; L
°R	degree Rankine	K
rd	rad	Gy
sb	stilb	lx
SCF	standard cubic foot	m^3
sq	square	exponential form
thm	therm	J
yd	yard	m

BIBLIOGRAPHY

1. The International Bureau of Weights and Measures, BIPM (Parc de Saint-Cloud, France) is described in Appendix X2 of Ref. 4. This bureau operates under the exclusive supervision of the International Committee for Weights and Measures (CIPM).
2. *Metric Editorial Guide (ANMC-78-1)*, latest ed., American National Metric Council, 5410 Grosvenor Lane, Bethesda, Md. 20814, 1981.
3. *SI Units and Recommendations for the Use of Their Multiples and of Certain Other Units (ISO 1000-1981)*, American National Standards Institute, 1430 Broadway, New York, 10018, 1981.
4. Based on *ASTM E380-89a (Standard Practice for Use of the International System of Units (SI))*, American Society for Testing and Materials, 1916 Race Street, Philadelphia, Pa. 19103, 1989.
5. *Fed. Reg.*, Dec. 10, 1976 (41 FR 36414).
6. For ANSI address, see Ref. 3.

R. P. Lukens
ASTM Committee E-43 on SI Practice

Continued

HELIUM GROUP

GASES

The helium-group gases are helium [*7440-59-7*], He; neon [*7440-01-9*], Ne; argon [*7440-37-1*], Ar; krypton [*7439-90-9*], Kr; xenon [*7440-63-3*], Xe; and radon [*14859-67-7*], Rn. These are all members of Group 18 (VIIIA) of the Periodic Table and are characterized by completely filled valence electron shells. Historically, they have been called the rare, noble, or inert gases. But although comparatively rare, krypton, xenon, and radon are not completely inert; all three form stable molecules with highly electronegative elements such as F, Cl, and O (see HELIUM GROUP, COMPOUNDS). Although inert enough, helium and argon are not rare; both are bulk items of commerce. At some future time, however, when helium-bearing natural gases have been depleted and the atmosphere becomes its only source, helium is expected to return to being a truly rare gas.

The helium-group gases were not isolated until the last decade of the nineteenth century making them the most recently discovered group of stable elements (1–3). While seeking more accurate density values for certain gases during the period 1882–1894, Rayleigh noted that the density of nitrogen isolated by removing oxygen from air was consistently about one-half percent greater than that of nitrogen obtained from chemical reactions. Working with Rayleigh, Ramsay treated atmospheric nitrogen with hot magnesium and obtained a residual gas, ca 1.25% of the nitrogen, that was completely unreactive. The gas had a relative density of 19.075 (O = 16) and exhibited spectral lines not seen before. In 1894 the discovery of an inert gas was announced and called argon from the Greek *argos*, inactive.

1

In 1868, within a decade of the development of the spectroscope, an orange-yellow line was observed in the sun's chromosphere that did not exactly coincide with the D-lines of sodium. This line was attributed to a new element which was named helium, from the Greek *helios*, the sun. In 1891 an inert gas isolated from the mineral uranite showed unusual spectral lines. In 1895 a similar gas was found in cleveite, another uranium mineral. This prominent yellow spectral line was then identified as that of helium, which to that time had been thought to exist only on the sun. In 1905 it was found that natural gas from a well near Dexter, Kansas contained nearly 2% helium (see GAS, NATURAL).

The existence of neon (Greek *neos*, new) was predicted, as was the existence of heavier members of the group. In 1898 krypton (Greek *kryptos*, hidden) was discovered by spectroscopic examination of the residue from a sample of liquid air. Neon was discovered in the same year. A month later, xenon (Greek *xenos*, strange) was isolated from the residue left after distillation of krypton.

Radon-220 [22481-48-7], ^{220}Rn, a decay product of thorium, was discovered by Owens and Rutherford in 1900. The more common radon-222, a decay product of radium, was discovered later in the same year and was isolated in 1902.

Helium-3 [14762-55-1], ^{3}He, has been known as a stable isotope since the middle 1930s and it was suspected that its properties were markedly different from the common isotope, helium-4. The development of nuclear fusion devices in the 1950s yielded workable quantities of pure helium-3 as a decay product from the large tritium inventory implicit in maintaining an arsenal of fusion weapons (see DEUTERIUM AND TRITIUM). Helium-3 is one of the very few stable materials where the only practical source is nuclear transmutation. The chronology of the isolation of the other stable isotopes of the helium-group gases has been summarized (4).

Occurrence

Helium, plentiful in the cosmos, is a product of the nuclear fusion reactions that are the prime source of stellar energy. The other members of the helium-group gases are thought to have been created like other heavier elements by further nuclear condensation reactions occurring at the extreme temperatures and densities found deep within stars and in supernovas.

On earth, the only practical sources of the stable helium-group gases are the atmosphere and certain helium-bearing natural gases. Faint concentrations of helium and argon are occluded in some minerals. The neon, krypton, xenon, and most of the argon isotopes were likely part of the original mass that condensed to form the earth. However, the earth's gravitational field is inadequate to prevent helium's escape from the atmosphere. Helium is being formed continuously on earth by α-decay of heavier elements such as uranium and thorium, the α-particle being simply a fully ionized helium atom. Thus, the atmospheric helium concentration represents a dynamic equilibrium between the gain of helium diffusing from the earth's crust and the loss of helium into space. The total terrestrial inventory of helium is estimated to be 4.9×10^{14} m^3 where the volume is measured throughout at 101.3 kPa (1 atm) absolute pressure and 15°C (5).

Argon-40 [7440-37-1] is created by the decay of potassium-40. The various isotopes of radon, all having short half-lives, are formed by the radioactive decay of radium, actinium, and thorium. Krypton and xenon are products of uranium and plutonium fission, and appreciable quantities of both are evolved during the reprocessing of spent fuel elements from nuclear reactors (qv) (see RADIOACTIVE ELEMENTS).

In the composition of dry air (Table 1), nine principal constituents are constant wherever free air is sampled. Five of these are the stable helium-group gases. Other trace impurities in air vary in concentration from place to place; among these is radon. The longest lived isotope, ^{222}Rn, has a half-life of 3.825 days which is short compared to the atmosphere's mixing time. The concentrations of several of the variable impurities are comparable to that of krypton and an order of magnitude greater than that of xenon. Also listed in Table 1 is the thermodynamic minimum work for isobaric, isothermal separation per mole of each component from air at 300 K (8). Practical separations most often involve cryogenic condensations and distillations, thus the normal boiling point of each component is also listed (see CRYOGENICS; NITROGEN; OXYGEN).

The principal source of helium is certain natural gas fields. The helium contents of more than 10,000 natural gases in various parts of the world have been measured (9). Helium concentrations of a few are listed in Table 2. In the United States, recovery of helium is economical only for helium-rich gases

Table 1. Components of Dry Air

Gas	Concentration,[a] μL/L	Minimum work of separation at 300 K, kJ/mol[b]	Normal boiling point, K
Principal constituents			
nitrogen	780,840 ± 40	1.68	77.35
oxygen	209,460 ± 20	6.11	90.19
argon	9,340 ± 10	14.14	87.28
carbon dioxide	300 ± 30	22.73	194.67[c]
neon	18.21 ± 0.04	29.72	27.09
helium-4	5.24 ± 0.05	32.82	4.22
krypton	1.14 ± 0.01	36.63	119.80
hydrogen	ca 0.05	38.68	20.27
xenon	0.087 ± 0.001	43.04	165.03
Variable impurities			
methane	1.7[d]		111.66
carbon monoxide	<1		81.70
sulfur dioxide	1		263.2
nitrogen dioxide	0.5		294.2
ozone	<0.1		161.8
nitrous oxide	<0.02		184.68
radon	6×10^{-14}		211

[a]Ref. 6.
[b]To convert J to cal, divide by 4.184.
[c]Sublimation temperature.
[d]Ref. 7.

Table 2. Helium Concentrations of Helium-Bearing Natural Gases[a]

Location	Field site	Concentration, vol %
United States		
Wyoming	Tip Top Field	0.4–0.8
San Juan, New Mexico	Beautiful Mountain	4.05
North Slope, Alaska	South Barroweast	2.54
Young, Texas	Young Regular	1.17
Grant, Kansas	Hugoton (Comprstation)	0.44
Canada		
Alberta	Worsley	0.53
Ontario	Norfolk	0.36
the Netherlands	Groningen	0.05
	DeWijk	0.05
Poland	Ostrow	0.4
North Sea, British sector	Indefatigable	0.05
West Germany, Niedersachsen	Apeldorn	0.12
CIS	Urengoi	0.055
	Orenburg	
Orange Free State	O.F.S. Goldfields	2.91
Algeria	Hassi R'Mel	0.19
Australia, North Territory	Palm Valley	0.21

[a] Refs. 9 and 10.

containing more than about 0.3 vol % helium. Most of the United States helium resources are located in the midcontinent and Rocky Mountain regions, and about 89% of the known United States supply is in the Hugoton field in Kansas, Oklahoma, and Texas; the Keyes field in Oklahoma; the Panhandle and Cliffside fields in Texas; and the Riley Ridge area in Wyoming (11).

Resources and Conservation. The availability of the helium-group gases from the atmosphere is unlikely to change. There are no environmental sinks for these practically inert materials, and quantities removed from the atmosphere are eventually returned.

Upon the United States entry into World War I in 1917, helium became a war material of first priority. Helium was sought to replace hydrogen as the lifting gas in lighter-than-air craft for military use. As a war material, helium became a government monopoly, was given a code name, and was shrouded in secrecy. By the war's end, quantities of helium had been produced, but none had reached combat.

In 1925, the first *Helium Conservation Act* made the U.S. Bureau of Mines responsible for all helium activities including the construction and operation of helium-separation plants, planning, and experimental work. Sales of helium to nongovernmental users were prohibited until 1937. Helium facilities were, therefore, in place at the beginning of World War II, and helium's availability as a lifting gas proved to be of great value in the antisubmarine campaigns. Following World War II, new technologies demanded ever-increasing amounts of helium. By the end of the 1950s, demand exceeded production capabilities of the Bureau of Mines' plants, and depletion of resources came into question (12).

In a move to conserve the helium being wasted in natural-gas combustion, the 1960 *Helium Act Amendments* allowed the government to enter long-term contracts for the purchase of crude helium recovered by private companies and to store helium. The program was to be supported by an increase in the price of Bureau of Mines helium. Over the next decade, however, helium usage increased more slowly than had been projected, reserves accumulated more rapidly than had been anticipated, and private producers took an increasing share of the market by selling pure helium at prices far below federal prices (13). Thus in 1971 the government canceled the entire helium conservation program. In the mid-1970s, the Bureau of Mines began accepting privately owned crude helium for storage in the Cliffside gasfield under long-term contracts. As of 1993, about 1.0×10^9 m^3 (35.3×10^9 ft^3) of helium were stored in the U.S. Cliffside field. Private industry had a seven-month supply of helium in government storage at Cliffside (11).

Helium continues to be a strategic material, but its importance has shifted from a simple lifting gas to a unique medium essential to high technology. The cost of separation from fuel gases and of long-term storage is great, and the probability of discovering new sources is uncertain at best. Yet the future cost of satisfying all helium requirements by extraction from the atmosphere is assuredly even higher. Estimates in 1991 (11) of the total U.S. helium reserves were about 12.8×10^9 m^3. Resources having a minimum helium content of 0.3 vol % measured at 6.9×10^9 m^3; another 0.9×10^9 m^3 has been indicated. Measured resources having helium contents of less than 0.39% were 1.1×10^9 m^3; another 3.9×10^9 m^3 have been indicated.

Physical Properties

Pure Elements. All of the helium-group elements are colorless, odorless, and tasteless gases at ambient temperature and atmospheric pressure. Chemically, they are nearly inert. A few stable chemical compounds are formed by radon, xenon, and krypton, but none has been reported for neon and helium (see HELIUM GROUP, COMPOUNDS). The helium-group elements are monoatomic and are considered to have perfect spherical symmetry. Because of the theoretical interest generated by this atomic simplicity, the physical properties of all the helium-group elements except radon have been well studied.

Some of the physical properties of the helium-group elements are summarized in Table 3. The values are those for the naturally occurring isotopic mixtures of each element except for helium which has separate ^{3}He and ^{4}He entries. These two stable helium isotopes exhibit large differences in many of their physical properties. Extensive reviews of physical properties are available (21,44–46). Detailed thermodynamic property correlations and tables have been published for helium (18,47,48), neon (19,48), and argon (18,20,48). Tabulations of the unstable isotopes of the helium-group elements and the corresponding decay modes are given in References 49 and 50.

Radon is the heaviest of the helium-group elements and the heaviest of the normal gaseous elements. It is strongly radioactive. The most common isotope, ^{222}Rn, has a half-life of 3.825 days (49). Radon's scarcity and radioactivity have severely limited the examination of its physical properties, and the values given

Table 3. Physical Properties of the Helium-Group Elements

Property	^{3}He	^{4}He	Ne	Ar	Kr	Xe	Rn	References
atomic number	2	2	10	18	36	54	86	
atomic weight	3.0160	4.0026	20.183	39.948	83.80	131.30	222	14–16
critical point								
temperature, K	3.324	5.2014	44.40	150.86	209.4	289.74	378	17–22
pressure, kPa[a]	116.4	227.5	2654	4898	5502	5840	6280	17–22
density, kg/m^3	41.3	69.64	483	535.7	908	1100		14,18–22
normal bp, K	3.1905	4.224	27.102	87.28	119.79	165.2	211	17,18,20,21, 23,24
triple point (tp)								
temperature, K			24.562	83.80	115.76	161.37	202	19,21,23–25
pressure, kPa[a]			43.37	68.90	73.15	81.66	70[b]	19,23–25
density, kg/m^3								
gas at 101.3 kPa,[a] 0°C	0.1347	0.17850	0.9000	1.7838	3.7493	5.8971	9.73	21,26
gas, normal bp	23.64	16.89	9.552	5.767	8.6	11		18–21,27
liquid, normal bp	58.9	125.0	1207	1393.9	2415	3057	4400	14,18,20, 21,24,28
liquid, tp			1247	1415	2451	3084		19,20,23,24
solid, tp			1444	1623	2826	3540		21,29,30
gas:liquid volume ratio[c]	437.4	700	1340	781	644	518	452	14,21
heat of vaporization, normal bp, J/mol[d]	25.48	81.70	1741	6469	9012	12640	18100	14,16,18 20,21,23,24
heat of fusion, tp, J/mol[d]			335	1183	1640	2313	3247	16,21,23,24
heat capacity, C_p, gas at 101.32 kPa,[a] 25°C, J/(mol·K)[d]	20.78	20.78	20.79	20.85	20.95	21.01	21[b]	18,20,26, 31,32
C_s, saturated liquid, normal bp,	16.74	18.12	37.24	45.6	44.9	44.56		19,21,28 33–35

6

Property								References
$J/(mol·K)$ [d]								
sonic velocity, gas at 101.32 kPa,[a] 0°C, m/s	1122[b]	973	433	307.8	213	168		18,21,34,36
thermal conductivity, mW/(m·K) gas, 101.32 kPa,[a] 0°C	163.6[b]	141.84	46.07	16.94	8.74	5.06		21,37,38
liquid, normal bp	21.3	31.4	129.7	121.3	88.3	73.2		14,21,39
viscosity gas, 101.32 kPa,[a] 25°C, Pa·s	17.2[b]	19.85	31.73	22.64	25.3	23.1	23.3	21,39
liquid, normal bp, mPa·s(=cP)	0.00161	0.0030	0.124	0.275	0.431	0.528		21,40–42
solubility in water, 20°C, mL/kg[e]		8.61	10.5	33.6	59.4	108.1	230	21
first ionization potential, eV		24.586	21.563	15.759	13.999	12.129	10.747	43
minimum excitation energy, eV		19.818	16.618	11.548	9.915	8.315	6.772	43
stable isotopic abundance, atomic %	1.3×10^{-4} [f]; 1.7×10^{-5} [g]	100	$^{20}\text{Ne}=90.5$; $^{21}\text{Ne}=0.27$; $^{22}\text{Ne}=9.23$	$^{36}\text{Ar}=0.337$; $^{38}\text{Ar}=0.063$; $^{40}\text{Ar}=99.60$	$^{78}\text{Kr}=0.35$; $^{80}\text{Kr}=2.27$; $^{82}\text{Kr}=11.56$; $^{83}\text{Kr}=11.55$; $^{84}\text{Kr}=56.90$; $^{86}\text{Kr}=17.37$	$^{124}\text{Xe}=0.096$; $^{126}\text{Xe}=0.090$; $^{128}\text{Xe}=1.92$; $^{129}\text{Xe}=26.44$; $^{130}\text{Xe}=4.08$; $^{131}\text{Xe}=21.18$; $^{132}\text{Xe}=26.89$; $^{134}\text{Xe}=10.44$; $^{136}\text{Xe}=8.87$	no stable isotopes	4

[a] To convert kPa to psi, multiply by 0.145.
[b] Estimated.
[c] Volume of gas at 101.32 kPa and 0°C equivalent to unit volume of liquid at normal bp.
[d] To convert J to cal, divide by 4.184.
[e] mL (101.32 kPa, 0°C) dissolved per kg water having partial pressure of pertinent gas of 101.32 kPa.
[f] In the atmosphere.
[g] In wells.

in Table 3 are much more uncertain than are the values listed for the other elements.

The physical properties of argon, krypton, and xenon are frequently selected as standard substances to which the properties of other substances are compared. Examples are the dipole moments, nonspherical shapes, quantum mechanical effects, etc. The principle of corresponding states asserts that the reduced properties of all substances are similar. The reduced properties are dimensionless ratios such as the ratio of a material's temperature to its critical temperature, density to critical density, pressure to critical pressure, etc. The classical, two-parameter law of corresponding states requires that if pure materials are compared at conditions where two of reduced properties are identical, then the values of any third reduced property also will be identical. Argon, krypton, xenon, and presumably radon all obey the two-parameter law of corresponding states particularly well, a fact that is further indication of their normal behavior.

Except for helium, all of the elements in Group 18 freeze into a face-centered cubic (fcc) crystal structure at normal pressure. Both helium isotopes assume this structure only at high pressures. The formation of a high pressure phase of solid xenon having electrical conductivity comparable to a metal has been reported at 33 GPa (330 kbar) and 32 K, and similar transformations by a band-overlap process have been predicted at 15 GPa (150 kbar) for radon and at 60 GPa (600 kbar) for krypton (51).

Quantum Mechanical Effects.   The very light gases show significant deviations from the classical law of corresponding states, especially at cryogenic temperatures. This anomalous behavior is caused by quantum mechanical effects that become increasingly significant with decreasing atomic weight. Only small quantitative effects can be observed in neon's properties; the effects in hydrogen are somewhat more pronounced; but these quantum mechanical effects have profound qualitative effects on the behavior of helium at low temperature. The liquid and solid phases of the two helium isotopes exhibit physical characteristics found in no other substances.

Solid Phases.   Quantum mechanical descriptions of solids do not allow the complete cessation of molecular motion even at absolute zero. Instead, a certain zero-point motion about the molecule's average position in the crystal lattice is required. The magnitude of this motion increases both with decreasing molecular weight and with decreasing van der Waal's attraction. The helium isotopes have the largest zero-point motion of any substance. This is manifested in several unique characteristics of condensed helium (52). The heliums are the only known substances that do not freeze under their own vapor pressure. All other materials have a triple point, that unique temperature and pressure at which the solid, liquid, and vapor phases coexist. Neither helium-3 nor helium-4 has a triple point. Under moderate pressure, both remain liquid to absolute zero.

Even at the lowest temperatures, a substantial pressure is required to solidify helium, and then the solid formed is one of the softest, most compressible known. The fluid–solid phase diagrams for both helium-3 and helium-4 are shown in Figure 1 (53). Both isotopes have three allotropic solid forms: an fcc structure at high pressures, an hcp structure at medium and low pressures, and a bcc structure over a narrow, low pressure range for helium-4 and over a

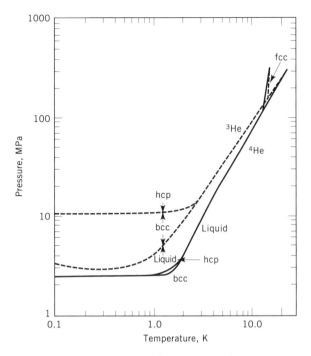

Fig. 1. Solid–liquid phase diagram for (- - -) ^{3}He and (—) ^{4}He where bcc = body-centered cubic, fcc = face-centered cubic, and hcp = hexagonal close-packed (53). To convert MPa to psi, multiply by 145.

somewhat larger range for helium-3. The melting pressure of helium-4 has been measured up to 24°C, where it is 11.5 GPa (115 kbar) (54).

The melting curves of both helium isotopes show a minimum. For helium-4, the minimum is at 2.53 MPa (25 atm) and 0.775 K (55) but the minimum is very shallow, only ca 0.7 kPa (0.1 psi) below the melting pressure at absolute zero. The minimum in the helium-3 melting curve, 0.32 K and 2.93 MPa (425 psia) (56), is much more distinct. It was predicted (57) from quantum mechanical arguments that at very low temperatures the entropy of liquid helium-3 should be less than that of the solid; hence, the melting curve slope should be negative, and the heat of solidification should also be negative. This is indeed the case, and adiabatic compression refrigerators based on the Pomeranchuk effect are used to produce temperatures in the millikelvin (mK or 0.001 K) range (see also REFRIGERATION AND REFRIGERANTS).

Liquid Helium-4. Quantum mechanics defines two fundamentally different types of particles: bosons, which have no unpaired quantum spins, and fermions, which do have unpaired spins. Bosons are governed by Bose-Einstein statistics which, at sufficiently low temperatures, allow the particles to collect into a low energy quantum level, the so-called Bose-Einstein condensation. Fermions, which include electrons, protons, and neutrons, are governed by Fermi-Dirac statistics which forbid any two particles to occupy exactly the same quantum state and thus forbid any analogue of Bose-Einstein condensation. Atoms may be thought of as assemblies of fermions only, but can behave as either fermions or bosons. If the total number of electrons, protons, and neutrons is odd, the atom is a fermion; if it is even, the atom is a boson.

Because the helium-4 atom contains an even number of fermions, it is a boson. When saturated liquid helium is cooled below 2.175 K, it undergoes what is generally recognized as the manifestation of a Bose-Einstein condensation (58). The liquid displays a striking and unique change of properties; it becomes a superfluid. A phase diagram for helium-4 is shown in Figure 2. Normal liquid helium-4 is termed helium I, and the superfluid liquid, helium II. The transition temperature varies with pressure from 2.172 K and 5.04 kPa (0.7304 psia) for the saturated liquid to 1.763 K and 3.01 MPa (437 psia) at the melting point (18,59). The superfluid transition is not first order; there is no heat of transition as such. Instead, the heat capacity of liquid helium-4 shows a sharp peak at the transition temperature. Because the peak somewhat resembles the lower case Greek letter lambda, the superfluid transition point is often called the lambda point of helium-4. The transition to the superfluid state can be clearly seen in experiments in which the temperature of a liquid helium-4 bath is reduced by continuously pumping away the helium vapors above the liquid. As the helium vapor pressure is reduced, the liquid boils vigorously until the transition temperature is reached. The liquid then abruptly becomes completely quiescent, although evolution of helium vapor from the liquid surface continues unabated.

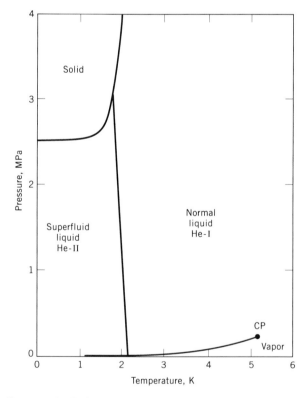

Fig. 2. Phase diagram for helium-4 where CP is the critical point. To convert MPa to psi, multiply by 145.

In many ways, helium II behaves as a liquid having a vanishingly small viscosity and a very high thermal conductivity. Detailed explanations of these properties involve complex quantum mechanical analyses (55,60,61). However, at least a phenomenological description of many of the characteristics of helium II can be gained from a two-fluid model. This model postulates that helium II is a mixture of two interpenetrating liquid components, one being similar to helium I, a reasonably normal liquid, and the other being a perfect superfluid, the concentration of the perfect zero to 0% at the lambda point. The converse is also assumed, namely, that any increase in superfluid concentration must be accompanied by a reduction in temperature.

There is assumed to be no interaction between the superfluid and normal components, thus the superfluid component can diffuse very rapidly to a heat source where it absorbs energy by reverting to the normal state. It thereby produces the very high effective thermal conductivity observed in helium II.

The viscosity of a liquid can be determined either from its rate of isothermal flow through a capillary or from the hydrodynamic drag it exerts on a rotating disk or cylinder (see RHEOLOGICAL MEASUREMENTS). Provided that the flow rate through the capillary is not too fast, such a determination indicates a vanishingly small viscosity for helium II. Viscosities of helium II determined from hydrodynamic drag experiments, however, have reasonable and finite values. The explanation based on the two-fluid model is that only the superfluid component flows through the capillary and that the normal component in the bulk liquid helium II surrounding a rotating cylinder produces the hydrodynamic drag.

When the superfluid component flows through a capillary connecting two reservoirs, the concentration of the superfluid component in the source reservoir decreases, and that in the receiving reservoir increases. When both reservoirs are thermally isolated, the temperature of the source reservoir increases and that of the receiving reservoir decreases. This behavior is consistent with the postulated relationship between superfluid component concentration and temperature. The converse effect, which may be thought of as the osmotic pressure of the superfluid component, also exists. If a reservoir of helium II held at constant temperature is connected by a fine capillary to another reservoir held at a higher temperature, the helium II flows from the cooler reservoir to the warmer one. A popular demonstration of this effect is the fountain experiment (55).

Superfluid helium can pass easily through openings so small that they cannot be detected by conventional leak detection methods. Such leaks, permeable only to helium II, are called superleaks. They can be a source of frustrating difficulties in the construction of apparatus for use with helium II.

Another unique phenomenon exhibited by liquid helium II is the Rollin film (62). All surfaces below the lambda point temperature that are connected to a helium II bath are covered with a very thin (several hundredths μm) mobile film of helium II. For example, if a container is dipped into a helium II bath, filled, and then raised above the bath, a film of liquid helium flows up the inner wall of the container, over the lip, down the outer wall, and drips from the bottom of the suspended container back into the helium II bath. Similarly, if the empty container is partially submerged in the helium II bath with its lip above the surface, the helium film flows up the outer wall of the container, over its lip, and into the container. This process continues until the level of liquid in the partially submerged container reaches that of the helium II bath.

Several oscillatory phenomena occur in helium II. Ordinary sound waves are propagated in a normal manner in helium II. To distinguish ordinary sound from the other wave-like phenomena, it is called first sound. In first sound, the superfluid and normal components oscillate exactly in phase, thus producing the density wave associated with normal sound. When the superfluid and normal components oscillate exactly out of phase, there is no periodic change in bulk liquid density, but there is a periodic change in the concentration of the superfluid component. This concentration difference is observable as a periodic change in temperature. Second sound, then, is a propagation of heat pulses that has all of the characteristics associated with wave propagation such as reflection, refraction, and interference. Third sound involves a simultaneous thermal and density wave propagation in the Rollin film in which free oscillation of the normal fluid component is inhibited by viscous wall effects. Fourth sound is the transmission of correlated thermal and density waves through densely packed beds of fine powders which effectively immobilize the normal fluid component.

Liquid Helium-3.   The helium-3 atom contains an odd number of fermions; thus it is itself a fermion. For many years, it was expected that liquid helium-3 might have relatively normal properties at even the lowest temperatures. This expectation was tempered, however, by quasisuperfluid behavior observed in another fermion system. The conduction electrons in metals are fermions, and at low temperatures, some metals become superconducting. Their resistance to direct current becomes identically zero (see SUPERCONDUCTING MATERIALS) which implies a superfluid behavior of the conduction electrons. In 1957 (63) superconductivity was explained by showing that a net attractive force between pairs of electrons could be created through their interactions with the lattice ions (63). These bound pairs formed a single quasiparticle having an even number of fermions which then behaved as a boson. This theory implied that, at low enough temperature, helium-3 atoms should also form bound pairs and show some form of superfluidity.

In 1971 a helium-3 adiabatic compression refrigerator was used to discover the superfluid transition at temperatures below 0.003 K (64). It was soon found that helium-3 has not one but three superfluid phases, and that these have properties quite different from superfluid helium-4 (65–67). The phases are magnetic, and many of the physical properties are anisotropic. A solid–liquid phase diagram showing the three primary variables of state-pressure, temperature, and magnetic field, is given in Figure 3. In the absence of a magnetic field, only two superfluid phases, A and B, exist; and there is a polycritical point (PCP) at which three liquid phases, A, B, and normal, coexist. As a magnetic field is applied to the liquid, indicated by the dashed plane in Figure 3, the polycritical point disappears; the A–B transition extends downward in pressure to the saturated liquid line, which is practically zero pressure; and a third superfluid phase, A1, interposes itself between the A and normal phases. A principal distinguishing feature of the three superfluid phases is their distinctive nuclear magnetic resonance characteristics (see MAGNETIC SPIN RESONANCE).

Mixtures.   A number of mixtures of the helium-group elements have been studied and their physical properties are found to show little deviation from

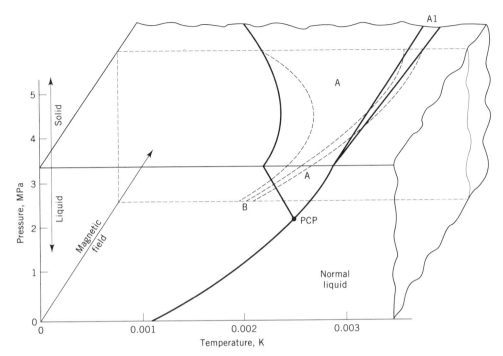

Fig. 3. Phase diagram for helium-3 where A, B, and A1 represent the three superfluid phases and PCP is the polycritical point. The dashed lines indicate the effect of an applied magnetic field. See text. To convert MPa to psi, multiply by 145.

ideal solution models. Data for mixtures of the helium-group elements with each other and with other low molecular weight materials are available (68). A similar collection of gas–solid data is also available (69).

The fundamental differences in the quantum mechanical character of the two helium isotopes created much interest in the properties of mixtures. Several reviews are available (70–72). Mixtures of isotopes of a single element usually behave quite ideally, but in the case of ^{3}He–^{4}He solutions, nonideality reaches the point of forming two immiscible liquid phases. The liquid-phase diagram for ^{3}He–^{4}He solutions at low pressure is shown as the solid curves in Figure 4. The solutions undergo the superfluid transition, but the transition temperature is depressed by increases of ^{3}He concentration. Below a solution critical point at 0.867 K and 67.5 mol % ^{3}He (71), two immiscible liquid phases can form. The ^{4}He-rich phase is superfluid and the ^{3}He-rich phase remains normal, at least to below 0.003 K. The solubility of ^{4}He in ^{3}He appears to approach zero as the temperature approaches zero, but the solubility of ^{3}He in ^{4}He does not; it remains finite (6.4 mol %) as absolute zero is approached (70).

Below about 0.5 K, the interactions between ^{3}He and ^{4}He in the super-fluid liquid phase becomes very small, and in many ways the ^{4}He component behaves as a mechanical vacuum to the diffusional motion of ^{3}He atoms. If ^{3}He is added to the normal phase or removed from the superfluid phase, equilibrium is restored by the transfer of ^{3}He from a concentrated phase to a dilute phase. The effective ^{3}He density is thereby decreased producing a heat-absorbing expansion analogous to the evaporation of ^{3}He. The ^{3}He density in the superfluid

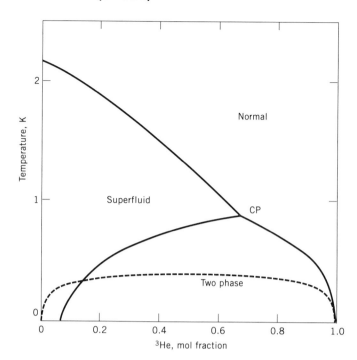

Fig. 4. Phase diagram for liquid and solid mixtures of ^{3}He and ^{4}He where CP is the critical point; (—), liquid; (---), solid.

phase, and hence its mass-transfer rate, is much greater than that in ^{3}He vapor at these low temperatures. Thus, the pseudoevaporative cooling effect can be sustained at practical rates down to very low temperatures in helium-dilution refrigerators (72).

Unusual behavior has also been observed in solid mixtures of ^{3}He and ^{4}He. In principle, all solid mixtures should separate as absolute zero is approached, but because of kinetic limitations, this equilibrium condition is almost never observed. However, because of high diffusivity resulting from the large zero-point motion in solid helium, this sort of separation takes place in a matter of hours in solid mixtures of ^{3}He and ^{4}He (53,61). The two-phase region for the solid mixture is outlined by the dashed curve in Figure 4. The two-phase dome is shallow, and its temperature maximum is 0.38 K.

Production

Helium is separated from helium-bearing natural gases usually, but not always, in the process of removing nitrogen to improve the fuel value of the gas. Thus, in a sense, the principal part of commercial helium is as a by-product. Argon, neon, krypton, and xenon, as well as small quantities of helium, are obtained from by-product streams that are concentrated during the separation of air to produce oxygen (qv) and nitrogen (qv). Radon is collected as a daughter product of the fission of radium, and helium-3 is the daughter product of tritium.

Helium from Natural Gas. Recovery of helium from a given natural-gas stream depends almost entirely on the total economic picture of the stream. In the United States, the lowest practical helium level that is recovered is most frequently 0.3 vol %, although helium is frequently ignored, and hence wasted, in streams containing somewhat high concentrations. In other parts of the world where political considerations sometimes interact with the economic, the use of helium concentrations lower than 0.3 vol % may be dictated.

Natural-gas components include water vapor, carbon dioxide (qv), sometimes hydrogen sulfide, heavier hydrocarbons (qv), methane, nitrogen, small amounts of argon, traces of neon and hydrogen, and helium. The production of pure helium from natural gas requires three basic processing steps (73).

First, is the removal of impurities. Water, carbon dioxide, and sulfides are removed by scrubbing with monoethanolamine (see ALKANOLAMINES) and diethylene glycol (see GLYCOLS), followed by drying with alumina. Then the natural gas is concentrated in helium as the higher boiling hydrocarbons are liquefied and collected. Crude helium, concentrated to perhaps 70% and containing nitrogen, argon, neon, and hydrogen, undergoes final purification at pressures up to 18.7 MPa (2700 psi). The crude material is chilled to 77 K in liquid nitrogen-cooled coils of a heat exchanger. Under the high pressure, the low temperature liquefies most of the remaining nitrogen and argon, allowing the helium together with last traces of nitrogen, neon, and hydrogen to separate. Evaporation of nitrogen reduces the temperature and nitrogen content of the helium before it passes into liquid-nitrogen-cooled adsorbers. Activated charcoal operating at liquid nitrogen temperatures or below is capable of adsorbing all nonhelium gases. Hence, passage through these adsorbers yields helium that exceeds 99.9999% in purity. Both concentration and purification steps require nitrogen refrigeration, which is obtained by expansion engines or turbines as well as by expansion valves.

A typical variation of the process suitable for low heating-value gas upgrade would operate with a 5.5 MPa (800 psig) feed followed by Joule-Thomson expansion to 2 MPa (300 psig) and double-column distillation to separate nitrogen from liquid hydrocarbons. This column would also contain a noncondensable gas stream containing helium. This process supplies its own refrigeration and eliminates the need for liquid nitrogen. The need for a 14 MPa (2000 psi) system is also eliminated if product helium is in liquefied form.

Relatively new methods for separating helium from natural gas use pressure swing adsorption (PSA) processes to recover helium at better than 99.99% purity. This type of process is probably less costly for the production of gaseous helium but might be uneconomical for liquefied helium production. The PSA process is widely used to produce specification pure helium from 85+ % crude helium in conjunction with cryogenic enrichment of the ca 50% helium raffinate.

Helium-Group Gases from the Atmosphere. Air contains 0.93% argon, lesser amounts of neon, helium, and krypton, and less of xenon (Table 1). Air is the only practical source in all of these elements except for helium. In the separation of air distillation–liquefaction process, the helium-group gases concentrate as noncondensable gases, ie, helium, neon, plus hydrogen, at liquid nitrogen temperatures; as a primary inert, ie, argon, that could reduce oxygen purity up to 5% if not removed; and as a trace inert, up to 10 ppm Kr–Xe also in the oxygen product, which may be ignored or removed for its own value (74,75).

Concentration of Rare Gas Crudes. The distillation of air is classically carried out in the double-column and auxiliary equipment of Figure 5. Dry,

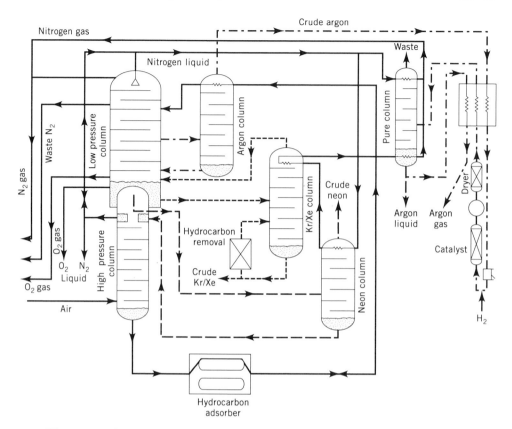

Fig. 5. Production of helium-group gases in a classical air-separation plant.

CO_2-free air, chilled to partial liquefaction by heat exchange, is introduced into the lower nitrogen or high pressure column. This unit is typically operated at 620–720 kPa (90–104 psi) pressure. As the air is introduced, partial condensation occurs; the nitrogen vapor distills up the column to become pure N_2 at the top, while the remaining liquid air having lost approximately half the original nitrogen and having as a consequence a higher boiling point, collects at the bottom. Condensation of nitrogen at the top provides the reflux necessary to rectify the gas streaming up as well as pure nitrogen liquid for reflux liquid to be supplied to the upper oxygen (or low pressure) column and product for removal to storage. Note that although the top of the lower (nitrogen) column and the bottom of the upper (oxygen) column are in thermal equilibrium; no gas or liquid can pass directly between them.

Because of their very low boiling points, helium, neon, and hydrogen are noncondensable under the conditions at the top of the nitrogen column, and they concentrate in the nitrogen gas there. Because they cut down on the rate of condensation of nitrogen and thereby reduce the thermal efficiency of the process, they must be withdrawn. The noncondensable stream withdrawn may have a neon, helium, or hydrogen content that varies from 1–12% (the balance being nitrogen), depending on plant design, objective of plant operation, and hence rate of withdrawal. For recovery, the noncondensable gases are first concentrated in the neon column shown in Figure 5, where a significant part of the nitrogen is liquefied and returned. The neon–helium crude is purified in further processing described below.

The oxygen-rich liquid (ca 40% O_2) that arrives at the bottom of the lower (high pressure) column is transferred by pressure differential to the upper column, known also as the oxygen (low pressure) column. En route to the upper column, the liquid passes through the hydrocarbon adsorber containing silica gel to remove unsaturated hydrocarbons and through a pressure dropping Joule-Thompson valve which gives added refrigeration. At its now lower temperature and pressure, this stream of rich liquid becomes the refrigerant in the crude argon column producing the reflux needed there. Rich liquid then enters the oxygen column somewhere near the middle at a pressure of ca 150 kPa (22 psia).

Of all the gases in the oxygen column, nitrogen is the most volatile and streams to the top where it is removed as a product, waste gas, or for recycle purposes. Oxygen flows to the bottom carrying dissolved traces of Kr and Xe which may concentrate to the sum of 10 ppm. These gases remain dissolved because of their much higher boiling points. Owing to differences in pressure between the two columns, heat flowing from nitrogen in the lower column causes the nitrogen there to condense, while on the upper side of the heat exchanger, the same heat causes the oxygen to boil (hence named the reboiler) to keep the rectification process active. Krypton and xenon may be tolerated as trace inert contaminants in the oxygen. To be recovered they must be substantially concentrated in the separate krypton–xenon column shown. Traces of hydrocarbons also concentrate with the heavy rare gases and, because the liquid is almost pure oxygen, it is necessary for safety to operate some continuous method of hydrocarbon removal. Removal is most frequently effected by adsorption, but it could also be achieved by catalytic combustion or by a combination of both. Flows must be adjusted to keep the concentration below the lower flammable limit but high enough to make removal efficient. In a succession of steps, the Kr–Xe crude may be concentrated to as much as a 95% mixture, the balance being small amounts of oxygen, argon, and trace amounts of other atmospheric contaminants.

This krypton–xenon mixture is usually sent to a different location for separation by distillation and further purification by catalytic and/or adsorptive processes.

Argon Separation and Purification. Argon, more volatile than oxygen and less volatile than nitrogen, concentrates in oxygen to perhaps 10% at an intermediate level in the low pressure column. From this point, which is determined by design, operational variables, and analysis, it is withdrawn into the argon concentration column, where the system, now simplified to two components, can be more effectively distilled. Beyond 98%, however, the 3 K difference in boiling points makes other means of oxygen removal practical. As indicated in Figure 5, the condenser of the argon column is conveniently refrigerated by rich liquid, condensing oxygen that, depleted of argon, returns to the column for further rectification.

Final purification of argon is readily accomplished by several methods. Purification by passage over heated active metals or by selective adsorption (76) is practiced. More commonly argon is purified by the addition of a small excess of hydrogen, catalytic combustion to water, and finally redistillation to remove both the excess hydrogen and any traces of nitrogen (see Fig. 5) (see EXHAUST CONTROL, INDUSTRIAL). With careful control, argon purities exceed 99.999%.

The removal of 2% oxygen requires more equipment and is costly in material usage and operating expenses. Industry is attempting to reduce these costs via several approaches. By improving the argon column the oxygen impurities can be reduced to <10 ppmv. This can be coupled with an improvement in the efficiency of the low pressure column to remove most of the nitrogen impurities (77,78). This allows the production of pure argon without the crude argon compressor, a Deoxo-type unit, and the pure argon distillation column.

Another variation would be to improve the argon column to produce argon containing considerably less than 1% by volume of oxygen that could be removed by catalysis. The need for the catalysis and reaction gases can be removed by using adsorbents to remove the nitrogen and oxygen impurities from the crude argon (79).

Argon also can be recovered from the spent gases remaining from air-consuming processes such as the production of hydrogen by reforming natural gas (see HYDROGEN) or from ammonia synthesis (see AMMONIA) in which argon is concentrated both in the production of the hydrogen and the nitrogen (80).

Neon–Helium Separation and Purification. As indicated earlier, neon, helium, and hydrogen do not liquefy in the high pressure (nitrogen) column because these condense at much lower temperatures than nitrogen. As withdrawn, the noncondensable stream has a neon–helium content that varies 1–12% in nitrogen, depending on the rate of withdrawal and elements of condenser design and plant operation.

In the neon column, the nitrogen content is reduced by passing the 620–720 kPa (90–104 psia) stream through a column under reflux with some colder, lower pressure nitrogen obtained from the top of the upper oxygen column. A composition typical of the resulting crude neon is 52% neon, 18% helium, 2% hydrogen, and balance nitrogen, although it varies with adjustments and plants. The crude neon may be vented to the atmosphere or processed in a neon separator and stored as compressed gas to await final purification. With only 18×10^{-6} m^3 (6.4×10^{-4} ft^3) neon and 5.5×10^{-6} m^3 (1.9×10^{-4} ft^3) of helium in each cubic meter (35 ft^3) of air entering the air separation process, the small quantities of crude collected in even a large air separation unit may be easily appreciated. It is sometimes desirable, therefore, to combine crudes collected from several air plants and to process them at a centralized location in specialized equipment.

Resolution of the noncondensable gas mix may be accomplished in a variety of ways, pure neon being the usual objective. In some processes, hydrogen is removed by catalytic oxidation, sufficient oxygen being introduced for the purpose; by diffusion through palladium; or by fractional adsorption on charcoal. Any water formed must be frozen out or removed by adsorptive drying. Most of the nitrogen is commonly removed by liquefaction from the pressurized crude in a liquid-nitrogen-cooled trap, followed by adsorption on liquid-nitrogen-cooled charcoal. Cold, well-activated charcoal completely adsorbs all gases other than neon and helium. Neon itself does adsorb on charcoal to some small extent, whereas helium adsorption is negligible. The final product of these steps, frequently called first-run neon, approximates 75% neon, 25% helium, and is free of all other gases. In older processes, this mixture was finally separated by differential adsorption on cold charcoal. With the increasing availability of neon crudes and very low temperature refrigeration processes, equipment in which the neon is condensed near its boiling point has come into increased use. As a result, most neon first appears as the liquid in purities that exceed 99.995%, the sole residual impurity being helium. At this point the uncondensed helium contains substantial amounts (up to 10%) of pure neon as its only diluent. Neon can be removed completely by additional fractional condensation followed by adsorption of neon on charcoal at temperatures below 77 K. In the United States, economics favor wasting most of this atmospheric helium. It remains of interest, however, because it contains about 10 times as much helium-3 as does helium from natural gas. Additionally, as helium-bearing natural gases become depleted, the atmosphere is expected to be the only remaining source. There has been much speculation about dedicated plants required to produce helium when natural-gas sources are exhausted, and some designs have been made. Only one plant is known to have been built exclusively for helium. That plant, put into operation in Germany in 1972 (81), obtains helium from ammonia synthesis gases.

Krypton–Xenon Purification and Separation. Highly concentrated Kr–Xe crudes are purified by removing residual hydrocarbons as described, then removing CO_2 and H_2O; batch distilling to remove completely other atmospheric components, and then to separate krypton and xenon. Inert atmospheric pollutants may survive the many steps of the purification process (82). In particular, tetrafluoromethane [75-73-0], CF_4, is sometimes found in pure krypton to the extent of a few parts per million (see FLUORINE COMPOUNDS, ORGANIC) indicating an atmospheric level of CF_4 on the order of 10^{-2} ppm. It has been postulated that CF_4 results from the reaction of carbon with flurospar, CaF_2, in the reducing gases of iron (qv) smelters. Removal of CF_4 from Kr can be accomplished by proprietary processes when its presence is detrimental to the intended use of the Kr. Crude Kr–Xe mixtures have also been found to contain varying levels of other halogenated compounds, oxides of nitrogen, and random impurities.

Krypton and xenon are recoverable from large air-separation units only when these plants have expensive additional processing equipment. Limited quantities are produced with much more effort. In European countries and the CIS, certain plants have been designed for the selective recovery of krypton and xenon from air. Several types of processes have been designed and constructed (83), but basic to all of them is the scrubbing of very large quantities of air using a relatively small quantity of liquid air. The krypton–xenon content of a large volume of air is thus concentrated and can be more readily and economically refined. Similarly, as for the argon and helium sources discussed above, air-consuming processes concentrate Kr–Xe and thus become attractive sources. Although many patents have been issued, only a few plants have been built. Purification details depend on the quality of the crude.

Krypton and Xenon from Nuclear Power Plants. Both xenon and krypton are products of the fission of uranium and plutonium. These gases are present in the spent fuel rods from nuclear power plants in the ratio 1 Kr:4 Xe. Recovered krypton contains ca 6% of the radioactive isotope Kr-85, with a 10.7 year half-life, but all radioactive xenon isotopes have short half-lives.

Separation of krypton and xenon from spent fuel rods should afford a source of xenon, technical usage of which is continuously growing (84). As of this writing, however, reprocessing of spent fuel rods is a political problem (see NUCLEAR REACTORS). Xenon from fission has a larger fraction of the heavier isotopes than xenon from the atmosphere and this may affect its usefulness in some applications.

Radon Separation.   Owing to its short half-life, radon is normally prepared close to the point of use in laboratory-scale apparatus. Radium salts are dissolved in water and the evolved gases periodically collected. The gas that contains radon, hydrogen, and oxygen is cooled to condense the radon, and the gaseous hydrogen and oxygen are pumped away.

Commercial Distribution

Commercially pure ($\geq$99.997%) helium is shipped directly from helium-purification plants located near the natural-gas supply to bulk users and secondary distribution points throughout the world. Commercially pure argon is produced at many large air-separation plants and is transported to bulk users up to several hundred kilometers away by truck, by railcar, and occasionally by dedicated gas pipeline (see PIPELINES). Normally, only crude grades of neon, krypton, and xenon are produced at air-separation plants. These are shipped to a central purification facility from which the pure materials, as well as smaller quantities and special grades of helium and argon, are then distributed. Radon is not distributed commercially.

The helium-group gases are distributed as gaseous and cryogenic liquid products in containers having capacities ranging from less than 0.001 m^3 (0.035 ft^3) to about 45,000 m^3 (1.6 $\times$ 10^6 ft^3).

Most gaseous products are distributed in forged steel or aluminum alloy cylinders at high pressure, typically 15 MPa (2200 psi), but ranging from 6 to 41 MPa (900–6000 psi). Small quantities of research-grade materials are still sold in glass flasks at atmospheric pressure. The (U.S.) Compressed Gas Association (CGA) specifies different cylinder valve connectors for different types of gases. For the helium-group gases the connectors used are CGA 100 for lecture bottles, CGA 580 for 21 MPa (3000 psig), CGA 680 for 24 MPa (3500 psig), and CGA 677 for 41 MPa (6000 psig) cylinders.

For distributing larger quantities of gaseous helium, argon, and occasionally neon, a number of large, horizontal, compressed gas cylinders are manifolded on truck semitrailers (called tube trailers) or railroad cars. Like individual cylinders, these serve both as transport containers and rental storage containers. Capacities of tube trailers range from about 300 to 5,000 m^3 (10,000–175,000 ft^3) of gas.

Bulk quantities of helium and argon and smaller quantities of neon are distributed as cryogenic liquids in insulated containers of liquid capacities of 10–56,000 L (2.5–14,800 gal). Although the cryogenic refrigeration of the liquid is sometimes utilized as for magnetic resonance imaging (MRI), another justification for this mode of distribution is the lower transportation costs resulting from reduced ratios of shipping container weight to product weight (see IMAGING TECHNOLOGY). This factor is particularly important for such a low density material as helium. For example (85), the weight of compressed-gas cylinders is about 45 times that of their helium content; on the other hand, the weight of a 500-L cryogenic liquid container is only about eight times the weight of the helium, and large liquid–helium semitrailers are only about four times heavier than their helium load.

Because of liquid helium's uniquely low temperature and small heat of vaporization, containers for its storage and transportation must be exceedingly well insulated. Some containers are insulated with only a fairly thick layer of very efficient insulation, but containers with the least heat leak use an inexpensive sacrificial cryogenic liquid, usually liquid nitrogen, to shield thermally the liquid helium contents.

Two types of insulation are used in liquid helium containers. Simple high vacuum insulation, exemplified by the common thermos bottle, consists of a well-evacuated space between highly reflective walls; however, its limited effectiveness restricts its use to liquid nitrogen-shielded containers. Highly effective, evacuated multilayer insulations are constructed from laminar arrangements of many radiation shields, metal foil or metal-coated plastic film, in a vacuum envelope. Thermal contact between shields is minimized either by interposing a low conductivity, fibrous separating membrane between shields or, in the case of coated plastic film, by crinkling the film so that adjacent layers touch only at a few, discrete, comparatively widely spaced points. A third type of cryogenic insulation, evacuated, low density, mineral powders, such as expanded perlite, is not often used in liquid helium containers but it is used for large, liquid argon vessels.

Most small liquid helium containers are unpressurized: heat leak slowly boils away the liquid, and the vapor is vented to the atmosphere. To prevent plugging of the vent lines with solidified air, check valves of some sort are included in the vent system. Containers used for air transportation are equipped with automatic venting valves that maintain a constant absolute pressure with the helium container in order to prevent liquid flash losses at the lower pressures of flight altitudes and to prevent the inhalation of air as the pressure increases during the aircraft's descent. Improved super insulation has removed the need for liquid nitrogen shielding from almost all small containers.

For shipping large (>5000 L) quantities of liquid helium over great distances, nonvented pressurized containers are used. At the purification plant these are almost filled with liquid helium at the atmospheric boiling point and sealed. As heat leaks into the container, no product is vented; instead the pressure is allowed to build up as the fluid temperature increases. Often the interior pressure and temperature are above the critical point upon arrival at the destination, and the contents are no longer a liquid but a cold, dense, supercritical fluid. However, by careful venting procedures, a substantial part of the contents can be recovered as liquid, and the vented vapor is compressed into cylinders. For deliveries involving only a week or so transit time, eg, deliveries with the continental United States, large trailers with heat leaks limited to about 30 W by evacuated multilayer insulation are used. For more extended shipments, eg, by freighter from the United States to Japan, which may involve nearly a month's transit time, up to 40,000-L capacity, truck-transportable containers are used. Their heat leak is held to ca 8 W by a combination of liquid–nitrogen shielding and evacuated multilayer insulation, and their liquid–nitrogen consumption is ca 40 L/d.

Liquid argon is distributed by truck from the point of production using large cryogenic tank trailers similar to those used for liquid oxygen and nitrogen. At each stop on the delivery route, liquid argon is transferred into fixed, well-insulated, cryogenic storage tanks called customer stations from which gaseous or liquid argon is withdrawn as needed (see CRYOGENICS; INSULATION, THERMAL).

Economic Aspects

The world production statistics for high purity helium are listed in Table 4. The peak in production during the 1960s reflects the large helium usage by the U.S. space program. Most of this production was by the U.S. Bureau of Mines. Private plants began significant production in 1963 and have taken an ever increasing share of the market. Since 1987, the market for U.S.-produced helium has grown at an average annual rate of 9%. Exports of helium started in the early 1960s, and became a significant fraction of total production only after the development of equipment and techniques for transoceanic shipment of bulk liquid helium. The foreign market made up about 31% of U.S. helium sales in 1991.

The U.S. production of argon is summarized in Table 5. Because argon is a by-product of air separation, its production is ca 1% that of air feed. Total 1988 United States consumption of neon, krypton, and xenon was 36,400, 6,800, and 1,200 m^3, respectively (88).

Comparative prices of helium-group gases in 1993 are listed in Table 6. The prices for Ne, Kr, and Xe dropped significantly from 1990 as a result of Russian exports of helium-group gases. Prices in 1990 were $120/$m^3$ for Ne, 700–800/m^3 for Kr, and $5,750–8,750/$m^3$ for xenon.

Table 4. Annual U.S. Production and Exports of High Purity Helium, 10^6 m^3 [a,b]

Years[c]	Production		Total	Exports
	Bureau of Mines	Private plants		
1960–1964	16.7	0.5	17.2	0.1
1965–1969	16.7	7.3	24.0	1.6
1970–1974	5.2	12.9	18.1	3.4
1975–1979	5.7	20.1	25.8	5.2
1980–1884	7.2	30.6	37.8	10.3
1985	11.3	42.1	53.4	12.4
1986	9.4	45.6	55.0	12.1
1987	7.4	54.5	61.9	14.0
1988	8.8	62.6	71.4	18.4
1989	9.7	70.2	79.9	22.1
1990	11.2	73.7	84.9	24.7
1991	9.4	78.7	88.1	27.1
1992	8.6	85.8	94.4	27.1
1993	9.0	90.0	99.0	30.7

[a]To convert 10^6 m^3 to 10^6 ft^3, multiply by 35.3.
[b]Reference 86.
[c]Values for time periods exceeding one year are average values.

Table 5. Annual U.S. Production of Argon[a]

Years[b]	Production, 10^6 m^{3c}
1960–1964	23.1
1965–1969	54.5
1970–1974	105.3
1975–1979	173.4
1980–1984	237.5
1985–1989	318.4
1990	364.3
1991	377.8
1992	394.4
1993	427.2

[a]Ref. 87.
[b]Values for time periods exceeding one year are average values.
[c]To convert 10^6 m^3 to 10^6 ft^3 multiply by 35.3.

Table 6. 1993 U.S. Prices of Helium-Group Gases[a]

Gas	Price,[b] $/m^3$
Bulk quantities	
helium	
gas	
U.S. Bureau of Mines	1.94
private	1.77
liquid	
private	2.30
argon	0.76
Laboratory quantities	
helium[c]	4.20–4.90[d]
	22.28–44.84[e]
neon	70
argon[c]	2.63–8.47
krypton	350
xenon	3500

[a]Ref. 88.
[b]To convert $/m^3$ to $/ft^3$, multiply by 0.0283.
[c]Varying purity levels.
[d]Industrial grade.
[e]Laboratory quality.

Specifications and Standards

In the United States, government agencies and trade organizations have established standards and specifications for helium and argon (Tables 7 and 8). Typical commercial purities for neon, krypton, and xenon are shown in Table 9. In general, all grades delivered from a given facility are prepared from a single product stream. The differences between grades lie in the care taken in subsequent handling, ie, the preparation of containers, elimination of intrusion of air and other gases, and the sampling and analytical efforts. These factors assure

Table 7. Selected Commercial Specifications for Helium Compressed Gas Association Grades[a]

Specification	M	N	P	G
minimum purity, %	99.995	99.997	99.999	99.9999
maximum impurities, μL/L				
total	50	30	10	1
water	9	3	1.5	
THC as CH_4[b]	5	1	0.5	
oxygen	3	3	1	
nitrogen + argon	14	5[c]	5	
neon	23	23	2	
hydrogen	37[d]	1	1	
CO/CO_2			0.5	

[a]Ref. 11.
[b]Total hydrocarbons (THC) reported as methane.
[c]Maximum 5 μL/L nitrogen only.
[d]Value noted is combined hydrogen, carbon monoxide, and dioxide.

Table 8. Selected Commercial Specifications for Argon

Specification	Compressed Gas Association grades[a]				
	B	C	D	E	F
minimum purity	99.996	99.997	99.998	99.999	99.9985
maximum impurity, μL/L					
total	40	30	20	10	15
water	14.3	10.5	3.5	1.5	1
THC as CH_4[b]	5[c]	3[c]	0.5	0.5	0.5
oxygen	7	5	2	1	2
nitrogen	15	20	10	5	10
hydrogen	1	1	1	1	1
CO_2			0.5	0.5	0.5[d]

[a]Ref. 89.
[b]Total hydrocarbons (THC) reported as methane.
[c]Value noted in combined total hydrocarbons and carbon dioxide.
[d]Value noted is combined carbon dioxide and carbon monoxide.

the quality of the gas but add to its cost. Because all other substances are solid at liquid helium temperatures, liquid helium filtered through a nominal 10-μm filter is accepted as pure helium.

Analysis

The most common techniques for measuring the concentrations of helium-group gases in mixtures are gas–solid chromatography (qv) (91) and mass spectrometry (qv) (92). The sensitivity of both techniques can be increased many-fold by concentrating the helium-group components by such methods as selective adsorption (qv) or removal of the noninert components using hot reactive metals (89) or other gettering means.

Table 9. Purities of Commercial Neon, Krypton, and Xenon[a]

Purity	Neon		Krypton		Xenon	
minimum purity, %	99.996	99.999	99.997	99.999	99.997	99.999
maximum impurity, μL/L						
total	40	10	30	10	30	10
helium	30	8				
xenon			20	5		
krypton					25	5
nitrogen	4	2	10	3	5	3
oxygen	1	0.5	2	1	1	0.5
CO/CO$_2$	0.5	0.5	2	2	1	1
THC as CH$_4$[b]	0.5	0.5	1	1	1	0.5
water	1	1	1	1	1	0.5

[a]Ref. 90.
[b]Total hydrocarbons (THC) reported as methane.

The determination of impurities in the helium-group gases is also accomplished by physical analytical methods and by conventional techniques for measuring the impurity in question (93), eg, galvanic sensors for oxygen, nondispersive infrared analysis for carbon dioxide, and electrolytic hygrometers for water.

Uses

In 1991, the primary domestic uses of helium (9) were magnetic resonance imaging (20%) and other cryogenics (10%), cryogenics (30%), welding (qv) (21%), and pressurizing and purging (11%). Minor uses included controlled atmospheres (5%), breathing mixtures (3%), leak detection (6%), and other uses such as lifting gas, heat transfer, etc.

The main uses for argon are in metallurgical applications and in electric lamps. Neon, krypton, and xenon, because of high costs, are limited to specialized uses in research, instrumentation, and electric lamps. There are no significant technical uses for radon.

Metallurgical. To prevent reaction with atmospheric oxygen and nitrogen, some metals must be shielded using an inert gas when heated or melted (94). Applications in metals processing account for most argon consumption and an important part of helium usage (see METALLURGY).

In gas tungsten-arc welding (95), also called tungsten inert gas (TIG) welding, the nonconsumable tungsten electrode, hot filler metal, and weld zone are protected by a continuous stream of helium, argon, or their mixtures. The arc is struck between the electrode and the workpiece. Helium shielding produces a hotter arc than does argon shielding. When the electrode is positive and argon or a predominantly argon mixture is used for shielding, a cleaning action occurs on the surface of the weld zone that does not occur with a shield of helium. Argon-shielded, alternating-current arcs produce the same surface-cleaning action, and the electrode has a higher current-carrying capacity than a continuously positive electrode. Helium shielding, however, decreases the stability of alternating-current arcs.

In gas metal-arc, also called metal inert gas (MIG) welding, the arc is struck between the workpiece and a metal electrode that is consumed as it is transferred in the form of molten droplets across the arc into the weld joint. Using a predominantly argon shield gas, the droplets are transferred with little radial dispersion. Using pure helium, on the other hand, the droplets are larger and have more of a tendency to spray radially from the arc's axis.

Argon, helium, and their mixtures with other gases are used as the working fluids in plasma arc devices for producing plasma jets with temperatures in excess of 50,000 K. These devices are used for cutting metals and for spray coating of refractory alloys and ceramics (qv) (see PLASMA TECHNOLOGY).

Large quantities of argon are used in the argon–oxygen decarburizing (AOD) process for stainless-steel production (96). A charge containing stainless and carbon-steel scrap and low cost, high carbon ferrochrome is melted in an arc furnace and then bottom-blown with an argon–oxygen mixture in a separate AOD vessel. The oxygen performs the necessary decarburizing, and the argon serves to reduce the partial pressure of the carbon monoxide reaction product, thereby selectively increasing the rate of carbon oxidation relative to that of chromium. Because of the arc furnace's lower materials costs, reduced chromium losses, and increased productivity, more than 90% of United States stainless-steel production is by the AOD process (86) (see STEEL).

Argon bubbles are frequently used to stir iron and steel ladles to prevent stratification, to help remove dissolved gases, to remove potential inclusion by flotation (qv), and to control temperatures. Argon is used as an inert blanket in melting, casting, and annealing certain alloys. Helium and argon environments are used for high temperature refining and fabrication of specialty materials such as zirconium, niobium, tantalum, titanium, uranium, thorium, plutonium, and reactor-grade graphite. Some furnace brazing, soldering, and powder-metal sintering operations (97) use helium and argon atmospheres either alone or in mixtures with hydrogen when a reducing atmosphere is required. Jets of argon can be used to atomize molten reactive metals to produce metal powders (98). In some metallurgical reactions, argon or helium is used as a carrier to transport products to and from the reaction zone and as an inert diluent to modify reaction rates. Typical applications are the Kroll process for making zirconium and titanium (99) and the submerged injection of pulverized reagents such as lime and calcium carbide into molten steel.

Low pressure argon is the usual medium for industrial sputtering of metals and other solid films (100) (see THIN FILMS, FILM FORMATION TECHNIQUES).

Laboratory and Scientific.   The helium-group gases are closely associated with precision measurement techniques. The international standard of length, the meter, is defined in terms of a spectral line of krypton-86. The boiling point of neon is a defining point on the international practical temperature scale (101). The vapor pressures of the two helium isotopes, helium-3 (17) and helium-4 (102), furnish an internationally recognized temperature scale below 5.2 K, and the vapor pressures of neon (16), argon (103), krypton (23), and xenon (24) afford convenient and reliable secondary temperature scales. Precision gas thermometers, one of the fundamental devices for defining temperatures, normally use helium as a fill gas because of its small deviations from ideal gas behavior (see TEMPERATURE MEASUREMENT). Quantum superconducting devices (SQUIDs) are used increasingly for precise and fundamental measurements (see SUPER-CONDUCTING MATERIALS), and the small, neon–helium, continuous-wave lasers (qv) are widely used for interferometric techniques and for precision alignments ranging from laboratory equipment to large tunnel bores.

Helium is the most commonly used carrier gas for gas–liquid and gas–solid chromatography (104). It is chemically inert and minimally dissolved or adsorbed on the column packing. Its low viscosity and density minimize pressure drop. But its primary advantage is that, because of its low molecular weight, its physical properties differ greatly from those of most materials being analyzed. Using a helium carrier thus yields maximum sensitivity for detectors based on changes in thermal conductivity, sonic velocity, density, etc. Argon is also used as a carrier gas, especially when analyzing samples containing hydrogen or in regions outside the United States where helium is more expensive. Thermal conductivity detectors, particularly, respond in a highly nonlinear fashion to hydrogen peaks in a helium carrier.

The helium leak detector is a common laboratory device for locating minute leaks in vacuum systems and other gas-tight devices. It is attached to the vacuum system under test; a helium stream is played on the suspected leak; and any leakage gas is passed into a mass spectrometer focused for the helium-4 peak. The lack of nearby mass peaks simplifies the spectrometer design; the low atmospheric background of helium yields high sensitivity; helium's inertness assures safety; and its high diffusivity and low adsorption make for fast response.

Detectors for ionizing radiation use the helium-group gases; mixtures containing neon, argon, and helium are used in Geiger-Muller counters; argon and xenon are used in proportional counters; and electron avalanche in liquid xenon (105) is the basis for very accurate photon detection (see PHOTODETECTORS). In neutron detectors, helium-3 has a high cross section to neutrons for the formation of tritium whose decay is then detected by a conventional counter (4). Krypton is frequently used to determine surface areas of fine solids by adsorption techniques. And helium and argon are often used as protective atmospheres when working with highly reactive materials.

Light Sources. In common incandescent lamps, the metal filament, usually tungsten, is protected by a glass bulb filled with a gas mixture, usually argon–nitrogen (106,107). Argon's higher molecular weight compared to nitrogen's helps reduce the rate of sublimation of the glowing tungsten filament, thereby prolonging its life, and argon's lower thermal conductivity reduces thermal losses, thereby improving the electrical efficiency. The arcing tendency of argon, however, necessitates the use of some (typically 12%) nitrogen in the fill gas. Krypton's still higher molecular weight and lower thermal conductivity makes it technically superior as a fill gas, but its higher cost has limited use to special-purpose lamps and to those parts of the world where electrical energy is especially expensive. Krypton-filled light bulbs are appearing on the U.S. consumer market.

Low pressure gas-discharge tubes consist of a sealed glass tube with an electrode at each end and a filling of gas at about 1 kPa (0.15 psia) absolute pressure. When a sufficiently high, alternating voltage is impressed across the tube, a low pressure plasma fills the tube bore and emits radiation at wavelengths dependent primarily on the composition of the fill gas. A familiar form of this light source is the neon sign used for advertising displays, a term that has been generalized to include any colored tubular lighting display regardless of the fill gas used. In clear glass, pure neon yields a red-orange light; helium produces a yellowish-white; argon emits a bluish color; helium–argon forms an orange tint; and neon–argon makes a deep lavender glow. By using these various filling gas mixtures in conjunction with colored glass tubes and sometimes a bit of mercury vapor, practically the entire spectrum of colors can be produced (106). Small, low pressure gas-discharge bulbs are widely used for indicator lights.

Neon, argon, and krypton are used to initiate and sustain the arc in metal-vapor discharge tubes, notably various mercury- and sodium-vapor lamps. The most familiar of these are the tubular fluorescent lamps in which a mercury-vapor discharge produces radiation that is converted to visible light by a phosphor coating on the tube wall (see LUMINESCENT MATERIALS). Most fluorescent tubes are filled to roughly 500 Pa (0.07 psia) absolute pressure with about a half-and-half mixture of argon and krypton. Krypton produces the greatest luminous efficiency, but it can be troublesome to start. Neon yields the greatest absolute light output, but it requires a higher starting voltage and tends to erode the electrodes by ion bombardment. Argon is used to start the arc and vaporize the mercury in high pressure mercury-arc lamps, and neon performs the analogous function in high pressure sodium-vapor lamps. Xenon, with or without mercury, is used at relatively high pressure in short-arc lamps which provide a highly concentrated light source.

Intense and brief light pulses are obtained by discharging a capacitor through xenon-filled flash lamps. Small lamps of this type are widely used for photography because light color approximates daylight, and the brief light pulse (about a millisecond duration) effectively freezes all motion. Stroboscope lamps are similar to flash lamps but are designed for repetitive discharges. Although xenon produces the greatest light intensity, neon is frequently used because of the contrast to daylight provided by its red color.

The helium-group gases are used throughout laser technology (108). From the first, large xenon flash lamps have been a common means for exciting pulsed lasers such as the ruby and neodymium glass devices. The neon–helium continuous lasers are widely used for optical applications that require neither high efficiency nor powers above a fraction of a watt. The simplicity and reliability derived from the use of a low pressure, glow discharge for excitation plus the perfect chemical stability of the lasing medium account for the popularity of these devices. Noble-gas ion lasers operate at very high power densities and over a wide frequency range (109). Carbon dioxide lasers are the most powerful and efficient of the common continuous lasers; the lasing gas is a carbon dioxide mixture which, among several other components, includes helium to increase gas cooling rates (110).

The efficiency of a helium–neon laser is improved by substituting helium-3 for helium-4, and its maximum gain curve can be shifted by varying the neon isotopic concentrations (4). More than 80 wavelengths have been reported for pulsed lasers and 24 for continuous-wave lasers using argon, krypton, and xenon lasing media (111) (see LASERS).

Electronic.   The largest use of helium-group gases in the electronics industry is for the manufacture of semiconducting devices. The starting materials, single crystals of ultrapure semiconductors (qv) such as silicon or germanium, are grown from melts using helium or argon as both a protective atmosphere and a heat-transfer medium. Only these very pure and inert gases can ensure the extremely high crystal purities required. These same inert gases are also used as diluents and carriers for the dopant gases such as arsine or phosphine used to form the n and p regions in the semiconducting devices.

Semiconducting devices have replaced many of the gas-filled electronic tubes, and this use for the helium-group gases continues to dwindle. Helium, neon, and argon are still used in certain high voltage switching and regulating tubes and for some lightning arrestors. Traces of krypton-85 are used in some of these tubes to furnish a continuous supply of gas ions which serve to stabilize the starting voltage.

Cryogenics. The usefulness of a substance as a low temperature working fluid is limited by its boiling point and freezing point. Cost and safety must also be considered. Krypton, xenon, and radon are seldom used as cryogenic fluids. Boiling argon is occasionally used for isothermal baths near the oxygen boiling point where oxygen might be too hazardous. Similarly, liquid-neon baths are sometimes used when the use of liquid hydrogen is judged to be too dangerous. However, both gaseous and liquid helium are quite widely used in cryogenic applications (see CRYOGENICS).

Gaseous helium is commonly used as the working fluid in closed-cycle cryogenic refrigerators because of chemical inertness, nearly ideal behavior at all but the lowest temperatures, high heat capacity per unit mass, low viscosity, and high thermal conductivity.

Below hydrogen's triple point temperature of 13.80 K, only helium exists as a dense fluid, and at lower temperatures, it is the only substance that retains appreciable heat capacity. Many cryogenic systems operate near the 4.22 K boiling point of helium-4. Lower temperatures can be maintained using vacuum pumps to lower the boiling pressure; however, once the superfluid transition occurs, the Rollin film phenomenon severely impedes the pumping process making it very difficult to attain temperatures much below 1 K by this technique (112). Helium-3 is more volatile than is helium-4, and in the range of interest, its use is not complicated by the Rollin film. By boiling liquid helium-3 under vacuum, temperatures down to ca 0.25 K can be attained (113). Helium-3 dilution refrigerators can be operated in either batch or continuous mode (58) and can reach temperatures as low as 0.005 K (114). Adiabatic helium-3 compression refrigerators based on the Pomeranchuk effect can produce temperatures below 0.002 K (37), the lowest temperatures presently attainable by mechanical means. Still lower temperatures are produced by various adiabatic demagnetization processes precooled with one of the helium refrigeration processes.

Several applications of cryogenic temperatures are based on properties that vary more or less smoothly with temperature. The electrical conductivity of pure metals increases as the temperature is lowered, allowing cryogenically cooled, normal metals (aluminum, copper, or sodium) to be used for electromagnets (115) and in electrical power distribution (116). The vapor pressures of all substances decrease with decreasing temperature; at the helium boiling point, only helium itself retains any substantial vapor pressure. Helium-cooled panels form cryopumps used for high capacity vacuum pumping systems or for smaller systems where very clean pumping is required (117). Even residual helium gas can be removed effectively by an adsorbent such as activated charcoal cooled to helium temperature. Large cryopumps are used to sustain vacuums in space-simulation chambers and in the neutral-hydrogen beam injectors and plasma chambers of experimental hydrogen fusion reactors (118,119). The thermal noise in electrical components decreases at low temperature. Many radio, and other radiation-detecting devices use neon or helium cooling to enhance their sensitivities (120).

Other applications of cryogenic temperatures involve a phenomenon that does not exist at normal temperatures. A number of metals and alloys abruptly change from normal electrical conductors into superconductors as they are cooled into the helium temperature range. In the superconducting state, all resistance to direct current disappears, and the resistance to alternating currents becomes very small. The current density and magnetic field that a superconductor can sustain increases as the material is further cooled below its transition temperature. Thus the lowest temperatures practical are desirable in most applications of superconductivity.

Large, helium-cooled superconducting electromagnets are used for bubble chambers, for bending and deflecting magnets in particle accelerators, for plasma confinement in fusion reactors, and for magnetohydrodynamic power generators (121) (see MAGNETOHYDRODYNAMICS). Experimental systems using superconducting coils to levitate high speed vehicles magnetically have been investigated (122). Superconducting windings have been built for motors and generators (123), and experimental superconducting cables have been built for direct- and alternating-current power transmission (121). Smaller superconducting magnets are used for magnetic separating devices (124) and in certain instruments requiring intense, stable, and uniform fields (see SUPERCONDUCTING MATERIALS; SEPARATION, MAGNETIC).

Because of the tonnage quantities of fluid involved (125), large superconducting machinery must be cooled with commercial helium. Most commonly, such machines are cooled by liquid helium boiling near atmospheric pressure. The principal limitations are necessary to provide channels for vapor transport and the film boiling instability if local heat fluxes exceed a critical value (46). Other cooling systems utilize forced-flow convention with single-phase, compressed fluid helium, often called supercritical helium, which does not suffer from a critical heat-flux limitation but which does have a definite total heat load limitation. A few large systems have been designed for cooling by free convection of compressed fluid helium, by gas–liquid forced convection, and by conduction through superfluid helium (126). Heat- and mass-transfer (qv) rates in bulk fluid helium have been concisely reviewed (46).

Nuclear Reactors.   The first generation of gas-cooled fission reactors used coolant temperatures of only ca 400°C which permitted the use of inexpensive carbon dioxide as the coolant gas. For improved thermal efficiency, second-generation reactors were designed to use gas-coolant temperatures up to 790°C, temperatures at which only helium offers the necessary chemical stability, inertness, high heat-transfer rates, low aerodynamic pressure losses, and low neutron cross section (^{4}He only) (127). Although gas-cooled reactors claim certain operating advantages and relative immunity from loss-of-coolant accidents, only one commercial power generating reactor of this type (Fort St. Vraine, Colorado) is in operation (128).

Different combinations of stable xenon isotopes have been sealed into each of the fuel elements in fission reactors as tags so that should one of the elements later develop a leak, it could be identified by analyzing the xenon isotope pattern in the reactor's cover gas (4). Historically, the sensitive helium mass spectrometer devices for leak detection were developed as a crucial part of building the gas-diffusion plant for uranium isotope separation at Oak Ridge, Tennessee (129), and helium leak detection equipment is still an essential tool in nuclear technology (see DIFFUSION SEPARATION METHODS).

Programs for the development of fusion power provide applications for the helium-group gases. The magnetic plasma confinement approaches require massive quantities of liquid helium to cool the large superconducting magnets and cryopumps (130). In the laser-ignited, inertial confinement approaches, xenon discharge lamps are a common means for energizing the large pulsed lasers. To improve energy yields, liquid helium is used to freeze the tritium fuel into a uniform, thin solid film on the inner surface of the glass microballoon targets (131) (see FUSION ENERGY).

In particle-track detectors, neon and helium are the common filling gases for spark chambers and streamer chambers (132), and liquid helium, neon, neon–hydrogen mixtures, and xenon have been used in bubble chambers (133). Xenon provides a very heavy target nucleus useful for detecting photons and neutral particles. Because of their similar physical properties, neon–hydrogen mixtures can be used in dual-fluid bubble chambers with liquid hydrogen (134). The liquid neon charge in the 4.6-m diameter bubble chamber at Fermilab (135) probably represents the greatest quantity of neon ever used for a single application.

Helium-3 ions are used as bombarding particles in activation analysis. These ions require less energy than protons, and their limited penetration permits surface analysis only (4). The concentrations of argon-40 accumulated from the radioactive decay of potassium-40 is used to determine the age of minerals and meteorites (4). Hermetically sealed electronic components have been tested for leaks by several hours exposure to a pressurized atmosphere containing a trace of krypton-85 followed by a rapid purge and inspection for residual radioactivity that possibly has leaked into the small containers (136). Krypton-85, because of its freedom from chemical holdup, is sometimes used as a tracer for movements of hydrological sediment flows (136) and for tracking the movement of large air masses (137).

Radioactive isotopes of the helium-group elements also pose some problems in nuclear fission technology. The off-gases from reactors and, more importantly, from nuclear fuel reprocessing plants, contain many hazardous radioactive materials. The chemically active ones can be removed and concentrated for storage by various chemical means, but these methods are ineffective for the radioisotopes of the helium-group elements. Xenon and krypton are principal products of uranium fission. All of xenon's unstable isotopes and most of krypton's have such short half-lives that they practically disappear after a reasonable detention period. Krypton-85, however, has a half-life of 10.76 yr, long enough to make detaining the whole gas stream impractical and short enough to account for its intense radioactivity. Various physicochemical processes for concentrating the krypton fraction for long-term storage have been described (138), but the only processes now operating in the United States rely on cryogenic condensation and adsorption techniques (139).

Aerospace. Between World Wars I and II, the primary use of helium was as the only safe, nonflammable light gas for lighter-than-air craft, but in the 1990s this use is very limited. Balloons and blimps are used only for sports and advertising purposes. Many unoccupied weather balloons and upper atmosphere research balloons still use helium, but these could use hydrogen with little hazard were it necessary. For safety, most toy carnival balloons are inflated with helium instead of hydrogen. From time to time, efforts are made to revive commercial lighter-than-air technology for such things as logging and cargo handling, but no such project has yet proved commercially successful.

In rocket and space vehicle systems, helium is used for purging cryogenic fuel and oxidizer tanks and for pressurizing the ullage space in cryogenic tanks either to provide pressure for direct liquid transfer or simply to provide net positive suction pressure for a transfer pump. Helium frequently is also used as the working fluid in pneumatic control systems. Cold helium has also been used to precool upper-stage, liquid-hydrogen rocket engines prior to launch in order to gain maximum propulsive efficiency during the first few moments of firing (117). Liquid helium and neon are extensively used to cool cryogenic radiation detectors and other specialized instruments used in the space exploration program (140).

Medical and Breathing Gases. The main physiological effect of the non-radioactive helium-group gases is a narcotic one. The so-called inert breathing gases which, although they are chemically unchanged by metabolic processes, exert prominent effects on some cellular functions (141). Of the several schemes for correlating narcotic potency with other gas properties, one of the earliest and still useful schemes is the Meyer-Overton hypothesis. This correlates narcotic potency, a gas's solubility in nerve tissue lipids which can be approximated by its solubility in light oils such as olive oil.

The narcotic potency and solubility in olive oil of several metabolically inert gases are listed in Table 10. The narcotic potency, ED_{50}, is expressed as the partial pressure of the gas in breathing mixtures required to produce a certain degree of anesthesia in 50% of the test animals. The solubilities are expressed as Bunsen coefficients, the volume of atmospheric pressure gas dissolved by an equal volume of liquid. The lipid solubility of xenon is about the same as that of nitrous oxide, a commonly used light anesthetic, and its narcotic potency is also about the same. As an anesthetic, zenon has the virtues of reasonable potency, nonflammability, chemical inertness, and easy elimination by the body, but its scarcity and great cost preclude its wide use for this purpose (see ANESTHETICS).

The main physiological application of the helium-group gases comes at the other extreme; because of its small (142) or nonexistant (143) narcotic effect, helium is used to replace nitrogen in breathing mixtures for deep sea divers. Nitrogen in compressed air creates narcosis noticeable at depths below ca 40 m and debilitating at around 80 m. Helium–oxygen breathing mixtures have permitted successful open sea dives to 350 m (144) and simulated dives in wet chambers to 490 m and in dry chambers to 610 m (145). In addition, the lower density and viscosity of helium–oxygen mixtures lessens the physical effort required to

Table 10. Narcotic Potency and Solubility in Olive Oil of Some Metabolically Inert Gases[a]

Gas	ED_{50},[b] kPa	Bunsen coefficient at 37°C
cyclopropane	4.6	7.15
dichlorodifluoromethane	26	5.1
ethylene	48	1.28
xenon	52	1.7
nitrous oxide	59	1.6
krypton	182	0.43
sulfur hexaflouride	189	0.25
methane	294	0.28
argon	1,280	0.14
nitrogen	1,820	0.067
helium	>16,500	0.015

[a]Ref. 142.
[b]ED_{50}, narcotic potency, is expressed as the partial pressure of a gas in breathing mixtures required to produce a certain degree of anesthesia in 50% of the test animals.

overcome the high breathing resistance of the heavier compressed gases. However, the much higher sonic velocity in helium mixtures radically changes vocal frequencies, thereby creating communications problems, and the high thermal conductivity of helium increases the rate of body heat loss to the extent that actively heated diving suits and breathing gas preheaters are often necessary (146). Sometimes, neon and neon–helium mixtures are also used in hyperbaric breathing mixtures.

The quantity of breathing gas consumed in deep dives is of both economical and logistical concern; at depths of 300 m, a reasonably active diver requires ca 1.8 m^3 (64 ft^3 at STP) of breathing gas per minute. In closed-cycle breathing systems, of both the self-contained and umbilical types, the helium is recirculated after carbon dioxide is removed and the oxygen replenished (147).

The low breathing resistance of helium–oxygen mixtures is of therapeutic advantage for patients suffering from asthma and other obstructive respiratory difficulties. The mixtures have also been used for hyperbaric therapy.

BIBLIOGRAPHY

"Helium-Group Gases" in *ECT* 1st ed., Vol. 7, pp. 398–419, by R. A. Cattell and H. P. Wheeler, Jr., (Helium), Petroleum and Natural Gas Branch, Bureau of Mines, U.S. Department of the Interior, and R. A. Jones and M. A. Dubs (Neon, Argon, Krypton, and Xenon), Linde Air Products Co., a division of Union Carbide and Carbon Corp.; "Helium" in *ECT* 1st ed., Suppl. 2, pp. 454–459, by H. P. Wheeler, Jr., Bureau of Mines, U.S. Department of the Interior; "Helium-Group Gases" in *ECT* 2nd ed., Vol. 10, pp. 862–888, by R. A. Hemstreet and B. S. Kirk, Central Research Laboratory, Air Reduction Co., Inc.; in *ECT* 3rd ed., Vol. 12, pp. 249–287, by B. S. Kirk and A. H. Taylor, Airco, Inc.

1. G. K. Horton, in M. L. Klein and J. A. Venables, eds., *Rare Gas Solids*, Vol. 1, Academic Press, Inc., New York, 1976, pp. 1–26.
2. R. J. Havlik, in G. A. Cook, ed., *Argon, Helium and the Rare Gases*, Vol. 1, Wiley-Interscience, New York, 1961, pp. 17–33.

3. M. W. Travers, *Discovery of the Rare Gases*, Edward Arnold Co., London, 1928.
4. C. R. Eck, in R. E. Stanley and A. A. Moghissi, eds., *Noble Gases*, CONF 730915 (Library of Congress 75-27055), *ERDA TIC*, 1973, pp. 580–586.
5. E. F. Hammel, *Adv. Cryog. Eng.* **25**, 810–821 (1980).
6. B. A. Mirtov, *Gaseous Composition of the Atmosphere and Its Analysis*, transl. *NASA TTF-145, OTS 64-11023*, U.S. Dept. of Commerce, Washington, D.C., 1964, p. 22.
7. Unpublished data, National Oceanic and Atmospheric Administration, U.S. Dept. of Commerce, Washington, D.C., 1991.
8. R. Barron, *Cryogenic Systems*, McGraw-Hill Book Co., Inc., New York, 1966, pp. 184–189.
9. J. E. Hamak and B. D. Gage, *Analysis of Natural Gases, 1991*, information circular 9318, U.S. Bureau of Mines, Washington, D.C., 1992; *Helium Resources of the United States, 1991*, information circular 9342, U.S. Bureau of Mines, Washington, D.C., 1993.
10. B. J. Moore, *Inf. Circ. U.S. Bureau of Mines*, information circular 8870, Washington, D.C., 1982.
11. W. D. Leachman, *Helium, 1992 Annual Report*, U.S. Bureau of Mines, Amarllo, Tex., 1993.
12. R. A. Markel, *Statement of Proposed Helium Energy Act of 1979 to Subcommittee on Energy and Power*, House Committee on Interstate and Foreign Commerce, Washington, D.C., 1979.
13. C. Seibel, *Helium–Child of the Sun*, University of Kansas Press, Lawrence, Ks., 1968.
14. D. P. Kelly and W. J. Hauback, *Comparative Properties of Helium-3 and Helium-4*, U.S. Atomic Energy Comm. Res. Dev. Rep. MLM-1161, Washington, D.C., July 1964.
15. A. E. Cameron and E. Wichers, *J. Am. Chem. Soc.* **84**, 4175 (1962).
16. W. T. Ziegler, G. N. Brown, and J. D. Garber, *Calculation of the Vapor Pressure and Heats of Vaporization and Sublimation of Liquids and Solids Below One Atmosphere Pressure. IX. Neon, Tech. Report No. 1, Contract No. CST-7973*, U.S. National Bureau of Standards, Boulder, Colo., 1970.
17. T. R. Roberts and co-workers, in C. J. Gorter, ed., *Progress in Low Temperature Physics*, Vol. 4, Wiley-Interscience, New York, 1964, pp. 480–514.
18. R. D. McCarty, *J. Phys. Chem. Ref Data* **2**, 923 (1973).
19. R. D. McCarty, *Interactive FORTRAN Programs for Micro Computers to Calculate the Thermo Physical Properties of Twelve Fluids [MIPROPS]*, NBS Technical Note 1097, U.S. Dept. of Commerce, Washington, D.C., 1986.
20. A. L. Gosman, R. D. McCarty, and J. G. Hust, *Natl. Stand. Ref Data Ser. Natl Bur. Stand.* **27** (1969).
21. A. C. Jenkins, in Ref. 2, pp. 391–394.
22. H. W. Habgood and W. G. Schneider, *Can. J. Chem.* **32**, 98 (1954).
23. W. T. Ziegler, D. W. Yarbrough, and J. C. Mullins, *Calculations of the Vapor Pressure and Heats of Vaporization and Sublimation of Liquids and Solids, Especially Below One Atmosphere Pressure, VI, Krypton, Tech. Rept. No. 1, Contract No. CST 1154*, National Standard Reference Data Program, U.S. National Bureau of Standards, Washington, D.C., 1964.
24. W. T. Ziegler, J. C. Mullins, and A. R. Berquist, *Calculation of the Vapor Pressure and Heats of Vaporization and Sublimation of Liquids and Solids Below One Atmosphere Pressure, VIII, Xenon, Tech. Rep. No. 3, Contract No. CST-1154*, National Standard Reference Data Program, U.S. National Bureau of Standards, Washington, D.C., 1966.
25. J. L. Tiggelman, C. Van Rijn, and M. Durieux, *Temp. Its Meas. Control Sci. Ind.* **4**, 137 (1972).
26. J. A. Beattie, in Ref. 2, pp. 251–312.
27. E. C. Kerr, *Phys. Rev.* **96**, 551 (1954).
28. G. Gladun and F. Menzel, *Cryogenics* **10**, 210 (1970).

29. K. Clusius, *Z. Phys. Chem. Abt. B* **31**, 459 (1936).
30. K. Clusius and K. Wiegand, *Z. Phys. Chem. Abt B* **46**, 1 (1940).
31. E. R. Grilly, *Cryogenics* **2**, 226 (1962).
32. S. S. Lestz, *J. Chem. Phys.* **38**, 2830 (1963).
33. L. Goldstein, *Phys. Rev.* **112**, 1465 (1958).
34. V. J. Johnson, ed., *A Compendium of the Properties of Materials at Low Temperature*, WADD Tech. Rept. 60-56, Part 1, July 1960, Part 2, Dec. 1961.
35. P. Flubacher, A. J. Leadbetter, and J. A. Morrison, *Proc. Phys. Soc.* (London) **78**, 1449 (1961).
36. D. H. Smith and R. G. Harlow, *Brit. J. Appl. Phys.* **14**, 102 (1963).
37. J. V. Gengers, W. T. Bolk, and C. J. Sigter, *Physica* **30**, 1018 (1964).
38. A. Michels and co-workers, *Physica* **22**, 121 (1956).
39. *Data Book*, Vol. 2, Thermophysical Properties Research Center, Purdue University, W. Lafayette, Ind., 1964, Chapt. 1.
40. S. Forster, *Cryogenics* **3**, 176 (1963).
41. J. P. Boon and G. Thomaes, *Physica* **29**, 208 (1963).
42. J. C. Legros and G. Thomaes, *Physica* **31**, 70.3 (1965).
43. W. F. Edgell, in Ref. 2, Chapt. 5, p. 99.
44. W. Braker and A. L. Mossman, eds., *The Matheson Unabridged Gas Data Book*, Matheson Gas Products, East Rutherford, N.J., 1974.
45. N. Marshall, transl., *Gas Encyclopedia*, Elsevier Scientific Pub. Co., Amsterdam, 1976.
46. D. S. Betts, *Cryogenics* **16**, 3 (1976).
47. R. D. McCarty and R. B. Stuart, in S. Gratch, ed., *Advances In Thermophysical Properties at Extreme Temperatures and Pressures*, American Society of Mechanical Engineers, New York, 1965, pp. 84–97.
48. R. D. McCarty, *Interactive FORTRAN IV Computer Programs for the Thermodynamic and Transport Properties of Selected Cryogens [Fluids Pack]*, NBS technical note 1025, U.S. Dept. of Commerce, Washington, D.C., 1980.
49. D. T. Goldman, in D. E. Gray, ed., *American Institute of Physics Handbook*, McGraw-Hill Book Co., Inc., New York, 1972, Sect. 8b.
50. R. L. Heath, in R. C. Weast, ed., *Handbook of Chemistry and Physics*, CRC Press, Cleveland, Ohio, 1975, pp. B252–B336.
51. D. A. Nelson, Jr., and A. L. Ruoff, *Phys. Rev. Lett.* **42**(6), 383 (1979).
52. B. Bertman and R. A. Guyer, *Sci. Am.* **217**, 85 (1967).
53. C. M. Varma and N. R. Werthamer, in K. H. Bennemann and J. B. Ketterson, eds., *The Physics of Liquid and Solid Helium*, Part 1, John Wiley & Sons, Inc., New York, 1976, p. 505.
54. J. M. Benson and J. P. Pinceaux, *Science* **206**, 1073 (1979).
55. W. E. Keller, *Helium-3 and Helium-4*, Plenum Press, New York, 1969, p. 364.
56. *Ibid.*, p. 360.
57. I. Pomeranchuk, *Zh. Eksp. Theor. Fiz.* **20**, 919 (1950).
58. Ref. 55, p. 16 ff.
59. K. F. Hwang and B. M. Khorana, *Metrologia* **12**, 61 (1976).
60. I. M. Khalatnikov, in Ref. 53, pp. 1–84.
61. J. Wilks, *The Properties of Liquid and Solid Helium*, Clarendon Press, Oxford, U.K., 1967.
62. W. H. Keesom, *Helium*, Elsevier, Amsterdam, 1942, p. 302.
63. J. Bardeen, L. N. Cooper, and J. R. Schrieffer, *Phys. Rev.* **108**, 1175 (1957).
64. D. D. Osheroff, R. C. Richardson, and D. M. Lee, *Phys. Rev. Lett.* **28**, 885 (1972).
65. N. D. Mermin and D. M. Lee, *Sci. Am.* **235**, 56–71 (1976).
66. D. M. Lee and R. C. Richardson, in Ref. 54, pp. 287–496.
67. J. C. Wheatley, in M. Krusius and M. Vuorio, eds., *Low Temperature Physics–1, T14*, Vol. 5, American Elsevier Publishing, New York, 1975, pp. 6–51.

68. M. J. Hiza, A. J. Kidnay, and R. C. Miller, *Equilibrium Properties of Fluid Mixtures: A Bibliography of Data on Fluids of Cryogenic Interest*, Plenum Press, New York, 1975.

69. R. L. Robinson, Jr. and M. J. Hiza, in K. D. Timmerhaus, ed., *Advances in Cryogenic Engineering*, Vol. 20, Plenum Press, New York, 1975, pp. 218–239.

70. J. C. Wheatley, in C. J. Gorter, ed., *Progress in Low Temperature Physics*, Vol. 4, North Holland Publishing Co., New York, 1970, pp. 77 ff.

71. G. Ahler, in Ref. 51, pp. 120–206.

72. J. C. Wheatley, in K. D. Timmerhaus, ed., *Advances in Cryogenic Engineering*, Vol. 15, Plenum Press, New York, 1970, pp. 415–421.

73. W. M. Deaton and P. V. Mullins, in R. H. Kropschot, B. W. Birmingham, and D. B. Mann, eds., *Technology of Liquid Helium*, National Bureau of Standards Monograph 111, U.S. Government Printing Office, Washington, D.C., 1968, pp. 1–32.

74. G. Kobelt and W. Otto, *Chem. Tech.* (Berlin) **19**, 449 (1967).

75. G. C. Handley, in Ref. 4, pp. 90–99.

76. G. Gnauck, *Chem. Tech.* (Berlin) **6**, 551 (1954).

77. U.S. Pat. 5,019,145 (May 28, 1991), W. Rohde and H. Cordman (to Linde Aktiengesellschaft).

78. U.S. Pat. 5,133,790 (July 28, 1992), J. R. Bianchi, D. P. Bonaquist, and R. A. Victor (to Union Carbide Corp.).

79. U.S. Pat. 5,159,816 (Nov. 3, 1992), K. W. Kovak, R. Agrawal, and J. C. Peterson (to Air Products and Chemicals, Inc.).

80. W. Otto, *Ind. Chem. Belg.* **32**, 1080 (1967).

81. M. Streich and P. Daimer, in *Cryotech '73*, IPC Science and Technology Press, Guildford, Sussex, UK, 1974, pp. 31–35.

82. C. McKinley, in Ref. 4, pp. 391–404.

83. M. Ruhemann, *The Separation of Gases*, Clarendon Press, London, 1949, pp. 225–229.

84. C. A. Rohrmann, *Isot. Radiat. Technol.* **8**, 253 (1970).

85. L. E. Scott and co-workers, in Ref. 71, pp. 155–195.

86. "Helium", in *Minerals Yearbook*, U.S. Bureau of Mines, U.S. Department of the Interior, Washington, D.C., 1961–1991.

87. *Current Industrial Reports: Industrial Gases*, U.S. Bureau of the Census, Washington, D.C., 1961–1993.

88. J. A. Cheney, private communication, Airco Industrial Gases Division, Murray Hill, N.J., 1983.

89. G. H. Cady and H. P. Cady, *Ind. Eng. Chem. Anal. Ed.* **17**, 760 (1945).

90. *Special Gases Data Sheet, Airco Rare and Specialty Gases*, Airco Industrial Gases Division, Murray Hill, N.J., 1991.

91. P. G. Jeffery and P. J. Kipping, *Gas Analysis by Gas Chromatography*, MacMillan, New York, 1964, pp. 102–106.

92. J. C. Newton, F. B. Stephens, and R. K. Stump, in Ref. 4, pp. 218–224.

93. G. Schmauch, in F. D. Snell and C. L. Hilton, eds., *Encyclopedia of Industrial Chemical Analysis*, Vol. 13, Wiley-Interscience, New York, 1971, pp. 286–309.

94. E. C. Nelson, in Ref. 2, pp. 609–633.

95. H. B. Cary, *Modern Welding Technology*, Prentice-Hall, Englewood Cliffs, N.J., 1979.

96. H. R. Miller, *Metall. Met. Form.* **44**, 392 (1977).

97. C. H. Simpson, *Chemicals from the Atmosphere*, Doubleday, New York, 1969, pp. 160–161.

98. H. R. Miller and F. W. Lunau, in Ref. 8–11.

99. C. T. Baroch, T. B. Kaczmarek, and J. F. Lenc, *U.S. Bur. Mines, Rep. Invest.* **5253** (1956).

100. L. Maissel, in L. Maissel and R. Glang, eds., *Handbook of Thin Film Technology*, McGraw-Hill Book Co., Inc., New York, 1970, Chapt. 4, pp. 4–2 ff.

101. C. R. Barber, *Metrologia* **5**(2), 35 (1969).

102. F. G. Brickwedde and co-workers, *Natl. Bur. Stand. U.S. Monogr.* **10** (1960).
103. W. T. Ziegler, J. C. Mullins, and B. S. Kirk, *Calculation of the Vapor Pressure and Heats of Vaporization and Sublimation of Liquids and Solids, Especially Below One Atmosphere Pressure II, Argon*, Tech. Report No. 2, Contract No. CST-7238, U.S. National Bureau of Standards, Boulder, Colo., June 1962.
104. G. Zweig and J. Sherma, *Handbook of Chromatography*, Vol. 2, CRC Press, Cleveland, Ohio, 1972, pp. 12–14.
105. *Chem. Eng. News* **50**, 36 (May 8, 1972).
106. A. C. Jenkins, in Ref. 2, pp. 685–724.
107. N. Booth, *Industrial Gases*, Pergamon Press, Oxford, U.K., 1973.
108. J. E. Geusic, W. B. Bridges, and J. I. Pankove, in I. P. Kaminow and A. E. Siegman, eds., *Laser Devices and Applications*, IEEE Press, New York, 1973, pp. 2–22.
109. W. B. Bridges and co-workers, in Ref. 108, pp. 94–107.
110. J. A. Beaulieu, in Ref. 108, pp. 108–115.
111. G. J. Linford, in Ref. 4, pp. 561–579.
112. J. G. Daunt, in R. W. Vance, ed., *Applications of Cryogenic Technology*, Vol. 4, Xyzyx Information Corp., 1972, pp. 365–371.
113. J. Daunt and E. Lerner, in K. D. Mendelssohn, ed., *International Cryogenic Engineering Conferences*, Vol. 5, IPC Science and Technology Press, Ltd., Guildford, Surrey, U.K., 1974, pp. 238–239.
114. T. O. Niinikoski, in Ref. 113, pp. 102–111.
115. S. S. Papell and R. C. Hendricks, in K. D. Timmerhaus and D. H. Weitzel, eds., *Advances in Cryogenic Engineering*, Vol. 21, Plenum Press, New York, 1975, pp. 278–281.
116. B. C. Belanger, in Ref. 69, pp. 1–22.
117. R. A. Kamper, R. H. Kropschot, and A. F. Schmidt, in Ref. 73, pp. 327–357.
118. R. F. Barron, in Ref. 69, pp. 402–410.
119. J. C. Boissin and J. J. Thibault, in Ref. 113, pp. 136–138.
120. J. W. Meyer and R. A. Kamper, in Ref. 73, pp. 300–323.
121. B. W. Birmingham and C. N. Smith, *Cryogenics* **16**, 59 (1976).
122. R. H. Borcherts, in *Cloud and Bubble Chambers*, Methuen and Co. Ltd., London, 1970, pp. 26–27.
123. J. L. Smith, Jr., G. L. Wilson, and J. L. Kirtley, Jr., *IEEE Trans. Magn.* **15**(1), 727 (1979).
124. J. A. Good and E. Cohen, *Cryogenics* **16**, 588 (1976).
125. S. W. Van Scives and co-workers, in Ref. 112, pp. 296–307.
126. A. J. Steel, S. Bruzzi, and M. E. Clarke, in Ref. 113, pp. 58–61.
127. C. Kuzmycz, *Cryog. Ind. Gases* **9**(4), 15 (1974).
128. *Nucl. News* **21**(3), 42 (1978).
129. S. Groueff, *Manhattan Project: The Untold Story of the Making of the Atomic Bomb*, Bantam Books, New York, 1967, pp. 211–219.
130. M. N. Wilson, in K. D. Timmerhaus, R. P. Reed, and A. F. Clarke, eds., *Advances in Cryogenic Engineering*, Vol. 24, Plenum Press, New York, 1978, pp. 1–16.
131. L. C. Pittenger, in K. D. Timmerhaus, ed., *Advances in Cryogenic Engineering*, Vol. 23, Plenum Press, New York, 1977, pp. 648–657.
132. *CERN Courier* **15**(7–8), 231 (1975).
133. C. Henderson, in Ref. 122.
134. Y. D. Aleshin and co-workers, *Cryogenics* **16**, 607 (1976).
135. *CERN Courier* **16**(2), 51 (1976).
136. Y. Wang, ed., *Handbook of Radioactive Nuclides*, CRC Press, Cleveland, Ohio, 1969, p. 515 ff.
137. G. J. Ferber, K. Telegadas, and M. E. Smith, *Atmospheric Environment*, Vol. 11, Pergamon Press, New York, 1977, pp. 379–385.

138. C. L. Bendixsen and J. A. Buckham, in Ref. 4, pp. 290–295.

139. R. W. Stuart, in Ref. 69, pp. 61–69.

140. J. W. Vorreiter, in *International Cryogenic Engineering Conferences*, Vol. 7, IPC Science and Technology Press, Guildford, Surrey, U.K., 1978, pp. 1–12.

141. C. J. Lambertson, in *Fundamentals of Hyperbaric Medicine*, Pub. No. 1298, National Academy of Sciences, Washington, D.C., 1966, pp. 3–11.

142. P. B. Bennett, S. Simon, and Y. Katz, in B. R. Fink, ed., *Progress in Anesthesiology*, Vol. 1, Raven Press, New York, 1975, pp. 367–402.

143. K. W. Miller, in Ref. 142, pp. 343–351.

144. H. Reuter, *Faceplate* **6**(2), 26 (1975).

145. *Faceplate* **4**(3), 20 (1973).

146. W. Spaur, *Faceplate* **5**(3), 8 (1974).

147. B. Barrett, *Faceplate* **5**(3), 8 (1974).

General References

G. A. Cook, ed., *Argon, Helium, and the Rare Gases*, Vols. 1–2, Wiley-Interscience, New York, 1961.

R. E. Stanley and A. A. Moghissi, eds., *Noble Gases*, CONF 730915 (Library of Congress 75-27055), ERDA TIC, 1973.

K. D. Timmerhaus and co-eds., *Advances in Cryogenic Engineering*, Vols. 1–38, Plenum Press, New York, 1960–1992.

International Cryogenic Engineering Conferences, Vols. 1–14, IPC Science and Technology Press and Butterworth, London, 1967–1993.

B. Elvers, S. Hawkins, and G. Schulz, eds., *Ullmann's Encyclopedia of Industrial Chemistry*, Vol. A17, VCH Publishers, Weinheim, Germany, 1991.

SHUEN-CHENG HWANG
WILLIAM R. WELTMER, JR.
The BOC Group, Inc.

COMPOUNDS

The helium-group (noble) gases had been characterized for many years as inert and incapable of forming compounds with other elements. Attempts to induce chemical reactivity led to assumptions of chemical inertness and inviolability of the noble-gas valence shell octet of electrons, the so-called octet rule (1). The discovery of helium-group gases (qv) reactivity was reported by Neil Bartlett (2) in 1962. Deep red-brown PtF_6 vapor in the presence of xenon gas at room temperature produced a yellow-orange compound then formulated as $Xe^+PtF_6^-$. Within a few months of this discovery, xenon tetrafluoride was prepared (3) as was xenon difluoride (4). A number of other xenon compounds, a krypton fluoride, and a radon fluoride followed (5).

The heavy gases, krypton, xenon, and radon, had been shown to react with fluorine (qv) and other powerful oxidants to form a number of stable products. As of this writing no stable compounds of the lighter gases, helium, neon, and argon, have been found. A complete survey of noble-gas compounds may be found elsewhere (5–12). Some of the properties of the fluorides, oxides, and oxofluorides are given in Table 1. Prior to the discovery of these compounds, hydrates and

Table 1. Physical and Thermodynamic Properties of Helium-Group Gas Fluorides, Oxofluorides, and Oxides

Compound	CAS Registry Number	Melting point, °C	Color	ΔH_{sub}, kJ/mol[a]	ΔH_f, kJ/mol[a]	ΔG_f, kJ/mol[a]	S, J/mol·K[a]	C_p, J/mol·K[a]	Density, g/cm³	References
KrF_2	[13773-81-4]	dec 25[b]	colorless	41	60.2				3.22	13–15
XeF_2	[13709-36-9]	129.03[c]	colorless	55.71	−162.8	−86.08	115.09	75.60	4.32	16–20
XeF_4	[13709-61-0]	117.10[c]	colorless	60.92	−267.1	−145.5	167.00	118.39	4.04	16,18,21,22
XeF_6	[13693-09-9]	49.48	colorless[d] or yellow-green[e,f]	59.12[g]	−338.2[g]	−169.0[g]	210.38[g]	171.59[g]	3.56[g,h]	16,23–25
$XeOF_2$	[13780-64-8]	dec ca 0[i]	yellow							26,27
$XeOF_4$	[13774-85-1]	−46.2	colorless		−25[j]				3.11[e,k]	28,29
XeO_2F_2	[13875-06-4]	30.8[i]	colorless		234[i]				4.10	29,30
XeO_3	[13776-58-4]	dec ca 25[i]	white	100[j]	402		287		4.55	29,31,32
XeO_4	[12340-14-6]	dec < 0[i]	yellow solid at −196°C		642					29,33

[a]To convert J to cal, divide by 4.184.
[b]Decomposes at ~ 10% h⁻¹.
[c]Triple point.
[d]Solid.
[e]Liquid.
[f]Vapor.
[g]Phase I.
[h]Phase II has a density of 3.71 g/cm³; Phase III, 3.82 g/cm³; and Phase IV, 3.73 g/cm³.
[i]Explosive.
[j]Estimated.
[k]Liquid at 22.5°C.

other clathrate compounds of the helium-group gases were known. A discussion of clathrates may be found in the literature (6,34) (see also INCLUSION COMPOUNDS).

Xenon Compounds

Halides. Xenon fluorides, xenon oxide tetrafluoride [13774-85-1], $XeOF_4$, and their complexes are the only thermodynamically stable xenon compounds. Xenon difluoride [13709-36-9], xenon tetrafluoride [13709-61-0], and xenon hexafluoride [13693-09-9] are colorless, crystalline solids which can be sublimed under vacuum at 25°C. The mean thermochemical bond energies are XeF_2, 132.3 ± 0.7 kJ/mol (31.6 ± 0.2 kcal/mol); XeF_4, 130.3 ± 0.5 kJ/mol (31.1 ± 0.1 kcal/mol); and XeF_6, 125.3 ± 0.7 kJ/mol (29.9 ± 0.2 kcal/mol) (16). Xenon hexafluoride is yellow-green as a liquid or gas. Reports of xenon octafluoride [17457-75-9], XeF_8, remain unsubstantiated. Xenon difluoride is a linear symmetrical molecule having a Xe—F distance of 197.73(15) pm (35), and XeF_4 is a square planar molecule (Xe—F, 195.3(4) pm) (36). Experimental evidence is consistent with a distorted octahedral structure for gaseous XeF_6 (37) arising from the presence of an extra pair of nonbonding electrons in the xenon valence shell. Solid XeF_6 exists in at least four phases which consist of tetrameric and hexameric rings of virtually undistorted square pyramidal XeF_5^+ cations linked together by fluoride ion bridges (24,38).

Xenon dichloride [13780-38-6], $XeCl_2$, and xenon(II) chlorofluoride [73378-52-6], XeClF, have been prepared by photochemical and electric discharge methods and have been examined at low temperatures by matrix-isolation techniques (39,40). The dichloride has a linear structure like that of XeF_2. Evidence for the existence of $XeCl_2$, $XeBr_2$, and xenon tetrachloride [14989-42-5], $XeCl_4$, has been obtained from Mössbauer studies (41,42). Owing to thermal chemical instabilities, no dihalide other than the binary fluorides has been prepared in macroscopic amounts.

Unstable monohalides of xenon ([16757-14-5], XeF, [55130-03-5], XeCl, [55130-04-6], XeBr, and [55130-05-7], XeI), have been produced in the gas phase by electron bombardment methods (43,44) and in solid matrices by gamma and ultraviolet irradiation methods (45,46). Although short-lived in the gas phase, these halides are of considerable importance as light-emitting species in gas lasers (qv).

Oxides. Two oxides of xenon are known: xenon trioxide [13776-58-4], XeO_3, and xenon tetroxide [12340-14-6], XeO_4 (Table 1). Xenon trioxide is most efficiently prepared by the hydrolysis of XeF_6 (47) or by the reaction of XeF_6 with $HOPOF_2$ (48). The XeO_3 molecule has a trigonal pyramidal shape Xe—O, 176(3) pm (49), and XeO_4 is tetrahedral with Xe—O, 173.6(2) pm (50). Xenon tetroxide is prepared by the interaction of concentrated sulfuric acid with sodium or barium perxenate, Na_4XeO_6, Ba_2XeO_6 (33). Both oxides are thermodynamically unstable, explosive solids which must be handled with the greatest care. On decomposing to the elements, solid XeO_3 and gaseous XeO_4 release 402 and 642 kJ/mol (96.1 and 153 kcal/mol), respectively (29). Xenon trioxide has a negligible vapor pressure at room temperature and readily dissolves in water to give stable solutions containing mainly molecular XeO_3 and xenic acid anion, $HXeO_4^-$, which is vanishingly small ($K \approx 3 \times 10^{-11}$ for $XeO_3(aq)$ + 2 $H_2O(1) \rightleftharpoons HXeO_4^-(aq) + H_3O^+(aq)$), except in basic solution ($K \approx 1.5 \times 10^3$ for $XeO_3(aq) + OH^-(aq) \rightleftharpoons HXeO_4^-(aq)$). Xenon tetroxide is volatile at 25°C, but frequently decomposes violently before this temperature is reached.

Oxofluorides. Xenon oxide tetrafluoride, $XeOF_4$, has a square pyramidal geometry; the oxygen is trans to the electron lone pair; the Xe—F distance is 170.3(15) pm; Xe—O, 190.0(5) pm (51). It is a volatile, colorless liquid. XeO_2F_2 has a trigonal, bipyramidal geometry; both oxygens and the lone pair of electrons are in the equatorial plane; the Xe—F distance is 171.4(4) pm; the Xe—O distance 189.9(3) pm (52). It is a colorless solid (Table 1). The compound, $XeOF_4$, is formed by the interaction of XeF_6 with an equimolar amount of water (53) or $NaNO_3$ (54). The reaction of stoichiometric amounts of N_2O_5 and $XeOF_4$ affords XeO_2F_2 (55). Xenon oxide difluoride [13780-64-8], $XeOF_2$, is formed as an unstable yellow solid by the interaction of XeF_4 with a stoichiometric amount of water (26,27). Two additional oxofluorides are known, XeO_3F_2 [15192-14-0] (56) and XeO_2F_4 [15195-51-4] (57). Both compounds are thermodynamically unstable, and little is known about XeO_2F_4.

Xenates and Perxenates. Alkali metal xenates of composition $MHXeO_4$· $1.5H_2O$, where M is sodium, potassium, rubidium, or cesium, have been prepared by freeze-drying mixtures of xenon trioxide and the corresponding metal hydroxides in 1:1 molar ratios. The xenates are unstable, explosive solids.

When XeF_6 is hydrolyzed in a strongly alkaline solution, part of the xenon is lost as gas, but a large fraction precipitates as a perxenate, XeO_6^{4-}, salt in which xenon is in the +8 oxidation state (7). Among the salts that have been prepared in this manner, or by alternative procedures, are $Na_4XeO_6·6H_2O$ [30970-85-5], $K_4XeO_6·9H_2O$ [60763-18-0], $Li_4XeO_6·2H_2O$ [34901-38-7], and $Ba_2XeO_6·1.5H_2O$ [15842-32-7]. The average Xe—O bond length is 185(2) pm (7). The salts are kinetically very stable, losing water gradually when heated; for example, $Na_4XeO_6·6H_2O$ dehydrates at 100°C and decomposes at 360°C. A number of transition-metal and actinide perxenates have been prepared but are not thoroughly characterized and include those of copper, lead, silver, zinc, thorium, and uranium. Perxenate solutions are powerful oxidants in aqueous solution and are capable of oxidizing iodate ion to periodate and manganous ion to permanganate. The aqueous chemistry of perxenates and other xenon compounds has been fully described (74).

Complex Salts and Molecular Adducts. The majority of the known complexes of xenon can be classified as cation or anion derivatives of binary fluorides, oxofluorides, and XeO_3. Gross geometries are listed in Table 2 along with those of the parent compounds. Although complex compounds derived from the interaction of a helium-group gas fluoride or oxofluoride with a strong fluoride ion acceptor are generally written as ionic formulations, the cations and anions interact with one another by means of weak covalent interactions of the cation with one or more fluorines on the anion, an interaction that is termed fluorine bridging (10).

Xenon difluoride behaves as a fluoride ion donor toward many metal pentafluorides to form complex salts containing the XeF^+ and $Xe_2F_3^+$ cations (10). In reactions with the pentafluorides of arsenic, antimony, and ruthenium, for example, it forms the salts $Xe_2F_3^+AsF_6^-$ [21308-45-2], $XeF^+AsF_6^-$

Table 2. Fluoro- and Oxofluoro-Cations and Anions of Xenon, Their Parent Compounds, and Geometries

Parent compound	Structure[a]	Cation(s)	Structure[a]	References	Anion(s)	Structure[a]	References
XeF_2	linear ($D_{\infty h}$)	XeF^+	linear ($C_{\infty v}$)	10	XeF_5^-	pentagonal plane (D_{5h})	61
		$FXe\text{-}\text{-}F\text{-}\text{-}XeF^+$	V-shape (C_{2v})[b]	10			
XeF_4	square plane (D_{4h})	XeF_3^+	T-shape (C_{2v})	10,58–60	$XeOF_3^-$	plane (C_{2v})	27
$XeOF_2$	T-shape (C_{2v})				XeF_7^-	geometry unknown	63,64
XeF_6	monocapped octahedron (C_{3v})	XeF_5^+	square pyramid (C_{4v})	10,11,63	XeF_8^{2-}	square antiprism (D_{4d})	63,64,66
		$F_5Xe\text{-}\text{-}F\text{-}\text{-}XeF_5^{+}$[b]		10,11,65	$XeOF_5^-$	pentagonal pyramid (C_{5v})[c]	69,70
					$(XeOF_4)_3F^-$[d]		70
$XeOF_4$	square pyramid (C_{4v})	$XeOF_3^+$	disphenoid (C_s)	67	$XeO_2F_3^-$	square pyramid (C_s)	27,55
XeO_2F_2	disphenoid (C_{2v})	XeO_2F^+	trigonal pyramid (C_{2v})	71–73			
		$FO_2Xe\text{-}\text{-}F\text{-}\text{-}XeO_2F^{+}$[b]		71			
XeO_3	trigonal pyramid (C_{3v})				XeO_3F^-	[e]	75

[a] Point group is given in parentheses.

[b] Cations that are mononuclear in xenon and the $F_5Xe\text{-}\text{-}F\text{-}\text{-}XeF_5^+$ ($Xe_2F_{11}^+$) cation interact with their fluoroanions through one or more fluorine bridges. Details of the structure and fluorine bridging in the $FO_2Xe\text{-}\text{-}F\text{-}\text{-}XeO_2F^+$ ($Xe_2O_4F_3^+$) cation are unknown, but the Xe--F--Xe arrangement is assumed to be bent as in the $FXe\text{-}\text{-}F\text{-}\text{-}XeF^+$ ($Xe_2F_3^+$) and $Xe_2F_{11}^+$ cations. The $Xe_2F_3^+$ cation forms no fluorine bridges with its fluoroanion. The XeF_5 groups in $Xe_2F_{11}^+$ have essentially square pyramidal geometries.

[c] Point symmetry determined by vibrational spectroscopy (68).

[d] Three $XeOF_4$ molecules, having essentially square pyramidal geometries, are coordinated through the xenon atoms to a single fluoride ion to give a trigonal pyramidal arrangement about the fluoride ion.

[e] The structure consists of open polymeric chains, $[XeO_3F^-]_n$, with two fluorine bridges to each xenon atom.

[26024-71-5], $Xe_2F_3^+SbF_6^-$ [12528-47-1], $XeF^+SbF_6^-$ [36539-18-1], [17679-45-7], $XeF^+Sb_2F_{11}^-$ [15364-10-0], [36539-19-2], $Xe_2F_3^+RuF_6^-$ [26297-25-6], $XeF^+RuF_6^-$ [22527-13-5], [26500-06-1], and $XeF^+Ru_2F_{11}^-$ [22527-14-6]. Bartlett's original compound, $Xe^+PtF_6^-$, is still not completely understood, but is thought to be of this general type. Adducts containing the weak fluoride ion acceptors $MoOF_4$ and WOF_4 are known in which XeF_2 interacts with the metal by formation of asymmetric Xe---F---M bridges (M = Mo or W), eg, F—Xe---F---$MoOF_4$, F—Xe---F---$MoOF_4(MoOF_4)_2$ (76,77). A crystal structure of $XeF_2 \cdot WOF_4$ [55888-48-7] exhibits the bond lengths, $Xe - F_{bridge}$, 204(3) pm; $W - F_{bridge}$, 218(3) pm; and $Xe - F_{terminal}$, 189(3) pm (78). Xenon difluoride also forms a related fluorine bridged adduct cation, F—Xe---F---$BrOF_2^+$ [77071-47-7], in the compound $XeF_2 \cdot BrOF_2^+AsF_6^-$ (79). The difluoride also forms a number of molecular complexes in which the molecular parameters are essentially the same as in the pure compound; these include $XeF_2 \cdot IF_5$ [21992-39-2] (80), $XeF_2 \cdot XeOF_4$ [34166-84-2] (81), and $XeF_2 \cdot XeF_4$ [12337-03-0] (82).

The only example of xenon in a fractional oxidation state, $+1/2$, is the bright emerald green paramagnetic dixenon cation, Xe_2^+ [12185-20-5]. Mixtures of xenon and fluorine gases react spontaneously with liquid antimony pentafluoride in the dark to form solutions of $XeF^+Sb_2F_{11}^-$, in which Xe_2^+ is formed as an intermediate product that is subsequently oxidized by fluorine to the XeF^+ cation (83). Spectroscopic studies have shown that xenon is oxidized at room temperature by solutions of XeF^+ in SbF_5 solvent to give the Xe_2^+ cation (84).

Xenon tetrafluoride is a much weaker fluoride ion donor and only forms stable complex salts with the strongest fluoride ion acceptors, eg, $XeF_3^+SbF_6^-$ [39797-63-2] (58), $XeF_3^+Sb_2F_{11}^-$ [35718-37-7] (59), and $XeF_3^+BiF_6^-$ [66121-33-3] (60). Xenon tetrafluoride has also been shown to behave as a weak fluoride ion acceptor toward alkali metal fluorides and the naked fluoride ion source $N(CH_3)_4^+F^-$ to give salts of the novel pentagonal planar XeF_5^- [133042-38-3] anion (61). Xenon oxide difluoride is a fluoride ion acceptor, forming the only other anion containing xenon in the +4 oxidation state; the $XeOF_3^-$ anion in $Cs^+XeOF_3^-$ [65014-02-0] (27).

Xenon hexafluoride is both a strong fluoride ion donor and acceptor. Examples of salts containing the XeF_5^+ cation are numerous (10,11). A representative crystal structure is that of $XeF_5^+PtF_6^-$ [18533-64-7] (62). There are several examples of salts that contain the fluoride bridged $Xe_2F_{11}^+$ cation that have a number of counter-anions in common with those of the XeF_5^+ salts (10,11). The structure of the $Xe_2F_{11}^+$ cation is exemplified in the crystal structure of $Xe_2F_{11}^+AuF_6^-$ [39043-77-1] (65). A number of salts in which XeF_6 behaves as a fluoride ion acceptor toward alkali metal fluorides are known which probably contain the XeF_7^- and XeF_8^{2-} anions (85). Several nonalkali metal salts have been shown by x-ray crystallography and vibrational spectroscopy to contain the anions XeF_7^- and XeF_8^{2-} and include $NF_4^+XeF_7^-$ [82963-12-0] (63,64), $[NF_4^+]_2XeF_8^{2-}$ [82963-15-3] (63,64), and $[NO^+]_2XeF_8^{2-}$ [17501-82-5] (66).

The oxofluorides of xenon(VI) exhibit analogous fluoride ion donor and acceptor properties. Salts of both the $XeOF_3^+$ and XeO_2F^+ cations are known, as well as a salt of the fluoride bridged cation $Xe_2O_4F_3^+$, and include $XeOF_3^+SbF_6^-$ [42861-25-6], $XeOF_3^+Sb_2F_{11}^-$ [15600-67-6], $XeO_2F^+Sb_2F_{11}^-$ [52078-91-8], $XeO_2F^+AsF_6^-$ [115117-21-0], and $Xe_2O_4F_3^+AsF_6^-$ [13875-06-4] (67,71–73).

Several alkali metal fluoride complexes with $XeOF_4$ are known, such as $3KF \cdot XeOF_4$ [12186-19-5], $3RbF \cdot 2XeOF_4$ [12186-23-1], $CsF \cdot XeOF_4$ [12191-01-4], and $CsF \cdot 3XeOF_4$. Structural studies show that the CsF complexes are best formulated as $Cs^+ XeOF_5^-$ [12191-01-4] (69,70) and $Cs^+(XeOF_4)_3F^-$ [76077-76-4] (70). The only complexes between XeO_2F_2 and a strong fluoride ion donor are the salts $Cs^+ XeO_2F_3^-$ [65014-03-1] and $[NO_2^+][XeO_2F_3^- \cdot nXeO_2F_2]$ [116025-38-8] (27,74).

Alkali metal fluoroxenates $KXeO_3F$ [23525-88-4, 27002-68-2], $RbXeO_3F$ [12434-32-1, 33572-57-5], and $CsXeO_3F$ [12443-46-8, 33572-58-6], which decompose above 200°C, and a chloroxenate, $CsXeO_3Cl$ [26283-13-6, 27002-67-1], which decomposes above 150°C, have been prepared by evaporating aqueous solutions of XeO_3 and the corresponding alkali metal fluorides and chlorides. The alkali metal fluoroxenates are the most stable solid oxygenated compounds of xenon(VI) known. X-ray crystallography shows that $KXeO_3F$ is best formulated as $nK^+[XeO_3F^-]_n$ in which each XeO_3 group is bonded to two fluorine atoms which bridge adjacent XeO_3 groups to give an open-chain polymeric structure (75). Similarly, the structures of the compounds $2.25\ MCl \cdot XeO_3$ (M = Rb, Cs) feature infinite chains of $[XeO_3Cl]^-$ units linked by nearly linear chlorine bridges (86). The $CsXeO_3Br$ compound is unstable even at room temperature (87).

Derivatives in Which Xenon is Bonded to Polyatomic Groups. *Xenon Bonded to Oxygen.* The greatest variety of polyatomic ligand groups bonded to xenon occur for xenon in its +2 oxidation state, and those bonded through oxygen are the most plentiful. Both mono- and disubstituted derivatives having the formulations FXeL and XeL_2 are known where L = $OTeF_5$, $OSeF_5$, OSO_2F, $OP(O)F_2$, $OClO_3$, ONO_2, $OC(O)CH_3$, $OC(O)CF_3$, OSO_2CH_3, OSO_2CF_3, and $OIOF_4$. With the exception of $OIOF_4$ and $OP(O)F_2$, the syntheses usually involve HF elimination reactions of the parent acid HL and XeF_2 (eg, $XeF_2 + x\ HL \longrightarrow F_{2-x}XeL_x + x\ HF$; $x = 1, 2$) (9,88).

The highly electronegative $-OTeF_5$ group is the only ligand known to completely replace fluorine in compounds of xenon in the +4 and +6 oxidation states. The derivative $Xe(OTeF_5)_4$ [66255-64-9] is the only known example of xenon in the +4 oxidation state bonded exclusively to oxygens (89–91); the average Xe–O bond length is 203.2(5) pm (92). The mixed derivatives $F_xXe(OTeF_5)_{4-x}$ ($x = 1-3$) result from the ligand redistribution reaction of XeF_4 and $Xe(OTeF_5)_4$ (93). Reaction of $Xe(OTeF_5)_4$ with SbF_5 results in the formation of the mixed cations $F_xXe(OTeF_5)_{3-x}^+$ ($x = 1, 2$) (94). The known $-OTeF_5$ derivatives of xenon in the +6 oxidation state include $Xe(OTeF_5)_6$ [68854-31-9] (90), which, unlike its fluorine analogue, XeF_6, is monomeric in the solid state; $O_2Xe(OTeF_5)_2$ (92,93); and $O=Xe(OTeF_5)_4$ [68854-32-0] (95). The mixed ligand derivatives, $O_2XeF(OTeF_5)$ [91002-54-9] (5) and $O=XeF_x(OTeF_5)_{4-x}$ ($x = 1-3$) (5,90,93), are also known, as well as the $O_2Xe(OTeF_5)^+$ [142533-95-7] (94) and $O=XeF_x(OTeF_5)_{3-x}^+$ ($x = 1, 2$) cations (94,95).

Xenon Bonded to Nitrogen. Several ligand groups form compounds containing xenon-nitrogen bonds (12). The first xenon-nitrogen bonded compound, $FXeN(SO_2F)_2$ [53719-78-1], was prepared by reaction of $HN(SO_2F)_2$ and XeF_2 at 0°C in CF_2Cl_2 solvent (96). Proof of Xe—N bonding was established by x-ray crystallography (97). The Xe—N and Xe—F bond distances are 220.0(3) pm and

196.7(3) pm, respectively, with trigonal planar coordination around the nitrogen atom. Compounds containing N–Xe–N linkages are also known, namely Xe[N(SO$_2$F)$_2$]$_2$ [85883-06-3] (98,99) and Xe[N(SO$_2$CF$_3$)$_2$]$_2$ [82113-64-2] (100). The fluoride ion donor properties of FXeN(SO$_2$F)$_2$ have been studied, leading to the characterization of salts of the XeN(SO$_2$F)$_2^+$ cation (98,101) and the fluorine-bridged (FO$_2$S)$_2$NXe---F---XeN(SO$_2$F)$_2^+$ cation (98,99,101).

The XeF$^+$ cation forms Lewis acid–base adduct cations containing N–Xe–F linkages with nitrogen bases that are resistant to oxidation by the strongly oxidizing XeF$^+$ cation having an estimated electron affinity of the XeF$^+$ cation of 10.9 eV (12). The thermally unstable colorless salt, HC≡NXeF$^+$AsF$_6^-$ [112144-23-7], is prepared by reaction of HC≡N with Xe$_2$F$_3^+$AsF$_6^-$ or XeF$^+$AsF$_6^-$ in anhydrous HF solvent or by reaction of HC≡NH$^+$AsF$_6^-$ with XeF$_2$ in BrF$_5$ solvent (102,103). Other nitriles, RC≡N, forming the white solids RC≡N–XeF$^+$AsF$_6^-$ (R = CH$_3$ [112144-25-9], CH$_2$F [112144-27-1], CF$_3$ [119127-10-5], C$_2$H$_5$ [112144-29-3], C$_2$F$_5$ [112144-31-7], C$_3$F$_7$ [112144-33-9], and C$_6$F$_5$ [112154-24-2]), have been prepared using methods similar to those used for the preparation of HC≡NXeF$^+$AsF$_6^-$ and generally are less thermally stable than HC≡NXeF$^+$AsF$_6^-$ (102,104). The fluoro(perfluoropyridine)xenon(II) cations 4-RC$_5$F$_4$N–XeF$^+$ (R = F [114481-54-8], CF$_3$ [114481-56-0]) and the s-C$_3$F$_3$N$_2$N–XeF$^+$ cation have been prepared as colorless AsF$_6^-$ salts and are planar cations in which the xenon atom is bonded to the ring through the lone pair of electrons on a nitrogen atom (105). The salt, s-C$_3$F$_3$N$_2$N–XeF$^+$AsF$_6^-$ [119127-04-7], is the only compound in this series that is stable at room temperature (104) and is prepared by the reaction of XeF$^+$AsF$_6^-$ with liquid s-trifluorotriazine, s-C$_3$F$_3$N$_3$, at room temperature.

Xenon Bonded to Carbon. A number of structurally well-characterized compounds containing Xe—C bonds are known. In all cases these occur as colorless salts of xenonium cations, R–Xe$^+$ where R is a fluorophenyl or alkynyl group. The formation of the pentafluorophenylxenon(II) cation, C$_6$F$_5$Xe$^+$ [121850-39-3] in CH$_2$Cl$_2$ (−30°C) and CH$_3$C≡N (0°C) solutions with the anions B(C$_6$F$_5$)$_3$F$^-$ [121850-40-6], B(C$_6$F$_5$)$_2$F$_2^-$ [123168-25-2], and B(C$_6$F$_5$)F$_3^-$ [124302-51-8] has been established (106–110). The salts are formed by the reaction of XeF$_2$ with the ligand transfer reagent B(C$_6$F$_5$)$_3$. The x-ray crystal structure of [CH$_3$C≡N—Xe—C$_6$F$_5$]$^+$[(C$_6$F$_5$)$_2$BF$_2$]$^-$ shows that the xenon atom is weakly coordinated to the nitrogen atom of a CH$_3$C≡N molecule, with a Xe—N bond length of 268.1(8) pm, and a Xe—C bond length of 209.2(8) pm (111).

The existence of the XeCH$_3^+$ [34176-86-8] cation has been established in the gas phase. The Xe—C bond energy of the XeCH$_3^+$ cation has been estimated to be 180 ± 33 kJ/mol (112) and more recently, 231 ± 10 kJ/mol (113) by ion cyclotron resonance. The compound Xe(CF$_3$)$_2$ [72599-34-9] is reported to be a waxy white solid having a half-life of ca 30 min at room temperature (114). The synthesis involved the addition of XeF$_2$ to a trifluoromethyl plasma, but the characterization of this compound is limited and has not been independently confirmed.

Krypton Compounds

Krypton Difluoride. Krypton difluoride [13773-81-4], KrF$_2$, is a colorless crystalline solid which can be sublimed under vacuum at 0°C but is thermodynamically unstable and slowly decomposes to the elements at ambient temperatures (Table 1). It can, however, be stored for indefinite periods of time at $-78°C$. The KrF_2 molecule has been shown, like XeF_2, to be linear in the gas phase, in the solid state, and in solution. The standard enthalpy of formation, derived from calorimetric measurements of the gaseous compound at 93°C, is 60.2 kJ/mol (14.4 kcal/mol) (15). Consistent with its thermodynamic instability, krypton difluoride is a powerful oxidative fluorinating agent, and is capable of oxidizing xenon to XeF_6 (115) and gold to AuF_5 (116). The heat of atomization for KrF_2 is only 97.8 kJ/mol (23.4 kcal/mol) (14,15) and is substantially less than that of F_2 at 157.7 ± 0.4 kJ/mol (37.7 ± 0.1 kcal/mol) (117), making it a better low temperature source of fluorine atoms and an aggressive fluorinating agent at even low temperatures. Although the first krypton compound to be prepared was described as the tetrafluoride, the properties ascribed to this material have been shown to be those of the difluoride. No other molecular fluoride of krypton is known, so that the chemistry of krypton is limited to that of krypton difluoride and its derivatives. Krypton monofluoride [34160-02-6], KrF, is short-lived in the gas phase and is an important species in excimer laser systems.

Complex Salts. The cationic species KrF^+ and $Kr_2F_3^+$ are formed in reactions of KrF_2 with strong and weak fluoride ion acceptors to give complex salts that are analogous to those of XeF_2. Salts of the former type are formed with the pentafluorides of Group V metals and those of platinum and gold: $KrF^+SbF_6^-$ [52708-44-8], [35289-40-8]; $KrF^+Sb_2F_{11}^-$ [35140-44-4], [39578-36-4]; $Kr_2F_3^+SbF_6^-$ [52708-43-7], [52721-22-9]; $KrF^+AsF_6^-$ [50859-36-4]; $Kr_2F_3^+AsF_6^-$ [52721-23-0]; $KrF^+Nb_2F_{11}^-$ [58815-73-9]; $KrF^+TaF_6^-$ [39438-53-4]; $KrF^+Ta_2F_{11}^-$ [58815-72-8]; $KrF^+PtF_6^-$ [52707-25-2]; and $KrF^+AuF_6^-$ [57583-94-5] (10,118,119). Unlike their xenon analogues, the majority of these salts decompose below room temperature, but three, $KrF^+AsF_6^-$, $KrF^+SbF_6^-$, and $KrF^+Sb_2F_{11}^-$, are moderately stable at room temperature. The KrF^+ cation ranks as the most powerful chemical oxidizer known (120) and is capable of oxidizing gaseous xenon to XeF_5^+, gaseous oxygen to O_2^+, NF_3 to NF_4^+, and chlorine, bromine, and iodine pentafluorides to the ClF_6^+, BrF_6^+, and IF_6^+ cations, respectively (118,121–124). Adducts with the weak fluoride ion acceptors $CrOF_4$, $MoOF_4$, and WOF_4 are known in which KrF_2 interacts with the metal center by formation of asymmetric Kr---F---M bridges (M = Cr, Mo, W), eg, F—Kr---F---MOF_4 [102110-05-4], [77744-88-8], [77744-91-3], F—Kr---F---$MoOF_4(MoOF_4)_2$ [77744-90-2] (77,125).

Other Derivatives. Despite the strong oxidizing properties of the KrF^+ cation, it has been shown to behave as a Lewis acid toward a limited number of Lewis bases at low temperatures. These bases are resistant to oxidation by the strongly oxidizing KrF^+ cation and, as in the case of the XeF^+ cation, they have first adiabatic ionization potentials that are greater than or comparable to the estimated electron affinity of the KrF^+ cation (13.2 eV) (12). The Lewis acid-base cations are all thermally unstable above ca $-40°C$ and consist of $HC{\equiv}N—KrF^+$ [117222-92-1], $F_3CC{\equiv}N—KrF^+$ [119127-12-7], $F_3CCF_2C{\equiv}N—KrF^+$ [119127-06-9], and n-$F_3CCF_2CF_2C{\equiv}N—KrF^+$ [119127-08-1], all having AsF_6^- as the counter anion (126,127). These cations comprise the only examples of krypton bonded to nitrogen. The compound $Kr(OTeF_5)_2$ [68854-33-1] provides the only reported example of a compound in which krypton is bonded to oxygen (128). The existence of the $KrCH_3^+$ cation [109282-51-1] has been established in the gas phase by ion cyclotron resonance spectroscopy (129). The Kr—C bond energy of the $KrCH_3^+$ cation has been estimated to be 199.6 ± 10.5 kJ/mol (47.7 ± 2.5 kcal/mol) and is considerably more stable than the Kr—F bonds of KrF_2 (mean thermochemical bond energy of 48.9 kJ/mol (11.7 kcal/mol)) and KrF^+ (~ 155 kJ/mol (~ 37 kcal/mol)). No compounds in which krypton is bonded to elements other than fluorine, oxygen, and nitrogen have been isolated. Indeed, the nonexistence of simple oxides or oxofluorides is consistent with the lack of a higher oxidation state of krypton.

Radon Compounds

Radon Fluoride. When a mixture of trace amounts of radon-222 and fluorine gas are heated to approximately 400°C, a nonvolatile fluoride is formed. The intense α-radiation of radon provides the activation, allowing radon in such quantities to react spontaneously with gaseous fluorine at room temperature and with liquid fluorine at -196°C. Radon is also oxidized by chlorine and bromine fluorides, IF_7 or NiF_6^{2-}, in HF to give stable solutions of radon fluoride. The products of these fluorination reactions have not been analyzed because of small masses and intense radioactivity. It has nevertheless been possible to deduce that radon difluoride [18976-85-7], RnF_2, is formed, as well as derivatives of the difluoride, by comparing reactions of radon with those of krypton and xenon. Electromigration and ion-exchange (qv) studies show that ionic radon is present in many of these solutions and is believed to be RnF^+ and Rn^{2+}. The chemical behavior of radon is similar to that of a metal fluoride and is consistent with its position in the Periodic Table as a metalloid element (130). The chemistry of radon has been fully described (130,131).

Complex Salts. Radon reacts at room temperature with solid oxidants, such as $O_2^+SbF_6^-$, $O_2^+Sb_2F_{11}^-$, $N_2F^+SbF_6^-$, and $BrF_2^+BiF_6^-$ to form nonvolatile complex salts which are believed to be $RnF^+SbF_6^-$ [73384-63-1], $RnF^+Sb_2F_{11}^-$ [53851-50-6], and $RnF^+BiF_6^-$ [73384-62-0], by analogy with xenon, which forms the well-characterized products, $XeF^+SbF_6^-$, $XeF^+Sb_2F_{11}^-$, and $XeF^+BiF_6^-$.

Prospects for Argon Compounds

Theoretical calculations indicate that argon difluoride should be unstable but that the ArF^+ cation should be stable in the presence of a suitable oxidatively resistant anion. The corresponding HeF^+ [12336-97-9] and NeF^+ [12518-02-4] cations are predicted to be unstable (132). Experimental evidence for ArF^+ [11089-94-4] in the gas phase has been obtained, leading to $D_o(ArF^+) \geq 1.655$ eV and confirming the instability of HeF^+ and NeF^+ in their electronic ground states (133). The electronegativity values assigned to the compound-forming helium-group gases (Kr, 2.9; Xe, 2.3; and Rn, 2.1) and those predicted for the gases which do not appear to form compounds (He, 5.2; Ne, 4.5; and Ar, 3.2) suggest that the values for argon and krypton are rather close, and that efforts to synthesize an ArF^+ salt are realistic. The ArF^+ cation (estimated electron affinity, 13.7 eV) is expected to be an even stronger oxidant than KrF^+ (~ 13.2 eV), making it an oxidizer of unprecedented strength (120,132). However, unlike KrF^+, which can only be synthesized from KrF_2, any synthesis of an ArF^+

salt cannot rely on the difluoride precursor, and the anion used to stabilize the ArF^+ cation must be capable of withstanding oxidation by ArF^+. Theoretical calculations also suggest that $HC{\equiv}N{-}ArF^+$ [124354-45-6] may be stable (134).

Methods of Preparation of Binary Fluorides

All helium-group gas chemistry originates with the binary fluorides. The amounts of the binary xenon fluorides suitable for synthetic work are generally prepared by heating mixtures of xenon and fluorine to 250–400°C in nickel or Monel vessels (135–137). Although all three fluorides coexist in equilibrium, suitable adjustments of temperature, pressure, and xenon:fluorine ratio can be made to yield primarily difluoride, tetrafluoride, or hexafluoride. Xenon difluoride can be prepared photochemically by exposing xenon and fluorine, contained in a Pyrex flask, either to direct sunlight or to ultraviolet light from a mercury arc (138–140). Other methods, including electric discharges, proton and electron beams, and γ-rays, have been employed for the preparation of xenon fluorides but are rarely used. Both XeF_2 and XeF_4 can be manipulated in glass vacuum systems, but XeF_6 must be handled in either fluorine-passivated metal or fluoroplastic vacuum systems. Safety measures, such as the use of protective glasses, face shields, and other personal protective covering, are essential for work involving tetrafluoride and hexafluoride which react with water to form explosive XeO_3; XeF_4 disproportionates in water according to the reaction, $6\ XeF_4(s) + 12\ H_2O(l) \longrightarrow 2\ XeO_3(s) + 4\ Xe(g) + 3\ O_2(g) + 24\ HF(aq)$.

Krypton difluoride cannot be synthesized by the standard high pressure–high temperature means used to prepare xenon fluorides because of the low thermal stability of KrF_2. There are three low temperature methods which have proven practical for the preparation of gram and greater amounts of KrF_2 (141–143). Radon fluoride is most conveniently prepared by reaction of radon gas with a liquid halogen fluoride (ClF, ClF_3, ClF_5, BrF_3, or IF_7) at room temperature (144,145).

Uses

Stable noble-gas compounds have no industrial uses as of this writing but are frequently utilized in laboratories as fluorinating and oxidizing agents. Xenon difluoride and xenon tetrafluoride are relatively mild oxidative fluorinating agents and have been used for the preparation of phosphorus, sulfur, tellurium, and silicon derivatives (146,147). Xenon difluoride has proven to be a versatile and stable fluorinating agent for use in synthetic organic chemistry (148). It has been used for the fluorination of alkenes, in fluorodecarboxylation, and for the fluorination of thioethers, aromatic, and aliphatic compounds. Xenon hexafluoride has been used to synthesize transition-metal fluorides and oxofluorides where the metal is in its highest oxidation state, eg, $Xe_2F_{11}^+AuF_6^-$ (65) and TcO_2F_3 (149). Krypton difluoride and its complex salts are extremely powerful oxidative fluorinating agents and can be used to oxidatively fluorinate gold, silver, and halogen fluorides to their highest oxidation states, eg, AuF_5, AgF_4^-, ClF_6^+, BrF_6^+. Krypton species can also be used to fluorinate oxocompounds of elements that are already in their highest attainable oxidation states as is exemplified by the fluorination of OsO_4 to cis-OsO_2F_4 by KrF_2 (150) and $XeOF_4$ to XeF_5^+ by KrF^+ (116). Aqueous solutions of sodium perxenate and of xenon trioxide are intermediate in oxidizing strength and are useful for analyzing manganese (151), and alcohols and carboxylic acids (152,153), respectively. Radon-222, an air contaminant in uranium mines, can be analyzed by means of oxidants that form nonvolatile radon salts (154).

A particularly important use for unstable noble-gas halides is as the gain medium in excimer lasers, which are increasingly being employed as high power sources of tunable laser light in the ultraviolet and visible spectral regions (155). The diatomic halides are used for this purpose because they are readily formed in their excited states by electron-beam pumping or by discharge pumping of suitable gas mixtures. Compared to other laser systems, high gains can be achieved because the ground states are generally dissociative. Consequently, these lasers provide the only sources of high energy pulses (up to several joules of typically 10 to 100 ns duration) in the ultraviolet. The laser systems include ArCl [54635-29-9] (175 nm), ArF [56617-31-3] (193 nm), KrCl [56617-29-9] (222 nm), KrF [34160-02-6] (249 nm), XeBr [55130-04-6] (282 nm), XeCl [55130-03-5] (308 nm), and XeF [16757-14-5] (350 nm). Of these, XeCl is especially useful because of the capability for long operating lifetimes. Most applications of excimer lasers are based on the removal of material. Unlike longer wavelength lasers, which heat and vaporize, excimer lasers remove material by ablation, which results in much finer detail and much less damage to surrounding areas. Applications include medical uses, processing of integrated solid-state devices, and use in photochemistry, isotope separation, mineral exploration, and studies of nuclear fusion. Medical applications include laser keratectomy and laser angioplasty.

BIBLIOGRAPHY

"Helium-Group Gases, Compounds" in *ECT* 2nd ed., Vol. 10, pp. 888–894, by H. H. Hyman. Argonne National Laboratory, in *ECT* 3rd ed., Vol. 12, pp. 288–297, by L. Stein, Argonne National Laboratory.

1. P. Laszlo and G. J. Schrobilgen, *Angew. Chem. Int. Ed. Engl.* **27**, 479 (1988).
2. N. Bartlett, *Proc. Chem. Soc.*, 218 (1962).
3. H. H. Claassen, H. Selig, and J. G. Malm, *J. Am. Chem. Soc.* **84**, 3593 (1962).
4. R. Hoppe, W. Dähne, H. Mattauch, and K. Rödder, *Angew. Chem. Int. Ed. Engl.* **1**, 599 (1962).
5. H. H. Hyman, ed., *Noble Gas Compounds*, University of Chicago Press, Chicago, 1963.
6. "Edelgasverbindungen" in *Gmelins Handbuch der anorganischen Chemie*, 8th ed., Main Supplement, Vol. 1, Verlag Chemie, Weinheim, Germany, 1970.
7. N. Bartlett and F. O. Sladky, in J. C. Bailar, Jr., H. J. Emeléus, R. Nyholm, and A. F. Trotman-Dickenson, eds., *Comprehensive Inorganic Chemistry*, Vol. 1, Pergamon Press, New York, 1973, pp. 213–330.
8. D. T. Hawkins, W. E. Falconer and N. Bartlett, *Noble Gas Compounds. A Bibliography, 1962–1976*, Plenum Press, Inc., New York, 1978.
9. K. Seppelt and D. Lentz, in S. J. Lippard, ed., *Progress in Inorganic Chemistry*, John Wiley & Sons, Inc., New York, 1982, pp. 167–202.

10. H. Selig and J. H. Holloway, in F. L. Boschke, ed., *Topics in Current Chemistry*, Vol. 124, Springer-Verlag, Berlin, 1984, pp. 33–90.
11. B. Žemva, *Croat. Chem. Acta* **61**, 163 (1988).
12. G. J. Schrobilgen, in R. D. Chambers, G. A. Olah, and G. K. S. Prakash, eds., *Synthetic Fluorine Chemistry*, John Wiley & Sons, Inc., New York, 1992, Chapt. 1, pp. 1–30.
13. S. Siegel and E. Gebert, *J. Am. Chem. Soc.* **86**, 3896 (1964).
14. S. R. Gunn, *J. Am. Chem. Soc.* **88**, 5924 (1966).
15. S. R. Gunn, *J. Phys. Chem.* **71**, 2934 (1967).
16. G. K. Johnson, J. G. Malm, and W. N. Hubbard, *J. Chem. Thermodynam.* **4**, 879 (1972).
17. D. W. Osborne, H. E. Flotow, J. G. Malm, *J. Chem. Phys.* **57**, 4670 (1972).
18. F. Schreiner, G. N. McDonald, and C. L. Chernick, *J. Phys. Chem.* **72**, 1162 (1968).
19. S. Siegel and E. Gebert, *J. Am. Chem. Soc.* **85**, 240 (1963).
20. P. A. Argon, G. M. Begun, H. A. Levy, A. A. Mason, G. Jones, and D. F. Smith, *Science* **139**, 842 (1963).
21. D. W. Osborne, F. Schreiner, H. E. Flotow, and J. G. Malm, *J. Chem. Phys.* **57**, 3401 (1972).
22. D. H. Templeton, A. Zalkin, J. D. Forrester, and S. M. Williamson, *J. Am. Chem. Soc.* **85**, 242 (1963).
23. F. Schreiner, D. W. Osborne, J. G. Malm, and G. N. McDonald, *J. Chem. Phys.* **51**, 4838 (1969).
24. P. A. Agron, C. K. Johnson, and H. A. Levy, *Inorg. Nucl. Chem. Lett.* **1**, 145 (1965).
25. R. D. Burbank and G. R. Jones, *Science* **168**, 248 (1970); R. D. Burbank and G. R. Jones, *Science* **171**, 485 (1971).
26. J. S. Ogden and J. J. Turner, *Chem. Commun.*, 693 (1966).
27. R. J. Gillespie and G. J. Schrobilgen, *J. Chem. Soc., Chem. Commun.*, 595 (1977).
28. H. Selig, *Inorg. Chem.* **5**, 183 (1966).
29. S. R. Gunn, *J. Am. Chem. Soc.* **87**, 2290 (1965).
30. S. W. Peterson, R. D. Willett, and J. L. Huston, *J. Chem. Phys.* **59**, 453 (1973).
31. S. R. Gunn, in H. H. Hyman, ed., *Noble Gas Compounds*, University of Chicago Press, Chicago and London, 1963, pp. 149–151.
32. G. Nagarajan, *Bull. Soc. Chim. Belges.* **73**, 665 (1964).
33. H. Selig, H. H. Claassen, C. L. Chernick, J. G. Malm, and J. L. Huston, *Science* **143**, 1322 (1963).
34. G. A. Cook, ed., *Argon, Helium and the Rare Gases*, Vol. 1, Interscience Publishers, New York, 1961, pp. 157–172.
35. S. Reichman and F. Schreiner, *J. Chem. Phys.* **51**, 2355 (1969).
36. J. H. Burns, P. A. Agron, and H. A. Levy, *Science* **139**, 1208 (1963).
37. H. H. Rupp and K. Seppelt, *Angew. Chem. Int. Ed. Engl.* **13**, 613 (1974).
38. R. D. Burbank and G. R. Jones, *J. Am. Chem. Soc.* **96**, 43 (1974); R. D. Burbank and G. R. Jones, *Science* **171**, 485 (1971); G. R. Jones, R. D. Burbank and W. E. Falconer, *J. Chem. Phys.* **52**, 6450 (1970); G. R. Jones, R. D. Burbank, and W. E. Falconer, *J. Chem. Phys.* **53**, 1605 (1970).
39. H. Meinert, *Z. Chem.* **6**, 71 (1966); H. Meinert, *Z. Chem.* **9**, 349 (1969).
40. W. F. Howard and L. Andrews, *J. Am. Chem. Soc.* **96**, 7864 (1974).
41. J. G. Perlow and H. Yoshida, *J. Chem. Phys.* **49**, 1474 (1968).
42. B. Jaselskis and J. P. Warriner, *Anal. Chem.* **38**, 563 (1966).
43. C. A. Brau and J. J. Ewing, *J. Chem. Phys.* **63**, 4640 (1975).
44. J. Tellinghausen, A. K. Hays, J. M. Hoffman, and G. C. Tisone, *J. Chem. Phys.* **65**, 4473 (1976).
45. J. R. Morton and W. E. Falconer, *J. Chem. Phys.* **39**, 427 (1963).

46. B. S. Ault and L. Andrews, *J. Chem. Phys.* **65**, 4192 (1976).
47. E. H. Appelman, *Inorg. Synth.* **11**, 211 (1968).
48. J. Foropoulos, Jr., and D. D. DesMarteau, *Inorg. Chem.* **21**, 2503 (1982).
49. D. H. Templeton, A. Zalkin, J. D. Forrester, and S. M. Williamson, *J. Am. Chem. Soc.* **85**, 817 (1963).
50. G. Gunderson, K. Hedberg, and J. Huston, *Acta Cryst.* **25**, 124 (1969).
51. J. Martins and E. B. Wilson, Jr., *J. Mol. Spectrosc.* **26**, 410 (1968).
52. S. W. Peterson, R. D. Willett, and J. L. Huston, *J. Chem. Phys.* **59**, 453 (1973).
53. J. Shamir, H. Selig, D. Samuel, and J. Reuben, *J. Am. Chem. Soc.* **87**, 2359 (1965).
54. K. O. Christe and W. W. Wilson, *Inorg. Chem.* **27**, 1296 (1988).
55. K. O. Christe and W. W. Wilson, *Inorg. Chem.* **27**, 3763 (1988).
56. J. L. Huston, *Inorg. Nucl. Chem. Lett.* **4**, 29 (1968).
57. J. L. Huston, *J. Am. Chem. Soc.* **93**, 5255 (1973).
58. P. Boldrini, R. J. Gillespie, P. Ireland, and G. J. Schrobilgen, *Inorg. Chem.* **13**, 1690 (1974).
59. D. E. McKee, A. Zalkin, and N. Bartlett, *Inorg. Chem.* **12**, 1713 (1973).
60. R. J. Gillespie, D. Martin, G. J. Schrobilgen, and D. R. Slim, *J. Chem. Soc., Dalton Trans.*, 2234 (1977).
61. K. O. Christe, E. C. Curtis, D. A. Dixon, H. P. Mercier, J. C. P. Sanders, and G. J. Schrobilgen, *J. Am. Chem. Soc.* **113**, 3351 (1991).
62. N. Bartlett, F. Einstein, D. F. Stewart, and J. Trotter, *J. Chem. Soc., Chem. Commun.*, 550 (1966).
63. K. O. Christe and W. W. Wilson, *Inorg. Chem.* **21**, 4113 (1982).
64. K. O. Christe, W. W. Wilson, R. D. Wilson, R. Bougon, and T. Bui Huy, *J. Fluorine Chem.* **23**, 399 (1983).
65. K. Leary, A. Zalkin, and N. Barlett, *Inorg. Chem.* **13**, 775 (1974); K. Leary, A. Zalkin, and N. Bartlett, *J. Chem. Soc., Chem. Commun.*, 131 (1973).
66. S. W. Peterson, J. H. Holloway, J. H. Coyle, and J. M. Williams, *Science (Washington, D.C.)* **173**, 1238 (1971).
67. H. P. A. Mercier, J. C. P. Sanders, G. J. Schrobilgen, and S. S. Tsai, *Inorg. Chem.* **32**, 386 (1993).
68. K. O. Christe, J. C. P. Sanders, G. J. Schrobilgen, and W. W. Wilson, unpublished results.
69. M. C. Waldman and H. Selig, *J. Inorg. Nucl. Chem.* **35**, 2173 (1973).
70. J. H. Holloway, V. Kaučič, D. Martin-Rovet, G. J. Schrobilgen, and H. Selig, *Inorg. Chem.* **24**, 678 (1985).
71. K. O. Christe and W. W. Wilson, *Inorg. Chem.* **27**, 2714 (1988).
72. R. J. Gillispie and G. J. Schrobilgen, *Inorg. Chem.* **13**, 2370 (1974).
73. R. J. Gillespie, B. Landa, and G. J. Schrobilgen, *Inorg. Chem.* **15**, 1256 (1976).
74. J. G. Malm and E. H. Appelman, *At. Energy Rev.* **7**, 3 (1969).
75. D. J. Hodgson and L. A. Ibers, *Inorg. Chem.* **8**, 326 (1969).
76. J. H. Holloway and G. J. Schrobilgen, *Inorg. Chem.* **19**, 2632 (1980).
77. J. H. Holloway and G. J. Schrobilgen, *Inorg. Chem.* **20**, 3363 (1981).
78. P. A. Tucker, P. A. Taylor, J. H. Holloway, D. R. Russell, *Acta Cryst.* **B31**, 906 (1975).
79. N. Keller and G. J. Schrobilgen, *Inorg. Chem.* **20**, 2118 (1981).
80. G. R. Jones, R. D. Burbank and N. Bartlett, *Inorg. Chem.* **9**, 2264 (1970).
81. M. Wechsberg and N. Barlett, *Z. Anorg. Allg. Chem.* **385**, 5 (1971).
82. J. H. Burns, R. D. Ellison and H. A. Levy, *Acta. Cryst.* **18**, 11 (1965).
83. L. Stein, *J. Fluorine Chem.* **20**, 65 (1982).
84. D. R. Brown, M. J. Clegg, A. J. Downs, R. C. Fowler, A. R. Minihan, J. R. Norris, and L. Stein, *Inorg. Chem.* **31**, 5041 (1992).
85. R. D. Peacock, H. Selig, and I. Sheft, *Proc. Chem. Soc.*, 285 (1964); R. D. Peacock, H. Selig, and I. Sheft, *J. Inorg. Nucl. Chem.* **28**, 2561 (1966).

86. R. D. Willett, S. W. Peterson, and B. A. Coyle, *J. Am. Chem. Soc.* **99**, 8202 (1977).
87. B. Jaselskis, J. L. Huston, and T. M. Spittler, *J. Am. Chem. Soc.* **91**, 1874 (1969).
88. R. G. Syvret and G. J. Schrobilgen, *Inorg. Chem.* **28**, 1564 (1989).
89. D. Lentz and K. Seppelt, *Angew. Chem. Int. Ed. Engl.* **17**, 356 (1978).
90. E. Jacob, D. Lentz, K. Seppelt, and A. Simon, *Z. Anorg. Allg. Chem.* **472**, 7 (1981).
91. T. Birchall, R. D. Myers, H. deWaard, and G. J. Schrobilgen, *Inorg. Chem.* **21**, 1068 (1982).
92. L. Turowsky and K. Seppelt, *Z. Anorg. Allg. Chem.* **609**, 153 (1992).
93. G. A. Schumacher and G. J. Schrobilgen, *Inorg. Chem.* **23**, 2923 (1984).
94. R. G. Syvret, K. M. Mitchell, J. C. P. Sanders, and G. J. Schrobilgen, *Inorg. Chem.* **31**, 3381 (1992).
95. D. Lentz and K. Seppelt, *Angew. Chem. Int. Ed. Engl.* **18**, 66 (1979).
96. R. D. LeBlond and D. D. DesMarteau, *J. Chem. Soc. Chem. Commun.*, 555 (1974).
97. J. F. Sawyer, G. J. Schrobilgen, and S. J. Sutherland, *Inorg. Chem.* **21**, 4064 (1982).
98. D. D. DesMarteau, R. D. LeBlond, S. F. Hossain, and D. Nothe, *J. Am. Chem. Soc.* **103**, 7734 (1981).
99. G. A. Schumacher and G. J. Schrobilgen, *Inorg. Chem.* **22**, 2178 (1983).
100. J. Foropoulos, Jr. and D. D. DesMarteau, *J. Am. Chem. Soc.* **104**, 4260 (1982).
101. R. Faggiani, D. K. Kennepohl, C. J. L. Lock, and G. J. Schrobilgen, *Inorg. Chem.* **25**, 563 (1986).
102. A. A. A. Emara and G. J. Schrobilgen, *J. Chem. Soc., Chem. Commun.*, 1644 (1987).
103. A. A. A. Emara and G. J. Schrobilgen, *Inorg. Chem.* **31**, 1323 (1992).
104. G. J. Schrobilgen, *J. Chem. Soc., Chem. Commun.*, 1506 (1988).
105. A. A. A. Emara and G. J. Schrobilgen, *J. Chem. Soc., Chem. Commun.*, 257 (1988).
106. H. J. Frohn and S. Jakobs, *J. Chem. Soc., Chem. Commun.*, 625 (1989).
107. H. J. Frohn, S. Jakobs, and C. Rossbach, *Eur. J. Solid State Inorg. Chem.* **29**, 729 (1992).
108. H. J. Frohn and A. Klose, *J. Fluorine Chem.* **64**, 201 (1993).
109. H. Butler, D. Naumann, and W. Tyrra, *Eur. J. Solid State Inorg. Chem.* **29**, 739 (1992).
110. D. Naumann and W. Tyrra, *J. Chem. Soc., Chem. Commun.*, 47 (1989).
111. H. J. Frohn, S. Jakobs and G. Henkel, *Angew. Chem. Int. Ed. Engl.* **28**, 1506 (1989).
112. D. Holtz and J. L. Beauchamp, *Science* **173**, 1238 (1971).
113. J. K. Hovey and T. B. McMahon, *J. Am. Chem. Soc.* **108**, 528 (1986).
114. L. J. Turbini, R. E. Aikman, and R. J. Lagow, *J. Am. Chem. Soc.* **101**, 5833 (1979).
115. B. Frlec and J. H. Holloway, *J. Chem. Soc., Chem. Commun.*, 370 (1973).
116. J. H. Holloway and G. J. Schrobilgen, *J. Chem. Soc., Chem. Commun.*, 623 (1975).
117. A. G. Sharpe, in V. Gutmann, ed., *Halogen Chemistry*, Vol. 1, Academic Press, New York, 1967, p. 1.
118. R. J. Gillespie and G. J. Schrobilgen, *Inorg. Chem.* **15**, 22 (1976).
119. B. Frlec and J. H. Holloway, *Inorg. Chem.* **15**, 1263 (1976).
120. K. O. Christe and D. A. Dixon, *J. Am. Chem. Soc.* **114**, 2978 (1992).
121. D. E. McKee, C. J. Adams, A. Zalkin, and N. Bartlett, *J. Chem. Soc., Chem. Commun.*, 26 (1973).
122. R. J. Gillespie and G. J. Schrobilgen, *Inorg. Chem.* **13**, 1230 (1974).
123. K. O. Christe and W. W. Wilson, *Inorg. Chem.* **22**, 3056 (1983).
124. K. O. Christe, W. W. Wilson, and R. D. Wilson, *Inorg. Chem.* **23**, 2058 (1984).
125. K. O. Christe, R. Bougon, and W. W. Wilson, *Inorg. Chem.* **25**, 2163 (1986).
126. G. J. Schrobilgen, *J. Chem. Soc., Chem. Commun.*, 1506 (1988).
127. G. J. Schrobilgen, *J. Chem. Soc., Chem. Commun.*, 863 (1988).
128. J. C. P. Sanders and G. J. Schrobilgen, *J. Chem. Soc., Chem. Commun.* 1576 (1989).
129. J. K. Hovey and T. B. McMahon, *J. Phys. Chem.* **91**, 4560 (1987).

130. L. Stein, *Radiochimica Acta*, **32**, 163 (1983); L. Stein, *J. Chem. Soc., Chem. Commun.*, 1631 (1985).
131. L. Stein, in *ACS Symposium Series*, Vol. 331, Chapt. 18, pp. 240–251.
132. G. Frenking, W. Koch, C. A. Deakyne, J. F. Liebman, and N. Bartlett, *J. Am. Chem. Soc.* **111**, 31 (1989).
133. J. Berkowitz and W. A. Chupka, *J. Chem. Phys.* **7**, 447 (1970).
134. M. W. Wong and L. Radom, *J. Chem. Soc., Chem. Commun.*, 719 (1989).
135. J. G. Malm and C. L. Chernick, *Inorg. Synth.* **8**, 254 (1966).
136. A. Šmalc, K. Lutar, and B. Žemva, *Inorg. Synth.* **29**, 4 (1992).
137. C. L. Chernick and J. G. Malm, *Inorg. Synth.* **8**, 258 (1966).
138. J. H. Holloway, *J. Chem. Educ.* **43**, 202 (1966).
139. J. L. Weeks and M. S. Matheson, *Inorg. Synth.* **8**, 260 (1966).
140. A. Šmalc and K. Lutar, *Inorg. Synth.* **29**, 1 (1992).
141. F. Schreiner, J. G. Malm, and J. C. Hindman, *J. Am. Chem. Soc.* **87**, 25 (1965).
142. A. Šmalc, K. Lutar, and B. Žemva, *Inorg. Syn.* **29**, 11 (1992).
143. V. N. Bezmel'nitsyn, V. A. Legasov, and B. B. Chaivanov, *Proc. Acad. Sci. USSR (Engl. trans.)* **235**, 365 (1977); *Dokl. Akad. Nauk SSSR* **235**, 96 (1977).
144. L. Stein, *J. Am. Chem. Soc.* **91**, 5396 (1969).
145. L. Stein, *Science* **175**, 1462 (1972).
146. M. J. Shaw, H. H. Hyman, and R. Filler, *J. Am. Chem. Soc.* **91**, 1563 (1969).
147. R. K. Marat and A. F. Janzen, *Can. J. Chem.* **55**, 3031 (1977).
148. J. A. Wilkinson, *Chem. Rev.* **92**, 505 (1992).
149. H. P. A. Mercier and G. J. Schrobilgen, *Inorg. Chem.* **32**, 145 (1993).
150. K. O. Christe and R. Bougon, *J. Chem. Soc., Chem. Commun.*, 1056 (1992); K. O. Christe and co-workers, *J. Am. Chem. Soc.* **115**, 11279 (1993).
151. R. W. Bane, *Analyst* **90**, 756 (1965).
152. B. Jaselskis and J. P. Warriner, *Anal. Chem.* **38**, 563 (1966).
153. B. Jaselskis and R. H. Krueger, *Talanta* **13**, 945 (1966).
154. L. Stein, J. A. Shearer, F. A. Hohorst, and F. Markum, *Report USBM H0252019*, U.S. Department of the Interior, Bureau of Mines, Washington, D.C., Jan. 14, 1977.
155. M. H. R. Hutchinson in L. F. Mollenauer, J. C. White, and C. R. Pollock, eds., *Topics in Applied Physics*, Vol. 59, Springer-Verlag, Berlin, 1992, pp. 19–56.

GARY J. SCHROBILGEN
J. MARC WHALEN
McMaster University

HELMHOLTZ (YOUNG-HELMHOLTZ). See COLOR PHOTOGRAPHY; PRINTING PROCESSES.

HEMATOLOGY. See AUTOMATED INSTRUMENTATION, HEMATOLOGY.

HEMICELLULOSE

Hemicellulose [*9034-32-6*] is the least utilized component of the biomass triad comprising cellulose (qv), lignin (qv), and hemicellulose. The term was originated by Schulze (1) and is used here to distinguish the noncellulosic polysaccharides of plant cell walls from those that are not part of the wall structure. Confusion arises because other hemicellulose definitions based on solvent extraction are often used in the literature (2–4). The term polyose is used in Europe to describe these noncellulosic polysaccharides from wood, whereas hemicellulose is used to describe the alkaline extracts from commercial pulps (4). The quantity of hemicellulose in different sources varies considerably as shown in Table 1.

The complex nature of hemicellulose components and their degradation products after commercial processing makes a logical system of nomenclature difficult (12). An arbitrary system of nomenclature has evolved in which source and history is included along with a generic classification based on IUPAC conventions. Thus *O*-acetyl 4-*O*-methyl-D-glucuronoxylan, arabinogalactan, etc, are generic terms in which the main components are enumerated, and the unique character of each polymer is highlighted by including a distinguishing aspect, such as aspen glucomannan, esparto xylan, or spruce kraft arabinoxylan.

Pure hemicellulose components are seldom extracted directly from their source. Extracts are a mixture of polysaccharides, lignin, and lignin–hemicellulose complexes (by chemical linkages and possibly physical interactions) (13) characteristic of their origin and the solvent employed. Hemicellulose has a lower degree of polymerization (DP) than cellulose (about 200 vs more

Table 1. Approximate Hemicellulose Content of Selected Vegetable Materials and Their Residues

Raw material	Cellulose, %	Hemicellulose,[a] %	Lignin, %	Pectic material, %	Extractives, %	Reference
algae (green)	20–40	20–50	0	30–50	[b]	5
bast fibers	80–95	5–20	0	trace	[b]	6
grasses[c]	[d]	25.1	4.3	[d]	[d]	7
corn cobs	41	36	6	3	14	8
cornstalks	29	28	3	trace	[b]	9
wheat straw	40	29	14	trace	[b]	10
temperate hardwood	3–47	25–35	16–24	trace	2–8	11
softwood	40–44	25–29	25–31	trace	1–5	11
chemical pulps	60–80	5–15	0–10	0	0	[e]

[a]Quantity can vary according to sampling and definition employed.
[b]Present, but quantities not reported.
[c]Average composition of grasses grown in temperature zone.
[d]Not reported in reference.
[e]See CELLULOSE.

than 10,000) and its lower limits have not been clearly defined. The extract may contain two or more polymers of similar composition but different structures (polydiversity) or of different distributions and amounts of branching or bonding in otherwise similar molecules (polydispersity) (2,14,15). If a single polymer is present, it may exhibit a spectrum of molecular weights (polymolecularity) which may exhibit a Gaussian or biased distribution. A pure hemicellulose component is one where polydiversity has been avoided and a degree of heterogeneity has been attained compatible with end use application.

The data in Table 2 illustrate the composition of the hydrolysates of arborescent plants. Most of the glucose comes from cellulose, and the remaining glucose and other sugars are derived chiefly from the hemicellulose components. The most common hemicellulose in angiosperms is composed of D-xylose [58-86-6] arranged in a linear manner. D-Mannose [3458-28-4] is derived from a glucomannan which is the most common hemicellulose in most gymnosperms. Both contain other sugars and exist in a variety of configurations and molecular weights. Specialized parts such as reaction wood, branch wood, cambium, bark, rays, vessels and resin canals, etc, contain these components and other exotic polysaccharides whose exact function in the plant is not understood (see CARBOHYDRATES; WOOD). This article concentrates primarily on the components of tracheids and fibers of arborescent plants.

The common hemicellulose components of arborescent plants are listed in Table 3. Xylans, arabinogalactans, and pectic substances are common to all while only traces (if at all) of glucomannans are found in the cell walls of bamboo.

Table 2. Composition of Hydrolysates of Vegetable Matter[a,b]

Component	Angiosperms	Gymnosperms
D-galactose	$1 \rightarrow 5$	$1 \rightarrow 20$
D-glucose	$45 \rightarrow 50$	$35 \rightarrow 40$
D-mannose	$0 \rightarrow 5$	$12 \rightarrow 20$
L-arabinose	$1 \rightarrow 3$	$2 \rightarrow 5$
D-xylose	$18 \rightarrow 25$	$8 \rightarrow 15$
uronic acids	$6 \rightarrow 10$	$3 \rightarrow 6$
klason lignin	$16 \rightarrow 25$	$25 \rightarrow 34$

[a] Percent of component.
[b] Derived from data in Ref. 11.

Table 3. Hemicellulose Components of Arborescent Plants

Gymnosperms	Dicotyledons	Monocotyledon (bamboo)
arabino-(4-O-methyl-glucurono)xylan	O-acetyl-(4-O-methyl-glucurono)xylan	arabino-(4-O-methyl-glucurono)xylan
O-acetylgalacto-glucomannan (0.1:1:3)	glucomannan	heteroxylans
O-acetylgalacto-glucomannan (1:1:3)	arabinogalactan	D-glucans
arabinogalactan	pectic substances	arabinogalactans
pectic substances	tension wood components	pectic substances
compression wood components		

Other polysaccharides are found in trace amounts in wood as well as in bark, growing tissues, and other specialized parts of trees.

Structures of Hemicellulose Components

Xylans. The polymers which give rise to the sugars shown in Table 2 vary in composition and structure from species to species but can be grouped into several families. The most plentiful of these is the xylan [*9014-63-5*] family (Fig. 1) which is characterized by a relatively short backbone of $\beta(1\rightarrow4)$-linked D-xylopyranosyl units with 4-*O*-methyl-D-glucuronopyranosyl units attached by $\alpha(1\rightarrow2)$ bonds to some of the anhydroxylose units (11), although linkages to the C-3 position have been reported (16). The acids occur in the native state as carboxyls, esters (17), or possibly as salts. Arabinofuranosyl units, mannose, rhamnose, glucuronic acid, galacturonic acid, and *O*-acetyl groups have also been found in xylan hydrolysates and may be a component in some instances. Glycosidic and ester linkages to lignin also occur.

This polymer can be extracted in near quantitative yield from many lignified hardwoods but not from others such as elm (*Ulmus americana*) (11). The number average degree of polymerization (DP_n) of hardwood xylan varies from about 100 to 200 units depending on the source and the extent of degradation during isolation. Some may possess a T or comb shape instead of a filiform structure (18). Some researchers believe the distribution of acid units on the chain is random (19); others disagree (20). A distribution of acid groups between molecules in a given extract may also occur (21). Hardwood xylans have *O*-acetyl groups on 7 out of 10 anhydroxylose units at the 2,3 position and at the 2 and 3 hydroxyl positions (22,23). It has been shown (24,25) that rhamnose and galacturonic acid units are part of the chain located toward the reducing end of birch wood xylan (*Betula verrucosa*). Since this terminal configuration is not present in xylans isolated from chlorite holocelluloses (26), some chlorite-labile intermediate may exist between it and the rest of the xylan molecule. Arabinose (27) and mannose units (28) are also reported in some xylan hydrolysates as terminal units. Many of the xylans of herbaceous plants and seed hairs (except for cotton which has none) are similar to those in arborescent plants (29). Removal of branch substituents from isolated xylans can result in crystallization of the residual chain (30). The xylans in other components of dicotyledons such as bark, leaf stems, etc, vary in uronic acid content and degree of polymerization from those in fibers. The greatest variation occurs in mature lateral roots of sugar maple (*Acer saccharum* Marsh.) where the main chain is composed of $\beta(1\rightarrow4)$-

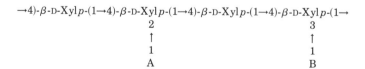

Fig. 1. A schematic representation of the xylan backbone of arborescent plants, where Xyl*p* = xylopyranosyl unit; A = α-(4-*O*)-methyl-D-glucuronopyranosyl) unit, sometimes an acetyl; and B = α-(L-arabinofuranosyl) unit, sometimes an acetyl.

linked D-glucopyranosyl and D-xylopyranosyl units (31) to which are attached branches of uronic acid and perhaps other structures.

Molecular weight studies of hardwood xylans are complicated by degradation during extraction (32) and conflicting theoretical considerations (33,34). Although hardwood xylan is polydisperse, its polymolecularity is very low (11) and shows a maximum at $DP_n = 200$. Other studies show solutions of xylan in water (35) and aqueous dimethyl sulfoxide (DMSO) (36), unlike the acetate dissolved in organic solvents (32), undergo changes in molecular associations with time and temperature. Branching inhibits the sorption of xylan from water and alkaline solutions onto cellulosic surfaces (37).

Xylan is plentiful in and easily extracted from monocotyledons. Its composition depends on the species involved and the fraction examined (3) and has a greater degree of polymolecularity, polydiversity, and polydispersity than wood xylan (14,38). Features of softwood xylans have been found in some xylans from bamboo (*Bambusa* sp.) (39), but most other fractions from stems and leaves are more complex (40). The DP of many nonarborescent monocotyledon xylans is generally lower than that found in wood, but it increases as the tissue matures (41,42). These complex xylans are referred to as heteroxylans and their composition and DP in the stalk differs from that in the leaves (43). Sorghum husk xylan (*Syrian granum*) is composed of three such polymers of varying composition and degrees of branching (44), the ratios of which can change with plant growth (45). Rice bran (*Oryza sativa*) (45) and wheat bran (*Tritium aestivum* L.) (46) yield a heteroxylan which contains more double-substituted xylose units and more complicated branches than are found in the polymer from the endosperm (47).

The xylan of gymnosperms is best isolated from holocellulose. It differs from that of dicotyledons by having a greater solubility in water, no acetyl substituents, a lower molecular weight (but of uncertain magnitude), greater substitution by L-arabinofuranosyl and 4-O-methyl-D-glucuronopyranosyl units (11), and a different distribution of branches along the chain. Uronic acids linked 1→3 to xylose have been reported occasionally (48), but most are linked by $\alpha(1 \to 2)$ bonds. There is random distribution of arabinofuranosyl units along the chain (49) and the distribution of uronic acids in softwood xylans often occurs in pairs. The isolation of two electrophoretically distinct xylans from slash pine (*Pinus elliottii* Engelm.) of identical chemical composition and structure suggests that the block formation of these xylans is dissimilar (50). Like hardwood xylans, many softwood xylans are terminated at the reducing end by rhamnose units and galacturonic acid (51).

The 4-O-methyl-D-glucuronoarabinoxylan in other parts of conifers differs slightly in composition and in quantity from that of tracheids. The gymnosperm *Ginkgo biloba* L. contains a polymer almost identical to that isolated from conifers (54). A softwood-like xylan is present in the primitive cinnamon fern (*Osmundia cinnamomea*) (55). The data in Table 4 illustrate the compositions of xylans from different sources. The molecular weight of undegraded softwood xylan is not known. A polymer isolated from Norway spruce (*Picea abies* (L.) Karst.) has many long branches, is probably degraded, and the DP_n of 128 is shifted toward the high end of the distribution spectrum (56).

The amount of softwood xylan sorbed from alkali at kraft cooking temperatures (100–170°C) is proportional to the quantity of hemicellulose present and

Table 4. A Comparison of Isolated Xylans From Different Sources

Source	Relative amounts of components in hydrolysate (xylose = 10)								Ref.
	Gal[a]	Glc[b]	Man[c]	Ara[d]	Rha[e]	GlcA[f]	MGlcA[g]	GalA[h]	
kenaf, *Hibiscus cannabinus*	0.2	0.07	0.02	0.02	0.07	0	1.4	0	28
birch, *Betula papyrifera*	0.02	0.4	0.07	0.2	trace	0	1.2	+	11
sorghum husk	0.5	5.5	trace	7.5	trace	0	1.5	−	44
esparto grass, *Stipa tenacissima*	0.5	0.9	0.05	1.4	trace	0	0	0	46
tobacco ECP,[i] *Nicotiana tabacum*	0.2	0.3	0.1	0.4	trace	2	0	0	52
softwood, average	trace	trace	0	1.3	trace	0	2	+	11
sapote gum, exudate from *Sapota achras*	0	0	0	4.5	0	1.7	1.8	0	53

[a]Galactose. [e]Rhamnose. [i]Extracellular polysaccharide from culture medium.
[b]Glucose. [f]Glucuronic acid.
[c]Mannose. [g]4-*O*-Methyl-D-glucuronic acid.
[d]Arabinose. [h]Galacturonic acid.

inversely to the extent of branching (57). At neutral pH, the presence of carboxyl groups inhibits sorption compared to a control, but no difference is observed when ionization is suppressed (58).

Glucomannans. Glucomannans [*11078-31-2*] are found in growing tissues as well as in the mature cells of many plants. They contain a chain of $\beta(1\rightarrow4)$-linked D-mannopyranosyl units (11) and D-glucopyranosyl units; the latter is randomly distributed in the case of spruce (*Picea mariana* (Mill.) B.S.P.) (59) (Fig. 2). The polymer is chiefly located in the secondary wall of fibers (60) and is often bonded to lignin (61). It undergoes noncovalent chain–chain aggregations with other polysaccharides (62). The principal variations in composition between species of plants are the ratio of components, the presence of acetyl groups, and the occurrence of terminal D-galactopyranosyl units.

Glucomannans in dicotyledon fibers are resistant to extraction even from holocellulose and consist of a short chain of glucose and mannose building blocks devoid of branches. Most glucomannans have a ratio of glucose to mannose

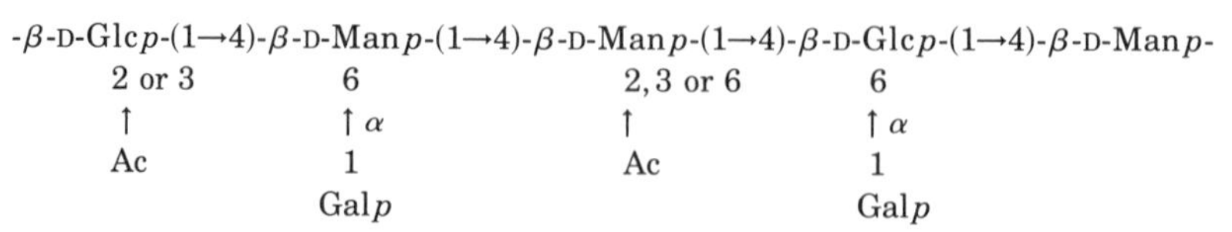

-β-D-Glc$p$-(1→4)-β-D-Man$p$-(1→4)-β-D-Man$p$-(1→4)-β-D-Glc$p$-(1→4)-β-D-Man$p$-
2 or 3   6   2, 3 or 6   6
↑   ↑ α ↑   ↑ α
Ac   1   Ac   1
                 Gal$p$ Gal$p$

Fig. 2. A simplified schematic representation of glucomannan, where Glc$p$ is the D-glucopyranosyl unit; Man$p$ is the D-mannopyranosyl unit; Gal$p$ is the D-galactopyranosyl unit; and Ac is the acetyl group.

(Glc/Man) of 1/1.5 to 2 (11) which represents a natural variation (Table 5). Those with a Glc/Man ratio of less than unity (68,69) may be contaminated with glucan. The isolation of a glucomannan in trace amounts by direct aqueous extraction of sugar maple wood (*Acer saccharum* Marsh.) with a ratio of components of Glc/Man = 1/4.6 illustrates the great variation in composition occurring in many natural products (70). Both glucose and mannose have been reported as nonreducing terminal groups. Galactoglucomannans are present in aspen bark (*Populus tremuloides* Michx.) (63), clover (*Trifolium* sp.) (64), in the midrib of tobacco leaves, and in cultured cells of tobacco (65,71), and have Glc/Man ratios characteristic of dicotyledons with the addition of terminal D-galactopyranosyl units attached to C-6 of the main chain.

Only traces of mannose occur in the hydrolysates of some monocotyledon stems (7,9) although glucomannans occur in the tubers and leaves of some monocotyledon species (72).

In gymnosperms, D-glucose and D-mannose components are linked by $\beta(1\rightarrow4)$ glycosidic bonds into a linear chain to which galactose units are attached by $\alpha(1\rightarrow6)$ glycosidic bonds (11). It is the primary hemicellulose component of the normal cell wall of most, but not all (73), conifers and in the native state contains O-acetyl units (74). Polydiversity is seen in the widely varying galactose content. The galactose-deficient galactoglucomannan (0.1:1:3) of conifers displays a relatively constant composition from source and species to species. Glc/Man ratios mostly range from 1:3 to 1:4 (11,75) and the slight variation observed in galactose content is probably within experimental error. The galactose content of the galactose-rich galactoglucomannans (1:1:3) exhibits a much greater variation from 11 to 27% (76–78). In some instances, polymers containing about 10% galactose content (0.5:1:3) can be fractionated further (75). Greater deviations in composition are reported for small fractions from Engelmann spruce (*Picea engelmannii* Parry) (79), Norway spruce (*Picea abies* (L.) Karst.) (80), and black spruce (75). It is possible that galactose to galactose linkages also exist in one fraction from Engelmann spruce (79).

The presence of galactose units increases the solubility of the polymer in water as well as that of its various derivatives in their appropriate solvents (11,80,81), allowing some to be extracted directly from milled wood (77). Acetyl groups such as galactose branches also confer solubility in water and DMSO (82–84). The naturally occurring acetyl groups are located along the chain in a random manner (85) on the C-2 and C-3 positions of the anhydromannose

Table 5. Comparison of Glucomannan Compositions from Various Sources

Source	Ratio of components in hydrolysate to glucose (=1.0)				
	Galactose	Mannose	Xylose	$[\alpha]_D$	Ref.
red maple	0	1.3		−31	11
aspen bark	0.5	1.3	trace	+10	63
red clover	0.25	1.1		−8.9	64
tobacco leaf midrib	0.5	2	0.5	−2	65
Engelmann spruce	1	3			66
	0.2	3			
ginkgo wood	1.0	1.7	0	−7	67
	0.2	3.6		−36	

units of the polymer from pine (86) and on the C-3 position of both glucose and mannose in the case of Parana pine (*Auraucaria angustifolia*) (84). Substitution at C-6 occurs in pine (*Pinus densiflora* S. and Z.) as well as at C-2 and C-3 (87).

Very little galactose-rich polymer can be isolated from coniferous bark, although a water-soluble component with a lesser galactose ratio was isolated (88–90). The bark of Lodgepole pine (*Pinus contorta* Dougl.) and Ginkgo xylem (as well as the shell of Ginkgo nut (90)) yields glucomannan extracts rich in glucose (91). Ray cells of normal red pine wood (*Pinus resinosa* Ait.) contain smaller amounts of galactose-rich glucomannan than tracheids. Both types of galactoglucomannans are claimed to be present in the cambium tissues of jack pine (*Pinus banksiana* Lamb.) before lignification takes place (92). The wide range of galactoglucomannan compositions from various sources is illustrated in Table 5.

The molecular weight of glucomannans is not as great as that of hardwood xylan (11), and when isolated from the same source can have different ratios of number to weight average molecular weights; being a Flory distribution in one instance (93) and not in another (94). No correlation exists between the viscosity of red pine (*Pinus resinosa* Ait.) glucomannan fractions, and their DP (95), and may reflect linking between lignin and glucomannan (60). The sorption of a ^{14}C-labeled white spruce (*Picea glauca* (Moench) Voss) glucomannan onto cellulose from water at temperatures from 5 to 40°C was found to occur more rapidly than a tritium labeled birch xylan (96). The apparent Arrhenius activation energy from initial sorption rates was 27.6 kJ/mol (6.6 kcal/mol) for the former and 41.8 kJ/mol (10 kcal/mol) for the latter. A galactoglucomannan (0.2:1:3) exhibited monolayer deposition onto a cellulose substrate from water when acetyl groups were present and multilayer deposition when deacetylated (97).

Galactans.   Hemicellulose galactans are found in the reaction wood of dicotyledons and gymnosperms. The tension wood of beech (*Fagus* sp.) contains a polymer composed of D-galactose, L-rhamnose, L-arabinose [5328-37-0] and D-xylose in the ratio of 62.4:22.2:9.7:5.7 (98). Galactopyranosyl units constitute the main chain; the other sugars as well as some galactose occur in the branches. Galacturonic acid and 4-*O*-methyl-D-glucuronic acid are also present. Identified fragments from partial acid hydrolysis resemble those from the gum of *Combretum leonense* (99). The degree of polymerization of 344 is much less than that of the gum and is skewed toward low molecular weight species. A less complicated, possibly linear β(1→3)-linked galactan has been found in *Rosa glauca* cells grown *in vitro* (100).

A polysaccharide is present in coniferous compression wood having a slightly branched main chain of (1→4)-linked β-D-galactopyranose residues with single-terminal β-D-galacturonic acid residues attached to C-6 of the main chain (101). It is best isolated from the chlorite liquor of holocellulose preparations of the compression wood of red spruce (*Picea rubens*), balsam fir (*Abies balsamea*), and tamarack (*Larix larcianan*). It is also found in the ray cells of normal and compression wood.

Glucans.   It has been shown (102,103) that the secondary phloem and ray cells of Scots pine (*Pinus sylvestris* L.), as well as many coniferous compression woods (101), contain a (1→3)-linked glucan (named laricinan) with a degree of polymerization of 174 to 205 (104). Its composition and insolubility after isolation resembles that of curdlan, but structural differences may exist. About 6 to 7% (1→4) linkages and traces of glucuronic and galacturonic acids are also present in the main chain as well as about eight branch points per average molecule.

Monocotyledon cell walls contain polysaccharides composed of a mixture of β-(1→3)- and β-(1→4)-linked glucopyranosyl residues which are commonly referred to as β-D-glucans (105). Only a few (1→3) linkages are present in this polymer and uronic acids have not been reported to be present. It is a significant hemicellulose component in bamboo stems and leaves (36) and constitutes up to 75% of the cell wall of barley (*Hordeum distichum* L.) endosperm (106). The sequencing of the linkages in the latter polymer indicates few contiguous β(1→3) linkages are present. The insertion of irregularly spaced β(1→3) linkages in a predominantly β(1→4)-linked structure loosens the otherwise rigid structure and allows it to assume a worm-like configuration (107).

Xyloglucans. These polysaccharides occur in trace amounts in tracheids and fibers of arborescent plants and in larger quantities in forage plants (7). They are also found in the cambial tissues of dicotyledons (108) and gymnosperms (66), the cultured cell walls of gymnosperms (109) and primary walls of cultured angiosperms cells (105), as well as the medium in which the cells were grown (110,111). Its sources, structures, and functions have been reviewed (112). They are similar in many respects to amyloid, a reserve carbohydrate in many seeds (113), but they possess additional components. Both polysaccharides give a characteristic blue-green color with iodine. The polymer consists of a cellulose-like backbone with D-xylosyl side chains linked by α(1→6) bonds to many of the glucosyl residues (114). There are many variations of this basic structure such as the presence of acetyl groups and terminal D-galactose, L-rhamnose, L-arabinose, and L-fucose units (114). The polymer has been detected in the cambial tissues of pine (*Pinus banksiana* Lamb.) (92), identified in aspen (*Populus tremuloides*) and linden (*Tilia americana* L.) cambial tissues (103), but has not been detected in the mature (one-year-old) tracheids of pine (92) suggesting it is a transient component of the primary wall which disappears on maturation.

The xyloglucan found in tamarind seeds has a greater molecular weight than that found in the cell walls but it does not contain L-fucose, acetyl, or pyruvyl substituents. The latter probably plays a unique role in the development and growth of plants (112,115). Xyloglucan is polydisperse, the number-average molecular weight of the polymer from the cambium tissues of poplar and linden is 62,000 Daltons, and the distribution shows a pronounced left-hand skewness (108). A similar distribution was found for tamarind amyloid although the molecular weight was much greater (1.4×10^6 Daltons) (116).

Both types of xyloglucans exhibit monolayer sorption onto cellulose (116) and tamarind xyloglucan exhibits maximum specific sorption onto cellulose less than that of coniferous xylan. By inference with other data, this is also less than that of glucomannan and hardwood xylan, but similar to many additives used in the paper industry.

Pectic Substances. Polymers rich in D-galacturonic acid have been isolated from pollen, fruit, seeds, bark, stalk, roots, aquatic plants, exudates, cultured tissue, and the media in which it is grown. They are not considered to be hemicellulose by some definitions (2), but many are cell wall components. Because of their susceptibility to degradation by enzymes, acids, alkalies, and oxidants during aging and isolation, research has concentrated on accessible polymers from cultured plant tissues (117). True pectins are characterized by the presence of an O-(α-D-galacturonopyranosyl)-(1→2)-L-rhamnopyranosyl linkage within the molecule but it is absent in many other pectin-like substances. At one extreme are simple galacturonans (118–120), whereas at the other is rhamnogalacturonan II which contains at least 10 different monosaccharide components, some unique to pectin, in the main chain or as a component of branches (121). Pectins of intermediate complexity contain alternate rhamnose and galacturonic acid units and are found in certain roots (122); others have branches of glucuronic acid linked to galacturonic acid. Rhamnogalacturonan I is most plentiful in the cell walls of cultured sycamore tissue (*Acer pseudoplatanus* L.) and has complex branches attached to half of the rhamnopyranosyl units (117). Acetyl and methoxyl groups are present in amounts which often vary with the state of development of the plant.

A rationalization of the complex behavior of pectins in solutions and gels with respect to their structure, solvation, and the presence of ions and other saccharides has been presented (123). The solution and sorption properties of gum tragacanth and the pectin isolated from the roots of *Hibiscus manihot* L. (Tororoaoi) contributes to their use in specialty paper manufacture (124–126).

Arabinan. This highly soluble polymer is found in the extracts of many fruits and seeds, in the boiling water extracts of pine wood (127), in the extracts of marshmallow roots (*Althaea officinalis*) (128), and aspen (63) and willow (*Salix alba* L.) (129) bark. Because arabinan can be isolated from mildly degraded pectin fractions, it is often difficult to determine whether it is a hemicellulose or a labile fragment of a larger polysaccharide and/or lignin complex. Arabinans have a complex structure composed almost entirely of 5-linked α-L-arabinofuranosyl units with similar residues linked to them at C-2 and/or C-3 and is soluble in 70% aqueous methanol solution.

Arabinogalactans. Arabinogalactan [9036-66-2] is found in association with proteins (as components of proteoglycans and glycoproteins) and free of such associations (130,131). It occurs in cell walls of plants and as an exudate in the pores and lumens of larch tracheids. The water-soluble extracts of Western (*Larix occidentalis* Null.) and Siberian (*L. sibirica*) larch heartwood are especially rich in this polymer. Although it is not a hemicellulose, its structure has been well investigated because of its availability (up to 20% of the heartwood), economic potential, detrimental effects on pulping processes, and because it probably resembles that polymer found in the cell wall very closely. Larch arabinogalactan is composed of two easily separated fractions (molecular weights 10,000 and 16,000) in which the ratio of galactose to arabinose is about 6:1. The polymer is highly branched with $\beta(1,3)$ and $\beta(1,6)$ linkages between D-galactose units (11). Research suggests that the main chain of the polymer from larch (like that of some Acacia gums) consists of blocks of about 12 D-galactopyranosyl units joined by (1→6) bonds which provide considerable flexibility to the molecule. Branches are composed of (1→6) linked galactopyranosyl units and 3- or 5-linked arabinofuranosyl residues terminated by galactose or arabinose. Glucuronic acid is a component of the arabinogalactan of many other conifers except Western larch. No clear line of structural demarcation exists between those coniferous arabinogalactans and exudates such as Acacia gums (120).

Glucuronomannans. Most glucuronomannans occur as components of certain plant gums and are characterized by the presence of an O-(β-D-glucurono-pyranosyl)-($1\rightarrow2$)-D-mannopyranosyl linkage in the molecule (65). Less complex glucuronomannans are found in the tissues of bracken (*Pteridium aquilinum*) and in tobacco leaf cells where they may be part of the cultured cell wall (132) and exude into the medium in which the cells are grown (133).

Isolation and Analysis

Techniques for the isolation of hemicellulose depend on the intended end use and whether it occurs in soluble waste material or is part of a solid matrix. Isolation is more difficult from solids as diminution of particle size and removal of undesired encrustants such as lignin is necessary to increase accessibility and destroy lignin hemicellulose bonds (13). Delignification techniques, except for those using ethanolamine, employ oxidants. Peroxides, peroxyacetic acid (134), and chlorine dioxide (1) have been used but the most common reagents are chlorine and acidified sodium chlorite (135). The former reagent is the least degradative (134), but it does not lend itself readily to large-scale preparations. The latter method requires less attention during preparation, can be conducted safely on a large scale, and can be manipulated to obviate the need for attrition (136).

Delignification extracts varying amounts of hemicellulose. Low reaction temperatures and (where possible) high salt concentrations minimize losses and concomitant chemical degradations such as oxidation and the effects of pH. Pectic substances and easily soluble arabinans, arabinogalactans, galactoglucomannans, xylans, and compression wood galactans are found in waste chlorite liquors (136,137). Carbonyl groups (excepting carboxyls) are frequently reduced with a suitable reagent before alkaline extractions are attempted to minimize β-elimination reactions (134). The use of an inert atmosphere during alkaline extraction prevents oxidation by oxygen.

Extraction of hemicellulose is a complex process that alters or degrades hemicellulose in some manner (11,138). Alkaline reagents that break hydrogen bonds are the most effective solvents but they de-esterify and initiate β-elimination reactions. Polar solvents such as DMSO and dimethylformamide are more specific and are used to extract partially acetylated polymers from milled wood or holocellulose (11,139). Solvent mixtures of increasing solvent power are employed in a sequential manner (138) and advantage is taken of the different behavior of various alkalies and alkaline complexes under different experimental conditions of extraction, concentration, and temperature (4,140). Some sequences for these elaborate extraction schemes have been summarized (138,139) and an experimenter should optimize them for the material involved and the desired end product (102).

The separation of the polysaccharide components utilizes their different solubilities, polar groups, extents of branching, molecular weights, and molecular flexibilities and may be accomplished batchwise or with easily automated column techniques such as column or high performance liquid chromatography. These procedures have been summarized in several reviews (3,141–143).

The increasing sophistication of analytical techniques coupled with suitable fractionation procedures has made the heterogeneity of hemicellulose components increasingly apparent (142). These techniques common to polymer chemistry include gas chromatography–mass spectroscopy, and proton and ^{13}C-nuclear magnetic resonance spectroscopy (144). Advances in functional group analysis and techniques for configurational and molecular studies (especially on the microscale) are frequent and the subject of numerous reviews (145). Molecular studies employ viscometry, osmometry, x-ray techniques, light scattering, and chromatographic and centrifugal techniques (11,146), as well as the use of optical rotatory dispersion and circular dichroism (4,140–146).

Pulping

The complex behavior of hemicellulose during pulping has been reviewed (147). When hemicellulose and lignin dissolve with the help of chemical transformations, fresh cellulosic surfaces are created and competition for deposition in these spaces arises between the dissolved components (148–150). Hemicellulose degradations also occur and are related to pH (acid hydrolysis, β-elimination reactions, redox reactions, etc (147)), and pyrolytic effects (151) to an extent dependent upon the time, temperature, and liquor composition of the cook (147,152). These reactions are rendered more complex because the cell wall controls the diffusion of the reactants and products into and out of the fiber so that hemicellulose may not be able to react or diffuse out of the fiber before the cook is completed. Under pulping conditions, the formation and cleavage of lignin–hemicellulose bonds (147) and possible carbohydrate–carbohydrate bonds (153) occurs complicating the nature of the product. The extent to which these competing reactions is accomplished is reflected in product composition and end use quality and is the subject of much empirical research (152).

Alkaline (kraft) processes are much more amenable to kinetic studies than acidic (sulfite) processes. After an initial rapid loss of easily soluble and accessible polysaccharides and lignin, hemicellulose losses proceed at a slower apparent first-order rate (154,155), dependent upon the temperature, time, and pH (liquor composition). Because of the changing nature of hemicellulose, the solubility and sorption characteristics are changed (156). Model experiments show isolated hemicellulose components behave in a similar manner and are influenced by the nature of the glycosidic bonds and the type of branching (157–162). Hydrolysis of kraft pulps at different yields shows the losses of lignin, and sugars (except mannose) proceed in a manner proportional to yield loss throughout the cook (154,163). Mannose losses from softwoods are greatest in the 80 to 90% pulp yield range and then become very resistant to removal. The quantity of hemicellulose remaining in pulp after processing is shown in Table 1. Those pulps with about 15% hemicellulose are usually used for paper manufacture, whereas those with 5% or less are used where a high cellulose content is required. The proportion of glucomannan is slightly greater in sulfite pulps, whereas the quantity of xylan is somewhat greater in kraft and soda pulps. These proportions can be altered slightly by changes in the cooking schedule.

Suitable pretreatment of wood before pulping alters this behavior of hemicellulose significantly. Saponification of the acetyl groups of softwood before sulfite cooking results in glucomannan retention in the final product (163,164). Those treatments that limit the peeling reaction during alkaline pulping processes (reductions with $NaBH_4$, H_2S, or oxidations with chlorite, polysulfide, anthraquinone, etc), can result in polysaccharide retention (165). The pretreatment of wood with mineral acid or liberated acids of wood at elevated temperatures (ie, 170°C for 30 min) diminishes the DP of hemicellulose components sufficiently that they will be consumed mostly during a subsequent alkaline cook (166). The resulting pulp behaves more like cotton cellulose in many industrial applications.

Figure 3 illustrates the transformations hemicellulose undergoes during commercial pulping operations as a result of acid hydrolysis or the many degradations induced by alkaline reagents. The changes shown are those that would occur if all were brought to completion after removal of most of the lignin, after washing with water, and before the pulps have undergone bleaching (147). Separation of these new compounds is best accomplished after destruction of lignin–carbohydrate bonds and the removal of lignin. Miscellaneous, labile, and accessible soluble polysaccharides are removed under these pulping conditions but they may be present together with incompletely transformed hemicellulose in high yield pulps. The hemicellulose components of hardwoods and bamboo follow the same pattern.

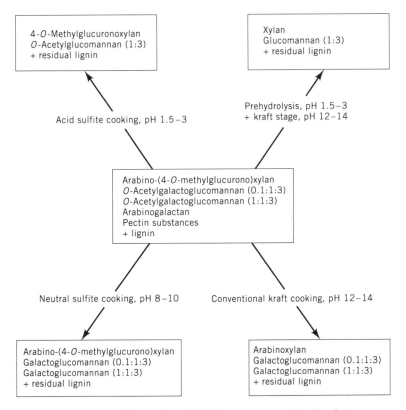

Fig. 3. The changes in coniferous hemicellulose as a result of pulping processes and before bleaching.

The Effect of Hemicellulose in Commercial Products

Hemicellulose components have an effect on the properties of products in which they are present. In the case of viscose manufacture much of the hemicellulose which remains in the product has only a marginal effect on strength properties and brightness and no effect on heat stability (167). It does contribute to swelling in yarn, and that remaining in the spent viscose liquor is harmful to filter presses. Increased clogging of spinnerets, low color index, and decreased yarn strength can also result when resin, cations, and hemicellulose are together in the steeping liquor (168).

When uronic acid and L-arabinofuranosyl branches are lost as a result of processing, the poor solubility of xylan acetate in organic solvents contributes adversely to cellulose acetate processing (169,170). The haze density (opaqueness) of solutions and solid products increases as hemicellulose components become less branched. Glucomannan in an acetate product contributes less haze than xylans but more than arabinoxylans. It is responsible for the false viscosity of cellulose acetate solutions derived from wood pulp. Glucomannan also causes filtration difficulties during processing, but the effects of xylans are unpredictable.

An understanding of the effect of hemicellulose on paper products is less clear because the formation of paper webs largely depends on their structure, which is influenced by many factors besides hemicellulose (171). Arabinogalactans, which are not sorbed onto cellulose, do not have an influence (172), whereas xylans and glucomannans and their survivors from the pulping processes do have an influence (173). The highly branched xyloglucan is sorbed but does little to improve the results of paper tests (116). Hemicellulose and related gums and mucilages help maintain a random dispersion of fibers in the furnish which results in more uniform and mechanically stronger paper webs. It also increases the rate at which pulp fibers respond to mechanical action (beating) and increases fiber bonding. The strong interfiber bonds formed after drying can alter paper structure and lead to losses in tearing strength and decreases in opacity if they are too extensive (171).

It has been proposed that differences in the tearing strengths of sulfite and kraft pulps are due to differences in the location of hemicelluloses in fiber walls (174). Other experiments suggest the location of hemicellulose in fiber walls does have an influence on tearing strength but not on tensile strength (175). A correlation of retained hemicellulose and operational variables during pulping processes has been carefully examined (176), but theoretical relationships between the parameters have not been established.

Applications

Hemicellulose and hemicellulose-like polysaccharides are beneficial components of foodstuffs because of their interactions between water and water-insoluble components (177). Endogenous polysaccharides are responsible for processing characteristics as well as texture and mouthfeel. Other properties such as gel formation, swelling of dough, fermentation, and optical properties are achieved using suitable treatments with enzymes and chemicals. Hemicellulose in spent liquor is mostly destroyed during modern alkaline pulping processes, and the residual fragments, unlike lignin, do not contribute greatly to the energy requirements for chemical recovery.

Hemicellulose constitutes a large reserve of raw material which can be converted to many products by chemical and biochemical means. Besides its inherent usefulness as a naturally occurring component of some manufactured products, it can be utilized either as a polymer or as the source of chemical intermediates. The former use is complicated since the mixture of polysaccharides and lignin in extracts requires special treatment if one component is to be isolated. The complexity of most hemicellulose extracts complicates the choice of use. As a result, the naturally occurring gums and mucilages are at a competitive advantage in the marketplace (178–180).

Larch arabinogalactan is easy to isolate and requires limited purification for many uses (181). The mixture of sugars, oligosaccharides, degraded hemicellulose, and lignin found in the liquors and condensates from Asplund-like and prehydrolysis pulping processes is used for binding and extending animal fodder. The sugars and oligosaccharides in waste sulfite liquor can be used for furfural production and the growth of yeast but no sugars survive in liquors from the predominant kraft processes. The quantities of xylan available for utilization greatly exceeds the quantity of arabinogalactan but its diverse nature complicates such use and the accumulation of the biomass necessary is expensive. Pectin is used in the food industry (182), and apart from specialty uses finds limited application in the paper industry (183).

Derivatives of hemicellulose components have properties similar to the cellulosic equivalents but modified by the effects of their lower molecular weight, more extensive branching, labile constituents, and more heterogeneous nature. Acetates, ethers, carboxymethylxylan (184), and xylan–poly(sodium acrylate) (185) have been prepared.

Reviews of the chemicals that can be derived from hemicellulose and arabinogalactan have been published (178–181). The xylan family has the greatest potential because of its widespread distribution. Furfural [98-01-1] is obtained from pentose by strong acid treatments but changes in pulping processes and farming practices make the collection of a suitable waste difficult. Xylose, xylitol, tartaric, and trioxyglutaric acids, methanol, butanol, and 2,3-butanediol can be manufactured on a large scale but only xylose [25990-60-7] and xylitol [87-99-0] have achieved temporary economic viability. There is little demand for mannose and mannose-derived chemicals because more convenient sources than softwoods exist. Nevertheless, several schemes have been developed to provide sweeteners, humectants, and unique compounds for starting chemical syntheses if market conditions improve.

BIBLIOGRAPHY

1. E. Schulze and E. Steiger, *Berichte* **20**, 290–294 (1887); **23**, 3110–3113 (1890).

2. G. O. Aspinall, in G. O. Aspinall, ed., *The Polysaccharides*, Vol. 2, Academic Press Inc., New York, 1983, Chapt. 1.

3. K. C. B. Wilkie, *Chemtech*, 306–319 (1983); *Adv. Carbohydr. Chem. Biochem.* **36**, 215–264 (1979).

4. D. Fengel and G. Wegener, in *Wood, Ultrastructure, Reactions*, DeGruyter, Berlin, 1983, Chapt. 5.

5. R. D. Preston, *The Physical Biology of Plant Cell Walls*, Chapman and Hall, London, 1974.

6. E. B. Cowling and T. K. Kirk, *Biotech. Bioeng. Symp.* **6**, 95 (1976).

7. F. E. Barton and co-workers, in E. J. Soltes, ed., *Wood and Agricultural Residues*, Academic Press, Inc., Harcourt Brace Jovanovich, New York, 1983.

8. K. F. Foley, *Chemical Properties and Uses of the Anderson's Corn Cob Products*, Anderson's brochure, Maumee, Ohio, 1978.

9. J. H. Sloneker, *Biotech. Bioneg. Symp.* **6**, 235–250 (1976).

10. L. E. Wise, R. C. Rittenhouse, and C. Garcia, *Tappi* **34**(4), 185–188 (1951).

11. T. E. Timell, *Adv. Carbohydr. Chem.* **19**, 247–302 (1964); **20**, 409–483 (1965).

12. C.-M. Kuo and T. E. Timell, *Svensk Papperstidn.* **72**(21), 703–716 (1969).

13. D. Fengel and G. Wegener, *1983 International Symposium on Wood and Pulping Chemistry (Japan)*, Vol. 4, Mar. 23–27, 1983, pp. 144–148.

14. J. S. G. Reid and K. C. B. Wilkie, *Phytochem.* **8**, 2045–2058 (1969); C. G. Fraser and K. C. B. Wilkie, *Phytochem.* **10**, 1539–1542 (1971).

15. A. G. McNeil and co-workers, *Ann. Rev. Biochem.* **53**, 625–663 (1984).

16. J. B. C. Correa, S. L. Gomes, and M. Gebara, *Carbohydr. Chem.* **60**, 337–343 (1978).

17. J. Comtat and J.-P. Joseleau, *Cell. Chem. Technol.* **7**, 653–657 (1973).

18. M. Zinbo and T. E. Timell, *Svensk Papperstidn.* **68**(19), 647–662 (1965); M. J. Song and T. E. Timell, *Cell. Chem. Technol.* **5**, 67–74 (1971).

19. K.-G. Roselle and S. Svensson, *Carbohydr. Res.* **42**, 297–304 (1975).

20. S. Kubackova and co-workers, *Carbohydr. Res.* **76**, 177–188 (1979).

21. C. P. J. Glaudemans and T. E. Timell, *Svensk Papperstidn.* **61**(1), 1–9 (1958).

22. B. Lindberg, K. G. Roselle, and S. Svensson, *Svensk Papperstidn.* **76**, 30–32 (1973).

23. S. Karcsonyi and co-workers, *Cell. Chem. Technol.* **17**, 637–645 (1983).

24. K.-G. Roselle and O. Samuelson, *Carbohydr. Res.* **42**, 297–304 (1975); M. H. Johansson and O. Samuelson, *Wood Sci. Technol.* **11**, 251 (1977); K. Shimizu and O. Samuelson, *Svensk Papperstidn.* **76**, 150–155 (1973).

25. F. Reicher, J. B. C. Correa, and P. A. J. Gorin, *Carbohydr. Res.* **135**, 129–140 (1984).

26. J. Y. Lee, in *1983 International Symposium on Wood and Pulping Chemistry (Japan)*, Vol. 4, Mar. 23–27, 1983, pp. 39–44.

27. C. S. Shulka, S. R. D. Guha, and J. S. Negi, *Indian Pulp Paper* **27**(9), 20–23 (1973).

28. L. Duckert, E. Byers, and N. S. Thompson, *Cell. Chem. Technol.* **22**, 29–37 (1988).

29. R. E. Lyons and T. E. Timell, *Cell. Chem. Technol.* **3**, 543–554 (1969).

30. R. H. Marchessault and co-workers, *J. Polymer Sci.* **51**(156), s66–s68 (1961).

31. P. C. Baradalaye and G. W. Hay, *Carbohydr. Res.* **37**(2), 339–350 (1974).

32. T. Koshijima and T. E. Timell, *Mokuzai Gakkaishi* **12**(4), 166–172 (1966).

33. R. G. LeBel and D. A. I. Goring, *J. Polymer. Sci., Part C*, (2), 29–48 (1963).

34. H. A. Swenson, C. A. Schmitt, and N. S. Thompson, *J. Polymer Sci., Part C*, (11), 243–252 (1965).

35. J. D. Blake and G. N. Richards, *Carbohydr. Res.* **18**, 11–21 (1970).

36. I. C. M. Dea and co-workers, *Carbohydr. Res.* **29**, 363–372 (1973).

37. J.-A. Hansson and N. Hartler, *Svensk Papperstidn.* **72**(17), 521–530 (1969).

38. K. C. B. Wilkie, *Adv. Carbohydr. Chem. Biochem.* **36**, 215–264 (1979).

39. E. Maekawa, *Wood Res. (Kyoto)* **59/60**, 153–179 (1976).

40. K. C. B. Wilkie and S. L. Woo, *Carbohydr. Res.* **49**, 399–409 (1976); **57**, 145–162 (1977).

41. U. Sharma and A. K. Mukherjee, *Carbohydr. Res.* **105**(2), 247–250 (1982).

42. J. P. Joseleau and F. Barnoud, *Phytochem.* **13**, 1155–1158 (1974).
43. A. J. Buchala and K. C. B. Wilkie, *Phytochem.* **13**, 1347–1351 (1974).
44. G. R. Woolard, E. B. Rathbone, and L. Novellie, *Carbohydr. Res.* **51**, 239–271 (1976); *Phytochem.* **16**, 957–959 (1977).
45. J. P. Joseleau and F. Barnoud, *Appl. Polymer Symp.*, (28), 983–992 (1976).
46. J. M. Brillouet and co-workers, *J. Agr. Food Chem.* **30**, 488–495 (1982).
47. N. Shibuya and T. Iwasaki, *Phytochem.* **24**(2), 285–289 (1985).
48. D. J. Brasch and L. E. Wise, *Tappi* **39**(11), 768–774 (1956).
49. J. M. Brillouet and J.-P. Joseleau, *Carbohydr. Res.* **159**, 109–126 (1987).
50. G. N. Richards and R. L. Whistler, *Carbohydr. Res.* **31**, 47–55 (1973).
51. T. Erickson, G. Peterson, and O. Samuelson, *Wood Sci. & Technol.* **11**, 219–223 (1973).
52. Y. Akiyama, S. Eda, and K. Kato, *Phytochem.* **23**(9), 2061–2062 (1984).
53. R. D. Lambert, E. E. Dickey, and N. S. Thompson, *Carbohydr. Res.* **6**, 43–51 (1968).
54. A. Jabbar Mian and T. E. Timell, *Svensk Papperstidn.* **64**(21) 769–774 (1960).
55. T. E. Timell, *Svensk Papperstidn.* **65**(4), 122–125 (1962).
56. M. Zinbo and T. E. Timell, *Svensk Papperstidn.* **70**(21), 695–701 (1967).
57. S. Yllner and B. Enstrom, *Svensk Papperstidn.* **59**(6), 229–232 (1956); **60**, 549 (1957); I. Croon and B. Enstrom, *Tappi* **44**(12), 870–874 (1961).
58. E. F. Walker, *Tappi* **48**(5), 298–303 (1965).
59. J. M. Vaughn and E. E. Dickey, *J. Org. Chem.* **29**, 715–718 (1964).
60. H. Meier, *Holzforschung* **13**, 177–182 (1959).
61. W. S. Linnell, N. S. Thompson, and H. A. Swenson, *Tappi* **49**(11), 491–493 (1966); W. S. Linnell and H. A. Swenson, *Tappi* **49**(11), 494–497 (1966).
62. I. C. M. Dea, in D. A. Brant, ed., *Solution Properties of Polysaccharides*, A.C.S. Series 150, Washington, D.C., 1981, pp. 439–454.
63. K.-S. Jiang and T. E. Timell, *Cell. Chem. & Technol.* **6**, 499–502, 503–505 (1972).
64. A. J. Buchala and H. Meier, *Carbohydr. Res.* **31**, 87–92 (1978).
65. Y. Akiyama and co-workers, *Phytochem.* **22**(5), 1177–1180 (1983).
66. A. R. Mills and T. E. Timell, *Can. J. Chem.* **41**, 1389–1395 (1963).
67. A. Jabbar Mian and T. E. Timell, *Svensk Papperstidn.* **63**(24), 884–888 (1960).
68. M. S. Dudkin, N. G. Shkanova, and M. A. Parfent'ova, *Khim. Prirod. Soed.* **9**(5), 598–601 (1973).
69. J.-Y. Lee, *J. Tappik* **13**(2), 3–9 (1981).
70. G. A. Adams, *Svensk Papperstidn.* **67**(3), 82–87 (1964).
71. S. Eda and co-workers, *Carbohydr. Res.* **131**, 105–118 (1984).
72. F. Smith and R. Montgomery, *The Chemistry of Plant Gums and Mucilages*, ACS Monograph Series, Reinhold Publishing Co., New York, 1959, pp. 48–49.
73. N. S. Thompson and co-workers, *Pulp & Paper Mag. Can.* **67**(12), T-541–T-551 (Dec. 1966).
74. T. Koshijima, *Mokuzai Gakkaishi* **6**(5), 194–198 (1960).
75. N. S. Thompson and O. A. Kaustinen, *Paperi Puu* **47**(4), 541–637 (1964).
76. J. K. Hamilton, E. V. Partlow, and N. S. Thompson, *J. Am. Chem. Soc.* **82**, 451–457 (1960).
77. G. A. Adams, *Tappi* **40**, 721–722 (1957).
78. T. E. Timell, *Tappi* **44**, 88–96 (1961).
79. J. K. Rogers and N. S. Thompson, *Svensk Papperstidn.* **72**(3), 61–67 (1969).
80. H. Meier, *Acta Chem. Scand.* **14**, 749–756 (1960).
81. R. J. Conca, J. K. Hamilton, and H. W. Kircher, *Tappi* **46**(11), 644–648 (1963).
82. H. Meier, *Acta Chem. Scand.* **15**, 1381–1385 (1961).
83. E. Hagglund, B. Lindberg, and J. McPherson, *Acta Chem. Scand.* **10**, 1160–1164 (1956).
84. G. Katz, *Tappi,* **48**(1), 34–41 (1965).

85. L. Kenne, K.-G. Rosell, and S. Swensson, *Carbohydr. Res.* **44**, 69–76 (1975).
86. B. Lindberg, K.-G. Rosell, and S. Svensson, *Svensk Papperstidn.* **76**, 383–384 (1983).
87. R. Tanaka and co-workers, *Mokuzai Gakkaishi* **31**(10), 859–861 (1985).
88. T. E. Timell, *Svensk Papperstidn.* **64**(18), 651–661 (1961); **64**(20), 744–747 (1961); **65**(21), 843–846 (1962).
89. V. Ramalingam and T. E. Timell, *Svensk Papperstidn.* **67**(12), 512–521 (1964).
90. E. Maekawa and K. Kitao, *Wood Res. (Kyoto)* **58**, 33–44 (1975).
91. G. C. Hoffmann and T. E. Timell, *Tappi* **55**(5), 733–736 (1972); **55**(6), 871–873 (1972); **53**(10), 1896–1899 (1970).
92. N. S. Thompson, R. E. Kremers, and O. A. Kaustinen, *Tappi* **51**(3), 127–131 (1968).
93. T. E. Timell and A. Tyminski, *J. Am. Chem. Soc.* **82**, 2823–2827 (1960).
94. B. Lindberg and H. Meier, *Svensk Papperstidn.* **60**(21), 785–790 (1957).
95. G. C. Hoffmann and T. E. Timell, *Tappi* **53**(10), 1896–1899 (1970).
96. D. W. Clayton and G. R. Phelps, *J. Polymer Sci. (Part C, Polymer Symposia)*, (11), 197–220 (1968).
97. K. B. Laffend and H. A. Swenson, *Tappi* **51**(3), 118–123, 141–143 (1968).
98. H. Meier, *Acta Chem. Scand.* **16**, 2275–2283 (1962).
99. G. O. Aspinall and V. P. Bhavanandan, *J. Chem. Soc.*, 2685–2693 (1965).
100. A. Mollard, F. Barnoud, and G. G. S. Dutton, *Physiol. Veg.* **14**(1), 101–108 (1976).
101. T. E. Timell, in *Compression Wood in Gymnsperms*, Vol. 1, Springer Verlag, New York, 1985, Chapt. 5.
102. G. C. Hoffmann and T. E. Timell, *Svensk Papperstidn.* **75**(4), 135–141 (1972); *Tappi* **55**(5), 733–736 (1972); *Tappi* **55**(6), 871 (1972).
103. Y.-L. Fu, P. J. Gutman, and T. E. Timell, *Cell. Chem. Technol.* **6**(5), 507–512 (1972); Y.-L. Fu and T. E. Timell, *Cell. Chem. Technol.* **6**(5), 513–515; **6**(5): 517–519 (1972).
104. G. C. Hoffmann and T. E. Timell, *Wood Sci Technol.* **4**, 159–162 (1970).
105. M. McNeil and co-workers, *Ann. Rev. Biochem.* **53**, 625–663 (1984).
106. J. R. Woodward, D. R. Phillips, and G. B. Fincher, *Carbohydr. Polymers* **3**, 143–156 (1983).
107. G. B. Fincher, *J. Inst. Brew.* **81**, 116 (1975).
108. B. W. Simpson and T. E. Timell, *Cellulose Chem. Technol.* **12**, 51–62 (1978).
109. D. Burke and co-workers, *Plant Physiol.* **54**, 109–115 (1974).
110. L. K. Nealey, *The Isolation, Characterization of a Xyloglucan from Suspension Cultured Loblolly Pine Cell Medium*, Ph.D. dissertation, Institute of Paper Chemistry, Appleton, Wis., 1987.
111. W. D. Bauer and co-workers, *Plant Physiol.* **51**, 174–187 (1974).
112. S. F. Fry, *J. Expt. Botany* **40**(210), 1–11 (1989).
113. P. Kooimann, *Rec. Traveaux Chim. Pays-Bas* **80**(8), 849–865 (1961); *Phytochem.* **6**, 1665–1673 (1967).
114. G. O. Aspinall, J. A. Molloy, and J. M. T. Craig, *Can. J. Biochem.* **47**, 1063–1070 (1969).
115. I. R. Siddiqui and P. J. Wood, *Carbohydr. Res.* **53**, 85–94 (1977); **50**, 97–107 (1976).
116. S. L. Molinarolo, *Sorption of Xyloglucan onto Cellulosic Fibers*, Ph.D. Dissertation, Institute of Paper Chemistry, Appleton, Wis., 1989.
117. J. M. Lau and co-workers, *Carbohydr. Res.* 137, 111–125 (1985); M. McNeil, A. G. Darvill, and P. Albersheim, *Plant Physiol.* **66**, 1128–1134 (1980).
118. V. Zitco and C. T. Bishop, *Can. J. Chem.* **44**, 1275–1282 (1966).
119. G. Chambat and J.-P. Joseleau, *Carbohydr. Res.* **85**, C10–C12 (1980).
120. G. O. Aspinall, in F. Loewus ed., *Biogenesis of Plant Cell Wall Polysaccharides*, Academic Press, Inc., New York, 1973.
121. D. Melton and co-workers, *Carbohydr. Res.* **146**, 279–395 (1986).
122. M. Tomoda and co-workers, *Carbohydr. Res.* **151**, 29–35 (1986).

123. D. A. Rees, *Carbohydr. Polymers* **2**, 254–263 (1982).
124. M. Mori and K. Kato, *Carbohydr. Res.* **91**, 49–58 (1981).
125. H. Ishikawa and co-workers, *Mokuzai Gakkaishi* **16**(4), 173–180 (1970).
126. J. W. Swanson, *Tappi* **44**(1), 142A–181A (1961).
127. A. J. Roudier and L. Eberhard, *Bull. Soc. Chim. France*, 460–464 (1965).
128. P. Capek and co-workers, *Carbohydr. Res.* **117**, 133–140 (1983).
129. S. Karacsonyi and co-workers, *Carbohydr. Res.* **44**, 285–290 (1975).
130. A. E. Clarke, R. L. Anderson, and B. A. Stone, *Phytochem.* **18**, 521–540 (1979).
131. Y. Akiyama and K. Kato, *Phytochem.* **20**(11), 2507–2510 (1981).
132. Y. Akiyama and co-workers, *Carb. Res.* **91**, 49–58 (1981); *Agric. Biol. Chem.* **48**(2), 403–407 (1984).
133. Y. Akiyama and co-workers, *Biol. Chem.* **48**(2), 403–407 (198).
134. B. Leopold, *Tappi* **44**(3), 230–232, 232–235 (1961).
135. L. E. Wise, M. Murphy, and A. A. D'Addieco, *Paper Trade J.* **122**(2), 35–43 (Jan. 10, 1946).
136. N. S. Thompson and O. A. Kaustinen, *Tappi* **47**(3), 157–163 (1964); **49**(2), 83–90 (1966); **53**(8), 1502–1505 (1970).
137. Y.-L. Fu, P. J. Gutman, and T. E. Timell, *Cell. Chem. Technol.* **6**(5), 507–512 (1972); Y.-L. Fu and T. E. Timell, *Cell. Chem. Technol.* **6**(5), 513–515, 517–519 (1972).
138. K. Ward and A. J. Morak, *Pure and Applied Chem.* **5**(1-2), 77–89 (1962).
139. E. Hagglund, B. Lindberg, and J. McPherson, *Acta Chem. Scand.* **10**, 1160–1164 (1956).
140. A. Beelik and co-workers, *Tappi* **50**(2), 78–81 (1967).
141. Ref. 2, Vol. 2, Chapt. 2.
142. A. M. Stephen, in Ref. 2, Vol. 2.
143. R. R. Selvendran and M. A. O'Neill, *Methods Biochem. Anal.*, **32**, 25–153 (1987).
144. M. McNeil and co-workers, *Methods Enzymol.* **83**, 3–45 (1982).
145. K. Shimizu, in N. S. Hon and N. Shiraishi, eds., *Wood and Cellulosic Chemistry*, Marcel Dekker, Inc., New York, 1991, Chapt. 5.
146. W. Banks and C. T. Greenwood, *Adv. Carbohydr. Chem.* **18**, 357–398 (1963).
147. D. Clayton and co-workers, "Chemistry of Alkaline Pulping," in *Pulp and Paper Manufacture*, 3rd ed., Vol. 5, Alkaline Pulping, The Joint Textbook Committee of the Paper Industry, TAPPI, CPPA, Technology Park Atlanta, Ga., 1989.
148. D. W. Clayton and G. R. Phelps, *J. Polymer Sci. Part C*, (11), 197–220 (1968).
149. J.-A. Hansson and N. Hartler, *Svensk Papperstidn.* **72**(17), 521–530 (1969).
150. S. Yllner and B. Enstrom, *Svensk Papperstidn.* **59**(6), 229–232 (1956); **60**, 549–554 (1957); I. Croon and B. Enstrom, *Tappi* **44**(12), 870–874 (1961).
151. N. S. Thompson and O. A. Kaustinen, *Tappi* **49**(12), 550–553 (1966).
152. S. A. Rydholm, *Pulping Processes*, Wiley-Interscience, New York, 1965.
153. J. W. McKinney, *Paper Trade J.* **122**(4), 58–62 (1946).
154. C. H. Matthews, *Svensk Papperstidn.* **77**(17), 629 (1974).
155. H. D. Wilder and E. J. Daleski, *Tappi* **47**, 270–275 (1964); **48**, 293–297 (1965).
156. A. Meller, *Holzforschung* **19**(4), 118–124 (1965); **22**(3), 88–92 (1968).
157. J. K. Hamilton and N. S. Thompson, *Pulp Paper Mag. Can.* **61**(4), T263–T272 (1960).
158. J.-A. Hansson and N. Hartler, *Holzforschung* **24**(2), 54–59 (1970).
159. R. L. Casebier and J. K. Hamilton, *Tappi* **48**(11), 664–669 (1965); **50**(9), 441–449 (1967).
160. Y. Z. Lai and K. V. Sarkanen, *J. Polymer Sci., Part C* **28**, 15 (1969).
161. R. A. Young, K. V. Sarkanen, and G. G. Allan, *Carbohydr. Res.* **21**, 111–122 (1972).
162. R. A. Young and L. Liss, *Cell. Chem. Technol.* **12**, 399–411 (1978).
163. S. Yllner, K. Ostberg, and L. Stockman, *Svensk Papperstidn.* **54**(21), 795–802 (1957).
164. G. E. Annergren and S. A. Rydholm, *Svensk Papperstidn.* **63**(18), 591–600 (1960).

165. G. E. Annergren and co-workers, *Svensk Papperstidn.* **64**(10), 386–393 (1961).

166. D. Brasch and K. W. Free, *Tappi* **47**, 186–189 (1964); **48**, 245–248 (1965).

167. E. Treiber, *Das Papier* **7**(12), 591–600 (1983).

168. I. Croon, H. Jonsen, and H.-G. Olofsson, *Svensk Papperstidn.* **71**(2), 40–45 (1968).

169. R. J. Conca, J. K. Hamilton, and H. W. Kircher, *Tappi* **46**(11), 644–648 (1963).

170. J. D. Wilson and R. S. Tabke, *Tappi Dissolving Pulp Conference*, Atlanta, Ga., Oct. 24–25, 1973, pp. 55–68; *Tappi* **57**(8), 77–80 (1983).

171. L. G. Cotrall, *World's Paper Trade Review*, **142**(5), 393, 394, 396, 406; (6) 445, 446, 448, 450, 454, 456 (July 29, Aug. 5, 1954).

172. G. K. Hunger, *Das Papier* **37**(12), 582–591 (1983).

173. J. O. Thompson, J. W. Swanson, and L. E. Wise, *Tappi* **36**(12), 534–541 (1953).

174. J. E. Luce, *Pulp Paper Mag. Can.* **65**(10), T419–T423 (1964).

175. N. S. Thompson and O. A. Kaustinen, "The Physics and Chemistry of Wood Pulp Fibers," *TAPPI Seminar*, Appleton, Wis., May 12–15, 1970, pp. 146–152; *Tappi* **53**(8), 1502–1505 (1970).

176. H. Makkonen, *Paperi ja Puu* **49**(6), 383–390 (1967); **49**(7), 437–440, 442–444, 446–450, 452–455 (1967).

177. W. Pilinik and F. M. Rombouts, *Carbohydr. Res.* **142**, 93–105 (1985).

178. R. L. Whistler and C.-C. Chen, in M. Lewin and I. S. Goldstein, eds., *Wood Structure and Composition*, Vol. 11, International Fiber Science and Technology Series, Marcel Dekker, Inc., New York, 1991, Chapt. 7.

179. N. S. Thompson, in Ref. 7.

180. N. S. Thompson, in I. S. Goldstein ed., *Organic Chemicals from Biomass*, CRC press, Boca Raton, Fla., 1981.

181. V. B. Ettling and M. F. Adams, *Tappi* **51**(3), 116–119 (1968).

182. J. K. Pedersen, in R. L. Davidson and M. Sittis, eds., *Handbook of Water-Soluble Gums and Resins*, McGraw-Hill Book Co., Inc., New York, Chapt. 15, 1962.

183. S. Annus, I. Hegedus, and L. Kobor, *Papiripar* **4**(4), 144–148 (1960).

184. J. Schmorak and G. A. Adams, *Tappi* **49**, 378–383 (1957).

185. J. Church, *J. Polymer Sci.* **5**(12), 3183–3192 (1967).

NORMAN S. THOMPSON
Consultant

HEPARIN. See BLOOD, COAGULANTS AND ANTICOAGULANTS.

HERBICIDES

An herbicide, in the broadest definition, is any agent which destroys or inhibits plant growth. Thus an herbicidal agent may be animal, ie, a home-gardener with a hoe or a grazing herbivore; vegetable, ie, a parasitic weed or one plant species competing successfully with another; or mineral, ie, chemicals with herbicidal activity. This discussion of herbicides specifically addresses chemicals which destroy or inhibit plant growth.

In the early twentieth century, the available weed killers or herbicides were inorganic acids and salts, eg, sulfuric acid, sodium chlorate, arsenicals, and copper sulfate. Until the 1932 introduction of the first organic chemical herbicide, 4,6-dinitro-o-cresol [534-52-1] (DNOC), weed control in fields and turf depended on these inorganic compounds and various combinations of surface tillage, mowing, chopping, hand-weeding, scorching, and burning of unwanted plants. Those time-honored but highly inefficient and labor-intensive methods were an essential part of agriculture because weeds successfully compete with crop plants for water, sunlight, and nutrients. Uncontrolled weed infestations drastically reduce crop yields and decrease crop, turf, timber, and forage quality. For example, the post-harvest presence of weed seeds reduces crop quality, ie, cocklebur in soybeans, wild mustard in canola, and red rice and Northern jointvetch in rice. Weeds also serve as alternative hosts for crop-infesting fungi and harbor insect pests such as whiteflies.

The introduction of DNOC was followed by the appearance in the 1940s of the substituted phenoxy acids, and in 1951 of the substituted ureas. The triazine family of herbicides appeared in and after 1955, and the bipyridiniums in 1960. In 1967, two new chemical herbicides (1) were registered under the Federal Insecticide, Fungicide, and Rodenticide Act (FIFRA). The number of new registrations swelled to 11 in 1975 and subsequently dropped to three in 1990 and two in 1991.

Herbicide Classes and Databases. Herbicides can be classified as selective and nonselective. Selective herbicides, like 2,4-D (2,4-dichlorophenoxyacetic acid), metolachlor [51218-45-2], and EPTC [759-94-4], are more effective against some types of plants than others, eg, broadleaved plants vs grasses. Glyphosate [1071-83-6] is representative of the nonselective herbicides used for total vegetable control.

The classes of herbicidally active toxophores are limited in number. Arbitrary classification by toxophore reveals eight generic herbicide groupings, ie, triazines, amides (haloacetanilides), carbamates, toluidines (dinitroanilines), ureas, plant growth hormones (phenoxy acids), diphenyl ethers, and miscellaneous unrelated compounds. Classification of commercial herbicides by chemical structure yields 10 related groupings with subgroups, ie, phenoxy alkanoic acids; bipyridiniums; benzonitriles with phthalic compounds; dinitroanilines; acid amides; carbamates; thiocarbamates; heterocyclic nitrogen compounds including triazines, pyridines, pyridazinones, sulfonylureas, and imidazoles; substituted ureas; and miscellaneous groupings that include halogenated aliphatic carboxylic acids, inorganics and organometallics, and derivatives of biologically important amino acids.

Herbicides are also sometimes classified according to mode of action, selectivity, registered uses, and toxicity. The ever-increasing importance of herbicides and other pesticides and agrochemicals to a wide range of users, regulators, and researchers has led to the development of multiple and extensive computer databases. The primary database resources contain collected information relevant to herbicides, and numerous resource publications are available to those needing information on the various aspects of herbicides (2).

Database	Type of information
Agribusiness	agricultural chemicals and finance; agribusiness companies, product development history, and government policies
Agricola	National Agricultural Library database; general coverage of U.S. agriculture
Agrochemicals handbook	active components of agrochemicals
BIOSIS previews (Biological Abstracts)	research literature in biological, biomedical, and life sciences
Chemical Abstracts search	chemical literature and applications
CAB Abstracts	general agricultural and biological information, including weed science, from U.S. sources
Claims/U.S. patent abstracts	patents issued by the U.S. Patent Office; chemical patent records from 1950 to present
Derwent world patents index	patent data from 30 patent-issuing authorities around the world; agricultural chemical patents from 1965 to present
EMMI	U.S. EPA Environmental Monitoring Methods Index, including regulatory lists, analytical methods, detection and regulatory limits
Enviroline	coverage of worldwide environmental information
Pollution abstracts	references to environmental literature, including pollutant sources and control
Toxline	National Library of Medicine toxicological information; references to pesticides, herbicides, environmental pollution, carcinogenic chemicals, food contamination, toxicological analyses
Toxnet	National Library of Medicine Toxicology Data Network, including Hazardous Substances Data Bank (HSDB), Registry of Toxic Effects on Chemical Substances (RTECS), Toxic Chemical Release Inventory (TRI), plus several other toxicological/carcinogenesis-related files

Herbicide Development

Examination of the various classified listings of herbicides provides insight into the processes and approaches that lead to the discovery of new pesticides. The four principal development approaches are random screening, imitative chemistry, testing natural products, and biorational development.

Random screening techniques have provided 50% of herbicides available (ca 1993). Chemicals that act on one biological screen frequently act on other pesticidal, plant growth regulator, or pharmaceutical screens. However, a steadily increasing number of chemicals must be tested to uncover a single compound with pesticidal potential. The time required for the discovery of one new compound with good herbicidal activity, which also meets tightening biological and environmental standards, is becoming progressively longer. On the average, more than 20,000 chemicals must be screened to produce one new product. The screening process is complicated further by large numbers of false positives and significant discrepancies between field and screening environments. Carefully balanced judgment of apparent activity must always take into account the level and specificity of biological action, the chemical and action novelty with respect to patenting requirements, and the potential and scope for commercial exploitation.

Historically, the discovery of one effective herbicide has led quickly to the preparation and screening of a family of imitative chemicals (3). Herbicide developers have traditionally used combinations of experience, art-based approaches, and intuitive applications of classical structure–activity relationships to imitate, increase, or make more selective the activity of the parent compound. This trial-and-error process depends on the costs and availabilities of appropriate starting materials, ease of synthesis of usually inactive intermediates, and alterations of parent compound chemical properties by stepwise addition of substituents that have been effective in the development of other pesticides, eg, halogens or substituted amino groups. The reason a particular imitative compound works is seldom understood, and other pesticidal applications are not readily predictable. Novices in this traditional, quite random, process require several years of training and experience in order to function productively.

Quantitative Structure–Activity Relationship Design. Increasing economic pressures toward more, better, and cheaper pesticides have led to the development and application of the Quantitative Structure–Activity Relationship (QSAR) paradigm and related experimental design principles for pesticides (4). Theoretically, quantitative determination of the relationships between chemical structure and biological and environmental properties of a molecule should permit the design of a novel molecule with exactly those properties considered ideal for the intended application. In 1964, using earlier studies of the relationships of chemical composition to biological activity (5), the use of multivariate linear regression analyses to develop the quantitative structure–activity rules required for successful QSAR application was outlined (6). Subsequently, QSAR has been refined, expanded, and successfully applied in the chemorational design of pesticidal compounds (3,7–11).

The correlations between chemical descriptors of molecular properties and biological activity, especially the activity of herbicides and/or plant growth regulators has been described (12). Several alternatives or improvements on the Hansch-Fujita QSAR system have been developed (13–15).

Some successful pesticide development programs have been based on existing empirical data concerning a specific biologically active moiety. For instance, the trifluoromethylated pyridinyl building block (16) has been used in the design and development from nitrofen [1836-75-5] of the diphenyl ether herbicide, fluazifop [69806-50-4]. The QSAR method also was fundamental to the design of amide and urea herbicides, such as the N-alkoxymethyl-N-(2,6-dialkylphenyl) analogues of chloroacetamide and selective post-emergence phenylureas and pyrazolium pesticides (17,18).

These chemorational techniques have generated great interest in, and high expectations for, the acceleration of development of innovative pesticides. However, many purportedly successful applications of QSAR procedures have relied on the qualitative insights traditionally associated with art-based pesticide development programs. Retrospective QSAR analyses have, however, been helpful in identifying the best compounds for specific uses (17). Chemorational techniques have also found some applications in the development of pesticides from natural product lead compounds, the best known examples being the synthetic pyrethroid insecticides (19) modeled on the plant natural product, pyrethrum.

Development Based on Natural Products.   Development, with or without the use of QSAR, of herbicides from natural products has been rare. Although a number of natural products have shown herbicidal activity, specificity has been lacking, the effective concentration has proven too high for practicality, or the cost of manufacture on a mass scale has been considered prohibitive. These characteristics have led to a variant of the lead-compound approach in which the microorganisms, usually fungi, that produce the biologically active natural products are applied to the weed in the form of mycoherbicides (20–22), ie, herbicides based on fungal weed pathogens. The principal advantage of mycoherbicides over conventional herbicides is the relative ease of registration. The principal disadvantage is very limited target specificity; a mycoherbicide is usually effective against only one weed species under rather specific environmental conditions. Future commercial development of mycoherbicides depends heavily on improvements in the formulations that are necessary for increased shelf-life and vitality of the living fungal spores that constitute the active ingredient of this type of pesticide (23).

Biorational Approaches to Development.   Rather than depending on chemorational approaches that rely on logical alterations of molecular structures, herbicide development programs have increasingly utilized biorational approaches (24). Biorational techniques include identification of a particular biochemical mechanism that, when inhibited, causes plant death. Molecules found to specifically inhibit an essential biochemical pathway in plants are chemically modified to produce a family of herbicides with greater potency and/or specificity than the original parent compound. For instance, studies of the stepwise mechanisms of photosynthesis have facilitated the development of the ureas, triazines, and uracils, as well as other herbicides that act through inhibition of photosynthesis (25). Glyphosate, an amino acid analogue (26), and the sulfonylureas (27) fatally inhibit essential amino acid production. Phosphinothricin [3557-44-5], the active herbicidal molecule in bialaphos and glufosinate, inhibits glutamine synthase and allows toxic levels of ammonia to accumulate in plant tissues (28). Substituted pyridazinones, dithiocarbamates, and a broad class of Graminaceae-selective herbicides that includes the oxyphenoxy propionic acids and cyclohexanediones inhibit lipid synthesis (29).

Biorational approaches have proven useful in the development of classes of herbicides which inhibit essential metabolic pathways common to all plants and thus are specific to plants and have low toxicity to mammalian species. Biorational herbicide development remains a high risk endeavor since promising high activities observed in the laboratory may be nullified by factors such as limitations in plant uptake and translocation, and the instability or inactivity of biochemical enzyme inhibitors under the harsher environmental conditions in the field. Despite these recognized drawbacks, biorational design of herbicides has shown sufficient potential to make the study of herbicide modes of action an important and growing research area.

Modes of Herbicide Action

Photosynthesis is the light-driven, membrane-localized electron/proton transport system by which plants, algae, and some bacteria convert the energy of a quanta of light into the phosphoryl group transfer potential of adenosine triphosphate (ATP) and the redox potential of nicotinamide adenine dinucleotide phosphate (NADPH) while oxidizing water to produce oxygen (30–32). In higher plants, the photosynthetic light energy conversion processes are localized in the thylakoid membranes of the grana of chloroplasts and the carbon-fixing processes occur in the stroma. Chloroplasts are chlorophyll-bearing, double membrane-bound organelles within photosynthetic plant cells; grana consist of stacks of thylakoids, vesicle-like structures that have internal spaces defined by a membrane and that are connected by unstacked stromal thylakoids. The thylakoid membranes contain the light-harvesting pigments and the electron- and proton-translocating components of both Photosystem I (PSI) and Photosystem II (PSII) of photosynthesis.

Traditionally, the electron and proton transport pathways of photosynthetic membranes (33) have been represented as a "Z" rotated 90° to the left with noncyclic electron flow from left to right and PSII on the left-most and PSI on the right-most vertical in that orientation (25,34). Other orientations and more complex graphical representations have been used to depict electron transport (29) or the sequence and redox midpoint potentials of the electron carriers. As elucidation of photosynthetic membrane architecture and electron pathways has progressed, PSI has come to be placed on the left as the "Z" convention is being abandoned. Figure 1 describes the orientation in the thylakoid membrane of the components of PSI and PSII with noncyclic electron flow from right to left.

Both PSI and PSII are necessary for photosynthesis, but the systems do not operate in the implied temporal sequence. There is also considerable pooling of electrons in intermediates between the two photosystems, and the indicated photoacts seldom occur in unison. The terms, PSI and PSII, have come to represent two distinct, but interacting reaction centers in photosynthetic membranes (36,37); the two centers are considered in combination with the proteins and electron-transfer processes specific to the separate centers.

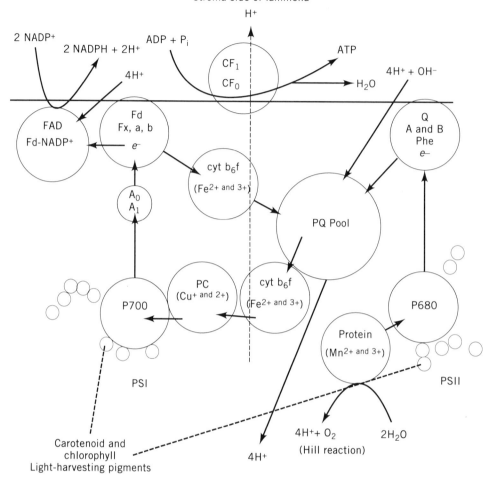

Stroma side of lammella

Fig. 1. An abbreviated schematic representation of photosynthetic electron flow in chloroplasts where PSII is photosystem II; PSI, photosystem I; Protein, Mn-containing oxygen evolving enzyme complex; P680, PSII reaction center chlorophyll; Phe, pheophytin a PSII e^- acceptor; Q_A, a bound PSII single e^- acceptor plastoquinone; Q_B, a bound PSII double e^- acceptor plastoquinone; PQ, membrane mobile reduced and oxidized plastoquinone; cyt, Fe-containing cytochrome e^- acceptor/donor; PC, copper-containing plastocyanin; P700, PSI reaction chlorophyll; A_0 and A_1, uncharacterized PSI e^- acceptors; Fx, a, b, FeS containing PSI e^- acceptors; Fd, ferredoxin; FAD, flavin adenine dinucleotide; Fd-NADP$^+$, nicotinamide adenine dinucleotide phosphate-reducing enzyme that contains ferredoxin; NADP$^+$ and NADPH, oxidized and reduced nicotinamide adenine dinucleotide phosphate, respectively; P_i, inorganic phosphate; ADP, adenine diphosphate; ATP, adenine triphosphate; and CF_1 and CF_0, ATP-synthesizing coupling factors. Light-harvesting pigments (smallest circles) transfer energy to reaction center chlorophyll molecules initiating a series of redox reactions that result in an electrochemical gradient across the thylakoid membrane. These reactions produce the reduced nucleotide and ATP required for anabolic metabolism in the plant. Solid arrows indicate the flow of H$^+$ and/or e^- among acceptor–donor compartments (larger circles). The dotted arrow indicates the flow of H$^+$ through the ATP synthetic coupling factors (35).

78

Photosystem I Inhibitors. Photosystem I is the reaction center or site in photosynthetic membranes of oxygen-evolving organisms at which light-activated electron transfers lead to reduction of the iron–sulfur, FeS^-, centers of ferredoxin (36) indicated as Fd in Figure 1. PSI cycling is ensured by electrons transferred from PSII or the cytochrome b_6f complex via the copper-containing plastocyanin (PC) which is the primary electron donor to P700, the specialized chlorophyll a molecule associated with PSI (25). From P700, electrons are transferred singly to FeS^- and thence to soluble ferredoxin. Ferredoxin nicotinamide adenine dinucleotide phosphate reductase transfers electrons to $NADP^+$ from soluble ferredoxin.

PSI transport processes include both this directional electron transfer to produce NADPH and a cyclic electron transfer that pumps protons into the stroma, resulting in the synthesis of ATP (36). Production of ATP by photophosphorylation can be inhibited (Table 1) by either uncouplers, eg, phenylhydrazones, carbanilates, diphenylamines, and ethane diamines (which dissipate the proton gradient necessary to drive ATP synthesis), or by inhibitors at ATP synthase, ie, coupling factor CF_1, nitrofen, and other chlorinated p-nitrodiphenyl ethers (37). Since ATP synthesis is not specific to oxygen-producing organisms, PSI inhibitors of the latter type are toxic to both plants and animals. Other herbicides, eg, oxyflurofen and other bipyridylium salts of p-nitro- or p-chlorodiphenyl ethers which act on PSI, have been reported to act by causing general destruction of the chloroplasts through membrane component peroxidation (37). Studies indicate that the diphenyl ethers, acifluorfen and oxyfluorfen, inhibit protoporphyrinogen oxidase, the penultimate enzyme in heme synthesis (38,39). These herbicidal compounds are discussed more fully in the section dealing specifically with membrane-active herbicides.

Electron Transport Between Photosystem I and Photosystem II Inhibitors. The interaction between PSI and PSII reaction centers (Fig. 1) depends on the thermodynamically favored transfer of electrons from low redox potential carriers to carriers of higher redox potential. This process serves to communicate reducing equivalents between the two photosystem complexes. Photosynthetic and respiratory membranes of both eukaryotes and prokaryotes contain structures that serve to oxidize low potential quinols while reducing high potential metalloproteins (40). In plant thylakoid membranes, this complex is usually referred to as the cytochrome b_6/f complex, or plastoquinol:plastocyanin oxidoreductase, which oxidizes plastoquinol reduced in PSII and reduces plastocyanin oxidized in PSI (25,41). Some diphenyl ethers, eg, 2,4-dinitrophenyl 2'-iodo-3'-methyl-4'-nitro-6'-isopropylphenyl ether [69311-70-2] (DNP-INT), and the quinone analogues, 2,5-dibromo-3-methyl-6-isopropyl-p-benzoquinone [29096-93-3] (DB-MIB) and 5-n-undecyl-6-hydroxy-4,7-dioxobenzothiazole [43152-58-5] (UHDBT) are presumed to interfere with the cytochrome b_6f complex by altering the Rieske Fe-S center (40,42). The high potential Fe-S protein with its characteristic epr spectrum was first discovered in the mitochondrial cytochrome bc_1 complex and is present in animal and plant cytochrome complexes that are active in quinol–cytochrome c oxidoreduction (40).

Photosystem II Inhibitors. The PSII complex usually is assumed to be that structural entity capable of light absorption, water oxidation, plastoquinone reduction, and generation of transmembrane charge asymmetry and the chemical

Table 1. Modes of Herbicide Action

Common name	CAS Registry Number	Chemical name	Molecular formula
		Photosystem I inhibitors	
acifluorfen	[50594-66-6]	5-[2-chloro-4-(trifluormethyl)phenoxy]-2-nitrobenzoic acid	$C_{14}H_{17}ClF_3NO_5$
nitrofen	[1836-75-5]	2,4-dichloro-1-(4-nitrophenoxy)benzene	$C_{12}H_7Cl_2NO_3$
oxyflurofen	[42874-03-3]	2-chloro-1-(3-ethoxy-4-nitrophenoxy)-4-(trifluoromethyl)benzene	$C_{15}H_{11}ClF_3NO_4$
		Photosystem II inhibitors	
atrazine	[1912-24-9]	6-chloro-N-ethyl-N'-isopropyl-1,3,5-triazine-2,4-diamine	$C_8H_{14}ClN_5$
metribuzin	[21087-64-9]	4-(amino-6-t-butyl-3-methylthio)-1,2,4-triazin-5(4H)-one	$C_8H_{14}N_4O_5$
diuron	[330-54-1]	N'-(3,4-dichlorophenyl)-N,N-dimethylurea	$C_9H_{10}Cl_2N_2O$
bromacil	[314-40-9]	5-bromo-6-methyl-3-sec-butyl-2,4(1H,3H)pyrimidinedione	$C_9H_{13}BrN_2O_2$
ioxynil	[132-66-1]	4-hydroxy-3,5-diiodobenzonitrile	$C_7H_3I_2NO$
dinoseb	[88-85-7]	2-sec-butyl-4,6-dinitrophenol	$C_{10}H_{12}N_2O_5$
bromoxynil	[1689-84-5]	3,5-dibromo-4-hydroxybenzonitrile	$C_7H_3Br_2NO$
dinitrocresol	[534-52-1]	2-methyl-4,6-dinitrophenol	$C_7H_6N_2O_5$
		Bleaching herbicides	
fluridone	[59756-60-4]	1-methyl-3-phenyl-5-(3-trifluoromethylphenyl)-4(1H)-pyridinone	$C_{19}H_{14}F_3NO$
flurochloridone	[61213-25-0]	3-chloro-4-(chloromethyl)-1-(3-trifluoromethylphenyl)-2-pyrrolidinone	$C_{12}H_{10}Cl_2F_3NO$
flurtamone	[96525-24-4]	5-(methyl)amino-2-phenyl-4-(3-trifluoromethylphenyl)-3(2H)-furanone	
S3442	[65261-98-5]	3-(2,5-dimethylphenoxy)-N-ethylbenzamide	$C_{17}H_{19}NO_2$
diflufenican	[83164-33-4]	N-(2,4-difluorophenyl)-2-(3-trifluoromethylphenoxy) nicotinamide	$C_{19}H_{11}F_5N_2O_2$
difunon	[7703-36-8]	5-(dimethylaminomethylene)-2-oxo-4-phenyl-2,5-dihydrofurane-carbo-nitrile	$C_{14}H_{12}N_2O_2$
norflurazon[a]	[27314-13-2]	4-chloro-5-(methylamino)-2-(3-trifluoromethylphenyl)-3(2H)-pyridazione	$C_{12}H_9ClF_3N_3O$
amitrole[b]	[61-82-5]	1H-1,2,4-triazol-3-amine	$C_2H_4N_4$
fluometuron	[2164-17-2]	N,N-dimethyl-N'-(3-trifluoromethylphenyl)urea	$C_{10}H_{11}F_3N_2O$
fomesafen	[72178-02-0]	5-[2-chloro-4-(trifluoromethyl)phenoxy]-N-(methylsulfonyl)-2-nitrobenzamide	$C_{15}H_{10}ClF_3N_2O_6S$

Common name	[CAS]	Chemical name	Formula
Chlorophyll biosynthesis inhibitors			
oxadiazon	[19666-30-9]	3-[2,4-dichloro-5-(1-methoxyethoxy)phenyl]-5-(1,1-dimethylethyl)-1,3,4-oxadiazol-2-(3H)-one	$C_{15}H_{18}Cl_2N_2O_3$
DTP	[58010-98-3]	1,3-dimethyl-4-(2,4-dichlorobenzoyl)-5-hydroxypyrazole	
MK-616	[39985-63-2]	N-(4-chlorophenyl)3,4,5,6-tetra-hydrophthalimide	$C_{14}H_{12}ClNO_2$
Lipid and wax synthesis inhibitors			
clethodim	[99129-21-2]	(E,E)-(±)-2-[1-[[3-chloro-2-propenyl)oxy]imino]propyl]-5-[2-(ethylthio) propyl]-3-hydroxy-2-cyclohexen-1-one	$C_{17}H_{26}ClNO_3S$
sethoxydim	[74051-80-2]	2-[1-(ethoxyimino)butyl]-5-[2-(ethylthio)propyl-3-hydroxy-2-cyclohexen-1-one	$C_{17}H_{29}NO_3S$
haloxyfop, methyl	[69806-40-2]	2-[4-[[3-chloro-5-trifluoromethyl)-2-pyridinyl]oxy]phenoxy]propanoic acid, methyl ether	$C_{16}H_{13}ClF_3NO_4$
tralkoxydim	[87820-88-0]		
fenoxaprop, ethyl	[82110-72-3]	ethyl (±)-2-[4-[(6-chloro-2-benzoxazoly)oxy]phenoxy]propanoate	$C_{18}H_{16}ClNO_5$
fluazifop, butyl	[69806-50-4] [79241-46-6][c]	butyl (±)-2-[4-[[5-(trifluoromethyl)-2-pyridinyl]oxy]phenoxy]propanoic acid	$C_{19}H_{20}F_3NO_4$
alachlor[d]	[15972-60-8]	2-chloro-N-(methoxymethyl)acetamide	$C_{14}H_{20}ClNO_2$
metolachlor[d]	[51218-45-2]	2-chloro-N-(2-ethyl-6-methyphenyl)-N-(methoxy-1-methyethyl)acetamide	$C_{15}H_{22}ClNO_2$
diclofop, methyl	[51338-27-3]	methyl (±)-2-[4-(2,4-dichlorophenoxy)phenoxy] propanoic acid	$C_{16}H_{14}Cl_2O_4$
CDEC	[95-06-7]	2-dichloroallyldiethyldithiocarbamate	$C_8H_{14}ClNS_2$
diallate	[2303-16-4]	S-(2,3-dichloro-2-propenyl)bis(1-methylethyl)carbamothioate	$C_{10}H_{17}Cl_2NOS$
EPTC	[759-94-4]	S-ethyl dipropyl carbamothioate	$C_9H_{19}NOS$
triallate	[2303-17-5]	S-(2,3,3-trichloro-2-propenyl)bis(1-methylethyl)carbamothioate	$C_{10}H_{16}Cl_3NOS$
metflurazon	[23576-23-0]	2-(3-trifluoromethylphenyl)-4-chloro-5-dimethylamino-3(2H)-pyridazinone	$C_{13}H_{11}ClF_3N_3O$
Inducers of damage to antioxidative systems			
paraquat	[4685-14-7]	1,1'-dimethyl-4,4'-bipyridinium ion	$C_{12}H_{14}N_2$
diquat	[2764-72-9]	6,7-dihydrodipyridol[1,2-α:2',1'-c]pyrazinedium ion	$C_{12}H_{12}N_2$
tridiphane	[58138-08-2]	2-(3,5-dichlorophenyl)-2-(2,2,2-trichloroethyl)oxirane	$C_{10}H_7Cl_5O$

Table 1. (Continued)

Common name	CAS Registry Number	Chemical name	Molecular formula
		Herbicidal inhibition of enzymes	
MAA	[124-58-3]	methylarsonic acid	CH_5AsO_3
MSMA	[2163-80-6]	monosodium salt of methylarsonic acid	$CH_5AsO_3 \cdot Na$
DSMA	[144-21-8]	disodium salt of methylarsonic acid	$CH_5AsO_3 \cdot 2Na$
AMA		ammonium methylarsonic acid	
cacodylic acid	[75-60-5]	dimethyl arsinic acid	$C_2H_7AsO_2$
glufosinate	[51276-47-2]	2-amino-4-(hydroxymethylphosphinyl)butanoic acid	$C_5H_{12}NO_4P$
glufosinate, ammonium	[77182-82-2]	ammonium-2-amino-4-(hydroxymethylphosphinyl)butanoic acid	$C_5H_{12}NO_4P \cdot H_3N$
		Amino acid and nucleotide biosynthesis inhibitors	
phaseolotoxin	[62249-77-8]	L-lysine, N^6-(aminoiminomethyl)-N^2-[N-[N^5-[[(amino sulfonyl)oxy]hydroxyphosphinyl]-L-ornithyl]-L-alanyl]	$C_{15}H_{33}N_8O_9PS$
glyphosate	[1071-83-6]	N-(phosphonomethyl)glycine	$C_3H_8NO_5P$
rhizobitoxine	[37658-95-0]	2-amino-4-(2-amino-3-hydroxypropoxy)-*trans*-3-butanoic acid	$C_7H_{14}N_2O_4$
chlorsulfuron	[64902-72-3]	2-chloro-N-[[4-methoxy-6-methyl-1,3,5-triazin-2-yl)amino]carbonyl]benzenesulfonamide	$C_{12}H_{12}ClN_5O_4S$
chlorimuron, ethyl	[90982-32-4]	2-[[[[4-chloro-6-methoxy-2-pyrimidinyl)amino]carbonyl]amino] sulfonyl]benzoic acid, ethyl ester	$C_{15}H_{15}ClN_4O_6S$
sulfometuron	[74222-97-2]	2-[[[[(4,6-dimethyl-2-pyrimidinyl)amino]carbonyl]amino] sulfonyl]benzoic acid, methyl ester	$C_{15}H_{16}N_4O_5S$
bensulfuron, methyl	[83055-99-6]	methyl-2-[[[[4,6-dimethoxy-2-pyrimidinyl)amino]carbonyl] amino]sulfonyl]benzoic acid, methyl ester	$C_{16}H_{18}N_4O_7S$
imazaquin	[81335-37-7]	2-(4,5-dihydro-4-methyl-4-isopropyl-5-oxo-1*H*-imidazol-2-yl)-3-quinolinecarboxylic acid	$C_{17}H_{17}N_3O_3$
imazapyr	[81334-34-1]	(±)-2-(4,5-dihydro-4-methyl-4-isopropyl-5-oxo-1*H*-imadazol-2-yl)-3-pyridinecarboxylic acid	$C_{13}H_{15}N_3O_3$

82

Name	CAS	Chemical name	Formula
imazethapyr	[81335-77-5]	2-(4,5-dihydro-4-methyl-4-isopropyl-5-oxo-1H-imadazol-2-yl)-5-ethyl-3-pyridinecarboxylic acid	$C_{15}H_{19}N_3O_3$
imazamethabenz	[81405-85-8]	(±)-2-(4,5-dihydro-4-methyl-4-isopropyl-5-oxo-1H-imadazol-2-yl)-4 (and 5)-methylbenzoic acid, methyl ester	$C_6H_{20}N_2O_3$

Cell division inhibitors

Name	CAS	Chemical name	Formula
trifluralin	[1582-09-8]	2,6-dinitro-N,N-dipropyl-4-(trifluoromethyl)benzenamine	$C_{13}H_{16}F_3N_3O_4$
oryzalin	[19044-88-3]	4-(dipropylamino)-3,5-dinitrobenzene-sulfonamide	$C_{12}H_{18}N_4O_6S$
pendimethalin	[40487-42-1]	N-(1-ethylpropyl)-3,4-dimethyl-2,6-dinitrobenzenamide	$C_{13}H_{19}N_3O_4$
nitralin	[4726-14-1]		
dinitramine	[29091-05-2]	N^3N^3-diethyl-2,4-dinitro-6-(trifluormethyl)-1,3-benzenediamine	$C_{11}H_{13}F_3N_4O_4$
asulam	[3337-71-1]	methyl 4-aminophenylsulfonylcarbamate	$C_8H_{10}N_2O_4S$
propham	[122-42-9]	isopropyl phenylcarbamate	$C_{10}H_{13}NO_2$
chloropropham	[101-21-3]	isopropyl 3-chlorphenylcarbamate	$C_{10}H_{12}ClNO_2$
barban	[101-27-9]	4-chloro-2-butynyl 3-chlorophenylcarbamate	$C_{11}H_9Cl_2NO_2$
butylate	[2008-41-5]	S-ethyl bis(2-isobutyl)carbamothioate	$C_{11}H_{23}NOS$
cycloate	[1134-23-2]	S-ethyl cyclohexylethylcarbamothioate	$C_{11}H_{21}NOS$
propachlor	[1918-16-7]	2-chloro-N-isopropyl-N-phenylacetamide	$C_{11}H_{14}ClNO$
DCPA	[709-98-9]	dimethyl 2,3,5,6-tetrachloro-1,4-benzenedicarboxylate	$C_9H_9Cl_2NO$
pronamide	[23950-58-5]	3,5-dichloro-(N-t-butyl-2-propynyl) benzamide	$C_{12}H_{11}Cl_2NO$
bensulfide	[741-58-2]	O,O-bis(isopropyl)-S-[2-[(phenyl sulfonyl)amino]ethyl]phosphorodithoate	$C_{11}H_{24}NO_4PS_3$
cinmethylin	[87818-31-3]	exo-1-methyl-4-isopropyl-2-[(2-methylphenyl)methoxy]-7-oxabicyclo[2.2.1]heptane	$C_{18}H_{26}O_2$

83

Table 1. (Continued)

Common name	CAS Registry Number	Chemical name	Molecular formula
		Plant growth regulator synthesis and function inhibitors	
naphthalene acetic acid	[86-87-3]	1-naphthaleneacetic acid	$C_{12}H_{10}O_2$
indolebutyric acid	[133-32-4]	1H-indole-3-butanoic acid	$C_{12}H_{13}NO_2$
2,4-D	[94-75-7]	2,4-dichlorophenoxyacetic acid	$C_8H_6Cl_2O_3$
2,4,5-T	[93-76-5]	2,4,5-trichlorophenoxyacetic acid	$C_8H_5Cl_3O_3$
MCPA	[94-74-6]	(4-chloro-2-methylphenoxy)acetic acid	$C_9H_9ClO_3$
dicamba	[1918-00-9]	3,6-dichloro-2-methoxybenzoic acid	$C_8H_6Cl_2O_3$
chloramben	[133-90-4]	3-amino-2,5-dichlorobenzoic acid	$C_7H_5Cl_2NO_2$
picloram	[1918-02-1]	4-amino-3,5,6-trichloro-2-pyridinecarboxylic acid	$C_6H_3Cl_3N_2O_2$
naptalam	[132-66-1]	N-napthylphthalamide	$C_{18}H_{13}NO_3$
TIBA	[88-82-4]	2,3,5-triiodobenzoic acid	$C_7H_3I_3O_2$
diclofop	[40843-25-2]	(±)-2-[4-(2,4-dichlorophenoxy)phenoxy] propanoic acid	$C_{15}H_{12}Cl_2O_4$
ethephon	[16672-87-0]	(2-chloroethyl) phosphonic acid	$C_2H_6ClO_3P$
tetcyclacis	[65245-23-0]	1-(4-chlorophenyl)-3a,4,4a,6a,7,7a-hexahydro-4,7-methano-1-H-[1,2]diazeto[3,4-f]benzotriazole	$C_{13}H_{12}ClN_5$
AMO-1618	[2438-53-1]	2-isopropyl-5-methyl-4-(trimethylammonium chloride)-phenyl-1-piperidiniumcarboxylate	$C_{19}H_{31}N_2O_2 \cdot Cl$
chlormequat chloride	[999-81-5]	(2-chloroethyl)-trimethylammonium chloride	$C_5H_{13}ClN \cdot Cl$
mepiquat chloride	[24307-26-4]	N,N-dimethylpiperdinium chloride	$C_7H_{16}N \cdot Cl$
ancymidol	[12771-68-5]	α-cyclopropyl-α-(4-methoxyphenyl)-5-pyrimidine methanol	$C_{15}H_{16}N_2O_2$
uniconazole	[83657-22-1]	(E)-1-(4-chlorophenyl)-4,4-dimethyl-2-(1,2,4-triazol-1-yl)-penten-3-ol	$C_{15}H_{18}ClN_3O$
paclobutrazol	[76738-62-0]	1-(4-chlorophenyl)-4,4-dimethyl-2-(1H-1,2,4-triazol-1-yl)-pentan-3-ol	$C_{15}H_{20}ClN_3O$
BAS 11100W	[80553-79-3]	1-phenoxy-3-(1H-1,2,4-triazol-1-yl)-4-hydroxy-5,5 dimethylhexane	$C_{16}H_{23}N_3O_2$

[a] Also effective as an inhibitor of lipid and wax synthesis.
[b] Primary mode of action is inhibition of amino acid synthesis.
[c] (R) isomer, commonly called fluazifop-P-butyl.
[d] Also general cell growth inhibitors.

potential of hydrogen ions (41). The typical PSII complex contains approximately a dozen different polypeptides; 200 chlorophyll a molecules; 100 chlorophyll b molecules; 50 carotenoid molecules; at least three different plastoquinones, ie, PQ, Q_A, and Q_B in Figure 1; one iron; two pheophytin a (Phe) molecules; one or more cytochrome b_6f molecules; four manganese ions; and varying numbers of chloride and calcium ions. The reaction center of PSII is more narrowly defined (43) as consisting of P680, a photo-oxidizable chlorophyll a which transfers an electron to the primary acceptor; pheophytin a; and, thence to the first quinone acceptor (Q_A), a plastoquinone. The so-called core of PSII is a set of five hydrophobic polypeptides (41), two of which form the reaction center that performs the primary photochemical charge separation of PSII, ie, the 47 and 43 kDa polypeptides (43–45). Since polypeptide molecular weights are only estimates and species-specific variations in molecular weights exist, this pair of polypeptides has been reported as 51 and 45 kDa (46). Evidence (ca 1993) favors the association of P680 with the larger polypeptide of this pair; the smaller polypeptide serving primarily as a light harvesting antenna for the reaction center. The plastoquinone Q_A is tightly bound to a PSII reaction center polypeptide, most likely the larger of the pair under discussion here (41,43).

The PSII complex contains two distinct plastoquinones that act in series. The first is the Q_A mentioned above; the second, Q_B, is reversibly associated with a 30–34 kDa polypeptide in the PSII core. This secondary quinone acceptor polypeptide is the most rapidly turned-over protein in thylakoid membranes (41,46). It serves as a two-electron gate and connects the single-electron transfer events of the reaction center with the pool of free plastoquinone in the membrane (25,41,46). The Q_B is probably the most studied protein in thylakoid membranes (46) since it is the binding site of many, if not most, PSII-inhibiting herbicides (46).

Many commercial herbicides inhibit electron flow on the reducing side of PSII (37,41,46). Compounds as chemically different as atrazine, metribuzin, diuron, bromacil, ioxynil, and dinoseb (see Table 1) all block electron transfer from Q_A to Q_B (46). Herbicidal PSII inhibitors, such as these and other triazines, ureas, pyrimidines, nitriles, and phenols, all appear to have the same site and mode of action; differences in activity are determined by the lipophilicity of the various side chains (41). These herbicides prevent electron transfer from Q_A to Q_B by displacing Q_B from its binding site on the Q_B polypeptide (46). This displacement from the proteinaceous receptor is competitive in the case of the phenylureas, triazines, pyridazinones, and biscarbamates (25). The herbicides cannot be reduced by Q_A, and electron transfer is blocked. The herbicides are competitive with plastoquinone which accepts electrons from Q_B (41), and schemes visualizing binding sites of herbicides to the Q_B protein have been presented (47,48). An allosteric action has been suggested for these herbicides (49,50). Conformational alterations of the Q_B protein through herbicide binding would affect both electron transport and binding of Q_B, plastoquinone, and additional herbicide molecules (41).

Polypeptide conformational changes through herbicide binding also have been suggested as the mode of action of the phenol PSII inhibitors, eg, dinitrocresol (DNOC), dinoseb, bromoxynil, and ioxynil (25) (Table 1). These phenolic herbicides also uncouple oxidative phosphorylation in PSI at high concentrations and have been classed as inhibitory uncouplers (51). Phenolic herbicides inhibit at the same site as ureas and triazines, but there is no common basic chemical structure, and the interactions at the thylakoid membrane receptor site are different (25). Structure–activity relationship studies of halogenated nitro- and dinitrophenols suggest that phenolic herbicidal activity is determined by steric parameters. Additional phenol analogues, eg, benzoquinones, napthoquinones, pyridones, quinolones, pyrones, dioxobenzthiazoles, and cyanoacrylates are potent inhibitors of PSII (25,47).

The extensive research effort that has increased understanding of PSII and the binding of herbicides to the various polypeptides in photosynthetic membranes has initiated molecular modeling studies aimed at new inhibitors and herbicides (52). QSAR techniques have been applied, usually *a posteriori*, in investigations of various chemical classes, ie, ureas, carbamates, and anilides; triazines; triazinones; pyridazinones, uracils, and pyrimidones; imidazoles and pyrazolones; cyanoacrylates; pyrrolones and pyridones; phenols and quinones; and phenylureas (18).

From the beginning, herbicide developmental research has been focused on PSII and, more recently, the different modes of inhibitor binding that confer specificity and efficacy (47,53). The characteristics and level of understanding of the water-oxidation and donor-side reactions in photosynthesis encourage this emphasis on the PSII plastoquinone–polypeptide complexes (41,45). However, there is some question whether inhibition of photosynthesis remains the best modern herbicide target. Indeed, in 1980, photosynthesis inhibitors represented 50% of the commercial herbicide market (51); that market share dropped to 30% in 1990. Decreasing tolerance for herbicides which persist in the soil is a contributing factor in this lessening use and acceptability. Other modes of action, particularly those relevant to membrane function or enzymatic activity, now show greater potential for producing new herbicides with desirably low application rates.

Bleaching Herbicides. Membrane-based modes of herbicidal action relevant to photosynthesis (37) include those of inhibitors of carotenoid biosynthesis, eg, norflurazon, difunon, *m*-phenoxybenzamines; inhibitors of chlorophyll biosynthesis, eg, oxadiazon, DTP or 1,3-dimethyl-4-(2,4-dichlorobenzoyl)-5-hydroxypyrazole [*19666-30-9*], MK-616 or *N*-(4-chlorophenyl)-3,4,5,6-tetrahydrophthalimide; and promoters of peroxidative destruction of membrane lipids, eg, bipyridyliums and diphenyl ethers. Bleaching herbicides can act at multiple sites in lipid metabolism and are reported to affect chloroplast pigments, ie, carotenoids and chlorophylls, by interfering with phytoene desaturation. This interference usually results in the accumulation of phytoene, a tetraterpene formed by the condensation of two molecules of geranylgeranyl pyrophosphate and the first carotene precursor of the carotenoid auxiliary pigments of photosynthesis.

There are three distinct groups of phytoene desaturase (dehydrogenase) inhibitors. The norflurazon class includes fluridone, flurochloridone, flurtamone, S3442, diflufenican, and difunon (Table 1). These compounds directly inhibit the conversion of phytoene to the colorless α- and β-carotenes which are the substrates for phytoene desaturase (PD), the enzyme which produces the colored carotenoid, ζ-carotene (37). Norflurazon (54) and fluridone (55) act as reversible noncompetitive PD inhibitors in cell-free systems.

A second class of herbicides primarily affects ζ-carotene desaturase. These herbicides are apparent feedback inhibitors of PD as well. This class of compounds includes dihydropyrones like LS 80707 [90936-96-2] (56) and 6-methylpyridines (57,58). The third class consists of the benzoylcyclohexane-diones, eg, 2-(4-chloro-2-nitrobenzoyl)-5,5-dimethyl-cyclohexane-1,3-dione. This class of atypical bleaching herbicides induces phytoene accumulation when applied either pre- or post-emergence. However, it does not inhibit phytoene de-saturase activity *in vitro* (59). Amitrole also has been considered a bleaching herbicide, though its main mode of action is inhibition of amino acid synthesis.

Quantitative structure–activity studies of the typical PD inhibitors from the norflurazon group have been used to elucidate the influence of various sub-stituents of herbicidal activity (60). Some herbicides known to inhibit PD inter-act with other targets, eg, fluometuron and fomesafen (37) also inhibit electron transport. Fluometuron is more potent as an inhibitor of electron flow *in vitro* through PSII than as an inhibitor of carotenoid formation. The action of fome-safen occurs more through initiation of peroxidative destruction of membranes than through inhibition of carotogenesis. Some phenylpyridazinones interfere with desaturation of linoleic acid to linolenic acid (29,37), probably through inter-action with other lipid desaturase enzymes, specifically δ-15-desaturase (29,61). The bleaching herbicide, fluorochloridone, inhibits PD *in vivo* and *in vitro* and also inhibits linolenic acid formation (37,62).

Structure–activity studies of the seven herbicide classes that inhibit PD suggest that all classes target the same enzyme (37). Chemically, these classes are phenoxybenzamides, phenoxynicotinamides, phenylpyridazinones, phenylpyrrolidinones, phenylfuranones, phenylpyridinones, and phenyltetrahy-dropyrimidones. Cross-resistance studies of *Synechococcus* mutants (63) indicate that inhibitors of PD bind to the enzyme in the same general region but not to the same amino acid residues. Computer modeling techniques have been used to define and compare four regions in PD-inhibitor molecules (64). Two to four of these regions were present in the different PD-inhibitors examined. The first steps have also been made toward employing QSAR in the construction of a model describing the general features of PD-inhibitors and in the characteriza-tion of PD through determination of PD enzymology, amino-acid composition, and genetic markers (65,66).

Chlorophyll Biosynthesis Inhibitors. Chemically, the chlorophylls are magnesium–porphyrin complexes in which the four central nitrogen atoms of the pyrrole rings are coordinated with a Mg^{2+} ion to form an extremely stable planar complex. Chlorophyll also has a long hydrophobic terpenoid side chain consisting of phytol [150-86-7], an alcohol which is esterified to a propionic acid residue in ring IV. Several herbicides are reported to inhibit chlorophyll biosyn-thesis, eg, oxadiazon, DTP, and MK-616 (see Table 1), but the target or targets of these compounds is unknown (37). Along with similar compounds, amitrole has been reported to induce accumulation of ζ-carotene and also to induce chlorophyll bleaching and inactivation of enzymes other than PD (67).

Lipid and Wax Synthesis Inhibitors. Lipids, primarily in the form of acyl lipids derived from long-chain fatty acids, are present in all plant organs and on leaf surfaces (68–70). In plant roots and shoots, acyl lipids such as phospho- or glycolipids are structural components of the essential biological membranes of cell compartmentation, enzymology, and bioenergetics. Acyl lipids are constituents of a large variety of different structures with different functions and are, therefore, promising potential target sites for herbicide action. The effects of herbicides on lipid metabolism have been reviewed (71–73).

Fatty acid synthesis in plants has been reviewed (69,74). The reactions that lead to the formation of fatty acids are roughly divided into three classes, ie, initial reactions, biosynthesis of 16:0 and 18:0 saturated fatty acids by fatty acid synthetase (FAS), and biosynthesis of 18:1, 18:2, and 18:3 unsaturated acids. The initial steps in fatty acid biosynthesis are those which produce acetyl-CoA and malonyl-CoA. The key enzyme in these steps is acetyl-CoA carboxylase (ACC), a multifunctional protein located in the chloroplasts (71). Clethodim is reported to inhibit ACC (2), as are sethoxydim, haloxyfop, and tralkoxydim (75,76) (Table 1). Fenoxaprop; fluazifop, butyl; and fluazifop-P-butyl also act through inhibition of fatty acid biosynthesis in sensitive plant species (2,77) (Table 1). These oxyphenoxy propionic acids and the analogous diclofop–methyl; clofop–isobutyl [51337-71-4]; haloxyfop–methyl; and fenthiaprop–ethyl [93921-16-5] all inhibit fatty acid synthesis *de novo* (78,79). Sethoxydim and alloxydim [55634-91-8], also inhibitors of fatty acid synthesis, are cyclohexanedione derivatives (29,80). The herbicidal activity of the cyclohexanediones is similar to the oxyphenoxy propionic acids, but alloxydim and sethoxydim also cause necrosis in meristematic regions and leaf chlorosis. These herbicides are selective against grasses (29,81). With some exceptions (81), cyclohexandiones do not inhibit incorporation of ^{14}C-acetate into chloroplast lipids of tolerant monocotyledonous, eg, *Poa annua*, *Festuca longifolia*, *F. rubra*, and *F. myuros* (76), and dicotyledonous, eg, pea, spinach, and tobacco species (82).

Incorporation inhibition of ^{14}C-acetate into lipids is also the most rapid and pronounced effect of the α-chloracetamides, alachlor, metazachlor, and metolachlor (83) (Table 1), suggesting that the acetate-incorporating steps of lipid synthesis are the site of action for these herbicides. The thiocarbamate, EPTC, inhibits three enzymes that produce acetyl-CoA, ie, chloroplastic acetyl-CoA synthetase and both the chloroplastic and mitochondrial pyruvate dehydrogenase complexes (84). These two classes of herbicides have somewhat parallel properties and modes of action, possibly through reaction with sulfhydryl groups of proteins and low molecular weight thiols such as glutathione (85). Further, both the α-chloracetamides and the thiocarbamates have been reported to inhibit the lipid metabolic pathways that lead to the formation of the very long-chain epicuticular waxes of leaves (29,68,86). Reductions in leaf cuticular waxes increase leaf wetability by decreasing water surface tension, thus increasing plant sensitivity to subsequently applied herbicides (29,68). It has been reported that dithiocarbamates, eg, diallate, CDEC, EPTC, and triallate, alter the structure, composition, or amount of wax (87,88). The alkanes and secondary alcohol components of the waxes were reported to be significantly affected by CDEC and EPTC (87). EPTC, diallate, and triallate also inhibit suberin formation (89). Suberin consists of relatively high amounts of dicarboxylic acids, phenolics, very long-chain fatty acids, and very long-chain alcohols (29).

Pyridazinone herbicidal activity depends on inhibition of multiple target sites in plants, eg, PSII, the Hill-reaction, and carotenoid biosynthesis, as well as changes in fatty acid composition (29,69). Pyridazinone-induced changes in fatty acid composition include increases in the 18:2/18:3 fatty acid ratios, suggesting inhibition of δ-15-desaturase (29,61). The 5-dimethylamino-substituted pyridazinones, eg, BAS 13 338 [3707-98-0] and metflurazon, have strong effects on the 18:2/18:3 ratio; some monomethylamino derivatives, eg, norflurazon, and structurally analogous phenylpyridazinones are also effective (90). Metflurazon and norflurazon increase the 18:2/18:3 ratio in galactolipids and phospholipids (29) in sensitive species. Metflurazon affects primarily the saturation of C-16:0, but it also inhibits desaturation of 18:2 to 18:3 when it is applied in the right concentrations to sensitive species (91).

The oxyphenoxypropionic acids and the cyclohexadiones are phytotoxic because they inhibit synthesis *de novo* of fatty acids (29,78,81). Inhibition of lipid synthesis could also produce the other physiological effects attributed to these herbicides, ie, membrane disruption (29) and chloroplast damage with accompanying decreases in chlorophyll, CO_2 fixation, and ATP production (29,92), and the disruption of mitochondrial function (29,81). However, the phenoxy propionic acids, cyclohexadiones, thiocarbamates, and chloroacetamides, classes of herbicides reported to affect lipid synthesis, also have been shown to have significant effects on the production and activity of plant growth regulators such as auxin (93,94) and gibberellins (95).

Radical Damage to Antioxidative Systems and Cellular Components Inducers. The herbicidal activities of many of the inhibitors of PSII are enhanced by light, eg, metflurazon, norflurazon, and fluridone (2,96), possibly through the mechanism of photooxidative destruction (97,98). Excitation of chlorophyll pigments leads initially to a singlet state chlorophyll, ^{1}Chl. If electron transport is inhibited and the excitation is unquenched, a lower energy triplet state, ^{3}Chl, is generated. This triplet state chlorophyll interacts with oxygen to produce $^1O_2$. If ^{3}Chl and $^1O_2$ are formed during normal photosynthesis, quenching occurs through the agency of carotenoids, membrane-bound α-tocopherol, and stromal radical scavengers like ascorbate, glutathione, polyamines, and flavanols (96). Herbicides that enhance the production and/or activity of toxic free radicals and singlet oxygen include the bipyridiniums, paraquat and diquat (97). The *p*-nitro- or *p*-chlorodiphenyl ethers (DPEs), acifluorfen–methyl, and oxyfluorfen (37,97) have been reported to be involved in the initiation of free-radical chain reactions with polyunsaturated fatty acid moieties of phospholipid molecules intrinsic to cell membranes. More recent reports indicate that the phytotoxic mechanism of DPE herbicides depends on inactivation of protoporphyrinogen oxidase and the subsequent accumulation of Protoporphyrinogen IX (Proto IX), a potent photosensitizer (99). Proto IX accumulates, and illumination then leads to the formation of singlet oxygen and lipid peroxides which result in loss of membrane integrity and cell death. The herbicidal activity of the bipyridiniums is also enhanced by both light and oxygen (97). The dicationic nature of the bipyridiniums allows easy reduction to a cation radical through electron donation from the terminal end of PSI. This diverts electrons from ferredoxin, the natural PSI acceptor, and leads to inhibition of $NADP^+$ reduction and CO_2 fixation. Electron flow from water with the photosynthetically driven release of O_2 in the chloroplasts permits reoxidation of the bipyridinium radical and formation of superoxide (97). Superoxide, $O_2^{\cdot}$, is generated as a normal part of chloroplast electron transport, and the radical is scavenged by superoxide dismutase (SOD) enzymes, ie, metalloproteins which catalyze the conversion of two superoxide radicals and two protons to hydrogen peroxide and water. Leaves of paraquat-tolerant plants are reported to have higher levels of SOD than do the leaves of sensitive species (37,100). The hydrogen peroxide produced by SOD is further converted to very active hydroxy radicals, probably by a metal-catalyzed Fenton-type reaction.

Light and photosynthetic electron transport convert DPEs into free radicals of undetermined structure. The radicals produced in the presence of the bipyridinium and DPE herbicides decrease leaf chlorophyll and carotenoid content and initiate general destruction of chloroplasts with concomitant formation of short-chain hydrocarbons from polyunsaturated fatty acids (37,97).

The effectiveness of herbicides that induce lipid peroxidation depends on the activity of the natural protective mechanisms which are based on antioxidants and radical scavengers, ie, α-tocopherol, the carotenoids, ascorbate, glutathione, polyamines, and the flavanols (37,97,101). Plant defenses against radical-inducing compounds include an antioxidative system consisting of ascorbate, α-tocopherol, and reduced glutathione (GSH), closely associated with SOD and catalase. Peroxidation by oxyfluorfen is counteracted by the appropriate ratio of ascorbate and α-tocopherol. Increased radical formation leads to higher production of antioxidants. For example, acifluorfen–sodium induces increased glutathione reductase activity, elevated levels of glutathione and ascorbate, and simultaneous increases in galactonolactone oxidase activity. Conjugation of metabolites and xenobiotics with GSH is catalyzed by GSH-transferase(s), the activity of which can be increased by low concentrations of chloroacetamides and other safeners, compounds which can protect selected crops from some herbicides. Tridiphane inhibits GSH transferases and may prevent conjugation of atrazine in some species (2,101).

The chemical mechanism of other herbicides also involves peroxidative destruction of polyunsaturated fatty acids by starter radicals (37). Fenton reactions produce alkoxy radicals which can split into alkyl radicals leading to hydrocarbon gases (102) or can initiate further radical destruction of other chloroplast components (37,97). Potent peroxy and alkoxy radicals and lipid hydroperoxides are formed (103). Lipid hydroperoxides also decompose to form cytotoxic malondialdehyde [542-78-9] (MDA), a compound often used as an index of lipid peroxidation (97,103). MDA, a significant 2-thiobarbituric reactant, can cause intra- and intermolecular cross-linking of sulfhydryl-containing proteins (97). Proteins can also be fragmented or modified by hydrogen peroxide in the presence of transition metals (97,104). The resulting hydroxyl radicals and the alkoxy radical intermediates from lipid peroxidation also attack proteins and individual amino acids (104), particularly histidine, cysteine/cystine, methionine, lysine, tyrosine, and tryptophan (97,103,105).

Herbicidal Inhibition of Enzymes. The list of known enzyme inhibitors contains five principal categories: group-specific reagents; substrate or ground-state analogues, ie, rapidly reversible inhibitors; affinity and photo-affinity labels; suicide substrate, or k_{cat}, inhibitors; and transition-state, or reaction-intermediate, analogues, ie, slowly reversible inhibitors (106).

The radical-generating herbicides, described above, that attack specific amino acid residues are examples of group-specific enzyme inhibitors. Substrate analogue enzyme inhibitors include the organoarsenicals, MAA and MSMA (see

Table 1). Arsenical pesticides, known since the time of Aristotle, have been widely used as herbicides since 1951 (107). Arsenite (As^{3+}) reacts with sulfhydryl groups in enzymes and other thiol groups (108). Arsenate (As^{5+}) replaces phosphate in essential metabolic phosphorylation reactions, eg, glyceraldehyde-3-phosphate dehydrogenase (109), and oxidative phosphorylation (110). The arsonic acid herbicides, MAA, MSMA, DSMA, and AMA, are not technically plant growth regulators since they act through enzyme systems to inhibit growth (107) and, thus, kill plants relatively slowly. Cacodylic acid and its sodium salt are used extensively as selective post-emergence herbicides in cotton and noncrop areas and orchards. Cacodylate is reported to be a nonspecific competitive inhibitor of adenine nucleotide deaminase (111) and may inhibit other enzymes as well (107).

Suicide substrate and reaction intermediate inhibitors promise the highest degree of specificity and have drawn increased attention (106). Heteroatom or radical replacement in reaction–intermediate analogues is a simple pesticide development strategy that offers potential for achieving extremely potent inhibition without high chemical reactivity. This simple design strategy may produce effective intermediate inhibitors for families of mechanistically related enzymes. For instance, substitution of a phosphorous for a carbon has produced potent inhibitors of metalloproteases, cytidine deaminase, and glutamine synthase (106,112–114).

In the formation of glutamine by glutamine synthase, ammonia attacks the carbonyl of a glutamate phosphate intermediate (106). In the potent herbicide, glufosinate, ie, phosphinothricin, the tetrahedral carbon of the enzymatic intermediate is replaced by phosphorus and the attacking ammonia is replaced by a methyl group (106). Primary ammonium assimilation is a plant-specific process (114–116), although glutamine synthase plays a role in recycling catabolically produced ammonia in both plants and animals. In plants, ammonium ions can be taken up by the plant or can originate through turnover of endogenous N-containing cell components. Elevated concentrations of ammonia are cytotoxic (117), and ammonium ions are usually reassimilated through the action of glutamine synthase which binds ammonium to glutamate (28).

Glutamine synthase is inhibited by glufosinate (114,118), methionine sulfoximine [15985-39-4] (MSX) (119), bialaphos [35597-43-4] (57), hydroxylysine [28902-93-4] (114), and tabtoxin [40957-90-2] (57). Bialaphos is a tripeptide, from *Streptomyces hydroscopicus*, that can split into two alanine molecules and glufosinate (120). Glufosinate is activated by light and conditions that promote photorespiration; it blocks photosynthesis while inducing high accumulations of ammonia (117,118,121,122). Depletion of glutamine in the presence of glufosinate leads to a shutdown of the oxidative C_2-carbon cycle. Intermediates of that cycle, phosphoglycolate and possibly glyoxylate, accumulate, shutting down photosynthesis (123). Tabtoxin is a dipeptide composed of threonine or serine and tabtoxinine (57); it is excreted by *Pseudomonas syringae* spp. Tabtoxinine [40957-88-8] also causes ammonia accumulation (124) through noncompetitive and irreversible binding to glutamine synthase (125). Other biologically derived inhibitors of glutamine synthase have been reported (126).

A carbonyl reagent, aminooxyacetate, inhibits pyridoxalphosphate-dependent enzymes such as decarboxylases and transaminases. This reagent is the basis for the herbicides, benzadox [5251-93-4], benzamidooxyacetic acid

[5251-93-4], and other lipophilic analogues (127). Benzadox decreases photosynthesis and inhibits both alanine and aspartate aminotransferase (128). Isonicotinic hydrazide [54-85-3] affects glycine–serine aminotransferase (129), and aminoacetonitrile [540-61-4] inhibits glycine decarboxylation in a manner similar to that of gulfosinate (130).

When ATP-synthase, specifically the plastidic coupling factor CF_0-CF_1, is inhibited, ATP formation by photophosphorylation is blocked. Inhibitors of this energy-transfer process prevent conversion into ATP of the electrochemical potential formed by electron transport during photosynthesis. One such inhibitor, dicyclohexylcarbodiimide [538-75-0] (DCCD), binds irreversibly to the F_0 part of the synthase, preventing transfer of protons to the F_1 portion (57). Two allelochemicals, phlorizin [60-81-1] and quercetin [117-39-5], inhibit ATP-synthase by competing with phosphate; chlorinated p-nitrodiphenyl ethers compete with adenosine diphosphate (ADP) (131). One of the most studied inhibitors of the plastidic coupling factor is tentoxin [28540-82-1], a cyclic tetrapeptide produced by *Alternaria alternata* f. *tenuis* (132). Tentoxin is plant-specific, able to pass through membranes, and highly active, binding to the catalytic site of the CF_1. Its mode of action may also include interference with transport through the plastid envelope and with transport polypeptides (57). Tentoxin is an example of a naturally occurring compound with potential as a model for synthetic analogues with herbicidal activity.

Amino Acid and Nucleotide Biosynthesis Inhibitors. The metabolism of amino acids is affected by both chemical herbicides and biogenic inhibitors, eg, bialaphos, phaseolotoxin, and rhizobitoxine (57). Phaseolotoxin, a tripeptide from *Pseudomonas syringae* pv. *phaseolicola*, causes increased ornithine accumulations through inhibition of ornithine carbamoyltransferase, an enzyme in the pathway from ornithine to citrulline, the precursor of arginine. Peptidase activity cleaves phaseolotoxin to form a sulfodiaminophosphinyl-L-ornithine (133) that binds irreversibly and covalently to ornithine carbamoyltransferase like a natural affinity label. Rhizobitoxine, formed by *Rhizobium japonicum* spp., inhibits β-cystathionase which cleaves β-cystathionine to produce homocysteine, the precursor in the pathway to methionine (134,135).

Herbicides also inhibit 5-enol-pyruvylshikimate synthase, a susceptible enzyme in the pathway to the aromatic amino acids, phenylalanine, tyrosine and tryptophan, and to the phenylpropanes. Acetolactate synthase, or acetohydroxy acid synthase, a key enzyme in the synthesis of the branched-chain amino acids isoleucine and valine, is also sensitive to some herbicides. Glyphosate (26), the sulfonylureas (136), and the imidazoles (137) all inhibit specific enzymes in amino acid synthesis pathways.

In plants and microorganisms, synthesis of aromatic amino acids, ie, p-aminobenzoic acid, and ubiquinone, proceeds by the shikimate pathway (26,138). In plants, this pathway also provides the precursors for indoleacetic acid [87-51-4] (IAA), a plant growth regulator, alkaloids, lignin, the flavonoids, and a wide variety of secondary metabolites (139). Some 20 enzyme-catalyzed reactions are involved in the production of the aromatic amino acids. However, herbicide mode of action studies have focused on the three enzymatic steps that convert shikimic acid to chorismic acid, the branch intermediate for a variety of metabolites. In the first step (140–142), shikimic acid is phosphorylated by shikimic kinase to form shikimate-5-phosphate which condenses with phosphoenolpyruvate (PEP) to form 3-phospho-5-*enol*-pyruvylshikimic acid (EPSP). This reaction is catalyzed by 3-phospho-5-*enol*-pyruvylshikimic synthase (EPSP synthase), the target site of glyphosate. The EPSP ring is oxidized with the loss of a phosphate group to give chorismic acid. Nanomolar levels of glyphosate inhibit only EPSP synthase. Glyphosate shows competitive inhibition with respect to PEP and noncompetitive inhibition with respect to shikimate-3-phosphate (142,143). Inhibition of aromatic amino acid synthesis by glyphosate exerts early effects on a broad range of plant processes, ie, ion uptake and transport, chlorophyll synthesis, photosynthetic CO_2 uptake, and protein and nucleic acid synthesis (144).

Two relatively new classes of herbicides, the sulfonylureas, eg, chlorsulfuron, sulfometuron, bensulfuron methyl, and chlorimuron ethyl; and the imidazolinones, eg, imazaquin, imazapyr, imazethapyr, and imazamethabenz, have totally different chemical structures but remarkably similar modes of action in plants (2,18,27) (Table 1). Both types of herbicides inhibit the same key enzyme, acetolactate synthase, in the biosynthetic pathway to branched-chain amino acids. The branched-chain amino acids, ie, valine, leucine, and isoleucine, are essential amino acids produced by microorganisms and plants only. Their synthesis proceeds from threonine and pyruvate through a common series of reactions (114,145). The first common reaction in the branched-chain pathway is the formation of acetohydroxy acids by acetolactate synthase, and alternatively, acetolactate synthase (ALS) or acetohydroxyacid synthase (AHAS) (114) and imidazolinone herbicides. ALS, a nonoxidative thiamine pyrophosphate-dependent decarboxylase, is found in a wide variety of plant species; compartmentation studies indicate that it is localized in the chloroplasts (146). Unlike bacteria, plants contain a single form of ALS, and resistance to sulfonylureas is inherited as a single Mendelian trait (114,147). Sulfonylurea herbicidal activity is very potent and rapid (114,148). Sulfonylureas inhibit ALS at levels in the nanomolar range (27). These compounds appear to bind tightly to FAD (flavin adenine dinucleotide; hydroxyethylthiamine pyrophosphate) in the ALS complex. Sulfonylureas interfere with the binding of pyruvate in the case of valine synthesis and 2-oxobutyrate in isoleucine and leucine synthesis (114,149). In addition to inhibiting ALS, sulfonylureas also indirectly inhibit plant-cell division and DNA synthesis (114,148). The α-ketobutyrate accumulating when ALS is inhibited may also be cytotoxic (150). QSAR techniques applied to sulfonylurea herbicides have elucidated the physicochemical factors that determine activities *in vivo* and *in vitro*, allowing modeling of the sulfonylurea binding site (27).

The imidazolinone herbicides selectively block branched-chain amino acid synthesis through inhibition of ALS (114,151). The structures of these compounds, however, are very different from those of the sulfonylureas that block the same enzyme. Based on the extraordinary activity of the pyridyl imidazolinones (18,152), the QSAR approach was used to examine imidazolinones (18) with the goal of identifying novel synthesis candidates of high predicted activity. Initial reports (153) indicated that inhibition of maize suspension cell cultures and seedlings by imazapyr was reversed by the addition of valine, leucine, and isoleucine and that these compounds are uncompetitive (154) or simple noncompetitive (24) inhibitors of ALS. Treatment with imidazolinones inhibited plant cell division and reduced root soluble protein, without significant effect on protein synthesis (151). Genetic evidence strengthens the hypothesis that ALS is the site of action of both the imidazolinones and the sulfonylureas (155,156).

Amitrole (3-amino-s-triazole) blocks histidine synthesis in bacteria by inhibiting imidazole glycerol phosphate dehydrase (156). In higher plants the active site is not known (157). In light-grown plants, amitrole increases free amino acids and decreases protein (158). The effect on protein synthesis may be indirectly caused by interferences with purine metabolism (159) and/or glycine–serine interconversion (158). Amitrole may also interfere with purine synthesis at the step in which formylglycineamidine ribotide is cyclized to form 4-aminoimidazole (160). Amitrole also inhibits catalase (161). Plastids from amitrole-treated plants grown in light are highly aberrant, containing few 70 S ribosomes and reduced amounts of protein and plastid DNA (162). Amitrole also inhibits ζ-desaturase activity (37).

Cell Division Inhibitors. The most common mode of action of soil-applied herbicides is growth inhibition, primarily through direct or indirect interference with cell division (163). Such growth inhibitory activity is the basis for most pre- or post-emergent herbicides intended to control germinating weed seeds. In germinating seeds, cell division occurs in the meristems of the root and the shoot. Meristematic cells go through a cycle (163,164) consisting of four discrete periods: G_1, gap 1; S, DNA synthesis; G_2, gap 2; and M, mitosis. The time to complete this cell cycle is species-dependent and ranges from 12 to 17 hours (165). The G_1 is sometimes called the pre-DNA synthesis period and the G_2 the post-DNA synthesis period. The period of DNA transcription is represented by S, and the G_1, S, and G_2 stages constitute the interphase of the cell cycle. Many interrelated biochemical reactions occur during these stages, and interruption of any of the metabolic events in the cell cycle halts DNA synthesis, as well as RNA, protein, and essential metabolite syntheses (166).

Mitosis, the physical division of individual G_2 cells into two complete cells, is the most studied stage of the cell cycle. Mitosis itself is divided into four stages: prophase, metaphase, anaphase, and telophase (163). The two key processes of mitosis are the physical movement of chromosomes from one cellular location to another during the metaphase and anaphase and the shuttling of Golgi vesicles into the cell plate region during cell wall formation during telophase. Chromosome movement requires the attachment of several hundred microtubules to the kinetochore of each chromosome. Other microtubules extend from one end of the cell to the other to complete the mitotic spindle apparatus. Microtubules may also provide the cell skeletal framework needed for orientation and movement of Golgi vesicles during formation of the new cell wall between the two daughter cells resulting from mitosis (163).

The influences of herbicides on cell division fall into two classes, ie, disruption of the mitotic sequence and inhibition of mitotic entry from interphase (G_1, S, G_2). If cell-cycle analyses indicate increases in abnormal mitotic figures, combined with decreases in one or more of the normal mitotic stages, the effect is upon mitosis. Mitotic effects usually involve the microtubules of the spindle apparatus in the form of spindle depolymerization, blocked tubulin synthesis, or inhibited microtubule polymerization (163). Alkaloids such as colchicine [64-86-8], vinblastine [865-21-4], and vincristine [57-22-7] disrupt microtubule function (164). Colchicine prevents microtubule formation and promotes disassembly of those already present. Vinblastine and vincristine also bind to free tubulin molecules, precipitating crystalline tubulin in the cytoplasm. The capacities of these drugs to interfere with mitotic spindles, blocking cell division, makes them useful in cancer treatment.

Those herbicides that block mitotic entry decrease or prevent the formation of mitotic figures in meristems. Amino acid, protein, RNA, DNA, and ATP synthesis and/or utilization can all arrest cell growth (163,166). Although not registered as herbicides, cycloheximide [66-81-9] inhibits mitotic entry by inhibiting protein synthesis (167); hydroxyurea [127-07-1] inhibits DNA synthesis (168); and actinomycin D [50-76-0] inhibits RNA synthesis (167).

The best understood of those herbicides inhibiting cell division are the dinitroanilines (see Table 1). Micromolar levels of these herbicides, eg, trifluralin, oryzalin, pendimethalin, nitralin, and dinitramine, act by disrupting mitosis (169–171) and producing aberrant mitotic figures in which there is no chromosome movement due to the absence or dysfunction of the spindle apparatus (172). Dinitroanilines are reported to bind to higher plant tubulin (173) and to prevent, in a concentration-dependent manner, the polymerization of higher plant tubulin in microtubles (173,174). Characteristically, dinitroaniline treatment induces swelling of the cell-elongation zone of the root-tip area (175). This hypertrophy is isotropic, suggesting that the microtubule orientation skeleton necessary for cell wall formation is absent or functioning improperly (176). Phosphorothioamidates, although not developed commercially as herbicides, also disrupt tubulin function and the mitotic sequence (177).

Various compounds in the N-phenylcarbamate class of herbicides, eg, propham, chlorpropham, asulam, barban, and carbetamide [16118-49-3], act through disruption of mitosis (178–180), inhibition of PSII electron transport, or uncoupling of oxidative phosphorylation and photophosphorylation (163). These compounds affect the microtubule organizing center, causing the formation of multipolar spindle configurations (181). The N-phenylcarbamates also cause root-tip swelling (182) and branching of the cell plate during cell-wall formation during telophase (180,182), and chromosome abnormalities like bridging between daughter chromosomes (178,179,182).

The growth inhibitory mechanism of the thiocarbamate herbicides, eg, EPTC, butylate, cycloate, diallate, and triallate, is not well defined. Cell elongation, rather than cell division, appears to be inhibited (183), although mitotic entry may be inhibited by diallate (184). Thiocarbamates have a greater effect on shoot than root tissue (163,184). The well-documented inhibition of lipid synthesis by thiocarbamates certainly contributes to the observed inhibitions of cell division and elongation. These compounds may also inhibit gibberellic acid synthesis (185).

Chloroacetamide herbicides, such as alachlor, allidochlor [93-71-0], metolachlor, and propachlor, are general growth inhibitors that inhibit both cell enlargement and mitotic entry (163,186). The mechanisms by which mitotic entry is prevented are not known nor are the biochemical causes of cell-enlargement inhibition. Chloroacetamide herbicides are alkylating agents that could inhibit the cell cycle at interphase through alkylation of essential enzymes (187).

The inhibitors of amino acid synthesis, sulfonylureas, imidazolinones, and glyphosate, were first recognized as general growth inhibitors that prevent mitotic entry (188,189). Whatever the mode of action, herbicides that inhibit amino acid synthesis also cause a rapid inhibition of cell growth, usually through inhibition of mitotic entry.

DCPA inhibits the growth of grass species by disrupting the mitotic sequence, probably at entry (190). DCPA influences spindle formation and function (181) and causes root-tip swelling (182) and brittle shoot tissue (191). It has been reported that DCPA, like colchicine and vinblastine, arrests mitosis at prometaphase and is associated with formation of polymorphic nuclei after mitotic arrest (192). Pronamide also inhibits root growth by disrupting the mitotic sequence in a manner similar to the effect of colchicine and the dinitroanilines (193,194). Cinmethylin and bensulide prevent mitotic entry by unknown mechanisms (194).

Before the germination process has begun, quiescent or dormant seeds are not sensitive to chemical herbicides, and many noxious weeds owe their persistence to seed dormancy survival strategies (195,196). Weed seeds lie dormant in the soil until specific conditions of environment and time since dispersal are met, and a portion of the seeds in the soil seed bank germinate. These germinated seedlings then become susceptible to herbicides. Other dormant seeds, with different germination requirements, remain in the soil and eventually germinate after earlier herbicide applications are no longer effective. Programs to control weed species, eg, *Avena fatua* L., or wild oats, surviving through seed dormancy and parasitic weeds, eg, *Striga* spp. or witchweed, that germinate only in the presence of host roots frequently include development of germination stimulants that break seed dormancy (196,197). The nonparasitic weed seedlings are then susceptible to pre-emergence herbicides such as those described as inhibiting cell division, an essential process in seed germination. Further, suicide germination in the absence of an obligate host kills the parasitic weed seedlings by starvation (196–198).

Plant Growth Regulator Synthesis and Function Inhibitors. In a broad sense, herbicides are exogenous plant growth regulators, ie, plant growth inhibitors (48). Endogenous plant growth regulators, which can both stimulate and retard growth, fall into five categories, ie, auxins, gibberellins, cytokinins, abscisic acid, and ethylene (199). In the past, these compounds have been called plant hormones or phytohormones because of similarities with animal hormones, ie, defined as organic compounds synthesized in one part of an organism and translocated to another part where, in very low concentrations, they cause a physiological response. However, significant differences between plant and animal physiology have led to the use of plant growth regulator (PGR) as the preferred descriptor for these compounds which are common to all higher plants and also for the synthetic compounds which are analogues, competitors, or antagonists of the natural PGR compounds.

The auxins include the first natural PGR recognized, indoleacetic acid (IAA). In plants, IAA is the chemical signal responsible for the first identified auxin-response, phototropism (199). Auxin mediation is observed in stem and root elongation, as well as in control of lateral bud development and fruit set. Auxin also stimulates intracellular ethylene production. Similar synthetic compounds that induce the physiological responses associated with IAA include naphthalene acetic acid, indolebutyric acid, 2,4-D, 2,4,5-T, and MCPA (see

Table 1). The last three compounds belong to the chlorophenoxy acid class of herbicides (2,199). Development of 2,4-D and its phenoxyalkanoic acid analogues occurred during World War II, and use of these herbicides in food production was readily accepted as a consequence of increased labor costs in the industrialized countries after that war (3,199). The structure–activity studies of the phenoxyacetic acids were the basis of the initial QSAR papers (5,6). Trace levels of 2,4-D stimulate elongation growth, as does IAA. Higher concentrations are needed to produce the inhibitory effects associated with these herbicides (72,200).

Other auxin-like herbicides (2,48) include the chlorobenzoic acids, eg, dicamba and chloramben, and miscellaneous compounds such as picloram, a substituted picolinic acid, and naptalam (see Table 1). Naptalam is not halogenated and is reported to function as an antiauxin, competitively blocking IAA action (199). TIBA is an antiauxin used in receptor site and other plant growth studies at the molecular level (201). Diclofop–methyl and diclofop are also potent, rapid inhibitors of auxin-stimulated response in monocots (93,94). Diclofop is reported to act as a proton ionophore, dissipating cell membrane potential and perturbing membrane functions.

All auxins, both natural IAA and the synthetics, stimulate ethylene production, and some reports have suggested that auxin effects may be due to increased ethylene concentration (199,202). Auxin-induced tissue elongation is apparently not directly affected by ethylene, but leaf epinasty, inhibition of stem, root, and leaf elongation, and floral senescence may involve ethylene to some degree. Lateral expansion or swelling is a rapid, easily recognizable physiological response to ethylene (202). This ethylene-induced change from elongation to isodiametric expansion is often accompanied by changes in basic cell wall structure, hormone balance, water relationships, metabolic pool sizes, intercellular localization of enzyme activities, and wall extensibility. Ethylene exposure frequently increases enzyme synthesis (199) and inhibits auxin-induction of enzymes in cell wall metabolism (203). Plant responses to ethylene are the basis for the growth-inhibiting uses of ethephon (Ethrel) which is applied to promote fruit ripening and abscission, and boll opening and defoliation in cotton (2). It has been suggested that an antagonism or buffering effect exists between ethylene and auxin (203–206), as well as between ethylene and gibberellic acid (207).

Abscisic acid [21293-29-8] (ABA), a sesquiterpenoid, is a natural plant growth inhibitor found in all higher plants (208). The three principal activities of ABA are inhibition of auxin-induced growth through plasmalemma charge alteration, inhibition of RNA synthesis, and inhibition of protein synthesis (199). ABA also interacts with other PGRs to control bud dormancy and induce flower, leaf, and fruit abscission (199). The biosynthetic pathway from farnesyl pyrophosphate to ABA is similar to those which produce sterols, carotenoids, and gibberellins. Several fungicides, eg, triadimefon [43121-43-3], and plant growth retardants, eg, tetcyclacis, that block sterol metabolism also block ABA biosynthesis (209,210).

The gibberellins (GAs), which constitute a large PGR class, occur in all higher plants and influence many plant growth processes, eg, induction of hydrolytic enzymes during seed germination, induction of flowering in some plants, stimulation of cell elongation and, to some degree, cell division (199). The history, occurrence, and chemistry of the GAs have been reviewed (211–213). Like the other natural PGRs, GAs are theoretically susceptible to influence by xenobiotics at the biosynthesis stage, during catabolism, during transport of the GAs or precursors, by alteration of the number and activity of GA receptors, and through modifications of the reactions induced by GAs.

Since GAs as diterpenes share many intermediates in the biosynthetic steps leading to other terpenoids, eg, cytokinins, ABA, sterols, and carotenoids, inhibitors of the mevalonate (MVA) pathway of terpene synthesis also inhibit GA synthesis (57). Biosynthesis of GAs progresses in three stages, ie, formation of *ent*-kaurene from MVA, oxidation of *ent*-kaurene to GA_{12}-aldehyde, and further oxidation of the GA_{12}-aldehyde to form the different GAs; more than 70 different GAs have been identified.

Compounds that slow cell division and shoot elongation without causing malformation are defined as growth retardants rather than herbicides (215,216). Growth retardants have value in the horticultural industry where short, compact flowering plants are desired and in cereal grain production where they serve as antilodging agents. Grain and fruit yields may also be increased when vegetative growth is retarded (211). Most commercial growth retardants act on GA biosynthesis. The onium growth retardants containing positively charged ammonium, phosphonium, or sulfonium moieties (2,217,218) interfere directly with the biosynthetic steps leading to *ent*-kaurene. AMO-1618, chlormequat chloride (CCC), mepiquat chloride, chlorphonium chloride, and some trimethyl iodides inhibit *ent*-kaurene synthetase A in several species and standard assay systems. However, CCC does not seem to lower GA contents of some higher plants (219,220). The dwarfing effects of these compounds can be reversed in some cases by exogenously applied GA, suggesting that other factors beyond inhibition of GA biosynthesis are involved in the action of these onium compounds. CCC, AMO 1618, and chlorphonium chloride also restrict synthesis of sterols and other terpenoids (221).

Ancymidol, a pyrimidine derivative, has found some commercial application since pyrimidines inhibit the oxidative reactions that produce *ent*-kaurenoic acid from *ent*-kaurene in the GA synthetic pathway (222,223). Other related pyrimidines and cytokinins have also been described as inhibiting GA synthesis (224). A second class of growth retardants, the norbornenodiazetine derivatives, eg, tetcyclacis, also inhibit the first three steps of *ent*-kaurene synthesis in higher plants (215). After the triazole fungicides, triadimefon [*43121-43-3*] and tridimenol [*55219-65-3*], were reported to inhibit shoot elongation (225), other triazoles were developed as growth retardants, eg, uniconazole (226), paclobutrazol (227), and BAS 11100W (228). The site of action of these triazole fungicides and plant growth retardants appears to involve the cytochrome P-450-containing monooxygenase(s) that catalyze the oxidation of *ent*-kaurene to *ent*-kaurenoic acid (229,230). This hypothesized mechanism may also extend to imidazoles, eg, 1-*n*-decylimidazole [*53529-02-1*] (231) and 4-pyridines which also retard plant growth (232). The noborenenodiazetine and triazole growth retardants are translocated in the xylem and are more active when applied via the roots or stems than when sprayed on the leaves (211,214).

Cytochrome P-450 is frequently the oxygenase which detoxifies xenobiotics, including herbicides. Blocking the metabolism of a herbicide increases the activity or delays the inactivation, thus increasing the effectiveness of such herbicides as chlortoluron [15545-48-9] and bentazon [25057-89-0] (233–235). Most of the GA-synthesis inhibitors characterized so far affect two segments of the complicated pathway from MVA to the many different GAs identified. The cyclization reactions that produce *ent*-kaurene are inhibited by the onium growth retardants, and the oxidations of *ent*-kaurene to *ent*-kaurenoic acid are sensitive to heterocyclic triazoles such as paclobutrazol and similar compounds. Other enzymes in the pathway are points for pathway disruption by as yet undeveloped GA biosynthesis inhibitors (236).

Cytokinins, first recognized as inducers of cell division, also evoke a diversity of other responses in plants (237). Root-produced cytokinins move to the shoot, interacting with other PGRs and factors to control both development and senescence. The biosynthesis and metabolism of cytokinins are quite complex, and the effects of herbicides on cytokinins are not well understood. Cytokinins also occur as component nucleosides in tRNA in plants, as well as animals and microorganisms. The lack of plant-specificity has made cytokinins less interesting than other PGRs to developers of herbicides. Some exogenous chemicals, including ABA, significantly modify cytokin in metabolism in plants.

Environmental Fate of Herbicides

Herbicide Fates in Plants. Beyond modes of action and structure–activity relationships, developers of new herbicides must also consider uptake by plants, translocation within the plant, and possible deactivation of herbicides by contact with soil. Some of these problematic factors can be addressed as part of the QSAR studies and during the screening process. Considerable attention is also being paid to the use of safeners (2,101) which protect the crop from herbicides that specifically target the weeds usually associated with that crop. Environmental protection and pesticide regulation concerns are the driving forces in the current efforts toward minimizing application rates, optimizing delivery through improved formulations and application equipment, and increasing target specificity. These research and development efforts include other important and related areas of interest to chemists, eg, the fate and detection of herbicides in the soil and ground and surface water.

Factors Affecting Environmental Fate. The fate of herbicides in the environment is influenced by many chemical, biological, and physical factors. The principal transport and dissipation pathways include sorption to organic and mineral soil and sediment constituents; transport to groundwater in the solution phase by mass flow and/or diffusion; transport to surface water in either the solution or sorbed phases; loss to the atmosphere through volatilization, with redeposition at a later time and location; transformation or mineralization by biological, chemical, or photochemical processes; and uptake by plant or animal species. These processes do not operate as isolated systems, but occur simultaneously and involve significant interaction and feedback. Although the environmental fates of most herbicides are controlled primarily by one or two of the outlined processes, all of these factors influence the fate to some extent. Each of the processes are discussed briefly to provide a basis for understanding their relative importance to an individual herbicide's environmental fate.

Sorption. The retention or sorption of a herbicide by soil, sediment, or aquifer material is one of the most important factors in deciding a particular chemical's environmental fate. The degree to which a herbicide is retained determines its concentration in the soil solution and thus the amount that is available to be leached. Sorption of a herbicide is dependent on the sum of the attractive and repulsive forces between the solid and solution (or vapor) phases. The majority of the interactions between the herbicide and the soil are electrostatic in nature and are therefore influenced by factors that affect the charge status of the system. Proposed sorption mechanisms would include cation and anion exchange, hydrogen bonding, ligand exchange, hydrophobic bonding, and nonpolar van der Waals interactions. Herbicide sorption is most often quantified through the use of a soil sorption, ie, distribution, coefficient (K_d), which is the ratio of the herbicide in the solid, or sorbed phase, to that in the solution phase (238,239).

The amount of herbicide sorbed by a given soil is influenced by properties of both the soil and the herbicide. Important properties related to the soil's retention ability include clay mineralogy, organic matter content, soil pH, and iron and aluminum oxide content. These properties, in turn, affect the soil's cation- and anion-exchange capacities (239). Important properties of a herbicide include charge, polarity, size, and flexibility (238). The charge of the herbicide is frequently influenced by the soil pH because many herbicides are ionizable. Sorption theory as it relates to herbicides and other organics has been reviewed (240,241).

Leaching. Leaching can be loosely defined as the transport of chemicals in the soil profile, as a result of the action of percolating water (242). Strictly speaking, this occurs through two distinct mechanisms, diffusion and mass flow. Diffusion results from random molecular motion and occurs from areas of high to low concentration. Movement of a herbicide through dispersion, which results from the effects of differential mixing and pore water velocities in the soil matrix, is usually included with the effects of diffusion. Transport by mass flow occurs through the movement of water in which the herbicide is dissolved or by the movement of suspended soil or sediment to which the herbicide is sorbed. Thus the total amount of a herbicide leached is the sum of that transported by diffusion, dispersion, and mass flow processes (243,244).

As previously stated, the degree to which a given herbicide is leached is significantly influenced by its sorption to soils and sediments. Therefore, factors of the soil and herbicide that affect sorption also have an influence on the degree to which the herbicide is leached. However, other factors related to the movement of water through the soil can influence this process. Important contributing factors include the amount and intensity of precipitation or irrigation, the soil texture, the tillage system, and the soil topography. Reviews are available on the theory and modeling of herbicide leaching (240,245–247) and on experimental techniques (241,242).

Runoff. Runoff can be defined as water and any dissolved or suspended matter it contains that leaves a plot, field, or small-cover watershed in surface drainage (244). Processes that influence herbicide leaching also influence the degree to which a herbicide is subject to runoff loss. Those factors that encourage the leaching of a herbicide generally reduce its loss in runoff and vice versa.

Specific factors known to influence the amount of herbicide lost to runoff include rainfall timing and intensity with respect to herbicide application time, herbicide application rate, herbicide solubility in water, terrain slope, vegetative cover, and soil texture (244,248). Factors relating to the herbicide's sorption, mobility, and persistence also influence its runoff susceptibility. The practical and theoretical aspects of herbicide runoff have been reviewed (249).

Volatilization. The susceptibility of a herbicide to loss through volatilization has received much attention, due in part to the realization that herbicides in the vapor phase may be transported large distances from the point of application. Volatilization losses can be as high as 80–90% of the total applied herbicide within several days of application. The processes that control the amount of herbicide volatilized are the evaporation of the herbicide from the solution or solid phase into the air, and dispersal and dilution of the resulting vapor into the atmosphere (250). These processes are influenced by many factors including herbicide application rate, wind velocity, temperature, soil moisture content, and the compound's sorption to soil organic and mineral surfaces. Properties of the herbicide that influence volatility include vapor pressure, water solubility, and chemical structure (251).

Degradation or Transformation. Degradation or transformation of a herbicide by soil microbes or by abiotic means has a significant influence not only on the herbicide's fate in the environment but also on the compound's efficacy. Herbicides that are readily degraded by soil microbes or other means may have a reduced environmental impact but may not be efficacious. Consider the phenomenon of herbicide-resistant soils. In these cases, repeated application of a given herbicide has led to a microbial population with an enhanced ability to degrade that herbicide (252,253). This results in a decrease or total loss of the ability of the herbicide to control the weed species in question in a cost-effective manner.

The degradation of a herbicide by soil microbes is primarily an enzymatic process in which cellular or extracellular enzymes break down the herbicide into smaller molecules that may be used by the organism as an energy or nutrient source. If the herbicide is used by the organism, the process is called catabolism. If the herbicide is degraded incidentally and not used by the organism, the process is called co-metabolism (254). Microbes also may influence the degradation of a herbicide through changes in soil pH or other soil properties. The degradation of herbicides by soil microbes is an extremely active area of research, and reviews have been published (254,255).

Degradation of a herbicide by abiotic means may be divided into chemical and photochemical pathways. Herbicides are subject to a wide array of chemical hydrolysis reactions with sorption often playing a key role in the process. Chloro-s-triazines are readily degraded by hydrolysis (256). The degradation of many other herbicide classes has been reviewed (257,258).

The photochemical degradation of herbicides is dependent on the ability of the herbicide to absorb light at a wavelength between 285 and 400 nm (259,260). Light below these wavelengths is generally absorbed by the earth's ozone layer and does not reach the surface. Light above 400 nm does not have sufficient energy to alter chemical bonds and thus does not photodegrade herbicides. Considerable work is being conducted to investigate the possibility of utilizing photochemical reactions to degrade waste herbicides. Examples of these approaches would include photocatalytic systems, eg, ultraviolet (uv) light plus a photo-catalyst; ozonation/uv light systems; and free-radical generating systems, eg, Fe^{2+}/H_2O_2 or Fe^{3+}/H_2O_2 with or without uv light. Specific examples of these systems are discussed in the sections on individual herbicide classes.

Plant or Animal Uptake.   Uptake and accumulation of a herbicide by a plant species is an important dissipation pathway for some herbicides. Indeed, for the target weed species, uptake of the herbicide is the desired outcome. The principal point of entry for soil-applied herbicides in seedling and adult plants is usually the root system, although this may be complicated by volatilization of the herbicide and subsequent absorption of the herbicide through above-ground plant organs (245). Foliage-applied herbicides must enter the plant through the leaf cuticle or the stomata. Successful penetration through the cuticle requires several complex steps and is significantly more involved than uptake through the plant root system (261). Distribution of the herbicide throughout the plant after uptake is primarily dictated by the mode of action of the herbicide. Some plants have the ability to detoxify a herbicide after uptake, eg, corn (*Zea mays*) cultivars which can hydrolyze atrazine or propazine to their nonphytotoxic hydroxy metabolites (262).

The extent to which a herbicide is accumulated in living organisms is defined as bioaccumulation. If a compound is transferred in the food chain, with a concurrent increase in concentration, it is called biomagnification (263). Uptake and accumulation of a herbicide by a living organism is controlled largely by the compound's solubility in water (264). Herbicides that tend to bioaccumulate and biomagnify are those compounds that are not readily soluble in water. Other properties that give an indication of a herbicide's tendency to bioaccumulate or biomagnify include the octanol–water partition coefficient (K_{ow}) and the soil sorption coefficient (K_d). The octanol–water partition coefficient indicates a herbicide's lipophilic nature and thus its tendency to accumulate in fatty tissue. The soil sorption coefficient relates the extent to which a compound will be retained by the soil and thus potentially be removed from the food chain. It should be noted, however, that even a herbicide which is strongly bound to the soil may be transported to an aquatic system by runoff and become available to bottom-feeding fish or benthic organisms (264). Individuals interested in the study of bioaccumulation and biomagnification indicators are referred to References 264 and 265.

Measurement of Environmental Fate.   *Water Quality Risk.*   Continued concern is expressed over the potential contamination of surface and groundwaters by agricultural chemicals. Herbicides have received much of this attention, due to their widespread use and the large total volume applied. However, this perceived threat to groundwater resources appears to be largely unfounded. A survey of private wells and public water well supplies in the United States has revealed that <1% contain herbicides at levels that would affect human or animal health (266). In addition, those sources that are contaminated can usually be attributed to point rather than nonpoint sources. A point source of contamination is readily located and thus more easily controlled and remediated, and is generally associated with industrial sources or municipal wastewater plants, although agricultural sources such as herbicide equipment rinsing stations also could be point sources. A nonpoint contamination source is one in which the exact source is unknown. They are typically diffuse, often of large areal extent (267), and are generally of agricultural origin. Nonpoint sources are generally treated by modifications in agricultural management practices. Typical modifications would include the use of alternative herbicide formulations, the splitting of the herbicide application in time, or the installation of vegetative buffer strips to trap runoff.

A re-evaluation of the water quality problem has revealed that surface water resources, rather than groundwater resources, are at higher risk of contamination from agricultural chemicals. It has been demonstrated that 94% of the herbicides reaching the Cedar River Basin in Iowa were transported in runoff, compared to 6% through groundwater flow (268). In addition, runoff had a higher herbicide concentration than groundwater sources. Additional information on the effects of herbicides on drinking water quality (269) and groundwater hydrology or remediation are available (270–272).

Carcinogenicity. The public health implications of drinking water contamination by herbicides are unclear. The levels that have been detected in groundwater are generally in the part per billion (ppb) or part per trillion (ppt) range and are below estimated acute toxicity levels. However, the long-term health effects of this exposure are generally unknown. Several studies have demonstrated that the mortality from some types of cancer is significantly higher in rural residents of many corn belt states (273). This trend is particularly evident in a study from Kansas involving 2,4-D exposure; however, factors other than 2,4-D exposure are also being considered. The U.S. Environmental Protection Agency (EPA) developed (ca 1993) a classification scheme in an attempt to further evaluate the carcinogenic potential of herbicides and pesticides (269). In this system, chemicals are placed in one of five groups, A–E, according to their carcinogenic potential, ranging from definite (A) human carcinogens to no evidence of carcinogenicity for humans (E). The principal difference between these groups is the amount of accumulated evidence demonstrating carcinogenic potential. There is sufficient evidence of a causal association between Group A compounds and cancer, but this is not the case with the other groups. Group B has been further divided into B1, for which a limited amount of epidemiological evidence indicating carcinogenicity is present, and B2, for which adequate evidence from animal studies is present. For those compounds in Group C, limited animal data suggesting carcinogenicity is present, but inadequate or no human data exist. There is inadequate animal and human data for Group D compounds. Finally, for Group E compounds, there is no evidence of carcinogenicity in at least two animal tests or in adequate animal and human studies (269).

This classification scheme is used in part in the determination and calculation of health advisory (HA) drinking water levels or carcinogenic risk estimates. The majority of herbicides in use in the United States (ca 1993) for which HAs have been issued fall into Group D, with a smaller percentage falling into Group C (269). This would indicate that there are insufficient data to classify the carcinogenic potential of many herbicides. This does not imply that chemical companies are not adequately testing herbicides. To the contrary, exhaustive toxilogical testing of a potential herbicide is required by the U.S. EPA before registration. The lack of data does indicate, however, that further testing will be required before the carcinogenic potential of many herbicides is known. Based on available HAs and the U.S. EPA classification scheme, acifluorfen, alachlor, amitrole, haloxyfop–methyl, lactofen, and oxadiazon have been listed as B2 carcinogens (274). Further information on carcinogenic risk assessment is available (275).

Analytical Methods. Since 1984, dramatic technical advances have been made in the analysis of trace organic chemicals in the environment. Indeed, these advances have been largely responsible for the increased public and governmental awareness of the wide distribution of herbicides in the environment. The ability to detect herbicides at ppb and ppt levels has resulted in the discovery of trace herbicide residues in many unexpected and unwanted areas. The realization that herbicides are being transported throughout the environment, albeit at extremely low levels, has caused much public and governmental concern. However, the public health implications remain unclear.

Numerous collections of herbicide analysis methods have been published (276–279). An increased emphasis has been placed on the first step in the environmental sampling process, that of obtaining a representative, uncontaminated sample. If this is to be accomplished, consideration must be made of such factors as sample size and location (280–283). After the sample has been obtained, it must be stored in such a way as to minimize degradation. This generally consists of refrigeration, possibly preceded by some type of drying (284).

Preparation of soil–sediment of water samples for herbicide analysis generally has consisted of solvent extraction of the sample, followed by cleanup of the extract through liquid–liquid or column chromatography, and finally, concentration through evaporation (285). This complex but necessary series of procedures is time-consuming and is responsible for the high cost of herbicide analyses. The advent of solid-phase extraction techniques in which the sample is simultaneously cleaned up and concentrated has condensed these steps and thus greatly simplified sample preparation (286).

Traditionally, herbicides have been analyzed by gas chromatography (gc) or spectrophotometric methods. The method of choice when accuracy and sensitivity are of the utmost importance is gc, especially when combined with mass spectrometry (287). However, several other methods are used for routine monitoring or screening purposes. High pressure liquid chromatography (hplc) provides detection limits that nearly rival gc and require significantly less sample preparation and cleanup (285,288). Advances in the 1980s have made thin-layer chromatography (tlc) a valuable tool in herbicide analysis (285,289). The combination of high performance tlc plates and scanning densitometers allows quantitative results to be obtained at detection limits that nearly rival hplc. Significant advances have been made in stationary phases for both hplc and tlc systems, including reverse-phase options. These have proven to be invaluable for herbicide analysis. Another analytical tool that has received much attention and shows great promise for routine analysis is enzyme immunoassay (eia) (290–292). This technique offers the advantages of a low cost analysis, few interferences, high specificity and sensitivity, and a minimal amount of sample preparation.

A mobility ranking based on soil thin-layer chromatography (stlc) is used to classify the herbicide leaching potential of various herbicides (293,294). The rankings range from I (immobile) to V (very mobile) with intermediate categories of II (low mobility), III (intermediate), and IV (mobile). This method is widely used and has been accepted for submission of leaching data for herbicide registration purposes by the U.S. EPA (242).

A comprehensive search (295) of the STORET water quality database, maintained by the U.S. EPA Office of Water, is used to evaluate the potential water quality implications of various herbicides. This database contains information on contamination of surface water (SW) and groundwater (GW) supplies. The data are provided to give a general impression of the occurrence of a given herbicide in SW and GW (269). The U.S. EPA scheme for categorizing a chemical's carcinogenic potential is used for herbicides for which healthy advisory information (HA) is available. The U.S. EPA is continually issuing HAs for various environmental contaminants; HAs available in Reference 269 were used in preparation of this article.

Herbicide Groups

Herbicides can be grouped according to common structural features. Sometimes the assignment is arbitrary when there are a multitude of functional groups, eg, acifluorfen which is a diphenyl ether (phenoxy compound) as well as a trifluoromethyl compound.

Phenoxyalkanoics.   The phenoxyalkanoic herbicide grouping is composed of two subgroups, the phenoxyacetic acids and the phenoxypropionic acids. The phenoxyacetic acid herbicides include some of the first commercially successful herbicides, eg, 2,4-D. They continue to be widely used for foliar control of broadleaf weeds. The more heavily functionalized phenoxypropionic acid herbicides are relatively new herbicides compared to the phenoxyacetic acids and are used primarily for selective control of grassy weeds in broadleaf crops (2,296,297).

The phenoxyalkanoic herbicides are acidic in nature and thus subject to some degree of ionization. The extent to which the herbicide ionizes is controlled by the acid dissociation constant (K_a) of the herbicide in question and the soil solution pH (238). The leaching potential is significantly influenced by these reactions.

The sorptive behavior of the phenoxyacetic acid herbicides has been investigated. Both the free acid and amine salt formulations of 2,4-D are minimally sorbed to the soil (298) and would be classified as mobile by stlc techniques (293). Some increase in sorption is expected in soils containing significant quantities of organic matter (2). The salt formulations of 2,4-D, 2,4-DB [94-82-6] (4-(2,4-dichlorophenoxy)-butanoic acid) and 2,4-DP [120-36-5] (2-(2,4-dichlorophenoxy)-propanoic acid), tend to be slightly more mobile than the free acids (299). No difference was noted in the runoff potentials of an amine salt and ester formulation of 2,4-D; however, the ester formulation resulted in significant volatilization losses (300). MCPA is weakly retained by soil and is also classified as mobile. The phenoxybenzenes, nitrofen and oxyfluorfen, were found to be strongly sorbed to organic soils and were only minimally leached (301).

Considerable research has been conducted on the breakdown of phenoxyacetic acids in soil. The decomposition of 2,4-D appears to be primarily a microbial process that occurs rapidly in surface soils under aerobic conditions and decreases with depth (302). MCPA is also degraded by microbial means although at a slower rate than 2,4-D (303). MCPA has also been shown to photodecompose rapidly, losing >80% of the initial herbicide application in six days of exposure (304). 2,4-D has been shown to be completely degraded in aerated solutions of hydrogen peroxide containing Fe^{3+}. This reaction can be further accelerated by irradiation with visible light containing a small ultraviolet (uv) component (305). Other related materials include bifenox [42576-02-3] and (4-(4-chloro-2-methyl-phenoxy)-butanic acid) MCPB [94-81-5].

The phenoxypropionic acids, eg, haloxyfop–methyl and fluazifop–butyl (Table 1) bear a variety of other functional groups and are not strongly sorbed to soils of widely varying constituents (297). Thus the leaching potential of these compounds is significant. Both compounds are rapidly hydrolyzed to the parent free acids in soil and gradually decomposed by microbial means (297,306). Diclofop–methyl (Table 1) is hydrolyzed in hydroalcoholic solutions in the presence of montmorillonite (307). Finally, mecoprop [7085-19-0] was degraded by several microbial communities (308) but not by the individual members, indicating a co-metabolic relationship. Other phenoxypropionic acid herbicides include dichlorprop [120-36-5] and quizalofop–ethyl [76578-14-8].

Considerable concern has been raised over the carcinogenic potential of 2,4-D. However, the World Health Organization (WHO) has evaluated the environmental health aspects of this chemical and concluded that 2,4-D posed an insignificant threat to the environment. They did indicate, however, that only limited data on toxicology in humans are available (309). An HA has been issued (269) for MCPA. It was found in 4 of 18 SW samples analyzed and in none of 118 GW samples (295), and has been placed in Group D for carcinogenic potential (269). EPA has published two gc methods for the analysis of the phenoxyalkanoic herbicides (277).

Bipyridiniums. The bipyridinium herbicides (Table 2), paraquat and diquat, are nonselective contact herbicides and crop desiccants. Diquat is also used as a general aquatic herbicide (2,296). Bipyridinium herbicides are organic cations and are retained in the soil complex via cation exchange. They are strongly sorbed to most soils and are not readily desorbed (332). Both paraquat and diquat are not readily leached (293).

Paraquat and diquat can both be degraded by photochemical means, and degrade quite rapidly under natural sunlight (333). Paraquat can be degraded by microbial means (335), and rapidly degraded by uv-ozonation (334); this process can be accelerated by the addition of a photosensitizer such as acetone. Finally, both paraquat and diquat have been demonstrated to dissipate rapidly in aquatic systems, paraquat taking a slightly longer time (336). Paraquat and diquat are much more toxic than most herbicides and ingestion of sufficient quantities can result in death if prompt medical treatment is not obtained (2).

Benzonitrile, Acetic Acid, and Phthalic Compounds. Benzonitrile herbicides (Table 2) are generally used for pre-emergence and post-emergence control of broadleaf weeds. Dichlobenil [1194-65-6] also controls grass weeds (299) and dichlobenil, endothall [145-73-3], and fenac [85-34-7] are used as aquatic herbicides (2). Most benzonitriles are selective in their control (296). Benzonitrile herbicides are acidic in nature, thus their environmental fate is influenced by changes in soil pH. Sorption of these herbicides is expected to increase with decreasing pH (337). This is the case with dicamba, which is minimally sorbed at near neutral pH but demonstrates a dramatic increase in sorption as the soil

Table 2. Environmental Health Advisories for Herbicides

Herbicide	Health advisories[a] SW	GW	Mobility[b]	Carcinogenic potential group[c]	Analytical methods[d]
Bipyridinium compounds					
diquatop			immobile		hplc
paraquat		0/843	immobile	E	hplc
Benzonitrile, acetic acid, and phthalic compounds					
chloramben	13/34	1/566	very mobile	D	gc[e]
DCPA	386/1995	12/982		D	gc[e]
dicamba	262/806	2/230	very mobile	D	gc[e]
dichlobenil			low		
endothall	0/3	0/604		D	gc
naptalam					uv
Dinitroaniline and derivatives					
benefin					gc
dinitramine					gc
dinoseb	1/89	0/1270		D	
fluchloralin					gc
oryzalin					uv
pendimethalin					gc
trifluralin	172/2047	1/507	immobile	C	ir
Acid amides					
alachlor					gc
bensulide			immobile[f]		hplc
diphenamide	0/3	0/678	intermediate	D	gc
metolachlor	2091/4161	13/596		C	gc
napropamide					
pronamide	20/391			C	gc[g]
propachlor	34/1690	2/99	intermediate	D	gc[h]
propanil			low		
Phenyl carbamates					
chloropropham			low		
karbutilate					hplc
propham	1/392	0/583	intermediate	D	hplc[i]
Thiocarbamates					
asulam					uv
butylate	91/836	2/152		D	gc, glc
EPTC					hplc
thiobencarb			relatively immobile		gc, glc
triallate					gc, glc
vernolate					hplc
Triazines					
ametryn	2/1190	24/560	intermediate	D	general[j]
atrazine	4123/10,942	343/3208	intermediate	C	gc
cyanazine	1708/5297	21/1821	intermediate[k]	D	ir
hexazinone			relatively immobile[l]	D	gc

Table 2. (Continued)

Herbicide	Health advisories[a] SW	Health advisories[a] GW	Mobility[b]	Carcinogenic potential group[c]	Analytical methods[d]
metribuzin	938/4651	0/416		D	general[j]
prometon	386/1419	36/746	intermediate	D	gc
prometryn			low		
propazine	33/1097	15/906	intermediate	C	general[j]
simazine	922/5873	202/2654	intermediate	C	gc
terbutryn					general[j]
Pyridines					
clopyralid			minimal[m]		
fluroxypyr			varied		
picloram	420/744	3/64	mobile[m]	D	general[n]
triclopyr			intermediate[o]		hplc
Pyridazinones					
norflurazon			low		
pyrazon					uv
Sulfonylureas					
chlorimuron, ethyl			mobile[p]		
chlorsulfuron			intermediate to very mobile[q]		hplc, gc[r]
metsulfuron, methyl					gc[r]
sulfometuron			mobile to very mobile[s]		
Imidazole compounds					
buthidazole					eia[t]
imazamethabenz					eia[t]
imazapyr					eia[t]
imazaquin			mobile to very mobile[u]		eia[t]
imazethapyr			immobile to mobile[u]		eia[t]
Other heterocyclic nitrogen derivatives					
amitrole			mobile		vis
bentazon			very mobile[v]	D	hplc
isoxaben			immobile[w]		
Ureas and uracils					
bromacil	0/3	0/841	mobile	C	glc
chloroxuron			immobile		glc
diuron	0/25	0/1337	low	D	ir
fluometuron	0/14	0/156	intermediate	D	uv
linuron					uv
tebuthiuron			intermediate to very mobile[x]	D	uv
terbacil				E	uv

Table 2. (*Continued*)

Herbicide	Health advisories[a] SW	Health advisories[a] GW	Mobility[b]	Carcinogenic potential group[c]	Analytical methods[d]
		Aliphatic–carboxylic			
dalapon	0/14	0/14	very mobile	D	ir
TCA			very mobile		
		Inorganics and metal organics			
AMS				D	titration
		Miscellaneous trifluoromethyl compounds			
acifluorfen				B$_2$	hplc
fluridone					gc^y
lactofen					hplc
		Amino acid analogues			
glufosinate,			intermediate^z		
glyphosate	0/6	0/98	immobile to low mobility$^{z'}$	D	hplc
		Other miscellaneous compounds			
cinmethylin					gc
ethofumesate					gc
tridiphane					

[a]Ref. 295 unless otherwise noted. SW = surface water; GW = ground water. Positive results/number of tests.
[b]Refs. 293 and 294. Mobility ranking based on soil thin-layer chromatography (stlc).
[c]Ref. 269. Group A, human carcinogen; Group B, probable human carcinogen; Group C, possible human carcinogen; Group D, not classifiable; Group E, no evidence of carcinogenicity for humans.
[d]Ref. 277 unless otherwise noted: gc = gas chromatography; hplc = high pressure liquid chromatography; ir = infrared spectroscopy; uv = ultraviolet spectroscopy; glc = gas–liquid chromatography; eia = enzyme immunoassay; vis = visible spectroscopy.
[e]Ref. 310. Gc for chlorinated pesticides can be used.
[f]Ref. 311.
[g]Ref. 312.
[h]Ref. 313.
[i]Ref. 314.
[j]General draft method for nitrogen- and phosphorus-containing pesticides[g].
[k]Ref. 315.
[l]Ref. 316.
[m]Mobility has been reported to be mobile (Refs. 293 and 317) and minimal (318) in different studies.
[n]General draft method for determination of chlorinated acids in water (310).
[o]Ref. 317.
[p]Ref. 319.
[q]Ref. 320.
[r]Ref. 321.
[s]Ref. 322.
[t]Ref. 323.
[u]Ref. 324. Mobility is a function of soil pH (325).
[v]Ref. 326.
[w]Ref. 327.
[x]Ref. 328.
[y]Ref. 329.
[z]Ref. 330.
[z']Ref. 331.

pH decreases (338). Endothall is also only minimally sorbed at a near neutral pH (339).

Benzonitrile herbicides tend to possess a high leaching potential; dichlobenil is an exception, due to its stronger sorption. The benzonitrile herbicides are also prone to volatilization losses (340) and off-site deposition (341).

Benzonitrile herbicides are readily degraded by soil microbes. Dicamba is rapidly degraded in soil (342) and water samples (343) by microbial means, and is metabolized by several species of soil bacteria (344). DCPA degrades very rapidly under optimum conditions (25°C) and slower at lower temperatures (345). Bromoxynil can be degraded by microbial (346) or photochemical means. The photodegradation of bromoxynil is highly pH-dependent, a decrease in degradation occurring at lower pH values (347). Endothall, which is widely used as an aquatic herbicide, is rapidly dissipated in water (348) and dissipates only slightly slower in soil (349). Dichlobenil is apparently degraded by a combination of microbial and photochemical processes (350). Finally, fenac is slowly degraded in water and soil by microbial processes (351).

HAs have been issued for chloramben, DCPA, dicamba, and endothall (269); health advisories have not been issued for the remaining benzonitrile herbicides (277).

Dinitroanilines and Derivatives.   Dinitroaniline herbicides are used principally for the selective, pre-emergence control of annual grasses and broadleaved weeds. They have little or no post-emergence activity. Oryzalin is used for selective weed control in flooded rice culture. In general, dinitroaniline herbicides are extremely prone to volatilization losses (299). For this reason, they should always be incorporated in the soil immediately after application; oryzalin and pendimethalin are exceptions to this statement. Dinitroaniline herbicides are nonionic and retained in soil primarily by hydrogen bonding to soil organic matter, or possibly through hydrophobic or van der Waals forces. The uptake of nonionic herbicides has been described as a chemical partition of the herbicide into soil organic matter (352,353). The reactions are governed, to a large extent, by the herbicide's polarity.

Considerable research has been conducted to investigate the soil sorption and mobility of dinitroaniline herbicides. In general, these herbicides are strongly sorbed by soil (354), and sorption has been correlated to both soil organic matter and clay content (355). Dinitroaniline herbicides are not readily leached in most soils (356), although leaching of trifluralin is enhanced by addition of surfactants (357).

Degradation of dinitroaniline herbicides has also been extensively investigated. Trifluralin undergoes rapid initial dissipation, primarily by volatilization, and then steady, but slower, dissipation through a combination of volatilization and degradation pathways (358,359). A similar dissipation pattern was noted for dinitramine and fluchloralin [33245-39-5] (360). Oryzalin and isopropalin [33820-53-0] also degrade and do not accumulate in field soils, even after repeated applications (361). Benefin [1861-40-1] shows a similar two-stage degradation pattern consisting of a rapid initial decomposition followed by a slower first-order breakdown (362). The degradation of benefin also proceeds faster in turf thatch than in soil (363) and under anaerobic, compared to aerobic, conditions (364). Pendimethalin has been rapidly degraded in grass clippings and compost (365), as well as under anaerobic conditions (366). The breakdown of pendimethalin in soil increases with temperature and, to a lesser extent, with moisture (367). Finally, several species of soil fungi have been isolated that can effectively degrade pendimethalin (368).

Health advisories have been issued for the phenol dinoseb and trifluralin; health advisories have not been issued for the remaining dinitroaniline herbicides, eg, profluralin [26399-36-0].

Acid Amides. The principal use of acid amide herbicides is the selective control of seedling grass and certain broadleaved weeds (299). The majority of acid amide herbicides are applied pre-emergence or pre-plant incorporated, except for propanil which is applied post-emergence (2). In general, the acid amide herbicides are not considered subject to large volatilization losses (2). However, under ideal conditions, eg, high soil moisture and low soil sorption, volatilization may be significant (250).

Acid amide herbicides are nonionic and moderately retained by soils. The sorption of several acid amide herbicides has been investigated (369). Acetochlor [34256-82-1] is sorbed more than either alachlor or metolachlor, which are similarly sorbed by a variety of soils. Sorption of all the herbicides is well correlated to soil organic matter content. In a field lysimeter study, metolachlor has been found to be more mobile and persistent than alachlor (370); diphenamid [957-51-7] and napropamide [15299-99-2] have been found to be more readily leached (356).

The breakdown of acid amide herbicides in soil has been extensively investigated. They do not appear to be persistent in the soil and most are readily metabolized by soil microbes. Alachlor degradation is well correlated with microbial biomass and respiration (371) although breakdown is slower in subsoils. Pretreatment of alachlor samples with uv light has been shown to accelerate the microbial breakdown process (372). The pretreatment process results in the dechlorination of the alachlor parent molecule and the formation of several intermediates. Pronamide is readily transformed in the soil via hydrolysis (373). Breakdown of metolachlor is also microbial in nature, though it is significantly longer lived in soil than alachlor, with a half life as long as 50 days. Preacclimation of the soil with metolachlor results in an enhanced rate of degradation (374). The isolation of soil microbes capable of degrading various acid amide herbicides has been reported, ie, several species that could transform, but not mineralize, metolachlor (375); microbial communities capable of metabolizing propachlor (376); one species that could metabolize propanil [709-98-8] (377); and finally, a microbial community which degraded diphenamid (378). The degradation process was accelerated when the microbes were preacclimated with the herbicide.

Health advisories have been issued (269) for diphenamid, metolachlor, pronamide, and propachlor. Other acid amide herbicides include butachlor [23184-66-9] and ethalfluralin [55283-68-6].

Phenylcarbamates. Phenylcarbamate herbicides represent one of two subgroups of carbamate herbicides, the phenylcarbamates and the thiocarbamates (299). Both groups are prone to volatilization losses; the thiocarbamates are particularly susceptible and should be soil-incorporated immediately after application (2). The carbamate herbicides are used, in general, for the selective pre-emergence control of grass and broadleaved weeds (299). Exceptions would include barban, desmedipham, and phenmedipham which are applied post-emergence.

Phenylcarbamate herbicides are nonionic and, in general, readily leached in soils (299). One notable exception is chlorpropham which is strongly sorbed to soils (379). Movement of karbutilate [4849-32-5] has been studied in several Texas rangeland soils and found to be greater in a loamy sand soil than in a clay loam soil (380), but more persistent in the clay loam soil. The phosphonate fosamine–ammonium [25954-13-6] readily degrades in soils, having a half-life of one week in the field and 10 days in the laboratory (381); applications of fosamine–ammonium to the soil do not adversely affect soil microbial populations (382). Finally, a bacterial strain has been isolated that can utilize chlorpropham as a sole carbon source (383). Other phenylcarbamate herbicides include desmedipham [13684-56-5] and phenmedipham [13684-63-4].

Thiocarbamates. Thiocarbamate herbicides are also nonionic. Diallate and triallate were strongly sorbed to both cation- and anion-exchange resins, but minimally to kaolinite or montmorillonite (354). This behavior suggests a physical rather than an ionic mechanism of attraction. The mobility of the thiocarbamate herbicides increases with increasing water solubility. The ranking of five thiocarbamate herbicides, in terms of leaching depth, is molinate [2212-67-1] > EPTC > vernolate [1929-77-7] > pebulate [1114-71-2] > cycloate (384). Thiobencarb [408-27-5] has been found to be relatively immobile in soil columns under saturated flow conditions (385).

The degradation of thiocarbamate herbicides has been extensively studied. Cycloate and EPTC are readily degraded in the air by reaction with OH and NO_3 radicals (386). This is an important observation, considering the volatile nature of the thiocarbamates. Thiobencarb is rapidly degraded in three Florida soils, exhibiting half-lives from 16–33 days (387). An increased breakdown of many of the thiocarbamate herbicides has been linked to previous application of thiocarbamates. This has generally been attributed to high enzymatic activity, resulting from an adaptation and acclimization of the microbial community to the herbicide in question. Accelerated breakdown of vernolate, EPTC, and butylate, but not of cycloate, was reported in vernolate-history soils (252). This trend also has been reported for butylate-history soils that exhibited accelerated breakdown of EPTC, but not of vernolate, pebulate, or cycloate (388). Finally, a strain of bacteria that degraded EPTC also degraded diallate (389). Metham–sodium [137-42-8] also is a thiocarbamate herbicide. A health advisory (HA) has been issued for butylate (269).

Triazines. Triazine herbicides are one of several herbicide groups that are heterocyclic nitrogen derivatives. Triazine herbicides include the chloro-, methylthio-, and methoxytriazines. They are used for the selective pre-emergence control and early post-emergence control of seedling grass and broadleaved weeds in cropland (299). In addition, some of the triazines, particularly atrazine, prometon [1610-18-0], and simazine [122-34-9], are used for the nonselective control of vegetation in noncropland (2). Simazine may be used for selective control of aquatic weeds (2).

The environmental fate of the triazines, particularly atrazine, has been extensively studied. This is due, in part, to their widespread use and the regularity with which they are found in GW and SW. Triazine herbicides are weak bases and can be protonated to form cationic species depending on the herbicide pK_a and the soil pH (337). Sorption of the triazines has been positively correlated with soil organic matter content, clay content, and soil cation-exchange capacity (CEC) (390,391). When five triazines were evaluated on 25 Missouri soils, prometryn [7287-19-6] was the most strongly sorbed, followed by prometon, simazine, atrazine, and propazine [139-40-2] (390).

Triazine herbicides are not readily volatilized. However, given ideal circumstances, volatilization losses may be significant. The tendency to volatilize varies among herbicides and is highly dependent on soil type and moisture conditions. A study of the volatilization potential of seven triazines reported the following ranking of decreasing volatilization losses: prometon ≥ trietazine > atrazine ≥ ametryn [834-12-8] ≥ prometryn > propazine ≥ simazine (392).

Triazine herbicides are subject to degradation by a wide variety of mechanisms, eg, the photocatalytic degradation of atrazine, simazine, trietazine, prometon, and prometryn (393). Degradation of all herbicides is rapid, although complete mineralization does not occur. Atrazine is also subject to degradation through chemical hydrolysis reactions (256). In general, the chlorotriazines appear to be more readily degraded through chemical hydrolysis reactions and the methoxy- and methylthiotriazines are more susceptible to microbial processes (299).

The microbiological degradation of the triazine herbicides has been thoroughly investigated. Although atrazine is thought to degrade primarily by abiotic means, several papers have demonstrated degradation of atrazine by soil bacteria (394) and soil fungi (395). Cyanazine [21725-46-2] has been found to degrade significantly faster than atrazine in soils by a combination of chemical and microbiological processes (396). Metribuzin degraded significantly faster in surface soils than in subsoils (397). The retarded breakdown of metribuzin in subsoils was attributed to lower microbial populations and activity. Finally, terbutryn [886-50-0] degradation has been found to be particularly sensitive to soil moisture content, degradation decreasing with increasing soil moisture content (398). Other triazine herbicides include dipropetryn [4147-51-7] and hexazinone [51235-04-2]. Health advisories have been issued for most of the triazine herbicides (269).

Pyridines and Pyridazinones. Pyridine herbicides are auxin-type herbicides generally used for selective control of broadleaved weeds in cropland, rangelands, and noncroplands (2,296). The pyridazinones are used primarily for the selective pre- and post-emergence control of seedling grass and broadleaved weeds in cotton and sugarbeets (299). The pyridines are slightly acidic in nature and the pyridazinones, slightly basic.

Pyridine herbicides are not strongly sorbed to soils and are readily leached. The mobility of fluroxypyr [69377-81-7] has been found to decrease with increasing incubation time (399); this is attributed to entrapment of the herbicide within the soil organic matter.

A study investigating the breakdown of clopyralid [1702-17-6] reported half-lives on different soils of approximately 2–7 weeks in a laboratory incubation (400); it was indicated that carryover was likely to occur in field soil. Picloram degrades and does not accumulate in field soil although low residue levels do persist for several years (401). The half-life for triclopyr [55335-06-3] is reported to be two weeks in two Canadian soils (402), and it has been shown to be rapidly degraded by aqueous photolysis (403).

Pyridazinone herbicides tend to be strongly sorbed in soils and do not leach readily. Norflurazon sorption increases as organic matter and clay contents increase (404), and it is subject to degradation through photolysis but only minimally through volatilization processes (404). Pyrazon [1698-60-8] sorption also has been shown to increase, and mobility to decrease, with increasing soil organic matter contents (405). The degradation of pyrazon appears to be a microbially mediated process directly related to soil organic matter content (406). Difenzoquat [43222-43-6] also is a pyridazinone herbicide.

Sulfonylureas.   Sulfonylurea herbicides are a relatively new class of herbicides generally used for selective pre- and post-emergence control of broadleaved weeds in croplands (2,296). In general, the sulfonylureas are applied in significantly lower amounts than most herbicides, and they tend to be more active against broadleaved species than grasses. Sulfometuron–methyl [74222-97-2] is used for broad-spectrum selective or nonselective weed control in noncroplands (296).

Sulfonylurea herbicides are weak acids and, in general, are not strongly sorbed to soils. Sorption of chlorsulfuron and metsulfuron–methyl is inversely related to soil pH (407) and is positively correlated to soil organic matter (408).

The degradation of sulfonylurea herbicides in soils appears to occur by two processes. The first pathway involves the acid-catalyzed hydrolysis of the urea function (409,410). This process is highly pH dependent, the rate increasing as pH decreases. Several herbicidally inactive fragments formed by this process may then be degraded by microbial means. A second pathway involves the direct microbial degradation of the sulfonylurea herbicide. This process generally occurs in conjunction with breakdown by chemical hydrolysis (411,412). Microbial degradation may be the dominant mechanism in neutral or alkaline soils where hydrolysis is minimal (411).

The EPA has not issued HAs for any of the sulfonylurea herbicides (269) and data on the occurrence of the sulfonylurea herbicides in SW or GW are not available. Additional sulfonylurea herbicides include bensulfuron [99283-01-9] and metsulfuron, methyl [74223-64-6].

Imidazoles.   Imidazole herbicides are generally used for selective pre- and post-emergence control of grass and broadleaved weeds in croplands. Buthidazole [55511-98-6] and imazapyr are used for broad-spectrum, nonselective weed control in noncroplands (2,296). Imidazole herbicides are amphoteric, possessing both acidic and basic functional groups (413). A notable exception is buthidazole which is nonionic in nature (2). At typical soil pH values, most of the imidazole herbicides exist as anions (324).

Sorption of imidazole herbicides has been shown to increase with decreasing pH. This is most likely due to protonation of the basic functional groups. Imazethapyr is more strongly sorbed to soils, and thus less mobile, than imazaquin (325). The classifications for imazethapyr range from immobile on a silty clay soil (pH 5) to mobile on a sandy loam soil (pH 7); the classifications for imazaquin range from low mobility to mobile for the same soils. Similar amounts of buthidazole leached in four soil columns of varying texture (413); however, the distribution of the herbicide within the columns was different. In the fine textured soils, a greater amount of the herbicide has been detected in the surface layers than in lower layers. In the sandy soils, the herbicide is uniformly distributed throughout the column.

The persistence of imidazole herbicides varies significantly with soil pH. Imazaquin persistence increases with decreasing soil pH (413). The increase is attributed to increased sorption and thus decreased availability for microbial degradation. Imazaquin and imazethapyr have both been shown to degrade primarily by microbial or enzymatic means (414). Degradation is faster in warm, moist soils than in cool, dry soils. Finally, imazaquin also undergoes significant photodecomposition when exposed to artificial uv light or sunlight (415).

No HAs have been issued for any of the imidazole herbicides (269) and data on the occurrence of the imidazole herbicides in SW or GW are not available.

Other Heterocyclic Nitrogen Derivative Herbicides. The herbicides in this group are heterocyclic nitrogen derivatives that do not readily fall into one of the previously discussed groups. They have a wide range of uses and properties. Most of these herbicides are used for selective, pre- and/or post-emergence weed control. Amitrole is used for post-emergence, nonselective weed control in noncroplands and also as an aquatic herbicide (2,296).

Bentazon [25057-89-0] is anionic in nature and is not significantly sorbed to any of 11 Illinois soils; its half-lives have been determined to range from 11 to 32 days in water and 5 days in soil (416). Isoxaben [8255-50-7] is a nonionic compound with low water solubility (417) which degrades in aqueous systems by photolysis (417). Amitrole is degraded by free-radical generating systems (418). Finally, methazole [20354-26-1] is strongly sorbed to soils, has a low leaching potential, and rapidly degrades in soils (419). A health advisory (HA) (269) has been issued for bentazon, though it was not found in sampling performed at two water supply stations (295).

Ureas and Uracils. Urea herbicides are generally used for selective pre-emergence and early post-emergence control of seedling grass and broadleaved weeds. Uracil herbicides are generally used for selective control of annual and perennial weed control in certain crops and for general weed control in noncrop areas. Bromacil, linuron [330-55-2], and tebuthiuron [34014-18-1] are used for the nonselective control of weeds in noncropland (2,296,299). Bromacil is also used in citrus crops, and linuron is used in sorghum and corn crops. Urea herbicides are nonionic and generally of low water solubility. The uracils are ionic herbicides that are not strongly sorbed to soils and readily leach (299).

The sorption of diuron and bromacil has been investigated on two Florida soils (420). Diuron is strongly sorbed to both soils, while bromacil was only weakly retained. Bromacil has been found to be very mobile in a related field study (421). Fenuron [101-41-8] and linuron are strongly sorbed to several Hawaiian soils, and the degree of sorption has been related to soil organic matter contents (422). The mobilities of the urea herbicides are directly related to herbicide water solubilities with mobility increasing with solubility. The mobility rankings for tebuthiuron and fluometuron may range from intermediate on a silt loam soil to very mobile on a sandy soil (328).

Urea and uracil herbicides tend to be persistent in soils and may carry over from one season to the next (299). However, there is significant variation between compounds. Bromacil is debrominated under anaerobic conditions but does not undergo further transformation (423), linuron is degraded in a field soil and does not accumulate or cause carryover problems (424), and terbacil [5902-51-2] is slowly degraded in a Russian soil by microbial means (425). The half-lives for this breakdown range from 76 to 2,475 days and are affected by several factors including moisture and temperature. Finally, tebuthiuron applied to rangeland has been shown to be phytotoxic after 615 days, and the estimated time for total dissipation of the herbicide is from 2.9 to 7.2 years (426).

HAs have been issued for bromacil, diuron, and fluometuron; no occurrence data are available for tebuthiuron or terbacil (295). Chloroxuron [1982-47-4], fenuron TCA [4482-55-7], and norea [18530-56-8] also are urea herbicides.

Aliphatic-Carboxylics.   There are only two herbicides present in this class, trichloroacetate [76-03-9] (TCA) and dalapon [75-99-0]. These are used primarily for the selective control of annual and perennial grass weeds in cropland and noncropland (2,299). Dalapon is also used as a selective aquatic herbicide (427). Dalapon and TCA are acidic in nature and are not strongly sorbed by soils. They are reported to be rapidly degraded in both soil and water by microbial processes (2,427). However, the breakdown of TCA occurs very slowly when incubated at 14–15°C in acidic soils (428). Liming not only accelerates this degradation but also increases the numbers of TCA-degrading bacteria. An HA has been issued for dalapon, but not TCA (269).

Metal Organics and Inorganics.   The metal organic herbicides are arsenicals used for the selective, post-emergence control of grass and broadleaved weeds in cropland and noncroplands. These herbicides are particularly useful for weed control in cotton and turf crops (2,296,294). Cacodylic acid is a contact herbicide used for nonselective weed control in cropland and noncropland (299). Ammonium sulfamate [7773-06-0] (AMS) is an inorganic herbicide used for control of woody plants and herbaceous perennials (2).

Arsenical herbicides are salts of methylarsonic acid, eg, calcium salt of methylarsonic acid [5902-95-4] (CMA), and are thus freely soluble in water (299). They are strongly sorbed to soils and not readily leached (2). The sorption of DSMA is greater on clay soils than on sandy soils (429). In addition, the amount sorbed is greater on kaolinite than on montmorillonite or vermiculite, indicating possible retention by exposed hydroxyl groups. Sorption of MSMA is also significantly higher on clay soils than on sandy soils (430), and MSMA is essentially immobile in field studies and not expected to leach. AMS is not retained in soils and is susceptible to leaching losses (431). Cacodylic acid and MSMA are both degraded in field soils and do not accumulate with repeated application (432). MSMA is degraded at a faster rate under flooded soil conditions than in soils at a moisture content less than field capacity (433). Finally, MSMA appears to be degraded, at least partially, by soil microbes (434). An HA has been issued for AMS, but not for any of the arsenical herbicides. A method for the analysis of the arsenicals by hplc is also available (435).

Miscellaneous Trifluoromethyl Compounds. The herbicides in this group are used for a wide variety of weed-control purposes. Acifluorfen, lactofen [77501-63-4], and oxyfluorfen are used for selective, pre-, and post-emergence weed control in croplands. Fluorochloridone is used for selective, pre-emergence weed control in cropland, and fluridone, fomesafen, and mefluidide [53780-34-0] are used for post-emergence control (296). Fluridone is also used as an aquatic herbicide (2).

Fluridone is a weak base with low water solubility. Sorption of fluridone increases with decreasing pH (436). Leaching of fluridone was not significant in field study, and the persistence has been determined to be less than 365 days. The degradation of fluridone appears to be microbial in nature, and accelerated breakdown of the herbicide occurs upon repeated applications (437). Fluorochloridone is shown to degrade by hydrolysis at pH 7 and 9, but not at lower pH. The half-lives for this reaction are 190 and 140 days for pH 7 and 9, respectively. Breakdown by photolysis occurs rapidly with a half-life of 4.3 days at pH 7 (438). An HA is available for acifluorfen.

Amino Acid Analogues. Amino acid analogue herbicides also control a large variety of weeds. Glyphosate and glufosinate are used for the broad-spectrum, nonselective control of grass and broadleaved weeds. Diethatyl [38725-95-0] is used for selective, pre-emergence control of grass and broadleaved weeds. Flamprop [58667-63-3] is used to control the growth of wild oats in wheat (2,296).

Glyphosate is zwitterionic and thus can be sorbed as an anion, cation, or zwitterion (245). Although the amount of glyphosate sorbed decreases with increasing soil pH (439), at the pH of typical agricultural soils glyphosate is strongly sorbed relatively immobile (440). The mobility classification varies from immobile on an acidic sandy clay loam soil to low mobility on an alkaline clay loam soil. The increase in mobility with increasing pH arises from a decrease in sorption (331).

Glyphosate is readily degraded by microbial means in most soils (331). A species of bacteria (*Pseudomonas* sp.) capable of degrading glyphosate has been isolated (441). Flamprop–methyl [52756-25-9] is transformed by a combination of chemical and microbial processes when incubated under aerobic conditions (442). The degree of transformation increases when the herbicide is incubated under flooded conditions. Finally, glufosinate is rapidly degraded by microbial means with half-lives ranging from 3 to 7 days (330). An HA is available for glyphosate. Diethatyl–ethyl [38727-55-8], and sulfosate [81591-81-3] are additional amino acid analogue herbicides.

Miscellaneous Other Herbicides. The herbicides in this group are not readily included in any of the preceding groups. Acrolein [107-02-8] (2-propenal) is used as a contact, aquatic herbicide. Sethoxydim, clethodim, and tridiphane are used for selective, post-emergence weed control. Cinmethylin and cloma-zone [81777-89-1] are used for selective pre-emergence control and ethofumesate [26225-79-6] for selective pre- and post-emergence weed control (2,296).

Cinmethylin has been found to resist leaching in several soils and is less mobile than metolachlor (443). Clomazone is a nonionic herbicide that is sorbed primarily to soil organic matter (444); it is rapidly dissipated in soil with half-lives ranging from 33 to 37 days (445). Clethodim is degraded by chemical hydrolysis and photolysis (446). Half-lives for these reactions range from 2.4 to 3.2 hours in aqueous solution, the rate increasing with decreasing pH. Finally, ethofumesate is strongly sorbed to soils and is subject to degradation via chemical hydrolysis (447). The sorption of ethofumesate is greater on dry soils. Cycloxydim [101205-02-1] also is a herbicide. Health advisories have not been issued for any of the aforementioned herbicides (269), and data on the occurrence of these herbicides in SW or GW are not available.

Economic Aspects

During the period from 1979 through 1991, the estimated U.S. total annual volume of herbicide usage increased somewhat from 254 million kg active ingredient (AI) to 285 million kg. Peak herbicide usage of 306 million kg occurred in 1984. During the years between 1979 and 1992, agricultural uses accounted for 76 to 81% of the herbicide applied in the United States. Combined U.S. government, industrial, and commercial herbicide usage during those years ranged from 14% in 1979, to 19% in 1986, and 17% in 1991. Home gardens and lawns received the remaining 4 to 5%.

Between 1979 and 1991, the amounts of herbicide applied in the United States have remained constant, but the expenditures on herbicides have increased 54%. Agricultural costs accounted for all of this increase and more, since herbicide user expenditures in the government/commercial and home sectors combined dropped 3 to 4% during that period. Increased weed control costs related to crop protection have also contributed to the 37% increase, since 1988, in total annual user expenditures for pesticides in general, ie, herbicides, fungicides, and insecticides. In the United States, agricultural uses (ca 1993) account for more than 67% of total pesticide user expenses and 75% of the quantity used annually. Herbicides are now the leading type of pesticides in terms of both user expenditures and volumes used (1).

Based on 1990–1991 estimates, the most used herbicides in the United States are, in descending order of usage, atrazine, alachlor, metolachlor, 2,4-D, trifluralin, cyanazine, EPTC, metham–sodium, glyphosate, and butylate (see Table 1). The estimated 1990–1991 usage ranged from 31.7 to 36.3 million kg AI, for atrazine widely used for control of broadleaf and grassy weeds in corn (maize), sorghum, rangeland, orchards, and turf, to 2.3 to 4.5 million kg, for butylate used for selective pre-emergence control of nutsedge, as well as of perennial and annual grasses.

Although the ratios have varied from year to year since 1979, the selective herbicides used in corn production have accounted for approximately 21% of herbicide use on a per crop basis (24). Herbicide use in soybean and cotton production combined account for ca 23% of the selective herbicide market. Graminicides, which selectively kill grasses, constitute 40% of the total market, leaving a market share of approximately 16% for the nonselective herbicides.

Innovative Weed Management Agents

Chemical, cultural, and mechanical weed control practices have been relatively successful in reducing yield losses from weeds (448). However, herbicide-resistant weed populations, soil erosion, pesticide persistence in the environment, and other problems associated with technologies used (ca 1993) to control weeds have raised concerns for the long-term efficacy and sustainability of herbicide-dependent crop production practices (449). These concerns, coupled with ever-increasing demands for food and fiber, contribute to the need for innovative weed management strategies (450).

Adoption by the agricultural community requires that an innovative weed management agent must be an effective control of the target species, be cost-effective, and be practical to employ. It must not interfere with crop production practices such as crop rotation or the use of other pesticides. Additionally, new weed-control agents cannot pose a significant threat to human health or the environment. Considerable costs are incurred in the development, registration, production, and marketing of weed control agents. These costs require that an herbicide have sufficient long-term market viability and market niche potential to justify these costs in time and money. The need for safe and effective methods of crop production in an environment that contains competitive weeds is becoming increasingly critical.

Weed Management Strategies. The paradigm that all noncrop plant populations in a field should be controlled, regardless of the actual impact on crop yield and quality, is not justifiable. The objective determination *a priori* of which plant populations require control and which do not, directly reduces the economic, environmental, and social costs associated with weed control and can be considered an innovative approach to weed management. For example, some noncrop plant populations do not significantly hinder production. In some developing areas of the world, producers have found uses for noncrop plants that would otherwise be considered weeds (451), and many weeds are both edible and nutritious (452). In aquaculture systems, certain highly problematic algal and bacterial weeds are also essential to the overall stability and productivity of the production system (453).

The immediate and total removal of weeds is often recommended. However, this recommendation may be based more on when control methods can most easily be applied, rather than on considerations of the optimal time for effective weed control (454). Controlling plants that are not actually problems or that are present at noncritical times is costly and may not truly benefit the producer. However, weeds that are present initially in very low numbers may require subsequent eradication if introduction of a new noxious species is to be prevented.

Managers of agroecosystems are being encouraged to manage weed populations at levels that are below their economic optimum thresholds (455), rather than attempting to eliminate or control all noncrop plants, regardless of their actual impact. Decisions concerning management of weed populations should be governed by both agroecological principles and site-specific considerations in the context of an overall integrated pest management program (451,456). However, the practical implementation of integrated pest management (IPM) programs can be difficult (457).

Nonchemical or traditional practices, such as weed seed removal, optimal crop seeding rates, crop selection, enhanced crop competitiveness, crop rotation, and mechanical weed control are all important components of an effective weed management program (458,459). In the context of modern intensive chemical herbicide application, nonchemical practices may even represent an innovative approach to weed management and should receive careful consideration.

Natural Products and Allelopathic Compounds as Herbicides. Approximately 60% of the registered herbicides are halogenated hydrocarbons. These compounds were discovered primarily by screening large numbers of chemically synthesized compounds for phytotoxic activity (460). The chemical synthesis *de novo* and bioscreening of large numbers of complex organic compounds are extremely costly and time consuming. In terms of yielding new chemical control agents, this approach is considered by many to have reached a point of diminishing returns (461). Additionally, there is growing concern that compounds that do not occur in nature may produce unanticipated health and environmental problems. However, plants, fungi, marine organisms, and certain bacteria produce a vast array of organic compounds, and many of these natural products exhibit biological activity (462–464). In nature, these compounds are produced in minute quantities and present interesting chemical problems in detection, identification, quantification, and production of active and stable analogues of these natural products. Although these compounds appear to be ecologically safe in naturally occurring amounts, the large quantities required for agricultural applications may cause environmental problems similar to those associated with chemical herbicides.

Natural products have exerted evolutionary pressure that has led ecological and biological systems to develop mechanisms that efficiently degrade or metabolize such organic compounds. Therefore, natural products may be less likely to accumulate in the environment than would metabolically resistant synthetic compounds. Although some natural products can be highly toxic, eg, aflatoxin in grains and cottonseed, and their safety cannot be assumed, there is great interest in the development potential of environmentally safe natural products and natural product derivatives that could control specific weeds and other pests.

Approximately 7000 naturally occurring secondary metabolites have been reported. Many of these compounds are difficult or impossible to synthesize chemically (465). If sufficient quantities of these natural products can be obtained, possibly through fermentation technology, their efficacy as commercial pest control agents can be evaluated more fully (460,466). Although there are difficulties associated with the direct commercialization of natural products as herbicides, the chemical alteration and optimization of natural products can still yield patentable and marketable control agents (467–469). A primary benefit of investigations of the biological activity of natural products may be the provision of leads to new classes of weed control agents (460).

Investigations of natural product chemistries have aided in the development of bialaphos, cinmethylin, picloram, glufosinate, and other important herbicides (448). Additional compounds may be found through investigations of natural products that cause plants and other organisms to undergo rapid physiological change, such as plant hormones and phytotoxins (132). Many plant hormones and phytotoxins are also produced by microorganisms. For example, it has been reported that the plant hormones, indole-3-acetic acid, gibberellins, ethylene, abscisic acid, and cytokinins, are produced by various microorganisms. Additionally, microorganisms have been reported to contain novel natural products that could provide basic structural templates for the development of new herbicides (470).

One route to the discovery of innovative control agents involves the search for compounds that affect interactions among plants and other organisms. The negative connotation of the term allelopathy refers to chemical interactions among plants that result in the suppression of other plant species (471). Although allelopathic compounds are often affected by microbial activity (472) and nonchemical interactions and competition among plants can complicate investigations of allelopathic interactions (471,473), identification of the causative suppressive compounds may lead to the discovery of novel control agents. In addition, new crop varieties that directly suppress weed growth with endogenous natural products could be developed (472,474).

In the early 1990s, many useful natural products, eg, fragrances, flavors, and potential natural pesticides, cannot be economically synthesized by chemical means (475). However, advances in biotechnology and fermentation technology, coupled with a desire for naturally derived compounds, show promise for the utilization of microorganisms in the commercial manufacture of natural products. A primary constraint in this approach is the limited availability of microbial strains to produce commercially exploitable amounts of the desired compounds. Such strains often require mutagenesis and extensive selection before they can be used on a commercial scale. Although this is a costly and primarily random process that may not yield a useful result, enhanced understanding of microbial physiology and genetics can greatly expedite the development of useful strains (475).

Biological systems produce an extremely wide variety of natural products. This ecological and genetic diversity offers researchers a vast index of compounds to search for innovative weed management agents.

Plant Pathogens and Insects as Control Agents.　Concerns about accumulations of chemical control agents in the environmental and food resources have also increased interest in microbial weed control agents (476). Controlling weeds with carefully screened plant pathogens offers several benefits, including a high degree of specificity for a given target weed, low potential for negative human health and environmental impact, inability to accumulate in the food chain, and other advantages (477,478). The high degree of host specificity may limit the market size for some biological control agents (477,479), but these biocontrol agents can be combined with chemical herbicides and other pathogens to increase the spectrum of weeds controlled (477). The marketing of biological control agents may also be constrained by slow expression of phytotoxicity, pathogen dependence on optimum environmental conditions, potential resistance of the weed towards the pathogen, and lack of formulation stability under field conditions and during preuse storage (480). These constraints can be addressed by genetic manipulation of selected pathogenic strains to produce more effective control agents (480,481) and by the investigation of the mechanisms of disease resistance in plants (482).

There are two principal approaches to the biological control of weeds (483–485). The first approach is referred to as classical or inoculative biological weed control. Plants that have been introduced to areas outside of their natural range often encounter fewer growth and seed dissemination constraints. This release from constraining factors can stimulate such migrant plants to become highly competitive and problematic weeds. The intent of classical biological weed control approaches is to manage introduced weed populations by introducing host-specific pathogens from the weed's native range, thus moderating the growth of weed populations by the reestablishment of an old association between host and pathogen populations in the expanded range (483,486). Just as a release from constraining factors can stimulate weed growth, release from hyperparasites, antagonists, fungivores, and other constraining factors in the newly expanded range of a plant pathogen can improve its effectiveness on target weeds in that expanded range (485). If an association can be established, the pathogen may become epiphytotic and require no further manipulations or repeated inoculations (487). This approach may be of particular benefit in developing nations where periodic reapplications of control agents may be difficult. The rich biodiversity in developing nations also provides a potential source for pathogen strains appropriate for biocontrol applications (488).

The long association of pathogen and host-plant in the host's native range, however, can contribute to the coevolution of polygenic resistance to the pathogen. This resistance can sometimes be overcome by the use of several and perhaps novel pathogens to form a new association in the expanded range of the weed. This approach takes full advantage of available biodiversity to overcome any polygenic resistance to biological control agents (489). Reviews of the concept of biological weed management are available (485,486,489,490).

An additional approach to biological weed control is referred to as the inundative or augment approach to biological weed management. This approach utilizes pathogenic propagules formulated as a weed control agent, eg, mycoherbicides. The mass-inoculation of pathogenic propagules in an effective formulation can enhance the dissemination and survival of the pathogens, overwhelm target weed resistance, and produce results similar to those achieved with chemical herbicides. Mycoherbicides often contain native pathogens that are active against native weeds and are thus highly selective against the target weed species (483,484,491).

Typically, mycoherbicides are developed by isolating useful pathogens from the environment, followed by the mass production of large quantities of pathogenic propagules. Once large amounts of propagules, ie, spores, are obtained, they are then combined with formulation components that increase the viability and longevity of the propagules in the formulation, as well as the ability of the propagules to withstand desiccation after application; these are all requirements for an overall increase in efficacy. Mycoherbicides may be applied using methods similar to those used to apply chemical herbicides and often require application on a repeated, periodic basis, as do many chemical herbicides.

Research concerning plant pathogen control agents has resulted in two commercially available mycoherbicides. The mycoherbicide Collego is a formulated product consisting of propagules of the fungus *Colletotrichum gloeosporioides*, and Devine is a formulated product containing the sexual spores of the oomycete, *Phytophthora palmivora* (477). Devine is used to control stranglevine (*Morrenia odorata*) in citrus, and Collego is used in northern jointvetch (*Aeschynomene virginica*) control in rice and soybean. In addition to these commercially available products, *Alternaria cassiae, Colletotrichum gloeosporioides clidemiae, C. gloeosporioides cuscutae, C. malvarum, C. coccodes, C. orbiculare, Phomopsis convolvulus, Cephalosporium diospyri, Cercospora rodmanii, Mycoletodiscus terrestris,* and *Chondrostereum purpureum* are under consideration as components of mycoherbicide agents (476,479,480). The plant pathogens, *Uromycladium tepperianum, Phaeoramularia* sp., *Entyloma ageratinae, E. compositarum, Phragmidium violaceum, Pucciunia carduorum, P. chondrillina, Ceratocystis fagacearum, Cactoblastus cactorum, Gleosporium lunatum,* and other species have also been investigated as weed control agents (459,476,485,492). Difficulties maintaining pathogen virulence until application, limitations in formulation and application technology, and various marketing factors have so far limited the commercial life of many mycoherbicides, including Devine which has been withdrawn from the market.

The isolation of potentially useful pathogens with appropriate host-specificity is a critical first step in the development of biological control agents. However, full commercialization requires that effective pathogenic propagules be mass-produced and that effective formulations be developed that enhance the stability, ease of application, and overall efficacy of mycoherbicides. If these constraints can be overcome, the use of biological weed management agents can expand more rapidly.

Control of Weed Seeds. Efforts to control parasites often focus on the most vulnerable stage in the life-cycle of the parasite, such as when the parasite is present and the host is absent. If weeds can be considered a type of parasite in cropping systems, then a point of vulnerability for weeds occurs after harvest and prior to the next crop planting. However, during this period weeds are usually present as seeds and/or other over-wintering storage structures. With the exception of a few soil fumigants that cannot be used over large areas, there are no agents available (ca 1993) for elimination of weed seed populations in the soil, ie, the soil seed bank. If agents that control weed seed germination could be applied prior to planting, interference from weeds would be prevented until reintroduction of weed propagules. Additionally, if a very large portion of the weed seed bank could be stimulated to germinate prior to planting, weeds could be controlled by a single cultivation or application of nonselective herbicide (493,494).

Efforts toward developing agents which destroy weed seeds in the soil are hindered by several factors. The soil seed bank can be very large with estimates ranging as high as 70,000–90,000 weed seeds per m^2 of the upper 15–25 cm of the soil (495). Seeds and other over-wintering propagules typically exhibit reduced metabolic activity, compared to growing plants. This lack of metabolic activity makes it difficult to render seeds nonviable with chemical agents that rely on the inhibition of metabolic pathways to produce a lethal effect, eg, most commercially available herbicides.

The soil seed bank consists of both dormant and nondormant seeds. Dormant seeds can remain viable in the soil for several years and in some cases as long as 50–100 years (493–495). These factors, as well as the technical difficulties associated with preventing seed production and distribution and removing seeds from the soil, limit the effectiveness of strategies to deplete the seed bank (494). Control efforts have focused on investigating factors, such as moisture, light, temperature, oxygen, and chemical germination stimulants, that effect dormancy and control weed seed germination (493,494). In addition, increased seed predation by insects and rodents has been considered a weed control strategy (496).

Development of agents that stimulate weed seed germination and/or attack weed seeds would have a profound impact on weed management. However, herbicide development programs in the early 1990s do not focus on identifying agents that are effective on weed propagules. A systematic search for compounds that render weed seeds nonviable or cause them to germinate simultaneously could provide important new weed management tools.

BIBLIOGRAPHY

"Weed Killers" in *ECT* 1st ed., Vol. 15, pp. 18–24, by D. E. H. Frear, Pennsylvania State University; in *ECT* 2nd ed., Vol. 22, pp. 174–220, by J. R. Plimmer, U.S. Department of Agriculture; "Herbicides: in *ECT* 3rd ed., Vol. 12, pp. 297–351, by J. R. Plimmer, U.S. Department of Agriculture.

1. A. Aspelin, A. H. Grube, and R. Torla, *Pesticide Industry Sales and Usage*, H-7503W, U.S. Environmental Protection Agency, (EPA), Office of Pesticide Programs, Washington, D.C., 1992.
2. *Herbicide Handbook*, 6th ed., Weed Science Society of America, Champaign, Ill., 1989.
3. W. Draber and T. Fujita, eds., *Rational Approaches to Structure, Activity, and Ecotoxicology of Agrochemicals*, CRC Press, Boca Raton, Fla., 1992.
4. E. L. Plummer, in Ref. 5, pp. 3–39.
5. R. M. Muir and C. Hansch, *Annu. Rev. Plant Physiol.* **6**, 157 (1955).
6. C. Hansch and T. Fujita, *J. Am. Chem. Soc.* **86**, 1616 (1964).
7. I. Takemoto and co-workers, *Pestic. Biochem. Physiol.*, **23**, 341 (1985).
8. D. R. Baker and co-workers, eds., *Synthesis and Chemistry of Agrochemicals*, ACS Symposium Series, **355**, American Chemical Society, Washington, D.C., 1987.
9. S. Creuzet, B. Gilquin, and J. M. Ducruet, *Z. Naturforsch.* **44c**, 435 (1989).
10. T. Sotomatsu and T. Fujita, *Quant. Struct.-Act. Relat.* **9**, 295 (1990).
11. D. R. Baker, J. G. Fenyes, and W. K. Moberg, eds., *Synthesis and Chemistry of Agrochemicals Part II*, ACS Symposium Series 443, American Chemical Society, Washington, D.C., 1991.
12. G. Sandmann and co-workers, *Pestic. Biochem. Physiol.* **42**, 1 (1992).
13. A. Verloop, W. Hoogenstraten, and J. Tipker, in E. J. Ariens, ed., *Drug Design*, Vol. 7, Academic Press, Inc., New York, 1976, p. 165.
14. H. Kubinyi, *Arzneim.-Forsch.* **26**, 1991 (1976).
15. H. Kubinyi and O. H. Kehrhahn, *Arzeim.-Forsch.* **28**, 598 (1978).
16. T. Haga, in Ref. 3, pp. 103–117.
17. C. Takayama and co-workers, in Ref. 3, pp. 315–330.
18. B. Cross and D. W. Ladner, in Ref. 3, pp. 331–355.
19. H. Yoshioka, in Ref. 3, pp. 185–217.
20. R. Hoagland, ed., *Microbes and Microbial Products as Herbicides*, ACS Symposium Series 439, American Chemical Society, Washington, D.C., 1990.
21. J. S. Bannon and co-workers, in Ref. 3, pp. 305–319.
22. G. E. Templeton, in Ref. 3, pp. 320–329.
23. W. J. Connick, Jr., in Ref. 20, pp. 241–250.

24. K. P. Parry, in A. D. Dodge, ed., *Herbicides and Plant Metabolism*, Cambridge University Press, New York, 1989, pp. 1–20.
25. J. J. S. van Rensen, in Ref. 2, pp. 21–36.
26. J. R. Coggins, in A. D. Dodge, ed., *Herbicides and Plant Metabolism*, Cambridge University Press, New York, 1989, pp. 97–112.
27. T. A. Andrea and co-workers, in Ref. 3, pp. 373–395.
28. M. Eto, in Ref. 3, pp. 147–161.
29. J. L. Harwood, S. M. Ridley, and K. A. Walker, in Ref. 2, pp. 73–96.
30. A. Trebst and M. Avron, eds., *Photosynthesis I: Photosynthetic Electron Transport and Photophosphorylation, Encyclopedia of Plant Physiology, N.S.*, Springer-Verlag, Berlin, 1977.
31. M. Gibbs and E. Latzko, eds., *Photosynthesis II: Photosynthetic Carbon Metabolism and Related Processes, Encyclopedia of Plant Physiology, N.S.*, Springer-Verlag, Berlin, 1979.
32. L. A. Staehelin and C. J. Arntzen, eds., *Photosynthesis III: Photosynthetic Membranes and Light Harvesting Systems, Encyclopedia of Plant Physiology, N.S.*, Springer-Verlag, Berlin, 1986.
33. R. Hill and F. Bendall, *Nature* **186**, 136 (1960).
34. M. F. Hipkins, in M. B. Wilkins, ed., *Advanced Plant Physiology*, Pitman, London, 1984, pp. 218–248.
35. F. B. Salisbury and C. W. Ross, *Plant Physiology*, 2nd ed., Wadsworth, Belmont, Calif., 1978, p. 133.
36. P. Sétif and P. Mathis, in Ref. 32, pp. 476–486.
37. G. Sandmann and P. Böger, in Ref. 32, pp. 595–602.
38. D. J. Gillham and A. D. Dodge, *Pestic. Sci.* **19**, 25 (1987).
39. G. C. Cruz and co-workers, *J. Biol. Chem.* **263**, 3835 (1988).
40. G. Hauska, in Ref. 32, pp. 496–507.
41. D. R. Ort, in Ref. 32, pp. 143–196.
42. P. Böger, *Z. Naturforsch.* **39c**, 468 (1984).
43. B. A. Diner, in Ref. 32, pp. 422–436.
44. E. L. Camm and B. R. Green, *Plant Physiol.* **66**, 428 (1980).
45. C. F. Yocum, in Ref. 32, pp. 437–446.
46. C. J. Arntzen and H. B. Pakrasi, in Ref. 32, pp. 457–475.
47. A. Trebst and W. Draber, *Photosynthesis Res.* **10**, 381 (1986).
48. O. L. Lange and co-workers, *Physiological Plant Ecology IV, Ecosystem Processes: Mineral Cycling, Productivity, and Man's Influence, Encyclopedia of Plant Physiology, N.S.*, Spriger-Verlag, Berlin, 1983.
49. U. Johanningmeier, E. Neumann, and W. Oettmeier, *J. Bioenerg. Biomembr.* **15**, 43 (1983).
50. G. Renger, *Physiol. Veg.* **24**, 509 (1986).
51. D. E. Moreland, *Annu. Rev. Plant Physiol.* **31**, 597 (1980).
52. A. Trebst and co-workers, in H. Frehse, ed., *Pesticide Chemistry, Advances in International Research, Development, and Legislation*, Proc. 7th Int. Congr. Pestic. Chem. (IUPAC), VCH, Weinheim, 1991, p. 111.
53. A. Trebst, *Z. Naturforsch.* **42c**, 742 (1987).
54. G. Sandmann, H. Linden, and P. Böger, *Z. Naturforsch.* **44c**, 787 (1989).
55. S. Kowlaczyk-Schröder and G. Sandmann, *Pestic. Biochem. Physiol.* **42**, 7 (1992).
56. G. Sandmann, P. M. Bramley, and P. Böger, *J. Pestic. Sci.* **10**, 19 (1985).
57. P. Böger and G. Sandmann, eds., *Target Sites of Herbicide Action*, CRC Press, Boca Raton, Fla., 1989.
58. M. P. Mayer and co-workers, *Pestic. Biochem. Physiol.* **34**, 111 (1989).

59. G. Sandmann, I. Kumita, and P. Böger, *Pestic. Sci.* **30**, 353 (1990).
60. W. J. Michaely and A. D. Gutman, in Ref. 8, pp. 54–64.
61. C. Willemot, *Plant Physiol.* **60**, 1 (1977).
62. J. B. St. John, *Pestic. Biochem. Physiol.* **23**, 13, (1985).
63. H. Linden and co-workers, *Pestic. Biochem. Physiol.* **35**, 46 (1990).
64. G. Mitchell and D. L. Bartlett, *7th International Congress of Pesticide Chemists, Abstr.* Vol. 1, Hamburg, 1990, p. 170.
65. G. Sandmann and S. Kowalczyk, *Biochem. Biophys. Res. Commun.* **163**, 916 (1989).
66. D. Chamovitz, I. Pecker, and F. Hirschberg, *Plant. Mol. Biol.* **16**, 967 (1991).
67. J. Feierabend, *Z. Naturforsch.* **39c**, 450 (1984).
68. P. K. Stumpf and E. E. Conn, eds., *The Biochemistry of Plants, Vol. 4, Lipids: Structure and Function*, Academic Press, Inc., New York, 1980.
69. S. Numa, ed., *Fatty Acid Metabolism and Its Regulation*, Elsevier/North Holland, Amsterdam, 1984.
70. P. K. Stumpf, W. J. Nes, and J. B. Mudd, eds., *Biochemistry of Plant Lipids: Structure and Function*, Plenum Press, New York, 1987.
71. D. E. Moreland, J. B. St. John, and F. D. Hess, eds., *Biochemical Responses Induced by Herbicides*, ACS Symposium Series **181**, American Chemical Society, Washington, D.C., 1982.
72. C. Fedtke, *Biochemistry and Physiology of Herbicide Action*, Springer-Verlag, New York, 1982.
73. S. O. Duke, ed., *Weed Physiology, Vol. II, Herbicide Physiology*, CRC Press, Boca Raton, Fla., 1985.
74. J. L. Harwood, *Annu. Rev. Plant Physiol. Plant Mol. Biol.* **39**, 101 (1988).
75. J. D. Burton and co-workers, *Biochem. Biophys. Res. Commun.* **148**, 1039 (1987).
76. I. Iwataki, in Ref. 3, Chapt. 16, pp. 397–426.
77. K. A. Walker, S. M. Ridley, and J. L. Harwood, *Biochem. J.* **254**, 307 (1988).
78. *Ibid.*, p. 811.
79. K. Kobek, H. Focke, and H. K. Lichtenthaler, *Z. Naturforsch.* **43c**, 47 (1988).
80. P. C. Kearney and D. Kaufmann, eds., *Herbicide Chemistry, Degradation and Mode of Action*, Vol. 3, Marcel Dekker, New York, 1988.
81. S. O. Duke and W. H. Kenyon, in Ref. 79, Chapt. 2, pp. 72–116.
82. H. K. Lichtenthaler and K. Kobek, *Z. Naturforsch.* **42c**, 1275 (1987).
83. K. Kobek and co-workers, *Physiol. Plant.* **72**, 492 (1988).
84. P. Zama and K. K. Hatzios, *Pestic. Biochem. Physiol.* **27**, 86 (1987).
85. R. E. Wilkinson and T. H. Oswald, *Pestic. Biochem. Physiol.* **28**, 38 (1987).
86. E. P. Fuerst, *Weed Technol.* **1**, 270 (1987).
87. D. G. Davis and K. E. Dusbabek, *Weed Sci.* **21**, 16 (1973).
88. G. Ezra, J. E. Gressel, and H. M. Flowers, *Pestic. Biochem. Physiol.* **19**, 225 (1983).
89. J. L. Harwood and N. J. Russell, *Lipids in Plants and Microbes*, Allen and Unwin, London, 1984.
90. F. A. Eder, *Z. Naturforsch.* **34c**, 1052 (1979).
91. A. O. Davies and J. L. Harwood, *J. Exp. Bot.* **34**, 1089 (1983).
92. J. W. Gronwald, *Weed Sci.* **34**, 196 (1986).
93. M. A. Shimabukuro, R. H. Shimabukuro, and W. C. Walsh, *Physiol. Plantarum* **56**, 444 (1982).
94. R. H. Shimabukuro, W. C. Walsh, and R. A. Hoerauf, *Plant Physiol.* **80**, 612 (1986).
95. R. E. Wilkinson, *Pestic. Biochem. Physiol.* **25**, 93 (1986).
96. P. G. Bartels and A. Hyde, *Plant Physiol.* **45**, 807 (1970).
97. K. J. Kunert and A. D. Dodge, in Ref. 57, Chapt. 3, pp. 45–63.
98. K. E. Pallett and A. D. Dodge, *Z. Naturforsch.* **39c**, 482 (1984).

99. D. A. Witkowski and B. P. Halling, *Plant Physiol.* **90**, 1239 (1989).

100. Y. Shaaltiel and J. Gressel, *Pestic. Biochem. Physiol.* **26**, 22 (1986).

101. P. Böger, in Ref. 57, Chapt. 10, pp. 247–282.

102. G. Sandmann and P. Böger, *Lipids* **17**, 35 (1982).

103. A. L. Tappel, in W. A. Prior, ed., *Free Radicals in Biology*, Vol. 4, Academic Press, Inc., New York, 1980, p. 1.

104. S. P. Wolff, A. Garner, and R. T. Dean, *TIBS* **11**, 27 (1986).

105. K. J. Kunert and co-workers, *Weed Sci.* **33**, 766 (1985).

106. J. V. Schloss, in Ref. 57, Chapt. 9, pp. 165–245.

107. E. A. Woolson, in Ref. 57, Chapt. 15, pp. 741–776.

108. R. M. Hochster and J. H. Quastel, eds., *Metabolic Inhibitors: Comprehensive Treatise*, Academic Press, Inc., New York, 1963.

109. O. Warburg and W. Christian, *Biochem. Z.* **303**, 40 (1939).

110. R. A. Mitchell and co-workers, *Biochemistry* **40**, 2049 (1971).

111. C. H. Huang and R. A. Mitchell, *Biochemistry* **11**, 2278 (1972).

112. J. A. Colanduoni and J. J. Villafranca, *Bioorg. Chem.* **14**, 163 (1986).

113. R. Manderscheid and A. Wild, *J. Plant Physiol.* **123**, 135 (1986).

114. T. B. Ray, in Ref. 57, Chapt. 6, pp. 105–125.

115. P. K. Stumpf and E. Conn, eds., *The Biochemistry of Plants*, Vol. 5, *Amino Acids and Derivatives*, Academic Press, Inc., New York, 1980.

116. D. A. Boulten and B. Parthier, eds., *Encyclopedia of Plant Physiology*, Vol. 14A, Springer-Verlag, New York, 1982.

117. H. Köcher, *Aspects Appl. Biol.* **4**, 227 (1983).

118. M. A. Acaster and P. D. J. Weitzman, *FEBS Lett.* **189**, 241 (1985).

119. J. J. Kaiser and O. A. M. Lewis, *New Phytol.* **85**, 235 (1980).

120. K. Tachibana, in R. Greenhalgh and T. R. Roberts, eds., *Pesticide Science and Biotechnology*, Blackwell Scientific, Oxford, 1987, p. 145.

121. A. Wild, H. Sauer, and W. Rühle, *Z. Naturforsch.* **42c**, 263 (1986).

122. *Ibid.*, p. 270.

123. R. M. Wallsgrove and co-workers, *Plant Physiol.* **83**, 155 (1987).

124. J. G. Turner and J. M. Debbage, *Physiol. Plant Pathol.* **20**, 223 (1982).

125. P. L. Langston-Unkefer, P. A. Macy, and R. D. Durbin, *Plant Physiol.* **76**, 71 (1984).

126. S. Omura and co-workers, *J. Antibiot.* **37**, 1324 (1984).

127. H. P. Fischer and D. Bellus, *Pestic. Sci.* **14**, 334 (1983).

128. H. Nakamoto, M. S. B. Ku, and G. E. Edwards, *Photosynth. Res.* **3**, 293 (1982).

129. G. Pitchard, W. Griffin, and C. P. Whittingham, *J. Exp. Bot.* **13**, 176 (1962).

130. E. Creach and C. R. Stewart, *Plant Physiol.* **70**, 1444 (1982).

131. H. Strotmann and S. Bickel-Sandkötter, *Annu. Rev. Plant Physiol.* **35**, 97 (1984).

132. J. Bennett, *Phytochemistry* **15**, 263 (1976).

133. R. E. Moore and co-workers, *Tetrahedron Lett.* **25**, 3931 (1984).

134. B. J. Miflin, ed., *The Biochemistry of Plants*, Vol. 5, Academic Press, Inc., New York, 1980.

135. J. Giovanelli, S. H. Mudd, and A. H. Datko, *Plant Physiol.* **78**, 555 (1985).

136. T. A. Amdrea and co-workers, in Ref. 3, Chapt. 15, pp. 373–396.

137. M. Los, in Ref. 120, p. 35.

138. E. E. Conn, *The Shikimic Acid Pathway*, Plenum Press, New York, 1986.

139. C. F. Van Sumere and P. J. Lea, eds., *The Biochemistry of Plant Phenolics*, Clarendon Press, Oxford, 1985.

140. C. C. Smart and co-workers, *J. Biol. Chem.* **260**, 16,338 (1985).

141. H. C. Steinrücken and co-workers, *Arch. Biochem. Biophys.* **244**, 169 (1986).

142. D. M. Mousdale and J. R. Coggins, *Planta* **163**, 241 (1985).
143. L. Comai and co-workers, *Nature* **317**, 741 (1985).
144. D. J. Cole, J. C. Caseley, and A. D. Dodge, *Weed Res.* **23**, 173 (1983).
145. J. Bryan, in Ref. 115, Chapt. 11.
146. B. Miflin, *Plant Physiol.* **54**, 550 (1974).
147. P. A. Hedin, J. J. Menn, and R. M. Hollingworth, eds., *Biotechnology for Crop Protection*, ACS Symposium Series 379, American Chemical Society, Washington, D.C., 1988.
148. T. Rost, *J. Plant Growth Reg.* **3**, 51 (1984).
149. R. C. Bray, P. C. Engel, and S. G. Mayhew, eds., *Flavins and Flavoproteins*, de Gruyter, Berlin, 1984.
150. T. K. VanDyk and R. A. LaRossa, *J. Bacteriol.* **165**, 386 (1986).
151. D. L. Shaner and M. L. Reider, *Pestic. Biochem. Physiol.* **25**, 248 (1986).
152. U.S. Pat. 4,798,619 (1989), M. Los.
153. D. L. Shaner and co-workers, *Proc. Br. Crop Protect. Conf.* **1**, 147 (1985).
154. D. L. Shaner, P. C. Anderson, and M. A. Stidham, *Plant Physiol.* **76**, 545 (1984).
155. L. G. Copping and P. Rodgers, eds., "Biotechnology and its Application to Agriculture," *British Crop Protection Conference*, Croydon, U.K., Monograph 32, 1985.
156. P. C. Kearney and D. D. Kaufman, eds., *Herbicides, Chemistry, Degradation, and Mode of Action*, Vol. 1, 2nd ed., Marcell Dekker, New York, 1975.
157. A. Wiater, T. Klopotowski, and G. Bagdasarian, *Acta Biochim. Pol.* **18**, 309 (1971).
158. C. G. McWhorter and J. L. Hilton, *Physiol. Plant* **20**, 30 (1967).
159. P. G. Bartels and F. T. Wolf, *Physiol. Plant.* **18**, 805 (1965).
160. D. Hulanicka, T. Klopotowski, and G. Bagdasarian, *Acta Biochim. Pol.* **16**, 127 (1969).
161. B. B. L. Agrawal and co-workers, *Fed. Proc.* **29**, 732 (1970).
162. P. G. Bartels and A. Hyde, *Plant Physiol.* **46**, 825 (1970).
163. F. D. Hess, in Ref. 57, Chapt. 5, pp. 85–103.
164. T. W. Goodwin and E. I. Mercer, *Introduction to Plant Biochemistry*, 2nd ed., Pergamon Press, Oxford, 1983.
165. T. L. Rost and E. M. Gifford, Jr., eds., *Mechanisms and Control of Cell Division*, Dowden, Hutchinson and Ross, Stroudsburg, Pa., 1977.
166. T. L. Rost, in Ref. 165, p. 111.
167. P. L. Webster and J. Van't Hof, *Am. J. Bot.* **57**, 130 (1970).
168. P. W. Barlow, *Planta* **88**, 215 (1969).
169. E. M. Lignowski and E. G. Scott, *Weed Sci.* **20**, 267 (1972).
170. F. D. Hess, in Ref. 71, p. 207.
171. E. A. Badir, M. Mousa, and M. A. Seehy, *Egypt J. Genet. Cytol.* **12**, 123 (1983).
172. A. S. Bajer and J. Molè-Bajer, *Ann. N.Y. Acad. Sci.* **466**, 767 (1986).
173. K. C. Vaughn, *Pestic. Biochem. Physiol.* **26**, 66 (1986).
174. J. W. Shay, ed., *Cell and Molecular Biology of the Cytoskeleton*, Plenum Press, New York, 1986.
175. A. E. Dowidar and A. El-Nahas, *Biol. Plant.* **20**, 1 (1978).
176. B. E. Struckmeyer, L. K. Binning, and R. G. Harvey, *Weed Sci.* **24**, 366 (1976).
177. L. C. Morejohn and D. E. Fosket, *Science* **224**, 874 (1984).
178. T. L. Rost and S. L. Morrison, *Cytologia* **49**, 61 (1984).
179. E. A. Badr, *Cytologia* **48**, 451 (1983).
180. L. Clayton and C. W. Lloyd, *Eur. J. Cell. Biol.* **34**, 248 (1984).
181. R. A. Cross and J. D. Pickett-Heaps, *J. Cell. Biol.* **63**, 84 (1974).
182. J. D. Holmsen and F. D. Hess, *J. Exp. Bot.* **36**, 1504 (1985).
183. J. D. Holmsen, *Proc. Calif. Weed Conf.* **32**, 33 (1980).

184. J. D. Banting, *Weed Sci.* **18**, 80 (1970).

185. R. E. Wilkinson and D. Ashley, *Weed Sci.* **27**, 270 (1979).

186. L. M. Deal and F. D. Hess, *Weed Sci.* **28**, 168 (1980).

187. J. E. McFarland and F. D. Hess, *Plant Physiol.* **75S**, 49 (1984).

188. T. B. Ray, *Pestic. Biochem. Physiol.* **17**, 10 (1982).

189. K. C. Vaughn and S. O. Duke, *Pestic. Biochem. Physiol.* **26**, 56 (1986).

190. J. D. Holmsen and F. D. Hess, *Weed Sci.* **32**, 732 (1984).

191. J. L. Anderson and B. Shaybany, *Weed Sci.* **20**, 434 (1972).

192. K. C. Vaughn, M. D. Marks, and D. P. Weeks, *Plant Physiol.* **83**, 956 (1987).

193. M. A. Vaughn and K. C. Vaughn, *Pestic. Biochem. Physiol.* **28**, 182 (1987).

194. M. H. El-Deek and F. D. Hess, *Weed Sci.* **34**, 684 (1986).

195. M. E. Foley, *Weed Sci.* **35**, 180 (1987).

196. J. M. Bradow, *Weed Sci.* **34**, 1 (1986).

197. J. M. Bradow and co-workers, *J. Plant Growth Regul.* **9**, 35 (1990).

198. J. M. Bradow, W. J. Connick, Jr., and A. B. Pepperman, *J. Plant Growth Regul.* **7** (1988).

199. F. B. Salisbury and C. W. Ross, *Plant Physiology*, 2nd ed., Wadsworth, Belmont, Calif., 1978.

200. M. A. Loos in Ref. 155, Chapt. 1, pp. 1–128.

201. P. H. Rubery, *Annu. Rev. Plant Physiol.* **32**, 569 (1981).

202. W. Eisinger, *Annu. Rev. Plant Physiol.* **34**, 225 (1983).

203. Y-S. Wong and G. A. Maclachlan, *Plant Physiol.* **65**, 222 (1980).

204. J. A. Sargent, A. V. Atack, and D. J. Osborne, *Planta* **115**, 213 (1974).

205. M. E. Terry, B. Rubenstein, and R. L. Jones, *Plant Physiol.* **68**, 538 (1981).

206. P. I. Pilet, ed., *Plant Growth Regulation*, Springer-Verlag, Heidelberg, 1977.

207. R. N. Stewart, M. Lieberman, and A. T. Kunishi, *Plant Physiol.* **54**, 1 (1974).

208. J. A. D. Zeevaart and R. A. Creelman, *Annu. Rev. Plant Physiol.* **39**, 439 (1988).

209. S. M. Norman and co-workers, *Plant Physiol.* **80**, 122 (1986).

210. M. Bopp, ed., *Plant Growth Substances*, Springer-Verlag, Berlin, 1986.

211. N. K. Asare-Boamah and co-workers, *Plant Cell Physiol.* **27**, 383 (1986).

212. N. Takahashi, ed., *Chemistry of Plant Hormones*, CRC Press, Boca Raton, Fla., 1986.

213. J. C. Graebe, *Annu. Rev. Plant Physiol.* **38**, 419 (1987).

214. R. P. Pharis and D. M. Reid, eds., *Encyclopedia of Plant Physiology: Hormonal Regulation of Development III, Role of Environmental Factors*, Vol. 11, Springer Verlag, Berlin, 1985.

215. W. Rademacher and J. Jung, *Proc. 13th Annu. Meet. Plant Growth Regulator Soc. Amer.*, 102 (1986).

216. R. Mehnenett and D. K. Lawrence, eds., *Biochemical Aspects of Synthetic and Naturally Occurring Plant Growth Regulators*, Br. Plant Growth Regulator Group Monograph 11, Wantage, 1984.

217. K. Y. Cho and co-workers, *Plant Cell Physiol.* **20**, 75 (1979).

218. T. K. Scott, ed., *Plant Regulation and World Agriculture*, Plenum Press, New York, 1979.

219. R. C. Coolbaugh, in Ref. 210, pp. 53–98.

220. R. P. Pharis and R. W. King, *Annu. Rev. Plant Physiol.* **36**, 517 (1985).

221. T. J. Douglas and L. G. Paleg, *J. Exp. Bot.* **32**, 59 (1981).

222. T. H. Thomas, ed. *Plant Growth Regulator Potential and Practice*, Brit. Crop Protect. Conf. Publ., Croydon, U.K., 1982, p. 117.

223. R. C. Coolbaugh, S. S. Hirano, and C. A. West, *Plant Physiol.* **62**, 571 (1978).

224. J. P. Sterrett and T. J. Tworkoski, *J. Am. Soc. Hort. Sci.* **112**, 341 (1987).

225. H. Buchenauer and E. Röhner, *Pestic. Biochem. Physiol.* **15**, 58 (1981).

226. K. Izumi and co-workers, *Plant Cell Physiol.* **26**, 821 (1985).
227. P. Hedden and J. E. Graebe, *J. Plant Growth Regul.* **4**, 111 (1985).
228. J. Jung and co-workers, *J. Agron., Crop Sci.* **158**, 324 (1987).
229. E. P. Hasson and C. A. West, *Plant Physiol.* **58**, 479 (1976).
230. R. C. Coolbaugh, S. S. Hirano, and C. A. West, *Plant Physiol.* **62**, 571 (1976).
231. K. Wada, T. Imai, and H. Yamashita, *Agric. Biol. Chem.* **45**, 1833 (1981).
232. W. Rademacher and co-workers, *Pestic. Sci.* **21**, 241 (1987).
233. C. Fedtke and A. Trebst, in *Pesticide Science and Biotechnology, Proc. 6th Int. Congr. Pestic. Chem.*, Blackwell Scientific, Oxford, 1987, p. 161.
234. D. J. Cole and W. J. Owen, *Plant Sci.* **50**, 13 (1987).
235. Z. Ekler and G. R. Stephenson, *1987 British Crop Protection Conference—Weeds*, Vol. 3, Brit. Crop Protect. Conf., Croydon, U.K., 1987, p. 1105.
236. A. Crozier and J. R. Hillman, eds., *The Biosynthesis and Metabolism of Plant Hormones*, Oxford University Press, Oxford, U.K., 1984.
237. D. S. Letham and L. M. S. Palni, *Annu. Rev. Plant Physiol.* **34**, 163 (1983).
238. R. E. Green, in W. D. Guenzi, ed., *Pesticides in Soil and Water*, Soil Science Society of America, Madison, Wis., 1974, pp. 3–37.
239. G. W. Bailey and J. L. White, *J. Agric. Food Chem.* **12**, 324 (1964).
240. P. S. C. Rao and R. E. Jessup, in *Chemical Mobility and Reactivity in Soil Systems*, Soil Science Society of America, Madison, Wis., 1983, pp. 183–201.
241. J. B. Weber and co-workers, in N. D. Camper, ed., *Research Methods in Weed Science*, Southern Weed Science Society, Champaign, Ill. 1986, pp. 155–188.
242. C. S. Helling and J. Dragun, in *Test Protocols for Environmental Fate and Movement of Toxicants*, Proc. Symp. Assoc. Off. Anal. Chemists, 94th Ann. Mtg., Washington, D.C., 1980, pp. 43–88.
243. J. Letey and W. J. Farmer, in Ref. 238, 1974, pp. 67–97.
244. R. J. Hance, in R. Grover, ed., *Environmental Chemistry of Herbicides*, Vol. 1, CRC Press, Boca Raton, Fla., 1980, pp. 1–19.
245. P. S. C. Rao, R. E. Jessup, and J. M. Davidson, in Ref. 244, 1988, pp. 20–43.
246. L. Luckner and W. M. Schestakow, *Migration Processes in the Soil and Groundwater Zone*, Lewis Publ., Chelsea, Mich., 1991.
247. J. B. Weber and C. T. Miller, in B. L. Sawhney and K. Brown, eds., *Reactions and Movement of Organic Chemicals in Soils*, Soil Science Society of America, Madison, Wis., 1989, pp. 305–334.
248. M. G. Merkle and R. W. Bovey, in Ref. 238, 1974, pp. 99–106.
249. R. D. Wauchope and D. G. DeCoursey, in Ref. 241, 1986, pp. 135–154.
250. A. W. Taylor and D. E. Glotfelty, in Ref. 3, 1988, pp. 90–129.
251. D. E. Glotfelty and C. J. Schomburg, in Ref. 247, 1989, pp. 181–207.
252. A. Tal, B. Rubin, and J. Katan, *Pestic. Sci.* **25**, 343 (1989).
253. F. W. Roeth and co-workers, *Weed Technol.* **3**, 24 (1989).
254. L. Torstensson, in R. J. Hance, ed., *Interaction Between Herbicides and the Soil*, Academic Press, Inc., London, 1980, pp. 159–178.
255. I. R. Hill and D. J. Arnold, in S. R. Hill and S. J. L. Wright, eds., *Pesticide Microbiology*, Academic Press, Inc., London, 1978, pp. 203–245.
256. H. D. Skipper, V. V. Volk, and R. Frech, *J. Agric. Food Chem.* **24**, 126 (1976).
257. N. L. Wolfe, M. E-S. Metwally, and A. E. Moftah, in Ref. 247, 1989, pp. 229–242.
258. A. E. Smith, in Ref. 245, 1988, pp. 172–200.
259. A. J. Cessna and D. C. G. Muir, in R. Grover and A. J. Cessna, eds., *Environmental Chemistry of Hericides*, Vol. 2, CRC Press, Boca Raton, Fla., 1991, pp. 199–263.
260. R. C. Miller, V. R. Hebert, and W. W. Miller, in Ref. 247, 1989, pp. 99–110.
261. M. D. Devine and W. H. V. Born, in Ref. 259, 1991, pp. 119–140.

262. K. K. Hatzios, in Ref. 259, 1991, pp. 141–185.

263. R. L. Metcalf, *Essays in Toxicol.* **5**, 17 (1974).

264. A. R. Isensee, in Ref. 259, 1991, pp. 187–197.

265. S. W. Karickhoff and D. S. Brown, *Determination of Octanol Water Distribution Coefficients, Water Solubilities, and Sediment/Water Partitions Coefficients for Hydrophobic Organic Pollutants*, EPA-600/4-79-032, report, EPA, Washington, D.C., 1979.

266. U.S. EPA, *National Pesticide Survey: Phase I Report, Office of Water, Office of Pesticides and Toxic Substances*, EPA 570/9-90-015, U.S. Govt. Printing Office, Washington, D.C., 1990.

267. C. M. Macal and B. J. Broomfield, in G. S. Tolley, D. Yaron, and G. C. Blomquiist, eds., *Environmental Policy*, Vol. 3, Water Quality, Ballinger Publ. Co., Cambridge, Mass., 1983, pp. 163–182.

268. P. J. Squillace and E. M. Thurman, *Environ. Sci. Technol.* **26**, 538 (1992).

269. U.S. EPA, *Drinking Water Health Advisory: Pesticides*, Lewis Pub., Chelsea, Mich., 1989.

270. E. K. Nyer, *Practical Techniques for Groundwater and Soil Remediation*, Lewis Publ., Chelsea, Mich., 1992.

271. D. W. Nelson and R. H. Dowdy, eds., *Methods for Ground Water Quality Studies*, University of Nebraska, Lincoln, 1988.

272. R. F. Carsel, in Ref. 247, 1989, pp. 439–445.

273. G. R. Halberg, *J. Soil Water Cons.* **41**, 357 (1986).

274. U.S. Environmental Protection Agency, *List of Chemicals Evaluated for Carcinogenic Potential*, internal memorandum, EPA, Washington, D.C., Feb. 1992.

275. U.S. EPA, *Fed. Reg.* **51**(185) 33992–34003 (1986).

276. A. E. Greenberg, L. S. Clesceri, and A. D. Eaton, *Standard Methods for the Examination of Water and Wastewater*, 18th ed, American Public Health Association, Washington, D.C., 1992.

277. U.S. EPA, *Manual of Chemical Methods for Pesticides and Devices*, 2nd ed., Assoc. Off. Anal. Chem., Arlington, Va., 1992.

278. U.S. EPA, *Methods for the Determination of Nonconventional Pesticides in Municipal and Industrial Wastewater*, EPA 821 RR-92-002, U.S. Govt. Printing Office, Washington, D.C., 1992.

279. U.S. EPA, *List of Lists, A Catalog of Analytes and Methods*, 21W-4005, US Govt. Printing Office, Washington, D.C., 1991.

280. L. H. Keith, *Environ. Sci. Technol.* **24**, 610 (1990).

281. L. H. Keith, *Principles of Environmental Sampling*, American Chemical Society, Washington, D.C., 1988.

282. L. H. Keith, *Practical Guide for Environmental Sampling and Analysis*, Lewis Publishers, Chelsea, Mich., 1990.

283. C. F. D'Elia, J. G. Sanders, and D. G. Capone, *Environ. Sci. Technol.* **23**, 768 (1989).

284. D. J. Munch and C. P. Frebis, *Environ. Sci. Technol.* **26**, 921 (1992).

285. J. D. Weete, in Ref. 241, 1986, pp. 219–245.

286. R. G. Nash, *J. Assoc. Off. Anal. Chem.* **73**, 438 (1990).

287. L. Q. Huang, *J. Assoc. Off. Anal. Chem.* **72**, 349 (1989).

288. M. D. Osselton and R. D. Snelling, *J. Chromatogr.* **368**, 265 (1986).

289. J. C. Touchstone and M. F. Dobbins, *Practice of Thin Layer Chromatography*, 2nd ed., John Wiley & Sons, Inc., New York, 1983.

290. R. O. Harrison and B. S. Ferguson, *Bull. Environ. Cont. Toxicol.* **46**, 283 (1991).

291. J. M. Van Emon and R. O. Mumma, *Immunochemical Methods for Environmental Analysis*, ACS Symposium Series 442, American Chemical Society, Washington, D.C., 1990.

292. M. Vanderlaan and co-workers, *Immunoassays for Trace Chemical Analysis: Monitoring Toxic Chemicals in Humans, Food and the Environment*, ACS Symposium Series 451, American Chemical Society, Washington, D.C., 1991.
293. C. S. Helling, *Soil Sci. Soc. Amer. Proc.* **35**, 737 (1971).
294. *Ibid.*, p. 743.
295. *STORET Water Quality File*, Office of Water, U.S. EPA, U.S. Govt. Printing Office, Washington, D.C., 1988.
296. W. T. Thomson, *Agricultural Chemicals, Book II Herbicides, 1989–1990 rev.*, Thomson Publishers, Fresno, Calif., 1989.
297. S. K. Rick, F. W. Slife, and W. L. Banwart, *Weed Sci.* **35**, 282 (1987).
298. R. Grover and A. E. Smith, *Can. J. Soil Sci.* **54**, 179 (1974).
299. W. P. Anderson, in Ref. 259, 1977, pp. 201–298.
300. R. G. Wilson and H. H. Cheng, *Weed Sci.* **24**, 461 (1976).
301. O. Fadayomi and G. F. Warren, *Weed Sci.* **25**, 97 (1977).
302. P. J. McCall, S. A. Vrona, and S. S. Kelley, *J. Agric Food Chem.* **29**, 100 (1981).
303. M. A. Loos, I. F. Schlosser, and W. R. Mapham, *Soil Biol. Biochem.* **11**, 377 (1979).
304. D. G. Crosby and J. B. Bowers, *J. Agric. Food Chem.* **33**, 569 (1985).
305. J. J. Pignatello, *Environ. Sci. Technol.* **26**, 944 (1992).
306. M. Negre and co-workers, *J Agric. Food Chem.* **36**, 1319 (1988).
307. A. Pusino and C. Gessa, *Pestic. Sci.* **30**, 211 (1990).
308. K-H. Oh and O. H. Tuovinen, *Bull. Environ. Contam. Toxicol.* **47**, 222 (1991).
309. WHO, *Environmental Health Criteria 84; 2,4-Dichlorophenoxyacetic Acid (2,4-D)-Environmental Aspects*, World Health Organization, Geneva, 1989.
310. U.S. EPA, Method 515.1–*Determination of Chlorinated Pesticide in Water by GC/ECD*, draft, Apr. 15, 1988; available from U.S. EPA Environmental Monitoring and Support Laboratory, Cincinnati, Ohio, 1988.
311. R. M. Menges and S. Tamez, *Weed. Sci.* **22**, 67 (1974).
312. U.S. EPA, Method 507–*Determination of Nitrogen- and Phosphorus Containing Pesticides in Ground Water by GC/NPD*, draft, Apr. 15, 1988; available from U.S. EPA Environmental Monitoring and Support Laboratory, Cincinnati, Ohio, 1988.
313. U.S. EPA, *Method 508-Determination of Chlorinated Pesticides in Ground Water by GC/ECD*, draft, Apr. 15, 1988, available from U.S. EPA Environmental Monitoring and Support Laboratory, Cincinnati, Ohio, 1988.
314. U.S. EPA, *Method 4–Determination of Pesticides in Ground Water by HPLC/UV*, draft, Jan. 1986; available from U.S. EPA Environmental Monitoring and Support Laboratory, Cincinnati, Ohio, 1986.
315. J. T. Majka and T. L. Lavy, *Weed Sci.* **25**, 401 (1977).
316. D. N. Roy and co-workers, *J. Agric. Food Chem.* **37**, 443 (1989).
317. J. R. Jotcham, D. W. Smith, and G. R. Stephenson, *Weed Technol.* **3**, 155 (1989).
318. R. W. Bovey and C. W. Richardson, *J. Environ. Qual.* **20**, 528 (1991).
319. A. J. Goetz and co-workers, *Weed Sci.* **37**, 428 (1989).
320. W. Mersie and C. L. Foy, *J. Agric. Food Chem.* **34**, 89 (1986).
321. E. G. Cotterill, *Pestic. Sci.* **34**, 291 (1992).
322. G. Wehtje and co-workers, *Weed Sci.* **35**, 858 (1987).
323. R. B. Wong and Z. H. Ahmed, *J. Agric. Food Chem.* **40**, 811 (1992).
324. A. J. Goetz and co-workers, *Weed Sci.* **34**, 788 (1986).
325. R. N. Stougaard, P. J. Shea, and A. R. Martin, *Weed Sci.* **38**, 67 (1990).
326. J. R. Abernathy and L. M. Wax, *Weed Sci.* **21**, 224 (1973).
327. P. Jamet and J.-C. Thoisy-Dur, *Bull. Environ. Contam. Toxicol.* **41**, 135 (1988).
328. S. S. Chang and J. F. Stritzke, *Weed Sci.* **25**, 184 (1977).
329. S. D. West, *J. Agric. Food Chem.* **26**, 644 (1978).

330. M. A. Gallina and G. R. Stephenson, *J. Agric. Food Chem.* **40**, 165 (1992).
331. P. Sprankle, W. F. Meggitt, and D. Penner, *Weed Sci.* **23**, 229 (1975).
332. M. A. Constenla and co-workers, *J. Agric. Food Chem.* **38**, 1985 (1990).
333. P. Slade, *Nature* **207**, 515 (1965).
334. P. C. Kearney and co-workers, *J. Agric. Food Chem.* **33**, 953 (1985).
335. R. J. G. Carr, R. F. Bilton, and T. Atkinson, *Appl. Environ. Microbiol.* **49**, 1290 (1985).
336. R. R. Yeo, *Weeds* **15**, 42 (1967).
337. S. B. Weed and J. B. Weber, in Ref. 238, 1974, pp. 39–66.
338. M. R. Murray and J. K. Hall, *J. Environ. Qual.* **18**, 51 (1989).
339. K. H. Reinert and J. H. Rodgers, Jr., *Bull. Environ. Contam. Toxicol.* **32**, 557 (1984).
340. R. Behrens and W. E. Lueschen, *Weed Sci.* **27**, 486 (1979).
341. L. J. Ross and co-workers, *J. Environ. Qual.* **19**, 715 (1990).
342. A. E. Smith, *J. Agric. Food Chem.* **22**, 601 (1974).
343. C. J. Scifres and co-workers, *J. Environ. Qual.* **2**, 306 (1973).
344. J. P. Krueger and co-workers, *J. Agric. Food Chem.* **37**, 534 (1989).
345. J. S. Choi and co-workers, *Agron. J.* **80**, 108 (1988).
346. K. E. Mcbride, J. W. Kenny, and D. M. Stalker, *Appl. Environ. Microbiol.* **52**, 325 (1986).
347. J. Kochany, G. G. Choudhry, and G. R. B. Webster, *Pestic. Sci.* **28**, 69 (1990).
348. P. Maini, *Pestic. Sci.* **34**, 45 (1992).
349. H. C. Sikka and C. P. Rice, *J. Agric. Food Chem.* **21**, 842 (1973).
350. C. P. Rice, H. C. Sikka, and R. S. Lynch, *J. Agric. Food Chem.* **22**, 533 (1974).
351. A. Rosenberg, *Bull. Environ. Contam. Toxicol.* **32**, 383 (1984).
352. C. T. Chiou, in Ref. 529, 1989, pp. 1–29.
353. J. J. Hasset and W. L. Banwart, in Ref. 247, 1989, pp. 31–44.
354. R. Grover, *Weed Sci.* **22**, 405 (1974).
355. C. J. Peter and J. B. Weber, *Weed Sci.* **33**, 861 (1985).
356. C-H. Wu and P. W. Santelmann, *Weed Sci.* **23**, 508 (1975).
357. E. Koren, *Weed Sci.* **20**, 230 (1972).
358. R. Grover and co-workers, *J. Environ. Qual.* **17**, 543 (1988).
359. C. C. Reyes and R. L. Zimdahl, *Weed Sci.* **37**, 604 (1989).
360. J. A. Poku and R. L. Zimdahl, *Weed Sci.* **28**, 650 (1980).
361. J. E. Nelson, W. F. Meggitt, and D. Penner, *Weed Sci.* **31**, 68 (1983).
362. R. L. Zimdahl and S. M. Gwynn, *Weed Sci.* **25**, 247 (1977).
363. K. A. Hurto, A. J. Turgeon, and M. A. Cole, *Weed Sci.* **27**, 154 (1979).
364. T. Golab and co-workers, *J. Agric. Food Chem.* **18**, 838 (1970).
365. C. R. Lemmon and H. M. Pylypiw, Jr., *Bull. Environ. Contam. Toxicol.* **48**, 409 (1992).
366. G. Kulshrestha and S. B. Singh, *Bull. Environ. Contam. Toxicol.* **48**, 269 (1992).
367. R. L. Zimdahl, P. Catizone, and A. C. Butcher, *Weed Sci.* **32**, 408 (1984).
368. A. S. Barua and co-workers, *Pestic. Sci.* **29**, 419 (1990).
369. J. B. Weber and C. J. Peter, *Weed Sci.* **30**, 14 (1982).
370. B. T. Bowman, *Environ. Toxicol. Chem.* **9**, 452 (1990).
371. A. Walker, Y.-H. Moon, and S. J. Welch, *Pestic. Sci.* **35**, 109 (1992).
372. C. J. Somich and co-workers, *J. Agric. Food Chem.* **36**, 1322 (1988).
373. R. Y. Yih, C. Swithenbank, and D. H. McRae, *Weeds* **18**, 604 (1970).
374. S.-Y. Liu, R. Zhang, and J.-M. Bollag, Biol. Fertil. Soils **5**, 276 (1988).
375. A. Saxena, R. Zhang, and J.-M. Bollag, *Appl. Environ. Micrbiol.* **53**, 390 (1987).
376. D. T. Villarreal, R. F. Turco, and A. Konopka, *Appl. Environ. Microbiol.* **57**, 2135 (1991).

377. A. Dahchour and co-workers, *Bull. Environ. Contam. Toxicol.* **36**, 556 (1986).
378. E. Avidov, N. Aharonson, and J. Katan, *Weed Sci.* **38**, 186 (1990).
379. C. I. Harris and T. J. Sheets, *Weeds* **13**, 215 (1965).
380. J. L. Mutz and co-workers, *Weed Sci.* **26**, 179 (1978).
381. J. C-Y. Han, *J. Agric. Food Chem.* **27**, 564 (1979).
382. J. C-Y. Han and R. L. Krause, *Soil Sci.* **128**, 23 (1979).
383. D. Vega, J. Bastide, and C. Coste, *Soil Biol. Biochem.* **4**, 541 (1985).
384. R. A. Gray and A. J. Weierich, *Weeds* **16**, 77 (1968).
385. M. P. Braverman and co-workers, *Weed Sci.* **38**, 607 (1990).
386. E. S. C. Kwok, R. Atkinson, and J. Arey, *Environ. Sci. Technol.* **26**, 1798 (1992).
387. M. P. Braverman and co-workers, *Weed Sci.* **38**, 583 (1990).
388. V. W. McCusker and co-workers, *Weed Sci.* **36**, 818 (1988).
389. R. M. Behki, *Soil Biol. Biochem.* **23**, 789 (1991).
390. R. E. Talbert and O. H. Fletchall, *Weeds* **13**, 46 (1965).
391. O. K. Borggaard and J. C. Streibig, *Acta Agric. Scand.* **38**, 293 (1988).
392. P. C. Kearney, T. J. Sheets, and J. W. Smith, *Weeds* **12**, 83 (1964).
393. E. Pelizzetti and co-workers, *Environ. Sci. Technol.* **24**, 1559 (1990).
394. R. M. Behki and S. U. Khan, *J. Agric. Food Chem.* **34**, 746 (1986).
395. D. D. Kaufman and J. Blake, *Soil Biol. Biochem.* **2**, 73 (1970).
396. M. R. Blumhorst and J. B. Weber, *J. Agric. Food Chem.* **40**, 894 (1992).
397. T. B. Moorman and S. S. Harper, *J. Environ. Qual.* **18**, 302 (1989).
398. C.-H. Wu, P. W. Santelmann, and J. M. Davidson, *Weed Sci.* **22**, 571 (1974).
399. R. G. Lehmann, J. R. Miller, and D. A. Laskowski, *Weed Res.* **30**, 383 (1990).
400. A. E. Smith and A. J. Aubin, *Bull. Environ. Contam. Toxicol.* **42**, 670 (1989).
401. J. D. Fryer, P. D. Smith, and J. W. Ludwig, *J. Environ. Qual.* **8**, 83 (1979).
402. G. R. Stephenson and co-workers, *J. Agric. Food Chem.* **38**, 584 (1990).
403. P. J. McCall and P. D. Gavit, *Environ. Toxicol. Chem.* **5**, 879 (1986).
404. C. W. Hubbs and T. L. Lavy, *Weed Sci.* **38**, 81 (1990).
405. D. T. Smith and W. F. Meggitt, *Weeds* **18**, 255 (1970).
406. *Ibid.*, p. 260.
407. K. Thirunarayanan, R. L. Zimdahl, and D. E. Smika, *Weed Sci.* **33**, 558 (1985).
408. A. Walker, E. G. Cotterill, and S. J. Welch, *Weed Res.* **29**, 281 (1989).
409. J. P. Cambon, S. Q. Zheng, and J. B. Gerap, *Weed Res.* **32**, 1 (1992).
410. J. Sabadie, *Weed Res.* **30**, 413 (1990).
411. M. M. Joshi, H. M. Brown, and J. A. Romesser, *Weed Sci.* **33**, 888 (1985).
412. T. P. Fuesler and M. K. Hanafey, *Weed Sci.* **38**, 256 (1990).
413. M. M. Loux and K. D. Reese, *Weed Sci.* **40**, 490 (1992).
414. J. R. Cantwell, R. A. Liebl, and F. W. Slife, *Weed Sci.* **37**, 815 (1989).
415. G. W. Basham and T. L. Lavy, *Weed Sci.* **35**, 865 (1987).
416. L. J. Ross and co-workers, *J. Environ. Qual.* **18**, 105 (1989).
417. A. Mamouni and co-workers, *Pestic. Sci.* **35**, 13 (1992).
418. J. R. Plimmer and co-workers, *J. Agric. Food Chem.* **15**, 996 (1967).
419. W. Koskinen, *Weed Sci.* **32**, 273 (1984).
420. A. K. Alva and M. Singh, *Bull. Environ. Contam. Toxicol.* **45**, 365 (1990).
421. K. N. Reddy, M. Singh, and A. K. Alva, *Bull. Environ. Contam. Toxicol.* **48**, 662 (1992).
422. H. W. Hilton and Q. H. Yuen, *J. Agric. Food Chem.* **11**, 230 (1963).
423. N. R. Adrian and J. M. Suflita, *Appl. Environ. Microbiol.* **56**, 292 (1990).
424. L. Mapplebeck and C. Waywell, *Weed Sci.* **31**, 8 (1983).
425. Y. Y. Spiridonov and co-workers, *Sov. Soil Sci.* **22**, 61 (1990).

426. W. E. Emmerich and co-workers, *J. Environ. Qual.* **13**, 382 (1984).

427. K. H. Reinert and J. H. Rodgers, *Rev. Environ. Contam. and Toxicol.* **98**, 61 (1987).

428. R. Lignell, H. Heinonen-Tanski, and A. Uusi-Rauva, *Acta Agric. Scand.* **34**, 3 (1984).

429. R. Dickens and A. E. Hiltbold, *Weed Sci.* **15**, 299 (1967).

430. A. E. Hiltbold, B. F. Hajek, and G. A. Buchanan, *Weed Sci.* **22**, 272 (1974).

431. A. S. Crafts, *Hilgardia* **16**, 483 (1945).

432. E. A. Woolson and R. Isensee, *Weed Sci.* **29**, 17 (1981).

433. K. H. Akkari, R. E. Frans, and T. L. Lavy, *Weed Sci.* **34**, 781 (1986).

434. D. W. Von Endt, P. C. Kearney, and D. D. Kaufman, *J. Agric. Food Chem.* **16**, 17 (1968).

435. R. Iadevaia, N. Aharonson, and E. A. Woolson, *J. Assoc. Off. Anal. Chem.* **63**, 742 (1980).

436. P. J. Shea and J. B. Weber, *Weed Sci.* **31**, 528 (1983).

437. J. Schroeder and P. A. Banks, *Weed Sci.* **34**, 612 (1986).

438. L. L. Chang, K.-S. Lee, and C. K. Tseng, *J. Agric. Food Chem.* **36**, 844 (1988).

439. J. S. McConnell and L. R. Hossner, *J. Agric. Food Chem.* **33**, 1075 (1985).

440. D. N. Roy and co-workers, *J. Agric. Food Chem.* **37** (1989).

441. G. S. Jacob and co-workers, *Appl. Environ. Microbiol.* **54**, 2953 (1988).

442. T. R. Roberts and M. E. Standen, *Pestic. Biochem. Physiol.* **9**, 322 (1978).

443. D. R. Wendt, J. B. Weber, and T. J. Monaco, *Proc. South. Weed Sci. Soc.* **40**, 391 (1987).

444. M. L. Loux, R. A. Liebl, and F. W. Slife, *Weed Sci.* **37**, 440 (1989).

445. E. R. Gallandt, P. K. Fay, and W. P. Inskeep, *Weed Technol.* **3**, 146 (1989).

446. L. N. Falb, D. C. Bridges, and A. E. Smith, Jr., *J. Agric. Food Chem.* **38**, 875 (1990).

447. D. McAuliffe and A. P. Appleby, *Weed Sci.* **32**, 468 (1984).

448. I. Yamaguchi, *Japan Pesticide Information* **50**, 17 (1987).

449. C. A. Edwards, in V. Delucchi, ed., *Integrated Pest Management, Protection Integee, Quo Vadis? An International Perspective*, Parasitis, Geneva, 1987, pp. 309–329.

450. M. Konnai, *Japan Pesticide Information* **50**, 10 (1987).

451. M. A. Altieri and M. Liebman, eds., *Weed Management in Agroecosystems: Ecological Approaches*, CRC Press, Boca Raton, Fla., 1988.

452. J. A. Duke, *Edible Weeds*, CRC Press, Boca Raton, Fla, 1992.

453. P. B. Johnsen and C. P. Dionigi, *J. Appl. Aquacult.* **2**(314) (1993).

454. R. L. Zimdahl, in Ref. 451, p. 145.

455. N. Jordan, *Weed Technol.* **6**, 184 (1992).

456. M. A. Altieri, *Agroecology, The Scientific Basis of Alternative Agriculture*, Division of Biological Control, University of California, Berkeley, 1983.

457. D. C. Thill and co-workers, *Weed Technol.* **5**, 648 (1991).

458. M. Barrett and W. W. Witt, *Energy in World Agriculture* **2**, 197 (1987).

459. R. W. Bovey, *Rev. Weed Sci.* **3**, 57 (1987).

460. S. O. Duke, *Rev. Weed Sci.* **2**, 15 (1986).

461. S. O. Duke, *Abstr. Pap. Am. Chem. Soc.*, 1–3 (1992).

462. J. A. Duke, *Handbook of Biologically Active Phytochemicals and Their Activities*, CRC Press, Boca Raton, Fla, 1992.

463. B. G. Isaac and co-workers, *J. Antibiotics* **44**, 729 (1991).

464. H. G. Cutler, *Biologically Active Natural Products, Potential Use in Agriculture*, ACS Symposium Series 380, American Chemical Society, Washington, D.C., 1988.

465. N. Porter, *Pestic. Sci.* **16**, 422 (1985).

466. P. A. Hedin, *Use of Natural Products in Pest Control, Developing Research Trends*, ACS Symposium Series 449, American Chemical Society, Washington, D.C., 1991, pp. 1–11.

467. S. O. Duke and J. Lydon, *Weed Technol.* **1**, 122 (1987).
468. T. Fujita, *Japan Pesticide Information* **50**, 12 (1987).
469. N. Takahashi, *Japan Pesticide Information* **50**, 18 (1987).
470. H. G. Cutler, *Crit. Rev. Plant Sci.* **6**, 323 (1988).
471. A. C. Thompson, ed., *The Chemistry of Allelopathy: Biochemical Interactions Among Plants*, ACS Symposium Series 268, American Chemical Society, Washington, D.C., 1985.
472. A. R. Putnam and C-S. Tang, *The Science of Allelopathy*, John Wiley & Sons, Inc., New York, 1986.
473. J. R. Qasem and T. A. Hill, *Weed Res.* **29**, 345 (1989).
474. A. R. Putnam and C-S. Tang, in Ref. 472, p. 1 (1986).
475. F. W. Welsh, W. D. Murray, and R. E. Williams, *Crit. Rev. Biotechnol.* **9**, 105 (1989).
476. R. Charudattan, *Fitopatol. Bras.* **15**, 13 (1990).
477. G. E. Templeton, R. J. Smith, and D. O. TeBeest, *Rev. Weed Sci.* **2**, 1 (1986).
478. G. E. Templeton, *Am. J. Alternative Agric.* **3**, 69 (1988).
479. M. P. Greaves and M. D. Macqueen, *Aspects Appl. Biol.* **17**, 417 (1988).
480. K. K. Hatzios, *Adv. Agron.* **41**, 325 (1987).
481. M. P. Greaves, J. A. Bailey, and J. A. Hargreaves, *Pestic. Sci.* **26**, 93 (1989).
482. R. E. Hoagland, ed., *Microbes and Microbial Products as Herbicides*, ACS Symposium Series 439, American Chemical Society, Washington, D.C., 1990, pp. 88–113.
483. A. J. Wapshere, E. S. Delfosse, and J. M. Cullen, *Crop Protection*, **8**, 227 (1989).
484. H. C. Evans, in R. A. Samson, J. M. Vlak, and D. Peters, eds., *Fundamental and Applied Aspects of Invertebrate Pathology*, Proc. 4th International Colloquium of Invertebrate Pathology, Veldhoven, Netherlands, 1986, pp. 475–480.
485. H. C. Evans, *Biocontrol News and Information* **8**, 7 (1987).
486. G. B. Dennill and H. M. T. Hokkanen, *Agric. Ecosyst. Environ.* **33**, 1 (1990).
487. D. O. TeBeest and G. E. Templeton, *Plant Dis.* **1**, 6 (1985).
488. M. J. Whitten and J. G. Oakeshott, *FAO Plant Protec. Bull.* **39**, 155 (1991).
489. D. Pimentel, *Crop Protection* **10**, 243 (1991).
490. J. M. Cullen and S. Hasan, *Phil. Trans. R. Soc. Lond. B* **318**, 213 (1988).
491. K. Mortensen, *Can. J. Plant Path.* **8**, 229 (1986).
492. M. J. Morris, *Agric. Ecosyst. Environ.* **37**, 239 (1991).
493. G. H. Egley, *Rev. Weed Sci.* **2**, 67 (1986).
494. R. B. Taylorson, *Rev. Weed Sci.* **3**, 135 (1987).
495. J. M. Baskin and C. C. Baskin, *Bioscience* **35**, 492 (1985).
496. G. E. Brust and G. J. House, *Am. J. Alternative Agric.* **3**, 19 (1988).

General References

Weed Control Manual, Meister Publishing Co., Willoughby, Ohio, 1992 and later updates.
Farm Chemicals Handbook '93, Meister Publishing Co., Willoughby, Ohio, 1993 and later updates.
Labels, books, and sales literature from herbicide manufacturers.
Herbicide Handbook, 6th ed., Weed Science Society of America, Champaign, Ill., 1989.

JUDITH M. BRADOW
CHRISTOPHER P. DIONIGI
RICHARD M. JOHNSON
SUHAD WOJKOWSKI
United States Department of Agriculture

HEXAMETHYL PHOSPHORIC TRIAMIDE.   See Phosphorus Compounds, Organic.

HEXANES.   See Hydrocarbons.

HIGH PERFORMANCE FIBERS

High performance fibers are generally characterized by remarkable unit tensile
strength and resistance to heat, flame, and chemical agents that normally
degrade conventional fibers. Applications include uses in the aerospace, biomedical, civil engineering, construction, protective apparel, geotextiles, and electronic areas.

For many years, plastics reinforced with polymer fibers have been utilized in
the manufacture of boats and sports cars. More recently, ultrahigh strength, high
modulus fibers have been invented and combined into composites whose strength
and stiffness on a specific basis are unmatched by conventional construction
materials. Composites are now replacing metals in such crucial applications
as aircraft and the space shuttle. The polymeric composites contain carbon
or aramid fibers several times stiffer, weight for weight, than steel embedded
in plastics. In composite materials (qv) the fibers support the load which is
distributed by the plastic which also prevents fatigue and failure.

In addition to their role in composites, high performance fibers are also
found in coated and laminated textile products, three-dimensional fabric structures, multifunctional property improvement, and intelligent or self-adaptive
materials.

In this article the preparation and properties of typical high performance
fibers are discussed, then their applications are classified and detailed.

Preparation and Properties

The principal classes of high performance fibers are derived from rigid-rod
polymers, gel spun fibers, modified carbon fibers, synthetic vitreous fibers, and
poly(phenylene sulfide) fibers.

RIGID-ROD POLYMERS

Rigid-rod polymers are often liquid crystalline polymers classified as lyotropic,
such as the aramids Nomex and Kevlar, or thermotropic liquid crystalline polymers, such as Vectran.

Liquid Crystallinity.   The liquid crystalline state is characterized by ori-
entationally ordered molecules. The molecules are characteristically rod- or

lathe-shaped and can exist in three principal structural arrangements: nematic, cholesteric, and smectic (see LIQUID CRYSTALLINE MATERIALS).

In the nematic phase, within volume elements of the macroscopic sample, the axes of the molecules are oriented on average in a specific direction in various domains. The centers of gravity of the molecules are arranged in a random fashion, and consequently no positional long range order exists. The molecules are arranged in essentially parallel arrays. Without the presence of an orienting magnetic or physical force, the molecules exist in random parallel arrays. When an orienting force is applied, these domains orient easily. The nematic phase is amenable to translational mobility of constituent molecules.

The cholesteric phase may be considered a modification of the nematic phase since its molecular structure is similar. The cholesteric phase is characterized by a continuous change in the direction of the long axes of the molecules in adjacent layers within the sample. This leads to a twist about an axis perpendicular to the long axes of the molecules. If the pitch of the helical structure is the same as a wavelength of visible light, selective reflection of monochromatic light can be observed in the form of iridescent colors.

In the smectic phase, the centers of gravity of the rod-like molecules are arranged in equidistant planes, ie, the ends of the molecules are correlated. The planes may move perpendicular to the layer normal, and within layers different arrangements of the molecules are possible. The long axes of the molecules may be parallel to the layer normal or tilted with respect to it. A two-dimensional short or long range order may exist within the smectic layers. The smectic modifications are labeled according to the arrangement of the molecules within the layers using the symbols A–H.

Poly(1,4-benzamide) (PBA) (1) was the first nonpeptide synthetic polymer reported to form a liquid crystalline solution. In order to obtain liquid crystalline solutions of poly(1,4-benzamide), it was first necessary to prepare the polymer in the proper solvent. Preparation of the polymer in N,N-dialkylamide solvents at low temperatures from p-aminobenzoyl chloride hydrochloride produces tractable PBA polymers with inherent viscosities of as much as 5 dL/g. In solvents such as N,N-dimethylacetamide [127-19-5] and N,N,N',N'-tetramethylurea [632-22-4] a coupled polymerization-spinning process in liquid crystalline solution has been developed. If polymerization is initiated at temperatures greater than 25°C, lower molecular weight polymer is formed. Above 25°C, chain termination by reaction of acid chloride chain ends with N,N-dialkylamide is significant. To obtain high molecular weights, a lithium base such as lithium hydride, lithium carbonate, or lithium hydroxide is added to the polymerization solution after the first 1–2 h of reaction time to neutralize the hydrogen chloride generated. As the reaction proceeds, the polymerization rate decreases because the increasing amounts of hydrogen chloride consequently produce fewer free-terminal amine groups.

When pure needle-like crystals of p-aminobenzoyl chloride are polymerized in a high temperature, nonsolvent process, or a low temperature, slurry process, polymer is obtained which maintains the needle-like appearance of monomer. PBA of inherent viscosity, 4.1 dL/g, has been obtained in a hexane slurry with pyridine as the acid acceptor. Therefore PBA of fiber-forming molecular weight can be prepared in the solid state.

In 1975, the synthesis of the first main-chain thermotropic polymers, three polyesters of 4,4′-dihydroxy-α,α′-dimethylbenzalazine with 6, 8, and 10 methylene groups in the aliphatic chain, was reported (2). Shortly thereafter, at the Tennessee Eastman Co. thermotropic polyesters were synthesized by the acidolysis of poly(ethylene terephthalate) by p-acetoxybenzoic acid (3). Copolymer compositions that contained 40–70 mol % of the oxybenzoyl unit formed anisotropic, turbid melts which were easily oriented.

Polyesters such as poly(p-phenylene terephthalate), which would be expected to form liquid crystalline phases, decompose at temperatures below the melting point. Three principal methods have been used for lowering the melting temperatures of thermotropic copolyesters: (1) the use of flexible groups as spacers to decouple the mesogenic units and reduce the axial ratio; (2) the use of unsymmetrical groups on mesogenic units; and (3) the copolymerization of rigid units with nonlinear, bent units which add a "kink" to the rod-like system.

According to patents obtained by Carborundum, Celanese, Du Pont, and Eastman (4) most industrial main-chain thermotropics are prepared by condensation polymerization involving transesterification. Hydroxy-substituted monomers are acetylated before polymerization by acetic anhydride in the presence of a suitable catalyst. The transesterification reactions involve acetylated diol, or monosubstituted hydroxybenzoic or hydroxynaphthoic acids (Vectran), and diacids. The polymerizations are carried out in an inert atmosphere to prevent oxidation. A stainless steel stirrer is utilized to improve mixing and to accelerate the release of the reaction by-products. The polymerizations are carried out at 50 to 80°C above the melting point of the highest melting monomer. After a low melt viscosity prepolymer is obtained, a vacuum is applied to remove the additional acetic acid and increase the molecular weight of the polymer. Finally, solid-state polymerization under reduced pressure or in nitrogen at a temperature of 10 to 30°C below the melting point may be utilized to increase the molecular weight. The heat treatment of spun fibers under these conditions leads to spectacular increases in tensile strength and modulus.

Researchers at Du Pont used hydroquinone asymmetrically substituted with chloro, methyl, or phenyl substituents and swivel or nonlinear bent substituted phenyl molecules such as 3,4′- or 4,4′-disubstituted diphenyl ether, sulfide, or ketone monomers. For example, poly(chloro-1,4-phenylene-trans-hexahydroterephthalate) and related copolymers were prepared in a melt polymerization process involving the reaction of molar equivalents of the diacetoxy derivatives of diphenols and hexahydroterephthalic acid (5). During polymerization, a phase transition from isotropic to anisotropic occurred soon after the rapid melting of the intermediates to form a clear, colorless liquid.

Also in 1972 (6), Carborudum researchers described a family of aromatic copolyesters which were recognized later to form liquid crystalline melts. The polymers are based on a bisphenol monomer. In 1976, in a patent assigned to Carborundum, a hydroxybenzoic acid–terephthalic acid–bisphenol system, modified and softened with isophthalic acid, was reported to be melt spinnable to produce fiber.

Industrial Lyotropic Liquid Crystalline Polymers (Aramid Fibers). The first polyaramid fiber (MPD-1) was based on poly(m-phenylene isophthalamide) [24938-60-1]. The fiber was not liquid crystalline but was the first aramid fiber to be commercialized by Du Pont under the trademark Nomex in 1967. The principal market niche for Nomex was as a heat-resistant material. Teijin also introduced a fiber (trademark Conex) based on MPD-1 in the early 1970s. Fenilon, also based on MPD-1, was produced in the former USSR for civilian, military, and space exploration applications. In 1970, Du Pont introduced an aramid fiber, Fiber B, for use in tires, which was probably based on polybenzamide PBA spun from an organic solvent. Fiber B had high strength and exceptionally high modulus. Another version of Fiber B, based on poly(p-phenylene terephthalamide) [24938-64-5] (PPT) was introduced in the 1970s. This version of Fiber B was spun from sulfuric acid and had a tensile strength approximately twice that of the Fiber B based on MPD-1. An even higher modulus fiber based on PPT, in which the modulus was increased by the drawing of the as-spun fiber, was introduced under the name PRD-49 for use in rigid composites. The undrawn and drawn fibers were later announced as Kevlar-29 and Kevlar-49, respectively. In 1975, Akzo of the Netherlands reported the commercialization of an aramid fiber, Twaron, based on PPT.

Nomex.   This fiber was commercialized for applications requiring unusually high thermal and flame resistance. Nomex fiber retains useful properties at temperatures as high as 370°C. Nomex has low flammability and has been found to be self-extinguishing when removed from the flame. On exposure to a flame, a Nomex fabric hardens, starts to melt, discolors, and chars thereby forming a protective coating. Therefore an outstanding characteristic is low smoke generation on burning. The limiting oxygen index (LOI) value (top down) for Nomex fabrics is 26.0. Nomex has a tga weight loss of 10% at 450°C and a use temperature of 370°C. Nomex has good to excellent strength, a tenacity of 0.42–0.51 N/tex (4.8–5.8 gf/den), good extendability, and a modulus greater than that of nylon-6,6. The density is 1.38 g/cm^3. Nomex is more difficult to dye than nylon, but the use of dye carriers (qv) allows dyeing to proceed at high temperatures with temperature-resistant basic dyes. The structure of Nomex may be represented as follows.

MPD-1 fibers may be obtained by the polymerization of isophthaloyl chloride and *m*-phenylenediamine in dimethylacetamide with 5% lithium chloride. The reactants must be very carefully dried since the presence of water would upset the stoichiometry and lead to low molecular weight products. Temperatures in the range of 0 to −40°C are desirable to avoid such side reactions as transamidation by the amide solvent and acylation of *m*-phenylenediamine by the amide solvent. Both reactions would lead to an imbalance in the stoichiometry and result in forming low molecular weight polymer. Fibers are dry spun directly from solution.

Kevlar.   In the 1970s, researchers at Du Pont reported that the processing of extended chain all para-aromatic polyamides from liquid crystalline solutions produced ultrahigh strength, ultrahigh modulus fibers. The greatly increased order and the long relaxation times in the liquid crystalline state compared to conventional systems led to fibers with highly oriented domains of polymer molecules. The most common lyotropic aramid fiber is poly(*p*-phenylene terephthalamide) (PPT) which is marketed as Kevlar by Du Pont. Aramid fiber is available from Akzo under the trade name Twaron. These fibers are used in body armor, cables, and composites for sports and space applications. Kevlar has the following structure:

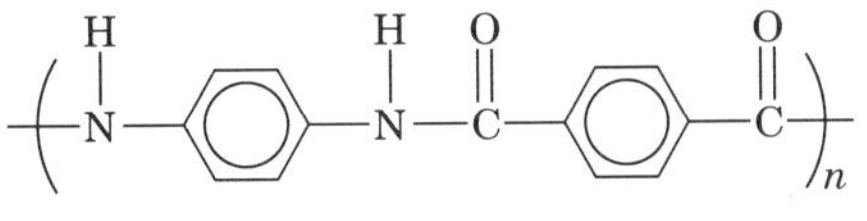

PPT of high molecular weight (inherent viscosity of 22 dL/g, corresponding to a molecular weight of 123,000) can be prepared by low temperature polymerization in various solvents (7). PPT is less soluble in amide solvents than PBA and the most successful polymerization solvents are a mixture of hexamethylphosphoramide [*680-31-9*] (HMPA) and *N*-methylpyrrolidinone [*872-50-4*] (NMP) or NMP-containing calcium chloride. These solvent systems yield fiber-forming polymer. As the molecular weight increases rapidly during the first few seconds of the polymerization, the critical concentration is exceeded and the solution develops the opalescence characteristic of the liquid crystalline state. The critical factors influencing the molecular weight include stoichiometry, solvent composition, temperature, and solids concentration. At low monomer concentrations, side reactions can occur between the acid chloride chain ends and the amide solvents. At higher solids concentrations, gelation acts to limit the development of high molecular weights. It is of critical importance to keep the initial temperature low in order to prevent the reaction of the amide solvents with the acid chloride groups.

The preparation of high molecular weight PPT in HMPA/NMP shows a strong dependence of inherent viscosity on reactant concentrations. In 2:1 (by volume) HMPA/NMP, the highest inherent viscosity polymer is obtained when each reactant is present in concentrations of ca 0.25 M; higher and lower concentrations result in the formation of polymer of lower inherent viscosities. A typical procedure is as follows: 1,4-phenylenediamine, HMPA, and NMP are added to an oven-dried resin kettle equipped with a stirrer and stirred for ca 15 min with cooling to −15°C, followed by the addition of powdered terephthaloyl chloride to the rapidly stirred solution. The reaction mixture changes to a thick, opalescent, paste-like gel in ca 5 min.

The manufacturing process utilizes continuous polymerization in order to minimize cost. A continuous stream of *p*-phenylenediamine solution is impinged on a continuous stream of molten terephthaloyl chloride. Volumetric control is easily achieved because both reactants are in the liquid state. Residence time in the mixing apparatus is on the order of 1 s. Next, the reactants enter a high shear, continuous screw mixer, in which the inherent viscosity of the polymer increases to 4–4.5 dL/g. The minimum inherent viscosity required for fiber spinning in sulfuric acid is 4 dL/g. The residence time is less than 15 s so the polymer solution which enters the third stage is still a fluid. The third stage is a high shear, twin-screw mixer with blades positioned for a number of recycle zones within the mixer, thereby achieving lower temperatures, higher residence times, and higher molecular weights.

An alternative polymerization process utilizes a slurry of calcium chloride in NMP as the polymerization medium. The solubility of calcium chloride is only 6% at 20°C; however, the salt continues to dissolve as conversion of monomers to polymer proceeds and calcium chloride/polyamide complexes are formed. Polymer molecular weight is further increased by the addition of N,N-dimethylaniline as an acid acceptor. This solvent system produces fiber-forming polymer of molecular weights comparable to that formed in HMPA/NMP.

Since PPT melts with decomposition at ca 560°C, melt spinning cannot be employed. Thus, solution spinning techniques must be used to prepare fibers. Although dry, wet, and dry jet-wet spinning methods have all been used to prepare fibers, ordinarily PPT is spun from nematic sulfuric acid solutions using the dry jet-wet spinning process with cold water as the coagulant. In the dry spinning process, a polymer solution is passed through a spinnerette followed by flash evaporation of the solvent in a heated chamber and subsequent winding of the fiber produced on a bobbin. In the wet spinning process, the polymer solution is passed through a spinnerette located in a coagulation bath. The fiber formed is then drawn to increase molecular orientation, tenacity, and modulus. In dry jet-wet spinning, the polymer solution is allowed to flow through a spinnerette into a separated coagulation bath. Therefore the temperatures of the spinnerette and coagulation baths may be independently controlled. The liquid crystalline nature of the PPT dopes and the dry jet-wet spinning technology are principally responsible for the development of commercial high performance Kevlar fibers.

Drawdown of the coagulated fiber is an essential element in high performance fiber technology. Under shear, the unoriented domains become oriented in the direction of stretch. In the fiber manufacturing process, the unoriented liquid crystalline domains are oriented in the spinnerette, followed by disorientation as the dope exits the capillary (causing die swell) and subsequent formation of a high degree of order during subsequent attenuation. Coagulation sets the high degree of orientational order achieved by stretching. Tenacity increases with increasing drawdown and inherent viscosity as well as decreasing air gap (between the spinnerette and coagulation bath). Modulus increases with increasing drawdown and total spinning strain.

In a typical commercial dry jet-wet spinning process, PPT polymer of inherent viscosity 6.0 dL/g is added to 99.7% sulfuric acid in a water-jacketed commercial mixer in a ratio of 46 g of polymer to 100 mL of acid. The mixture is sealed in a vacuum of 68.5–76 mL of mercury. Mixing takes place for two hours at temperatures of 77 to 85°C. The dope is then transferred to a glass-lined, water-jacketed kettle at 90°C. Any air or bubbles caused by the transfer are removed under vacuum for about 30 min. The dope is then pumped through a heated (90°C) transfer line to an electrically heated spinning block with an associated gear pump. The gear pump then meters the dope through a heated (80°C) 1.25 cm diameter spinnerette containing 100 holes of 51 μm diameter. The dope is extruded from the spinnerette at a velocity of ca 63 m/min vertically through a 0.5-cm layer of air (air gap) into water at a temperature of 1°C. The yarn is wound on a bobbin under a 50°C water spray. The bobbin is then submerged in 0.1 N NaHCO$_3$ solution and then further extracted with water at 70°C.

Dramatic increases in the mechanical properties of aramid fibers are observed on heat treatment under tension. Tenacity and modulus increase exponentially with increasing temperature (and draw ratios) of wet spun fibers at temperatures of ca 360°C (the glass-transition temperature, T_g) to 550°C (the melting temperature). Heat treatment of dry jet-wet spun yarns under tension show substantial increases in modulus at temperatures greater than 200°C; the already high values of tenacity remain essentially unchanged. An intermediate modulus, high tenacity dry jet-wet spun yarn is thus converted into high modulus, high strength fiber.

Because the inherent viscosities of the heated yarns remain constant, it is postulated that the changes are physical. Yarns with as-spun moduli of 8.8–88 N/tex (100–1000 gf/den) may be obtained directly by dry jet-wet spinning. Yarns with as-spun tenacities of greater than 1.8 N/tex (20 gf/den) are obtained by dry jet-wet spinning. Kevlar-29 has a tenacity of ca 2.5 N/tex (28 gf/den) and a specific modulus of ca 41 N/tex (464 gf/den). Kevlar-49 has a tenacity of ca 2.5 N/tex (28 gf/den) and a specific modulus of ca 86 N/tex (980 gf/den). A relatively new fiber, Kevlar-149, is the highest tensile modulus aramid fiber currently available. Its specific modulus is ca 126 N/tex (1430 gf/den) and tenacity ca 2.3 N/tex (26 gf/den).

The crystal structure of PPT is pseudo-orthorhombic (essentially monoclinic) with $a = 0.785/nm$; $b = 0.515/nm$; c (fiber axis) $= 1.28/nm$; and $\partial = 90°$. The molecules are arranged in parallel hydrogen-bonded sheets. There are two chains in a unit cell and the theoretical crystal density is 1.48 g/cm^3. The observed fiber density is 1.45 g/cm^3. An interesting property of the dry jet-wet spun fibers is the lateral crystalline order. Based on electron microscopy studies of peeled sections of Kevlar-49, the supramolecular structure consists of radially oriented crystallites. The fiber contains a pleated structure along the fiber axis, with a periodicity of 500–600 nm.

Technora. In 1985, Teijin Ltd. introduced Technora fiber, previously known as HM-50, into the high performance fiber market. Technora is based on the 1:1 copolyterephthalamide of 3,4′-diaminodiphenyl ether and *p*-phenylenediamine (8). Technora is a wholly aromatic copolyamide of PPT, modified with a crankshaft-shaped comonomer, which results in the formation of isotropic solutions that then become anisotropic during the shear alignment during spinning. The polymer is synthesized by the low temperature polymerization of *p*-phenylenediamine, 3,4′-diaminophenyl ether, and terephthaloyl chloride in an amide solvent containing a small amount of an alkali salt. Calcium chloride or lithium chloride is used as the alkali salt. The solvents used are hexamethylphosphoramide (HMPA), *N*-methyl-2-pyrrolidinone (NMP), and dimethylacetamide (DMAc). The structure of Technora is as follows:

The polymerization is carried out at temperatures of 0–80°C in 1–5 h at a solids concentration of 6–12%. The polymerization is terminated by neutralizing agents such as calcium hydroxide, calcium oxide, calcium carbonate, or lithium hydroxide. Inherent viscosities of 2–4 dL/g are obtained at 3,4'-diaminodiphenyl ether contents of 35–50 mol %. Because of the introduction of nonlinearity into the PPT chain by the inclusion of 3,4'-diaminodiphenyl ether kinks, the copolymer shows improved tractability and may be wet or dry jet-wet spun from the polymerization solvent. The fibers are best coagulated in an aqueous equilibrium bath containing less than 50 vol % of polymerization solvent and from 35 to 50% of calcium chloride or magnesium chloride.

The copolymer fiber shows a high degree of drawability. The spun fibers of the copolymer were highly drawn over a wide range of conditions to produce fibers with tensile properties comparable to PPT fibers spun from liquid crystalline dopes. There is a strong correlation between draw ratio and tenacity. Typical tenacity and tensile modulus values of 2.2 N/tex (25 gf/den) and 50 N/tex (570 gf/den), respectively, have been reported for Technora fiber (8).

Heterocyclic Rigid-Rod Polymers. *PBO and PBZT.* PBZ, a family of *p*-phenylene–heterocyclic rigid-rod and extended chain polymers includes poly(*p*-phenylene-2,6-benzobisthiazole) [69794-31-6] (*trans*-PBZT) and poly(*p*-phenylene-2,6-benzobisoxazole) [60871-72-9] (*cis*-PBO). PBZT and PBO were initially prepared at the Air Force Materials Laboratory at Wright-Patterson Air Force Base, Dayton, Ohio. PBZT was prepared by the reaction of 2,5-diamino-1,4-benzenedithiol dihydrochloride with terephthalic acid in polyphosphoric acid (PPA) and PBO by the reaction of 4,6-diamino-1,3-benzenediol dihydrochloride with terephthalic acid in PPA.

Although the crystal structures of the 2,6-diphenyl-*cis*- and *trans*-benzobisoxazole compounds have colinear exocyclic bonds with the coplanar condensed rings, and the phenyl rings coplanar with the heterocycles, the central ring of the 2,6-diphenyl-*cis*-benzobisthiazole system is bent. The exocyclic bonds of the 2,6-diphenyl-*cis*-benzobisthiazole system are bent out of linearity. The central, condensed ring system of 2,6-diphenyl-*trans*-benzobisthiazole is planar with the exocyclic bonds showing a deviation of only 0.06 nm from colinearity. The phenyl rings of 2,6-diphenyl-*trans*-bisthiazole deviate from planarity with a dihedral angle of ca 23°. The phenylene rings in the *trans*-PBZT polymers are coplanar with the central condensed heterocyclic ring system. The structures of PBO and PBZT are as follows:

PBO   PBZT

The early syntheses of *cis*-PBO and *trans*-PBZT were conducted at polymer concentrations of 3 wt % or less. Since these isotropic solutions had high bulk viscosities, polymerizations had to be carried out at low solids concentrations to maintain tractability. When the concentration of *trans*-PBZT was raised to 5 to

10 wt %, nematic solutions were formed and polymers with intrinsic viscosities as high as 31 dL/g were obtained. Initially, the formation of *trans*-PBZT solutions of concentrations greater than 10% caused foaming problems during the polymerization and low molecular weights. The discovery of the P_2O_5 adjustment method was the breakthrough that resulted in the production of nematic spinnable dopes.

The P_2O_5 adjustment method involves adding P_2O_5 to the PPA polymerization solvent to maintain an effective PPA composition as the PPA acts as solvent, catalyst, and dehydrating agent. PPA acts as the solvent for monomer, oligomers, and polymer. PPA also activates the functional groups for polymerization and removes the water of condensation. Also, P_2O_5 is added at the end of the polymerization to achieve the viscosity necessary for spinning. At the end of the polymerization process, the P_2O_5 content must be greater than 82% to keep all the components in solution and less than 84% to give a solution of the proper viscosity for spinning. The temperatures of the PBO and PBZT polymerizations are raised in steps from 100 to 200°C to avoid decomposition of monomers.

These rigid-rod polymers are spun using the dry jet-wet spinning technique also used for the spinning of aramid dopes. The solution is extruded under heat and pressure through a single or multihole spinnerette and an air gap into a coagulation bath, followed by washing, drying, and heat setting. The ordered polymers have been spun in PPA and methanesulfonic acid. Water, dilute phosphoric acid, methanol, and ammonia have been used as coagulants. Heat treatment involves temperatures of 500–700°C with residence times on the order of a few seconds to several minutes. The nematic PPA solution formed in the polymerization may be used as the spinning dope. The typical molecular weight range used to spin fibers is 50,000–100,000 daltons.

In thermogravimetric analyses (9) of the ordered polymers, the extrapolated onset of degradation of PBO and PBZT is reported to be 620°C in air. The extrapolated onset of degradation of PBO in helium is over 700°C. In isothermal aging studies in air at 343°C, PBO and PBZT retain ca 90% of the weight after 200 hours. At 371°C in air, PBO and PBZT retain ca 78% and 71% of the original weight, respectively. PBZ polymers degrade without the observation of crystalline melting points or glass-transition temperatures.

A *cis*-PBO fiber is currently in the market development stage at the Dow Chemical Co. The Air Force Materials Laboratory developed these ordered polymers for aerospace applications. PBO fibers have the highest reported tensile modulus of any known polymeric fiber, 280–360 GPa ($41–52 \times 10^6$ psi). PBO and PBZT are among the most radiation-resistant polymers. Although the compressive strengths of PBO and PBZT are approximately an order of magnitude less than the tensile strengths, alloys of these fibers with high compressive strength fibers can be produced. The polymers are now being evaluated for other applications such as nonlinear optics. Possible PBO applications include reinforcing fibers in composites, multilayer circuit boards, athletic equipment, marine applications, woven fabrics, and fire-resistant fibers (10).

Polybenzimidazole (PBI) Fibers. Poly(2,2'-(m-phenylene)-5,5'-bisbenzimidazole) [25734-65-0] is a textile fiber marketed by Hoechst-Celanese (11) which does not form liquid crystalline solutions due to its bent meta backbone monomeric component. PBI has excellent resistance to high temperature and chemicals.

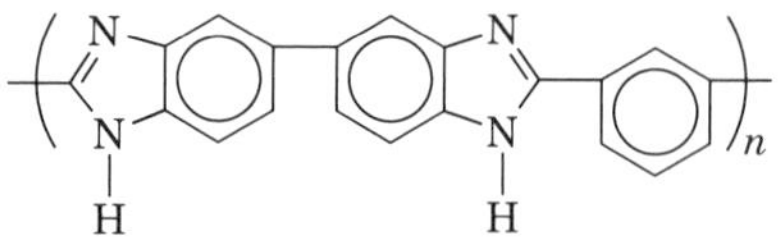

PBI

PBI is being marketed as a replacement for asbestos and as a high temperature filtration fabric with excellent textile apparel properties. The synthesis of wholly aromatic polybenzimidazoles with improved thermal stabilities was reported in 1961 (12). The Non-Metallic Materials and Manufacturing Technology Division of the U.S. Air Force Materials Laboratory, Wright-Patterson Air Force Base, awarded a contract to the Narmco Research and Development Division of the Whittaker Corp. for development of these materials into high temperature adhesives and laminates.

Poly[2,2'-(m-phenylene)-5,5'-bisbenzimidazole] was chosen as the most promising candidate for further development as a fibrous material. Under the terms of an Air Force contract, Du Pont was able to spin fibers from both dimethylsulfoxide and dimethylacetamide solutions to form relatively strong, thermally stable fibers. In 1963, an Air Force contract was awarded to Celanese Research Co. for the development of a manufacturing process for the scale-up of PBI production. PBI fiber of tenacities 0.31–0.44 N/tex (3.5–5.0 gf/den) were produced in sufficient quantity for large-scale evaluation. The fiber was discovered to have a soft hand in addition to possessing a high degree of non-flammability. In the limited oxygen index (LOI) test, the concentration of oxygen required for sustained, steady-state burning was 41%. A new development program was started at Celanese with funding from NASA and the Air Force to develop a flight suit material, fabrics for fatigues worn in space capsules, and utility equipment such as ropes and bungee cords.

Further field tests demonstrated that in spite of the excellent thermal and fire resistance, shrinking of the fabrics occurred above the glass-transition temperature which might expose the wearer to flames. Based on the results obtained in an Air Force contract at Dynatech Co., the Celanese Research Co. developed a two-stage process which reduced the shrinkage from 50 to 6%. The process was also amenable to on-line processing. The sulfonated derivative is the fiber being marketed by the Celanese Corp. Some potential end uses include replacement of asbestos, thermal and chemical safety apparel, and stack gas filter bags.

Development efforts at Celanese Research Co. established solid-state polymerization as the most practical process for engineering scale-up. Homogeneous solution polymerization of PBI in polyphosphoric acid was eliminated because of the need to work with low solid compositions (in the range of 3–5%) during the precipitation, neutralization, and washing steps required for isolation of the product.

In the first stage of the engineering scale-up process, a 189 L oil-jacketed, stainless steel reactor is charged with diphenyl isophthalate (DPIP) and 3,3′,4,4′-tetraaminobiphenyl (TAB). The reactor is deoxygenated by alternative application of vacuum and filling with nitrogen three times, followed by agitation and heating to 250°C under a stream of nitrogen, followed by heating at 290°C for 1.5–3.0 h in the absence of agitation before cooling. In the second stage of the process, the polymer obtained in three to four runs is ground to 0.84 mm (20 mesh) and charged into a 38 L oil-heated stainless steel reactor for a final heating step with agitation at 370–390°C for 3–4 h. Initially, large amounts of foam were produced in the first stage of the process. Foam reduction involves the addition of 10–20% by weight of an organic additive such as diphenyl ether. At the lower temperature of the first stage, the additive acts to prevent foaming and the additive is then removed at the higher temperatures involved in the second stage. Although the foam volume is significantly reduced, the additive residues are removed only with considerable difficulty.

The spinning process used to produce PBI fibers is dry spinning. The preferred solvent for dry spinning of PBI is dimethylacetamide (DMAc). The powdered polymer is dissolved in DMAc at high temperatures (ca 250°C) to form ca 23% wt/wt concentration spinning dopes. The spinning dope is fed by a metering pump through a spinnerette (following filtering) into a countercurrent of hot nitrogen gas in the spinning column. Nitrogen gas is used to prevent oxidation of the oxidatively sensitive filaments formed as the hot gas evaporates the DMAc. The filaments pass to a godet roll and then onto a winder. Washing of the fiber takes place on perforated bobbins to remove lithium chloride stabilizer and residual solvent. The fiber is drawn to achieve improved mechanical properties by passing it from feed rolls to draw rolls through an oven set at temperatures greater than 400°C while under a positive nitrogen pressure. Acid treatment to minimize shrinkage involves the use of aqueous sulfuric acid to produce an acid salt followed by heat treatment to form sulfonic acid groups. If all the imidazole rings were substituted, the final stabilized product would contain 8% sulfur; however, the level of sulfur ordinarily obtained (ca 6%) is sufficient for the required improvement in dimensional stability.

Typical properties of stabilized PBI are a tenacity of 0.27 N/tex (3.1 gf/den), a fiber breaking elongation of 30%, an initial modulus of 3.9 N/tex (45 gf/den), a density of 1.43 g/cm^3, and a moisture regain of 15% (at 21°C and 65% relative humidity).

Solution dyeing of PBI is necessary because the glass-transition temperature (T_g) of PBI is greater than 400°C, and as a result dye molecules only slowly diffuse into the PBI fiber structure. Since the pigments are added to the spinning dope, the pigments must be capable of withstanding the high temperatures used in the various fiber-forming processes.

Industrial Thermotropic LCPs.   Vectran, poly(6-hydroxy-2-naphthoic acid-*co*-4-hydroxybenzoic acid) [*81843-52-9*], is currently the only thermotropic fiber which is commercially available (13). Vectran is synthesized by the melt acidolysis of *p*-acetoxybenzoic acid and 6-acetoxy-2-naphthoic acid.

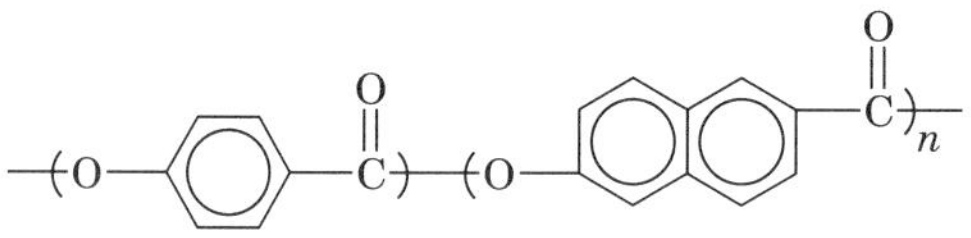

$$-\left(O-\bigcirc-\overset{\overset{O}{\|}}{C}\right)-\left(O-\bigcirc\bigcirc-\overset{\overset{O}{\|}}{C}\right)_{n}$$

First, p-hydroxybenzoic acid (HBA) and 6-hydroxy-2-naphthoic acid (HNA) are acetylated to produce the low melting acetate esters which are molten at 200°C. In an inert gas, the two monomers are melted together at 200°C. The temperature is raised to 250–280°C and acetic acid is collected for one-half to three hours. The temperature is raised to 280–340°C and additional acetic acid is removed in vacuum for a period of 10 min to 1 h. The opalescent polymer melt produced is extruded through a spinning jet, followed by melt drawdown. The use of the parallel offset monomer, acetylated HNA, results in the formation of a series of random copolyesters of different compositions, many of which fall within the commercially acceptable melting range of 250–310°C. Characteristically, these nematic melts show the persistence of orientational order under the influence of elongational flow fields which result in low melt viscosities under typical fiber formation conditions even at high molecular weights.

Axial orientation develops quickly during melt drawdown with a concomitant increase in fiber modulus. At a drawdown ratio of ca 10, the fiber achieves a maximum modulus in the range of 44.1 to 61.7 N/tex (500 to 700 gf/den). Neither cold drawing nor annealing led to additional increases in modulus. The high level of mechanical properties is the result of the comparative ease of axial orientation of the nematic phase leading to a highly oriented rod-like fiber structure. This is borne out by x-ray fiber analysis which shows well-defined meridonal maxima characteristic of highly oriented parallel arrays of polymer chains with poor lateral spacing.

Heat treatment of the as-spun fibers results in an increase in tenacity but no attendant increase in modulus. Typically, the as-spun fibers are heat treated in an inert environment at temperatures 10–20°C below the melting point for from 10 min to several hours. There is a corresponding increase in chemical resistance and melting temperatures, presumably due to increases in molecular weight rather than improvements in structural perfection. This is in agreement with x-ray fiber diagram results which show no increase in orientation of mesophases during the heat treatment process. Vectran HS fibers are reported to have typical tensile strength and modulus values of 2 N/tex (23 gf/den) and 46 N/tex (525 gf/den), respectively. The melting point and density are reported to be 330°C and 1.4 g/cm^3. The fibers have excellent chemical resistance except for their resistance to alkali.

GEL SPUN FIBERS

In the mid-1970s it was discovered at the Dutch States Mines Co. (DSM) that through an ingenious new method of gel spinning ultrahigh molecular weight polyethylene it was possible to produce fibers having twice the tenacity of Kevlar, which was then considered to be the strongest known fiber (14). The discovery was important not only because of the exciting 3.8 N/tex

(44 gf/den) strengths these new fibers displayed, but also because it clearly demonstrated that factors other than monomer polarity were critical in controlling fiber performance characteristics. These high performance polyethylene fibers (HPPE) produced by the DSM subsidiary company, Stamicarbon, are called Dyneema and those produced by the AlliedSignal Corp. in the United States are sold under the trade name of Spectra 1000. The commercial products have somewhat lower strengths than the laboratory fibers but still are in the high 2.6 N/tex (30 gf/den) range.

Process. In the gel spinning process, 1–8% solutions of polyethylene are prepared by dissolving polymer of molecular weights of 1–4 million in hot hydrocarbon liquids such as decalin, melted waxes, or mineral oils at ca 150°C. These hot solutions are then screw extruded through spinnerettes having holes of 0.5–2.0 mm diameter and an L/D ratio of 25 to control the viscoelastic flow properties of the fluid. The fibers are spun into a cooling bath which yields disoriented highly crystalline gel fibers of sufficient stability to be wound onto a first godet at several m/min. These gel fibers are then processed in solvents at about 50°C to remove the hydrocarbons. The solvent-free gels are then stretched in progressively hotter zones at temperatures from 120 to 160°C with an overall final windup/extrusion speed of about 1000/1 or whatever is required to give the final desired strengths of 1.7 to 3.5 N/tex (20–40 gf/den) (15).

The patent literature indicates that the AlliedSignal process uses lower boiling solvents such as chlorofluorocarbons as the cooling/extraction baths (16), whereas the processes of Stamicarbon indicate the use of decalin solvent followed by cooling and slow removal of the decalin in successively hotter chambers while stretching (17).

Properties. Fiber property comparisons for the different products are given in Table 1.

The attributes of HPPE fibers include high strength; high abrasion resistance; high uv stability as compared to other synthetics; high resistance to acids, alkali, organic chemicals, and solvents; and low density. Disadvantages are a low melting point of about 150°C, which means performance is limited to no more than 120°C; difficult processing; and poor surface adhesion properties.

It is difficult to process HPPE staple fibers mechanically because of so-called married fibers which are bundles of 4–6 fibers that firmly adhere to each other and resist separation by conventional processing. Although HPPE fibers like to adhere to each other, they exhibit poor adhesion to other materials.

It is possible to modify HPPE to overcome the poor adhesion of the fiber surfaces by using corona discharge in an oxygen atmosphere previously developed for polyolefin films or by the addition of fillers to the polymer solution prior to spinning. The melting point of HPPE fibers embedded in polymer matrices is increased by about 8°C (18,19). Temperature performance can also be enhanced

Table 1. Properties of Commercial HPPE Fibers

Fiber	Tenacity, N/tex[a]	Initial modulus, N/tex[a]	Elongation at break, %
Dyneema	1.01–3.57	57–128	3–7
Spectra 1000	3.4–3.57	162–171	3–7

[a]To convert N/tex to gf/den, multiply by 11.33.

by wrapping the HPPE fibers with other fire-resistant or fire-retardant fibers (20). Other ultrahigh molecular weight polymers have also been spun via the gel spinning process. These include polypropylene, polyacrylonitrile, poly(vinyl alcohol), and nylon-6. However, the property improvements in these cases evidently have not warranted commercialization.

ELONGATABLE CARBONACEOUS FIBER

It is difficult to weave or knit regular carbon fiber. For any fiber to be considered as a satisfactory textile fiber it should have an elongation of at least 3% and preferably more in the range of 5–8%. The extreme brittleness, high modulus, and low elongation of standard carbon fibers restricts them to be woven only on a special Dornier Rigid Rapier type of loom. To overcome these drawbacks, an exciting new modification of carbon fiber technology was developed; by using less stringent carbonizing conditions and only partially carbonizing the precursor fibers, improved textile fiber properties have been achieved (Fig. 1) (21).

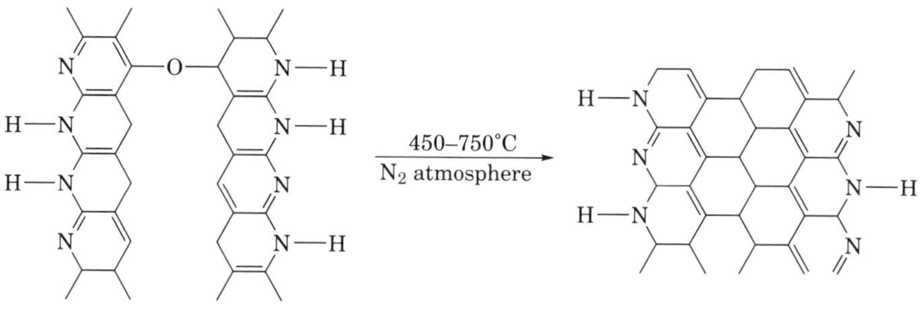

Fig. 1.   The structure of Dow EDF fiber. EDF = elongatable Dow fiber; OPF = oxidized PAN fiber.

Process.   Any standard precursor material can be used, but the preferred material is wet spun Courtaulds special acrylic fiber (SAF), oxidized by RK Carbon Fibers Co. to form 6K Panox B oxidized polyacrylonitrile (PAN) fiber (OPF). This OPF is treated in a nitrogen atmosphere at 450–750°C, preferably 525–595°C, to give fibers having between 69–70% C, 19% N; density less than 2.5 g/mL; and a specific resistivity under 10^{10} ohm·cm. If crimp is desired, the fibers are first knit into a sock before heat treating and then de-knit. Controlled carbonization of precursor filaments results in a linear Dow fiber (LDF), whereas controlled carbonization of knit precursor fibers results in a curly carbonaceous fiber (EDF). At higher carbonizing temperatures of 1000–1400°C the fibers become electrically conductive (22).

Properties.   Unlike regular carbon fibers, these new products do not conduct electricity, but do exhibit good textile processing properties and possess exceptional ignition-resistant, flame-retardant, and even fire blocking properties. The limiting oxygen index (LOI) defines the percentage of oxygen necessary in an oxygen/nitrogen mixture before a material supports combustion. Typical LOI values for various fibers are given in Table 2.

Table 2. Limiting Oxygen Values of Fibers

Fiber	LOI, %	Fiber	LOI, %
polyethylene	17	polyimide	37
polystyrene	19	rigid PVC	40–44
cotton	20	oxidized PAN	40+
nylon	20	Hoechst-Celanese PBI	41
polycarbonate	22	Dow's EDF	45–55
Du Pont Nomex	26	Phillips' PPS	44
PPO	26	graphite	55
polysulfone	30	PTFE	95

Previous results with ignition resistant (IR) blends, where such fibers as aramids or PBI are used as the high LOI fibers, show that they need at least 65% and typically 85% fiber content to pass the vertical burn test for lightweight nonwoven batting. In contrast only 7–20% of either the Dow EDF or LDF mixed with flammable natural and synthetic fibers allow the blends to pass such tests while still retaining most of the base natural or synthetic fiber properties. Blends of 50/50 EDF/polyester also passed the stringent FAA airlines ignition resistance tests with zero flame length and no after-burn, whereas other blends of 65% LOI fiber/40% synthetic blends gave burn lengths of 20 cm and 15 second after-burn, clearly demonstrating the superiority of the lower level carbonaceous fiber as a flame blocker (22,23). Such nonwoven batting has exceptional thermal and sound insulation properties and has been successfully tested by the U.S. Navy for pilot's arctic wear.

VITREOUS FIBERS

Man-made vitreous fibers (MMVF) comprise a number of glass and specialty glass fibers and also refractory ceramic fibers. The vitreous state in glass is somewhat analogous to the amorphous state in polymers. However, unlike organic polymers, it is not desirable to achieve the crystalline state in glass. Glasses are produced from glass-forming compounds such as SiO_2, P_2O_5, etc, which are mixed with other intermediate oxides such as Al_2O_3, TiO_2, or ZnO, and modifiers or fluxes like MgO, Li_2O, BaO, CaO, Na_2O, and K_2O (see GLASS).

The purpose of the fluxes is to break down the SiO_2 network so that the molten glass has the proper viscosity characteristics to allow it to cool to the desired vitreous state. Glasses with large fractions of noncross-linking monovalent alkaline fluxes allow the melts to form at lower temperatures but correspondingly have lower chemical resistance. For example, sodium silicate glasses with larger amounts of Na_2O are sold as water solutions (water glass).

A wide range of glass compositions is available to suit many textile fiber needs; the three most common glass compositions are referred to as E, S, and AR glasses. AR glass is a special glass with higher contents of Zr_2O designed to resist the calcium hydroxide in the cementitious products where it is used. S glass is a magnesium–aluminum–silicate cross-linked glass used where high mechanical strength or higher application temperatures are desired. E glass is a member of the calcium–aluminum–silicate family containing less than 2% alkali (see composition in ASTM specification D578-89a) and is the predominant glass used to make textile and continuous filament fibers.

Glass fibers < 3 μm are to be avoided because these are classed as respirable fibers which can enter and damage lung passages. Most glass fiber products have sufficient fiber lengths to prevent lung entry even if their diameters are < 3 μm.

Manufacture.   Vitreous fibers are produced by several processes (24).

Continuous Drawing Process.   Textile glass filaments are made by a process different from that used for making discontinuous fibers, but literally parallel to a standard organic polymer melt spinning operation that does not employ a screw extruder. Premelted glass or glass marbles are fed into an electrically heated furnace called a bushing which contains 204 platinum nozzles (or multiples thereof). The exiting glass filaments are drawn down into the desired diameters, water sprayed, coated with a sizing, and the multiple filaments are collected as bundles of strands which are then wound onto a suitable cone.

Rotary Process.   This process is much like the making of cotton candy except that molten glass is used in place of molten sugar. The melted glass is dropped into a rotating spinner with sidewall perforations and the exiting glass filaments are drawn to a fine diameter by the centrifugal force. These fibers are collected and coated with a protective spray containing either lubricants, binders, or antistatic and wetting agents. Other versions of the rotary process are (*1*) the wheel centrifugal process where molten glass is cascaded over spinning wheels and the formed fibers are stretched and broken by variations in the wheel speeds prior to being collected, and (*2*) the Downey process where molten glass is dropped onto a centrifuge wheel and then exits into a stream of high velocity air much like the melt blown process for making textile nonwovens.

Flame Attenuation.   This process closely resembles the continuous drawing process except that the melted strand is not wound onto a cone. Rather, the exiting strand is blown at right angles with a high velocity gas burner so as to remelt and reform the glass as small fibers, which are collected as a mat onto a moving belt. A modification of this process simply uses the high velocity flame at right angles to a dropping melted stream of glass to fibrillate the mass into minute fibers.

Properties.   Glass fibers made from various compositions have softening points in the range 650–970°C. Fiber length and diameter distributions are significant factors in determining thermal and acoustical insulation properties. Slag wool and rock wool fibers are prepared from the slag from pig iron blast furnaces. They contain significant amounts of iron oxides and have a glass-transition temperature of 760–870°C. Slag and rock wool fibers are used to prevent fires from spreading. At temperatures above 850°C, these fibers partially devitrify and form polycrystalline material that melts at 1225–1360°C, which is high enough to contain the fires for several hours. Seventy percent of the slag wool in the United States is used for ceiling tiles.

Refractory Ceramic Fibers (RCF).   These MMVF materials constitute only about 1% of the vitreous fiber market but have exceptional high temperature performance characteristics. They are produced by using high percentages of Al_2O_3 about 50/50 with SiO_2 as is or modified with other oxides like ZrO_2 or by using Kaolin clay which has similar high amounts of Al_2O_3. Different compositions result in modifying end use temperatures from about 1050°C or higher for the kaolin-based products to 1425°C and above for the zirconium-containing materials. At temperatures above 1000°C these ceramic fibers tend to devitrify and partially crystallize. Specially prepared ceramic fibers are used to protect space vehicles on re-entry and can withstand temperatures above 1250°C (see ABLATIVE MATERIALS; REFRACTORY FIBERS).

SULFAR FIBERS

Ryton fibers are high performance products developed by Phillips Petroleum Co. by reaction of p-dichlorobenzene with sodium sulfide in the presence of a polar solvent according to the following:

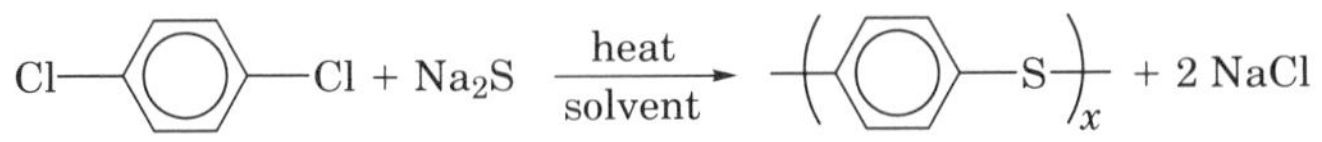

$$\text{Cl}-\!\!\bigcirc\!\!-\text{Cl} + \text{Na}_2\text{S} \xrightarrow[\text{solvent}]{\text{heat}} {\left(\!\!-\!\!\bigcirc\!\!-\text{S}\!\!-\!\!\right)}_x + 2\,\text{NaCl}$$

Because the chemical structure of poly(phenylene sulfide) [9016-75-5] (PPS) does not fall into any of the standard polymer classes, the Federal Trade Commission granted the fiber the new generic name of Sulfar. The fiber has excellent chemical and high temperature performance properties (see SULFUR-CONTAINING POLYMERS).

Manufacture. The virgin polymer as originally prepared by the reaction shown is an off-white solid powder of modest 18,000 molecular weight. The molecular weight can, however, be increased in the solid state by heating in air at a temperature near its melting point. The mechanism for this is a combination of chain extension and cross-linking. Reaction improvements utilize an alkali metal carboxylate modifier which eliminates the need for a post-cure and directly gives a high molecular weight polymer having an inherent viscosity of 0.28 and a melt flow of 250. This polymer is suitable for producing fibers via melt spinning (25,26).

Properties. As prepared, the polymer is not soluble in any known solvents below 200°C and has limited solubility in selected aromatics, halogenated aromatics, and heterocyclic liquids above this temperature. The properties of Ryton staple fibers are in the range of most textile fibers and not in the range of the high tenacity or high modulus fibers such as the aramids. The density of the fiber is 1.37 g/cm³ which is about the same as polyester. However, its melting temperature of 285°C is intermediate between most common melt spun fibers (230–260°C) and Vectran thermotropic fiber (330°C). PPS fibers have a T_g of 83°C and a crystallinity of about 60%.

The main advantage of Ryton is not any particular excelling property, but rather the ability to retain its standard properties under adverse conditions. For example, it retains essentially 100% of its strength after exposure to most organic solvents even at the boil and also retains 100% of its strength after exposure to most strong acids and alkalies even after a week at 93°C. Only concentrated nitric acid and concentrated sulfuric acid cause fiber degradation after a one week exposure at 93°C. At lower acid concentrations of 50% or in

10% sodium hydroxide, Sulfar PPS fibers show no strength loss after one week at 93°C. Other high performance and regular textile fibers, such as Nomex, PBI, polyester, acrylic, and glass fibers, all display essentially total strength loss under the same conditions. In standard thermal exposure tests, PPS fibers retain essentially 100% of their strength and 50% of their elongation after 54 days at 200°C (27).

Applications

Commercial high performance fibers and high technology textile products are becoming an increasingly important segment of fiber and textile consumption worldwide. Although breakdown of the numerous applications by weight and/or economic value is impractical, one review indicates that high technology textile uses will account for 50% of all worldwide fiber consumption by the year 2000 compared to 10–15% in 1990 (28). In some instances various technologies and concepts are combined or refined to produce a textile product for the desired application(s). Thus, sophistication and enhancement of properties may be introduced at the fiber, yarn, and/or fabric levels.

STRUCTURE/PROPERTY CLASSIFICATION

The relationship between structure and properties of textile or fibrous substrates and their applications is one method of classifying nontraditional or high technology textiles. At the fiber level, the distinguishing high performance characteristics are high tenacity/strength fibers, hollow fibers, very fine or microtex (microdenier) fibers (hollow or nonhollow), fibers with unique porosities, bicomponent and biconstituent fibers, and fibers with superior resistance to extreme heat, flame, and/or chemical agents (Table 3). At the fabric or product level, the classes may be described as coated and laminated fabrics, composites and fiber-reinforced materials, three-dimensional fabric structures, and fabrics containing polymers or structural features that impart multifunctional properties or allow the fibrous substrate to act as an intelligent material. Although some of these fiber and fabric characteristics also apply to conventional textile uses and products, many

Table 3. Classification of High Performance Fibers and High Technology Textiles by Property

Property	Fiber types	Applications
high tenacity and high modulus	aramids, gel spun polyethylene, polyarylate	tires, antiballistic, ropes, optical cables
resistant to heat and flame	aramids, PPS, PEEK, PBI, polyimides, EDF	protective clothing for various applications
resistance to chemical agents	PPS, fluorocarbon, polyolefins	filters, geotextiles, marine applications
microtex and hollow fibers	most synthetics and regenerated fibers	filtration, leisure, insulation, biomedical
intricate shapes and porosities	most synthetics and regenerated fibers	fashion, fragrances, antimicrobial, fiber optics, specialty wipes

of these concepts have evolved from the production and use of high performance fibers and products.

Fiber Properties.   *High Strength Fibers.*   Super fibers or fibers with very high tenacities and Young's moduli have been defined as those with a tenacity of at least 2.5 GPa (255 kgf/mm^2) and a modulus of at least 55 GPa (5600 kgf/mm^2) (29). Fibers meeting these criteria are aramids such as Kevlar and Twaron, gel spun polyethylene such as Dyneema and Spectra, and various carbon fibers and aromatic liquid–crystalline polyesters such as Vectran. Representative applications are for antiballistic clothing, building materials, aerospace, and as reinforcing material in composites for various applications.

Heat-Resistant Fibers.   Inherently flame- and heat-resistant fibers have other criteria for performance in addition to high tenacity and modulus. These fibers must be suitable for protective clothing or for use as a material in a particular application such as firefighters' uniforms, race car drivers apparel for protection from hot metals and gas explosions, and as components in commercial and military aircraft. Dimensional stability and strength retention on exposure to intense heat sources as well as a limiting oxygen index (LOI) above 30 are essential for most of these applications. Fiber types and blends meeting these criteria are various aramids such as Nomex and Kevlar, polybenzimidazole (PBI), poly(phenylene sulfide) (PPS) such as Ryton, Dow's EDF, and wool blends with these various inherently flame-resistant fibers.

Chemically Resistant Fibers.   Fibers with excellent chemical resistance to corrosive and/or chemical warfare agents or extreme pH conditions (eg, very acidic or very alkaline) were initially used for protective clothing. However, applications for filtration of gases and liquids in numerous industrial facilities are now the more important. For example, PPS is suitable for use in filter fabrics for coal-fired boilers because of its outstanding chemical and heat resistance to acidic flue gases and its excellent durability under these end use conditions. Many high tenacity fibers are also chemically inert or relatively unaffected under a variety of conditions. Aramids, gel spun polyethylene, polypropylene, fluorocarbon, and carbon fibers meet these criteria and have been used or are being considered for applications where chemical resistance is important.

Fine and Hollow Fibers.   Controlling and designing the geometry, fineness or denier, and porosities of fibers (and occasionally of yarns) have led to novel and high technology textile products for diverse applications. Hollow fibers derived from regenerated cellulose or from synthetic fibers have been used in the development of artificial body organs such as the kidney, pancreas, and lung (see HOLLOW-FIBER MEMBRANES). Hollow fibers have also frequently been employed to increase the insulation value of garments due to the benefits of the air trapped inside the fiber cavity. A variety of ultrafine fibers, ranging in tex (denier) from as little as 0.0011 (0.01) up to 0.011 (0.1) have been commercialized (primarily in Japan) to impart various surface characteristics that change fabric hand and appearance. Because spinning ultrafine fibers directly is not technically feasible, such fibers are produced by spinning bicomponent or biconstituent polymer mixtures, highly stretching them to form ultrafine deniers, and extracting or otherwise removing the undesired matrix carrier material to release the desired ultrafine fibers (Fig. 2). Controlling the porosity of fibers has been advantageously used to release antimicrobial agents at different rates for

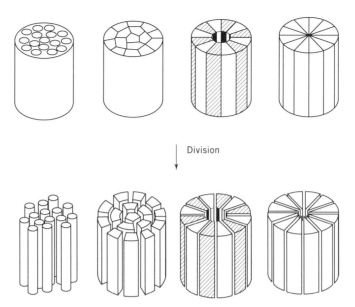

Fig. 2.   Ultrafine fibers are produced by spinning bicomponent or biconstituent polymer mixtures, highly stretching such products to ultrafine deniers, and extracting or otherwise removing the undesired matrix carrier to release the desired ultrafine fibers (30). For example, spinning polyester islands in a matrix of polystyrene and then, after stretching, dissolving the polystyrene to leave the polyester fibers; cospinning polyester with polyamides, then stretching, removing the polyester by caustic treatment, leaving 100% nylon ultrafine fibers.

specific applications. Novel core/sheath yarns that control selective distribution of both components are being evaluated for numerous applications (20).

·**Product Types.**   *Composites.*   Various composite materials have evolved over the years as a significant class of high performance textile products. The prototype composite is carbon fiber with an epoxy resin matrix for structural aircraft components and other aerospace and military applications. Carbon fiber composites are also used in various leisure and sporting items such as golf clubs, tennis rackets, and lightweight bicycle frames. However, other types of applications and composites are also entering the marketplace. For example, short cellulose fiber/rubber composites are used for hoses, belting, and pneumatic tire components.

Coated and laminated fabrics have progressed from relatively impermeable or nonporous structures to coated and membrane-laminated fabrics that are microporous yet waterproof. The advent of GoreTex in the late 1970s, a microporous polytetrafluoroethylene membrane that can be laminated to fabrics, has led to numerous other coated and laminated textile products. These products are now used in applications for thermal comfort, biomedical devices such as degradable bone plates, and geotextiles/geomembranes for pond liners and confinement of hazardous waste (see GEOTEXTILES).

3-D Structures.   Three-dimensional textile structures have been developed primarily for architectural, civil engineering, and aerospace applications. Such structures fall into four different categories: nonwoven orthogonals (straight, continuous fibers arranged in three directions with no interlacing); multilayer woven (multiple layers interlaced at selected points); multilayer knitted or stitched (multiple layers of yarns stitched in diverse or desired directions); and three-dimensional braids (intertwining of multiple yarns so that a particular yarn follows a path taking it completely across the structure several times) (31).

Multifunctional Materials. Multifunctional property improvement by binding of specific polymers to fibers and fibrous products has been extensively investigated and reviewed (32). With poly(ethylene glycol) as the bound polymer, functional and aesthetic property improvements include thermal comfort, liquid absorbency/repellancy, increased wear life, soil release, resistance to static charge, antimicrobial activity, and resiliency. Numerous applications such as sportswear/ski wear, protective clothing for health care workers, durable and nondurable hygienic items, work uniforms, and space suits are being commercialized and evaluated.

Smart Materials. The field of intelligent or self-adaptive materials is in its infancy, particularly with regard to textile products. Conceptually, any material that is responsive to one or more external stimuli (heat, force, light, moisture, electrical current) and that responds to such exposure by changing shape or related characteristics is classified as an intelligent material. Synthetic gels that act like artificial muscle and fabrics that self-repair to avoid tensile and other forms of mechanical failure are examples of ongoing research and uses (33).

CLASSIFICATION BY TYPES OF APPLICATION

Another way to classify high performance fibers and high technology textile materials or products is by types of applications. A scheme of 10 main categories has been adopted (Table 4) and is similar to several classification schemes previously reported (28).

Transportation. High performance fibers and high technology textile products have many applications in the transportation area. Composites are increasingly used as structural materials in aircraft components such as horizontal stabilizers, fins, landing gear doors, fan blades, and nose spin cones. In addition to carbon and glass fibers in composites, aramid and polyimide fibers are also used in conjunction with epoxy resins. Safety requirements by the U.S. Federal Aeronautics Administration (FAA) have led to the development of flame- and heat-resistant seals and structural components in civilian aircraft cabins. Wool blend fabrics containing aramids, poly(phenylene sulfide), EDF, and other inherently flame-resistant fibers and fabrics containing only these highly heat- and flame-resistant fibers are the types most frequently used in these applications.

The introduction of air bags into automobiles represents a new and enormous market for high performance textiles. Polyamide-coated fabrics are primarily used because of the high strength of the polyamide, but there have been marketing and technical studies that indicate the feasibility of using high tenacity polyester as the base fabric. Performance requirements for air bags include tear; tensile, seam, and bursting strength; dimensional stability; resistance to puncture, abrasion, and buckling; flame resistance; and resistance to delami-

Table 4. Classification of High Performance and High Technology Textiles by Types of Application

Application class	Subcategories	Examples of specific products
transportation	civilian and military aircraft, traffic	components of aircraft, air bags for vehicles, seats in planes and cars
manufacturing	electronics, information and communication	optical fibers for telecommunications and computers, printed circuit boards, industrial filters and belts
agriculture and forestry	horticulture, erosion control, barriers	greenhouse covers, control of drainage, land nets, cultivation of plants
civil engineering and construction	geotextiles and geomembranes, architectural	road reinforcement, pond liners and dams, fabric roofs, soil stabilization
fishery and marine	pollution containment, breeding of aquatic life, components for industrial and leisure marine equipment and vehicles	corrosion-resistant composites, conveyers, floating breakwaters, screens for fish breeding and pollution control, speedboat components, artificial floating islands
protective clothing	chemical/environmental, security biological, heat and flame	firefighters' uniforms, bulletproof vests, protection from diseases and toxic agents
sports and leisure	sporting goods and vehicles, spas and pools	golf clubs, tennis rackets, snowmobiles, bicycle frames, spa and pool parts, ski wear and sportswear, luxury apparel
biomedical and health care	devices, artificial organs, sutures, wound care, prostheses	kidneys and artificial limbs, bioimplants, dressings for burns, hydrogel composites
defense and aerospace	chemical/biological protection, camouflage, components for weapons	chemical/biological warfare protection composites for armor and other weapons, space suits and materials for space travel
energy use and conservation	insulation, containment of hazardous waste, shields from high energy sources	all types of insulation, waste containers, electromagnetic shielding

158

nation (34). It has been estimated that by the end of the century more than 41×10^6 m of coated fabrics (qv) will be used in the United States alone to make automobile air bags (35).

Manufacturing. The use of advanced textile materials in manufacturing is a diverse and expanding market. Optical fibers (qv) are being increasingly used in telecommunications, computers, cable television, and to facilitate process control in the nuclear, petrochemical and chemical, and food industries. Although most optical fibers are glass or of related inorganic composition, there are optical fibers for special applications comprised of poly(methyl methacrylate), polystyrene, and polycarbonate that are coextruded with fluorinated acrylate polymers. Polyacetal, ie, poly(oxymethylene) fiber has also been used as a reinforcing material for optical fibers. Basic fiber properties required for an optical fiber are (1) a structure with a high refractive index core and a low refractive index cladding (sheath bonded to core under high temperature and pressure); (2) fiber with a low attenuation or low power loss of light over distance; and (3) fiber with low dispersion or pulse broadening as light travels down the fiber (36).

Electronics and electrical insulation are other manufacturing areas in which advanced fibers are extensively used. Manufacturing of computer components such as printed circuit boards and printed wiring boards require use of reinforced fibers that have good dielectric properties, thermal and dimensional stability, chemical resistance, and low moisture regain. For electrical insulation applications, fiber requirements are high dielectric strength, low power loss, and good thermal and chemical resistance. Although inorganic fibers are extensively used for both electronic and electrical applications, aramid fibers are also now used for both types of applications. A variety of natural and synthetic fibers are also used for various electrical applications (37).

A third area that has rapidly expanded in the manufacturing sector is the demand for clean room working garments. The primary electronics and computer industries have the most stringent requirements for cleanliness in their manufacturing environments; clean room garments are also used by personnel in the pharmaceutical, food, precision engineering, and biotechnology/biomedical industries. Requirements for clean room garments and the increasingly stringent standards for both particle size and number of acceptable particles have been critically reviewed and discussed (38). Required properties for garments in clean rooms include fabrics or materials that are dustproof, antistatic, durable to laundering and sterilization procedures, and that are comfortable. Garments have been designed with ability to filter air by mechanical or electrostatic concepts and collect fine particles. Microtex polyester fibers are most frequently used to make these clean room garments. Diameters of particles to be removed have progressed from about 1 μm in the early 1970s to only 0.1 μm in the 1990s. Industry standards for the corresponding degree of cleanliness (particles/m^3) have decreased from 35,000 to 35 (10,000 to 1 per ft^3). Thermal comfort of many of the garments is unacceptable due to their low moisture content, but some efforts are being made to develop microporous coatings to provide both comfort and function required for this application.

The fourth area of manufacturing for advanced fibrous materials is that of filtration of liquids and gases. Filters have diverse requirements for different industries and environments to which they are exposed. The use of cellulose acetate and aramid hollow fibers to desalinate water by a reverse osmosis process is an example of one of the earlier applications for filtration. Important fiber properties for such filtration are hydrolytic, oxidative, and biological stability over wide pH ranges. Separation of gases by ultrafiltration (qv) for a variety of industries is another example of the importance of advanced fibrous substrates. Because of the diversity of exposure to extreme temperature changes and corrosive chemical agents, many of the high temperature, high tenacity, and inert fibers are replacing more conventional types of fibers for such filtration. Poly(phenylene sulfide), polysulfone, aramids, polyimide, PEEK, fluorocarbon, and related fibers are examples of high performance fibers effective for gas and liquid filtration under extreme and rapidly changing environmental conditions.

Agriculture and Forestry.   Agricultural and silvicultural applications for high performance textile materials range from geotextile and geomembrane materials to horticultural uses to facilitate the growth, yield, and production of edible and ornamental plants. Geotextile structures, either permanent or biodegradable, are inexpensive alternatives to more expensive, conventional concrete or other materials for prevention and control of gully formation. Also, drainage and soil erosion control on farm acreage have been facilitated by geotextiles. Wind fences and other geotextile barriers have also been advantageously used.

If the geotextile is above ground, it should be stable to sunlight, air pollution, rot, and mildew-producing fungi besides having mechanical integrity and dimensional stability. If the textile is below ground, it should be dimensionally stable and inert to most chemical agents and have good long-term permeability so that it does not become easily clogged with soil particles.

Other examples of agricultural applications are flexible and lightweight silos constructed from geotextiles that hold grain, manure, or any other agricultural commodity of interest. Polypropylene fiber is used primarily in most geotextile (agricultural, civil engineering, and other applications) structures because of its excellent inertness to chemical and biological agents and its dimensional stability over long periods of use. Polyester and glass fibers are also used to some extent in geotextiles.

Horticultural applications include use of greenhouse thermal screens, row-crop and turf covers, conveyer belts to process agricultural products, and other similar items. High performance fibers are not normally used in these applications, but high strength fibers are preferable for conveyer belts. Environmentally inert low cost fibers such as polypropylene are used for many of the outdoor horticultural applications.

Civil Engineering and Construction.   These textile applications can be classified as civil engineering applications (also commonly referred to as geotextiles (qv)) and architectural applications. High performance fibers are not currently used to any great extent in these applications due to their high unit cost. However, mechanical and environmental requirements for such applications demand that even conventional fibers withstand challenging structural demands and sustained long-term performance outdoors. Polypropylene fibers have captured a sizable portion of the geotextile market because of their excellent dimensional stability and chemical inertness. Polyester and glass fibers are two other types of fibers used in geotextile applications. Reinforcement of highways and roadbeds is by far the principal use of geotextiles. Their primary function is to serve as a soil stabilizer by spreading the impact load of vehicles over a larger area thus prolonging the useful life of roadways. As in agricultural applications, geotextiles are also used to prevent or minimize soil erosion and facilitate drainage. These materials have been utilized in important projects such as reinforcement of dikes in Holland and to construct dams in the United States. Mechanisms for geotextile failure have been critically evaluated and discussed (39). The types of failure must be evaluated relative to the function of the geotextile. Failure modes include piping of soils through or clogging of the geotextile, reduced tensile resisting force, deformation of the fabric, reduced resistance to puncture, and reduced in-plane flow to liquids.

Most architectural fabrics are usually flexible composites comprised of glass fibers coated with fluorocarbons to resist wind, mechanical forces, and outdoor environmental degradation. The airport terminal in Saudi Arabia, and the roofs for the Hubert Humphrey Dome in Minneapolis and the Tokyo Dome Stadium are a few examples of the successful use of architectural fabrics.

Fishery and Marine.   Marine applications for textiles include (1) those concerned with fishing and fishing industries, (2) leisure items and components, and (3) industrial, including control of various types of environmental pollution and influences (40,41). The earliest use of textiles in fishery and marine applications were for high strength nets, lines, and ropes. Conventional natural and synthetic fibers are being replaced by high performance fibers such as aramids, gel spun polyethylene, and polyacetal. However, cost is still a factor because the latter fibers are still much more expensive than the conventional fibers. More advanced applications for fishing and marine industries include the construction of artificial seaweed beds comprised of synthetic fiber membranes that facilitate fish breeding.

Advanced composites and fiber-reinforced materials are used in sailcloth, speedboat, and other types of boat components, and leisure and commercial fishing gear. Aramid and polyethylene fibers are currently used in conveyer belts to collect valuable offshore minerals such as cobalt, uranium, and manganese. Construction of oil-adsorbing fences made of high performance fabrics is being evaluated in Japan as well as the construction of other pollution control textile materials for maritime use. For most marine uses, the textile materials must be resistant to biodeterioration and to a variety of aqueous pollutants and environmental conditions.

Protective Clothing.   Protection against adverse chemical and biological environments, from heat and/or flame, and antiballistic protection are the principal applications of protective clothing. Several high performance fibers are finding increasing use in many protective clothing applications due to their high strength, resistance to cutting and firearms, and their ability to withstand temperature extremes and corrosive chemical agents. Aramids such as Kevlar and polyethylene fibers such as Spectra are used in many bulletproof vests and other types of security garments.

Aramids are also used in firefighters' uniforms for their excellent flame resistance. The new Dow Curlon EDF fibers have exceptional flame resistance both by themselves and in combination with other flammable fibers. At only 20% by weight, Curlon has been shown to totally prevent the flaming of polyester and other similar flammable materials. This property along with Curlon's good textile processibility will undoubtedly lead to many new applications in the protective garment area (21–23). Garments derived from blends of polybenzimidazole (PBI) and aramid fibers have excellent chemical resistance and a low propensity to severe burns as shown by tests in a gas pit where temperatures could rise as high as 1100°C (42).

There are distinct performance differences for garments that are resistant to flame alone compared to those that are resistant to both heat buildup and flammability. Appropriate tests have been devised that measure the thermal protective performance of fabrics when exposed to a radiant heat source. Sophisticated constructions against biohazards need further development to produce materials that are both thermally comfortable and impermeable to blood-borne pathogens and other deleterious microorganisms. The most promising approaches are materials that are laminated or coated fabrics that are permeable to vapor but impermeable to liquids. Statistics included in recent OSHA standards for protection against blood-borne pathogens (eg, hepatitis and AIDS viruses) estimate that currently close to six million persons in the United States (health care workers and many other occupations) require protective clothing and other types of safeguards against these biohazards (43).

Sports and Leisure. The application of high technology fiber properties and complex structures for sports and leisure are identifiable in three areas: components for sporting goods, boats, and other rigid structures; high technology sportswear and ski wear; and luxury apparel. Although most components for sporting goods (such as tennis rackets, golf clubs, lightweight bicycle frames, spa components, speedboat components) are usually carbon-fiber reinforced composites, the high performance aramids and polyethylene are now being used to manufacture items such as bicycle helmets and other sports items. For these end uses, impact and overall tensile strength, dimensional stability, and resistance to specific environmental agents are required. High technology sportswear and ski wear evolved with the advent of GoreTex. GoreTex was the first garment comprised of a microporous polytetrafluoroethylene film laminated to a polyamide fabric. The microporosity of the film allowed the garment to be breathable or allow passage of water vapor yet be waterproof or resistant to liquid penetration. Since that time, a variety of microporous materials or related structures have been developed in the United States and Japan. Another example for high technology sportswear is the latent heat and high water absorbency of a cross-linked poly(ethylene glycol) coating on fabrics that makes it possible for the garment to be thermally adaptable in hot and cold environments (32). Also, microencapsulation (qv) of zirconium carbide into polyamide or polyester fibers produces a ski wear garment (called Solar-α) that absorbs sunlight and converts it into heat.

Numerous high touch fibers have been produced and commercialized (primarily in Japan) for luxury apparel. Conceptually, the geometry and fineness of the fibers are carefully modified and controlled to produce fabrics with desirable sensual responses of touch, sight, sound, comfort, and even odor (30,44). There are several techniques for producing ultrafine fibers and this has led to the production and commercialization of materials such as Ultrasuede, artificial leather and silk, fabrics with a peach-fuzz sensation, perfumed hosiery, fabrics that change color due to microencapsulation of cholesteric liquid crystals, and numerous other luxury and novelty textile products.

Biomedical and Health Care. This area is an extremely diverse and expanding market for high performance fiber concepts and their reduction to the desired end use performance. Although the complexity of the human body and its physiological functions are difficult to mimic with artificial materials, significant progress has been made in the introduction of biomaterials and related health care products. The numerous uses of fibers in medicine have been critically and concisely reviewed, including nonimplantable items (such as surgical dressings, gowns, drapes, pressure bandages), extra corporeal devices (components of dialyzers and oxygenators, catheters) and implants (sutures, vascular grafts, artificial organs, fibers and fabrics for surgical reinforcement, optical fibers for medical procedures, orthopedic devices) (45). In addition to having the desired mechanical and functional integrity for the particular biomedical application, the fibrous product must be nonallergenic, usually nonthrombogenic, noncarcinogenic, sterilizable, and biocompatible. Nonimplant items are usually derived from conventional fibers and fabrics, except for clean room garments. Examples of extra corporeal devices or components are regenerated hollow cellulose fibers for the artificial kidneys or dialyzers and hollow fibers or flat-sheet fibrous membranes derived from polypropylene, polytetrafluoroethylene, and various coated fabrics in blood oxygenation devices.

Implants generally can be classified as those that are designed to be bioabsorbable (eg, sutures) and those that are designed for more permanent function (nonabsorbable sutures, vascular grafts, and artificial organs). Bioabsorbable materials have been critically reviewed with emphasis on sutures and related fibrous products (46). Bioabsorbable sutures were originally derived from catgut or collagen, but now a variety of poly(glycolic acid) homo- and copolymers as well as poly(dioxanones) are used. Hydrogel fibrous composites have become important as wound dressings and as implants. Many of these hydrogel composites are poly(ethylene oxide) homopolymers and/or copolymers of poly(acrylic acid), polyacrylamide, or other related structures (47). Nonabsorbable sutures and reinforced structures for vascular grafts and artificial organs are derived from high tenacity polyamide, polypropylene, polyester, polytetrafluoroethylene, and various coated fibers and fabrics. Optical fibers for endoscopy and angioplasty procedures are usually composed of glass or poly(methyl methacrylate). In addition to these classifications, the production of biomimetic fibers, ie, fibrous materials that are specifically designed to mimic physiological and other anatomical functions of living organisms, has been described (48). Applications include artificial cells, nerves and muscles, biosensors, and artificial bones and teeth.

Defense and Aerospace. Military applications for high performance fibers and high technology textiles generally parallel those discussed earlier for civilian aircraft and protective clothing. Protective clothing is usually designed to be impermeable to chemical and biological warfare agents such as nerve and mustard gas. Design of protective clothing for these purposes is usually classified information, but it is reasonable to assume that the clothing assembly consists of microporous membranes and elaborate filtration devices. Another military application is the design of combat uniforms that are dyed in such a manner that they provide effective camouflage and are not detectable by night surveillance devices. Not surprisingly such materials have been adapted to civilian use by hunters to give better camouflage against deer and turkeys that see uv reflections. Printing of aramid fabrics for this purpose has been described (49).

Requirements for space suits are more complex and frequently involve garments that can circulate water and/or air through the fibrous assembly. Laminated and/or coated garments with specific requirements to pressure, radiation, temperature, and humidity are more structurally complex as a textile product relative to the types of fibers used in this aerospace fabrication.

Fiber composites and three-dimensional textile structures are frequently used in advanced military weapons and aircraft such as extended use armor tracks, materials for engine parts (nose spin cones and fan blades), and primary and secondary structural components of aircraft (main wings, horizontal tail and stabilizer, speed brakes, high performance gaskets, fins, and ailerons). Although many of these materials are carbon fiber–epoxy resin composites or glass fiber–epoxy resin composites, there is an increase in the use of aramid, polyimide, and polyetheretherketone (PEEK) fibers because of their excellent strength and thermal/oxidative and chemical stabilities.

Energy Use and Conservation.   A variety of materials are needed for high performance thermal insulation, particularly as components of nuclear reactors. Replacements for asbestos fibers are needed for components such as reactor core flooring, plumbing, and packaging. The fibers must be very resistant to high temperatures with outstanding dimensional stability and resistance to compression.

Filters for nuclear fuel reprocessing and replacement of fibrous composites to reduce or prevent radiation leaks are two other areas of application. Advanced fiber composites, aramids, and high tenacity polyethylene for specific applications are types of high performance fibers and structures being considered (38). Metal-coated fibers and fabrics have been effective in electromagnetic shielding applications. High performance fabrics have also been used for containment of hazardous waste in addition to being used for conservation of energy and improving the safety of nuclear power plants (50). The same considerations that apply to geotextiles for other applications also apply to geotextiles used for waste containment. However, there are additional performance requirements such as excellent resistance to municipal waste leachate. Thus, polypropylene and to a lesser extent polyethylene fibers are used in the construction of geotextiles for this type of application.

BIBLIOGRAPHY

1. S. L. Kwolek and co-workers, *Macromolecules* **10**, 1390 (1977).
2. A. Roviello and A. Sirigu, *J. Polym. Sci. Polym. Lett. Ed.* **13**, 455 (1975).
3. W. J. Jackson and H. F. Kuhfuss, *J. Polym. Sci. Polym. Chem. Ed.* **14**, 2043 (1976).
4. J.-I. Jin and co-workers, *Brit. Polym. J.*, 132 (1980).

5. S. L. Kwolek and R. R. Luise, *Macromolecules* **19**, 1789 (1986).
6. U.S. Pat. 3,637,595 (1972), S. G. Cottis, J. Economy, and B. E. Nowak (to Carborundum).
7. M. Jaffe and R. S. Jones, in M. Lewin and J. Preston, eds., *Handbook of Fiber Science and Technology*, Vol. 3, Part C, 1983, p. 349.
8. S. Ozawa and K. Matsuda, in Ref. 7, Part B, p. 1.
9. L. R. Denny, I. J. Goldfarb, and E. J. Soloski, *Materials Research Society Symposium Proceedings* **134**, 395 (1989).
10. S. Kumar, *Indian J. Fibre and Textile Res.* **16**, 52 (1991).
11. A. B. Conciatori, A. Buckley, and D. E. Stuetz, in Ref. 7.
12. U.S. Pat. 3,174,974 (Mar. 23, 1965), H. Vogel and C. S. Marvel (to the University of Illinois).
13. M. Jaffe, G. W. Calundann, and H.-N. Yoon, in Ref. 7, Part B, p. 83.
14. U.S. Pat. 4,137,394 (1976), C. E. Meihuizen, A. J. Pennings, and A. Zwijnenburg.
15. *Textiltechnik*, (4), 175 (1985).
16. U.S. Pat. 4,403,012 (1983), (to Allied Corp.).
17. Eur. Pat. 77,590 (1983), (to Stamicarbon, BV); Ger. Pat. 3,023,726 (1981), (to Stamicarbon, BV).
18. Eur. Pat. 144,997 (1985), (to Stamicarbon, BV).
19. Eur. Pat. 55,001 (1992), (to Stamicarbon, BV).
20. A. P. S. Sawhney and co-workers, *Text. Res. J.* **62**(1), 21 (1992); U.S. Pat. 4,976,096, P. S. Sawhney.
21. U.S. Pat. 4,837,076 (1989), F. P. McCullough and D. M. Hall.
22. F. P. McCullough and B. C. Goswami, "Novel Fibers and Their Ignition Behavior," *Hi-Tech Conference*, Clemson University, S.C., July 21, 1993.
23. U.S. Pat. 4,879,168; 4,943,478; 4,950,533; 4,950,540; 4,980,233; 4,997,716; 4,999,236; 5,024,877; F. P. McCullough and co-workers.
24. W. Eastes, ed., *Man-made Vitreous Fibers*, TIMA Inc., Mar. 1993.
25. H. W. Hill, Jr., *I&EC Prod. Res. Dev.* **18**, 252 (Dec. 1979).
26. U.S. Pat. 3,919,177 (Nov. 11, 1975), R. W. Campbell.
27. J. G. Scruggs and J. O. Reed, *Handbook of Fiber Science and Technology*, Vol. 3, Marcel Dekker, Inc., New York, 1985, pp. 341–342.
28. K. Matsumoto, *Jpn. Textile News* **456**, 59 (Nov. 1992).
29. T. Hongu and G. O. Phillips, *New Fibres*, Ellis Horwood, West Sussex, U.K., 1990, Chapt. 2.
30. M. Takahashi, in L. Rebenfeld, ed., *Science and Technology of Fibers and Related Materials*, Wiley-Interscience, New York, 1991, p. 38.
31. P. S. Tung and S. Jayaraman, in T. L. Vigo and A. F. Turbak, eds., *High-Tech Fibrous Materials*, American Chemical Society, Washington, D.C., 1991, Chapt. 4.
32. T. L. Vigo and J. S. Bruno, *Proc. NASA Tech. Conf. 2002*, **2**, 307 (1993).
33. K. Kajiwara and S. B. Ross-Murphy, *Nature* **355**, 208 (Jan. 16, 1992).
34. *Chemiefasern/Textil-industrie* **39**(5), T146/E68 (May 1989).
35. *High Performance Textiles*, 12 (Jan. 1991).
36. E. L. Barish and B. Tariyal, in M. Lewin and J. Preston, eds., *High Technology Fibers, Part C*, Marcel Dekker, Inc., New York, 1993, Chapt. 1.
37. A. K. Dinghra and H. G. Lauterbach, in J. Kroschwitz, ed., *Encyclopedia of Polymer Science and Engineering*, Vol. 6, John Wiley & Sons, Inc., New York, 1986, p. 756.
38. Ref. 29, Chapt. 6.
39. C. J. Sprague and G. W. Davis, in Ref. 31, Chapt. 20.
40. *Jpn. Textile News* **391**, 48 (June 1987).
41. M. Issi, *Jpn. Textile News* **456**, 63 (Nov. 1992).
42. N. Fahl and M. Faile, in Ref. 31, Chapt. 14.

43. T. L. Vigo, in M. Raheel, ed., *Protective Clothing*, Marcel Dekker, Inc., New York, in press, Chapt. 9.
44. Ref. 29, Chapt. 3.
45. D. Lyman, in Ref. 31, Chapt. 8.
46. Y. Ikada, in Ref. 7, Part B, Chapt. 8.
47. P. H. Corkhill, C. J. Hamilton, and B. J. Tighe, *Biomaterials* **10**(1), 3 (1989).
48. Ref. 29, Chapt. 4.
49. J. D. Hodge and E. A. Dodgson, in Ref. 34, Chapt. 17.
50. S. D. Menhoff, J. W. Stenborg, and M. J. Rodgers, in Ref. 34, Chapt. 23.

General References

References 31 and 37.
M. Lewin and J. Preston, eds., *High Technology Fibers, Part A*, Marcel Dekker, Inc., New York, 1985, 397 pp.
M. Lewin and J. Preston, eds., *High Technology Fibers, Part B*, Marcel Dekker, Inc., New York, 1989, 332 pp.
M. Lewin and J. Preston, eds., *High Technology Fibers, Part C*, Marcel Dekker, Inc., New York, 1993, 376 pp.
T. L. Vigo and B. J. Kinzig, eds., *Composite Applications: The Role of Matrix, Fiber, and Interface*, VCH Publishers, New York, 1992, 407 pp.
T. L. Vigo and A. F. Turbak, eds., *High-Tech Fibrous Materials: Composites, Biomedical Materials, Protective Clothing, and Geotextiles*, American Chemical Society, Washington, D.C., 1991, 398 pp.

MALCOLM POLK
Georgia Institute of Technology

TYRONE L.VIGO
U.S. Department of Agriculture

ALBIN F. TURBAK
Consultant

HIGH PRESSURE TECHNOLOGY

During the nineteenth century the growth of thermodynamics and the development of the kinetic theory marked the beginning of an era in which the physical sciences were given a quantitative foundation. In the laboratory, extensive researches were carried out to determine the effects of pressure and temperature on the rates of chemical reactions and to measure the physical properties of matter. Work on the critical properties of carbon dioxide and on the continuity of state by van der Waals provided the stimulus for accurate measurements on the compressibility of gases and liquids at what, in 1885, was a surprisingly high pressure of 300 MPa (~3,000 atm or 43,500 psi). This pressure was not exceeded until about 1912.

High Pressure in the Chemical Industry. The use of high pressure in industry may be traced to early efforts to liquefy the so-called permanent gases using a combination of pressure and low temperature. At about the same time the chemical industry was becoming involved in high pressure processes. The discovery of mauveine in 1856 led to the development of the synthetic dye industry which was well established, particularly in Germany, by the end of the century. Some of the intermediate compounds required for the production of dyes were produced, in autoclaves, at pressures of 5–8 MPa (725–1160 psi).

A pressure process for synthesizing ammonia from nitrogen and hydrogen was patented in 1881, and a modification using pressures up to 10 MPa (1450 psi) was described in 1901. However, it was not until 1904 that the full significance of pressure as a means of increasing the yield of ammonia was properly appreciated. It was soon apparent that autoclaves designed for batch processing could not be used to handle large volumes of gas continuously at pressures up to 20 MPa (2900 psi) and temperatures in the region of 500°C. The modern form of reactor designed to withstand what were then abnormal working conditions was soon developed. Between 1910 and 1913 the Badische Analin and Soda Fabrik (BASF) constructed the first commercial ammonia plant at Oppau. Since then many variations have been made to the original process primarily in catalyst composition, in type of feedstock and method of generating synthesis gas, and in operating conditions. In 1921 a modified process which operated at 100 MPa (14,500 psi) was produced, but this was never popular and to this day most ammonia plants have been designed to operate at pressures in the range 15–35 MPa (2200–5100 psi) (see AMMONIA).

Other important chemical processes such as the synthesis of methanol and urea were developed in Germany. By 1923 BASF was manufacturing methyl alcohol by the catalytic reduction of carbon monoxide by hydrogen at 20–30 MPa and 390–425°C. A large number of so-called high pressure plants which were in operation throughout the world have been replaced by the low pressure process, 5–10 MPa, developed by ICI in 1966 (see METHANOL). In all urea synthesis plants NH_3 reacts with CO_2 to form ammonium carbamate which is simultaneously dehydrated to give urea. This synthesis, first carried out at the Leuna works of IG Farben, is now performed in reactors in which the pressure ranges from 13–30 MPa and the temperature from 180–200°C depending on the process (see UREA AND UREA DERIVATIVES).

Imperial Chemical Industries (ICI) operated a coal hydrogenation plant at a pressure of 20 MPa (2900 psi) and a temperature of 400–500°C to produce liquid hydrocarbon fuel from 1935 to the outbreak of World War II. As many as 12 such plants operated in Germany during World War II to make the country less dependent on petroleum from natural sources but the process was discontinued when hostilities ceased (see COAL CONVERSION PROCESSES, LIQUEFACTION). Currently the Fisher-Tropsch process is being used at the Sasol plants in South Africa to convert synthesis gas into largely aliphatic hydrocarbons at 10–20 MPa and about 400°C to supply 70% of the fuel needed for transportation.

Following the discovery of polyethylene by ICI in the early 1930s, work went ahead first on the design of a pilot plant and then of a commercial plant, commissioned late in 1939, for the continuous production of polymer at a pressure of 150 MPa (22,000 psi) and a temperature of 170°C. Because the operating pressure was about four times that employed in the majority of the ammonia and methanol plants in use at the time the engineering problems in the design of the plant were considerable. Furthermore, process design proved difficult because of the need to remove the high heat of polymerization. By the mid-1940s a number of alternative processes had been designed by E. I. du Pont de Nemours & Co., Inc. and Union Carbide in the United States and BASF in Germany. These processes all operated at high pressure but differed in the way in which the heat of polymerization was removed. ICI and Du Pont used stirred reactors, known as autoclaves, in which the heat of reaction was used to raise the temperature of the incoming stream of ethylene containing small quantities of initiator. Union Carbide and BASF developed tubular reactors in which the heat of polymerization is removed by coolant circulating through jackets surrounding the reactor.

The two main types of process for the production of low density polyethylene (LDPE), the stirred autoclave and the tubular reactor, have been modified in many ways to enlarge the range of products that can be made and to increase conversion. The introduction of low pressure processes for the manufacture of high density polyethylene (HDPE) in the late 1950s and those for the manufacture of linear low density polyethylene (LLDPE) in the years following 1968 have reduced the dependence of the plastics industry on high pressure processes for some types of resin, but as long as there is a need for LDPE and copolymers such as ethylene–vinyl acetate (EVA), high pressure processes will continue to be of importance. It is estimated that the current annual production of LDPE throughout the world exceeds 15×10^6 t (see OLEFIN POLYMERS).

Other Industrial Applications.   High pressures are used industrially for many other specialized applications. Apart from mechanical uses in which hydraulic pressure is used to supply power or to generate liquid jets for mining minerals or cutting metal sheets and fabrics, most of these other operations are batch processes. For example, metallurgical applications include isostatic compaction, hot isostatic compaction (HIP), and the hydrostatic extrusion of metals. Other applications such as the hydrothermal synthesis of quartz (see SILICA, SYNTHETIC QUARTZ), or the synthesis of industrial diamonds involve changing the phase of a substance under pressure. In the case of the synthesis of diamonds, conditions of 6 GPa (870,000 psi) and 1500°C are used (see CARBON, DIAMOND–SYNTHETIC).

Very high dynamic pressures can be produced in material by shock waves generated by exploding charges adjacent to the material; this technique is used for welding, forming, and metal cutting operations (see METALLIC COATINGS).

Technological Aspects. The many and varied industrial applications of high pressure make it necessary to be selective; here consideration is confined to the techniques used in the design and safe operation of continuous chemical processes operating at pressures above about 20 MPa (2900 psi). Authorities responsible for national pressure vessel and piping codes give detailed design procedures for vessels and piping systems, but until recently little consideration was given to providing guidance for the design of plants to operate at pressures above 70 MPa (10,000 psi). Since the 1940s the design of the various processes for the manufacture of LDPE has had to be based on first principles and modified in the light of experience. For reasons of industrial secrecy, very little was published about the detailed design of the equipment used, particularly the compressors, which were of vital importance in determining the capacity and operating pressure of the reactors. In the 1990s there are continuous processes working at 350 MPa (51,000 psi) and there is no reason why, if required, this pressure could not in favorable circumstances be increased.

Design of Thick-Walled Cylinders

Loss of containment of chemical process vessels operating at high pressures is a rare occurrence, and when it happens it is likely to have been caused either by gross deformation or fatigue cracking. Gross deformation may result from the inability of pressure relieving devices fitted to the vessel to reduce the pressure generated within the vessel by a deflagration or rapid decomposition with sufficient speed. The resulting overpressure may rupture the vessel, particularly if its strength is weakened as a result of high temperature produced simultaneously. Alternatively the deformation may be confined to a critical part of the vessel, such as a removable closure, and result in leakage.

Repetitive loading of high pressure components by repeated application or generation of pressure, or as a result of mechanical vibration, may cause fatigue cracks to propagate through the wall thickness of the component so that leakage ensues. However, if the material of construction lacks toughness the crack may propagate only a small distance through the wall thickness before fast fracture intervenes and the component fails catastrophically.

Much information is available on the deformation and fatigue behavior of simple thick-walled hollow cylinders; the more important aspects necessary for an understanding of the techniques used in the design of modern chemical process plant are given in this section. However, it must be remembered that most process vessels are not simple hollow cylinders; they are provided with removable closures, pipe connections, side holes, etc. The shape of these features is such that it is difficult to obtain an analytical solution to the stresses induced in them by the application of internal pressure. Since the 1970s great progress has been made in the use of finite element techniques to analyze the stresses in fittings, such as screwed closures, and in components where the stress distribution is modified by the presence of cracks.

Elastic Behavior.   In the following discussion of the equations relevant to the design of thick-walled hollow cylinders, it should be assumed that the material of which the cylinder is made is isotropic and that the cylinder is long and initially free from stress. It may be shown (1,2) that if a cylinder of inner radius, r_i, and outer radius, r_o, is subjected to a uniform internal pressure, P_i, the principal stresses in the radial and tangential directions, σ_r and σ_t, at any radius r, such that $r_o > r > r_i$, are given by

$$\sigma_r = \frac{P_i}{k^2 - 1}\left(1 - \frac{r_o^2}{r^2}\right) \tag{1}$$

$$\sigma_t = \frac{P_i}{k^2 - 1}\left(1 + \frac{r_o^2}{r^2}\right) \tag{2}$$

where the diameter ratio $k = r_o/r_i$.

The third principal stress, that in the axial direction, σ_z (Fig. 1), depends on whether the cylinder wall resists the force acting on closures attached to the ends of the cylinder or whether the force is opposed by some external support. The former method is the most common, the cylinder being fitted with removable closures attached by nuts and bolts, studs, or other coupling devices, so that the cylinder is said to be sealed under closed-end conditions (Fig. 2a). The second method makes use of one or two floating heads which allow the cylinder to expand or contract freely in the axial direction. The force on the heads arising from the pressure inside the vessel is then resisted by some external support and in this case the vessel is said to be sealed under open-end conditions (Fig. 2b). This arrangement has seldom been used for pressure vessels in the chemical industry, but is sometimes employed in the autofrettage process of strengthening components and to seal vessels used for isostatic compaction.

The stress in the axial direction of a cylinder sealed under closed-end conditions is given by

$$\sigma_z = \frac{P_i}{k^2 - 1} = \frac{\sigma_t + \sigma_r}{2} \tag{3}$$

Under open-end conditions, neglecting the small frictional force between the bore of the cylinder and the sealing rings,

$$\sigma_z = 0 \tag{4}$$

All cylinders analyzed in this article are sealed under closed-end conditions unless stated otherwise.

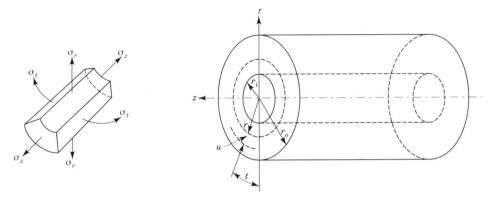

Fig. 1. Principal stresses acting on small element of cylinder wall at radius, r, when the cylinder is stressed elastically by internal pressure, P_i (3).

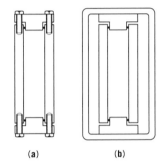

(a)   (b)

Fig. 2.   (a) Closed-end condition; (b) open-end condition (4).

Equations 1 to 3 enable the stresses which exist at any point across the wall thickness of a cylindrical shell to be calculated when the material is stressed elastically by applying an internal pressure. The principal stresses cannot be used to determine how thick a shell must be to withstand a particular pressure until a criterion of elastic failure is defined in terms of some limiting combination of the principal stresses.

Yield Pressure.   Figure 3 shows the pressure-expansion curve of a thick-walled cylinder made of a ductile material. In this figure, the strain in the tangential direction at the outside surface of the cylinder has been plotted against the internal pressure as the latter is slowly increased. From O to A the cylinder behaves elastically and if the pressure is released it will return to its original dimensions. At A the material at the bore reaches its elastic limit, and the corresponding pressure is known as the yield pressure, P_y. As the pressure is further increased, plastic deformation progresses through the wall thickness until somewhere between A and C the cylinder becomes completely plastic. At C a condition of instability is reached and localized bulging takes place. Finally the cylinder ruptures at some slightly lower pressure, D.

Early in the twentieth century it was considered prudent to design the shells of pressure vessels so that they operated within the elastic range of the

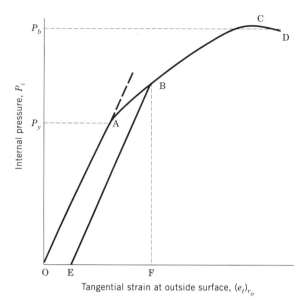

Fig. 3. Pressure expansion curve for thick-walled cylinder made of ductile material (5). See text.

material at all times. To that end designers arranged for the working pressure to be less than the yield pressure and the ratio of the latter to the former was regarded as the factor of safety with respect to yielding. This philosophy continued with the introduction of the pressure vessel codes but, as is shown later, the factor of safety with respect to yielding is expressed as the ratio of the yield strength of the material to the maximum permitted design stress generated by the internal pressure.

The state of stress in a cylinder subjected to an internal pressure has been shown to be equivalent to a simple shear stress, τ, which varies across the wall thickness in accordance with equation 5 together with a superimposed uniform (triaxial) tensile stress (6).

$$\tau = \frac{\sigma_t - \sigma_r}{2} = \frac{k^2}{(k^2 - 1)} \left(\frac{r_i}{r}\right)^2 P_i \tag{5}$$

The maximum shear stress occurs at the bore and is given by

$$\tau = \frac{k^2}{k^2 - 1} P_i \tag{6}$$

If it is assumed that uniform tensile stress, like uniform compressive stress (7), has no significant effect on yield, then the yield pressure of a cylinder subjected solely to an internal pressure may be calculated from

$$P_y = \tau_y \frac{(k^2 - 1)}{k^2} \tag{7}$$

where τ_y is the yield strength of the material in torsion.

Equation 7 predicts the correct yield pressure only if the material is isotropic, the cylinder free from residual stress prior to the application of pressure, and sufficiently long, eg, more than five diameters, for there to be no end effects.

Although a torsion test is simple to carry out, it is not commonly accepted as an integral part of a material specification; furthermore, few torsion data exist in handbooks. If, as is usually the case, the design needs to be based on tensile data, then a criterion of elastic failure has to be invoked, and this introduces some uncertainty in the calculated yield pressure (8).

Criteria of Elastic Failure. Of the criteria of elastic failure which have been formulated, the two most important for ductile materials are the maximum shear stress criterion and the shear strain energy criterion. According to the former criterion, $\tau_y = \sigma_y/2$ and from equation 7

$$P_y = \frac{\sigma_y}{2} \frac{(k^2 - 1)}{k^2} \tag{8}$$

For the gun steels often used in the construction of high pressure vessels, the latter criterion is favored (9) according to which, $\tau_y = \sigma_y/\sqrt{3}$. Hence from equation 7

$$P_y = \frac{\sigma_y}{\sqrt{3}} \frac{(k^2 - 1)}{k^2} \tag{9}$$

It may be seen from equation 7 that even if the wall of an initially stress-free cylinder is infinitely thick, $k = \infty$, the yield pressure cannot exceed τ_y. The wall thickness of the cylinder is not used very efficiently since, provided the pressure is sufficiently high, the inner layers may be on the point of yielding whereas the outer are comparatively lightly stressed. Furthermore, the thicker the wall the more inefficiently is the strength of the steel utilized. The yield strength of steels which can be forged and uniformly heat treated is such that the yield pressure of a cylinder having a bore of, eg, 350 mm is restricted to about 200 MPa (29,000 psi). To achieve a higher yield pressure steels having a greater yield stress would be required. Alternatively, the inherent strength of the steel may be used more effectively by prestressing cylinders to ensure a more uniform stress distribution under load.

Partially Plastic Thick-Walled Cylinders. As the internal pressure is increased above the yield pressure, P_y, plastic deformation penetrates the wall of the cylinder so that the inner layers are stressed plastically while the outer ones remain elastic. A rigorous analysis of the stresses and strains in a partially plastic thick-walled cylinder made of a material which work hardens is very complicated. However, if it is assumed that the material yields at a constant value of the yield shear stress (Fig. 4a), that the elastic–plastic boundary is cylindrical and concentric with the bore of the cylinder (Fig. 4b), and that the axial stress is the mean of the tangential and radial stresses, then it may be shown (10) that

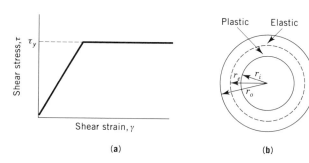

Fig. 4. (a) Shear stress diagram for elastic ideal plastic material; (b) partially plastic thick-walled cylinder.

the internal pressure, P_e, needed to take the boundary to any radius r_e such that $r_i < r_e < r_o$ is given by

$$P_e = \tau_y \left\{ 1 - \left(\frac{r_e}{r_o} \right)^2 + \ln\left(\frac{r_e}{r_i} \right)^2 \right\} \tag{10}$$

The resulting shear stress distribution while the internal pressure is acting shows the extent of yielding at radius r.

in the plastic region $r_i < r < r_e$, $\tau = \tau_y$ (11)

in the elastic region $r_e < r < r_o$, $\tau = \tau_y(r_e/r)^2$ (12)

If it is assumed that yield and subsequent plastic flow of the material occurs in accordance with the maximum shear stress criterion, then $\sigma_y/2$ may be substituted for τ_y in the above and subsequent equations. For the shear strain energy criterion it may be assumed, as a first approximation, that the corresponding value is $\sigma_y/\sqrt{3}$. Errors in this assumption have been discussed (11).

Assume pressure, P_e, needed to take the elastic–plastic boundary to radius r_e corresponds to point B (see Fig. 3). Then provided the cylinder unloads elastically when the internal pressure is removed, ie, unloading path BE is parallel to OA, the residual shear stress distribution is as follows.

$$\text{when } r_i < r < r_e, \qquad \tau = \tau_y - P_e \frac{k^2}{(k^2 - 1)} \left(\frac{r_i}{r} \right)^2 \tag{13}$$

$$\text{when } r_e < r < r_o, \qquad \tau = \tau_y \left(\frac{r_e}{r} \right)^2 - P_e \frac{k^2}{(k^2 - 1)} \left(\frac{r_i}{r} \right)^2 \tag{14}$$

The effect of subjecting a thick-walled cylinder to a pressure greater than the yield pressure and then releasing the pressure is to put the material adjacent to the bore of the cylinder in compression while the outer layers remain in

tension. On subsequent repressurization the cylinder will, to a first approximation, retrace the unloading path BE (see Fig. 3) so that the cylinder withstands
elastically a pressure equal to that applied originally.

Collapse and Bursting Pressure. If the pressure is sufficiently large to
push the plastic–elastic boundary to the outer surface of the cylinder so that
the fibers at that surface yield, then there is nothing to restrain the wall, and
the cylinder is said to collapse. With an ideal material which does not work
harden the collapse pressure, P_c, sometimes called the full plastic flow pressure,
the full overstrain pressure or the full thickness yield pressure, would be the
bursting pressure of the cylinder. It is given by equation 10 when $r_e = r_o$ thus

$$P_c = 2\tau_y \ln k \tag{15}$$

Little error is introduced using the idealized stress–strain diagram
(Fig. 4a) to estimate the stresses and strains in partially plastic cylinders since
many steels used in the construction of pressure vessels have a flat top to their
stress–strain curve in the region where the plastic strain is relatively small.
However, this is not true for large deformations, particularly if the material work
hardens, when the pressure can usually be increased above that corresponding
to the collapse pressure before the cylinder bursts.

An approximate procedure for estimating the stresses and strains in a
partially plastic cylinder, which uses actual stress–strain data for the material
and takes account for dimensional changes, has been devised (12,13). More
importantly, this procedure, which is based on shear stress–strain data obtained
in a torsion test, may also be used to estimate the bursting pressure. Tension data
may be used but the method is less accurate, particularly at large strains, where
the specimen may have necked (8,14). It is assumed that the stress system is one
of simple shear and that the material is incompressible, the internal pressure
corresponding to any given bore strain, and hence the pressure expansion curve,
may be computed (15). Figure 5 shows the calculated pressure expansion curves
for cylinders of various diameter ratios made of Ni–Cr–Mo steel EN25 (BS970-
826M31) (16). At the peak pressure the decrease in strength of the cylinder
due to wall thinning is just counterbalanced by the increase in strength due
to work hardening of the material of which the cylinder is made. It has been
observed that if the pressure in a cylinder made of a ductile material is increased
slowly, bursting may not occur at the peak pressure and that the cylinder may
continue to expand at a reduced pressure prior to bursting. The peak pressure
reached during a burst test is often known as the ultimate bursting pressure (16).

The validity of Manning's method (12,13) for predicting the pressure–expansion curve of thick-walled cylinders up to their ultimate bursting
pressure has been experimentally confirmed, at room temperature, for a 0.15%
C steel, a 0.3% C steel and a Ni–Cr–Mo steel EN25 (BS970-826M31) and
on two alloy steels EN25 and EN40 (BS970-722M24) and a carbon steel EN3
(BS970-M26) up to temperatures of 370°C (18). In addition, ICI has provided
data on 3.5% NCMV steel cylinders having a diameter ratio of 1.61 at 20°C
and 350°C (15). Based on some of the earlier tests it was suggested (14,16)

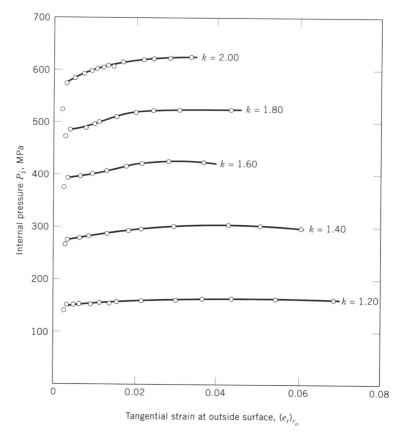

Fig. 5. Internal pressure expansion curves for cylinders of EN25 (16,17); $k = r_o/r_i$. To convert MPa to psi, multiply by 145.

that the following equation might be used to estimate the bursting pressure of thick-walled cylinders.

$$P_b = 2\sigma_u \frac{(k - 1)}{(k + 1)} \tag{16}$$

Bursting tests have been carried out on nearly a hundred thick-walled cylinders made of carbon, low alloy, and stainless steels, together with some nonferrous materials. The diameter ratio of the cylinders varied from 1.75 to 5.86, and some tests were carried out at 660°C. An analysis of the results (19) showed that 90% of the cylinders burst within ±15% of the value given by equation 17.

$$P_b = \frac{\sigma_y}{\sqrt{3}} \ln k \left(2 - \frac{\sigma_y}{\sigma_u} \right) \tag{17}$$

Plots of the bursting pressures of the Ni–Cr–Mo cylinders (EN 25) vs k derived from equations 16 and 17 show that neither equation is in such good agreement with the experimental results as is the curve derived from Manning's theory. Similar conclusions have been reached for cylinders made of other materials which have been tested (16). Manning's analytical procedure may be programmed for computation and, although torsion tests are not as commonly specified as tension tests, they are not difficult or expensive to carry out (20).

Prestressed Cylinders. Two considerations of importance in the design of a thick-walled cylinder to operate at high pressures are the initial yield pressure and the bursting pressure. Whereas the yield pressure approaches the yield shear stress of steel asymptotically as the diameter ratio increases, the bursting pressure continues to increase monotonically; hence the margin between the initial yield and bursting pressure gets larger as the diameter ratio increases. Unless the tensile strength of the steel is changed, nothing can be done to affect the bursting pressure of a cylinder of given diameter ratio. On the other hand, the pressure to cause initial yielding of the bore of the cylinder can be raised by developing compressive stresses at the bore prior to the application of pressure, a technique known as prestressing.

The potential benefits of prestressing may be seen from Figure 6, in which curve AB represents the distribution of shear stress across the wall of an initially unstressed cylinder generated by an internal pressure P. If the cylinder could be prestressed in such a way that the stress distribution prior to pressure, P, being applied were to be given by curve CD, then the resultant distribution of shear stress when the cylinder is under load would be represented by line EF obtained from the algebraic summation of curves AB and CD. One effect of the prestressing would be to produce a more uniform resultant stress across the wall thickness at the design pressure; another would be to lower the maximum shear stress at the bore. Figure 6 shows that a prestressed cylinder behaves elastically provided the stress at C induced by prestressing, or that at E induced by the working pressure, do not exceed the yield strength of the material of construction. If it is assumed that the shear yield stress of the material in compression is the same as that in tension, then the maximum internal pressure which a prestressed cylinder, having a diameter ratio $k > 2.2$, could withstand elastically would be twice as large as that which an initially stress-free cylinder of the same radius ratio could withstand.

A more important effect of prestressing is its effect on the mean stress at the bore of the cylinder when an internal pressure is applied. It may be seen from Figure 6 that when an initially stress-free cylinder is subjected to an internal pressure, the shear stress at the bore of the cylinder increases from O to A. On the other hand, when a prestressed cylinder of the same dimensions is subjected to the same internal pressure, the shear stress at the bore changes from C to E. Although the range of shear stress is the same in the two cases (distance OA = CE), the mean shear stress in the prestressed cylinder, represented by point G, is smaller than that for the initially stress-free cylinder represented by point H. This reduction in the mean shear stress increases the fatigue strength of components subjected to repeated internal pressure.

In practice a uniform distribution of the working stress cannot be achieved, but it may be approached by various methods of construction such as compound shrinkage, tape winding, and autofrettage which have their origin in the design of ordnance.

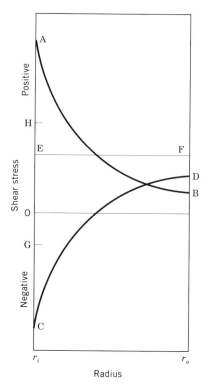

Fig. 6. An idealized example of prestressing (21). See text.

Autofrettage.   The process of overstraining a thick-walled cylinder by applying an internal pressure in excess of the yield pressure, so as to develop a
favorable residual stress distribution to increase the range of internal pressure
which the cylinder can withstand elastically, is known as autofrettage. It is a
French word meaning self-hooping, first used in 1906 to describe a procedure
for strengthening gun barrels. However, it was not until World War II that the
process was widely used for this purpose, and it now plays an important role
in the design of high pressure plants. Since the 1960s, the value of autofrettage
in increasing the initial yield pressure of a plain cylinder has been almost completely overshadowed by its widespread use in increasing the fatigue strength of
components, some of which are of complex shape.
    The pressure expansion curve of a thick-walled cylinder undergoing auto-
frettage is shown diagrammatically in Figure 7. If the autofrettage pressure
represented by point B is released, the unloading curve is more or less straight
and parallel to the elastic line OA. The permanent strain at the outside surface is represented by distance OC. If the pressure is now reapplied, the loading
curve is initially parallel to the elastic line and the pressure corresponding
to point D is sometimes known as the reyield pressure. When the pressure
approaches the maximum value of the previous loading, the stress–strain curve
bends over and continues as if the process has not been interrupted at B.
This behavior is typical when the degree of overstrain is small; however, if
the pressure is increased to point F, such that more of the wall thickness

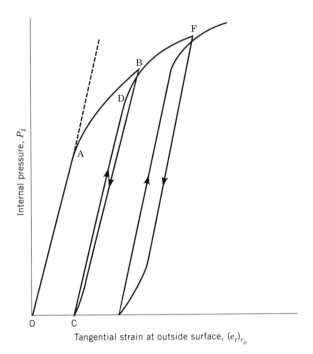

Fig. 7.   Pressure expansion curve of a thick-walled cylinder undergoing autofrettage (22).

is overstrained, then the slope of the unloading curve decreases as the inner layers yield in compression. The decrease in the yield strength of the material in compression, resulting from plastic deformation in tension, is known as the Bauschinger effect, and the assumption that the cylinder unloads elastically from the autofrettage pressure may lead to a higher estimate of the residual stresses than is realized in practice.

The pressure external expansion relationship for a thick-walled cylinder and the subsequent distribution of the residual stresses when the autofrettage pressure is removed may be calculated from material properties obtained from torsion tests on a small number of thin tubes (23). Using this procedure, the residual stresses in cylinders made of EN25 induced by autofrettage have been calculated and compared with those measured experimentally, using the Sachs boring-out technique, and with those calculated on the assumption that the cylinders unloaded elastically. It was found that with cylinders having a diameter ratio of about 2.4, autofrettaged to take the plastic–elastic boundary to the geometric mean radius, the residual stress at the bore of the cylinder was only about 65% of that expected had the release been elastic.

Quenched and tempered low alloy steels, often used for vessels, have a low strain hardening coefficient and a significant Bauschinger effect. Although other methods have been proposed to allow for the Bauschinger effect (24–27), when calculating the reyield pressure and the residual stresses in autofrettaged plain cylinders, none has yet achieved widespread acceptance. In addition to the Sachs boring-out technique, a number of other experimental methods, some nondestructive, are available for determining residual stress distributions (25).

Low temperature heat treatment, usually in the range 250–300°C, was given after the autofrettage of guns as it was thought to reduce the hysteresis loops, shown in Figure 7, and restore elasticity (28). Although tests (23) have shown that it does have a slightly beneficial effect, the problems of controlling it on all but the smallest of specimens have cast doubts on its value for long pipes and large vessels. At higher temperatures residual stresses induced by autofrettage may be relaxed unless the component is made of creep-resistant material (29).

The autofrettage process is usually controlled by measuring the strain in the tangential direction on the outside surface of the component while the pressure is acting and again after it has been removed. If the component is long, as with a length of tube, it may be necessary to record the strain at a number of points along the length. Since the autofrettage pressure may be two or three times the working pressure of the component, special arrangements may be needed to withstand the end load generated by the autofrettage pressure. Precautions and techniques used in the autofrettage of components for reciprocating compressors at pressures up to 1.3 GPa (189,000 psi) have been described (30) and the freezing pressures of pressure transmitting oils at room temperature have been measured (31).

Autofrettage of small bore tubes can be effected by pushing highly polished and generously lubricated oversized drifts or balls through the bore. The process can be controlled by measuring the tangential strain at the outside surface of the tube, and, if a number of passes are used, a very good bore finish is achieved. Residual stresses developed in tubes ($k = 2.5$) 300 mm long, having a bore diameter of 27 mm made of 2.5% Ni–Cr–Mo steel (EN26) as a result of forcing oversized tungsten carbide balls through the bore, have been measured experimentally (29).

Compound Shrinkage.   In its simplest form (Fig. 8a) compound shrinkage consists of machining the inner radius of an outer component I, $(r_i)_I$, so that it is smaller than the outer radius of an inner component II, $(r_o)_{II}$. The difference between the two is known as the radial interference δ. To assemble the cylinders, outer component I is heated and/or inner component II cooled so that the outer component can be slipped over the inner as shown in Figure 8b. When the temperature of the assembly returns to ambient, a compressive stress (pressure) is generated across the interface which simultaneously compresses the inner and expands the outer component and, in so doing, displaces radius $(r_i)_I$ by U_I and radius $(r_o)_{II}$ by U_{II}. Unfortunately, it is difficult to carry out this operation without setting up stresses in the axial direction (32).

The residual shear stress distribution in the assembled cylinders, prior to the application of internal pressure, may be calculated, from pressure P^*, generated across the interface. The resulting shear stress distribution in the compound cylinder, when subjected to an internal pressure P_i, may be calculated from the sum of the residual stress distribution and that which would have been generated elastically in a simple cylinder of the same overall radius ratio as that of the compound cylinder.

In a correctly designed and assembled compound cylinder the stresses are all within the elastic range; however, if the interference is too large the inner component can yield in compression or the outer component in tension. On the

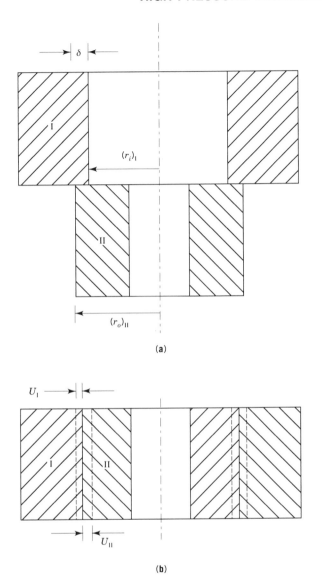

Fig. 8. Compound cylinders (**a**) before and (**b**) after assembly. Not to scale.

other hand, if the internal pressure is too high the inner or outer component can yield in tension.

The radial interference, δ, necessary to achieve pressure P^* may be calculated from the radial displacements U_I and U_{II} generated during assembly, assuming that the shrinkage is carried out without generating an axial stress in either component.

It may be shown (33) that when the inner surface of a cylinder made of n components of the same material is subjected to an internal pressure, the bore of each component experiences the same shear stress provided all components have the same diameter ratio. For these optimum conditions,

$$k_1 = k_2 = k_3 \ldots = k_n = k_o^{1/n} \tag{18}$$

The yield pressure of the multicomponent cylinder is given by

$$P_y = \tau_y n \frac{(k_o^{2/n} - 1)}{k_o^{2/n}} \tag{19}$$

and plotted in Figure 9 in dimensionless form against the overall diameter ratio k_o with the number of components, n, as parameter. It might appear possible from Figure 9 to design a cylinder to withstand very high internal pressures if both the number of components and the overall diameter ratio k_o were large enough. However, this cannot be done because the residual shear stress at the bore of the vessel after assembly would be so large that the material would yield. If it is assumed that the shear yield stress is the same when the material is stressed in compression as it is when stressed in tension, then the maximum internal pressure which the multicomponent cylinder can withstand without yielding is given by

$$p_y = \frac{2\tau_y(k_o^2 - 1)}{k_o^2} \tag{20}$$

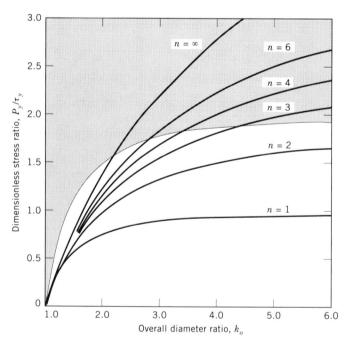

Fig. 9. Yield pressure of multicomponent vessels designed for optimum conditions (34).

Provided the design is such that it can be represented by coordinates which fall within the unshaded area, then the residual stress will not exceed the yield strength of the material. When the cylinder is built up of n components of the same material, it can be shown (35,36) that the interference per unit radius required for all cylinder mating operations is given by

$$\frac{\sigma}{r} = \frac{2P_i}{nE} \tag{21}$$

where σ is the interference of the components in contact at radius r, P_i is the internal pressure, n the number of components, and E the modulus of elasticity.

To utilize fully the strength of the inner member it should be stressed from the yield point in compression to the yield point in tension. From Figure 9 it is seen that if the members are initially stress-free at least three must be employed to make this possible.

In practice compound shrinkage is often used to prestress a high strength or corrosion-resistant liner. The optimum radius ratios of components of different yield strengths have been shown (37,38) to be

$$\frac{\tau_{y_1}}{k_1^2} = \frac{\tau_{y_2}}{k_2^2} = \frac{\tau_{y_3}}{k_3^2} = \text{etc} \tag{22}$$

The maximum internal pressure, subject to the avoidance of reversed yielding, is then given by

$$P = \tau_{y_1} \frac{k_1^2 - 1}{k_1^2} + \tau_{y_2} \frac{k_2^2 - 1}{k_2^2} + \ldots \tau_{y_n} \frac{k_n^2 - 1}{k_n^2} \tag{23}$$

If all the components have the same elastic constants, the condition for reversed yielding is the same as that given by equation 20.

This optimum condition is designed to ensure that the bores of all components yield at the same time. If the cylinder is subjected to fatigue conditions, it has been suggested (39) that a better design criterion would arrange for the maximum normal stress, which controls fatigue crack propagation, to be the same in each component.

When the components are shrunk together there is a limitation on the temperature difference that can be employed. The upper temperature must not exceed the tempering temperature of the outer members and the lower temperature is limited by the dew point of the atmosphere, otherwise water condenses on the cooled component. It is also necessary to have a clearance of about 1 mm/m diameter to facilitate assembly. It is difficult to carry out the shrinkage operation without compressive axial stresses being generated in those components which are cooled, and tensile axial stresses being generated in those components which are heated, prior to assembly. During use the components in tension tend to contract in length; those in compression expand so as to relax the axial stresses. The resulting strains are unimportant unless, as with some compressor components, the ends of the compound assembly are lapped to provide sealing surfaces. If a compound vessel is to be used under conditions where the heat flux changes rapidly over the length of the vessel, such as during start-up and shut-down of a reactor, it is important to anchor the inner component to the outer at one point along their length so as to avoid relative movement of the components as a result of thermal ratcheting. The difficulties of shrinkage may be reduced, at least in the case of relatively small components, if the components are tapered (using an angle less than the angle of friction) and pushed together in a press.

Probably the largest compound vessels built were two triple-wall vessels, each having a bore diameter of 782 mm and a length of 3048 mm designed for a pressure of 207 MPa (30,000 psi). These vessels were used by Union Carbide Co. for isostatic compaction; unfortunately the first failed at the root of the internal thread of the outer component which was required to withstand the end load (40). A disadvantage of compound shrinkage is that, unless the vessel is sealed under open-end conditions, the end load on the closures has to be resisted by one of the components, which means that the axial stress in that component is high.

Although compound shrinkage has occasionally been used in the manufacture of pressure vessels for the polymerization of ethylene, its main application in the chemical industry has been in the construction of components, such as packing cups for reciprocating compressors, which are subjected to fatigue conditions or to the manufacture of compressor cylinders with wear-resistant liners made of brittle materials. Since these liners are made of hard materials such as tungsten carbide, which have poor tensile properties, it is necessary to ensure that they are never subjected to tensile stress. The differences in the modulus of elasticity and in Poisson's ratio between tungsten carbide and steel must be taken into account when calculating the required interference. The coefficient of thermal expansion of tungsten carbide is approximately 7×10^6 per °C compared with 12×10^6 per °C for steel; provided the shrinkage requirement can be kept below about 2 mm/m, a worn or damaged liner can usually be made to drop out of its outer mantle by heating the assembly through about 450°C.

Composite Vessels.   *Wire-Wound Vessels.*   Wire-winding was used extensively in the manufacture of large guns, but is seldom employed for chemical process vessels mainly because of the difficulty of providing the vessel with sufficient strength in the axial direction. However, this problem does not arise if, as in the case of some isostatic presses, the vessel is sealed under open-end conditions when the advantages of prestressing by wire winding can be realized (41,42).

Tape-Wound Vessels.   In 1937 it was suggested (43) that lack of axial strength in wire-wound vessels might be overcome if the vessels were reinforced with interlocking tape, and in 1938 IG Farben produced the first Wickelkorper or tape-wound vessel. In this process a forged or welded inner tube was set up in a special lathe and spiral grooves were cut in the outer surface to match tongues in the tape (Fig. 10**a**). Before wrapping was commenced on the lathe, one end of the tape was welded to one of the dummy ends of the inner tube. As the tape unwound from a spool, it was heated electrically to 800–900°C and immediately after winding cooled first by jets of compressed air then by cold water. After the first layer of tape had been completely wound the end was welded to the other dummy end. The beginning of the next layer was then welded to the first layer so that it overlapped the joints of the underlying layer (Fig. 10**b**). Succeeding layers were similarly wrapped until the required wall thickness, up to 380 mm, was obtained. The theory of compound shrinkage shows that the residual contact pressures between 10 or more components are sufficiently small that they could be achieved by the tension generated as a result of the contraction of the tape and the wrapping tension. Flanges were screwed or shrunk on the layers, after which the dummy ends were cut off and the vessel machined to its proper length, as shown in Figure 10**c**. Tried methods for attaching cooling/heating coils to the outside surface of tape wound vessels are available.

The theory of tape-wound reinforcement and the calculated residual and working stress distributions in the radial and tangential directions in a typical vessel used for coal hydrogenation have been presented (45,46). By the end of World War II over 200 vessels for service at 30–70 MPa (4350–10,000 psi) had been made, the largest being 18 m long and 1.2 m internal diameter. Since then vessels have been manufactured for use as pulsation dampeners in LDPE plants, some of which have been in operation at pressures of 276 MPa (40,000 psi) since the 1950s (46). Few tape-wound vessels similar to those described have been made outside Germany (47).

Multilayer Construction. An alternative method of construction developed since 1931, first by A. O. Smith Corp. and subsequently used in Europe, consists of wrapping and welding successive layers of thin steel plate round an inner cylinder (48,49). The original objective was to produce a thick-wall from much thinner material, thereby easing the task of ensuring sound material throughout. Later, multilayer vessels were found to have greater elastic strength than was expected because of the residual stresses produced by a combination of wrapping tension and weld contraction and are reported to be used at pressures up to 250 MPa (36,000 psi) (50). Current practice assumes that the residual stresses are generated by the transverse weld shrinkage in the longitudinal seams but analyses of the problem (50–52) show that many other factors are involved. The ends of the layer built shell are scarfed so that closures of conventional design

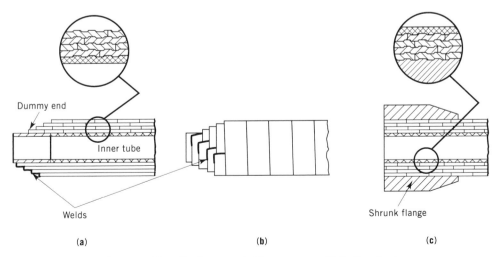

(a)   (b)   (c)

Fig. 10. Detail of tape-winding process (44). See text.

may be welded on to them. The relief provided by the movement of the individual layers prevents the development of internal stress; hence heat treatment of the circumferential welds which may be as much as 400 mm thick is not required.

The main advantages of multilayer construction are that it (1), reduces the risk of fragmentation; (2), resists the growth of flaws from one layer to another; and (3), enables leakage of the inner layer to be detected if the outer layers are provided with vent holes (53).

Multilayer vessels burst at about the same pressure as initially unstressed vessels of the same diameter ratio made of the same material (54). A bursting test was carried out on a special multilayer vessel manufactured by Krupp which consisted of an inner layer 5 mm thick, three middle layers 10 mm thick which had not been welded along their longitudinal seams, and an outer shell 5 mm thick. Friction between the layers was such that the bursting pressure was three-quarters that of a normal vessel of the same materials and dimensions (55). The bursting pressure of a model vessel of 307 mm bore diameter, made of seven 3-mm thick layers in which the longitudinal seams of the inner five layers were staggered over 60° and tack welded at a few points to keep the layers in contact, was only 4% below that calculated for an equivalent monobloc vessel (56). Similar experiments, including fatigue tests, have been carried out by other workers on model vessels (52,57).

Helically Coiled Vessels. These were developed in the 1970s in Japan, China, and the former Soviet Union and are produced by winding long strips of metal plate 900–1800 mm wide around a core tube at an angle of 15–30° by means of pressure and guide rolls, using a procedure similar to the winding of a coil spring. After the vessel is fabricated, outer wrapper sheets are used to enclose the helically coiled courses, the spring effect of which greatly improves the axial strength of the vessel (58).

Effect of Temperature on Design. In gas separation processes high pressure equipment is needed to operate at temperatures considerably below atmospheric, whereas some heterogeneous gas reactions have to be carried out at high temperatures to make them economically feasible. Both high and low temperatures have an effect on the mechanical properties of metals. In general, the ductile properties, and in particular the toughness and impact strength, of most low alloy steels, decrease sharply as the temperature is reduced and care has to be taken over the choice of materials if low temperature embrittlement is to be avoided. On the other hand, the yield and tensile strength of steels decrease as the temperature increases and allowance has to be made for this in estimating the static strength of a thick-walled cylinder at temperatures above ambient. Above about 350°C, creep starts to become an important factor with Ni–Cr–Mo steels of the sort used for high pressure applications, and the stresses in the wall of the vessel and its deformation are no longer independent of time. In addition to the effect of temperature on mechanical properties, temperature gradients, generated in the walls of vessels as a result of applied heat or of heat liberated by exothermic reactions proceeding within the vessel, cause thermal stresses which may need to be considered when estimating the stresses in a thick-walled cylinder subjected to both internal pressure and heat flux.

Thermal Stresses. When the wall of a cylindrical pressure vessel is subjected to a temperature gradient, every part expands in accordance with the thermal coefficient of linear expansion of the steel. Those parts of the cylinder at a lower temperature resist the expansion of those parts at a higher temperature, so setting up thermal stresses. To estimate the transient thermal stresses which arise during start-up or shutdown of continuous processes or as a result of process interruptions, it is necessary to know the temperature across the wall thickness as a function of radius and time. Techniques for evaluating transient thermal stresses are available (59) but here only steady-state thermal stresses are considered. The steady-state thermal stresses in the radial, tangential, and axial directions at a point sufficiently far away from the ends of the cylinder for there to be no end effects are as follows:

$$\sigma_r = -\frac{\beta}{2}\left[\frac{k^2}{k^2 - 1}\left(1 - \frac{r_i^2}{r^2}\right) - \frac{\ln(r/r_i)}{\ln k}\right] \tag{24}$$

$$\sigma_t = -\frac{\beta}{2}\left[\frac{k^2}{k^2 - 1}\left(1 + \frac{r_i^2}{r^2}\right) - \frac{\ln(r/r_i)}{\ln k} - \frac{1}{\ln k}\right] \tag{25}$$

$$\sigma_z = -\beta\left[\frac{k^2}{k^2 - 1} - \frac{\ln(r/r_i)}{\ln k} - \frac{1}{2\ln k}\right] \tag{26}$$

where $\beta = \alpha E \Delta T/(1 - v)$; $\Delta T = T_i - T_o$ is the steady-state temperature difference between the inner radius r_i and the outer radius r_o; α is the coefficient of thermal expansion; E, Young's modulus; and v, Poisson's ratio.

In the derivation of equations 24–26 (60) it is assumed that the cylinder is made of a material which is isotropic and initially stress-free, the temperature does not vary along the length of the cylinder, and that the effect of temperature on the coefficient of thermal expansion and Young's modulus may be neglected. Furthermore, it is assumed that the temperatures everywhere in the cylinder are low enough for there to be no relaxation of the stresses as a result of creep.

Figure 11 shows the thermal stresses in an austenitic stainless steel pipe having a diameter ratio of 2, when subjected to a temperature gradient of 100°C arising on the one hand from internal and on the other from external heating. The thermal stresses in the axial direction, unlike those induced by internal pressure, are not constant across the wall thickness. Furthermore the axial stress is not always the intermediate principal stress, so that care must be taken in evaluating the maximum shear stress. The tangential stress at the bore may be compressive or tensile, depending on whether heat flows from the inner to the outer wall or vice versa. Thus, when a vessel is heated internally, the thermal stresses oppose the pressure stresses, whereas when it is heated externally they augment them. This is illustrated in Figure 12 where the principal thermal stresses shown in the previous figures have been combined to give the distribution of the shear stress across the wall thickness, for the case of internal heating and that of external heating. Superimposed on this figure is the distribution of shear stress, neglecting the thermal stress present which arises from an internal pressure of 138 MPa (20,000 psi), together with the resulting working stress given by the algebraic summation of the thermal and pressure stresses.

Seldom is the temperature difference across the wall thickness of an item of equipment known. Since large temperature gradients may occur in the boundary

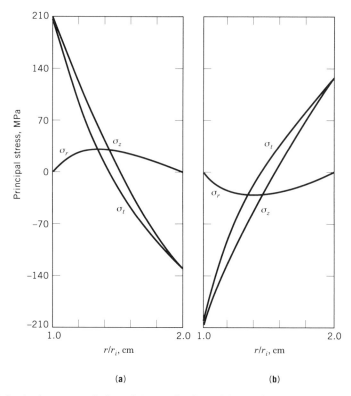

Fig. 11. Principal stresses induced in a cylinder of $k = 2$ by a temperature gradient of 100°C: (**a**) maximum temperature at outside surface; (**b**) maximum temperature at inside surface (61). To convert MPa to psi, multiply by 145.

layers adjacent to the metal surfaces, the temperature difference across the wall should not be estimated from the temperatures of the fluids on each side of the wall, but from the heat flux using equation 27

$$\Delta T = Q \ln k / 2\pi\lambda \qquad (27)$$

where Q is the rate of flow or heat from the cylinder per unit length and λ is the thermal conductivity of the metal.

Creep of Thick-walled Cylinders. The design of relatively thick-walled pressure vessels for operation at elevated temperatures where creep cannot be ignored is of interest to the oil, chemical, and power industries. In steam power plants, pressures of 35 MPa (5000 psi) and 650°C are used. Quartz crystals are grown hydrothermally, using a batch process, in vessels operating at a temperature of 340–400°C and a pressure of 170 MPa (25,000 psi). In general, in the chemical industry creep is not a problem provided the wall temperature of vessels made of Ni–Cr–Mo steel is below 350°C.

The stress system within the wall of a pressure vessel under creep conditions does not lend itself to simple analysis because of the continual redistribution of stress which occurs as creep takes place. Consequently, the calculation of the rate of deformation of a thick-walled cylinder subjected to internal

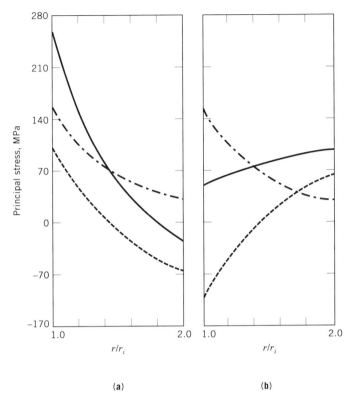

Fig. 12. Pressure and temperature stresses in a cylinder, $k = 2$, subjected to a steady temperature gradient of 100°C and an internal pressure of 138 MPa (20,000 psi): (**a**) maximum temperature at outside surface; (**b**) maximum temperature at inside surface (62). (-----), Thermal stress; (-- -- --), stress due to pressure; (——), combined pressure and thermal stress.

pressure undergoing creep poses many problems (63). From results of pressure creep tests (64) on cylinders, of diameter ratio 2, made of 0.19% carbon steel at 450°C, it was concluded (65) that a previously outlined procedure (66,67) provides a satisfactory correlation between tension and thick-walled creep data in the secondary region of creep. Subsequent (68) pressure creep tests were made on cylinders of 2.5% Ni–Cr–Mo steel (EN25) having a diameter ratio of 1.67 at pressures of 309–340 MPa (45–49,000 psi) and the results correlated (69,70). Before vessels can be designed with confidence to operate at high pressures in the creep range, more reliable experimental data against which the various theories may be tested is required.

Creep Rupture. The results from creep rupture tests on tubes under internal pressure at elevated temperatures (71,72) may be correlated by equation 16, in which σ_u is replaced by σ_{cr} the tensile creep rupture stress after time t at temperature T.

$$P_b = 2\sigma_{cr} \frac{(k-1)}{(k+1)} \tag{28}$$

Short-term (100 h) internal pressure creep rupture tests of small bore cylinders having diameter ratios of 3 to 4, made of a number of different refractory metals, have been investigated at temperatures in the range 900–1300°C (73).

Internal Heating. In many high pressure gas reactors, where for economic reasons it is essential to operate at a high temperature, special provision is made by heat exchange or direct cooling to keep the inner walls at a temperature low enough to minimize creep. One way of doing this is to allow the cool gas entering the reactor to flow between the inner surface of the vessel and an insulated layer on the outside of the reaction chamber. Thus the outer vessel, which has to withstand the internal pressure, is at a relatively low temperature, while the inner reaction chamber, which is maintained at a high temperature, is not subjected to an unbalanced pressure.

With a batch process, such as hot isostatic compaction (HIP), heat exchange as used in a continuous reactor is not possible, and it is common practice to provide a furnace within the pressure vessel which is thermally insulated to ensure that the temperature of the vessel does not rise above 300°C. Most HIP operations involve gas pressures in the range 70–200 MPa (10–29,000 psi) and temperatures of 1250–2000°C, occasionally 2250°C (74). The pressure vessel may have a bore diameter from 27 to 1524 mm (75) and is nearly always provided with threaded closures sealed with O-rings made of elastomer provided the temperature is low enough.

In the 1950s, serious problems were experienced in the design of furnaces (qv) to prevent unrestrained gas convection currents which gave rise to very large temperature gradients (76,77) and made furnace control difficult. Figure 13**a** represents convection currents around an electrically heated furnace tube in an enclosure pressurized by gas. To achieve a uniform temperature over the central section of tube shown in Figure 13**b**, the furnace was wound in four sections so that the power input to each section could be varied independently. Covers were fitted to the top and bottom of the tube to prevent gas flowing through it, and the charge within the furnace was surrounded by fine dense alumina powder of low permeability. Convection currents on the outside of the tube were largely prevented by a series of impermeable baffles, the volume between the baffles being filled with alumina powder. The performance of the furnace at temperatures up to 1250°C and helium gas pressures of 100 MPa (14,500 psi) was satisfactory (79).

During the 1960s, improvements in furnace design made possible the replacement of helium by less expensive argon. A convection furnace, widely used since about 1976, heats the dense gas and transfers heat by natural convection to the workpiece above. A liner is used to provide a path for the gas to flow (Fig. 14**a**) and circulation helps to achieve a uniform temperature throughout the work area. In the forced circulation furnace (see Fig. 14**b**) high heating and cooling rates may be achieved. Convection furnaces are favored because, for a given vessel diameter, the work cavity is larger and the heating elements are less susceptible to damage during the loading/unloading process.

The workpiece volume of production HIP equipment ranges from 0.05 to 2.5 m^3 but seldom exceeds 1.5 m^3, and the most common materials for the elements of the furnaces are graphite, molybdenum, and Ni–Cr alloy (75). To ensure that the temperature of the vessel is sufficiently low to enable O-rings

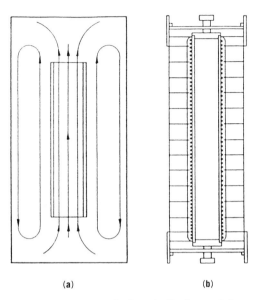

(a)   (b)

Fig. 13. (**a**) Convection currents around electrically heated furnace tube; (**b**) furnace construction to minimize convection (78).

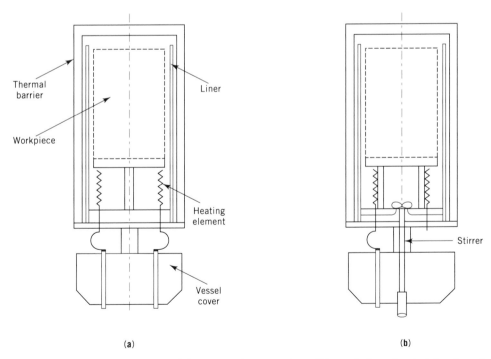

(a)   (b)

Fig. 14. (**a**) Workpiece heated by natural convection; (**b**) workpiece heated by forced convection (80).

made of elastomer to be used to seal the upper and lower closures, it is necessary to cool the vessel wall. This can be done by pumping coolant through helical passages of a liner shrunk into the bore of the vessel or through cooling coils attached to the outside of the vessel. The need, for process reasons, to use argon containing oxygen in HIP furnaces for super-conductive materials has generated a need for oxygen compatible furnaces, vessels, seals, and pumps (79).

Fatigue. Fatigue, or the failure of metals under repeated application of a stress insufficient to cause failure on the first application, was recognized in the middle of the nineteenth century. However, it was not until the mercury lute compressors, used by ICI in their process for polymerizing ethylene, failed unexpectedly after only several months of operation that engineers became aware that fatigue data, which relate to simple kinds of loading, eg, reversed bending, may not be directly applicable to the design of equipment subjected to repetitions of internal pressure (81).

After World War II, at the University of Bristol, a machine was developed (usually known as a Bristol-type machine) to study the fatigue strength of tubular specimens subjected to a pulsating internal pressure generated by reciprocating a plunger within the specimen, the internal space being filled with oil (82,83). The early machines were capable of speeds in excess of 15 Hz which enabled them to generate 10^7 cycles in a week of continuous operation. With specimens of the size shown in Figure 15, the maximum cyclic pressure, which could be varied by changing the stroke of the plunger, was limited to 350 MPa (51,000 psi). However, both specimen bore size and maximum cyclic pressure have been increased as similar machines have been built and developed at a number of facilities in the U.K. and United States. One of the latest is designed for operation at pressures up to 700 MPa (101,000 psi) at speeds up to 16 Hz (85).

In Figure 16 the maximum shear stress at the bore of cylinders made of a 2.5% Ni–Cr–Mo steel EN25 T, generated by repeated internal pressure, has been plotted against the number of cycles to failure for cylinders, of six different diameter ratios in the range 1.2 to 3.0 (83). After machining, the bore of each specimen was honed to a smoothness of 0.025–0.1 μm or better, before being stress relieved in vacuo to remove the residual compressive stresses generated by the honing process. Tests discontinued after 10^7 cycles are indicated on Figure 16 by an arrow. The results for each diameter ratio show considerable scatter, which

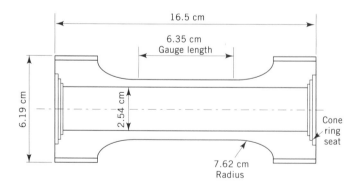

Fig. 15. 2.54 cm (1 in.) bore diameter fatigue test specimen (84).

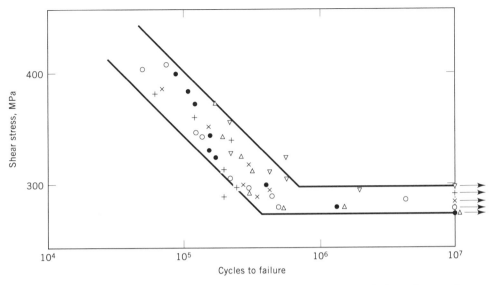

Fig. 16. Maximum shear stress at bore vs number of cycles to failure for specimens of EN25 T steel at various k values. $+$, $k = 1.2$; $\bigcirc$, $k = 1.4$; $\times$, $k = 1.6$; $\triangle$, $k = 2.0$; $\triangledown$, $k = 3.0$ (83). To convert MPa to psi, multiply by 145.

is normal for all fatigue tests, but there is a well-defined fatigue limit at a bore shear stress of about 270–300 MPa (39,000–43,000 psi). A cylinder made of the same material, having the same mechanical properties and bore surface finish designed so that the shear stress at the bore is less than the fatigue limit, should under conditions of repeated internal pressure last indefinitely. In practice, because of the statistical nature of fatigue and the differences between the test samples and an actual cylinder, the shear stress at the bore has to be limited to, eg, 0.3–0.5 of that at the fatigue limit to increase the probability of the cylinder lasting indefinitely.

Similar correlations, based on the shear stress at the bore of the cylinder, were found with the following ductile materials tested (82,83,86–88): mild steel, EN25 T, EN25 V, EN40 S, EN56 C, 18-8-Ti stainless steel, and maraging steel. Cylinders of titanium (82), beryllium copper (86), and an aluminium alloy (82) failed to give consistent results (86). There was no clear indication of a "knee," ie, fatigue limit, with these materials and the endurance limit or cyclic stress to cause failure in a defined number of cycles, such as 10^7, was reported.

The composition of the materials tested, their heat treatment, and their static and fatigue properties have been reviewed (89). Because fatigue results for cylinders of different diameter ratios subjected to repeated internal pressure can be correlated on the basis of the maximum bore shear stress, it might have been expected that this stress at the fatigue limit would correspond to the fatigue limit for the material in repeated torsion. However, it was found that the limit for the pressure tests was much lower than expected. Attempts to relate the limiting shear stress at the bore of the cylinder to the fatigue or endurance limit of the material in other forms of loading, eg, torsion, were not successful, and it was necessary to adopt an empirical approach. With the exception of

EN56 and titanium, the ratio of the range of shear stress at the bore at the fatigue limit of 10^7 cycles to the ultimate stress $\Delta\tau_{r_ifl}/\sigma_u$ lies between 0.36 and 0.27. Thus it appears that the limiting maximum shear stress which can be endured indefinitely at the bore of a plain cylinder is about one third of the ultimate tensile stress at least up to tensile strengths of 1 GPa (145,000 psi). If the material is markedly anisotropic the ratio $\Delta\tau_{r_ifl}/\sigma_u$ may be smaller (90).

A number of workers (90,91) concluded that the reason the fatigue results under pulsating pressure were lower than expected was due to the ingress of oil into incipient microcracks at the bore surface. There is a good correlation (eq. 29) between the range of repeated internal pressure ΔP, which must not be exceeded if fatigue initiation is to be avoided, and the range of repeated stress for the material of construction at the fatigue limit in uniaxial tension $\Delta\sigma_{fl}$ (91).

$$\Delta P = \frac{\Delta\sigma_{fl}(k^2 - 1)}{2k^2} \tag{29}$$

The uniaxial specimen should be cut so that its axis is tangential to the cylinder.

Nature of Failure.   Fatigue cracks in a thick-walled cylinder of ductile material generated by pulsating internal pressure are very different from those produced by static loading in which the pressure is gradually increased to the bursting point. In the latter case, fracture nearly always takes a spiral course across the wall thickness (maximum shear stress trajectory), and is preceded by considerable plastic deformation. On the other hand, with a plain cylinder, a fatigue crack initiates and propagates from inclusions intersecting the bore surface and usually spreads outward in all directions in the axial–radial plane so that its shape is approximately semicircular. If the cylinder is made of ductile material, the crack usually continues to propagate across the wall thickness until it intersects the outside surface and leakage ensues. The crack at the outside surface is so fine that it is difficult to detect with the naked eye. The fracture surfaces, revealed when the tube is broken open, are found to be smooth with a rippled appearance characteristic of fatigue. This type of behavior is sometimes known as leak before break. On the other hand, if the material lacks toughness, the propagation of the fatigue crack may be interrupted part way through the wall by the intervention of fast fracture, resulting in what is sometimes known as the break before leak mode of failure.

The life of a component, as measured in a fatigue test, is the number of cycles needed to initiate a crack and cause it to propagate across the wall until it intersects the outside surface or until fast fracture intervenes.

Mean Stress.   The fatigue strength of plain bored cylinders (Fig. 15) was established using repeated pressure cycles, in which the pressure increased to a specified maximum value and then dropped almost to zero, thus making the mean stress half the maximum. Using a test machine in which both the mean stress and the range of stress could be varied, it was found, in the life range 10^6 to 10^7 cycles, that the fatigue strength of thick-walled cylinders, $k = 1.8$, of EN25 depended not only on the maximum range of bore shear stress, but also on the associated mean shear stress (92) or the mean normal stress on the plane of maximum shear (93). Assuming the critical stress to be the mean shear stress at the bore, then as this increases, the fatigue strength, as measured by the maximum bore shear stress to cause failure in 10^6 to 10^7 cycles, decreases as shown in Figure 17. The reversed stress condition (zero mean stress) was achieved using specimens around which an outer mantle was shrunk. The fatigue strength of autofrettaged cylinders is increased relative to that of a nonautofrettaged cylinder, as a result of the reduction in the mean shear stress brought about by overstrain. However, the increase is not as large as expected because of the Bauschinger effect. It has been suggested that residual stresses induced by autofrettage are significantly altered by cyclic loading (92), but this is not thought to be so at temperatures low enough for there to be no relaxation of residual stress due to creep.

One of the effects of autofrettage is to raise the mean stress on the outside surface of a thick-wall cylinder, subjected to a pulsating internal pressure, above that for a similar nonautofrettaged cylinder. If there are design features, such as threads, grooves, or severe metallurgical defects, to concentrate the stress at the outside surface, a fatigue crack may propagate from that surface to the bore

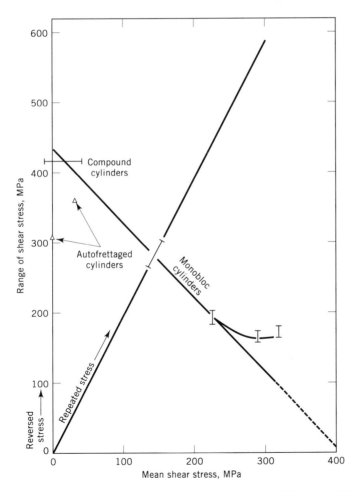

Fig. 17. Effect of mean shear stress on the fatigue strength of EN25 for life of 10^7 cycles (92). To convert MPa to psi, multiply by 145.

surface leading to a reduction in fatigue life. The effect of diameter ratio, degree of autofrettage, stress concentration, and internal pressure on the fatigue life of cylinders subjected to a repeated internal pressure on a Bristol machine have been measured (94). The specimens were made of AISI 4340, heat treated to give a nominal yield strength of 1100 MPa (160,000 psi), and the autofrettage was carried out mechanically using the swage method. A combination of shot peening and a reduction of the stress concentration of notches on the outside surface has been shown to be very effective in increasing the fatigue life of autofrettaged cylinders, of a design of interest to ordnance engineers, when subjected to hydrostatic fatigue loading (95).

Cross-Bores. The results of tests (87,88) carried out on specimens with a small radial cross-bore made of EN25 T which were stress-relieved indicate that the fatigue limit is at a shear stress of 136 MPa, as compared with 288 MPa for a cylinder without a cross-bore. This gives an actual fatigue strength reduction factor of 2.1 compared with a theoretical shear stress concentration value of 2.5 (89). The benefit of autofrettaging cross-bore specimens of the same design is shown in Figure 18, in which the specimens were autofrettaged at various pressures and subsequently given a low temperature heat treatment at 250°C for one hour (86).

A considerable reduction in stress concentration could be achieved by using a cross-bore which is elliptical in cross-section, provided the major axis of the ellipse is normal to the axis of the main cylinder. A more practical method of achieving the same effect is to have an offset radial hole whose axis is parallel to a radius but not coincident with it (97,98). Whenever possible the sharp edges at the intersection of the main bore with the cross bore are removed and smooth rounded corners produced so as to reduce the stress raising effects.

Short Life Tests. To cause thick-walled cylinders to fail in less than 10^4 cycles, larger internal pressures than those generated by the Bristol machine are required and low cycle, high stress fatigue studies are usually carried out by

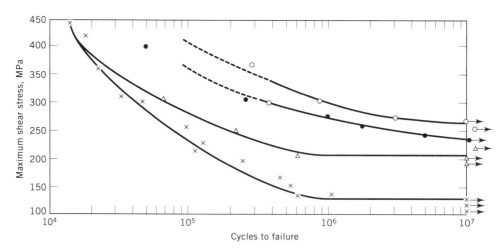

Fig. 18. Repeated pressure tests on autofrettaged EN25 steel cylinders, $k = 2.25$, with cross-bores (86,96). △, Autofrettaged at 324 MPa; ●, at 552 MPa; ○, at 665 MPa; ×, normal cross-bore cylinders. To convert MPa to psi, multiply by 145.

piping pressure from an intensifier to the component or components under test (99). When the set pressure is reached, a valve is opened and the pressure drops to near atmospheric after which the cycle is repeated. Alternatively, the intensifier is used as a jack to reciprocate a ram within an oil-filled space (88,100). About 10–20 cycles per min can be achieved using an hydraulic drive, but pressures of 1 GPa (145,000 psi) can be achieved in specimens of small volume (99) or 300 MPa (43,500 psi) in vessels having a bore of 300 mm (100,101). The hazards and safeguards of fatigue tests on large vessels have been reported (102).

Short life tests have been carried out at Watervliet Arsenal (99) on autofrettaged cylinders, autofrettaged cylinders with subsequent low temperature heat treatment, and nonautofrettaged cylinders made of AISI 4340 steel with a nominal ultimate tensile strength of 1100 MPa. The cylinders which had an internal finish varying from 0.4 to >3 μm (16 to 125 micro inches) were tested in the open-end condition. Tests on cylinders with diameter ratios of 1.4–2.0 showed that autofrettage improves the fatigue life at lower pressures but had no effect at the higher pressures where the lives were very small. Further results obtained in the range $10^3 - 10^5$ cycles on smooth and rifled bore autofrettaged and nonautofrettaged cylinders with diameter ratios varying between 1.2 and 2 made of material roughly equivalent to AISI 4330 steel have been reported (103).

Increased Tensile Strength. Because the limiting maximum shear stress that can be endured by a thick-walled cylinder indefinitely is about one-third of the ultimate tensile strength, it might be thought that increasing the tensile strength would be a good way to increase the fatigue strength. Fatigue tests on cross-bored cylinders of EN25 in various states of hardness (104) show that the fatigue limit is raised as the ultimate tensile strength increases, but at much higher pressures and shorter lives the higher strength cylinders survive for fewer cycles than those of lower strength. Most of the failed cylinders of the two highest tensile strengths suffered fast fracture of increasing severity as the pressure increased, whereas those cylinders having the lowest tensile strength all failed as a result of fatigue cracks with no fast fracture. For fatigue cracks associated with the cross-bore configuration a small increase in fracture toughness from 77 to 88 MPa$\sqrt{m}$ is sufficient to ensure that the fatigue crack penetrates the wall before fast fracture intervenes. For long plane fronted cracks of the sort found in rifled gun barrels, material with a much higher fracture toughness is needed to prevent fast fracture.

Surface Treatments. Any surface treatment which sets up compressive stresses in the bore layers of a cylinder increases its endurance to high cycle fatigue. Such treatments include peening, nitriding (83), honing, etc. Nitriding produces a very hard case which is also helpful in preventing the initiation of fatigue cracks, providing the component does not contain areas of stress concentration which would result in strains sufficiently large to cause the case to crack when the component is stressed. Finishing operations which might leave scratches on the bore surface should be avoided, particularly if the scratches are parallel to the axis of the cylinder.

Cumulative Damage. Pressure vessels may be subjected to a variety of stress cycles during service; some of these cycles have amplitudes below the fatigue (endurance) limit of the material and some have amplitudes various amounts above it. The simplest and most commonly used method for evaluating the cumulative effect of these various cycles is a linear damage relationship in which it is assumed that, if N_1 cycles would produce failure at a given stress level, then n_1 cycles at the same stress level would use up fraction n_1/N_1 of the total life. Failure occurs when

$$n_1/N_1 + n_2/N_2 + \ldots + n_i/N_i = 1.0 \tag{30}$$

The linear damage rule takes no account of the order in which the stress cycles are applied.

Fracture Mechanics. Linear elastic fracture mechanics (qv) (LEFM) can be applied only to the propagation and fracture stages of fatigue failure. LEFM is based on a definition of the stress close to a crack tip in terms of a stress intensification factor K, for which the simplest general relationship is

$$K = Y\sigma\sqrt{\pi b} \tag{31}$$

where σ is the nominal stress normal to the crack plane, b is the crack depth, and Y is a constant containing material and crack shape parameters.

The use of the single parameter, K, to define the stress field at the crack tip is justified for brittle materials, but its extension to ductile materials is based on the assumption that although some plasticity may occur at the tip the surrounding linear elastic stress field is the controlling parameter.

One of the most important applications of LEFM is to estimate the critical crack or defect size which causes fast fracture to occur. This occurs when the value of K in a structure becomes equal to the plain strain fracture toughness, K_{IC}, of the material; the critical crack size, for a given stress and fracture toughness, is then given by equation 31.

Lack of accepted stress intensity factors for internally pressurized components has, until recently, limited this application. The factors are a function of the size and shape of both cracks and high pressure components as well as modes of loading (91). Stress intensity factors can be derived analytically for some simple geometries, but most require the application of advanced numerical methods (105–107). Alternatively they may be determined experimentally (108).

Standard procedures for fracture toughness testing of materials give reproducible values of minimum fracture toughness, but little guidance is available for incorporating these values into the design process (109). A design philosophy for the avoidance of unstable fracture in high pressure components based on LEFM has been given (91); for all practical combinations of yield stress and fracture toughness an LEFM analysis proves conservative except in the case of very thick-walled components.

This belief is supported by the results of burst tests on tubular specimens made of AISI 4335 tempered to various fracture toughness levels from 65–120 MPa√m and precracked to a radial depth of half the wall thickness. For the lowest tempering temperature it was found that the burst pressure, at the onset of fast fracture, was 22% higher than that expected on the basis of a careful LEFM analysis, whereas at the highest tempering temperature the discrepancy was 100%. Apparently the conventional requirements for plane strain fracture

toughness testing are not adequate for internally pressurized thick cylinders because of the enhanced plasticity generated by the pressure acting at the tip of
the crack.

Another important application of LEFM is the rate of growth of a fatigue
crack under cyclic loading. This is also controlled by the stress intensity factor
through an equation of the following form (110):

$$\frac{db}{dn} = C(\Delta K)^m \tag{32}$$

where db/dn is the crack growth/cycle or crack growth rate, ΔK is the range
of stress intensity, and C and m are material constants. Knowledge of crack
growth rates in components, in which it is difficult to prevent crack initiation,
are of value in establishing inspection frequencies to ensure that the crack is
not allowed to reach critical size.

Irrespective of whether unstable fracture or subcritical crack growth is being considered, reliable values for the stress intensity factors are required. Those
for single semielliptical or straight fronted cracks located on the inside or outside
surfaces of a nonautofrettaged cylinder are straightforward to calculate. Multiple flat fronted cracks of equal depth, multiple unequal flat fronted cracks, and
multiple flat fronted cracks with one crack deeper than others have been studied
(111). Cracks initiating in complex regions of a vessel, such as blind end corners,
cross-bore corners, or the root of threads in a threaded closure, are difficult to
calculate. In addition, the problem becomes much more complicated when residual stress fields created accidentally or deliberately, as with autofrettage, need
to be considered.

One aspect of pressure vessel design which has received considerable attention in recent years is the design of threaded closures where, due to the high
stress concentration at the root of the first active thread, a fatigue crack may
quickly initiate and propagate in the radial–circumferential plane. Stress intensity factors for this type of crack are difficult to compute (112,113), and more
geometries need to be examined before the factors can be used with confidence.

Another possible mode of failure for this type of vessel is the growth of
a fatigue crack in the radial–axial plane, which may grow from a pre-existing
defect at the bore surface or initiate in the usual way. Because the growth of such
a crack to a critical size can represent a significant portion of the fatigue life, it
would be advantageous to be able to estimate this. The parameters which need
to be considered are cylinder size, and diameter ratio, initial crack depth and
shape, the allowable crack depth, the cyclic internal pressure, and the residual
stress field induced by autofrettage (114).

Design of Removable Closures

PRESSURE VESSELS

Closures usually consist of three elements: a cover to the opening in the vessel,
a coupling device holding the cover in position against the internal pressure, and
a sealing ring or gasket between the cover and the vessel. The sealing ring or

gasket is nearly always made of a softer material than the vessel and end cover so that when it is tightened to make an initial pressure tight seal, it deforms and follows the irregularities in the mating surfaces closely. Thus deformation is almost entirely confined to the sealing ring which may be replaced as necessary.

An alternative method of making a pressure-tight seal between a vessel and its cover is to develop such a high compressive stress between the two components that they yield along a narrow, axisymmetrical, circular band of contact. By this means the asperities on the mating surfaces are smoothed out and what is usually described as a metal-to-metal seal is formed. This technique makes it difficult to refurbish the mating surfaces of large vessels and it is almost entirely confined to fittings used for connecting small-bore pipelines and to sealing adjacent packing cups or valve components in reciprocating compressors.

Sealing Rings.   Gaskets or sealing rings may be of the compression type or the self-sealing type. Figures 19**a** and **b** shows two compression-type gaskets in which the initial stress across the jointing faces decreases as the internal pressure in the vessel increases. To a first approximation these joints remain pressure tight as long as the compressive stress along the sealing paths on each face of the gasket is greater than the pressure to be resisted. Thus the maximum pressure that this type of gasket is capable of withstanding is determined by the initial compressive stress that can be developed across the sealing surfaces. This imposes severe limitations on the pressure which can be sealed even when, as in Figure 19, the gaskets are fully confined.

Self-sealing types of ring in which the stress across the jointing faces is automatically maintained at a higher value than the pressure to be sealed are almost exclusively used for chemical process vessels designed for pressures above about 20 MPa (2900 psi). These rings are sometimes known as unsupported area or Bridgman seals, following Bridgman's development of the unsupported area principle (116) in the 1920s to seal pressures of 1200 MPa (174,000 psi) in laboratory equipment.

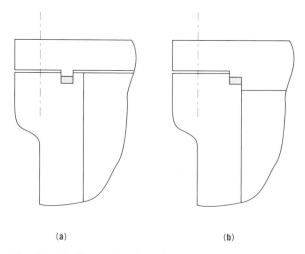

(a)   (b)

Fig. 19. Fully confined gasket joints (115). See text.

Lens Ring. The lens ring joint in Figure 20**a** illustrates the principle by which self-sealing is achieved; the ring, usually of steel, seats between conical surfaces and receives no lateral support, except that provided by the line contacts with the body of the vessel and its cover. Once an initial seal has been formed, the pressure on the inside surface of a correctly proportioned ring causes a compressive stress to be developed across the sealing surfaces which is always greater than the internal pressure. There is a small axial component of the load across these surfaces which increases that generated by the internal pressure and has to be resisted by the coupling device. If the included angle of the conical surfaces is increased the ring may yield and collapse inwardly when the joint is tightened initially. In the case of the conically shaped D-ring shown in Figure 20**b**, this may be prevented by providing a spigot on the underside of the end cover to support the ring. Note the use of a soft material interposed between the ring and the mating surfaces to facilitate the making of an initial pressure tight seal and the grooves to ensure that the internal pressure has unrestricted access to the inside surface of the ring.

Delta Ring. This triangular section ring (Fig. 21) is used extensively in the United States. To prevent it from collapsing during initial tightening, it is located in grooves in the body of the vessel and the end cover in such a way that the combined depth of the grooves is slightly less than the height of the ring. A relatively small coupling force is required to make the initial seal, and contact between the flanges prevents the ring from being crushed when this force is exceeded; hence the initial sealing stress is determined by the geometry of the ring and grooves. Under load the inner face of the ring is subjected to the internal pressure and the self-sealing effect comes into play. Another ring which makes use of the self-sealing principle is the metal lip sealing ring shown in Figure 22**a**. On initial tightening the inward movement of the ring is prevented by the reinforcement profvided by the rim, which is compressed between the head and body of the vessel.

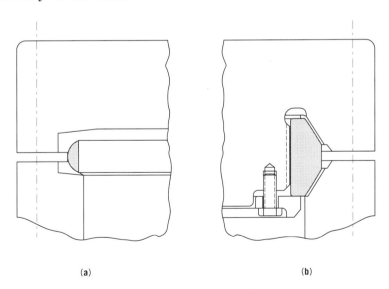

(a)   (b)

Fig. 20. Lens rings (117). See text.

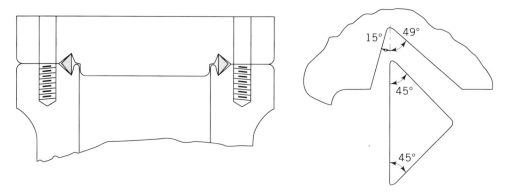

Fig. 21. Delta ring (118). See text.

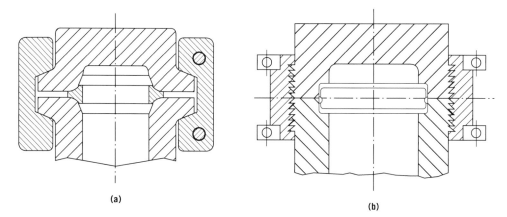

(a)

(b)

Fig. 22. (**a**) Metal lip sealing ring (119); (**b**) wave ring (120).

Wave Ring. The ring shown in Figure 22**b** has two crests which are slightly greater in diameter than the recesses in the vessel and end cover. To make an initial seal the ring has to be forced or sprung into position; this is often done by cooling it and allowing it to expand in the recesses. The ring, usually made of hardened steel, is highly finished and polished, as are the recesses in each of the mating components. To avoid damage when fitting the ring, the entrance to each recess is provided with a lead-in and the ring is coated with suitable lubricant or flashed with a soft metal to facilitate making an initial pressure-tight seal. There is no axial component of the force generated by the stress across the seating surfaces, and the self-sealing property of the joint is fully effective. Unlike the lens or metal lip sealing rings the wave ring accommodates itself to the radial expansion of the joint without relative movement at the lines of contact and provides accurate location of the end cover relative to the vessel.

O-Ring. Although these are not used to seal vessels for continuous chemical processes, they are widely used in other applications such as isostatic pressing. Figure 23 shows an elastomeric O-ring in its undeformed state, backed up by a chamfer or miter ring to prevent the extrusion of the elastomer into the gap between the head and body of the vessel when the internal pressure is applied.

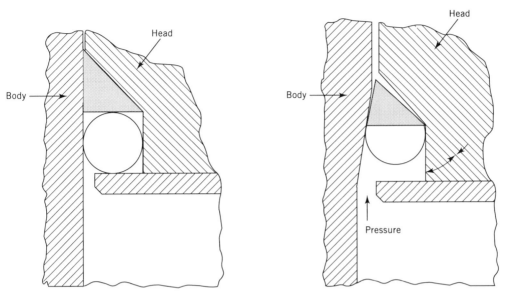

Fig. 23. O-ring backed up by miter ring to seal vessel (121).

For the seal to work the antiextrusion ring must have a high yield strength to enable it to maintain its shape under maximum pressure conditions. Provided the geometry of the recess is such that sufficient initial squeeze is applied to the O-ring and it is backed up by an antiextrusion ring, it continues to work to very high pressures. The maximum operating temperature for elastomeric materials is limited to about 200°C; at higher temperatures PTFE may be employed provided some device, such as a spring, is used to give the polymer the required elasticity.

Bridgman Seal. With the exception of the lens ring, the initial stress across the sealing surfaces of all the self-sealing gaskets described is governed by the geometry of the closure. Should the joint leak it cannot be tightened more and usually has to be remade. With the Bridgman-type closure shown in Figure 24, the functions of taking the end load and making an initial pressure-tight seal can be separated. Nuts, A (Fig. 24), enable the vessel to be tightened easily without heavy sledging of the large nuts, B, required to resist the end load. The high stresses set up in the gasket may deform the inner surface of the vessel, making it difficult to remove the end plug, and provision is usually made for its extraction by means of jack bolts, C.

Coupling Devices. Usually the coupling device has to withstand not only the stress produced by the internal pressure acting on the cover, but that produced by the initial tightening of the seal. With some self-sealing gaskets such as the wave ring, the only stress which has to be considered is the former but with others, such as the delta and the lens rings, there is a small wedging effect in addition to the initial tightening up stress.

Bolted Flanges. Probably the oldest method of coupling is the use of bolted flanges; the flange may be integrally forged with the body of the vessel or alternatively welded or screwed to it. In designing a flanged closure, the aim

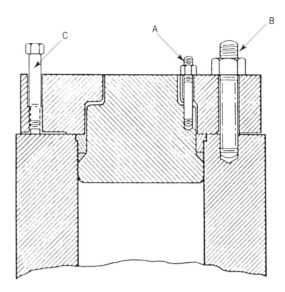

Fig. 24. Bridgman-type closure (122). See text.

is to reduce the diameter of the bolt circle so that the coupling action of the nut and bolt is as near vertically above the gasket or line seating as possible, thereby reducing the required thickness of the end cover as well as reducing the overall diameter of the closure. A limit is imposed by the necessity of providing adequate room to tighten the nuts.

Screwed Plugs.   Vessels for isostatic pressing, made in the mid-1950s, were provided with screwed plug closures. Isostatic pressing, whether carried out hot or cold, is a batch operation, and these closures are compact, inexpensive to manufacture, and can be opened and closed easily. Because the thread on the screwed plug is in compression and that on the vessel in tension, it is nonuniformly loaded along its length. The stress acting at the root of the first engaged thread may be 10 times that at the thread remote from the seal. One of the earliest failures occurred in a vessel, having a bore of ca 450 mm, in which the crack propagated, from the first thread, across the wall of the vessel in the radial tangential plane as a result of axial bending (123). This failure led to Sopwith's analysis (124) of the distribution of load in screw fastenings being used to predict the effect of changing many of the screwed plug parameters. Photoelastic tests on model vessels, in which both the thread and vessel parameters were changed, showed that the experimentally measured peak stresses at the root of the first thread were higher than those predicted (123).

With the help of photoelastic (125) and fatigue (126) studies on models of specific screwed plug closures, many design changes have been proposed. For example, different thread forms such as acme and buttress have been used. The root radii of the threads have been increased and the undercut adjacent to the first active thread modified to reduce stress concentration; the length of the thread has been increased to obtain a more even distribution of load, the angle and pitch of the threads have been varied, and the threads have been ground to introduce residual compressive stresses. A novel arrangement, known as a resilient thread (127), makes use of a spiral wound spring to take the place of the protruding portion of the threads, and fit closely into opposing grooves in both the vessel and closure.

The development of finite element methods, since the late 1960s, has made possible the exploration of a wide range of variables relevant to the design of screwed plug closures. Work (126,128–132) on the stress at the root of the first loaded thread, where most failures occur, and the load distribution along the thread length has led to the conclusions that the load carried by the first three threads decreases considerably as the number of active threads increases to 20, and the load carried by the second thread, f_2, is approximately 75% of the load on the first thread, f_1, and that on the third thread, f_3, about 60% of f_1, that on the first thread, regardless of the number of threads.

Empirical equations have been proposed (133) which enable a combination of thread and vessel parameters to be chosen to minimize the stresses at the root of the first three active threads. The load distribution along the thread is sensitive to machining tolerances and temperature differentials (134) and it is clear that threaded vessels should not be used for high cyclic service without rigorous and frequent inspections (135).

Quick Opening Devices. Breech block, tapered or interrupted thread, or pinned closures are often used when an end cover has to be removed quickly, as with some isostatic presses (126,136), or to enable the end cover to be removed easily after the vessel has been heated to high temperatures.

Vickers-Anderson Coupling. Whenever possible it is desirable to take the end load on the outside of the vessel, since this is less highly stressed than the inside. One way in which this may be done is with the Vickers-Anderson coupling. The end cover is secured by means of a collar which is usually made in three segments drawn together by means of tangential bolts (Fig. 25a). The joint may be sealed by any of the self-sealing gaskets previously described and in Figure 22a a metal lip sealing ring is shown. The contact pressure to effect the initial seal is obtained by machining a shoulder with an external conical face on to the body of the vessel and the end cover and clamping them together with the collar, which is counterbored with corresponding conical surfaces. The angle of the cones should be less than the angle of friction of the mating metallic surfaces, so that the tangential bolts do not carry any stresses; however, for reasons of safety the bolts are usually designed to withstand the separating forces which would be generated in the absence of friction. The stresses in the critical regions of a closure of Vickers-Anderson type, in which eight hydraulically operated clamp jaws were used to secure the end cover of an isostatic compaction vessel having a bore of 610 mm at pressure up to 138 MPa (20,000 psi), have been estimated (123,139).

Buttress-Shaped Grooves. The cones of the Vickers-Anderson coupling may be replaced by a series of parallel grooves of buttress section machined on the outside surface of the vessel and its end cover, which mate with similar ones counterbored on the inside of a split collar as shown in Figure 22b. This arrangement is only possible for a ring such as the wave ring which requires no initial end load to make a pressure-tight seal. Instead of the three component split collar shown in Figure 25a, two half laps may be used as in Figure 25b. This closure has formed the basis of the commonly adopted design for the largest

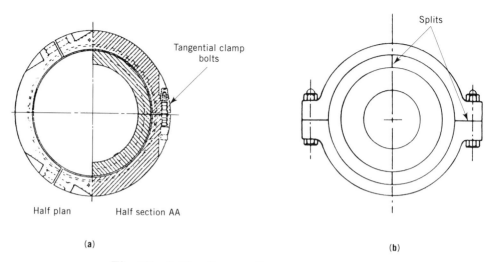

Half plan   Half section AA

(a)   (b)

Fig. 25.   Split collar coupling (137,138). See text.

autoclaves used for LDPE production, although there are variations in the number and size of the buttress grooves employed. Some vessels use a split collar to engage with a buttress-shaped thread and so overcome the problem of ensuring that the buttress grooves are of uniform pitch. Seals for vessels commonly used in the German chemical industry at pressures of 30 to 400 MPa (4350–58,000 psi) have been reviewed (140).

Yokes.   The need to couple the end cover to the body of the vessel may be avoided if yokes, external to the vessel, are used to resist the load arising from the internal pressure acting on the closures. However the necessity to move the vessel out of the yoke and remove one of the closures to gain access to the inside of the vessel limits its use for chemical process equipment. Yokes may be pinned, welded, bolted, or wire wound. Both the vessel and yoke may be wire wound (136).

End Covers.   Generally it is desirable to have as few openings as possible in the walls of a vessel, and to this end it is customary to accommodate ancillary equipment, such as stirrer glands, pipe connections, and small fittings for measuring instruments, in the more lightly stressed end cover.

TUBES

Tubes having a wide range of bore sizes are required to operate at pressures in the region of 150–300 MPa (22–44,000 psi). Small bore tubes, say 3–15 mm dia, are used to supply lubricant to the packing cups of secondary compressors, initiator to reactors, etc, while larger bore tubes, say 25–75 mm bore dia, are used to connect compressors to reactors and for the construction of coolers and tubular reactors.

Union Connectors.   Collar and cone-type connectors (Figs. 26**a** and **b**) are nearly always used to connect small bore tubes at pressures up to about 400 MPa (140). The included angle of the end of the tube is 55–57° and that of the seat 60–61° so that when the gland nut is tightened into the connector a metal-to-

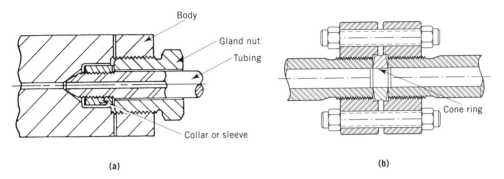

Fig. 26. Joints for tubes. See text.

metal seal is formed along a narrow band of contact between the end of the tube and the connector. Union connectors for small bore tubes at pressures of 500 MPa and above usually employ lens rings.

Loose Screwed Flanges. The ends of tubes having a bore greater than about 12 mm are usually threaded and coupled together with loose screwed flanges designed in accordance with the ASME procedure. Cone rings, lens rings, or metal lip sealing rings may be used to make the seal between the ends of adjacent lengths of tube. For ease of assembly the cone ring in Figure 26**b** (140) is located in a recess in the flange which is concentric with the outside diameter of the tube. However, since the bore of the tube may not be concentric with the outside diameter, it is necessary to establish a sealing diameter at the end of the tube, which is larger than the tube bore, concentric with the outside diameter, and blended into the bore. In spite of this preparation the cone ring requires less machining on the ends of the tubes than a lens or metal lip sealing ring. To achieve optimum performance with a cone ring or lens ring, the stud bolts which provide the coupling force to effect the initial pressure-tight seal are often tightened hydraulically. This is not necessary with the metal lip sealing ring because the ends of the pipe are forced into contact with the rim of the ring and the joint cannot be overtightened. Unlike the cone ring or lens ring the metal lip sealing ring does not suffer if the tube is subjected to a bending stress. On the other hand, it may be difficult to spring the pipes sufficiently far apart to replace a metal lip sealing ring should it be necessary. Seals for pipes commonly used in the German chemical industry at pressures of 30 to 400 MPa (4350–58,000 psi) have been reviewed (140).

Manufacture of Pressure Vessels

Steels for Forged Vessels. The choice of steel for early forged pressure vessel bodies and closures was influenced by ordnance practice. In the U.K. vessels built during the 1950s were made of a 2.5% Ni–Cr–Mo steel EN25 (BS970-826M31), a gun steel which could be heat treated to achieve uniform strength throughout the wall thickness while retaining adequate ductility and transverse impact strength. As more highly stressed vessels were needed to minimize the size of forging for a given reaction volume, the Ni–Cr–Mo–V steels found favor because increased strength could be achieved without loss of ductile and impact properties. It is unlikely that 3.5% Ni−Cr−Mo−V steel will be replaced (141) for the production of large forgings, and much information is available on the mechanical properties of forgings of this material.

In the United States similar gun steels, 2300 series nickel steels and Ni−Cr−Mo−V steels, are widely used for forged pressure vessels. ASTM specification A508 first issued in 1964 covers quenched and tempered vacuum-treated carbon and low alloy steels for pressure vessels (142), but it was not until 1975 that the ASTM issued specification A723 to cover three grades of Ni−Cr−Mo−V steel for pressure vessels, containing approximately 2, 3, and 4% Ni, in six classes of tensile strength from 795 to 1310 MPa (143). Tensile, impact, and axial fatigue properties have been reported on large forgings of A723 in three directions (144).

The effects of variations in composition, cleanliness, structure, and mechanical properties of electroslag remelted (ESR) 35NiCrMoV12.5 steel have been reported. This steel which lies between Grades 2 and 3 of ASTM specification A723 is widely used in Europe (145).

Forging Process. The prime reason for forging is to shape the component, but it also improves the properties of the steel (qv). The introduction of large electric arc furnaces and the development of liquid steel degassing techniques in the late 1950s made possible the manufacture of large forgings which were less susceptible to hydrogen flakes or hairline cracks. These days ingots are relatively free from porosity; the final portion of the steel to freeze, which is high in carbon and sulfur, is contained within the head which, like the bottom of the ingot, is discarded prior to forging. Suitable forging techniques can greatly lessen the differences in the ductile properties of the steel in the longitudinal and transverse directions by breaking up inclusions and working the steel in more than one direction.

One such technique, once widely used in the U.K., is hollow forging. This consists of heating to forging temperature an ingot, the core of which has been removed, and forging it on a massive mandrel so that the metal is squeezed out circumferentially and worked in the direction to resist internal pressure. However, the process cannot be carried out unless the final minimum internal diameter is at least 300 mm and, even then, only if the overall length is less than about 2000 mm (146). Hollow forging longer cylinders may require several heats, with the result that parts of the forging are heated to forging temperature and not reworked. This results in local grain growth which is difficult to refine by heat treatment and which may lead to a degradation of the mechanical properties.

A more satisfactory procedure, which ensures work in all three directions, involves upsetting the ingot in the axial direction followed by drawing down and shaping to size. This enables the whole forging to be hot worked thereby reducing grain size and ensuring that it is relatively uniform. As a consequence of the upsetting treatment, better transverse properties are achieved in the final heat treatment.

Gun steels are generally used in a quenched and tempered condition which means heating the forging to about 850−920°C, followed by quenching in oil or, with 3.5% Ni−Cr−Mo−V steel, water. The steel is then extremely hard and brittle, but by tempering, ie, heating again usually in the range 500−650°C, the tensile strength is reduced, whereas the ductility and impact strength are improved. The mechanical properties of the forging are determined from test specimens cut from prolongations on the ends of the forging, after it has been demonstrated that heat treatment has been carried out satisfactorily. After preliminary surface machining, ultrasonic inspection is used to ensure that no harmful internal cracks or concentrations of nonmetallic inclusions are present (141). A second ultrasonic inspection, made before finish machining but after final heat treatment, enables the forging to be examined before cross-bores, grooves, etc, which would interfere with the examination, have been machined. After finish-machining all surfaces are subjected to a magnetic particle examination before the vessel is assembled and pressure tested. On completion of pressure testing the vessel is dismantled prior to a final magnetic particle and ultrasonic check.

Remelting Processes. Vacuum arc remelting (VAR), which came into prominence in the early 1960s, provides material almost completely free of oxide inclusions and in which the crystal size and the directional properties are both remarkably uniform and dissolved hydrogen, oxygen, and nitrogen are greatly reduced. The remelting process consists of preparing, from a cropped ingot, a rough machined solid cylinder, which is progressively fused by an electric arc under high vacuum, after which the material resolidifies to form a new ingot. To manufacture stirred reactors for polyethylene production from VAR steel would require a remelted ingot larger than that readily available and, since satisfactory reactors have been forged using stream degassed steel, the process has not been widely used for such vessels. However, VAR steel is often specified for components to work under severe conditions, such as those which form part of high pressure reciprocating compressors.

An alternative process is electroslag remelting (ESR). More oxide inclusions are found in ESR steel than in VAR steel, but their size and distribution are such that they normally have no noticeable adverse effect on properties (141).

Design of Forged Vessels. Out of concern for public safety, organizations within many industrially developed countries have been given the responsibility of issuing and maintaining codes of practice for the design, construction, and testing of unfired pressure vessels. In the United States it is the American Society of Mechanical Engineers (ASME); in the U.K. it is the British Standards Institution (BSI).

ASME Boiler and Pressure Vessel Code. The ASME Boiler and Pressure Vessel Code is published in 11 sections. Section VIII, which is concerned with rules for the design of unfired pressure vessels, was first published in 1925 and since 1968 it has been issued in two parts, Division 1 (147) and Division 2 (148), the latter being known as the Alternative Rules.

Division 1. Below the creep range, design stresses are based on one-fourth of the tensile strength or two-thirds of the yield, or 0.2% proof stress. Design procedures are given for typical vessel components under both internal pressure and external pressure. No specific requirements are given for the assessment of fatigue and thermal stresses.

Division 2. With the advent of higher design pressures the ASME recognized the need for alternative rules permitting thinner walls with adequate safety factors. Division 2 provides for these alternative rules; it is more restrictive in both materials and methods of analysis, but it makes use of higher allowable stresses than does Division 1. The maximum allowable stresses were increased from one-fourth to one-third of the ultimate tensile stress or two-thirds of the yield stress, whichever is least for materials at any temperature. Division 2 requires an analysis of combined stress, stress concentration factors, fatigue stresses, and thermal stress. The same type of materials are covered as in Division 1.

Limitations of ASME Code.   In its present form Section VIII of the code contains disclaimers for its application to high pressure. Division 1 requires additional consideration for pressures above 20 MPa (3000 psi); although Division 2 is not restricted to a particular pressure, in practice, some additions to or deviations from the rules may be necessary to meet the design principles and construction practice required for high pressure vessels. The limitations of Division 2 (*149*) are as follows: (*1*) restriction of the maximum allowable stress to one-third of the ultimate tensile strength does not permit utilization of the inherent strength of the steel to the extent that has been reliably demonstrated to be reasonable and prudent; (*2*) materials of higher ultimate tensile strength than 932 MPa (135,000 psi) and of higher yield strength than 828 MPa (120,000 psi) are currently not listed in the Code; (*3*) prestressing produced by compound shrinkage, autofrettage, or other means cannot be taken into account when assessing the fatigue strength of a vessel.

In 1979 because neither Division of Section VIII was suited to the design of high pressure vessels, the ASME Board on Pressure Technology, Codes & Standards approved the establishment of a Special Working Group for high pressure vessels under the ASME Subcommittee on Pressure Vessels (Section VIII). The main design criteria, which are likely to be incorporated in a new Division of Section VIII, have been set out (*149*).

Design Criteria.   Traditionally the yield pressure has been regarded as an important design criterion because it is the largest pressure to which an initially stress-free cylinder may be subjected without the cylinder suffering any permanent deformation when the pressure is removed. Customarily, calculation of the yield pressure has been based on measurements of the tensile rather than the shear yield strength of material of construction. If it is assumed that the material yields in accordance with the shear strain energy criterion of failure, then the yield pressure is given by equation 9. From this equation, even for an infinitely thick-walled vessel ($k = \infty$), which is free from stress prior to the application of pressure, the yield pressure is limited to $\sigma_y \sqrt{3}$. If a factor of safety against yield of, eg, 2.5 is used, then for a vessel to withstand a pressure of 200 MPa (29,000 psi) a material with a tensile yield of 866 MPa (125,600 psi) would be required for an infinitely thick-walled vessel, or 1155 MPa (167,500 psi) for a vessel with $k = 2$.

Thus, to increase the yield pressure, the designer is tempted to use an excessively high strength steel with much lower transverse impact strength. Manning, who was responsible for the design of the vessels used by ICI in their LDPE process, realized that a low transverse impact strength was undesirable in that it could lead to what he referred to as brittle fracture, now known as fast fracture. Consequently he decided to adopt a factor of safety of 2.5 against bursting based on room temperature data and a factor of safety of 1.5 against yield, with an additional requirement that the transverse impact strength as determined by an Izod test should exceed 34 J (25 ft·lbf). It was required that the vessel be proof tested to the yield pressure.

To a first approximation the materials normally used for high pressure vessels behave in an ideal elastic–plastic manner and the collapse pressure is given by equation 15. If the factor of safety of 2.5 based on the bursting pressure is applied to the collapse pressure, then for a design pressure of 200 MPa (29,000 psi) and $k = 2$ the tensile yield stress, based on the assumption that the material yields in accordance with the shear strain energy criterion, would be 625 MPa (90,600 psi) which is much less than the value of 1155 MPa quoted for a factor of safety of 2.5, based on the yield pressure. In reality the bursting pressure is a little higher than the collapse pressure as most high strength steels exhibit some strain hardening.

Figure 27 shows the initial yield, collapse, and bursting pressure for EN25 used in the early autoclaves. Adopting Manning's proposals gives the two lower curves which show that, up to a radius ratio of about 1.7, the design is dictated by the lower curve based on bursting pressure and above this radius ratio the curve based on yield pressure is critical. For a vessel of this material with a design pressure of 200 MPa a radius ratio of 1.8 is needed. At 300°C the bursting strength is reduced by 10% compared with atmospheric temperature tests (18). So the actual factors of safety against yield and bursting of 1.5 and 2.5 adopted by Manning were, under operating conditions, somewhat reduced.

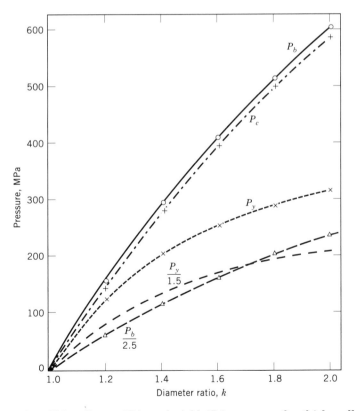

Fig. 27. Bursting (P_b), collapse (P_c), and yield (P_y) pressure for thick-walled vessel of EN25 (138). To convert MPa to psi, multiply by 145.

For an impact strength of 34 J (25 ft·lbf) the equivalent fracture toughness (150) is approximately 120 MPa$\sqrt{m}$. The fracture toughness dictates the critical size of crack above which fast fracture intervenes, so the smaller its value the smaller the critical crack and hence the greater significance of the transverse impact requirement specified by Manning.

These design rules have been extensively used by ICI since the 1940s and as far as is known there have been no catastrophic failures (151). More recently it has been proposed (152) that the factor of safety with respect to yield be reduced to 1.25 and the factor with respect to bursting to 2.25. If the initial yield and bursting pressures are based on data at the operating temperature, which might be as high as 300°C, then it is suggested that these factors can be further reduced to 1.0 and 2.0, respectively. At present, static design of high pressure vessels is based almost exclusively on the estimated bursting or collapse pressure of the vessel together with the fracture toughness properties of the material of construction.

Autoclaves of 250 L capacity built during the 1950s were made of 2.5% Ni–Cr–Mo steel having an ultimate tensile strength (UTS) of 965 MPa. Later, as the required volume increased, 3.5% Ni–Cr–Mo–V steel having a UTS of 1112 MPa was used. Probably the largest vessels currently in use are about 1.25 m external diameter and up to 10 m long with a nominal output of 150,000 t/yr (151).

Manufacture of Tubing

In the early polyethylene plants built by ICI, high pressure tube up to 3 m long was made from 12–14% chromium steel to BS 970 EN56C (now grade 420837). These tubes were made from bored bar by a number of cold reducing operations. Subsequently, tubes were made in material equivalent to EN56C by several companies in the United States using a cold pilgering process. Tubing made by tempering cold reduced bored round bars had a transverse Charpy impact strength of less than 5 J at −17.8°C and failed in a break-before-leak manner (153).

In the late 1950s the Timken Co. began to produce tubes 10 m long in AISI 4300 series steel. The process starts by piercing a billet, which is then hot rolled or hot Assel elongated to reduce the outside diameter to the required size. This tubing may then be finished, heat treated, and sold as hot worked tubing. Alternatively, the hot worked tube may become the hollow for subsequent cold reduction by cold pilgering. After cold pilgering the tube is finished, heat treated, and sold as cold reduced or Rotorolled tubing. A fine bore finish can be achieved by honing.

Defects such as hot tears or laps, quench cracks, localized overheating during stress relief, and corrosion may occur during the tubemaking process (154). Magnetic particle, ultrasonic, and visual inspection techniques are used to ensure that relatively few tubes enter service with significant defects.

Design Criteria for High Pressure Piping.   Since 1984 the Chemical Plant and Petroleum Refinery Piping Code, ASME B31.3, has included rules for the design of high pressure piping systems. If the new rules are to be used, all the requirements in Chapter IX must be met. In the context of the code, high pressure is considered to be pressure in excess of that allowed by the appropriate ANSI B16.5 Class 2500 flange ratings, about 45 MPa (6500 psi) at room temperature. Reference must be made to the latest edition of the code for full details of the requirements which have been reviewed (155); only those aspects to ensure that the piping has adequate static and fatigue strength and that it is made of material with acceptable toughness are considered here.

Static Strength. For a straight pipe the internal design pressure, P, is given by

$$P = \frac{S}{1.155} \ln\left(\frac{d + 2T}{d + 2c}\right) \tag{33}$$

where $S = 2\sigma_y/3$ is the basic allowable stress for the material of construction, d = inside diameter of the pipe, c = sum of mechanical allowances (thread or groove depth plus corrosion and erosion allowances), and T = the pipe wall thickness.

If the sum of the mechanical allowances, c, is neglected, then it may be shown from equation 15 that the pressure given by equation 33 is half the collapse pressure of a cylinder made of an elastic ideal plastic material which yields in accordance with the shear stress energy criterion at a constant value of shear yield stress $\tau_y = \sigma_y\sqrt{3}$.

Fatigue Strength. Alternative approaches to fatigue design are under consideration by the ASME Special Working Group on High Pressure Vessels for inclusion in the High Pressure Vessel Code which is under development. Other than this writing the allowable amplitude of alternating stress is determined from the ASME Boiler and Pressure Vessel Code. Credit is not allowed for favorable compressive mean stresses such as those induced by autofrettage or shot-peening, unless it can be demonstrated by testing or successful service experience that the design is adequate and that it is consistent with the design criteria in Chapter IX.

Fracture Toughness. In an effort to ensure that all piping components have sufficient toughness to resist fast fracture in the presence of a through-thickness crack, minimum Charpy V-notch impact values have been specified in the code. The values range from 20–95 J (15–70 ft·lbf) depending on the wall thickness of the component, the specified minimum yield strength of the material, the direction in which the test specimen is cut, and the number of tests carried out. For example, the average of three Charpy V-notch impact values on specimens, cut in the transverse direction from a tube having a bore of 25 mm ($k = 2.625$), made of steel having a minimum specified yield strength of 965 MPa, is required to be 40.7 J (30 ft·lbf) at the lowest metal temperature at which the component will be subjected to a stress greater than 41 MPa (5945 psi).

Low Density Polyethylene Process

Autoclave Process. When exothermic polymerization of ethylene is carried out in a stirred reactor, it is difficult to transfer much heat through the wall, and cold ethylene is passed continuously into the polymerizing mixture to maintain a constant temperature in the range 200–300°C and a pressure of

150–300 MPa (22,000–43,500 psi). Maximum conversion of ethylene to polymer is determined by the temperature difference between the incoming gas and the reacting mixture, and the need to achieve high conversion and throughput requires the reaction temperature to be set fairly close to the temperature at which ethylene decomposes into carbon, hydrogen, and methane. The decomposition is strongly exothermic and, once started, it proceeds very rapidly and might cause a pressure increase sufficient to rupture the vessel were it not for the pressure relief provided. Furthermore, if the hot products of the decomposition reaction are not cooled sufficiently before being vented, they may explode when they come into contact with the atmosphere and produce what is known as an aerial decomposition.

Stirrer.   Initiator may be introduced into the cold ethylene feed or directly into the reactor at one or more points. To control the reaction temperature closely and avoid hot spots caused by nonuniform dispersion of the initiator, which might result in a decomposition, it is necessary to achieve good mixing of the polymerizing fluid from one end of the reactor to the other and rapid mixing of the cold ethylene with this fluid. The turbulence necessary to accomplish this requires a high stirrer speed with considerable power input. Stirrers which are prone to attract a buildup of polymer need to be avoided, not only because they become unbalanced and vibrate making fatigue failure more likely, but also because the polymer deposits undergo graft polymerization which degrades the product.

A rotary stirrer driven by an induction motor housed within the vessel was an integral part of the ICI process. Figure 28 shows a general arrangement of the 250 L autoclave introduced in 1943 in which the stirrer motor housing was coupled to the reaction vessel, both being designed for the same pressure. The shaft from the motor passed through the stem connecting the two components and was coupled to the stirrer (not shown), which extended over the length of the reaction space. Cool ethylene entered the top of the motor housing and passed through the gap between the rotor and stator down the stem containing the shaft into the reaction space. Development of a satisfactory stirrer involved extensive flow visualization and mathematical modeling; from an engineering point of view, many of the early problems stemmed from decompositions caused by local overheating or breakage of stirrer bearings.

The use of an internal motor has found wide acceptance even for the very large autoclaves now in service, although Du Pont has used an external motor with a shaft passing through high pressure glands (157).

Heating and Cooling.   The body of the autoclave is heated by steam or other fluid to bring it up to reaction temperature and enable the process to be brought on stream as quickly as possible. The heating jackets may be of the separate compartmental clamp on type or they may be welded directly to the outside of the vessel. End covers to the autoclave may have machined passages to allow them to be heated. Once the reaction is established cooling fluid may be passed through the jackets to remove some of the heat of polymerization.

Protection Against Overpressure.   A decomposition propagating through the large mass of ethylene contained in a stirred reactor leads to a high rate of rise of temperature and pressure. To protect the reactor from excessive pressure, without having to resort to a large vent area, requires the pressure relieving

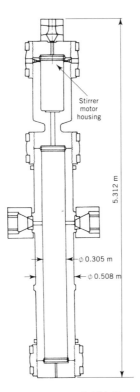

Fig. 28.   General arrangement of 250-L autoclave (ICI) (156).

device to act quickly and open at an overpressure which is as small as practicable. The early autoclaves used by ICI were protected by hydraulically balanced safety valves which were held closed by constant oil pressure. As soon as the pressure in the vessel increased significantly above the operating pressure, an oil relief valve lifted allowing the safety valve to open. In the case of the 250-L autoclave shown in Figure 28, the safety valves were housed in the recesses located round that part of the vessel provided with the cross-bore (156). Although a bursting disk was fitted in the oil system to accelerate the opening of the safety valve, it was still relatively slow in operation and was replaced by precisely machined bursting caps which, when the pressure significantly exceeded the operating pressure, failed in tension (158). The end of the bursting cap which becomes detached is caught in a dead-ended hole in-line with the exhaust port. Other types of bursting and shear element can also be used, provided the element which fails as a result of overpressure is located as close to the bore of the vessel as possible and that the material of construction does not creep significantly at the operating temperature. It may be necessary to provide a number of bursting elements to secure the required vent area with a large autoclave. The steel surrounding the exhaust ports through which the products of the decomposition reaction pass at sonic velocity when the pressure relieving device opens would be subjected to severe decarburization and cracking were it not for the replaceable sleeves fitted in the ports.

The autoclave is not the only component of an LDPE plant which may be exposed to a decomposition. Local hot spots in a secondary compressor may initiate a decomposition reaction; consequently it is necessary to protect these units from serious overpressure by pressure relieving devices and to release the products of the decomposition reactions safely. The problem of the aerial decomposition referred to earlier has been largely overcome by rapidly quenching the decomposition products as they enter the vent stack.

Tubular Process. Tubular reactors vary in length and may be as much as 1000 m long. They are constructed by joining together lengths of high pressure tubing, ~10 m long, with a bore diameter in the range 25–75 mm and a radius ratio of about 2.5. To reduce the ground space required, the reactor is usually of zigzag or serpentine construction, adjacent straight lengths of tubing being joined by return bends or loops. The tubes are jacketed so that they can be heated or cooled by passing suitable fluids through the jackets. The first part of the reactor, which may be a quarter of the total length, is used to preheat ethylene to 100–200°C. A small amount of initiator is then injected continuously into the reactor and the heat of polymerization raises the temperature to 250–300°C. After the temperature of the polymerizing ethylene has been reduced by transferring heat through the wall of the jacketed tube, more initiator may be introduced at a second initiation point, and after further cooling yet more at a third point. Multiple point initiation serves to increase the percentage conversion of ethylene to polyethylene but it increases the length of the reactor. Further increases in conversion may be obtained if side streams of cold ethylene are introduced into the reactor, but in this event it may be necessary to use tubes of a number of different bore sizes so as to maintain the velocity through each section of the reactor approximately constant.

As the percentage conversion increases, the polymer has a greater tendency to form a second phase on the bore surface of the reactor where it reduces heat transfer and makes temperature control of the polymerizing mixture difficult. To increase the flow velocity through the reactor and slough the polymer from the wall, the pressure at the exit to the reactor is rapidly reduced from its maximum operating value of ~300 MPa to ~200 MPa (29,000 psi) every few minutes. The depth and frequency of the so-called bump cycle varies widely from plant to plant. In addition, there is a superimposed small high frequency cyclic pressure from the compressors; hence fatigue is of primary consideration in the design of a tubular reactor.

Tubing. Results of a large number of fatigue tests on tubing in the as received (pressure tested) and autofrettaged states and also a few tests on pipe bends have been reported. These tests were carried out at high mean pressure with a superimposed cyclic pressure to simulate the conditions in LDPE tubular reactors with periodic let-down in pressure (159). At a mean pressure of 207 MPa (30,000 psi) autofrettaged tubes withstood a range of cyclic pressure at the fatigue limit in excess of 310 MPa (45,000 psi), which means that under the conditions similar to those of the test, such tubes can withstand being cycled from 52 to 362 MPa (7500–52,500 psi) indefinitely (160).

Tests cannot be used to estimate actual service life because laboratory test conditions, particularly in the absence of transient and steady state temperature gradients through the tube wall, are not identical to those used in the

LDPE process. Nevertheless a comparison between components A and B in the laboratory, coupled with a knowledge of how component A behaves in the plant, will usually enable an estimate to be made of how B is likely to perform under the same operating conditions. Comparative fatigue tests have proved to be a valuable tool in assessing the effect of heat treatment, anisotropy, surface finish, degree of autofrettage, etc, in the development of tubes and components for the LDPE tubular process.

Autofrettage. To achieve the required fatigue strength the tubes have to be autofrettaged to induce favorable residual stresses. The degree of autofrettage used by different manufacturers varies from that needed to take the elastic–plastic boundary a small distance through the wall to a pressure which may be 75% of the burst pressure of the tube. When high autofrettage pressures are used, the process is usually controlled by measuring the strain on the outside surface at, eg, three points along the length of the tube. Provided the variation in tensile strength along the length of each tube is small, this procedure ensures that each tube is subjected to a similar degree of autofrettage irrespective of the strength of the individual tubes. Tubes may be hot or cold bent provided the radius of the tube center line is not less than 10 times the nominal tube outside diameter. Any heat treatment required after cold bending is carried out before autofrettage.

Relaxation of the residual stresses induced by autofrettage at 720 MPa (104,400 psi) in reactor tubes ($k = 2.4$), of AISI 4333 M6 at a uniform temperature of 300°C has been studied and it was concluded, on the basis of creep tests for 10,000 h, that after 5.7 years 60% of the original stress would remain (161).

Heating and Cooling Jackets. The tubes of a tubular reactor are surrounded by heating or cooling jackets which are interconnected. In some early designs the jacket was fixed to the tube at one end and a gland was provided at the other to allow for thermal expansion. To obviate inevitable leaks at the moving gland the modern trend is to fix a flexible jacket at each end, either by welding to the outside of the tube or by welding to rings which are shrunk on to the outside surface of the tube.

With many reactors only the straight lengths of tube are water jacketed; however, processes are now available for continuously bending jackets round preformed high pressure bends so that no welding is required except at the jacket to tube ends. Care is needed in the design of the supports to the reactor; otherwise thermal expansion may induce large bending stresses in the tubes. At the same time, it is necessary to ensure that the reactor does not vibrate under the influence of the cyclic pressures and generate axial bending stresses.

Side Inlets. Some early tubular reactors were fitted with T-blocks, to provide side inlets for pressure and temperature measurement, initiator injection, etc. A large number of these forged blocks were required; furthermore, they were heavy and expensive because they had to be large enough to take the stud bolts needed to join the block to the adjacent tubes. An alternative arrangement is to put the cross-bore through the sealing ring (162). Unfortunately the fatigue strength of the cross-bored sealing ring, which is subjected to an axial compressive stress in the direction of the main bore, is less than that of a cross-bored block in which the axial stress is tensile, and special precautions have to be taken to ensure that the fatigue strength of the cross-bored sealing ring is adequate.

Protection Against Overpressure.   The problem of protecting a tubular reactor from high pressures generated by decompositions is not as serious as with a stirred autoclave because the much smaller mass of ethylene contained in the tube leads to a slower rate of rise of pressure. The first line of defence against overpressure is usually one or more externally controlled dump valves (162) such as that shown in Figure 29 for an operating pressure of 320 MPa (52,000 psi) fitted to a tube of 60 mm bore. In addition rupture disks are usually provided at various points along the length of the reactor, but these cannot be set to burst at a pressure close to the operating pressure. This is because the temperature is not constant along the length of the reactor and because it may be necessary to replace the rupture disks on a preventative basis, if failure due to fatigue caused by the periodic let-down cycle is to be avoided. Rupture disks are often housed in Y blocks with a generously radiused intersection to avoid fatigue (163). All the fittings such as that shown in Figure 29 and the blocks required to house the bursting disks must be designed so that the dead space is as small as possible, otherwise the space itself may become a source of decompositions.

If the speed with which ethylene is passing through a tube is comparable to the speed with which the decomposition reaction travels through the ethylene, then one or other of the fronts where the decomposition is occurring will be stationary relative to the tube. Under these conditions the tube will be heated to a very high temperature rapidly and fail at a pressure much lower than the burst pressure of the tube at ambient temperature.

Initiator Pumps.   Initiator, usually a mixture of organic peroxides dissolved in solvent, may be injected into the reactor through a sparger using hydraulically driven intensifiers or multicylinder mechanically driven pumps. With intensifiers in which the piston is relatively slow moving, it is essential to ensure that the flow of initiator is not interrupted at the end of each stroke. In principle, while initiator is being discharged from intensifier 1 the piston in intensifier 2 has to precompress the liquid to the discharge pressure, so that when intensifier 1 has completed its stroke intensifier 2 can take over without any interruption to the flow. Unfortunately, some of the control systems used to synchronize the operation of the intensifiers, which may be single or double-acting, are not easy to maintain and an interruption to the flow may result in a decomposition or loss of reaction at the sparger fed by that pump.

Many initiators attack steels of the AISI 4300 series and the barrels of the intensifiers, which are usually of compound construction to resist fatigue, have an inner liner of AISI 410 or austenitic stainless steel. The associated small bore pipework and fittings used to transfer the initiator to the sparger are usually made of cold worked austenitic stainless steel. The required pumping capacity varies considerably from one process to another, but an initiator flow rate 0.5 L/min is more than sufficient to supply a single injection point in a reactor nominally rated for 40 t/d of polyethylene.

The biggest problem associated with high pressure reciprocating pumps is that of satisfactorily sealing the plungers. To avoid excessive rate of wear a liquid film must be maintained between the plunger and the packing and, if this cannot be done by the fluid being handled, separate lubrication must be arranged. The plungers in the intensifiers are usually sealed using stationary chevron or U-type packings in which the liquid under pressure deflects the limbs of the U to provide

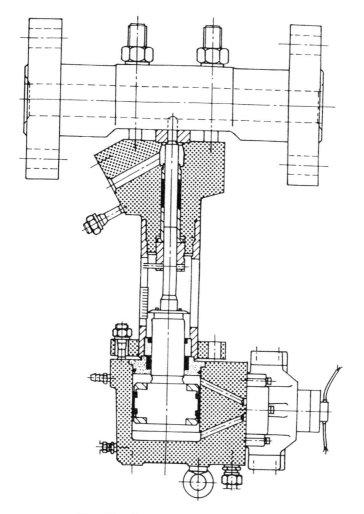

Fig. 29. Dump valve (Uhde) (162).

the seal. Packing rings are often made of filled polytetrafluoroethylene (PTFE), which has low friction coefficient and good resistance to attack from solvents, such as benzene, toluene, hexane, etc, used to dissolve the organic peroxides. The rings held between suitably shaped steel rings are compressed by a plug being screwed into a recess at the end of the cylinder. Some peroxides deposit solid material during the compression process, which is difficult to control; this has an adverse effect on the life of the packing and the operation of the valves which are usually of the ball or double ball type. The valves may be accommodated in the pump body itself or more usually in a separate head attached to the body.

Because of the problems of securing an uninterrupted flow, some manu-facturers have developed mechanically driven pumps of three or more throws. Some of these pumps avoided the use of conventional chevron-type packing by arranging for a stationary lubricated sleeve or bushing to be hydraulically tight-ened round the plunger so as to effect a satisfactory seal. Some of the difficulties in developing metallic lip clearance seals for high pressure reciprocating pumps operating on oil have been described (164) (see PUMPS).

Reciprocating Compressors.   Prior to 1895, when Linde developed his air liquefaction apparatus, none of the chemical processes used industrially required pressures much in excess of 1 MPa (145 psi) and the need for a continuous supply of air at 20 MPa provided the impetus for the development of reciprocating compressors. The introduction of ammonia, methanol, and urea processes in the early part of the twentieth century, and the need to take advantage of the economy of scale in ammonia plants, led to a threefold increase in the power required for compression from 1920 to 1940. The development of reciprocating compressors was not easy; little was known about the effects of cycles of fluctuating pressure on the behavior of the limited number of materials of construction which were then available, and failures of high pressure cylinders, valves, and gland packings occurred frequently.

Although the advantages of centrifugal compressors, one of which was the delivery of oil-free gas, had been recognized since 1926 (165) two problems had to be solved before they could be used in ammonia synthesis plants. First, the discharge pressure had to be raised, and second, the throughput of the compressor had to be decreased. It was not until 30 years later that a centrifugal machine was used in an ammonia plant; it operated at about 15 MPa and was followed by a single-stage reciprocating compressor to raise the gas to the final pressure. Further development, pioneered by Clarke Brothers in the United States, resulted in centrifugal machines capable of compressing synthesis gas to 30–40 MPa being used in all new single-stream plants having an output of 600 t/d or more. By the early 1960s only methanol and urea plants made use of reciprocating compressors. The development, by ICI in the late 1960s, of the supported copper oxide catalyst made possible the synthesis of methanol at pressures in the range 5 to 10 MPa with the results that centrifugal compressors could be used for all but the smallest plants.

The discovery of polyethylene created a need for compressors capable of compressing ethylene to about 150 MPa (22,000 psi). Initially the type of compressor used for this duty was based on a technique used in the laboratory (166), whereby ethylene was compressed from 25 (3600 psi) to about 150 MPa by an oscillating mercury piston contained within a U-tube, the movement of the mercury needed to compress the gas being produced by high pressure oil acting on one side of the mercury column. Although it accomplished its purpose by ensuring a steady production of polyethylene during the war years, it was not until World War II that any real progress was made in the development of reciprocating compressors suitable for pressures of 150 MPa.

GHH or Maschinenfabric Esslingen, as it was then, produced the reciprocating compressor used by BASF in its first polyethylene plant in 1942. GHH probably manufactured the first commercially available compressor in 1948, followed by Burckhardt in 1951. With the licensing of the LDPE process and the rapid development of the market for polyethylene, both ICI and its licensees gave up their attempts to develop their own compressors and sought to collaborate with selected compressor makers in the development, first, of machines capable of generating pressures of 150 MPa and later of larger machines for higher pressures. Feedback of operating experience from users provided technical know-how which enabled the makers to establish a virtual monopoly in the manufacture of these compressors. Each manufacturer developed machines which differed from those of his competitors, and users were reluctant to introduce alternative designs into their plant, with the result that comparatively little has been disclosed about the detailed design and performance of the different machines.

However, some of the more important techniques in high pressure engineering which made possible the increase in throughput of reciprocating compressors from 4–5 t/h at a pressure of 125 MPa (18,000 psi) in the early 1950s to 120 t/h at a pressure of 350 MPa (51,000 psi) can be considered.

Hypercompressors. In an LDPE plant a primary compressor, usually of two stages, is used to raise the pressure of ethylene to about 25–30 MPa and a secondary compressor, often referred to as a hypercompressor, is used to increase it to 150–315 MPa (22,000–45,700 psi). The thermodynamic properties of ethylene are such that the secondary compressor requires only two stages and this results in a large pressure difference between the second stage suction and discharge pressures.

The principal problems which had to be solved in the development of secondary compressors for higher operating pressures and throughputs were restriction of ethylene leakage past the plunger or piston to an acceptable level without jeopardizing the life of the elements used to seal the plunger or piston; and the avoidance of fatigue failure of high pressure components such as the cylinder and valves. In a compressor cylinder the stress fluctuates cyclically between a minimum and maximum corresponding to the suction and discharge pressures. A compressor running at 5 Hz completes 10^7 cycles in about 23 days of continuous operation; hence design must be based on high cycle fatigue data.

In addition to these mechanical problems there are two aspects of the compression process which relate specifically to ethylene. First, there is a tendency for small amounts of low molecular weight polymer to be formed and, second, the gas may decompose into carbon, hydrogen, and methane if it becomes overheated during compression. Cavities in which the gas can collect and form polymer, which hardens with time or in which the gas can become hot, need to be avoided.

Sealing Arrangements. Moving or stationary friction seals are used to make an acceptable seal between the reciprocating piston or plunger (the terms are used interchangeably), and the stationary cylinder. Moving seals make use of piston or sealing rings which are attached to the piston and reciprocate within the cylinder, whereas stationary seals make use of sealing elements through which the plunger moves. The former is known as the ring on piston, and the latter as the packed plunger arrangement.

Ring on Piston. Figure 30 shows a secondary cylinder of an Ingersoll Rand (Dresser Rand) compressor built in the 1960s in which the plunger is sealed with a ring on piston arrangement (167). Figure 31 shows the details of two typical piston ring assemblies used by Burckhardt, which consist of pairs of sealing rings with each of the rings covering the slot in the other, and an expander ring behind the pair, which also seals the gaps in the radial direction (168). Special grades of cast iron, bronze, or a combination of both are used for the rings; cast iron or steel are employed for the expander. Tungsten carbide has been found to

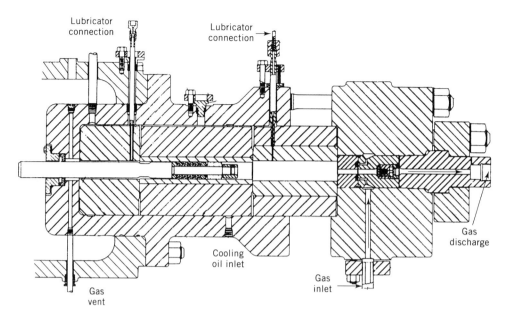

Fig. 30.   Secondary cylinder with piston rings. Courtesy of Dresser Rand (167).

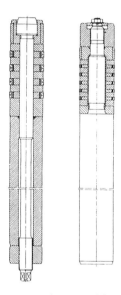

Fig. 31.   High pressure pistons with piston rings (168).

be the most satisfactory material for cylinder liners in which sealing is effected by piston rings of the type described. However, because this material is weak in tension, it is essential to prestress the liner in compression by shrinking it into an outer mantle. To ensure that prestressing is not reduced as a result of the relative thermal expansion between the tungsten carbide liner and the outer steel cylinder into which it is shrunk, it is necessary to cool the cylinder as shown in Figure 30.

Lubrication of piston rings in a hard liner is achieved by injecting oil into the suction stream or outer end sleeve (see Fig. 30) but this is not always satisfactory because compressed ethylene is a good solvent for mineral oils and very little oil reaches rings other than the first one or two. The temptation to use larger amounts of lubricant has to be resisted as the oil usually has an adverse effect on the properties of the product. Furthermore, if the product is to be used for food packaging, the lubricant has to be acceptable to the health authorities, and special lubricants with less satisfactory lubricating properties must be used in as small a quantity as possible (168). Fortunately, low molecular weight polymers carried by the recycle gas are reasonably good lubricants and supplement lubricant injected into the suction gas; on the other hand, too much polymer of higher molecular weight causes the rings to stick in their grooves. Maintaining conditions for good lubrication of piston rings is difficult to achieve. Experience has shown that rings in the assembly do not share the pressure difference across the plunger equally; the largest pressure drop is taken by one or two rings, not necessarily those nearest the upstream pressure. When these, as a result of wear, are no longer able to hold the pressure the sealing is taken over by other rings. There is no simple rule for the number of rings to seal a given pressure, and fewer than 10 have been found adequate to seal pressures of 250 MPa (36,000 psi).

Packed Plunger. With a packed plunger, lubricant is injected directly into the sealing elements. This has a number of advantages; first, all the oil is used in lubricating the moving surfaces at critical locations; second, unlike the piston ring arrangement, there is no direct contact between lubricant and ethylene, and this reduces both the amount of lubricant required and the contamination of the compressed gas.

A typical secondary compressor cylinder with packed plunger manufactured by Dresser Rand is shown in Figure 32. Initially steel plungers were either nitrided or flame plated but most of these have been replaced by plungers of tungsten carbide available up to 300 mm diameter. Misalignment between the axis of a tungsten carbide plunger and the axis of motion of the thrust block attached to the cross-head may lead to fracture of the plunger, as a result of bending stresses, unless precautions are taken (169,170). Sealing rings, not shown in Figure 32, are located in the recesses in the packing cups which are of compound construction. Metallic self-adjusting sealing rings of the type shown in Figure 33 are widely used. They are usually assembled in pairs, the actual rings which are tangentially split into three or six pieces being covered by a three piece radially cut section. Both, usually made of bronze with high lead content, are kept closed by surrounding garter rings. The optimum number of sealing elements is 4–5 (168). On the suction stroke, gas at discharge pressure, trapped in the packing cup recesses, flows toward the cylinder as the pressure drops to suction value. This backflow may result in ring and garter spring damage in those rings nearest the pressure side of the packing assembly unless a breaker ring, such as that shown in Figure 33, is used to minimize backflow. The pressure distribution along the length of such packings has been examined (172). Lubricant is supplied directly to the sealing rings by quills which pass through packing cups; each quill and its associated check valve is provided with lubricant from a separate pump. Lubricants used (subject to agreement by the health authorities) include

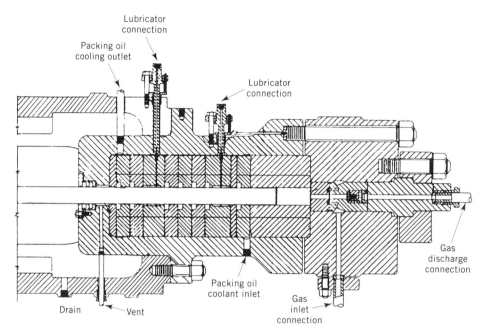

Fig. 32. Secondary cylinder with packed plunger. Courtesy of Dresser Rand (167).

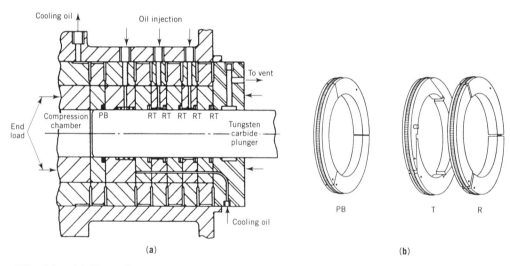

Fig. 33. (**a**) Typical secondary cylinder with packing cups (171); (**b**) stationary sealing rings for packing cups (France Products Division, Garloc Inc.). PB, pressure break; T, tangential ring; R, radial ring.

mineral oil mixed with various percentages of polybutene, polybutene with no additives, and polyglycol. By restricting plunger velocities to 2–3 m/s cylinder lives of nearly 70,000 h, with an end of life leakage of ethylene past the plunger of about 200 kg/h at a delivery pressure of 290 MPa (42,000 psi), have been achieved (151).

The packing cups, high pressure cylinder, and cylinder head are held together by tensioned bolts so that the axial stress across the carefully ground and lapped mating surfaces, finished to about 0.2 μm, is sufficiently high to prevent leakage. Duplex or triplex shrink fit construction is used to overcome the stress concentration effect of the cross-bores needed for the lubrication quills and the axial holes needed for the supply of coolant oil. Although fatigue failure in the radial–axial plane is rare, cups may fail from fatigue as a result of fretting induced by bending stresses. An assessment has been made of the stresses in packing cups at the junction of their mating surfaces and the critical value of the axial load to prevent fretting in secondary cylinders designed by Nuovo Pignone (173). Techniques used by the same manufacturer to autofrettage various components of the packed plunger design to improve their fatigue life have been described (30).

The disadvantage of the packed plunger design lies in the much larger joint diameters of the packing cups and other static cylinder parts which require higher closing forces than a comparable piston ring design. The cylinder bolts of cylinders of large bore need to be pretensioned to 10 or more times the maximum plunger load, and various hydraulic tensioning devices have been developed for this purpose. In spite of this, the superiority of the packed plunger design for secondary compressors has been generally accepted, although many primary compressors still use pistons fitted with rings.

With both the packed plunger and ring on piston designs it is essential that the piston or plunger be accurately centered if the seals are to be effective, and to this end guide rings are provided within the packing cup assembly (see Fig. 32) and at the connection between the piston and the driving rod. At the low pressure end of the cylinder, additional seals allow gas which has escaped past the sealing rings or piston rings to be collected.

The rate of ethylene leakage past the plunger is measured for each secondary cylinder on a periodic basis, and these measurements provide valuable information on the degradation of the friction seals. The decision as to when to change a cylinder is usually based on economic considerations relating to loss of product resulting from leakage and that resulting from downtime caused by a cylinder change. The maximum rate of leakage acceptable before a cylinder change varies from plant to plant and with the demand for product.

The time the plant is down for a cylinder change may be reduced, and cylinder life increased, if the cylinder with worn packings is removed from the compressor in its entirety and replaced with one which has been assembled with care in a clean environment. Alternatively, with some designs it is possible to replace the packing cups and valve assembly as a complete cartridge.

Pressure Wrapped Cylinder. The pressure wrapped, sometimes called wrap around or fluid ring, cylinder aims to overcome fatigue failure of the packing cups by ensuring that they are at all times exposed to compressive stresses by subjecting their outer diameter to gas at discharge pressure. Thus, simple one piece packing cups, instead of duplex or triplex designs, can be used; however, the outer cylinder must be designed to withstand the more or less constant discharge pressure which acts on its bore. Figure 34 shows a pressure wrapped cylinder in which packing cups, A, outer end sleeve, B, and valve assembly, C, are all subjected to the discharge pressure which acts on the

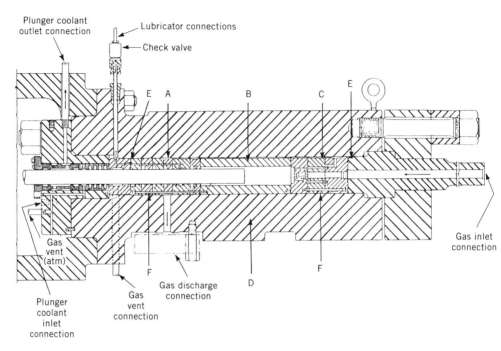

Fig. 34. Pressure wrapped secondary cylinder. Courtesy of Dresser Rand.

bore of outer cylinder, D, between static seals, E. To facilitate assembly, the packing cups are held together and joined to the outer end sleeve by tie rods, F. One of the problems in handling ethylene at high pressures is that polymer formed during the compression process may pass through the discharge valve and be deposited in the gap between the outside of the packing cups and the inside of the cylinder. This polymer hardens with age and may make it difficult to withdraw the valve, outer end sleeve, and packing cup assemblies from the cylinder. More importantly the design must be such that, if the annulus is partially blocked, the resulting transverse load cannot force one or more of the packing cups out of alignment and damage or break the carbide plunger.

So as not to decrease the strength of the outer cylinder by introducing cross-bores, in the section subjected to the discharge pressure, lubricant may be supplied to quills at the low pressure end of the cylinder and then through axial drilled holes to the appropriate packing cup. Lubricant has to be supplied at a pressure greater than the discharge pressure and fatigue failure of check valves and lubrication quills sometimes proved to be a problem until the check valves were enclosed within the cylinder. The discharge gas passed around the packing cups contributes little cooling, and additional cooling is provided by passing oil through passageways drilled in the packing cups in the axial direction.

Valve Design.   The reliability of the suction and discharge valves is crucial to the performance of the compressor. Initially, compressor manufacturers designed cylinder heads containing radial suction and discharge valves, so that it was possible to pull a complete piston with its rings through the cylinder head without disconnecting the suction and discharge lines or removing the valves. With this arrangement fatigue proved to be a serious problem which the design, shown in Figure 35, attempted to overcome. The well radiused intersection of the

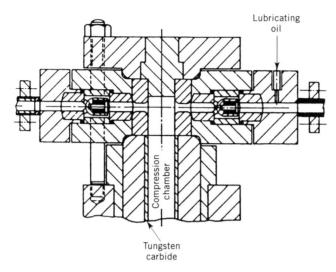

Fig. 35. Precompressed cylinder head with radial inlet and outlet valves (174).

gas passages with the main bore was located in a small forged core shrunk into a heavy flange and compressed axially by the cover plate so that it was subjected essentially to all-round compression (168). The head made use of the single poppet valves shown in Figure 36**a** in which the lapped mating surfaces of the three components were loaded in compression to produce a pressure-tight seal.

As the difference between the suction and discharge pressure increased, it became necessary to remove the pressure fluctuations from the area of the cross-bores and several ways of doing this have been described (168). In the arrangement shown in Figure 36**b**, the suction and discharge valves are arranged in line and the entire valve body is subjected to suction pressure on the outside; only the gas passage leading to the discharge valve is subject to cyclic pressure. Separation of suction and delivery pressure is assured by the sealing ring and the entire valve is pressed on the end of the cylinder liner by the difference of pressure and the precompressed belville washers. The components of this valve assembly can be autofrettaged to increase their fatigue resistance. Other manufacturers subject the outside of their axial valves to discharge pressure, which gives added strength, and may use multipoppet valves or plate valves instead of single poppet valves. Figure 36**b** shows an axial plate valve in which the discharge pressure acts on the outside surface of the components; such a valve might be used in the pressure wrapped cylinder shown in Figure 34.

Driving Mechanism.   The first essential is to ensure axiality of motion of the high pressure plunger, so as to avoid side loading of the packings and plungers which would jeopardize packing life and might result in damage or fracture of the plunger. The second requirement is to reduce the cyclic fluctuation of torque.

Basically, a vertical mechanism (Fig. 37**a**) was adopted from the design of large diesel engines and this, together with a horizontal mechanism (Fig. 37**b**), was used for some of the earliest secondary compressors. Since there is only one working stroke in two, both mechanisms suffer from large cyclic torque fluctuations. The mechanisms in Figure 37**c** and **d** give a more uniform torque diagram and the load on the small-end bearing reverses which aids their lubrication. The horizontal machine, unlike the vertical, has excellent accessibility to all parts of

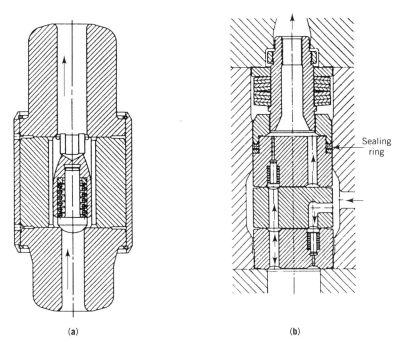

Fig. 36. (**a**) Single poppet valve; (**b**) combined inlet–outlet axial plate valve. Courtesy of Dresser Rand (169).

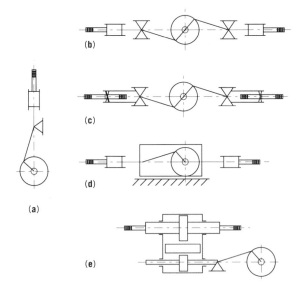

Fig. 37. Driving mechanisms used in secondary compressors (174). See text.

the drive mechanism as well as to the valves and glands; furthermore, layout of the pipework and auxiliary plant such as coolers, pulsation dampers, and separators is easier. The mechanism in Figure 37**c** which makes use of a yoke or secondary cross-head to which the plungers of an opposed pair of cylinders are attached, was adopted by Dresser Rand. The inboard cylinders are not as accessible as the outboard, but alignment is good. The arrangement has the added advantage that it is easy to incorporate a distance piece to prevent ethylene entering the crankcase, an important safety consideration, but it suffers because of its great overall length. The most commonly adopted design is that of Figure 37**d** in which the connecting rod is connected to a cross-head which surrounds the crankshaft. The cross-head is connected to a second cross-head on the opposed high pressure cylinder. With this design the cross-head guides can be arranged in the base of the machine as is done by Burckhardt and GHH or in the plane of the center-line of the plungers which is the design adopted by Nuovo Pignone (175).

Capacity. As a consequence of the limitation on plunger velocity, compressors are usually slow speed with a short stroke. At the same time constructional problems limit the bore size and large capacities have to be achieved by multiplying the number of cylinders. A range of compressors with different strokes and bore diameters, based on the driving mechanism of Figure 37**d** and a module of two cylinders, has been designed by both Nuovo Pignone and Burckhardt with up to six modules, that is twelve cylinders. For example, a machine with 10 cylinders operating between a suction pressure of 25 MPa (3600 psi) and a delivery pressure of 315 MPa (45,700 psi) would have a throughput of 118,400 kg/h and require a power input of 15.2 MW (151).

It was concluded in 1975 that centrifugal secondary compressors with a delivery pressure of 200–250 MPa (29,000–36,000 psi) would be feasible for ethylene capacities of more than 100,000 kg/h. However, in order to replace the primary compressor with a centrifugal machine, it would be necessary to increase the throughput to 200,000–250,000 kg/h (176). Rotary machines have yet to be used in LDPE plants.

Variable Flow Rate. Conventional variable clearance volume and valve lifting devices are impracticable at high pressures and, should it be necessary to vary the flow rate, use has to be made of variable speed electric drives or magnetic clutches. Integral steam and gas engines have been used and Burckhardt (168) developed an hydraulic drive to provide an integrated variable capacity machine, but its efficiency is less than that of a straight mechanical drive.

Operational and Maintenance Problems. Stud Tightening. A large number of screwed fastenings are used to secure the component parts of a reciprocating compressor, eg, cylinder, cylinder head, tie rods, drive rods, etc. Each of the threaded components is subjected to an alternating load, and pretensioning forms the best and simplest way of increasing the fatigue strength of the fastening. Pretensioning serves to increase the mean stress in the bolt but reduces the amplitude of the cyclic stress by an amount determined by the relative stiffness of the bolt and the parts being joined together (177). The pretension may be 80% or more of the yield stress of the bolt material based on core area. A number of methods of controlling the pretension in bolts may be used, of which measurement of the axial stretch by means of a dial indicator is the most common. In a bolted joint about 95% of the torque applied by wrenching is needed to overcome friction and less than 5% becomes useful preload. The torque needed for preload is a function of the cube of the nominal bolt diameter; hence, it becomes very difficult to preload large diameter threads properly unless some form of hydraulic stud tensioning is used.

The pretension in the bolts gradually becomes less as the contact surfaces settle down, as a result of the surface roughness peaks being smoothed out during subsequent loading cycles, and routine checking of tightness is desirable. Loss of pretension is more rapid with machine cut and ground threads than with rolled threads.

Pulsation Dampening.   At the compressor discharge, the peak to peak pressure variation as a percentage of the mean pressure may be as much as 15–18%. It is seldom reduced using volume bottles because the required volume is large; hence, steps have to be taken to minimize the potentially damaging effects. Pressure pulsations induce vibrations in the downstream piping, particularly if it contains numerous bends, and it is necessary to restrain the movement of the pipe to limit the stresses, especially in bending, to a safe level. Mechanical resonance occurs if the frequency of the pressure pulsations, or one of the harmonics, coincides with the natural frequency of vibration of part of the piping system. In this case selection of the correct distances between supports or the use of vibration dampers, in areas where supports cannot be used, may be necessary to reduce the amplitude of the vibration to acceptable levels. Acoustic resonance gives rise to larger pressure pulses than would otherwise be expected and, when designing a complicated piping system to operate over a range of conditions, is difficult to predict. Analogue simulation shows how the potential for resonance conditions can be minimized by altering the pipe layout and by inserting restrictions at appropriate points.

Axial bending induced by mechanical vibration may cause a pipe to fail by fatigue cracking in a transverse direction, usually at the last engaged thread behind a screwed flange. To avoid this mode of failure, which is particularly hazardous as it leads to full bore leakage, it is essential not only to reduce the amplitude of vibration of the pipe between supports to a safe level but also to monitor it periodically in the suction, delivery and interstage piping to ensure that it remains so. Advice on this subject has been given both by users (178) and by manufacturers of compressors (179,180).

Nonlubricated Compressors.   In processes such as hot isostatic compaction, gas purity is very important and diaphragm compressors, nonlubricated reciprocating compressors, or liquid pumps, are often used. In the operation of a diaphragm compressor, oil, above a reciprocating piston, flexes a metal diaphragm in a lenticular cavity, allowing gas on the other side to be sucked in and discharged through inlet and outlet valves housed in recesses (181). The diaphragm compressor is not able to handle large flow rates; the displacement of a single head ranges from 1.5×10^{-4} to 0.1 m^3/s. Each stage is capable of a compression ratio of about 15 to 1 by virtue of the large cooling area surrounding the diaphragm; hence, fewer stages are required than would be the case for a reciprocating compressor. The weakest element is the diaphragm, which is subjected to high cyclic stresses and is easily damaged by small solid particles which may enter with the gas. Triple diaphragms with leak detection monitors ensure that oil cannot enter the gas stream, but diaphragm life is from 200–500 h depending on gas cleanliness and service conditions (182). With a batch process this may be acceptable; the alternative is to use a nonlubricated piston compressor. These are electrically or hydraulically driven, single or double stage and rated up to 200 MPa (29,000 psi). The drive fluid in the hydraulically driven compressor is segregated from the process gas and avoids contamination. Seal life is about 500–1000 h, depending on service conditions (182).

Safety and Testing

The engineering problems involved in the development of a safe and cost effective chemical plant such as that for the manufacture of polyethylene depend on the skill and experience of the craftsmen involved. It is common experience that high pressure work requires special skills and abilities and, even more importantly, a dogged perseverance. In particular, it requires the ability to work with high strength and frequently intractable materials which are more difficult to machine to high accuracy and surface finish. Great attention is needed to minute details in order to achieve reliability as, for instance, in the lapping of the mating surfaces of packing cups for reciprocating compressors and the installation of the sealing rings and plunger. Cleanliness in the assembly of valves and catalyst pumps, for example, is essential for their successful operation.

Reliability and safety depends not only on the craftsmen but also on the quality of the materials used in construction. Very large forgings in high strength steel alloy required for autoclaves needed considerable developments in the casting of the original ingot and its subsequent forging to achieve the high quality demanded, particularly in relation to defects. This improvement in the production of large forgings coupled with metallurgical development work has been necessary in the tube manufacturing process to produce the high quality of tubing required and the larger diameter tubing demanded as a result of the increased size of plant.

The safe operation of high pressure plant of the type described here necessitates a suitable in-service inspection program to ensure that the equipment remains within acceptable design limits. Nondestructive testing plays an important role, and dye penetrant, magnetic particle, eddy-current, and ultrasonic techniques have been widely used to detect flaws or fatigue cracks in high pressure components. A realistic approach to inspection and maintenance of high pressure equipment in polyethylene plants and the compromise which is often necessary between unreasonably rigorous standards and those dictated only by plant production requirements have been discussed (183). The need for adequate records and the extent to which in-service programs have complemented and taken account of more fundamental investigative work have been stressed (183). The concept of a structural integrity program introduced by Du Pont, as a result of its experience with high pressure operations has been discussed (184). This program is aimed at improving the general awareness of those designing and maintaining plants with unusual hazards and standards, as well as providing specific technical control in areas such as failure analysis and inspection.

BIBLIOGRAPHY

"Pressure Techniques (Compressors)" in *ECT* 1st ed., Vol. 11, pp. 115–123, by E. I. Case and J. Charls, Jr., Worthington Corp.; "Pumps and Compressors (Compressors)" in *ECT*

2nd ed., Vol. 12, pp. 728–762, by J. F. Julian and J. F. Hendricks, Washington Corp.; "High Pressure Technology" in *ECT* 3rd ed., Vol. 12, pp. 352–416, by I. L. Spain, Colorado State University.

1. W. R. D. Manning and S. Labrow, *High Pressure Engineering*, Leonard Hill, London, 1971, p. 13.
2. J. H. Faupel and F. E. Fisher, *Engineering Design*, McGraw Hill Book Co., Inc., New York, 1982, p. 230.
3. K. E. Bett and D. M. Newitt, in H. W. Cremer and T. Davies, eds., *The Design of Pressure Vessels*, Vol. 5, *Chemical Engineering Fracture*, Butterworth, London, 1958, p. 199.
4. Ref. 3, p. 200.
5. Ref. 3, p. 205.
6. Ref. 1, p. 18.
7. B. Crossland, *Proc. Inst. Mech. Engrs.* **168**, 935 (1954).
8. B. Crossland, *Welding Research Council Bulletin No. 94*, (1964).
9. J. L. M. Morrison, *Proc. Inst. Mech. Engrs.* **159**, 81 (1948).
10. Ref. 1, p. 41.
11. B. Avitzur, *Int. J. Press. Ves. Piping* **38**, 147 (1989).
12. W. R. D. Manning, *Engineering* **159**, 101, 183 (1945).
13. W. R. D. Manning, *Engineering* **169**, 479, 509, 562 (1950).
14. B. Crossland, S. M. Jorgenson, and J. A. Bones, *TASME* **81**, 95 (1959).
15. B. Crossland and S. A. Gaydon, *PVP* **61**, 167 (1982).
16. B. Crossland and J. A. Bones, *Proc. Inst. Mech. Engrs.* **172**, 777 (1958).
17. Ref. 3, p. 249.
18. B. Crossland and W. F. K. Kerr, *High Temp-High Press.* **1**, 133 (1969).
19. J. H. Faupel, *Trans. ASME* **78**, 1031 (1956).
20. B. Crossland, *Proc. Inst. Mech. Engrs., part 3A*, **180**, 243 (1966).
21. Ref. 3, p. 218.
22. Ref. 3, p. 241.
23. G. L. Franklin and J. L. M. Morrison, *Proc. Inst. Mech. Engrs.* **174**, 947 (1960).
24. D. P. Kendall, *PVP* **125**, 17 (1987).
25. A. Stacey and G. A. Webster, *Int. J. Pres. Ves. Piping* **31**, 205 (1988).
26. P. C. T. Chen, *PVP* **110**, 61 (1986).
27. A. Chaaban, K. Leung, and D. J. Burns, *PVP* **110**, 55 (1986).
28. A. E. Macrae, *Overstrain of Metals and its Application to the Autofrettage Process of Cylinder and Gun Construction*, HMSO, London, 1930.
29. D. Brown and W. J. Skelton, *PVP* **110**, 61 (1986).
30. E. Giacomelli, P. Pinzauti, and S. Corsi, *PVP* **61**, 63 (1982).
31. A. W. Birks, in *Proceedings of the 2nd International Conference on High Pressure Engineering*, University of Sussex, U.K., 1975.
32. B. Crossland and D. J. Burns, *Proc. Inst. Mech. Engrs.* **175**, 1083 (1962).
33. Ref. 1, p. 55.
34. Ref. 3, p. 224.
35. W. R. D. Manning, *Engineering Lond.* **163**, 349 (1947).
36. W. R. D. Manning, *Engineering Lond.* **170**, 464 (1950).
37. R. A. Strub, *Trans. Am. Soc. Mech. Engrs.* **75**, 73 (1953).
38. S. J. Becker and L. Mollick, *J. Eng. Ind., Trans. Am. Soc. Mech. Engrs.* **82**, 136 (1960).
39. J. A. Knapp and P. S. J. Crofton, *PVP* **110**, 21 (1986).
40. H. A. Photo, in Ref. 31.
41. A. H. Ghosn and M. Sabbaghian, *PVP* **192**, 9 (1990).
42. P. J. James, ed., *Isostatic Pressing Technology*, Applied Science Publishers, London, 1983, p. 214.

43. J. Schierenbeck, *Brennstoff-Chemie* **31**, 375 (1950).
44. Ref. 3, p. 231.
45. E. Seibel and S. Schwaigerer, *Chem.-Ing.-Tech.* **24**, 199 (1952).
46. E. Karl, *Chem. Eng. Prog.* **68**(11), 56 (1972).
47. Anon., *Engineering Lond.* **187**, 155 (1949).
48. T. M. Jasper and C. M. Scudder, *Trans. Amer. Inst. Chem. Engrs.* **37**, 885 (1941).
49. T. M. Jasper, *Chem. Eng. Prog.* **52**, 521 (1956).
50. Z. Xiao-Qin, W. Jing-Sheng, and C. Guo-Li, *PVP* **165**, 47 (1989).
51. K. Boudjelida, A. H. Ghosn, and M. Sabbaghian, *J. Press. Vessel Tech.* **113**, 459 (1991).
52. Yu Wang, *PVP* **110**, 75 (1986).
53. J. S. McCabe and E. W. Rothrock, *Mech. Eng.* **93**, 34 (Mar. 1971).
54. W. R. D. Manning, *Proc. Inst. Mech. Engrs.* **156**, 362 (1947).
55. K. Opitz, *Chem. Ing. Tech.* **51**, 5, 398 (1979).
56. Y. Wang, A. I. Soler, and G. L. Chen, *J. Press. Vessel Tech.* **112**, 410 (1990).
57. H. C. Rauschnplat, *PVP* **57**, 57 (1982).
58. P. S. Huang and G. Zhu, *J. Press. Vessel Tech.* **114**, 94 (1992).
59. Z. Zudans, T. C. Yen, and W. H. Steigelmann, *Thermal Stress Techniques*, Elsevier Publishing Co., New York, 1965.
60. S. Timoshenko and J. N. Goodier, *Theory of Elasticity*, McGraw-Hill Book Co., Inc., London, 1951.
61. Ref. 3, p. 256.
62. Ref. 3, p. 257.
63. W. J. Skelton and B. Crossland, *Proc. Inst. Mech. Engrs.* **182**(pt. 3C), 25 (1967–1968).
64. *Ibid.*, p. 139.
65. *Ibid.*, p. 159.
66. C. R. Soderberg, *Trans. Amer. Soc. Mech. Engrs.* **63**, 737 (1941).
67. F. H. Norton and C. R. Soderberg, *Trans. Amer. Soc. Mech. Engrs.* **64**, 769 (1942).
68. W. J. Skelton, A. Salim, and R. G. Patton, in Ref. 24.
69. R. W. Bailey, *Proc. Inst. Mech. Engrs.* **164**, 324 (1951).
70. *Ibid.*, p. 425.
71. W. B. Carlson and D. Duval, *Engineering, Lond.* **193**, 829 (1962).
72. A. Chitty and D. Duval, *Proc. Inst. Mech. Engrs.* **178**(pt. 3A), 4-1 (1963–1964).
73. D. W. Williams and P. G. Harris, *Proc. Inst. Mech. Engrs.* **182**(pt. 3C), 166 (1967–1968).
74. P. Snowden, *Proc. Inst. Mech. Engrs.* **182**(pt. 3C), 283 (1967–1968).
75. F. X. Zimmerman and W. H. Walker, in P. J. James, ed., *Isostatic Pressing Technology*, Applied Science Publishers, London, 1983, Chapt. 7, pp. 183–201.
76. K. E. Bett and G. Saville, *AIChE Chem. E. Symp. Series*, (2), 71 (1965).
77. K. E. Bett, G. Saville, and M. Brown, *Proceedings 3rd International Conference on High Pressure*, Aviemore, Scotland, 1970, Institute of Mechanical Engineers, 1971.
78. Ref. 74, p. 285.
79. J. B. Toops and E. Enroth, *PVP* **148**, 15 (1988).
80. P. J. James, ed., *Isostatic Pressing Technology*, Applied Science Publishers, London, 1983, p. 189.
81. Ref. 1, p. 108.
82. J. L. M. Morrison, B. Crossland, and J. S. C. Parry, *Proc. Inst. Mech. Engrs.* **170**, 697 (1956).
83. J. L. M. Morrison, B. Crossland, and J. S. C. Parry, *Proc. Inst. Mech. Engrs.* **174**, 95 (1960); Ref. 1, p. 115.
84. H. L. D. Pugh, ed., *Mechanical Behaviour of Materials Under Pressure*, Applied Science Publishers, London, 1971, Chapt. 7, p. 316.

85. D. Brown and W. J. Skelton, *PVP* **61**, 89 (1982).
86. J. S. C. Parry, *Proc. Inst. Mech. Engrs.* **180**(pt. 1), 387 (1965–1966).
87. J. L. M. Morrison, B. Crossland, and J. S. C. Parry, *J. Mech. Eng. Sci.* **1**, 207 (1959).
88. B. A. Austin and B. Crossland, *Proc. Inst. Mech. Engrs.* **180**, 134 (1965).
89. Ref. 84, pp. 299–353.
90. G. H. Haslam, *High Temp.-High Press.* **1**, 705 (1969).
91. P. S. J. Crofton and W. A. Lees, *PVP* **61**, 115 (1982).
92. D. J. Burns and J. S. C. Parry, *Proc. Inst. Mech. Engrs.* **182**(pt. 3C), 72 (1967).
93. P. M. Jones and B. Tomkins, *Proc. Inst. Mech. Engrs.* **182**(pt. 3C), 311 (1967).
94. J. A. Kapp and P. S. J. Crofton, *PVP* **125**, 81 (1982).
95. S. Tauscher, *PVP* **125**, 73 (1982).
96. Ref. 84, p. 327.
97. B. N. Cole, *J. Mech. Eng. Sci.* **11**, 151 (1969).
98. T. E. Davidson, B. B. Brown, and D. P. Kendall, in Ref. 24.
99. T. E. Davidson, E. Eisenstadt, and A. N. Reiner, *J. Basic Eng. Trans. ASME* **85**(2), 555 (1963).
100. B. Crossland and B. A. Austin, *Proc. Inst. Mech. Engrs.* **180**(3A), 118 (1956).
101. B. Crossland and co-workers, *Proc. 4th Int. Conf. Pressure Vessel Tech.* **2**, 375 (1980).
102. B. B. Brown, *PVP* **125**, 9 (1982).
103. B. A. Austin, A. N. Reiner, and T. E. Davidson, *Proc. Inst. Mech. Engrs.* **182**(pt. 3C), 91 (1967).
104. B. Crossland and W. J. Skelton, *Proc. Inst. Mech. Engrs.* **182**(pt. 3C), 106 (1967).
105. C. L. Tan and R. T. Fenner, *J. Strain Anal.* **13**, 213 (1978).
106. C. L. Tan and R. T. Fenner, *Proc. R. Soc. Lond.* **A369**, 243 (1979).
107. C. L. Tan and R. T. Fenner, *Int. J. Fracture* **16**, 233 (1980).
108. H. Price and B. A. Austin, *PVP* **61**, 135 (1982).
109. C. E. Turner, *High Temp.-High Press.* **6**, 1 (1974).
110. P. C. Paris, *Trans. ASME, J. Basic Eng.* **85**, 528 (1953).
111. G. Hartwig, in J. I. Kroschwitz, ed., *Encyclopedia of Polymer Science and Engineering*, John Wiley & Sons, Inc., New York, 1986, p. 458.
112. A. Chaaban and M. Jutras, *PVP* **165**, 9 (1989).
113. A. Chaaban and M. Jutras, *PVP* **148**, 35 (1988).
114. G. Bing-Liang, L. Ting-Xin, and L. Tian-Xiang, *PVP* **148**, 19 (1988).
115. Ref. 3, p. 285.
116. Ref. 1, p. 281.
117. Ref. 3, p. 286.
118. Ref. 3, p. 288.
119. Ref. 1, p. 115
120. Ref. 138, p. 242.
121. Ref. 80, p. 133
122. Ref. 3, p. 288.
123. E. G. Warnke, *Proc. Inst. Mech. Engrs.* **182**(pt. 3C), 47 (1967–1968).
124. D. G. Sopwith, *Proc. Inst. Mech. Engrs.* **159**, 373, 395 (1948).
125. A. Kuske, W. Steinchen, and J. Zech, in Ref. 31.
126. E. H. Perez, J. G. Sloan, and K. J. Kelleher, *PVP* **125**, 53 (1987).
127. D. M. Fryer and C. W. Smith, in Ref. 31.
128. B. Kenny and E. A. Patterson, *Exp. Mech.* **25**, 208 (1985).
129. D. J. Burns and co-workers, *PVP*, **125**, 63 (1987).
130. A. Chaaban and M. Jutras, *PVP* **148**, 35 (1988).
131. R. G. Fasiczka, *PVP* **148**, 139 (1988).
132. A. Chaaban and U. Muzzo, *PVP* **192**, 23 (1990).
133. M. Jutras and A. Chaaban, *PVP* **165**, 57 (1989).

134. R. G. Fasiczka, *PVP* **192**, 29 (1990).
135. C. B. Boyer, *Hot Isostatic Pressure Systems Failures and Accident History*, Battelle Memorial Institute, Columbus, Ohio, 1987.
136. E. L. Danfelt, W. J. O'Donnell, and E. L. Westermann, *PVP* **148**, 119 (1988).
137. Ref. 3, p. 291.
138. B. Crossland and co-workers, *Proc. Inst. Mech. Engrs.* **200**(A4), 240 (1986).
139. B. W. Rolfe, *Proc. Inst. Mech. Engrs.* **182**(pt. 3C), 239 (1967–1968).
140. E. Karl, *PVP* **48**, 37 (1981); M. D. Biggs, *PVP* **48**, 9 (1981).
141. E. H. Watson, *Am. Inst. Chem. Eng.* **2**, 35 (1974).
142. *ASTM Specification A 508/A 508M-88b*, Philadelphis, Pa.
143. *ASTM Specification A 723/A 723M-88*, Philadelphia, Pa.
144. A. K. Khare, *PVP* **114**, 47 (1986).
145. V. Placania, D. Hengerer, and H. J. Mueller-Aue, *PVP* **114**, 33 (1986).
146. Ref. 1, p. 139.
147. *ASME Boiler and Pressure Vessel Code, Section 8, Rules for Construction of Pressure Vessels*, ASME, New York, 1989.
148. *ASME Boiler and Pressure Vessel Code, Section 8, Rules for Construction of Pressure Vessels, Alternative Rules*, ASME, New York, 1989.
149. G. J. Mraz, *J. Press. Vessel Tech.* **109**, 257 (1987).
150. J. G. Logan and B. Crossland, *Proceedings Conference on Practical Applications of Fracture Mechanics to Pressure Vessel Technology*, Institution of Mechanical Engineers, London, 1971, p. 148.
151. Ref. 138, p. 237.
152. H. Ford, E. M. Watson, and B. Crossland, *J. Press. Vessel Tech.* **103**, 133 (1980).
153. W. T. Hughes, *PVP* **192**, 33 (1990).
154. H. I. Burrier, *PVP* **48**, 53 (1981).
155. J. R. Sims, *PVP* **110**, 35 (1986).
156. Ref. 138, p. 243.
157. U.S. Pat. 2,772,103 (1952), E. Strub.
158. Ref. 138, p. 244.
159. H. Ford and co-workers, *Chem. Eng. Prog.* **68**, 77 (1972).
160. J. Rogan in Ref. 31.
161. J. Bognar and co-workers, *PVP* **148**, 95 (1981).
162. Ref. 138, p. 247.
163. J. E. Aller, *PVP* **192**, 35 (1990).
164. M. J. Grey and W. J. Skelton, *PVP* **61**, 103 (1982).
165. J. B. Allen, *Chem & Proc. Eng.*, 493 (Sept. 1965).
166. Ref. 1, p. 208.
167. Ref. 1, pp. 245 and 246.
168. C. Mantile, *Proc. Inst. Mech. Engrs.* **184**(pt. 3R), 1 (1969–1970).
169. J. S. Prentice, S. E. Smith, and L. S. Virtue, *Am. Inst. Chem. Engrs.* **2**, 1 (1974).
170. B. W. Sander, *Am. Inst. Chem. Engrs.* **3**, 47 (1978).
171. Ref. 138, p. 249.
172. A. Traversari and E. Giacomelli, in Ref. 31, pp. 161–173.
173. A. Traversari, M. Ceccherini, and A. Del Puglia, *PVP* **48**, 81 (1981).
174. Ref. 138, p. 250.
175. A. Traversari and P. Beni, *Am. Inst. Chem. Engrs.* **2**, 8 (1974).
176. C. Matile and R. A. Strub, *Sulzer Tech. Rev.* **57**, 1 (1975).
177. Ref. 1, p. 184.
178. A. K. Gardner and W. T. Hughes, *PVP* **125**, 1 (1987).
179. J. R. Olivier and K. Scheuber, *Am. Inst. Chem. Engrs.* **3**, 37 (1978).
180. M. Michelini, *Am. Inst. Chem. Engrs.* **3**, 41 (1978).

181. Ref. 1, p. 202.
182. Ref. 42, pp. 191–193.
183. A. K. Gardner, K. B. King, and R. J. Cooper, *PVP* **48**, 151 (1981).
184. W. T. Hughes, *PVP* **98-8**, 193 (1985).

General References

P. W. Bridgman, *The Physics of High Pressure*, G. Bell & Sons Ltd., London, 1931.

D. M. Newitt, *The Design of High Pressure Plant and the Properties of Fluids at High Pressures*, Clarenden Press, Oxford, 1940.

E. W. Comings, *High Pressure Technology*, McGraw-Hill Book Co., Inc., New York, 1956.

S. D. Hamann, *Physico-Chemical Effects of Pressure*, Butterworths, London, 1957.

K. E. Bett and D. M. Newitt, in H. W. Cremer and T. Davies, eds., *The Design of High Pressure Vessels*, Vol. 5, Chemical Engineering Practice, Butterworths, London, 1958, pp. 196–298.

H. Tongue, *The Design and Construction of High Pressure Chemical Plant*, 2nd ed., Chapman & Hall Ltd., London, 1959.

D. S. Tsiklis, *Handbook of Techniques in High Pressure Research and Engineering*, trans. A. Bobrowsky, Plenum Press, Inc., New York, 1959.

R. H. Wentorf, ed., *Modern Very High Pressure Techniques*, Butterworths, London, 1962.

R. S. Bradley, ed., *High Pressure Physics and Chemistry*, Vols. 1 and 2, Academic Press Inc., London, 1963.

W. Paul and D. M. Warschauer, eds., *Solids Under Pressure*, McGraw-Hill Book Co. Inc., New York, 1963.

K. E. Weale, *Chemical Reactions at High Pressures*, E. & F. N. Spon Ltd., London, 1967.

H. H. Buchter, *Apparate und Armatieren der Chemischen Hochdrucktechnik Konstruktion, Berechnung und Herestellung*, Springer-Verlag, Berlin, 1967.

W. R. D. Manning and S. Labrow, *High Pressure Engineering*, Leonard Hill, London, 1971.

H. Ll. D. Pugh, ed., *Mechanical Behaviour of Materials Under Pressure*, Applied Science Publishers Ltd., London, 1971.

G. C. Ulmer, *Research Techniques for High Pressure High Temperature*, Springer-Verlag, Berlin, 1971.

B. A. Sykes and D. Brown, *A Review of the Technology of High Pressure Systems*, Institute of Gas Engineers, London, 1975.

I. L. Spain and J. P. Paauwe, *High Pressure Technology*, Marcel Dekker, Inc., New York, 1977.

R. S. Dadson, S. L. Lewis, and G. N. Peggs, *The Pressure Balance*, HMSO, London, 1982.

J. H. Faupel and F. E. Fisher, *Engineering Design*, McGraw Hill Book Co., Inc., New York, 1982.

J. F. Harvey, *Theory and Design of Pressure Vessels*, Van Nostrand Reinhold Co. Inc., New York, 1985.

H. H. Bednar, *Pressure Vessel Design Handbook* 2nd ed., Van Nostrand Reinhold Co. Inc., New York, 1986.

W. F. Sherman and A. A. Stadtmuller, *Experimental Techniques in High-Pressure Research*, John Wiley & Sons Ltd., Chichester, 1987.

W. Cross, *History of the ASME Boiler and Pressure Vessel Code*, ASME, New York, 1989.

K. E. Bett
University of London

HIGH TEMPERATURE ALLOYS

High temperature alloys are those combinations of metals that are used specifically for their heat resisting properties. Physical properties such as melting temperatures, elastic modulus, density, and thermal conductivity of the elemental metals that serve as the bases for most high temperature alloys are listed in Table 1. Contrary to expectations, melting point is not the primary indicator of adequate high temperature strength. For example, nickel, which has the lowest melting point of any element in Table 1, is the choice for the most severe high temperature structural applications in air. There are several reasons for this, including the lack of an allotropic phase transformation in nickel below its melting point, its high tolerance for alloying elements without causing a phase change from the close packed face-centered cubic (fcc) crystal structure, and the ability to produce a very stable precipitate, γ'-(Ni$_3$Al), that is the primary source of high temperature strengthening in nickel-base superalloys (see NICKEL AND NICKEL ALLOYS).

Density is a particularly important characteristic of alloys used in rotating machinery, because centrifugal stresses increase with density. Densities of the various metals in Table 1 range from 6.1 to 19.3 g/cm^3. Those of iron, nickel, and cobalt-base superalloys fall in the range 7–8.5 g/cm^3. Those alloys which contain the heavier elements, ie, molybdenum, tantalum, or tungsten, have correspondingly high densities.

Thermal expansion coefficients are important for mating components, as well as in the development of thermal stresses, which vary directly with the expansion coefficients. In general, thermal expansion coefficients are inversely proportional to melting point. This is readily seen from Table 1 and Figure 1a (1). Thus molybdenum, tungsten, and tantalum have the lowest thermal expansion coefficients of the metals listed. Nickel alloys tend to have the highest thermal expansion and thermal conductivity coefficients of the iron, nickel, and cobalt-based superalloys. Thermal conductivity data for several high temperature materials are shown in Table 1 and in Figure 1b (1). High thermal conductivity results in reduced thermal gradients and, consequently, lower stresses.

The bcc refractory metals (Table 1) are usually categorized by location in the Periodic Table rather than by melting point or physical properties. The Group 5 (VB) elements, vanadium, niobium, and tantalum, all are ductile at ambient and cryogenic temperatures, provided that contamination by interstitial solutes is avoided. In contrast, the unalloyed Group 6 (VIB) metals, tungsten, chromium, and molybdenum, are generally classified as brittle at low temperatures, although some ductility may be displayed by single crystals or zone refined polycrystals of molybdenum and tungsten. Alloys of the VIB metals having appreciable room temperature ductility can be made if rhenium is added in large quantities. Also, the preparation of a fibered microstructure, eg, cold or warm worked, has a very beneficial effect on ductility.

One common characteristic held by the bcc refractory metals is very poor resistance to oxidizing atmospheres. All of these metals readily form nonprotective, low melting oxides, so that the atmospheres in which they are used must

Table 1. Physical Properties of High Temperature Metals

Metal	Crystal structure		Melting point, °C	Density, g/cm^3	Thermal expansion coefficient at RT, 10^6/°C	Thermal conductivity at RT, W/(m·K)[a]	Young's modulus, GPa[b]
	22°C	1000°C					
Co	hcp	fcc	1495	8.85	13.8	69.0	206.8
Ni	fcc	fcc	1453	8.90	13.3	92	206.8
Fe	bcc	fcc	1537	7.87	11.76	75	196.5
Cr	bcc	bcc	1890	7.2	6.2	66.9	248.2
Nb	bcc	bcc	2468	8.6	7.1	52.3	103.4
W	bcc	bcc	3410	19.3	4.5	20.1	344.7
Ta	bcc	bcc	2996	16.6	6.6	54	186.2
V	bcc	bcc	1900	6.1	9.7	31	125
Mo	bcc	bcc	2610	10.22	5.4	146	324.1

[a]To convert W/(m·K) to cal/(cm·s·°C), multiply by 2.39×10^{-3}.
[b]To convert GPa to psi, multiply by 1.45×10^5.

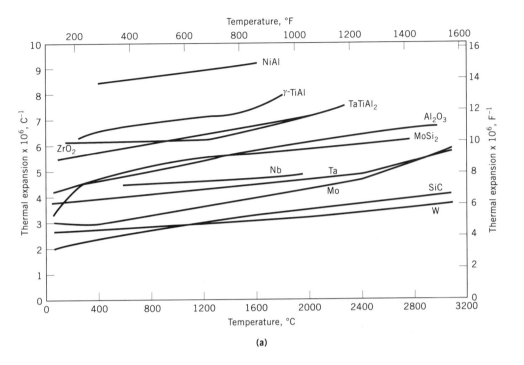

(a)

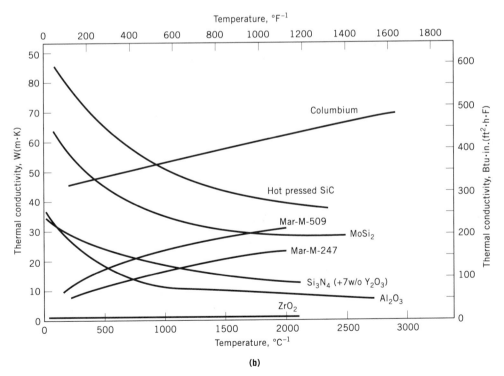

(b)

Fig. 1. Thermal properties of high temperature materials: (**a**), expansion coefficient; (**b**), thermal conductivity (1). Mar-M-247 has an Ni base; Mar-M-509, a Co base.

be inert, eg, helium or a vacuum, or they need to be coated. The iron, cobalt, and nickel-base superalloys, on the other hand, have sufficient oxidation resistance without having surface coatings to permit operation in oxidizing atmospheres at much higher temperatures than the Group VB and VIB metals. This oxidation and hot corrosion resistance is conferred chiefly by chromium additions, and, in the case of nickel alloys, from aluminum in solid solution. Nevertheless, under severe operating conditions, eg, as turbine blades in aircraft engines, even the surfaces of nickel and cobalt-base superalloys must be coated with oxidation-resistant compounds or alloys such as NiAl, CoCrAlY, or PtAl.

Mechanical Behavior

Creep Rupture. Metals and their alloys lose appreciable strength at elevated temperatures. For most materials, the ultimate tensile and yield strengths fall off regularly as the temperature increases, as illustrated in Figure 2 (2). The exceptions are some intermetallics, eg, nickel aluminide(3:1) [*12003-81-5*], Ni$_3$Al, and alloys containing a large volume fraction of Ni$_3$Al, eg, single-crystal alloys PWA 1480. Not only is temperature important, but time also influences mechanical behavior at elevated temperatures. Stresses well below the yield strength can cause gradual deformation and eventual fracture if sustained long enough. For this reason, the standard tensile test is inadequate for providing design data at elevated temperatures, and creep–rupture tests must be conducted, in which the time-dependent deformation and fracture are determined from periodic measurements under a fixed stress or load.

Time-dependent deformation under stress is known as creep, and the curve showing the deformation as a function of time is known as a creep curve. Creep curves obtained for different materials, temperatures, and stresses have certain common features, illustrated in Figure 3 for a constant load situation. After

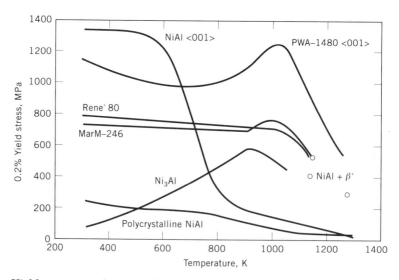

Fig. 2. Yield stress as a function of temperature for NiAl, Ni$_3$Al, and several commercial superalloys where <001> is the parallel-to-tensile axis for single crystals and (○) are data points for NiAl + β' (2). See text. To convert MPa to psi, multiply by 145.

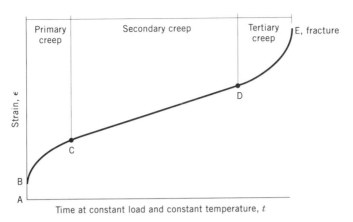

Fig. 3. The idealized creep curve where the distance from A to B represents the initial elongation on loading. See text.

the instantaneous extension on loading, there is a period of primary or transient creep during which the rate of creep continually diminishes. During this stage the material strain hardens, so that resistance to deformation continually increases as the deformation increases. The secondary stage of creep, or steady-state creep, represents a running balance between strengthening by work hardening and weakening by recovery, so that the rate of creep remains essentially constant. The tertiary stage is the period of accelerating creep that precedes fracture. This increase in creep rate is often caused by an increase in stress that accompanies the decrease in cross-sectional area of the specimen, which can result from either the decrease of the diameter of the specimen as it elongates, or to the formation of cracks, which can also act as stress-raisers. In other cases, the accelerating creep rate results from changes in the metallurgical structure, such as recrystallization, precipitation, or re-solution of second phases that originally contributed to alloy strength. Were the stress to be maintained constant during the test, rather than the load, the creep curve would be significantly different. Most notably, the tertiary stage of creep would be eliminated.

Both the initial elongation on loading and the amount of primary creep increase as the temperature or the stress is raised. At low stresses for a given temperature, or at low temperatures for a given stress, the curves are characterized by prolonged secondary or constant rate periods of creep. When either the stress or the temperature is low, fracture does not occur unless the test is continued for a very long time. In fact, creep rate and rupture time are generally inversely related as demonstrated through the empirical relationship

$$\log t_r + m \log \dot{\epsilon}_s = B \tag{1}$$

where t_r = rupture life, h; $\dot{\epsilon}_s$ = steady-state creep rate, s^{-1}; and m and B are constants (3). As the stress or temperature is increased, the minimum creep rate, ie, the slope of the creep curve in the secondary stage increases; at the same time the duration of secondary creep becomes much shorter, until at high stresses or temperatures it may hardly be discerned in the creep curve. Under the latter conditions the creep curve can best be described as consisting of an initial decelerating rate of creep followed by an accelerating rate of creep leading to fracture. Finally, both the amount of creep strain during the third stage of creep and the final elongation at fracture generally increase with stress and temperature, ie, as the time to fracture decreases.

Stress–rupture tests are similar to creep tests except that the test is carried out to failure. Higher loads are used in the stress–rupture tests, thus creep rates are higher. In creep tests the total strain is often very small, whereas the total strain in a stress–rupture test may approach 50%. Simpler strain measuring devices, such as dial gauges, may therefore be used in stress–rupture tests, and provision of multiple testing units is more likely. The basic data obtained from a stress–rupture test is time to failure at a given temperature and nominal stress level. Elongation measurements vs time give minimum creep rate. Rupture life or time to reach a fixed strain vs stress generally are plotted on a log–log scale.

For many applications, the life of a part is well beyond the time that laboratory tests are practical. The creep deformation during very long times is often estimated as the product of minimum creep rate and life. Minimum creep rates can be established with sufficient accuracy for this calculation by running a test for between 1000 and 10,000 h, ie, between 42 days and about 14 months, and these rates can be used to estimate the deformation that would result after continuous running for 100,000 h (about $11\frac{1}{2}$ yr), or more. In order for such a calculation to have meaning, the stress and temperatures must be low enough for the part neither to fracture nor to enter the third stage of creep. This can be guarded against by carefully extrapolating the creep–rupture lifetime curves to lower stresses, and making sure that the rupture life at the design stress is well beyond the service life. There are several standard extrapolation techniques to relate stress, temperature, and rupture life (or time to a specified strain). The best known of these, the Larson-Miller parameter (P_{LM}) (4), relates the stress, σ, to the time to rupture, t_R (in hours), as follows:

$$\sigma = f(P_{LM}) = f[T(C + \log t_R)]. \tag{2}$$

where C is a constant of the order of 20 which is derivable from experiments, and temperature, T, is expressed in Rankine. All extrapolation procedures should be used with caution.

Relaxation. Relaxation is also associated with creep at high temperatures. If a specimen is stretched or compressed and is then held over a period of time at a high temperature with its ends in fixed positions, the stresses within the specimen gradually diminish. After sufficient time has elapsed, the tensile or compressive stresses may relax to only a fraction of the original values. Creep and relaxation are thus complementary processes, and an alloy having high creep strength generally resists relaxation. Creep refers to the changes in strain with time while under stress; relaxation refers to the changes in internal stress with time while at constant strain.

Relaxation is an important example of a creep phenomenon encountered in practice. Bolts, studs, flanges, and springs of all kinds are subject to relaxation when used at high temperatures. These members are loaded to a stress that must be maintained for proper functioning. If relaxation occurs, the stress decreases. Thus bolts can become loose so that bolted joints develop leaks after operation at elevated temperatures.

Fatigue. Engineering components often experience repeated cycles of load or deflection during their service lives. Under repetitive loading most metallic materials fracture at stresses well below their ultimate tensile strengths, by a process known as fatigue. The actual lifetime of the part depends on service conditions, eg, magnitude of stress or strain, temperature, environment, surface condition of the part, as well as on the microstructure.

Resistance to fatigue fracture is an important consideration in selecting materials for many high temperature applications, most notably in rotating machinery such as gas or steam turbines. Generally, two classifications of fatigue behavior are made, depending on the parameter being reversed, stress, or strain. Under strain controlled cycling, where appreciable plastic strains are involved in each cycle, the process is known as low cycle fatigue. Cycles to failure, N_f, and the range of plastic strain per cycle, $\Delta \epsilon_p$, are related through the Coffin-Manson equation (5,6):

$$N_f^b \Delta \epsilon_p = C \tag{3}$$

The constant, C, is proportional to the ductility of the material in tension; the exponent, b, is near 0.5 for most materials over a wide temperature range. This equation applies usually in the range $1-10^5$ cycles, and typical data are shown in Figure **4a** (5). The exponent rises when creep or environmental interactions affect fatigue behavior.

When constant stress (S) amplitudes are encountered, the process is known as high cycle fatigue, because failure generally occurs only when N exceeds 10^4 cycles. Data from high cycle fatigue tests are reported in the form of an S vs N curve, as shown in Figure **4b** (7).

The fatigue resistance of metals and alloys frequently parallels the tensile properties, thus high temperatures generally diminish both strength and fatigue resistance. However, at elevated temperatures an added complication is provided by changes in the frequency of load or strain reversal. At low test frequencies, more time is afforded for either creep or deleterious environmental interactions to occur, and the number of cycles to failure for a given stress or strain level decreases. In vacuum, the environmental interaction is avoided. At ordinary temperatures, in contrast, frequency is not a significant variable. Another change often brought about by high temperatures is a different crack appearance brought about by a shift to a new fatigue damage mechanism.

High temperature materials which exhibit the greatest resistance to high cycle fatigue on a strength basis, ie, fatigue limit/tensile strength vs N_f, are composite materials and dispersion strengthened alloys. Precipitation hardened alloys, on the other hand, usually demonstrate very poor fatigue resistance relative to the tensile strengths, perhaps because the precipitates become unstable as a result of repeated cycling. Therefore, nickel-base alloys strengthened by γ'-precipitates exhibit inferior fatigue resistance, relative to nickel-base composite

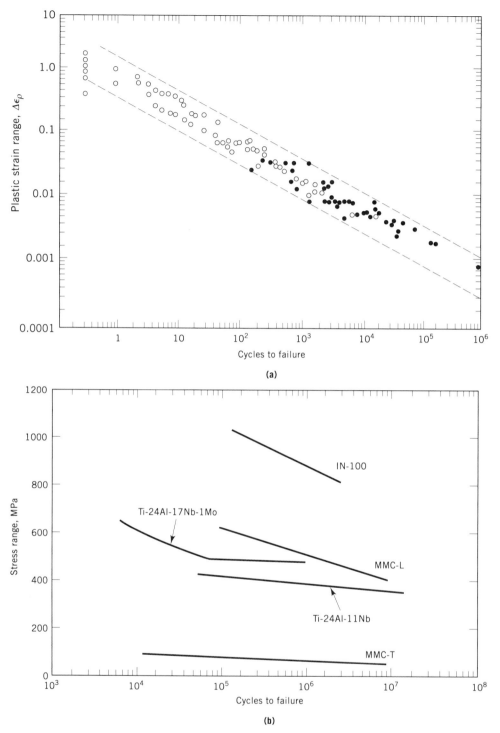

Fig. 4. Fatigue data for high temperature alloys: (**a**) $\Delta\epsilon_p$ vs cycles to failure for various alloys tested under strain control. (○) = testing at RT in air; (●) = elevated temperatures in a vacuum or in argon (5). (**b**) High cycle fatigue behavior of α_2-Ti$_3$Al and its composites (MMC-L is longitudinal, MMC-T is transverse orientation) compared to IN-100 at 650°C, in air, $\nu = 0.2$ Hz, R (ratio of minimum to maximum stress) $\simeq 0$ (7). To convert MPa to psi, multiply by 145.

materials of the same tensile strength. A more detailed account of fatigue resistance of metallic materials is available (8,9).

The rate of crack growth is often a more useful parameter than is fatigue life. Fracture mechanics (qv) techniques have been widely applied to the crack growth behavior of high temperature alloys, as typified by the data of Figure 5 (7). The relationship between crack growth rate, da/dN, and stress intensity range, ΔK, is given by

$$da/dN = C\Delta K^m \tag{4}$$

over most of the life of the part, C and m are experimental constants, ΔK is given approximately by the product of the stress range, $\Delta\sigma$, and the square root of the crack length, a. Crack growth rates may be influenced by microstructure, although the effects are much less than that achieved by changing temperature. Crack growth rates tend to increase sharply for many structural alloys between 400 and 1000°C (10).

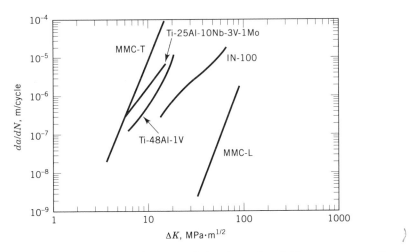

Fig. 5.   Comparison of crack growth rates of titanium aluminides, α_2 composites, and IN-100; tests at 650°C, R = 0.1, ν = 0.2 Hz except for the composite for which ν = 3.33 Hz (7). See Figure 4 for term definitions.

Strengthening Mechanisms

Solid-Solution Strengthening. *Tensile Strength.*   Commercial high temperature alloys generally contain substantial alloying additions in solid solution to provide strength, creep resistance, or resistance to surface degradation. In addition, the stronger alloys may contain elements which, after suitable heat treatment or thermal-mechanical processing, result in the formation of small coherent particles of an intermetallic compound or of a carbide. Many alloying elements partition to some degree between both the matrix and the precipitate, resulting in considerable alloying of both phases. In the case of solid-solution strengthening, yielding may be discussed in terms of the effects of solutes on various physical or crystallographic properties, for example, lattice parameter and elastic modulus.

Both solid-solution hardening and precipitation hardening can be accounted for by internal strains generated by inserting either solute atoms or particles in an elastic matrix (11). The degree of elastic misfit, δ, produced by the difference, Δa, between the lattice parameter, a_o, of the pure matrix and, a, the lattice parameter of the solute atom is given by

$$\delta = \frac{1}{c} \frac{\Delta a}{a_o} \tag{6}$$

where c is the solute concentration and $\Delta a = a_o - a$.

Whereas a linear relation between flow stress and lattice-parameter change is obeyed for any single solute element in nickel, the change in yield stress for various solutes in nickel is not a single-valued function of the lattice parameter, but depends directly on the position of the solute in the Periodic Table (12). For the same lattice strains, the larger the valency difference between solute and solvent, the greater the hardening. The strengthening influence of alloying elements persists to temperatures at least as high as 815°C. Valency effects may be explained by modulus differences between the various alloys (13). Alternatively, the effects of valency may be felt through the decrease in stacking fault energy (SFE) of fcc alloys having increasing electron to atom ratio (14).

For nickel, cobalt, and iron-base alloys the amount of solute, particularly tungsten or molybdenum, intentionally added for strengthening by lattice or modulus misfit is generally limited by the instability of the alloy to unwanted σ-phase formation. However, the Group 5(VB) bcc metals rely on additions of the Group 6(VIB) metals Mo and W for solid-solution strengthening.

Creep Resistance. Principles of alloy design for creep resistance are not unlike those for yield or tensile strength. Creep rate is generally reduced when solutes of higher modulus and melting point than the base metal are utilized. Low stacking fault energy also has been associated with high resistance to creep (15). However, perhaps the principal determinant of creep resistance is a parameter which expresses the rate at which atomic species can migrate in the alloy. The second-stage creep rate, $\dot{\epsilon}$, can be described by an Arrhenius-type equation:

$$\dot{\epsilon} = K(\sigma/G)^n \exp(-Q_c/RT) \tag{7}$$

where Q_c is the activation energy for creep, K is a constant, σ is stress, G is shear modulus, and n is an empirical constant. The activation energy for creep, Q_c, is often independent of stress or strain, but can increase substantially with temperature. Between temperatures of $0.5\,T_m$ and T_m, where T_m is the melting point, Q_c is nearly constant, and for many crystalline solids has been shown to equal the activation energy for self-diffusion, Q_D (16). Diffusivity is higher in the more open bcc structure relative to close packed fcc or hcp structures. Lowered diffusivity in close packed alloys accounts for the superior creep resistance of austenitic steels relative to ferritic steels (17), and of alpha (hcp) titanium alloys relative to beta (bcc) alloys (see STEEL).

In summary, the addition of soluble alloying elements generally improves the creep resistance of a metal at a given temperature, unless the melting point is greatly depressed by the solute. Elements that retard recovery and recrystallization during service are effective in improving creep strength, as are solutes having high elastic moduli. For maximum creep resistance, nonclose packed crystal structures should be avoided, and solutes should depress the stacking fault energy of the base metal.

Short-Range Order. Concentrated solid solutions are likely to exhibit appreciable short-range order, in which case there is a preponderance of un-like nearest neighbor atoms surrounding each lattice site. Short-range or-der strengthening should provide an athermal increment to flow stress (18). However, the degree of short-range order increases with decreasing annealing temperature. Consequently the short-range order component of flow stress is sensitive to thermal history. Short-range order hardening may contribute to flow stress when solute content approaches 20 wt % as, for example, with chromium in many wrought nickel and cobalt-base alloys.

Grain Boundary Strengthening. Grain boundaries are strong barriers to plastic flow at low and intermediate temperatures, provided that there is a significant lattice misorientation between the adjacent grains. The yield stress, σ_y, in terms of grain diameter, $2d$, is given by the well-known Hall-Petch relation (19,20):

$$\sigma_y = \sigma_i + k_y d^{1/2} \tag{8}$$

where k_y is a constant representing the strength of grain boundary obstacles. At temperatures $>0.5\ T_m$ grain boundaries no longer are significant barriers to plastic flow, as these can either migrate normal to their length (grain growth) or shear parallel to their length (grain boundary sliding). Moreover, whereas the path of fracture of most pure metals at room temperature is transcrystalline, at high temperatures the boundaries themselves tend to fracture. Given these circumstances, fine grained metals should be preferred for low and intermediate temperature use; coarse grains are preferable for high temperature service. However, the effect of grain size on primary and secondary creep is usually not large when compared to that of stress or temperature, within the normal range of grain size (0.01–1 mm diameter) found in polycrystalline alloys (21). Moreover, fine grained metals produced by a working operation tend to have higher fatigue resistance than do coarse grained, cast metals. Therefore, it is often necessary to compromise between advantages and disadvantages of different grain sizes in order to produce an optimum microstructure. Also, there are circumstances where an elongated grain structure aligned parallel to the stress axis is preferable for creep resistance. This is because of a reduction in the number of transverse grain boundaries as well as the occurrence of a preferred orientation.

Cold Working. Cold working, ie, rolling, forging, or similar processes, car-ried out near room temperature, is an important means of imparting strength to metals for low temperature service. However, when a cold worked part is used at elevated temperature, annealing of the heavily dislocated structure occurs, leading eventually to recovery or recrystallization. Because the lattice vacancies produced by cold working tend to accelerate atomic rearrangements during annealing, cold working actually can be detrimental to high temperature strength. An optimum amount of prior strain exists for maximum strengthening at elevated temperatures, and this amount depends on both the material and the experimental conditions. As a practical matter, cold working is applied principally to the Group VIB refractory metals, for both higher strength and improved ductility, and to zirconium alloys, which tend to operate at low temperatures. Thermal–mechanical processing of superalloys, in such a manner that stable dislocation substructures are produced, also has been employed for improved intermediate temperature strength (22).

Precipitation Hardening. With the exception of ferritic steels, which can be hardened either by the martensitic transformation or by eutectoid decomposition, most heat-treatable alloys are of the precipitation-hardening type. During heat treatment of these alloys, a controlled dispersion of submicroscopic particles is formed in the microstructure. The final properties depend on the manner in which particles are dispersed, and on particle size and stability. Because precipitation-hardening alloys can retain strength at temperatures above those at which martensitic steels become unstable, these alloys become an important, in fact pre-eminent, class of high temperature materials.

Strengthening by second-phase particles may be by direct or indirect means, or both. Direct strengthening occurs when, during deformation, particles act as barriers to dislocation motion. The particles then must be penetrated by dislocations, or the latter must bypass the particles by climb or cross-slip. Indirect strengthening occurs when particles help to stabilize a worked microstructure, or produce a columnar microstructure resistant to grain boundary sliding as, for example, is the case of thermal–mechanical processing of dispersoid-containing metals.

The specific means by which direct strengthening occurs are varied and complex. Precipitate particles which are coherent with the matrix lattice, ie, all planes and directions in particle and matrix are parallel, and have no discrete interface, cause strains in the lattice known as coherency strains. Hardening increases are proportional to the misfit, ie, the difference in lattice spacing between precipitate and matrix, but very large misfits cannot be accommodated without loss of coherency. If the particle possesses long-range order, eg, γ'-(Ni_3Al) in nickel-base alloys, then the resistance of shear of the particles increases with increasing energy to disorder the particle. The very high strength of nickel-base alloys hardened by γ' arises from a combination of order and misfit strengthening, coupled with the very high resistance of the particles to coarsening at elevated temperatures. Other factors which may influence strengthening are particle radius, volume fraction, and interparticle spacing. The quantities are, of course, interrelated, spacing being inversely proportional to volume fraction for constant particle radius. Maximum strengthening is achieved when particles are small and closely spaced.

Once the precipitates grow beyond a critical size they lose coherency and then, in order for deformation to continue, dislocations must avoid the particles by a process known as Orowan bowing (23). This mechanism applies also to alloys strengthened by inert dispersoids. In this case a dislocation bends between adjacent particles until the loop becomes unstable, at which point it is released for further plastic deformation, leaving a portion behind, looped around the particles. The smaller the interparticle spacing, the greater the strengthening.

Dispersion Strengthening. Dispersion-hardened alloys are strengthened by particles of a second phase that are dispersed in the microstructures by methods other than heat treatment. Although similar in some respects to precipitation-hardened alloys, dispersion-hardened alloys differ in the types of secondary phases, the means for dispersing them, and the mechanism of strengthening. The highest temperature at which precipitation and dispersion-hardened alloys retain useful strength is limited to the temperature where the second-phase particles become unstable. Precipitation-hardened alloys are more restricted in maximum use temperature because, in order to be capable of heat treatment, the second phase must dissolve into the matrix at some temperature below the melting point of the alloy. Even at temperatures below the solvus, finely precipitated particles tend to agglomerate and overage, and the matrix may undergo recrystallization and grain growth. The second phase in dispersion-hardened alloys, on the other hand, can be completely insoluble in the matrix up to (and even above) the melting point of the matrix and it must, therefore, be dispersed by some means other than heat treatment. Once dispersed, however, the particles are more stable and can retain the strengthening action to higher temperatures than is possible with precipitation-hardened alloys. These particles also tend to stabilize the matrix and make it more resistant to recrystallization. Creep–rupture strength generally is improved as the interparticle spacing becomes less and as the volume fraction of the dispersed phase becomes greater, although optimum values may exist. Oxides, eg, Y_2O_3, ThO_2, and Al_2O_3, are among the most stable compounds at elevated temperatures. These have large negative free energies of formation and high melting points and, accordingly, have been most widely studied and used as dispersion-hardening agents. Powders containing the dispersed phase are produced by a variety of means, are consolidated by powder metallurgy techniques, and then either extruded or fabricated by other techniques (see METALLURGY, POWDER).

Composite Strengthening. An alternative strengthening method which holds great promise for producing advanced high temperature alloys involves the incorporation of fibers or lamellae of a strong, often brittle phase, in a relatively weak, ductile, metallic matrix. This technique has been commercially exploited for polymer-matrix, glass-reinforced materials such as fiber glass, but has not yet been applied commercially to high temperature systems using a metal matrix (see COMPOSITE MATERIALS; GLASS, POLYMER-MODIFIED). The principal benefit arising from composite strengthening is the very high strength achievable at extremely high fractions of the melting point. In this respect composite and dispersion strengthening offer substantially identical benefits insofar as tensile or creep strength and fatigue resistance are concerned.

Two approaches have been taken to produce metal-matrix composites (qv): incorporation of fibers into a matrix by mechanical means; and *in situ* preparation of a two-phase fibrous or lamellar material by controlled solidification or heat treatment. The principles of strengthening for alloys prepared by the former technique are well established (24), primarily because yielding and even fracture of these materials occurs while the reinforcing phase is elastically deformed. Under these conditions both strength and modulus increase linearly with volume fraction of reinforcement. However, the deformation of *in situ*, ie, eutectic, eutectoid, peritectic, or peritectoid, composites usually involves some plastic deformation of the reinforcing phase, and this presents many complexities in analysis and prediction of properties.

Summary of Strengthening Methods. In practice, few alloys are strengthened by only one or merely a few of the mechanisms described, as shown in Table 2. For example, all precipitation-hardened alloys must have solutes which may produce significant solid-solution strengthening. Nickel-base superalloys, which are strengthened primarily by γ'-precipitates, also are hardened by aluminum and molybdenum, tungsten or rhenium in solid solution. Cobalt-base alloys rely on solid-solution strengthening as well as carbide precipitation for high temperature strength. Fine grain size usually accompanies dispersion strengthening, both because of the powder metallurgical techniques involved and because particles tend to pin grain boundaries. In the mechanical alloying process (25) precipitation and dispersion strengthening are combined in a single alloy. Although it is difficult to assess how the various strengthening mechanisms interact, it is often assumed that disparate strengthening mechanisms are additive. In a qualitative sense such approximations can be useful, but it should be recognized that the actual strength of a high temperature alloy can rarely be predicted with any degree of confidence.

Table 2. Strengthening Mechanisms in High Temperature Alloys

Alloys	Primary strengthening	Secondary strengthening
superalloys		
Ni-base	precipitation of γ'	solid solution
Co-base	solid solution	precipitation of carbides
bcc refractory metals	cold working	solid solution, precipitation
directionally solidified eutectics	composite	solid solution, precipitation
dispersion-strengthened alloys	dispersion	solid solution, grain size
intermetallics	long-range order	solid solution, precipitation

Surface Stability

Oxidation. Immense progress in technology has imposed ever-increasing demands on the mechanical and chemical properties, in particular the oxidation and scaling resistance, of metallic materials. The terms metal oxidation and scaling can be used to describe the attack of a metal or an alloy by aggressive gases such as oxygen, sulfur, the halogens, or water vapor. This attack can take place under a variety of circumstances, varying from the mild oxidizing conditions which may exist at room temperature in air to the severe conditions imposed by hot furnace gases contacting metallic surfaces. Especially stringent requirements are placed on the scaling resistance of components where condensed molten salts and oxidizing atmospheres simultaneously exist. The attack under such conditions is commonly referred to as hot corrosion, ie, a particularly severe form of corrosion (see CORROSION AND CORROSION CONTROL).

Industrial materials without sufficient scaling resistance frequently fail after a short period of time as a result of rapid oxidation or hot corrosion, in conjunction with severe spalling owing to poor adherence of the scale to the metallic component. As a result, the permissible limits of metal loss often are exceeded and expensive, and premature replacement of parts is required. Extensive efforts are made to develop alloys which are not simply strong at elevated temperatures but which also possess the adequate surface stability.

Oxide Scale Morphology. The scale morphology is dependent on the conditions of reaction, the time of oxidation, the composition of the corrosive medium, and the type and composition of the particular alloy involved (26). The most desirable scale morphology is the nonporous (compact) oxide. If a scale is porous throughout its entire cross-section, then the chemical composition and structure of the reaction products are of no essential significance to the course of oxidation. The oxidation rate is then determined by the chemical reaction at the metal–scale interface because the oxidant gas penetrates at a sufficient rate through fissures in the scale. Such course of oxidation automatically excludes the possibility of using such a material at high temperatures. Fortunately, most materials do not behave in such a manner. Complex alloys may form two or more layers differing in either composition or microstructure or both. In order to maintain good oxidation resistance at least one of the layers must be compact and preferably be a slow growing oxide.

The oxidation of most modern alloys is dependent on the formation of a compact protective film of a slow growing chemically stable oxide such as chromium(III) oxide [1308-38-9], Cr_2O_3, alumina [1344-28-1], Al_2O_3, or silica [7631-86-9], SiO_2. These oxides grow much more slowly than do the oxides of cobalt, iron, or nickel (26). The growth of the former oxides is particularly slow because the native defect concentration responsible for oxidation during parabolic oxide growth is extremely low. Design of corrosion-resistant materials is thus based on maintaining a compact protective scale so as to induce a slow growing, highly coherent oxide. Those alloys which depend on the formation of a protective Cr_2O_3 film are limited to temperatures below 1000°C. Above this temperature volatile CrO_3 and, in the presence of water vapor, $CrO_2(OH)_2$, are formed above the Cr_2O_3, and rapid vaporization of the protective Cr_2O_3 results.

The oxidation behavior of multicomponent γ'-strengthened alloys can be estimated by considering the NiCrAl content of the alloy (26). For typical NiCrAl-base alloys, such as U-500, at 713°C the spinel $Ni(Cr,Al)_2O_4$, NiO, and Cr_2O_3 are reported to form initially during transient oxidation (27). Thereafter both Cr_2O_3 and Al_2O_3 are present as internal oxides. The proportions in which the elements are present in both the external and internal scales is determined by the alloy composition. The critical parameter in such alloys is the Cr–Al ratio, which determines the composition of the protective scale. The oxidation kinetics of such alloys are generally quite complex and may depend on the alloy additions, particularly the niobium, molybdenum, and tungsten content, some of which have an adverse effect on the oxidation resistance of the nickel-base superalloys. In general, good cyclic oxidation resistance is associated with Al_2O_3 and/or $NiAl_2O_4$ formation (26). Al_2O_3 protective scales appear to be the most advantageous of the protective oxides on iron, nickel, and cobalt alloys. However, in Fe–Al alloys, poor ductility limits the amount of aluminum which can be used. Furthermore, in the presence of a molten salt and an oxidizing atmosphere rapid degradation may ensue in the temperature range of 850–900°C.

Because of the great difference in coefficient of thermal expansion between the protective oxides and the base alloys, severe spallation of the oxide may result, leaving base metal in contact with oxidizing environment. When that occurs, rapid oxidation ensues. After a number of thermal cycles, base metal adjacent to the protective scale can become depleted in those elements used to form protective oxides, particularly Cr and Al. In this case a faster growing oxide forms and enhanced degradation of the alloy ensues. Despite such drawbacks, there are no alternatives to the formation of protective Cr_2O_3 or Al_2O_3 scales on superalloys. A comprehensive review of the oxidation behavior of nickel-base superalloys is available (26).

Hot Corrosion. Hot corrosion is an accelerated form of oxidation that arises from the presence not only of an oxidizing gas, but also of a molten salt on the component surface. The molten salt interacts with the protective oxide so as to render the oxide nonprotective. Most commonly, hot corrosion is associated with the condensation of a thin molten film of sodium sulfate [7757-82-6], Na_2SO_4, on superalloys commonly used in components for gas turbines, particularly first-stage turbine blades and vanes. Other examples of hot corrosion have been identified in energy conversion systems, particularly coal gasifiers and direct coal combustors. In these cases the salt originates from alkali impurities in the coal which condense on the internal components, thereby initiating the hot corrosion (see COAL CONVERSION PROCESSES).

The deposition of molten Na_2SO_4 in gas turbines is believed to be related to the reaction between the residual sulfur in fuel and sodium which may be contained either in the fuel or the intake air. The sodium in the air is normally present as an aerosol of sea salt. Salt concentrations of over 0.01 ppm in the intake air may be necessary to initiate hot corrosion.

The morphology of the hot corrosion reaction also is different from simple oxidation in that a form of concurrent oxidation and sulfidation commonly occurs. It has been observed that a layer of sulfide particles forms beneath a region of porous oxide (28). In service cases the sulfides are nearly always chromium rich; however, some nickel sulfides have been observed. The extent of the internal sulfide layer can vary considerably. At times the sulfide layer appears to be virtually absent or as a very thin band of fine discrete sulfide particles. Under different conditions a large band of interconnected sulfide particles may form. For most materials, the outer porous oxide consists of the simple oxide of the base metal, ie, either NiO or CoO. Beneath this zone a region of mixed oxide, sulfide, and pure metal may exist. In such cases the oxide may be a spinel, Cr_2O_3 (for chromia formers), or Al_2O_3 (for alumina formers). The sulfides again are usually chromium rich, whereas the metal is generally rather pure discontinuous fragments of the base metal.

For most alloys, the corrosion rate displays a maximum at 850–900°C, and decreases very rapidly at temperatures up to 1000°C (28), again strongly suggesting that a molten salt is necessary in order to initiate hot corrosion.

It is generally conceded that the chromium content is the most important factor in hot corrosion resistance, as may be seen in Figure 6 (29). For this reason cobalt alloys, which generally contain 20% or more chromium, display better hot corrosion resistance than nickel alloys, which typically contain 8–15% Cr. Much of the disagreement concerning the effects of other elements is perhaps a result of interactive effects within the alloy scale and salt. Some elements, which can produce beneficial effects over certain concentration ranges, are deleterious over others. For example in comparing the corrosion of a Ni–20%Cr–6%Al alloy and

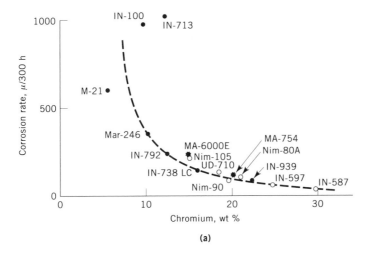

(a)

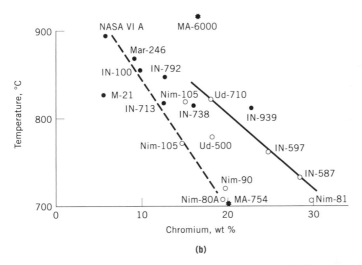

(b)

Fig. 6. Properties of the ($*$) oxide dispersion strengthening (ODS) materials MA-6000 and MA-754 and conventional ($\bullet$) cast and ($\circ$) wrought superalloys as a function of chromium content, where (–––) represents Al_2O_3 and (——) Cr_2O_3 formers. Testing conditions: 300 h, 850°C, 0.11 MPa, extra-light fuel oil (0.3–0.4% S), 15 ppm Na, 5 ppm V (29). (**a**) Corrosion rate and (**b**), temperature capability at $\sigma = 200$ MPa, for 10^4 h. To convert MPa to psi, multiply by 145.

a Ni–30%Cr alloy, Al is beneficial initially, but then becomes deleterious after 100 hot corrosion cycles (30). Molybdenum and tungsten additions are considered to be detrimental to the hot corrosion resistance of nickel- and cobalt-based alloys. The mechanisms of hot corrosion depend on both alloy content and gas composition. A full description of these mechanisms appears elsewhere (31).

Coatings. It is common practice to apply some type of protective coating to extend the surface stability of superalloy or refractory metal components (32). Superalloy coatings provide an aluminum reservoir for growth of protective Al_2O_3 scales and inhibit further oxidation. Coating microstructures can be varied from the relatively brittle high aluminum content intermetallic matrix phase, eg, CoAl or NiAl, to the relatively ductile low aluminum content metal plus intermetallic-type structures, eg, Co + CoAl. During cyclic oxidation and hot corrosion, excessive spallation occurs. Coatings have been developed, particularly Co–25%Cr–2%Al–0.1%Y and Ni–15%Cr–6%Al–0.1%Y, that exhibit excellent resistance to thermal cycling. These overlay coating compositions are based on the knowledge that aluminum is required to form protective Al_2O_3, and chromium is necessary to enhance Al_2O_3 formation and to improve hot corrosion resistance further. Excellent resistance to spallation is generally attributed to the addition of yttrium. A compact adhesive $Al_2O_3/CoAl_2O_4$ scale is formed having a low parabolic growth rate, thereby protecting the underlying base alloy. Both $CoAl_2O_4$ and $NiAl_2O_4$ spinels grow rather slowly when compared with most oxides. Such coatings degrade after a period of time, because in addition to attack of the protective oxide by spallation, erosion, and chemical means, inward diffusion of aluminum and chromium occurs. Therefore, although coatings can improve the oxidation resistance, degradation can occur and oxidation of the alloy ensues. NiCoAlY compositions offer superior elevated temperature oxidation resistance and diffusional stability on nickel-base superalloys. Additions of cobalt to NiCrAlY enhance hot corrosion resistance and improve coating ductility. CoCrAlY coatings provide superior protection in a hot corrosion environment.

In the case of refractory metals, coatings generally are silicides, applied by pack cementation or slurry processes. Typical silicide compositions are Si–20Cr–20Fe for niobium alloys and $MoSi_2$ for molybdenum alloys (see MOLYBDENUM AND MOLYBDENUM ALLOYS; NIOBIUM AND NIOBIUM COMPOUNDS; REFRACTORY COATINGS) (33).

Specific Alloy Systems

Plain Carbon and Low Alloy Steels. For the purposes herein plain carbon and low alloy steels include those containing up to 10% chromium and 1.5% molybdenum, plus small amounts of other alloying elements. These steels are generally cheaper and easier to fabricate than the more highly alloyed steels, and are the most widely used class of alloys within their serviceable temperature range. Figure 7 shows relaxation strengths of these steels and some nickel-base alloys at elevated temperatures (34).

Of the common alloying elements in steel (qv), molybdenum is the most effective in increasing creep–rupture strength, and the carbon–molybdenum steels generally have more than twice the creep–rupture strength of plain carbon steel at the same temperature (34). The most commonly used steels for

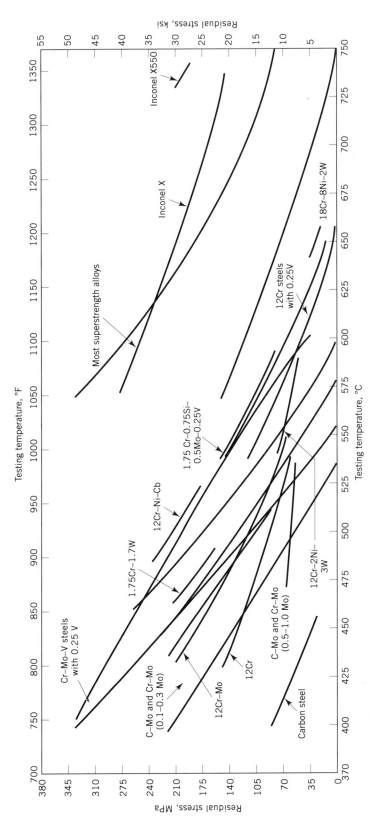

Fig. 7. Comparison of relaxation strengths at 1000 h as measured by residual stress of various steels and nickel alloys (34). To convert MPa to psi, multiply by 145.

high temperature service contain from 0.5 to 1.5% molybdenum. Carbon–molybdenum steels are about equivalent to plain carbon steels in metallurgical stability and resistance to corrosion and oxidation, and are therefore used where greater strength is needed, although plain carbon steel would otherwise be acceptable. The straight carbon–molybdenum steels should not be used continuously above 470°C because of graphitization, ie, the breakdown of the cementite present in the pearlitic structure to form ferrite and carbon in the form of graphite flakes.

Chromium is the most effective addition to improve the resistance of steels to corrosion and oxidation at elevated temperatures, and the chromium–molybdenum steels are an important class of alloys for use in steam (qv) power plants, petroleum (qv) refineries, and chemical-process equipment. The chromium content in these steels varies from 0.5 to 10%. As a group, the low carbon chromium–molybdenum steels have similar creep–rupture strengths, regardless of the chromium content, but corrosion and oxidation resistance increase progressively with chromium content. Carbon content is usually about 0.15% but may be higher in bolting steels and hot-work die steels. Molybdenum content is usually between 0.5 and 1.5%; it increases creep–rupture strength and prevents temper embrittlement at the higher chromium contents. In the modified steels, silicon is added to improve oxidation resistance, titanium and vanadium to stabilize the carbides to higher temperatures, and nickel to reduce notch sensitivity. Most of the chromium–molybdenum steels are used in the annealed or in the normalized and tempered condition; some of the modified grades have better properties in the quench and tempered condition.

Stainless Steels. *Standard Wrought Steels.* Steels containing 11% and more of chromium are classed as stainless steels. The prime characteristics are corrosion and oxidation resistance, which increase as the chromium content is increased. Three groups of wrought stainless steels, series 200, 300, and 400, have composition limits that have been standardized by the American Iron and Steel Institute (AISI) (see STEEL). Figure 8 compares the creep–rupture strengths of the standard austenitic stainless steels that are most commonly used at elevated temperatures (35). Compositions of these steels are listed in Table 3.

Steels in the AISI 400 series contain a minimum of 11.5% chromium and usually not more than 2.5% of any other alloying element; these steels are either hardenable (martensitic) or nonhardenable, depending principally on chromium content. Whereas these steels resist oxidation up to temperatures as high as 1150°C, they are not particularly strong above 700°C. Steels in the AISI 300 series contain a minimum of 16% chromium and 6% nickel; the relative amounts of these elements are balanced to give an austenitic structure. These steels cannot be strengthened by heat treatment, but can be strain-hardened by cold work.

Ferritic stainless steels depend on chromium for high temperature corrosion resistance. A Cr_2O_3 scale may form on an alloy above 600°C when the chromium content is ca 13 wt % (36,37). This scale has excellent protective properties and occurs in the form of a very thin layer containing up to 2 wt % iron. At chromium contents above 19 wt % the metal loss owing to oxidation at 950°C is quite small. Such alloys also are quite resistant to attack by water vapor at 600°C

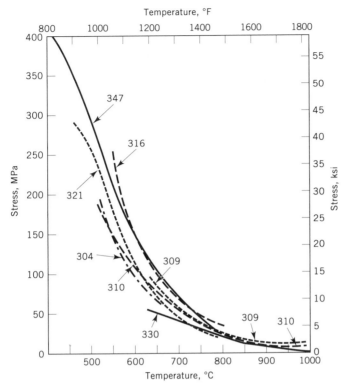

Fig. 8. Stress–rupture curves for annealed H-grade austenitic stainless steels. AISI numbers are given (see Table 3). Rupture in 10,000 h (35). To convert MPa to psi, multiply by 145.

Table 3. Compositions of Stainless Steels, wt %

AISI number	UNS number	C	Mn	Si	Cr	Ni	Other[a,b]
201	S20100	0.15	6.50	0.75	17.0	4.50	0.20[c]
202	S20200	0.15	8.75	0.75	18.0	5.00	0.20[c]
205	S20500	0.12	15.0	0.50	17.0	1.75	0.35[c]
304	S30400	0.06	1.50	0.75	19.0	10.0	
309	S30900	0.16	1.50	0.75	23.0	13.5	
310	S31000	0.20	1.50	1.00	25.0	20.5	
316	S31600	0.06	1.50	0.75	17.0	12.0	2.50[d]
321	S32100	0.06	1.50	0.75	18.0	10.5	0.50[e]
330	S33000	0.08	2.0	1.0	18.0	35.0	
347	S34700	0.06	1.50	0.75	18.0	11.0	1.00[f]
410	S41000	0.12	0.75	0.75	12.5		
430	S43000	0.10	0.75	0.50	16.0	0.30	
446	S44600	0.30	1.00	0.75	25.0		0.20[c]

[a]S and P are held below 0.03 and 0.04 wt %, respectively.
[b]The balance is Fe.
[c]Nitrogen. [d]Molybdenum. [e]Titanium. [f]Niobium.

(38). Isothermal oxidation resistance for some ferritic stainless steels has been reported after 10,000 h at 815°C (39). Grades 410 and 430, with 11.5–13.5 wt % Cr and 14–18 wt % Cr, respectively, behaved significantly better than type 409 which has a chromium content of 11 wt %.

AISI 321 and 347 are stainless steels that contain titanium and niobium in order to stabilize the carbides (qv). These metals prevent intergranular precipitation of carbides during service above 480°C, which can otherwise render the stainless steels susceptible to intergranular corrosion. Grades such as AISI 316 and 317 contain 2–4% of molybdenum, which increases their creep–rupture strength appreciably. In the AISI 200 series, chromium–manganese austenitic stainless steels the nickel content is reduced in comparison to the AISI 300 series.

Stainless Alloy Castings.   Although the stainless steels usually are specified in the wrought condition, a number of iron–chromium, iron–chromium–nickel, and nickel or cobalt-base alloys are produced as castings. Castings are classified as heat resistant when utilized in applications at 650°C or higher. Examples of Fe–Ni–Cr heat resisting castings are HC (equivalent to wrought AISI type 446), HH (AISI type 309), and HD (AISI type 327). These alloys are used in metallurgical furnaces, oil-refinery furnaces, power plant equipment, gas turbines, and in the manufacture of glass and synthetic rubber. Iron–chromium castings containing 10–30% Cr are useful chiefly for oxidation resistance, whereas iron alloys containing more than 18% Cr and more than 7% Ni have superior strength and ductility. Other iron-base alloys containing more than 10% Cr and more than 25% Ni are used in both reducing and oxidizing atmospheres.

Modified Stainless Steels.   The 12%-chromium ferritic superalloys are a group of proprietary steels that are essentially modifications of AISI 403 stainless steel. Examples are Crucible 422, Lapelloy (AISI 619), and Jessop-H46. The modifications include adding up to several percent of molybdenum and/or tungsten to stiffen the matrix, and up to 0.5% of niobium and vanadium to improve the dispersion and stability of the carbides. Up to 2% nickel, copper, and aluminum also may be present in these steels. The modified steels have a substantially greater creep–rupture strength than the standard AISI 403 stainless steel, and about the same level of corrosion and oxidation resistance. Like the standard composition, the 12%-chromium ferritic superalloys are normally used in the quenched and tempered condition. Typical applications include high temperature bolts, blades for jet-engine compressors and for high temperature steam turbines, compressor and turbine disks for jet engines, boiler, superheater, and reheater tubes and valve parts. These steels are available in most wrought forms.

The precipitation-hardening stainless steels are proprietary grades hardened by both the martensitic transformation and precipitation hardening. These contain higher amounts of chromium (16–17%) and nickel (4–7%) than the 12%-chromium ferritic alloys, so that the martensitic transformation is very sluggish and is easily depressed to below room temperature by raising the austenitizing temperature. Up to about 3% molybdenum and 0.25% niobium may be added, as well as several percent of aluminum or copper to promote precipitation hardening. These steels are normally used at lower temperatures than the 12%-chromium ferritic superalloys.

The highly alloyed austenitic stainless steels are proprietary modifications of the standard AISI 316 stainless steel. These have higher creep–rupture strengths than the standard steels, yet retain the good corrosion resistance and forming characteristics of the standard austenitic stainless steels.

Nickel-Base Superalloys. *Conventional Wrought and Cast Alloys.* The nickel-base superalloys are the most complex in composition and microstructures and, in many respects, the most successful high temperature alloys. Development commenced in the late 1930s with the need for aircraft gas turbine component materials that were stronger than the then-available austenitic stainless steels. The earliest superalloys were wrought, ie, fabricated to final size by a mechanical working operation. The earliest superalloy (Nimonic 75) was produced by adding 0.3%Ti and 0.1%C to an oxidation resistant solid solution 80%Ni–20%Cr (Nichrome) base. Higher engine speeds and turbine inlet temperatures were the impetus for succeeding modifications, first by adding aluminum and titanium to produce γ' strengthening (Nimonic 80), and later by adding cobalt to raise the volume fraction of γ' and to improve workability (Nimonic 90). Later alloys have incorporated higher aluminum plus titanium contents, as well as molybdenum for solid-solution strengthening (Nimonics 115 and 120). The compositions of these and other wrought nickel-base alloys are listed in Table 4. Also included are compositions of some powder processed alloys.

A series of nickel–chromium–iron alloys based on the solid solution Inconel 600 alloy (see Table 4) was developed, initially depending on aluminum and titanium for γ' strengthening. To this was added 1% Nb to form the well-known Inconel X-750 alloy. Other early wrought alloys developed in the United States included Waspaloy and M-252, which utilized molybdenum for solid-solution strengthening and carbide formation in addition to the γ' strengthening produced by aluminum and titanium. Both alloys, used initially as turbine blades, are used in other gas turbine components, such as turbine disks. Later, stronger wrought alloys, including Udimet 500, Udimet 700, and René 41, all strengthened principally by γ', became commercial. Hastelloy X, a solid-solution and carbide-strengthened alloy, is used in sheet form for less highly stressed parts requiring high oxidation resistance.

Alloys developed by processing through the investment casting process had higher strength and design flexibility, which led to many further advances through air cooling. Among the most notable of these very strong alloys developed in the 1960s were Inconel 713 and a low carbon version, 713LC, as well as IN-100, B-1900, and MarM-200 (Table 5). The cast alloys tended to contain less chromium, which was replaced by molybdenum, tungsten, and tantalum, while retaining high volume fractions (to 60%) of γ'.

Alloy development in the former Soviet Union has produced alloys having strengths equivalent to IN-100 and Mar-M-200. Alloys developed in the United States and United Kingdom are also widely used in French aircraft engines.

All of the alloys listed in Tables 4 and 5 are austenitic, ie, fcc. Apart from γ' and solid-solution strengthening, many alloys benefit from the presence of carbides, carbonitrides, and borides. Generally the cubic MC-type monocarbides, which tend to form in the melt, are large and widely spaced, and do not contribute to strengthening. However, the formation, distribution, and solid-state reactions of $M_{23}C_6$ carbides are very important because of their role in improving grain boundary rupture strength together with their association with harmful intermetallic compounds which can precipitate on them. The general characteristics

Table 4. Chemical Composition of Wrought and Powder Nickel-Base Superalloys, wt %

Alloy designation	UNS number	Ni	Cr	Co	Mo	W	Al	Ti	Fe	Mn	Si	C	B	Other
AF-115			10.7	15	28	5.9	3.8	3.9				0.05	0.02	1.7Nb, 0.75Hf, 0.05Zr
Astroloy		55.1	15.0	17.0	5.25		4.0	3.5				0.06	0.030	
Inconel alloy 600	N06600	76.6	15.8						7.2	0.20	0.20	0.04		
Inconel alloy 601		60.7	23.0				1.35		14.1	0.50	0.25	0.05		
Inconel alloy 625	N06625	61.1	22.0		9.0		0.2	0.2	3.0	0.15	0.30	0.05		4.0Nb
Inconel alloy X-750	N07750	73.0	15.0				0.8	2.5	6.8	0.70	0.30	0.04		0.9Nb
IN-102	N06102	67.9	15.0		3.0	3.0	0.4	0.6	7.0			0.06	0.005	3.0Nb, 0.03Zr, 0.02Mg
IN-587		47.2	28.5	20.0			1.2	2.3				0.05	0.003	0.7Nb, 0.05Zr
IN-597		48.4	24.5	20.0	1.5		1.5	3.0				0.05	0.012	0.5Zr, 0.02Mg, 1.0 Nb
IN-853		74.6	20.0				1.5	2.5				0.05	0.007	0.07Zr, 1.3Y$_2$O$_3$
M-252		55.2	20.0	10.0	10.0		1.0	2.6		0.50	0.50	0.15	0.005	
Merl-76			12.4	18.5	3.2		5	4.3				0.025	0.02	1.4Nb, 0.75Hf, 0.06Zr
Nimonic alloy 80A	N02080	74.7	19.5	1.1			1.3	2.5		0.10	0.70	0.06		
Nimonic alloy 90	N07090	57.4	19.5	18.0			1.4	2.4		0.50	0.70	0.07		
Nimonic alloy 105		53.3	14.5	20.0	5.0		1.2	4.5		0.50	0.70	0.20		
Nimonic alloy 115		57.3	15.0	15.0	3.5		5.0	4.0				0.15		
Nimonic alloy PK.33		55.9	18.5	14.0	7.0		2.0	2.0	0.25	0.10	0.15	0.05	0.030	
Nimonic alloy 120		63.8	12.5	10.0	5.7		4.5	3.5				0.04		
Nimonic alloy 942		49.5	12.5	1.0	6.0		0.6	3.9	27.5			0.03		
RA-333	N06333	45.0	25.5	3.0	3.0	3.0			18.0	1.50	1.20	0.05		3.5Ta, 0.05Zr
René 41	N07041	55.3	19.0	11.0	10.0		1.5	3.1				0.09	0.005	
René 95		61.3	14.0	8.0	3.5	3.5	3.5	2.5				0.15	0.010	
TD Nickel		98.0												2.0ThO$_2$
TD NiC		78.0	20.0											2.0ThO$_2$
Udimet 500	N07500	53.6	18.0	18.5	4.0		2.9	2.9				0.08	0.006	0.05Zr
Udimet 520		56.9	19.0	12.0	6.0	1.0	2.0	3.0				0.05	0.005	
Udimet 700		53.4	15.0	18.5	5.2		4.3	3.5				0.08	0.030	
Udimet 710		54.9	18.0	15.0	3.0	1.5	2.5	5.0				0.07	0.020	
Unitemp AF2-IDA	N07012	59.5	12.0	10.0	3.0	6.0	4.6	3.0				0.32	0.015	1.5Ta, 0.10Zr

Table 5. Chemical Composition of Cast Nickel-Base Superalloys, wt %

Alloy designation	NI	Cr	Co	Mo	W	Ta	Nb	Al	Ti	C	B	Zr	Other
Alloy 713C	74	12.5		4.2			2.0	6.1	0.8	0.12	0.012	0.10	
B-1900	64	8.0	10.0	6.0		4.0		6.0	1.0	0.10	0.015	0.10	
IN-100/René 100	60	10/9.5	15.0	3.0				5.5	4.7/4.2	0.18	0.014	0.06	1.0V
IN-162	73	10.0		4.0	2.0	2.0	1.0	6.5	1.0	0.12	0.020	0.10	
IN-731	67	9.5	10.0	2.5				5.5	4.6	0.18	0.015	0.06	
IN-738	61	16.0	8.5	1.7	2.6	1.7	0.9	3.4	3.4	0.17	0.010	0.10	
IN-792	61	12.4	9.0	1.9	3.8	3.9		3.1	4.5	0.12	0.020	0.10	
Mar-M 200	60	9.0	10.0		12.0		1.0	5.0	2.0	0.15	0.015	0.05	
Mar-M 200(DS)	60	9.0	10.0		12.0		1.0	5.0	2.0	0.13	0.015	0.05	
Mar-M 246	60	9.0	10.0	2.5	10.0	1.5		5.5	1.5	0.15	0.015	0.05	
Mar-M 421	61	15.8	9.5	2.0	3.8		2.0	4.3	1.8	0.15	0.015	0.05	
Mar-M 432	50	15.5	20.0		3.0	2.0	2.0	2.8	4.3	0.15	0.015	0.05	
NX188(DS)	74			18.0				8.0		0.04			
René 77	58	14.6	15.0	4.2				4.3	3.3	0.07	0.016	0.04	
René 80	60	14.0	9.5	4.0	4.0			3.0	5.0	0.17	0.015	0.03	
Taz-8A	68	6.0		4.0	4.0	8.0	2.5	6.0		0.12	0.004	1.00	
TRW-NASA VIA	61	6.1	7.5	2.0	5.8	9.0	0.5	5.4	1.0	0.13	0.020	0.13	0.5Re, 0.411f
Udimet 500	52	18.0	19.0	4.2				3.0	3.0	0.07	0.007	0.05	
Waz-20(DS)	72				20.0			6.5		0.20		1.50	
IN-939	48	22.5	19.0		2.0	1.4	1.0	1.9	3.7		0.010		

of temperature-induced changes in strength and ductility of nickel-base alloys strengthened by γ' appear in Figure 9 (40). Yield and tensile strengths are nearly constant between room temperature and about 800°C, and often, depending on volume fraction of γ', rise to a peak between 700 and 800°C. Accompanying the peak in flow stress is a characteristic ductility minimum. The peak in flow stress is a consequence of the presence of a large volume fraction of γ' which, unlike conventional nickel alloys, tends to strengthen anomalously with increasing temperature. In general, wrought nickel-base alloys are not as strong as cast alloys of similar composition, nor as capable of resisting high temperatures (41). The strongest alloys contain up to 70% by volume of γ' produced by precipitation. The development of alloys with large volume fractions of γ' appears to have had a considerably greater impact on stress capability than has the trend from wrought to cast. Also, the achievement of higher temperature capability has come about in part at the expense of chromium content. Low chromium alloys, however, generally require coatings.

An undesirable feature of the most highly alloyed superalloys is the tendency to develop unwanted phases such as σ and μ. Sigma (σ) phase, a plate-like intermetallic compound of two or more transition metals, eg, Cr_xFe_y or $(CrMo)_x(NiCo)_y$ where x and y can vary from 1 to 7, may precipitate from alloys containing a high refractory metal content, eg, IN-100. There is a critical temperature range, centered around 800°C, for the precipitation of σ, and precipitation leads to a decrease in rupture properties. Low temperature ductility also is adversely affected. The recognition that σ and other topologically close

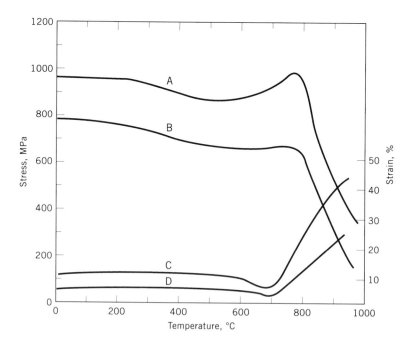

Fig. 9. Effect of temperature on strength and ductility of a nickel-base superalloy, IN-939, showing A, tensile strength; B, 0.2% proof stress; C, reduction in area; and D, elongation (40). To convert MPa to psi, multiply by 145.

packed (tcp) phases are electron compounds where precipitation from solution can be predicted by knowledge of the average electron vacancy number of the alloy matrix was the basis of the Phacomp system for predicting safe alloy compositions (42). In brief, the total concentration of elements having high electron vacancy number (Cr, Mo, W, Mn) must be limited to avoid σ-phase precipitation, either during alloy processing or in service. Computer programs derived from the principles of electron vacancy numbers and phase stability are used in engineering specifications for superalloys utilized in aircraft engines and industrial gas turbines.

The temperature capability of nickel-base alloys has been improved markedly by processing techniques such as vacuum arc melting. This process avoids the loss of titanium and aluminum by oxidation and keeps gaseous element contaminants down to acceptable levels. The vacuum induction furnace, having capacity to 45,500 kg, provides optimum capabilities for close compositional control and removal of deleterious gases and other elements. Air induction melting is still used for some high temperature alloys, but there is little opportunity for refining during melting.

Vacuum arc remelting often is used to develop optimum solidification structures. Electroslag remelting, which utilizes a molten pool of slag in which the electrode is immersed, yields a cleaner and purer material.

Cleanliness is critical in modern superalloys because of the role of inclusions in initiating fatigue cracks and fracture. Several refining processes to produce cleaner superalloys have been introduced, including the use of ceramic-foam filters in conjunction with vacuum induction melting, the reintroduction of electroslag remelting, and the development of electron-beam cold-hearth refining (43). Directional solidification (DS), in which heat withdrawal is made to occur parallel to the ingot axis, has been introduced to produce large columnar grains parallel to that axis. The technique, which resembles the Bridgeman method for growing single crystals, produces increased ductility at intermediate temperatures (760°C), improved rupture strength in thin sections, and improved low cycle fatigue life. MarM-200 (Table 5) is the principal alloy to which the DS process has been applied. Compositional control also can be utilized for improved castability and properties, as in the case of hafnium additions to MarM-200, or the replacement of carbon by 0.1% boron in some newer casting superalloys.

The best elevated temperature properties of nickel-base alloys are obtained by an adaptation of the DS process to produce single crystals (44). Single-crystal turbine blades, introduced into aircraft engines by Pratt and Whitney Aircraft in the 1970s, as of this writing are utilized by all engine manufacturers. Some typical compositions are listed in Table 6. These alloys do not require grain boundary strengtheners such as carbon, boron, or zirconium, thereby reducing segregation and enhancing incipient (local) melting temperature. This in turn allows more leeway in fully solutionizing alloys (dissolving all precipitates) prior to a final aging treatment to bring out the desired distribution of γ'-strengthening precipitate.

Several other composition trends are readily apparent in Table 6: 8–12% Cr and 2–12% Ta were combined to maximize strength (higher chromium contents reduce strength) in first-generation alloys; Re was added to the second generation because it is perhaps the most effective strengthener of all with respect

Table 6. Compositions and Densities of Superalloy Single Crystals

Alloy[a]	Cr	Co	Mo	W	Ta	Al	Ti	Hf	Other	Density, g/cm^3
					First-generation alloys					
PWA-1480	10	5		4	12	5.0	1.5			8.70
René N-4	9	8	2	6	4	3.7	4.2		0.5[b]	8.56
SRR-99	8	5		10	3	5.5	2.2			8.56
RR-2000	10	15	2			5.5	4.0		1[c]	7.87
AM1	8	6	3	6	9	5.2	1.2			8.59
AM3	8	6	2	5	4	6.0	2.0			8.25
CMSX-2	8	5	0.6	8	6	5.6	1.0			8.56
CMSX-3	8	5	0.6	8	6	5.6	1.0	0.1		8.56
CMSX-6	10	5	3		2	4.8	4.7	0.1		7.98
AF-56	12	8	2	4	5	3.4	4.2			8.25
					Second-generation alloys					
CMSX-4	6.5	9	0.6	6	6.5	5.6	1.0	0.1	3[d]	8.70
PWA-1484	5	10	2	6	9	5.6		0.1	3[d]	8.95
SC-180	5	10	2	5	8.5	5.2	1.0	0.1	3[d]	8.84
MC2	8	5	2	8	6	5.0	1.5			8.63

[a]Balance in all alloys is nickel.
[b]Columbium.
[c]Vanadium.
[d]Rhenium.

to creep. In addition, rhenium slows growth of fine γ' during high temperature service. However, rhenium is expensive so that methods must be developed to recover rhenium in casting gates and risers and in scrap (45) (see RECYCLE, NONFERROUS METALS). A sharp increase in performance has been achieved with single-crystal turbine airfoils, which as of this writing are in the third generation of development (46). Creep and thermal fatigue resistance are superior to those of conventionally cast or directionally solidified (DS) alloys. Further, oxidation resistance using vacuum plasma deposited NiCoCrAlY coatings modified with hafnium and silicon is outstanding (46). The increased temperature capability more than compensates for the higher costs associated with single-crystal production. Single-crystal casting defects such as included grains, microporosity, and freckling (a chain of equiaxed grains caused by segregation of elements in the liquid state) must be avoided. Rejection of blades containing such defects contributes further to the cost of the process.

The coarse grains developed by conventional casting processes usually are deleterious to fatigue life. For parts such as turbine disks that are life limited by fatigue rather than creep, fine grains are produced by powder metallurgical techniques. Typical disk alloys include Astroloy, IN-100, René 95, and Merl 76 (Table 4). Consolidation of parts by hot isostatic pressing (HIPing), which has now supplanted more conventional canning and extrusion of powders, may also be applied to cast parts to heal casting defects. The Gatorizing process (Pratt and Whitney Aircraft) involves working of fine grained parts produced from powders in a temperature range where the material is superplastic. This permits working of very strong alloys such as IN-100, which ordinarily would be used in the cast condition, with little applied force.

Some of the superalloys can be welded by arc melting processes, as well as by resistance and electron-beam techniques. Alloys having low γ' contents are readily weldable. However, with increasing volume fractions of γ', welding becomes more difficult. René 41 and Waspaloy are borderline cases. These can be welded relatively easily, but may crack during post-weld heat treating (47). Cast alloys having high aluminum and titanium contents exhibit low ductility at all test temperatures and generally crack during welding.

It is possible to decrease the metal temperature of turbine blades and vanes by cooling them using high pressure air diverted from the compressor or, conversely, to allow higher turbine inlet temperatures by means of insulating oxide thermal barrier coatings (32). A layer 75–125 μm thick of plasma-sprayed Ni–16Cr–5Al–0.6Y to which is bonded a plasma sprayed zirconia–yttria overlay 125–380 μm thick has been used. The ceramic layer insulates the metallic substrate from higher surface temperatures than it otherwise might tolerate. These thermal barrier coatings may be of value also in turbines operated for electric power generation (qv). The best ZrO_2-based coating under severe sulfidation conditions contain 6–8% Y_2O_3, applied over an alloy containing 0.15Y; approximately the same composition was found to be best for aircraft turbines (32).

Low Expansion Alloys. Binary Fe–Ni alloys as well as several alloys of the type Fe–Ni–X, where X = Cr or Co, are utilized for their low thermal expansion coefficients over a limited temperature range. Other elements also may be added to provide altered mechanical or physical properties. Common trade names include Invar (64%Fe–36%Ni), Elinvar (52%Fe–36%Ni–12%Cr) and super Invar (63%Fe–32%Ni–5%Co). These alloys, which have many commercial applications, are typically used at low (25–500°C) temperatures. Exceptions are automotive pistons and components of gas turbines. These alloys are useful to about 650°C while retaining low coefficients of thermal expansion. Alloys 903, 907, and 909, based on 42%Fe–38%Ni–13%Co and having varying amounts of niobium, titanium, and aluminum, are examples of such alloys (2).

Oxide Dispersion Strengthened Alloys. Oxide dispersion strengthening (ODS) was first applied to tungsten in 1919, and was later reported for aluminum-base alloys in 1949. The first commercial nickel-base dispersion strengthened alloy, TDNi (Ni–2%ThO_2), dating from 1962, exhibited high strength in the range 980–1200°C but was subject to severe attack in oxidizing environments. Improved static oxidation resistance was achieved in Ni–20Cr–2ThO_2 type alloys, but dynamic oxidation resistance suffered, owing to the volatilization of chromium oxide. The latest class of ODS alloys to be developed, based on Ni, Cr, and Al, relies on an Al_2O_3 protective scale for dynamic oxidation resistance; Y_2O_3 is the dispersoid in each of these alloys (Table 7).

The inadequate low temperature strength displayed by many dispersion strengthened alloys led to attempted combinations of dispersion, and γ'-hardening through the mechanical alloying technique, ie, simultaneous melding of all constituents: master alloy, solutes, and oxide dispersoids, in a special high energy ball mill. At low temperatures γ'-hardening is achieved and above 1000°C strength parallels that of TD–Ni sheet. Mechanical properties of ODS

Table 7. Nominal Compositions of Oxide Dispersion Strengthened Alloys, wt %

Alloy	Cr	Y_2O_3	Al	Ti	C	Ta	Mo	W	Zr	B
Inconel MA-754[a]	20	0.6	0.3	0.5	0.05					
Incoloy MA-956[b]	20	0.5	4.5	0.5						
MA-6000[a]	15	1.1	4.5	2.5		2.0	2.0	4.0	0.1	0.01
MA-760[a]	20	0.95	6.0		0.05		2.0	3.5	0.15	0.01
MA-758[a]	30	0.6	0.3	0.5	0.05					

[a]The balance is nickel.
[b]The balance is iron.

alloys are outlined in Table 8. All are available as bars. MA-754 forged airfoils have been available commercially since the late 1970s. Stress–rupture properties of ODS alloys have been compared with those of several other alloys (48). MA-6000, which displays the highest strength at 1093°C, is stronger than directionally solidified MarM-200+Hf at temperatures above 870°C, while manifesting good oxidation resistance at 1093°C. It is approximately equal to that of conventionally cast alloys such as IN-100 and 713LC. Sulfidation resistance is comparable to that of the cast Ni-base alloys IN-792 and IN-738. Fusion welding can only be carried out in regions of low stress, because the uniform dispersion of oxides is disrupted by melting. Nonfusion techniques are needed to obtain tensile and stress–rupture properties near those of the parent alloy.

Applications. The principal applications of nickel-base superalloys are in gas turbines, where they are utilized as blades, disks, and sheet metal parts. Aircraft gas turbines utilized in both commercial and military service depend upon superalloys for parts exposed to peak metal temperatures in excess of 1000°C. Typical gas turbine engines produced in the United States in 1990 utilized nickel and cobalt-base superalloys for 46% of total engine weight (41). However, programs for future aerospace propulsion systems emphasize the need for lightweight materials having greater heat resistance. For such applications, intermetallics matrix composites and ceramic composites are expected to be needed.

Power sources for vehicles in space similarly must incorporate superalloys into their design. For example, high temperature alloys have been used extensively in the NASA space shuttle main engine. A high pressure oxidizer turbopump, driven by a hot gas turbine having inlet temperature of up to 850°C, delivers liquid oxygen at a rate of over 450 kg/s. The turbine is constructed primarily of nickel and cobalt-base alloys, with highly stressed parts cooled by liquid hydrogen. MarM-246 (Hf-modified) nozzles, Waspaloy turbine disks and shaft, and Inconel 718 high pressure main pump impeller are typical superalloy parts. Nickel-base alloys of the Hastelloy series are utilized for their resistance to extreme corrosion environments such as nitric and hydrofluoric acids.

Several nickel alloys have usable strength at elevated temperatures even in the absence of significant quantities of strengthening precipitates. Inco alloy G-3 is Ni–22 wt %Cr–20%Fe–2%Cu. Its low carbon content is beneficial for preventing sensitization and intergranular corrosion of weld heat affected zones. This alloy is used for flue-gas scrubbers and for handling phosphoric and sulfuric acids. Inconel 600 has good oxidation resistance coupled with low corrosion

Table 8. Material Properties of Oxide Dispersion Strengthened Superalloys[a]

Property	Inconel MA-754[b]	Incoloy MA-956[c]	Inconel MA-6000[b]
Mechanical properties			
ultimate tensile strength, MPa[d]			
at 21°C, longitudinal	965	645	1295
at 540°C, longitudinal	760	370	1155
at 1095°C			
longitudinal	148	91	222
transverse	131	90	177
yield strength, 0.2% offset, MPa[d]			
at 21°C, longitudinal	585	555	1285
at 540°C, longitudinal	515	285	1010
at 1095°C			
longitudinal	134	84.8	192
transverse	121	82.7	170
elongation, %			
at 21°C, longitudinal	21	10	4
at 540°C, longitudinal	19	20	6
at 1095°C			
longitudinal	12.5	3.5	9.0
transverse	3.5	4.0	2.0
reduction in area at 1095°C, %			
longitudinal	24		31.0
transverse	1.5		1.0
1000-h rupture strength, MPa			
at 650°C	255	110	
at 980°C	130	65	185
Physical properties			
melting range, °C		1480	
specific heat capacity at 21°C, J/(kg·K)[e]		469	
thermal conductivity at 21°C, W/(m·K)		10.9	
mean coefficient of thermal expansion at			
538°C, 10^{-6}/K		11.3	
electrical resistivity, nΩ·m		1310	

[a]Ref. 35.
[b]Bar.
[c]Sheet.
[d]To convert MPa to psi, multiply by 145.
[e]To convert J to cal, divide by 4.184.

resistance. It is used for furnace components, as piping in nuclear plants, and in chemical and food processing (qv) industries. Inconel 601 has even higher strength and oxidation resistance owing to the addition of 1–1.7 wt % Al. Typical applications are in heat treating equipment, petrochemical equipment, and gas turbine components.

Other Ni–Cr–Fe alloys having higher Fe than Cr contents include Incoloy 800 and Incoloy 825 and Inco alloy 330. Incoloy 800 displays excellent oxidation and carburization resistance, and resists corrosion by many aqueous environments. It is used for process piping, heat exchangers, heating element sheaths, and nuclear steam-generator tubing. Alloy 330, which contains Si for enhanced oxidation resistance, is used for furnace and heat treating parts.

Iron–Nickel Base Superalloys

Iron–nickel base superalloys were developed primarily from the stainless steels. In the United States, these alloys included 19-9 DL, 16-25-6, and A-286 (Table 9). Later, higher nickel contents were employed to take advantage of the superior oxidation resistance of nickel and the beneficial effects of γ'-forming elements. All iron–nickel base superalloys rely on solid solution hardening to some extent. Hastelloy X, more accurately classified as nickel-base, N-155, and Inconel 625 also exhibit some precipitation strengthening owing to the presence of small amounts of carbon, aluminum, or titanium, but are used primarily in low stress applications to about 1100°C where oxidation resistance is important. Other iron–nickel superalloys, widely used for turbine wheels, are strengthened additionally either by ordered fcc γ'-Ni$_3$Al (A-286, V-57, Incoloy 901) or by ordered body-centered tetragonal (bct) γ'-Ni$_3$Nb (Inconel 706, Inconel 718). Finally, a group of alloys achieves strengthening from carbides, nitrides, and carbonitrides (16-25-6, CRMD series). The temperature capabilities of the iron–nickel alloys varies widely: γ' or carbide-strengthened alloys can be used to 815°C; γ'-strengthened alloys are used for high stress applications to 650°C. The stress–rupture capabilities of several alloys in this group are summarized in Figure 10, which shows clearly the lack of any appreciable strength above about 800°C except for the ODS alloy MA-956 (49).

Because iron–nickel alloys tend to contain large amounts of ferrite stabilizers such as chromium and molybdenum, the minimum nickel content required to maintain a fcc matrix is about 25 wt %. High iron contents lower cost, increase fabricability, and tend to raise the melting point, at the expense of poorer oxidation resistance than nickel-base alloys. Chromium is added for surface protection and solid-solution strengthening of gamma. Molybdenum also is added for solid-solution strengthening, but is present also in carbides and γ'. Small quantities of boron or zirconium are added to improve workability and stress–rupture properties, and carbon is useful as a deoxidant and to provide MC carbides to help refine grain size during hot working. Finally, ductilizing effects may be realized with small addition of magnesium, calcium, and certain rare-earth elements. Iron–nickel alloys are used extensively in aircraft gas turbines and in the space shuttle main engine.

Cobalt-Base Superalloys. Cobalt-base superalloys are used principally where operating metal temperatures range from 650 to 1000°C and stresses are relatively low. Strengthened primarily by carbide precipitation and solid-solution effects, these alloys are widely used as forgings and castings for nozzle vanes in gas turbine engines, because of good thermal shock and hot corrosion resistance, and in sheet metal assemblies, such as combustion chamber liners, tail pipes, and afterburners. However, rotating parts such as turbine blades and disks, are more likely to be made from nickel-base alloys, because of the latter's superior strength at low and intermediate temperatures. Some of the industrial uses of

Table 9. Iron–Nickel-Base Alloys, wt %

Alloy	UNS number	Ni	Cr	Co	Mo	Al	Ti	Fe	Mn	Si	C	B	Other
Hastelloy X	N06002	47.3	22.0	1.5	9.0			18.5	0.5	0.5	0.1		0.6W
Incoloy 901	N09901	42.5	12.5		5.7	0.2	2.8	36.0	0.10	0.10	0.05	0.015	
A-286	S66286	26.0	15.0		1.3	0.2	2.0	53.6	1.35	0.50	0.05	0.015	
Discaloy	S66720	26.0	13.5		2.7	0.1	1.7	54.3	0.90	0.80	0.04	0.005	
N-155	R30155	20.0	21.0	20.0	3.0			30.3	1.50	0.50	0.15		2.5W, 1.0Nb
V-57		27.0	14.8		1.25	0.25	3.0	52.0	0.35	0.75	0.08	0.010	0.50V
Inconel 706	N09706	41.5	16.0	0.5	0.5	0.2	1.75	40.8	0.18	0.18	0.03		2.9Nb
Pyromet 860		43.0	12.6	4.0	6.0	1.25	3.0	30.0	0.05	0.05	0.05	0.010	
Inconel 718	N07718	53.0	18.6		3.1	0.4	0.9	18.5	0.2	0.3	0.04		5.0Nb
16-25-6		25.5	16.2	6.25							0.1		0.15N

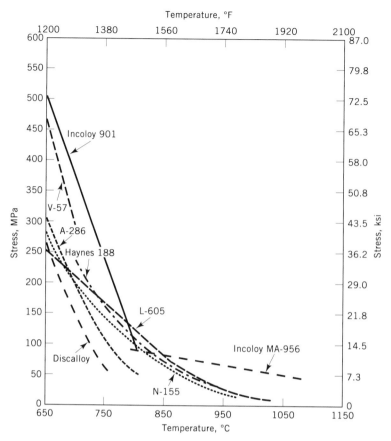

Fig. 10. 1000 h stress–rupture curves of wrought cobalt-base (Haynes 188 and L-605) and wrought iron-base superalloys (49). To convert MPa to psi, multiply by 145.

cobalt-base alloys include grates for heat-treating furnaces, quenching baskets, pouring funnels for molten copper, skids for slab reheating furnaces, and other foundry and metalworking operations where the prime requisites are resistance to oxidation at elevated temperatures, comparability with slags, and resistance to thermal and mechanical shocks. In addition, the wrought grades are used for high temperature springs and fasteners. Cast alloys where carbon exceeds 1 wt % and chromium contents are in the range 26–32% are used for cutting tools and wear-resistant facings, where high hardness and abrasion resistance at elevated temperatures are the prime requirements (see TOOL MATERIALS). Cobalt alloys also are used extensively in valves, pumps (qv), nozzles, and mixers for the chemical process industries. These alloys generally also contain significant amounts of tungsten, iron, nickel, and sometimes molybdenum. A prominent example is Stellite 6, which has 30% Cr, 2.5% Ni, 3% Fe, 1.5% Mo, 4% W, 1.4% Mn, and 1% C.

Tables 10 and 11 list typical compositions of cast and wrought cobalt-base alloys, respectively. Stress–rupture properties of two wrought cobalt alloys,

Table 10. Compositions of Wrought Cobalt-Base Alloys, wt %

Alloy	UNS number	Ni	Cr	Co	W	Fe	Mn	Si	C	Other	Density, g/cm³
CM-7		15.0	20.0	48.0	15.0				0.10	0.5Al, 1.3Ti	9.05
Haynes 188	R30188	22.0	22.0	39.2	14.0	1.5	0.75	0.40	0.10	0.08La	9.13
L-605	R30605	10.0	20.0	52.9	15.0		1.50	0.50	0.10		9.13
MarM-918		20.0	20.0	52.0					0.05	7.5Ta	8.86
S-816	R30816	20.0	20.0	42.0	4.0	4.0	1.2	0.40	0.38	4.0Mo, 4.0Nb	8.59
TD Co		20.0	18.0	60.0						2.0ThO₂	8.61
UMCO-50			28.0	51.0		21.0			0.10		8.06

Table 11. Compositions of Cast Cobalt-Base Alloys, wt %

Alloy	Ni	Cr	Co	W	Ta	Nb	Al	Ti	C	Zr	Other	Density, g/cm³
AR-13	1.0	21.0	58.0	11.0		2.0	3.5		0.45		0.1Y, 2.5Fe, 0.5Mn	8.43
AR-213		19.0	66.0	4.7	6.5		3.5		0.18	0.15	0.1Y	8.51
AR-215		19.0	64.0	4.5	7.5		4.3		0.35	0.13	0.17Y	8.47
FSX-414	10.0	29.0	52.0	7.5					0.25		1.0Fe, 0.010B	8.30
MarM-302		21.5	58.0	10.0	9.0				0.85	0.20	0.005B	9.21
MarM-322		21.5	61.0	9.0	4.5			0.75	1.00	2.25		8.91
MarM-509	10.0	23.5	55.0	7.0	3.5			0.20	0.60	0.50		8.85
NASA CoWRe		3.0	68.0	25.0				1.00	0.40	1.00	2.0Re	9.59
WI-52		21.0	63.0	11.0		2.0			0.45		2.0Fe, 0.25Mn, 0.25Si	8.88
X-40/X-45	10.5	25.5	54.0	7.5					0.50/0.25		0.75Mn, 0.75Si	8.60

272

Haynes 188 and L605, are compared to those of iron–nickel alloys in Figure 10 (49). The cobalt alloys generally are inferior in strength to the strongest cast nickel-base superalloys. Tensile strengths at low and intermediate temperatures are particularly deficient for the cobalt alloys.

Cobalt-base superalloys are not normally heat treated, except for stress relief following machining or welding, prior to being placed in service. No stable intermetallic compound having the beneficial effects of γ' in nickel-base alloys has been found in cobalt-base systems. As a result of the reliance predominantly on solid-solution hardening for strength, cobalt-base alloys tend to be useful above 1100°C, where γ'-strengthening becomes ineffective in nickel-base alloys, and at temperatures on the order of 815°C in relatively low stress applications. The intervening temperature range is dominated by nickel-base alloys.

Older cobalt-base superalloys (HAS-31, WI-52) do not benefit appreciably from processing under vacuum, and thereby offer a considerable savings in costs. However, newer alloys such as the MarM series having substantial quantities of tantalum, and the AiResist alloys (AR-13, 213, 215), containing 0.1% yttrium, do require vacuum melting. A limited amount of work on powder processing indicates that reduced segregation, improved workability, and superior fatigue resistance may result for selected cobalt alloys (Table 11).

Cobalt-base alloys generally rely on chromium for high temperature corrosion resistance, and most contain at least 20–25% Cr to form protective Cr_2O_3. Only minor amounts of CoO and $CoCr_2O_4$ form. Whereas chromium is generally recognized as the most beneficial element for high temperature oxidation resistance, yttrium has been shown to significantly reduce the oxidation rate and improve scale adherence of a pure Co–20 wt %Cr alloy between 900 and 1200°C (50). Similar results have been noticed in steels and with additions of lanthanum and cerium. Co–30Cr alloys are far superior to Co–10Cr alloys even to temperatures of 1200°C (51), and oxidation resistance of Co–30Cr is further improved by additions of zirconium, cerium, aluminum, and boron, whereas tantalum appears to be of some benefit to Co–10Cr alloys.

Cobalt alloys may find application in a fluidized-bed process for the direct combustion of coal (qv). CoCrAlY-coated Haynes 188 has proven to be one of the most resistant materials to a fireside corrosion process encountered in tubes connected the fluidized-bed combustor to a steam turbine.

Refractory Metals and Their Alloys.   Many elements which could be called refractory are found in the Periodic Table, but those which have received the most attention for potential structural applications are the bcc metals, tantalum, molybdenum, vanadium, niobium, and tungsten, all of which melt at about 2000°C. Chromium is sometimes included in this group, but its lack of ductility and toughness at room temperature, together with its reaction with nitrogen in air at high temperatures, has resulted in termination of efforts to develop Cr-base alloys. Molybdenum and tungsten also suffer from low temperature brittleness, but this problem can be circumvented by alloying or thermomechanical working. Other high melting metals have comparatively poor high temperature strength, and many, eg, the platinum-group metals (qv), are too rare and expensive for widespread use. Figure 11 (52) compares some of the unalloyed refractory metals on the basis of ultimate tensile strengths at elevated temperatures. Commercially available alloys (53,54) are

Alloy designation	Nominal compositions, wt %
unalloyed niobium	Nb–0.030O–0.01C–0.03N
Nb–1Zr	Nb–1Zr
WC-103	Nb–10Hf–1Ti
FS-85	Nb–27Ta–10W–1Zr
SCB-291	Nb–10W–10Ta
B-88	Nb–28W–2Hf–0.07C
WC-129Y	Nb–10W–10Hf–0.24Y
D-43	Nb–10W–1Zr–0.1C
unalloyed tantalum	Ta–0.015O–0.01C–0.01N
Ta–10W	Ta–10W
T-111	Ta–8W–2Hf
T-222	Ta–10W–2.5Hf–0.01C
Astar 811C	Ta–8W–1Re–1Hf–0.025C
unalloyed molybdenum	Mo–0.04C–0.003O–0.001N
Mo-TZM	Mo–0.5Ti–0.1Zr–0.03C
Mo–42Re	Mo–42Re
Mo–50Re	Mo–50Re
unalloyed tungsten	W–0.01C–0.006O–0.005N
W–3Re	W–3Re
W–5Re	W–5Re
W–25Re	W–25Re
W–0.3Hf–0.025C	W–0.3Hf–0.025C
W–4Re–0.3Hf–0.025C	W–4Re–0.3Hf–0.025C
W–24Re–0.3Hf–0.025C	W–24Re–0.3Hf–0.025C

Rupture strength data for several of these alloys also are available (53).

Nitrogen and carbon are the most potent solutes to obtain high strength in refractory metals (55). Particularly effective are carbides and carbonitrides of hafnium in tungsten, niobium, and tantalum alloys, and carbides of titanium and zirconium in molybdenum alloys.

The technology of molybdenum is well advanced, although niobium also has received considerable attention because of its good fabricability. Tungsten has a highly developed metallurgical technology because of its vital use as lamp and electron tube filaments. The low temperature ductility and corrosion resistance of tantalum make it attractive for cryogenic applications and for chemical equipment, but it has received only limited use at elevated temperatures (see CRYOGENICS).

Because of high melting temperatures, the Group 6 (VIB) refractory metals and tantalum all were first fabricated by powder metallurgy techniques. Since about 1950, arc melting of refractory metals in vacuum or inert atmospheres by the consumable electrode technique has been used commercially to produce large ingots and billets. Extrusion, forging, and sheet-rolling technologies have advanced rapidly so that many of the refractory metal alloys are now available in various mill forms. For example, ingots of molybdenum over 30 cm in diameter

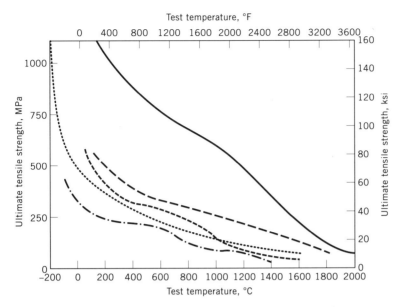

Fig. 11. Test temperature vs ultimate tensile strength for pure refractory metals (52): (——) rhenium; (– – –) tungsten; (- - - -) molybdenum; (·····) tantalum; and (– · – ·) niobium. To convert MPa to psi, multiply by 145.

are produced and converted to billets, forgings, sheets, tubes, bars, and special shapes. Electron-beam melting, plasma-arc spraying, fused-salt electroplating, and vapor deposition are among the specialized techniques being used to produce and fabricate the refractory metals.

In many high temperature applications in the electrical and electronics industry, the refractory metals are protected by a vacuum or an inert gas, so that oxidation is not a problem. However, for most other high temperature applications, poor oxidation resistance has limited use. The oxides of the refractory metals, rather than existing as tight, protective barriers, suffer from porosity at moderately elevated temperatures, volatility at higher temperature, and spalling of the oxide scales away from the substrate, especially at corners and edges. For comparison, the temperatures at which the oxides of the three most widely used refractory metals become significantly volatile are approximately 788°C for MoO_3, 900°C for WO_3, and 1275°C for Nb_2O_5. Bare molybdenum and tungsten literally go up in smoke if used in air at temperatures around 1100°C. Molybdenum oxidizes to MoO_2 and MoO_3. The latter melts at 797°C, thus allowing rapid transport through the oxide at higher temperatures. At 1100°C the vapor pressure of MoO_3 is approximately 100 kPa (1 atm) in still air, and significant vapor pressures of the other two species are observed. The detrimental effect of molybdenum additions of the oxidation resistance of alloys has been noted. Linear kinetics was observed for tungsten in all cases within the temperature regime of 750–1050°C. Although a large number of tungsten oxides are stable, only WO_2 and WO_3 were found in the high temperature regime. Both tantalum and niobium oxidize rapidly to form porous oxides. Only Nb_2O_5 is readily identified in porous, cracked scales formed on pure niobium in air at high temperature. The scale grows by anion diffusion inward; after a certain scale thickness is reached, the scale cracks and becomes porous. The process is then repeated many times.

In addition to oxidation itself, gas diffusion into the base metal can be more damaging than the actual loss of metal from the surface. Thus the loss in mechanical properties owing to diffusion of oxygen into niobium makes it more difficult to protect niobium against oxidation damage than molybdenum, even though molybdenum has less resistance to normal oxidation effects than niobium.

Alloy research in the 1950s concentrated on improving the oxidation resistance of the refractory metals. Efforts in the 1960s concentrated on developing oxidation-resistant coatings. The most successful coatings are disilicides of the base metal, 2–5 mm thick, usually applied by a high temperature pack cementation process. Aluminide coatings also have been applied successfully. The aluminide and silicide coatings form thin, impervious layers of alumina and silica that protect against oxidation at temperatures from 1100 to 1500°C, depending on the substrate. In the case of molybdenum alloys, temperatures of less than 1093°C are required in order for useful life to exceed 10^4 h (53). At reduced oxygen pressures, such as are encountered during aerodynamic heating by reentry of aerospace vehicles, silicide coatings break down and are not as protective as under normal atmospheric pressures of oxygen. Other coating systems have been based on noble metals and Ni–Cr alloys (56).

Joining is another problem. Fusion welding in inert atmospheres often develops recrystallized structures in the heat-affected zones, so that welded parts lose strength and are embrittled. Brazing can overcome these difficulties, but the brazed parts are limited to the maximum service temperatures of the filler metal, and brittle intermetallic compounds are formed in some cases. Special joining techniques being used to help overcome these deficiencies include electron-beam and solid-phase welding, and the development of special brazing materials.

Molybdenum.   Molybdenum is the most readily available and widely utilized refractory metal. Most engineering applications of this metal utilize the high melting temperature, high strength and stiffness, resistance to corrosion in many environments, or high thermal and electrical conductivity. However, molybdenum must be coated if exposed to air above 535°C. The melting point of 2610°C, over 1000°C higher than for most high temperature superalloys, permits molybdenum to be used in inert atmosphere furnace equipment. Furnace hardware and heat shields also perform well under extreme temperature conditions. The high thermal and electrical conductivity of molybdenum, as well as its inertness to molten glasses, permits it to be used for electrical heating or heat booster electrodes in commercial continuous glass making operations. Molybdenum also is used in a wide range of electronic and thermionic devices as well as crystal growing devices, x-ray tubes, magnetism and thyristors, and resistance weld electrodes. Other characteristics of molybdenum are its low thermal expansion, high stiffness, and the ability to take a high surface finish. Molybdenum is useful for high temperature laser mirror components such as those to be used in fusion power systems (see FUSION ENERGY; LASERS).

The most widely used molybdenum alloy for high temperature applications is TZM, containing 0.5% Ti and 0.08% Zr. The tensile strength of this alloy is substantially higher than that of unalloyed molybdenum (57). Whereas unalloyed molybdenum is strengthened only by cold work, TZM molybdenum relies also on solid-solution strengthening by small quantities of zirconium, titanium, and carbon, as well as dispersion strengthening by precipitation of complex Mo–Ti–Zr carbide particles. The principal applications for TZM are at service temperatures of 1095°C and above. The alloy is used for cores and inserts of die casting equipment for aluminum, zinc, and copper alloys as well as steels. Die casting of an aluminum carburetor for the automotive industry has been a particularly successful application of TZM.

TZM also is utilized in hot work tool applications, eg, tools for hot extrusion, hot rolling, and hot forging. TZM dies are used for the isothermal forging process known as Gatorizing (Pratt and Whitney Aircraft) for large turbine disks. TZM also has been employed for hot gas valves and seals for high temperature gas systems. A modified TZM alloy, called TZC Mo (1.25%Ti–0.3%Zr–0.15%C), has been developed for higher strength at elevated temperatures.

A solid solution Mo–30%W alloy, which was developed originally for rocket nozzle erosion resistance (see ABLATIVE MATERIALS), has been found to have excellent resistance to chemical attack by molten zinc or zinc vapor. This alloy, therefore, is used as critical elements in pumps for liquid zinc, for example as impeller shafts and impellers in centrifugal impeller-type pumps. Valves for precision melting of zinc also have been fabricated from Mo–30%W.

Two experimental alloys, having 1.2% Hf, 0.1% C and 0.6% Zr, 0.1% C, are promising for high temperature applications beyond the capabilities of the existing TZM and TZC alloys. Strength of these alloys at elevated temperatures is superior to that of TZM by at least 25%, and the recrystallization temperature is at least 275°C higher than for TZM. Creep–rupture resistance of the Mo–Hf–C alloy at 1370°C also is substantially higher than for TZM.

Chromium.   Chromium offers a number of potential advantages for high temperature applications, including oxidation resistance, low density, and a melting point over 400°C higher than that of nickel. Nevertheless, all efforts to use chromium have been hindered by its extreme brittleness (58).

Nitrogen was identified as a primary embrittling impurity in chromium in the early 1950s. However, even pure chromium single crystals are known to be brittle at ambient temperatures, thereby presenting a significant problem in developing ductile chromium alloys. Nevertheless, several promising approaches to improve ductility and/or strength have been reported. Several ductile Cr–MgO alloys made by a powder process have been developed in the United States, the MgO improving oxidation resistance as well. Alloy development in the United States also has combined both solid-solution strengthening by tungsten or molybdenum with precipitate strengthening by carbides, and small amounts of yttrium or yttrium and lanthanum for solute scavenging and slightly improved oxidation resistance (58). Other chromium alloys having improved strength but only marginally increased ductility have been developed in Australia and the former Soviet Union. In all cases there is a trade-off between high temperature strength and low temperature ductility.

Niobium and Tantalum.   Niobium is the lightest of the four most widely utilized bcc refractory metals and, when pure, is tough and ductile at room temperature. Although alloys have been developed having useful strength above 1000°C, their oxidation resistance is so poor that uses in oxygen bearing environments require a protective coating. Two classes of niobium alloys having improved oxidation resistance are (*1*) Nb–Ti–X, where X = W, Mo, or V, and some Nb–Al–V alloys which are moderately better in oxidation resistance but lack adequate strength for high temperature structural applications; (*2*) much improved in oxidation resistance but brittle are Nb–Cr–Al, Nb–Fe–Al, and Nb–Ti–Cr. Elements that substantially improve oxidation resistance include zirconium, aluminum, chromium, molybdenum, and titanium (59).

In spite of the poor oxidation resistance of most commercial niobium alloys, some gas turbine components, rocket components, thermal shields, fasteners, and bellows have been fabricated from alloys such as C-103, C-129Y, and Cb-752. In all cases of high temperature exposure these parts are coated with a silicide. The influence of temperature on stress–rupture strength of three of the strongest niobium alloys is shown in Figure 12 (31). These alloys are coated with Si–20Cr–20Fe. Other variants contain hafnium silicide, which raises the melting temperature of the coating. By comparison, Nb–1Zr is much weaker.

Niobium has sufficient ductility at room temperature to allow conventional forming practices. Most niobium alloys display good formability, weldability, and moderate strength. However, alloys require protective coatings at temperatures above 425°C in oxidizing atmospheres. Substitutional solutes most often added to niobium for creep resistance are tungsten, molybdenum, and tantalum. However, the reactive metals zirconium and hafnium often are added, together with carbon and nitrogen, to produce fine precipitates which further contribute to creep strength. The Nb–1%Zr alloy is used extensively in nuclear systems containing liquid metals operating in the 980–1200°C range because of its low thermal neutron cross section, moderate strength, and excellent fabricability. C-103 has been utilized for rocket components requiring moderate strength from 1093–1370°C.

Tantalum alloys also are readily fabricable, have high melting points, and relatively good mechanical properties. However, high cost and high density are

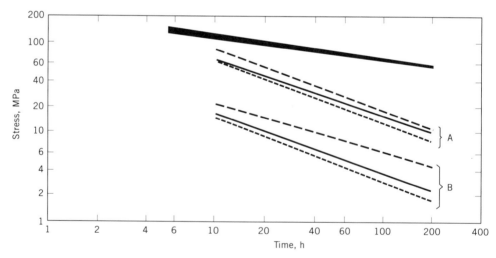

Fig. 12. Effect of temperature on the stress–rupture properties of three niobium alloys coated with a silicide at A, 1205°C, and B, 1315°C. (■) represents all three alloys at 1095°C. (——) C-129Y; (– – –) Cb-752; and (----) FS-85. Stress levels are based on material thickness prior to application of coating (31). To convert MPa to psi, multiply by 145.

negative factors that have slowed development. Strengthening of tantalum by substitutional solutes generally parallels trends in niobium. However, tungsten is present in all commercial tantalum-base alloys, because W is a more potent strengthener than molybdenum. The Ta–10%W alloy is widely used for aerospace parts. Maximum use temperature is about 2480°C in aerospace applications (31). T-111 and T-222 are higher strength modifications of Ta–10W and have comparable fabricability. However, products of Ta alloys are limited to Ta–W binaries owing to the demand for corrosion resistance (60). Protective coatings are required for tantalum and its alloys above 480°C.

Vanadium. The mechanical properties of unalloyed vanadium resemble those of the other Group 5 (VB) elements, niobium and tantalum. All are ductile at ambient and cryogenic temperatures, and therefore are easily fabricable by standard cold working techniques. However, interstitial impurities have a significant effect on strength and ductility. Embrittlement by hydrogen is the most notable single impurity effect.

Principal vanadium alloy development programs for possible applications as breeder reactor fuel element cladding (see NUCLEAR REACTORS) were conducted both in the United States and Germany during the 1950s and 1960s. As of this writing, however, there is no commercial production of either vanadium or its alloys in the United States, owing in part to the lack of sufficient market. Nevertheless, there has been considerable interest in vanadium-base alloys for the first wall in fusion reactors. Vanadium alloys offer significant advantages over iron-base alloys in high temperature strength, corrosion resistance to liquid lithium, and resistance to radiation damage (51).

Research has focused on V–Cr–Ti–Si alloys, having 0.15 wt % Cr, 0–20% Ti, and less than 1% Si. Titanium additions above 3 wt % drastically reduce the magnitude of radiation-induced swelling. The alloy of greatest interest is V–5Ti–5Cr (31).

Tungsten. Only three tungsten alloy systems are produced commercially: $W–ThO_2$, W–Mo, and W–Re (61). The use of tungsten for filaments in incandescent lamps dates to about 1904; its use for anodes in x-ray tubes to the 1930s. A more recent use for tungsten is as cathodes in magnetron power tubes for generating microwaves for use in radar, communications, microwave, diathermy, and electronic cooling (see MICROWAVE TECHNOLOGY). Industrial furnaces use tungsten for heating elements, furnace boats, firing trays, and vaporizing coils. Welding electrodes of unalloyed tungsten or tungsten plus thoria or zirconia are used for tungsten inert gas (TIG) and atomic hydrogen welding. Tungsten also is used for electrical contacts and disks for ignition systems and semiconductor applications. In the aerospace industries, tungsten infiltrated with silver is used in rocket nozzles. Tungsten, as well as the other Group VIB metals, can be alloyed with 20–40% rhenium to improve ductility. However, the high cost of rhenium does not permit widespread commercial exploitation of this ductility effect, the origin of which seems to be based on carbide redistribution in the presence of rhenium.

The brittle to ductile transition (BTD) of recrystallized tungsten is above 205°C; however, this can be lowered to below room temperature by heavy warm or cold working. Annealing subsequent to working progressively restores brittle behavior. Tungsten is very notch sensitive; therefore, removal of surface defects can substantially decrease the BDT temperature. Lowering impurity content is another important factor in improving low temperature ductility.

Both molybdenum and tungsten can be worked in air without ductility loss. All refractory metals can be made into tubing by extrusion, and most refractory metals, except chromium, are available as wire. Tungsten wires were attempted as fiber reinforcement for experimental nickel-base composites.

New Materials and Processes. *Aligned Eutectics.* Metal-matrix composites (qv) produced by directional solidification of eutectic or near eutectic compositions, have been considered prime candidates for application in advanced gas turbine engines. Several eutectic composites were identified in the 1980s which promised significant improvements over conventional nickel and cobalt-base superalloys in terms of tensile strength, impact strength, creep–rupture properties, and fatigue resistance. Successful growth was demonstrated of either solid or hollow blades of several advanced eutectic alloys having no loss of alignment, and some alloys successfully passed engine tests (62,63).

Table 13 is a representative list of nickel and cobalt-base eutectics for which mechanical properties data are available. In most eutectics the matrix phase is ductile and the reinforcement is brittle or semibrittle, but this is not invariably so. The strongest of the alloys listed in Table 13 exhibit ultimate tensile strengths of 1300–1550 MPa. Appreciable ductility can be attained in many fibrous eutectics even when the fibers themselves are quite brittle. However, some lamellar eutectics, notably $\gamma/\gamma'-\delta$, reveal little plastic deformation prior to fracture.

The tensile strengths of nickel-base eutectic alloys containing aluminum can be increased by precipitation of the ordered γ'-(Ni_3Al) phase from solution. Such aging treatments can produce increased strength over a wide range of test temperatures, but may also reduce ductility, as in the γ/γ'–Mo system. In addition, precipitation hardening has been successfully applied to Cotac, leading to improvements in stress–rupture and fatigue properties.

Aligned eutectics exhibit excellent longitudinal creep and stress–rupture properties (64). At low stresses and high temperatures several eutectic alloys exhibit stress–rupture resistance superior to that of the superalloys. Additional strong eutectics have been reported (65), but the simultaneous development of superalloy single crystals in the 1970s and 1980s resulted in widespread use of the latter. Work on eutectics was halted around 1985, but efforts were then renewed on eutectics containing one or more intermetallic phases, eg, Nb_3Al or NiAl.

Oxide and Fiber-Reinforced Superalloys. Apart from the aligned eutectics, two other approaches promise to improve the maximum service temperature of turbine blade materials: oxide dispersion strengthened superalloys (ODSS), and oxidation-resistant alloys reinforced with refractory metal fibers. The properties of some FRS alloys are compared with those of superalloys in Figure 13 (66). When the density-compensated stress–rupture resistance of these advanced alloys is compared with that of MarM-200 + Hf, a commercial directionally solidified superalloy to which hafnium is added, the strength of the superalloy can be approximately doubled by the advanced alloys at temperatures in excess of 1000°C. However, a detailed cost estimate for manufacturing turbine blades of each of the four types of alloys indicates that a substantial cost penalty

Table 13. High Temperature Eutectic Alloy Compositions

Alloy	v_f^a	Ni^b	Co	Cr	Al	Nb	Ta	C	Other
Fibrous materials									
Nitac 13		69	3.3	4.4	5.4		8.1	0.54	3.1W, 6.2Re, 5.6V
Cotac 741	0.10	bal	10	10	5	4.7	14	0.5	10W
Cotac	0.10	10	65	10			13	1	
Cotac 3 or 33^c	0.10	10	56	20			12	1	
Cotac 50B3W		9.5	59	15.7			12	0.77	3W
γ'/γ–Mo (AG-15)	0.26	65.5			8.1				26.4Mo
Co, Cr–(Co,Cr)$_7$C$_3$	0.30		56.6	41				2.4	
Cotac 744		bal	10	4	6	3.8		0.46	10W, 2Mo
Lamellar materials									
γ'/γ–δ (6% Cr)	0.37	71.5		6	2.5	20			
γ–δ	0.26	66.7				23.3			
γ–γ'–Ni$_3$Ta		67.6			3.7		28.7		
Ni–Ni$_3$Ta		63					37		
γ'–δ	0.44	bal			4.4	23.4			
γ–β	0.4	43			9.6				37.4Fe

$^a v_f$ = volume fraction of reinforcing phase.
^{b} bal = balance of material.
^{c} 1300°C, 2 h; 1000°C, 24 h, air cooled.

281

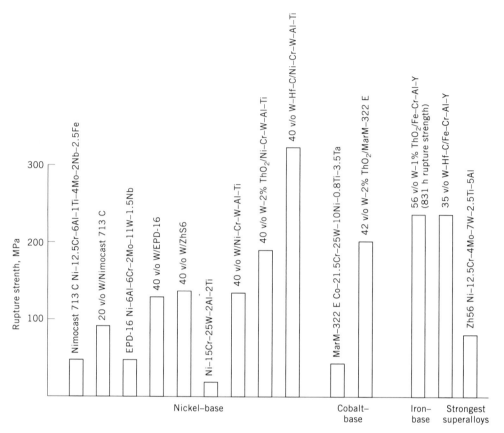

Fig. 13. Comparison of 100 h rupture strength at 1093°C for composites and super-alloys (66). To convert MPa to psi, multiply by 145.

is assumed with aligned eutectics. Dispersion-strengthened alloys and fiber-reinforced superalloys (FRS), however, may be more competitive with conventional alloys, especially because some fiber-reinforced superalloys may not require coatings.

A more extensive comparison of many potential turbine blade materials is available (67). The refractory metals and a ceramic, silicon nitride, provide a much higher value of 100 h stress–rupture life, normalized by density, than any of the cobalt- or nickel-base alloys. Several intermetallics and intermetallic matrix composites, eg, alloyed Nb_3Al and $MoSi_2$–SiC composites, also show very high creep resistance at 1100°C (68). Nevertheless, the superalloys are expected to continue to dominate high temperature alloy technology for some time.

Intermetallic Compounds and Other Ordered Phases. Intermetallic compounds and other ordered phases offer several very attractive features for high temperature applications. Within a long-range ordered lattice, most diffusion-controlled processes are slowed, as is reflected by high activation energies. Consequently, such processes as creep, recrystallization, and oxidation can be markedly hindered by long-range order. Many attempts have been made to exploit these features, particularly in aluminides of nickel, titanium, and iron (Table 14) (69), because of the excellent resistance to further oxidation offered by a surface film of Al_2O_3. Nevertheless, attempts to develop high temperature alloys based on these aluminides have been hindered by brittleness and lack of fabricability. Poor ductility also implies lack of toughness and poor low cycle fatigue and thermal shock resistance extending from room temperature to extremely high temperatures. Numerous solutions to the ductility problem have been achieved, however, based on compositional or microstructural changes as well as the recognition that moisture in air is a common cause of embrittlement (70).

Titanium Aluminides. Two classes of titanium aluminides have been studied extensively since the 1970s: those based on hcp Ti_3Al (alpha 2) and those based on tetragonal TiAl (gamma). Both classes of alloys offer lower density, superior oxidation resistance, and higher operating temperatures than conventional titanium alloys (Table 15) (71). However, these alloys still cannot compete with nickel and cobalt-base superalloys, also listed in Table 15. Two serious problems remain: limited room temperature toughness and ductility as well as insufficient oxidation resistance. In addition, fatigue crack growth rates are very high, with a high exponent, m, in the crack growth relation (eq. 4). Both classes of titanium aluminides exhibit the best properties when reinforced with small amounts of second phase. Thus Ti–Al–Nb alloys, which have received most attention in the alpha 2 class, contain small amounts of bcc beta phase, whereas most gamma alloys are reinforced with small amounts of alpha 2. Increasing amounts of beta phase in alpha 2 lead to improved strength and low temperature ductility, but creep strength is adversely affected. Compositions and room temperature mechanical properties of specific alpha 2 alloys may be found in the literature (71,72). Many different microstructures are obtainable by heat treatment in alpha 2 alloys, and properties vary appreciably with thermal

Table 14. Properties of Nickel, Iron, and Titanium Aluminides[a]

Alloy	Crystal structure[b]	Critical ordering temperature, T_c °C	°F	Mp, °C	Material density, g/cm³	Young's modulus, GPa[c]
Ni_3Al	$L1_2$ (fcc)	1390	2535	1390	7.50	179
NiAl	B2 (bcc)	1640	2985	1640	5.86	294
Fe_3Al	DO_3 (bcc)	540	1000	1540	6.72	141
	B2 (bcc)	760	1400	1540		
FeAl	B2 (bcc)	1250	2280	1250	5.56	261
Ti_3Al	DO_{19} (hcp)	1100	2010	1600	4.2	145
TiAl	$L1_0$ (tetragonal)	1460	2660	1460	3.91	176
$TiAl_3$	DO_{22} (tetragonal)	1350	2460	1350	3.4	

[a]Ref. 69.
[b]All structures are ordered.
[c]To convert GPa to psi, multiply by 145,000.

Table 15. Properties of Titanium Aluminides, Titanium-Base Conventional Alloys, and Superalloys[a]

Property	Ti-Base	Ti$_3$Al-Base	TiAl-Base	Superalloys
structure	hcp/bcc	DO$_{19}$	L1$_0$	fcc/L1$_2$
density, g/cm^3	4.5	4.1–4.7	3.7–3.9	8.3
modulus, GPa[b]	95–115	110–145	160–180	206
yield strength, MPa[c]	380–1150	700–990	400–650	
tensile strength, MPa[c]	480–1200	800–1140	450–800	
ductility, %				
at RT	10–25	2–10	1–4	3–5
at HT[d]	12–50	10–20	10–60	10–20
fracture toughness,				
MPa·m$^{1/2}$	high	13–30	10–20	25
creep limit, °C	600	760	1000	1090
oxidation, °C	600	650	900	1090

[a]Refs. 2, 7, 10, 11, and 71.
[b]To convert GPa to psi, multiply by 145,000.
[c]To convert MPa to psi, multiply by 145.
[d]HT = high temperature.

history. Most research on gamma alloys has been centered on Al contents in the range 47–49 atomic %, because the highest (of the order of 2–4%) ductilities are achieved in this range, especially when a lamellar structure is produced during processing. Although absolute strengths of gamma alloys at elevated temperatures are lower than those of nickel-base superalloys, specific strengths (normalized by density) are quite high (73). Potential applications for gamma alloys include industrial and aircraft gas turbine hardware as well as automotive turbochargers. Both titanium aluminides also are candidate materials, either in monolithic form or as the matrix for composites, for hypersonic vehicles as well as for the Integrated High Performance Turbine Engine Technology (IHPTET) project of the U.S. Air Force and the High Speed Civil Transport project funded by NASA.

    Nickel Aluminides. Two nickel aluminides, fcc Ni$_3$Al and CsCl structure NiAl, have received widespread attention for high temperature applications. The strongest nickel-base superalloys contain up to 70 vol % Ni$_3$Al in a fcc matrix, but single-phase Ni$_3$Al is not nearly as strong or creep resistant. Binary Ni$_3$Al actually increases in strength with increasing temperature, reaching a peak at about 600°C (see Fig. 2) (2). Many elements are soluble in this compound, thereby offering wide opportunities for structural alloy development. Accordingly, a series of Ni$_3$Al-base alloys has been developed at Oak Ridge National Laboratory and elsewhere for potential use as fasteners, diesel engine parts, and furnace hardware. These alloys contain both boron and chromium to improve low temperature and elevated temperature ductility, respectively. Other alloying elements include zirconium and molybdenum. The ductilizing effect of boron set off a wave of research when it was discovered in Japan in 1979 (74). Boron is only effective as a ductilizer in nickel-rich compositions. Its influence appears to result primarily from suppression of the embrittling effects of moisture in air (75). The role of chromium, on the other hand, is to reduce the embrittling effects of oxygen at temperatures above about 500°C (76).

In spite of extreme brittleness and lack of high temperature strength of NiAl (Fig. 1), much effort has been directed at alloy development for aircraft turbine blade applications, especially at General Electric Co. This is a consequence of excellent oxidation resistance, low density, and high thermal conductivity compared to superalloys. Alloyed single crystals, strengthened by substitutional solutes as well as precipitates, display stress rupture properties comparable to nickel-base alloys such as René 80 (Fig. 14) (77). NiAl composites reinforced with ceramic fibers or particles also are being evaluated (78). However, low toughness and high notch sensitivity near room temperature in both single crystals and polycrystals remain as serious impediments to structural applications.

Iron Aluminides. Two iron aluminides, FeAl and Fe$_3$Al, have potential applications which can exploit good oxidation and corrosion resistance, coupled with very low material costs. Although considered for many years to be brittle, it has been established that both alloys are ductile when moisture or hydrogen is excluded from the atmosphere. Both intermetallics are cubic; FeAl has a CsCl structure and Fe$_3$Al a bcc-like DO$_3$ structure. Because of the relatively open crystal structures, together with an order–disorder transition temperature of about 550°C for Fe$_3$Al, neither alloy could be expected to rival nickel aluminides or conventional superalloys at high temperatures. Nevertheless, the U.S. Department of Energy has funded many studies of these alloy systems for potential use in energy generation systems where ferrous alloys and stainless steels are used (79).

Other Intermetallics. Several other low density, oxidation-resistant intermetallics have been studied, including Al$_3$X alloys where X = Nb, Ta, Ti, Zr or Hf, and several silicides such as MoSi$_2$ and Cr$_3$Si. As in the case of other

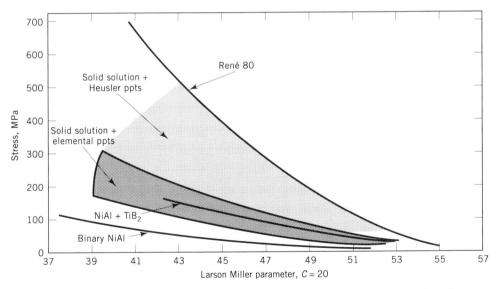

Fig. 14. Stress–rupture properties of NiAl alloys and composites compared with superalloy René 80 (77). The Heusler precipitates (ppts) = 50 Ni–25 Al–25X (at. %), where X = a Group 4 (IVB) or Group 5 (VB) element such as Hf. To convert MPa to psi, multiply by 145.

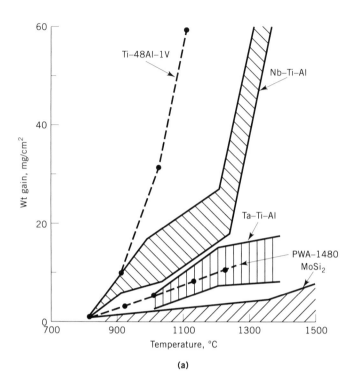

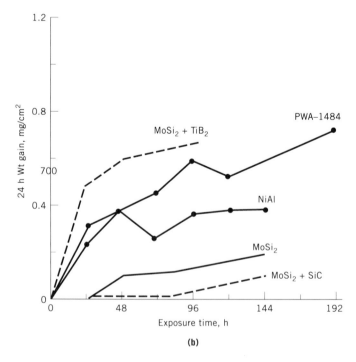

Fig. 15. Comparison of MoSi$_2$ and other intermetallics and superalloys: (**a**), isothermal; and (**b**), cyclic oxidation (1200°C) behavior (80).

intermetallics, these compounds suffer from extreme brittleness at low temperatures. Attempts have been made to ductilize the Al_3X compounds, which crystallize as tetragonal DO_{22} or DO_{23} structures, by alloying with transition metals to produce a cubic structure. Slight improvements in ductility have been reported, eg, in Fe or Mn-modified Al_3Ti. Much attention has been devoted to $MoSi_2$ because of its outstanding oxidation resistance (Fig. 15) (80) and low density. Significant increases in high temperature strength have been achieved by alloying with tungsten and by reinforcement with ceramic fibers such as SiC, Al_2O_3, and ZrO_2 (81). These improvements in toughness have not been appreciable, however. Niobium and tantalum reinforcements have been shown to substantially improve low temperature toughness, but at the expense of high temperature strength and oxidation resistance. Further, ductile reinforcements are only effective when in the form of continuous aligned fibers or tubes. Research on other silicides as structural materials has been relatively sparse, although silicides have been used successfully as coatings for refractory metals for many years (see REFRACTORY COATINGS). Several silicides, eg, $MoSi_2$ and Ti_5Si_3, have melting points in excess of 2000°C, and most have densities lower than those of superalloys. Research on this class of materials is likely to increase, especially with respect to improving toughness at low and intermediate temperatures.

The physical and mechanical properties of several refractory metal beryllides have also been studied. The most extensively examined include $NbBe_2$, $NbBe_7$, NiBe, and $TiBe_2$. Low toughness is a problem. There is, however, considerable interest in developing refractory beryllide fibers having a high thermal expansion coefficient for use as reinforcements of brittle intermetallics such as NiAl.

BIBLIOGRAPHY

"High Temperature Alloys" in *ECT* 2nd ed., Vol. 11, pp. 6–34, by F. J. Clauss, Lockheed Missiles & Space Co.; in *ECT* 3rd ed., Vol. 12, pp. 417–458, by N. S. Stoloff and S. R. Shatynski, Rensselaer Polytechnic Institute.

1. A. K. Vasudevan and J. J. Petrovic, *Mat. Sci. and Eng.* **155A**, 1–17 (1992).
2. R. Noebe, in S. D. Antolovich and co-workers, eds., *Superalloys '92*, TMS, Warrendale, Pa., 1992, pp. 341–350.
3. F. C. Monkman and N. J. Grant. *Proc. ASTM* **56**, 593 (1956).
4. F. R. Larson and J. Miller, *Trans. ASME* **74**, 765–771 (1952).
5. L. F. Coffin, *Trans. ASME* **76**, 931 (1954).
6. S. S. Manson, *Exp. Mech.* **5**, 193 (1965).
7. J. M. Larsen and co-workers, in S. H. Whang and co-workers, eds., *High Temperature Aluminides and Intermetallics*, TMS, Warrendale, Pa., 1990, pp. 521–556.
8. S. D. Antolovich and B. Lerch, in J. K. Tien and T. Caulfield, eds., *Superalloys, Supercomposites and Superceramics*, Academic Press, Inc., San Diego, Calif., 1989, pp. 363–411.
9. N. S. Stoloff and D. J. Duquette, *Crit. Rev. Solid State Sci.*, **4**, 615–687 (1974).
10. R. Viswanathan, in *Damage Mechanisms and Life Assessment of High Temperature Components*, ASM, Materials Park, Ohio, 1989.
11. N. F. Mott and F. R. N. Nabarro, *Report of Conference on Strength of Solids*, Physical Society, 1948, pp. 1–19.
12. R. M. N. Pelloux and N. J. Grant, *Trans. Met. Soc. AIME* **218**, 232–237 (1960).

13. R. L. Fleischer, *The Strengthening of Metals*, Reinhold Publishing Co., New York, 1964, p. 93.

14. B. E. P. Beeston, I. L. Dillamore, and R. E. Smallman, *Met. Sci. J.* **2**, 12–14 (1968).

15. C. R. Barrett and O. D. Sherby, *Trans. Met. Soc. AIME* **230**, 1322–1327 (1964).

16. J. Weertman, *Trans. AIME* **61**, 681–694 (1968).

17. O. D. Sherby and J. L. Lytton, *Trans. AIME* **206**, 928–930 (1956).

18. J. C. Fisher, *Acta Met.* **2**, 9–10 (1954).

19. E. O. Hall, *Proc. Roy. Soc. (London)* **B64**, 747–753 (1951).

20. N. J. Petch, in *Fracture, Proceedings Swampscott Conference*, John Wiley & Sons, Inc., New York, 1959, p. 54.

21. F. Garafolo, *Fundamentals of Creep and Creep-Rupture in Metals*, MacMillan, New York, 1965, p. 27.

22. J. M. Oblak and W. A. Owczarski, *Met. Trans.* **3**, 617–626 (1972).

23. E. Orowan, in *Symposium on Internal Stresses in Metals*, Institute of Metals, London, 1948, pp. 451–453.

24. K. K. Chawla, *Composite Materials Science and Engineering*, Springer-Verlag, New York, 1987.

25. J. S. Benjamin, *Met. Trans.* **1**, 2943–2951 (1970).

26. J. L. Smialek and G. H. Meier, in C. T. Sims, N. S. Stoloff, and W. C. Hagel, eds., *Superalloys II*, John Wiley & Sons, Inc., New York, 1987, pp. 293–320.

27. G. E. Wasielewski and R. A. Rapp, in C. T. Sims and W. C. Hagel, eds., *The Superalloys*, John Wiley & Sons, Inc., New York, 1972, pp. 287–316.

28. J. Stringer, *Ann. Rev. Mat. Sci.* **7**, 477 (1977).

29. G. H. Gessinger, *Powder Metallurgy of Superalloys*, Butterworths, London, 1984, p. 282.

30. G. S. Giggins and F. S. Pettit, *Pratt and Whitney Aircraft*, report no. FR-11545, East Hartford, Conn., Sept. 1978.

31. F. S. Pettit and C. S. Giggins, in Ref. 26, pp. 327–358.

32. J. H. Wood and E. H. Goldman, in Ref. 26, pp. 359–384.

33. S. Geradi, in *Metals Handbook*, Vol. 2, 10th ed., ASM, Materials Park, Ohio, 1990, pp. 565–571.

34. J. W. Freeman and H. Voorhees, *Relaxation Properties of Steels and Superstrength Alloys at Elevated Temperatures*, ASTM STP 187, ASTM, Philadelphia, Pa., 1956, p. 7; *Metals Handbook*, Vol. 1, 10th ed., ASM, Materials Park, Ohio, p. 631.

35. *Metals Handbook*, Vol. 1, 10th ed., ASM, Materials Park, Ohio, 1990, pp. 934, 976.

36. W. Smeltzer, *Trans. Canad. Inst. Met. Min.* **65**, p. 366 (1967).

37. M. Brabers and C. Birchenall, *Corrosion* **14**, 79 (1958).

38. I. A. Rohrig, R. M. Van Duzer, and C. H. Fellows, *Trans. ASME* **66**, 277 (1944).

39. C. A. Barrett, in *Proceedings of Conference on Environmental Degradation of Engineering Materials*, 1977, p. 319.

40. G. W. Shaw, *Metal Prog.* **115**, 67 (1979).

41. C. T. Sims, N. S. Stoloff, and W. C. Hagel, eds., in Ref. 26, pp. 591–593.

42. C. T. Sims, in Ref. 27, pp. 259–284.

43. J. K. Tien, G. E. Vignoul, and M. W. Kopp, *Mat. Sci. and Eng.* **A143**, 43–49 (1991).

44. D. N. Duhl, in Ref. 26, pp. 189–214.

45. G. S. Hoppin III, and W. P. Danesi, in Ref. 26, pp. 549–561.

46. M. Gell and co-workers, *J. Metals* **39**(7), 11–15 (1987).

47. G. K. Bouse and J. R. Mihalisin, in Ref. 8, pp. 99–148.

48. J. K. Tien and E. Jacobs, in Ref. 8, pp. 285–300.

49. C. T. Sims, N. S. Stoloff, and W. C. Hagel, eds., in Ref. 26, pp. 594–578.

50. A. M. Beltran, *Cobalt* **46**, 3 (1970).

51. A. Davin, D. Coutsouradis, and L. Habraken, *Cobalt* **35**, 67 (1967).

52. J. B. Lambert and J. J. Rausch, in Ref. 33, pp. 557–565.
53. R. W. Buckman, Jr., in J. L. Walter, M. R. Jackson, and C. T. Sims, eds., *Alloying*, ASM, Materials Park, Ohio, 1988, pp. 419–445.
54. ASTM B391, ASTM B364, ASTM B387, and ASTM B410, American Society for Testing and Materials, Philadelphia, Pa., 1993.
55. B. A. Wilcox, in I. Machlin, R. T. Begley, and E. D. Weisert, eds., *Refractory Metal Alloys*, Plenum Press, New York, 1968, pp. 1–34.
56. J. Wadsworth, *Int. Mat. Rev.* **33**, 1988, pp. 131–150.
57. R. W. Burman, *J. Metals* **29**(12), 9–14 (1977).
58. W. D. Klopp, in Ref. 27, pp. 175–196.
59. E. Loria, *J. Metals* **39**(8), 22–26 (1987).
60. R. W. Buckman, Jr., in *Proceedings MRS Symposium on High Temperature Silicides and Refractory Alloys*, Pittsburgh, Pa., Nov.–Dec. 1993, pp. 329–339.
61. W. A. Johnson, in Ref. 35, pp. 577–588.
62. M. Rabinovitch and J. M. Hauser, in J. L. Walter and co-workers, eds., *Proceedings of the Conference on In-Situ Composites III*, Ginn Custom Publishers, Lexington, Mass., 1979, pp. 246–247.
63. C. A. Bruch and co-workers, in Ref. 62, pp. 258–267.
64. E. R. Thompson and F. D. Lemkey, *Directionally Solidified Eutectic Superalloys*, report M110544-2, United Aircraft Research Lab., East Hartford, Conn., Apr. 1973.
65. N. S. Stoloff, in Ref. 62, pp. 357–377.
66. D. W. Petrasek and co-workers, in Ref. 8, pp. 625–670.
67. D. M. Dimiduk, M. G. Mendiretta, and P. R. Subramanian, in R. Darolia and co-workers, eds., *Structural Intermetallics*, TMS, Warrendale, Pa., 1993, pp. 614–630.
68. D. L. Anton and D. M. Shah, in C. T. Liu and co-workers, eds., *MRS Symp. Proc.* **133**, 361–371 (1989).
69. C. T. Liu and J. O. Stiegler, in Ref. 33, pp. 913–942.
70. C. T. Liu, in C. T. Liu, R. W. Cahn, and G. Sauthoff, eds., *Ordered Intermetallics-Physical Metallurgy and Mechanical Behavior*, Kluwer Academic Publishers, Dordrecht, Netherlands, 1992, pp. 321–333.
71. Y.-W. Kim and F. H. Froes, in Ref. 7, pp. 465–492.
72. F. H. Froes, C. Suryanarayana, and D. Eliezer, *J. Mater. Sci.* **27**, 5113 (1992).
73. Y.-W. Kim, *J. Metals* **41**(7), 24–30 (1989).
74. K. Aoki and O. Izumi, *J. Japan Inst. Met.* **43**, 1190 (1979).
75. E. P. George, C. T. Liu, and D. P. Pope, *Scripta Metall.* **30**, 37–42 (1994).
76. C. T. Liu and V. K. Sikka, *J. of Metals* **38**(5), 19 (1986).
77. R. Darolia, *J. Metals* **43**(3), 44–49 (1991).
78. D. E. Alman and N. S. Stoloff, *Int. J. Powder Met.* **27**, 22–41 (1991).
79. V. K. Sikka, in Ref. 67, pp. 483–491.
80. J. Cook, R. Mahapatra, E. W. Lee, and A. Khan, reprinted from Ref. 1, p. 13.
81. K. Sadananda and C. R. Zeng, *Mat. Sci. and Eng.* **A170**, 199–213 (1993).

NORMAN S. STOLOFF
Rensselaer Polytechnic Institute

HIGH TEMPERATURE COMPOSITES.   See COMPOSITE MATERIALS.

HISTAMINE AND HISTAMINE ANTAGONISTS

The history of histamine [51-45-6], $C_5H_9N_3$ (**1**), and the development of antihistamines have been reviewed (1,2). Histamine was the first to be characterized of a series of biogenic amines that are released in the inflammatory process (Fig. 1). As early as 1910, it was shown that histamine caused constriction of isolated guinea pig ileum and, subsequently, it was found that histamine induced a shock-like syndrome. In 1927 the presence of histamine in normal tissues was demonstrated. Attempts to reduce histamine manifestations led to the report, in 1933, that certain phenolic ethers inhibited histamine action (3). Toxicity precluded clinical use. In 1942 phenbenzamine [961-71-7] (Antergan), $C_{17}H_{22}N_2$, was the first antihistamine to be successfully used in humans (4).

In 1966, the name H_1 was proposed (5) for receptors blocked by the at that time known antihistamines. It was also speculated that the other actions of histamine were likely to be mediated by other histamine receptors. The existence of the H_2 receptor was accepted in 1972 (6) and the H_3 receptor was recognized in rat brain in 1983 (7). H_3 receptors in the brain appear to be involved in the

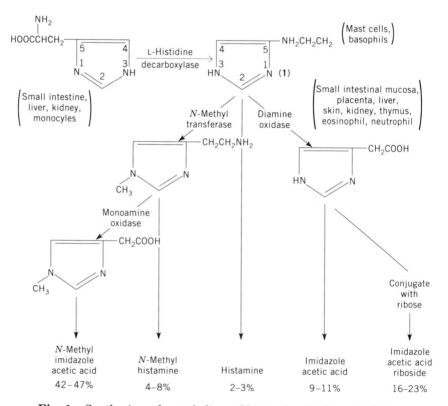

Fig. 1. Synthesis and metabolism of histamine (**1**) from histidine.

feedback control of both histamine synthesis and release, whereas release of various other neurotransmitters, eg, serotinin (5-HT), dopamine, noradrenaline, and acetylcholine, is also modulated (8) (see NEUROREGULATORS). H_3 receptor effects have also been demonstrated in various peripheral tissues and as of this writing H_3 agonists and antagonists are undergoing intensive study for therapeutic applications.

Histamine Synthesis, Metabolism, and Distribution

The synthesis and disposition of histamine is well described both in allergy textbooks (9,10) and in review articles (11).

Synthesis. Histamine [51-45-6], 2-(4-imidazolyl)ethylamine (1) is formed by decarboxylation of histidine by the enzyme L-histidine decarboxylase (Fig. 1). Most histamine is stored preformed in cytoplasmic granules of mast cells and basophils. In humans mast cells are found in the loose connective tissue of all organs, especially around blood and lymphatic vessels and nerves. These cells are most abundant in the organs expressing allergic diseases: the skin, respiratory tract, and gastrointestinal tract.

Histamine Release. Antigen cross-linked immunoglobulin E (IgE) is the classic mast cell secretagogue in allergic diseases, but other physiologic mediators may also be involved, eg, opioids (qv) or neuropeptides. Basophil degranulation is caused by histamine-releasing factors, which are produced by most inflammatory cells, eg, neutrophils, platelets, and eosinophils. Several of the cytokines also cause basophil degranulation. Mast cells may also form a functional link between sensory nerve terminals and blood vessels, as these seem to be in close proximity to sensory nerve terminals. In addition, several sensory neuropeptides have histamine-releasing activity, suggesting a possible role of histamine in sensory nervous system functions. Thus histamine could play a special role in reinforcing the inflammatory response (see IMMUNOTHERAPEUTIC AGENTS).

Histamine in the Blood. After its release, histamine diffuses rapidly into the blood stream and surrounding tissues (12). Histamine appears in blood within 2.5 min after its release, peaks at 5 min, and returns to baseline levels by 15 to 30 min. In humans, the diurnal mean of plasma histamine levels is 0.13 ng/g. In urine, elevations of histamine or metabolites are more prolonged than plasma elevations. Consequently, abnormalities are more easily detected by urinary histamine assay. About one-half of the histamine in normal blood is in basophils, one-third in eosinophils, and one-seventh in neutrophils; the remainder is distributed among all the other blood components. Increases in blood histamine levels occur in several pathological conditions, but provide no positive evidence for the involvement of histamine in the pathogenesis of various diseases (13). In asthmatic patients blood histamine levels appear to be higher during asthmatic attacks but are close to normal during intermissions (see ANTIASTHMATIC AGENTS).

Metabolism. Metabolism of histamine occurs via two principal enzymatic pathways (Fig. 1). Most (50 to 70%) histamine is metabolized to N-methylhistamine by N-methyltransferase, and some is metabolized further by monoamine oxidase to N-methylimidazoleacetic acid and excreted in the urine.

The remaining 30 to 40% of histamine is metabolized to imidazoleacetic acid by diamine oxidase, also called histaminase. Only 2 to 3% of histamine is excreted unchanged in the urine.

Histamine in the Brain and the Histamine Neurone System.   There is evidence that histamine functions as a neurotransmitter or a neuromodulator in the brain (14). In the mammalian brain, several thousand cell bodies are located in the posterior basal hypothalamus, comprising a functional unit, the tubero-mamillary nucleus. The latter sends nerve fibers widely and unevenly to various regions from the olfactory bulb to the spinal cord. In the brain, histamine is related to functions such as the regulation of neuroendocrine and cardiovascular systems, thermoregulation, the circadian rhythm of sleep–wakefulness, behavior, vestibular function, cerebral vascular regulation, and antinociception and analgesia.

Histamine in the Cardiovascular System.   It has been known for many years that histamine is present in sympathetic nerves and has a distribution within the heart that parallels that of norepinephrine (see EPINEPHRINE AND NOREPINEPHRINE). A physiological role for cardiac histamine as a modulator of sympathetic responses is highly plausible (15). A pool of histamine in rat heart located neither in mast cells nor in sympathetic nerves has been demonstrated. The turnover of this metabolically active pool of histamine appears to be maintained by normal sympathetic activity.

Histamine Receptors

The actions of histamine are mediated through at least three distinct receptors defined pharmacologically by the actions of the respective agonists and antagonists. Reviews have been published (16,17).

The H_1 Receptor and its Ligands.   The H_1 receptor mediates most of the important histamine effects in allergic diseases (11). These include smooth muscle contraction, increased vascular permeability, pruritus, prostaglandin generation, decreased atrioventricular node conduction time with resultant tachycardia, activation of vagal reflexes, and increased cyclic guanosine monophosphate (cGMP) production. The mean density of cerebral H_1 receptors is about 100 fmol/mg membrane protein, in the same range as that of receptors of other neurotransmitters. In the human brain the highest density is found in the neocortex and various limbic structures.

H_1 Receptor Agonists.   In the histamine molecule there are two principal structural elements: an imidazole moiety and an ethylamine side chain (16). Only the N^π-position is absolutely necessary for H_1 agonism. The imidazole ring can be replaced, eg, 2-pyridylethylamine, 2-thiazolylethylamine, or substituted at the 2-position. 2-Methylhistamine is often used as a selective H_1 agonist; however, larger substituents are not allowed unless a phenyl ring is used. 2-Phenylhistamine analogues appear to be very selective H_1 receptor agonists. In this series of compounds both full and partial agonists and even H_1 antagonists are encountered, representing useful tools for studies of pharmacology and structure–activity relationships. Methylation of the amine group is allowed, but this change does not lead to selective compounds.

H₁ Receptor Antagonists.   The classical H_1 receptor antagonists are reversible, competitive, dose-dependent inhibitors of the action of histamine on H_1 receptors. Histamine H_1 antagonists are usually divided into two classes: the first-generation or classical H_1 antagonists and the second-generation H_1 antagonists. The main distinction between the first- and second-generation drugs is the absence of sedative and anticholinergic side effects in the latter.

Classical H₁ Receptor Antagonists.   Since the initial discovery of an antihistaminic compound in 1933, a very high number of potent compounds were synthesized. Several reviews are available (16,18–20).

The classical histamine H_1 receptor antagonists are structurally very similar, all being substituted ethylamines (**2**). Table 1 presents a member of each structural subclass. Presumably, the ethylamine core is needed to accomplish nonfruitful binding to the high affinity conformation of the receptor. The other substituents modify the physical chemistry of the compounds. A disubstituted terminal amino group, often a dimethylamino group, is connected to an atom X via a short carbon chain. The chain can be saturated, unsaturated, branched, or part of a ring system, whereas X can be an oxygen, nitrogen, or carbon atom or a polyatomic group; X links the side chain to the aromatic region. This aromatic part generally contains two aromatic rings, eg, phenyl, thiophene, 2-pyridyl, which may eventually be fused to form tricyclic derivatives. Based on the type of X-linkage and the type of aromatic region, the classical H_1 receptor antagonists can be subdivided into six classes: aminoalkylethers, ethylenediamines, alkylamines, piperazines, phenothiaxines, and piperidines.

The prototype aminoalkylether or ethanolamine is diphenhydramine (Benadryl). Clemastine [*15686-51-8*], $C_{21}H_{26}ClNO$, differs from it in having one phenyl ring chlorinated, a methylated benzhydryl carbon, and a three-carbon amino side chain, partly incorporated into a pyrrolidinyl ring system. Setastine [*64294-95-7*], $C_{22}H_{28}ClNO$, which has been launched in Hungary, differs from clemastine only in the type of heterocyclic ring.

The prototype of a pure ethylenediamine is tripelennamine; antazoline [*91-75-8*], $C_{17}H_{19}N_3$, belongs to the same family of compounds. Several well-known alkylamines in addition to chlorpheniramine are known. Dexchlorpheniramine maleate [*2438-32-6*] and triprolidine monohydrochloride monohydrate [*6138-79-0*], an alkenyl derivative, are two examples.

Replacement of the oxygen bridge atom of diphenhydramine by nitrogen and incorporation of the two nitrogens into a piperazine ring leads to cyclizines. Hydroxyzine is a well-known piperazine derivative. Cinnarizine [*298-57-7*], $C_{26}H_{28}N_2$, and oxatomide [*60607-34-3*], $C_{27}H_{30}N_4O$, both structurally related to cyclizine [*82-92-8*], differ only by having respectively a cinnamyl or a propylbenzimidazolone *N*-substituent instead of a methyl group. Oxatomide also has mast cell stabilizing actions.

There are two types of tricyclic H_1 antagonists: the phenothiazine and piperidine derivatives. Promethazine is the prototype molecule of the phenothiazine derivatives. Mequitazine [*29216-28-2*], $C_{20}H_{22}N_2S$, a quinuclidinylmethyl derivative, is less sedative than promethazine. Cyproheptadine and pizotifen [*5189-11-7*], $C_{19}H_{21}NS \cdot C_4H_6O_5$, are piperidine derivatives. Though antihistamines, they are mainly used because of their antiserotonergic properties. Azatadine maleate [*3978-86-7*], $C_{20}H_{22}N_2$, is a pyridyl variant of

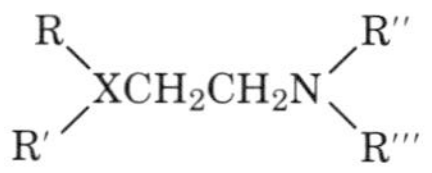

Table 1. Classical H₁ Antagonists

$$\underset{R'}{\overset{R}{\diagdown}}XCH_2CH_2N\underset{R'''}{\overset{R''}{\diagdown}}$$

(2)

Compound	CAS Registry Number	Molecular formula	Structure
		Aminoalkylether	
diphenhydramine	[58-73-1]	$C_{17}H_{21}NO$	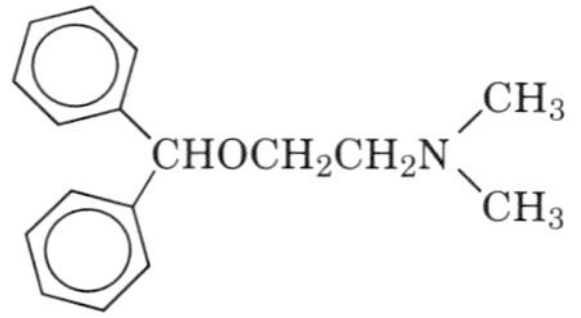
		Ethylenediamine	
tripelennamine	[91-81-6]	$C_{16}H_{18}N_3$	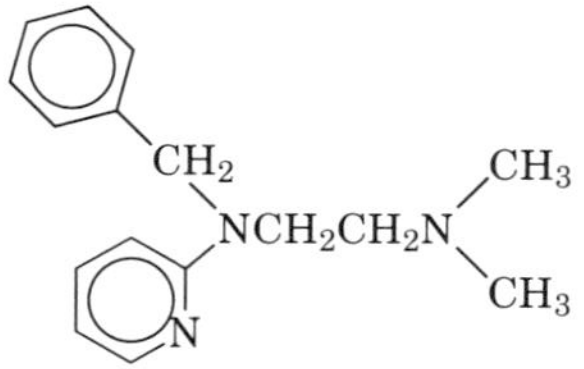
		Alkylamine	
chlorpheniramine	[132-22-9]	$C_{16}H_{19}ClN_2$	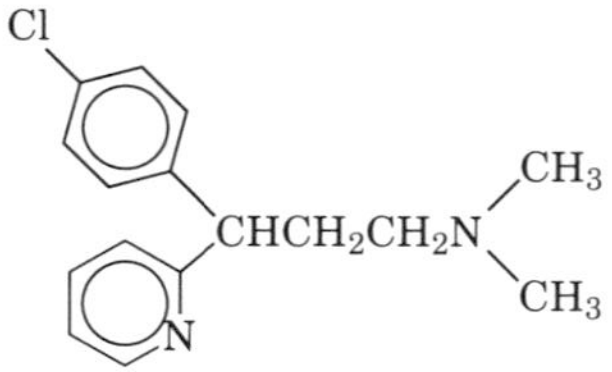

Piperazine

hydroxyzine [68-88-2] $C_{21}H_{27}ClN_2O_2$

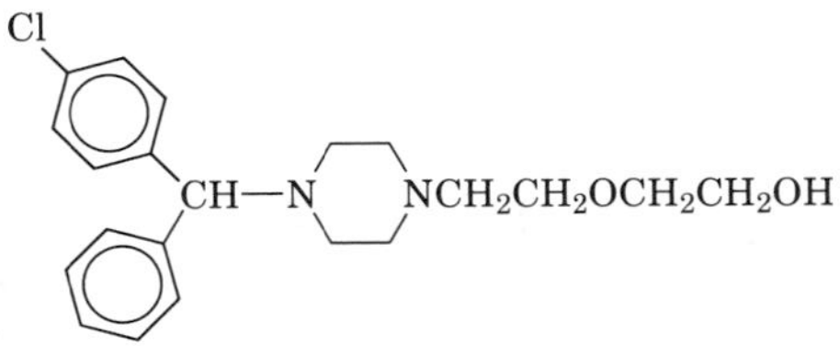

NCH$_2$CH$_2$OCH$_2$CH$_2$OH

Phenothiazine

promethazine [60-87-7] $C_{17}H_{20}N_2S$

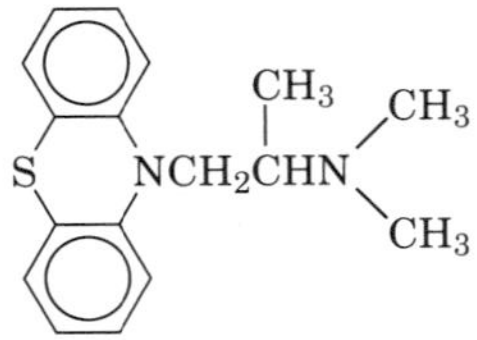

Piperidine

cyproheptadine [129-03-3] $C_{21}H_{21}N$

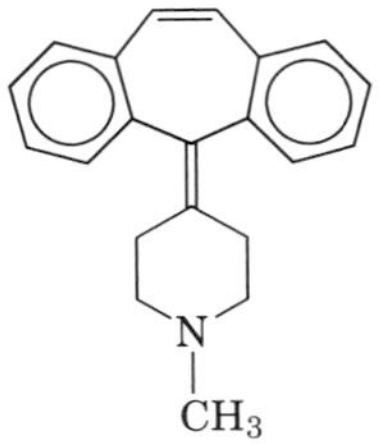

295

cyproheptadine. The compound mianserin hydrochloride [21535-47-7], $C_{18}H_{21}ClN_2$, may structurally be regarded as a double-bridged analogue of phenbenzamine. Ketotifen [34580-14-8], $C_{19}H_{19}NOS$, is also a piperidine derivative; like oxatomide, it has mast cell stabilizing actions besides its antihistamine properties.

Second-Generation H_1 Receptor Antagonists.   Because of undesirable side effects caused by classical H_1 receptor antagonists, drugs having enhanced clinical effectiveness and a reduced side-effect profile have been sought. The main progress has been in the development of antihistamines that, unlike the classical H_1 antagonists, do not cause sedation. Thus the term second-generation antihistamines usually refers to those nonsedating antihistamines. The reduced ability to penetrate the central nervous system (CNS) can generally be explained by physicochemical properties. Several reviews address these new antihistamines (16,21,22). Table 2 shows some of the second-generation H_1 antagonists.

Several antihistamines have been derived from classical structures that do not penetrate into the brain. Most of these compounds are modifications of the classical H_1 receptor antagonists. Terfenadine (**9**) was studied as part of a butyrophenone program for new antipsychotics (see PSYCHOPHARMACOLOGICAL AGENTS). Whereas it was ineffective in the models used because it did not cross the blood brain barrier readily, terfenadine was then found to be an effective H_1 receptor antagonist having little or no central activity. Oxidative metabolization of hydroxyzine afforded the active metabolite cetirizine (**4**). Owing to the presence of a carboxy group, cetirizine has a reduced access to the CNS as compared to the parent compound. An analogue of cetirizine, pibaxizine (**8**) is being developed.

Modification of the classical H_1 antagonist azatadine also resulted in a potent nonsedating antihistamine, loratadine (**7**). *Ex vivo* binding studies in animals demonstrated a selectivity for occupancy of lung H_1 receptors compared with cerebral cortex H_1 receptors, which might partially explain the observed lack of significant CNS activity. Acrivastine (**3**) is closely related to the potent alkylamine H_1 receptor antagonist triprolidine.

Ebastine (**5**) is a more recent H_1 antagonist which structurally may be considered a hybrid between diphenylpyraline [147-20-6], $C_{19}H_{23}NO$ (**7**), a classical antihistamine, and terfenadine (**9**). Epinastine (**5**), structurally related to the antidepressant and sedative H_1 receptor antagonist mianserin [24219-97-4], $C_{18}H_{20}N_2$, has reduced CNS penetration probably because of its different physicochemical properties.

Some of the second-generation H_1 antagonists have nonclassical structures. The phthalazinone derivative azelastine (**11**)combines potent H_1 receptor antagonism with moderate to weak serotonergic, leukotriene C_4 and D_4, and platelet activating factor (PAF) receptor antagonism. Besides an oral formulation, a nasal spray has also been developed. A somewhat unusual side effect of azelastine is taste alteration. Astemizole (**10**), a benzimidazole derivative, is a potent H_1 receptor antagonist having moderate serotonin- and α1-adrenergic blocking activity. The most striking property is its extremely slow rate of dissociation from H_1 receptors *in vitro*. A unique feature of astemizole is the presence of a guanidino function, although incorporated in a benzimidazole ring system. This property may contribute to the compound's failure to penetrate the CNS. Other benzimidazole derivatives being developed include emedastine (**12**) and mizolastine (**14**).

Table 2. Second-Generation H$_1$ Antagonists

Compound	CAS Registry Number	Molecular formula	Structure	Structure number
			Classical structures	
acrivastine	[87848-99-5]	C$_{22}$H$_{24}$N$_2$O$_2$	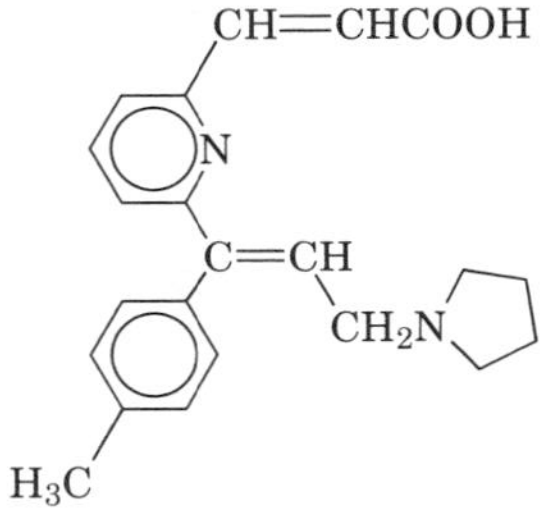	(3)
cetirizine	[83881-51-0]	C$_{21}$H$_{25}$ClN$_2$O$_3$	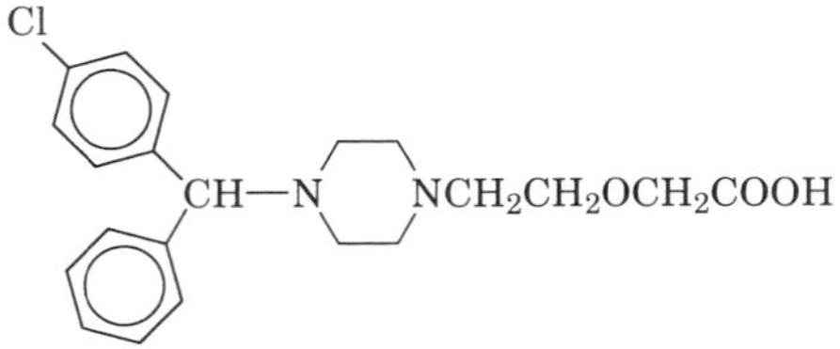	(4)
ebastine	[90729-43-4]	C$_{32}$H$_{39}$NO$_2$	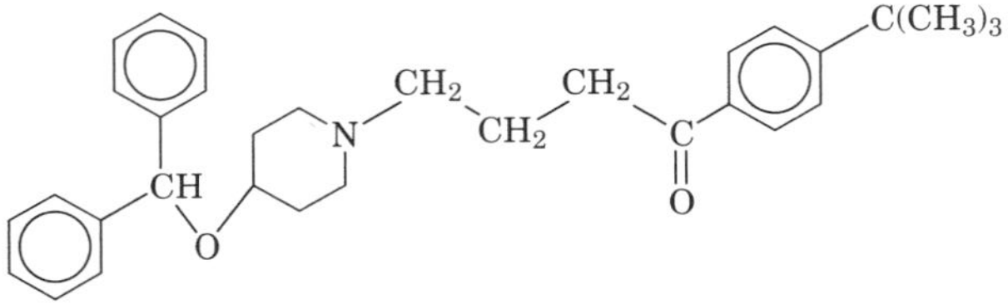	(5)
epinastine	[80012-43-7]	C$_{16}$H$_{15}$N$_3$	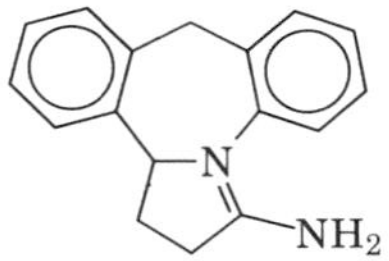	(6)

297

Table 2. (*Continued*)

Compound	CAS Registry Number	Molecular formula	Structure	Structure number
loratadine	[79794-75-5]	$C_{22}H_{23}N_2O_2Cl$		(**7**)
pibaxizine	[82227-39-2]	$C_{24}H_{29}NO_4$		(**8**)
terfenadine	[50679-08-8]	$C_{32}H_{41}NO_2$		(**9**)

Nonclassical structures

astemizole   [68844-77-9] $C_{28}H_{31}FN_4O$ (10)

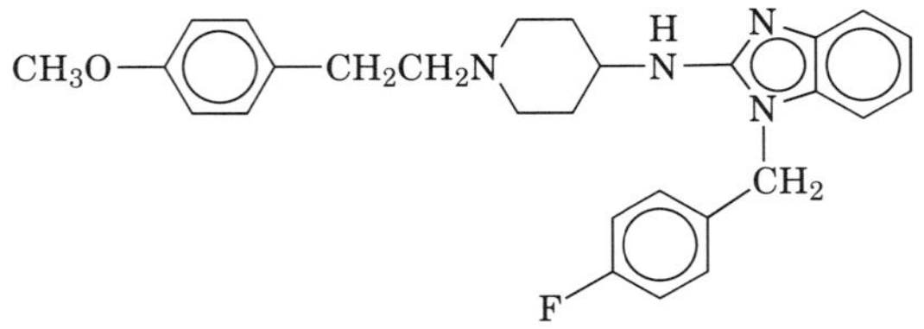

azelastine   [58581-89-8] $C_{22}H_{24}ClN_3O$ (11)

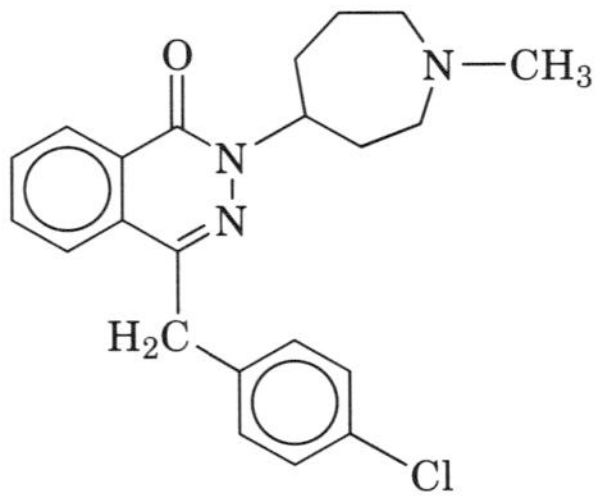

emedastine   [87233-61-2] $C_{17}H_{26}N_4O$ (12)

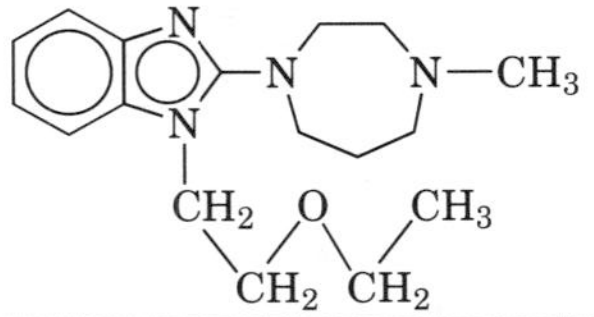

299

Table 2. (*Continued*)

Compound	CAS Registry Number	Molecular formula	Structure	Structure number
levocabastine	[79516-68-0]	$C_{26}H_{29}FN_2O_2$	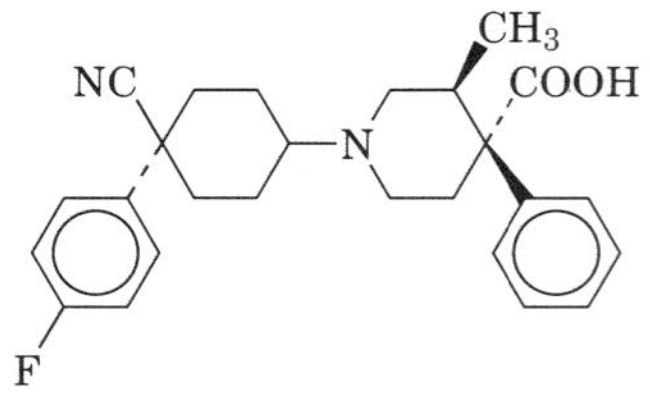	(**13**)
mizolastine	[108612-45-9]	$C_{24}H_{25}FN_6O$	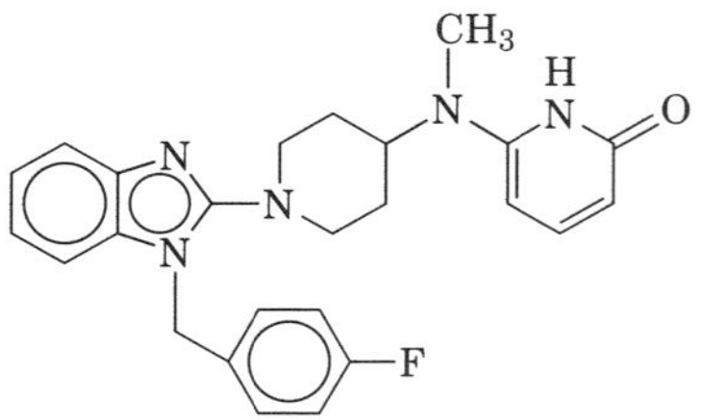	(**14**)

300

Astemizole (**10**) has further been modified into a series of 4-phenylcyclo-hexylamine compounds, resulting in the synthesis of cabastine, for example. Cabastine is a highly active compound and its geometric isomers are also active, demonstrating the stereoselectivity of histamine H_1 receptors toward chiral ligands. The $3S$, $4R$-levo antipode of cabastine was the most active, and therefore this isomer, levocabastine (**13**), has been chosen for further development. Because of high potency, levocabastine has been developed for topical application such as eye drops and nasal spray.

Compounds have also been described (23) in which H_1 receptor antagonism is combined with histamine H_2 receptor agonism. These are primarily derivatives of the potent H_2 receptor agonist impromidine. Also, compounds having other dual activity are being developed, eg, Schering 37370, an antagonist of PAF and H_1 receptors and AL-3264 (Dainippon), a 5-lipoxygenase inhibitor with H_1 blocking activity.

The H_2 Receptor and its Ligands. The discovery of H_2 receptors and antagonists occurred in 1972. This topic, including therapeutic implications, is considered in more detail elsewhere (see GASTROINTESTINAL AGENTS).

H_2 Receptors. The H_2 receptor mediates effects, through an increase in cyclic adenosine monophosphate (cAMP), such as gastric acid secretion; relaxation of airway smooth muscle and of pulmonary vessels; increased lower airway mucus secretion; esophageal contraction; inhibition of basophil, but not mast cell histamine release; inhibition of neutrophil activation; and induction of suppressor T cells. There is no evidence that the H_2 receptor causes significant modulation of lung function in the healthy human subject or in the asthmatic, except for some weak relaxing effect (24). In the brain, the H_2 receptors are distributed widely and mainly in association with neurones.

Combined H_1/H_2 receptor stimulation by histamine is responsible for vasodilation-related symptoms, such as hypotension, flushing, and headache, as well as for tachycardia stimulated indirectly through vasodilation and catecholamine secretion.

Histamine H_2 Receptor Agonists. Extended reviews on histamine H_2-active compounds have been published (16,25). Structural requirements of histamine as an H_2 agonist are considered to be the protonated side-chain nitrogen atom and the ability of the imidazole amidine system to undergo a tautomeric shift. 4-Methylhistamine is often used as a selective H_2 agonist. Larger substituents are not allowed. Methylation of the amine group is allowed but leads to nonselective analogues.

H_2 agonists are divided in three chemical classes, ie, analogues of histamine (**1**), dimaprit (**16**), and impromidine (**17**) (Table 3). In histamine analogues, substitution of the imidazole ring results in many cases in a sharp decrease in H_2 agonistic activity. Dimaprit, a highly selective H_2 agonist, is a simple nonaromatic compound. The optimal chain length in dimaprit for H_2 agonistic activity corresponds to three CH_2 groups, which is contrary to histamine having two CH_2 groups as optimal chain length. Comparison of the structures of histamine and dimaprit have also led to the development of a new class of H_2 agonists (26,27). The thiazole amthamine (**15**) appears to be a selective and rather potent H_2 agonist. It is remarkable in that in this compound the tautomeric shift, suggested to be essential for H_2 receptor activation, is impossible. Impromidine, a

Table 3. H₂ Agonists and Antagonists

Compound	CAS Registry Number	Molecular formula	Structure	Structure number
H₂ agonists				
amthamine	[142437-67-0]	$C_6H_{11}N_3S$	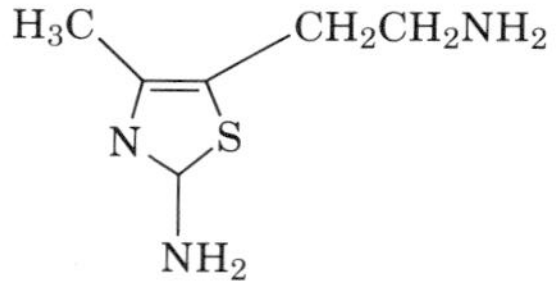 H_3C, $CH_2CH_2NH_2$ thiazole ring with NH_2	(15)
dimaprit	[65119-89-3]	$C_{16}H_{15}N_3S$	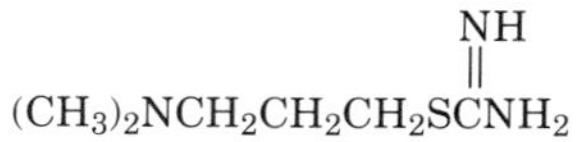 $(CH_3)_2NCH_2CH_2CH_2SCNH_2$, $=NH$	(16)
impromidine	[55273-05-7]	$C_{14}H_{23}N_7S$	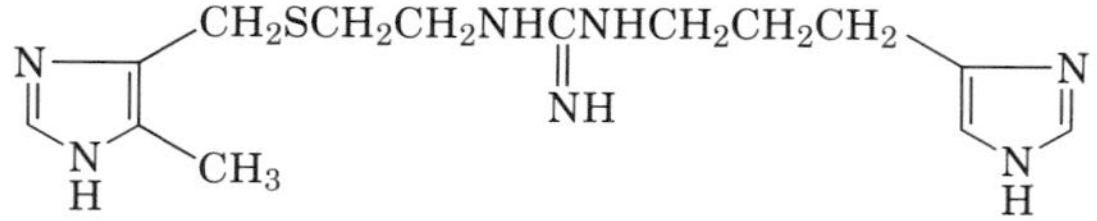 $CH_2SCH_2CH_2NHCNHCH_2CH_2CH_2$, $=NH$, imidazole rings with CH_3	(17)
H₂ antagonists				
cimetidine	[51484-61-9]	$C_{10}H_{16}N_6S$	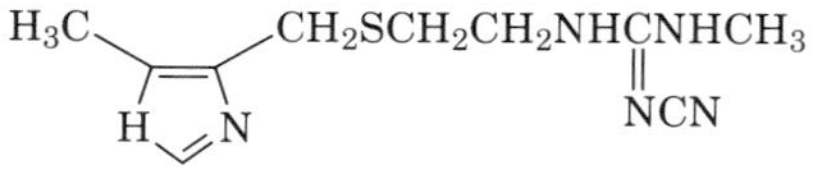 $CH_2SCH_2CH_2NHCNHCH_3$, $=NCN$, H_3C imidazole	(18)
famotidine	[76824-35-6]	$C_8H_{15}N_7O_2S_3$	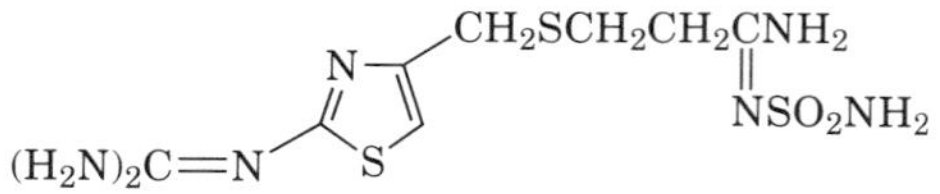 $CH_2SCH_2CH_2CNH_2$, $=NSO_2NH_2$, thiazole, $(H_2N)_2C=N$	(19)
ranitidine	[66357-35-5]	$C_{13}H_{22}N_4O_3S$	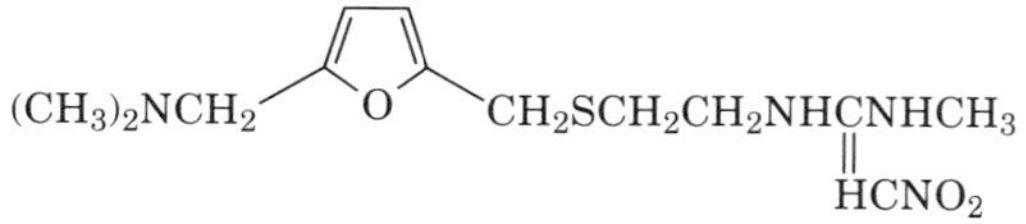 $(CH_3)_2NCH_2$ furan $CH_2SCH_2CH_2NHCNHCH_3$, $=HCNO_2$	(20)

highly potent H_2 agonist, is a substituted guanidine. It is much more potent than histamine, which appears to be the result of an increased affinity of the compound for H_2 receptors rather than an increased efficacy. Other studies have shown that impromidine is an antagonist of both H_1 and H_3 receptors.

From a therapeutic point of view, selective H_2 agonists may become useful in the treatment of heart failure and catecholamine-insensitive cardiomyopathy, but only if compounds become available that do not stimulate gastric acid secretion or cause other unforeseen problems.

Histamine H_2 Receptor Antagonists. In 1972 a new class of histamine antagonists was described that was capable of antagonizing histamine-induced gastric acid secretion (6). The H_2 antagonists are divided into five structural classes, some of which are shown in Table 3. A more complete review can be found in Reference 25.

The H_2 antagonists are very potent compounds. In the early 1990s research for new drugs has moved away from potency, because profound inhibition of gastric acid secretion has been found to result in severe gastric damage, owing to the induction of elevation of gastrin levels. Since their discovery, the H_2 antagonists have quickly earned their place in the treatment of peptic ulcer disease and esophageal reflux (28,29).

The design of H_2 antagonists has been based principally on the structure of histamine (**1**). Burimamide [*34970-69-9*], $C_9H_{16}N_4S$, the first selective H_2 receptor antagonist, was poorly absorbed after oral administration, had insufficient potency, and was rather toxic. Metiamide [*34839-70-8*] was more potent but caused bone marrow suppression, probably because of the presence of a thiourea group. Cimetidine (**18**), a nonthiourea congener of metiamide, was first marketed in the United Kingdom in 1976 and in the United States in 1982. Ranitidine (**20**), in which the imidazole ring of cimetidine was replaced by a dimethylaminomethylfuryl ring system, became available in 1983. Famotidine (**19**) was introduced in 1986.

For control of gastric acid secretion, the H_2 antagonists have encountered competition from the potent H^+K^+-ATPase inhibitors such as omeprazole [*73590-58-6*], $C_{17}H_{19}N_3O_3$ (29), which reduce acid secretion better than the H_2 antagonists.

The H_3 Receptor and its Ligands. The first evidence for the existence of a third histamine receptor subtype was published in 1983 (7) and great advances have been made since then. The moderately active H_2 antagonist burimamide proved to be a rather potent H_3 receptor antagonist, whereas the H_2 receptor agonist impromidine (**17**) (see Table 3) is also an active H_3 antagonist (7) (Table 4). Soon after, R-(α)methylhistamine (**23**) and thioperamide (**27**) were described as highly potent and selective ligands for this new receptor subtype (30). These compounds are still valuable tools for receptor identification as H_3 receptor agonist and antagonist, respectively. Structural requirements for ligands of the H_3 receptor are very different from those for H_1 and H_2 receptors (8,31).

H_3 Receptors. Originally described as an autoreceptor in rat brain (7), the H_3 receptor has since been reported to modulate the release of a variety of neurotransmitters and can be regarded as a general regulatory mechanism (8,31). It is responsible for presynaptic inhibition of neurotransmitter release in CNS, and inhibition of cholinergic transmission and microvascular leakage in

Table 4. H₃ Agonists and Antagonists

Compound	CAS Registry Number	Molecular formula	Structure	Structure number
			H₃ agonists	
imetit	[102203-18-9]	$C_6H_{10}N_4S$	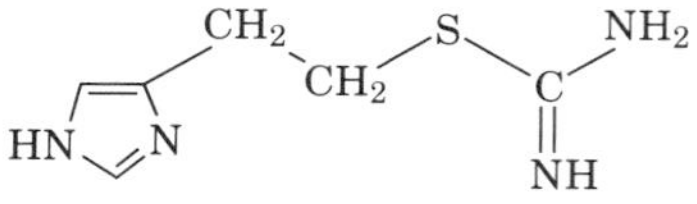	(**21**)
immepip	[151070-83-6]	$C_9H_{15}N_3$	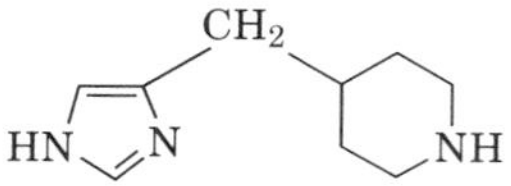	(**22**)
R-(α)-methylhistamine	[6986-90-9]	$C_6H_{11}N_3$	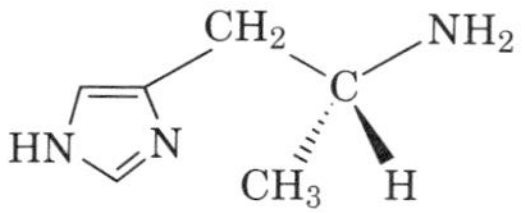	(**23**)
R-(α), *S*-(β)-methyl-histamine	[127607-86-7]	$C_7H_{13}N_3$	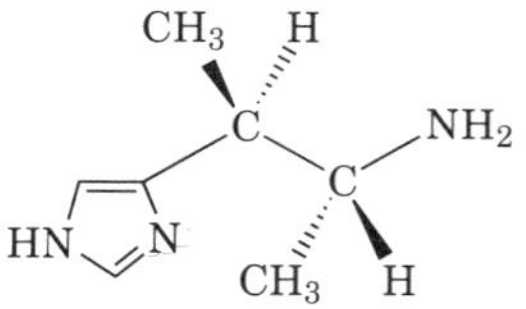	(**24**)
			H₃ antagonists	
betahistine	[5638-76-6]	$C_8H_{12}N_2$	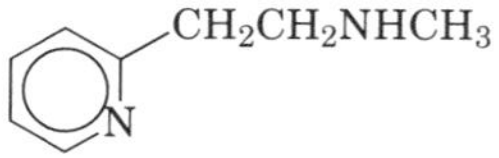	(**25**)

304

clobenpropit [145231-45-4] C$_{14}$H$_{17}$N$_4$SCl (26)

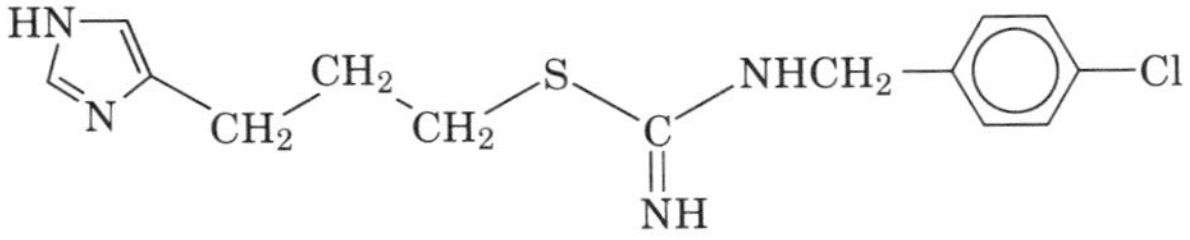

thioperamide [106243-16-7] C$_{15}$H$_{24}$N$_4$S (27)

305

airways, as well as inhibition of gastric acid secretion. Moreover, the localization of this receptor subtype is not restricted to brain areas. Also, in several peripheral tissues, eg, airways, gastrointestinal (GI) tract, and heart, H_3 receptors seem to be present and exert important (patho)physiological actions (8).

Histamine H_3 Receptor Agonists.   In contrast to the development of selective agonists for the H_1 and H_2 receptor, potent agonists for the H_3 receptor can be obtained by simple modification of the histamine molecule. Histamine itself is already a rather potent agonist of the H_3 receptor, although it is of course not very selective. Its high affinity for the H_3 receptor in comparison with its affinity for the H_1 and H_2 receptor is striking. Modification of the histamine molecule is usually not well tolerated. Substitution or replacement of the imidazole moiety is not beneficial for H_3 receptor activity (8,31). As for the H_2 receptor, both nitrogen atoms of the imidazole ring appear to be essential for agonistic activity (see Table 4).

Methylation of the side chain at the α-position leads to the potent and selective H_3 agonist R-(α)-methylhistamine (30). Yet branching the side chain at the β-position does not affect the H_3 activity, whereas α-substitution using larger substituents or two methyl groups leads to a dramatic loss of activity (8,32). Methylation at both the α- and the β-position is also allowed and results in the potent H_3 agonist R-(α),S-(β)-methylhistamine (**24**) (8,32). Methylation of the amine function is acceptable, although no selective agents are obtained because both N^α-methylhistamine and N^α,N^α-dimethylhistamine are also effective H_1 and H_2 agonists (8). Substitution of the amine function using larger substituents is not acceptable for potent H_3 receptor agonism.

The amine function of histamine can be replaced by other basic groups. Whereas replacement with a guanidino group results in a partial agonist, an isothioureum group appears to be suitable for agonism. The isothioureum analogue of histamine, imetit (**21**), is one of the most potent H_3 receptor agonists known (8). Moreover, it is highly selective for the H_3 receptor and is active both *in vitro* and *in vivo*. Incorporating the amine function of histamine in a piperidine ring results in immepip (**22**), another potent and selective H_3 receptor agonist. Also, this compound is highly selective for the H_3 receptor (33).

H_3 Receptor Antagonists.   Already in 1983 it was noticed that several H_2 receptor ligands, both agonists, eg, impromidine (**17**) and dimaprit (**16**), and antagonists, eg, burimamide, were relatively potent H_3 antagonists. The H_3 antagonistic properties are, however, unrelated to the effects at the H_2 receptor. Burimamide is approximately 100-fold more active than tiotidine [69014-14-8] (base) at the H_3 receptor, whereas the reverse is observed for the H_2 receptor. H_1 antagonists do not show any appreciable effect at the H_3 receptor. In 1987 thioperamide (**27**) was introduced as the first selective and potent H_3 antagonist (30). Moreover, also for the therapeutically used betahistine (**25**), a moderate H_3 antagonism was reported. As of this writing, there are no clear ideas about the mechanism of action of betahistine, but these observations have led to the suggestion that perhaps the H_3 antagonistic effects could be involved (34). Within a series of imetit derivatives it was observed that methylation of the isothioureum moiety resulted in antagonistic properties. N-Substitution of the isothioureum group with aromatic groups further increased the antagonistic activity. Finally, lengthening of the side chain resulted in very potent H_3 receptor antagonists

(8). Clobenpropit (**26**), which has a chlorobenzyl substituent at the isothioureum group, is approximately 10-fold more potent than thioperamide and as of this writing is the most potent H_3 antagonist known.

Uses of Histamine Receptor Ligands

H_1 Antihistamine Treatment in Allergic Diseases. H_1-receptor antagonists are used for the symptomatic treatment of several allergic diseases where histamine release from mast cells is induced via immunological or nonimmunological mechanisms (9–11) (see also IMMUNOTHERAPEUTIC AGENTS).

Allergic Seasonal or Perennial Rhinoconjunctivitis. Histamine can cause all pathologic features of allergic rhinitis (35–37), with the exception of late-phase inflammatory reactions. Pruritus is caused by stimulation of H_1 receptors on sensory nerve endings; prostaglandins (qv) may also contribute. Sneezing, like pruritus, is an H_1-mediated neural reflex and can also be mediated by eicosanoids. Mucosal edema, which manifests as nasal obstruction, can be caused by H_1 stimulation as well as by eicosanoids and kinins. Increased vascular permeability is H_1 mediated. Nasal mucus secretion can be mediated by histamine both directly and indirectly through muscarinic discharge and by eicosanoids. Late-phase reactions, which manifest as nasal congestion and hyperirritability, are not mediated by histamine, but rather by inflammatory and chemotactic factors such as eicosanoids. In view of the excellent response to H_1 antagonists experienced by most patients with allergic rhinitis, histamine is a likely primary mediator of this disease. In numerous studies in this indication, H_1-receptor antagonists have proven to be extremely useful in ameliorating sneezing, nasal discharge, itchy nose and eyes, tearing, and eye redness. However, they are rather ineffective in relieving nasal congestion. Hence, decongestants such as pseudoephedrine or phenylpropanolamine have been added to H_1 antihistamines in order to provide relief of congestion.

Urticaria. The action of histamine on the skin is manifested by the classical Lewis triple response: oedema (wheal), erythema (owing to direct vasodilation, redness) which spreads because of axon reflex (flare), and pruritus or pain (38,39). Histamine, acting through its H_1 receptor, can mediate all three pathologic components of urticaria. The mechanism by which H_1 receptors mediate pruritus is indirect and involves stimulation of sensory nerve endings. Histamine-induced cutaneous vasodilation and flushing are probably partially mediated by neurohormones; other vasoactive mediators, eg, bradykinin and prostaglandins, can contribute. The partial failure of H_1 antihistamines in the treatment of urticaria may result from the involvement of other mediators or from the presence of H_2 as well as H_1 receptors in the skin. Direct vasodilation seems to involve H_2 receptors. H_3 receptors have not yet been found in human skin. Nevertheless, H_1 antihistamines remain the primary therapy for urticaria; these reduce pruritus and the number, size, and duration of urticarial lesions.

Asthma. In asthma, bronchospasm and mucosal edema can be caused by H_1 receptor stimulation. H_2 and possibly H_1 activation may be minor causes of mucus secretion. However, other mediators, such as leukotrienes, prostaglandins, bradykinin, and PAF, may be more important in asthma than histamine. Airway inflammation is not stimulated by histamine but can be caused by inflammatory factors, such as chemotactic factors. Mucus glycoprotein secretion can be induced by H_2 receptor activation, whereas increased movement of interstitial fluid into the airway lumen can be mediated by H_1 receptors. Clinical trials using classical H_1 antagonists for the treatment of asthma have yielded disappointing results. The drug concentrations achieved in the lungs with the available H_1 antagonists may not be high enough for an effective H_1 receptor blockade. H_1 antihistamines provide some relief from seasonal or chronic asthma when taken over weeks or months by patients with mild asthma, but these certainly are not drugs of first choice (40,41). High dose nonsedating H_1 antagonists, however, deserve further study as potential agents in the treatment of asthma.

Anaphylaxis.   All the symptoms of anaphylaxis can be reproduced by histamine. Vascular permeability manifests as angioedema, and laryngeal and intestinal edema; vasodilation leads to flushing and headache; smooth muscle contraction results in wheezing, abdominal cramping, and diarrhea. Tachycardia, often in the form of palpitations, can be caused by decreased atrioventricular node conduction time and indirectly by histamine-induced vasodilation and resultant catecholamine secretion. Reduced peripheral vascular resistance is responsible for syncope. Mucus secretion manifests as rhinorrhea and bronchorrhea. In acute anaphylaxis, treatment of first choice is epinephrine; H_1 receptor antagonists are a useful adjunctive treatment for control of pruritus, rhinorrhea, and some other symptoms (21).

Atopic Dermatitis.   The mechanism of itching associated with atopic dermatitis remains unknown, but histamine is almost certainly involved to some extent as histamine concentrations are increased in the skin and in the plasma of patients with this disorder (39,42). Second-generation H_1 receptor antagonists, unlike first-generation H_1 receptor antagonists, have not been uniformly found to be effective in relieving itching in atopic dermatitis, which may be related to the absence of a sedative effect (43).

Clinical Efficacy and Side-Effects of H_1 Antihistamines.   *Clinical Efficacy.* It is evident from the mechanism of action of antihistamines and the etiology of allergic diseases that antihistamines in no sense achieve a cure of the patient's allergy. After the administration of a therapeutic dose, a temporal blockade of the effects of histamine is obtained. Whereas classical antihistamines needed at least twice daily administration, for most of the more recently introduced agents administration once daily is sufficient.

Nevertheless, although the nonsedating H_1 antihistamines have substantially improved the acceptability and clinical efficacy of this class of compounds, these do not provide complete relief; eye disease responds less well than nasal disease, of the rhinitis symptoms nasal congestion responds poorly, breakthrough symptoms occur at high pollen counts, and only some 70% of patients report excellent to good treatment responses. Considerable research therefore still continues in the H_1 antihistamine field. New antihistamines are continually being introduced.

Side Effects.   The classical H_1 antihistamines are not very specific, and several compounds have some degree of anticholinergic activity, eg, diphenhydramine; serotonin antagonism, eg, cyproheptadine; or adreno-receptor blocking activity, eg, promethazine. Anticholinergic effects can present side effects such as dry mouth, blurred vision, and urine retention, whereas the anticholinergic action of some antihistamines is probably the reason for effectiveness in motion sickness (19). Interactions in the brain with noradrenergic, serotonergic, and dopaminergic uptake systems may play a role in behavioral effects of H_1 antihistamines.

At therapeutic doses, the classical H_1-receptor antagonists generally produce sedation. This usually unwanted effect is probably caused by the H_1-receptor blockade in the CNS. Among the conventional antihistamines, the aminoalkylethers, eg, diphenhydramine, and phenothiazines, eg, promethazine, have the most prominent sedative effects (promethazine is widely used as a hypnotic agent), whereas the alkylamines seem to cause less CNS depression. Sedation is often transient, owing to the development of tolerance, but may be sufficient to persuade the patient to discontinue treatment. Patients vary considerably in susceptibility to the sedating effects of antihistamines, but on average sedation using classical H_1 antagonists has been reported in some 20% of users.

The second-generation H_1 antihistamines, such as terfenadine (**9**), astemizole (**10**), cetirizine (**4**), and loratadine (**7**), generally present few side effects and, in particular, are considered not to cause sedation, mainly because of reduced ability to penetrate the CNS. However, terfenadine (**9**) and astemizole (**10**), the two most widely used nonsedating antihistamines available in the United States, have been associated with prolongation of the QT-interval in the electrocardiogram (ECG) and ventricular arrhythmias. The pharmacodynamic mechanism is unclear, but may be related to inhibition of the delayed potassium rectifier current (44). The QT prolongation generally occurs at higher than therapeutic plasma levels. This can be the case with overdoses or in combination with drugs that impair hepatic metabolism, eg, azole antimycotics (see ANTIPARASITIC AGENTS) and macrolide antibiotics (qv). It remains to be seen whether this side effect is particular to terfenadine and astemizole or whether it may be a class effect of antihistamines which may also become apparent with the more recently introduced antihistamines after more widespread exposure.

Topical Application.   Azelastine (**11**) and levocabastine (**13**) have been developed for topical application (45). The topical antihistamines address the preference of some patients for a local treatment and allow administration of drug directly to the site required. The advantage of this therapeutic approach is likely to be in the speed of onset of symptom relief. In contrast to earlier reports of sensitization with older antihistamines locally applied to the skin (46), sensitization has not been reported with local application to the nose or eyes.

Selective H_3 Agonists and Antagonists.   No clear therapeutic indications have been reported for H_3 receptor ligands. Yet, with the development of highly potent and selective agonists and antagonists, insights in the role of histamine H_3 receptors in various (patho)physiological processes have been obtained. An interesting option for therapeutic application of H_3 agonists would be in asthmatic diseases (8). H_3 receptors are known to be present in guinea pig and human airways, where they can inhibit nonadrenergic/noncholinergic (NANC) and cholinergic neurotransmission (8). Moreover, H_3 agonists can inhibit *in vivo* the microvascular leakage often associated with neurogenic airway inflammation. Using sensitized guinea pigs (47), thioperamide (**27**) was shown to enhance the allergen-induced bronchoconstriction in ovalbumine-sensitized animals. This bronchoconstricition resulted from H_1 receptor stimulation after histamine release from mast cells. These data indicate that H_3 receptors might be present on mast cells and could regulate the release of histamine. Previously, the H_3 receptor was indeed suggested to be involved in the regulation of histamine synthesis in lung tissue (30). Therapeutic application of H_3 receptor agonists could also be envisaged for gastrointestinal disorders. H_3 receptors are present throughout the GI tract and inhibit again the NANC and cholinergic neurotransmission (8). These properties could become useful for the treatment of diarrhea. Moreover, H_3 receptors also control gastric acid release, although some marked species dependence is noticed (8). However, application of H_3 agonists in this area does not seem to be probable; the H_2 antagonists and the proton pump inhibitors serve quite well.

Because histamine in the CNS is important for the regulation of sleep/wakefulness, applications in this area could be found. In cats, the H_3 receptor has been shown to affect the sleep pattern, and it has also been implicated in this respect in rats and mice (8).

Economic Aspects

The sales of antagonists of H_1 receptors, eg, diphenhydramine, terfenadine, and astemizole, used in the treatment of allergic diseases, represent 1% of the overall pharmaceutical market, ie, $1.7 billion (U.S.). H_2 antagonists, eg, cimetidine and ranitidine, are effective in peptic ulcer disease and esophageal reflux. Sales represent 3.5% of the world market, ie, $6 billion (U.S.). H_3 agonists or antagonists have not yet found a clear indication.

BIBLIOGRAPHY

"Histamine and Antihistamine Agents" in *ECT* 1st ed., Vol. 7, pp. 469–475, by R. W. Fleming and G. Rieveschl, Jr., Parke Davis & Co.; in *ECT* 2nd ed., Vol. 11, pp. 35–44, by R. W. Fleming and J. M. Grisar, The Wm. S. Merrell Co., Division of Richardson-Merrell, Inc.; "Histamine and Histamine Antagonists" in *ECT* 3rd ed., Vol. 12, pp. 481–491, by R. W. Fleming, Warner-Lambert Co., and J. M. Grisar, Merrell National Laboratories.

1. M. B. Emanuel, *Drugs of Today* **22**(1), 39 (1986).
2. P. S. Norman, *J. All. Clin. Immunol.* **76**, 366 (1985).
3. E. Fourneau and D. Bovet, *Arch. Intl. Pharmacodyn. Ther.* **46**, 178 (1933).
4. B. N. Halpern, *Arch. Intl. Pharmacodyn. Ther.* **68**, 339 (1942).
5. A. S. F. Ash and H. O. Schild, *Brit. J. Pharmacol.* **27**, 427 (1966).
6. J. W. Black and co-workers, *Nature* **236**, 385 (1972).
7. J.-M. Arrang, M. Garbarg, and J.-C. Schwartz, *Nature* **302**, 832 (1983).
8. R. Leurs and H. Timmerman, *Progress in Drug Research* **39**, 127 (1992).
9. E. Middleton, Jr. and co-workers, eds., Allergy, Principles and Practice, 3rd ed., The C. V. Mosby Co., St. Louis, Mo., 1988.
10. B. Uvnäs, ed., Histamine and Histamine Antagonists, Springer-Verlag, Berlin, 1991.
11. M. V. White, *J. All. Clin. Immunol.* **86**, 599 (1990).
12. K. Tasaka, in Ref. 10.
13. J. F. Porter and R. G. Mitchell, *Physiol. Rev.* **52**, 361 (1972).
14. J.-C. Schwartz and co-workers, in Ref. 10.

15. R. Levi, L. E. Rubin, and S. S. Gross, in Ref. 10.
16. R. Leurs, H. Van der Goot, and H. Timmerman, *Adv. Drug Res.* **20**, 217 (1991).
17. A. F. Casy, in Ref. 10.
18. D. S. Pearlman, *Drugs* **12**, 258 (1976).
19. J. P. Trzeciakowski, N. Mendelsohn, and R. Levi, in Ref. 9, p. 715.
20. M. H. Loveless and M. Dworin, *Bull. of the New York Acad. of Med.* **25**, 473 (1949).
21. F. E. R. Simons and K. J. Simons, *Ann. All.* **66**, 5–19 (1991).
22. M. M.-L. Janssens, ed., *Clin. Rev. All.* **11**(1), 1 (1993).
23. G. J. Sterk and co-workers, *Eur. J. Med. Chem.* **22**, 491 (1987).
24. J. C. Foreman, in Ref. 10.
25. H. van der Goot, A. Bast, and H. Timmerman, in Ref. 10, Vol. 97.
26. E. E. J. Haaksma and co-workers, *J. Mol. Graphics* **10**, 79 (1992).
27. J. C. Eriks and co-workers, *J. Med. Chem.* **35**, 3239 (1992).
28. M. Deakin and J. G. Williams, *Drugs* **44**(5), 709 (1992).
29. R. D. Shamburek and M. L. Schubert, *Gastroenterol. Clin. North Am.* **21**(3), 527 (1992).
30. J.-M. Arrang and co-workers, *Nature* **327**, 117 (1987).
31. H. Timmerman, *J. Med. Chem.* **33**, 4 (1990).
32. R. Lipp and co-workers, in H. Timmerman and H. van der Goot, eds., *New Perspectives in Histamine Research*, Birkhäuser Verlag, Basel, 1991, p. 227.
33. R. C. Vollinga and co-workers, *J. Med. Chem.* **37**, 332 (1994).
34. J.-M. Arrang, J.-C. Schwartz, and W. Schunack, *Eur. J. Pharmacol.* **117**, 109 (1985).
35. H. M. Druce and M. A. Kaliner, *J. Am. Med. Assoc.* **259**, 260 (1988).
36. J. Bousquet, P. Chanez, and F. B. Michel, *Resp. Med.* **84** (Suppl. A), 11 (1990).
37. R. M. Naclerio, *New Engl. J. Med.* **325**(12), 860 (1991).
38. A. Kobza-Black, *Skin Pharmacol.* **5**(1), 21 (1992).
39. C. Advenier and C. Queille-Roussel, *Drugs* **38**(4), 634 (1989).
40. M. Kaliner, *Chest* **87** (Suppl.), 2S (1985).
41. S. T. Holgate and J. P. Finnerty, *J. All. Clin. Immunol.* **83**, 537 (1989).
42. H. Behrendt and J. Ring, *Clin. Exp. All.* **20** (Suppl. 4), 25 (1990).
43. L. Krause and S. Shuster, *Br. Med. J.* **287**, 1199 (1983).
44. J. Kemp, *Ann. All.* **69**, 276 (1992).
45. R. L. Mabry, *Southern Med. J.* **85**(2), 149 (1992).
46. M. M. Moske and W. L. Peterson, *J. Invest. Dermatol.* **14**, 1 (1950).
47. M. Ichinose and P. J. Barnes, *J. Allergy Clin. Immunol.* **86**, 491 (1990).

MONIQUE M.-L. JANSSENS
Janssen Research Foundation

HENDRIK TIMMERMAN
ROBERT LEURS
Leiden/Amsterdam Center for Drug Research

HOLLOW-FIBER MEMBRANES

The development of hollow-fiber membrane technology has been greatly inspired by intensive research and development of reverse-osmosis membranes during the 1960s. Du Pont's pioneer aramid polymer device was commercialized in 1969, followed by a cellulose triacetate polymer developed by the Dow Chemical Co. and Toyobo (Japan). The excellent mass-transfer properties conferred by the hollow-fiber configuration soon led to numerous applications (1). Commercial applications have been established in the medical field (see BLOOD FRACTIONATION), in water reclamation (purification and desalination) (see WATER, SUPPLY AND DESALINATION), in gas separations and pervaporation; other applications are in various stages of development. A hollow-fiber membrane is a capillary having an inside diameter of >25 μm and an outside diameter <1 mm and whose wall functions as a semipermeable membrane. The fibers can be employed singly or grouped into a bundle which may contain tens of thousands of fibers and up to several million fibers as in reverse osmosis (Fig. 1). In most cases, hollow fibers are used as cylindrical membranes that permit selective exchange of materials across their walls. However, they can also be used as containers to effect the controlled release of a specific material (2), or as reactors to chemically modify a permeate as it diffuses through a chemically activated hollow-fiber wall, eg, loaded with immobilized enzyme (see ENZYMES).

Hollow-fiber membranes, therefore, may be divided into two categories: (1) open hollow fibers (Figs. 2a and 2b) where a gas or liquid permeates across the fiber wall, while flow of the lumen medium gas or liquid is not restricted, and (2) loaded fibers (Fig. 2c) where the lumen is filled with an immobilized solid, liquid, or gas. The open hollow fiber has two basic geometries: the first is a loop of fiber or a closed bundle contained in a pressurized vessel. Gas or liquid

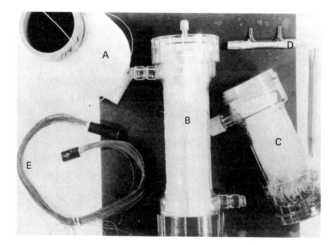

Fig. 1. A, hollow-fiber spool; B, hollow-fiber cartridge employed in hemodialysis; C, cartridge identical to item B demonstrating high packing density; D, hollow-fiber assembly employed for tissue cell growth; E, hollow-fiber bundle potted at its ends to be inserted into a cartridge or employed in a situation that requires mechanical flexibility.

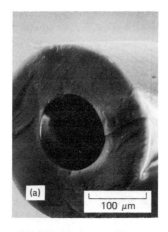

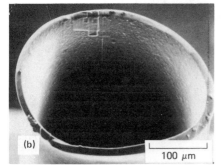 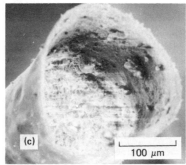

Fig. 2. (**a**) Thick-walled hollow fiber for high pressure desalination; (**b**) thin-walled acrylic hollow fiber; (**c**) sorbent-filled fiber. Courtesy of I. Cabasso.

passes through the small diameter fiber wall and exits via the open fiber ends. In the second type, fibers are open at both ends. The feed fluid can be circulated on the inside or outside of the relatively large diameter fibers. These so-called large capillary (spaghetti) fibers are used in microfiltration, ultrafiltration (qv), pervaporation, and some low pressure (<1035 kPa $= 10$ atm) gas applications.

In open fibers the fiber wall may be a permselective membrane, and uses include dialysis, ultrafiltration, reverse osmosis, Donnan exchange (dialysis), osmotic pumping, pervaporation, gaseous separation, and stream filtration. Alternatively, the fiber wall may act as a catalytic reactor and immobilization of catalyst and enzyme in the wall entity may occur. Loaded fibers are used as sorbents, and in ion exchange and controlled release. Special uses of hollow fibers include tissue-culture growth, heat exchangers, and others.

Hollow fibers offer three primary advantages over flat-sheet or tubular membranes. First, hollow fibers exhibit higher productivity per unit volume; second, they are self-supporting; and third, high recovery in individual units can be tolerated. The high productivity is derived from a high packing density and a large surface area. Since surface area-to-volume ratio varies inversely with fiber diameter, a 0.04 m^3 (1.5 ft^3) membrane device can easily accommodate 575 m^2 (6200 ft^2) of effective membrane area in hollow-fiber form (90 μm in diameter), compared to about 30 m^2 (330 ft^2) of spiral wound flat-sheet membrane and about 5 m^2 (50 ft^2) of membrane in a tubular configuration. Because they are self-supporting, the hollow-fiber membranes greatly simplify the hardware for fabrication of a membrane permeator. Whereas flat-sheet membranes employed in ultrafiltration or reverse-osmosis modules must be assembled with spacers, porous supports, or both, a bundle of hollow fibers can simply be potted into a standard size tube of plastic or metal, as shown in Figure 1. The primary disadvantage of the hollow-fiber unit as compared to the other membrane configurations is its sensitivity to fouling and plugging by particulate matter due to a relatively low free space between fibers. In commercial applications this problem is greatly lessened by designing systems with good pretreatment of feeds to the hollow fiber devices.

Hollow fibers can be prepared from almost any spinnable material. The fiber can be spun directly as a membrane or as a substrate which is post-treated to achieve desired membrane characteristics. Analogous fibers have been spun in the textile industry and are employed for the production of high bulk, low density fabrics. The technology employed in the fabrication of synthetic fibers applies also to the spinning of hollow-fiber membranes from natural and synthetic polymers.

Properties

Morphology. The desired fiber-wall morphology frequently dictates the spinning method. The basic morphologies are isotropic, dense, or porous; and asymmetric (anisotropic), having a tight surface (interior or exterior) extending from a highly porous wall structure (Fig. 3). The tight surface can be a dense, selective skin, permitting only diffusive transport, or a porous skin, allowing viscous flow of the permeate as in conventional ultrafiltration (Fig. 4) or reverse osmosis. Membrane-separation technology is achieved by use of these basic morphologies. The semipermeability of the porous morphology is based essentially on the spatial cross section of the permeating species, ie, small molecules exhibit a higher permeation rate through the fiber wall. The semiper-

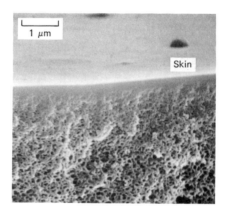

Fig. 3. Anisotropic hollow-fiber morphology exhibiting a dense skin. Courtesy of I. Cabasso.

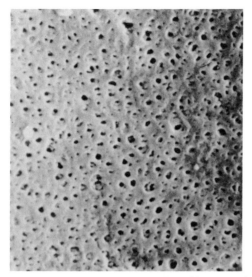

Fig. 4. Surface of a polysulfone ultrafiltration hollow-fiber membrane spun with poly-(vinylpyrrolidinone) (3). Surface pore diameter is 0.2–0.4 μm.

meability or anisotropic morphology of the dense membrane, which exhibits a dense skin, is obtained through a solution–diffusion mechanism. The permeating species chemically interacts with the dense polymer matrix and selectively dissolves in it, resulting in diffusive mass transport along a chemical potential gradient. Thus the dense membrane may exhibit semipermeability toward the large molecules with which it interacts, whereas the smaller, noninteracting species do not permeate. This is well demonstrated in the pervaporation process, where one-stage separation of toluene, as the permeating species, from its mixture with hexane or pentane can be accomplished employing alloys of cellulose acetate–polyphosphonate for the hollow fibers (4).

The asymmetric configuration is of special value. In the early 1960s, the development of the asymmetric membranes (5) exhibiting a dense, ultrathin skin on a porous structure provided momentum to the progress of membrane-separation technology. The rationale behind this development is that the transport rate through a dense membrane is inversely proportional to the membrane thickness, and membrane permselectivity is independent of thickness. Thus membranes with this structure permit high transport rates, yet can yield excellent separation. High manufacturing standards of reproducibility and quality control are required to maintain the integrity of the separation process. In addition, mechanical integrity problems associated with ultrathin membranes are obviated by use of asymmetric morphologies.

Mechanical Considerations and Fiber Dimensions. The hollow fiber is self-supporting, and is actually a thick wall cylinder. The ratio of outside to inside diameter in some reverse-osmosis applications is about 2 to 1 thus providing the strength to withstand high operating pressures, commercially up to 10,000 kPa (96 atm), without collapsing. A hollow fiber that is exposed to external pressure would exhibit a collapse pressure P_c that depends on the inner and outer fiber radii (IR, OR) and the Young's modulus E and Poisson ratio v of the material. The approximate relationship is given by the expression

$$P_c = \frac{2E}{(1 - v^2)} [(OR - IR)/(OR + IR)]^3 \tag{1}$$

For most hydraulic pressure-driven processes (eg, reverse osmosis), dense membranes in hollow-fiber configuration can be employed only if the internal diameters of the fibers are kept within the order of magnitude of the fiber-wall thickness. The asymmetric hollow fiber has to have a high elastic modulus to prevent catastrophic collapse of the filament. The yield-stress $\sigma\gamma$ of the fiber material, operating under hydraulic pressure, can be related to the fiber collapse pressure to yield a more realistic estimate of plastic collapse:

$$P_c = \sigma\gamma(OR - IR)/OR \tag{2}$$

For the asymmetric membranes, progressive yield can cause a loss of production rate due to compaction of the matrix in a prolonged operation. The pressure at which hollow-fiber compaction is initiated, P_γ, can be approximated:

$$P_\gamma = \sigma\gamma \, ln\,(OR/IR) \tag{3}$$

This value is taken into account when planning hollow-fiber dimensions. A partial account of these considerations can be found in References 6 and 7. In practical applications, ie, reverse osmosis, membrane compaction with time is experimentally derived as a function of the polymeric material at given temperatures and pressures (8).

   When the operation of the hollow-fiber membrane is to be reversed, and permeation from the bore to outer zone is required, circumferential stress and pressure drop along the fiber capillary (bore) must be considered in the design of the fiber unit. The circumferential stress, S_c, is expressed as

$$S_c = \frac{P_b IR}{(OR - IR)} \tag{4}$$

where P_b is the bore fluid pressure (9). Knowing the relationship between S_c, the applied pressure, and the membrane radius of curvature thickness, one can determine the minimum yield stress or ultimate tensile strength required to prevent failure or massive distension under maximum anticipated transmembrane pressure.

   The relationship between the bore fluid pressure drop, ΔP_b, and its flow rate is defined by Poiseuille's law:

$$\Delta P_b = K \frac{\eta L J}{\pi (2IR)^4} \tag{5}$$

(for laminar flow) where K, η, L, and J are dimensional constant, viscosity, fiber length, and flow rate, respectively (10).

Fiber dimensions have been studied for hemodialysis. When blood is circulated through the fiber lumen (*in vivo*), a significant reduction in apparent blood viscosity may occur if the flow-path diameter is below 100 μm (11). Therefore, current dialyzers use fibers with internal diameters of 180–250 μm to obtain the maximum surface area within a safe range (see DIALYSIS). The relationship between the fiber cross section and the blood cells is shown in Figure 5. In many industrial applications, where the bore fluid is dialyzed under elevated pressure (> 200 kPa or 2 atm), fibers may burst at points of imperfection. Failure of this nature is especially likely for asymmetric fibers that display a large number of macrovoids within the walls.

10 20 30 40 50 μm

Fig. 5.   Cross section of blood-clogged hemofiltration acrylic hollow-fiber membrane. The spheroids are red blood cells. Courtesy of I. Cabasso.

Spinning

In preparation of permselective hollow-fiber membranes, morphology must be controlled to obtain desired mechanical and transport properties. Fiber fabrication is performed without a casting surface. Therefore, in the moving, unsupported thread line, the nascent hollow-fiber membrane must establish mechanical integrity in a very short time.

Various common principles of hollow-fiber spinning processes are illustrated in Figure 6, which describes the dry-jet wet method. In this scheme, the spinning dope, consisting of a viscous, degassed, and filtered polymer solution (20–40% polymer by weight) is pumped into a coaxial tube, jet spinneret. The polymer content is usually close to the precipitation point. The thread line emerging from the spinneret is quickly stabilized by an internal quench (coagulating) medium (nonsolvent liquid or gas) as it emerges from the jet orifice.

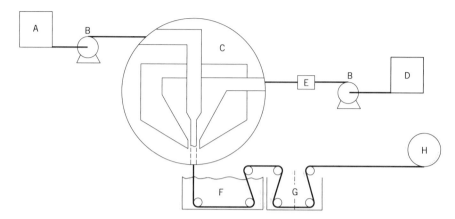

Fig. 6. Schematic of dry-jet wet spinning employing tube-in-orifice spinneret: A, bore injection medium (liquid, gas, or suspended solids); B, pump; C, spinneret; D, polymer spinning solution; E, micrometer (μm) "dope" filter; F, coagulation or cooling bath; G, quench bath; and H, collection spool.

The nascent hollow thread is further stabilized in a quench bath. At this point, the fiber has sufficient mechanical integrity to pass over guides and rollers under moderate tension. In most commercial production lines, the spinning rate is 10–1000 m/min and is governed by the spinning method, dope compositions, and the morphological and dimensional requirements. Therefore, dope compositions and spinning conditions are sought that result in establishment of the hollow fiber immediately on emergence from the orifice. Residual quenching liquid and solvent are usually removed by some sort of a washing step prior to use.

There are three conventional synthetic fiber spinning methods that can be applied to the production of hollow-fiber membranes: in melt spinning, a polymer melt is extruded into a cooler atmosphere which induces phase transition and controlled solidification of the nascent fiber; in solution (wet) spinning, the spinning dope, consisting of the polymer(s) predissolved in a volatile solvent mixture, is spun into a liquid coagulating bath (12). A combination of these first two methods is applied for hollow-fiber fabrication in the dry-jet wet-spinning technique, in which the spinneret is positioned above a coagulation bath (Fig. 6). In this process, all three mechanisms of formation (temperature gradient, solvent evaporation, and solvent–nonsolvent exchange) can be combined.

Spinnerets.   In all methods, a tubular cross section is formed by delivering the spinning dope through an extrusion orifice. Four schemes of spinneret nozzle cross sections are shown in Figure 7: (1) the segmented-arc design, which has a C-shaped orifice, is suitable for melt spinning. The extrudate rapidly coalesces to complete the annular configuration. There is no need for gas injection to prevent collapse of the hollow fiber, because the gas is drawn through the unwelded gaps. (2) The plug-in-orifice design simply provides the extrudate with the annular shape; however, a gas supply is usually required to prevent collapse. (3) The tube-in-orifice jet design is the most versatile combination since it can be applied to all three spinning techniques mentioned above. Gas, liquid, or suspended solids can be delivered through the inner tube to maintain the annular structure, and/or to control coagulation of the fiber bore, as well as to encapsulate gases, liquids, or

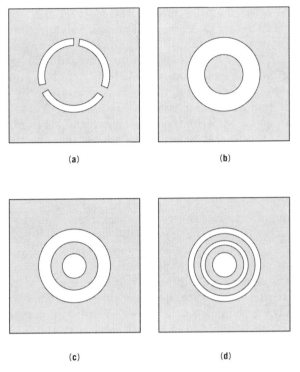

(a)   (b)

(c)   (d)

Fig. 7. Spinneret faces: (**a**) segmented-arc design; (**b**) plug-in-orifice design; (**c**) tube-in-orifice; (**d**) multiannular design.

solids to form the so-called loaded fibers. (4) The multiannular design is employed in spinning of multilayer fiber walls, or for entrapping (encapsulating) activated species in a composite hollow-fiber wall. In all cases the gas used is usually inert (nitrogen) so no chemical interactions occur between gas and organic polymer(s).

The spinnerets in a production line assembly consist of multiple groupings of orifices; thus the number of fibers from a single spinning cell can vary. All three spinning techniques mentioned above commonly employ spinning lines that carry as many as 250 fibers (sometimes referred to as a rope) drawn from a single multiorifice spinneret.

The multiorifice spinnerets require a high degree of precision in design and manufacture. The main problems encountered are the delivery of identical quantities of dope to each orifice and the instantaneous self-adjustment of the spinneret's internal pressure if an orifice plugs during spinning.

In most spinning processes, the fiber emerging from the spinneret is drawn down to a desired dimension before complete solidification. In some production lines, a laser beam is utilized to monitor the dimensions of emerging thread; any alteration is recorded; and if required, the dope delivery rate, take-up rate, and internal-injection medium delivery rate are adjusted automatically.

Melt Spinning. Melt spinning is the simplest spinning process. The spinning rate can be fast (over 1000 m/min), and fiber reproducibility is high. The polymer is extruded through the outer capillary of the spinneret. Melt spinning is often limited by the thermal stability of the polymer melt. Thermal degradation, cross-linking, and other modifications of the polymer melt can easily occur at high temperatures. To obviate these difficulties, polymers are sometimes melt spun with volatile or extractable plasticizers (as much as 50 wt %), which are blended with the polymer prior to spinning. For example, grades of cellulose triacetate, whose melting points are 292–314°C, can be spun at 150–230°C if premixed with sulfolane, dimethyl sulfoxide, glycerol, diethylene glycol, caprolactam, 2-pyrrolidinone, or other compounds (see FIBERS, CELLULOSE ESTERS). Melt-spun fibers can be stretched as they leave the spinneret, facilitating the production of very thin fibers, a significant advantage over solution spinning.

Other important operating parameters include plasticizer changes, gas type and pressure, gap length between spinneret and quench bath, line speed, and rope tension. All variables must be carefully controlled to obtain a hollow fiber of desired characteristics.

A common modification of the melt-spinning technique involves using dopes containing polymer-diluent pairs compatible in the melt but incompatible at lower temperatures. Thus phase separation follows phase transition from melt to solid. Examples of such dope compositions are poly(2,6-dimethyl-*p*-phenylene oxide) (PPO) with caprolactam (qv) and cellulose esters with dimethyl phthalate–glycerol mixtures (6). In such spinning, the gelled fiber walls consist of two phases: a continuous polymer matrix and solid or liquid domains of the diluent. Leaching of the latter with an appropriate solvent (water or aqueous solution) produces a porous matrix. This aspect is discussed in further detail below. Conventional melt-spinning methods tend to yield isotropic (dense or porous) wall structures. Asymmetric structures result, in most cases, from rapid loss of the diluent by evaporation, or by interaction with the quench medium when the nascent fiber is extruded into a gas flow-through chamber, or into an aqueous cooling bath. The fibers have low void volume and hence low permeabilities. The latter is overcome by use of high packing densities (very large surface areas) in a bundle. Polymers not soluble in common solvents have an advantage using this process.

Solution (Wet) Spinning. The key to the dry spinning of hollow-fiber membranes is the utilization of volatile solvents that boil below 40°C. Wet spinning techniques yielding cellulose acetate hollow-fiber membranes have been reported by Chemical System, Inc. (13). In this process, methyl formate (bp 30°C) and propylene oxide (bp 35°C) are employed as solvents. Such highly volatile solvents are necessary because the extruded fiber must be self-supporting over a considerable drop height (6–7 m) in the spinning cell, and completely dry after minimal festooning to maintain lumen potency during take-up. The spinneret generally consists of two concentric capillaries, the outer capillary having a diameter of approximately 400 μm and the central capillary having an outer diameter of approximately 200 μm and an inner diameter of 100 μm. Polymer solution is forced through the outer capillary while gas (usually nitrogen) or liquid is forced through the inner one. The rate at which the core fluid is injected into the fibers relative to the flow of polymer solution governs the ultimate wall thickness of the fiber (14).

Fibers spun by this method may be isotropic or asymmetric, with dense or porous walls, depending on the dope composition. An isotropic porous membrane results from spinning solutions at the point of incipient gelation. The dope mixture comprises a polymer, a solvent, and a nonsolvent, which are spun into an evaporative column. Because of the rapid evaporation of the solvent component, the spinning dope solidifies almost immediately upon emergence from the spinneret in contact with the gas phase. The amount of time between the solution's exit from the spinneret and its entrance into the coagulation bath has been found to be a critical variable. Asymmetric fibers result from an inherently more compatible solvent/nonsolvent composition, ie, a composition containing lower nonsolvent concentrations. The nature of the exterior skin (dense or porous) of the fiber is also controlled by the dope composition.

Wet- and Dry-Jet Wet Spinning. Although melt and solution (wet) spinning are convenient methods for obtaining dense isotropic morphologies, wet- and dry-jet wet methods can be used to obtain almost every known membrane morphology. The desired fiber morphology dictates the selection of the solvents, coagulants (interior and exterior), and the location of the spinneret, as well as other specific parameters. In most production lines, an aqueous solution serves as an exterior coagulant. The spinning dope, therefore, is composed of water-soluble solvents that are eventually extracted from the nascent fiber through the exterior zone of the fiber, although coagulant is often injected through the bore of the emerging fiber as well (Fig. 6). Thus a skin can be formed either on the external fiber surface or on the bore surface. The spinning rate in this method is relatively low (up to 100 m/min, but usually 15–50 m/min).

Many parameters are involved in the dry-jet wet technique, and these interact during the extrusion/coagulation steps. The principal variables are more or less the same for all spinning techniques and include dope composition, dope viscosity, spinning temperature, dope-pumping rate, composition of coagulants (interior and exterior), spinneret distance from the coagulation bath, interior-medium flow rate, coagulation temperature, and fiber draw rate (take-up rate). For example, the relationships among fiber dimensions, fiber morphology, fiber properties, and spinning variables are given in the literature for polysulfone hollow fibers spun from a dope composition of polysulfone/poly(vinylpyrrolidinone) (PVP) in dimethylacetamide (DMA) (3,7,15). The composition of the bore fluid and the rate of its delivery are important parameters for the initiation of fiber formation. The dope emerging from the orifice immediately interacts with the interior medium (which can be either gas or liquid), and the pressure within the nascent bore determines the initial filament diameter. For some dope compositions, the bore fluid contains dry nitrogen or mineral oil, whose only function is to maintain the annular configuration of the fiber as it is drawn down and coagulated from the exterior zone. Instant coagulation at the orifice can take place if the interior fluid is a strong coagulant. In such an event, an internal skin is formed and the fiber draw-down rate is relatively low.

To increase the viscosity of a spinning solution, a coagulant-soluble additive polymer may be included in the dope composition. For example, polysulfone cannot be spun adequately from low viscosity solution mixtures, ca 13–16 wt % of a polysulfone in dimethylformamide (DMF). However, if PVP is added to the polysulfone spinning dope, the solution viscosity is increased, and the viscosity required to maintain the falling lumen configuration is obtained. Then the PVP, a water-soluble polymer, dissolves in the coagulation bath, leaving behind a porous hydrophobic polysulfone matrix; the wall porosity and structure of the fiber can be controlled by varying the polysulfone/PVP/DMF composition (3,7,15) (Fig. 4).

The interplay between exterior and interior coagulation results in the range of fiber wall morphologies desired for various applications. Two underlying principles are (*1*) the nascent fiber wall facing a strong coagulant forms a skin and (*2*) the nature of the skin, degree of porosity, macrovoid formation, and other morphological traits chiefly depend on the dope composition. For example, cellulose acetate spun from acetone as solvent (25 wt % polymer) forms a dense wall if coagulated in water. If the solvent mixture consists of 60/40 acetone/formamide (formamide is a poor solvent for cellulose acetate), a skin resting on a porous structure is formed upon coagulation with water. If the bore fluid contains 70/30 water/glycerol, and the fiber is spun into a water coagulation bath, a skin forms in the exterior zone. If the coagulants are reversed, the skin forms on the interior bore wall. A cellulose acetate dope composition consisting of 40/60 acetone/formamide is at the point of incipient gelation. Therefore, in the dry-jet wet-spinning method, if the injected bore fluid is 60/40 glycerol/water, the nascent fiber may gel before contact with the water. In this case, the gelled matrix is plasticized by the glycerol, and the fiber can be drawn down from the spinneret orifice to give an isotropic porous morphology as shown in Figure 8.

Macrovoids.   Hollow-fiber membranes that are solution-spun by the foregoing methods can exhibit large voids in conical, droplet, or lobe configurations. These voids may extend through the entire fiber cross section (Fig. 9). The voids, in general, result from fast coagulation of a spinning solution that is relatively low in either polymer concentration or viscosity. Fast coagulation does not allow for a gradual exchange of solvent with nonsolvent at the nascent-membrane surface. In most cases, the voids occur when the coagulant intrudes into the ex-

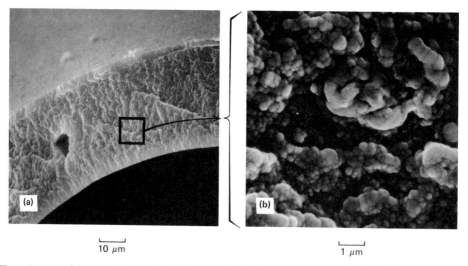

└──┘
10 μm

└──┘
1 μm

Fig. 8.   (**a**) Cross section of cellulose acetate isotropic porous hollow-fiber dry-jet wet spun at incipient gelation point of dope mixture; (**b**) magnification at the central cross section of (**a**). Courtesy of I. Cabasso.

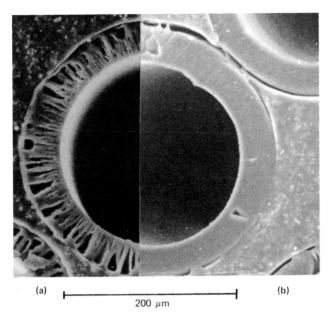

(a)　|——————————————————|　(b)

200 μm

Fig. 9. Cross sections of acrylonitrile hollow fibers spun by the dry-jet wet-spinning method. The two fibers were spun from the same dope composition into a water bath and were internally coagulated by aqueous solutions. Section (**a**) was spun with strong coagulation bore fluids resulting in tight interior skin with macrovoids extending to the exterior zone, and (**b**) was spun with a moderate bore coagulant, slowing the gelation rate of the nascent fiber. The relatively few macrovoids seen in (**b**) are in reverse direction to those in (**a**), and were caused by the water (a strong coagulant) in the quenching bath. Courtesy of I. Cabasso.

trudate at an early stage of gelation. The gross inward contraction of the surface at subsequent stages causes the surface to seal around the intrusion, forming a tight surface skin. The use of a less severe quenching medium, on the other hand, yields a macrovoid-free hollow fiber (16).

The presence of macrovoids in hollow-fiber membranes is a serious drawback since it increases the fragility of the fiber and limits its ability to withstand hydraulic pressures. Such fibers have lower elongation and tensile strength.

Fiber Treatment

Treatment In-line. The coagulated fiber on the moving threadline may be subjected to cooling (for melt spinning, or for dry-jet wet spinning conducted at high temperatures), washing to remove trace solvents and dope additives, swelling with diluted solvent and/or plasticizers, stretching between godets, and heat-treating (annealing) to consolidate its morphology and impose transport properties (such as closing the skin pores of an asymmetric hollow-fiber membrane for reverse-osmosis applications). In the continuous processing of hollow fibers, these steps add little to costs but are fundamental to achieving the desired functionality of the product.

Post-Treatment of Hollow Fibers.   End use of the hollow-fiber membrane dictates the type of post-treatment, if any. There are three main categories: fibers that are spun, fibers that will be chemically or physically modified, and fibers that will serve as a porous matrix for support of another (active) polymer deposited (or entrapped) upon (or within) its walls. There is no theoretical impediment to the inclusion of all conventional treatments in the spinning line: photochemical cross-linking, fluorination, and antiplasticizers have been successful. Fibers that are wet or melt spun with a plasticizer additive often need to be kept plasticized (or wet) to retain their morphology and membrane properties. This is true for many cellulosic, cellulose–ester, and polyamide fibers, as well as for fibers spun from other synthetic materials. However, dried (by gas) fibers, although unsuitable for one application, eg, reverse osmosis, can be good for another, eg, gas separation. The fibers may be plasticized in line; often, however, the hollow-fiber spool is subsequently submerged in a plasticizing medium (batch treatment). The plasticizing agents most commonly used are glycerol, poly(ethylene glycol), Triton X-100 (alkylphenoxylpolyethoxyethanol), and related compounds. Formaldehyde or sodium *meta*-bisulfite is often added to protect the fiber from biological fouling. In some cases fiber stability is enhanced by adding a soluble divalent cation, eg, magnesium ion, to a storage solution.

Fiber Modification.   Chemical modification of the fiber is usually a separate operation. The largest such commercial processing is the deacetylation of cellulose acetate hollow fibers, which converts them into regenerated cellulose hollow fibers employed in hemodialysis. The Cordis-Dow (Miami, Fla.) hemodialyzer shown in Figure 1 is prepared from such fibers. The modification is an alkaline hydrolysis in which bundles of fibers are hydrolyzed, thoroughly washed, neutralized, and plasticized with glycerol in preparation for cutting and encasing into the dialysis cartridge. Another example is provided by Du Pont (17) where a strong oxidizing agent is employed to increase the porosity of a reverse osmosis hollow fiber. The resultant product finds application in concentrating fruit juices.

Composite Hollow-Fiber Membrane.   Composite membranes consist of highly porous substrates, having minimum resistance to the permeates, which support ultrathin semipermeable membranes. The appeal of this concept is that it combines the properties of two or more different materials to yield a desired product. It has a special value in membrane technology where compounds that have the required semipermeabilities cannot be extruded into hollow fibers but can be deposited on the interior or exterior surface of a fiber. Composite hollow fibers, consisting of a polysulfone support matrix coated with cross-linked polyethyleneimine (PEI) or furan resin, are shown in Figure 10. The thickness of the deposited dense layers (resting on a porous, asymmetric polysulfone fiber) is 0.1–1 μm. Such fibers were developed for desalination of saline waters (brackish water and seawater) using the reverse-osmosis process (see WATER, SUPPLY AND DESALINATION). The principal fabrication difficulty is in maintaining continuity of the ultrathin layer, which requires adequate methods of deposition. The production scheme for a composite hollow-fiber membrane, consisting of polysulfone coated by PEI that is cross-linked *in situ* (on the exterior surface of the fiber), is shown in Figure 11. The furan-resin-coated polysulfone hollow fiber is produced by passing the fiber through a furfuryl alcohol solution followed by treatment

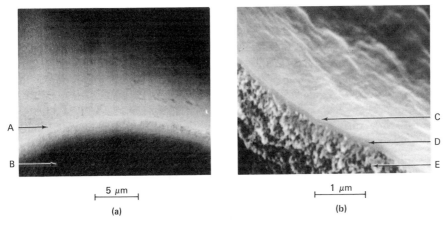

Fig. 10. Composite hollow-fiber membranes: (**a**) polysulfone hollow fiber coated with furan resin. A and B denote furan resin surface and porous support, respectively; (**b**) cross section of composite hollow fiber (PEI/TDI coated on polysulfone matrix). C, D, and E denote tightly cross-linked surface, "gutter" gel layer, and porous support, respectively. Both fibers were developed for reverse osmosis application (15).

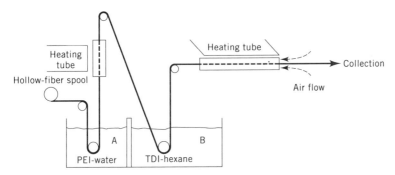

Fig. 11. Composite hollow-fiber production scheme (PEI = polyethyleneimine; TDI = toluene 2,4-diisocyanate). Anisotropic (porous skin) polysulfone hollow fiber is rolled into bath A and is lifted vertically (to avoid droplet formation) into a heating tube. The fiber is then passed through bath B and is annealed in a ventilated heating tube (110°C).

with sulfuric acid (18,19), to effect *in situ* cationic polymerization on the exterior surface as shown in Figure 10**a** (see MEMBRANE TECHNOLOGY).

Although these composite fibers were developed for reverse osmosis their acceptance in the desalination industry has been limited due to insufficient selectivity and oxidative stability. The concept, however, is extremely viable; composite membrane flat films made from interfacial polymerization (20) have gained wide industry approval. Hollow fibers using this technique to give equivalent properties and life, yet to be developed, should be market tested during the 1990s.

Interpenetrated Wall Matrix. Ion-exchange hollow fibers can be produced by polymerizing an ionic monomer within the porous wall matrix of a hollow fiber. For example, 4-vinylpyridine has been polymerized in a porous wall matrix of polyacrylonitrile (PAN) hollow fiber (21), and monomers containing sulfonic acid moieties have been polymerized in the wall matrix of polysulfone to yield ion-exchange hollow fibers (employed in Donnan-type dialysis). Requirements of such a fabrication are (*1*) the monomers should not dissolve or plasticize the polymer from which the fibers are made; (*2*) the heat generated during the polymerization and contraction prior to the formation of new interpenetrating polymer should be minimized; and (*3*) the polymerization should not occur within the lumen (and hence cause plugging of the fiber). The fabrication of such fibers is accomplished by forcing the monomers into the matrix under pressure while maintaining a flow of gas or liquid in the bore. High charge densities can be obtained by cross-linking the polymer network; for example, by employing sulfonated phenol–formaldehyde as the ionic species, highly cross-linked resin within the fiber wall is obtained (22). Drawbacks of such fibers are brittleness.

Fiber Handling and Unit Assembly. Most hollow fibers can be collected on spools by winding machines analogous to those used in the textile industry. Individual or multifilaments can be crosswound, or may be wound in a simple parallel arrangement (for highly plasticized, or large ID fibers, where cross-winding intersections may weaken the structure). Subsequent handling of the filament depends on the intended use of the hollow fiber.

Assembling and potting (cementing together) of hollow-fiber bundles, as shown in Figure 1, require great care and precision technology. The potting agent must be compatible with the function assigned to the fiber, as well as with the fiber material. For example, the potting materials employed in a hemodialysis cartridge (Fig. 12) must be blood-compatible and nontoxic, and adhere to the exterior surface of the fibers as well as to the fiber-housing unit. Another factor important in the selection of a potting agent is its surface tension (ability to wet fibers yet not excessively wick). Commonly employed potting agents include epoxy resins (qv), polyurethanes, and silicone rubbers. A potting agent is used in a liquid form that is eventually polymerized and cured in bulk (without solvent). In general, the potting agent must not shrink or evolve excess heat when cured; it must not penetrate the fiber, plug the bore, wick on the fiber walls, or damage any ultrathin coating. It must be hard enough after curing so as not to creep under pressure (crucial for ultrafiltration, gas separation, and reverse osmosis) and to be capable of further mechanical machining, ie, placed in a lathe and cut to open hollow-fiber bores for reverse osmotic flow.

For the high pressure reverse-osmosis units, epoxy resins which can withstand elevated hydraulic pressures (>10,000 kPa, 1,450 psi) are used as potting agents. Composite polysulfone hollow fibers have been potted with an epoxy resin sandwiched between two layers of silicone rubber. The rubber has low adhesion to the fiber, but protects it from epoxy wicking and breakage near the potting fixation spots and permits cutting of the fibers while they are fixed in the rubbery medium. In large commercial reverse-osmosis (RO) units for desalination of seawater and brackish water (eg, Permasep, Hollosep), the fibers are assembled in a U shape (Fig. 13) and are epoxy-potted. Winding of RO hollow-fiber membranes into a permeator is described in References 6, 17, and 23. Various types of blood-compatible polyurethanes are available for hemodialysis potting; others that resist attack by solvents are available for liquid-mixture separation by pervaporation. A difficulty sometimes encountered with ion-exchange hollow fiber is

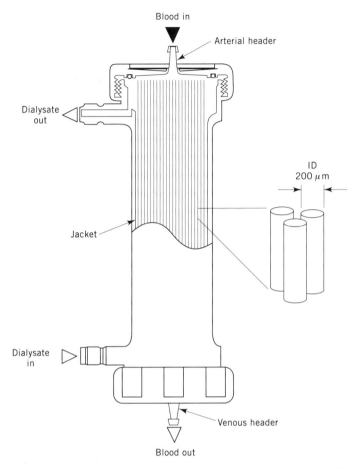

Fig. 12. Schematic of hollow-fiber membrane cartridge employed for blood dialysis. Courtesy of Cordis-Dow.

their tendency to undergo dimensional changes when wetted. Since most potting agents require dry potting conditions, the adhesive bond may fail after several wet–dry cycles. To circumvent this problem, chemical treatment is employed to neutralize the ion-exchange sites at the ends of the bundles.

A useful technique commonly employed in manufacturing dialyzers is centrifugal potting, in which the potting agent is introduced to the ends of the pre-assembled rotating cartridge. The potting agent cures while the centrifugal forces assure bubble-free, maximum potting density. The hemodialysis units shown in Figures 1 and 12 contain thousands of fibers that were potted in this manner.

In general there are three main types of hollow-fiber flow configurations. In the most common, for reverse osmosis and ultrafiltration, the feed enters outside the fiber; permeate is inside the fibers and flow is countercurrent. In the second, for large diameter fibers, where the feed has a high loading of particulates, the feed is through the fiber bore; permeate is outside the fiber; flow is usually countercurrent. In cross flow, shell side feed is prevalent as with microfiltration. Gas permeation uses all three flow patterns (24).

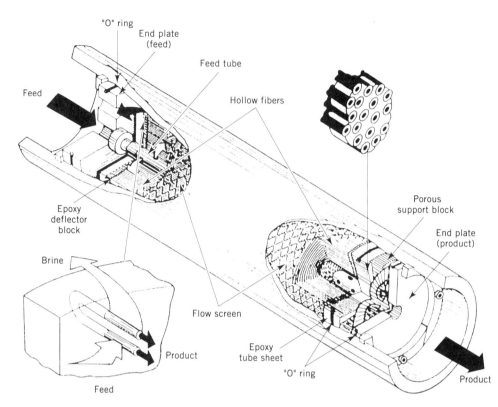

Fig. 13. Du Pont hollow fine fiber Permasep permeator.

Materials

The components employed in spinning-dope formulations must be consistent in every batch preparation, because numerous parameters are involved in the spinning process. Thus stringent criteria are imposed on the selection of components to be used in each spinning operation. The components are rigorously tested for purity, molecular weight, molecular weight distribution, chemical composition, viscoelastic properties, and other specific parameters that might influence hollow-fiber production and final membrane properties. This often requires close cooperation between the producers of the polymer and the hollow fiber manufacturers.

Cellulose. Cellulosic hollow fibers are produced chiefly by either of two methods: (1) wet spinning of a dope mixture containing cellulose dissolved in cuprammonium solution, eg, Cuprophan fibers, or (2) deacetylating cellulose acetate fiber to produce a regenerated cellulose fiber. The cuprammonium process involves mixing and dissolving cellulose at a low temperature. The dope solution can then be stored for prolonged periods if kept in an oxygen-free container. In the spinning process, the water-swollen, coagulated filament gel is thoroughly washed, and the copper is separated from the also regenerated cellulose fiber by acid extraction (see FIBERS, REGENERATED CELLULOSICS). The process has been perfected to the point of virtually full recovery of ammonia and copper residue.

This recovery rate has reduced the cost of manufacturing the fiber, and this factor plus reusability enables the hollow-fiber artificial kidney to dominate the hemodialysis market, though the initial cost is considered somewhat high.

The cellulosic fibers produced by both processes are used primarily for hemodialysis. The blood to be cleaned is circulated through the fiber bore (200–300 μm ID) in a cartridge such as that shown in Figure 12. Metabolic waste, consisting of low molecular weight components, eg, urea and uric acid, rapidly diffuses through the cellulosic fiber wall (thickness of 16–25 μm) which contains 40–60 wt % water. The pore radii generally are smaller than 6 nm. Surface areas range from 0.3 to 3.0 m^2, depending on the model. Manufacturers adjust the surface area by varying the length of the fibers (150–300 mm) and the number of fibers (3,000 to 20,000) employed (25).

The Cuprophan hollow fiber is manufactured with a bore fluid (isopropyl myristate) that has to be removed before the fiber is packed into a dialyzer. In general, the fiber bundles are drained, washed with alcohol, plasticized with glycerol, and sterilized with ethylene oxide. The hollow fibers are interwoven with thread and grouped into bundles. Other sterilization techniques for dialyzers include stream autoclaving, gamma ray, and formaldehyde. The Cuprophan hollow-fiber (and flat-sheet) membranes are often thinner than the cellulose acetate hollow-fiber membranes. They are produced by Enka Glanzstoff AG (Wuppertal, Germany) and Cordis-Dow. Three such bundles are contained in the annular space between two stainless steel tubes and encapsulated with polyurethane at each end (the tubesheets). The inner tube acts as a conduit feeding the shell side of the module, the flow exiting at one end of the outer tube. Two such submodules of 28 cm effective fiber length are joined in series giving 22.5 m^2 membrane area (24).

Gas separation using ethylcellulose hollow fiber has also recently become important. A/G Technology is leading this effort. Fluoroaceylated ethylcellulose is reported to have good gas permeation and blood compatibility (26).

Cellulose Ester.   Among the cellulose esters, cellulose acetate [9004-35-7] and cellulose triacetate [9012-09-3] have drawn the most attention. Both polymers have been developed commercially for desalination of brackish water and seawater. When dried the polymers are suitable for gas separation. Commonly, these hollow fibers have an asymmetric structure with a dense skin at the outer surface. Reverse osmosis permeators were manufactured by Dow Chemical (Dowex) until the mid-1980s and consisted of cellulose triacetate (CTA) fibers comprising two types of membrane: asymmetric CTA fat hollow fiber (250 μm OD, 90 μm ID) that is melt-spun into a cooling-leaching bath, and an isotropic melt-spun dense fiber (90 μm OD, 35 μm ID). The first is designed for desalination of brackish water, and the second for seawater. In the 1990s the only CTA hollow fiber manufacturer is Toyobo (Japan) whose fibers have dimension of 70 μm ID by 165 μm OD. The advantage of CTA over cellulose acetate, which has a degree of acetylation in the range of 2.3 to 2.7 per glucose unit, is that the latter is more vulnerable to biodegradation and hydrolysis (narrow pH operating range of 5–7). An early historical case description on the production and performance of cellulose acetate (CA) fibers for reverse osmosis is given in References 6 and 27. Since then CA hollow fibers have had limited commercial use (see also REVERSE OSMOSIS). A cellulose acetate unit, such as that shown in Figure 12, is also manufactured for hemofiltration by Cordis-Dow. The difference between hemodialysis and hemofiltration lies essentially in the transport mechanism; the first is based on diffusive transport, and the second on hydraulic transport, as shown in Figure 14. Cellulose acetate butyrate [9004-36-8] is another membrane polymer with low industrial acceptance.

One report (13) describes the procedure for spinning dry asymmetric cellulose acetate fiber with a bore skin. Such fibers are spun in a modified dry-spinning process in which a volatile liquid (methyl formate) is used as the cellulose acetate solvent. The bore coagulating liquid is isopropyl alcohol, which is subsequently removed. The advantages of these dry fibers over most cellulose acetate membranes are that they can be stored dry, they are wet–dry reversible, they can be sterilized and packed dry, and they are ready for use without removal of preservatives.

Polysulfone. Polysulfone is a commercial polymer that is a product of bisphenol A and 1,1'-sulfonylbis (4-chlorobenzene) (see POLYMERS CONTAINING SULFUR). This high strength thermoplastic can be spun into hollow fiber by melt- or dry-jet wet-spinning techniques. The polymer is hydrophobic, but is soluble in water-miscible solvents such as DMF. This fiber has been investigated as a matrix component for the preparation of composite hollow-fiber membranes (3,7,15). Composite hollow-fiber membranes consisting of polysulfone fibers coated by semipermeable ultrathin furan-resin membranes (19) are being manufactured by FRL Corp. (Dedham, Mass.). Various companies (Amicon; Romicon; and Berghof GmBH, Germany) manufacture an essentially asymmetric polysulfone hollow fiber with a core skin for industrial and medical ultrafiltration. Since polysulfone

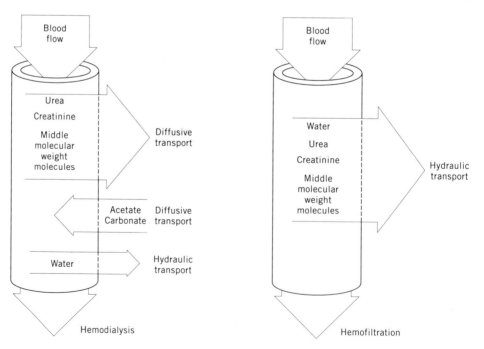

Fig. 14. Mass transfer across hemodialysis and hemofiltration hollow-fiber membranes.

is a hydrophobic material, wetting agents (eg, glycerol, dodecyl sodium sulfate) are often incorporated into the porous walls of the dry fiber after manufacturing. Polysulfone hollow fibers by Fresenius and Amicon employed in hemofiltration display exceptionally high interaction with blood and plasma proteins. Membrane thickness is in the 40 to 70 μm range. For gas separation, membranes can be made into defect-free thin skin structures by coating the membrane in a post-treatment with a dilute solution of silicone rubber. Since the silicone rubber is thin, its permeability is high, thus sealing defects in the base membrane without materially affecting the permeability of the membrane. The Monsanto Prism separator embodies this concept. Encased ultrafiltration size fibers have found use in laboratories as a bioreactor. The molecular weight cut-off of 10,000 daltons is advantageous (28). Separation of cells from fermenter broths, purification, and concentration of antibiotic products and enzymes are other limited use areas (29).

Poly(methyl methacrylate). Poly(methyl methacrylate) (PMMA) is a hydrophobic polymer that is soluble in some water-miscible solvents, such as dimethyl sulfoxide and DMF (see METHACRYLIC POLYMERS). Membranes that are produced by casting atactic PMMA into a water gelation bath have poor mechanical and transport properties. However, solution compositions consisting of isotactic and syndiotactic PMMA mixtures form thermoreversible sol–gel stereotopic complexes that can be cast or spun into hollow fibers with controlled wall porosity and improved mechanical strength (30,31). Such dope solutions are spun hot into a subambient (<30°C) atmosphere; there, the nascent fiber forms a gel and is subsequently passed through water to effect an exchange of water for solvent. Since matrix formation is established during gelation, the microstructure of the fiber wall does not collapse when submerged in water, and water-swollen fiber (ca 65 wt % water content for dope composition spun with 20 wt % polymer) is obtained. Toray Industries, Inc. (Japan) has developed a PMMA hollow-fiber membrane. Use of this polymer has found some application as a hollow-fiber artificial kidney. Membrane thickness is in the 30–40 μm range (25).

Polyamide. Nylon hollow fibers are produced by Du Pont, Berghof GmbH, and many others. The development of hollow fiber initially from nylon-6 or nylon-6,6 was a natural extension of technology established in the textile industry (see POLYAMIDES, FIBERS). These materials were aimed toward the desalination of brackish water employing high pressure reverse osmosis. Fiber dimensions were 50–60 μm OD and 25–30 μm ID. Hydraulic permeability through these aliphatic nylon derivatives was very low. The second generation of asymmetric polyamide hollow-fiber membranes developed for high pressure reverse osmosis consist of derivatives of aromatic polyamides (aramids) with improved water permeability and water (brackish and seawaters) separation. They are the largest consumers of hollow fibers. The fibers are spun from a solution of inorganic salts and DMA while a nitrogen stream is maintained through the nascent fiber bore. The extrusion is carried into a high temperature nitrogen atmosphere, resulting in solvent evaporation, and skin establishment in the outer zone is annealed. These fibers must be stored wet to retain the asymmetric morphology essential for high hydraulic permeability.

Du Pont, the world's leading supplier of hollow fibers, has introduced two types of hollow-fiber permeators for desalination and water reclamation: the B-10 Permasep permeator for desalination of seawater at up to 8300 kPa (up to 1200 psig), and the B-9 Permasep module for low pressure brackish water operations (1400 to 2800 kPa, 200 to 400 psig) (32). The fiber dimensions for the B-9 permeator are 91 μm OD and 44 μm ID, and for B-10 module 95 μm OD and 42 μm. In comparison to the smaller diameter fibers mentioned above, Berghof GmbH has developed an ultrafiltration fiber (aromatic polyamide) by the dry-jet wet-spinning method. A variety of asymmetric fibers with bore skin having large inside diameters (600–1500 μm) and a molecular weight cut-off of 2,000–50,000 are being manufactured for industrial and medical ultrafiltration processes. Fibers with dimensions of about 200 μm (ID) have been developed for hemofiltration, and are manufactured in cartridges 0.1 to 0.3 m long containing 1 to 2 m^2 of membrane area. Gambio in Sweden is a leading supplier.

The literature notes many other polyamide hollow fibers, none of which have achieved significant commercial success. Included in this category are such polymers (some of which are cross-linked) as piperazinamides, hydrazine, substituted acrylamide, and modified and grafted nylons.

Other Nitrogen-Containing Polymers. Polymeric membranes which are nitrogen based appear to be preferred for their overall combination of properties; high selectivity, good flux rate, longevity, and relative stability to oxygen and other chemicals. Other than the polyamides an example would be polybenzimidazole. Polybenzimidazole polymers have been spun chiefly from poly[2,2'-(*m*-phenylene)-5,5'-bibenzimidazole] (PBI) (33,34) (see POLYIMIDES). Asymmetric fibers with outside skins have been spun from a solution consisting of 20–25 wt % PBI in DMA by the dry-jet wet-spinning technique. An inert hydrocarbon liquid is used as the bore fluid, and the fiber is annealed in-line at elevated temperatures (140–180°C) in ethylene glycol. The PBI fibers, which have excellent chemical resistivity and mechanical properties have not realized their full potential because PBI is still an expensive polymer (see HIGH PERFORMANCE FIBERS). There is limited use in reverse osmosis and gas separations.

Another example is polyacrylonitrile (PAN). Hollow fibers consisting of polyacrylonitrile and its copolymers have been developed by Gulf South Research Institute (Figs. 2b and 9) and are manufactured among others by Hospal and Asahi Medical Corp. Commercially available acrylonitrile copolymers are soluble in water-miscible solvents such as DMA and DMSO, and therefore can be spun by the dry-jet wet-spinning technique with a water coagulant. These fibers are used for pervaporation ultrafiltration and hemofiltration, and as porous substrates (interpenetrated networks) for the production of ion-exchange hollow fibers (21) (see ACRYLONITRILE POLYMERS). In dialyzers these membranes have received a lot of attention since the 1970s. Problems associated with their costs and high permeability to water have limited use. Additional nitrogen-containing polymers for which hollow fibers have been made include polyimides, polyoxadiazole, nitrogen-substituted aromatics (26), polyurea, and polyethylenimine.

Glass and Inorganic Hollow-Fiber Membranes. The development of porous, glass hollow fibers for membrane applications is reported in Reference 35 (see also GLASS). The glass material provides higher dimensional and chemical stability than most polymeric materials, but the fibers have the disadvantages of brittleness and low permeabilities. The glass is essentially composed of the system, $Na_2O–B_2O_3–SiO_2$. In the processing, an induced microphase separation results in rich alkali borate domains. When leached with mineral acids, the fibers are left with a porous silica matrix (>96% silica). Fibers having diameters of 0.3 mm ID are manufactured by Schott for ultrafiltration. In general, the economics of the production are poor and use is limited.

Inorganic membranes (29,36) are generally more stable than their polymeric counterparts. Mechanical property data have not been definitive for good comparisons. Industrially, tube bundle and honeycomb constructions predominate with surface areas 20 to 200 m². Cross-flow is generally the preferred mode of operation. Packing densities are greater than 1000 m²/m³. Porous ceramics, sintered metal, and metal oxides on porous carbon support (for microfiltration and ultrafiltration) are the most common materials. Use of glass and inorganic membranes has been held back by the difficulty in developing tube sheet seals which prevent mixing of feed with permeate. Among the current solutions are epoxies, Teflon, Viton, polyester, and silicone rubber.

Others. Hollow-fiber membranes have been spun from various other polymers, including polyesters, silicone rubber, silicone rubber/polycarbonate copolymers, ceramics, porous metallics, and polyphosphazenes for gas separations. Ultrafilters and reverse osmosis hollow-fiber membranes have been made from polyfuran poly(vinylidene fluoride), polyethylene, and poly(vinyl alcohol). A patent by Du Pont in 1971 (37) identified a number of noncrystalline (glassy) polymer materials with high intrinsic permeabilities, providing by melt spinning the potential for gas separations using hollow fibers and for composite hollow-fiber membranes; as such, *in situ* plasma polymerization on the surface of porous hollow fibers has been reported (38). There are numerous compounds that can be plasma-deposited on a porous fiber to yield a multitude of classes of composite membranes. Such fibers will become more readily available as plasma polymerization technology progresses (see PLASMA TECHNOLOGY). Ethylene–vinyl alcohol fibers are under study as hemofiltration membranes.

Sorbent Fibers

Filled Fibers. Interest in the encapsulation of specific active materials (eg, activated charcoal, enzymes, drugs) led to the development of encapsulation spinning, usually employing a wet- or dry wet-spinning process. In the encapsulation process, the filling ingredient is suspended or dissolved in the bore liquid medium (usually a coagulant) which is injected through the internal orifice during the spinning process (Fig. 6). Subsequent quenching and washing in a water bath are employed. Reports on the spinning of polysulfone, cellulose acetate, and polyacrylonitrile fibers filled with activated charcoal, enzymes, drugs, ion exchangers, and phosphate sorbents are described in References 39 and 40. The spinning conditions and materials for such fibers must not reduce the activity of the encapsulated ingredient. Ion-exchange beads encapsulated in a lumen of cellulose acetate hollow fiber (developed at Gulf South Research Institute) are shown in Figure 2c (see MICROENCAPSULATION).

The rationale for the development of such fibers is demonstrated by their application in the medical field, notably hemoperfusion, where cartridges loaded with activated charcoal-filled hollow fiber contact blood. Low molecular weight body wastes diffuse through the fiber walls and are absorbed in the fiber core. In such processes, the blood does not contact the active sorbent directly, but faces the nontoxic, blood compatible membrane (see CONTROLLED RELEASE TECH-NOLOGY, PHARMACEUTICAL). Other uses include waste industrial applications as general as chromates and phosphates and as specific as radioactive/nuclear materials.

Hollow Fiber with Sorbent Walls. A cellulose sorbent and dialyzing membrane hollow fiber was reported in 1977 by Enka Glanzstoff AG (41). This hollow fiber, with an inside diameter of about 300 μm, has a double-layer wall. The inner wall consists of Cuprophan cellulose and is very thin, approximately 8 μm. The outer wall, which is ca 40-μm thick, consists mainly of sorbent substance bonded by cellulose. The advantage of such a fiber is that it combines the principles of hemodialysis with those of hemoperfusion. Two such fibers have been made: one with activated carbon in the fiber wall, and one with aluminum oxide, which is a phosphate binder (also see DIALYSIS).

Future Prospects

Membrane science began emerging as an independent technology only in the late 1960s, and its concepts still are being defined. Intra- and interprocess competitive pressures are the driving force. Many developments evolved from fundamental studies that were government-sponsored. However, in the 1990s as it has become a feasible technology in the industrial complex, investment and resources grow mostly corporate instigated. Currently, hollow-fiber membrane units are used in water desalination as the large volume area; they are well established in hemodialysis and the potential in gas separation, microfiltration, and pervaporation is very large. Hollow-fiber technology seems to be one step behind flat-sheet membrane technology, chiefly because the laboratory casting of the latter is simple and does not require special equipment. Nevertheless, the hollow-fiber configurations always appear to be competitive.

Hollow-fiber membranes are subjected to extensive studies for gaseous separation (eg, CO_2, H_2, O_2, N_2, H_2S, CO, CH_4), where the capillary configuration has an advantage over the spiral-wound flat film (42) and plate-and-frame devices. Such fibers achieved first niche commercial prominence in such medical purposes as membrane oxygenators. Commercializations and development activities are now occurring rapidly at a number of corporations including A/G Technology, Dow Chemical Co., Du Pont, Monsanto, Perma Pure, Toyobo, Ube Industries, and Union Carbide. For high pressure applications it appears glassy rigid polymers as polysulfone, polycarbonate polyaramid, and polyamide are preferred. Sintered inorganics, ie, iron, nickel, aluminas, and carbides are involving much attention. Glassy polymers have an amphorous polymeric material that is below its softening or glass-transition temperature under the conditions of use. This concept is opposed to a rubber polymer which is employed above its glass-transition temperature. The rigidity of the glassy polymers offer better selectivity than the rubbery polymers (24).

Another significant area of development and commercialization is pervaporation. These membranes are dense, rather than porous structures. Generally asymmetric composite constructions are employed with the ultrathin membranes on an open support. Key economic variables for commercial viability require high productivity, excellent selectivity, life expectancy, and low internal pressure drops. Hollow fibers inherently can satisfy these needs vs flat films but problems with temperature variations with length have to be considered. Some common membrane materials are poly(vinyl alcohol), silicones, cellulose acetates, polysulfones, polyacrylic acid, polyetheramides, polyolefins, ion-exchange resins, and combinations of these. Acceptance is being gained for use in separation and recovery of liquid mixtures, ie, dehydration of ethanol, isopropyl alcohol, and ethylene glycol.

Supported liquid membranes in the configuration of porous hollow-fiber matrices (29,43) consist of an organic solution of an extracting agent (carrier), absorbed on a thin microporous support. The supported liquid membrane separates the aqueous solutions initially containing the permeating ions (feed solution) from the aqueous solution initially free of these ions. In this process, metal ions are pumped across the membrane, from a low to a highly concentrated solution, via coupling of metal ion to another ion. In such a process, the recovery of metals (eg, copper, gold, silver, platinum, or uranium) from dilute leach solution is possible. Within the same concept is the recently developed hollow fiber containing liquid membrane (24) for gas and liquid separations. In this process, thousands of microporous hydrophobic hollow fibers are packed into a permeator shell filled with the aqueous solution that acts as the membrane. The fibers are present in two distinct sets, feed set and the sweep set, with the ends of each set being separated. The feed gas mixture flows through the lumen of the feed hollow fibers and is swept into the second hollow fibers via an aqueous nonwetting membrane liquid solution. The liquid membrane concept is experimental at this time.

Considerable research and development effort is being placed on a chlorine-resistant membrane that will maintain permeability and selectivity over considerable time periods (years). This polymer activity is not limited to hollow fibers, but the thick assymetric skin of hollow-fiber construction might offer an advantage in resolving the end use need as opposed to the ultrathin flat-sheet composite membranes.

When pressurized liquid is used to separate micrometer-size particles from fluids, the process is called microfiltration. Generally particle sizes are from 0.02 to 10 μm. Thus compared to ultrafiltration and reverse osmosis, fluxes and pore sizes are large, osmotic pressure low, and pressures moderate. Two types of microfiltration processes exist, crossflow and deadend. Commercially, the former is growing at the expense of the latter. Many polymeric materials are used, eg, polysulfone, polyproplyene, nylon, PTFE, and cellulose derivatives. The key to industrial success appears to be mainly commercial factors of cost per unit permeate flow, reliability, life, and on-stream time.

Sorbent fibers were developed in the late 1970s, particularly by California Institute of Technology and Gulf South Research Institute. The concept of encapsulation within a hollow fiber, gas, liquid, suspended solid, catalyst, or others, has potential. For example, Massachusetts Institute of Technology is researching this field in conjunction with developing an artificial pancreas (44). Similarly, University of Minnesota, Cellex Biosciences, Regenerex, and Baylor College of Medicine are looking at artificial livers using hollow fibers (45).

Progress, however, in the 1980s and 1990s has been slow in medicine, agriculture, waste recovery, and other fields mostly due to low value-in-use and high development and acceptance costs.

A full listing of all U.S. patents issued between February 1970 through February 1981 is given in Reference 26. Similar related material on membranes, ultrafiltration, and reverse osmosis can be found in References 46–49.

BIBLIOGRAPHY

"Hollow-Fiber Membranes" in *ECT* 3rd ed., Vol. 12, pp. 492–517, by I. Cabasso, Gulf South Research Institute.

1. H. I. Mahon and B. J. Lipps, in N. M. Bikales, ed., *Encyclopedia of Polymer Science and Technology*, Vol. 15, Interscience Publishers, a division of John Wiley & Sons, Inc., New York, 1971, p. 258.
2. R. W. Baker and H. K. Lonsdale, in A. C. Tanquary and R. E. Lacey, eds., *Controlled Release of Biologically Active Agents*, Plenum Press, Inc., New York, 1974.
3. I. Cabasso and co-workers, *J. Appl. Polym Sci.* **21**, 1883 (1977).
4. I. Cabasso and I. Leon, *Boston AIChE 80th National Meeting*, microfiche no. 32, 1975.
5. S. Loeb and S. Sourirajan, *Adv. Chem. Ser.* **38**, 117 (1962).
6. T. A. Orafino, in S. Sourirajan, ed., *Reverse Osmosis and Synthetic Membranes*, National Research Council, Ottawa, Ontario, Canada, 1977, p. 313.
7. I. Cabasso, E. Klein, and J. K. Smith, *J. Appl. Polym. Sci.* **20**, 2377 (1976).
8. *Permasep Products Engineering Manual*, E. I. du Pont de Nemours & Co., Inc., Wilmington, Del., 1992.
9. F. L. Singer, *Strength of Materials*, 2nd ed., Harper & Row, New York, 1962, p. 17.
10. K. H. Keller, *Fluid Mechanics and Mass Transfer in Artificial Organs*, American Society of Artificial Internal Organs, Apr., 1973.
11. R. H. Haynes, *Am. J. Physiol.* **198**, 1193 (1960).
12. L. Rebenfeld, in N. M. Bikales, ed., *Encyclopedia of Polymer Science and Technology*, Vol. 6, Interscience Publishers, a division of John Wiley & Sons, Inc., New York, 1967, p. 505.
13. U.S. Pat. 4,035,459 (July 12, 1977), R. Kesting (to Chemical Systems, Inc.).
14. D. R. Lloyd, ed., "Materials Science of Synthetic Membranes," *ACS Symposium Series 269*, ACS, Washington, D.C., 1985.
15. I. Cabasso, E. Klein, and J. K. Smith, *J. Appl. Polym. Sci.* **21**, 165 (1977).
16. I. Cabasso, in A. Cooper, ed., *Ultrafiltration*, Plenum Press, New York, 1980.
17. U.S. Pat. 4,938,872 (July 3, 1990), J. W. Strantz and W. J. Brehm (to E. I. du Pont de Nemours & Co., Inc.).
18. I. Cabasso and A. P. Tamvakis, *J. Appl. Polym. Sci.* **23**, 1509 (1979).
19. A. F. Allegrezza, Jr. and co-workers, *Desalination* **20**, 87 (1977).
20. U.S. Pat. 4,277,344 (July 7, 1981), John E. Cadotte (to Filmtec Corp.).
21. A. Rembaum and E. Selegny, eds., *Polyelectrolytes and Their Applications*, Vol. 2, Reidl Publishing Co., Boston, 1975.
22. E. Klein and co-workers, *Polym. Lett.* **13**, 45 (1975).
23. U.S. Pat. 3,690,465 (Oct. 15, 1970), P. R. McGinnis and G. J. O'Brien (to E. I. du Pont de Nemours & Co., Inc.).
24. W. S. W. Ho and K. K. Sirkar, eds., *Membrane Handbook*, Von Nostrand Reinhold, New York, 1992.
25. M. A. Newberry, *Textbook of Hemodialysis for Patient Care Personnel*, CC Thomas, Springfield, Ill., 1989.

26. Jeanette Scott, ed., *Hollow Fibers Manufacture and Applications, Chemical Technology Review No. 194*, Noyes Data Corp., Park Ridge, N.J., 1981.
27. T. A. Orofino, *Progress Report No. 549*, Office of Saline Water Research and Development, U.S. Dept. of the Interior, Washington, D.C., May 1970.
28. J. C. Gomez and co-workers, *Progress in Membrane Biotechnology*, Birkhauser Verlag, Boston, 1991.
29. L. Cecille and J. C. Toussaint, *Future Industrial Prospects of Membrane Processes*, Elsevier Applied Science, London, 1989.
30. Y. Sakai and co-workers, *Preprint, 178th Meeting ACS, Division of Colloid and Surface Chemistry*, Washington, D.C., Sept. 1979.
31. Y. Sakai and H. Tanzawa, *J. Appl. Poly. Sci.* **22**, 1805 (1978).
32. V. P. Caracciolo, N. W. Rosenblatt, and V. J. Tomsic, in Ref. 6, p. 343.
33. F. S. Model and L. A. Lee, in H. K. Lonsdale and H. E. Podall, eds., *Reverse Osmosis Membrane Research*, Plenum Press, New York, 1972, p. 285.
34. F. S. Model, H. J. Davis, and J. E. Poist, in Ref. 6, p. 23.
35. R. Schnabel and W. Vaulont, *Desalination* **24**, 249 (1978).
36. R. R. Bhave, *Inorganic Membranes Synthesis, Characteristics and Applications*, Van Nostrand Reinhold, New York, 1991.
37. U.S. Pat. 3,567,632 (Mar. 1971) J. W. Richter and H. H. Hoehn (to E. I. du Pont de Nemours & Co., Inc.).
38. H. Yasuda, in Ref. 6, p. 263.
39. E. Klein and co-workers, *Trans. Am. Soc. Artif. Intern. Organs* **24**, 127 (1978).
40. U.S. Pat. 3,875,008 (Apr. 1, 1975), M. Yoshino, Y. Hashino, and M. Morishita (to Asahi Kasei Kogoyo Kabushiki Kaisha).
41. *Cuprophan Technical Information Bulletin No. 12.1 Enka AG*, Product Group, Wuppertal, Germany, 1977.
42. D. L. MacLean and T. E. Graham, *Chem. Eng.*, 54 (Feb. 25, 1980).
43. W. C. Babcock and co-workers, *Symposium on Separation Science and Technology for Energy Applications*, sponsored by DOE and Oak Ridge National Laboratory, Gatlinburg, Tenn., Oct. 1979, p. 46.
44. *Chem. Eng. Prog.*, 13 (Nov. 1993).
45. *In Vivo, the Business & Medicine Report*, 17 (Nov. 1993).
46. J. Scott, ed., *Membrane and Ultrafiltration Technology 1980, Recent Advances, Chemical Technology Review No. 147*, Noyes Data Corp., Park Ridge, N.J., 1980.
47. J. Scott, ed., *Desalination of Seawater by Reverse Osmosis, Pollution Technology Review No. 75*, Noyes Data Corp., Park Ridge, N.J., 1981.
48. S. Torrey, ed., *Membrane and Ultrafiltration Technology, Developments Since 1981*, Noyes Data Corp., Park Ridge, N.J., 1984.
49. R. W. Baker and co-workers, *Membrane Separation Systems, Recent Developments and Future Directions*, Noyes Data Corp., Park Ridge, N.J., 1991.

IRVING MOCH, JR.
E. I. du Pont de Nemours & Co., Inc.

HOLMIUM. See LANTHANIDES.

HOLOGRAPHY

Holography involves an image recording technique whereby the complete wave information (optical or other wave phenomena) emanating from a three-dimensional scene is captured in a suitable material and a reconstruction step where the information is replayed to reconstruct the true three-dimensional character of the recorded scene. The three-dimensional character of scenes reconstructed with a hologram is illustrated by photographs of the reconstruction from a single hologram taken from different viewing angles as shown in Figure 1. Each picture captures a unique perspective of the true scene as if the original scene were being viewed, to the extent that a properly prepared holographic portrait can seem very much as if the person were being viewed directly. In fact, a properly prepared hologram can exactly duplicate a particular wavefront at a specified wavelength. This is radically distinct from conventional photography where the film records only the intensity fluctuations of a flat projection (image) of the original scene with complete disregard of the relative phase distribution which stems from the wave nature of light. As implied by the name hologram which literally means "whole record," a hologram contains the complete intensity and phase information associated with a given scene.

In optical holography, the phase information carried by an object wave to be recorded is transformed into a complex interference pattern by combining it with a mutually coherent reference beam. The resulting nonuniform intensity distribution is then recorded with either standard photographic film or other specialized materials for holography. This indirect method of capturing both the intensity and phase information is necessary because the extremely high frequencies of light make direct sensing and recording impossible. This is in sharp contrast to acoustic and microwave phenomena where direct coherent sensing and recording are performed routinely with coherently phased array sonar and radar.

With the advent of new materials including polymers and photorefractive media, which require less handling complexity, relatively recent applications involving the storage and manipulation of information have been introduced. Significant advances in optoelectronics such as the introduction of compact semiconductor and solid-state lasers have also helped to fuel such new applications. This article describes a variety of such applications in addition to display holography which is more familiar and widespread. Before addressing the applications topics, however, the basic concepts of holographic recording and reconstruction must be addressed as well as the variety of materials and recording/reconstruction geometries that are currently available. There have been excellent discussions of the history behind the invention and development of holography which by itself is both fascinating and enlightening (1,2).

The Holographic Principle

Holography involves the recording of the mutual interference pattern due to two mutually coherent optical fields. A generic holographic recording experiment is shown in Figure 2: an expanded and collimated laser beam is split into two paths, with one falling directly on the holographic material and the other scattering off

(a)

(b)

(c)

Fig. 1. Reconstructions from a hologram. (a) Normal viewing angle; (b) viewing from left side; (c) viewing from right side. Courtesy of Mr. Tae Jin Kim.

339

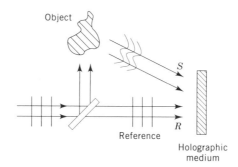

Fig. 2. Holographic recording system. S is the object wave and R the plane wave reference.

an object to be collected on the same material. The first is called the reference and the second the object. Although variations of this basic arrangement exist, its simplicity is most suitable for the present discussion. The intensity distribution falling on the film is given by equation 1, where S is the object wave amplitude and R represents the plane wave reference.

$$I \propto |S + R|^2 = |S|^2 + |R|^2 + SR^* + RS^* \tag{1}$$

Film and other holographic recording media respond to the overall exposure or total optical energy per unit area deposited during the exposure time τ given by $\epsilon = \tau I$. An assumed linear relationship between the exposure and the transmittance of the film after exposure and development is given by equation 2, where t_0 is the average transmittance of the film and κ is a constant that depends on the material and processing.

$$t = t_0 + \kappa\tau(|S|^2 + |R|^2 + SR^* + RS^*) \tag{2}$$

The last two terms in the transmittance expression enable the holographic reconstruction process.

Reconstruction of the object wave is achieved by illumination of the developed hologram with the reference wave as shown in Figure 3a. The diffracted wave amplitude from the hologram is given by equation 3, where the first term represents the attenuated reference wave after passage through the hologram.

$$A_{\text{diff}} \propto tR = [t_0 + \kappa\tau(|S|^2 + |R|^2)]R + \kappa\tau|R|^2S + \kappa\tau R^2S^* \tag{3}$$

The second term represents a virtual image of the original object signal which can be viewed by an observer looking at the hologram along the original object signal direction. The last term represents an unfocused wave which carries the complex conjugate of the signal amplitude emerging at a distinct angle from the hologram as shown in Figure 3a. This term can be brought to a focus by reconstructing the hologram with a plane wave traveling in exactly the opposite direction as the original reference wave as shown in Figure 3b. If $R = e^{ikx}$ represents the

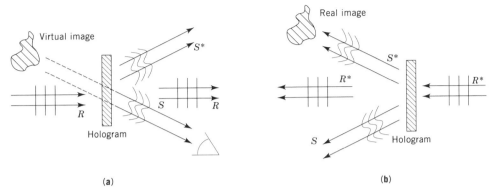

Fig. 3. Holographic reconstruction. (**a**) Reconstruction of virtual image of object; (**b**) reconstruction of real image of object.

original plane wave reference, then $R^* = e^{-ikx}$ represents a wave traveling in the opposite case. In such a case, the diffracted wave amplitude is given by equation 4, where the last term represents a real image formed precisely at the location of the original object.

$$A_{\text{diff}} \propto tR^* = [t_0 + \kappa\tau(|S|^2 + |R|^2)]R^* + \kappa\tau R^{*2}S + \kappa\tau|R|^2S^* \qquad (4)$$

Yet another variation in the reconstruction is to use the object wave to reconstruct the reference wave. This will be pursued further in a later section that describes the use of holography for pattern recognition.

Hologram Varieties

The rich variety in the types of holograms stems from the specifics of how the interference patterns are recorded. Although a more complete classification of the various hologram types is possible, the most important classification parameters are the material perturbation (amplitude or phase), material thickness (thin or thick), and the recording format (transmission or reflection). Most holograms can be classified using a combination of these three parameters. In describing the basic hologram types, only the simplest planar gratings recorded by two intersecting plane waves are considered. These results can be generalized using the principles of coherent optics theory to address more complex problems involving nonplanar waves.

Material Perturbation. Exposure of a suitable holographic material to recording waves must induce perturbations of its optical transmission properties to record holograms. The perturbations may include a complex set of physical effects (3), but in most cases they can be lumped into one of two categories: (1) absorption perturbations and (2) phase perturbations. A unit amplitude plane wave at wavelength λ passing through a homogenous material of thickness T, refractive index n, and absorption constant α emerges with amplitude A, where the first multiplicative term describes the effect of material absorption and the second represents the imposed change in phase of the wave.

$$A = e^{-\alpha T} e^{i\frac{2\pi}{\lambda} nT} \tag{5}$$

Holographic information can be recorded by spatially modulating either (or a combination of) the absorption or phase (index or thickness change).

In choosing a material for a particular application, the relative merits of each candidate must be considered. The material sensitivity, which can be loosely defined as the energy density that is required to induce a perturbation in either absorption or phase that represents a significant portion of the operating range, is an important characteristic. Holographic resolution defined as the number of line pairs per millimeter that the material can support is another important consideration. These parameters along with other less prominent but still important quantifiers are used to describe holographic materials in Table 1. In the following discussion the physical mechanisms that enable holographic recording in the materials are presented.

Absorption Materials.   Materials that respond to incident exposure by absorption perturbations include photographic film and photochromic glasses/crystals (see PHOTOGRAPHY; CHROMOGENIC MATERIALS, PHOTOCHROMIC). The ideal absorption material would respond with an amplitude transmittance (the first multiplicative factor in eq. 5) that is linearly proportional to the recording exposure. All materials exhibit such a linearity only over a limited range of exposures as illustrated by the exposure response curve shown in Figure 4. The nonlinearities are manifested as spurious gratings and result in reconstruction distortions.

Photographic film consists of fine grains of silver halide compounds and sensitizing agents imbedded in gelatin. The film is coated on a substrate which is a transparent plate or sheet (glass or plastic). The typical thickness of the coatings range from 5 to 15 μm and the size of the photosensitive grains are typically less than 0.1 μm. Exposure to light and subsequent wet processing converts the silver halide into metallic silver which modifies the absorption. Although the wet processing that is required to develop the holograms is inconvenient, this medium is still the most popular because of its high sensitivity (0.3–10 μJ/cm^2 to effect 50% change in transmission), commercial availability, capability of sensitizing to various wavelengths, and high resolution (1000–3000 line pairs/mm). An additional feature is that by a bleaching process, the developed absorption hologram can be converted into a phase hologram where the absorptive silver is changed into a transparent compound.

Phase Materials.   Phase holograms can be recorded in a large variety of materials, the most popular of which are dichromated gelatin, photopolymers, thermoplastic materials, and photorefractive crystals. Dichromated gelatin and some photopolymers require wet processing, and thermoplastic materials require heat processing. Photorefractive crystals are unique in that they are considered to be real-time materials and require no after-exposure processing.

Dichromated gelatin (DCG) as a holographic recording medium offers a large index modulation capability, low absorption/scattering, and high

Table 1. Characteristics of Holographic Materials

Material[a]	Reference	Sensitivity, (J/cm^{2b})	Resolution, lines/mm	Wavelength, nm	Processing method
silver halide emulsion[c]	4	$3.5 \times 10^{-7} - 5 \times 10^{-4}$	1000–5000	450–700	wet/bleach
DCG[d]	2	1.5×10^{-2}	5000	355–700[e]	wet
photopolymer	5,6	$80{-}100 \times 10^{-3}$	5000	400–700	uv/heat
thermoplastic	2,4	10^{-5}	750–1500[e]	350–700	heat
photorefractors					
semiconductors	7	5×10^{-5}	1000[f]	700–1300	none
sillenites	8	1.5×10^{-2}	1000[f]	400–700	none
ferroelectrics	9–11	$10^{-2} - 10^{f}$	5000[f]	400–700	none

[a]Subject to phase perturbations unless otherwise noted.
[b]To convert J/cm^2 to dyn/cm, multiply by 10^7.
[c]Also absorption perturbations.
[d]DCG = dichromated gelatin.
[e]$\lambda > 550$ nm requires sensitizing dyes.
[f]Due to the wide variability of material parameters, these figures represent only rough estimates.

343

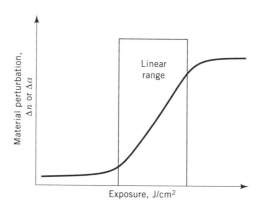

Fig. 4. Material exposure response curve.

resolution, and is a popular material for display holograms. The only drawbacks are that it requires wet chemical processing, plates must be fabricated since they are not commercially available in ready-to-use form, and the developed plates must be sealed as the material is hygroscopic. A DCG plate consists of a gelatin layer (typically $1-15$ μm thick) with a small amount of ammonium dichromate $((NH_4)_2Cr_2O_7)$ spun on a glass substrate. Standard photographic plates can also be converted into a DCG plate by a well-known process (12). Exposure to light ($\lambda = 488$ nm for DCG without sensitizing dyes) causes the gelatin to harden with unexposed areas remaining relatively soft. Wet processing produces an index modulation through the introduction of vacuoles (voids whose sizes are smaller than the wavelength) (13) with the resulting index variation being as high as 0.08 (12).

Certain polymers doped with appropriate dyes can also be used to create phase holograms. Typically, exposure to light causes local polymerization of the material causing spatial variations in the monomer to polymer ratio, which after appropriate processing leads to index perturbations that can be almost as high as that achieved in DCG . Polaroid's DMP128 (14) and Du Pont's Omnidex (4,5,15) materials are among the most popular. Photopolymers can be spun or rolled onto substrates or cast into molds, depending on the particulars of the material. The Polaroid material requires wet processing while the Du Pont material is developed after exposure by uv light and heat treatment.

Holograms can be recorded in certain thermoplastic materials in the form of surface relief gratings. The so-called photothermoplastics (6) usually consist of a thermoplastic on top of a photoconductor layer, all deposited on a glass substrate. The medium is prepared by uniformly applying a static charge on the top layer (thermoplastic) usually through corona discharge methods. Exposure to a holographic interference pattern then causes a relaxation of the static charge where the intensity is high and thus imprints a spatially varying electric field across the thermoplastic layer. Heating the entire structure causes the thermoplastic to physically deform in response to the intensity of the local electric field, thus creating a surface relief pattern which can be read out holographically. The medium can then be reused by heating to erase the recorded gratings. Complete systems to record holograms using thermoplastics are available (6).

The photorefractive effect can be used to record holograms in a variety of materials which exhibit a combination of impurity levels in the band structure, charge transport mechanisms, and the linear electrooptic (Pockel's) effect. This includes semiconductors (7), sillenites (8), ferroelectrics (9–11), and most recently, polymers (16) and exotic multiple quantum well semiconductor structures (17). In such materials, the holographic interference pattern causes a spatially nonuniform excitation of charges out of the impurity trap levels into the conduction (or valence for holes) band where they move under the influence of diffusion and drift to be retrapped in relatively darker regions. The multistep process is repeated until a steady state is reached yielding a grating in the form of variations in the space charge field. The space charge electric field in turn modulates the local refractive index via the linear electrooptic effect resulting in a phase hologram. The time required to reach the steady state is approximately inversely proportional to the exposure intensity, and the proportionality constant depends on the material that is used. At modest intensity levels of 1 W/cm^2, the time constant can range from many seconds in the slowest ferroelectrics to microseconds in the fastest semiconductors. Recent developments in these materials have yielded thermal and electrical techniques of fixing such holograms so that they are impervious to further exposures of light (18–22), making the materials useful for not only real-time applications but also for long term storage.

Material Thickness. Holograms recorded in a material whose physical thickness is small when compared with the grating spacing is considered a thin hologram; the effect on an incident optical wave is characterized by a spatially varying transmittance function. The output of such a hologram can be determined by multiplying the input field with the transmittance of the hologram and analyzing the resultant diffraction using the principles of coherent optics theory. The most prominent features of thin gratings are lack of strong angular selection in reconstruction and limited diffraction efficiency. A quantity that is often used to delineate the boundary between thin and thick holograms (thick implies that angular/wavelength selectivity effects are exhibited) is where n_0 is the average refractive index of the hologram, Λ_g is the grating spacing, and T is the physical thickness of the hologram.

$$Q = \frac{2\pi\lambda T}{n_0 \Lambda_g^2}$$

Values of $Q < 1$ imply a thin hologram and $Q > 1$ imply a thick hologram (23).

Thin Holograms. A thin amplitude grating (recorded in, for example, silver halide film) formed by two plane waves at wavelength λ_w, intersecting at a half angle θ_w as illustrated in Figure 5a, can be characterized by the transmittance t, where Λ_g is the grating spacing and $t_0 = t_1 = 0.5$ represents the ideal case of full modulation depth.

$$t_a = t_0 + t_1 \cos(K_g x), \qquad K_g = \frac{2\pi}{\Lambda_g} = \frac{4\pi}{\lambda_w}\sin\theta_w \qquad (6)$$

A uniform plane wave at wavelength λ, incident at an angle θ as shown in Figure 5b, has a complex amplitude given by equation 7 at the hologram plane.

$$A = e^{i\frac{2\pi}{\lambda}\sin\theta x} \qquad (7)$$

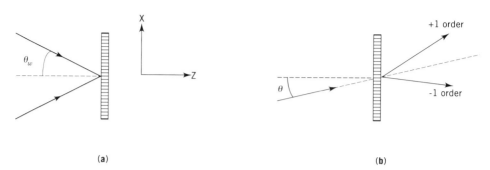

(a) (b)

Fig. 5. Uniform plane wave grating where (**a**) is the recording, and (**b**) the readout.

The diffracted field from the hologram is then given by the amplitude function as shown in equation 8.

$$A_{\text{diff}} = At_a = t_0 A + \tfrac{1}{2} t_1 e^{i\alpha_+ x} e^{i\beta_+ z} + \tfrac{1}{2} t_1 e^{i\alpha_- x} e^{i\beta_- z},$$

$$\alpha_\pm = 2\pi\left(\frac{\sin\theta}{\lambda} \pm \frac{1}{\Lambda_g}\right) = 2\pi\left(\frac{\sin\theta}{\lambda} \pm \frac{2\sin\theta_w}{\lambda_w}\right), \qquad \beta_\pm = \sqrt{\left(\frac{2\pi}{\lambda}\right)^2 - \alpha_\pm^2} \qquad (8)$$

This is easily derived by expanding equation 6 in terms of complex exponentials. The first term represents the undeflected passage of the readout beam whereas the second and third terms represent the waves diffracted by the hologram. Although there is a range of parameters θ, λ, λ_w, for which one or both diffracted orders are exponentially attenuated and thus do not propagate, such a thin grating can be reconstructed from a wide range of angles and wavelengths, albeit with some distortions in the case of information carrying holograms. The maximum efficiency with which energy is diverted into the diffracted orders is $\eta = 1/16$ for the ideal case of full modulation grating ($t_0 = t_1 = 1/2$).

If the index of refraction of a thin material were modulated in lieu of its absorption, the resultant transmittance function for a grating prepared as in the absorption case is given by equation 9 where n is the average index of the thin film, Δn is the amplitude of the index perturbation, and T is the thickness of the film.

$$t_\phi = e^{ia\cos K_g x} \qquad a = \frac{2\pi}{\lambda}\Delta nT \qquad (9)$$

The diffracted amplitude from illuminating such a grating with a unit plane wave normal to the surface is easily calculated again by resolving equation 9 into complex exponentials (as in eq. 10) where $J_m(a)$ is the mth Bessel function.

$$A_{\text{diff}} = \sum_{m=-\infty}^{\infty} J_m(a)\,(-i)^m e^{imK_g x} e^{i\sqrt{k_0^2 - m^2 K_g^2}\,z} \qquad k_0 = \frac{2\pi}{\lambda} \qquad (10)$$

The chief factor that distinguishes phase and absorption gratings is the appearance of multiple diffraction orders. Although the expansion has an infinite number of terms, the diffraction amplitude is zero beyond a certain number due again to evanescent waves. Each of the first diffraction orders (corresponding to $m = \pm1$) has a diffraction efficiency of $[J_{\pm1}(a)]^2$ whose maximum value is approximately equal to 0.34. Thus the maximum diffraction efficiency from a thin sinusoidal phase grating is 34%. By using optimized blazed phase profiles, the diffraction efficiency into the first order can approach unity as is done for diffraction gratings for spectroscopy (1).

Thick Holograms.   When the recording medium is significantly thicker than the grating spacing, the recorded hologram is said to be a volume hologram and is marked by high diffraction efficiencies and strong angular–wavelength selectivity during reconstruction. The interference patterns are recorded as either absorption or index gratings throughout the volume of the material. The two basic configurations for recording volume holograms are transmission and reflection geometries as illustrated in Figure 6. The distinction between the two is simply that in recording transmission gratings, the two interacting waves enter from the same side of the medium, whereas the recording waves enter from opposite faces to prepare the reflection hologram. The resultant grating planes are nominally perpendicular to the hologram face in the transmission geometry and they are parallel in the reflection case.

Because the losses tend to be excessive in volume absorption gratings, volume holograms are realized in most cases using index (phase) gratings. The maximum achievable diffraction efficiency (theoretical value assuming ideal parameters) from an absorption transmission grating is 3.7 and 7.2% from an absorption reflection hologram (1). Much higher diffraction efficiencies can be obtained using phase volume holograms; basic expressions for estimating them are given below.

The most distinctive feature of thick holograms is the angular and wavelength selective reconstruction. With increasing thickness of the hologram, such effects (also known as Bragg effects which are analogous to the angularly selective x-ray diffraction in solid-state physics) become more pronounced. The Bragg effect also suppresses higher diffraction orders which are prevalent in thin-phase gratings. Maximum diffraction efficiency is achieved when the reading beam matches the angle and wavelength of one of the beams used to write the grating. Small deviations $\Delta\theta$ and $\Delta\lambda$ from the Bragg angle and wavelength yield rapid decrease in the diffraction efficiency. A measure of the total detuning from the

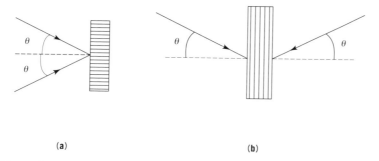

(a)   (b)

Fig. 6.   Volume hologram recording where (**a**) is the transmission, and (**b**) the reflection.

optimum angle and/or wavelength is given by the phase mismatch function $\Delta\phi$ (24), where T is the thickness of the grating, Λ_g is the grating spacing, n_0 is the average refractive index in the hologram, θ is the half-angle defined in Figure 6, and $\psi = \pi/2$ for transmission holograms and 0 for reflection holograms.

$$\Delta\phi = \frac{\pi T}{2\Lambda_g \cos\theta}\left[2\Delta\theta \sin(\psi - \theta) - \frac{\Delta\lambda\pi}{\Lambda_g n_0}\right] \qquad (11)$$

The diffraction efficiency for the transmission geometry is given by equation 12 (24) where n_1 is the amplitude of the index grating.

$$\eta_{\text{trans}} = \frac{\sin^2\sqrt{\phi^2 + (\Delta\phi)^2}}{1 + \left(\dfrac{\phi}{\Delta\phi}\right)^2} \qquad \phi = \frac{\pi n_1 T}{\lambda \cos\theta} \qquad (12)$$

For the reflection geometry shown in Figure 6b, the diffraction efficiency is given by equation 13.

$$\eta_{\text{refl}} = \frac{1}{1 + \dfrac{1 - (\Delta\phi/\phi)^2}{\sinh^2\sqrt{\phi^2 - (\Delta\phi)^2}}} \qquad (13)$$

For both geometries the diffraction efficiency approaches unity in value for $\Delta\phi = 0$ with the transmission hologram exhibiting a periodic behavior (24,25) efficiency as a function of the grating strength ϕ, whereas the reflection efficiency exponentially approaches unity.

Computer-Generated Holograms.   Instead of physically recording an object wave by mixing it with a reference on a suitable recording medium, holograms can be prepared in the following two steps. First, the desired object wave distribution can be calculated across the plane where the holographic plate would have been positioned. The object can be real or fictitious since a computer is used to simulate the effects of diffraction. Next, a transparency is produced using computer-controlled photolithography, which can be as simple as a laser printer with proper photographic demagnification, that yields a reconstruction of the object when properly illuminated. Because of the flexibility introduced by the computer in the calculation stage, either off-axis or on-axis holograms with high efficiency and quality can be prepared (26). Although the former variety is done quite routinely by optical means, preparation of on-axis holograms optically introduces undesirable noise terms which severely degrade the reconstruction quality. Such is not the case for computer-generated on-axis holograms because the physical limitations of the recording geometry are removed.

In principle, masks with complex valued transmittance functions can be prepared using a composite of two masks, with one implementing only the amplitude perturbations and the other having only a phase distribution via surface deformations; usually one method or the other is used. The binary detour-phase method (27) and its variants (28,29) yield masks which have a collection of transparent holes in an opaque background where the size of the holes encode the amplitude and whose spatial position within some predefined limits encode the phase. Because of the binary nature of the masks thus produced, such holograms typically yield higher order diffraction and diffract only a small percentage of the light into the useful first order.

Kinoforms are phase-only masks which are prepared by setting the amplitude values to unity and changing only the phase across the hologram (30). The kinoform has the advantage that proper preparation can lead to holograms which can diffract almost all of the light into the useful first order.

By using electrically addressed spatial light modulators (31) in place of photolithographic masks, the preparation of holograms can be made essentially real-time with the reconstruction changing dynamically as fast as the computer calculations and the spatial light modulator response allow (32). Typically, however, the limited resolution in such devices leads to a sequentially synthesized reconstruction where the entire object field is divided into smaller units and a sequence of reconstructions performed rapidly appears as a large field reconstruction to the observer. An impressive real-time holographic display system using a Connection Machine (a parallel supercomputer) for the calculations has been demonstrated at MIT (33).

Display Holograms

Display holography is by far the most familiar and well-developed holographic application to date with several museums dedicated to holographic art. These holograms range from extremely simple ornaments to large color portrait holograms with strikingly real appearance. The one practical feature of all display holograms is that they be observable with white light so that potentially dangerous laser beams need not be used.

Reflection Hologram. Perhaps the simplest such hologram is a reflection hologram (34). A laser beam is passed through a holographic plate to scatter off a nearby object as shown in Figure 7. The beam that passed through the plate mixes with the backscattered waves to form the holographic fringes which are recorded. When developed, the plate can be illuminated with an incoherent point source of light and the observer sees the object with full parallax through the plate. Because the hologram was made with a single wavelength laser beam, the reconstruction, as dictated by the Bragg effect, will be primarily at one wavelength, although, depending on the spatial information that is recorded, various colors may be observed across the reconstruction. As the observer changes his viewing angle, a corresponding shift in the color is noticeable. This geometry has been widely commercialized for ornaments and jewelry.

Achromatic Hologram. Transmission holograms present a challenge for white light viewing since the dispersive nature of holographic gratings result in the various colors smearing the reconstruction unless special care is taken. The simplest realization of an achromatic transmission hologram is to use a lens to form an image of the object on the holographic plate (35). A collimated white light beam illuminating the developed hologram then reconstructs the object where the various color components diverge due to grating dispersion. When the observer focuses on the holographic plate, however, the diverging colors come

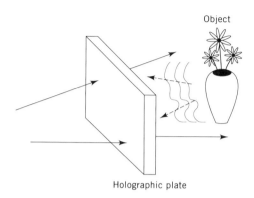

Fig. 7. Reflection display hologram (34).

back into registration and form an achromatic reconstruction. This technique, however, works well only for objects with very limited depth since points on the object whose image points are not precisely at the plane of the holographic plate result in color smearing. A better technique is to use a diffraction grating whose spatial frequency is equal to the average spatial frequency of the transmission hologram to precompensate the white light illumination (36). If the +1 diffraction order from the grating is used to illuminate the hologram to yield a −1 order, the dispersive action of the hologram is compensated by the opposite sign dispersion from the grating. The grating must be sufficiently close to the hologram to minimize color separation, while the unwanted diffracted orders from the grating must be suppressed.

Rainbow Hologram. For the reasons just described, the transmission hologram that is most popular in display applications is the rainbow hologram (37) which achieves object monochromatic reconstruction under white light illumination with depth not possible using other methods. The only drawbacks are that vertical parallax (with respect to the observer) is sacrificed, and the recording process requires either a two-step procedure (37) or a complex one-step apparatus (38). The first drawback does not represent a serious problem since vertical parallax is not appreciated as much as horizontal by human observers. Instead of seeing vertical parallax when the observer moves the viewing angle up and down, the reconstruction changes in color. The second problem has been solved for the most part by process optimization, and large beautiful holograms with striking detail and depth have been produced (39).

True Color Holograms. To record holograms in full color (of objects under white light illumination) as viewed by an observer, three suitably chosen separate laser wavelengths are required. The three colors are chosen such that when mixed in proper proportions, the entire color range of the object to be recorded is covered. The most easily obtained combination which yields good results is the He−Ne laser which emits red light (λ = 633 nm) used with an argon ion laser emitting both green (λ = 515 nm) and blue (λ = 488 nm) light. A wide range of wavelengths can also be obtained with a dye laser system.

The method for recording color holograms is essentially a simple procedure whereby three separate holograms, each formed at a different wavelength, are prepared in the same medium. Readout using the three lasers yields a superposition of three separate reconstructions whose colors are mixed by the observer's eye to yield the desired true colors. Recording the three holograms in a common medium is important for color registration, and care must be taken to keep the object fixed with respect to the holographic plate between the exposures lest color smearing result during the reconstruction.

The simultaneous reconstruction using three different colors yields cross-talk noise components in addition to the desired color reconstruction since a given color will access not only the hologram that was recorded at the same wavelength but also the other holograms. Volume holograms have been used to effectively suppress such cross-talk (40) via the Bragg effect which allows efficient reconstruction only for the particular color and angle that was used during recording.

Composite Holograms.   For applications where lasers cannot be used for the recording process, for reasons of safety or convenience, composite holograms (41) can be prepared from regular photographs of an object taken from a number of different perspectives (along an arc centered on the object). A series of contiguous, narrow strip holograms are then exposed using these photographic slides as the object so that each narrow strip presents a slightly different view to the observer. When properly illuminated, the composite hologram presents a view which changes with the observation angle. Animation effects can be included by moving the object between each photographic exposure. The composite hologram prepared with such a set of pictures then presents a view which appears to move with the observation angle.

HOLOGRAPHIC APPLICATIONS

Holographic Interferometry.   An important industrial application of holography involves the interference of reconstructed and real object waves to measure surface and internal stress and deformations in solid objects (42). The measurement method is nondestructive and can yield highly accurate results. In holographic interferometry, a hologram of the object to be tested is recorded. After it is developed, the hologram is replaced in exactly the same position in which it was recorded. When the object is illuminated, as during the recording phase, and the hologram is illuminated to yield a reconstruction, the virtual image from the hologram coincides with the waves emanating from the object. The two waves interfere, and slight changes in the object's position or shape from its recording conditions result in shifts and deformations in the interference fringes. When fringe pattern is properly interpreted, it yields important information about the detailed changes in the object's shape. A variant of this procedure using time-averaging techniques has been used to study spatiotemporal oscillation modes of vibrating surfaces (43,44).

Holographic Optical Elements.   Although holography has served niche roles in the fabrication of gratings and filters for spectroscopy (45) and the manufacture of distributed feedback semiconductor lasers (46), a wider role is being filled by holographic optical elements to replace or improve traditional imaging components (47,48). By using patterned grating structures which can be produced optically or by means of computer controlled lithography/etching techniques, diffractive elements can be produced to implement a wide range of imaging functions ranging from simple lenses to complex heads-up display systems for modern avionics. Because of rapid advances in lithography/etching techniques developed mainly for the production of electronic circuitry, complex grating structures can be prepared to implement diffractive lenses and beam splitters whose refractive counterparts would be difficult to fabricate. Moreover, the diffractive lenses are much thinner and lighter than their refractive counterparts. The production of master molds of such structures can lead to the mass production of such elements using embossing techniques which are already being used to prepare holographic insignia on various credit cards. Diffractive structures can also be prepared on the surface of refractive lenses to cancel various aberrations including chromatic and spherical aberrations (49).

Holographic Data Storage.   Advances in holographic materials and device technologies such as spatial light modulators and detector arrays are fueling developments in the use of holography for information storage and retrieval (see INFORMATION STORAGE MATERIALS, OPTICAL). For reasons of compactness and storage capacity, the most favorable approach uses a volume hologram in which data is multiplexed using either reference angle (50–54) or wavelength variations (55,56). As opposed to other long-term storage media such as magnetic or optical disks, data in holographic systems are usually organized as pages with each page recorded and retrieved as one entity. In the angular multiplexing scheme, for example, a spatial light modulator is programmed with the data to be recorded (as a spatially varying transmittance function). The object beam passes through the spatial light modulator (also known as the page composer) to acquire the data and is recorded in the holographic medium (thick) with a plane wave reference beam. The next page of data to be recorded is then impressed into the spatial light modulator and a new hologram is recorded in the same medium but with a reference wave whose direction is slightly detuned from the first. After a series of such holograms are exposed, the hologram can be interrogated to yield any particular page of information by illuminating it with the corresponding reference wave. The ability to record and independently access any particular page is a manifestation of the Bragg effect described earlier. Another multiplexing technique uses the wavelength of a tunable laser source to record and access information in a reflection hologram geometry. Instead of the angular orientation of the reference wave being varied, the wavelength of the laser is carefully tuned to record and access data which is also another manifestation of the Bragg effect. Experiments using photorefractive crystals as the recording medium have demonstrated the storage and recall of 5000 pages of information, with each page of information consisting of 320×200 pixels (51). The information can be reconstructed with high fidelity (leading to low error rates) as evidenced by the pictures shown in Figure 8 (51).

Holographic Pattern Recognition.   Whereas the holographic memory application uses the reconstruction of the object beam by illuminating the prepared crystal with a properly oriented reference wave (for the angularly multiplexed case), an object beam can be used to illuminate the crystal to reconstruct the reference waves (57–59). If, as illustrated in Figure 9, a series of patterns are recorded holographically in a crystal using reference waves multiplexed in one orientation, the illumination of the crystal with an arbitrary object pattern yields

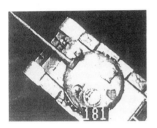

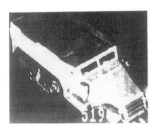

Fig. 8. Holographic data storage experimental results photographs of holographic reconstructions from a multiplexed hologram (51). Courtesy of Dr. Fai Mok.

an array of reconstructed reference waves which are focused to an array of spots using a lens. The amplitude value of each focused spot is proportional to the degree of correlation between the input object beam and the object pattern that corresponds to that particular reference wave. By locating the highest valued intensity spot, the stored object that is closest to the input object can be

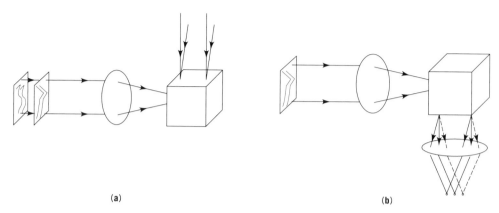

(a)   (b)

Fig. 9. Holographic pattern recognition system. (**a**) Recording an angularly multiplexed hologram; (**b**) forming correlation outputs using arbitrary input object pattern.

quickly identified. This holographic technique yields an extremely rapid method of comparing patterns since the output information is produced in parallel.

Coherent Optics Theory

The principles of coherent optics theory are essential to study and problem-solving in holography (60). The propagation and diffraction of light are described by Maxwell's equations in terms of the amplitudes of the associated electromagnetic field. For the current purpose, discussions can be limited to deal solely with monochromatic light (single wavelength) which can be realized approximately with lasers. By assuming then that the electric field amplitude of the light of frequency ν follows the form of equation 14,

$$E = Re\{E(x)e^{-i2\pi\nu t}\} \tag{14}$$

Maxwell's equations can be combined (61) to describe the propagation of light in free space, yielding the following scalar wave equation:

$$\nabla^2 E(\boldsymbol{x}) + k_0^2 E(\boldsymbol{x}) = 0 \tag{15}$$

where E is the complex electric field amplitude, c is the speed of light in vacuum, and $k_0 = \nu/c = 2\pi/\lambda$ is essentially the photon momentum at wavelength λ. The simplest solution to this equation is the uniform plane wave traveling in an arbitrary direction denoted by the vector $\mathbf{k}$:

$$E(\boldsymbol{x}) = e^{i\mathbf{k}\cdot x} \tag{16}$$

where the propagation vector is required to satisfy $|\mathbf{k}| = k_0$.

Any field amplitude distribution and associated propagation effects can be described equivalently by a superposition of plane waves of appropriate amplitude and direction provided that every component plane wave satisfies equation 16. If, for example, an optical field amplitude given by the function

$$a(x,y) = \frac{1}{2\pi} \int \int_{-\infty}^{+\infty} A(k_x, k_y) e^{i(k_x x + k_y y)} \, dk_x \, dk_y \tag{17}$$

is imposed across the input plane as in Figure 10 where $A(k_x, k_y)$ is the Fourier transform of $a(x, y)$, then the solution describing the effects of propagating along a distance z is given by the amplitude distribution

$$\hat{a}(x,y,z) = \int \int_{-\infty}^{+\infty} A(k_x, k_y) e^{i(k_x x + k_y y)} e^{iz\sqrt{k_0^2 - k_x^2 - k_y^2}} \, dk_x \, dk_y \tag{18}$$

If $a(x,y)$ is spatially bandlimited in extent so that

$$A(k_x, k_y) = 0, \quad \text{for } \sqrt{k_x^2 + k_y^2} > \frac{2\pi}{\Delta} \tag{19}$$

where Δ essentially measures the finest spatial feature in the amplitude distribution $a(x, y)$, the exact diffraction integral can be reduced to the familiar Fresnel form as long as the associated assumption relating Δ, λ, and z is met:

$$\hat{a}(x,y,z) = \frac{\pi e^{ik_0 z}}{i\lambda z} \int \int_{-\infty}^{+\infty} a(x', y') e^{i\frac{ik_0}{2z}[(x'-x)^2 + (y'-y)^2]} \, dx' \, dy' \tag{20}$$

provided that

$$\frac{z\lambda^3}{2\Delta^4} < 1$$

This expression is the main tool used in describing diffraction effects associated with Fourier optics. Holographic techniques and effects can, likewise, be approached similarly by describing first the plane wave case which can then be generalized to address more complex distribution problems by using the same superposition principle.

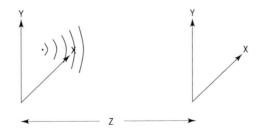

Fig. 10. Diffraction analysis coordinate system.

BIBLIOGRAPHY

"Holography" in *ECT* 3rd ed., Vol. 12, pp. 518–538, by W. T. Cathey, University of Colorado.

1. P. Hariharan, *Optical Holography: Principles, Techniques, and Applications*, Cambridge University Press, Cambridge, 1984.
2. R. J. Collier, C. B. Burkhardt, and L. H. Lin, *Optical Holography*, Academic Press, Inc., New York, 1971.
3. H. M. Smith, ed., *Holographic Materials, Topics in Applied Physics*, Vol. 20, Springer-Verlag, Berlin, 1977.
4. A. M. Weber, *SPIE OE/LASE Conference Proceedings*, V.1212-04, Los Angeles, Jan. 1990.
5. S. Zager and A. M. Weber, *SPIE Conference Proceedings*, V.1461, San Jose, Calif., Feb. 1991, pp. 58–67.
6. *Newport Corporation 1992 Catalog*, Irvine, Calif.
7. M. B. Klein, *Opt. Lett.* **9**, 350 (1984).
8. Ph. Refregier and co-workers, *Opt. Eng.* **49**, 45 (1985).
9. A. M. Glass, *Opt. Eng.* **17**, 470 (1978).
10. W. J. Burke and co-workers, *Opt. Eng.* **17**, 308 (1978).
11. M. H. Garrett and co-workers, *Opt. Lett.* **17**, 103 (1992).
12. B. J. Chang and C. D. Leonard, *Appl. Opt.* **18**, 2407 (1979).
13. S. K. Case and R. Alferness, *Appl. Phys.* **10**, 41 (1970).
14. *Polaroid DMP128 Holographic Material*, Cambridge, Mass.
15. *Du Pont Omnidex Holographic Material*, Wilimington, Del.
16. C. A. Walsh and W. E. Moerner, *J. Opt. Soc. Am.* **9**, 1642 (1992).
17. Q. N. Wang and D. D. Nolte, *Appl. Phys. Lett.* **59**, 256 (1991).
18. J. J. Amodei and D. L. Staebler, *Appl. Phys. Lett.* **18**, 540 (1971).
19. F. Micheron and G. Bismuth, *Appl. Phys. Lett.* **23**, 71 (1973).
20. Y. Qiao and co-workers, *Opt. Lett.* **18**, 1004 (1993).
21. A. Kewitch and co-workers, *Opt. Lett.* **18**, 1262 (1993).
22. D. Kirillov and J. Feinberg, *Opt. Lett.* **16**, 1520 (1991).
23. W. R. Klein and B. D. Cook, *IEEE Trans. Sonics Ultrason.* **SU-14**, 123 (1967).
24. H. Kogelnik, *Bell. Syst. Tech. J.* **48**, 2909 (1969).
25. J. Hong and co-workers, *Opt. Lett.* **15**, 344 (1990).
26. W. H. Lee, in E. Wolf, ed., *Progress in Optics*, Vol. 16, North-Holland, Amsterdam, the Netherlands, 1978, p. 121.
27. A. W. Lohmann and D. P. Paris, *Appl. Opt.* **6**, 1739 (1967).
28. W. H. Lee, *Appl. Opt.* **13**, 1677 (1974).
29. C. B. Burckhardt, *Appl. Opt.* **9**, 1949 (1970).
30. L. B. Lesem, P. M. Hirsch, and J. A. Jordan, Jr., *IBM. J. Res. Dev.* **13**, 150 (1969).
31. J. A. Neff, R. Athale, and S. H. Lee, *Proc. IEEE* **78**, 826 (1990).
32. F. Mok and co-workers, *Opt. Lett.* **11**, 748 (1986).
33. S. A. Benton, *Proceedings SPIE Institute on Holography, paper #5*, SPIE, Bellingham, Wash., 1991.
34. Yu N. Denisyuk, *Opt. Spectrosc.* **15**, 279 (1963).
35. L. Rosen, *Appl. Phys. Lett.* **9**, 337 (1966).
36. S. A. Benton, *J. Opt. Soc. Am.* **68**, 1441 (1978).
37. S. A. Benton, *J. Opt. Soc. Am.* **59**, 1545 (1969).
38. H. Chen and F. T. S. Yu, *Opt. Lett.* **2**, 85 (1978).
39. S. A. Benton, *Opt. Eng.* **19**, 686 (1980).
40. K. S. Pennington and L. H. Lin, *Appl. Phys. Lett.* **7**, 56 (1965).
41. J. T. McCrickerd and N. George, *Appl. Phys. Lett.* **12**, 10 (1968).

42. N. Abramson, *The Making and Evaluation of Holograms*, Academic Press, Inc., London, 1980.
43. N. H. Abramson and H. Bjelkhagen, *Appl. Opt.* **11**, 2792 (1973).
44. C. C. Aleksoff, *Appl. Opt.* **10**, 1329 (1971).
45. J. M. Tedesco and co-workers, *Anal. Chem.* **65**(9), 441 (1993).
46. G. P. Agrawal and N. K. Dutta, *Long-Wavelength Semiconductor Lasers*, Van Nostrand Reinhold, New York, 1986.
47. S. H. Lee, *Opt. Photon. News* **16**, 18 (1990).
48. J. N. Latta, *Appl. Opt.* **11**, 1686 (1972).
49. M. W. Farn, M. B. Stern, and W. Veldkamp, *Opt. Photon. News* **2**, 20 (1991).
50. F. H. Mok, M. C. Tackitt, and H. M. Stoll, *Opt. Lett.* **16**, 605 (1991).
51. F. H. Mok, *Opt. Lett.* **18**, 915 (1993).
52. S. Tao, D. R. Selviah, and J. E. Midwinter, *Opt. Lett.* **18**, 912 (1993).
53. K. Blotekjaer, *Appl. Opt.* **18**, 57 (1979).
54. D. L. Staebler and co-workers, *Appl. Phys. Lett.* **26**, 182 (1975).
55. P. J. VanHeerden, *Appl. Opt.* **2**, 393 (1963).
56. G. A. Rakuljic, V. Leyva, and A. Yariv, *Opt. Lett.* **17**, 1471 (1992).
57. D. Psaltis, D. Brady, and K. Wagner, *Appl. Opt.* **27**, 1752 (1988).
58. D. Psaltis, C. H. Park, and J. Hong, *Neural Networks* **1**, 149 (1988).
59. J. Hong and D. Psaltis, in G. P. Agrawal and R. Boyd, eds., *Contemporary Nonlinear Optics*, Academic Press, Inc., Boston, 1992.
60. J. W. Goodman, *Fourier Optics*, McGraw-Hill Book Co., Inc., San Francisco, 1968.
61. M. Born and E. Wolf, *Principles of Optics*, Pergamon Press, Oxford, U.K., 1980.

JOHN H. HONG
Rockwell International Science Center

HORMONES

SURVEY

The word hormone is derived from the Greek *hormaein*, meaning to set in motion or to excite. It was used initially to define the activity of secretin [*1393-25-5*] (1), a gastrointestinal polypeptide released into the blood by the duodenal mucosa to stimulate pancreatic acinar cells to release bicarbonate and water.

Vertebrate Hormones

The term hormone is used to denote a chemical substance, released from a cell into the extracellular fluid in low quantities, which acts on a target cell to produce a response. Hormones are classified on the basis of chemical structure; most hormones are polypeptides, steroids (qv), or derived from single amino acids (qv) (Table 1).

Polypeptide hormones are synthesized as part of a larger precursor molecule or prohormone. Cleavage of the prohormone by specific cellular enzymes, ie, peptidases, produces the secreted form of the hormone. In some cases, multiple bioactive hormones are produced from a single prohormone. In the anterior pituitary gland (see HORMONES, ANTERIOR PITUITARY HORMONES), both adrenocorticotropic hormones (ACTH) and the endogenous opiate hormone, β-endorphin, are synthesized from a common prohormone (2) (see OPIOIDS, ENDOGENOUS). In the adrenal medulla, five to seven copies of another opiate hormone, methionine–enkephalin (Met-enkephalin), and one copy of leucine–enkephalin (Leu-enkephalin) are synthesized from each precursor molecule (3).

Steroids are synthetic products of cholesterol [57-88-5]. The chemical structure of a steroid hormone is determined by sequential enzymatic processing of the cholesterol molecule. Steroid products differ among steroid-secreting glands because of differences in enzyme processing, eg, the production of estrogen by the ovary requires enzymatic steps that do not occur in the adrenal cortex.

Amino acid-derived hormones include the catecholamines, epinephrine and norepinephrine (qv), and the thyroid hormones, thyroxine and triiodothyronine (see THYROID AND ANTITHYROID PREPARATIONS). Catecholamines are synthesized from the amino acid tyrosine by a series of enzymatic reactions that include hydroxylations, decarboxylations, and methylations. Thyroid hormones also are derived from tyrosine; iodination of the tyrosine residues on a large protein backbone results in the production of active hormone.

Mechanisms of Action.   Biologically effective concentrations of hormones range between 10^{-7} to 10^{-12} M. Although enzymes and vitamins (qv) are also effective in small amounts in producing cellular responses, a fundamental defining characteristic of a hormone is that it binds to a stereospecific cellular receptor to activate a response. Binding to a receptor protein activates an intracellular transduction process that mediates the hormone action. Hydrophilic molecules, eg, polypeptide hormones and catecholamines, bind to membrane receptors. Common intracellular transducers for these hormones include cyclic AMP, calcium, and phosphatidyl inositides. In contrast, lipophilic molecules, eg, steroid and thyroid hormones, readily diffuse through the cell membrane, enter the nucleus, and bind to receptors that activate or inactivate specific genes. Termination of the hormone action occurs after metabolism of the hormone or the hormone–receptor complex.

Effects and Secretion.   An endocrine gland is defined classically as a tissue consisting of hormone-secreting cells that synthesize products for release into the blood. The pattern of release is reflected by changes in plasma

Table 1. Hormones in Vertebrates

Tissue of origin/hormone	CAS Registry Number	Chemical nature	Site of action	Effect
Adrenal				
adrenal cortex				
aldosterone	[52-39-1]	steroid	kidney	electrolyte and water metabolism
cortisol	[50-23-7]	steroid	multiple tissues	protein, lipid, and carbohydrate metabolism; cardiovascular stability; immune responses
corticosterone	[50-22-6]			
adrenal medulla				
epinephrine	[51-43-4]	catecholamine	cardiac muscle	increase heart rate
			skeletal muscle	vasodilation
			skin and kidney	vasoconstriction
			liver	vasodilation and glycogenolysis
			adipose tissue	lipolysis
			intestinal smooth muscle	relaxation
norepinephrine	[51-41-2]	catecholamine	cardiac muscle	increase heart rate
			skeletal muscle, skin, and kidney	vasoconstriction
			liver	vasoconstriction and glycogenolysis
			adipose tissue	lipolysis
			intestinal smooth muscle	relaxation
Leu-enkephalin, Met-enkephalin	[59141-40-1][a]	polypeptide	multiple tissues	endogenous opiates
Cardiovascular tissue				
endothelial cells				
endothelin	[116243-73-3]	polypeptide	vascular smooth muscle	vasoconstriction
heart				
atrial natriuretic hormone (ANH)	[9088-07-7]	protein	kidney	increased sodium excretion and decreased renin secretion
			vascular smooth muscle	vasodilation
			adrenal cortex	decreased aldosterone secretion

Table 1. (Continued)

Tissue of origin/hormone	CAS Registry Number	Chemical nature	Site of action	Effect
Gastrointestinal (GI) tract				
gastrin	[53989-98-0, 60748-06-3, 60748-07-4]	polypeptide	stomach	increased acid secretion
secretin	[1393-25-5]	polypeptide	pancreas	increased water and biocarbonate secretion
cholecystokinin (CCK)	[9011-97-6]	polypeptide	gallbladder	contraction
			pancreas	increased enzyme secretion
motilin	[52906-92-0]	polypeptide	gastrointestinal tract	smooth muscle contraction
neurotensin	[39379-15-2]	polypeptide	stomach	decreased acid secretion and emptying
			pancreas	increased bicarbonate secretion
			intestine	increased motor activity
peptide tyrosine tyrosine (PYY)	[106388-42-5]	polypeptide	stomach	decreased acid secretion and emptying
somatostatin	[38916-34-6]	polypeptide	GI tract	decreased GI hormone secretion; decreased adsorption of nutrients; decreased gastric emptying and gall bladder contraction
Gonadal tissue				
corpus luteum				
progesterone	[57-83-0]	steroid	uterus	proliferation and vascularization of the endometrium; preparation for ovum implantation and maintenance of pregnancy
			mammary glands	alveolar development
relaxin	[9002-69-1]	protein	uterus	cervical softening
ovary				
estrone	[53-16-7]	steroid	uterus	endometrial proliferation
estradiol	[5750-28-2]	steroid	ovary	increased cell division and follicle growth
			mammary glands	duct development; development of secondary sex characteristics
inhibin	[57285-09-3]	protein	pituitary	inhibits FSH
activin	[114949-22-3]	protein	pituitary	stimulates FSH

testis				
testosterone	[58-22-0]	steroid	accessory sex organs	maturation and normal function; development of secondary sex characteristics
inhibin	[57285-09-3]	protein	pituitary	inhibits FSH
activin	[114949-22-3]	protein	pituitary	stimulates FSH
Hypothalamic/brain hormones				
corticotropin-releasing hormone (CRH)	[9015-71-8]	polypeptide	pituitary	release of ACTH and β-endorphin
dopamine	[51-61-6]	catecholamine	pituitary	inhibition of prolactin
gonadotropin-releasing hormone (GnRH)	[9034-40-6]	polypeptide	pituitary	release of LH and FSH
growth hormone-releasing hormone (GRH)	[9034-39-3]	polypeptide	pituitary	release of growth hormone
somatostatin	[38916-34-6]	polypeptide	pituitary	inhibition of growth hormone and TSH
thyrotropin-releasing hormone (TRH)	[9015-91-2]	tripeptide	pituitary	release of TSH and prolactin
vasopressin	[11000-17-2]	polypeptide	pituitary	release of ACTH and β-endorphin
			Pituitary	
anterior pituitary (adenohypophysis)				
adrenocorticotropic hormone (ACTH)	[9002-07-2]	polypeptide	adrenal	secretion of adrenocortical steroids
β-endorphin	[60118-07-2]	polypeptide	multiple tissues	endogenous opiate
follicle-stimulating hormone (FSH)	[9002-68-0]	glycoprotein	gonads	steroid and peptide secretion

Table 1. (Continued)

Tissue of origin/hormone	CAS Registry Number	Chemical nature	Site of action	Effect
γ-melanocyte-stimulating hormone (γ-MSH)	[72711-43-4]	polypeptide	adrenal	potentiates response to ACTH
growth hormone	[9002-72-6]	protein	multiple tissues	growth of bone and muscle; metabolism of carbohydrate and lipid; anabolic effect on mineral metabolism
luteinizing hormone	[9002-67-9]	glycoprotein	gonads ovary testes	steroid secretion ovulation development of interstitial tissue
prolactin	[9002-62-4]	protein	mammary gland corpus luteum	proliferation; milk secretion
thyroid-stimulating hormone (TSH)	[9002-71-5]	glycoprotein	thyroid	development and functional activity growth and secretion
intermediate pituitary (pars intermedia)				
β-endorphin	[60118-07-2]	polypeptide	multiple tissues	endogenous opiate
β-lipotropin	[37199-43-2]	polypeptide	adipose tissue	lipolysis
α-melanocyte-stimulating hormone (α-MSH)	[9002-79-3]	polypeptide	melanophores	skin pigmentation
γ-melanocyte-stimulating hormone (γ-MSH)	[72711-43-4]	polypeptide	adrenal	potentiates responses to ACTH
posterior pituitary (neurohypophysis)				
oxytocin	[50-56-6]	polypeptide	uterus mammary gland	contraction milk ejection
vasopressin	[11000-17-2]	polypeptide	kidney vascular smooth muscle liver	water reabsorption vasoconstriction glycogenolysis

	CAS Registry Number[a]		Target tissue	Function
parathyroid				
parathyroid hormone	[52232-67-4]	polypeptide	kidney, bone, and GI tract	mobilization of calcium and phosphorus
pineal				
melatonin	[73-31-4]	indole	skin reproductive tissue	inhibits pigmentation regulates function
thyroid				
thyroxine	[51-48-9]	amino acid derived	multiple tissues	increases metabolic rate and oxygen consumption
triiodothyronine	[6893-02-3]			
calcitonin	[9007-12-9]	polypeptide	bone and kidney	mobilization of calcium and phosphorus
Various organs				
kidney				
erythropoietin	[11096-26-7]	glycoprotein	bone marrow	erythrocyte production
renin	[9015-94-5]		adrenal	aldosterone secretion
angiotensin II (AII)	[11128-99-7]	polypeptide	vascular smooth muscle brain	vasoconstriction drinking
pancreas				
glucagon	[9007-92-5]	polypeptide	liver	glycogenolysis and gluconeogenesis
insulin	[9004-10-8]	polypeptide	multiple tissues	carbohydrate utilization lipogenesis
pancreatic peptide	[59763-91-6]	polypeptide	adipose tissue	function unknown
somatostatin	[38916-34-6]	polypeptide		inhibition of all pancreatic hormones
placenta				
chorionic gonadotropin	[9002-61-3]	glycoprotein	corpus luteum	maintenance of function
placental lactogen	[9035-59-5]	polypeptide	material tissues	insulin-like effects
relaxin	[9002-69-1]	protein		cervical softening

[a]CAS Registry Number for enkephalin.

363

hormone concentration. Analysis of hormone concentration is performed using radioimmunoassay based on competition of a hormone for binding to a specific antibody, or bioassay based on measurement of a specific biological response (see IMMUNOASSAY). Most hormones are secreted episodically, showing minute-to-minute changes in the circulation. Also, rhythms of hormone secretion are common, having periodicities that vary from hours to days. For example, circadian rhythms, having a periodicity of approximately 24 h, have been described for many hormones, including adrenal corticosteroid hormones, pituitary thyrotropic hormone, and pineal melatonin. Both growth hormone and prolactin secreted from the anterior pituitary show daily maxima related to the sleep–wake cycle. Pituitary gonadotropins and gonadal steroids essential for reproductive functions demonstrate cycles ranging over periods of days.

The biological effect of a hormone traditionally has been defined both by the change in the physiology of an organism after endocrine gland ablation and by the reversal of the effect by replacement with glandular extracts. For example, removal of the pancreas, ie, pancreatectomy, results in hyperglycemia, or elevated plasma glucose, and injection of pancreatic extracts results in the reestablishment of euglycemia, ie, normal plasma glucose. This experimental approach has led to the identification of pancreatic insulin as a hormone essential for the utilization of carbohydrates (see INSULIN AND OTHER ANTIDIABETIC AGENTS). Using a similar approach, the anterior pituitary gland has been characterized as the master gland. Its removal impairs the function of many other glands and is deleterious to the general well-being of the organism. The anterior pituitary secretes multiple hormones that are tropic, ie, maintains secretory function, and trophic, ie, maintains growth, for endocrine glands. Thus adrenocorticotropic hormone (ACTH) directly stimulates the growth and secretion of the adrenal cortex; thyrotropic hormone, ie, thyroid-stimulating hormone (TSH), stimulates the thyroid gland; and gonadotropic hormones, ie, luteinizing hormone (LH) and follicle-stimulating hormone (FSH), stimulate the gonads. Maintenance of nonendocrine cell growth and metabolism is dependent on the anterior pituitary gland through its secretion of growth hormone (GH). Prolactin is required for proliferation and milk secretion by the mammary gland in mammals and for water and electrolyte metabolism in lower vertebrates.

The secretion of anterior pituitary hormones is controlled by stimulatory and inhibitory hormones secreted from nerve cells found in a region of the brain called the hypothalamus. The hypothalamic neurons secrete hormones into the blood that flows directly to the anterior pituitary gland. The brain hormones (see HORMONES, BRAIN OLIGOPEPTIDES) include corticotropin-releasing hormone (CRH), which stimulates the secretion of ACTH; thyrotropin-releasing hormone (TRH), which stimulates the secretion of TSH; gonadotropin-releasing hormone (GnRH), which stimulates FSH and LH; growth hormone-releasing hormone (GRH), which stimulates GH; somatostatin, which inhibits GH release; and the catecholamine dopamine, which inhibits the secretion of prolactin from the anterior pituitary.

Control of secretion of anterior pituitary hormones also includes inhibition by hormones produced by target organs. For example, CRH stimulates the anterior pituitary to secrete ACTH, which in turn stimulates the adrenal cortex to secrete corticosteroids. Corticosteroids then feed back to inhibit the secretion of ACTH. Feedback mechanisms are important for the control of most hormones. For example, insulin (qv) secretion from the pancreas increases in response to increased blood glucose resulting from ingestion of a meal. Insulin increases tissue uptake and metabolism of glucose, which lowers blood glucose and in turn reduces insulin secretion.

Endocrine Pathology. The health of an organism is dependent on the maintenance of hormone concentrations within a normal physiological range. Pathology occurs when hormone concentrations are higher or lower than normal for extended periods. Excess hormone concentrations can result from overproduction by endocrine cells, eg, hypercorticoidism owing to excess secretion of adrenal cortical hormones, or from reduced hormone metabolism. Endocrine cell tumors have been described that secrete excess amounts of polypeptide hormones, steroid hormones, or catecholamines. Hormone deficiency can result from reduced hormone production, eg, hypocorticoidism, or from increased metabolic clearance of a hormone from the circulation. Pathology also can result when adequate hormone concentrations are present, but the hormone receptor on target cells does not bind normally, eg, in certain forms of diabetes mellitus, insulin concentration are adequate, but insulin receptors on target cells are reduced. Thus target cells are unable to respond appropriately (4).

Endocrinologists use hormones or hormone analogues to treat endocrine disease. Potent analogues of some hormones have been synthesized that antagonize the action of the natural hormone. Their use is effective in treating endocrine problems related to excess or inappropriate hormone production, eg, hypertension, ie, elevated blood pressure, can result from excess constriction of vascular smooth muscle induced by high circulating concentrations of angiotensin II. Treatment with 1-sar-8-ala-angiotensin II (saralasin [*34273-10-4*]), an inactive angiotensin II analogue, lowers blood pressure (5) (see CARDIOVASCULAR AGENTS). Saralasin binds to the angiotensin receptor and prevents angiotensin II activity. To offset endocrine gland removal or reduced function, replacement therapy is performed. In most cases, synthetic hormones are administered; steroid hormones and some peptide hormones can be obtained in pure form. The availability of protein hormones for therapeutic or experimental use has been increased greatly through the application of genetic engineering (qv) techniques. The deoxyribonucleic acid (DNA) sequence of the human gene coding for a specific protein hormone is cloned. When inserted into rapidly replicating bacteria, enormous quantities of mammalian hormones can be produced. This approach has generated highly purified recombinant human hormones, including insulin and growth hormone.

New Perspectives in Endocrinology. The field of endocrinology has as its primary focus the study of the structure and function of endocrine glands and their secretory products. However, findings (ca 1994) have fostered a broader scope for endocrinology. Hormone secretion from cells that exist outside the classic endocrine glands can occur. The ability to synthesize and secrete hormones has been demonstrated for nonglandular tissue, including neurons, ie, cells identified in the central and peripheral nervous systems, and leukocytes (see NEUROREGULATORS). To identify cells that synthesize hormones, molecular biological approaches have been used to measure the messenger ribonucleic acid (mRNA) that encodes for a specific hormone or for an enzyme required for hormone synthesis.

Transport in the blood is no longer a requisite for a hormonal response. Responses can occur after release of hormones into the interstitial fluid with binding to receptors in nearby cells, called paracrine control, or binding to receptors on the cell that released the hormone, called autocrine control. A class of hormones shown to be synthesized by the tissue in which they act or to act in the local cellular environment are the prostaglandins (qv). These ubiquitous compounds are derived from arachidonic acid [506-32-1] which is stored in the cell membranes as part of phospholipids. Prostaglandins bind to specific cellular receptors and act as important modulators of cell activity in many tissues.

Cells synthesize and secrete multiple hormones. For example, some cells in the anterior pituitary have the capacity to secrete both polypeptide hormones, FSH and LH. Cells also can secrete different chemical classes of hormone, eg, cells in the adrenal medulla synthesize and secrete the amine-derived catecholamines, norepinephrine and epinephrine, as well as the polypeptide hormones, the enkephalins.

A hormone can have multiple biological effects that are conferred by binding to receptors on specific target cells. Angiotensin II stimulates the secretion of the steroid hormone aldosterone by binding to receptors in the adrenocortical cell and the constriction of blood vessels by binding to receptors on vascular smooth muscle cells; it also activates the drinking response by binding to neuronal receptors in the brain. A broader view of what constitutes a hormone is exemplified by the discovery that identical peptides are present as secretory products in neurons of the peripheral and central nervous system, as well as glandular cells in the gut, pancreas, and other tissues. For example, somatostatin, originally identified as a hypothalamic factor that affects anterior pituitary secretion, is synthesized in the gut and pancreas and contributes to functional control in those tissues. Many peptides have been localized both in the gastrointestinal (GI) tract and in the brain. These are listed in Table 2 with some of their GI and brain effects. In addition to affecting local cell activity in the nervous system and in nonneural tissue, many of these peptides also have been shown to act after secretion into the circulation.

Nontraditional Hormones. Novel hormones identified in cardiovascular tissue have profound effects on maintenance of blood pressure and blood volume in mammals. Atrial natriuretic hormone (ANH) is a polypeptide hormone secreted from the atria of the heart. When the cardiac atrium is stretched by increased blood volume, secretion of ANH is stimulated; ANH in turn increases salt and water excretion and reduces blood pressure (6). Endothelin is a polypeptide hormone secreted by endothelial cells throughout the vasculature. Although endothelin is released into the circulation, it acts locally in a paracrine fashion to constrict adjacent vascular smooth muscle and increase blood pressure (7).

Two protein hormones, inhibin and activin, have been identified in gonadal tissue. Inhibin has been isolated from ovarian follicular fluid and found to inhibit pituitary secretion of FSH. Inhibin is a glycoprotein heterodimer consisting of two disulfide-linked subunits, α and β; two types of β-subunit, β_A and β_B, exist in follicular fluid. Control of inhibin secretion involves a feedback relationship in

Table 2. Brain–Gut Peptide Hormones in Vertebrates

Hormone	CAS Registry Number	Effects[a]	
		Brain	GI tract
bombesin	[74815-57-9]	motor activity (I); body temperature and feeding (D)	gastrin secretion (I)
calcitonin gene-related peptide (CGRP)	[83652-28-2]	motor activity and feeding (D)	acid secretion (D)
cholecystokinin (CCK)	[9011-97-6]	motor activity and feeding (D)	enzyme secretion (I)
leu- and met-enkephalin	[59141-40-1]	body temperature (D)	smooth muscle contraction (I); GI secretion (D)
motilin	[52906-92-0]	drinking (D)	smooth muscle contraction (I)
neuropeptide Y	[82785-45-3]	feeding and drinking (I)	gut motility (D)
neurotensin	[39379-15-2]	motor activity and body temperature (D)	acid secretion and emptying (D)
secretin	[1393-25-5]	motor activity (D)	H$_2$O and bicarbonate secretion (I)
somatostatin	[38916-34-6]	memory (I); aggressive behavior and body temperature (D)	hormone secretion, nutrient adsorption, gastric emptying, and gall bladder contraction (D)
substance P	[33507-63-0]	motor activity (I); aggressive behavior (D)	intestinal motility and exocrine pancreatic secretion (I)
vasoactive intestinal peptide (VIP)	[37221-79-7]	motor activity (D)	intestinal H$_2$O and Cl$^-$ secretion and relaxation of esophageal sphincter (I)

[a](I), increases; (D), decreases.

which circulating FSH stimulates inhibin secretion, which in turn reduces the secretion of FSH (8). Both the homo- and the heterodimers of the β-subunits of inhibin promote the secretion of FSH and thus have been termed activins. Activin is secreted by the ovary and the testes into the circulation. In addition, both inhibin and activin have intragonadal autocrine and paracrine effects that influence gonadal steroidogenesis (9).

Erythroid differentiation factor (EDF) is a protein isolated originally from the culture fluid of a human leukemia cell line; it induces the proliferation and differentiation of hematopoietic progenitor cells (10). Interestingly, the sequence of the EDF mRNA is identical to that of the β_A subunit of activin, and inhibin (11). Both activin and EDF are composed of two β_A subunits, and their biological activity is similar, eg, activin can augment the proliferation and differentiation of erythroid progenitor cells and EDF can stimulate secretion of FSH from anterior pituitary cells. These findings provide additional support for the concept that one hormone can have multiple biological effects. In this instance, the same hormone has been given two different names because it expressed specific biological activities in two different physiological systems.

Insect and Plant Hormones

Insect Hormones.   Insects and crustaceans must shed their exoskeleton in order to grow. This molting process is called ecdysis and is initiated by a steroid hormone called ecdysone [3604-87-3], which is secreted by the Y-organ located at the base of the antennae. Secretion is under negative control by molt-inhibiting hormones secreted by a ganglion, the X-organ. To initiate release of ecdysone and molting, the nervous system inhibits the release of the peptide, molt-inhibiting hormone. In insects, a molt inhibitor is not involved. Instead, the corpus cardiaca of the brain produces prothoracotropic hormone [61583-57-1], which acts on the prothoracic gland to secrete ecdysone. Another hormone, juvenile hormone [23314-84-3], secreted by the corpora allata in insects, prevents development from the larval to the adult stage; when juvenile hormone concentrations decrease, metamorphosis is initiated. Multiple juvenile hormones have been isolated and all are derivatives of methyl-10,11-epoxytridecadienoate, a sesquiterpenoid compound (see INSECT CONTROL TECHNOLOGY).

Plant Hormones.   Plant hormones are organic substances, active in small (<1 μM) amounts, which are formed in one part of a plant and usually translocated to other sites to induce specific biochemical or morphological responses. Auxins derived from the amino acid tryptophan induce elongation in shoot cells; the principal naive auxin in higher plants is indole-3-acetic acid [87-51-4]. Another class of hormones, the gibberellins, are four-ring structures that occur as 19-carbon or 20-carbon, mono-, di-, or tricarboxylic acids. Originally isolated from fungi, gibberellins are natural products of higher plants and an intact ring structure is essential for their activity in stimulating cell division and cell elongation. Cytokinins are isopentenyl adenine derivatives that promote cell division. Kinetin [523-79-1], the first cytokinin studied, has been isolated from autoclaved herring sperm DNA and identified as 6-furfurylaminepurine. The first natural plant cytokinin, zeatin [1637-39-4], has been isolated from immature corn kernels and identified as 6-(4-hydroxy-3-methyl-*trans*-2-butenylamino) purine. Growth-inhibiting hormones are important for inducing periods of plant dormancy; abscisic acid [21293-29-8] is a natural growth-inhibiting hormone. Plants also synthesize brassinosteroids that contain the steroid nucleus, are active in concentrations lower than those of other plant hormones, ie, pM, and act alone or synergise with other plant hormones to affect plant growth (see GROWTH REGULATORS, PLANTS).

Pherohormones

Pherohormones, or pheromones, are interorganismal hormones that transmit information between members of a species. Insect behavior is affected by different pheromones used as sex attractants or as chemical markers of food sources. The chemical structure of pheromones varies. These may be small molecules that rapidly disperse and serve to signal alarm, eg, 4-methyl-3-heptanone [6137-11-7] in ants, or larger, less volatile compounds that persist for longer periods and function as sex attractants, eg, 3,13-octadecadien-1-ol [66410-24-0], [66410-28-4]. There is evidence for olfactory communication in a variety of mammalian species, including humans. In male and female rodents, different volatile constituents in urine induce aggressive behavior, delay or accelerate puberty, or attract the opposite gender. Originally isolated from porcine testicles, 16-androstenes are C-19 steroids that are volatile, have a pronounced odor, and have been implicated as sex attractants in some mammals. It has not been established whether C-19 steroids are sex attractants in all mammals. However, these compounds, recently isolated from the axillae of human males, may be produced by bacterial metabolism of a native androgen to function as human pheromones.

BIBLIOGRAPHY

"Hormones, Survey" in *ECT* 1st ed., Vol. 7, pp. 475–481, by O. Kamm, Parke Davis & Co.; in *ECT* 2nd ed., Vol. 11, pp. 45–52, by J. Fried, University of Chicago; in *ECT* 3rd ed., Vol. 12, pp. 538–545, by C. Rivier and J. Rivier, Salk Institute.

1. W. M. Bayliss and E. H. Starling, *J. Physiol.* **28**, 325 (1902).
2. B. A. Eipper and R. E. Mains, *Endocrine Revs.* **1**, 1–27 (1980).
3. R. V. Lewis and co-workers, *Science* **208**, 1459 (1980).
4. C. R. Kahn and co-workers, *N. Engl. J. Med.* **294**, 739 (1976).
5. M. L. Horne and co-workers, *Kidney Int.* **15**, s115 (1979).
6. B. J. Ballermann and B. M. Brenner, *Circ. Res.* **58**, 619–630 (1986).
7. G. M. Rubanyi and L. H. Parker Botelho, *FASEB J.* **5**, 2713–2720 (1991).
8. F. H. De Jong, *Physiol. Revs.* **68**, 555–607 (1988).
9. S.-Y. Ying, in L. Martini and W. F. Ganong, eds., *Frontiers in Neuroendocrinology,* Vol. 10, Raven Press, Ltd., New York, Chapt. 8, pp. 167–184, 1988.
10. Y. Eto and co-workers, *Biochem. Biophys. Res. Commun.* **142**, 1095–1097 (1987).
11. M. Murata and co-workers, *Proc. Nat. Acad. Sci. U.S.A.* **85**, 2434–2436 (1988).

General References

K. L. Becker, *Principles and Practice of Endocrinology and Metabolism*, J. B. Lippincott Co., Philadelphia, Pa., 1990.
D. T. Krieger, *Endocrine Rhythms*, Raven Press, New York, 1979.

H. Laufer and R. G. H. Downer, *Endocrinology of Selected Invertebrate Types*, Alan R. Liss, Inc., New York, 1988.

D. W. MacDonald, D. Muller-Schwarze, and S. E. Natynczuk, *Chemical Signals in Vertebrates 5*, Oxford University Press, Oxford, U.K., 1990.

J. B. Martin and S. Reichlin, in *Clinical Neuroendocrinology*, 2nd ed., F.A. Davis Co., Philadelphia, Pa., 1987, Chapt. 18, pp. 559–605.

T. C. Moore, *Biochemistry and Physiology of Plant Hormones*, 2nd ed., Springer-Verlag, New York, 1989.

G. D. Prestwich and G. J. Blomquist, *Pheromone Biochemistry*, Academic Press, Inc., Orlando, Fla., 1987.

WILLIAM C. ENGELAND
University of Minnesota

ANTERIOR PITUITARY HORMONES

The hormones of the anterior pituitary gland play a significant role in the maintenance of normal health and body function. This master gland produces hormones involved in the regulation of somatic growth, metabolic rate, carbohydrate and lipid metabolism, lactation, reproduction, and response to stress. Eleven anterior pituitary hormones have been extensively characterized at the protein and genomic levels. These hormones fall into three classic categories, ie, pro-opiomelanocortin-related (POMC-related) hormones, proteins structurally related to prolactin and growth hormone (PRL/GH-related), and glycoproteins. Structural similarities exist within each biochemical category, and the structural data presented represent listings in the Protein Identification Resource (PIR) database, compiled by the National Biomedical Research Foundation, Washington, D.C. Table 1 provides summary information on the principal biologic roles of each of these hormones. The National Hormone and Pituitary Program at the NIH and the USDA Animal Hormone Program provide anterior pituitary hormones free of charge for nonprofit research purposes. All principal hormones of the anterior pituitary gland are available commercially as material purified from anterior pituitary gland extracts, synthetic peptides, or proteins produced by the expression of recombinant DNA. Excellent listings of commercial sources of anterior pituitary hormones are available (1,2).

Pro-Opiomelanocortin-Derived Hormones

A single parent gene codes for the 267 amino acid glycoprotein which contains the sequences for adrenocorticotropic hormone (ACTH), melanocyte-stimulating hormone (MSH), endorphin, and lipotropin (LPH) (3). This precursor has been named pro-opiomelanocortin [66796-54-1] (POMC), and the biosynthesis and structure have been reviewed (4,5). The locations of the POMC-derived bioactive peptides relative to the precursor molecule are shown in Figure 1. Two disulfide bonds exist in the proximity of the N-terminus of the precursor, but there are no disulfide bonds in the bioactive peptides derived from the parent molecule. Glycosylation sites exist at Thr-45 and Asn-65. There also is evidence that

Table 1. Hormones of the Anterior Pituitary and Their Functions

Hormone	Acronym	CAS Registry Number	Number of amino acids[a]	Principal function
Pro-opiomelanocortin-derived peptides				
adreno-corticotropin	ACTH	[9002-60-2]	39	stimulates cortex of adrenal gland to produce steroid hormones, ie, glucocorticoids
β-endorphin		[59887-17-1]	31	functions as neurotransmitter; exerts opiate-like analgesia
lipotropin	LPH	[9035-55-6]		mobilizes fat; precursor for β-endorphin
β-LPH		[37199-43-2]	91	
γ-LPH		[60893-02-9]	58	
melanotropin	MSH	[9002-79-3]		darkening of the skin, ie, pigmentation
α-MSH		[37213-49-3]	13	
β-MSH		[9034-42-8]	22	
Prolactin/growth hormone-related peptides				
prolactin		[9002-62-4]	199	supports lactation
growth hormone	GH	[9002-72-6]	191	stimulates body growth; anabolism
Glycoproteins				
follicle-stimulating hormone	FSH	[9002-68-0]	210	supports maturation of ovarian follicles and sperm
luteinizing hormone	LH	[9002-67-9]	204	induces ovulation; maintains testicular function
thyroid-stimulating hormone	TSH	[9002-71-5]	211	stimulates thyroid hormone production and thyroid gland growth

[a]From human amino acid sequence (PIR).

Asp-29 of ACTH is glycosylated in the rat and mouse (6,7). The biological significance of POMC glycosylation is unknown, but it has been postulated to stabilize the prohormone (8).

Adrenocorticotropic Hormone. Adrenocorticotropic hormone [9002-60-2] is required for maintenance and function of the adrenal cortex, which secretes potent steroid hormones regulating metabolism, ie, glucocorticoids. Insufficient secretion of the steroid hormones of the adrenal cortex, as in Addison's disease, may result in insulin hypersensitivity, inability to maintain blood sugar levels during food deprivation, hypotension, general weakness, fatigue, and psychological disturbances. Glucocorticoids regulate their own production by suppressing the synthesis and secretion of POMC-derived hormones, ie, negative

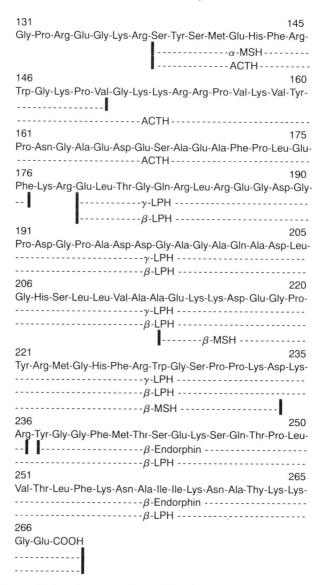

Fig. 1. Locations of bioactive peptides within human pro-opiomelanocortin for sites 131–267.

feedback. Removal of the pituitary gland induces a pronounced physical atrophy of the adrenal cortex, which accompanies loss of function including a reduction in steroid hormone secretion. The size and functional integrity of the adrenal gland can be restored by the administration of ACTH. The sequence of the first 24 N-terminal amino acids of ACTH cosyntropin [*16960-16-0*], confers full bioacitvity at the adrenal gland and is highly conserved among species. Across species, the ACTH molecule differs slightly in structure and retains similar biologic function. It is a single-chain peptide, 39 amino acids in length (Fig. 1). Adrenocorticotropic hormone is not generally used for the treatment of adrenal insufficiency, because the adrenal steroids can be easily synthesized and administered. Clinically, ACTH is used for diagnostic testing of adrenal function. Commercial preparations of ACTH are available from Armour Pharmaceutical, Ciba Pharmaceutical, Hoechst-Roussel, Organon Inc., and Wyeth-Ayerst Laboratories.

Melanotropins.   Two melanotropins, α-MSH and β-MSH, result from post-translational processing of POMC in the pituitary gland. The α-MSH peptide is comprised of the first 13 amino acids of ACTH, except that serine 1 is acetylated and valine 13 is amidated (Fig. 1). The cleavage of lipotropin (LPH) generates β-MSH (Fig. 1). In lower vertebrates, these hormones induce changes in melanin-containing cellular organelles which rapidly alter body pigmentation. While the physiologic role of MSH in mammals is questionable, MSH-secreting tumors and injections of MSH cause hyperpigmentation in humans. Darkening of the skin also occurs clinically with inadequate glucocorticoid secretion, which removes the negative feedback on POMC-derived hormone production and increases MSH secretion. α-MSH acts as a neurotransmitter within the central nervous system, where it may be involved in regulation of body temperature (9). Solid-phase synthesis is used to produce MSH for use in biomedical research. No other commercial applications exist.

β-Endorphin.   A peptide corresponding to the 31 C-terminal amino acids of β-LPH was first discovered in camel pituitary tissue (10). This substance is β-endorphin, which exerts a potent analgesic effect by binding to cell surface receptors in the central nervous system. The sequence of β-endorphin is well conserved across species for the first 25 N-terminal amino acids. Opiates derived from plant sources, eg, heroin, morphine, opium, etc, exert their actions by interacting with the β-endorphin receptor. On a molar basis, this peptide has approximately five times the potency of morphine. Both β-endorphin and ACTH are cosecreted from the pituitary gland. Whereas the physiologic importance of β-endorphin release into the systemic circulation is not certain, this molecule clearly has been shown to be an important neurotransmitter within the central nervous system. Endorphin has been invaluable as a research tool, but has not been clinically useful due to the availability of plant-derived opiates.

Lipotropin.   Lipotropin, first isolated from sheep pituitary glands (11,12) in the course of research directed at optimizing ACTH purification techniques, has been determined to be distinct from the other known pituitary hormones and possesses lipolytic activity. The location of the two lipotropins, β-LPH and γ-LPH, within the POMC protein sequence is shown in Figure 1. Lipotropin is important as a precursor for β-endorphin, but the biological significance of its fat-mobilizing properties is unresolved. In the early 1990s LPH is used only for basic research.

Prolactin/Growth Hormone-Related Hormones

Prolactin.   In 1928, bovine pituitary extract was shown to induce lactation in rabbits (13). Soon afterward, lactogenic activity was found in pituitary extracts from a variety of mammalian and nonmammalian species. The protein responsible for this activity has been appropriately named prolactin, although it is sometimes referred to as lactogenic hormone. The best-documented function of prolactin is to stimulate milk synthesis and secretion; however, a variety of other functions may exist. Advances in immunophysiology indicate an important role for this hormone in regulating immune function (14). In rodents, but not other species, prolactin acts to support the function of the corpus luteum, ie, the ovarian tissue which secretes hormones needed to maintain pregnancy. Prolactin also may be involved in the regulation of salt–water balance, growth, development, and metabolism (15).

The primary structure of prolactin consists of 199 amino acids in a linear sequence (see HORMONES, ANTERIOR PITUITARY-LIKE HORMONES). Similar to growth hormone (16), the tertiary structure of prolactin is thought to be arranged in a bundle of four α-helices (17). Detailed reviews of the structure–function relationships of prolactin among a variety of species have been published (17,18). Only one gene for prolactin appears to exist (19). Although classically placed in the category of simple protein hormones, prolactin can be glycosylated. Carbohydrate attachment occurs at Asn-31, where the consensus glycosylation sequence Asn–X–Ser is found.

The abundance of glycosylated prolactin varies between species, but can be significant; in the pig, 50% of prolactin can be in the glycosylated form (20). The physiologic importance of glycosylated prolactin is uncertain, but depending on the bioassay used for potency testing, glycosylated prolactin may be less (21) or more (22) active than the nonglycosylated hormone. In addition to 23 kD prolactin, 21 and 25 kD forms of this hormone exist which may be generated from alternative mRNA splicing. Variants of prolactin also arise from proteolytic cleavage, phosphorylation, deamidation, sulfation, and polymerization. The biopotencies of these variants relative to 23 kD prolactin vary with different assay systems. While exceptions exist, prolactin variants generally exert equal or lower potency than the 23 kD form of the hormone; a summary of relative potencies is available (20). The N-terminal portion of prolactin appears to contain sequences which confer bioactivity, as the 16 kD N-terminal product of enzymatic cleavage retains potency (23). Site-directed mutagenesis of bovine prolactin has been used to produce hormone variants lacking disulfide bridges between amino acids 4–11, 191–199, or 58–174 (24). Mutant prolactin lacking the large (sites 58–174), but not small disulfide loops has a greatly reduced bioactivity. Prolactin is produced for research purposes. Clinical applications may develop as prolactin's role as an immune system modulator is elucidated (14).

Growth Hormone. The growth-promoting property of pituitary extracts was evident to early endocrinologists. Their observations have been confirmed upon the isolation of growth hormone (GH), also known as somatotropin, from bovine pituitary glands (25). While variable potencies have been observed when GH from one species is administered to another, this hormone is obviously a powerful stimulant of body growth. The lack of endogenous GH production is a cause of dwarfism in humans which can be overcome by clinical replacement therapy. Growth hormone produced from recombinant DNA is the commercial source of hormone for the clinical treatment of GH deficiency. Much of the growth-stimulating action of GH is mediated by insulin-like growth factors, originally called somatomedins, which are produced primarily in the liver in response to GH (26). Growth hormone also stimulates the production of insulin-like growth factors which have autocrine actions in a variety of non-hepatic target tissues (27).

In addition to its growth-stimulating properties, GH exerts potent metabolic actions. It induces a decrease in blood urea and amino acid levels; this action may be secondary to an increase in the rate of amino acid uptake and protein synthesis in peripheral tissues. Growth hormone antagonizes the blood sugar-lowering effect of insulin by suppressing glucose uptake and utilization, and promotes body leanness by decreasing fat synthesis and stimulating adipose breakdown. As a result of these desirable metabolic effects, recombinant GH preparations may find significant commercial applications as performance stimulators in domestic animals. Research has shown that growth hormone enhances milk production in lactating animals and improves feed efficiency, growth rate, and carcass lean meat-to-fat ratio in growing animals (see GROWTH REGULATORS, ANIMAL). Interestingly, there is an increasing body of evidence to support the role of growth hormone as a stimulator of immune function, with the ability to reverse the aging-related decrease in thymus size (28).

Growth hormone from one species is not necessarily active when administered to other species. Bovine GH [66419-50-9], for example, is active in sheep and rodents, but not in primates. Growth hormone from primates, but not other species, exerts prolactin-like activity, which reflects its ability to bind to the prolactin receptor (18). The evidence supporting the ability of growth hormone to create a desirable physiological profile, ie, high protein–water and low fat body composition, and to enhance immunity has created a high degree of research interest in the pharmaceutical industry. U.S. Food and Drug Administration (FDA) approval has been granted for the use of a recombinant bovine GH preparation, produced by Monsanto Co., U.S.A., in lactating dairy cattle to increase milk production. Eli Lilly and Company, The Upjohn Company, and American Cyanamid Company also have interests in the commercial application of recombinant bovine GH. Recombinant porcine GH [9061-23-8] preparations from several companies, eg, The Upjohn Company, SmithKline Beecham Animal Health, Pitman-Moore, Inc., Monsanto Company, and American Cyanamid Company, are being evaluated for commercial use. Recombinant human GH for clinical use is marketed under such names as Protropin (Genentech), Umatrope (Eli Lilly), Genotropin (Sumitomo), and Somatonorm (Kabi-Vitrum) by a variety of pharmaceutical companies. A listing of additional suppliers is available (2).

Human chromosome 17 contains two genes that code for slightly different forms of GH (29). The products of these two gene copies differ in 12 amino acids, spread along the length of the molecule. These isohormones have been termed GH-normal (GH-N) and GH-variant [109675-94-7] (GH-V). GH-N is produced primarily in the pituitary gland, while GH-V is expressed in placental tissue (see HORMONES, ANTERIOR PITUITARY-LIKE HORMONES). As for prolactin, variants of the principal 22 kD form of pituitary GH exist; alternative mRNA splicing gives rise to a 20 kD form of GH, with residues 32–46 of 22 kD GH deleted. The 20 kD GH apparently does not bind the GH receptor as well as 22 kD GH, but exerts similar biological potency (16). Deamidated and acylated GH have been described which possess unaltered biopotency (16). A cluster of four α-helices comprises the tertiary structure of GH (16). Growth hormones' relation to prolactin and placental lactogen is evident, particularly in the preservation of two disulfide bonds which give rise to the so-called long and short loops (18). The synthesis of bovine GH from site-directed mutations of the encoding DNA has shown that the long rather than the short loop is needed for hormone secretion (30).

Full reduction and carbamidomethylation of the disulfide bonds of human GH does not affect bioactivity (31). However, there may be species differences in the role of the disulfide bonds on hormone action; for example, reduction followed by aminoethylation of porcine GH reduces bioactivity (32). Numerous studies have sought to determine the structural components of GH required for receptor binding. Site-directed mutagenesis shows that the N-terminus of GH, perhaps between residues 8 and 18, is required for receptor binding activity (33). On the other hand, the C-terminus also has been recognized to be important for the expression of full potency (18). The solution of the crystal structure of the GH-receptor complex has unified the results of earlier studies by showing that a number of sites over the length of GH are involved in the binding of one hormone molecule to two GH receptors (34).

Glycoprotein Hormones

There are three pituitary glycoprotein hormones, ie, luteinizing hormone (LH), follicle-stimulating hormone (FSH), and thyroid-stimulating hormone (TSH). Luteinizing hormone and FSH control significant aspects of reproduction. These hormones are referred to as gonadotropins, owing to their trophic and stimulatory effects on gonadal tissues. Follicle-stimulating hormone is needed for maturation of the ovarian follicles, for maintaining the size of the ovary, and for stimulating the production of female sex steroid hormones, ie, estrogens. In the male, FSH is necessary for sperm production. Luteinizing hormone induces ovulation, ie, release of the egg, from the mature follicle. In the male, LH induces the synthesis of the male sex steroid hormones, called androgens (see HORMONES, SEX HORMONES). Thyroid-stimulating hormone is necessary for the proper development, growth, and function of the thyroid gland. Most importantly, TSH is responsible for maintaining thyroid hormone production. Inadequate levels of thyroid hormone give rise to a clinical profile characterized by sluggish mental processes, slow growth, sensitivity to cold, low basal metabolic rate, cardiac insufficiency, general weakness, and susceptibility to infection.

Luteinizing hormone, FSH, and TSH are comprised of two dissimilar, non-covalently bound α and β subunits. The α subunit is coded by a single gene (35–37). The β subunits differ in structure and confer specificity of biologic action. There is no significant bioactivity without $\alpha-\beta$ subunit association. Both subunits possess N-linked oligosaccharide chains which are found at consensus glycosylation sequences, ie, Asn–X–Thr, as generalized in Figure 2. Depending on the species and the hormone, both carbohydrate chains can terminate in sialic acid, sulfate, or a combination of both. A variety of complex glycosylation variants can occur, giving rise to closely related isoforms of LH, FSH, and TSH (38–41). There is a convincing body of evidence showing the importance of glycosylation on the biopotency of these hormones. Terminal sialic acid moieties create acidic, low pI, glycoprotein isoforms (42–44). Sialic acids prolong

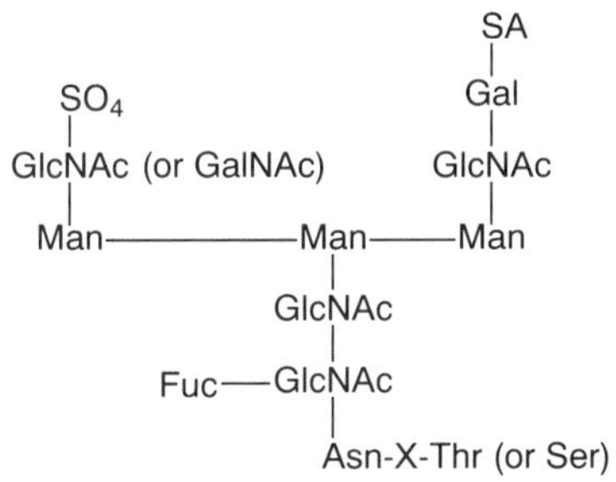

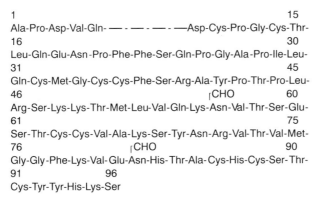

Fig. 2. Generalized structure of N-linked oligosaccharide of the pituitary glycoprotein hormones. SO_4=sulfate, GlcNAc=N-acetylglucosamine, GalNAc=N-acetylgalactosamine, Man=mannose, Fuc=fucose, SA=sialic (neuraminic) acid.

hormone half-life in the circulation (45,46), which confers significant *in vivo* potency (46,47). Glycoprotein subunits have no free sulfhydryl moieties. Technical limitations have hindered the complete assignment of disulfide bridges, but the following assignments are likely correct: $\alpha 11-35$, $\alpha 14-36$, $\beta 93-100$, $\beta 26-110$, and perhaps $\beta 23-72$ (48). A disulfide bond between $\beta 38-57$ also has been reported for human chorionic gonadotropin, a placental LH-like hormone (49), and ovine LH [*9002-67-9*] (50).

Relative to other species, human glycoprotein α subunit has a 4 amino acid deletion between positions 6 and 9. A deletion spanning positions 1 and 6 exists in the porcine α subunit. A comparison of the primary structures of glycoprotein hormone subunits among species has been reviewed (48,51). The α subunit possesses two carbohydrate moieties, Asn-56 and 82 (Fig. 3). The carbohydrate component of the α subunit is essential for post-receptor signal transduction. Deglycosylation of the α subunit increases receptor binding potency, but virtually eliminates the bioactivity of LH (52). Sites of the α subunit which participate in receptor binding may lie between amino acid positions 30–55 and 80–96 (48).

One glycosylation site exists on the β subunits of human LH [*53664-53-2*] and TSH [*64365-92-0*], ie, Asn-30 (Fig. 4). In some species, Asn-13 of LH-β is

```
1                                                15
Ala-Pro-Asp-Val-Gln- — - — - — - —Asp-Cys-Pro-Gly-Cys-Thr-
16                                               30
Leu-Gln-Glu-Asn-Pro-Phe-Phe-Ser-Gln-Pro-Gly-Ala-Pro-Ile-Leu-
31                                               45
Gln-Cys-Met-Gly-Cys-Cys-Phe-Ser-Arg-Ala-Tyr-Pro-Thr-Pro-Leu-
46                                 ⌈CHO          60
Arg-Ser-Lys-Lys-Thr-Met-Leu-Val-Gln-Lys-Asn-Val-Thr-Ser-Glu-
61                                               75
Ser-Thr-Cys-Cys-Val-Ala-Lys-Ser-Tyr-Asn-Arg-Val-Thr-Val-Met-
76            ⌈CHO                                90
Gly-Gly-Phe-Lys-Val-Glu-Asn-His-Thr-Ala-Cys-His-Cys-Ser-Thr-
91             96
Cys-Tyr-Tyr-His-Lys-Ser
```

Fig. 3. Human LH, FSH, and TSH α subunit [*69431-84-1*]. Amino acid numbering is relative to maximum homology between species (48). Note the 4 amino acid deletion in human α subunit between positions 6 and 9. Consensus glycosylation sites are at Asn-56 and 82. CHO = carbohydrate chain.

```
                                                              (FSHβ)
                  1                                           ┌CHO  15
LHβ       Ser-Arg-Glu-Pro-Leu-Arg-Pro-Trp-Cys-His-Pro-Ile-Asn-Ala-Ile-
FSHβ                                          Asn-Ser- * -Glu-Leu- * -Asn-Ile-Thr-
TSHβ                                          Phe- * -Ile- * -Thr-Glu-Tyr-Thr-


                  16                                          29 ┌CHO
LHβ       Leu-Ala-Val-Gln-Lys-Glu-Gly-Cys-Pro-Val-Cys-Ile-Thr-Val-Asn-
FSHβ      Ile- * -Ile-Glu- * - * -Glu- * -Arg-Phe- * -Leu- * -Ile- * -
TSHβ      Met-His-Ile-Glu-Arg-Arg-Glu- * -Ala-Tyr- * -Leu- * -Ile- * -


                  31                                                    45
LHβ       Thr-Thr-Ile-Cys-Ala-Gly-Tyr-Cys-Pro-Thr-Met- - - - -Met-Arg-Val-
FSHβ       * - * -Trp- * - * - * - * - * - * -Tyr- * -Arg-Asp-Leu-Val-Tyr-
TSHβ       * - * - * - * - * - * - * - * -Met- * -Arg- - - - -Asp-Ile-Asn-


                  46                                                    57
LHβ       Leu-Gln-Ala-Val-Leu-Pro- - - - - - - - -Pro-Leu-Pro-Gln- - - - -Val-Cys-
FSHβ      Lys-Asn-Pro-Ala-Arg- * - - - - - - - - - -Lys-Ile-Gln-Lys- - - - -Thr- * -
TSHβ      Gly-Lys-Leu-Phe- * - * -Lys-Tyr-Ala- * -Ser- * -Asp- * - * -


                  58                                                    72
LHβ       Thr-Tyr-Arg-Asp-Val-Arg-Phe-Glu-Ser-Ile-Arg-Leu-Pro-Gly-Cys-
FSHβ       * -Phe-Lys-Glu-Leu-Val-Tyr- * -Thr-Val- * -Val- * - * - * - *
TSHβ       * - * - * - * -Phe-Ile-Tyr-Arg-Thr-Val-Glu-Ile- * - * - * - *


                  73                                                    87
LHβ       Pro-Arg-Gly-Val-Asp-Pro-Val-Val-Ser-Phe-Pro-Val-Ala-Leu-Ser-
FSHβ      Ala-His-His-Ala- * -Ser-Leu-Tyr-Thr-Tyr- * - * - * - * -Thr-Gln-
TSHβ       * -Leu-His- * -Ala- * -Tyr-Phe- * -Tyr- * - * - * - * - * - *


                  88                                                    102
LHβ       Cys-Arg-Cys-Gly-Pro-Cys-Arg-Arg-Ser-Thr-Ser-Asp-Cys-Gly-Gly-
FSHβ       * -His- * - * -Lys- * -Asp-Ser-Asp-Ser-Thr- * - * -Thr-Val-
TSHβ       * -Lys- * - * -Lys- * -Asn-Thr-Asp-Tyr- * - * - * -Ile-His-


                  103                                                   117
LHβ        Pro-Lys-Asp-His-Pro-Leu-Thr-Cys-Asp-Pro-Gln-His-Ser-Gly
FSHβ      Arg-Gly-Leu-Gly- * -Ser-Tyr- * -Ser-Phe-Gly-Glu-Met-Lys-Gln-
TSHβ      Glu-Ala-Ile-Lys-Thr-Asn-Tyr- * -Thr-Lys-Pro-Gln-Lys-Ser-Tyr-


                  118                          124
FSHβ      Tyr-Pro-Thr-Ala-Leu-Ser-Tyr
TSHβ      Leu-Val-Gly-Phe-Ser-Val
```

Fig. 4. Human LH, FSH, and TSH β subunits. Amino acid numbering is relative to maximum homology between the three subunits. Consensus glycosylation sites are at Asn-13 and 30. Note that carbohydrate attachment at Asn-13 only occurs in human FSHβ. * indicates same amino acid as LHβ. CHO = carbohydrate chain (see Fig. 3).

glycosylated (48). FSH-β subunit [58857-12-8] is glycosylated at two sites, Asn-13 and 30. Based on interactions of synthetic peptides with the LH receptor, loops formed by β 93–100 and β 38–57 may be essential for hormone bioactivity (48). Highly conserved sequences between residues 31–37 have been implicated in the formation of the α–β subunit dimer (48), which is absolutely necessary for the expression of bioactivity.

Preparations of FSH, usually containing equal or greater quantities of LH, are used for superovulation, ie, the production of a large number of ova, in human medicine and domestic animal production. Schering-Plough, U.S.A., and Vetipharm, Canada, market FSH for use in domestic animals. Preparations of FSH used in human fertility control are produced by two European companies, Serono and Organon; Serono produces the only FSH/LH approved for use in the United States. Luteinizing hormone does not have significant commercial value in the induction of ovulation in women since human chorionic gonadotropin (hCG) (see HORMONES, ANTERIOR PITUITARY-LIKE HORMONES) exerts potent LH activity, is resistant to biological degradation, and is readily available. A preparation of porcine LH [9061-23-8] is produced by Vetipharm, Canada; however, many superovulation procedures use gonadotropin-releasing hormone (GnRH) analogues rather than LH to induce ovulation. Gonadotropin-releasing hormone is a hypothalamic peptide which stimulates the pituitary to release LH and FSH. The use of recombinant DNA technology for the production of LH and FSH is being explored, and commercial preparations (ca 1993) of LH and FSH are derived from natural sources.

Thyroid-stimulating hormone can be used clinically to test thyroid function but has not found practical application in the treatment of human thyroid insufficiency. Direct replacement therapy with thyroid hormone is easy and effective, owing to a simple molecular structure. TSH has been used in the veterinary treatment of hypothyroidism, and preparations of TSH are produced by Cooper Animal Health, Inc. and Armour Pharmaceuticals.

Names are necessary to report factually on available data; however, the U.S. Department of Agriculture neither guarantees nor warrants the standard of the product, and the use of the name by the U.S. Department of Agriculture implies no approval of the product to the exclusion of others that may also be suitable.

BIBLIOGRAPHY

"Anterior–Pituitary Hormones" in *ECT* 1st ed., Vol. 7, pp. 481–489, by C. H. Li, University of California; in *ECT* 2nd ed., Vol. 11, pp. 52–64, by H. Papkoff and C. H. Li, The Hormone Research Laboratory, University of California; in *ECT* 3rd ed., pp. 546–556, by C. H. Li, University of California, San Francisco.

1. W. D. Linscott, *Linscott's Directory of Immunological and Biological Reagents*, 7th ed., Linscott's Directory, Santa Rosa, Calif., 1992–1993, 239 pp.
2. *The Merck Index*, 11th ed., Merck & Co., Inc., Rahway, N.J., 1989.
3. P. L. Whitfeld, P. H. Seeburg, and J. Shine, *DNA* 1, 133 (1982).
4. M. Lamacz and co-workers, *Arch. Int. Physiol. Biochim. Biophys.* 99, 205 (1991).
5. P. L. Loh, *Biochem. Pharmacol.* 44, 843 (1992).
6. J. Drouin and H. M. Goodman, *Nature* 288, 610 (1980).
7. B. A. Eipper and R. E. Mains, *Endocr. Rev.* 1, 1 (1980).
8. Y. P. Loh and H. Gainer, *Mol. Cell Endocrinol.* 20, 35 (1980).
9. W. C. Clark and J. M. Lipton, *Pharmacol. Ther.* 22, 249 (1983).
10. C. H. Li and D. Chung, *Proc. Nat. Acad. Sci. USA* 73, 1145 (1976).
11. C. H. Li, *Nature (London)* 201, 924 (1964).
12. Y. Birk and C. H. Li, *J. Biol. Chem.* 239, 1048 (1964).
13. P. Stricker and F. Grueter, *C. R. Seances Soc. Biol. Paris* 99, 1978 (1928).

14. L. Matera, G. Bellone, and A. Cesaro, *Adv. Neuroimmunol.* **1**, 158 (1991).
15. C. S. Nicoll, *Fed. Proc.* **39**, 2563 (1980).
16. G. Bauman, *Endocr. Rev.* **12**, 424 (1991).
17. A. I. Rubinshtein and Y. A. Pankov, *Mol. Biol. (USSR)* **24**, 1562 (1990).
18. C. S. Nicoll, G. L. Mayer, and S. M. Russell, *Endocr. Rev.* **7**, 169 (1986).
19. W. L. Miller and N. L. Eberhardt, *Endocr. Rev.* **4**, 97 (1983).
20. Y. N. Sinha, *Trends Endocrinol. Metab.* **3**, 100 (1992).
21. E. Markoff and co-workers, *Endocrinology* **123**, 1303 (1988).
22. Y. A. Pankov and V. Y. Butnev, *Int. J. Peptide Protein Res.* **28**, 113 (1986).
23. C. Clapp and co-workers, *Endocrinology* **122**, 2892 (1988).
24. D. N. Luck and co-workers, *Protein Eng.* **5**, 559 (1992).
25. C. H. Li and H. M. Evans, *Science* **99**, 183 (1944).
26. B. T. Rudd, *Ann. Clin. Biochem.* **28**, 542 (1991).
27. J. M. P. Holly and J. A. H. Wass, *J. Endocrinol.* **122**, 611 (1989).
28. K. W. Kelley, *Biochem. Pharmacol.* **38**, 705 (1989).
29. G. S. Barsh, P. H. Seeburg, and R. E. Gelinas, *Nucleic Acids Res.* **11**, 3939 (1983).
30. Z. C. Chen and co-workers, *Mol. Endocrinol.* **5**, 598 (1992).
31. J. S. Dixon and C. H. Li, *Science* **154**, 785 (1966).
32. D. F. Nutting and co-workers, *Biochim. Biophys. Acta* **200**, 601 (1970).
33. L. Binder and co-workers, *Mol. Endocrinol.* **3**, 923 (1989).
34. A. M. de Vos, M. Ultsch and A. A. Kossiakoff, *Science* **255**, 306 (1992).
35. M. Boothby and co-workers, *J. Biol. Chem.* **256**, 5121 (1981).
36. J. C. Fiddes and H. M. Goodman, *J. Mol. Appl. Genet.* **1**, 3 (1981).
37. R. G. Goodwin and co-workers, *Nucleic Acids Res.* **11**, 6873 (1983).
38. H. E. Grotjan, Jr., in B. A. Keel and H. E. Grotjan, Jr., eds., *Microheterogeneity of Glycoprotein Hormones*, CRC Press, Boca Raton, Fla., 1989, pp. 23–52.
39. B. A. Keel, in Ref. 38, p. 203.
40. B. A. Keel and H. E. Grotjan, Jr., in Ref. 38, p. 185.
41. D. M. Robertson, in Ref. 38, p. 107.
42. R. L. Matteri and H. Papkoff, *Biol. Reprod.* **36**, 261 (1987).
43. R. L. Matteri and H. Papkoff, *Biol. Reprod.* **38**, 324 (1988).
44. R. L. Matteri and H. Papkoff, in Ref. 38, p. 185.
45. W. F. P. Blum and D. Gupta, *J. Endocrinol.* **105**, 29 (1985).
46. A. B. Galway and co-workers, *Endocrinology* **127**, 93 (1990).
47. R. L. Matteri, P. K. Warikoo, and B. D. Bavister, *Am. J. Primatol.* **27**, 205 (1992).
48. R. J. Ryan and co-workers, *Recent Prog. Hormone Res.* **43**, 383 (1987).
49. T. Mise and O. P. Bahl, *J. Biol. Chem.* **256**, 6587 (1981).
50. S. Tsunasawa and co-workers, *Biochim. Biophys. Acta* **492**, 340 (1977).
51. D. N. Ward and co-workers, in Ref. 38, p. 1.
52. M. R. Sairam and G. N. Bhargavi, *Science* **229**, 65 (1985).

ROBERT L. MATTERI
U.S. Department of Agriculture

ANTERIOR PITUITARY-LIKE HORMONES

Hormones with structural and functional similarities to the hormones of the anterior pituitary gland (see HORMONES, ANTERIOR PITUITARY HORMONES) fall into two main categories, ie, proteins of the prolactin/growth hormone (PRL/GH) family, and glycoproteins. These hormones are produced in the placenta by chorionic tissue, and so are normally found in the female during pregnancy. Placental lactogens, members of the PRL/GH family, support growth and development of mammary tissue in preparation for lactation. Chorionic gonadotropin (CG) is similar in structure to luteinizing hormone (LH) and supports the maintenance of early pregnancy. Chorionic gonadotropin [9002-70-4] is found in humans (hCG) and lower primates, as well as in horses (eCG). Additional members of the PRL/GH family of hormones with unknown biologic function have been discovered. Commercial sources are provided herein when available. Additional information on primary structure, glycosylation, and disulfide bonding relating to the bioactivity of PRL/GH-related hormones and glycoprotein hormones is available (see HORMONES, ANTERIOR PITUITARY HORMONES).

Prolactin/Growth Hormone Family

Chorionic Somatomammotropin. Three genes encode human chorionic somatomammotropin [11085-36-2] (hCS). These are located within a cluster of genes on human chromosome 17 which code for pituitary growth hormone [12629-01-5] (GH-N), placental GH [109675-94-7] (GH-V), and three hCS molecules, ie, hCS-A, hCS-B, and hCS-V (1–3), also referred to as human placental lactogens. All of these molecules are closely related to GH in structure (Fig. 1). Placental lactogens also exist in rodents and ruminants; however, these hormones are more closely related to prolactin than GH.

The amino acid sequences of hCS-A, hCS-B, and hCS-V are shown in relation to GH in Figure 1. The sequence of hCS-V is predicted from the DNA coding sequence and apparently does not possess amino acids 8–55 relative to GH and the other hCS molecules. It is not certain whether hCS-V is expressed or what function it may have. Human CS-A and hCS-B share approximately 85% identity with GH and also possess the disulfide bonds between Cys 53–165 and Cys 182–189 which produce the long and short S–S loops characteristic of the PRL/GH family.

Chorionic somatomammotropin is detectable in maternal serum by immunoassay (qv) at about two weeks of pregnancy. Its secretion increases thereafter to very high levels, 20 μg/mL serum, prior to delivery (4). Human CS does not bind the GH receptor significantly, even though there is high sequence homology between hCS and GH. Existing structure–function data, particularly from hydrophobicity evaluations, suggest that four amino acid substitutions relative to GH, ie, Asp-104, Asp-109, Asp-110, and His-112, may result in the lack of growth-promoting activity of hCS (5). Human CS behaves much like prolactin in a number of assay systems. Accordingly, hCS is thought to contribute to the development of mammary tissue in preparation for postnatal nursing. Some of

```
        1                                                          15
GH-N:Phe-Pro-Thr-Ile-Pro-Leu-Ser-Arg-Leu-Phe-Asp-Asn-Ala-Met-Leu-
GH-V:  * - * - * - * - * - * - * - * - * - * - * - * - * - * - * -
CS-A: Val-Gln- * -Val- * - * - * - * - * - * - * -His- * - * - * -
CS-B: Val-Gln- * -Val- * - * - * - * - * - * - * -His- * - * - * -
CS-V: Val-Gln- * -Val- * - * - * - * - * - * -Lys-Glu- * - * - * -

        16                                                         30
GH-N:Arg-Ala-His-Arg-Leu-His-Gln-Leu-Ala-Phe-Asp-Thr-Tyr-Gln-Glu-
GH-V:  * - * -Arg- * - * -Tyr- * - * - * -Tyr- * - * - * - * - * -
CS-A: Gln- * - * - * -Ala- * - * - * - * -Ile- * - * - * - * - * -
CS-B: Gln- * - * - * -Ala- * - * - * - * -Ile- * - * - * - * - * -
CS-V: Gln- * - * - * -Ala- * - * - * - * -Ile- * - * - * - * - * -

        31                                                         45
GH-N:Phe-Glu-Glu-Ala-Tyr-Ile-Pro-Lys-Glu-Gln-Lys-Tyr-Ser-Phe-Leu-
GH-V:  * - * - * - * - * - * -Leu- * - * - * - * - * - * - * - * -
CS-A:  * - * - * -Thr- * - * - * - * - * - * - * - * - * - * - * -
CS-B:  * - * - * -Thr- * - * - * - * -Asp- * - * - * - * - * - * -
CS-V:  * -Ile-Ser-Ser-Trp-Gly-Met --------------------------------

        46                                                         60
GH-N:Gln-Asn-Pro-Gln-Thr-Ser-Leu-Cys-Phe-Ser-Glu-Ser-Ile-Pro-Thr-
GH-V:  * - * - * - * - * - * - * - * - * - * - * - * - * - * - * -
CS-A: His-Asp-Ser- * - * - * -Phe- * - * - * -Asp- * - * - * - * -
CS-B: His-Asp-Ser- * - * - * -Phe- * - * - * -Asp- * - * - * - * -
CS-V: ------------------------------------Asp- * - * - * - * -

        61                                                         75
GH-N:Pro-Ser-Asn-Arg-Glu-Glu-Thr-Gln-Gln-Lys-Ser-Asn-Leu-Glu-Leu-
GH-V:  * - * - * - * - * -Val-Lys- * - * - * - * - * - * - * - * -
CS-A:  * - * - * -Met- * - * - * - * - * - * - * - * - * - * - * -
CS-B:  * - * - * -Met- * - * - * - * - * - * - * - * - * - * - * -
CS-V: Ser- * -Asn-Met- * - * - * - * - * - * - * - * - * - * - * -

        76                                                         90
GH-N:Leu-Arg-Ile-Ser-Leu-Leu-Leu-Ile-Gln-Ser-Trp-Leu-Glu-Pro-Val-
GH-V:  * - * - * - * - * - * - * - * - * - * - * - * - * - * - * -
CS-A:  * - * - * - * - * - * - * - * - * - * - * - * - * - * - * -
CS-B:  * - * - * - * - * - * - * - * -Glu- * - * - * - * - * - * -
CS-V:  * -His- * - * - * - * - * - * -Glu- * -Arg- * - * - * - * -

        91                                                         105
GH-N:Gln-Phe-Leu-Arg-Ser-Val-Phe-Ala-Asn-Ser-Leu-Val-Tyr-Gly-Ala-
GH-V:  * -Leu- * - * - * - * - * - * - * - * - * - * - * - * - * -
CS-A: Arg- * - * - * - * -Met- * - * - * -Asn- * - * - * -Asp-Thr-
CS-B:  * - * - * - * - * -Met- * - * - * -Asn- * - * - * -Asp-Thr-
CS-V: Arg- * - * - * - * -Thr- * -Thr- * -Asn- * - * - * -Asp-Thr-

        106                                                        120
GH-N:Ser-Asp-Ser-Asn-Val-Tyr-Asp-Leu-Leu-Lys-Asp-Leu-Glu-Glu-Gly-
GH-V:  * - * - * - * - * - * - * - * -His- * - * - * - * - * - * -
CS-A:  * - * - * -Asp-Asp- * -His- * - * - * - * - * - * - * - * -
CS-B:  * - * - * -Asp-Asp- * -His- * - * - * - * - * - * - * - * -
CS-V:  * - * - * -Asp-Asp- * -His- * - * - * - * - * - * - * - * -
```

Fig. 1. Primary structure of the human growth hormone family of proteins. GH-N = pituitary GH; GH-V = placental GH variant; CS = choriosomatomammotropin; * = same amino acid as GH-N. Disulfide bonds connect Cys molecules at positions 53–165 and 182–189. Asn-140 of GH-V is underlined to indicate an N-linked glycosylation site.

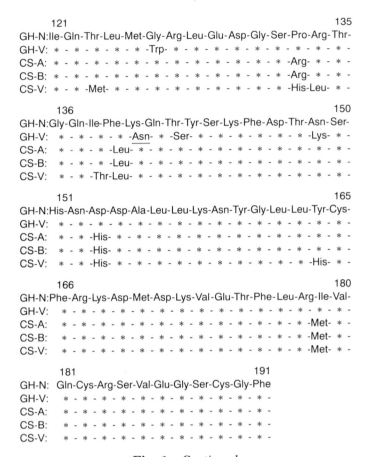

```
         121                                               135
GH-N:Ile-Gln-Thr-Leu-Met-Gly-Arg-Leu-Glu-Asp-Gly-Ser-Pro-Arg-Thr-
GH-V: * - * - * - * - * -Trp- * - * - * - * - * - * - * - * -
CS-A: * - * - * - * - * - * - * - * - * - * - * - * -Arg- * - * -
CS-B: * - * - * - * - * - * - * - * - * - * - * - * -Arg- * - * -
CS-V: * - * -Met- * - * - * - * - * - * - * - * - * -His-Leu- * -

         136                                               150
GH-N:Gly-Gln-Ile-Phe-Lys-Gln-Thr-Tyr-Ser-Lys-Phe-Asp-Thr-Asn-Ser-
GH-V: * - * - * - * -Asn- * -Ser- * - * - * - * - * - * -Lys- * -
CS-A: * - * - * - * -Leu- * - * - * - * - * - * - * - * - * - * -
CS-B: * - * - * - * -Leu- * - * - * - * - * - * - * - * - * - * -
CS-V: * - * -Thr-Leu- * - * - * - * - * - * - * - * - * - * - * -

         151                                               165
GH-N:His-Asn-Asp-Asp-Ala-Leu-Leu-Lys-Asn-Tyr-Gly-Leu-Leu-Tyr-Cys-
GH-V: * - * - * - * - * - * - * - * - * - * - * - * - * - * - * -
CS-A: * - * - * -His- * - * - * - * - * - * - * - * - * - * - * -
CS-B: * - * - * -His- * - * - * - * - * - * - * - * - * - * - * -
CS-V: * - * - * -His- * - * - * - * - * - * - * - * - * -His- * -

         166                                               180
GH-N:Phe-Arg-Lys-Asp-Met-Asp-Lys-Val-Glu-Thr-Phe-Leu-Arg-Ile-Val-
GH-V:  * - * - * - * - * - * - * - * - * - * - * - * - * - * - * -
CS-A:  * - * - * - * - * - * - * - * - * - * - * - * - * -Met- * -
CS-B:  * - * - * - * - * - * - * - * - * - * - * - * - * -Met- * -
CS-V:  * - * - * - * - * - * - * - * - * - * - * - * - * -Met- * -

         181                               191
GH-N: Gln-Cys-Arg-Ser-Val-Glu-Gly-Ser-Cys-Gly-Phe
GH-V:  * - * - * - * - * - * - * - * - * - * - * -
CS-A:  * - * - * - * - * - * - * - * - * - * - * -
CS-B:  * - * - * - * - * - * - * - * - * - * - * -
CS-V:  * - * - * - * - * - * - * - * - * - * - * -
```

Fig. 1. *Continued*

the metabolic effects of GH may be conserved in hCS. Human CS is thought to contribute to the commonly observed resistance to insulin in the mother, which serves to divert carbohydrate energy sources to the developing fetus. Human CS is available for research purposes from a variety of sources (6), but does not have clinical applications.

Placental Growth Hormone. The gene for placental growth hormone (GH-V, V = variant) is found within the GH/CS gene cluster. GH-V is expressed in the placenta, but not in the anterior pituitary. The structure of GH-V differs from pituitary GH (GH-N, N = normal), by only 12 substitutions dispersed along the length of the molecule (Fig. 1). Similarly to GH-N (see HORMONES, ANTERIOR PITUITARY HORMONES), GH-V possesses both long and short disulfide loops. A consensus glycosylation site is utilized for carbohydrate attachment during post-translational processing at Asn-140 (7). A frameshift in transcription gives rise to a 230 amino acid variant of GH-V which does not possess the glycosylation site and has a dissimilar carboxy terminus relative to GH-V. Interestingly, this form of GH-V is not secreted, but is believed to be membrane-bound (2).

Antibodies have been generated which produce immunoassays that discriminate between GH-V and GH-N. These assay systems have shown that the

```
        1                                                             15
bPRL:                      Thr-Pro-Val-Cys-Pro-Asn-Gly-Pro-Gly-Asn-Cys-
bPL:   Val-Glu-Asp-Tyr-Ala--*--Tyr--*--Lys- -*--Gln--*---*---*---*--
oPL:   Gln-Ala-Gln-His-Pro--*--Tyr--*--Arg--*--Gln--*----*---Lys--*--
bGH:                                                          Ala-Phe-Pro-

        16                                                            30
bPRL:Gln-Val-Ser-Leu-Arg-Asp-Leu-Phe-Asp-Arg-Ala-Val-Met-Val-Ser-
bPL:   Arg-Ile-Pro--*--Gln-Ser--*-- -*--Glu--*---*--Thr-Leu--*--Ala-
oPL:   -*--Ile-Pro--*--Gln-Ser--*----*---*----*---*--Thr-Thr--*--Ala-
bGH:   Ala-Met--*----*---Ser-Gly--*----*--Ala-Asn--*---*--Leu-Arg-Ala-

        31                                                            45
bPRL:His-Tyr-Ile-His-Asp-Leu-Ser-Ser-Glu-Met-Phe-Asn-Glu-Phe-Asp-
bPL:   Ser-Asn-Asn-Tyr-Arg- -*--Ala-Arg--*---*----*---*----*---*--Asn-
oPL:   Asn--*--Asn-Ser-Lys--*--Ala-Gly--*----*--Val--*--Arg--*----*--
bGH:   Gln-His-Leu--*- -Gln--*--Ala-Ala-Asp-Thr--*- -Lys--*---*--Glu-

        46                                                            60
bPRL:Lys-Arg-Tyr-Ala-Gln- Gly-Lys-Gly-Phe-Ile------------Thr-Met-
bPL:   -*--Gln-Phe-Gly-Glu--*----*--Asn--*--Thr------------Ser-Lys-
oPL:   Glu-Gln--*--Gly--*----*--Ile-Asn-Ser-Glu------------Ser-Lys-
bGH:   Arg-Thr- -*- -Ile-Pro-Glu-Gly-Gln-Arg-Tyr-Ser-Ile-Gln-Asn- Thr-

        61                                                            75
bPRL:Ala-Leu-Asn-Ser-Cys-His-Thr-Ser-Ser-Leu-Pro-Thr-Pro-Glu-Asp-
bPL:   Phe-Ile--*---*---*---*---*--Glu-Phe-Met-Thr- -*----*--Asn-Asn-
oPL:   Val-Ile--*--------*---*---*---*---*--Ile-Thr- -*----*--Asn-Ser-
bGH:   Gln-Val-Ala-Phe--*--Phe-Ser-Glu-Thr-Ile--*--Ala--*--Thr-Gly-

        76                                                            90
bPRL:Lys-Glu-Gln-Ala-Gln-Gln-Thr-His-His-Glu-Val-Leu-Met-Ser-Leu-
bPL:   -*---*--Ala--*--Ala-Asn--*--Glu-Asp--*--Ala--*--Leu-Arg--*--
oPL:   -*--Ala-Glu--*--Ile-Asn--*--Glu-Asp-Lys-Ile--*--Phe-Lys--*--
bGH:   -*--Asn-Glu--*----*--Lys-Ser-Asp-Leu-Glu- -*--Leu-Arg-Ile-

        91                                                           105
bPRL:Ile-Leu-Gly-Leu-Leu-Arg-Ser-Trp-Asn-Asp-Pro-Leu-Tyr-His-Leu-
bPL:   Val-Ile-Ser--*----*--His--*----*--Asp-Glu--*----*--His-Gln-Ala-
oPL:   Val-Ile-Ser--*----*--His--*----*--Asp-Glu--*----*--His--*--Ala-
bGH:   Ser--*--Leu--*--Ile-Gln--*----*--Leu-Gly--*----*--Gln-Phe--*- -

        106                                                          120
bPRL:Val-Thr-Glu-Val-Arg-Gly-Met---------Lys-Gly-Ala-Pro-Asp-Ala-
bPL:   -*---*---*---Leu-Leu-His-Arg---------Asn--*---*--Ser-Pro-Asp-
oPL:   -*---*---*---Leu-Ala-Asn-Ser-----------*---*--Thr-Ser-Pro--*--
bGH:   Ser-Arg-Val-Phe-Thr-Asn-Ser-Leu-Val-Phe--*--Thr-Ser-Asp-Arg-

        121                                                          135
bPRL:Ile-Leu-Ser-Arg-Ala-Ile-Glu-Ile-Glu-Glu-Glu-Asn-Lys-Arg-Leu-
bPL:   -*---*--Ala--*----*---Lys--*----*----*--Asp-Lys-Thr- -*--Val--*--
oPL:   Leu--*--Thr-Lys--*--Gln--*----*----*--Lys--*--Lys-Ala--*--Val--*--
bGH:   Val-Tyr-Glu-Lys-Leu-Lys-Asp-Leu--*----*--Gly-Ile-Leu-Ala--*--
```

Fig. 2. Primary structure of bPRL, bPL, oPL, and bGH. Amino acid numbers are relative to maximum homology among hormones (12). * = same amino acid as bPRL. Disulfide bonds: Cys 8–15 (not in bGH), 65–183, 200–208. Asn-53 of bPL is underlined to indicate an N-linked glycosylation site.

```
          136                                          150
bPRL:Leu-Glu-Gly-Met-Glu-Met-Ile-Phe-Gly-Gln-Val-Ile-Pro-Gly-Ala-
bPL:  -*---*---*--Val--*---*---*--Gln-Lys-Arg--*--His--*---*--Glu-
oPL:  Val-Asp--*--Val--*--Val--*--Gln-Lys-Arg-Ile-His--*---*--Glu-
bGH:  Met-Arg-Glu-Leu--*--Asp-Gly-Thr-Pro-Arg-Ala----------*--Gln-

          151                                          165
bPRL:Lys-Glu-Thr-Glu-Pro-Tyr-Pro-Val-Trp-Ser-Gly-Leu-Pro-Ser-Leu-
bPL:  -*--Lys-Asn--*---*---*--*---*---*----*--Glu-Lys-Ser--*---*--
oPL:  -*------Asn--*---*---*---*--*---*---*--*--Glu-Gln-Ser--*---*--
bGH:  Ile-Leu-Lys-Gln-Thr--*--Asp-Lys-Phe-Asp-Thr-Asn-Met---------

          166                                          180
bPRL:Gln-Thr-Lys-Asp-Glu-Asp-Ala-Arg-Tyr-Ser-Ala-Phe-Tyr-Asn-Leu-
bPL:  Thr-Ala-Asp--*---*---*--*--Val--*--Gln-Thr--*---*---*---*--Arg-Met-
oPL:  Thr-Ser-Gln--*---*--Asn-Val--*--Arg-Val--*---*---*--*--Arg--*--
bGH:  -----------Arg-Ser--*--Asp-Ala-Leu-Leu-Lys-Asn--*--Gly--*--

          181                                          195
bPRL:Leu-His-Cys-Leu-Arg-Arg-Asp-Ser-Ser-Lys-Ile-Asp-Thr-Tyr-Leu-
bPL:  Phe--*---*---*--His--*---*---*---*---*---*---*--Ser--*---*--Ile-
oPL:  Phe--*---*---*--His--*---*---*---*---*---*----*--Tyr--*---*---*--
bGH:  -*--Ser--*--Phe--*--Lys--*--Leu-His--*--Thr-Glu--*---*--*---*--

          196                                          210
bPRL: Lys-Leu-Leu-Asn-Cys-Arg-Ile-Ile-Tyr-Asn-Asn-Asn-Cys
bPL:  Asn--*---*--Lys--*---*------------Phe-Thr-Pro--*-
oPL:  Arg-Ile--*--Lys--*----*-----------Leu-Thr-Ser--*--Glu-Thr
bGH:  Arg-Val-Met-Lys--*----*--Arg-Phe-Gly-Glu-Ala-Ser--*--Ala-Phe
```

Fig. 2. *Continued*

secretion of GH-V becomes elevated at about three weeks of pregnancy and increases to approximately 15 ng/mL near term (8). The physiological role of GH-V is uncertain. Genetic deficiency of GH-V does not adversely affect pregnancy or fetal development (9). GH-V is a potent growth-stimulator but possesses considerably less lactogenic activity than GH-N (10). There are no clinical applications (ca 1993) for GH-V.

Placental Lactogens. The placentae of rodents and ruminants produce prolactin-related molecules called placental lactogens (11). Placental lactogens are not found in all species. Prolactin-like activity has not been demonstrated in placental extracts from rabbits, pigs, horses, or animals of the order Carnivora. Ovine [127497-22-7] (oPL) and bovine [116669-02-4] (bPL) placental lactogens have been studied extensively (12). The structures of bovine prolactin [56832-36-1] (bPRL), bPL, oPL, and bovine GH [66419-50-9] (bGH) are given in Figure 2. Bovine and ovine PL share about 50% homology with bPRL and approximately 20% homology with bGH. A short N-terminal disulfide loop exists in bPL and oPL, reflecting their relatedness to prolactin. As with other hormones of the PRL/GH family, bPL and oPL possess the long and short C-terminal loops formed by disulfide bridges. Bovine PL, but not oPL, possesses both O- and N-linked oligosaccharide moieties (13). Enzymatic deglycosylation of bPL may increase receptor binding potency, but does not appear to influence bioactivity *in vitro* (14). The carbohydrate moieties of bPL could serve to prolong the hormone's circulating half-life (15). Only the glycosylated form of bPL appears to be secreted (12).

The ruminant PLs are able to bind both GH and PRL receptors and exert lactogenic and somatogenic actions (12). As described for hCS, ruminant PLs are thought to contribute to mammary development and to alter maternal metabolism to support fetal nutrition. Evidence also suggests that ruminant PL may directly regulate fetal growth. Receptors which show greater affinity for oPL than for ovine growth hormone [37267-05-03] (oGH) and ovine prolactin [12585-34-1] (oPRL) have been identified in fetal ovine liver (16). The synthesis of hepatic glycogen, an important carbohydrate energy source, is stimulated by oPL (17). Maternal food deprivation induces a reduction of approximately 70% in PL binding to the fetal liver, perhaps contributing to the accompanying intrauterine growth retardation (18). The presence of bPL receptors in bovine endometrium (19,20) suggests that bPL could regulate the function of maternal tissues which support pregnancy, fetal growth, and development.

There are no commercial uses for ruminant PLs. However, bPL has been shown to stimulate insulin-like growth factor I and somatic growth in rats (12), suggesting potential applications as performance stimulators in domestic animal production. Milk yield and feed intake are stimulated by the administration of bPL (21) (see GROWTH REGULATORS, ANIMAL). The development of recombinant bPL for use in food animals is being investigated by Monsanto Co. (15). Limited availability of oPL has restricted opportunities for research into the pharmacologic value of this hormone. The development of a high yield expression system for oPL is in progress (22).

A review of the synthesis, structure, and potential functions of rodent PLs is available (11). These hormones have no commercial importance per se, but may prove useful in studies of biology and structure–function relationships essential for future practical applications of bPL and oPL. Two placental lactogens, PL-I [144591-44-6] and PL-II [121631-23-0], are produced in the rat. The production of PL-I, which is dominant at midgestation, is replaced by PL-II later in pregnancy. PL-I, but not PL-II, is glycosylated. Rodent PLs all demonstrate lactogenic activity, but do not bind GH receptors. Prolactin-like molecules support the existence and function of the corpus luteum in rodents. The maintenance of corpus luteum function, which is critical for the continuation of pregnancy, is probably mediated by PLs in mid- and late gestation.

Prolactin-Like Proteins.   A number of prolactin-like proteins (PLPs), which are distinct from the PLs, have been identified in ruminants and rodents (11,23). Several cDNA transcripts coding for PLPs in cattle have been identified (23). These transcripts code for proteins which possess about 40% sequence homology with bovine PRL; 60% if conservative substitutions are considered. Three glycosylated PLPs, ie, PLP-A, -B, and -C, are produced during pregnancy in the rat (11). Two additional prolactin-related molecules have been identified in the mouse (24,25), ie, proliferin [92769-12-5] (PLF) and PLF-related protein [98724-27-7]. These are not found in other rodents and may be unique to the mouse. The functional roles of PLPs remain to be determined.

Glycoprotein Family

Human Chorionic Gonadotropin.   Human CG (hCG) is produced by syncitiotrophoblast cells of the placenta. The secretion of hCG from chorionic cells begins about 10 days after fertilization. The detection of this early rise in hCG in urine forms the basis for a variety of nonprescription home pregnancy tests. Maximum production of CG occurs at approximately 70 days of gestation and declines rapidly, reaching relatively low levels in the serum by the second trimester. The secretion of hCG early in pregnancy prolongs the lifespan of the corpus luteum, an ovarian structure which secretes the steroid hormones progesterone and estradiol, needed to maintain pregnancy. After the first two weeks, the placenta is able to produce sufficient steroid hormones to maintain pregnancy.

Human CG is purified readily from urine collected during the first trimester of pregnancy and is very similar in structure to pituitary luteinizing hormone (LH). Like LH, hCG is comprised of noncovalently bound α- [56832-30-5] and β- [78690-52-6] subunits. The subunits are glycosylated during post-translational processing (26). Both subunits must be in association for the expression of biological activity. The subunits dissociate readily in the presence of chaotropic salts such as urea and guanidine–HCl. The α-subunit of hCG and the three pituitary glycoprotein hormones, ie, LH, follicle-stimulating hormone (FSH), and thyroid-stimulating hormone (TSH) (27), are coded by the same gene and therefore have identical amino acid sequences; structural information on the α-subunit is available (see HORMONES, ANTERIOR PITUITARY HORMONES). The β-subunit of hCG (hCGβ) is coded by a cluster of seven gene copies, some of which may be nonexpressed pseudogenes (28,29). At least two hCGβ genes are expressed and one occurs in tandem with the gene for LHβ (30). As for the pituitary glycoprotein hormones, the β-subunit of hCG confers the type of biologic action which is expressed, eg, the combination of TSHα and hCGβ would produce bioactive hCG. Human CGβ contains a carboxy terminal peptide extension of 30 amino acids relative to hLHβ [53664-53-2], which contains sites for O-linked carbohydrate attachment (Fig. 3). On a weight basis, hCG is comprised of approximately one-third carbohydrate, ie, 13.8% hexose, 10.8% hexosamines, and 9.6% sialic acid. The carbohydrate component of the α-subunit is necessary for the expression of bioactivity (32). Selective elimination of N-linked glycosylation sites by site-directed mutagenesis has revealed that the absence of the oligosaccharide at Asn-52α does not significantly affect receptor binding, yet reduces the ability of hCG to activate its target cell (33). Removal of the other carbohydrates at Asn-78α, Asn-13β, or Asn-30β has minor effects on *in vitro* bioactivity (33). The high content of sialic acid confers a prolonged biological half-life which promotes biological activity *in vivo*. The complex and varied carbohydrate structures of hCG have been reviewed (34).

Human CG has considerable commercial value in clinical fertility control due to its LH-like bioactivity and natural resistance to biological degradation. The injection of hCG is used to induce ovulation following treatment with high doses of follicle-stimulating hormone (FSH) to stimulate the maturation of multiple ova. Significant producers of hCG for clinical usage are Serono (Italy) and Organon (Sweden). Human CG is available for research purposes from many suppliers (6,35).

Equine Chorionic Gonadotropin. Equine CG (eCG) is produced by trophoblast-derived structures in the endometrium known as endometrial cups. The original name of this hormone was pregnant mare serum gonadotropin

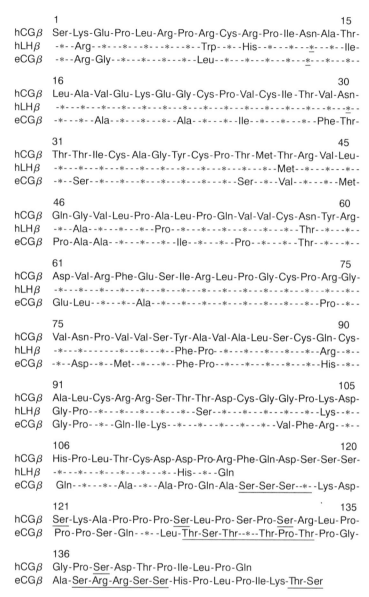

Fig. 3. Human CG, hLH, and equine CG (eCG) β-subunits. Amino acid numbering is relative to maximum homology between the three subunits. Consensus glycosylation sites are at Asn-13 and 30. * = same amino acid as hCGβ. Underlined Asn residues indicate attachment of N-linked carbohydrate chains. Serines at positions 121, 127, 132, and 138 of hCGβ are underlined to indicate sites of O-linked carbohydrate attachment. Residues 115–118, 127–133, 137–141, 148–149 of eCGβ are underlined to highlight probable locations of O-linked oligosaccharides attached to Ser or Thr (31).

(PMSG). Whereas eCG is the correct scientific name for this hormone, it is referred to by many as PMSG. Equine CG is secreted at high levels during the first trimester of the mare's pregnancy, ie, days 40–130. The high concentration of serum eCG is conducive to hormone extraction and purification. The purification of eCG is readily achieved by precipitation in metaphosphoric acid and ethanol, followed by ion-exchange and size-exclusion chromatography (36,37).

Like hCG, eCG is a placental gonadotropic hormone with close structural similarity to pituitary luteinizing hormone. Equine CG is comprised of noncovalently bound α- [112326-58-6] and β- [111092-61-6] subunits, which are glycosylated. Free subunits are not biologically active. Unlike hCGβ, which has a different genetic code than that for hLHβ, only one gene which codes for both eCGβ and eLHβ appears to exist (38). Accordingly, the amino acid sequence of eCGβ and eLHβ is identical (31,39). Equine CG possesses an extremely high carbohydrate content, ie, 38–45% by weight, with N-linked oligosaccharides at Asn-30β, Asn-56α, Asn-82α, and O-linked oligosaccharides within the last 35 C-terminal amino acids (30,39). The structures of eCG carbohydrate moieties appear to be varied and highly complex; review of this topic offers further detail (34). The approximate contents of hexose, hexosamine, and sialic acid are 15, 14, and 12%, respectively (40). The high content of sialic acid imparts a low isoelectric point, pI, of approximately 2, and a prolonged circulating half-life (41). A half-life of six days has been reported in geldings. In laboratory species, half-lives of 24 hours have been reported.

In nonequine species, eCG exerts both LH and FSH bioactivity (42). This is reflected in the ability of eCG to bind both gonadotropin receptors. Figure 4

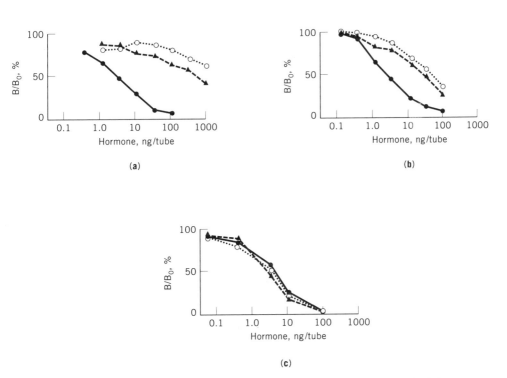

Fig. 4. FSH receptor-binding potencies of equine FSH ($\bullet$), eCG purified from pregnant mare's serum ($\circ$), and endometrial cups ($\blacktriangle$). Receptor-binding in cell membrane fractions, B/B$_0$ from (**a**) horse, (**b**) calf, and (**c**) rodent testes (40). Courtesy of Butterworth-Heinemann.

shows the ability of eCG purified from pregnant mare serum and endometrial cups to bind with rat, bovine, and equine FSH receptors (40). The structural component of eCG responsible for the expression of FSH activity is contained within the β-subunit, but still remains to be determined. Analysis of eLH, which is identical to eCG (38), has shown that chemical removal of the carboxy-terminal peptide does not eliminate FSH bioactivity (43). The prolonged biological half-life and FSH bioactivity make eCG extremely valuable commercially. This hormone is widely used internationally for the induction of superovulation in nonequine livestock. In the United States, Food and Drug Administration licensing has been issued only for the use of eCG in swine, as PG-600 (Intervet, Holland). Equine CG is readily available for research purposes from many suppliers (6,35).

Names are necessary to report factually on available data; however, the U.S. Department of Agriculture neither guarantees nor warrants the standard of the product, and the use of the name by the U.S. Department of Agriculture implies no approval of the product to the exclusion of others that may also be suitable.

BIBLIOGRAPHY

"Anterior–Pituitary-Like Hormones" in *ECT* 1st ed., Vol. 7, pp. 489–491, by P. A. Katzman, St. Louis University; in *ECT* 2nd ed., Vol. 11, pp. 64–68, by A. Albert, Mayo Clinic; in *ECT* 3rd ed., Vol. 12, pp. 557–563, by H. Papkoff, University of California, San Francisco.

1. G. S. Barsh, P. H. Seeburg, and R. E. Gelinas, *Nucleic Acids Res.* **11**, 3939 (1983).
2. G. Bauman, *Endocr. Rev.* **12**, 424 (1991).
3. W. L. Miller and N. L. Eberhardt, *Endocr. Rev.* **4**, 97 (1983).
4. H. M. Goodman in V. B. Mountcastle, ed., *Medical Physiology*, 13th ed. C. V. Mosby, St. Louis, Mo., 1974, p. 1741.
5. C. S. Nicoll, G. L. Mayer, and S. M. Russell, *Endocr. Rev.* **7**, 169 (1986).
6. W. D. Linscott, *Linscott's Directory of Immunological and Biological Reagents*, 7th ed., Linscott's Directory, Santa Rosa, Calif., 1992–1993, 239 pp.
7. J. Ray and co-workers, *Endocrinology* **125**, 566 (1989).
8. F. Frankenne and co-workers, *J. Clin. Endocrinol. Metab.* **66**, 1171 (1988).
9. P. Simon and co-workers, *Human Genetics* **74**, 235 (1986).
10. J. N. Macleod and co-workers, *Endocrinology* **128**, 1298 (1991).
11. M. J. Soares and co-workers, *Endocrine Rev.* **12**, 402 (1991).
12. J. C. Byatt and co-workers, *J. Anim. Sci.* **70**, 2911 (1992).
13. K. Shimomura and R. D. Bremel, *Mol. Endocrinol.* **2**, 845 (1988).
14. J. C. Byatt and co-workers, *Endocrinology* **4**, 231 (1990).
15. J. C. Byatt and co-workers, *J. Endocrinol.* **132**, 185 (1992).
16. M. Freemark, M. Comer, and S. Handwerger, *Am. J. Physiol.* **251**, E328 (1986).
17. M. Freemark and S. Handwerger, *Endocrinology* **112**, 402 (1983).
18. M. Freemark and co-workers, *Endocrinology* **125**, 1504 (1989).
19. S. S. Galosy and co-workers, *Mol. Cell. Endocrinol.* **78**, 229 (1991).
20. M. A. Kessler, T. M. Duello, and L. A. Schuler, *Endocrinology* **129**, 1885 (1991).
21. J. C. Byatt and co-workers, *J. Dairy Sci.* **75**, 1216 (1992).
22. R. V. Anthony, University of Colorado, Fort Collins, personal communication, 1993.
23. M. A. Kessler and co-workers, *Biochemistry* **28**, 5154 (1989).

24. D. I. H. Linzer and co-workers, *Proc. Nat. Acad. Sci. USA* **82**, 4356 (1985).
25. P. Colosi and co-workers, *Mol. Endocrinol* **2**, 579 (1988).
26. R. O. Hussa, *Endocrine Rev.* **1**, 268 (1980).
27. J. C. Fiddes and H. M. Goodman, *J. Mol. Appl. Genet.* **1**, 3 (1981).
28. W. R. Boorstein, N. C. Vamvakopoulos, and J. C. Fiddes, *Nature (London)* **300**, 419 (1982).
29. K. Talmadge, W. R. Boorstein, and J. C. Fiddes, *DNA* **2**, 279 (1983).
30. R. J. Ryan and co-workers, *Recent Prog. Hormone Res.* **43**, 383 (1987).
31. H. Sugino and co-workers, *J. Biol. Chem.* **262**, 8603 (1987).
32. M. R. Sairam and G. N. Bhargavi, *Science* **229**, 65 (1985).
33. M. M. Matzuk and I. Boime, *Biol. Reprod.* **40**, 48 (1989).
34. H. E. Grotjan, Jr., in B. A. Keel and H. E. Grotjan, eds., *Microheterogeneity of Glycoprotein Hormones*, CRC Press, Boca Raton, Fl., 1989, p. 23.
35. *Merck Index*, 11th ed., Merck & Co., Inc., Rahway, N.J., 1989.
36. D. Gospodarowicz and H. Papkoff, *Endocrinology* **80**, 699 (1967).
37. D. Gospodarowicz, *Endocrinology* **91**, 101 (1972).
38. G. B. Sherman and co-workers, *Mol. Endocrinol.* **6**, 951 (1992).
39. G. R. Bousfield and co-workers, *J. Biol. Chem.* **262**, 8610 (1987).
40. R. L. Matteri and co-workers, *Dom. Anim. Endocrinol.* **3**, 39 (1986).
41. O. J. Ginther, ed., in *Reproductive Biology of the Mare: Basic and Applied Aspects*, 2nd ed., Equi Services, Cross Plains, Wis., 1992, p. 41.
42. G. R. Bousfield and D. N. Ward, *Biochim. Biophys. Acta* **885**, 327 (1986).
43. G. R. Bousfield, W. K. Liu, and D. N. Ward, *Endocrinology* **124**, 379 (1989).

ROBERT L. MATTERI
U.S. Department of Agriculture

POSTERIOR PITUITARY HORMONES

The posterior lobe of the pituitary, ie, the neurohypophysis, is under direct nervous control (1), unlike most other endocrine organs. The hormones stored in this gland are formed in hypothalamic nerve cells but pass through nerve stalks into the posterior pituitary. As early as 1895 it was found that pituitrin [50-57-7], an extract of the posterior lobe, raises blood pressure when injected (2), and that Pitocin [50-56-6] (Parke-Davis) causes contractions of smooth muscle, especially in the uterus (3). Isolation of the active materials involved in these extracts is the result of work from several laboratories. Several highly active posterior pituitary extracts have been discovered (4), and it has been determined that their biological activities result from peptide hormones, ie, low molecular weight substances not covalently linked to proteins (gv) (5).

The principal hormones of the human posterior pituitary include the two nonapeptides, oxytocin [50-56-6] and arginine vasopressin [11000-17-2] (antidiuretic hormone, ADH). Many other hormones, including opioid peptides (see OPIOIDS, ENDOGENOUS), cholecystokinin [9011-97-6] (CCK) (see HORMONES, BRAIN OLIGOPEPTIDES), and gastrointestinal peptides, also have been located in mammalian neurohypophysis (6), but are usually found in much lower concentrations (7). Studies have demonstrated that oxytocin and vasopressin are synthesized in other human organs, both centrally and peripherally, and there is considerable evidence for their role as neurotransmitters (see NEUROREGU-LATORS) (8).

Oxytocin, although found in large quantities in both males and females, primarily functions in contracting the uterus during childbirth and releasing milk from mammary tissue. The fowl-pressor effect, ie, elevation of chicken blood pressure following iv administration of peptide, is a third action. It is fundamentally pharmacological, but an assay based on the effect often has been useful for rapid screening of potential active analogues. Arginine vasopressin (AVP) promotes water readsorption by the kidney, ie, the antidiuretic effect, and contraction of the smooth muscle of vascular tissue, ie, the pressor effect. Additional biological activities based on the behavioral properties of both hormones represent areas of intense research interest (9–11).

The structure of oxytocin was the first of the peptide hormones to be elucidated and synthesized by conventional methods of solution synthesis and fragment condensation (5,12); structure elucidation and synthesis of AVP followed (Fig. 1) (13). At least twelve related nonapeptides have been isolated from various vertebrate species (8,14). An additional three analogues have been described among the invertebrates. Oxytocin seems ubiquitous among higher mammals, although lysine vasopressin [50-57-7] (lysipressin, [Lys-8]VP, LVP), a generally less potent analogue, is found in place of arginine vasopressin, [Arg-8]VP or AVP, in the pig and related species.

Structure of the Principal Neurohypophyseal Hormones

The primary structures of oxytocin, the vasopressins, and some of the other natural neurohypophyseal analogues are given in Table 1. Each compound contains nine amino acids and each has a disulfide bridge between cysteine residues at positions 1 and 6. All are amidated at their C-terminal glycine residue. Variations among naturally occurring nonapeptides are found primarily in positions 3, 4, and 8. The relative activities of these compounds, when contrasted with their phyletic distributions and structural variations, suggest an evolutionary relationship among the various peptides (14). A common link may be arginine vasotocin, which is found in lower vertebrates as well as in the most complex mammals.

The wide-ranging differences in biological activities among peptide congeners that have small structural variations, eg, oxytocin and aspartocin [4117-65-1], are remarkable; in determining their conformation–activity parameters (15,16) and their evolutionary significance (17–19), such variations have made the neurohypophyseal hormones very valuable. In general, oxytocin and analogues that bind well to oxytocic receptors all contain tyrosine at the 2 position and a hydrophobic residue, eg, Leu, Ile, Val, or Gln, at the 8 position. In contrast, vasopressin and other analogues that bind to the vasopressin receptor classes are characterized by a Phe at the 3 position and basic residues, ie, Arg or Lys, at the 8 position.

Other, more recently described, naturally occurring neurohypophyseal hormone analogues include phenypressin [30635-27-9] ([Phe-2]-arginine vasopressin), found in kangaroos and other macropodidae (20), and conopressins found in *Conus* snails (21). *Conus* snails are representative of invertebrate

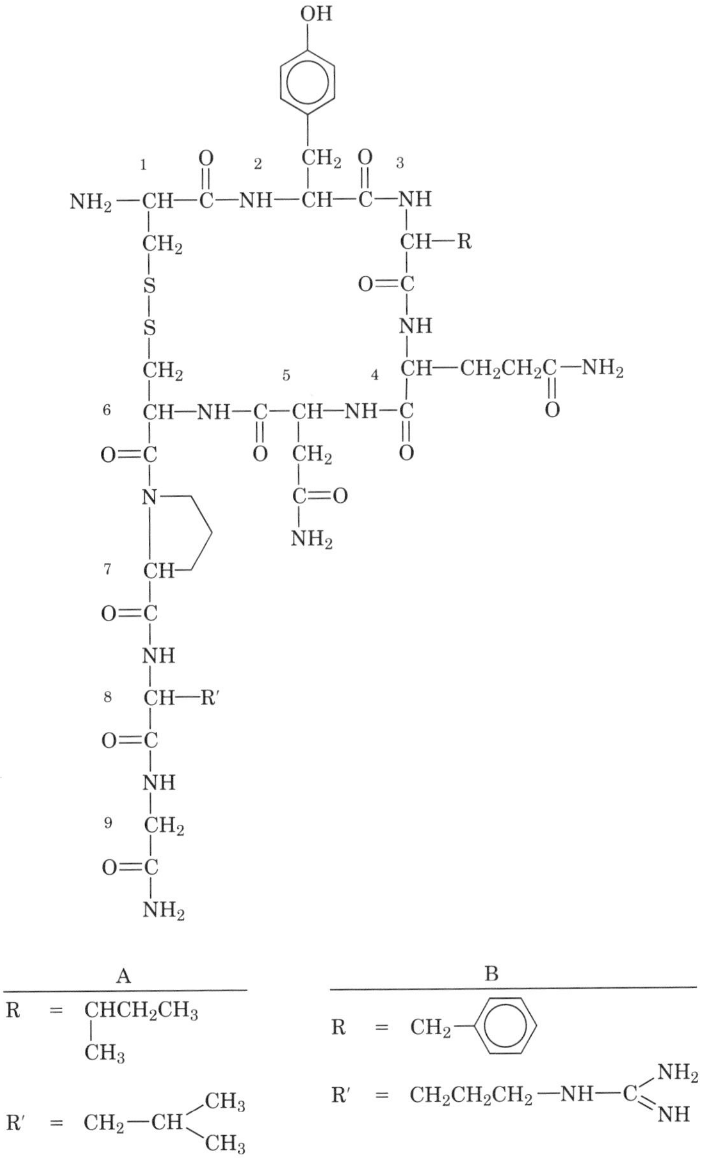

Fig. 1. Structure of oxytocin and arginine vasopressin. Numbers indicate approximate location of amino acids sequences found in Table 1. A, oxytocin; B, arginine vasopressin.

classes that diverged from vertebrates over 700×10^6 years ago and also represent rich sources of many other biologically active peptides such as ion-channel blockers. In the opossum, an American marsupial, four hormones are found, ie, oxytocin, mesotocin [362-39-0], lysine vasopressin, and arginine vasopressin.

The structural similarities between oxytocin and vasopressin analogues can be contrasted to their distinct physiological roles and even their separate receptor

Table 1. Structures of the Neurohypophyseal Hormone-Like Peptides[a,b]

	1 2 3 4 5 6 7 8 9		
	Cys—X—Y—Z—Asn—Cys—Pro—W—Gly—NH$_2$		
Species	Peptide	X, Y, Z	W
	Oxytocin-like peptides		
tetrapods	oxytocin	Tyr–Ile–Gln	Leu
	mesotocin	Tyr–Ile–Gln	Ile
bony fishes	isotocin	Tyr–Ile–Ser	Ile
cartilaginous fishes			
rays	glumitocin	Tyr–Ile–Ser	Gln
spiny dogfish	valitocin	Tyr–Ile–Ser	Val
	aspartocin	Tyr–Ile–Asn	Leu
spotted dogfish	asvatocin	Tyr–Ile–Asn	Val
	phasvatocin	Tyr–Phe–Asn	Val
octopus	cephalotocin	Tyr–Phe–Arg	Ile
	Vasopressin-like peptides		
mammals	arginine vasopressin	Tyr–Phe–Gln	Arg
pig, marsupials	lysine vasopressin[c]	Tyr–Phe–Gln	Lys
macropodids	phenypressin	Phe–Phe–Gln	Lys
nonmammalian vertebrates	arginine vasotocin	Phe–Phe–Gln	Arg
invertebrates	locust suboesophagial ganglia peptide	Leu–Ile–Thr	Arg
Conus striatus	arginine conopressin	Ile–Ile–Thr	Arg
Conus geographicus	lysine conopressin	Phe–Ile–Arg	Lys

[a]Ref. 14. Courtesy of Elsevier Science.
[b]Numbers represent the position of the amino acids relative to the N-terminus.
[c]Lysipressin.

second messengers. Oxytocin seems to have evolved from the more primitive isotocin [550-21-0], found in fish, to mesotocin, found in amphibians and reptiles, to oxytocin, found in higher mammals. In contrast, vasotocin is found in fish, amphibians, and reptiles, but evolved to vasopressin in mammals. Because there are separate oxytocin receptors in mammals, the presence of vasotocin, which has significant oxytocic activity, could have proven problematic and led to this evolutionary change. Vasotocin is an antidiuretic in amphibians and reptiles but is a diuretic in freshwater fishes, including lungfish (14).

Conformation and Structure–Activity Relationships

Hundreds of analogues of both oxytocin and vasopressin have been synthesized, and the conformations of many of these have been studied (22,23). Several large structural compilations are available. Conformational and topographic features of these analogues have been studied by nmr (24–26), raman spectroscopy (27), and circular dichroism (28). These methods, supplemented by conformational calculations, have provided a reasonably clear understanding of the structures of the neurohypophyseal hormones in various solutions (16). Newer techniques in nmr spectroscopy have improved the quality and quantity of structural information regarding peptide hormones (see MAGNETIC SPIN RESONANCE). For example the presence of a minor amount of a *cis*-6-7-amide bond in oxytocin was revealed by one- and two-dimensional proton and carbon spectroscopies; as of this writing the physiological relevance of this finding is not apparent (29).

Disulfide bridging and hydrogen bonds greatly limit the number of possible structures. The Urry-Walter proposed conformation (30) is the representative and, for the most part, biologically relevant model (16). Highly active analogues of oxytocin have been synthesized, based partly on this structure. A similar and key feature in oxytocin and LVP is the presence of a β-turn, involving residues 2–5, which is stabilized by a hydrogen bond from the 2-tyrosine CO to the 5-asparagine NH (31). The 5-asparagine of oxytocin also may contribute to structural rigidity by a second hydrogen bond from its side chain to the oxytocin tail -(Pro-Leu-Gly-NH$_2$); in contrast to a more compact structure when dissolved in organic solvents, eg, dimethyl sulfoxide, the tail is free to move in aqueous media (32). An x-ray structure (Fig. 2) of the potent oxytocin analogue, deaminooxytocin [*113-78-0*], provides additional support for this model (33) and has been used to guide both new agonist and antagonist design (34).

In water, the aromatic rings of vasopressin in residues 2 and 3 are believed to be in close proximity. However, the 2-tyrosine residue of oxytocin is folded over the 20-membered ring in the proposed biologically active conformer. The [2-D-tyrosine]AVP analogue, which can maintain a hydrophobic interaction with phenylalanine pointing away from the ring, retains significant antidiuretic activity, whereas the corresponding [2-D-tyrosine]oxytocin derivative is practically devoid of uterotonic potency (35). General findings support the concept that neurohypophyseal hormones are relatively flexible peptides which have various binding and active elements that can, in theory, be modified to form either superagonists or hormone inhibitors. A relevant example is a bicyclic oxytocin analogue formed by cyclizing a Glu^4 to a Lys^8. This lactam bridge converts a weak agonist, ie, the noncyclized form, to a more rigid and fairly potent oxytocin antagonist (36). Equally significant, however, is the report of linear analogues of vasopressin (37) that have been found to be surprisingly potent antagonists toward vasopressin receptor subclasses.

Synthesis of Posterior Pituitary Hormones and Their Analogues

Agonists. Both oxytocin and the vasopressins have been synthesized by numerous routes using both rapid-solution (38) and solid-phase methods (39). The latter route has facilitated the syntheses of hundreds of analogues which are extremely useful in establishing cogent structure–activity relationships (23,34,40,41). In general, large structural variations, which alter overall geometries, prove destructive to activity. Subtle variations, however, have led to compounds with retained and even markedly enhanced hormonal activities; several of these are summarized in Tables 2 and 3. In some cases, certain variations, eg, deaminooxytocin and carba analogues which substitute methylene groups for the disulfide bridge, apparently retard biodegradation, whereas others, eg, 7-thiaproline- [*59095-56-6*] and 7-dehydroproline-oxytocin, appear to increase the hormones' affinities to their respective receptors. A particularly potent

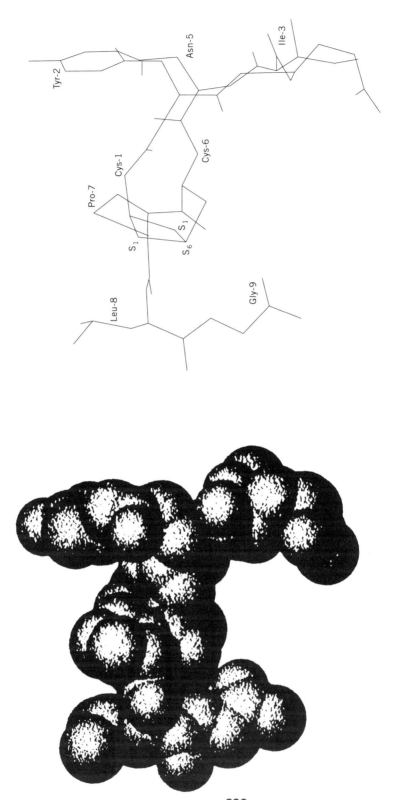

Fig. 2. X-ray structure of deaminooxytocin: (**a**) space-filling model; (**b**) equivalent stick model. Numbers and amino acids refer to positions indicated in Table 1. S_1—S_6 bond is indicated (Fig. 1) (33).

Table 2. Synthetic Analogues of Oxytocin With High Potency or Selectivity

Peptide	Biological activities[a]			O/A ratio[b,c]	Reference
	Oxytocic	Milk ejection	Antidiuretic		
oxytocin	520	474	4	130	23
1-deamino-oxytocin	803	541	19	42	23
[Thr-4]-oxytocin	923	543	0.9	1026	41
[Thr-4,Gly-7]-oxytocin	166	802	0.002	83,000	41

[a]Biological activities are in USP units/mg.
[b]O/A ratio = oxytocic/antidiuretic ratio.
[c]In the absence of Mg^{2+}.

Table 3. Synthetic Analogues of Vasopressin With High Potency or Selectivity

Peptide	Biological activities[a]		A/P ratio[b]	Reference
	Antidiuretic activity	Vasopressor activity		
arginine vasopressin	503	487	1	34
1-deamino[D-Arg]-8-vasopressin	1200	0.39	3000	41
1-deamino-Val-4, Arg-8	1230	antagonist	infinite	41

[a]Biological activities are in USP units/mg.
[b]A/P ratio = antidiuretic/pressor activity ratio.

synthetic analogue is the vasopressin derivative, [1-deamino,2-phenylalanine,7-(3,4-dehydroproline)]arginine vasopressin [66185-31-3]; its antidiuretic potency is 13,000 ± 1,250 IU/mg as compared to 503 ± 53 IU/mg for authentic AVP (42). In addition, this analogue has negligible pressor activity and thus exhibits the dissociated biological activity which frequently is a goal of synthetic analogue studies.

Hormone analogues have been tested that have nonparallel dose–response curves and that yield analogues with unusually high activities at low doses. This characteristic was exploited in the synthesis of the AVP analogue, [1-deamino,4-asparagine,8-D-arginine]vasopressin [65919-02-0] and yielded a hybrid having almost pure antidiuretic activity of ca 11,000 IU/mg (43).

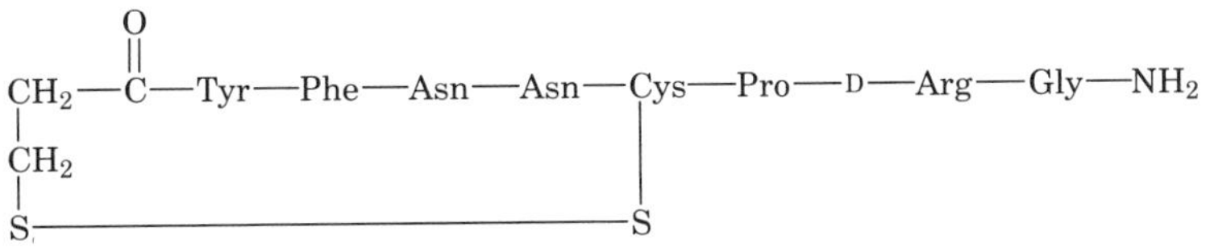

Antagonists. Another goal of structure–function studies of peptide hormones is the design of antagonists or hormone inhibitors that may have potential clinical usage (44). Modifications of the N-terminus of oxytocin and the vasopressins have yielded structures having significant *in vitro* inhibitory activities. As shown in Table 4, synthetic inhibitors of oxytocin contain variations in the 1 and 2 positions, eg, 1,1-dialkylmercaptopropionic acid and 2-O-alkyltyrosine. Conformationally restricted structures, especially near the N-terminus, yield some of the most potent inhibitory analogues (45,46). Analogues of vasopressin have been synthesized that are either potent pressor antagonists, such as [Mcpr-1,Tyr(CH_3)-2]AVP, or that have potent antiantidiuretic activity, ie, [Mcpr-1,Phe-2,Ile-4,Ala-9]AVP. Similarly, oxytocin analogues have been found having impressive antioxytocic activity but limited antipressor activity, eg, [deamino-1,cyclo(Glu-4, Lys-8)]oxytocin (36).

Two fundamental questions have emerged from these studies, ie, to what extent are agonists and antagonists binding similarly or differently to the respective receptors, and can inhibitory compounds be developed that are active *in vivo* in humans as well as *in vitro*. An oxytocin antagonist that can block premature uterine contractions presents a promising example of the clinical utility of such structures (47). Both linear as well as bicyclic modifications of these hormones also have provided new antagonist structures.

Other potential oxytocin antagonists are being developed using leads from naturally occurring, nonpeptide structures, such as an extract from *Streptomyces* (48). A selective nonpeptide vasopressin V_1 receptor antagonist also has been found (49).

Degradation. Both oxytocin and vasopressin have short plasma half-lives in humans; oxytocin has a half-life of several minutes (50,51); and ^{125}I vasopressin has a half-life in humans of 24 min (52). A pattern of hormone metabolism in various species, including identification of sites vulnerable to enzymic cleavage, has been developed by using radioactively labeled oxytocin and

Table 4. Synthetic Analogues of Oxytocin and Vasopressin Having Antagonist Activities

Analogue[a]	Activity, pA_2[b]		Ref.
	Antioxytocic	Antipressor[c]	
Oxytocin			
[Mcpr-1]oxytocin	7.61	weak	34
[penicillamine-1]oxytocin	6.86	0	45
[deamino,cyclo(Glu-4, Lys-8)]-oxytocin	8.74	6.3	36
[Mcpr-1,Tyr(CH_3)-2,Orn-8]-oxytocin	8.52	7.96	34
Vasopressin			
[Mcpr-1,Tyr(CH_3)-2]AVP	8.13	8.62	44
[Mcpr-1,Tyr(CH_2CH_3)-2,Val-4]AVP	7.88	8.16	44
[Mcpr-1,Phe-2,Ile-4,Ala-9]AVP		7.71	44

[a]Mcpr = 3-mercapto-3,3-cyclopentamethylene propionic acid.
[b]$pA_2 = -\log A_2$, where A_2 refers to concentration required to cause 50% inhibition of agonist response.
[c]Antiantidiuretic activity for [Mcpr-1,Tyr(CH_2CH_3)-2, Val-4]AVP = 7.57; for [Mcpr-1,Phe-2, Ile-4,Ala-9]AVP = 8.38.

vasopressin analogues and by detecting hormone fragments (see RADIOACTIVE TRACERS) (53). Cleavage sites of oxytocin include the 8–9 bond, yielding glycin-amide, and the 7–8 bond by virtue of post-proline cleaving enzyme (50), yielding the dipeptide Leu–Gly–NH$_2$ (54). In addition, oxytocin may be degraded in the testes by disulfide reduction (30%) and by cleavage of the 1–2 bond (70%) (55). A similar pattern of enzymatic cleavage for vasopressin has been reported to have four main sites (56):

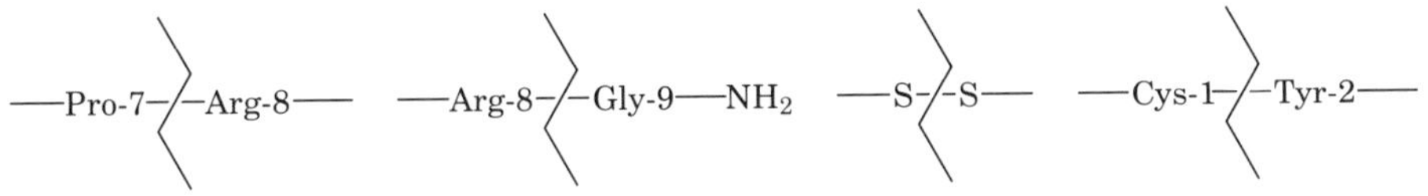

$$—Pro\text{-}7 \diagdown Arg\text{-}8— \quad —Arg\text{-}8 \diagdown Gly\text{-}9—NH_2 \quad —S \diagdown S— \quad —Cys\text{-}1 \diagdown Tyr\text{-}2—$$

In the placenta, the aminopeptidase oxytocinase, ie, cysteine aminopeptidase, is a principal catalyst for oxytocin hydrolysis and prevents premature uter-ine contractions.

Neurophysins, Hypothalamic Hormones, and Hormone Processing

The neurophysins represent a group of medium-sized (9000–10,000 mol wt) proteins found in the posterior pituitary that act as carriers of oxytocin and the vasopressins (57,58). Although the pattern is species-variant, the two principal forms found in higher mammals are neurophysin I (NP-I) and II (NP-II) (59); these appear to be associated with vasopressin and oxytocin, respectively. According to the concept of one neurophysin–one hormone, the carrier proteins and their respective hormones are synthesized as linear pep-tide chains on human chromosome 20 (60) and then cleaved and cyclized to yield NP-I and vasopressin and NP-II and oxytocin (61,62). The posterior pitu-itary hormones and their carrier proteins are synthesized in the supraoptic and paraventricular neurons of the hypothalamus (63,64) as part of larger preprohor-mone sequences, ie, prepro-oxytocin and prepro-vasopressin. These precursors have been verified from nucleotide sequence analysis of complementary deoxyri-bonucleic acid (cDNA) as well as by more traditional peptide sequencing studies (65). Following initial translation of the prepro sequences, the N-terminal sig-nal peptide region is cleaved to yield pro-oxytocin or pro-vasopressin. These prohormones are then further processed in the neurosecretory vesicles to yield three components, ie, the peptide hormone, the neurophysin, and a C-terminal glycopeptide having an N-terminal glycosylation site (Asn-Ala-Thr) (66). As for the neurophysins, no clear biological role has been established for either of the glycopeptides. Figure 3 shows a representation of the processing steps for rat and bovine preprovasopressins (14).

     In humans, the hypothalamic-derived protein and the hormone noncovalent complexes are packaged in neurosecretory granules, then migrate along axons at a rate of 1–4 mm/h until they reach the posterior pituitary where they are stored prior to release into the bloodstream by exocytosis (67). Considerable evi-dence suggests that posterior pituitary hormones function as neurotransmitters (68); vasopressin acts on the anterior pituitary to release adrenocorticotropic

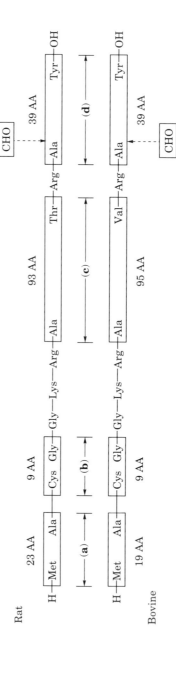

Fig. 3. Processing steps of rat and bovine prepro-vasopressin leading to the hormone vasopressin and its carrier protein, neurophysin (14). (**a**) Putative signal peptide; (**b**) vasopressin; (**c**) neurophysin; (**d**) glycopeptide. CHO = carbohydrate. Courtesy of Elsevier Science.

hormone [*9002-60-2*] (ACTH) (69) as well as on traditional target tissues such as kidneys. Both hormones promote other important central nervous system (CNS) effects (9,70).

The nature of the hormone–NP link is unclear. Although the binding is quite specific, it is sufficiently weak, ie, binding constant $10^{-5}-10^{-6}$ M (71), as to cast doubt upon a protective role. However, no compelling evidence for a biological function for the neurophysins, other than as hormone carriers, has been elucidated (59,72).

Sequences and partial sequences of several of the neurophysins, including those in rat, cod, and sheep, have been determined (73,74). Most neurophysins contain 90–95 individual amino acid residues. Considerable sequence homology exists among the various neurophysin classes and between vertebrate species, especially in the region between 10 and 76. The molecules contain seven disulfide bridges, and 13 or 14 cysteine residues are found within the highly conserved 10–76 region. Some repetition within the peptide chain suggests a possible second binding site, but most evidence is supportive of a single strong binding site for oxytocin and vasopressin under physiological conditions. The principal site for attachment of oxytocin to neurophysin appears to be mainly via the N-terminal residues (59).

Oxytocin and Vasopressin Receptors. The actions of oxytocin and vasopressin are mediated through their interactions with receptors. Different receptor types as well as different second messenger responses help explain their diverse activities in spite of the hormones' structural similarities. Thus oxytocin has at least one separate receptor and vasopressin has been shown to have two principal receptor types, V_1 and V_2. Subclasses of these receptors have been demonstrated, and species differences further complicate experimental analysis. It is apparent that both oxytocin and V_1 receptors function through the GP/1 phospholipase C complex (75), while the V_2 receptors activate cyclic AMP (76).

The ultimate goal of structure–function studies is complete understanding of the hormone's interaction with its receptor(s). Evidence continues to support an oxytocin model in which residues 2 and 5, ie, Tyr and Asn, are vital for interaction with the uterine smooth muscle receptor (77–79). In this model, the residues at positions 3, 4, 7, and 8 are at the corners of two β-turns and, as binding elements, in principle can be modified to provide greater affinity with the receptor. Early studies examined the binding of oxytocin and vasopressin analogues to porcine (78) and bovine (79) receptors. The correlation found between activation of adenylate cyclase from renal membrane preparations and the known relative activities or affinities of the preparations is good evidence for the suggested mechanisms and site of action for these hormones.

V_1 receptors, found in vascular smooth muscles (80) and in the liver (81), function by means of a calcium-dependent pathway. V_2 receptors, found in the kidney (82), modulate the antidiuretic response of vasopressin analogues through a cyclic adenosine monophosphate (AMP)-dependent pathway. With the development of potent and reasonably selective agonists and antagonists of vasopressin, the elucidation of additional physiological roles for VP has been shown (70). Similarly, through the use of oxytocin superagonists and inhibitors, a variety of behavioral properties affecting maternal, sexual, and social behavior have been discerned (83).

Several human receptors for the neurohypophyseal hormones have been cloned and the sequences elucidated. The human V_2 receptor for antidiuretic hormone presumably contains 371 amino acids and seven transmembrane segments and activates cyclic AMP (76). The oxytocin receptor is a classic G-protein-coupled type of receptor with a proposed membrane topography also involving seven transmembrane components (84). A schematic representation of the oxytocin receptor structure within the membrane is shown in Figure 4 (85).

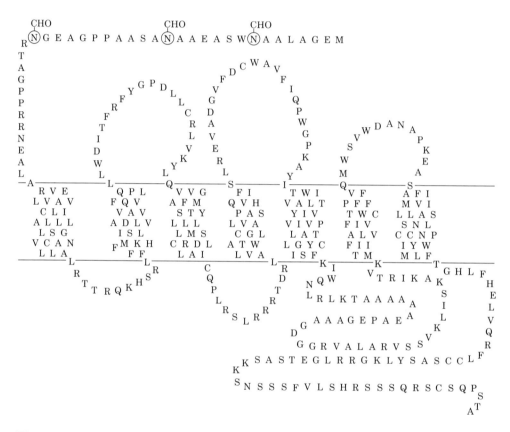

Fig. 4. Representation of proposed topography of human oxytocin receptor (85) where CHO represents carbohydrate. Courtesy of Elsevier Science.

Uses

Oxytocin has been used widely to induce labor, although its suitability has been questioned; prostaglandin E_2 may be a practical alternative (86) (see PROSTAGLANDINS). Among side effects cited with oxytocin use are water intoxication as a result of the natural natriuretic effect of oxytocin, venospasm, and increased incidence of neonatal jaundice. Oxytocin also is used clinically to facilitate milk ejection in nursing mothers. It is available for intravenous, oral, or nasal modes of administration, ie, it is supplied as Pitocin (Parke-Davis) for intravenous use and as its citrate for buccal administration, and is marketed as

Syntocinon (Sandoz) for injection or as a nasal spray. Syntocinon is one of the first examples of a solid-phase-derived synthetic polypeptide licensed for clinical use. The nasal spray is administered 2–3 min before nursing to induce contraction of the breast myoepithelial elements, thereby making milk more readily available in the larger mammary ducts. A number of synthetic oxytocin analogues are also finding clinical use, especially in Europe under such trade names as Sandopart, carbetocin, and methyloxytocin, and in Japan as Cargutocin (Statocin) (87).

Hyposecretion by the hypothalamic supraoptic nuclei or injury to the posterior pituitary can give rise to diabetes insipidus; a gene deletion in the rat also has been shown to lead to diabetes insipidus (88). The symptoms of the disease can be controlled by chronic administration of antidiuretic hormone (ADH). Vasopressin infusion also has been used in the management of gastrointestinal hemorrhage (see GASTROINTESTINAL AGENTS). A vasopressin analogue, desamino-8-D-arginine vasopressin [16679-58-6] (desmopressin, DDAVP), has prolonged biological activity and a high antidiuretic:pressor ratio which has led to its use in clinical trials as a substitute for AVP (89,90). In addition, it has proven useful in hemophilia as a hemostatic agent, reducing the need for blood and plasma products (91).

Vasopressin is supplied as Pitressin Tannate (Parke-Davis) for use as an antidiuretic agent administered intramuscularly (see DIURETIC AGENTS). Lysine vasopressin is marketed in nasal spray form as an antidiuretic under the trademark Diapid (Sandoz).

Vasopressin and Memory. Several hormones, eg, ACTH, vasopressin, and the catecholamines, may be involved in memory retention or consolidation (9,11,85) (see ANTIAGING AGENTS; MEMORY ENHANCING DRUGS). LVP and AVP and analogues significantly reduce the effect of puromycin-induced amnesia in mice (92), and vasopressin is present in several regions of the brain (93). The utility and long-lasting effect of vasopressin on retention of conditioned avoidance response has been demonstrated (94). These memory and learning effects are probably not related to the primary action of the vasopressins, since similar activities are seen using an octapeptide fragment, ie, desglycinamide LVP [32472-43-8], isolated from porcine pituitary and inactive in pressor and antidiuretic assays (95). While there is not universal agreement regarding the role of vasopressin and memory (86), a variety of clinical studies, including ones designed to test the effects of vasopressin analogues on memory retention in both the young and the elderly, continue to be performed (87).

BIBLIOGRAPHY

"Posterior-Pituitary Hormones" under "Hormones" in *ECT* 1st ed., Vol 7, pp. 491–495, by O. Kamm, Parke-Davis & Co.; in *ECT* 2nd ed., Vol. 11, pp. 68–77, by M. Bodanszky, The Squibb Institute for Medical Research; in *ECT* 3rd ed., Vol. 12, pp. 563–574, by A. F. Spatola, University of Louisville.

1. K. Lederis in E. Knobil and W. H. Sawyer, eds., *Handbook of Physiology*, Vol. 4, American Physiological Society, Waverly Press, Baltimore, Md., 1974, Sect. 7, pt. 1, pp. 81–130.

2. G. Oliver and E. A. Schäfer, *J. Physiol. London* **18**, 277 (1895).

3. I. Ott and J. C. Scott, *Proc. Soc. Exptl. Biol. Med.* **8**, 48 (1910).

4. O. Kamm, *Science,* **67**, 199 (1928).

5. V. du Vigneaud and co-workers, *J. Am. Chem. Soc.* **75**, 4879 (1953).

6. C. A. Bondy and co-workers, *Cell. Mol. Neurobiol.* **9**, 427–428 (1989).

7. T. L. O'Donohue and J. Z. Kiss, in S. Udenfriend and J. Meienhofer, eds., *The Peptides,* Vol. 8, Academic Press, Orlando, Fla., 1987, pp. 27–40.

8. J. J. Dreifuss, *Arch. Int. Physiol. Biochim.* **97**, 1–14 (1989).

9. P. L. Hoffman in Ref. 7, pp. 239–295.

10. R. Walter, *Ann. N.Y. Acad. Sci.* **248** (1975).

11. G. Fehm-Wolfsdorf and J. Born, *Peptides* **12**, 1399–1406 (1991).

12. V. du Vigneaud, H. C. Lawler, and E. A. Popenon, *J. Am. Chem. Soc.* **75**, 4880 (1953).

13. V. du Vigneaud, M. F. Bartlett, and A. Johl, *J. Am. Chem. Soc.* **79**, 5572 (1957).

14. R. Acher, *Regul. Peptides* **45**, 1 (1993).

15. V. J. Hruby and co-workers, *J. Am. Chem. Soc.* **101**, 202 (1979).

16. J. C. Hempel in Ref. 7, pp. 209–237.

17. R. Acher in E. Schoffeniels, ed., *Molecular Evolution,* Vol. II, North Holland, Amsterdam, the Netherlands, 1971, p. 43.

18. R. Acher, *Biochemistry* **70**, 1197–1207 (1988).

19. N. M. Sherwood and D. B. Parker, *J. Exp. Zool. Supp.* **4**, 63–71 (1990).

20. M. T. Chauvet and co-workers, *Nature* **287**, 640 (1980).

21. L. J. Cruz and co-workers, *J. Biol. Chem.* **262**, 15821–15824 (1987).

22. B. Berde and R. A. Boissonas, in B. Berde, ed., *Handbook of Experimental Pharmacology,* Vol. 23, Springer-Verlag, New York, 1968, pp. 807–808.

23. K. Jost, M. Lebl, and F. Brtnik, eds., *Handbook of Neurohypophyseal Hormone Analogs,* Vol. I and II, CRC Press, Boca Raton, Fla., 1987.

24. E. M. Krauss and D. Cowburn, *Biochemistry* **20**, 671 (1981).

25. R. Deslauriers and co-workers, *J. Am. Chem. Soc.,* **100**, 3912 (1978).

26. A. I. R. Brewster and V. J. Hruby, *Proc. Nat. Acad. Sci. U.S.A.* **70**, 3806 (1973); C. A. Biocelli, A. F. Bradbury, and J. Feeney, *J. Chem. Soc., Perkin II,* 477 (1977); D. W. Urry, M. Ohnishi, and R. Walter, *Proc. Nat. Acad. Sci., U.S.A.* **66**, 111 (1970); I. C. P. Smith and R. Deslauriers, *Recent Prog. Horm. Res.* **33**, 301 (1977); A. I. R. Brewster, V. J. Hruby, A. F. Spatola, and F. A. Bovey, *Biochemistry* **12**, 1643 (1973).

27. F. R. Maxfield and H. A. Scheraga, *Biochemistry* **16**, 4443 (1977); V. J. Hruby, K. K. Deb, and V. J. Hruby, *J. Biol. Chem.* **253**, 6060 (1978).

28. I. Fric, M. Kodicek, K. Jost, and K. Blaha, in H. Nanson and H.-D. Jakubke, eds., *Peptides, 1972,* North Holland, Amsterdam, 1973, pp. 318–323; A. T. Tu and co-workers, *J. Biol. Chem.* **254**, 3272 (1979).

29. C. K. Larive, L. Guerra, and D. L. Rabenstein, *J. Am. Chem. Soc.* **114**, 7331 (1992).

30. D. W. Urry and R. Walter, *Proc. Nat. Acad. Sci. U.S.A.* **68**, 956 (1971).

31. R. Walter and co-workers, *Disturbances in Body Fluid Osmolality,* American Physiological Society, Bethesda, Md., 1977, pp. 1–36.

32. J. A. Glasel, V. J. Hruby, J. F. McKelvy, and A. F. Spatola, *J. Mol. Biol.* **79**, 555 (1973); R. Deslauriers, I. C. P. Smith, and R. Walter, *J. Am. Chem. Soc.* **96**, 2289 (1974).

33. S. P. Wood and co-workers, *Science* **232**, 633–636 (1986).

34. V. J. Hruby and C. W. Smith in Ref. 7, 77–207.

35. V. J. Hruby and co-workers, *J. Am. Chem. Soc.* **101**, 2717 (1979).

36. P. S. Hill and co-workers, *J. Am. Chem. Soc.* **112**, 3110–3113 (1990).

37. M. Manning and co-workers, *Nature* **329**, 839–840 (1987).

38. M. Bodanszky and V. du Vigneaud, *J. Am. Chem. Soc.* **81**, 1258 (1959).

39. M. Manning, *J. Am. Chem. Soc.* **90**, 1348 (1968).

40. D. H. Schlesinger, ed., *Neurohypophyseal Peptide Hormones and Other Biologically Active Peptides, Developments in Endocrinology*, Vol. 13, Elsevier Science, New York, 1981.
41. M. Manning and W. H. Sawyer, *J. Receptor Res.* **13**, 195–214 (1993).
42. C. W. Smith and R. Walter, *Science* **199**, 297 (1978).
43. M. Zaoral, J. Kolc, and F. Sorm, *Coll. Czech. Chem. Comm.* **32**, 1250 (1967).
44. M. Manning and W. H. Sawyer, *J. Lab. Clin. Med.* **114**, 617 (1989).
45. M. Manning, W. Y. Chan, and W. H. Sawyer, *Regul. Peptides* **45**, 279 (1993).
46. R. J. Vavrek and co-workers, *J. Med. Chem.* **15**, 123 (1972).
47. M. Akerlund and co-workers, *Brit. J. Obstet. Gynecol.* **94**, 1040 (1987).
48. M. G. Bock and co-workers, *J. Med. Chem.* **33**, 2321–2323 (1990).
49. Y. Yamamura and co-workers, *Science* **252**, 572 (1991).
50. R. Walter, *Biochim. Biophys. Acta* **422**, 138 (1976).
51. R. Walter and T. Yoshimoto, *Biochem.* **17**, 4139 (1978).
52. G. Bauman and J. F. Dingman, *J. Clin. Invest.* **57**, 1109 (1976).
53. K. Jost, in Ref. 23, Vol. I, Pt. II, pp. 15–30.
54. J. Lowbridge and co-workers, *J. Med. Chem.* **22**, 565 (1979).
55. K. Wojcik and E. Brzezinska-Slebodzinska, *Bull. Acad. Pol. Sci. Ser. Sci. Biol.* **23**, 577 (1975).
56. R. Walter and W. H. Simmons, in A. M. Moses and L. Share, eds., *Neurohypophysis*, Basel, 1977, pp. 167–188.
57. M. J. Brownstein, J. J. Russell, and H. Gainer, *Science* **207**, 373 (1980).
58. M. T. Chauvet and co-workers, *Proc. Nat. Acad. Sci., U.S.A.* **80**, 2839 (1983).
59. E. Breslow, *Ann. Rev. Biochem.* **48**, 251 (1979).
60. N. E. Simpson, *J. Med. Genet.* **12**, 798–804 (1988).
61. H. Sachs, in *Handbook of Neurochemistry*, Vol 4, Plenum Press, New York, 1977, pp. 373–428.
62. H. Gainer, Y. Sarne, and M. J. Brownstein, *Science* **195**, 1354 (1977).
63. R. A. Reaves, Jr., and H. N. Hayward, *Cell Tissue Res.* **196**, 117 (1979).
64. K. Diericks and F. Vandesande, *Cell Tissue Res.* **196**, 203 (1979).
65. D. Richter and R. Ivell, in H. Imura, ed., *The Pituitary Gland*, Raven Press, New York, 1985, pp. 127–148.
66. H. Gainer, *Progr. Brain Res.* **60**, 205 (1983).
67. W. W. Douglas, J. Nagasawa, and R. Schulz in H. Heller and K. Lederis, eds., *Subcellular Organization and Function in Endocrine Tissues*, University Press, Cambridge, Mass., 1971, pp. 353–377.
68. S. H. Snyder and R. B. Innes, *Ann. Rev. Biochem.* **48**, 755 (1979).
69. E. A. Zimmerman and co-workers, *Ann. N.Y. Acad. Sci.* **297**, 405 (1977).
70. G. F. Koob and co-workers, in R. W. Schrier, ed., *Vasopressin*, Raven Press, New York, 1985.
71. R. Deslauriers, I. C. Smith, G. L. Stahl, and R. Walter, *Int. J. Pept. Protein Res.* **13**, 78 (1979); E. Breslaw and R. Walter, *Mol. Pharmacol* **8**, 75 (1972); J. A. Glasel, V. J. Hruby, J. F. McKelvy, and A. F. Spatola, *J. Biol. Chem.* **251**, 2929 (1976).
72. E. Breslow, *Regul. Peptides* **45**, 15 (1993).
73. D. H. Schlesinger, J. K. Audhya, and R. Walter, *J. Biol. Chem.* **253**, 5019 (1978).
74. R. Acher and J. Chauvet, *Biochimie* **70**, 1197–1207 (1988).
75. S. Jard, J. Elands, A. Schmidt, and C. Barberis, in H. Imura and K. Shizume, eds, *Progress in Endocrinology*, Elsevier, Amsterdam, the Netherlands, 1988, pp. 1183–1188.
76. M. Birnbaumer and co-workers, *Nature* **357**, 333–335 (1992).
77. R. Walter and co-workers, *Proc. Nat. Acad. Sci. U.S.A.* **68**, 1355 (1971).
78. R. Walter and co-workers, *Disturbances in Body Fluid Osmolality*, American Physiology Society, Bethesda, Md., 1977, pp. 1–36.

79. C. W. Smith, *Dev. Endocrinol.* **13**, 23 (1981).
80. E. L. Schiffrin and J. Genest, *Endocrinology* **113**, 409 (1983).
81. B. Cantau and co-workers, *J. Recept. Res.* **1**, 137 (1980).
82. G. Guillon and co-workers, *Eur. J. Pharmacol.* **85**, 291 (1982).
83. D. Gantter and D. Pfaff, eds., *Neurobiology of Oxytocin*, Springer-Verlag, Berlin, 1986.
84. T. Kimura and co-workers, *Nature* **356**, 526 (1992).
85. T. Kimura and co-workers, *Regul. Peptides* **45**, 73 (1993).
86. R. Dantzer and R. M. Bluthe, *Regul. Peptides* **45**, 121 (1993).
87. B. E. Weekwith and co-workers, *Ann. N.Y. Acad. Sci.* **579**, 215 (1990).
88. H. Schmale and D. Richter, *Nature* **308**, 705 (1984).
89. M. L. Forsling, *Antidiuretic Hormone*, Vol. 3, Eden Press, Montreal, Canada, 1978.
90. A. G. Robinson and J. G. Verbalis, in P. Czernichow and A. G. Robinson, eds., *Diabetes Insipidus*, Karger, Basel, Switzerland, 1985, pp. 293–303.
91. J.-L. David, *Regul. Peptides* **45**, 311 (1993).
92. D. DeWied, *Nature* **232**, 58 (1971).
93. D. DeWied, *Progr. Brain Res.* **60**, 155–167 (1983).
94. D. De Wied, *Life Sci.* **19**, 685 (1976); D. De Wied, B. Bohus, and V. Wimersma-Greidanus, *Brain Res.* **85**, 152 (1975).
95. M. P. Vawter and K. F. Green, *Physiol. Behav.* **25**, 851–854 (1980).

General References

B. Berde, *Handb. Exp. Pharmacol.* **23**, 1 (1968).
A. W. Cowley, Jr., J. F. Liard, and D. A. Ausiello, eds., *Vasopressin: Cellular and Integrative Functions*, Raven Press, New York, 1988.
M. L. Forsling, *Antidiuretic Hormone*, Vol. 1, Eden Press, Montreal, Canada, 1976.
M. L. Forsling, *Antidiuretic Hormone*, Vol. 2, Eden Press, Montreal, Canada, 1977.
M. L. Forsling, *Antidiuretic Hormone*, Vol. 3, Eden Press, Montreal, Canada, 1978.
D. M. Gash and G. Boer, eds., *Vasopressin: Principles and Properties*, Plenum, New York, 1987.
J. R. Roberts, *Oxytocin*, Vol. I, Eden Press, Montreal, Canada, 1977.
D. Ganter and D. Pfaff, eds., *Neurobiology of Oxytocin*, Springer-Verlag, Berlin, 1986.
S. Melmed and R. J. Robbins, eds., *Molecular and Clinical Advances in Pituitary Disorders*, Blackwell Scientific Publications, St. Louis, Mo., 1991.
A. T. Hagler and co-workers, *Science* **227**, 1309 (1985).
V. J. Hruby, *Trends Pharmacol. Sci.* **8**, 336–339 (1987).
S. Yoshida and L. Share, eds., *Recent Progress in Posterior Pituitary Hormones*, Elsevier Amsterdam, the Netherlands, 1988.

ARNO F. SPATOLA
University of Louisville

HUMAN GROWTH HORMONE

Human growth hormone [*12629-01-5*] (hGH), also known as somatotropin, is a protein hormone produced and secreted by the somatotropic cells of the anterior pituitary (see HORMONES, ANTERIOR PITUITARY HORMONES). Secretion is regulated by a releasing factor, ie, the growth hormone-releasing hormone, and by an inhibitory factor, somatostatin. Human growth hormone plays a key role in somatic growth through its effects on the metabolism of proteins (qv), carbohy-drates (qv), and lipids. Human growth hormone exerts its biological effects either through direct action of the hormone at the target tissue or indirectly through the action of a second class of peptide hormones, the somatomedins, also known as insulin-like growth factors, which are produced primarily in the liver in re-sponse to hGH binding to specific receptors there (see also INSULIN AND OTHER ANTIDIABETIC DRUGS).

It was known as early as the 1920s that the pituitary gland contained a factor that could promote growth, but not until 1956 (1) was the factor isolated. hGH is extremely abundant in the pituitary, representing as much as 10% of this gland's dry weight, or 8 mg of hormone per gland (2).

Two well-known pathological conditions are the result of an excess or a deficiency of this hormone. The condition in which the body produces an excess of hGH is known as acromegaly or giantism. The condition in which too little is produced is dwarfism.

Human growth hormone, isolated and purified from the pituitary glands of human cadavers, was first administered to humans in 1958. In 1963, the National Institute of Arthritis, Metabolism and Digestive Diseases formed a single collection program called the National Pituitary Agency (NPA). It was the responsibility of this agency to collect human pituitary glands in the United States, organize the extraction of hGH, and distribute it within the country for hypopituitary children and for use in other studies. Pituitary-source hGH from this program was used clinically until the mid-1980s, when hGH of recombinant deoxyribonucleic acid (DNA) origin was approved for sale by the U.S. Food and Drug Administration (see GENETIC ENGINEERING). One factor that led to the decline in use of pituitary-derived hGH and expedited the acceptance of the biosynthetic hGH was the diagnosis in several recipients of pituitary hGH of Creutzfeldt-Jakob disease, a severe neurological disease thought to be caused by a slow virus that was not effectively removed or deactivated in early preparations of pituitary hGH. This problem was eliminated with the use of biosynthetic hGH.

Chemical and Physical Properties

Human growth hormone is a single polypeptide chain of 191 amino acids (qv) having two disulfide bonds, one between Cys-53 and Cys-165, forming a large loop in the molecule, and the other between Cys-182 and Cys-189, forming a small loop near the C-terminus. The structure of hGH is shown in Figure 1; molecular mass is 22,125; the empirical formula is $C_{990}H_{1529}N_{262}O_{300}S_7$.

Purified hGH is a white amorphous powder in its lyophilized form. It is readily soluble (concentrations >10 mg/mL) in dilute aqueous buffers at pH values above 7.2. The isoelectric point is 5.2 (3) and the generally accepted value for the extinction coefficient at 280 nm is 17,700 $(M \cdot cm)^{-1}$ (4), although other values have been reported. The equilibrium denaturation of hGH has been studied. A two-state mechanism is indicated with a Gibbs free energy of unfolding of 60.7 ± 4 kJ/mol (14.5 ± 1 kcal/mol) (5).

In solution, hGH exists predominantly as a monomer, with a small fraction as a dimer and higher molecular weight oligomers. Under certain

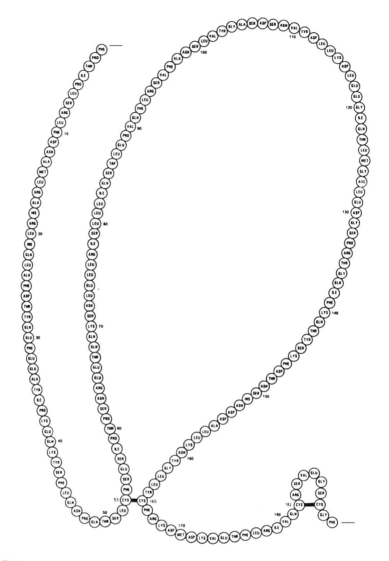

Fig. 1. Primary structure of hGH showing the amino acid sequence and the disulfide bonds.

conditions, hGH can be induced to form larger amounts of dimers, trimers, and higher oligomers although the conditions leading to self-association are not well understood.

As of this writing the crystal structure of hGH has not been solved. A considerable amount of information is known about its tertiary structure, however. hGH is a globular protein having approximately 55% α-helical character, a value determined independently in studies using circular dichroism (6) and Raman spectroscopy (7). A protein of a high degree of homology and structural similarity is porcine growth hormone, the crystal structure of which has been solved (8). This structure shows four antiparallel helices connected by strands of polypeptide having little structure. A crystal structure of hGH complexed with the extracellular domain of its receptor has been reported (9). In this complex, the structure of hGH is readily obvious and shows the four-helical bundle connected in an up−up−down−down fashion rather than the more common up−down−up−down. This work has convincingly shown that hGH possesses two binding sites for its receptor and that the hormone functions by dimerizing its receptor on the cell membrane. A more detailed discussion of the molecular mechanisms by which hGH functions can be found in the literature (10).

hGH Derivatives. Several derivatives of hGH are known. These derivatives include naturally occurring derivatives, variants, and metabolic products, degradation products primarily of biosynthetic hGH, and engineered derivatives of hGH produced through genetic methods.

Methionyl hGH. The first form of hGH to be produced through recombinant DNA technology was actually a derivative of hGH having one additional methionine residue at its N-terminus (11). Although technology has advanced to the stage where natural sequence hGH can easily be produced, as of this writing this derivative, referred to as methionyl hGH, is still produced commercially.

20-K hGH. A natural, structural variant of hGH called 20-K hGH has been reported to occur in the pituitary as well as in the bloodstream (12,13). This variant, which lacks the 15 amino acid residues from Glu-32 to Gln-46, arises from an alternative splicing of the messenger ribonucleic acid (mRNA) (14). This variant shares many but not all of the biological properties of hGH.

Acetylated hGH. A form of hGH that is acetylated at the N-terminus has been isolated and identified (15). It is not clear if acylation serves a regulatory role or is simply an artifact of the purification.

Size Isomers. In solution, hGH is a mixture of monomer, dimer, and higher molecular weight oligomers. Furthermore, there are aggregated forms of hGH found in both the pituitary and in the circulation (16,17). The dimeric forms of hGH have been the most carefully studied and there appear to be at least three distinct types of dimer: a disulfide dimer connected through interchain disulfide bonds (8); a covalent or irreversible dimer that is detected on sodium dodecylsulfate- (SDS-)polyacrylamide gels (see ELECTROSEPARATIONS, ELECTROPHORESIS) and is not a disulfide dimer (19,20); and a noncovalent dimer which is easily dissociated into monomeric hGH by treatment with agents that disrupt hydrophobic interactions in proteins (21). In addition, hGH forms a dimeric complex with Zn^{2+} (22). Scatchard analysis has revealed that two Zn^{2+} ions associate per hGH dimer in a cooperative fashion, and this Zn^{2+}-hGH dimeric complex was found to be more stable to denaturation than monomeric hGH (22). Because Zn^{2+} concentrations are high in hGH secretory granules of the anterior pituitary (23), it has been speculated that formation of a Zn^{2+}-hGH dimeric complex may be physiologically important in the storage and release of hGH from these granules (22).

Proteolytically Cleaved Two-Chain Forms. There are numerous derivatives of hGH that arise from proteolytic modifications of the molecule. The primary pathway for the metabolism of hGH involves proteolysis. The region of hGH around residues 130−150 is extremely susceptible to proteolysis and several derivatives of hGH having nicks or deletions in this region have been described (2). This region is in the large loop of hGH (see Fig. 1), and cleavage of a peptide bond there results in the generation of two chains that are connected through the disulfide bond at Cys-53 and Cys-165. Many of these two-chain forms are reported to have increased biological activity (24).

In addition to the naturally occurring proteolytically cleaved two-chain derivatives, there are many examples of derivatives generated artificially through the use of enzymes. The enzymes trypsin and subtilisin, as well as others, have been used to modify hGH at various points throughout the molecule (25,26). One such derivative, called two-chain anabolic protein (2-CAP), was formed through the controlled proteolysis of hGH using trypsin (27). 2-CAP was found to have biological properties very distinct from those of the intact hGH molecule. Whereas the growth-promoting activity of hGH was largely retained, most of the effects on carbohydrate metabolism were abolished.

Desamido hGH. Asparagine and glutamine residues in proteins are susceptible to deamidation reactions under appropriate conditions. Pituitary hGH has been shown to undergo this type of reaction, resulting in conversion of Asn-152 to aspartic acid and also, to a lesser extent, conversion of Gln-137 to glutamic acid (28). Deamidated hGH has been shown to have an altered susceptibility to proteolysis with the enzyme subtilisin suggesting that deamidation may have physiological significance in directing proteolytic cleavage of hGH. Biosynthetic hGH is known to degrade under certain storage conditions, resulting in deamidation at a different asparagine (Asn-149). This is the primary site of deamidation, but deamidation at Asn-152 is also seen (29). Deamidation at Gln-137 has not been reported in biosynthetic hGH.

Sulfoxide hGH. Methionine residues in proteins are susceptible to oxidation primarily to the sulfoxide. Both pituitary-derived and biosynthetic hGH undergo sulfoxidations at Met-14 and Met-125 (29). Oxidation at Met-170 has also been reported in pituitary but not biosynthetic hGH. Both desamido hGH and Met-14 sulfoxide hGH exhibit full biological activity (29).

Truncated Forms. Truncated forms of hGH have been produced, either through the actions of enzymes or by genetic methods. 2-CAP, generated by the controlled actions of the trypsin, has the first eight residues at the N-terminus of hGH removed. Other truncated versions of hGH have been produced by modification of the gene before expression in a suitable host. The first 13 residues have been removed to yield a derivative having distinctive biological properties (30). In this latter case the polypeptide chain is not cleaved.

Bound to Binding Proteins. In the circulation, it is estimated that approximately 50% of the hGH is present in a bound form in a noncovalent complex with a specific, high affinity growth hormone binding protein. Another 5% is bound to a second low affinity binding protein. The high affinity binding protein (GHBP) is a single-chain glycoprotein of ca 60,000 mol wt which binds hGH in a 1:1 stoichiometry (31). Although the human GHBP has not been sequenced, it is likely to be analogous to the situation in the rabbit where the GHBP is a fragment of the hepatic growth hormone receptor encompassing the extracellular domain (32). The GHBP is thought to play a role in influencing the *in vivo* kinetics of hGH, increasing the serum half-life and acting as a circulating reservoir of hGH that can dampen the oscillations caused by the pulsatile secretion of hGH by the pituitary.

Point Mutations.   Since the advent of recombinant DNA technology, a number of researchers have used point mutation techniques either to delete one or more residues within the hGH molecule or systematically to change from one amino acid to another to probe hGH structure/function relationships (33).

Biological Properties

Human growth hormone is a very complex molecule biologically. Several diverse biological activities such as anabolic, insulin-like, diabetogenic, and lactogenic activities have been ascribed to hGH, which also appears to promote water and salt retention. An in-depth discussion of these activities may be found in several excellent reviews available in the literature (34–36).

Uses

Human growth hormone, used as a human pharmaceutical, is approved for only one indication in the United States, treatment of growth failure owing to hGH deficiency, a condition known as pituitary dwarfism. However, clinical trials are underway to test its efficacy in Turner's syndrome, burns, wound healing, cachexia, osteoporosis, constitutional growth delay, aging, malnutrition, and obesity.

Pituitary Dwarfism.   Pituitary dwarfism is a condition characterized by an inability to produce or secrete normal levels of endogenous hGH. The condition results in reduced heights of individuals afflicted with the condition and has been treated by intramuscular or subcutaneous injection of hGH. Pituitary hGH was used prior to the approval of biosynthetic hGH. If treatment is initiated early enough, the patient can attain a final adult height well within the normal range.

Turner's Syndrome.   Turner's syndrome is a genetic disorder of females characterized by short stature, nonfunctioning ovaries, and failure to develop secondary sexual characteristics. Several clinical trials in the United States, Europe, and Japan have demonstrated that hGH can accelerate short-term growth, accompanied by a relatively modest advancement of skeletal maturation. Human growth hormone has been approved for this condition in Japan and in parts of Europe. Approval is expected in the United States.

Other Uses.   Other uses of hGH, eg, for burns, wound healing, cachexia, osteoporosis, aging, malnutrition, and obesity, are being investigated. These uses are in various stages of development and trials are being carried out by several different pharmaceutical companies.

Manufacture and Processing

Human growth hormone was originally manufactured by isolation of the natural product from human pituitaries and subsequent purification of the protein. Since 1985, manufacture of hGH has been almost exclusively by recombinant DNA technology.

Natural Product hGH.   In 1944 the preparation of a highly purified growth hormone from bovine pituitary glands was reported (37). Subsequently, growth hormones derived from animal pituitaries were found to be ineffective in humans; the existence of specificity among species for growth hormone was thus established.

Numerous methods for the extraction and purification of hGH have been developed since this hormone was first isolated in 1956 (1). Initially, the only source was the pituitary gland of cadavers. The sooner after death that the glands were removed and frozen, the higher the yield and quality of the resulting hGH preparation. After thawing, the glands were used directly for extraction of the hGH or an acetone powder was prepared for later extraction. Shortages of hGH led to development of a process to extract hGH from embalmed tissues (38). Yields using embalmed glands were only 20–25% of those when fresh-frozen glands were used, but the greater availability of embalmed glands compensated for the losses. It was found that if formalin was removed from embalmed pituitary glands with large volumes of cold acetone, appreciable quantities of hGH could be isolated from these glands.

Extraction of hGH from pituitary glands has been accomplished by a variety of procedures with or without protease inhibitors. One of the first commercial procedures used glacial acetic acid at 70°C to extract the hGH from the glands (39).

Following extraction from the gland the crude hGH was purified by a number of different procedures, including precipitation steps and different types of chromatography (qv). hGH could be separated from most of the other pituitary hormones by simple size exclusion chromatography on resins like Sephadex G-100. Cation- and anion-exchange steps were introduced into the purification schemes of a number of laboratory processes. Higher quality material was produced, but in significantly lower yields. The addition of proteinase inhibitors like diisopropylfluorophosphate (DFP) before loading the material onto the anion-exchange column dramatically improved the hGH yield. Purification procedures for pituitary-derived hGH began to appear which included ion-exchange steps sometimes containing $6\,M$ urea in the running buffer, ultrafiltration steps, and size exclusion steps. However, the need for increased purity was always modulated by the short supply of the starting material and the need for reasonably high yields in any purification process used for commercial purposes. Besides procedures to remove contaminating proteins from the hGH preparation, procedures were developed using $6\,M$ urea and ultrafiltration to alleviate any potential contamination by slow viruses (40).

One of the goals of the early purification procedures was to produce hGH in the highest yield possible. Preparations of hGH produced in high yield were found to be satisfactory in clinical investigation, but were by no means electrophoretically homogeneous, and some were even brown in color (41). Antibodies appeared in the serum of patients treated with preparations of the order of 50% purity, and the immunogenicity was attributable to inactive components. Recognition that highly purified preparations of hGH could overcome many of the complications arising in the clinical treatment was the driving force to develop more highly purified hGH.

Recombinant hGH. Introduction of recombinant DNA technology meant an unlimited supply of hGH could be produced in a number of different systems.

Escherichia coli (*E. coli*) was the first heterologous host used and originally produced Met-hGH, authentic hGH with the initiating methionine attached (11). This was followed by production in *E. coli* of precursors of hGH that could be cleaved *in vitro* to yield natural sequence hGH. High level expression of hGH in *E. coli* can lead to deposition of the hormone in inclusion bodies. This mode of production offers both advantages and disadvantages, from a production point of view. It is generally quite easy to isolate the inclusion bodies by lysis of the cell and collection of the inclusion bodies by centrifugation. The isolated inclusion bodies are greatly enriched in hGH. Alternative methods of production in *E. coli* allow the hGH produced to be exported to either the periplasmic space of the bacterium (42) or to the culture medium (43). Periplasmic production of hGH generally leads to a fully soluble form of hGH that is properly folded and which can be released from the cell by osmotic lysis of the *E. coli* outer membrane (42).

If the hGH is exported to the culture medium the product can easily be collected by removal of the cells from the culture medium by centrifugation. Purification of hGH from the culture medium is facilitated by low amounts of contaminating proteins present. In fact, it has been shown that hGH can be purified on a laboratory scale by a single purification step on a reversed-phase hplc column (43). Mammalian cells growing in tissue culture have also been used as hosts to produce hGH, which is exported into the culture media (44).

The availability of large amounts of hGH produced by recombinant DNA technology made it possible to investigate other forms of chromatographic purifications such as reversed-phase hplc (rp-hplc) (45), immunoadsorbant chromatography, and chromatofocusing (46). However, the classical forms of purification such as anion-exchange and size exclusion chromatography produce hGH from a recombinant source with greater than 90% purity (42).

Economic Aspects

Human growth hormone is one of the largest selling therapeutic proteins produced by recombinant DNA technology. Annual worldwide sales increased from $130,000,000 in 1987 to $575,000,000 in 1992 (47). Upon approval of additional indications, the sales of hGH are expected to increase even more.

hGH preparations produced and marketed in the United States include Humatrope from Eli Lilly and Co. (Indianapolis, Ind.) and Protropin from Genentech, Inc. (South San Francisco, Calif.). Protropin differs slightly in structure from natural sequence hGH, having an additional methionine residue at the N-terminus. Both Humatrope and Protropin are marketed as vials containing 5 mg of hGH plus additional excipients, ie, glycine and mannitol, in a lyophilized form. A companion diluent vial is provided, allowing the patient to reconstitute the product to the desired concentration prior to administration of the dose. Additional products are expected to become available in the United States during 1994, or thereafter. Outside the United States many commercial preparations are available in a variety of package sizes. In some cases the label quantity is expressed in international units (IUs) rather than milligrams. Package sizes of 2, 4, 12, and 16 IU/vial are available. A conversion factor of 3 IU/mg is generally accepted when calculating dosages.

Specifications, Standards, and Quality Control

Specifications for hGH products are defined by the governmental licensing authorities, eg, the U.S. Food and Drug Administration. Draft monographs for hGH have been prepared by both the *United States Pharmacopia* and the *European Pharmacopeia* commissions and should be formally adopted by 1995. These specifications are suitable for biosynthetic hGH. The much less purified pituitary-derived hGH has virtually disappeared from commercial production. An international reference standard for pituitary-derived hGH (lot 80/505) has been used for calibration, particularly for bioassay purposes. A highly purified biosynthetic hGH standard (lot 88/624) has been prepared and should be formally adopted by 1995, or before.

Quality control (qv) of hGH preparations includes proof of identity, potency, and purity. Purity of hGH must be carefully determined. The result can depend greatly on the specific assay used. The predominant chemical variant in hGH at the time of manufacture is desamido hGH (29), wherein the asparagine at position 149 has been converted to aspartic acid. Desamido hGH levels are generally in the range of 2–5% at the time of manufacture, and can increase slightly over the shelf-life of the product. A second chemical derivative present in hGH preparations is hGH methionine sulfoxide (29). Methionine sulfoxide forms of hGH are generally not present at significant levels at the time of manufacture, but typically increase to a level of 3–6% over the shelf-life of the product. Both the desamido and methionine sulfoxide forms of hGH show significant biological activity, and hence the formation of these derivatives during storage has little effect on the potency of the product. The specification for total chemical variants in hGH preparations at the time of expiration is not more than 12%, in the draft USP monograph.

In addition to chemical derivatives, purity of hGH must also be established with respect to physically associated forms. The hydrophobically linked, noncovalent dimer of hGH found to exhibit relatively low biological activity (21) is present at a level of 1–2% in most hGH preparations at the time of manufacture. In addition, both noncovalent and covalent forms of hGH dimer and higher order aggregates may form during storage of the product. Typical dimer contents of hGH preparations are in the range of 2–5%, with a product specification of not more than 6%, at time of product expiration.

Although biosynthetic hGH products are of very high purity relative to pituitary-derived products, additional evidence of purity with respect to removal of DNA and proteins from the host-cell line is required. Host-cell protein levels are routinely determined and controlled to a level of not more than 10 ppm. DNA levels are demonstrated to be less than 100 pg/dose, or not more than 40 pg/mg, based on a maximal 2.5-mg dose.

Analytical Test Methods and Storage

Proof of identity of hGH preparations is provided by several different test methods conducted in parallel. A single test method is not considered to be sufficient evidence of identity for a molecule as complex as hGH. The most powerful assay method for identification is peptide mapping, wherein an hGH solution is treated with trypsin at neutral pH for a few hours, and the resulting peptides separated by reversed-phase high performance liquid chromatography (rp-hplc). Because trypsin cleaves on the C-terminal side of each arginine and lysine with high specificity, the resulting rp-hplc chromatogram is highly reproducible and provides a fingerprint from which identity of the test material can be deduced (48). A second method for identification of hGH also involves a highly selective rp-hplc procedure, which is capable of resolving hGH from close structural variants such as deamidated forms, methionine sulfoxides, and N-terminal derivatives (49). This method resolves hGH from *N*-methionyl hGH sold commercially as Protropin.

Potency of hGH preparations is quantitatively determined, in terms of mass per vial, by one or more chromatographic procedures (50). Biopotency is calculated from the mass-based potency using a conversion factor, typically 3 IU/mg. Traditionally a bioactivity assay using hypophysectomized rats has been used to determine potency; however, the imprecision of this assay has resulted in its use only as a semiquantitative indicator of bioactivity (1), sometimes referred to as a bioidentity test.

Chemical derivatives of hGH can be determined by a variety of chromatographic procedures. The rp-hplc procedure described for identity confirmation is also capable of determining the levels of desamido hGH and hGH methionine sulfoxide (49). Ion-exchange hplc can also be used to determine the level of desamido hGH (49). Dimeric and aggregated forms of hGH can be determined by size-exclusion hplc, using nondenaturing conditions (21). However, denaturing conditions used in rp-hplc or in sodium dodecylsulfate polyacrylamide gel electrophoresis (sds-page) cause conversion of noncovalent dimers and aggregates to monomeric hGH; thus such techniques are not suitable for this purpose.

Trace contaminants such as host cell proteins (HCPs) and DNA are determined by more specialized techniques. Host cell proteins are generally determined using an immunochemical assay, in which an antibody preparation, raised against a mixture of the HCPs, is used to selectively detect the total level of HCPs in the product. DNA can be determined using a labeled mixture, or probe, of complimentary DNA from the host cell.

Freeze-dried hGH products are quite stable under refrigerated conditions, eg, a 24-month shelf-life is typical at this temperature. However, solutions of hGH at neutral pH readily deamidate, and storage of the reconstituted product is limited to a few weeks under refrigerated conditions, although the biological activity is relatively unaffected even after prolonged storage. Reconstituted solutions of hGH should not be frozen, because denaturation and subsequent aggregation may result.

Metabolism and Disposition

The pharmacokinetics of hGH have been evaluated in animals and humans. After intravenous administration, the elimination of hGH is described by first-order kinetics with a serum half-life of 12–30 min in both animals and humans (52,53). Traditionally, intramuscular (im) injection has been the method of choice for delivery of hGH. This has given way to subcutaneous (sc) administration as the preferred route of delivery, owing to increased patient compliance as a result of lessened local discomfort (54). In humans, absorption of exogenous hGH appears to be more rapid from the im site, with a time to maximum concentration of 2–3 h compared to 4–6 h after sc administration. The disappearance phase from serum has been reported to range from 12–20 and 20–24 h after im and sc administration, respectively (55,56). The plasma kinetics of hGH in humans has been best predicted by animal studies using the nonhuman primate and less so by rodent species. In general, no significant differences have been observed in the pharmacokinetics or biological activities of recombinant natural sequence hGH, recombinant N-methionyl–hGH, or pituitary-derived material in humans (52,57).

The principal organs involved in the peripheral clearance of hGH from the plasma are the kidney and liver. hGH is cleared via glomerular filtration at the kidney and by a receptor-mediated mechanism at the liver (58,59). In animal models, derivatives of hGH such as the 20,000 mol wt variant, oligomeric forms, and hGH complexed with GH-binding protein have been shown to be cleared from the serum at significantly lower rates than 22,000 mol wt hGH (60–62). The prolonged plasma half-life of these derivatives probably reflects a combination of decreased receptor affinity and size constraints on glomerular filtration.

Information regarding the metabolic fate of hGH in humans and animals is fragmentary. hGH has been demonstrated to be stable to proteolytic degradation in the plasma or serum from animals and humans (63). Thus it appears unlikely that intravascular degradation of hGH is relevant *in vivo*. Although a mechanism has not been established, hGH fragments formed after receptor-mediated uptake may be recycled into the circulation. *In vitro*, proteolytic processing of hGH in the large disulfide loop region along with truncation of the N-terminus resulted in hGH metabolites having masses ranging from 22,143 to 16,002 (64). These metabolites were formed by the sequential action of a chymotrypsin-like protease and a carboxypeptidase (64,65). In addition, membrane fractions from target tissue, ie, liver, have been shown by sds-page to convert hGH to a 15,000 mol wt fragment (66). The molecular weight and the cleavage site sensitivity in these metabolites are consistent with that of endogenous pituitary-derived variants of hGH, suggesting potential physiological or pharmacological relevance. In this regard, immunoreactive hGH forms having apparent molecular weights of 12,000, 16,000, and 30,000 have been observed in human serum (67). Although the source of this material has not been elucidated, these molecules are not detectable in plasma after a secretory stimulus, suggesting their formation by peripheral tissues. At this time, the actual concentration of hGH fragments in plasma and the precise nature of this immunoreactive material has not been established. Application of sensitive and accurate methods of protein analysis such as electrospray mass spectrometry (qv) should help resolve some of the deficiencies in the information available.

BIBLIOGRAPHY

1. C. H. Li and H. Papkoff, *Science* **124**, 1293 (1956).
2. U. J. Lewis and co-workers, *Recent Progr. Horm. Res.* **36**, 477 (1980).
3. B. C. Hummel and co-workers, *Endocrinology* **97**, 855 (1975).

4. Preliminary report on "Collaborative Study on the Proposed International Standard for Somatropin," National Institute for Biological Standards and Control, London, Dec. 21, 1992.
5. D. N. Brems, P. L. Brown, and G. W. Becker, *J. Biol. Chem.* **265**, 5504 (1990).
6. T. A. Bewley and C. H. Li, *Biochemistry* **11**, 884 (1972).
7. L. G. Tensmeyer, *J. Mol. Struct.* **175**, 73 (1988).
8. S. S. Abdel-Meguid and co-workers, *Proc. Nat. Acad. Sci. USA* **84**, 6434 (1987).
9. A. M. de Vos, M. Ultsch, and A. A. Kossiakoff, *Science* **255**, 306–312 (1992).
10. J. A. Wells and co-workers, *Recent Progr. Horm. Res.* **48**, 253 (1993).
11. D. V. Goeddel and co-workers, *Nature* **281**, 544 (1979).
12. U. J. Lewis and co-workers, *J. Biol. Chem.* **253**, 2679 (1978).
13. U. J. Lewis, L. F. Bonewald, and L. J. Lewis, *Biochem. Biophys. Res. Comm.* **92**, 511 (1980).
14. F. M. DeNoto, D. D. Moore, and H. M. Goodman, *Nucleic Acids Res.* **9**, 3719 (1981).
15. U. J. Lewis and co-workers, *Endocrinology* **104**, 1256 (1979).
16. M. W. Stolar, K. Amburn, and G. Baumann, *J. Clin. Endocrinol. Metab.* **59**, 212 (1984).
17. M. W. Stolar and G. Baumann, *Metabolism* **35**, 75 (1986).
18. U. J. Lewis and co-workers, *J. Biol. Chem.* **252**, 3697 (1977).
19. T. A. Bewley and C. H. Li, *Adv. Enzymol. Relat. Areas Mol. Biol.* **42**, 73 (1975).
20. N. D. Jones and co-workers, *Biotechnology* **5**, 499 (1987).
21. G. W. Becker and co-workers, *Biotechnol. Appl. Biochem.* **9**, 478 (1987).
22. B. C. Cunningham, M. G. Mulkerrin, and J. A. Wells, *Science* **253**, 545 (1991).
23. O. Thorlacius-Ussing, *Neuroendocrinology* **45**, 233 (1987).
24. R. N. P. Singh and co-workers, *Endocrinology* **94**, 883 (1974).
25. U. J. Lewis and co-workers, *Endocrinology* **101**, 1587 (1977).
26. L. Graff, C. H. Li, and M. D. Jibson, *J. Biol. Chem.* **257**, 2365 (1982).
27. G. W. Becker and C. J. Shaar, abstr. no. 342, 71st Annual Meeting, The Endocrine Society, Seattle, Wash., June 1989.
28. U. J. Lewis and co-workers, *J. Biol. Chem.* **256**, 11645 (1981).
29. G. W. Becker and co-workers, *Biotechnol. Appl. Biochem.* **10**, 326 (1988).
30. A. Gertler and co-workers, *Endocrinology* **118**, 720 (1986).
31. G. Baumann, *Acta Paediatr. Scand.* **367** (Suppl.), 142 (1990).
32. S. A. Spencer and co-workers, *J. Biol. Chem.* **263**, 7862 (1988).
33. S. Nishikawa and co-workers, *Protein Eng.* **3**, 49 (1989).
34. R. K. Chawla, J. S. Parks, and D. Rudman, *Ann. Rev. Med.* **35**, 519 (1983).
35. M. B. Davidson, *Endocr. Rev.* **8**, 115 (1987).
36. J. L. Kostyo, in L. E. Underwood, ed., *Human Growth Hormone: Progress and Challenges*, Marcel Dekker, New York 1988, p. 63.
37. C. H. Li and H. M. Evans, *Science* **99**, 183 (1944).
38. J. B. Mills and co-workers, *J. Clin. Endocrinol. Metab.* **29**, 1456 (1969).
39. M. S. Raben, *Science* **125**, 883 (1957).
40. M. Pocchiari and co-workers, *Horm. Res.* **35**, 161 (1991).
41. R. A. Reisfeld and co-workers, *Nature* **197**, 1206 (1963).
42. H. M. Hsiung, N. G. Mayne, and G. W. Becker, *Biotechnology* **4**, 991 (1986).
43. H. M. Hsiung and co-workers, *Biotechnology* **7**, 267 (1989).
44. S. Lefort and P. Ferrara, *J. Chromatogr.* **361**, 209 (1986).
45. B. S. Welinder, H. H. Sørensen, and B. Hansen, *J. Chromatogr.* **398**, 309 (1987).
46. S. Niimi and co-workers, *Chem. Pharm. Bull.* **10**, 4221 (1987).
47. A. Klausner, *Biotechnology* **11**, S35 (1993).
48. G. W. Becker and H. M. Hsiung, *FEBS Lett.* **204**, 145 (1986).
49. R. M. Riggin, G. K. Dorulla and D. J. Miner, *Anal. Biochem.* **167**, 199 (1987).
50. R. M. Riggin and co-workers, *J. Chromatogr.* **435**, 307 (1988).

51. W. Marx, M. E. Simpson, and H. M. Evans, *Endocrinology* **30**, (1942).
52. J. A. Moore and co-workers, *Endocrinology* **122**, 2920 (1988).
53. C. M. Hendricks and co-workers, *J. Clin. Endocrinol. Metal.* **60**, 864 (1985).
54. D. M. Wison and co-workers, *Pediatrics* **76**, 361 (1985).
55. K. Albertsson-Wikland, O. Westphal, and U. Westgren, *Acta Paediatr. Scand.* **75**, 89 (1986).
56. J. O. Jorgensen and co-workers, *Acta Endocrinol. (Copenhagen)* **116**, 381 (1987).
57. K. D. Jorgensson and co-workers, *Pharmacol. Toxicol.* **63**, 129 (1988).
58. F. A. Carone and co-workers, *Kidney Int.* **16**, 271 (1979).
59. M.-C. Postel-Vinay, C. Kayser, and B. Debuquois, *Endocrinol. Rev.* **111**, 244 (1982).
60. G. Baumann, K. D. Amburn, and T. A. Buchanan, *J. Clin. Endocrinol. Metab.* **64**, 657 (1987).
61. G. Baumann, M. W. Stolar, and T. A. Buchanan, *Endocrinology* **119**, 1497 (1985).
62. G. Baumann, M. W. Stolar, and T. A. Buchanan, *Endocrinology* **117**, 1309 (1985).
63. G. Baumann, *J. Clin. Endocrinol. Metab.* **43**, 222 (1976).
64. V. J. Wroblewski, R. E. Kaiser, and G. W. Becker, *Pharm. Res.* **10**, 1106 (1993).
65. V. J. Wroblewski, M. Masnyk, and G. W. Becker, *Endocrinology* **129**, 465 (1991).
66. J. M. Schepper and co-workers, *J. Biol. Chem.* **259**, 12945 (1984).
67. G. Baumann, M. W. Stolar, and K. Amburn, *J. Clin. Endocrinol. Metab.* **60**, 1216 (1985).

GERALD W. BECKER
WARREN C. MACKELLAR
RALPH M. RIGGIN
VICTOR J. WROBLEWSKI
Lilly Research Laboratories

BRAIN OLIGOPEPTIDES

Neuropeptides represent the most numerous of the classes of neurotransmitters and neuromodulators (see NEUROREGULATORS). There are at least 50 oligopeptides known to exist in the brain (1,2), and the total list of oligopeptide neurotransmitter candidates may eventually be significantly greater. The clinical impact of these peptides on brain function, however, has been limited because most peptides are not metabolically stable, may not cross the blood brain barrier, or there may be no specific antagonists available to test the normal functioning of a peptide system. The development of novel analogues which satisfy many of these concerns represents a significant potential. Central nervous system (CNS) disorders involving Alzheimer's disease, appetite control, depression, and analgesia are all potentially affected by drugs acting at specific neuropeptide systems. The cloning and sequencing of the receptors for these peptides and modeling of their binding sites should be important components in the design of such drugs.

Despite their relatively large number, brain oligopeptides represent the latest class of substances to be identified as neurotransmitter candidates. Although historically there has been great interest in the identification and purification of pituitary hormone-releasing factors from hypothalamic extracts (3–6), most of that interest has been focused not on their possible roles as neurotransmitters but instead on their endocrine functions. The first genuine neuropeptide to be isolated from the brain was the 11-amino acid peptide substance P, first identified in 1931 in equine intestine and brain extracts (7) and much later identified in the brain (8). The subsequent discovery that substance P [33507-63-0] plays a crucial role in the processing of pain information (9) clearly establishes its role in brain function. The field of neuropeptides exploded upon identification from porcine brain of the two pentapeptides, methionine enkephalin [58569-55-4] and leucine enkephalin [58822-25-6] as endogenous opioid peptides (see OPIOIDS, ENDOGENOUS) (10). Not only did this discovery provide the key finding that neuropeptides may have significant pharmacological actions, but the existence of specific opioid antagonists, such as naloxone [465-65-6], meant that the normal function of such neuropeptide systems in the brain could be determined by their pharmacological blockade. The synthesis of peptide antagonists has greatly aided in the development of knowledge of other neuropeptide systems as well.

Neuropeptides represent potentially the most diverse group of neurotransmitters, and their potential roles in normal and diseased brain function are wide-ranging. Many oligopeptides identified in the central nervous system exert both gastrointestinal and behavioral effects and also modify pituitary function (11–16). Within the brain, almost every possible behavioral action has been associated with some type of neuropeptide. Moreover, many neuropeptides have been associated with several CNS disease states. Opioid peptides and substance P have been associated with pain states, neurotensin [39379-15-2] has been associated with schizophrenia and the actions of psychotropic drugs, and corticotropin-releasing factor [9015-75-8] (CRF) has been associated with clinical depression. All neuropeptides are distributed in discrete areas of the brain and many are strategically located to mediate many of these effects. One important aspect of neuropeptide localization is colocalization, ie, the finding that many neuropeptides are stored and released along with other peptide and nonpeptide neurotransmitters (17–20). For this reason, neuropeptides have often been termed neuromodulators because they may act to modulate the effects of other neurotransmitters.

Many neurotransmitters are remarkably potent, both *in vivo* as well as in binding to their respective receptors. Many peptides exhibit K_D, dissociation constant, values at their receptors at less than 1 nM, which is an affinity several orders of magnitude higher than that observed for many classical nonpeptide neurotransmitters. Most neuropeptide receptors belong to the superfamily of G-protein-coupled receptors. Thus the immediate biological response in the cell to the binding of the peptide to its receptor is the production of an intracellular second messenger system, usually either cyclic adenosine monophosphate (cAMP) or phosphoinositide turnover.

Analysis. Neuropeptides are assayed either by bioassays, including receptor binding assays, which measure the biologically active material as opposed to inactive precursors, or by immunological assays, eg, radioimmunoassays or immunocytochemical assays, which utilize specific antibodies generated against native peptides or against fragments of peptides (see IMMUNOASSAYS). Although the latter assays have predominated since the early 1970s because of their relative simplicity and specificity, these are limited by the specificity of the antibodies used, and also may provide false positive values based on cross-reactivity of antibodies against inactive precursors. For localization of neuropeptides, the technique of immunocytochemical localization (21) has been most useful, although this methodology may suffer from a precise identification of the peptides being measured. Another important technique is *in situ* localization of messenger ribonucleic acid (mRNA) products. The biosynthesis of neuropeptides generally occurs through breakdown from high molecular protein precursors, and these precursors can be identified by specific probes directed against their mRNA sequences. Such probes can be extremely specific for their respective mRNA sequences and this method provides an extremely specific view of a particular peptidergic system. A disadvantage is that the *in situ* technique detects only mRNA levels of inactive precursors, and the resulting localization may not match that of the final active peptide fragments.

Synthesis and Biosynthesis

Synthesis. In contrast to pituitary hormones, which usually can be obtained in pure form only after extraction from animal tissues, brain oligopeptides are readily available because of their small size. The synthetic replica represents the most economical and readily accessible source for the oligopeptides. Two techniques are available for laboratory synthesis of oligopeptides, ie, solution chemistry and solid-phase peptide synthesis (SPPS). These methodologies have been reviewed (22). In both methods, synthesis involves assembly of protected peptide chains, deprotection, purification, and characterization. However, the solid-phase method, pioneered by Merrifield, dominates the field of peptide chemistry (23). In SPPS, the C-terminal amino acid of the desired peptide is attached to a polymeric solid support. The addition of amino acids (qv) requires a number of relatively simple steps that are easily automated. Therefore, SPPS contains a number of advantages compared to the solution approach, including fewer solubility problems, use of less specialized chemistry, potential for automation, and requirement of relatively less skilled operators (22). Additionally, intermediates are not isolated and purified, and therefore the steps can be carried out more rapidly. Moreover, the SPPS method has been shown to proceed without racemization, whereas in fragment synthesis there is always a potential for racemization. Solution synthesis provides peptides of relatively higher purity; however, the addition of hplc methodologies allows for pure peptide products from SPPS as well.

Several improvements have been made in the original SPPS methodology which have extended the general usefulness of this technique and enabled routine synthesis of peptides 50 amino acids in length (22). Longer peptides also have been reported using this technique. Although the original SPPS technique allowed for synthesis of only one peptide at a time, more recent developments have allowed multiple concurrent syntheses. For example, a method synthesizing peptides permanently bound to solid supports has been reported (24). Such peptides then undergo enzyme immunosuppressant assays (elisa). Another variation of SPPS has been described in which the solid support is contained within solvent-permeable packets. This is often referred to as the T-bag method (25) and it allows concurrent synthesis of extremely large numbers of different peptide sequences (26). It has become useful (ca 1993) in construction of extremely large peptide libraries.

The use of SPPS allows synthesis of both native oligopeptides and important synthetic analogues. Many of these fragments or substituted analogues have been found to be equal to, or more potent than, the parent molecule and to exhibit long-acting and antagonistic activities.

Biosynthesis. The biosynthesis of neuropeptides is much more complex and involves the multistep process of transcription of specific mRNA from specific genes, formation of a high molecular weight protein product by translation, post-translational processing of the protein precursor to allow for proper packaging within the cell, and final enzymatic cleavage to produce the active peptide product (2). The latter processing step is summarized by the schemes in Figure 1, which illustrates the precursors of several oligopeptides, including CRF, β-endorphin/ACTH/MSH, tachykinins, enkephalin, CGRP, and dynorphin. An example of this process is the ACTH/MSH/endorphin precursor proopiomelanocortin [78321-08-1] (POMC). Synthesized in the pituitary, this 31,000 mol wt protein is heavily glycosylated in several species before it is cleaved into at least three different classes of active peptide, ie, ACTH, β-endorphin, and three different variants of MSH (see Fig. 1). Like most precursors, the active peptides from POMC are hydrolyzed from the precursor by trypsin-like enzymes which cleave between double-basic amino acid residues, including Lys–Lys, Arg–Arg, Lys–Arg, and Arg–Lys. These cleavages result in at

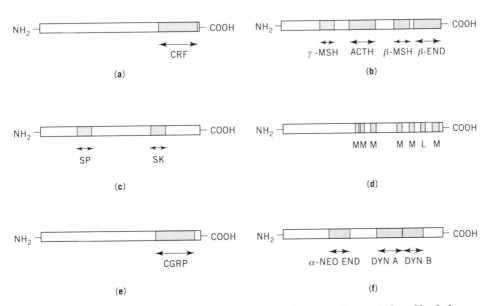

Fig. 1. Schematic drawing of precursors for selected brain oligopeptides. Shaded areas represent the location of sequences of active peptide products which are normally cleaved by trypsin-like enzymes acting on double-basic amino acid residues. Precursors are not necessarily drawn to scale. (**a**) CRF precursor; (**b**) proopiomelanocortin (POMC); (**c**) β-protachykinin; (**d**) proenkephalin A; (**e**) CGRP precursor; (**f**) preprodynorphin, ie, preproenkephalin B. Terms are defined in Table 1.

least four active oligopeptide products from POMC, including γ-MSH, ACTH, β-MSH, and β-endorphin (Fig. 1).

Peptide biosynthesis is affected by regulation of mRNA synthesis (27). For example, levels of some neuropeptides are increased by cyclic AMP, which stimulates gene expression through a specific promoter element known as the cyclic AMP responsive element (CRE). A CRE was first identified with the nucleotide sequence TGACGTCA in the somatostatin gene (28). Other neuropeptide genes contain homologous CRE sites. In addition, the AP-1 site on many peptide genes is activated by the product of the proto-oncogene c-*jun* and probably mediates protein kinase C-activated synthesis (27).

Individual Brain Oligopeptides

Table 1 lists neuropeptides, along with their common acronyms, CAS Registry Numbers, and molecular weights. The amino acid sequences of these neuropeptides are known. Although many peptides contain free amino- and carboxy-terminals, some peptides contain altered amino acids and termini. Particularly common is the presence of an amidated carboxy-terminal. In every case, this is caused by enzymatic decarboxylation of a carboxy-terminal glycine residue by glycine-converting enzyme. Other amino acids are also altered enzymatically, eg, in cholecystokinin, Tyr^{27} is sulfated. In thyrotropin-releasing factor (TRF),

Table 1. Names and Properties of Selected Brain Oligopeptides

Peptide[a]	Acronym[b]	CAS Registry Number	Species	Molecular weight
adrenocorticotrophic hormone	ACTH	[9002-60-2]	human	4538
angiotensin				
angiotensin I		[9041-90-1]	human	1296
angiotensin II		[11128-99-7]	human	1046
bombesin		[31362-50-2]		1619
bradykinin		[58-82-2]		1060
calcitonin gene-related peptide	CGRP	[83652-28-2]	human	3787
cholecystokinin	CCK			
CCK-33		[93443-27-7]	human	3950
CCK-8				1149
dynorphins				
dynorphin-A	DYN A	[88161-22-2]	porcine	2146
dynorphin-B	DYN B	[85006-82-2]	porcine	1570
α-neo-endorphin	α-NEO END	[69671-17-6]	porcine	1228
endorphins				
β-endorphin	β-END	[59887-17-1]	human	3463
α-endorphin		[59004-96-5]		1745
enkephalins				
Leu-enkephalin	L	[58822-25-6]		555
Met-enkephalin	M	[58569-55-4]		573
galanin		[88813-36-9]	human	3156
gastrin		[9002-76-0]	human	2097

Table 1. (*continued*)

Peptide[a]	Acronym	CAS Registry Number	Species	Molecular weight
hypothalamic-releasing factors				
corticotropin-releasing factor	CRF[c]	[9015-71-8]	human	4754
growth hormone-releasing factor	GRF[d]	[83930-13-6]	human	5037
luteinizing hormone-releasing hormone	LHRH[e]	[9034-40-6]		1182
somatostatin	SRIF[f]	[51110-01-1]		1637
thyrotropin-releasing factor	TRF[g]	[24305-27-9]		362
melanocyte-stimulating hormone	MSH			
α-MSH		[37213-49-3]		1664
β-MSH		[9034-42-8]	human	2659
γ-MSH		[72711-43-4]		1570
neurotensin		[39379-15-2]		1672
neuropeptide Y	NPY	[82785-45-3]	human	4269
peptide YY		[106388-42-5]	human	4307
secretin		[1393-25-5]	human	3038
tachykinins				
substance K	SK	[102577-24-2]		1133
neurokinin B[h]		[102577-19-5]		1210
substance P	SP	[33507-63-0]		1347
vasoactive intestinal peptide	VIP	[40077-57-4]	human	3324
vasopressin	ADH	[11000-17-2]		1083

[a] Alternative nomenclature for hypothalamic releasing factors are indicated.
[b] See Figure 1.
[c] Corticotropin-releasing factor (CRF) = corticoliberin.
[d] Growth hormone-releasing factor (GRF) = growth hormone-releasing hormone (GHRH) = somatoliberin.
[e] Luteinizing hormone-releasing hormone (LHRH) = gonadotropin-releasing hormone (GnRH) = gonadoliberin-luteinizing hormone-releasing factor (LRF) = luliberin.
[f] Somatostatin (SS) = somatotropin-releasing inhibiting factor (SRIF).
[g] Thyrotropin-releasing factor (TRF) = thyrotropin-releasing hormone (TRH) = thyroliberin.
[h] Neurokinin B = neuromedin K.

the amino-terminal Gln residue has been converted into a pyro-Glu. Many synthetic analogues also contain altered amino acids, including D-amino acids. Such alterations may increase potencies, change pharmacological properties, eg, antagonists vs agonists, or increase stability to enzymatic breakdown.

A summary of the properties of some of these neuropeptides is available herein; other neuropeptides have been reviewed (1,2,19,29–32).

Tachykinins. The group of peptides known as tachykinins include substance P, substance K or neurokinin A, and neuromedin K, ie, neurokinin B, as well as a number of nonmammalian peptides. All members of this family contain the conserved carboxy-terminal sequence -Phe-X-Gly-Leu-Met-NH$_2$, where X is an aromatic, ie, Phe or Tyr, or branched aliphatic, eg, Val or Ile, amino acid. In general, this C-terminal sequence is crucial for tachykinin activity (33);

in fact, both the methionineamide and the C-terminal amide are crucial for activity. The nature of the X residue in this sequence determines pharmacological identity (34,35); thus the substance P group contains an aromatic residue in this position, while the substance K group contains an aliphatic residue (33).

Substance P was the first neuropeptide to be discovered. It was originally detected in the acid–alcohol extracts of equine brain and intestine (7). Its role in the brain was confirmed by isolation and sequence of substance P from bovine hypothalamus (8,36). Although other members of the tachykinin family have been identified in nonmammalian species (37), it was assumed for years that substance P was the only tachykinin in the brain. That assumption was invalidated in the early 1980s when substance K and neuromedian K were isolated from brain extracts (38–41). The identification of these other tachykinins in the brain not only broadened the potential role of this family as neurotransmitters, but also confounded many previous localization studies, because it is likely that some of the antisera used in earlier studies may have cross-reacted with these other tachykinins (33). Nevertheless, the highest density of substance P in the brain is found in the *substantia nigra* (42,43), with relatively large levels also observed in hypothalamus. In several areas, substance K and substance P are colocalized (44), which would be predicted based on their common precursors.

Biosynthesis. Two closely related genes encode the three mammalian tachykinins. The preprotachykinin A gene encodes both substance P and substance K, while the preprotachykinin B gene encodes neuromedin K (45–47). The active sequences are flanked by the usual double-basic amino acid residues, and the carboxy-terminal amino acid is a glycine residue which is decarboxylated to an amide. As with most neuropeptide precursors, intermediates in peptide processing can be detected, but their biological activities are not clear (ca 1994).

Biological Activities. Substance P was first isolated for its hypotensive action and stimulation of rabbit jejunum contraction and later for following its ability to produce salivation in rats. It also stimulates glucagon secretion and produces hyperglycemia in rats, stimulates smooth muscle contraction in the guinea pig vas deferens and ileum, and elevates growth hormone and prolactin secretion in humans (48). In the brain, one of the most commonly discussed roles of substance P is the mediation of pain information, where this peptide acts as one of the primary neurotransmitters in the sensory afferent fibers which conduct pain information (49,50). Thus compounds which increase substance P release, eg, capsaisin, tend to increase pain, while potential substance P antagonists are sought-after analogues for analgesic purposes.

Analogues. The smallest sequence possessing most of the substance P spectrum of activity and high potency is the hexapeptide C-terminus. There are a number of analogues of substance P which are potent antagonists at substance P receptors; these include [D-Arg^1, D-Phe^5, D-Trp^{7,9}, Leu^{11}]-substance P (51), [Arg^4, Gly^5, D-Trp^7, Asp(OC$_4$H$_9{}^t$)$_2^{10}$]-substance P (52), and [Arg^6, D-Trp^{7,9}, CH$_3$-Phe^8]-substance P 6-11 (53).

Endorphins, Enkephalins, and Dynorphins. The discovery of Met- and Leu-enkephalin as the first opioid peptides in 1975 (10) was not only a breakthrough in the understanding of endogenous opioids, but also represented the beginning of the explosion of neuropeptide discoveries that would highlight the next 10 years (54). In 1994 it is clear that three separate opioid peptide families exist, each representing different gene products, ie, enkephalins, β-endorphin, and dynorphin. All of these peptides share the common sequence Tyr-Gly-Gly-Phe-X, where X is either Leu or Met. The shortest members of this family, the enkephalins, are pentapeptides. β-Endorphin contains 31 amino acids, and also the sequence of Met-enkephalin. Dynorphin A contains 17 amino acids, and the sequence of Leu-enkephalin. Both of the longer peptides, especially dynorphin, are relatively basic in nature, with a number of Lys and Arg residues. The enkephalin sequence is crucial for the biological activity of the longer peptides; in all members of this family, the N-terminal Tyr is also crucial for activity.

These three peptide groups exhibit different localization patterns in the brain. β-Endorphin cell bodies are mostly localized to the arcuate nucleus of the hypothalamus (55). Fibers from this area project to many different areas of the brain. However, pituitary is the principal organ of POMC synthesis, where it exists in anterior and intermediate lobes (56). In the anterior lobe, β-endorphin exists in corticotrophic cells, along with ACTH, whereas in the intermediate lobe it exists in melanotropic cells together with MSH. Thus the processing of the POMC precursor varies in different cells. In contrast to β-endorphin, enkephalins are found in cell bodies in many different brain areas (57). In some of these areas, enkephalins form local circuits and are located in interneurons, while in other areas enkephalin cell bodies form long projections. Moreover, enkephalins are colocalized with a large number of other peptide and nonpeptide neurotransmitters, including acetylcholine [51-84-3], epinephrine [51-43-4], norepinephrine [51-41-2], serotonin [50-67-9], γ-aminobutyric acid [56-12-2] (GABA), and substance P (57). Like enkephalin, dynorphin peptides also are widely distributed throughout the brain, although their presence in the supraoptic and paraventricular nuclei of the hypothalamus are particularly well known (57).

Several peptides are related in different ways to these classical opioid peptides. FMRFamide (Phe-Met-Arg-Phe-NH$_2$) contains the first four amino acids of enkephalin and is active in various invertebrates (58); FMRFamide-related peptides also have been located in the mammalian brain. Although these peptides share structural similarities with enkephalins, they do not have appreciable affinities for opioid receptors. Casomorpins, originally isolated from milk, do not contain the enkephalin sequence, but do bind to opioid receptors (59). Another group of nonenkephalin-like opioid peptides are derived from dermorphin [77614-16-5], an unusual D-amino acid-containing peptide from frog skin (60). One derivative of dermorphin, termed deltorphin, is extremely specific for delta-opioid receptors (61).

Biosynthesis. Three separate genes encode the opioid peptides (Fig. 1). Enkephalin is derived from preproenkephalin A, which contains six copies of Met-enkephalin and extended peptides, and one copy of Leu-enkephalin (62–66). β-Endorphin is one of the many products of POMC, and represents the N-terminal 31 amino acids of β-lipotropin (67,68). Three different dynorphin peptides are derived from the third opioid gene, preproenkephalin B, or preprodynorphin (69). The dynorphin peptides include dynorphin A, dynorphin B, and α-neo-endorphin.

Biological Activities. All opioid peptides elicit a number of morphinomimetic activities following intracerebroventricular injection, eg, analgesia and catatonia, and exert numerous behavioral effects (54). *In vitro*, they decrease the amplitude of muscle contractions induced electrically in the guinea pig ileum and in the mouse vas deferens. β-Endorphin also stimulates prolactin and growth hormone secretion *in vivo* (70). All of these are reversed by the opiate antagonist, naloxone [465-65-6]. Although they are not as selective as some of the synthetic analogues, the natural peptides preferentially bind to different types of opioid receptor. Both Met- and Leu-enkephalin preferentially bind to delta-opioid receptors, whereas dynorphin peptides bind to kappa-receptors. On the other hand, β-endorphin is relatively nonselective between mu-, delta-, and kappa-opioid receptors (54).

Analogues. Hundreds of enkephalin analogues have been synthesized in an effort to find a nonaddictive opioid. Some analogues have been synthesized as metabolically stable agonists for longer durations of action; these include [D-Ala^2, Met^5]-enkephalinamide (71) and [D-Ala^2-N-CH$_3$-Phe^4, Met(O)^5-ol]-enkephalin (72). The latter peptide is about 30,000 and 1,000 times more potent when injected intracerebroventricularly than Met-enkephalin and morphine [57-27-2], respectively, and 23 times as active as β-endorphin. Other analogues have high selectivity for different receptor types. DAMGO [78123-71-4] (sometimes abbreviated DAGO, [D-Ala^2-N-CH$_3$-Phe^4, Gly^5-ol]-enkephalin), is selective for mu-receptors (73), whereas DSLET ([D-Ser^2, Leu^5, Thr^6]-enkephalin) and DPDPE ([D-Pen^{2,5}]-enkephalin) are selective for delta-receptors (74,75).

The opioid peptides are unique from a pharmacological point of view because there historically exists a number of specific nonpeptide antagonists, typified by naloxone. Nevertheless, there are several peptide antagonists available for these receptors, particularly for delta-opioid receptors. In 1994 the most selective peptide antagonist for delta-receptors is Tyr-Tic-Phe-Phe-OH (TIPP), where Tic is tetrahydroisoquinoline-3-carboxylic acid (76). Finally, several somatostatin analogues exhibit reasonable affinities at various opioid receptors. One example is CTOP (Cys^2-Tyr^3-Orn^8-Pen^7 amide) somatostatin analogue [103429-31-8], which is a potent and highly selective mu receptor antagonist (77). This finding is particularly interesting because the cloned delta-opioid receptor has significant sequence homology with the cloned somatostatin receptor (78).

Corticotropin-Releasing Factor. Corticotropin-releasing factor (CRF) stimulates the release of ACTH. However, because ACTH, MSH, and β-endorphin all arise from the same precursor, POMC, CRF regulates the release of these peptides as well. Therefore, CRF plays a significant regulatory role in brain and hormonal function. CRF activity was first identified in the hypothalamus in the 1950s (3,4), but difficulties in the bioassay for this material precluded isolation and sequencing of this peptide until 1981 when the sequence of the 41-amino acid peptide CRF was reported (79,80). As shown in Table 1, CRF is one of the largest of the brain oligopeptides and activity appears to reside in several areas of its sequence, although much of its potency lies in the C-terminal 27 amino acids (81).

CRF is localized in several brain areas. The best studied structure is the parvocellular region of the periventricular nucleus of the hypothalamus, where fibers project to the median eminence. Other areas containing CRF include amygdala, lateral hypothalamus, central gray area, dorsal tegmentum, *locus coereleus*, parabrachial nucleus, dorsal vagal complex, and inferior olive (31).

CRF is colocalized with a number of other peptides, including enkephalin and vasopressin (82,83).

Biosynthesis.   CRF is derived from a precursor of 196 amino acids (84,85). This gene contains one copy of CRF, which is flanked by double basic amino acids. The amino acid sequence of the CRF precursor suggests that it may arise from proteins related to POMC and neurophysins (31). The CRF precursor contains a cAMP responsive element which allows stimulation of mRNA synthesis when intracellular levels of cAMP are increased (86).

Biological Activity.   The principal function of CRF is to regulate the release of POMC-derived peptides, especially ACTH. Because the role of ACTH is to release glucocorticoids, CRF is a significant regulator of pituitary–adrenal function. CRF also regulates several aspects of the immune system (31). Perhaps most interestingly, from the point of view of brain function, is the suggestion that CRF plays a principal role in the etiology of CNS depression. In this view, CRF causes many of the known signs of clinical depression, including decreased appetite, decreased libido, and insomnia (32). Such findings suggest that specific CRF antagonists may be valuable tools in the treatment of depression.

Analogues.   There is a definite lack of practical synthetic CRF analogues (ca 1994). The only CRF antagonist available is α-helical CRF 9-41 (87). Although this compound acts as an antagonist by binding to the CRF receptor and not producing a biological response, it is relatively weak and its role as a clinical tool is limited by the fact that it does not cross the blood brain barrier (32).

Somatostatin.   Somatostatin is somatotropin-releasing inhibiting factor (SRIF or SS). Somatostatin was isolated from ovine hypothalamic extracts because of its characteristic inhibition of the spontaneous release of growth hormone by cultured pituitary cells (88). It also inhibits a large number of other neurotransmitters and hormones, including thyroid stimulating hormone [9002-71-5] (TSH), prolactin [9002-62-4], insulin [9004-10-8], glucagon [9007-92-5], acetylcholine, gastrin [9002-76-0], gastric acid, and other digestive enzymes.

Somatostatin is widely distributed throughout the brain and the gastrointestinal tract; it also is found in mammalian plasma (89), in the rat retina (90), and in the human adrenal medulla (91). In the brain, the highest levels are found in the amygdala and hypothalamus, followed by several areas of frontal cortex (29). Significant loss of somatostatin occurs in the frontal cortex of Alzheimer's patients (92,93), and behavioral studies in rats suggest that this peptide may play a role in mediating recognition and memory (94).

Biosynthesis.   Somatostatin exists in longer forms in several biological tissues (95,96). One of the longer forms, which has been isolated from porcine intestine, has been characterized as a 28-amino acid peptide (97). Somatostatin is derived from a precursor containing 116 amino acids (98,99). The precursor contains one copy of the somatostatin tetradecapeptide, which is contained within the sequence of the 28-amino acid peptide at the carboxy-terminal end of the precursor. The 28-amino acid somatostatin is preceded by a single Arg residue, while somatostatin 1-14 is preceded by a pair of basic residues.

Biological Activities and Analogues.   Somatostatin exerts some neurotropic actions, eg, as a tranquilizer and as a spontaneous motor activity depressor. It also lengthens barbiturate anesthesia time and induces sedation and hypothermia. These actions are consistent with the strong association between somatostatin and GABA; in the primate cerebral cortex, 90–95% of somatostatin-positive cells also contain GABA (100).

Substitution of Trp^8 by D-Trp^8 increased the potency of somatostatin (101). Most other substitutions, however, are deleterious to biological activity. Cyclic octapeptide analogues of somatostatin retain high potency; one of them, CTOP, is a potent mu-opioid antagonist.

Hypothalamic-Releasing Peptides. *Gonadoliberin.* Gonadoliberin (LRF) has been isolated from porcine (102) and ovine (103) hypothalamic extracts. Its highest concentration is found in the hypothalamus; however, it also has been reported to be present in extrahypothalamic central nervous regions, blood, urine, and placenta (104–108). LRF acutely stimulates lutenizing hormone (LH) and follicle-stimulating hormone (FSH) secretion (see HORMONES, ANTERIOR PITUITARY HORMONES). Additionally, there is evidence that LRF can act on the central nervous system to modulate sexual behavior (109). Paradoxically, however, its long-term administration is associated with antigonadal effects which include termination of pregnancy, decreased gonadal weights, and lowered steroid secretion, possibly through desensitization at pituitary and gonadal levels and through alterations in steroidogenesis (110).

Potent and long-acting analogues of LRF have been designed. These peptides generally have a D-amino acid at the 6-position and a modified C-terminus. Specific LRF antagonists also have been designed. Although the first antagonist to LRF, ie, [des-His^2]-LRF, was very weak, more recent analogues are considerably more potent; these include [D-Phe^2, Pro^3, D-Phe^6]-LRF and [D-$pGlu^1$, D-Phe^2, D-$Trp^{3,6}$]-LRF (111).

Thyroliberin. Thyroliberin (TRF) originally was purified from alcohol–acid extracts of ovine (112) and porcine (113) hypothalamic extracts. Its greatest hypothalamic concentration is found in the mammalian median eminence; immunoreactive TRF has been found throughout the central nervous system as well as in the blood, urine, cerebrospinal fluid, and endocrine pancreas (114). The principal function of TRF is to stimulate thyroid-stimulating hormone (TSH) release, but it also releases prolactin and growth hormone under specific conditions (see HORMONES, HUMAN GROWTH HORMONE). TRF has been reported to alleviate depressive symptoms, to reverse the duration of anesthesia and hypothermia induced by a number of substances, and to increase spontaneous motor activity. It is derived from a precursor that contains five separate copies of this decapeptide, flanked as usual by double basic residues (115). After cleavage from the precursor, the N-terminal Gln is converted to Glu to produce the active peptide product.

Several hundred analogues to TRF have been synthesized. Some of them, like 1-CH_3-S-Dio-His-Pro-NH_2 are more potent than TRF (116); others are conformationally restricted and equally as potent as TRF (117).

Neurotensin. This hormone has been isolated and characterized from acid–acetone extracts of bovine hypothalamus (118) on the basis of its hypotensive activity. Immunoreactive neurotensin is present in mammalian gut and is distributed throughout the central nervous system; its highest concentration is in the hypothalamus and in the substantia gelatinosa of the spinal cord (119). Its overall brain distribution is not unlike that of enkephalin (1).

Biological Activities and Analogues. The many pharmacological actions of neurotensin include hypotension, increased vascular permeability, hyperglycemia, increased intestinal motility, and inhibition of gastric acid secretion (120). In the brain, it produces analgesia at remarkably low doses (121).

The smallest sequence possessing most of the neurotensin spectrum of activities and its high potency is the hexapeptide C-terminus (1). [D-Trp^{11}]-Neurotensin acts like a neurotensin antagonist in perfused heart preparations, but acts like a full agonist in guinea pig atria and rat stomach strips (122).

Calcitonin Gene-Related Peptide. Most peptides are discovered in a classical way, ie, they are first identified by their biological or receptor-binding activity, then they are isolated and purified from extracts and finally sequenced. The advent of molecular cloning, however, has revolutionized this approach. Several neuropeptides have been isolated by the fact they are derived from precursors closely related to known peptide precursors. An example of this approach is calcitonin gene-related peptide (CGRP), the existence of which was not even suspected before an analysis of the calcitonin precursor gene revealed that alternative splicing of mRNA would give rise to a structurally related sequence (123). By generating antibodies to this sequence, it has been possible to isolate CGRP from the brain and determine its localization (123). Unlike calcitonin itself, which is present primarily in the thyroid gland, CGRP is present throughout the nervous system. It contains 36 amino acids having a C-terminal amide.

CGRP has a wide distribution in the nervous system (19) and was the first peptide to be localized to motoneurons (124). It is also found in primary sensory neurons where it is colocalized with substance P (125). CGRP is derived from a precursor structurally related to the calcitonin precursor. The latter precursor produces two products, calcitonin itself and katacalcin, while the CGRP precursor produces one copy of CGRP (123). Like other peptides, CGRP is cleaved from its precursor by tryptic breakdown between double basic amino acid residues.

Biological Activity and Analogues. The colocalization of CGRP with substance P in primary sensory afferents indicates an important relationship between these two peptides. CGRP promotes release (126) and inhibits degradation of substance P (127), thus increasing its effects. Like substance P, CGRP is a vasodilator (124). Finally, CGRP may have trophic actions on motoneurons.

Relatively few CGRP analogues are available (ca 1994). One interesting analogue is CGRP 8-37, which appears to be an antagonist at CGRP receptors and an agonist at calcitonin receptors (128,129).

Cholecystokinin. Cholecystokinin (CCK) was first isolated from duodenal mucosa as a substance that produced contractions of the gallbladder (130). It is a 3-amino acid peptide that shares the same carboxy-terminal pentapeptide sequence with gastrin. In fact, its existence in the brain was first determined with gastrin antibodies (131) and this cross-reactivity has produced confusion regarding the relative localization of these two peptides in the brain. Authentic gastrin is known to be primarily localized to the magnocellular nuclei of the hypothalamus; all other gastrin-like immunoreactivity in the brain is actually CCK (132). There are relatively high quantities of CCK in the brain, eg, it has been calculated that the human brain contains 1–2 mg of CCK, compared to microgram quantities of many other neuropeptides (1). One unusual aspect of CCK is the presence of a sulfated Tyr residue (see Table 1) which is added to the peptide during post-translational processing.

CCK has been detected in two principal forms, ie, the traditional 33-amino acid peptide, and an octapeptide CCK-8. The intestine produces mainly CCK-33 (133) and the brain produces mainly CCK-8 (132). The CCK precursor contains one copy of CCK-33 (133,134); this peptide is flanked on both ends with double basic residues, whereas CCK-8 is formed from CCK-33 by cleavage of a single basic residue.

Biological Activity and Analogues.   In addition to its actions on the gastrointestinal tract, CCK in the brain has been associated with control of eating behavior (1). Like somatostatin, CCK has a strong relationship with GABA, ie, 90–95% of CCK immunoreactive cells colocalize with GABA cells (100). However, CCK and somatostatin do not coexist in the same GABA-containing cells (100). There is also colocalization of CCK with dopamine (135), which correlates with biochemical studies suggesting that CCK is a neuromodulator of dopamine function (29).

The CCK system shares one property with the opioid system, ie, the existence of selective nonpeptide antagonists. These include asperlicine, a natural benzodiazepine (136), and Devazepide (L-364,718; MK-329) (137). Selective, potent peptide antagonists for CCK, eg, Cl-988 and PD 134308, have been developed that may be useful as anxiolytics and as drugs which increase the analgesic effect of morphine but at the same time prevent morphine tolerance (138) (see HYPNOTICS, SEDATIVES, ANTICONVULSANTS, AND ANXIOLYTICS).

BIBLIOGRAPHY

"Brain Oligopeptides" under "Hormones" in *ECT* 3rd ed., Vol. 12, pp. 603–617, by J. Rivier and C. Rivier, Salk Institute.

1. S. H. Snyder and R. B. Innis, *Ann. Rev. Biochem.* **48**, 755 (1979).
2. D. R. Lynch and S. H. Snyder, *Ann. Rev. Biochem.* **55**, 773 (1986).
3. M. Saffran and A. V. Schally, *Can. J. Biochem. Physiol.* **33**, 408 (1955).
4. R. Guillemin and B. Rosenberg, *Endocrinology* **57**, 599 (1955).
5. J. Boler and co-workers, *Biochem. Biophys. Res. Commun.* **37**, 705 (1969).
6. R. Burgus and co-workers, *C.R. Acad. Sci (Paris)* **269**, 1870 (1969).
7. U. S. von Euler and J. H. Gaddum, *J. Physiol.* **72**, 74 (1931).
8. M. M. Chang, S. E. Leeman, and H. D. Niall, *Nature New Biol.* **232**, 86 (1971).
9. T. M. Jessel and L. L. Ivesen, *Nature* **268**, 549 (1977).
10. J. Hughes and co-workers, *Nature* **258**, 577 (1975).
11. W. Vale, C. Rivier, and M. Brown, *Ann. Rev. Physiol.* **39**, 473 (1977).
12. D. T. Krieger and A. S. Lotta, *Science* **205**, 365 (1979).
13. A. J. Prange and co-workers, in L. L. Iversen, S. D. Iversen, and S. H. Snyder, eds., *Handbook of Psychopharmacology*, Vol. 13, Plenum Press, New York, 1978, p. 1.
14. A.G.E. Pearse, *Nature* **262**, 92 (1976).
15. L. L. Iversen, *Sci. Am.* **241**, 134 (1979).
16. R. Guillemin, *Science* **202**, 390 (1978).
17. V. Chan-Palay and S. Palay, eds., *Coexistence of Neuroactive Substances in Neurons*, John Wiley & Sons, Inc., New York, 1984.
18. A. Cuello, ed., *Co-Transmission*, Macmillan, London, 1982.

19. T. Hokfelt, *Neuron* **7**, 867 (1991).
20. B. Meister and co-workers, *Neuroendocrinology* **48**, 516 (1988).
21. A. Coons, in J. Danielli, ed., *General Cytochemical Methods*, New York, Academic Press, Inc., 1958, p. 399.
22. S. B. H. Kent, *Ann. Rev. Biochem.* **57**, 957 (1988).
23. R. B. Merrifield, *J. Am. Chem. Soc.* **85**, 2149 (1963).
24. H. M. Geysen, R. H. Meloen, and S. J. Barteling, *Proc. Nat. Acad. Sci. USA* **81**, 3998 (1984).
25. R. A. Houghten, *Proc. Nat. Acad. Sci. USA* **82**, 5131 (1985).
26. R. A. Houghten and co-workers, *Nature* **354**, 84 (1991).
27. R. H. Goodman, *Ann. Rev. Neurosci.* **13**, 111 (1990).
28. M. R. Montminy and co-workers, *Proc. Nat. Acad. Sci. USA* **83**, 6682 (1986).
29. M. Hayashi, *Prog. Neurobiol.* **38**, 231 (1991).
30. F. L. Strand and co-workers, *Physiol. Rev.* **71**, 1017 (1991).
31. M. J. Owens and C. B. Nemeroff, *Pharmacol. Rev.* **42**, 425 (1991).
32. C. B. Nemeroff, *Neuropsychopharmacol.* **6**, 69 (1992).
33. J. E. Maggio, *Ann. Rev. Neurosci.* **11**, 13 (1988).
34. V. Erspamer, *Trends Neurosci.* **4**, 267 (1981).
35. C.-M. Lee and co-workers, *Naunyn-Schmeideberg's Arch. Pharmacol.* **318**, 281 (1982).
36. M. M. Chang and S. E. Leeman, *J. Biol. Chem.* **245**, 4784 (1970).
37. V. Erspamer and A. Anastasi, *Experientia* **18**, 58 (1962).
38. J. E. Maggio and co-workers, in *Substance P—Dublin 1983*, Poole Press, Dublin, 1983, p. 20.
39. S. Kimura and co-workers, *Proc. Jpn. Acad. Ser. B* **59**, 101 (1983).
40. K. Kangawa and co-workers, *Biochem. Biophys. Res. Commun.* **114**, 533 (1983).
41. N. Minamino and co-workers, *Neuropeptides* **4**, 157 (1984).
42. R. A. Nicoll, C. Schenker, and S. E. Leeman, *Ann. Rev. Neurosci.* **3**, 227 (1980).
43. N. Aronin, M. Difiglia, and S. E. Leeman, in D. T. Krieger, M. J. Brownstein, and J. B. Martin, eds., *Brain Peptides*, John Wiley & Sons, Inc., New York, p. 783.
44. A. Y. Deutsch and co-workers, *Peptides* **6**(Suppl. 2), 113 (1985).
45. H. Kotani and co-workers, *Proc. Nat. Acad. Sci. USA* **83**, 7074 (1986).
46. H. Nawa and co-workers, *Nature* **306**, 32 (1983).
47. H. Nawa, H. Kotani, and S. Nakanishi, *Nature* **312**, 729 (1984).
48. C. Rivier, J. Rivier, and W. Vale, *Endocrinology* **102**, 519 (1978).
49. T. Jessel, *J. Physiol. (London)* **270**, 56 (1977).
50. C. Lamotte, C. B. Pert, and S. H. Snyder, *Brain Res.* **112**, 407 (1976).
51. P. J. Woll and E. Rozengurt, *Proc. Nat. Acad. Sci. USA* **85**, 1859 (1988).
52. C. Kitada, *Peptide Chem.* **1983**, 297 (1983).
53. R. Laufer and co-workers, *Proc. Nat. Acad. Sci. USA* **82**, 7444 (1985).
54. H. Akil and co-workers, *Ann. Rev. Neurosci.* **7**, 223 (1984).
55. H. Khatchaturian and co-workers, *Handbook Chem. Neuroanat.* **4**, 216, 1985.
56. F. E. Bloom and co-workers, *Life Sci.* **20**, 43 (1977).
57. H. Khatchaturian, M. K. H. Schaefer, and M. E. Lewis, *Handbook Exp. Pharmacol.* **104**(1), 471 (1993).
58. D. A. Price and M. J. Greenberg, *Science* **197**, 670 (1977).
59. H. Teschemaker, *Handbook Exp. Pharmacol.* **104**(1), 471 (1993).
60. P. C. Montecucchi, R. de Castiglione, and V. Erspamer, *Int. J. Pept. Prot. Res.* **17**, 275 (1981).
61. G. Kreil and co-workers, *Eur. J. Pharmacol.* **162**, 123 (1989).
62. R. V. Lewis and co-workers, *Biochem. Biophys. Res. Commun.* **89**, 822 (1979).
63. S. Udenfriend and D. L. Kilpatrick, *Arch. Biochem. Biophys.* **221**, 309 (1983).
64. S. Kimura and co-workers, *Proc. Nat. Acad. Sci. USA* **77**, 1681 (1980).

65. U. Gubler and co-workers, *Nature* **295**, 206 (1982).
66. M. Noda and co-workers, *Nature* **295**, 202 (1982).
67. B. A. Eipper and R. E. Mains, *Endocrin. Rev.* **1**, 1 (1980).
68. S. Nakanishi and co-workers, *Nature* **278**, 423 (1979).
69. S. Horikawa and co-workers, *Nature* **306**, 611 (1983).
70. C. Rivier and co-workers, *Endocrinology* **100**, 238 (1977).
71. C. B. Pert, J. K. Chang, and B. T. W. Fong, *Science* **194**, 330 (1976).
72. D. Roemer and co-workers, *Nature* **268**, 547 (1977).
73. B. K. Handa and co-workers, *Eur. J. Pharmacol.* **70**, 531 (1981).
74. G. Gacel, M. C. Fournie-Zaluski, and B. P. Roques, *FEBS Lett.* **118**, 245 (1980).
75. H. I. Mosberg and co-workers, *Proc. Nat. Acad. Sci. USA* **80**, 5871 (1983).
76. P. W. Schiller and co-workers, *Proc. Nat. Acad. Sci. USA* **89**, 11871 (1992).
77. R. Mauer and co-workers, *Proc. Nat. Acad. Sci. USA* **79**, 4815 (1982).
78. C. J. Evans and co-workers, *Science* **258**, 1952 (1992).
79. W. Vale, J. Speiss, and J. Rivier, *Science* **213**, 1394 (1981).
80. J. Speiss and co-workers, *Proc. Nat. Acad. Sci. USA* **78**, 6517 (1981).
81. G. Aguilera and co-workers, *J. Biol. Chem.* **258**, 8039 (1983).
82. K. A. Roth, E. Weber, and J. Barchas, *Life Sci.* **31**, 1857 (1982).
83. T. Hokfelt and co-workers, *Proc. Nat. Acad. Sci. USA* **80**, 895 (1983).
84. Y. Furutani and co-workers, *Nature* **301**, 537 (1983).
85. S. Shibahara and co-workers, *EMBO J.* **5**, 775 (1983).
86. A. F. Seasholtz, R. C. Thompson, and J. O. Douglass, *Mol. Endocrinol.* **88**, 1311 (1988).
87. J. Rivier, C. Rivier, and W. Vale, *Science* **224**, 889 (1984).
88. P. Brazeau and co-workers, *Science*, **179**, 77 (1973).
89. J. M. Conlon and co-workers, *FEBS Lett.* **94**, 327 (1978).
90. O. P. Rostad, M. J. Brownstein, and J. B. Martin, *Proc. Nat. Acad. Sci. USA* **76**, 3019 (1979).
91. J. M. Lundberg and co-workers, *Proc. Nat. Acad. Sci. USA* **76**, 4079 (1979).
92. P. Davies, R. Katzman, and R. D. Terry, *Nature* **288**, 279 (1980).
93. M. N. Rossor and co-workers, *Neurosci. Lett.* **20**, 373 (1980).
94. V. Haroutunian and co-workers, *Brain Res.* **403**, 234 (1987).
95. B. D. Noe, D. J. Fletcher, and J. Spiess, *Diabetes* **28**, 724 (1979).
96. J. Spiess and W. Vale, *Metabolism* **27**, 1175 (1978).
97. L. Pradayrol and co-workers, *FEBS Lett.* **109**, 55 (1980).
98. L. P. Shen, R. L. Pictet, and W. J. Rutter, *Proc. Nat. Acad. Sci. USA* **79**, 4575 (1982).
99. G. H. Travis and J. G. Sutcliffe, *Proc. Nat. Acad. Sci. USA* **85**, 1696 (1988).
100. S. H. C. Hendry and co-workers, *Proc. Nat. Acad. Sci. USA* **81**, 6526 (1984).
101. J. Rivier, M. Brown, and W. Vale, *Biochem. Biophys. Res. Commun.* **65**, 746 (1976).
102. H. Matsuo and co-workers, *Biochem. Biophys. Res. Commun.* **43**, 1374 (1971).
103. R. Burgus and co-workers, *Proc. Nat. Acad. Sci.* **69**, 278 (1972).
104. W. Vale, *Neurosci. Res. Prog. Bull.* **16**, 521 (1979).
105. H. A. Jonas and co-workers, *Endocrinology* **96**, 384 (1975).
106. A. E. Bolton, *J. Endocrinol.* **63**, 255 (1974).
107. J. M. Gibbons, M. Mitnick, and V. Chieff, *Am. J. Obstetr. Gynecol.* **121**, 127 (1975).
108. T. M. Siler-Khodr and G. S. Khodr, *Am. J. Obstetr. Gynecol.* **130**, 216 (1978).
109. R. L. Moss, *Ann. Rev. Physiol.* **41**, 617 (1979).
110. C. Rivier, J. Rivier, and W. Vale, *Endocrinology* **103**, 2299 (1978).
111. J. Humphries and co-workers, *J. Med. Chem.* **21**, 120 (1978).
112. R. Burgus and co-workers, *Nature* **226**, 326 (1970).
113. R. M. G. Nair and co-workers, *Biochemistry* **9**, 1103 (1970).
114. J. E. Morley and co-workers, *Endocrinology* **104**, 137 (1979).

115. R. M. Lechman and co-workers, *Science* **231**, 159 (1986).
116. M. Suzuki and co-workers, *J. Med. Chem.* **33**, 2130 (1990).
117. B. Donzel and co-workers, *Nature* **256**, 750 (1975).
118. R. Carraway and S. E. Leeman, *J. Biol. Chem.* **248**, 6854 (1973).
119. G. R. Uhl, M. J. Kuhar, and S. H. Snyder, *Proc. Nat. Acad. Sci. USA* **74**, 4059, 1977.
120. J. Dolais-Kitabgi and co-workers, *Endocrinology* **105**, 256 (1979).
121. B. V. Clinesschmidt and J. C. McGuffin, *Eur. J. Pharmacol.* **46**, 395 (1977).
122. R. Quirion and co-workers, *Eur. J. Pharmacol.* **61**, 309 (1980).
123. M. G. Rosenfeld and co-workers, *Nature* **304**, 129 (1983).
124. S. J. Gibson and co-workers, *J. Neurosci.* **4**, 3101 (1984).
125. S. D. Brain and co-workers, *Nature* **313**, 54 (1985).
126. R. Oku and co-workers, *Brain Res.* **475**, 356 (1987).
127. P. Le Greves and co-workers, *Eur. J. Pharmacol.* **115**, 309 (1985).
128. S.-P. Han, L. Naes, and T. C. Westfall, *Biochem. Bipophys. Res. Commun.* **168**, 786 (1990).
129. T. Dennis and co-workers, *J. Pharmacol. Exp. Ther.* **254**, 123 (1990).
130. A. C. Ivy and co-workers, *Am. J. Physiol.* **91**, 336 (1929).
131. G. J. Dockray, *Nature* **264**, 568 (1977).
132. P. D. Marley, J. F. Rehfeld, and P. C. Emson, *J. Neurochem.* **42**, 1523 (1984).
133. U. Gubler and co-workers, *Proc. Nat. Acad. Sci. USA* **81**, 726 (1984).
134. R. J. Deschenes and co-workers, *Proc. Nat. Acad. Sci. USA* **81**, 726 (1984).
135. T. Hokfelt and co-workers, *Nature* **285**, 476 (1980).
136. R. Wang and R. Schoenfeld, eds., *Cholecystokinin Antagonists*, Alan R. Liss, Inc., New York.
137. B. E. Evans and co-authors, *Proc. Nat. Acad. Sci. USA* **83**, 4918 (1986).
138. J. Hughes and co-workers, *Proc. Nat. Acad. Sci. USA* **87**, 6728 (1990).

Steven R. Childers
Wake Forest University

SEX HORMONES

Progestins are a class of steroids (qv), named for their progestational effects, which are essential for the initiation and continuation of pregnancy. The primary progestin in humans is progesterone [57-83-0] (**1**), a hormone and intermediate in the synthetic pathways of estrogens, androgens, and corticosteroids. Progesterone is synthesized from pregnenolone [145-13-1], $C_{21}H_{32}O_2$ (**2**), which is derived from the side-chain cleavage of cholesterol [57-88-5] (**3**) (Fig. 1). This cleavage is the rate-limiting step in the biosynthesis. Pregnenolone is converted to progesterone by a two-enzyme system, ie, it is dehydrogenated to an intermediate 5-ene-3-oxosteroid by the mitochondrial enzyme 5-ene-3β-ol-dehydrogenase (EC 1.1.1.51), and this intermediate is irreversibly converted to progesterone by a 5-4 isomerase (EC 5.3.3.1) (**1**). Numerous reviews on the biosynthesis of progesterone can be found (2–4).

The primary sources of progesterone in women are the corpora lutea of the ovary and the placenta. Ovarian production of progesterone varies over the menstrual cycle. Baseline levels are 0.75–4 mg/d during the follicular phase, ie, days 1–14 of a typical 28-d cycle. After ovulation and formation of the corpus luteum,

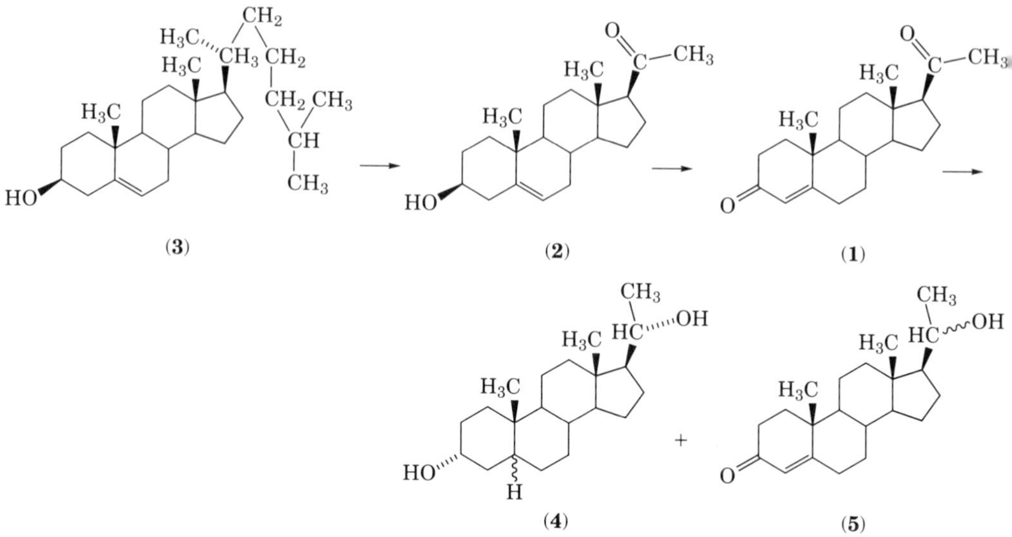

Fig. 1. Compounds involved in the biosynthesis and metabolism of progesterone (**1**).

production rises reaching peak levels of 15–50 mg/d midway through the luteal phase, ie, days 15–28 of the cycle (5–7). The plasma levels of progesterone reflect these differing rates of synthesis. Follicular levels are <0.2–2.0 ng/mL and luteal levels range from 9.5–22 ng/mL (8,9). Peak levels of plasma progesterone persist for only 4–6 days, then rapidly decline to initial baseline levels. This drop in plasma progesterone triggers the onset of menses in 1–2 days. During pregnancy, placental production of progesterone exceeds that of the ovary by nearly 10-fold and the resulting plasma progesterone concentrations reach 120–200 ng/mL near term. In men and ovariectomized women, progesterone can be detected in the plasma at concentrations of 0.3–0.4 ng/mL (5–7).

Other naturally occurring progestins such as pregnanediol (5α- [566-58-5] or 5β-pregnane-3α,20α-diol [80-92-2]) (**4**) and 20-dihydroprogesterone (20α- [145-14-2] or 20β-dihydroprogesterone [145-15-3]) (**5**) can result from progesterone metabolism in steroid-responsive tissues (Fig. 1) (10–12). These progestins have much weaker biological activity than progesterone and their physiological significance remains unclear. Progesterone has been used as a therapeutic agent, but is rapidly metabolized and has little or no oral activity and a short duration of action. In humans, it is metabolized primarily by the liver into the biologically inactive pregnanediol. Reviews of the physical characteristics, pharmacokinetics, and metabolism of progesterone can be found in several references (4,13).

All of the progestational agents in clinical use (ca 1994) are synthetic steroidal progestins. Since the discovery of ethisterone [434-03-7] (**6**), the first progestin having reasonable oral activity, an enormous number of progestins has been synthesized. As a class, progestins have been identified by their biological activity, rather than their basic chemical structure.

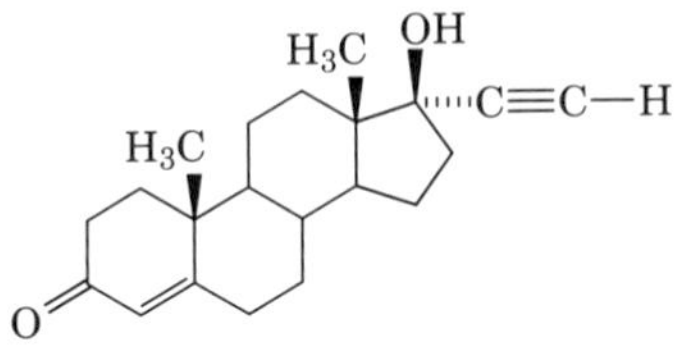

(6)

Most progestins prescribed in the early 1990s were developed a number of years prior. However, developments in the use of progestins have led to several newer progestins being brought to the marketplace (ca 1994). In addition to newer progestin agonists, potent progestin antagonists are available and may eventually replace progestin agonists for many therapeutic uses. Steroids, once given only orally or by injection, can be administered in sustained release formulations such as depot injection, vaginal silicone ring, or subcutaneous silicone rods (see CONTROLLED RELEASE TECHNOLOGY, PHARMACEUTICAL; DRUG DELIVERY SYSTEMS).

The many advances in molecular biology have aided in understanding the actions of progestins at the receptor and genomic level. The interaction of hormone and receptor is an active area of research, the results of which are the progestin agonists and antagonists being developed. As the effects of progestins at the molecular level are better understood, precise assays for progestin action have been devised. Progestin agonists and antagonists are being evaluated for the ability to stimulate the transcription of specific genes in cell culture.

Synthesis of Steroidal Agonists

Progestins are derivatives of a planar tetracyclic structure and fall into one of four structural classes: 5α-pregnanes (7), 5α-androstanes (8), estranes (9), and gonanes (10) (Fig. 2). Substituents that lie above the plane of the ring system are assigned a β-stereochemical configuration, indicated by a bold wedge. Substituents lying below the ring system are termed α-substituents and are joined to the ring system by a dashed line. Angular methyl groups at C-10 and C-13 have the β-configuration and are usually indicated by solid bonds only. When there are no methyl groups in these positions the structures are referred to as 19-nor and 18-nor compounds, respectively. Tertiary hydrogen atoms at C-5, C-8, C-9, C-14, and C-17 are generally not indicated unless their stereochemistry differs from (8) (14).

Extracts of corpora lutea were known in the early twentieth century to inhibit ovulation in animals. Pure progesterone (3), the active component of the extracts, was isolated in 1934 and its structure reported (15). Several problems limited its use and drove efforts to develop progesterone analogues, ie, it was only available in small quantities from animal sources, was not orally active, and was discovered to cause androgenic side effects.

The isolation of progesterone from animal sources has never been commercially viable, although enough has been purified to allow early pharmacological

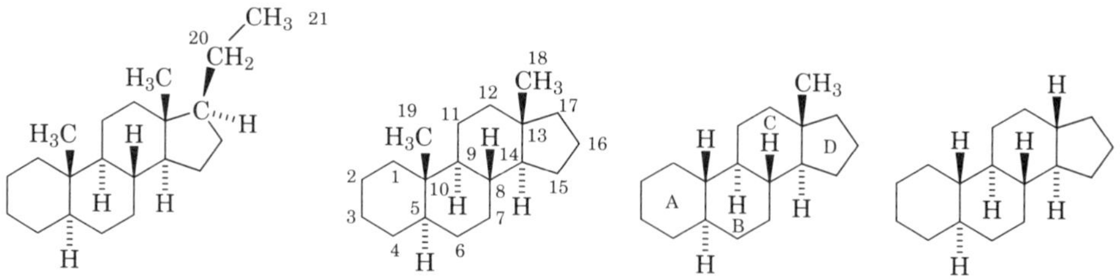

Fig. 2. The four structural classes of progesterone. Carbon atoms are numbered as shown in (**7**) and (**8**) and the rings are referred to as detailed in (**9**).

studies. Progress in improving the biological profile of progestins, therefore, depended on finding a source of steroid precursors. Three methods of obtaining steroids have been developed for the commercial preparation of progestins, ie, chemical degradation of steroids isolated from plants, microbial degradation of steroids isolated from plants, and total chemical synthesis. Advantages of using naturally occurring steroids as raw materials include availability of large quantities and presence of the required tetracyclic ring structure possessing the correct stereochemical orientation. Chemical degradation methodology was developed first. The discovery of mutant strains of bacteria that incompletely degrade steroids complemented the chemical degradation routes. Microbial degradation allowed the use of plant materials not amenable to chemical modification and the synthesis of precursors not easily produced chemically. The total synthesis of steroids from materials that do not contain the steroid nucleus allow the synthesis of steroids containing structural features not found naturally. Industrial processes (ca 1994) for the synthesis of progestins make use of all three methods (16,17). Most progestins are prepared from a few steroidal precursors.

Pregnanes. The first process to produce progesterone in quantity (15,18) used diosgenin [512-04-9] (**11**) isolated from the Mexican yam, a species of *Dioscorea*, as the starting material (Fig. 3). Treatment of (**11**) with hot acetic anhydride in the presence of a catalyst, eg, *p*-toluenesulfonic acid, provides a dihydrofuran. Oxidation of the dihydrofuran with chromium trioxide cleaves the side chain and results in the desired C-20 ketone (**12**). Treatment of this ester with acetic anhydride gives 16-dehydropregnenolone acetate [979-02-2] (**13**) and selective reduction of the conjugated ketone by catalytic hydrogenation provides pregnenolone acetate [1778-02-5] (**14**). Hydrolysis of the acetate and oxidation with aluminum tri-*tert*-butoxide, ie, Oppenauer oxidation, accompanied by *in situ* migration of the double bond, gives progesterone (**1**). Several modifications in the extraction and degradation process have appeared in the literature (19,20). Extensive surveys have been made of plant materials in an effort to find the optimal raw material for processing (18,19); diosgenin isolated from *Dioscorea composita*, barbasco root, has been found to be the preferred source (21,22).

It is possible to leave the sapogenin side chain intact while making modifications in the A- and B-rings of the steroid. Conversion of diosgenin to its

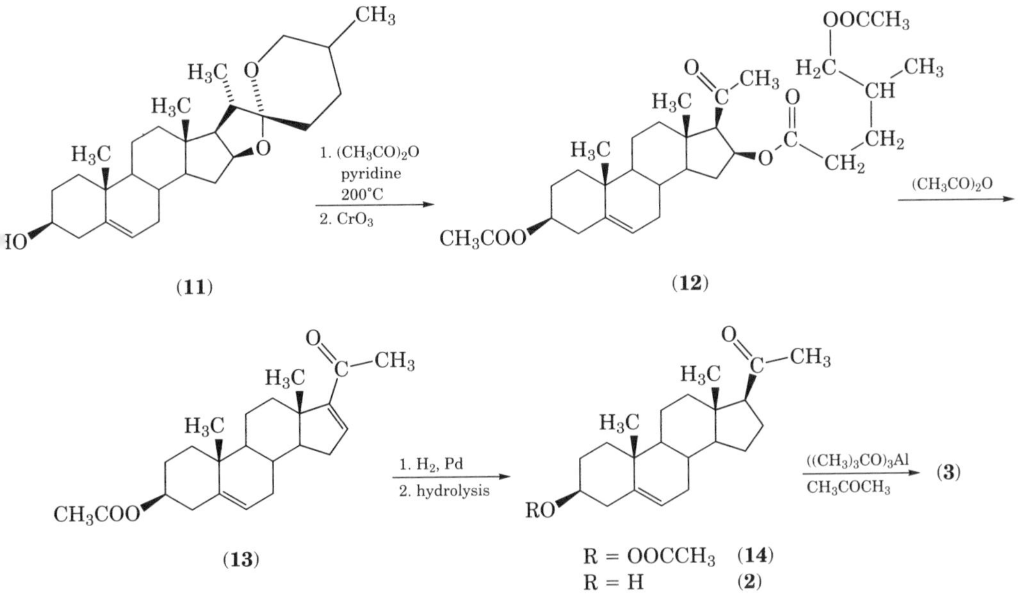

Fig. 3. First degradation process used to produce progesterone.

tosylate, followed by solvolysis, induces the *i*-steroid rearrangement. Oxidation using chromium trioxide–pyridine provides a ketone which upon reaction with methyl magnesium iodide gives a mixture of alcohols that are thus dehydrated under acidic conditions. After degradation of the side chain, the mixture is carried on to progestins, such as melengestrol acetate [2919-66-6], $C_{25}H_{32}O_4$, and medrogestone [977-79-7], $C_{23}H_{32}O_2$ (23).

At one time, the processing of diosgenin (**11**) provided most of the supply of steroid raw materials (16). The development of microbial degradation processes and total synthesis, together with a rapid rise in the price of diosgenin in the 1970s, has led to the increased use of other processes (16,24). Nevertheless, diosgenin remains an important source of steroid raw materials, particularly in countries with indigenous sources (19).

Soybeans provide another rich source of plant steroids. A process has been developed by Upjohn to isolate and purify one of the components, stigmasterol [83-48-7] (**15**) (25). Stigmasterol contains a double bond in its side chain making it amenable to chemical degradation. Treatment of stigmasterol under Oppenauer oxidation conditions results in the oxidation of the A-ring alcohol and migration of the double bond. Ozonolysis of the side chain leads to aldehyde (**16**). Formation of the enamine and further degradation through ozonolysis provides progesterone (**1**) (25).

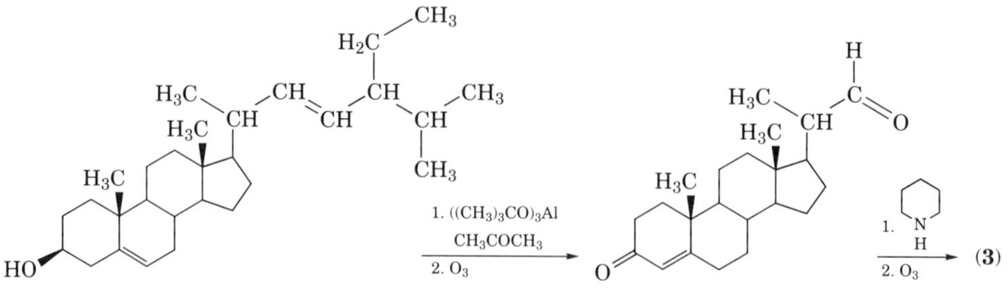

(15)   (16)   (3)

Androstanes. Androstanes, ie, pregnanes without the C-17 side chain, are important intermediates in the synthesis of progestins. One approach to the synthesis of androstanes is simply side-chain degradation of pregnanes. 16-Dehydropregnenolone acetate (**13**), an intermediate in the diosgenin to progesterone scheme, may be converted to androstenedione [63-05-8] (**17**). 16-Dehydropregnenolone acetate (**13**) is converted to its oxime. Treatment of the oxime with a Lewis acid induces a Beckmann rearrangement providing an enamide; hydrolysis and Oppenauer oxidation of the resulting alcohol forms androstenedione (**17**) (22).

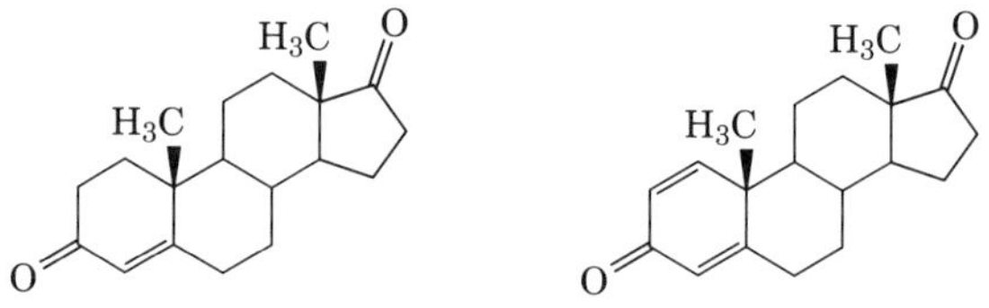

(17)   (18)

Microbial processes for transforming steroids such as cholesterol (**1**) or sitosterol, directly into androstanes are available (26). Neither steroid can be used easily as a substrate in the chemical degradation pathways owing to their relatively inert side chains. As a result sitosterol was an unused by-product of the stigmasterol (**15**) extraction from soybeans until it was discovered that it can be transformed into androstenedione (**17**) and androstadienedione [897-06-03] (**18**) by *Mycobacterium fortuitum* mutants (25–27). Cholesterol (**1**) has been degraded with *Mycobacterium phlei*, with added nickel sulfate, to androstadienedione (28). The use of microorganisms to degrade steroids continues to be examined as a source of raw materials (29).

Methods for reconstructing the C-17 side chain from intermediates such as androstendione have been developed. These are used when commercial considerations favor the production of androstanes, but pregnanes are the ultimate goal (25,30).

Estranes. Estranes lack the C-19 methyl group found in pregnanes and androstanes and are commonly referred to as 19-norsteroids. As a class,

they represent many of the oral contraceptives on the market as of 1994 (see
CONTRACEPTIVES). Interest in 19-norsteroids resulted from the observation that
19-norprogesterone [472-54-8] (**19**) possesses high oral progestational activity
(31). Small quantities of norprogesterone were produced from strophanthidin
[66-28-4], $C_{23}H_{32}O_6$, through a complex multistep procedure. The final oily product probably does not possess the same stereochemistry as progesterone at C-14
and C-17; it is known that small changes in steroid structure often result in
loss of activity (32). The Birch reduction, a method developed for the reduction of aromatic rings to dihydrobenzenes, provides an improved route to 19-
norprogesterone (33). Application of the Birch reduction allows the synthesis of
stereochemically pure, crystalline 19-norprogesterone in sufficient quantities to
be fully evaluated (32). It has proven to be four to eight times as biologically
active as progesterone (**1**) (34).

(**19**)   (**20**)

Interest in 19-norsteroids, and the expedient route provided to them by
the Birch reduction, has led to the need for A-ring aromatic steroids. However,
none of the plant steroids available in commercial quantities contained aromatic
A-rings (16). The principal difficulty in transforming available plant steroids
into aromatic A-rings is the presence of the angular methyl group at the A–B
ring junction. The methyl group is unactivated, making its removal relatively
difficult. *In vivo* removal is accomplished enzymatically and is a key step in
the biosynthetic conversion of androgens to estrogens (35). Loss of the methyl
group and subsequent aromatization of the A-ring through chemical methods has
been accomplished (36). Passing a solution of androstadienedione (**18**) in mineral
oil through a hot tube at approximately 500°C yields estrone [53-16-7] (**20**) in
reasonable amounts (37).

Several alternative methods followed this early work. In one, aromatization is effected by treating the ketal of androstadienedione with the
radical anion obtained from lithium and diphenyl in refluxing tetrahydrofuran. Diphenylmethane is added to quench the methyllithium produced from
the departing methyl group. An acidic work-up assures hydrolysis of the ketal,
providing estrone (**20**) in good yield (38). In another method, elimination of the
unactivated 19-methyl group is accomplished via intramolecular functionalization of C-19 (39,40). Figure 4 illustrates one route applicable to a wide variety
of 5-ene-3-ols. The acetate (**21**) is converted to bromohydrin (**22**). Irradiation of
(**22**) in the presence of lead tetraacetate, iodine, and calcium carbonate provides
the C-19 functionalized cyclic ether (**23**). Treatment of (**23**) with zinc in acetic

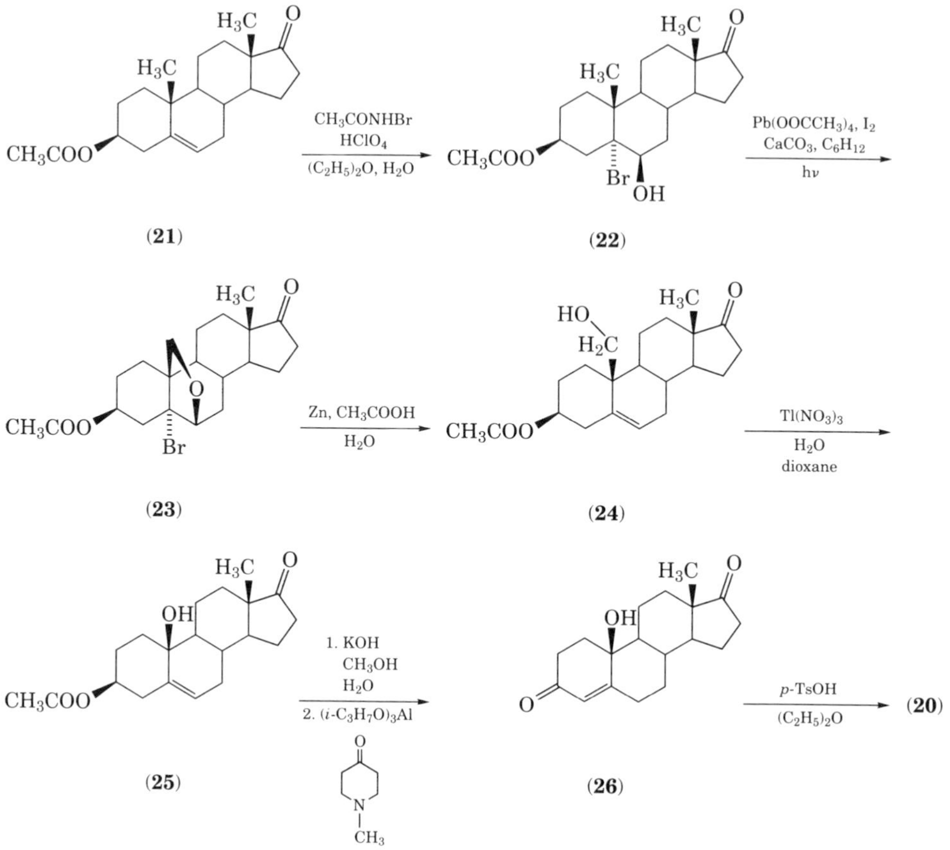

Fig. 4. Synthesis of estrone (**20**) through elimination of unactivated 19-methyl group via intramolecular functionalization of C-19. p-TsOH = p-toluenesulfonic acid.

acid/water provides the 19-hydroxy derivative (**24**). This intermediate can be converted to a 19-nor derivative in several ways (40–42). In one method, the conversion of (**24**) to estrone (**20**) is accomplished by treatment of (**24**) with thallium nitrate to provide the acetate (**25**), which is saponified and oxidized to (**26**) using modified Oppenauer conditions (43). Enone (**26**) is treated with p-toluenesulfonic acid to give estrone (**20**) (42).

The preparation of estranes via 19-hydroxy steroids has also been accomplished microbially. 19-Hydroxycholesterol, prepared chemically (39), is incubated with *Nicardia restrictus* to provide estrone (**20**) directly (44). Similarly, 19-hydroxyandrost-4-en-3,17-dione is converted to estrone (**20**) (45).

A direct, one-step aromatization of 19-substituted steroids has appeared in the literature, ie, cholesterol (**1**), dehydroisoandrosterone, androsterone, progesterone (**3**), and testosterone react with an electrophilic ruthenium complex, Cp*Ru$^+$, where Cp* represents η^5-cyclopentadienyl, obtained through protonation of [Cp*Ru(OCH$_3$)]$_2$ using triflic acid, to provide estrone (**20**) directly (46).

Gonanes. Progress at dissociating progestagenic from androgenic activity was made with the discovery of 13β-ethyl substituted gonane derivatives, originally prepared by total synthesis. Methodology has since developed that allows these compounds to be prepared from the natural product intermediates discussed in the preceding sections. Compounds of this type include levonorgestrel [797-63-7], norgestrel [6533-00-2], gestodene [60282-87-3], desogestrel [54024-22-5], and norgestimate [35189-28-7].

17-Ethynyl Steroids. Although 17-ethynyl steroids do not represent a separate skeletal class, they are significant in the development of an orally active progestin. Progress toward orally active progestins has been spurred by the observation that ethynyl groups, introduced at the 17α-position of estradiol and testosterone, result in orally active compounds (47). Ethynylestradiol [57-63-6] (**27**) is the estrogenic component of most oral contraceptives (ca 1994).

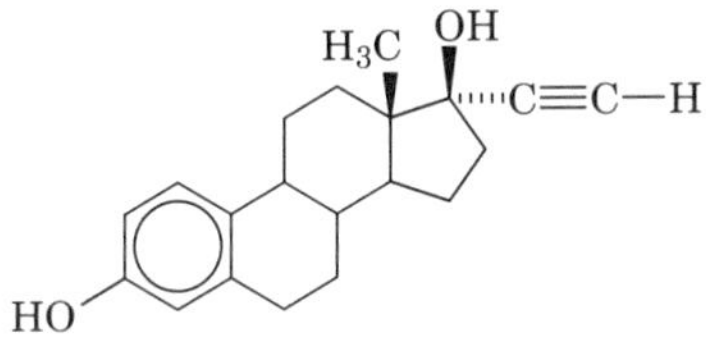

(**27**)

Ethynyltestosterone (ethisterone) (**6**) was the first marketed orally active progestin and has been used to treat gynecological disorders. Introducing an ethynyl substitution into a 19-norsteroid has led to the synthesis of norethindrone [68-22-4] (**28**) and norethynodrel [68-23-5] (**29**), the first two progestins used as oral contraceptives (32,48). Preparation of norethindrone is outlined in Figure 5 (32). Treatment of the monomethyl ether of estradiol (**30**) to Birch reduction conditions, followed by hydrolytic work-up, provides 19-nortestosterone (**31**). Chromium trioxide oxidation affords enone (**32**). Reaction of (**32**) with ethyl orthoformate, under controlled conditions, results in the selective protection of the A-ring ketone (**33**). Addition of the ethynyl group, followed by cleavage of the enol ether with aqueous mineral acid gives norethindrone (**28**). The route to norethynodrel (**29**) is shown in Figure 6. Estrone (**20**) is treated with dimethylsulfate to form the methyl ether, and with lithium aluminum hydride to reduce the C-17 ketone. The product (**34**) undergoes a Birch reduction to provide (**35**). The enol ether (**35**) is oxidized to (**36**). The enol ether formed in the Birch reduction (49) is used as the protecting group during the ethynylation. Subsequent mild hydrolysis maintains the double bond out of conjugation to provide norethynodrel (**29**) (48). Under more vigorous hydrolytic conditions, ie, mineral acids, the double bond migrates to provide norethindrone (**28**) (32).

Agonists and Antiprogestins

Steroidal Agonists. A comprehensive list of steroids possessing progestational activity has been compiled (50). Many drugs were originally prepared from optically active natural products and contain one enantiomer. In general they are stable, white to off-white crystalline solids. Because of possible instability

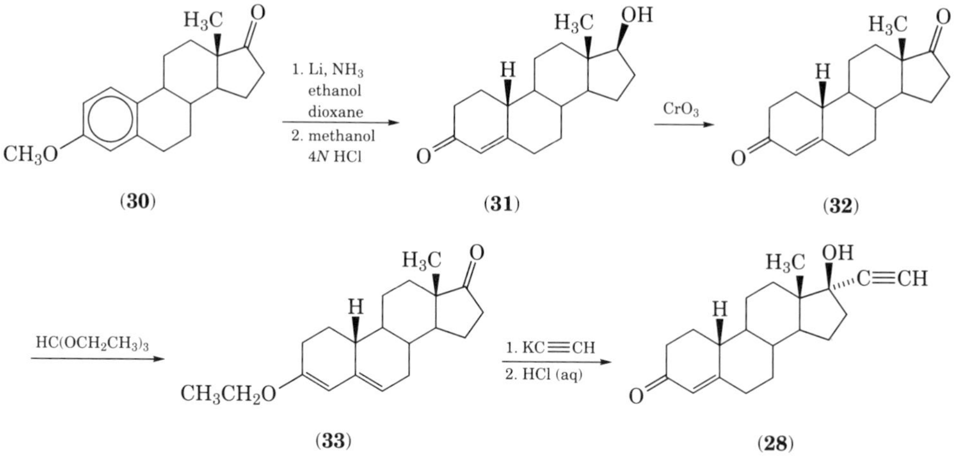

Fig. 5.   Preparation of norethindrone (28).

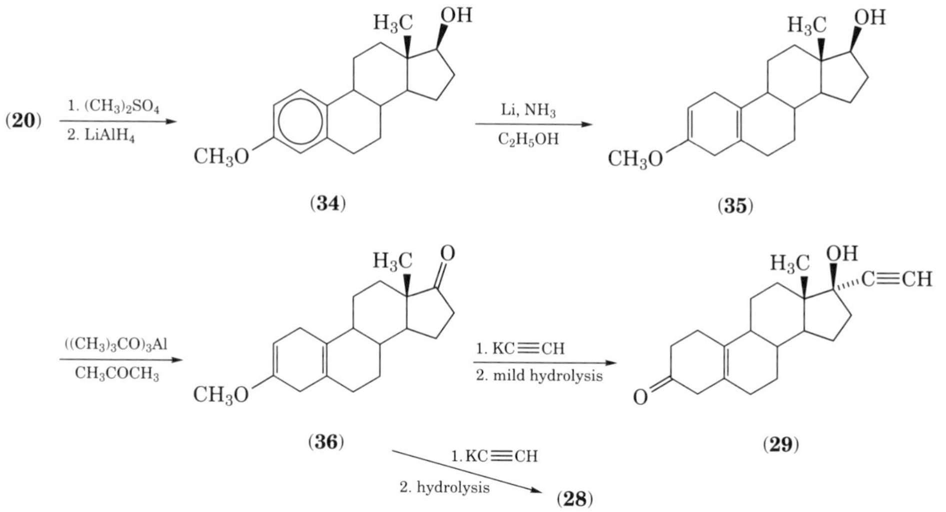

Fig. 6.   Preparation of norethynodrel (29).

of the A-ring, it has been recommended that many of them be stored protected from light (51). Table 1 lists data for a variety of steroidal progestin agonists. A list of progestins used in the United States and Europe is given in Table 2. General references for the analysis of steroids are also available (54).

   Allylestrenol.   Allylestrenol (37), which has been used to treat cases of habitual abortion (55), can be recrystallized from ether/petroleum ether (56). It is soluble in acetone, ethanol, ether, and chloroform and practically insoluble in water (57). The uv and ir spectra have been reported (58). Allylestrenol is sensitive to oxidizing agents (57).

   Allylestrenol (37) is prepared from (32), an intermediate in the synthesis of norethindrone. Treatment of (32) with ethanedithiol and catalytic boron

Table 1. Physical Properties of Progestins[a]

Common name	CAS Registry Number	Structure number	Molecular formula	Mol wt	Melting point, °C	$[\alpha]_D^b$
allylestrenol	[432-60-0]	(37)	$C_{21}H_{32}O$	300.483	79.5–80	+39[c]
chlormadinone acetate	[302-22-7]		$C_{23}H_{29}ClO_4$	404.932	211–212	+8
cyproterone acetate	[427-51-0]	(41)	$C_{24}H_{29}ClO_4$	416.943	200–201	
desogestrel	[54024-22-5]	(44)	$C_{22}H_{30}O$	310.478	109–110	+55
ethynodiol diacetate	[297-76-7]	(53)	$C_{24}H_{32}O_4$	384.514	129–132, 126–127	−72.5
gestodene	[60282-87-3]	(54)	$C_{21}H_{26}O_2$	310.435	198	−185.7
hydroxyprogesterone	[68-96-2]	(39)	$C_{21}H_{30}O_3$	330.466	221	+97
hydroxyprogesterone acetate	[17308-02-0]	(42)	$C_{23}H_{32}O_4$	372.503	246.5	+76.4
hydroxyprogesterone caproate	[630-56-8]	(59)	$C_{27}H_{40}O_4$	428.611	119–121	+61
levonorgestrel	[797-63-7]	(−)-(65)	$C_{21}H_{28}O_2$	312.451	238–242	−32.4
lynestrenol	[52-76-6]	(73)	$C_{20}H_{28}O$	284.441	162–164	−13
medroxyprogesterone acetate	[71-58-9]	(74)	$C_{24}H_{34}O_4$	386.530	205–209	+66
megestrol acetate	[595-33-5]	(79)	$C_{24}H_{32}O_4$	384.514	214–216	+5
norgestrel	[6533-00-2]	(65)	$C_{21}H_{28}O_2$	312.451	205–207	
norethindrone	[68-22-4]	(28)	$C_{20}H_{26}O_2$	298.424	203–204	−31.7
norethindrone acetate	[51-98-9]	(80)	$C_{22}H_{28}O_3$	340.461	161–163	−33
norgestimate	[35189-28-7]	(82)	$C_{23}H_{31}NO_3$	369.25	214–218	+41
progesterone	[57-83-0]	(3)	$C_{21}H_{30}O_2$	314.476	121–122, 127–131	+192
promegestone	[34184-77-5]	(83)	$C_{22}H_{30}O_2$	326.478	152	−262[d]

[a] Ref. 50, unless noted.
[b] $[\alpha]_D$ = specific optical rotation for the sodium D line.
[c] Ref. 52.
[d] Ref. 53.

Table 2. Progestins Marketed in the United States and Europe for Contraceptive and Noncontraceptive Uses

Progestin	CAS Registry Number	Primary use[a] C	Primary use[a] NC	Progestin	CAS Registry Number	Primary use[a] C	Primary use[a] NC
algestone acetophenide	[595-77-7] [24351-94-3]	+		lynestrenol	[52-76-6]	+	+
allylestrenol	[432-60-0]		+	medrogestone	[977-79-7]		+
chlormadinone	[1961-77-9]	+	+	medroxyprogesterone acetate	[71-58-9]	+	+
chlormadinone acetate	[302-22-7]	+	+	megestrol acetate	[595-33-5]		+
cyproterone acetate	[427-51-0]		+	milbolerone			+
demegestone	[10116-22-0]	+		nomegestrol acetate			+
desogestrel	[54024-22-5]	+		norethindrone	[68-22-4]	+	+
dydrogesterone	[152-62-5]		+	norethindrone acetate	[51-98-9]	+	+
ethisterone	[434-03-7]		+	norethindrone enanthate	[3836-23-5]	+	+
ethynodiol diacetate	[297-76-7]	+		norethynodrel	[68-23-5]	+	
gestodene	[60282-87-3]	+		norgestimate	[35189-28-7]	+	+
gestrinone	[16320-04-0]		+	norgestrel	[6533-00-2]	+	
gestronol hexanoate	[1253-28-7]		+	norgestrienone	[848-21-5]	+	+
hydroxyprogesterone hexanoate	[630-56-8]		+	progesterone	[57-83-0]	+	+
levonorgestrel	[797-63-7]	+	+	promegestone	[34184-77-5]	+	+

[a]C = contraceptive; NC = noncontraceptive.

444

trifluoride provides a thioketal. Reduction with sodium in liquid ammonia results in the desired reductive elimination of the thioketal along with reduction of the 17-keto group. Oxidation of this alcohol with chromic acid in acetone followed by addition of allyl magnesium bromide, completes the synthesis (52).

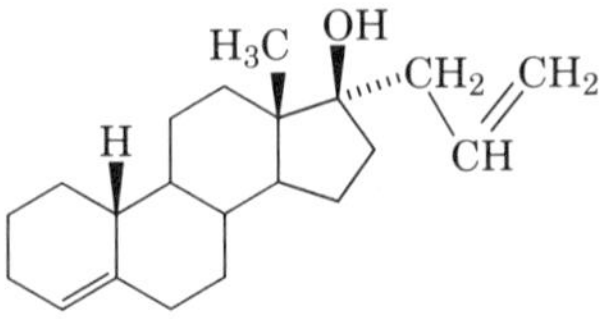

(37)

Chlormadinone Acetate. Chlormadinone acetate is used in combination with ethynylestradiol or mestranol [72-33-3], $C_{21}H_{26}O_2$, in the treatment of menstrual disorders and as an oral contraceptive. It has been reported to be slightly estrogenic (59). It may be recrystallized from acetone/ether (60), methanol, or ether (61) and is soluble in chloroform, benzene, acetone, ethyl acetate, and dimethylformamide, and insoluble in water and hexane (62). The uv and ir spectrum for chlormadinone acetate, along with other chromatographic data, have been summarized (62).

Chlormadinone (**38**) is prepared from hydroxyprogesterone (**39**) by epoxidation and treatment of the epoxide with hydrochloric acid to provide the chloroalkene (**40**). Oxidation, ie, dehydrogenation, with chloranil, provides chlormadinone (**38**), which may be acetylated to provide chlormadinone acetate (63,64).

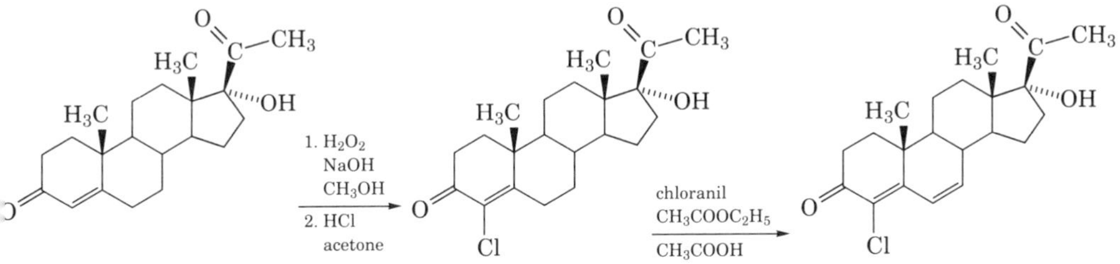

(39)   (40)   (38)

Cyproterone Acetate. Cyproterone, the free alcohol of cyproterone acetate, is an antiandrogen. Cyproterone acetate (**41**) is both an antiandrogen and a progestin. It is used in prostatic carcinoma treatment and for the control of libido. In females, it is used in conjunction with ethynylestradiol for the control of acne and hirsutism (65). Cyproterone acetate may be recrystallized from diisopropyl ether (66). It may be synthesized from 17-acetoxyprogesterone (**42**). The triene (67) reacts with diazomethane to give a pyrazoline; pyrolysis, to effect loss of nitrogen, provides a cyclopropane (**43**); and oxidation with perbenzoic acid leads to an epoxide. Treatment with hydrochloric acid results in both the desired opening of the epoxide and the undesired cleavage of the cyclopropane. The cyclopropane is re-formed through treatment with collidine to afford cyproterone acetate (**41**) (68–70).

(**41**)   (**42**)   (**43**)

Desogestrel.   Desogestrel is used in oral contraceptives. It is a white powder and can be recrystallized from pentane (71). It is practically insoluble in water and its crystal structure has been reported (72).

The structural features that distinguish desogestrel (**44**) are the absence of the 3-keto group, the presence of an 11-alkylidene, and the 13-ethyl substitution. Figure 7 outlines a route to desogestrel (**44**) that addresses the synthesis of these features. In this synthesis, the key step is the intramolecular oxidation of bis-ketal (**45**) to the lactone (**46**). This allows the transformation of a naturally occurring methyl group into the synthetically derived ethyl group. Treatment of (**45**) with lead tetraacetate, iodine, and azobisisobutyronitrile [78-67-1] (AIBN) as the radical initiator provides the lactone (**46**). Grignard reaction with methyl magnesium bromide gives a nearly quantitative yield of a solution phase equilibrium of (**47**) and (**48**). Wolff-Kishner reduction of the mixture gives (**49**). Oxidation with chromium trioxide/pyridine and Wittig reaction of the resulting 11-ketone provides (**50**). Removal of the oxo-ketals and selective protection of the 3-ketone of (**51**) as a thioketal yields (**52**). The synthesis of desogestrel (**44**) is completed by reaction with potassium acetylide and reductive removal, using Na/NH$_3$ of the thioketal (73). The introduction of 11-alkylidene groups in various steroid skeletons has been discussed (17,74). An alternative to the thioketalization/reduction step has been described; ie, 17-β-hydroxy-4-androsten-3-ones are reduced to the corresponding 17-β-hydroxy-4-androstenes by treatment with sodium borohydride and trifluoroacetic or trichloroacetic acid (75).

Ethynodiol Diacetate.   Ethynodiol diacetate has been used alone, and in combination with an estrogen, as an oral contraceptive and to treat disorders associated with progesterone deficiency (76). It may be crystallized from aqueous methanol (77) and is soluble in chloroform, ether, and ethanol; sparingly soluble in fixed oils; and insoluble in water (76). Extensive spectral and chromatographic data have been compiled (78).

Ethynodiol diacetate (**53**) is prepared by reduction of the 3-oxo group of norethindrone (**28**) with lithium tributoxyaluminum hydride, followed by acylation with acetic anhydride–pyridine (78,79). It has been reported that higher yields can be obtained in the reduction step by using triethylaminoaluminum hydride (80).

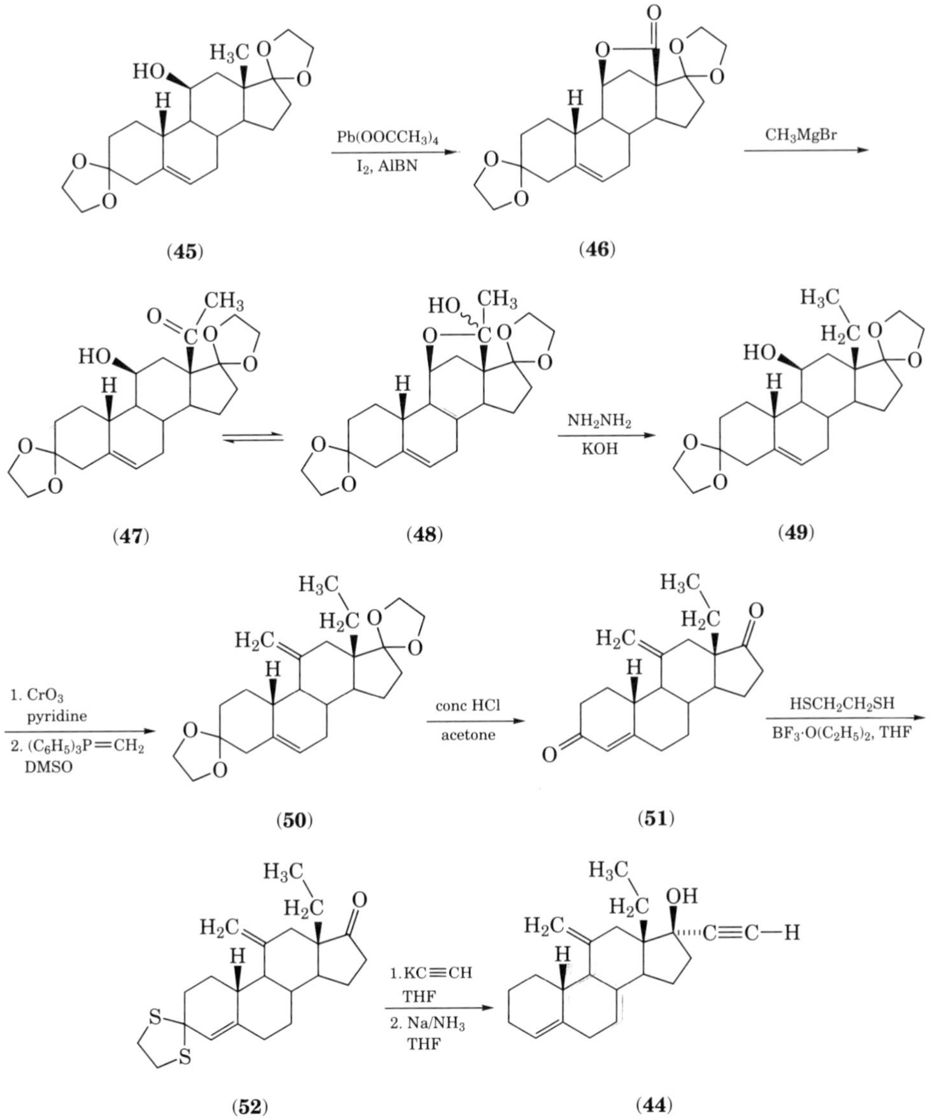

Fig. 7. Synthesis of desogestrel (**44**).

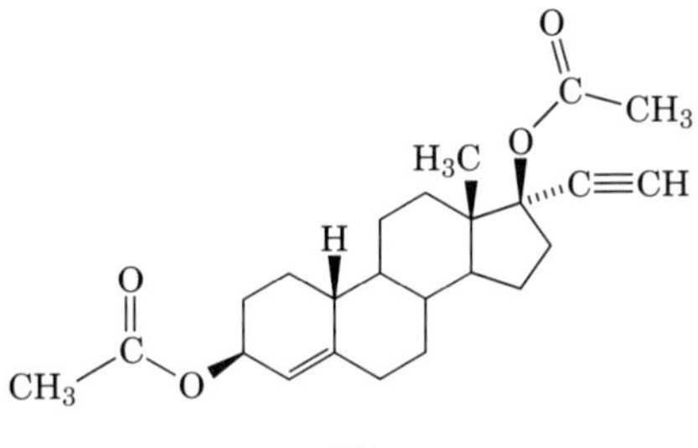

447

Gestodene.   Gestodene (**54**), along with norgestimate and desogestrel, are the progestin components of the third-generation oral contraceptives (see CONTRACEPTIVES). It may be crystallized from hexane/acetone (81) or ethyl acetate (82), and its crystal structure (83) and other spectral data have been reported (84).

Gestodene has been prepared in several ways (85). The route that provides the highest yield is shown in Figure 8. Microbial oxidation of (**55**) with *Penicillium raistrickii* results in the 15-alcohol (**56**). Protection of the alcohol as the acetate (**57**) and protection of the ketone as a dienolether provides (**58**). In a one-pot procedure (**58**) is treated with lithium acetylide and subjected to a hydrolytic work-up to provide gestodene (**54**) (86).

Hydroxyprogesterone Acetate and Caproate.   Hydroxyprogesterone (**39**) itself does not possess progestational activity, but its esters are active progestins. Hydroxyprogesterone caproate (**59**) is used mainly to treat threatened or habitual abortion (87). Although other esters are known (88), the caproate is the most widely used. The ester group of hydroxyprogesterone caproate is resistant to hydrolysis (89).

Hydroxyprogesterone is a colorless crystalline solid which may be recrystallized from acetone or ethanol (90) and is practically insoluble in ethanol and water (87). Hydroxyprogesterone acetate can be recrystallized from chloroform/methanol (90). Hydroxyprogesterone caproate, ie, hexanoate, can be recrystallized from isopropyl ether or methanol (90) and is soluble in chloroform, ethanol, ether, benzene, and fixed oils, and insoluble in water (91). An extensive profile of hydroxyprogesterone caproate, including spectral and analytical data, has been compiled (89). The crystal structures of hydroxyprogesterone and hydroxyprogesterone acetate have been discussed (92).

Hydroxyprogesterone (**39**) has been prepared from (**13**). Treatment of (**13**) in alkaline hydrogen peroxide provided an epoxide. Opening of the epoxide is accomplished by reaction with hydrogen bromide in acetic acid. Reductive

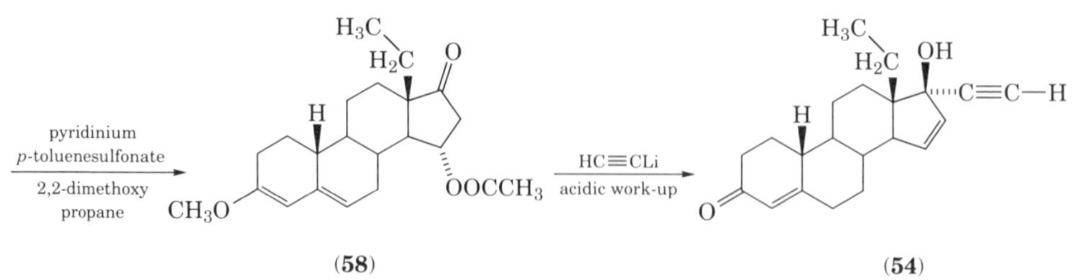

Fig. 8.   Preparation of gestodene (**54**).

removal of the bromide is performed in the presence of ammonium acetate, which prevents reduction of the double bond. The reaction mixture is heated with formic acid to provide the formate which is acylated. Subsequent Oppenauer oxidation (43) provides hydroxyprogesterone acetate (42). Saponification of the acetate with potassium hydroxide in methanol provides hydroxyprogesterone (39). This method is efficient in that direct oxidation of the formate avoids a deprotection step (93). Hydroxyprogesterone acetate (42) and hydroxyprogesterone caproate (59) also have been prepared from hydroxyprogesterone (39) (94).

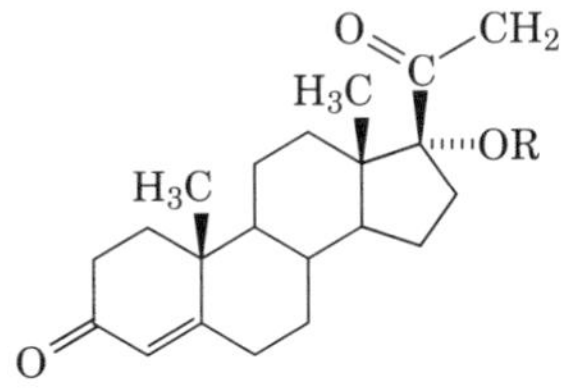

$$R = H \qquad\qquad (39)$$
$$R = COCH_3 \qquad\quad (42)$$
$$R = CO(CH_2)_4CH_3 \quad (59)$$

Levonorgestrel and Norgestrel. Both of these compounds are used in oral contraceptives. Norgestrel (65) is a racemic mixture; the biologically active isomer is known as levonorgestrel or (−)-norgestrel (65). Norgestrel can be recrystallized from ethyl acetate or methanol (95), and levonorgestrel can be recrystallized from chloroform–methanol (95). Both compounds are soluble in chloroform, slightly soluble in ethanol, and practically insoluble in water (96). Extensive physical, spectral, and analytical properties have been compiled (97).

Compound (60) in Figure 9, obtained by the reaction of 6-methoxytetralone and vinyl magnesium bromide, is condensed with 2-ethyl-1,3-cyclopentanedione (61). Cyclization of (62) to (63) is effected with hydrochloric acid in ethanol. Reduction of the D-ring double bond, 17-keto group, and finally the B-ring double bond provides (64). Birch reduction and Oppenauer oxidation, followed by addition of acetylide, completes the synthesis of norgestrel (65) (98). Early syntheses of levonorgestrel relied on chiral separation to obtain optically active material (98). Later, chiral syntheses were developed (99,100). One chiral synthesis is shown in Figure 9 (85). The optical activity is obtained through an asymmetric aldol and subsequent dehydration of the indanedione (66) (101). The chiral hydrindane (67) reacts with paraformaldehyde and benzenesulfinic acid to provide (68), which is hydrogenated to form sulfone (69). Sulfone reacts with the anion formed from (70) to provide the crude product (71), which is cyclized, hydrolyzed, and decarboxylated, resulting in (72). Hydrogenation, followed by treatment with acid to effect ketal cleavage and cyclization, leads to (55), which can be further converted to levonorgestrel (−)-(65) by ethynylation (86).

Lynestrenol. Lynestrenol (73) has been used in oral contraceptives and to treat menstrual disorders. It is converted *in vivo* to its active metabolite norethindrone (102,103). It can be recrystallized from methanol, and is soluble in ethanol, ether, chloroform, and acetone, and insoluble in water (102). The crystal structure (104) and other spectral and analytical data have been reported for lynestrenol (62).

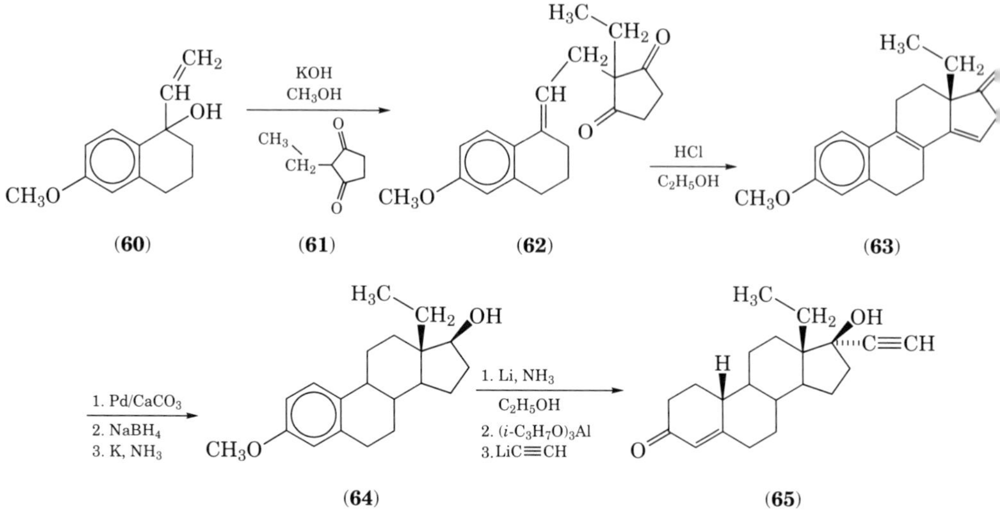

(a)

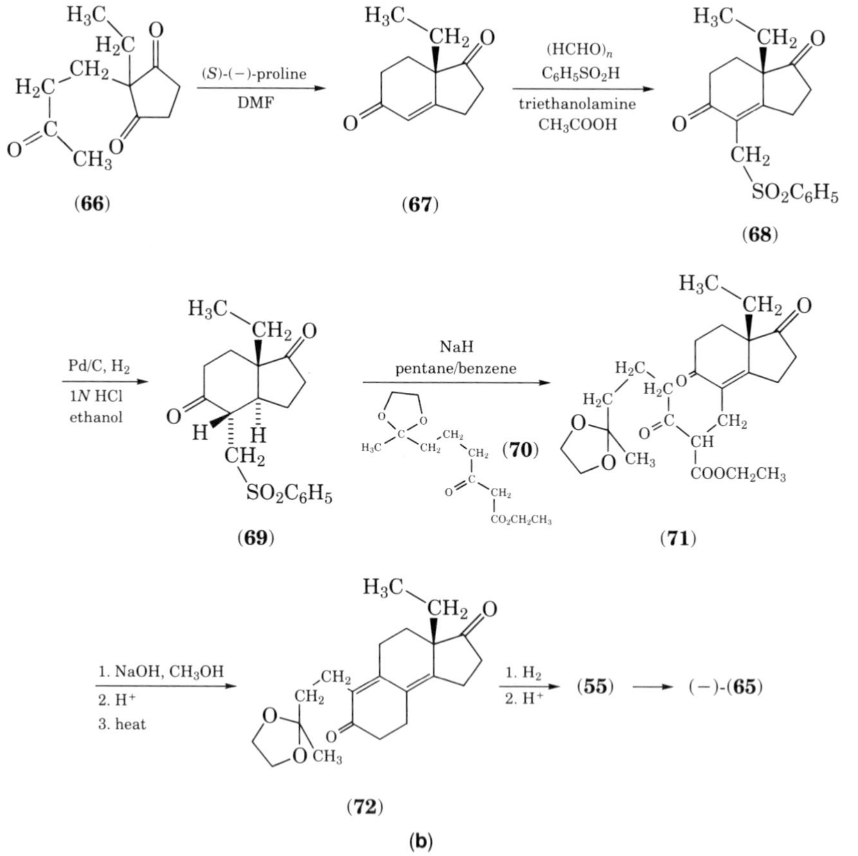

(b)

Fig. 9. Synthesis of (**a**) norgestrel and (**b**) levonorgestrel (−)-(**65**).

Lynestrenol is the des-3-oxo derivative of norethindrone (**28**). It has been prepared through a similar synthetic pathway as allylestrenol (**37**) (52), ie, addition of potassium acetylide, rather than allyl magnesium bromide, affords lynestrenol (**73**). Lynestrenol is also available from norethindrone (**28**). Reduction of the 3-keto group is accomplished by treating norethindrone (**28**) with sodium borohydride in the presence of trifluoro- or trichloroacetic acid (75).

(**73**)

Medroxyprogesterone Acetate. Medroxyprogesterone acetate (**74**), ie, Depo-Provera (Upjohn), is used as an injectable, long-acting contraceptive. It is also used in the treatment of menstrual disorders, endometriosis, and in the palliative treatment of hormone-responsive malignant neoplasms. Medroxyprogesterone acetate can be recrystallized from methanol (105). It is soluble in chloroform, acetone, and dioxane; sparingly soluble in ethanol and methanol; slightly soluble in ether; and insoluble in water (102). Its crystal structure (92) and other spectral and analytical data have been reported (62).

Medroxyprogesterone acetate (**74**) is structurally related to and has been prepared from hydroxyprogesterone (**39**) (Fig. 10). Formation of the bis-ketal accomplishes the protection of the ketones and the required migration of the double bond. Epoxidation with peracetic acid produces a mixture of epoxides (**75**), with α predominating. Treatment of the α-epoxide with methyl magnesium bromide results in diaxial opening of the epoxide. Deprotection of the ketones provides (**76**), which is dehydrated to (**77**) by treatment with dilute sodium hydroxide in pyridine. Upon treatment with gaseous hydrochloric acid in chloroform, the 6β-methyl in (**77**) is epimerized to the more stable 6α-methyl (**78**). Acylation of (**78**) completes the synthesis of medroxyprogesterone acetate (**74**) (106).

Megestrol Acetate. This compound is used outside the United States as an oral contraceptive. In the United States, it is used for the palliative treatment of breast cancer and endometrial cancer, or as an adjunct to other therapies. Its use has been associated with an increased appetite and food intake and has been evaluated in the treatment of anorexia and cachexia (107).

Megestrol acetate can be recrystallized from aqueous methanol (108). It is soluble in acetone, chloroform, and ethanol; slightly soluble in ether and fixed oils; and insoluble in water (107). Additional spectral and physical data have been published (62).

Megestrol acetate (**79**) is structurally related to progesterone (**1**). It has been prepared from medroxyprogesterone acetate (**74**) by chloranil-mediated dehydrogenation. It also has been prepared from hydroxyprogesterone acetate (**42**) via 6-methylenation and double-bond migration (109,110).

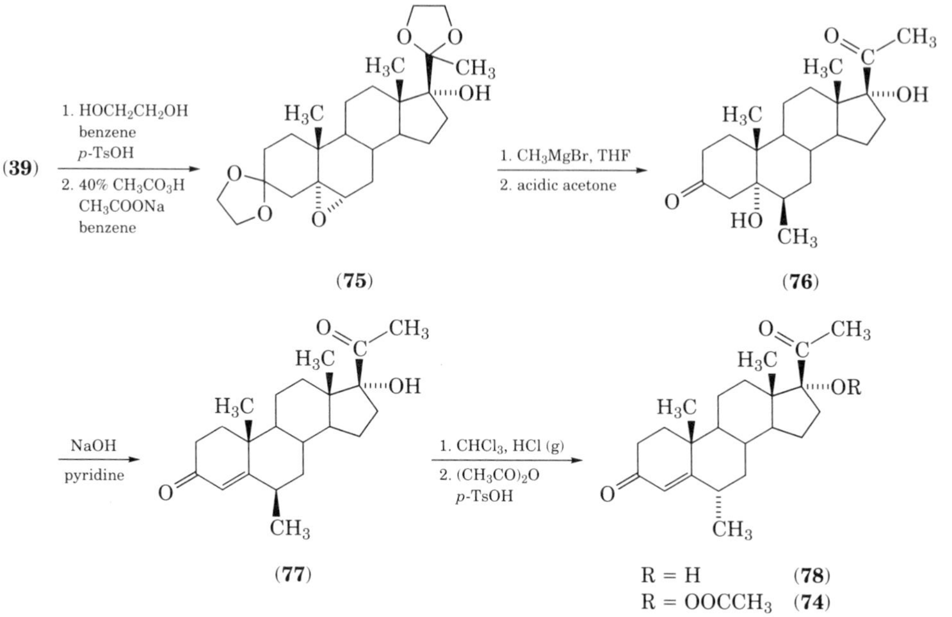

Fig. 10. Preparation of medroxyprogesterone acetate (**74**) and medroxyprogesterone (**78**) from hydroxyprogesterone (**39**). p-TsOH = p-toluenesulfonic acid.

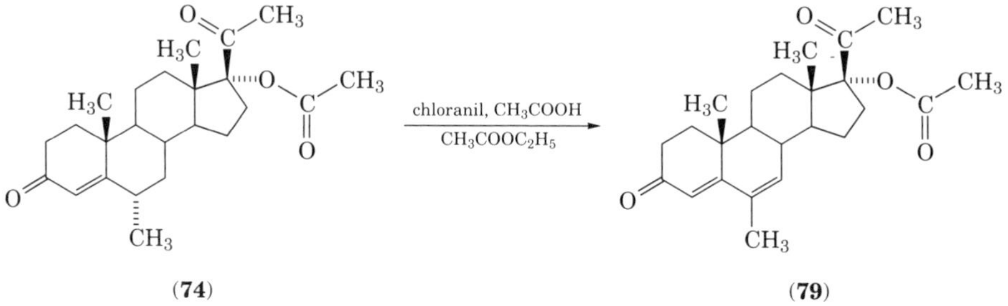

Norethindrone and Norethindrone Acetate. Norethindrone (**28**) and its acetate or enanthate are used in oral contraceptives. Norethindrone may be recrystallized from ethyl acetate (111). It is soluble in acetone, chloroform, dioxane, ethanol, and pyridine; slightly soluble in ether, and insoluble in water (112,113). Its crystal structure has been reported (114), and extensive analytical and spectral data have been compiled (115). Norethindrone acetate can be recrystallized from methylene chloride/hexane (111). It is soluble in acetone, chloroform, dioxane, ethanol, and ether, and insoluble in water (112). Data for identification have been reported (113). The preparation of norethindrone (**28**) has been described (see Fig. 5). Norethindrone acetate (**80**) is prepared by the acylation of norethindrone. Norethindrone esters has been described; ie, norethindrone, an appropriate acid, and trifluoroacetic anhydride have been shown to provide a wide variety of norethindrone esters including the acetate (**80**) and enanthate (**81**) (116).

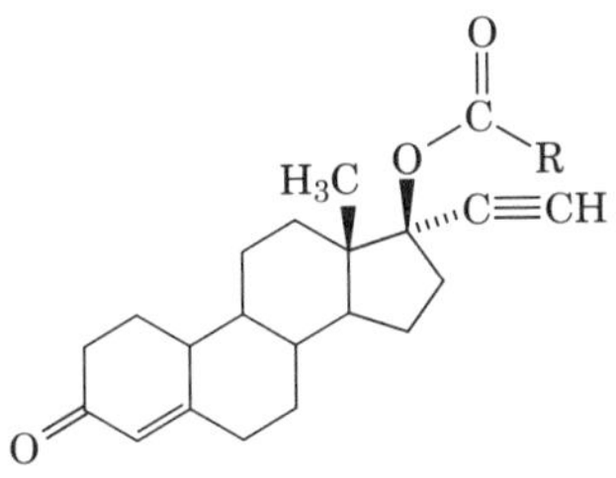

R = CH₃　　　　(80)
R = (CH₂)₅CH₃　(81)

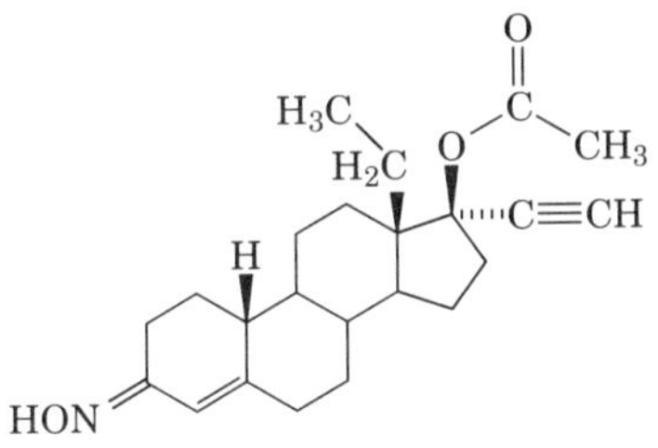

(82)

Norgestimate. Norgestimate (**82**) is used in oral contraceptives. It can be recrystallized from methylene chloride (117) and is the 17-acetoxy-3-oxime derivative of levonorgestrel. It may be prepared from levonorgestrel by bis-acylation, hydrolysis of the enol acetate, and oximation (118,119). In an alternative route, the A-ring is formed through cyclization to provide an enamine. Hydrolysis and formation of the oxime affords norgestimate (**82**) (120).

Progesterone. Progesterone (**3**) is not orally active. Although seldom used clinically, it can be administered as an intramuscular injection, pessaries, or suppositories in the treatment of menstrual disorders and habitual abortion (121). Progesterone can be recrystallized from dilute alcohol and exists in two crystalline forms (122). It is soluble in chloroform and ethanol; sparingly soluble in acetone, dioxane, ether, and fixed oils; and practically insoluble in water (121). Two syntheses of progesterone (**3**) are described in Figure 3.

Promegestone. Promegestone (**83**) or R5020, is an antineoplastic agent and is used in the treatment of menstrual disorders. Tritiated R5020 is widely used as a radioligand in progestin-binding studies (see RADIOACTIVE TRACERS) (123). Promegestone can be recrystallized from isopropyl ether, and is soluble in acetone and benzene and insoluble in water (53). In the synthesis of promegestone (**83**) (124), an enone is reduced with lithium in ammonia and the intermediate enolate is quenched with methyl iodide (125). The resulting compound is then subjected to Birch reduction and the intermediate enol ether is hydrolyzed under mild conditions to maintain the double bond out of conjugation. The resulting alcohol is oxidized, dehydrohalogenated, selectively protected then alkylated and hydrolyzed, to afford promegestone (**83**) (124–127).

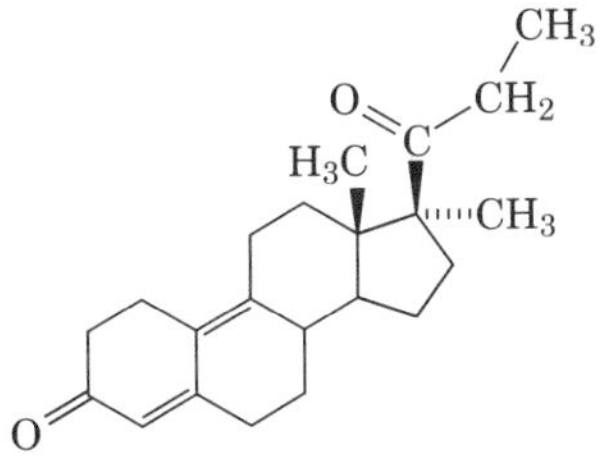

(**83**)

Steroidal Antiprogestins.   A significant breakthrough in the progestin antagonist area was the discovery of the 11β-substituted progestin antagonists. The first of these compounds to be used clinically was RU 38486, ie, RU 486 or mifespristone [84371-65-3] (**84**). It has proven to be a potent antiprogestin, but has the liability of also displaying antiglucocorticoid activity. Although this activity has been exploited in the treatment of Cushing's syndrome, for antiprogestational applications, a compound with reduced antiglucocorticoid activity is desirable. Synthetic changes have been made in an effort to improve the antiprogestational activity of RU 38486 while minimizing antiglucocorticoid activity (128–130).

The methodology used in the preparation of RU 486 (**84**) and other 11β-steroids is shown. Conjugate addition of a cuprate reagent to the α,β-unsaturated epoxide (**85**) provides the 11β-substituted steroid (**86**) stereospecifically (131). Subsequent steps lead to the synthesis of RU 486 (**84**).

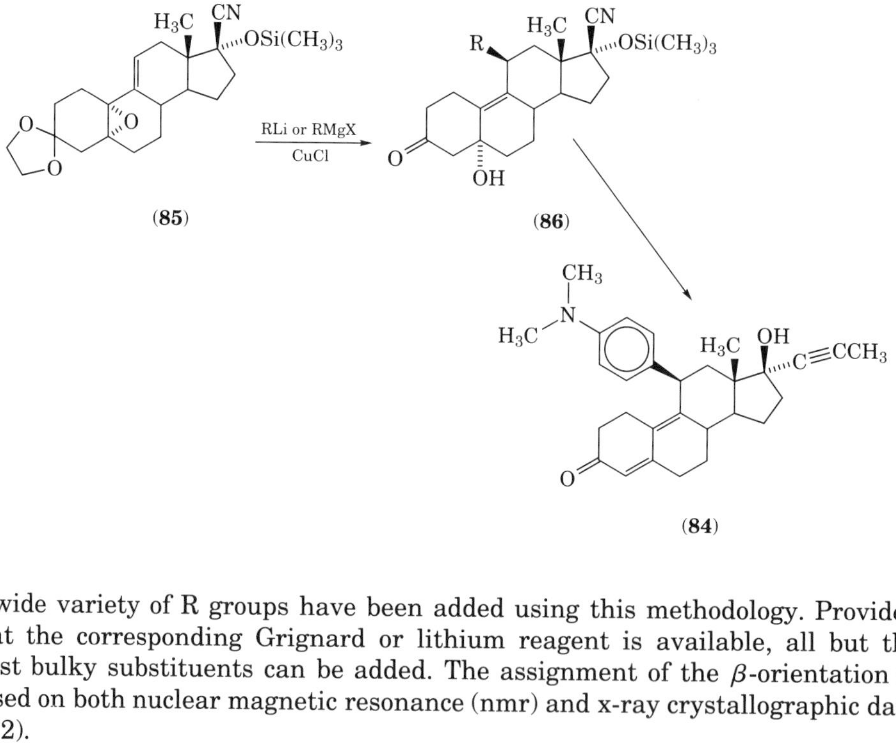

A wide variety of R groups have been added using this methodology. Provided that the corresponding Grignard or lithium reagent is available, all but the most bulky substituents can be added. The assignment of the β-orientation is based on both nuclear magnetic resonance (nmr) and x-ray crystallographic data (132).

A series of compounds with an epimeric methyl at C-13 has been prepared. A modification of the Teutsch route starting from a deconjugated ketone is used in the synthesis (132,133). Dehydrogenation (126) provides the dienone, which is protected as a ketal. 2,2-Dimethyl-1,3-propanediol is used to protect the ketone because it provides a more crystalline product than is obtained with ethylene glycol (134). Selective epoxidation, addition of the 11β-phenyl substituent, and oxidation provide the substrate for epimerization. Irradiation of this substrate leads to the C-13 epimer. It is preferable to epimerize at this stage, because the stereo- and regioselective epoxidation step used in the synthesis of RU 486 fails in the C-13 epimeric series. In contrast to the natural series, alkynylation of the C-13 epimer occurs less selectively and predominantly from the β-face (135). The alkyne is reduced and treated with acid to effect hydrolysis and dehydration of the C-5 alcohol to afford onapristone [96346-61-1] (ZK 98299) (**87**). On a technical scale, a modification that involved epoxidation followed by epimerization was reported to be the preferred pathway (133). Onapristone has been evaluated clinically. Compounds of this type have been shown to retain antiprogestational effects and to have reduced antiglucocorticoid activity (136).

Introduction of a 6β-methyl substituent, but not a 6α-methyl, into an 11β-phenyl steroid causes a reduction in affinity for the glucocorticoid receptor while maintaining affinity for the progesterone receptor (137). The starting material used is the hydroxy-steroid (**45**) (see Fig. 7). This is epoxidized and treated with methyl magnesium chloride to provide the desired 6β-methyl product in high yield. This material is then carried through using the chemistry of Teutsch (131) to introduce the 11β-phenyl substituent. Other chemistry (138) is used to introduce the 17α-substitution that ultimately will be converted into the 17-spiro compounds. ORG 31710 (**88**) and (**89**) have an affinity for the progesterone receptor similar to RU 486, whereas (**90**) has less affinity. All compounds have a lower affinity for the glucocorticoid receptor than RU 486 (**84**) (137).

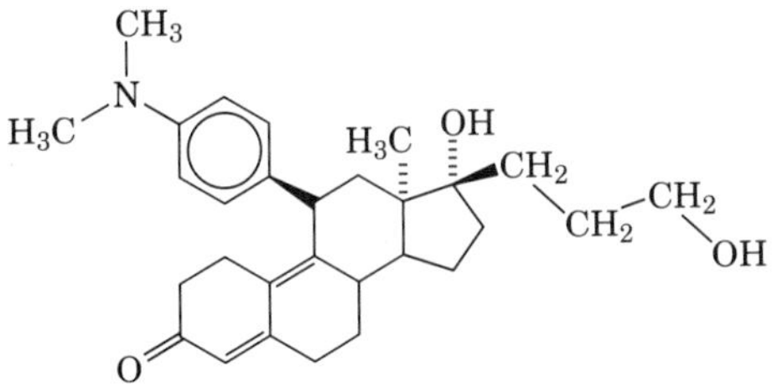

(**87**)

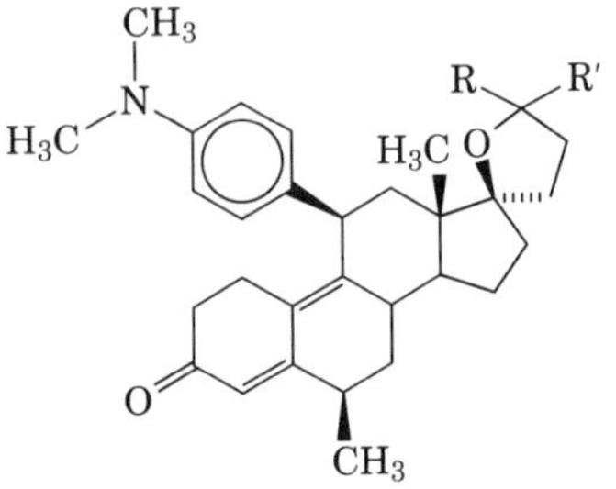

R, R' = H (**88**)
R = H, R'=OH (**89**)
R, R' = R=O (**90**)

Nonsteroidal Agonists and Antagonists. There are no nonsteroidal progestin agonists or antagonists in clinical use. This is in contrast to the wide variety of known nonsteroidal estrogens and androgens (139). Nonsteroidal compounds having affinity for the progesterone receptor or activity typical of such compounds have been reported. A series of acylanilides (**91**), which are structurally related to the antiandrogen flutamide [13311-84-7], has been reported to possess potent progestational activity and no antiandrogenic activity (140).

(**91**)

A closely related series has been reported to be antiandrogenic, and to possess some progestational and antiprogestational activity (141). Other compounds have been reported that either have affinity for the progesterone receptor or demonstrate progestational activity or antiprogestational activity (142–148). The antihormonal action of compounds with modified steroid ring structures has been reviewed (149). Potential mechanisms of action and a chemical classification of nonsteroidal hormone antagonists have been discussed (139). The relative binding affinities for the progesterone receptor of a large number of steroidal and nonsteroidal compounds have been compiled (143).

Structure–Activity Studies

An understanding of the ligand–receptor interaction is crucial in the effort to develop more potent and more selective ligands. Ideally, the structure of the receptor and the activity of a variety of ligands are known, and conclusions can be drawn about structure–activity relationships (SAR). Although the amino acid sequence of the progesterone receptor is known, the three-dimensional structure remains unsolved (150). In the absence of the purified progestin receptor, or even of the hormone-binding domain fragment, several techniques have been used to gain insight into ligand–receptor interactions (151). Techniques designed to identify which amino acids are important for ligand binding have been developed, including site-directed mutagenesis (152,153), affinity labeling (154–158), and reactions of specific amino acids with chemical reagents (159).

Another approach has been to use binding data and structures of ligands to develop a three-dimensional map of the receptor. Extensive work has been done in this area. The picture of the binding pocket that has emerged consists of a hydrophobic core with polar regions able to form hydrogen bonds on each end. There is also a hydrophobic region above and below the plane of the steroid. The large but narrowly defined hydrophobic region above the plane of the steroid is associated with agonist versus antagonist activity. Except for specific polar interactions, steroid–ligand binding is generally considered to be hydrophobic in nature (160–163). Antibodies have been raised to steroids in an effort to examine how steroids interact with receptors (164).

The prediction of protein structure through computational techniques has been applied to steroid hormone receptors. Hydrophobic cluster analysis (HCA) has been used to develop a model of the hormone binding domain of steroid receptors. Using HCA as a first approximation, the hormone-binding domains are proposed to be similar to a cleaved portion of human α_1-antitrypsin, the crystal structure of which is known (151). Another analysis has predicted that the hormone-binding domains of steroid receptors are similar to the substilisin-like serine proteases (165). Both models await further validation through comparison with biological data and, hopefully, the eventual solution of the three-dimensional structure of the progestin receptor.

Modes of Structural Studies. Determining the three-dimensional structure of relatively small steroids has been accomplished through x-ray crystallography and nuclear magnetic resonance (nmr) spectroscopy. If the structure is unknown or theoretical, or the compound is difficult to crystallize, molecular modeling techniques may be employed to predict the preferred conformation. This is a much easier task for the relatively small ligands, as compared to large receptors. From these techniques geometrical, conformational, and electronic information may be obtained. Many studies have focused on the conformation of the A-ring, as it is believed to be important in receptor binding (161); however, each method has its limitations. The structure of a steroid in its natural environment may not be the same as that obtained by any of these methods due to crystal packing forces, solvent effects, or an error in minimizing the structure.

Nuclear Magnetic Resonance Spectroscopy. The data from nmr studies complement information obtained by x-ray crystallography in that nmr is performed in the solution phase (see MAGNETIC SPIN RESONANCE). Selected studies have used nmr to draw conclusions about steroid structure or SAR. In a study of three 19-nor-4-en-3-ones, the normal A-ring conformation has been shown to predominate (166). An nmr study of 10 4-ene-3-one steroids has found that the normal $1\alpha,2\beta$-half-chair conformation predominates in all but one 2β-acetoxy-substituted steroid. A-Ring conformations have been found to vary less in solution than in the solid phase (167). The A- and D-ring conformations of 17 5α-androstanes and one pregnane have been discussed based on observed nmr data and calculated minimized structures. The A-ring sofa and D-ring half-chair/envelope intermediate conformations have been found to be preferred in solution (168).

X-Ray Crystallography. Structural data exists on more than one thousand steroids. Comparisons of the conformations obtained by x-ray crystallography have been made to the conformations obtained by other methods (161). Several studies use x-ray crystal structures in the study of progestins.

Crystal structures of phenylseleno-substituted progesterone analogues support the theory that a hydrogen-bond donor is present in the receptor toward O-20 in the β-region of C-16; eg, 21-(phenylseleno)progesterone bound with moderate affinity to the progesterone receptor, but 17α-(phenylseleno)progesterone bound with very little affinity. From the crystal structures, it has been observed that in the 21-substituted progesterone, the C-16−C17−C20−O20 orientation is closer to that of other compounds known to bind to the progesterone receptor (169). An inverted $1\beta,2\alpha$-half-chair A-ring conformation has been proposed to be the preferred conformation for progestin binding and activity (161). Steroids with affinity for the progestin receptor are able to form hydrogen bonds in the A-ring region. Based on crystal structure data, this hydrogen-bonding region is predicted to be located closer to the C-2 side than to the C-4 side of the ketone and above the β-face of the steroid (163,170).

Molecular Modeling. Molecular modeling (qv) techniques have been used to predict the conformation of progestins. Combined with computer graphics techniques, such tools allow visualization of a three-dimensional structure. Steric, hydrophobic, and electronic information can be incorporated into the model. The conformation of the progesterone side chain determined from x-ray crystallographic data has been found to be identical to that predicted from molecular mechanics calculations. Calculations using older force-field techniques have led to a discrepancy between the x-ray and calculated structures (161). When the tendency of a carbonyl group to eclipse neighboring aliphatic bonds is considered, a better correlation is obtained (171). A semi-empirical, all-valance-electron molecular orbital method successfully predicts the x-ray crystal structure of progesterone and these results imply the structures of other steroids can be calculated (172). A study of 4-ene-3-ones in the androstane, estrane, and estradiene classes has been reported. To exclude the possibility of crystal packing forces, structures are determined by molecular mechanics calculations. 10-Methyl group steric hindrance and/or steroid backbone flexibility play an important role in ligand binding (173).

Quantitative Structure–Activity Relationships. Many quantitative structure–activity relationship (QSAR) studies of progestins have appeared in the literature and an extensive review of this work is available (174). QSAR studies attempt to correlate electronic, steric, and/or hydrophobic properties to progestational activity or receptor binding affinity. A review focusing on the problems associated with QSAR of steroids has been published (175).

Examination by ^{13}C nmr of eight analogues of norethindrone, where structures varied at the C-11, C-18, and $\Delta^{15,16}$ position, suggests a correlation between the ^{13}C-resonance of C-17 and the relative binding affinity (RBA) to the progesterone receptor and to the progestin/androgen RBA ratios (176). A nonlinear relationship has been found between the lipophilicity of esters of norethindrone and levonorgestrel (177). The lipophilicity is expressed in terms of high performance liquid chromatography (hplc) retention times. It has been proposed that the rate of release and dissolution of the esters, as well as partitioning through biological membranes, not *in vivo* ester hydrolysis, are the rate-determining steps in obtaining biologically active drug. A pilot study has attempted to find a relationship between the calculated lipophilicity and steric effects of a 13β-substituent and progestational potency. The activity of the 13β-ethyl and 13β-acetyl pregnanes have been measured by the subcutaneous Clauberg assay (178). Binding energies of progesterone analogues, calculated from their binding affinities for the rabbit uterine progesterone receptor, have been used to draw conclusions about receptor–ligand interactions. Binding is attributed to hydrogen bonds involving the C-3 and C-20 carbonyls, and van der Waals' interactions at C-2, C-4, C-7, C-9, C-12, C-18, and C-19. A greater distance between the ligand and the receptor occurs at C-6, C-11, C-14, C-15, C-16, and C-21. An extensive discussion of structure and binding affinity correlations has been presented (179). A successful QSAR has been obtained for affinity to the progesterone receptor for 55 progesterone derivatives using the minimal steric, ie, topological, method (180).

Two correspondence analysis methods have been described (151,181). Using the minimum spanning tree method, many test compounds have been found to separate into four main branches corresponding to favorable androgen receptor (AR) binding, progesterone receptor (PR) binding, glucocorticoid receptor (GR) binding, and a nonspecific branch with AR/PR and GR binding. The favored PR branch is characterized by 19-norpregnanes or sultine structures. A related method, that of correspondence factoral analysis, also has been developed (181,182). This method allows representation of compounds and biological variables on the same two-dimensional graph. The results agree with those found using the minimum spanning tree method, ie, the same compounds show the highest specificities for the tree receptors (151). A comparison of the binding of steroids to the progestin and androgen receptors using the comparative molecular field analysis (CoMFA) technique has been published. Using relative binding affinities from the literature, 48 steroids have been analyzed. Differences in the electrostatic requirements in the C-17 region have been noted. The progestin model is found to be more predictive than the androgen model (183).

Mechanism of Action of Progestins and Antiprogestins

Like all steroid hormones, progestins exert their biological effects by altering the rate at which certain genes are expressed in hormone-responsive cells. Steroids bind in a competitive and reversible fashion to hormone-specific receptor proteins which act as transcription factors regulating ribonucleic acid (RNA) synthesis within the cell nucleus. For the progestin receptor, the binding of steroid is an absolute requirement for the receptor to interact with deoxyribonucleic acid (DNA) and initiate gene transcription in intact cells (184,185). The progestin receptor has two isoforms, A and B, having molecular weights of approximately 94,000 and 120,000, respectively. The isoforms are produced from the same gene but differ structurally; the B form is 164 amino acids longer at the amino-terminus than the A form (186). The isoforms show similar DNA and hormone-binding affinities (187,188). The entire amino acid sequence for the human progestin receptor is known and has been cloned (150). The progestin receptor is a member of the same steroid hormone receptor superfamily as the glucocorticoid, mineralcorticoid, estrogen, vitamin D_3, and thyroid receptors. There is considerable structural homology between the members of this receptor superfamily. The structure of the steroid receptors, including the progestin receptor, can be divided into several functional domains, ie, the progestin-binding domain at the carboxyl terminus, the hinge region, the DNA-binding domain, and the variable region at the amino-terminus (189). The structure of the progestin receptor has been under intense study and many reviews are available (190–192).

The binding of hormone to the receptor triggers a number of changes in the receptor that allow the hormone-receptor complex to bind to DNA and stimulate transcription. The occurrence of this ligand-induced transcriptional activity is termed receptor activation. The binding of progestin to the receptor induces the phosphorylation of serine residues in the receptor (193,194), the release of heat shock proteins normally associated with the hormone-free receptor (195), and the dimerization of activated receptors (196). Because there are A and B isoforms of the progestin receptor, three species of dimers may exist: A:A, A:B, and B:B. All three dimers can bind to the same specific DNA sequences, known as progesterone response elements (196). However, there is evidence that the three dimeric forms may differentially regulate target genes (197,198). To study the A and B isoforms in isolation, molecular biological techniques have been used to produce cells which express either the A isoform, the B isoform, or some known ratio of the two. In some cell types, the B isoform but not the A isoform activates a specific gene promoter (198). When the A isoform is coexpressed in the same cells as the B isoform, the transcriptional activity of the B isoform is repressed in direct proportion to the amount of A isoform present. Apparently, the A isoform has dual activities, acting both to promote and repress the expression of certain genes. The A and B isoforms show additional differences in their response to progestin antagonists. The A isoform is not transcriptionally activated by progestin antagonists, but only by progestin agonists. In contrast, the B isoform, in the absence of the A isoform, is activated to induce gene transcription by both progestin agonists and antagonists (199,200). The binding of progestin antagonists induces changes in receptor structure and phosphorylation different from those due to progestin agonist binding (201–203). A monoclonal antibody directed to the carboxyl terminus of the human progestin receptor can differentiate between agonist- and antagonist-receptor complexes (204). It is these ligand-induced changes in receptor structure which are believed to be the basis for the biological effects of progestin agonists and antagonists.

   RU 486.   The most well-studied progestin antagonist is RU 486 (**84**). RU 486 is a potent antiprogestin and antiglucocorticoid which has been used in the research laboratory for exploring receptor structure and in the clinic for terminating first-trimester pregnancies. The progestin receptor binding characteristics of RU 486 have been described (205,206). It binds to both A and B progestin receptor isoforms with high affinity and can induce receptor activation. RU 486 does not, however, bind to the progestin receptors of all species. Human but not chick progestin receptors recognize RU 486 (207). The species differences seen with RU 486 receptor binding result from as little as a single amino acid difference in the structures of the receptor. The substitution of a cysteine at position 575 with a glycine in the chick receptor allows RU 486 to bind to the modified receptor and antagonize agonist-induced gene transcription (208). Substitution with other amino acids containing side chains, ie, methionine or leucine, at position 575 produces receptors with the same properties as the native chick receptor, suggesting that the presence of an amino acid side chain at this critical position sterically hinders RU 486 binding.

   RU 486 blocks the actions of agonists in two ways. The antagonist binds to and competes for the same receptor as the agonists. In addition, after receptor binding has occurred, the antagonist-receptor complex binds to and competes for the same progesterone response elements on the DNA as the agonist-receptor complex (209). By allowing receptor-DNA binding, RU 486 may actually induce the transcription of some genes. This may explain some of the agonistic effects of RU 486 that have been reported. RU 486 alone does not alter insulin receptor concentrations in breast cancer cells, but can partially block progestin-induced increases in insulin receptors (see INSULIN AND OTHER ANTIDIABETIC DRUGS). In contrast, progestin suppression of breast cell growth is not blocked by RU 486. Further, RU 486 alone suppresses cell growth as effectively as the progestin agonist, R5020 (205). RU 486 can induce gene transcription when bound to either the A or B isoform of the human progestin receptor (199,200). In rats, RU 486 has some weak agonistic effects on sexual behavior (210). In women, RU 486 alone induces the same secretory changes in the uterine endometrium as progesterone; administered together, RU 486 blocks the effects of progesterone (211).

Classifications of Progestin Antagonists. Progestin antagonists can be classified based on their ability to form hormone-receptor complexes capable of binding to progesterone response elements (212). Type I antagonists bind to the receptor but do not induce the receptor activation necessary for DNA binding. Type II antagonists, such as RU 486, bind to the receptor and induce the binding of receptor to DNA. Type I antagonists may be viewed as pure antagonists because, owing to their blockade of receptor-DNA binding, no gene transcription is possible. However, type II antagonists may induce transcription of some genes. Presently there are several type I and type II antagonists being used for studying the progestin receptor. Type I antagonists include ZK 98299 [96349-61-1] (87) (209,213); type II antagonists such as RU 486 include ZK 112993 [105114-63-4] (92) (214), ORG 31710 [118968-41-5] (88), and ORG 31806 [123916-70-1] (93) (215).

(92)   (93)

Regulation of Progestin Receptors. The biological response of a tissue to a steroid hormone depends on the number of steroid receptors in the tissue. This makes regulation of receptor concentration an issue in determining the effectiveness of a steroid (216,217). Progestin receptors can be regulated by a number of hormones. Estrogens increase the levels of the progestin receptor in most target tissues and cells (218). Progestin receptor messenger RNA (mRNA) levels rise several hours prior to any increase in receptor concentration indicating that the effects of estrogen are due to an increase in progestin receptor synthesis (219,220). In contrast, progestins decrease the number of progestin receptors (221), increasing the rate of receptor degradation and decreasing the rate of receptor synthesis (222). Whereas estrogens and progestins are the primary regulators of progestin receptor concentrations, other factors and hormones may also play a role. Insulin, insulin-like growth factor, and epidermal growth factor can increase progestin-receptor concentrations in cultured breast and uterine cells (223,224). Luteinizing hormone and follicle stimulating hormone, the pituitary hormones responsible for ovulation, induce progestin receptors in cultured ovarian granulosa cells (225).

Assay Methods for Evaluating Progestin Agonists and Antagonists. *Progestin-Binding Assays.* The initial evaluation of a progestin often begins with the determination of its relative binding affinity for the progestin receptor.

Compounds can be directly compared based on their ability to compete with a radiolabeled progestin for a receptor binding site. Classically, progestin receptor studies have been performed using a crude receptor preparation made from the uteri of estrogen-primed rabbits. The assay itself involves combining the receptor preparation with a ^{3}H- or ^{125}I-labeled progestin and varying concentrations of the test compound. After incubating under conditions allowing equilibrium to be reached, the receptor-bound radioligand is separated from the free radioligand and is quantitated. The relative binding affinity is calculated from the displacement curve and is often expressed as either an IC_{50} value or as the ratio of the IC_{50} of progesterone, or another standard, versus the test compound IC_{50}. There are many detailed descriptions of the assay methodology (226–228). Differences between species, and even tissues within the same species, have been noted in progestin binding (229,230). However, the relative binding affinities of progestins in rabbit uterine preparations show a reasonable correlation with binding affinities measured in human tissues (143,231). In addition, the relative binding affinities of progestins are related to and can predict their relative biological activity *in vitro* and *in vivo* (232–234).

The utility of receptor binding assays in identifying active compounds and in predicting biological activity is not as clear with progestin antagonists. The potent progestin antagonist RU 486 shows a high degree of species specificity with regard to its receptor-binding affinity. It binds with a high affinity to the progestin receptors in human, rat, rabbit, and calf tissues (205,235–237) but does not bind to progestin receptors in chick oviduct (207) or hamster uterus (238). Receptor-binding assays cannot differentiate agonists from antagonists, nor can they predict the biological actions of mixed agonists/antagonists. An example of such a compound is RU 486 which antagonizes the effects of progestins and yet by itself can demonstrate agonistic properties *in vitro* and *in vivo* (205,211).

Progestin Effects In Vitro.   To quantitate hormone action in a defined system, several cell lines have been used to evaluate the biological activity of progestins. The proliferation of human breast cells has become one of the most commonly used *in vitro* assays of progestin action. Normal breast cells obtained from mammoplasties and breast tumor-derived cells respond to progestins with a concentration-dependent reduction in their rate of growth (239,240). Under different experimental conditions, a progestin-induced increase in breast cancer cell proliferation also has been reported (241,242). Possible explanations for these discrepancies include differences in the cell lines used between laboratories, the confounding estrogenic and androgenic properties of the progestational steroids used, and the presence of phenol red in the media (243–245). Human breast cancer cells synthesize a 250,000 mol wt protein in response to progestin stimulation. In further studies, this protein has been found to be fatty acid synthetase and has been shown to be produced in a dose-response fashion following progestin treatment (246). Another progestin-induced protein found in human breast cells is a 48,000 mol wt protein secreted from T47D breast cells as a specific response to progestin treatment (247).

Rather than use the synthesis of an endogenous protein as an index of progestin activity, a newer assay methodology has become available which uses cells modified or transfected to produce a foreign protein in response to progestins. To perform this transactivation assay, cells containing progestin receptors are transiently transfected, ie, made to take up a DNA construct comprising the gene for an easily quantitated enzyme (the reporter gene) linked to a hormonally responsive promoter (see GENETIC ENGINEERING). The reporter enzymes commonly used are chloramphenicol acetyltransferase [9014-00-0] or luciferase [9040-07-7], neither of which is found in mammalian cells. Addition of a progestin agonist to the transfected cells activates the endogenous progestin receptor which binds to the hormonally responsive promoter, triggers transcription of the reporter gene, and causes the synthesis of the reporter enzyme (248,249). Progestin antagonists also can be identified with this assay based on ability to block transactivation by the agonist R5020. Another method for exploring the action of antagonists is to measure gene activation in a hormone-dependent, cell-free system (250). In this system, the effect of an agonist or antagonist can be explored by measuring gene expression under controlled, well-defined conditions.

Progestin Effects In Vivo. A variety of methods has been devised to evaluate and compare progestins *in vivo*. Perhaps the best known is the Clauberg assay in which the end point is the glandular proliferation of the uterine epithelium in rabbits (251). Intact, immature rabbits are given estradiol for several days to prime the uterus; the progestin is then given orally or by subcutaneous injection for several days. The uterus is removed and histological sections are prepared for analysis. In response to a progestin, the endometrium shows a characteristic pattern of proliferation which is related to the dose and potency of the progestin. The progestin-induced proliferation of the endometrium is quantitated using the McPhail index (252). This index grades the histological changes in the endometrium on a 1 to 4 scale with +4 as the maximum proliferative response. A variation on this method is the McGinty test in which spayed adult rabbits have the drug administered directly into the uterine horn (182). The direct application of the progestin to the target tissue makes this a more sensitive test for progestational activity. Using these methods there can be problems quantifying the results. The dose–response curves for many synthetic progestins are not parallel with progesterone. Also, those compounds that produce shallow, nonparallel dose-response curves do not produce the maximal McPhail ratings (253).

There are other whole animal models to evaluate progestins based on their reproductive effects. Using rats or mice, progestational activity can be determined by measuring uterine deciduoma formation (251). Deciduoma formation is a proliferative response of the uterine stroma to a physical trauma which, in the case of this assay, is usually a silk suture sewn through the uterine wall. The magnitude of decidual reaction, quantitated by uterine weight or histology, occurs in direct proportion to the dose or potency of the progestin used to pretreat the animal. Progestins can also be compared based on their ability to inhibit parturition in rats when given near the time of birth (254). The maintenance of pregnancy in rats following ovariectomy is also a model of progestational activity (251).

In addition to histological changes, the effects of progestins on the endometrium can be quantitated by measuring the production of several proteins. Carbonic anhydrase, measured in rabbit uterine homogenates, increases in a dose–response fashion after progestin treatment (251,255). The synthesis of uteroglobin is also stimulated by progestins in the rabbit (256). After five days of progesterone treatment, the concentration of uteroglobin in the uterine luminal fluid is increased nearly 3000-fold (257). Estrogens and androgens, however, can also stimulate uterine uteroglobin levels (256).

Effects of Progestins on Various Target Tissues. Progestins exert a variety of direct and indirect effects on all tissues and organs within the reproductive system. Organs that contain progestin receptors, and thus can respond directly to progestins, include the uterus, cervix, vagina, breast, ovary, brain, and pituitary gland (258,259). Receptors have also been identified in rat thymus (260), human lymphocytes (261), and bone cells (262,263). Several factors are responsible for progestin effects including the dose and potency of the drug itself, the duration of treatment, the route of administration, and the ability of the drug to affect progestin-receptor number and estrogen levels.

The uterus is exceptionally sensitive to the effects of estrogens and progestins. Estrogens stimulate the proliferation and vascularization of the uterine endometrium (see HORMONES, ESTROGENS AND ANTIESTROGENS). Progestins serve to suppress the stimulatory effects of estrogen on uterine growth, an effect termed antiestrogenic. Progestins blunt the responsiveness of the uterus to estrogen by reducing estrogen receptor levels in the uterus (264,265). In addition, progestins affect estrogen metabolism within the uterus by inducing 17β-hydroxysteroid dehydrogenase which converts estradiol to the inactive estrone (266). Morphologically, the uterine effects of progestins can be seen as a reduction in the number and size of endometrial glands and a decrease in the proliferation of both glandular and stromal cells (253,267,268). Progestins also act on the uterine myometrium reducing the frequency of contractions (269). In nonpregnant women, menstrual bleeding is the most noticeable effect of progestins on the lower reproductive tract. After being transformed from a proliferative to a secretory state by the action of progesterone or a synthetic progestin, the uterine endometrium requires constant steroidal stimulation or it regresses and is lost in menses. Intermenstrual bleeding, or breakthrough bleeding, is a common side effect with the use of synthetic progestins, as in oral contraceptives. Elsewhere within the lower reproductive tract, progestins have other effects. The quantity of cervical mucus is lessened and the mucus is more viscous under the influence of progestins (253). The motility of the oviduct is affected by progestins which can act to delay the transport of ova from the ovary to the uterus (270).

The breast is another tissue sensitive to estrogens and progestins. In the presence of estrogens, which presumably increase progestin receptors in the breast, progestins stimulate ductal branching and lobuloalveolar development (271,272). While acting to promote mammary gland growth in concert with estrogens, progestins suppress milk production (268). The actions of progestins on breast cells in culture have been examined in a number of studies with conflicting results as to whether cell proliferation is stimulated or inhibited (201).

The effects of progestins are not limited to the reproductive system. Progestins exert direct actions on bone osteoblast cells, acting as a bone-forming hormone (273). Within the brain, progestins modulate neuroendocrine function, mood, and behavior via intracellular nuclear receptors (274,275). On the membrane of some neurons however, naturally occurring A-ring-reduced metabolites of progesterone bind to the chloride ion-GABA$_A$ receptor complex. The principal neuroactive metabolite of progesterone is 3α-hydroxy-5α-pregnan-30-

one [*516-54-1*] (allopregnanolone). Allopregnanolone and similar A-ring-reduced pregnanes potentiate GABA effects at these receptors. These steroids mimic the effects of the benzodiazepines, changing chloride ion conductance and producing sedative and hypnotic behavioral effects (276,277). Neuroactive steroids can be therapeutically useful as anticonvulsants, anxiolytics, or anesthetics (qv) (see also HYPNOTICS, SEDATIVES, ANTICONVULSANTS, AND ANXIOLYTICS).

Progestins have indirect effects on other nonreproductive systems and organs, effects not necessarily involving progestin receptors. Some synthetic progestins, especially medroxyprogesterone acetate (MPA) and other acetoxypro-gesterone derivatives, have sufficient glucocorticoid-like activity to lower adreno-corticotrophin hormone (ACTH) and cortisol levels via a negative feedback effect (278). Progesterone has antimineralcorticoid activity, and some other progestins, eg, gestodene, show surprisingly high affinity for the mineralcorticoid receptor (279,280). The hepatic production of sex hormone-binding globulin [*111566-18-8*] (SHBG) is lowered by some progestins, eg, MPA or norgestrel, but not others (280). It is not uncommon for progestins, especially 19-nortestosterone deriva-tives, to bind to SHBG, as do androgens (279,280). Many progestins show vary-ing degrees of affinity for the androgen receptor and produce androgenic effects *in vivo*. Gonane-based progestins tend to be more androgenic than estrane-based progestins (281). High blood glucose levels owing to insulin insensitivity occur during pregnancy, a time of high endogenous progesterone secretion. Depending on dose and potency, synthetic progestins can produce similar effects, impairing glucose tolerance and increasing insulin resistance (282).

There has been great interest in understanding the impact of synthetic progestins on the cardiovascular system and the risk of coronary heart disease in women (283). Coronary heart disease has been linked with high plasma lev-els of total cholesterol and low density lipoprotein cholesterol (LDL) and low plasma levels of high density lipoprotein cholesterol (HDL). Estrogens tend to in-crease plasma levels of cardiac-protective HDL and decrease LDL levels, whereas progestins have opposite effects (284). At clinically used doses, norgestrel and levonorgestrel have a greater effect on HDL and LDL levels than estrane deriva-tives such as norethindrone (285,286). The newer gonane-based progestins, eg, desogestrel, gestodene, and norgestimate, have minimal or even favorable ef-fects on the LDL/HDL ratio (287).

Therapeutic Uses of Progestins

Progestins Used for Contraception. The primary therapeutic use of pro-gestins is in combination with estrogens for contraception. Estrogen-containing contraceptives work primarily by inhibiting pituitary gonadotropin secretion and thus ovulation (288). Progestins are given with estrogens not so much as to increase their contraceptive effectiveness, but to prevent uncontrolled, estrogen-driven proliferation of the uterine endometrium. Synthetic progestins have a number of effects on the reproductive system, many of which by them-selves provide some degree of contraception. Progestins counter the effects of estrogens on the uterine endometrium, making it less suitable for implantation (253,264–268). Progestins block sperm passage into the uterus by thickening the cervical mucus (253). The movement of the egg through the fallopian tubes is also slowed by progestins (270). The combined effects of progestins on the uterus, cervix, and fallopian tubes are sufficient to achieve contraception as evidenced by the use of progestin-alone contraceptives.

As of 1994, there were approximately 47 progestin-containing contraceptive drug formulations sold in the United States for use as oral contraceptives (Table 3). In addition, there are three nonoral contraceptive formulations containing progestins; ie, one injectable (Depo-Provera), one as an intrauterine device (IUD) (Progestasert), and one implantable (Norplant). Of the oral formulations, all but two also contain an estrogen component, ethynylestradiol.

Progestins are formulated in a variety of doses. Oral contraceptives are available in a monophasic formulation, where the doses of progestin and estrogen are constant through the 21–28 day dosing regimen, and in a triphasic formulation, where the doses of progestin and estrogen are varied in three stages to more accurately mimic the normal changes in steroid levels during the menstrual cycle. The triphasic regimen was designed to reduce the total monthly dose of progestin and thus reduce side effects but still provide effective contraception (289). A number of studies have shown that this approach to lowering the total progestin dose does lessen the negative changes in plasma lipids, androgenic effects, and glucose tolerance that occur owing to the progestin component in oral contraceptives (290,291).

Contraception can also be achieved using progestins administered in a fashion to provide sustained release. Depo-Provera (Upjohn) is a microcrystalline solution of MPA designed to provide contraception for three months after a single administration. Each intramuscular injection contains 150 mg MPA, a dose appropriate for all body weights. Plasma MPA concentrations peak at 1–7 ng/mL within three weeks and fall to undetectable levels 120–200 days after injection (292). The Norplant system consists of six capsules made of Silastic (dimethylsiloxane–methylvinylsiloxane copolymer), each containing 36 mg of levonorgestrel. When inserted subcutaneously, the Norplant capsules provide contraception for five years. During Norplant development, levonorgestrel was chosen as the most suitable progestin owing to its slow release rate, approximately 10% of original load per year (293). Initially 85 μg of levonorgestrel is released per day producing blood levels of 0.25–0.3 ng/mL. The rate of release slowly falls over time, reaching 50 μg/d by nine months, 35 μg/d by 18 months, and 30 μg/d thereafter (294). Unlike oral contraceptives which show dramatic hepatic metabolism following absorption from the gastrointestinal tract ie, the first-pass effect, the release of progestins from Depo-Provera and Norplant avoids this initial hepatic metabolism. Other methods of progestin administration now being investigated include progestin-containing IUDs (Progestasert), progestin-filled Silastic vaginal rings (295), and injectable microcapsules of poly(lactide-co-glycolide) containing norethindrone (296).

Noncontraceptive Uses of Progestins.   Progestins have other therapeutic uses aside from contraception. Hormone-dependent tumors and cysts involving reproductive tissues respond to progestins. Megestrol acetate and MPA are the two most commonly used progestins to treat breast cancer (297). The progestin doses necessary for effective cancer therapy are very high, reaching 1000–2000 mg/d. These high concentrations of progestins may work through

Table 3. Progestin Formulations Marketed in the United States[a]

Progestin	Doses, mg	Ethynylestradiol component, μg	Trade name	Manufacturer
Contraceptive uses				
desogestrel	0.15	30	Desogen,	Organon
			Ortho-Cept	Ortho
ethynodiol diacetate	1.0	35	Demulen 1/35	Searle
	1.0	50	Demulen 1/50	Searle
levonorgestrel	0.15	30	Nordette	Wyeth-Ayerst
	0.05/0.075/ 0.125	30/40/30	Triphasil	Wyeth-Ayerst
	6×36[b]		Norplant	Wyeth-Ayerst
medroxyprogesterone acetate (MPA)	150[c]		Depo-Provera	Upjohn
norethindrone	0.35		Micronor,	Ortho
			Nor-Q D	Syntex
	0.4	35	Ovcon 35	Mead Johnson
	1.0	50	Ovcon 50	Mead Johnson
	1.0	50	Ortho Novum 1/50	Ortho
			Norinyl 1 + 50,	Syntex
	0.5	35	Ortho Novum	Ortho
			10/11, Brevicon	Syntex
	1.0	35	Ortho Novum 1/35,	Ortho
			Norinyl 1 + 35	Syntex
	0.5/0.75/1.0	35	Ortho Novum 7/7/7	Ortho
	0.5/1.0/0.5	35	Tri-Norinyl	Syntex
norethindrone acetate	1.0	20	Loestrin 1/20	Parke-Davis
	1.5	30	Loestrin 1.5/30	Parke-Davis
norgestimate	0.25	35	Ortho-Cyclen	Ortho
	0.18/0.215/ 0.25	35	Ortho Tri-Cyclen	Ortho
norgestrel	0.075		Ovrette	Wyeth-Ayerst
	0.5	50	Ovral	Wyeth-Ayerst
	0.3	30	Lo/Ovral	Wyeth-Ayerst
progesterone	38[d]		Progestasert	Alza
Noncontraceptive uses				
medroxyprogesterone acetate	2.5, 5, 10[e]		Amen	Carnrick
			Provera	Upjohn
			Cycrin	ESI-Pharma
norethindrone acetate	5[e]		Aygestin	Wyeth-Ayerst
			Norluate	Parke-Davis

[a]Oral formulations unless noted.
[b]Implant.
[c]Injection.
[d]Intrauterine device.
[e]Doses available.

several modes of action, including inhibition of estrogen-induced growth by both lowering estrogen receptors and increasing the oxidative activity of 17β-hydroxysteroid dehydrogenase which converts estradiol to less active estrone (297), and directly inhibiting the proliferation of breast cancer cells, as has been shown *in vitro* (258). Benign breast masses, ie, fibrocystic breast cysts, can be treated with progestins at doses found in oral contraceptives (298). The uterine endometrium is very sensitive to progestins and therefore abnormal growth of the endometrium, whether benign or malignant, often responds to progestin treatment. Endometrial cancer, endometriosis, and endometrial hyperplasia all show reduced growth and regression during progestin treatment (299,300). For the treatment of endometrial adenocarcinomas, MPA is given at a dose of 1 g/w orally (301). This dose produces therapeutically effective serum levels of approximately 90 ng/mL within one week. Part of the benefit of MPA treatment is believed to be its ability to make the tumor more sensitive to radiation therapy (302). In disease states involving excess androgen production, ie, polycystic ovary syndrome or excess androgen of adrenal origin, progestins can be used therapeutically to suppress the synthesis of androgenic steroids at the ovary and adrenal gland (303,304). For this purpose progestins having low androgenicity are the logical choices, ie, not levonorgestrel.

Progestins are commonly given in combination with estrogen in hormone replacement therapy following menopause or ovariectomy. The primary function of the progestin component is to reduce the likelihood of endometrial hyperplasia and endometrial carcinoma (305). The usual course of treatment calls for 12–13 days of progestin per month with continuous daily estrogen administration. To protect the endometrium, doses of 10–20 mg dydrogesterone, 10 mg MPA, 1 mg norethindrone, or 150 μg levonorgestrel are effective (306,307). By themselves, however, progestins have beneficial effects on the events associated with menopause. The vasomotor symptoms of menopause, ie, hot flushes, are decreased with MPA administered as either a 150-mg depot injection or 10 mg/d orally (308,309). Osteoporosis occurs in post-menopausal women primarily because of the loss of estrogen. In the absence of estrogens the rate of bone resorption exceeds the rate of bone formation and a steady loss of bone mass occurs. Estrogen replacement slows bone resorption but does not cause the replacement of lost bone (310,311). There is mounting evidence that progestins can prevent bone resorption and may also induce bone formation (273). *In vitro*, progestins stimulate the growth and activity of osteoblasts, the cells which lay down new bone (312). MPA, norethindrone, and progesterone have been reported to exert positive effects on bone mass in animal models and women, either stimulating bone formation and/or suppressing bone resorption (273,313–315). The drug most commonly used off-label is MPA (Provera, see Table 3), prescribed with estrogen for the prevention of post-menopausal osteoporosis (316).

Pharmacokinetics and Metabolism of Therapeutic Progestins. *Norethindrone.* Norethindrone (**28**) is rapidly and completely absorbed after oral administration (317). There is a significant loss of norethindrone owing to a first-pass effect. Approximately 36% of an oral dose is lost as a result of metabolism within the intestinal wall and liver (318). In women, 60.8% of norethindrone in the blood is bound to albumin, 35.5% to sex hormone binding globulin (SHBG), and 3.7% is free (319). The half-life for elimination has been measured in a number of studies in which the values ranged from 3.4 to 13.4 hours, thereby giving an average of approximately eight hours (320). Using single daily dosing, steady-state levels are reached in 5–10 days. Norethindrone is quickly distributed to all tissues, accumulating in, and being metabolized by, the liver. The primary metabolic pathway is reduction of the α,β-unsaturated oxo group in the A-ring producing four stereoisomeric tetrahydro-reduced metabolites ($3\alpha,5\beta$; $3\beta,5\beta$; $3\alpha,5\alpha$; $3\beta,5\alpha$) (**94**). Metabolites are conjugated with glucuronic acid and excreted in the urine. Sulfate conjugation and hydroxylation also occur to a lesser degree (321–323). The most widely used method to measure norethindrone and some of its metabolites is radioimmunoassay (320,324) but a gc/ms method has been described (325).

(**94**)

Levonorgestrel.   Levonorgestrel $(-)$-(**65**) is 100% orally absorbed with no first-pass effect. Peak plasma levels are reached between 0.5 and 2 hours (326,327). Plasma levels of levonorgestrel decline in a biphasic fashion; the initial phase has a half-life of 50 to 180 minutes followed by a second phase with a half-life of 10 to 26 hours. In plasma, approximately 98% of levonorgestrel circulates bind; ~50% to albumin, 48% to SHBG, and 2% free (328). The metabolic pathways of levonorgestrel are A-ring reduction and or hydroxylation in the liver. Secretion is in the form of sulfate and glucuronide conjugates in both urine and feces. All four tetrahydro-metabolites of levonorgestrel can be detected in the urine with the $3\alpha,5\beta$-isomer as the principal compound (329). The analytical method used for measuring levonorgestrel is radioimmunoassay (326).

Norgestimate.   Norgestimate (**82**) is completely and rapidly absorbed after oral administration with peak plasma levels reached within two hours (330). Over the course of two weeks following administration of ^{14}C-norgestimate, 35–49% is excreted in the urine and 16–49% in the feces (330,331). The urine does not contain any intact norgestimate or any metabolite retaining the oxime group. The principal metabolites of norgestimate are levonorgestrel $(-)$-(**65**), 3-ketonorgestimate (**95**), and $3\alpha,5\beta$-levonorgestrel and trihydroxy compounds; minor metabolites include 16β-hydroxylevonorgestrel, 2α-hydroxylevonorgestrel, and 3,16-dihydroxy-5α-tetrahydrolevonorgestrel (331). Based on these active metabolites, the argument has been raised that norgestimate is a prodrug for levonorgestrel. Several studies have refuted this possibility. Norgestimate itself binds to the progestin receptor and is biologically active when administered directly into the uterus in a McGinty test (332). Norgestimate is rapidly deacetylated to 17-deacetylnorgestimate (**96**) by the human gastric mucosa and liver *in vitro*. 17-Deacetylnorgestimate shows a pharmacological profile similar to that of norgestimate but different from that of the levonorgestrel-based metabolites (333). Norgestimate and 17-deacetylnorgestimate can be measured in plasma by either radioimmunoassay or hplc (330).

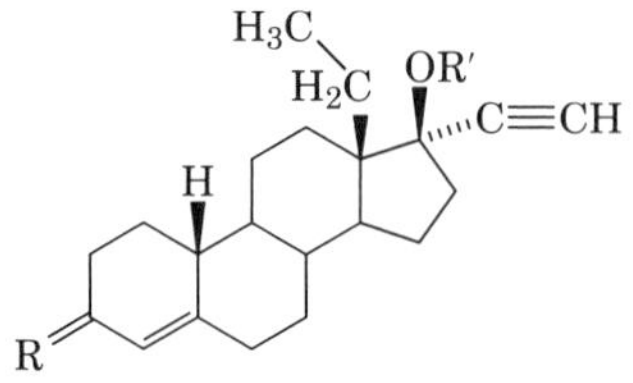

$$R = O; \quad R' = COCH_3 \quad (95)$$
$$R = NOH; \quad R' = H \quad\quad (96)$$

Gestodene.   Gestodene (**54**) is completely absorbed after oral administration with no first-pass metabolism (334). The half-life of elimination is approximately 20 hours. Repeated daily dosing results in steady-state serum levels after about 10 days (335). These levels are approximately four to five times those following a single dose, due to an increase in serum SHBG. The vast majority of gestodene in the serum is protein-bound. At steady-state conditions, about 75% of gestodene is bound to SHBG and 24% to albumin. In contrast, gestodene is 50% SHBG-bound and 48% albumin-bound after a single dose (336,337). Regardless of the frequency of administration, less than 1–2% of circulating gestodene is free and biologically active. The plasma clearance of gestodene is biphasic. The first half-life is 1.5 h; the second, 18 h (338). The metabolites of gestodene are highly water soluble and are excreted equally in the urine and feces following conjugation by the liver. Gestodene undergoes extensive reduction and oxidation in the liver. The metabolites measured in urine are mainly hydroxylated derivatives of A-ring-reduced gestodene. Hydroxylation occurs at several positions including C-1, C-6, and C-11 (338). Gestodene is measured in plasma by radioimmunoassay (334).

Desogestrel.   Desogestrel (**44**) is rapidly and nearly completely converted to the biologically active metabolite, 3-ketodesogestrel [*54048-10-1*] (**97**) (339,340). The oral administration of equal doses of desogestrel or 3-ketodesogestrel produces nearly identical serum concentrations of 3-ketodesogestrel (341). The affinity of 3-ketodesogestrel for the progestin receptor in human uterine myometrium is approximately six times greater than that of progesterone and 57 times greater than that of desogestrel (342). 3-Ketodesogestrel binds extensively to albumin and SHBG; only approximately 2.5% of 3-ketodesogestrel is free in circulation (343). Four metabolites of desogestrel (**44**) have been identified in *in vitro* studies; 3α- and 3β-hydroxydesogestrel, 3α-hydroxy-5α-desogestrel, and 3-ketodesogestrel (**97**). The half-life of elimination from the plasma for 3α-hydroxydesogestrel is 16 hours (340). The 3α- and 3β-hydroxydesogestrel metabolites are believed to be intermediates in the conversion of desogestrel to 3-ketodesogestrel (344). Human intestinal mucosa as well as liver can perform this conversion (344,345). Desogestrel and 3-ketodesogestrel can be measured by radioimmunoassay or hplc (339,345).

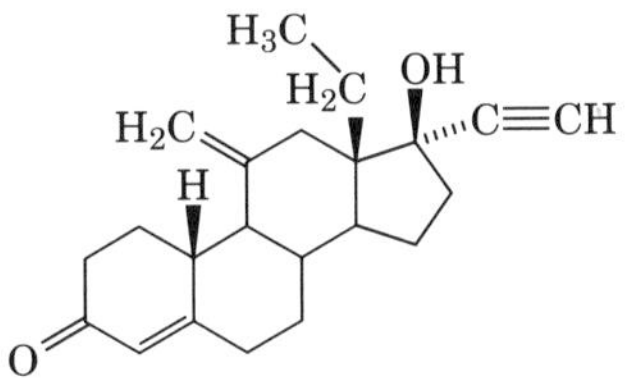

(**97**)

Medroxyprogesterone Acetate.   Accurate pharmacokinetic and metabolism studies on MPA have been difficult because the radioimmunoassays employed cannot differentiate between MPA and its metabolites (346). Comparison of MPA plasma levels assayed by hplc and radioimmunoassay show that radioimmunoassay may overestimate intact MPA concentrations by about fivefold (347). However, values of the mean elimination half-life of MPA were similar, being 33.8 and 39.7 hours when measured by hplc and radioimmunoassay, respectively (347). Approximately 94% of MPA in the blood is bound to albumin (348). When taken orally, MPA is rapidly absorbed with little or no first-pass metabolism (13). Peak serum levels are reached after three hours. Steady state occurs after three days of daily administration (349). The pharmacokinetics of MPA when administered in a depot formulation have been described (350).

Mifepristone.   After oral administration, peak plasma levels of mifepristone (**84**) (RU 486) are reached in one hour and over 95% was bound to plasma proteins (351,352). The plasma half-life of RU 486 is approximately 24 h (352,353). In humans, monodemethylated (**98**), didemethylated (**99**), and alcoholic nondemethylated (**100**) metabolites of RU 486 have been identified (351). These metabolites show some progestin-binding affinity, approximately five to ten times lower than that of RU 486 itself. RU 486 and its metabolites can be measured by radioimmunoassay and hplc (353,354).

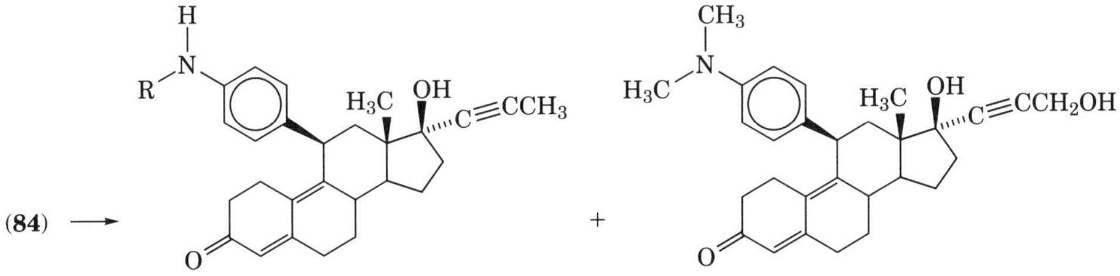

(**84**) ⟶

+

R = CH₃ (**98**)
R = H   (**99**)

(**100**)

BIBLIOGRAPHY

"Sex Hormones" under "Hormones" in *ECT* 1st ed., Vol. 7, pp. 513–536, by H. B. MacPhillamy, Ciba Pharmaceutical Products; in *ECT* 2nd ed., Vol. 11, pp. 92–127, by R. I. Dorfman, Syntex Corp.; in *ECT* 3rd ed., Vol. 12, pp. 618–657, by V. Petrow, consultant.

1. *Enzyme Nomenclature*, Academic Press, Inc., Orlando, Fla., 1984.
2. D. B. Gower, in H. L. Makin, ed., *Biochemistry of Steroid Hormones*, 2nd ed., Blackwell Scientific Publications, Oxford, U.K., 1984, p. 230.
3. C. R. Jefcoate, B. C. McNamara, I. Artemenko, and T. Yamazaki, *J. Steroid Biochem. Molec. Biol.* **43**, 751 (1992); F. Labrie, J. Simard, V. Luu-The, A. Belanger, and G. Pelletier, *J. Steroid Biochem. Molec. Biol.* **43**, 805 (1992).
4. F. A. Kincl, in M. Tausk, ed., *Pharmacology of the Endocrine System and Related Drugs: Progesterone, Progestational Drugs and Antifertility Agents*, Vol. 1, Pergamon Press, Ltd., Oxford, U.K., 1971, p. 13.
5. M. B. Aufrère and H. Benson, *J. Pharm. Sci.* **65**, 783 (1976).
6. T. J. Lin and co-workers, *J. Clin. Endocrinol. Metab.* **34**, 287 (1972).
7. A. Riondel, J. F. Tait, S. A. S. Tait, M. Gut, and B. Little, *J. Clin. Endocrinol. Metab.* **25**, 229 (1965).
8. D. R. Mishell and co-workers, *Am. J. Obstet. Gynecol.* **111**,60 (1971).
9. J. D. Neill, E. D. B. Johansson, J. K. Datta, and E. Knobil, *J. Clin. Endocrinol. Metab.* **27**, 1167 (1967).
10. A. F. Clark, in R. Hobkirk, ed., *Steroid Biochemistry*, Vol. 1, CRC Press, Boca Raton, Fla., 1979, p. 1.
11. C. W. Hooker and T. R. Forbes, *Endocrinology* **45**, 71 (1949).
12. W. Voigt, E. P. Fernandez, and S. L. Hsia, *J. Biol. Chem.* **245**, 5594 (1970).
13. H. Kuhl, *Maturitas*, **12**, 171 (1990).
14. R. A. Hill, D. N. Kirk, H. L. J. Makin, and G. M. Murphy, eds., *Dictionary of Steroids*, Chapman and Hall, London, 1991, p. xiv.
15. L. F. Feiser and M. Feiser, *Steroids*, Reinhold, New York, 1959.
16. M. H. Rabinowitz and C. Djerassi, *J. Am. Chem. Soc.* **114**,304 (1992).
17. E. M. Groen-Piotrowska and M. B. Groen, *Recl. Trav. Chim. Pays-Bas* **112**, 627 (1993).
18. R. E. Marker and co-workers, *J. Am. Chem. Soc.* **69**, 2167 (1947).
19. H. Herzog and E. P. Oliveto, *Steroids* **57**, 617 (1992).
20. I. V. Mícovíc, M. D. Ivanovic, and D. M. Piatak, *Synthesis*, 591 (1990).
21. H. Ringold, in M. J. Pramik, ed., *Norethindrone: the First Three Decades*, Syntex Laboratories, Palo Alto, Calif., 1978, p. 1.
22. G. Rosenkranz, *Steroids* **57**, 409 (1992).
23. D. Lednicer and L. A. Mitscher, *The Organic Chemistry of Drug Synthesis*, Vol. 1, John Wiley & Sons, Inc., New York, 1977, p. 182.
24. R. Wiechert, *Angew. Chem. Int. Ed. Engl.* **16**, 506 (1977).
25. J. A. Hogg, *Steroids* **57**, 593 (1992).
26. W. J. Marsheck, S. Kraychy, and R. D. Muir, *Appl. Microbiol.* **23**, 72 (1972).
27. M. G. Wovcha, F. J. Antosz, J. C. Knight, L. A. Kominek, and T. R. Pyke, *Biochim. Biophys. Acta* **531**, 308 (1978).
28. R. Wiechert, *Angew. Chem. Int. Ed. Engl.* **9**, 321 (1970).
29. K. C. Wang, L.-H. Young, Y. Wang, and S.-S. Lee, *Tetrahedron Lett.* **31**,1283 (1990).
30. D. A. Livingston, J. E. Petre, and C. L. Bergh, *J. Am. Chem. Soc.* **112**,6449 (1990).
31. M. Ehrenstein, *J. Org. Chem.* **9**, 435 (1944).
32. C. Djerassi, *Steroids* **57**, 631 (1992).
33. A. J. Birch, *J. Chem. Soc.*, 367 (1950).
34. W. W. Tullner and R. Hertz, *J. Clin. Endocrinol. Metab.* **12**,916 (1952).
35. P. A. Cole and C. H. Robinson, *J. Am. Chem Soc.* **113**,8130 (1991).
36. H. H. Inhoffen, *Angew. Chem.* **59**, 207 (1947).
37. E. B. Hershberg, M. Rubin, and E. Schwenk, *J. Org. Chem.* **15**, 292 (1950).

38. H. L. Dryden, G. M. Webber, and J. Wieczorek, *J. Am. Chem. Soc.* **86**, 742 (1964).
39. J. Kalvoda, K. Heusler, H. Ueberwasser, G. Anner, and A. Wettstein, *Helv. Chim. Acta* **4**, 1361 (1963).
40. T. B. Windholz and M. Windholz, *Angew. Chem. Inter. Ed. Eng.* **3**, 353 (1964).
41. T. Terasawa and T. Okada, *Tetrahedron* **42**, 537 (1986).
42. P. Kocovsky and R. S. Baines, *Tetrahedron Lett.* **34**, 6139 (1993).
43. R. Reich and J. F. W. Keana, *Synth. Commun.* **2**, 323 (1972).
44. C. J. Sih and K. C. Wang, *J. Am. Chem. Soc.* **87**, 1387 (1965).
45. C. J. Sih and A. M. Rahim, *J. Pharm. Sci.* **52**, 1075 (1963).
46. F. Urbanos and co-workers, *J. Am. Chem. Soc.* **115**, 3484 (1993).
47. H. H. Inhoffen, W. Logemann, W. Hohlweg, and A. Serini, *Chem. Ber.* **71**, 1024 (1938).
48. F. B. Colton, *Steroids* **57**, 624 (1992).
49. A. L. Wilds and N. A. Nelson, *J. Am. Chem. Soc.* **75**, 5366 (1953).
50. Ref. 14, p. 424.
51. R. E. Graham, P. A. Williams, and C. T. Kenner, *J. Pharmaceut. Sci.* **59**, 1152 (1970).
52. M. S. de Winter, C. M. Siegmann, and S. A. Szpilfogel, *Chem. Ind.*, 905 (1959).
53. S. Budavari, ed., *The Merck Index*, 11th ed., Merck & Co., Rahway, N.J., 1989, p. 1237.
54. S. Görög, ed., *Steroid Analysis in the Pharmaceutical Industry*, Ellis Horwood, Chichester, U.K., 1989.
55. J. E. F. Reynolds, ed., *Martindale, The Extra Pharmacopoeia*, 30th ed., The Pharmaceutical Press, London, 1993, p. 1178.
56. Ref. 14, p. 10.
57. Ref. 53 p. 49.
58. A. C. Moffat, ed., *Clarke's Isolation and Identification of Drugs*, 2nd ed., The Pharmaceutical Press, London, 1986, p. 328.
59. Ref. 55, p. 1179.
60. Ref. 14, p. 106.
61. Ref. 53, p. 325.
62. J. R. Prous, *Drugs Today* **4**, 4 (1968).
63. K. Brückner, B. Hampel, and U. Johnsen, *Chem. Ber.* **94**, 1225 (1961).
64. R. W. Draper, *J. Chem. Soc. Perkin Trans. I*, 2787 (1983).
65. Ref. 55, p. 1180.
66. Ref. 53, p. 435.
67. S. Kaufman, J. Pataki, G. Rosenkranz, J. Romo, and C. Djerassi, *J. Am. Chem. Soc.* **72**, 4531 (1950).
68. R. Wiechert and co-workers, *Chem. Ber.* **99**, 1118 (1966).
69. D. Lednicer and L. A. Mitscher **2**(23), 166 (1980).
70. E. L. Shapiro, T. L. Popper, L. Weber, R. Neri, and H. L. Herzog, *J. Med. Chem.* **12**, 631 (1969).
71. M. Sittig, *Pharmaceutical Manufacturing Encyclopediea*, Vol. 1, 2nd ed., Noyes Publications, Park Ridge, N.J., 1988, p. 445.
72. T. C. van Soest, L. A. Van Dijck, and F. J. Zeelen, *Recl. Trav. Chim. Pays-Bas* **99**, 323 (1980).
73. M. J. van den Heuvel, C. W. Van Bokoven, H. P. De Jongh, and F. J. Zeelin, *Recl. Trav. Chim. Pays-Bas* **107**, 331 (1988).
74. A. J. van den Broek, C. van Bokhover, P. M. J. Hobbelen, and J. Leemhuis, *Recl. Trav. Chim. Pays-Bas* **94**, 35 (1975).
75. Ger. Pat. 3,909,770 (Mar. 21, 1989), E. Winterfeldt, U. Tilstamm, H. Hofmeister, and H. Laurent (to Schering A. G.).
76. Ref. 55, p. 1184.

77. Ref. 53, p. 606.
78. E. P. K. Lau and J. L. Sutter, in K. Florey, ed., *Analytical Profiles of Drug Substances*, Vol. 3, Academic Press, Inc., New York, 1974, p. 253.
79. P. D. Klimstra and F. B. Colton, *Steroids* **10**, 411 (1967).
80. S. Cacchi, B. Giannoli, and D. Misiti, *Synthesis*, 728 (1974).
81. Ref. 53, p. 691.
82. Ref. 14, p. 457.
83. E. Eckle and co-workers, *Liebigs Ann. Chem.*, 199 (1988).
84. G. Cleve, E. Frost, G.-A. Hoyer, D. Rosenberg, and A. Seeger, *Arzneim. -Forsch.* **36**, 784 (1986).
85. G. Sauer, U. Eder, G. Haffer, G. Neef, and R. Wiechert, *Angew. Chem. Int. Ed. Eng.* **14**, 417 (1975).
86. H. Hofmeister, K. Annen, H. Laurent, K. Petzoldt, and R. Wiechert, *Arzneim. - Forsch.* **36**, 781 (1986).
87. C. Dollery, ed., *Therapeutic Drugs*, Vol. 1, Churchill Livingstone, Edinburgh, U.K., 1991, p. H77.
88. Ref. 14, p. 607.
89. K. Florey, in Ref. 78, Vol. 4, 1975, p. 209.
90. Ref. 53, p. 769.
91. Ref. 55, p. 1185.
92. W. L. Duax and P. D. Strong, *Steroids* **34**, 501 (1979).
93. H. J. Ringold, B. Löken, G. Rosenkranz, and F. Sondheimer, *J. Am. Chem. Soc.* **78**, 816 (1956).
94. U.S. Pat. 2,753,360 (July 3, 1956), E. Kaspar (to Schering AG).
95. Ref. 53, p. 1060.
96. A. M. Sopirak and L. F. Cullen, in Ref. 89, p. 294.
97. Ref. 55, p. 1190.
98. H. Smith and co-workers, *J. Chem. Soc.*, 4472 (1964).
99. H. Baier, G. Dürner, and G. Quinkert, *Helv. Chim. Acta* **68**, 1054 (1985).
100. *Synform* **3**, 19 (1985).
101. Z. G. Hajos and D. R. Parrish, *J. Org. Chem.* **39**, 1615 (1974).
102. Ref. 55, p. 1186.
103. R. E. Ranney, *J. Toxicol. Environ. Health* **3**, 139 (1977).
104. D. C. Rohrer, J. C. Lauffenburger, W. L. Duax, and F. J. Zeelen, *Cryst. Struct. Commun.* **5**, 539 (1976).
105. Ref. 53, p. 909.
106. J. C. Babcock and co-workers, *J. Am. Chem. Soc.* **80**, 2904 (1958).
107. Ref. 55, p. 1187.
108. Ref. 14, p. 571.
109. K. Annen, H. Hofmeister, H. Laurent, and R. Wiechert, *Synthesis*, 34 (1982).
110. Z. Guo, G. Peng, L. Wei, and Q. Liu, *Yiyao Gongye*, 16 (1984).
111. Ref. 53, p. 1058.
112. Ref. 55, p. 1189.
113. Ref. 58, p. 823.
114. J.-P. Mornon, G. Lepicard, and M. J. Delettré, *Compt. Rend. Hebd. Seances Acad. Sci. Sect. C* **282**, 387 (1976).
115. A. P. Schroff and E. S. Moyer, in Ref. 89, p. 268.
116. S. L. Leung, R. Karunanithy, G. Becket, and S. H. Yeo, *Steroids* **46**, 639 (1985).
117. Ref. 53, p. 1060.
118. Brit. Pat. 1,123,104 (Aug. 14, 1968), (to Ortho Pharmaceutical Corp.).
119. Ger. Pat. 2,633,210 (Feb. 17, 1977), A. P. Shroff (to Ortho Pharmaceutical Corp.).
120. Can. Pat. 1,122,592 (Apr. 27, 1982), J. Warnant and J. Jolly (to Roussel-UCLAF).

121. Ref. 55, p. 1194.
122. Ref. 53, p. 1234.
123. *Drugs Future* **8**, 562 (1983).
124. Ger. Pat. 2,107,835 (Sept. 2, 1971), K. Warnant and A. Farcilli (to Roussel-UCLAF).
125. M. J. Weiss, R. E. Schaub, J. F. Poletto, G. R. Allen, and C. J. Coscia, *Chem. Ind.*, 118 (1963).
126. M. Perelman, E. Farkas, E. J. Fornefeld, R. J. Kraay, and R. T. Rapala, *J. Am. Chem. Soc.* **82**, 2402 (1960).
127. P. J. Roberts, *Drugs Future* **3**, 469 (1978).
128. H. J. Kloosterboer, G. H. J. Deckers, M. J. van der Heuvel, and H. J. J. Loozen, *J. Steroid Biochem.* **31**, 567 (1988).
129. A. Cleve, E. Ottow, G. Neef, and R. Wiechert, *Tetrahedron* **49**, 2217 (1993).
130. C. E. Cook, M. C. Wani, Y.-W. Lee, P. A. Fail, and V. Petrow, *Life Sci.* **52**, 155 (1992).
131. A. Bélanger, D. Philibert, and G. Teutsch, *Steroids* **37**, 361 (1981).
132. G. Teutsch, in M. K. Agarwal, ed., *Adrenal Steroid Antagonism*, Walter de Gruyter & Co., Berlin, Germany, 1984, p. 43.
133. G. Neef, in E. Mutschler and E. Winterfeldt, eds., *Trends in Medicinal Chemistry*, VCH Publishers, New York, 1987, p. 565.
134. R. Rohde, G. Neef, G. Sauer, and R. Wiechert, *Tetrahedron Lett.* **26**, 2069 (1985).
135. R. Wiechert and G. Neef, *J. Steroid Biochem.* **27**, 851 (1987).
136. H. Michna, Y. Nishino, G. Neef, W. L. McGuire, and M. R. Schneider, *J. Steroid Biochem. Molec. Biol.* **41**, 339 (1992).
137. M. J. van den Heuvel and M. B. Groen, *Recl. Trav. Chim. Pays-Bas* **112**, 107 (1993).
138. G. E. Arth, H. Schwam, L. H. Sarett, and M. Glitzer, *J. Med. Chem.* **6**, 617 (1963).
139. J.-F. Miquel and J. Gilbert, *J. Steroid Biochem.* **31**, 525 (1988).
140. Eur. Pat. 253,500 (Jan. 20, 1988), H. Tucker (to Imperial Chemical Industries).
141. Eur. Pat. 253,503 (Jan. 20, 1988), H. Tucker (to Imperial Chemical Industries).
142. I. Hutton, ed., *Pharmaprojects*, Vol. 14, PJB Publications, Richmond, U.K., 1993, p. a640.
143. M. S. Neelima and A. P. Bhaduri, in E. Jucker, ed., *Progress in Drug Research*, Vol. 30, Birkhaüser Verlag, Basel, Germany, 1986, p. 151.
144. A. Agnihotri and co-workers, *Exp. Clin. Endocrinol.* **88**, 185 (1986).
145. S. K. Saxena, M. Seth, A. P. Bhaduri, and M. K. Sahib, *J. Steroid Biochem.* **18**, 303 (1983).
146. R. N. Iyer and R. Gopalchari, *Ind. J. Pharmacy* **31**, 49 (1969).
147. U.S. Pat. 3,760,007 (Sept. 18, 1973), M. Steinman (to Schering Corp.).
148. P. Gentili, *Farmaco Ed. Sci.* **31**, 572 (1976).
149. L. Starka, R. Hampl, and A. Kasal, in M. K. Agarwal, ed., *Receptor Mediated Antisteroid Action*, Walter de Gruyter, Berlin, Germany, 1987, p. 17.
150. M. Misrahi and co-workers, *Biochem. Biophys. Res. Commun.* **143**, 740 (1987).
151. T. Ojasoo, J.-C. Doré, J.-P. Mornon, and J.-P. Raynaud, in M. Bohl and W. L. Duax, eds., *Molecular Structure and Biological Activity of Steroids*, CRC Press, Boca Raton, Fla., 1992, p. 157.
152. R. L. Miesfeld, *Crit. Rev. Biochem. Molec. Biol.* **24**, 101 (1989).
153. A. Guiochon-Mantel and co-workers, *J. Steroid Biochem. Molec. Biol.* **41**, 209 (1992).
154. F. Sweet and G. L. Murdock, *Endocrine Rev.* **8**, 154 (1987).
155. P. R. Kym, K. E. Carlson, and J. A. Katzenellenbogen, *J. Med. Chem.* **36**, 1111 (1993).
156. G. Teutsch, M. Klich, F. Bouchoux, E. Cerede, and D. Philibert, *Steroids* **59**, 22 (1994).
157. D. J. Lamb, P. E. Kima, and D. W. Bullock, *Biochemistry* **25**, 6319 (1986).
158. P.-E. Strömstedt, A. Berkenstam, H. Jörnvall, J.-A. Gustafsson, and J. Carstedt-Duke, *J. Biol. Chem.* **265**, 12973 (1990).

159. M. E. Baker and L. S. Terry, *Steroids* **42**, 3121 (1983).
160. M. K. Agarwal, *Biochem. Pharmacol.* **43**, 2299 (1992).
161. W. L. Duax and J. F. Griffin, in Ref. 151, p. 1.
162. G. Teutsch, M. Gaillard-Moguilewshy, G. Lemoine, F. Nique, and D. Philibert, *Biochem. Soc. Transact.* **9**, 901 (1991).
163. T. Ojasoo, J. P. Raynaud, and J. P. Mornon, in C. Hansch, ed., *Comprehensive Medicinal Chemistry*, Vol. 3, Pergamon Press, Oxford, U.K., 1990, p. 1175.
164. J. H. Arevalo, E. A. Stura, M. J. Taussig, and I. A. Wilson, *J. Mol. Biol.* **231**, 103 (1993).
165. R. A. Goldstein, J. A. Katzenellenbogen, Z. A. Luthey-Schulten, D. A. Seielstad, and P. G. Wolynes, *Proc. Natl. Acad. Sci. USA*, **90**, 9949 (1993).
166. T. C. Wong and co-workers, *J. Chem. Soc. Perkin Trans. II*, 765 (1988).
167. K. Marat, J. F. Templeton, and V. P. S. Kumar, *Magn. Reson. Chem.* **25**, 25 (1987).
168. H.-J. Schneider, U. Buchheit, N. Becker, G. Schmidt, and U. Siehl, *J. Am. Chem. Soc.* **197**, 7027 (1985).
169. E. Surcouf, G. Lepicard, J. P. Mornon, T. Ojasoo, and J. P. Raynaud, *J. Med. Chem.* **26**, 1320 (1983).
170. J. Delettré, J. P. Mornon, G. Lepicard, T. Ojasoo, and J. P. Raynaud, *J. Steroid Biochem.* **13**, 45 (1980).
171. S. Profeta, P. A. Kollman, and M. E. Wolff, *J. Am. Chem. Soc.* **104**, 3745 (1982).
172. J. M. Adelantado and W. G. Richards, *J. Chem. Soc. Perkin Trans. II*, 1253 (1986).
173. M. Bohl, G. Kaufmann, M. Hübner, G. Reck, and R.-G. Kretschmer, *J. Steroid Biochem.* **23**, 895 (1985).
174. M. Bohl, in Ref. 151, p. 91.
175. F. J. Zeelen, *Quant. Struct.-Act. Relat.* **5**, 131 (1986).
176. H.-O. Hoppen and P. Hammann, *Acta Endocrinologica* **115**, 406 (1987).
177. R. Karunanithy, S. L. Leung, and S. H. Yeo, *J. Pharm. Biochem. Anal.* **5**, 597 (1987).
178. K. H. Schönemann, N. P. van Vliet, and F. J. Zeelen, *Steroids* **45**, 297 (1985).
179. D. H. Seeley, W.-Y. Wang, and H. A. Salhanick, *J. Biol. Chem.* **257**, 13359 (1982).
180. Z. Simon and M. Bohl, *Quant. Struct.-Act. Relat.* **11**, 23 (1992).
181. T. Ojasoo, J. C. Doré, J. Gilbert, and J. P. Raynaud, *J. Med. Chem.* **31**, 1160 (1988).
182. D. A. McGinty, L. P. Anderson, and N. B. McCullough, *Endocrinology* **24**, 829 (1939).
183. D. A. Loughney and C. F. Schwender, *J. Computer-Aided Molecul. Des.* **6**, 569 (1992).
184. C. A. Beck, N. L. Weigel, and D. P. Edwards, *Molec. Endocrinol.* **6**, 607 (1992).
185. M. K. Bagchi, S. Y. Tsai, M. J. Tsai, and B. W. O'Malley, *Mol. Cell. Biol.* **11**, 4998 (1991).
186. K. B. Horwitz, M. D. Francis, and L. L. Wei, *DNA* **4**, 451 (1985).
187. B. A. Lessey, P. S. Alexander, and K. B. Horwitz, *Endocrinology* **112**, 1267 (1983).
188. K. Christensen and co-workers, *Molec. Endocrinol.* **5**, 1755 (1991).
189. R. M. Evans, *Science* **240**, 889 (1988).
190. P. J. Godowski and D. Picard, *Biochem. Pharmacol.* **38**, 3135 (1989).
191. R. L. Miesfeld, *Crit. Rev. Biochem. Molec. Biol.* **24**, 101 (1989).
192. L. P. Freedman, *Endocrine Rev.* **13**, 129 (1992).
193. P. L. Sheridan, R. M. Evans, and K. B. Horwitz, *J. Biol. Chem.* **264**, 6520 (1989).
194. F. Logeat, M. Le Cunff, R. Pamphile, and E. Milgrom, *Biochem. Biophys. Res. Commun.* **131**, 421 (1985).
195. D. F. Smith, L. E. Faber, and D. O. Toft, *J. Biol. Chem.* **265**, 3996 (1990).
196. A. M. DeMarzo, C. A. Beck, S. A. Oñate, and D. P. Edwards, *Proc. Natl. Acad. Sci. USA* **88**, 72 (1991).
197. L. Tora, H. Gronemeyer, B. Turcotte, M.-P. Gaub, and P. Chambon, *Nature* **333**, 185 (1988).
198. E. Vegeto and co-workers, *Molec. Endocrinol.* **7**, 1241 (1993).

199. L. Tung, M. K. Mohamed, J. P. Hoeffler, G. S. Takimoto, and K. B. Horwitz, *Molec. Endocrinol.* **7**, 1256 (1993).
200. M. E. Meyer and co-workers, *EMBO J.* **9**, 3923 (1990).
201. K. B. Horwitz, *Endocrine Rev.* **13**, 146 (1992).
202. G. F. Allen and co-workers, *J. Biol. Chem.* **267**, 19513 (1992).
203. D. El-Ashry, S. A. Oñate, S. K. Nordeen, and D. P. Edwards, *Molec. Endocrinol.* **3**, 1545 (1989).
204. N. L. Weigel and co-workers, *Molec. Endocrinol.* **6**, 1585 (1992).
205. K. B. Horwitz, *Endocrinology* **116**, 2236 (1985).
206. M. Kalimi, in Ref. 149, p. 121.
207. V. K. Moudgil, G. Lombardo, C. Hurd, N. Eliezer, and M. K. Agarwal, *Biochim. Biophys. Acta* **889**, 192 (1986).
208. B. Benhamou and co-workers, *Science* **255**, 206 (1992).
209. A. Guiochon-Mantel and co-workers, *Nature* **336**, 695 (1988).
210. A. M. Etgen and I. Vathy, in M. K. Agarwal, ed., *Antihormones in Health and Disease*, Karger, Basel, Switzerland, 1991, p. 45.
211. A. Gravanis and co-workers, *J. Clin. Endocrinol. Metab.* **60**, 156 (1985).
212. L. Klein-Hitpass, A. C. B. Cato, D. Henderson, and G. U. Ryffel, *Nucleic. Acids Res.* **19**, 1227 (1991).
213. R. Nath, A. Bhakta, and V. K. Mougil, *Arch. Biochem. Biophys.* **292**, 303 (1992).
214. C. A. Beck, N. L. Weigel, M. L. Moyer, S. K. Nordeen, and D. P. Edwards, *Proceed. Natl. Acad. Sci. USA* **90**, 4441 (1993).
215. T. Mizutani, A. Bhakta, H. J. Kloosterboer, and V. K. Moudgil, *J. Steroid Biochem. Mol. Biol.* **42**, 695 (1992).
216. M. Beato, *Cell* **56**, 335 (1989).
217. K. R. Yamamoto, *Ann. Rev. Genet.* **19**, 209 (1985).
218. W. W. Leavitt, T. J. Chen, Y. S. Do, B. D. Carlton, and T. C. Allen, in B. W. O'Malley and L. Birnbaumer, eds., *Receptors and Hormone Action*, Academic Press, Inc., New York, 1978, p. 157.
219. A. H. Ree and co-workers, *Endocrinology* **124**, 2577 (1989).
220. A. M. Nardulli, G. L. Greene, B. W. O'Malley, and B. S. Katzenellenbogen, *Endocrinology* **122**, 935 (1988).
221. C. L. Clarke, *Molec. Cell. Endocrinol.* **70**, C29 (1990).
222. A. M. Nardulli and B. S. Katzenellenbogen, *Endocrinology* **122**, 1532 (1988).
223. B. S. Katzenellenbogen and M. J. Norman, *Endocrinology* **126**, 891 (1990).
224. C. Sumida, F. Lecerf, and J. R. Pasqualini, *Endocrinology* **122**, 3 (1988).
225. S. Hild-Petito, R. L. Stouffer, and R. M. Brenner, *Endocrinology* **123**, 2896 (1988).
226. R. E. Leake and F. Habib, in B. Green and R. E. Leake, eds., *Steroid Hormones: A Practical Approach*, IRL Press, Oxford, U.K., 1987, p. 67.
227. M. A. Blankenstein and E. Mulder, in B. A. Cooke, R. J. B. King, and H. J. van der Molen, eds., *Hormones and Their Actions, Part I*, Elsevier Sciences Publishers BV, Amsterdam, 1988, p. 49.
228. B. C. Goverde, H. J. Kloosterboer, and A. H. W. M. Schuurs, in Ref. 54, p. 156.
229. F. J. Zeelen, *Trends Pharm. Sci.* **4**, 520 (1983).
230. X. Zhang, G. M. Stone, and B. G. Miller, *Reproduct. Fertil. Dev.* **1**, 223 (1989).
231. V. Isomaa, *Biochem. Biophys. Acta* **675**, 9 (1981).
232. K. Kontula and co-workers, *Acta Endocrinol.* **78**, 574 (1975).
233. S. S. Shapiro, R. D. Dyer, and A. E. Colás, *Am. J. Obstet. Gynecol.* **132**, 549 (1978).
234. C. H. Spilman, J. W. Wilks, and J. A. Campbell, *J. Steroid Biochem.* **22**, 289 (1985).
235. V. K. Moudgil and C. Hurd, *Biochemistry* **26**, 4993 (1987).
236. O. Heikinheimo and co-workers, *J. Steroid Biochem.* **26**, 279 (1987).
237. J. R. Schreiber, A. J. Hsueh, and E. E. Baulieu, *Contraception* **28**, 77 (1983).

238. G. O. Gray and W. W. Leavitt, *J. Steroid Biochem.* **28**, 493 (1987).
239. K. B. Horwitz and G. R. Freidenberg, *Cancer Res.* **45**, 167 (1985).
240. R. L. Sutherland, R. E. Hall, G. Y. N. Pang, E. A. Musgrove, and C. L. Clarke, *Cancer Res.* **48**, 5084 (1988).
241. J. R. Hissom and M. R. Moore, *Biochem. Biophys. Res. Commun.* **145**, 706 (1987).
242. A. Manni, C. Wright, B. Badger, L. Demers, and M. Bartholomew, *Cancer Res.* **48**, 3058 (1988).
243. R. Poulin, D. Baker, D. Poirier, and F. Labrie, *Breast Can. Res. Treat.* **17**, 197 (1991).
244. L. Markiewicz, R. B. Hochberg, and E. Gurpide, *J. Steroid Biochem. Molec. Biol.* **41**, 53 (1992).
245. V. C. Jordan, M. H. Jeng, W. H. Catherino, and C. J. Parker, *Cancer* **71**(suppl), 1501 (1993).
246. D. Chalbos, M. Chambon, G. Ailhaud, and H. Rochefort, *J. Biol. Chem.* **262**, 9923 (1987).
247. D. Chalbos and H. Rochefort, *J. Biol. Chem.* **259**, 1231 (1984).
248. A. C. B. Cato, R. Miksicek, G. Schütz, J. Arnemann, and M. Beato, *EMBO J.* **5**, 2237 (1986).
249. A. C. B. Cato, D. Henderson, and H. Ponta, *EMBO J.* **6**, 363 (1987).
250. M. K. Bagchi, M.-J. Tsai, B. W. O'Malley, and S. Y. Tsai, *Endocrine Rev.* **13**, 525 (1992).
251. Z. S. Madjerek, in Ref. 4, p. 389.
252. M. K. McPhail, *J. Physiol.* **82**, 145 (1934).
253. R. A. Edgren, in J. W. Goldzieher and K. Fortherby, eds., *Pharmacology of the Contraceptive Steroids*, Raven Press, New York, 1994, p. 81.
254. R. A. Edgren and D. L. Peterson, *Proc. Soc. Exp. Biol. Med.* **123**, 867 (1966).
255. G. Pincus and G. Bialy, *Rec. Prog. Horm. Res.* **19**, 201 (1963).
256. O. A. Jänne, V. V. Isomaa, T. K. Torkkeli, H. E. Isotalo, and H. T. Kopu, in C. W. Bardin, E. Milgröm, and P. Mauvais-Jarvis, eds., *Progesterone and Progestins*, Raven Press, New York, 1983, p. 33.
257. O. Jänne and co-workers, in M. Beato, ed., *Steroid Induced Uterine Proteins*, Elsevier, Amsterdam, 1980, p. 319.
258. C. L. Clarke and R. S. Sutherland, *Endocrine Rev.* **11**, 266 (1990).
259. N. J. MacLusky and B. S. McEwen, *J. Clin. Endocrinol. Metab.* **106**, 192 (1980).
260. P. T. Pearce, B. A. K. Khalid, and J. W. Funder, *Endocrinology*, **113**, 1287 (1983).
261. J. Szekeres-Bartho, D. Philibert, and G. Chaouat, *Am. J. Reprod. Immunol.* **23**, 42 (1990).
262. M. C. Etienne and co-workers, *Eur. J. Cancer* **26**, 807 (1990).
263. E. F. Eriksen and co-workers, *Science* **241**, 84 (1988).
264. R. M. Brenner, J. A. Resko, and N. B. West, *Endocrinology* **95**, 1094 (1974).
265. J. H. Clark, Z. Paszko, and E. J. Peck, Jr., *Endocrinology* **100**, 91 (1977).
266. E. Gurpide, *Pediatrics* **62**, 1114 (1978).
267. E. Johannisson, B.-M. Landgren, and E. Diczfalusy, *Contraception* **25**, 13 (1982).
268. E. Johannisson and I. Brosens, in Ref. 253, p. 211.
269. L. P. Bengtsson, in Ref. 4, p. 487.
270. E. M. Coutinho, H. Maia, and R. X. da Costa, *Int. J. Fertil.* **18**, 161 (1973).
271. H. Vorherr, *Am. J. Obstet. Gynecol.* **154**, 161 (1986).
272. L.-M. Houdebine and co-workers, in Ref. 256, p. 297.
273. J. C. Prior, *Endocrine Rev.* **11**, 386 (1990).
274. V. D. Ramirez, K. Kim, and D. Dluzen, *Rec. Prog. Horm. Res.* **41**, 421 (1985).
275. D. Keefe, in Ref. 253, p. 283.
276. S. M. Paul and R. H. Purdy, *FASEB J.* **6**, 2311 (1992).
277. S. I. Deutsch, J. Mastropaolo, and A. Hitri, *Clin. Neuropharmacol.* **15**, 352 (1992).
278. L. Hellman and co-workers, *J. Clin. Endocrinol. Metab.* **42**, 912 (1976).

279. G. Hoppe, *Contraception* **37**, 493 (1988).
280. S. M. Petak and E. Steinberger, in Ref. 253, p. 233.
281. C. W. Bardin, in Ref. 256, p. 135.
282. K. Elkind-Hirch and J. W. Goldzieher, in Ref. 253, p. 345.
283. R. T. Burkman, *Semin. Reproduct. Endocrinol.* **7**, 224 (1989).
284. R. H. Knopp, *J. Reproduct. Med.* **31**, 913 (1986).
285. P. Wahl and co-workers, *Engl. J. Med.* **308**, 862 (1983).
286. P. G. Brooks, *J. Reproduct. Med.* **29**, 539 (1984).
287. S. O. Skouby and K. R. Peterson, *Int. J. Fert.* **36**(suppl. 1), 32 (1991).
288. J. W. Goldzieher, in Ref. 253, p. 185.
289. U. Lachnit-Fixson, *The Development of a New Triphasic Oral Contraceptive*, MTP Press, Lancaster, U.K., 1980, p. 23.
290. D. C. Foster, *Semin. Reproduct. Endocrinol.* **7**, 205 (1989).
291. D. Crook, I. F. Godsland, and V. Wynn, *Int. J. Fertil.* **36**(suppl.), 38 (1991).
292. *Physician's Desk Reference*, 48th ed., Medical Economics Data Production, Montvale, N.J., 1994, p. 2414.
293. S. Segal, *Am. J. Obstet. Gynecol.* **157**, 1090 (1987).
294. Ref. 292, p. 2564.
295. D. M. Potts and J. B. Smith, *Int. J. Fertil.* **36**(suppl.), 57 (1991).
296. F. M. Primiero and G. Benagiano, in Ref. 253, p. 153.
297. R. J. Santen, A. Manni, H. Harvey, and C. Redmond, *Endocrine Rev.* **11**, 221 (1990).
298. H. Vorherr, in E. J. Quilligan, ed., *Current Therapy in Obstetrics and Gynecology*, 3rd ed., W. B. Saunders, Philadelphia, Pa., 1990, p. 367.
299. J. M. Goldberg, in Ref. 298, p. 44.
300. H. R. K. Barber, in Ref. 298, p. 41.
301. J. Bonte, J. M. Decoster, and G. Billiet, *Gynecol. Oncol.* **6**, 60 (1978).
302. J. Bonte, J. M. Decoster, and P. Ide, *Cancer* **25**, 907 (1970).
303. R. A. Lobo, in Ref. 298, p. 3.
304. R. Cragun and R. J. Chang, in Ref. 298, p. 95.
305. R. D. Gambrell, Jr., *Int. J. Fertil.* **31**, 112 (1986).
306. M. I. Whitehead, P. T. Townsend, J. Pryse-Davies, T. A. Ryder, and R. J. B. King, *N. Engl. J. Med.* **305**, 1599 (1981).
307. G. Lane, N. C. Siddle, T. A. Ryder, J. Pryse-Davies, and R. J. B. King, *Br. J. Obstet. Gynaecol.* **93**, 55 (1986).
308. J. L. Bullock, F. M. Massey, and R. D. Gambrell, Jr., *Obstet. Gynecol.* **46**, 165 (1975).
309. B. H. Albrecht, I. Schiff, D. Tulchinsky, and K. J. Ryan, *Am. J. Obstet. Gynecol.* **139**, 631 (1981).
310. W. A. Peck, in H. J. Armbrecht, R. M. Coe, and N. Wongsurawat, eds., *Endocrine Function and Aging*, Springer-Verlag, New York, 1989, p. 67.
311. J. C. Stevenson, *Obstet. Gynecol.* **75**(suppl.), 36S (1990).
312. L. L. Wei, M. W. Leach, R. S. Miner, and L. M. Demers, *Biochem. Biophys. Res. Commun.* **195**, 525 (1993).
313. E. I. Barengolts and co-workers, *J. Bone Min. Res.* **5**, 1143 (1990).
314. J. C. Gallagher, W. T. Kable, and D. Goldgar, *Am. J. Med.* **90**, 171 (1991).
315. S. G. McNeeley, Jr., J. S. Schinfeld, T. G. Stovall, F. W. Ling, and B. H. Buxton, *Int. J. Gynecol. Obstet.* **34**, 253 (1991).
316. *FDC Rep.* **55**, T&G 4 (1993).
317. Y. E. Shi, C. H. He, J. Gu, and K. Fotherby, *Contraception* **35**, 465 (1987).
318. D. J. Back and co-workers, *Clin. Pharm. Therap.* **24**, 439 (1978).
319. G. L. Hammond, P. L. A. Lähteenmäki, P. Lähteenmäki, and T. Luukkainen, *J. Steroid Biochem.* **17**, 375 (1982).
320. K. Fotherby, in Ref. 253, p. 99.
321. D. S. Layne, T. Golab, K. Arai, and G. Pincus, *Biochem. Pharm.* **905**, B 12 (1963).

322. S. Kamyab, K. Fotherby, and A. I. Klopper, *J. Endocrinol.* **41**, 263 (1968).
323. E. Gerhards, W. Hecker, H. Hitze, B. Nieuweboer, and O. Bellmann, *Acta Endocrinol.* **68**, 219 (1971).
324. J. W. Munson, in Ref. 54, p. 371.
325. L. Siekmann, A. Siekmann, and H. Breuer, *Biomed. Mass. Spectr.* **7**, 511 (1980).
326. D. J. Back and co-workers, *Contraception* **23**, 229 (1981).
327. M. L. 'E. Orme, D. J. Back, and A. M. Breckenridge, *Clin. Pharmacokinet.* **8**, 95 (1983).
328. G. L. Hammond, M. S. Langley, P. A. Robinson, S. Nummi, and L. Lund, *Fertil. Steril.* **42**, 44 (1984).
329. D. C. Dejongh and co-workers, *Steroids* **11**, 649 (1968).
330. H. S. Weintraub, L. S. Abrams, J. E. Patrick, and J. L. McGuire, *J. Pharm. Sci.* **67**, 1406 (1978).
331. K. B. Alton, N. S. Hetyei, C. Shaw, and J. E. Patrick, *Contraception* **29**, 19 (1984).
332. J. Killinger, D. W. Hahn, A. Phillips, N. S. Hetyei, and J. L. McGuire, *Contraception* **32**, 311 (1985).
333. J. L. McGuire and co-workers, *Am. J. Obstet. Gynecol.* **163**, 2127 (1990).
334. U. Taüber, J. W. Tack, and H. Matthes, *Contraception* **40**, 461 (1989).
335. H. Kuhl, C. Jung-Hoffmann, and F. Heidt, *Contraception* **38**, 477 (1988).
336. Q. G. Li and M. Hümpel, *J. Steroid Biochem.* **35**, 319 (1990).
337. U. Tauber, W. Kuhnz, and M. Hümpel, *Am. J. Obstet. Gynecol.* **163**, 1414 (1990).
338. B. Düsterberg, J. Tack, W. Krause, and M. Hümpel, in M. Elstein, ed., *Gestodene*, Parthenon, Carnforth, U.K., 1987, p. 35.
339. M. A. Shaw, D. J. Back, A. M. Cowie, and M. L. 'E. Orme, *J. Steroid Biochem.* **22**, 111 (1985).
340. L. Viinikka and co-workers, *Eur. J. Clin. Pharm.* **15**, 349 (1979).
341. H. G. Hasenack, A. M. G. Bosch, and K. Käär, *Contraception* **33**, 591 (1986).
342. E. W. Bergink, A. D. Hamburger, E. de Jager, and J. van der Vies, *J. Steroid Biochem.* **14**, 175 (1981).
343. W. Kuhnz, M. Pfeffer, and G. Al-Yacoub, *J. Steroid Biochem.* **35**, 313 (1990).
344. S. Madden, D. J. Back, and M. L. 'E. Orme, *J. Steroid Biochem.* **35**, 281 (1990).
345. S. Madden, D. J. Back, C. A. Martin, and M. L. 'E. Orme, *Br. J. Clin. Pharm.* **27**, 295 (1989).
346. H. Aldercreutz, P. B. Erikson, and M. S. Christensen, *J. Pharm. Biomed. Anal.* **1**, 153 (1983).
347. G. P. Mould, J. Read, D. Edwards, and A. Bye, *J. Pharm. Biomed. Anal.* **7**, 119 (1989).
348. M. Mathrubutham and K. Fortherby, *J. Steroid Biochem.* **14**, 783 (1981).
349. A. Wikström, B. Green, and E. D. B. Johansson, *Acta Obstet. Gynecol. Scand.* **63**, 163 (1984).
350. U. Goebelsmann, in A. T. Gregoire and R. P. Blye, eds., *Contraceptive Steroids: Pharmacology and Safety*, Plenum Press, New York, 1986, p. 67.
351. P. Lähteenmäki and co-workers, *J. Steroid Biochem.* **27**, 859 (1987).
352. S. Kawai and co-workers, *J. Pharm. Exp. Therap.* **241**, 401 (1987).
353. J. Salmon and M. Mouren, in E. E. Baulieu and S. J. Segal, eds., *The Antiprogestin Steroid RU486 and Human Fertility Control*, Plenum Press, New York, 1985, p. 99.
354. O. Heikinheimo, M. Tevilin, D. Shoupe, H. Croxatto, and P. Lähteenmäki, *Contraception* **34**, 613 (1986).

JOSEPH W. GUNNET
LISA A. DIXON
The R. W. Johnson Pharmaceutical Research Institute

ESTROGENS AND ANTIESTROGENS

Estrogens are a group of naturally occurring steroid sex hormones which are characterized by their ability to induce estrus in the female mammal. They are derivatives of the planar tetracyclic structure estra-1,3,5(10)-trien-3-ol [53-63-4] (1), and the three principal estrogens in humans are estrone [56-16-7] (E$_1$) (2), estradiol [50-28-2] (E$_2$) (3), and estriol [50-27-1] (E$_3$) (4).

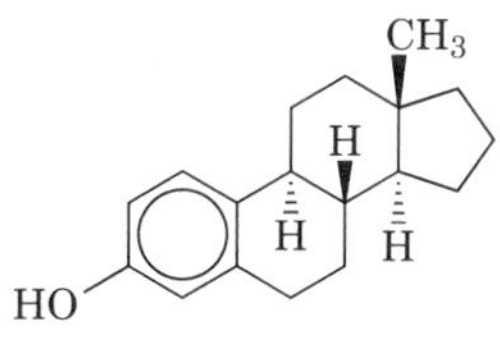

(1) 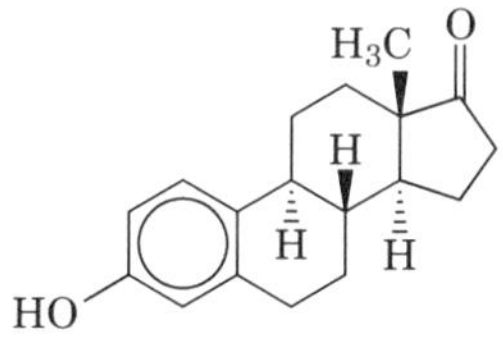　　　　　　　　　　(2)

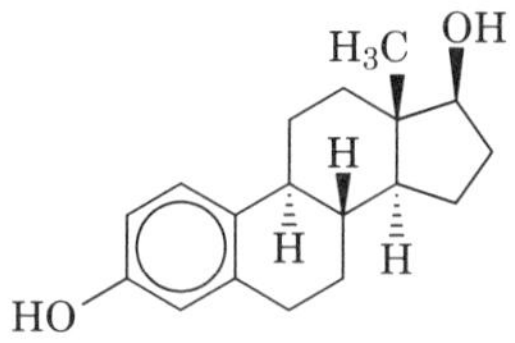

(3) 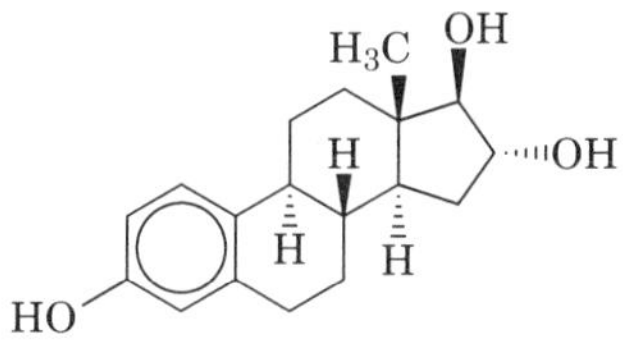　　　　　　　　　　(4)

The two synthetic steroidal estrogens which have attained the greatest degree of therapeutic use are ethinyl estradiol [57-63-6] (EE) (5) and its 3-methyl ether, mestranol [72-33-3] (6). In contrast to the naturally occurring estrone derivatives, these acetylenic analogues are orally active and are the main estrogenic components of combination oral contraceptives (see CONTRACEPTIVES) and certain estrogen replacement products.

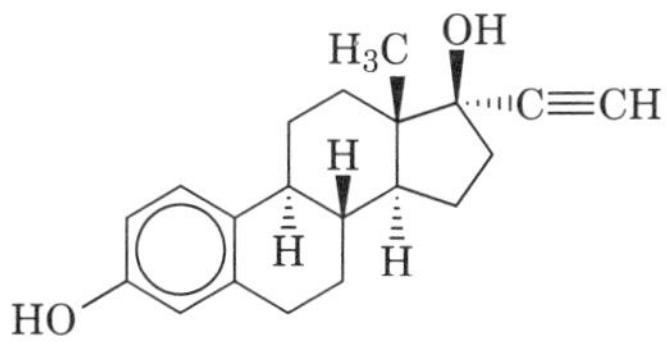

(5) 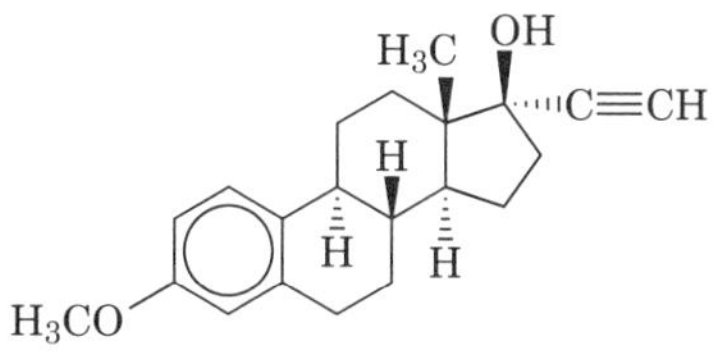　　　　　　　　　　(6)

Diethylstilbestrol [56-53-1] (DES) (7), which was first synthesized in the 1930s, is the most widely studied nonsteroidal estrogen and has been extensively reviewed. It is an extremely potent estrogen, possessing four times the oral potency of estradiol (3), but carcinogenicity problems have limited its use.

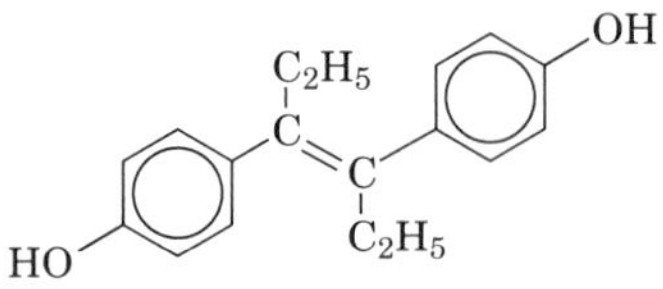

(**7**)

In the 1980s little progress occurred in the introduction of new estrogens, either steroidal or nonsteroidal, into human medicine. However, there has been a resurgence of interest in the development of novel steroidal estrogens for specific medical uses in the 1990s and a significant advancement in the understanding of the mechanism of estrogen action at the molecular level. For instance, the estrogen receptor has been characterized as consisting of three major domains, one of them being a DNA-binding domain which specifically binds to estrogen response elements in the target genes. The estrogen receptor will be discussed in more detail in the pharmacology section of this article. Further elucidation of the nature of this steroid-receptor complex will supply the groundwork for the development of new estrogens with unique and tissue-specific therapeutic effects via a structure-based approach to their design and discovery.

For the purposes of this article, antiestrogens are compounds that counteract the biological activity of estrogens at the receptor level. In the late 1970s, there were no steroidal antiestrogens in widespread clinical use. Clomiphene [*911-45-5*] (**8**) and tamoxifen [*10540-29-1*] (**9**) were nonsteroidal antiestrogens that had been employed for the treatment of female infertility and breast cancer, respectively.

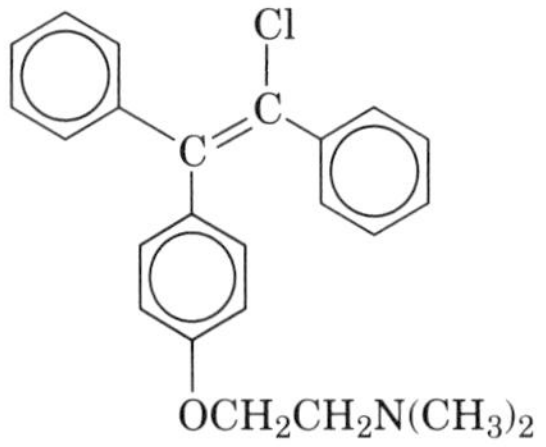

(**8**) 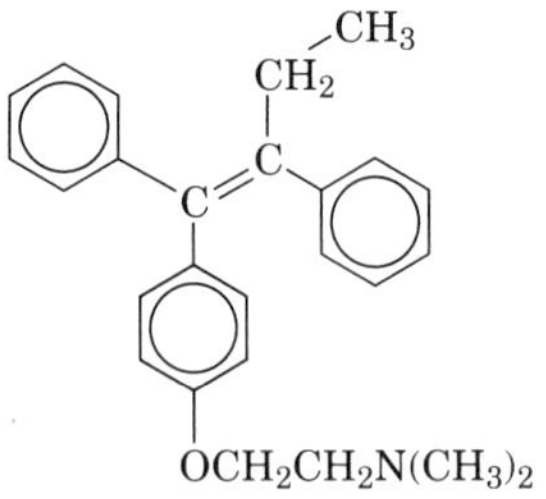   (**9**)

This article highlights the progress in the development of steroidal antiestrogens such as ICI 182780, compounds with greatly reduced partial estrogenic activity, and selected nonsteroidal antiestrogens classified as triarylethylenes (TAEs), chromenes, benzofurans, benzothiophenes, carbocyclic TAEs, diphenylmethanes, diphenylethanes, and indoles. Representative studies which have attempted to delineate the mechanism of action of estrogens and antiestrogens based on binding studies, molecular modeling strategies, or the like have also been highlighted. An overview of the pharmacology of estrogens is presented with a particular emphasis on receptor-mediated molecular events. Therapeutically, estrogens are used for contraception, hormone replacement therapy, and chemotherapy.

Chemistry

Steroidal and Nonsteroidal Estrogens.   Modification of the basic steroid skeleton and the nature of the functional groups in the B, C, and D rings while maintaining the phenolic A-ring has continued to be a primary approach in the development of new estrogens with unique biological profiles.

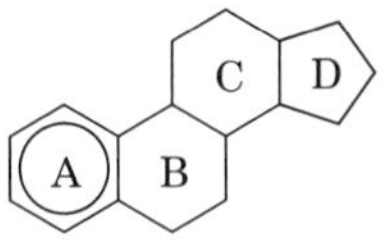

A series of patents from Schering AG has described the synthesis of various 14–17 α- and β-ethano-bridged estratriene compounds represented by compounds [116229-12-0] (**10**), [130693-81-1] (**11**), and [135768-83-1] (**12**). All are reported to be potent estrogens (1–3). A series of 17-halomethylene estratrienes represented by compound [123651-64-9] (**13**) have also been reported by Schering AG to be potent estrogens (4).

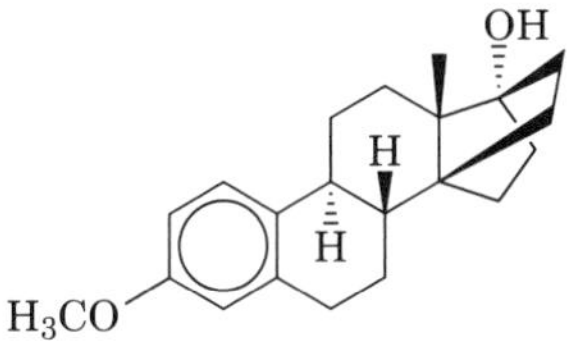

(10)

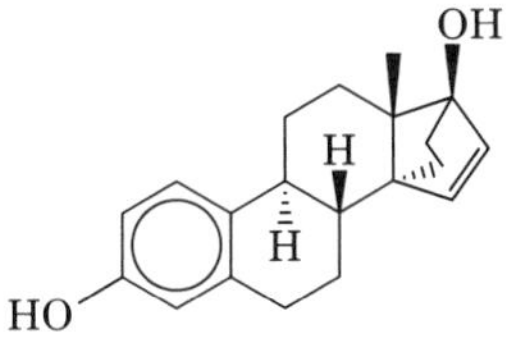

(11)

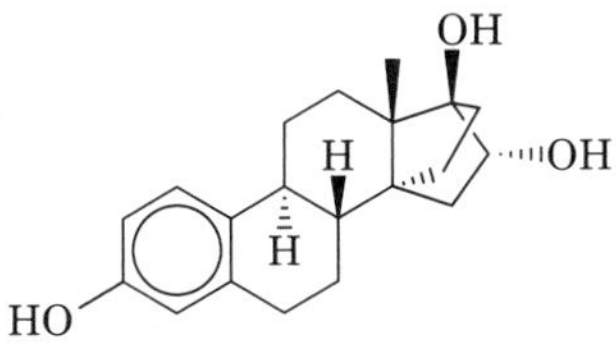

(12)

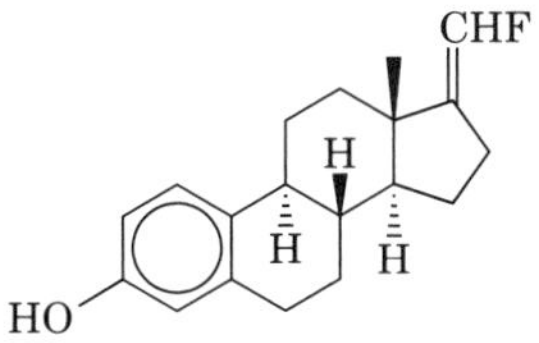

(13)

11β-Nitrate Esters.   It was discovered in the late 1930s that introduction of a 17α-alkynyl group in estradiol gave orally active estrogens such as EE (ethinyl estradiol), which have been widely employed with synthetic progestins (pharmaceutical agents which have effects similar to progesterone) in oral contraceptives.

The 17-alkynyl group is more resistant to metabolism in the liver than are the naturally occurring estrogens, and attention has refocused on the search for new orally active super estrogens which could be used at lower doses and theoretically reduce the metabolic burden. As was the case with the hormone antagonists, introduction of functionality at the 11-position of the estrane skeleton has resulted in the generation of orally active potent estrogen agonists. Agonists are drugs that stimulate activity at cell receptors normally stimulated by naturally occurring substances. Various estrane derivatives have been converted with ceric ammonium nitrate selectively and efficiently to the corresponding $9\alpha,11\beta$-hydroxy nitrate esters which were then deoxygenated at C-9 with triethylsilane–boron trifluoride etherate to yield the desired 11β-nitrate esters (5); standard transformations then gave the 7α-methyl target compounds such as (14). The results of uterotropic and post-coital antifertility activities in rats and estrogen withdrawal bleeding studies in monkeys for a similar series of estradiol analogues and ethinyl estradiol analogues were determined (6). The estrogenic potency in rats of the estradiol derivatives CDB-3280 and CDB-1357 [108887-25-8] when taken orally were 0.63 and 14.41 times the potency of EE (5), respectively, whereas the corresponding ethinyl estradiol derivatives, CDB-3294 and CDB-3322, were surprisingly only 5.09 and 2.0 times the potency of EE, respectively. Post-coital activity in the rat was on the same order as estrogenic potency, with the order being CDB-1357 $(ED_{100} = (5\ \mu g/kg)/d$ for each of 5 days) > CDB-3294 $(ED_{100} = (<10\ \mu g/kg)/d$ for each of 5 days) > CDB-3322 $(ED_{100} = (<20\ \mu g/kg)/d$ for each of 5 days) > CDB-3280 $(ED_{100} = (100\ \mu g/kg)/d$ for each of 5 days).

(14)

where R' = R= H for CDB-3280   R' = C≡CH; R = H for CDB-3294
R' = H; R = CH$_3$ for CDB-1357   R' = C≡CH; R = CH$_3$ for CDB-3322

C-11 functionalized estrone derivatives are synthetically available by the following process (5): treatment of 7α-methylestrone acetate [36014-09-2] (15) with four equivalents of ceric ammonium nitrate in 90% acetic acid provides the 9α-hydroxy-11β-nitrate ester in good yield. Subsequent deoxygenation of the C-9 benzylic position of the nitrate ester with retention of configuration is effected utilizing triethylsilane and boron trifluoride etherate to produce compound (16). Reduction with NaBH$_4$ affords the key target (14). A nearly identical approach has been used to prepare (16) which was converted to (14) and also ethinylated at C-17 with acetylene and potassium t-butoxide to give CDB-3322 (6).

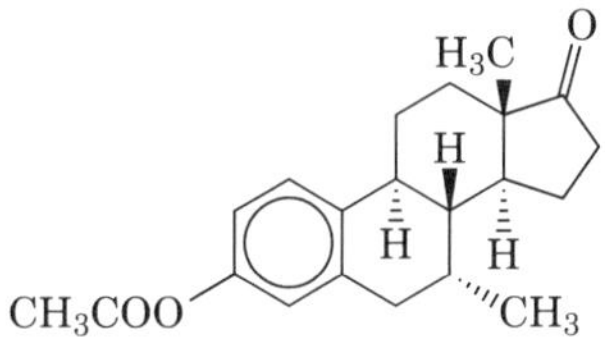

(**15**) 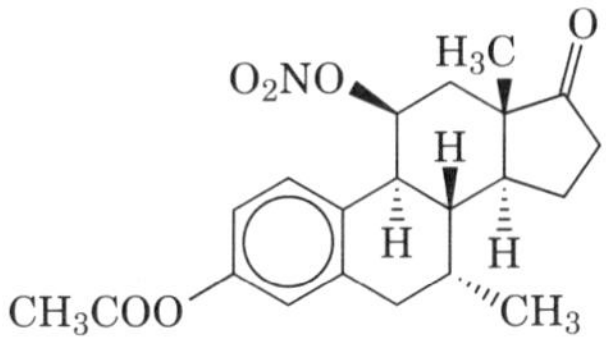   (**16**)

WS-7528 [*132147-69-4*] (**17**), a nonsteroidal estrogen, is an isoflavone which has been isolated from *Streptomyces* sp. No. 7528 and is an estrogen agonist. It inhibits [3*H*]-estradiol binding to its receptor in rat uterine cytosol at an inhibitor concentration of 5.7 n*M* for 50% of the rats tested (IC$_{50}$). It also induces the growth of estrogen-dependent human breast cancer cell line MCF-7 (7).

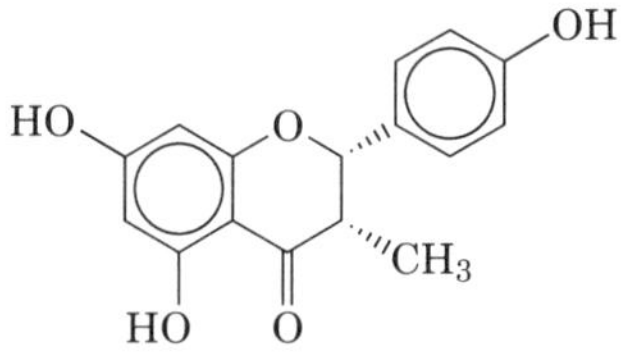

(**17**)

Steroidal Antiestrogens.   The balance of estrogenic and antiestrogenic activity expressed by the nonsteroidal antiestrogens varies widely across species, target organs, cells, and genes, depending on which indicator of response is measured (8). Tamoxifen (**9**), Nolvadex (ICI 46474), is a synthetic nonsteroidal antiestrogen which has been used for the control of hormone-sensitive breast cancer (see CHEMOTHERAPEUTICS, ANTICANCER). However, tamoxifen is a partial estrogen agonist and as such has estrogen-like stimulatory activity on the uterus, vagina, mammary glands, and the pituitary–ovarian axis in animals. Attempts to synthesize nonsteroidal antiestrogens devoid of partial estrogen agonist activity have met with limited success. The emphasis has returned to the synthesis of steroidal antiestrogens. The ICI compounds, ICI 164384 [*98007-99-9*] (**18**) and ICI 182780 [*129453-61-8*] (**19**) are examples of "pure" antagonists.

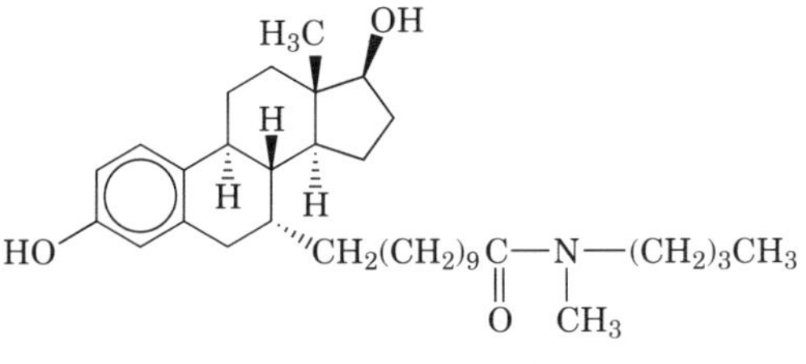

(**18**) 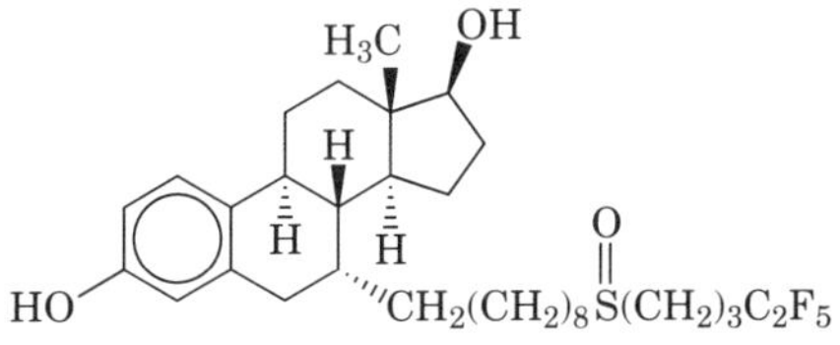   (**19**)

ICI 164384 is an antiestrogenic agent with a greater affinity for the rat uterus estrogen receptor than tamoxifen (0.19 and 0.025, respectively, relative to estradiol = 1), which dose-dependently inhibits estradiol-induced growth of ZR-71-1 human breast cancer cells *in vitro*; it is devoid of estrogenic activity in the rat and mouse uterus (9). Further exploration of the structure–activity relationships of the 7α-alkylamide analogues of 17β-estradiol led to the synthesis of ICI 182780, a compound with a fivefold increase of intrinsic potency compared to ICI 164384, measured by receptor binding and cell growth inhibition in human breast cancer cells. More importantly, ICI 182780 is 10-fold more potent *in vivo* and represents a candidate for further clinical evaluation (10). Biological activity in this series is confined to the 7α-isomers (11,12).

Based on the potent antagonist activity of the above ICI compounds and the observations that halogenation of the 16α-position of the D-ring in the steroid nucleus often leads to compounds with increased affinity for the estrogen receptor, a series of 7α-undecanamide-substituted 17β-estradiols with 16-halogen substituents were synthesized (13,14). These 16-halo-7-alkylamide antiestrogens, characterized by EM-139 [*131811-54-6*] (**20**) (13) and EM-170 [*131811-55-7*] (**21**, R = *n*-butyl) (14), demonstrated potent and pure antagonistic activity *in vivo* in screens where tamoxifen exhibited estrogenic activity and was only a weak partial antiestrogen.

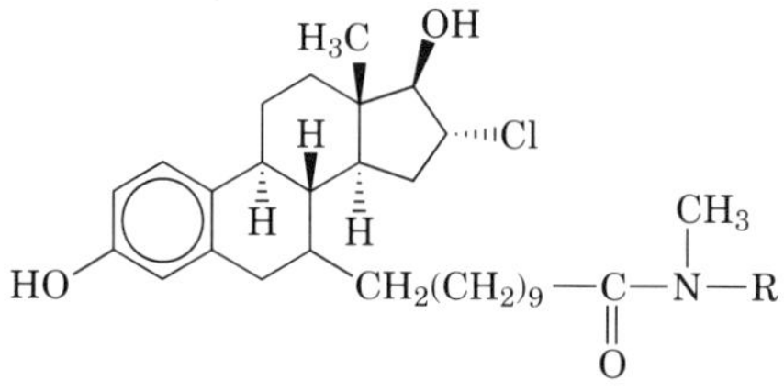

(**20**) 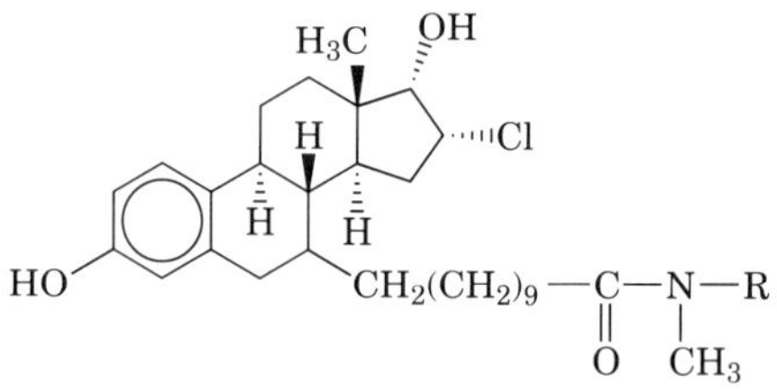   (**21**)

The synthesis of EM-139 and related 16α-halosteroids has been carried out utilizing an enol acetate (**22**) as a key intermediate; synthesis of (**22**) is shown in Figure 1. Stereospecific chlorination of (**22**) is effected with *tert*-butyl hypochlorite in acetone buffered with sodium acetate, acetic acid, and water to yield the 16α-chloro ketone amide. Reduction of the 17-keto group with lithium aluminum hydride gives a mixture of epimers from which EM-139 (**20**) is isolated by chromatography. In addition, the enol diacetate amide (**22**) is stereospecifically brominated with bromine in acetic acid to give the corresponding 16α-bromo ketone amide. The 16α-iodo analogue is synthesized from the above bromo derivative by treatment with sodium iodide in refluxing butanone under equilibrium conditions.

Related 7-substituted 19-norsteroids have appeared in the patent literature as potential antiestrogens including the 14α,17α-ethanoestratriene [*134514-24-2*] (**25**) (Schering AG) and the 7-dimethylaminoethoxyphenyl analogue [*119286-92-9*] (**26**) (Roussel-Uclaf). Compound (**25**) is reported to be antiestrogenic without the partial agonistic effects when given orally. It inhibited an estradiol-induced increase in uterine weight by 53 and 92%, respectively, and vaginal

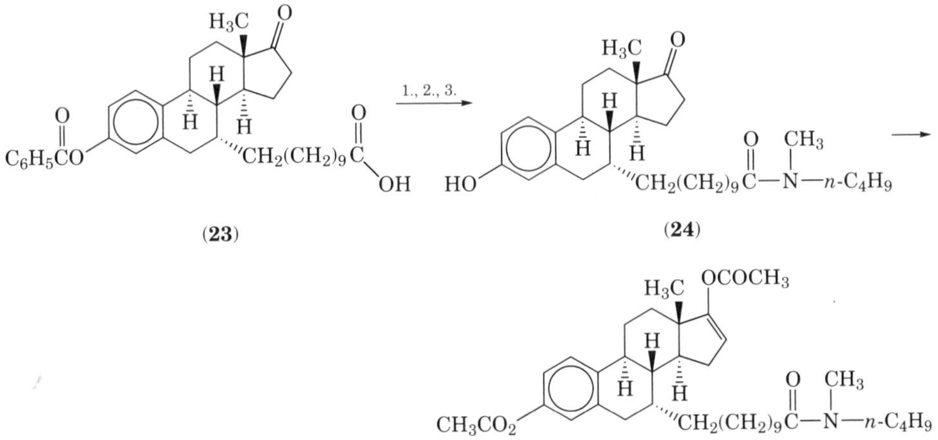

Fig. 1. Treatment of 11-[3′-(benzoyloxy)-17′-oxoestra-1′,3′,5′(10′)-trien-7′α-yl]undecanoic acid [55592-11-5] (**23**) in $CH_2Cl_2/(C_4H_9)_3N$ with 1. isobutyl chloroformate followed by 2. methylbutylamine gave the protected *N*-butyl-*N*-methyl undecanamide which was 3. deprotected with aqueous NaOH to give the estrone derivative [98013-89-9] (**24**). The enol diacetate amide [131811-72-8] (**22**) was then prepared under standard conditions with isopropenyl acetate and *p*-toluenesulfonic acid.

weight by 13 and 92% when administered at 3 and 30 (mg/kg)/d to adult ovariectomized rats (15). Compound (**26**) is a potent inhibitor of the growth of MCF-7 mammary tumor cells ($IC_{50} = 0.1$ n*M*) and thus has potential utility on hormone-dependent carcinomas (16).

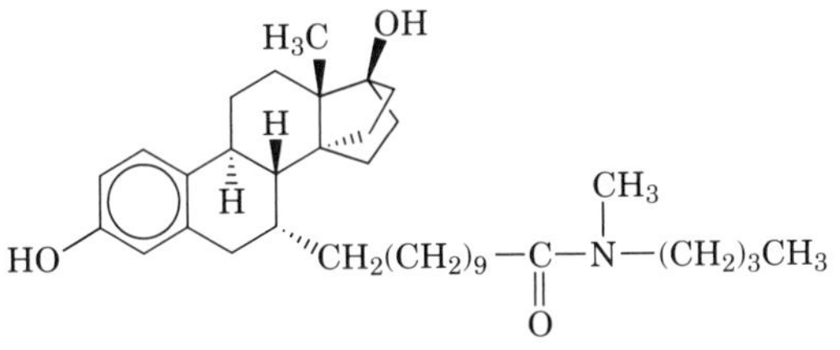

The 11-position of estradiol analogues has been a fruitful site of exploration in the development of hormone antagonists, eg, the antiprogestin RU 486 (see CONTRACEPTIVES). The Roussel group has also uncovered novel antiestrogens by investigating various substituents at the 11β-position. The bis-dimethylaminophenyl compound [123955-65-7] (**27**) and the phenoxyoctanamide [134411-59-9] (**28**) are examples of such an approach. The dimethylaminophenyl analogue (**27**) is also, like (**26**), effective against MCF-7 mammary tumor cells, but is much less potent ($IC_{50} = 500$ n*M*) (17). The importance of the long alkyl side chain for potency is evident as compound (**28**) ($IC_{50} = 0.006$ n*M*) is extremely potent and has potential use in a variety of indications (18).

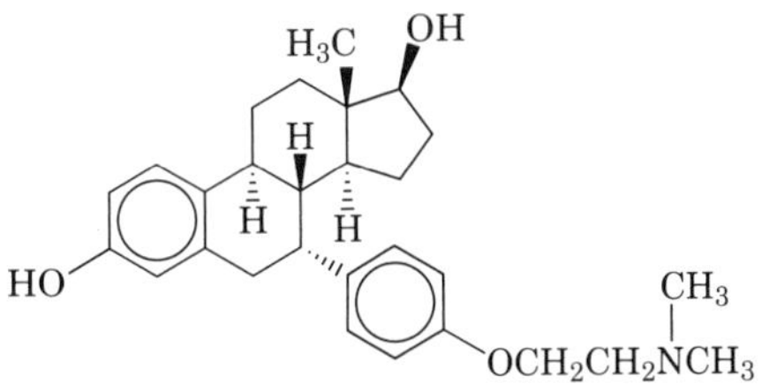

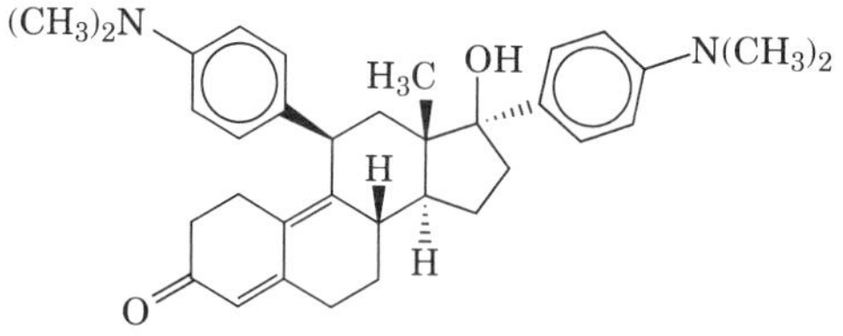

(27)

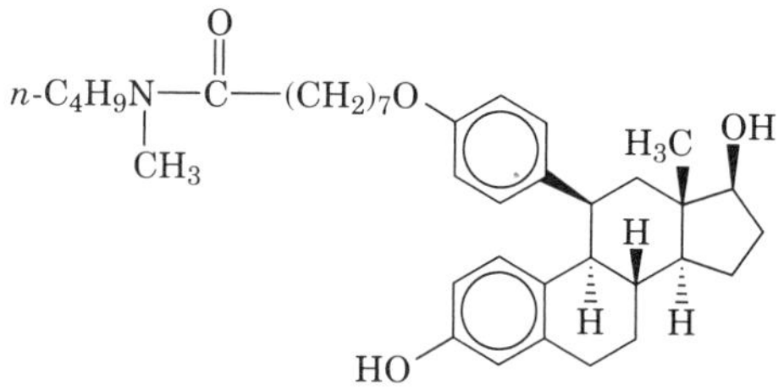

(28)

Nonsteroidal Antiestrogens. Clomiphene (**8**) and tamoxifen (**9**) which have been reviewed previously, are prime examples of nonsteroidal antiestrogens which have gained some degree of medical use in the treatment of female infertility and breast cancer, respectively (see CHEMOTHERAPEUTICS, ANTICANCER). However, the first generation of nonsteroidal antiestrogens all demonstrated some degree of agonist activity. Synthetic nonsteroidal antiestrogens are bound intracellularly by two high affinity saturable binding sites, the estrogen receptor and the antiestrogen-binding site (AEBS) (19,20). The physiological significance of the AEBS site(s) is still an open question. The search for compounds with improved specificity for each of these receptors has continued and representative examples of selected structural types are listed below. The first structural type is triarylethylenes (TAEs). Toremifene [*89778-26-7*] (**29**) is a chlorinated analogue of tamoxifen which is indicated for the treatment of post-menopausal breast cancer (21).

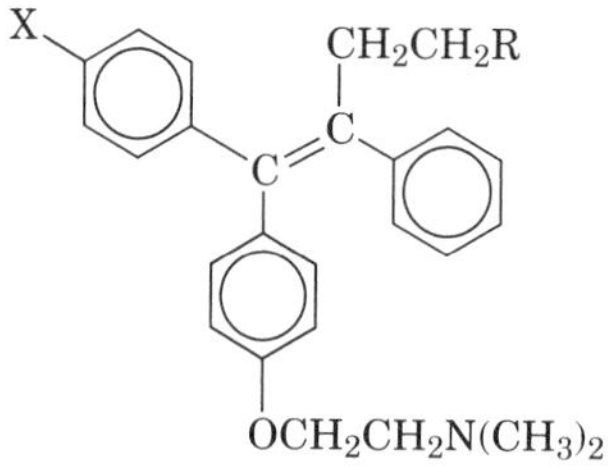

(**9**), R = X = H
(**29**), R = Cl; X = H
(**30**), R = H; X = I
(**31**), R = H; X = Br

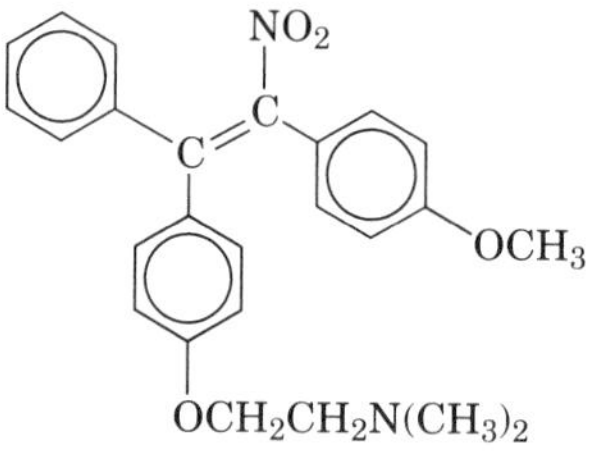

(**32**)

4-Iodotamoxifen [*116057-66-0*] (**30**) and 4-bromotamoxifen [*116057-69-3*] (**31**) have been reported in the patent literature (22) as more potent antiestrogens than tamoxifen. When the iodine is radioisotopic, compound (**30**) has potential utility for radiotherapy of breast cancer or as an imaging agent for diagnosis. However, its relatively low affinity for the estrogen receptor limits its practical usefulness. Nitromifene [*76313-96-7*] (CI 628) (**32**) (23) is the prototype of a series of nitrostilbenes with antiestrogen activity. Structure–activity relationship studies of a series of 32 nitromifene analogues indicated that replacement of the 4-OCH$_3$ with a 3-CF$_3$ moiety gave the analogue of highest affinity for the AEBS in both rat liver and MCF-7 human breast cancer cell preparations (24). Metabolites resulting from transformation of the pyrrolidine ring and of the nitro group were synthesized and the estrogen receptor affinities and estrogen agonist and antagonist properties evaluated vs nitromifene (25). Additional TAEs which have been reported are (**33**)–(**36**). Compound (**33**) is TAT 59 [*115767-74-3*] (26). Compounds (**34**) [*109517-76-2*], (**35**) [*115767-77-6*], and (**36**) [*129612-87-9*] are from Taiho (27–29). Compound (**36**) is structure (**33**) wherein the phosphate has been hydrolyzed to the phenolic OH.

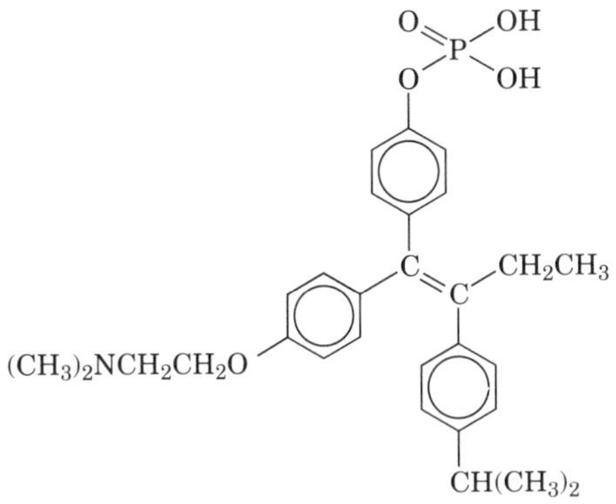

(33)

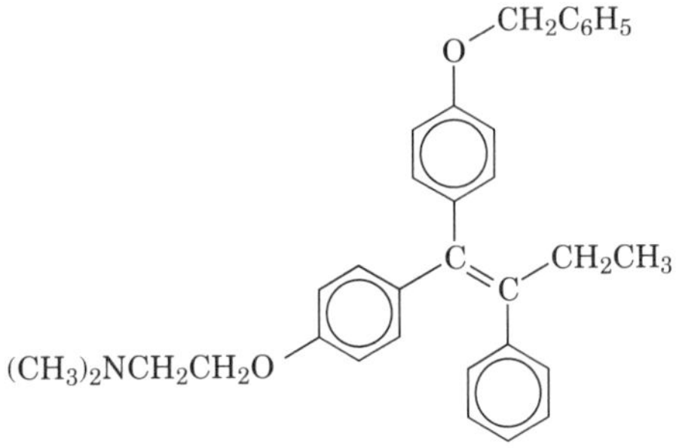

(34)

A structurally related series of phenylhydrazones resulted in the selection of compound A-007 [*2675-35-6*] (DEKK-TEC) (**37**) for the treatment of hormone-dependent tumors. A-007 is an antiestrogen that, in contrast to tamoxifen, demonstrated inhibitory activity both in the presence and absence of estradiol in ZR-75-1 estrogen-dependent human breast cancer cells, and afforded more protection than tamoxifen in the 7,12-dimethylbenz[*a*]anthracene [*57-97-6*] (DMBA) rat breast cancer model (30).

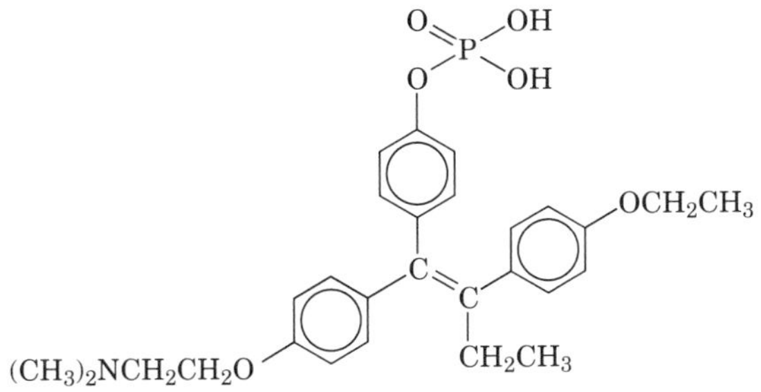

(35)

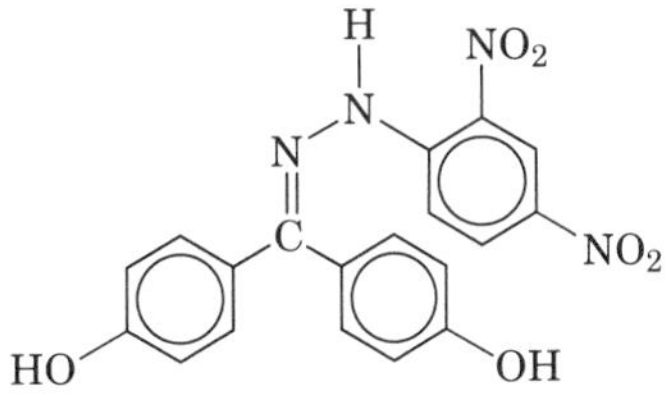

(37)

Another structural type is chromenes. Centchroman [*31477-60-8*] (**38**) is a pyrrolidinoethoxyphenyl chromane which is a potent antiestrogen with weak estrogenic activity. In India, it is used as a weekly contraceptive pill based on its reported ability to inhibit the uterine preparation for the attachment of the fertilized ova to the wall of the uterus (see CONTRACEPTIVES) (31).

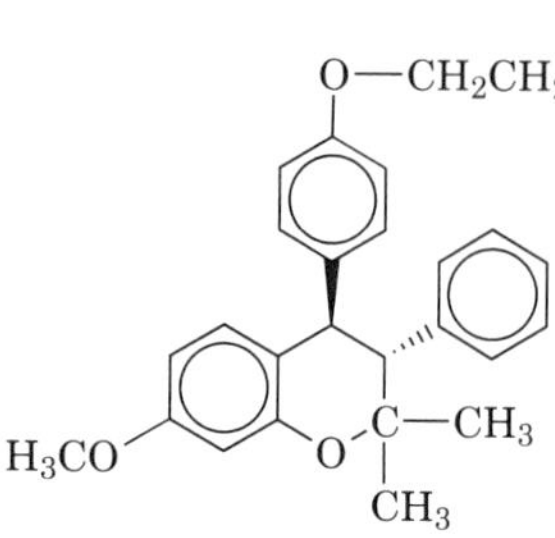

(38)

Related benzopyran derivatives include the compound CDRI-85/287 [130064-18-5] (39) from the Central Drug Research Institute (32). Analogues such as the pyrrolidinoethoxyphenyl have also been evaluated (33). An alternative series of basic ethers of 3-(p-halophenyl)-4-arylchrom-3-enes (40, X = F [128040-44-8], Cl, Br), has been synthesized and all found to be selective ligands for AEBS in vitro (34).

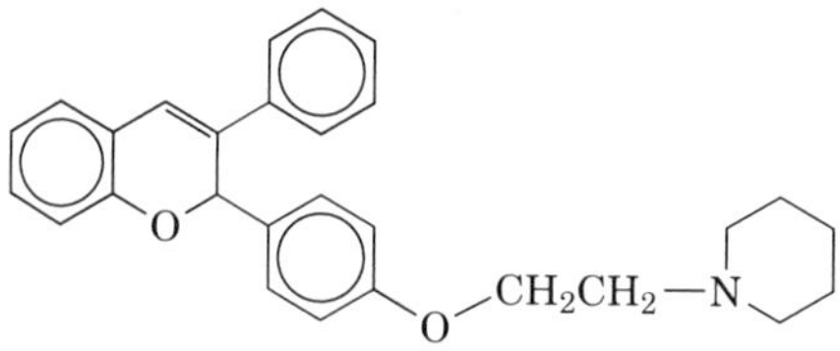

(39)

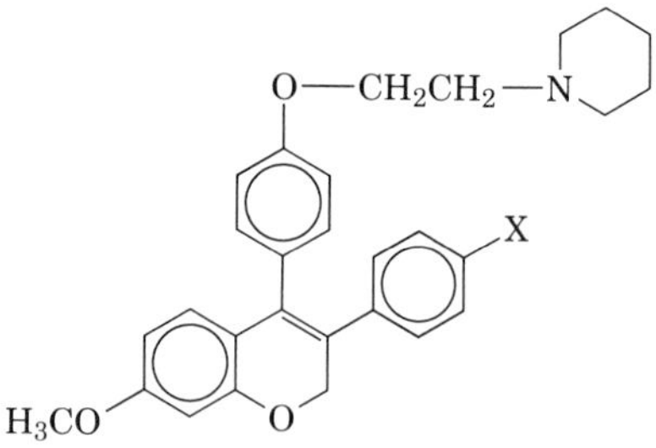

(40)

In an extension of the initial studies with chromenes related to (40), the synthesis of the five-membered benzofuran ring analogues such as the morpholino derivative (41) [139276-13-4] was carried out (34). In general, this series of benzofurans are ligands for AEBS and display no significant interaction with the estrogen receptor. Based on a comparison with a number of other reported antiestrogens, the high binding affinity for AEBS is postulated to be attributable to the following structural features: (1) the presence of a triarylethylene moiety in a fixed heterocyclic oxygen ring system, (2) the presence of an (alkylamino)-ethoxy side chain at the 4'-position of the 3-phenyl ring, and (3) the presence of an additional methylene group at the C-2 position of the benzofuran ring.

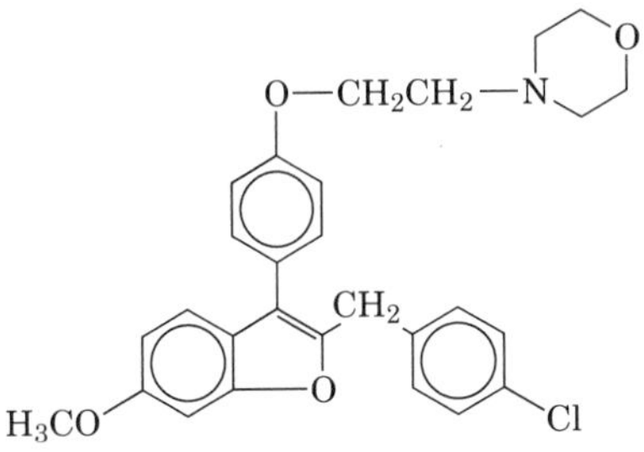

(41) 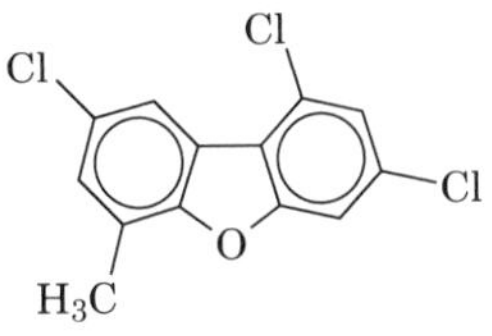   (42)

1,3,8-Trichloro-6-methyldibenzofuran [*118174-38-2*] (MCDF) (**42**) has been described as an antiestrogen which induces a broad spectrum of antiestrogenic responses in the female rat and in human breast cancer cells which are mediated through the aryl hydrocarbon (Ah) receptor. In contrast to tamoxifen, MCDF does not bind directly to the estrogen receptor. Neither tamoxifen or 17β-estradiol bind to the Ah receptor. 2,3,7,8-Tetrachlorodibenzo-*p*-dioxin [*1746-01-6*] (TCDD) was the initial lead in this structural series. However, although MCDF is 500 times less potent than TCDD as an antiestrogen, it is 10,000 to 100,000 times less toxic than TCDD for the traditional Ah receptor-mediated toxic responses (35).

A series of benzothiophenes characterized by LY 117018 [*63676-25-5*] (**43**, $n = 1$) and LY 156758 [*82640-04-8*] (**43**, $n = 2$) was synthesized with the aim of producing estrogen antagonists with minimal intrinsic estrogenicity using trioxifene [*63619-84-1*] (**44**, $n = 1$, R = OCH$_3$, R' = H) as the structural prototype (20,36). Each of these benzothiophenes has an affinity for the estrogen receptor approximately equal to estradiol and approximately 100 times that of tamoxifen. In addition, LY 117018 was 100–1000 times as potent as tamoxifen in suppressing MCF-7 human breast cancer cell growth (see CHEMOTHERAPEUTICS, ANTICANCER) and neither compound induced characteristics in these cells believed to be linked to residual estrogenicity.

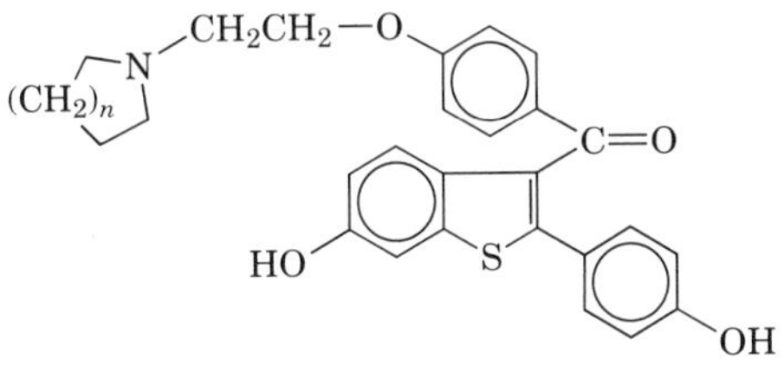

(43)

The results on the chromenes and benzothiophenes led to a further extension of the structure–activity relationship of the carbocyclic prototype trioxifene. As a class, these are called carbocyclic triarylethylenes. Dihydronaphthalene derivatives with a hydroxy moiety at R' and the corresponding benzo[*a*]fluorenes ([*138666-17-8*], R = H; [*138630-67-8*], R = OH) (**45**) were synthesized and evaluated for their antiestrogenic properties (37). Both the dihydronaphthalenes and the benzo[a]fluorenes had potent antiproliferative effects on MCF-7 cells in culture, but their binding to estrogen receptors was similar to that of other TAEs. The inhibitory effects of (**44**, R = H, OH, or OCH$_3$; R' = H or OH) and (**45**, R = H or OH) could not be reversed completely by estradiol, indicating interaction with receptors other than the estrogen receptor or a nonspecific toxic effect. Studies of these analogues with calmodulin, a putative target for TAEs, indicated only a weak interaction.

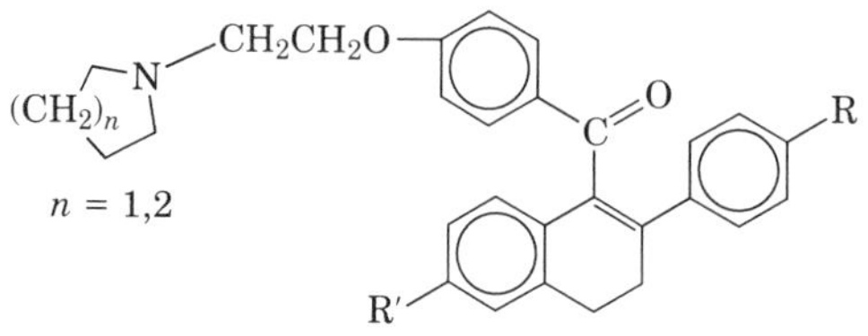

(**44**) 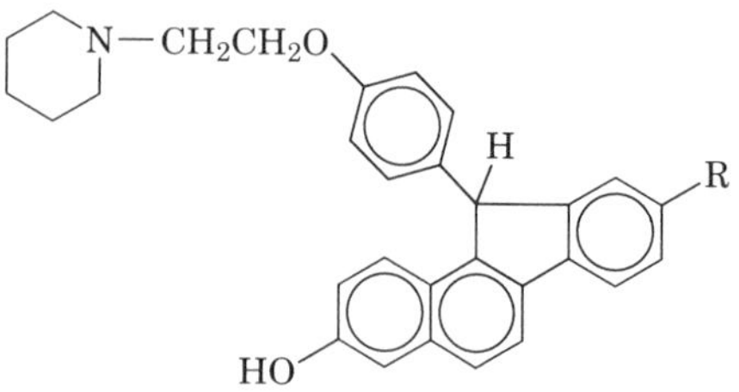   (**45**)

The synthesis of compounds of general structure (**44**) and (**45**) is carried out as shown in Figures 2 and 3. Nafoxidine [1845-11-0] (**52**) (38), the prototype dihydronaphthalene lacking an acyl group at C-1, was one of the first compounds found to have higher affinity at the AEBS than at the estrogen receptor (20).

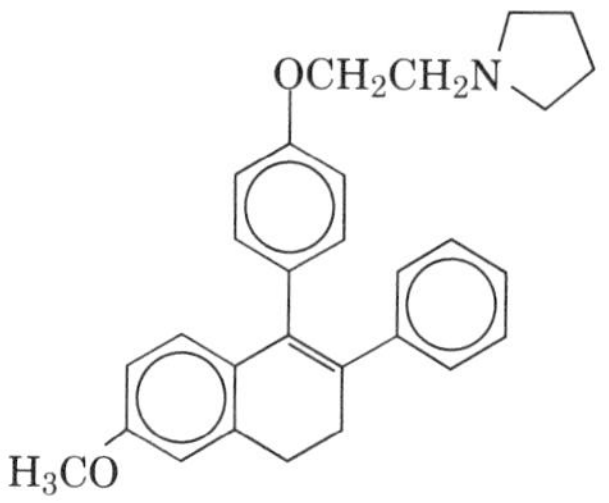

(**52**)

Another group of antiestrogens are diphenylmethanes and diphenyl-ethanes. CGS-20267 [112809-51-5] (**53**, Ciba-Geigy) is a potent aromatase inhibitor (IC$_{50}$ = 11 nM) with potential utility in the treatment of estrogen-

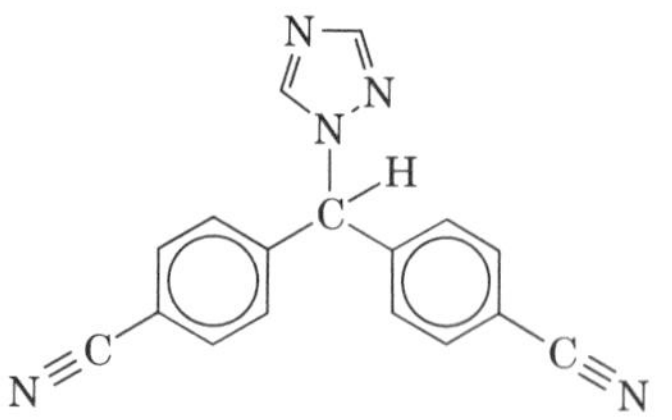

(**53**) 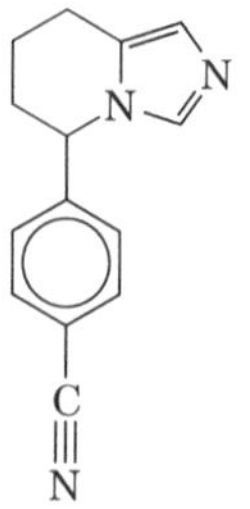   (**54**)

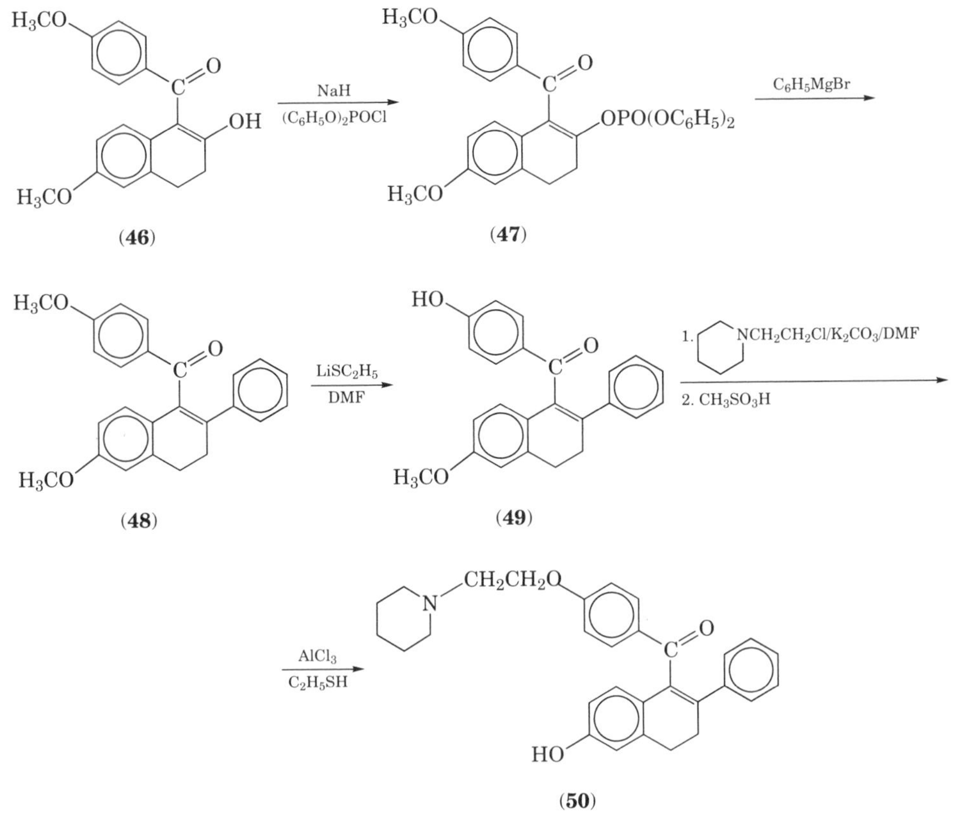

Fig. 2. The highly enolized diketone [*138630-68-9*] (**46**) is synthesized by acylation of the sodium salt of 6-methoxy-2-tetralone. The enol is then converted to the enol phosphate, tentatively assigned structure [*138630-69-0*] (**47**), with sodium hydride followed by diphenyl chlorophosphate. Addition of Grignard reagents, eg, phenylmagnesium bromide, to (**47**) gives dihydronaphthalene [*138630-70-3*] (**48**). Compound (**48**) is selectively demethylated with lithium thioethylate/DMF to remove the methoxy group para to the carbonyl group in good yield, producing monophenol [*138630-73-6*] (**49**). The basic piperidinylethoxy amine side chain is introduced using conventional conditions, and the remaining methoxy group is cleaved by the use of aluminum trichloride–ethyl mercaptan to give dihydronaphthalenes of general structure [*138630-65-6*] (**50**).

dependent disease; it was derived from CGS-16949 A [*102676-96-0*] (fadrozole HCl) (**54**) which is in clinical trials for breast cancer (39). Compounds (**53**) and (**54**) most likely exert their antagonist effects by inhibition of estrogen biosynthesis. D-18954 [*96826-17-4*] (**55**) is reported to be an antineoplastic agent which inhibits DMBA-induced mammary carcinoma in rats (74% remission at 5 mg/kg) and is active on transplantable Dunning R 3327 prostate carcinoma of rats owing to its antiestrogenic activity (40). The *gem*-dichlorocyclopropane [*126987-63-1*] (**56**) was selected from a series of cyclopropane analogues which again demonstrated greater antitumor activity when compared to tamoxifen in MCF-7 human breast cell proliferation assays *in vitro* (see CHEMOTHERAPEUTICS, ANTICANCER) (41).

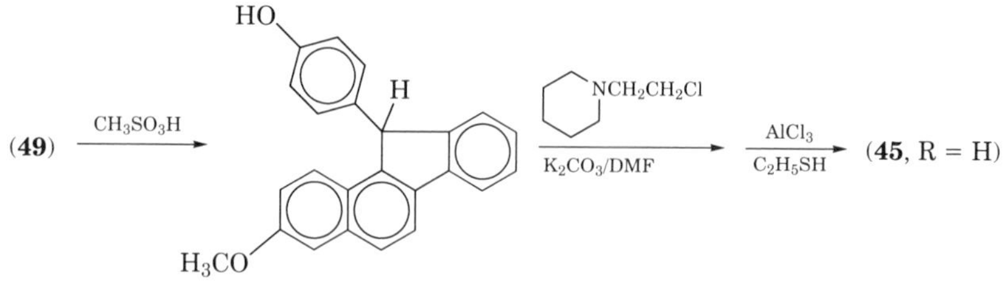

Fig. 3. After exposure of 1-acyl-2-aryl-3,4-dihydronaphthalene derivatives of structure (**49**) to strong acidic conditions, cyclization of the 2-aryl moiety onto the carbonyl group occurs rapidly and is accompanied by rapid dehydration of the carbinol intermediate to provide the 11-aryl-11*H*-benzo[a]fluorene compounds of structure [*138630-78-1*] (**51**). The corresponding analogues containing the aminoethoxy side chain [*138630-80-5*], $C_{31}H_{31}NO_2$, are then prepared by treatment with *N*-(2-chloroethyl)piperidine HCl/K_2CO_3 in refluxing DMF. Cleavage of the remaining methoxy group is again accomplished with aluminum trichloride–ethyl mercaptan, as described in Figure 2.

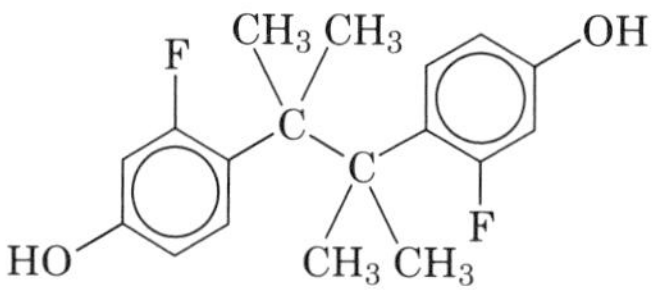

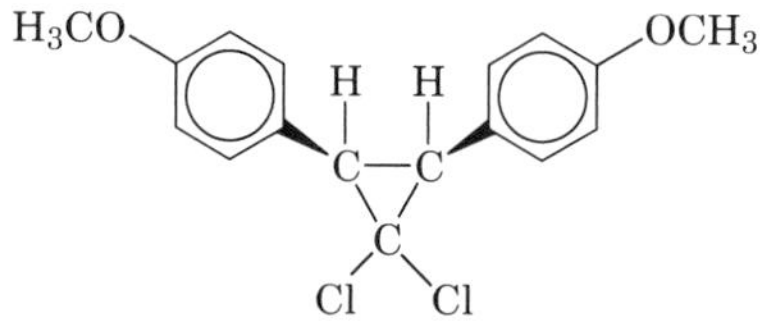

Other compounds of this general class which have been found to have antiestrogenic properties include the cytochrome P-450 inhibitor, SKF 525A [*302-33-0*] (**57**) (24); *N*,*N*-diethyl-2-[(4-phenylmethyl)phenoxy]ethanamine [*98774-23-3*] (DPPE) (**58**) (42); *t*-Butylphenoxyethyl diethylamine [*57586-10-4*] (BPEA) (**59**) (43); and cyclofenil [*110042-18-7*] (**60**, R = C_2H_5) (24) analogues.

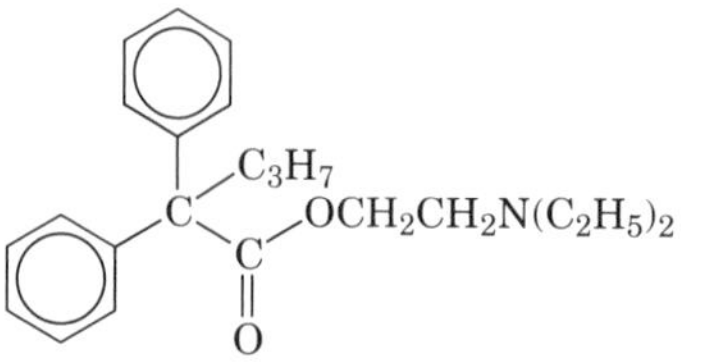

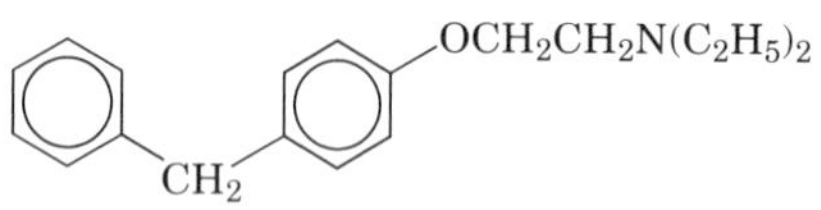

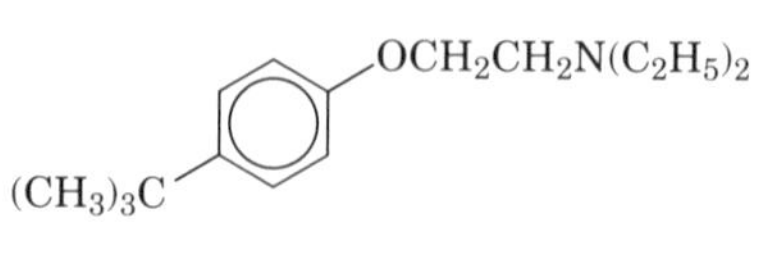

ZK-119010 [*127457-69-8*] (Schering AG) (**61**), a pyrrolidinohexyl indole derivative, is the prototype from this series of antiestrogens, with high affinity for and competitive inhibition of calf uterus [3*H*]-17β-estradiol receptors. ZK-119010 (**61**) inhibits hormone-related tumor growth, especially of human mammary carcinoma cells (MCF-7) and was reported to be superior to ICI-164384 (**18**) and tamoxifen (**9**) in antiuterine potency when evaluated in rats and mice (44). The indolo–isoquinoline [*123312-72-5*] (**62**) is reported to be a DNA-intercalating agent with high affinity for steroid hormone receptors and has anticancer properties (45).

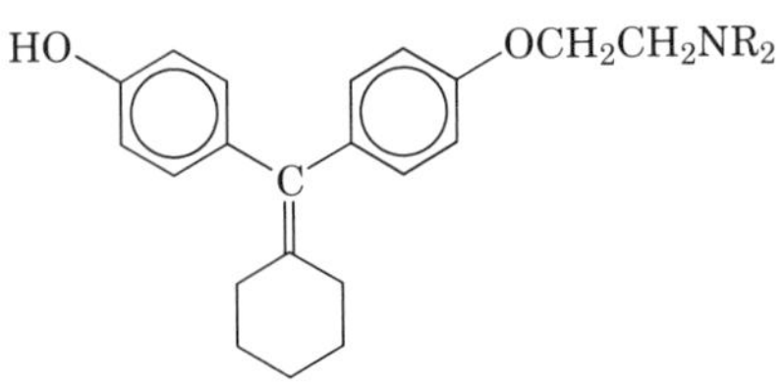

(**60**)

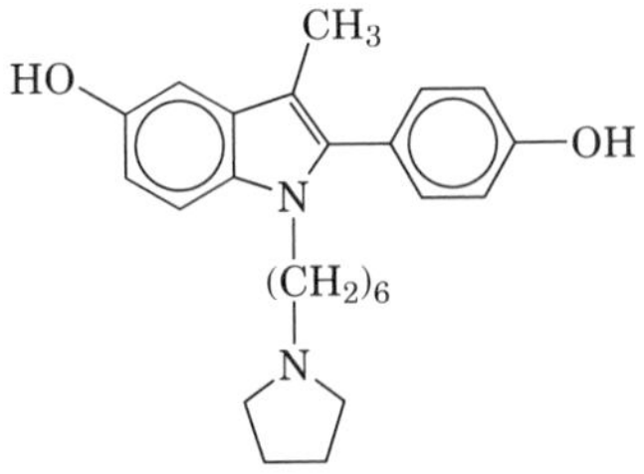

(**61**)

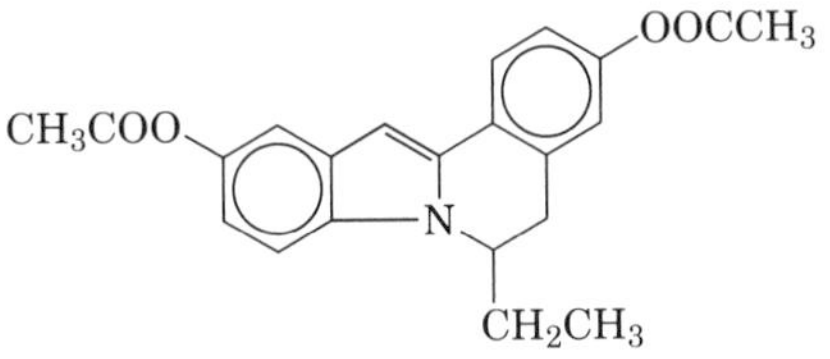

(**62**)

The area of nonsteroidal antiestrogens along with other classes of non-steroidal antagonists of sex-steroid hormone action has been reviewed to 1986, and these compounds have been grouped by chemical structure as a basis of classification rather than any biochemical or biological test system utilized to assess antagonist activity (46).

Modeling and Crystallographic Studies of Estrogen Agonists and Antagonists

There have been a number of studies that have attempted to describe the mechanism of action of estrogen agonists and antagonists at the molecular level based on binding studies, molecular modeling strategies, conformational analyses, and similar strategies. This is an area where limited progress has been achieved owing primarily to the lack of potent specific antagonists which totally block the action of endogenous and exogenous steroids.

The nature of inducible hormone effects and ligand structure remains of intense interest as the relationship between ligand–receptor binding and subsequent post-binding events such as induction of gene expression continues to be explored. One of the models proposed for the mechanism of estrogen and anti-estrogen action is based on the structure of the human estrogen receptor (ER) cloned from complementary DNA (cDNA) libraries prepared from the MCF-7 breast cancer cell line which encodes a 595 amino acid protein of 66 kiloDaltons (47,48). In this model, estrogen promotes dimerization of the ER which binds tightly to the estrogen response element (ERE), resulting in transcription via TAF-2, a ligand-inducible transcriptional activating factor and TAF-1, a constitutive transcriptional activating factor. Tamoxifen and other related antiestrogens promote dimerization and DNA binding and appear to act only through TAF-1, whereas "pure" steroidal antiestrogens such as ICI 164384 (**18**) activate neither TAF-1 or TAF-2. The interaction of tamoxifen (**9**) with the calcium-binding protein calmodulin was investigated by using computerized molecular modeling methods (49). In addition, six analogues of tamoxifen were also examined, exploring various portions of the molecule. These analogues included side-chain variants (diethylamino, piperazino, and pyrrolidino in place of dimethylamino), introduction of aromatic substitution in the 4-position in the phenyl ring cis to the ethyl substituent (iodo and hydroxyl), and replacement of the ethyl with a methyl group. The interaction models for tamoxifen with calmodulin support the observations that were obtained experimentally, which indicated that the ethyl group is essential for potency. The model also predicts the poor binding properties of the piperazino and C-methyl analogues; however, it does not account for the superior binding affinity of the iodo analogue which is observed experimentally. X-ray crystallographic data on estradiol, tamoxifen, and other estrogens and antiestrogens, combined with energy minimization of the observed crystal conformations, suggested that estrogen receptor binding is primarily the result of a tight fit only at the A-ring of the steroid. The most potent antagonists have phenolic rings capable of promoting binding but do not elicit a response due to a lack of the appropriate functional group elsewhere in the molecule (ie, hydrogen bond donor) or are sterically hindered in some way to prevent events subsequent to binding from taking place (50,51). Conformational analyses on a recently reported estrogen, bis(4-hydroxphenyl)[2-phenoxysulfonyl)-phenyl] methane, and related analogues showed that the spatial orientation of the phenyl sulfonate ring was a critically important determinant of receptor binding (52).

Pharmacology

Mechanism of Estrogen Action. *Estrogen Receptor and Receptor-Mediated Molecular Events.* Estrogen molecules are lipophilic and diffuse through the plasma membrane of all cells. The steroid ligands encounter their specific receptors only within target cells. In cells lacking estrogen receptors, estrogens are not readily retained and exit the cell. Estrogen receptors appear in target tissues before ovary maturation (53,54), and the concentration of estrogen receptor in the uterus correlates with the level of estrogen in the blood (55). The number of estrogen-binding sites in the uterus changes during the menstrual cycle from a minimum of 1000 sites per cell during estrus (ovulation), increasing to 3500 sites per cell and reaching a maximum at proestrus, which is immediately prior to the next cycle, of 5000 sites per cell. It has been reported that aging is associated with a decrease of estrogen receptor concentration (56). However, there is no decrease in binding affinity for estrogens. The affinity, specificity, and large concentration of estrogen receptors in cells allow a low concentration of estrogen to produce biological responses (57).

The estrogen receptors are large protein molecules that are mainly local-ized in the cell nucleus (58). The receptor contains three principal domains: an estrogen-binding site at the C-terminal region, a DNA-binding domain in the middle of the protein molecule which is capable of binding to specific regions within the target genes called estrogen response elements, and a modulating domain at the N-terminus (59). Recent evidence showed that the estrogen re-ceptor is bound to specific estrogen response elements of a variety of genes with or without estrogen (60).

Estrogen exerts hormonal effects by first binding with high affinity to the receptor to form an estrogen receptor complex. The hormone binding then in-duces conformational changes in the steroid-binding domain and other domains of the receptor, a process which is referred to as receptor activation or trans-formation. The transformed or activated estrogen receptor complex alters the interaction with target genes, leading to an increase of the affinity to DNA and other nuclear components, such as the nuclear matrix, nuclear proteins, as well as alteration of structure of a complex of nucleic acids and proteins which sur-round the genes. As a result, estrogen induces changes in the activity of tran-scription machinery associated with the target genes, which in turn modulates the expression of target genes and further regulates cell function, growth, or differentiation. Reviews with more detailed references on this topic have been published (61–64).

Cell- or Tissue-Specific Effects of Estrogen.   Out of the trillions of cells in the human body, only special cell types or tissues elicit biochemical or physiologic responses to estrogen. These target cells possess estrogen receptors, although the receptor concentration varies between cell types. The responses to estrogen of target cells are also divergent. The mechanism for the cell- or tissue-specificity of estrogen response is still under active investigation. However, several ele-ments are considered to modulate the nature of cell- or tissue-specific effects of estrogen following estrogen receptor binding, eg, specific target genes and the associated gene network in each type of cell; cell-specific regulating factors for transcription of the genes, such as regulatory sequences, transcriptional factors, and local chromatin conformation of the genes and nuclear matrix; and the rela-tive response of the cell to other synergistic molecules of estrogen action (64,65).

Current Concept of Estrogen Agonist, Antagonist, and Cell- or Tissue-Specific Response of Estrogen.   The presence of estrogen-receptor and estrogen-target genes containing estrogen response elements as well as appropriate transcription machinery determine the cell or tissue's competence to respond to estrogen. Factors such as the cell- or tissue-specific array of target genes, the associated transcription regulatory factors, the chromatin structure, and nuclear matrix surrounding the genes allow different target cells to elicit diverse cell-specific responses to a certain ligand. The molecular features of different ligands affect the stability of ligand-receptor binding and the interaction of ligand-receptor complex with DNA and transcription machinery which in turn governs the nature of induced estrogen responses, ie, active or latent, strong or weak, being agonist in one tissue and being antagonist in another tissue (10,66–68).

Biosynthesis.   Natural estrogens are produced by steroidogenesis in vari-ous tissues. The ovary is the primary source of the hormone in nonpregnant women (69). Estradiol (**3**) is the most potent and primary product of the ovary, although the organ also produces estrone (**2**). The estrogens are ultimately formed from either androstenedione or testosterone as immediate precursors. The key reaction is the aromatization of the A-ring to yield a phenolic hydroxyl at C-3 (70). Pathological conditions, such as hirsutism and virilism, are thought to be caused by a defect of the aromatization reaction. During pregnancy the placenta produces large amounts of estrogens, especially estriol (**4**). Other tissues such as liver, adipose tissue, skeleton muscle, and hypothalamus are also sources of estrogens where androgens are converted to estrone (71). In post-menopausal women, peripheral aromatization of adrenal androgens to estrone is the principal source of estrogen (72). Because significant extraglandular estrogen production occurs in adipose tissue, estrogen production is greater in obese than in thin post-menopausal women, and total estrogen production in the massively obese may be as great as, or greater than, in premenopausal women (73).

Metabolism and Distribution. Estrogens are readily absorbed through the gastrointestinal (GI) tract. During this process the unconjugated estrogens are converted primarily to estrone. Therefore, estrogens, if taken orally, cause increased serum estrone levels. Both endogenous and exogenous estrogens are metabolized similarly (74). Maximal serum estrogen levels after oral ingestion are reached in 4–6 h (75).

Inactivation of estrogen in the body is carried out mainly in the liver. A certain proportion of the estrogen reaching that organ is secreted into the bile, which is then reabsorbed from the intestine. Degradation of estradiol and estrone in the liver is mainly through conversion to less active products such as estriol and numerous other estrogens, through oxidation to nonestrogenic substances, and through conjugation with sulfuric and glucuronic acids. The end products are secreted by the kidney. A portion of estriol is metabolized and excreted in the gallbladder, and the majority is cleared in the urine. Many of the synthetic derivatives of steroidal estrogens, conjugated estrogens, and the nonsteroidal estrogens are orally active and inactivated slowly. The course of metabolism of ethinyl estradiol is different. The insertion of an ethinyl group at the 17α-position allows estrogen to be absorbed efficiently and inhibits intestinal or hepatic metabolism, which accounts for its high intrinsic potency (76).

The absorption of estrogen is efficient through other modes of delivery (transdermal, vaginal, nasal, or intramuscular) (77). Various vehicles by which estrogen is administered give different rates of absorption. Injected estrogens are rapidly absorbed and lead to a fast increase in blood concentration. However, esterification or polymerization slows the process. Vaginal, transdermal, or intramuscular administration of estradiol results in higher levels in plasma of estradiol than estrone, because the intestinal metabolism of oral estrogens is bypassed (78–80).

Circulating estrogens are tightly conjugated with sex hormone-binding globulin and weakly bound to albumin. The concentration of sex hormone-binding globulin in blood is regulated by hormones. Plasma sex hormone-binding globulin levels are increased by estrogens and pregnancy and decreased by testosterone administration (81). The physiological role of plasma protein binding of estrogen is still not fully understood. Binding is not necessary for estrogen transport, because estrogens are sufficiently water-soluble at physiological concentrations. Besides, only free estrogens are available to the tissues. It is suggested that the principal function of sex hormone-binding proteins may be to ensure uniform hormone distribution among all the cells of target tissue (82).

Pharmacological Effects. Three principal natural estrogens (E_1, E_2, and E_3) are produced by the ovary and play important roles in the development and support of female reproduction (65). The normal menstrual cycle is modulated by a variety of ovarian steroids and peptides, among which estrogen is a primary regulator. Cells in the anterior hypothalamus secrete gonadotropin-releasing hormone (GnRH, also known as luteinizing hormone releasing hormone, LHRH) in a pulsating manner. LHRH stimulates the gonadotrope cells of the anterior pituitary to release both luteinizing hormone (LH) and follicle-stimulating hormone (FSH). These pituitary gonadotropic hormones induce maturation of the oocyte (an ovarian cell which produces an ovum) and stimulate the ovarian follicles to synthesize estrogen and progestins as well as the peptide hormone inhibin. Estrogen and progestins directly inhibit LHRH and pituitary gonadotropin secretion and inhibit further stimulation of the ovary (83). The effect of estrogen on pituitary gonadotropin is biphasic. Early in the cycle when the concentration of estrogen is relatively low (20–60 pg/mL), estrogen inhibits LH secretion. The increasing concentration of estrogen in midcycle (200 pg/mL) leads to an increase of the frequency and amplitude of the LH pulse. The LH surge leads to ovulation of the dominant follicle and coincides with a transient decrease in estrogen levels. Ovulation typically occurs on day 14 of the typical 28-d cycle. The period after ovulation is the luteal phase, which is characterized by a decreasing LH pulse and a rise in progesterone concentration (83,84).

Estrogens coordinate the systemic response during the ovulatory cycle, including the growth and maintenance of the reproductive tract, pituitary, breasts, and other tissues. Estrogens are also responsible for maturation of the skeleton and development of female secondary sex characteristics when females enter puberty. The other important functions of estrogens include modulation of many metabolic processes (76).

Estrogens stimulate cellular proliferation, induce RNA and protein synthesis of uterine endometrium and the fibrous connective tissue framework for ovaries, and increase the size of the cells. This effect leads to the growth and regeneration of the endometrial layer and spiral arterioles, and increase in the number and size of endometrial glands. Under the influence of estrogen, vaginal mucosa becomes thicker, as cervical mucus becomes thinner (85,86).

Breast development is initiated by estrogens with both ductal and stromal growth, resulting in breast enlargement. Estrogens also promote body hair and female distribution of fat in the breasts, buttocks, and thighs. Estrogens modulate bone growth in a biphasic manner. At low dosages, estrogens help to maintain bone growth primarily by inhibiting bone resorption. At high dosages, the hormones stimulate closure of the shafts of the long bone. The positive effects of estrogens on salt and water retention usually cause edema and decrease of bowel motility. Estrogens also stimulate the synthesis and secretion of prolactin in pituitary lactotrophic cells (87).

Estrogens have an influence on hepatic metabolism. The production of sex hormone-binding globulin, thyroxine-binding globulin, blood-clotting factors (VII to X) and plasminogen in the liver is stimulated by estrogens. Estrogens promote the production of high density lipoprotein (HDL), especially HDL2 and its apolipoproteins A1 and A2, in liver. Contrarily, estrogens inhibit the hepatic formation of low density lipoprotein (LDL). These effects vary with different types and doses of estrogens and the route of administration (88,89).

Therapeutic Uses

Estrogens, along with progestins, are used to suppress ovulation for fertility control in the form of contraceptives (see CONTRACEPTIVES). Estrogens are applied to treat gonadal failure, to induce and maintain secondary sexual characteristics due to inadequate production of ovarian steroids in hypogonadal individuals. Estrogens are also indicated for hormone replacement therapy in postmenopausal women, for the preparation of the endometrium of hypogonadal women before donor egg and embryo transfer, and for treatment of breast cancer (see CHEMOTHERAPEUTICS, ANTICANCER).

Contraceptive Agents. Oral contraceptives consist of estrogens and progestins in combination or progestin alone. Both hormones act primarily to inhibit the production of gonadotropins, follicle-stimulating hormone (FSH), and luteinizing hormone (LH) in the pituitary. As a result, the midcycle surge of LH is suppressed and ovulation is prevented. Measurements of circulating FSH and LH show that estrogen–progestin combinations suppress both gonadotropins. Experience with the most common types of oral contraceptives shows them to be 99–100% effective (90).

The estrogen component consists of either ethinyl estradiol (**5**), a potent synthetic estrogen, or mestranol (**6**), the 3-methyl ether of ethinyl estradiol, a less potent compound. Mestranol is demethylated to ethinyl estradiol in the liver, which accounts for the majority of its estrogenic activity. Ethinyl estradiol is metabolized and conjugated in the liver very slowly, which enhances the oral activity (91). Peak levels of ethinyl estradiol in plasma are reached 1 h after oral administration, followed by an initial rapid decline and a second, slower phase of decline. Up to 60% of an oral dose is excreted in urine after 24 h (92). Early oral contraceptive formulations contained up to 100–150 μg of estrogen. Numerous studies were carried out to find the lowest doses of estrogen and progestogen, a hormone whose effects mimic those of progesterone, consistent with efficacy and having the least amount of side effects. By the early 1990s the majority of oral contraceptives contained only 30–35 μg of estrogen. These low dose products are commonly referred to as second-generation oral contraceptives, or are termed "low dose" (see CONTRACEPTIVES).

There were reports that the appearance of male characteristics, androgenicity, associated with the progestational component of some of the original progestogens was related to changes in lipid metabolism (93,94). New, pharmacologically more selective progestogens with no or less undesirable adverse effects have been investigated. Using medicinal chemistry approaches, companies like Ortho Pharmaceutical Corp., Organon, and Schering AG have dissociated the androgenicity and progestational activity of steroidal progestogens. More selective progestogens such as norgestimate, desogestrel, and gestodene emerged from this research (95–98). In combination with ethinyl estradiol, these new progestogens compose the third generation of oral contraceptives. The usage of these contraceptives is expected to continue to grow in the 1990s and beyond.

Estrogens alone have only been used as short-term "emergency" contraimplantative agents. Under high dose estrogen treatment, the endometrium becomes nonreceptive to the blastocyst, thus preventing implantation. Therefore, large doses of estrogens can prevent pregnancy when given within 48 h following unprotected coitus. The "morning-after pill" is effective; however, its use is associated with side effects such as nausea, vomiting, and menstrual disturbances. It is recommended only in case of rape, incest, failure of a barrier method, or unprotected intercourse (99). The use of 50 μg of ethinyl estradiol and 0.5 mg of norgestrel is equally effective and results in fewer side effects (100). Detailed reviews on contraceptive agents have been published (see CONTRACEPTIVES) (92,101,102).

Hormone Replacement Therapy. *Estrogen Deficiency.* Treatment with cyclic estrogen has been proven to benefit young women with estrogen deficiency, caused by primary ovarian failure or by hypogonadotropic hypogonodolism resulting from luteinizing hormone-releasing hormone (LH-RH) deficiency or hypopituitarism (103,104). The therapy should be initiated at the time of expected puberty for the promotion and maintenance of female sexual characteristics. The breasts and endometrium only respond to a high dosage of ethinyl estradiol. Low dosage estrogen therapy is used to promote the growth of the long bones. Estrogen is usually prescribed in a cyclic fashion with an initial dose of 0.3 mg/d of conjugated estrogens until growth ceases, at which time the daily estrogen dose is increased to 0.625–1.25 mg to augment breast development. Adding a progestogen at the time of the first breakthrough bleeding episode could induce cyclical withdrawal bleeding (105,106).

Menopause. The depletion of functional ovarian follicles leads to the natural cessation of menses called menopause (107). After menopause, ovarian production of estradiol falls to less than 20 μg/d from a mean value of 220 μg/d during the normal menstrual cycle (108,109). This causes a deficiency of estrogen and progesterone in post-menopausal women. However, the menopause does not result in total cessation of estrogen production. Androstenedione is continuously aromatized to estrone in peripheral tissues such as fat, liver, the skeletal muscles, hypothalamus, and hair follicles. Because estrone is a weak estrogen, most menopausal women are hypoestrogenic.

Estrogen replacement is successful in treating the symptoms in menopause (vasomotor symptoms and vaginal atrophy) and has been reported to reduce the risk of osteoporosis and atherosclerotic heart disease (110). A recent study suggests that estrogen replacement treatment prevents increases in abdominal fat that occur after menopause (111). For long-term use estrogens should be given in the minimally effective dose (0.625 mg conjugated estrogen, 0.625 mg estrone sulfate, or 1 mg of micronized estradiol) (112,113). The hormones are given orally or in estrogen-containing vaginal cream or transdermal estradiol patches. Progestogens given cyclically (10–14 d/mo) are recommended to avoid withdrawal bleeding (114) and to minimize the risk of endometrial cancer (115,116). Estrogens are most effective for the relief of vasomotor symptoms. The symptoms of genitourinary atrophy improve with estrogen replacement therapy of all forms, and it has been suggested that estrogens could be given before atrophy occurs.

Osteoporosis. Bone is constantly renewed to meet the stresses imposed upon it and to ensure proper calcium homeostasis. The process involves the removal of bone (resorption) by the osteoclasts and formation of bone by the osteoblasts. It is believed that estrogens may induce production of cytokines (nonantibody proteins) in osteoblasts or other progenitors of bone cells. In turn, the hormone inhibits the osteoclastic resorption and as a result, reduces the net bone loss (117,118). The administration of estrogens daily (and of progestogens intermittently if the woman has a uterus) is an effective method of preventing bone loss, slowing the progress of osteoporosis, and reducing the risk of fractures of the hip, radius, and vertebrae in postmenopausal women (119,120). Case-control and other retrospective studies suggest that the long-term use of estrogen in post-menopausal women, starting soon after the menopause, reduces the risk of subsequent vertebral fracture by up to 90% and of Colles' (wrist) and hip fractures by up to 50%, especially if therapy is started within 5 yr of the last menses (120–122). The use of hormone replacement therapy to treat established osteoporosis by preventing further bone resorption has been reported (123,124). The minimal effective doses of oral estrogen that reduce the loss of bone and the incidence of fracture are 0.625 mg/d of conjugated equine estrogens, 0.625 mg/d of estrone sulfate, or 0.02 mg/d of ethinyl estradiol. Tamoxifen (**9**), a partial-antiestrogen in certain tissues, acts as an estrogen agonist on trabecular bone, which supports connective tissue, causing decreases in the usual high levels of bone resorption and a net preservation of bone mineral mass (125–127). Detailed reviews on this subject have been published (128–130).

Cardiovascular System. Concerns about negative cardiovascular effects associated with the use of estrogen originated from the use of oral contraceptives when the steroid dose was much higher than that utilized in current low dose preparations. Recent data show that estrogens have positive effects on the cardiovascular system in post-menopausal women receiving estrogen replacement therapy (89,131,132). Post-menopausal estrogen treatment reduced the risk for angiographically significant coronary artery disease and prolonged survival when coronary artery disease was present, but had less effect in the absence of coronary artery disease. Epidemiological studies have shown that estrogen treatment after menopause significantly reduces the mortality rate from cardiovascular disease (89).

One theory is that the protective effect by estrogens against atherosclerosis is partially a result of hormonal effects on circulating lipid levels. Estrogen replacement therapy changes the overall lipid profiles of older women (133). The rise in high density lipoprotein (HDL) cholesterol and the lowering of low density lipoprotein (LDL) cholesterol may provide a protective effect from atherosclerotic heart disease (88) (see CARDIOVASCULAR AGENTS). Table 1 lists preparations of estrogens for hormone replacement therapy (134).

Chemotherapy. *Prostate Cancer.* Estrogen has an inhibitory effect on the prostate in addition to its suppression of gonadotropin secretion by the pituitary. The three- and five-year survival rates in prostate cancer patients with metastatic disease improved when treated with DES (**7**) alone or along with castration. However, DES does not improve the survival rates in patients whose carcinoma is confined to the prostate. Small doses of DES (1 mg/d) appear to retard prostate cancer growth and could reduce the cardiovascular complications associated with larger doses (5 mg/d) (135) (see CHEMOTHERAPEUTICS, ANTICANCER).

Breast Cancer. Although the mechanism of action is unknown, estrogens (DES, conjugated estrogens, and ethinyl estradiol) have been used in the

Table 1. Preparations of Estrogens[a]

Preparation	Form[b]	Trade name	Manufacturer
estrone aqueous suspension	I, 2–5 mg/mL	Estrone Aqueous	Wyeth-Ayerst
		Bestrone	Bluco
		Estrone-A	Kay Pharmaceuticals
		Estronol	Central
		Theelin aqueous	Parke-Davis
		Estrone 5	Keene Pharmaceuticals
		Kestrone-5	Hyrex
		Theogen	Jones-Western
		Estragyn 5	Clint Pharmaceuticals
estrogenic substance aqueous suspension	I, 2–5 mg/mL	Estrogenic Substance Aqueous	Wyeth-Ayerst
		Estrofol	Reid-Provident
		Estroject-2	Mayrand
		Estromone	Endo
		Gynogen	Forest Pharmaceuticals
		Hormogen-A	Mallard
		Kestrin Aqueous	Hyrex
		Unigen	Vortech
		Wehgen	Hauck
micronized estradiol	T, 1–2 mg	Estrace	Mead Johnson Laboratories
	C, 0.1 mg/g	Estrace	Mead Johnson Laboratories
	Tr, 0.05–0.1 mg/d	Estraderm	Ciba-Geigy
estradiol cypionate in oil	I, 1–5 mg	Depo-Estradiol	Upjohn
		Depostra	Tennessee Pharmaceuticals
		Depgyhogen	Forest Pharmaceuticals
		Depogen	Hyrex
		Dura-Estrin	Hauck
		E-lonate	Tunex
		Estra-D	Seatrace
		Estro-Cyp	Keene Pharmaceuticals
		Estrofem	Pasadena Research
		Estroject-LA	Mayrand
		Estronol-LA	Central
		Estragyn LA 5	Clint Pharmaceuticals
		Estro-L.A.	Hauser
		E-Cypionate	Legere
		Estro-Span C	Primedics
		Esdinate	Shoals Pharmaceuticals
estradiol valerate in oil	I, 10–40 mg	Delestrogen	Mead Johnson Laboratories
		Dioval, XX, 40	Keene Pharmaceuticals

503

Table 1. (*Continued*)

Preparation	Form[b]	Trade name	Manufacturer
		Duragen-10, 20, 40	Hauck
		Estradiol L.A., 20, 40	Vortech
		Estroval-10	Reid-Provident
		Feminate-10, 20, 40	Jones-Western
		Gynogen L.A. 10, 20	Forest Pharmaceuticals
		Valergen-10, 20, 40	Hyrex
		L.A.E. 20	Seatrace
		Clinagen LA 40	Clint Pharmaceuticals
		Deladiol-40	Dunhall
		Menaval-20	Legere
		Medidiol 10	Med-Tek Pharmaceuticals
		Estra-L	Pasadena Research
		Estro-Span	Primedics
		Repository Hormone	Rugby
		Esdival-10	Shoals Pharmaceuticals
polyestradiol phosphate	I, 40 mg	Estradurin	Wyeth-Ayerst
conjugated estrogens[c]	T, 0.3–2.5 mg	Premarin	Wyeth-Ayerst
		Progens	Major
		Estrocon	Savage Laboratories
	P, 25 mg	Premarin Intravenous	Wyeth-Ayerst
	VC, 0.625 mg/g	Premarin Cream	Wyeth-Ayerst
esterified estrogens[d]	T, 0.3–2.5 mg	Estratab	Solvay Pharmaceuticals
		Menest	Smith Kline Beecham
estropipate piperazine	T, 0.625–5 mg	Ogen	Abbott
estrone sulfate	VC, 1.5 mg/g	Ogen	Abbott
ethinyl estradiol	T, 0.02–0.5 mg	Estinyl	Schering
		Feminone	Upjohn
quinestrol	T, 100 μg	Estrovis	Parke-Davis
diethylstilbestrol (DES)	T, 1–5 mg, entric coated; 0.1–5 mg	Diethylstilbestrol	Lilly
	VS, 1–0.5 mg		Lilly
chlorotrianisene	Ca, 12–72 mg	Tace	Marion Merrell Dow
dienestrol	VC, 0.1 mg/g	Ortho Dienestrol	Ortho-McNill
		DV	Marion Merrell Dow
		Estraguard	Reid-Provident

[a]Ref. 134.
[b]I = injection; T = tablet; C = cream; Tr = transdermal, continuous delivery, P = parenteral; VC = vaginal cream; VS = vaginal suppositories; Ca = capsule.
[c]Contain 50–65% sodium estrone sulfate and 20–35% sodium equillin sulfate.
[d]Contain 75–80% sodium estrone sulfate and 6–15% sodium equillin sulfate.

treatment of advanced metastatic breast cancer in post-menopausal women (see CHEMOTHERAPEUTICS, ANTICANCER). However the estrogen responsiveness is not predictable. Some patients with breast cancer show tumor regression when treated with pharmacological doses of estrogens (136,137). Remission rates of 30–37% have been reported when estrogen is used as the initial therapy. In a randomized trial, estrogen gave a 29% remission rate. The duration of response to estrogen is relatively long-lived in most series. Remission rates of less than 10% have been reported when estrogens are used in patients who have relapsed from other therapy. A more recent clinical report concluded that DES remains a useful, active agent in the management of advanced breast cancer in post-menopausal women, even in patients with tumors unresponsive to other endocrine therapy (138).

Triphenylethylene derivatives, such as tamoxifen (**9**) or clomiphene (**8**), are long-acting antiestrogens. They act by binding to estrogen receptors and interfering with estrogen function. Both are mixed agonists–antagonists of estrogen action, which partially mimic the response of estrogen in some cell types, yet block estrogen action in others (139).

Tamoxifen is useful in the treatment of estrogen-receptor-positive breast cancer (140,141). The antiestrogen induces a therapeutic response in about a third of men and women with breast cancer. Tamoxifen acts as a cell cycle phase-specific which controls a biochemical process critical in cycle regulation of breast cancer cells. The mechanism of the antiestrogen action is not clear yet. *In vitro* studies show that tamoxifen stimulates hormone-dependent breast cancer cells to secrete the peptide transforming growth factor β (TGF-β), which can inhibit the growth of estrogen receptor-negative breast cancer cells which are negative for estrogen receptors (142–144).

Tamoxifen is converted to a number of well-characterized metabolites in patients. The majority of these metabolites exhibit antiestrogenic effects. The primary metabolite of tamoxifen is N-desmethyltamoxifen which has low affinity for the estrogen receptor. 4-Hydroxytamoxifen is a minor metabolite, but binds to the estrogen receptor with an affinity 25–50 times that of tamoxifen and equal to that of estradiol (145,146). Therefore, although 4-hydroxytamoxifen is found at lower levels in serum than N-desmethyltamoxifen, its potency is 1250 times that of N-desmethyltamoxifen. It suggests that 4-hydroxytamoxifen is primarily responsible for the observed antiestrogenic effect. The metabolites are secreted largely in the bile as conjugates. The initial blood half-life of tamoxifen is less than 8 h. The secondary half-life is about 7 d. It takes 16 weeks to reach steady state (147–149). At a dose of 20 mg twice a day, tamoxifen levels range from 285–310 mg/L, N-desmethyltamoxifen levels from 462–481 mg/L, and 4-hydroxytamoxifen levels from 6–7 mg/L. The relationship between the response and the dose of tamoxifen is minimal. However, increasing dose has been reported to result in a second remission after relapse (150–152). Animal carcinogenicity studies and human studies suggest that tamoxifen treatment of preclinical or early breast cancer (asymptomatic) has a major beneficial effect. With long-term therapy, a majority of preclinical breast-cancer clones should be suppressed.

Adverse Effects. The side effects of estrogens vary according to the type and dosage of the estrogen and if progestins are co-administrated. The most common side effect is nausea. However, it does not usually interfere with eating and cause weight loss. Increasing the dosage may further induce anorexia and vomiting. Breast enlargement, enlargement of endometrial tissue, and intermenstrual bleeding are other common side effects. Low dose estrogens, such as multiphasic oral contraceptives which contain $30-35$ μg of estrogen and new versions of progestogen, significantly reduce intermenstrual bleeding (153).

The DES Story. Diethylstilbestrol (DES), a potent estrogen, was in widespread use in midcentury for maintenance of pregnancies (154,155). It has been estimated that up to 4.5×10^6 children born during this period were exposed to DES *in utero* (156). It was not until the 1970s that an increase in cases of a rare form of vaginal adenocarcinoma was found in girls who had been exposed to DES *in utero*, a link between prenatal exposure to DES and this rare cancer was suggested (157,158). The annual incidence rate of genital cancer (per 10×10^6) for 16-yr-old females born in 1955 was 17.8 as compared to 0 for 16-yr-old females born in 1949 (158). In addition to the malignant growth (neoplasia), benign anomalies were also reported in women exposed to DES *in utero* (159–161). Approximately $0.14-1.4$ per thousand of the exposed offspring developed clear cell adenocarcinoma of the vagina and cervix, peaking at age 19 and decreasing to lower levels by age 30. Functional abnormalities include an increased incidence of spontaneous abortion, premature labor, and ectopic pregnancies. In male adults, the reproductive tracts were adversely affected by prenatal exposure to DES. Decreased fertility, abnormalities in quantity and quality of sperm, and epididymal cysts were present more often in men who were exposed to DES *in utero* compared with the unexposed controls (160,161). Since there have been no reports correlating DES treatment during pregnancy with genital carcinogenesis in the mother, it is believed that the teratogenic effects of DES are exerted only during the first 18 weeks of pregnancy, when the fetal urogenital tract is differentiating and appears to be sensitive to the harmful effects of DES (54).

Breast Cancer. Epidemiological studies of breast cancer have suggested that in the presence of other cocarcinogens such as virus, chemicals, and radiation, estrogen could induce breast cancer (162–164). The causation of breast cancer also includes genetic factors. A history of breast cancer in a first-degree relative elevates a woman's risk of contracting breast cancer more than twofold (165,166). Other predictors of risk include age, age at menarche and menopause, age at first pregnancy, geographic area of residence, dietary factors, and smoking habit. However, knowledge of the specific hormones involved in the carcinogenesis and their relative roles remains elusive. The results of many retrospective analyses and designed studies on the risk of breast cancer in various groups of estrogen users are conflicting (167–172).

Endometrial Cancer. Clinical, biological, and epidemiological data indicate that exogenous estrogens are linked to endometrial cancer (173,174). Increased risk (two- to eightfold) has been found in women under estrogen replacement therapy (175–179). The incidence depends on duration of the hormone treatment. Every type of estrogen that has been investigated has shown this relationship, including conjugated equine estrogens, ethinyl estradiol, and DES. The endometrial cancer associated with estrogen users is a less aggressive form, to which the relatively lower death rate from this disease can be attributed. The opposed estrogen treatment (estrogen–progesterone in sequence) reduces the frequency of hyperplasia (tissue enlargement) and atypical hyperplasia which is associated with unopposed estrogen treatment. Studies on combination oral contraceptives clearly show a 50–60% protection against endometrial cancer. The protective effect may be absent among the obese, long-term estrogen users, and women who have borne more than one child (180,181).

Cervical Cancer. Studies of cervical cancer found two principal risk factors for this malignancy: a large number of different sexual partners and an early age at first intercourse. However, the cervix is an endocrine target organ, and therefore its neoplastic potential may depend on hormonal influence. A series of case-control and follow-up studies have linked "high dose" oral contraceptive use to cervical intraepithelial neoplasia and frankly invasive cervical carcinoma (182–184). These studies report a risk that rises with the duration of use to approximately twofold among long-term users of "high dose" contraceptives. A systematic follow-up of women who were exposed to diethylstilbestrol has revealed an increased incidence of cervical intraepithelial neoplasia among these women compared with women unexposed to the drug (185).

Liver and Gallbladder. High dosages of oral estrogens have been reported to increase the risk for jaundice, cholestatic hepatitis, gallstones, and hepatic vein blood clots. Estrogens promote the development of hepatic neoplasms associated with increased hepatic cell regenerative activity (186,187).

Estrogens change the hepatic production of many proteins and metabolites, which can alter the rate of metabolism and excretion of other hormones and drugs, thus further influencing the interaction between these compounds and the hepatocytes (188). For example, estrogen enhances the synthesis of carrier proteins corticosteroid-binding globulin (CBG), thyroxine-binding globulin (TBG), sex hormone-binding globulin (SHBG), transferrin, and ceruloplasmin, which influences the results of laboratory tests used to determine the levels of substances bound to these proteins.

BIBLIOGRAPHY

"Synthetic Estrogens" under "Hormones" in *ECT* 1st ed., Vol. 7, pp. 536–547, by P. F. Dreisbach, Calco Chemical Division, American Cyanamid Company; "Nonsteroidal Estrogens" under "Hormones" in *ECT* 2nd ed., Vol. 11, pp. 127–141, by R. I. Dorfman, Syntex Research Division of Syntex Corporation; "Nonsteroidal Estrogens" under "Hormones" in *ECT* 3rd ed., Vol. 12, pp. 658–691, by G. C. Crawley, Imperial Chemical Industries Limited.

1. U.S. Pat. 4,789,671 (Dec. 6, 1988), J. R. Bull, R. I. Thomson, H. Laurent, H. Schroder, and R. Wiechert (to Schering AG).
2. Ger. Offen. 3,838,779 (May 17, 1990), G. Kirsch, G. Neef, H. Laurent, R. Wiechert, S. Beier, W. Elger, and J. R. Bull (to Schering AG).
3. Ger. Offen. 3,939,893; 3,939,894 (June 6, 1991), S. Beier, J. R. Bull, W. Elger, P. Esperling, G. Kirsch, H. Laurent, G. Neef, and R. Wiechert (to Schering AG).
4. Ger. Offen. 3,741,800 (June 15, 1989), P. Jungblut, R. Wiechert, and R. Bohlmann (to Schering AG).
5. R. H. Peters and co-workers, *J. Med. Chem.* **32**, 2306 (1989).
6. H. K. Kim, in B. K. Kim, E. B. Lee, C. K. Kim, and Y. N. Han, eds., *Advances in New Drug Development* (Invited Lectures of the International Congress of New Drug

Development, Seoul, South Korea, August 18, 1991) The Pharmaceutical Society of Korea, Seoul, 1991, pp. 547–555.

7. O. Nakayama and co-workers, *J. Antibiot.* **43**, 1394 (1990); H. Kondo and co-workers, *J. Antibiot.* **43**, 1533 (1990).

8. A. E. Wakeling and J. Bowler, *Biochem. Soc. Trans.* **19**, 899 (1991).

9. P. J. Weatherill and co-workers, *J. Steroid Biochem.* **30**, 263 (1988).

10. A. E. Wakeling, M. Dukes, and J. Bowler, *Cancer Res.* **51**, 3867 (1991).

11. A. E. Wakeling and J. Bowler, *J. Steroid Biochem.* **30**, 141 (1988).

12. A. E. Wakeling and J. Bowler, *J. Endocrinol.* **112**, R7 (1988).

13. C. Levesque and co-workers, *J. Med. Chem.* **34**, 1624 (1991).

14. Eur. Pat. 367576 (May 9, 1990), F. Labrie (to Endorecherche).

15. Ger. Offen. 3,925,507 (Jan. 31, 1991), R. Bohlmann, D. Henderson, H. Kunzer, Y. Nishino, M. Schneider, R. Wiechert, and H. Kuenzer (to Schering AG).

16. U.S. Pat. 4,874,754 (Oct. 17, 1989), M. Bouton, L. Nedelec, F. Nique, and D. Philibert (to Roussel-Uclaf).

17. U.S. Pat. 4,943,566 (July 24, 1990), F. Nique, D. Philibert, M. Moquilewsk, M. Bouton, and L. Nedelec (to Roussel-Uclaf).

18. Eur. Pat. 384842 (Aug. 29, 1990), A. Claussner, L. Nedelec, D. Philibert, and P. Vandevelde (to Roussel-Uclaf).

19. R. L. Sutherland and co-workers, *Nature* **288**, 273 (1980).

20. C. B. Lazier and B. V. Bapat, *J. Steroid Biochem.* **31**(4B), 665 (1988).

21. R. Valavaara and co-workers, *Eur. J. Cancer Clin. Oncol.* **24**, 785 (1988).

22. Eur. Pat. 260066 (Mar. 16, 1988), R. McCague (to National Research and Development Corp.).

23. *Drugs Fut.* **8**, 673 (1983).

24. C. K. W. Watts and R. L. Sutherland, *Mol. Pharmacol.* **31**, 541 (1987).

25. P. C. Ruenitz, J. R. Bagley, and C. M. Mokler, *J. Med. Chem.* **26**, 1701 (1983).

26. Jpn. Pat. 88503618 (April 7, 1988) T. Asao and co-workers (to Taiho).

27. Jpn. Pat. 87099352 (May 8, 1987) T. Asao and co-workers (to Taiho).

28. U.S. Pat. 4,897,503 (Jan. 30, 1990) T. Asao, S. Takeda, and Y. Sugimoto (to Taiho).

29. T. Asao, *16th Int. Cong. Chemother.* (June 11–16, Jerusalem), 248 (1989).

30. L. R. Morgan and co-workers, *Proc. Am. Assoc. Cancer Res.* **31**, Abst. 2629 (1990); *Drugs Fut.* **17**, 369 (1992).

31. N. C. Misra and co-workers, *Int. J. Cancer* **43**, 781 (1989).

32. A. P. Sharma and co-workers, *J. Med. Chem.*, **33**, 3216 (1990).

33. A. Saeed and co-workers, *J. Med. Chem.* **33**, 3210 (1990).

34. C. C. Teo and co-workers, *J. Med. Chem.* **35**, 1330 (1992).

35. B. Astroff and co-workers, *Mol. Pharmacol.* **33**, 231 (1988).

36. C. D. Jones and co-workers, *J. Med. Chem.* **27**, 1057 (1984).

37. C. D. Jones and co-workers, *J. Med. Chem.* **35**, 931 (1992).

38. *Drugs Fut.* **3**, 211 (1978).

39. M. Lang and co-workers, *Proc. Amer. Assoc. Cancer Res.* **31**, Abst. 1305 (1990).

40. R. W. Hartmann and co-workers, *J. Med. Chem.* **28**, 1295 (1985).

41. M. B. Hossain and co-workers, *Acta Crystallogr. Sect. B* **B47**, 511 (1991).

42. L. J. Brandes and M. W. Hermonat, *Biochem. Biophys. Res. Commun.* **123**, 724 (1984).

43. Y. Y. Sheen, D. M. Simpson, and B. S. Katzenellenbogen, *Endocrinology* **117**, 561 (1985).

44. M. R. Schneider, *Eur. J. Cancer* **115**, Abst. TH 11 (1989).

45. S. von Angerer and co-workers, *6th NCI-EORTC Symp. New Drugs Cancer Ther.* (March 7–10, Amsterdam) Abst. 060 (1989).

46. J. F. Miquel and J. Gilbert, *J. Steroid Biochem.* **31**(4B), 525 (1988).

47. S. Green and co-workers, *Nature* **320**, 134 (1986).
48. K. J. Edwards, C. A. Laughton, and S. Neidle, *J. Med. Chem.* **35**, 2753 (1992).
49. S. Green, *J. Steroid Biochem. Molec. Biol.* **37**, 747 (1990).
50. W. L. Duax and co-workers, *J. Steroid Biochem.* **31**(4B), 481 (1988).
51. W. L. Duax and J. F. Griffin, *J. Steroid Biochem.* **27**, 271 (1987).
52. R. D. Bindal, J. T. Golab, and J. A. Katzenellenbogen, *J. Amer. Chem. Soc.* **112**, 7861 (1990).
53. C. Holdregger and D. Keefer, *Am. J. Anat.* **177**, 285 (1986).
54. T. L. Greco, T. M. Duello, and J. Gorski, *Endocr. Rev.* **14**, 59 (1993).
55. J. H. Clark, J. Anderson, and E. J. Peck, Jr., *Science* **176**, 528 (1972).
56. G. S. Roth and G. D. Hess, *Mech. Ageing Dev.* **20**, 175 (1982).
57. A. Barbea and J. Gorsic, *Biochemistry* **9**, 1899 (1970).
58. W. V. Welshons, M. S. Lieberman, and J. Gorski, *Nature* **122**, 1165 (1984).
59. V. Kumar, S. Green, and G. Stack, *Cell* **51**, 941 (1987).
60. F. E. Murdoch and co-workers, *Biochemistry* **29**, 8377 (1990).
61. M. Beato, *Cell* **56**, 335 (1989).
62. R. M. Evans, *Science* **240**, 889 (1988).
63. J. Gorski and co-workers, *Biol. Reprod.* **48**, 8 (1993).
64. J. H. Clark, W. T. Schrader, and B. W. O'Malley, in J. D. Wilson and D. W. Foster, eds., *Textbook of Endocrinology*, 8th ed., W. B. Saunders, Philadelphia, Pa., 1991, p. 9.
65. J. E. Griffin and S. R. Ojeda, eds., *Textbook of Endocrine Physiology*, 2nd ed. Oxford University Press, New York, 1992.
66. L. Martin, in J. A. McLachlan, ed., *Estrogen in the Environment*, Elsevier North Holland Inc., New York, 1980, p. 103.
67. K. L. Kelner, H. J. Kirchick, and E. J. Peck, Jr., *Endocrinology* **111**, 1986 (1982).
68. S. W. Curtis and K. S. Korach, *Mol. Endocrinol.* **5**, 959 (1991).
69. A. J. W. Hsueh, E. Y. Adashi, and P. B. C. Jones, *Endocr. Rev.* **5**, 76 (1984).
70. E. R. Simpson, *Mol. Cell Endocrinol.* **13**, 213 (1979).
71. P. K. Siiteri and P. C. MacDonald, in R. O. Greep, and E. B. Astwood, eds., *Handbook of Physiology, Section 7: Endocrinology*, Vol. II, American Physiology Society, Williams & Wilkins, Baltimore, Md., 1973, p. 615.
72. D. L. Hemsell, J. M. Grodin, and P. F. Brenner, *J. Clin. Endocrinol. Metab.* **38**, 476 (1974).
73. C. D. Edman and P. C. MacDonald, *Am. J. Obstet. Gynecol.* **130**, 456 (1978).
74. K. Fotherby and F. James, *Adv. Steroid Biochem. Pharmacol.* **3**, 67 (1972).
75. D. E. Englund and E. D. B. Johansson, *Brit. J. Obstet. Gynecol.* **85**, 957 (1978).
76. G. I. Zatuchni, in K. L. Becker, ed., *Principles and Practice of Endocrinology and Metabolism*, Lippincott, Philadelphia, Pa., 1990, p. 868.
77. L. A. Rigg, B. Milanes, and B. Villanueva, *J. Clin. Endocrinol. Metab.* **45**, 1261 (1977).
78. M. S. Power, L. Schenkel, and P. E. Darley, *Am. J. Obstet. Gynecol.* **152**, 1099 (1985).
79. I. Schiff, D. Tulchinsky, and K. J. Ryan, *Fertil. Steril.* **28**, 1063 (1977).
80. S. Hass and co-workers, *Obstet. Gynecol.* **71**, 671 (1988).
81. W. Rosner, *Endocr. Rev.* **11**, 80 (1990).
82. W. M. Pardridge, *Amer. J. Physiol.* **252**, E157 (1987).
83. R. T. Scott, Jr., and G. D. Hodhen, *Clin. Obstet. Gynecol.* **33**, 551 (1990).
84. D. L. Healy and co-workers, *Fertil. Steril.* **41**, 114 (1984).
85. R. W. Rebar, D. Kenigsberg, and G. D. Hodgen, in K. L. Becker, ed., *Principles and Practice of Endocrinology and Metabolism*, Lippincott, Philadelphia, Pa., 1990, p. 788.
86. A. R. LaBarbera and R. W. Rebar, *Clin. Obstet. Gynecol.* **33**, 576 (1990).

87. P. A. Lee, *Semin. Reprod. Endocrinol.* **6**, 13 (1988).
88. J. Ginsburg, in H. Burger and M. Boulet, eds., *A Portrait of the Menopause*, Parthenon Publishing Group, Carnforth, Lancs, U.K., 1991, p. 45.
89. J. M. Sullivan and co-workers, *Arch. Intern. Med.* **150**, 2557 (1990).
90. *Oral Contraceptives*, Population Reports, Series A, No. 6, Population Information Program, The Johns Hopkins University, Baltimore, Md. 1982, p. A190.
91. U. J. Gospard, *Amer. J. Obstet. Gynecol.* **157**, 1029S (1987).
92. L. E. DeLia and M. G. Emery, *Clin. Obstet. Gynecol.* **24**, 879 (1981).
93. K. Fotherby, *Brit. J. Family Planning* **11**, 86 (1985).
94. G. Silfverstolpe and co-workers, *Acta Obstet. Gynecol. Scand.* **88** (Suppl), 89 (1979).
95. A. Phillips and co-workers, *Contraception* **36**, (1987).
96. A. Phillips and co-workers, *Contraception* **41**, 399 (1990).
97. H. J. Kloosterboer, C. A. Von Knoordegraaf, and E. W. Turpijn, *Contraception* **38**, 325 (1988).
98. D. W. Hahn, A. Phillips, and J. L. McQuire, in *Aktuelle Aspekte der Hormonalen Kontrazeption*, P. J. Keller, ed., Karger, New York, 1991, p. 46.
99. T. A. Grillo and R. Hatcher, *Bull. Woodruff Med. Cent.* **11**, 183 (1980).
100. A. Yuzpe, R. P. Smith, and A. W. Rademaker, *Feretil. Steril.* **37**, 508 (1982).
101. C.-R. Garcia and D. L. Rosenfeld, *Human Fertility: The Regulation of Reproduction*, F. A. Davis Company, Philadelphia, Pa., 1977, p. 59.
102. L. Mastroiani, Jr., P. J. Donaldson, and T. T. Kane, *Developing New Contraceptives*, National Academic Press, Washington, D.C., 1990, p. 11.
103. M. R. Soules, *Pediatr. Clin. North Amer.* **43**, 1083 (1987).
104. D. T. Baird, in L. J. LeGroot and co-workers, eds., *Endocrinology*, 2nd ed., W. B. Saunders, Philadelphia, Pa., 1989, p. 1950.
105. J. L. Ross and co-workers, *N. Engl. J. Med.* **309**, 1104 (1983).
106. J. L. Ross and co-workers, *J. Pediatr.* **109**, 950 (1986).
107. M. J. Odom, B. R. Carr, and P. C. MacDonald, in R. Andres, E. Bierman, and J. P. Blass, eds., *Principles of Geriatric Medicine and Gerontology*, 2nd ed., McGraw-Hill, Book Co., Inc., New York, 1990, p. 777.
108. D. T. Baird and I. S. Fraser, *J. Clin. Endocrinol. Metab.* **38**, 1009 (1974).
109. C. Longcope, *Amer. J. Obstet. Gynecol.* **111**, 778 (1971).
110. G. A. Colditz and co-workers, *JAMA* **264**, 2648 (1990).
111. J. Haarbo and co-workers, *Metabolism* **40**, 1323 (1991).
112. B. Ettinger and co-workers, *Amer. J. Obstet. Gynecol.* **166**, 479 (1992).
113. Consensus Development Conference, *Brit. Med. J.* **295**, 914 (1987).
114. A. T. Leather, M. Savvas, and J. W. W. Studd, *Obstet. Gynecol.* **78**, 1008 (1991).
115. I. Persson and co-workers, *Brit. Med. J.* **298**, 147 (1989).
116. L. F. Voigt and co-workers, *Lancet* **338**, 274 (1991).
117. M. C. Horowitz, *Science* **260**, 626 (1993).
118. R. L. Jilka and co-workers, *Science* **257**, 88 (1992).
119. K. Hunt, *Brit. J. Hosp. Med.* **38**, 450 (1987).
120. D. M. Hart, in J. O. Drife and J. W. W. Studd, eds., *HRT and Osteoporosis*, Springer-Verlag, London, 1990, p. 263.
121. N. S. Weiss, C. L. Ure, and J. H. Ballard, *N. Engl. J. Med.* **303**, 1195 (1980).
122. B. Ettinger, H. K. Genant, and C. E. Cann, *Ann. Intern. Med.* **102**, 319 (1985).
123. E. G. Lufkin and co-workers, *Ann. Intern. Med.* **117**, 1 (1992).
124. R. Lindsay and J. F. Tohme, *Obstet. Gynecol.* **76**, 290 (1990).
125. R. R. Love and co-workers, *Breast Cancer Res. Treat.* **12**, 297 (1988).
126. T. Fornander, L. E. Rutqvist, and H. E. Sjorberg, *Proc. Am. Soc. Clin. Oncol.* **8**, 21 (A-77) (1989).
127. S. Turken and co-workers, *J. Nat. Cancer Inst.* **81**, 1086 (1989).

128. R. Lindsay, *Lancet* **341**, 801 (1993).
129. J. D. Drife and J. W. W. Studd, *HRT and Osteoporosis*, Springer-Verlag, London, 1990, p. 251.
130. G. A. Greendale, K. J. Carlson, and I. Schiff, *J. Gener. Intern. Med.* **5**, 464 (1990).
131. J. M. Sullivan and R. A. Lobo, *Amer. J. Obstet. Gynecol.* **168**, 2006 (1993).
132. G. A. Colditz and co-workers, *N. Engl. J. Med.* **316**, 1105 (1987).
133. T. L. Bush and co-workers, *Circulation* **75**, 1102 (1987).
134. *Red Book*, Medical Economics Data, Montvale, N.J., 1993.
135. D. P. Byer, *Cancer* **32**, 1126 (1973).
136. B. J. Kennedy, *Surg. Gynecol. Obstet.* **113**, 635 (1965).
137. B. J. Kennedy, *Cancer* **24**, 1345 (1969).
138. M. J. Boyer and M. H. Tattersall, *Med. Pediatr. Oncol.* **18**, 317 (1990).
139. R. R. Love, *Prevent. Med.* **20**, 64 (1991).
140. V. C. Jordan and C. S. Murphy, *Endocr. Rev.* **11**, 578 (1990).
141. J. S. Patterson, L. A. Buttersby, and D. G. Edward, *Rev. Endocr. Related Cancer* **9**, 563 (1982).
142. C. Knabbe and co-workers, *Cell* **48**, 417 (1987).
143. B. Dickson and co-workers, *J. Steroid Biochem. Mol. Biol.* **43**, 69 (1992).
144. R. Clarke, R. B. Dickson, and M. E. Lippman, *Critical Rev. Oncol. Hematol.* **12**, 1 (1992).
145. C. Fabian, L. Tilzer, and L. Sternson, *Biopharmaceut. Drug Dispos.* **2**, 281 (1981).
146. R. I. Nicholson and co-workers, *Eur. J. Cancer* **15**, 317 (1979).
147. J. M. Fromson, S. Pearson, and S. Bramah, *Xenobiotica* **3**, 711 (1973).
148. C. Fabian and co-workers, *Cancer* **48**, 876 (1981).
149. B. M. Sherman and co-workers, *J. Clin. Invest.* **64**, 398 (1979).
150. J. V. Kemp and co-workers, *Biochem. Pharmacol.* **32**, 2045 (1983).
151. P. Daniel and co-workers, *Eur. J. Cancer Clin. Oncol.* **17**, 1183 (1981).
152. J. S. Patterson and co-workers, in H. T. Mourison and T. Palshoff, eds., *Breast Cancer—Experimental and Clinical Aspects*, Pergamon, Oxford, UK, 1980, pp. 89.
153. B. R. Carr and J. E. Griffin, in Ref. 64, p. 1007.
154. O. W. Smith, *Amer. J. Obstet. Gynecol.* **56**, 821 (1948).
155. O. W. Smith and G. V. S. Smith, *N. Engl. J. Med.* **241**, 562 (1949).
156. J. M. Reinisch and W. G. Karow, *Arch. Sex. Behav.* **6**, 257 (1977).
157. A. L. Herbst, H. Ulfelder, and D. C. Poskanzer, *N. Engl. J. Med.* **284**, 878 (1971).
158. A. L. Herbst and co-workers, *Amer. J. Obstet. Gynecol.* **128**, 43 (1977).
159. A. L. Herbst and co-workers, *N. Engl. J. Med.* **292**, 334 (1975).
160. W. B. Gill, G. F. B. Schumacher, and M. Bibbo, *J. Reprod. Med.* **16**, 147 (1976).
161. M. Bibbo and co-workers, *J. Obstet. Gynecol.* **49**, 1 (1977).
162. W. R. Miller and T. J. Anderson, in J. J. Studd and M. I. Whitehead eds., *The Menopause*, Blackwell Scientific Publishers, Oxford, UK, 1988, p. 241.
163. J. L. Kelsey and co-workers, *J. Natl. Cancer Inst.* **67**, 327 (1981).
164. A. B. Miller, *Cancer Res.* **38**, 3985 (1978).
165. B. L. Weber and J. E. Garber, *JAMA* **270**, 1602 (1993).
166. M. L. Slattery and R. A. Kerber, *JAMA* **270**, 1563 (1993).
167. R. K. Ross and co-workers, *JAMA* **243**, 1635 (1980).
168. R. Hoover and co-workers, *J. Natl. Cancer Inst.* **67**, 815 (1981).
169. L. A. Brinton, R. Hoover, and J. F. Fraumeni, Jr., *Brit. J. Cancer* **54**, 825 (1986).
170. B. S. Hulka and co-workers, *Amer. J. Obstet. Gynecol.* **143**, 638 (1982).
171. P. A. Wingo and co-workers, *JAMA* **257**, 209 (1987).
172. K. Hunt and co-workers, *Brit. J. Obstet. Gynecol.* **94**, 620 (1987).
173. L. A. Brinton, *Cancer Detect. Prev.* **7**, 159 (1984).
174. R. D. Gambrel, *Clin. Obstet. Gynecol.* **130**, 695 (1986).

175. A. Paganini-Hill, R. K. Ross, and B. E. Henderson, *Br. J. Cancer* **59**, 445 (1989).
176. J. L. Kelsey and co-workers, *Amer. J. Epidemiol.* **116**, 333 (1982).
177. S. Shapiro and co-workers, *N. Engl. J. Med.* **303**, 485 (1980).
178. B. S. Hulka and co-workers, *Amer. J. Obstet. Gynecol.* **137**, 92 (1980).
179. C. M. F. Antunes and co-workers, *N. Engl. J. Med.* **300**, 9 (1979).
180. D. W. Sturdee and co-workers, *Brit. Med. J.* **1**, 1575 (1978).
181. M. I. Whitehead and co-workers, *J. Reprod. Med.* **27** (Suppl), 539 (1982).
182. L. A. Brinton and co-workers, *Int. J. Cancer* **38**, 339 (1986).
183. M. P. Vessey and co-workers, *Lancet* **2**, 930 (1983).
184. S. H. Swan and W. L. Brown, *Amer. J. Obstet. Gynecol.* **139**, 52 (1980).
185. S. J. Robboy and co-workers, *JAMA* **252**, 2979 (1984).
186. I. R. Wanless and A. Medline, *Lab. Invest.* **46**, 313 (1982).
187. L. E. Porter, D. H. Thiel, and P. K. Eagon, *Semin. Liver Dis.* **7**, 24 (1987).
188. J. Dickerson, R. Bressler, and C. D. Christian, *Contraception* **22**, 597 (1980).

J. Z. GUO
D. W. HAHN
M. P. WACHTER
R. W. Johnson Pharmaceutical Research Institute

HORMONES, PEPTIDE HORMONES.   See HORMONES, ANTERIOR PI-
TUITARY; ANTERIOR PITUITARY-LIKE; POSTERIOR PITUITARY; BRAIN OLIGOPEP-
TIDES; PEPTIDES.

HPLC.   See CHROMATOGRAPHY.

HUMIDIFICATION.   See AIR CONDITIONING; SIMULTANEOUS HEAT AND MASS
TRANSER.

HYDANTOIN AND ITS DERIVATIVES

Hydantoin (**1**) is an accepted name for 2,4-imidazolidinedione [*461-72-3*]. This ring system rarely occurs in nature, although some natural products with hydantoin substructures are known. For example, allantoin [*97-59-6*] (**2**) is a constituent of urine and axinohydantoin [*125143-99-9*] (**3**) is an antitumor component of Indo-Pacific marine sponges (1). Hydantoin itself was synthesized by Baeyer in 1861, although its structure was not assigned correctly until 1870 by Strecker. A huge number of derivatives have been prepared since then, among which several 5,5-disubstituted derivatives have found use in medicine. Other applications of hydantoin derivatives include their use as synthetic and analytical reagents, and a variety of sanitary, agricultural, industrial, and other uses. The chemistry of hydantoins has been reviewed (2–5). A review of thiohydantoins is also available (6).

(1)   (2)   (3)

Physical Properties

Hydantoins are crystalline solids with high melting points, particularly those compounds in which nitrogen is unsubstituted, because this allows intermolecular association by hydrogen bonds (Table 1). Hydantoins are weak acids, which dissociate at the imidic N-3—H atom because this allows more efficient delocalization of the negative charge than ionization at N-1. Definite proof for the nonacidic character of the latter position comes from the fact that 3-substituted hydantoins show no appreciable ionization (7).

Several structure–acidity relationships have been established for hydantoin derivatives. Thus, ionization is known to be unaffected by alkyl

Table 1. Melting Points of Some Hydantoin Derivatives[a]

R	R''	Mp,°C
H	H	214–220[b,c]
CH_3	H	178–182.5[d]
C_6H_5	H	295–299
C_6H_5	CH_3	216–217
CH_2CH_3	CH_3	138
H	CH_2CH_3	94
CH_2CH_3	H	198

[a] R' = phenyl (C_6H_5) unless otherwise noted.
[b] R' = H.
[c] Literature values vary, depending on the rate of heating.
[d] R' = CH_3.

substituents at N-1 and at C-5; for example, pK_a of hydantoin is 9.0 and that of 1-methylhydantoin is 9.1 (8). However, aryl and other electron-withdrawing groups can considerably enhance the acidity of hydantoins; thus 5,5-diphenylhydantoin [57-41-0] (phenytoin) has a pK_a of 8.12 (9), and 1-benzenesulfonyl-5,5-diphenylhydantoin [21413-28-5] has a pK_a of 4.89 (10). Introduction of an arylmethylene side chain at C-5 increases the acidity of the N-1 hydrogen, making it measurable (11). This is due to delocalization of the negative charge at N-1 into the C-5 substituent.

Solvent variation can greatly affect the acidity of hydantoins. Although two different standard states are employed for the pK_a scale and therefore care must be exercised when comparing absolute acidity constants measured in water and other solvents like dimethyl sulfoxide (DMSO), the huge difference in pK_a values, eg, 9.0 in water and 15.0 in DMSO (12) in the case of hydantoin itself, indicates that water provides a better stabilization for the hydantoin anion and hence an increased acidity when compared to DMSO.

2-Thiohydantoin [503-87-7] (pK_a 8.5) is a slightly stronger acid than hydantoin (pK_a 9.0). 4-Thiohydantoins appear to be weaker acids (4).

Spectral Properties.   Hydantoin derivatives show weak absorption in the uv-visible region, unless a part of the molecule other than the imidazolidinedione ring behaves as a chromophore (13); however, pK_a values have been determined by spectrophotometry in favorable cases (14). Absorption of uv by thiohydantoins is more intense, and the two bands observed have been attributed to $n \rightarrow \pi^*$ and $n \rightarrow \sigma^*$ transitions of the thiocarbonyl group (15,16). Several pK_a values of thiohydantoins have been determined by uv-visible spectrophotometry (16).

Infrared spectra of hydantoins show two characteristic carbonyl absorptions at about 1720 and 1780 cm^{-1}. Assignment of these bands has been controversial, and the low frequency signal has been attributed both to C-4 and C-2 carbonyls. In an alternative interpretation, these bands have been assigned (17) to symmetrical and asymmetrical vibrations of a coupled carbonyl system similar to the one found in imides. The relative intensities of both bands can be used as a criterion for structural studies on hydantoins, such as hydrogen bonding (18) and the absence of oxo–enol tautomerism in hydantoins (19). Bands due to N—H stretching vibrations give rise to complex absorption patterns in the 3000–3500 cm^{-1} region, depending on the degree of molecular association by hydrogen bonds (18,20,21).

[1]H-nmr chemical shifts of N-1—H and N-3—H signals have been used as a criterion for distinguishing between N-1-substituted and N-3-substituted hydantoin derivatives (22). They can often be related to electronic properties, and thus good linear correlations have been found between the shifts of N—H and Hammett parameters of the substituents attached to the aryl group of 5-arylmethylenehydantoins (23).

[13]C-nmr data have been recorded and assigned for a great number of hydantoin derivatives (24). As in the case of [1]H-nmr, useful correlations between chemical shifts and electronic parameters have been found. For example, Hammett constants of substituents in the aromatic portion of the molecule correlate well to chemical shifts of C-5 and C-α in 5-arylmethylenehydantoins (23). Comparison between ^{13}C-nmr spectra of hydantoins and those of their conjugate bases has been used for the calculation of their pK_a values (12,25). ^{15}N-nmr

spectra of hydantoins and their thio analogues have been studied (26). The ^{15}N-nmr chemical shifts show a linear correlation with the frequencies of the N—H stretching vibrations in the infrared spectra.

Luminescence spectra of hydantoin have been compared (27) with those of related heterocycles, represented by the following structure. The spectra are more sensitive to variation in the exocyclic heteroatoms (Y) rather than the endocyclic ones (Z).

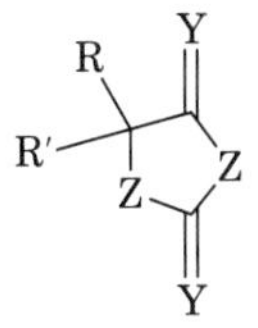

$$Y = O, S; Z = O, NH$$

Mass spectral fragmentation patterns of alkyl and phenyl hydantoins have been investigated by means of labeling techniques (28–30), and similar studies have also been carried out for thiohydantoins (31,32). In all cases, breakdown of the hydantoin ring occurs by α-fission at C-4 with concomitant loss of carbon monoxide and an isocyanate molecule. In the case of aryl derivatives, the ease of formation of Ar—NCO is related to the electronic properties of the aryl ring substituents (33). Mass spectrometry has been used for identification of the phenylthiohydantoin derivatives formed from amino acids during peptide sequence determination by the Edman method (34).

Chemical Properties

Hydantoins can react with electrophiles at both nitrogen atoms and at C-5. The electrophilic carbonyl groups can be attacked by nucleophiles, leading to hydrolysis of the ring or to partial or total reduction of the carbonyl system. Other reactions are possible, including photochemical cleavage of the ring.

Reactions at Nitrogen. The imide proton N-3—H is more acidic than N-1—H and hence this position is more reactive toward electrophiles in a basic medium. Thus hydantoins can be selectively monoalkylated at N-3 by treatment with alkyl halides in the presence of alkoxides (2,4). The mono-*N*-substituted derivatives (**5**) can be alkylated at N-1 under harsher conditions, involving the use of sodium hydride in dimethylformamide (35) to yield derivatives (**6**). Preparation of N-1 monoalkylated derivatives requires previous protection of the imide nitrogen as an aminomethyl derivative (36). Hydantoins with an increased acidity at N-1—H, such as 5-arylmethylene derivatives, can be easily monoalkylated at N-3, but dialkylation is also possible under mild conditions.

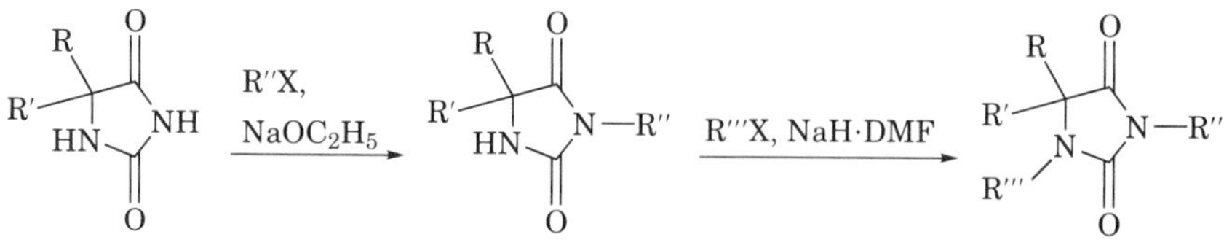

(4)   (5)   (6)

Similarly, hydantoins can be arylated at N-3. For example, treatment of 5,5-diphenylhydantoin (4), $R = R' = C_6H_5$ (phenytoin), with p-tolyllead triacetate in the presence of sodium hydride and a catalytic amount of copper(II) acetate (37) gives compound (7).

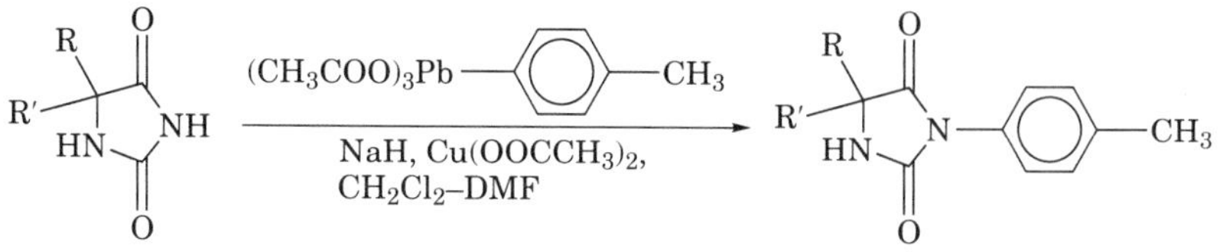

(4)   (7)

Other reactions that show preference for the acidic N-3—H group include Mannich aminomethylation by treatment with formaldehyde and an amine (38) to yield compound (8), reaction with ethyleneimine (39) to give (9), and Michael-type additions (40) such as the one with acrylonitrile to give (10):

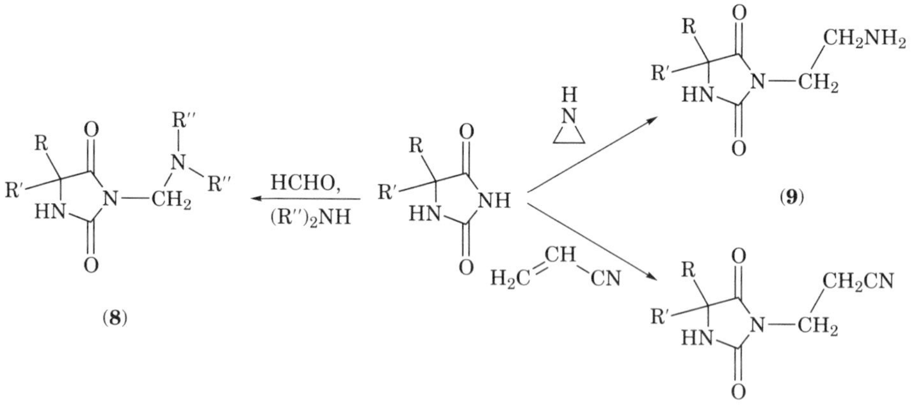

(8)   (9)

(10)

Several electrophiles, such as acetic anhydride, nitric acid or alternative nitrating agents, such as ammonium nitrate in trifluoroacetic anhydride (41), or sodium hypochlorite, react at N-1, which is followed by reaction at N-3 under suitable conditions. In the case of acetic anhydride, the reaction can take place exclusively at N-3 if N-1 is hindered; this fact has served as a criterion for studying the stereochemistry of 5-spirohydantoin derivatives (42,43).

Reactions at C-5.   The C-5 atom of hydantoins can be considered as an active methylene group, and therefore is a suitable position for base-catalyzed condensation reactions with aldehydes (44). 2-Thiohydantoins give the reaction more readily than their oxygen counterparts:

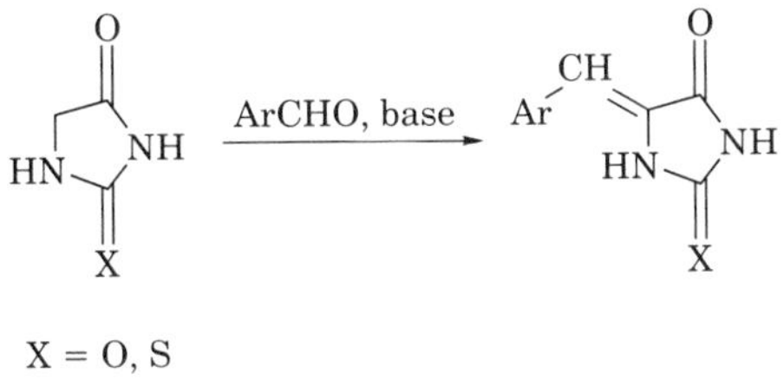

X = O, S

Some of the 5-arylmethylene derivatives thus obtained are useful synthetic intermediates, as shown in Figure 1.

Similarly, the reaction with bromine yields the corresponding 5-bromo derivatives, which are suitable substrates for nucleophilic displacement or dehydrohalogenation reactions. Typical nucleophiles (Nu) are H_2O, NH_3, RNH_2, and sodium barbiturates:

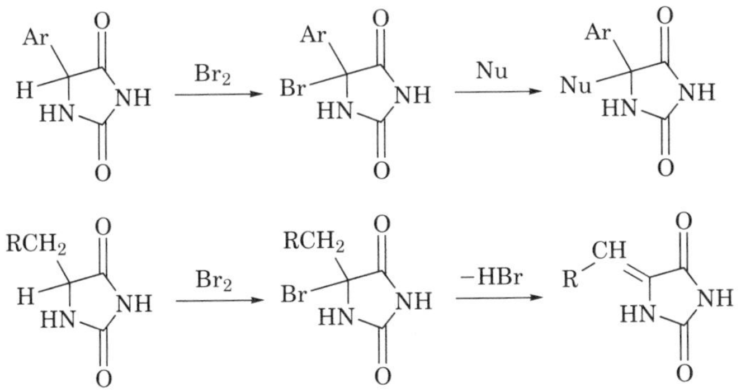

Hydrolysis.   Although hydantoins can be hydrolyzed under strongly acidic conditions, the most common method consists of heating in an alkaline medium

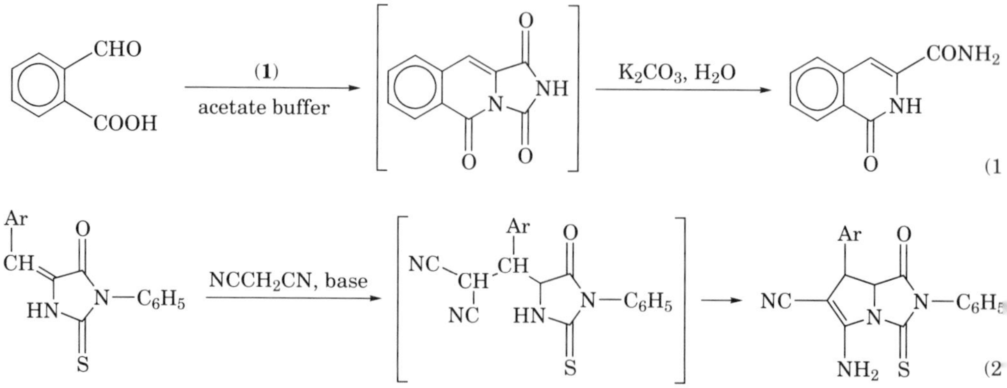

Fig. 1. 5-Substituted hydantoins as synthetic intermediates. Equation 1 (45); equation 2 (46).

to give intermediate ureido acids (the so-called hydantoic acids), which are finally hydrolyzed to α-amino acids.

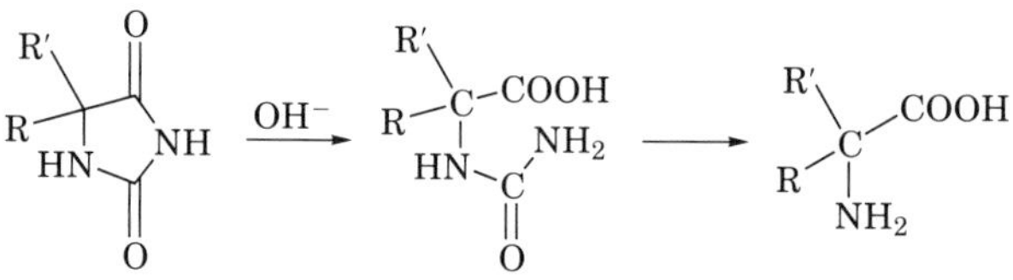

Preparation of amino acids (qv) by this sequence has traditionally been one of the main applications of hydantoins, particularly where other synthetic methods fail. Much of the research effort has been directed toward the development of enzymatic methods for the stereoselective hydrolysis of hydantoins (47). Both D-specific and L-specific hydantoinases (ie, enzymes capable of hydrolyzing hydantoins) are known, thus allowing the preparation of D- and L-amino acids. The use of these enzymes is illustrated in Figure 2 by the synthesis of D-tyrosine [556-02-5], a structural fragment of the antibiotic Amoxicillin (48).

  Reduction. Lithium aluminium hydride is the most common reagent for the reduction of hydantoins. The structure of the products obtained varies, depending on reaction conditions and on the nature of the substituents at C-5, N-1, and N-3. Room temperature reductions (49,50) usually yield 4-hydroxy-2-imidazolidinones (**11**) or the corresponding dehydration products (**12**). Reflux conditions (51,52) give 2-imidazolidinones (**13**), imidazoles (**14**), or imidazolidines (**15**).

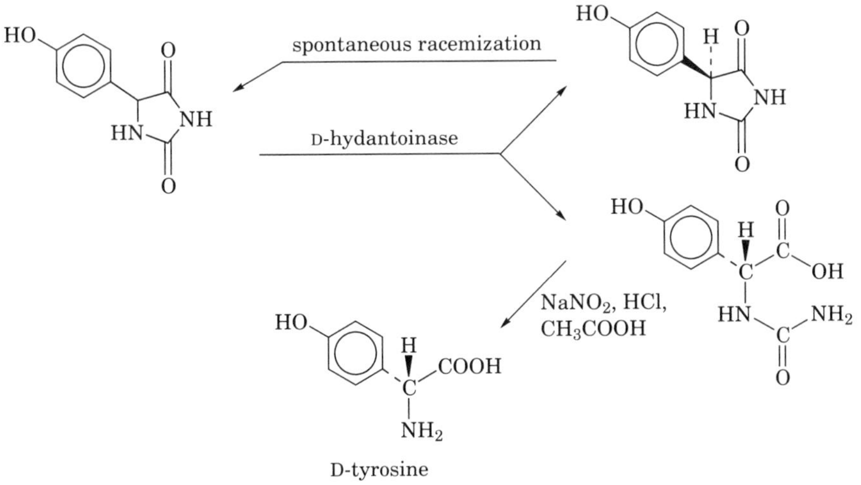

Fig. 2. Enzymatic synthesis of D-tyrosine from (±)5-(4-hydroxyphenyl)hydantoin [54832-24-5].

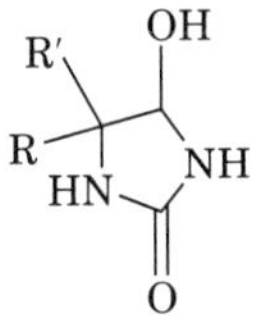

(11) 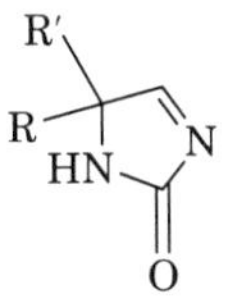　　　(12) 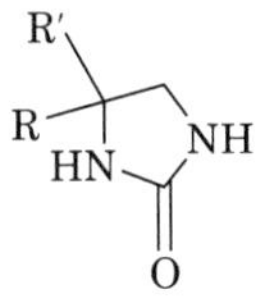　　　(13) 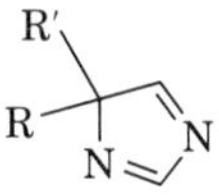　　　(14) 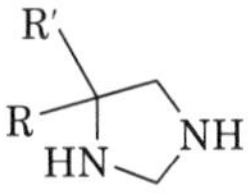　　　(15)

Reaction products (11) can rearrange under suitable conditions (53) to condensed 2(3H)-imidazolidinones such as (16):

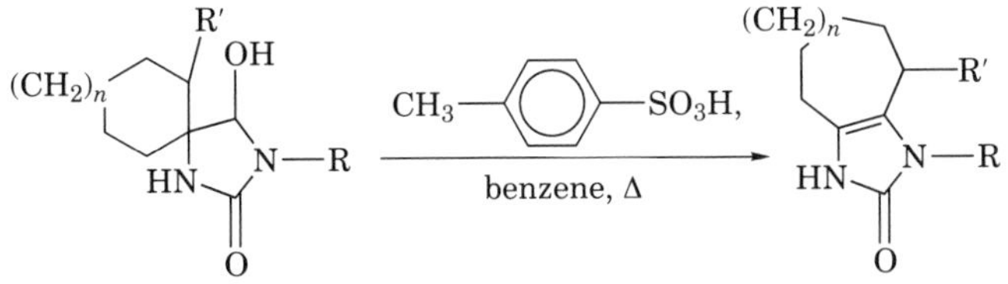

(16)

Compound (12) can also be obtained by dissolving metal reactions through treatment with an excess of lithium–liquid ammonia in *tert*-butyl alcohol (54).

Photochemical Reactions. Hydantoins and other heterocycles containing an NCO group suffer ring cleavage under photochemical conditions. Thus irradiation of 5,5-dimethylhydantoin [77-71-4] in methanol in the absence of benzophenone and oxygen leads to oxidative fission of the C-4–C-5 bond (55):

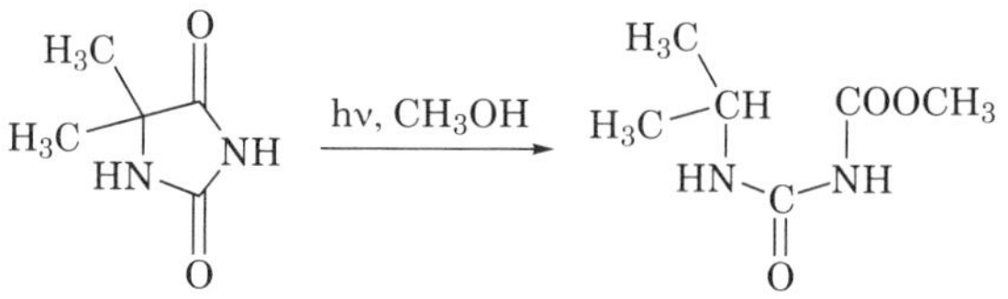

On the other hand, when a similar photoreaction is carried out on hydantoin or its 5-monosubstituted derivatives in the presence of benzophenone, the hydrogen atom at C-5—H is abstracted and the resulting radical couples with that of benzophenone (56):

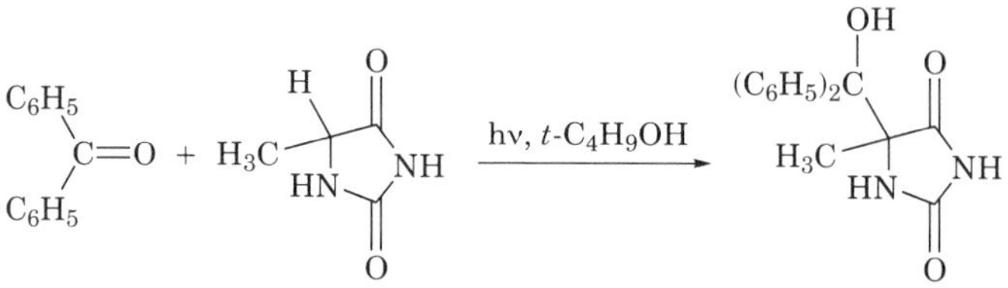

Miscellaneous Reactions.   Some hydantoin derivatives can serve as precursors of carbonium–immonium electrophiles (57). 5-Alkoxyhydantoins are useful precursors of dienophiles (**17**), which undergo Diels-Alder cycloadditions under thermal conditions or in the presence of acid catalysis (58). The pyridine ring of Streptonigrine has been constructed on the basis of this reaction (59).

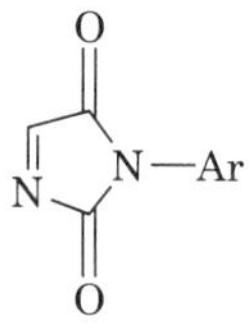

(**17**)

Synthesis

Synthesis From α-Amino Acids and Related Compounds.   Addition of cyanates, isocyanates, and urea derivatives to α-amino acids yields hydantoin precursors. This method is called the Read synthesis (2), and can be considered as the reverse of hydantoin hydrolysis. Thus the reaction of α-amino acids with alkaline cyanates affords hydantoic acids, which cyclize to hydantoins in an acidic medium.

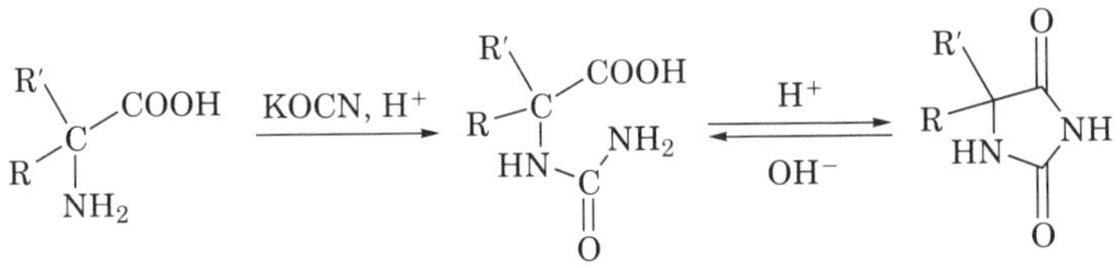

In a modification of the original method, Read (60) replaced α-amino acids with α-amino nitriles. This reaction is sometimes known as Strecker hydantoin synthesis, the term referring to the reaction employed for the synthesis of the α-amino nitrile from an aldehyde or ketone. The cyclization intermediate (18) has been isolated in some cases (61), and is involved in a pH-controlled equilibrium with the corresponding ureide.

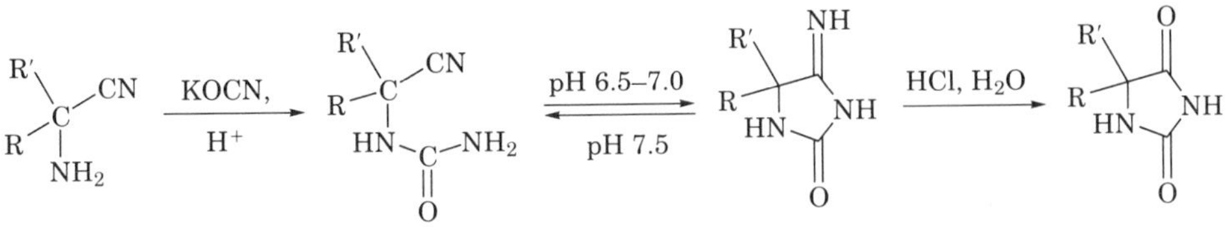

(18)

Chlorosulfonyl isocyanate is an excellent alternative to alkaline cyanates in the preparation of hydantoins from sterically hindered or labile amino nitriles (62). Imino derivatives similar to (18) can also be obtained by addition of isonitriles to imines followed by treatment with a cyanate (63).

Substitution of alkaline cyanates by isocyanates allows the preparation of 3-substituted hydantoins, both from amino acids (64) and amino nitriles (65). The related reaction between α-amino acids and phenyl isothiocyanate to yield 5-substituted 3-phenyl-2-thiohydantoins has been used for the analytical characterization of amino acids, and is the basis of the Edman method for the sequential degradation of peptides with concomitant identification of the N-terminal amino acid.

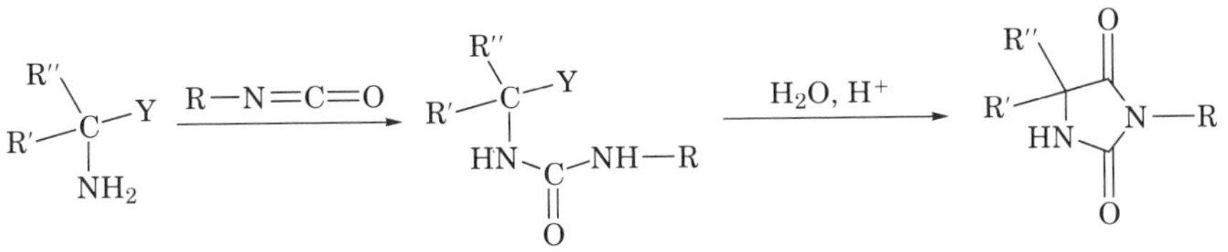

Y = CN, COOH

A related reaction sequence, which proceeds through a Curtius rearrangement, allows the transformation of α-cyano acids into hydantoins (66):

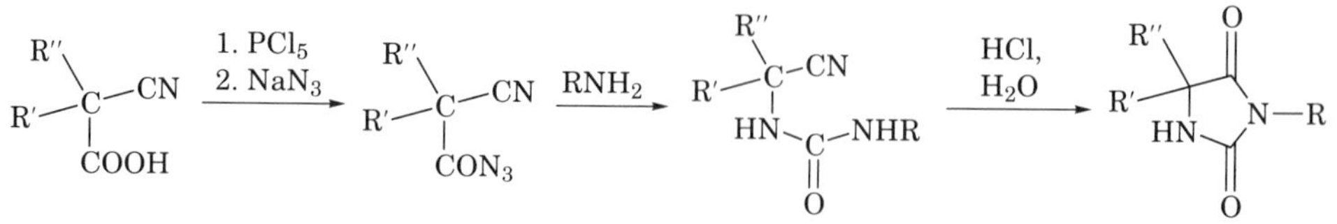

A variety of α-amino acid derivatives, including the acids themselves, halides, esters, and amides can be transformed into hydantoins by condensation with urea (67). α-Hydroxy acids and their nitriles give a similar reaction (68):

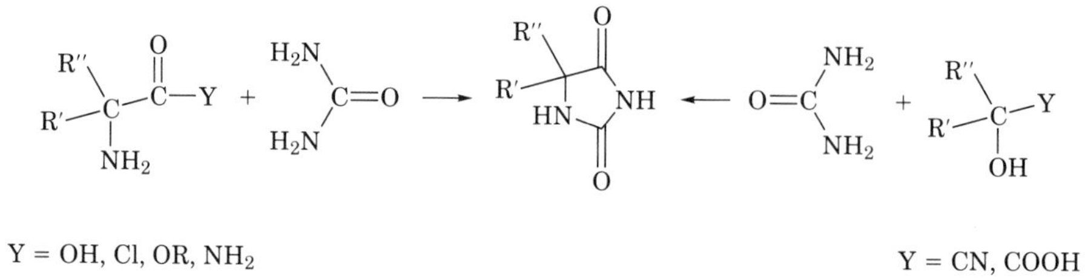

Y = OH, Cl, OR, NH_2

Y = CN, COOH

Synthesis From Aldehydes and Ketones. Treatment of aldehydes and ketones with potassium cyanide and ammonium carbonate gives hydantoins in a one-pot procedure (Bucherer-Bergs reaction) that proceeds through a complex mechanism (69). Some derivatives, like oximes, semicarbazones, thiosemicarbazones, and others, are also suitable starting materials. The Bucherer-Bergs and Read hydantoin syntheses give epimeric products when applied to cycloalkanones, which is of importance in the stereoselective synthesis of amino acids (69,70).

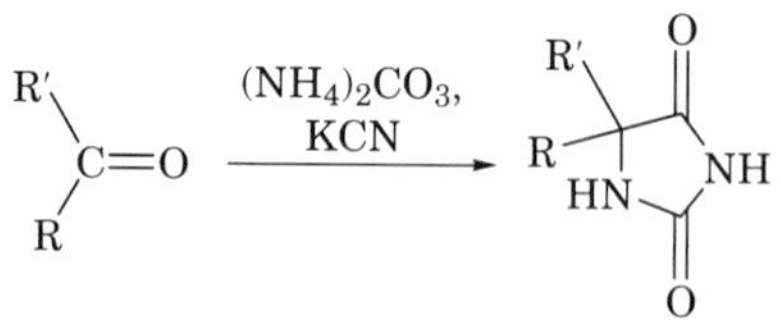

Treatment of α-dicarbonyl compounds with urea in a basic medium is a good method for the synthesis of 5,5-disubstituted hydantoins. This reaction is particularly useful for 5,5-diaryl derivatives, difficult to obtain by the Bucherer-Bergs method. It involves a benzylic rearrangement of the starting compound to an α-hydroxy acid, followed by cyclization of the latter:

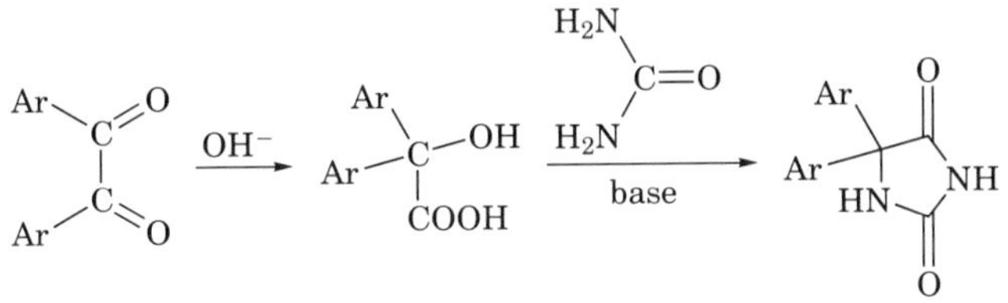

Synthesis From Thiohydantoins.　A modification (71) of the Bucherer-Bergs reaction consisting of treatment of an aldehyde or ketone with carbon disulfide, ammonium chloride, and sodium cyanide affords 2,4-dithiohydantoins (**19**). 4-Thiohydantoins (**20**) are available from reaction of amino nitriles with carbon disulfide (72). Compounds (**19**) and (**20**) can be transformed into hydantoins.

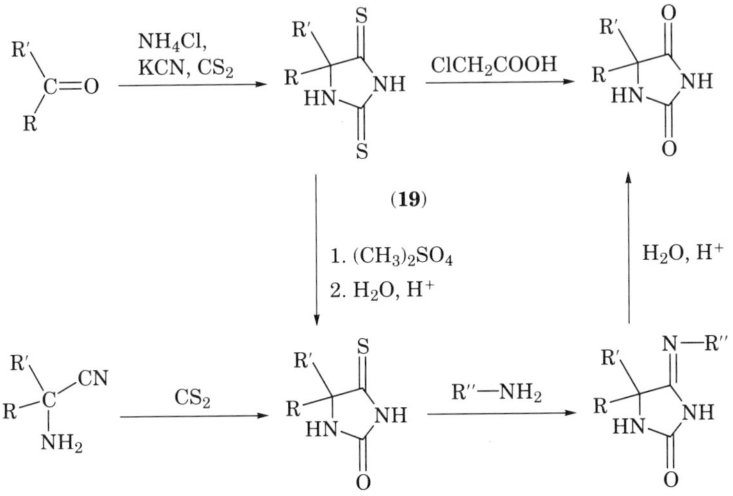

Analytical and Test Methods

Hydantoin itself can be detected in small concentrations in the presence of other NH-containing compounds by paper chromatography followed by detection with a mercury acetate–diphenylcarbazone spray reagent. A variety of analytical reactions has been developed for 5,5-disubstituted hydantoins, due to their medicinal interest. These reactions are best exemplified by reference to the assays used for 5,5-diphenylhydantoin (73–78), most of which are based on their cyclic ureide structure. Identity tests include the following: (*1*) the Zwikker reaction, consisting of the formation of a colored complex on treatment with cobalt(II) salts in the presence of an amine; (*2*) formation of colored copper complexes; and (*3*) precipitation on addition of silver(I) species, due to formation of insoluble salts at N_3.

An acidimetric quantitative determination is based on treatment of the hydantoin with silver nitrate and pyridine in aqueous solution. Complexation of the silver ion at N-3 liberates a proton, and the pyridinium ions thus formed are titrated using phenolphthalein as an indicator. In a different approach, the acidity of N-3—H is directly determined by neutralization with tetrabutylammonium hydroxide or sodium methoxide in dimethylformamide.

A review published in 1984 (79) discusses some of the methods employed for the determination of phenytoin in biological fluids, including thermal methods, spectrophotometry, luminescence techniques, polarography, immunoassay, and chromatographic methods. More recent and sophisticated approaches include positive and negative ion mass spectrometry (80), combined gas chromatography–mass spectrometry (81), and ftir immunoassay (82).

Health and Safety Factors (Toxicology)

The acute toxicity of hydantoin derivatives seems to be low. Most studies on long-term toxicity of hydantoins deal with phenytoin, due to the wide use of this compound as an anticonvulsant (83). Long-term toxic effects of phenytoin include folate deficiency due to impaired folate absorption, hypocalcemia and osteomalacia, alterations of carbohydrate metabolism, gingival hyperplasia, and teratogenic effects. These are grouped under the term fetal hydantoin syndrome, which is associated with continued use of hydantoins during the early stages of pregnancy, and consists of mild retardation of physical and mental indexes, dysmorphic faces and, occasionally, cleft palate, cleft lip, and cardiac defects.

Formation of cyanide by degradation of hydantoin derivatives used as antiseptics for water treatment has been described (84), and this fact might have toxicological relevance.

Applications

Halogenated Hydantoins.   Halogenation has been achieved by use of a variety of halogenating reagents (2,85). These derivatives are employed as reagents in synthesis and analysis and also as disinfectants and biocides in water treatment. In synthesis, 3-bromohydantoins are equivalents of N-bromosuccinimide and other N-bromoamides, acting as a source of halogen for selective allylic and benzylic bromination (86). Among other reactions, both 3-bromo- and 3-iodohydantoins are very convenient reagents for halolactonization of unsaturated carboxylic acids (87):

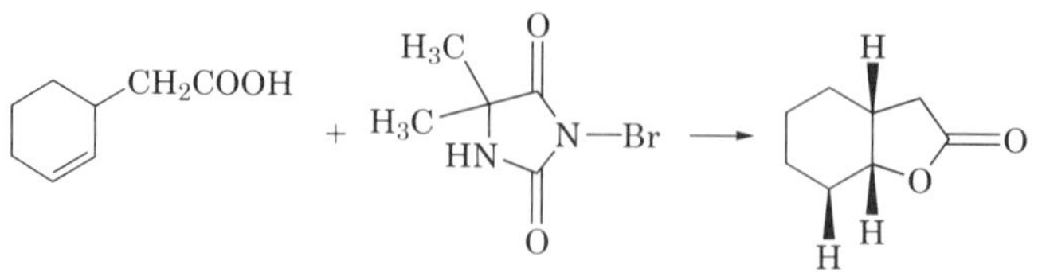

In analysis, 1,3-dihalohydantoins can be used as oxidimetric titrants for determining hydrazine and phenylhydrazine, thiourea, hydroquinone, ascorbic acid, quinoline derivatives such as primaquine, chloroquine, amodiaquine, quinidine and quinine, phenothiazines such as chlorpromazine, Sb(III), As(III), Sn(II), Fe(II), Tl(I), Fe(CN)$_6^{4-}$, I$^-$, SCN$^-$, Ti(III), etc (88–91). Other hydantoins and thiohydantoins have been used as analytical reagents for heavy metals because of their capacity for complex formation (92–94).

Halohydantoins, particularly 1,3-dichloro-5,5-dimethylhydantoin [118-52-5] (commercially available under the name of Dichlorantin) and 1-bromo-3-chloro-5,5-dimethylhydantoin [16079-88-2], ensure high disinfecting and bactericidal effects in water-purification plants, because they are capable of producing hypohalite ions (see CHLORAMINES AND BROMAMINES). They can also be used in sanitizing and bleaching toilet bowls, as disinfectants for dental appliances, for automatic dishwashers, as resin stabilizers, etc. Agglomerated halohydantoins releasing active halogen at a controlled rate have been patented (95–97).

N-**Methylolhydantoins.** 1,3-Bis(hydroxymethyl)-5,5-dimethylhydantoin [6440-58-0] is used extensively as a preservative in cosmetic and industrial applications, and carries EPA registration for the industrial segment. It is available in solid and in aqueous solution forms, including low free formaldehyde versions of the latter. A related derivative, 1,3-bis(hydroxyethyl)-5,5-dimethylhydantoin [26850-24-8], is used in the manufacture of high temperature polyesters, polyurethanes, and coatings, offering improved heat resistance, uv stability, flexibility, and adhesion.

1-Hydroxymethyl-5,5-dimethylhydantoin [116-25-6] is used as an odorless donor of formaldehyde for adhesive applications. Under other reaction conditions 5,5-dimethylhydantoin–formaldehyde resins are obtained, which are described in the patent literature as useful additives for several purposes.

Epoxy Resins. Urethane and ester-extended hydantoin epoxy resins cured with several compounds seem to have better properties than the previous ones (98). These resins are prepared from hydantoins such as (**21**) (99,100).

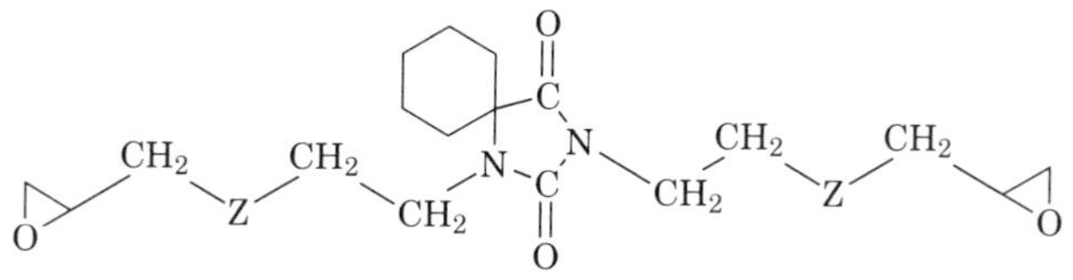

(**21**)

Adding amines to coating compounds containing other polymers of hydantoin derivatives permits thermal curing of the coating compounds, which are useful as electrical insulators of wires under a broad range of conditions without loss of coating flexibility (101).

5-Substituted Hydantoins. 5-Methylhydantoin [616-03-5] has been selected from several structures as a formaldehyde scavenger for color photosensitive materials and water-thinned inks and coatings (102,103).

5,5-Dimethylhydantoins. Some 5,5-dimethylhydantoin derivatives such as (**22**) have been patented as color photographic couplers (104):

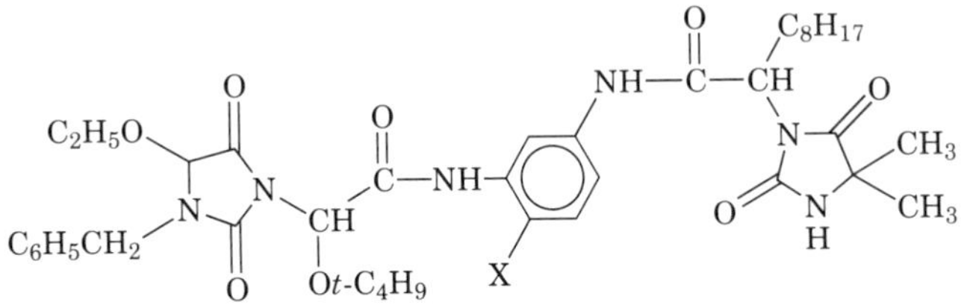

(**22**)

Although the hydantoin ring itself does not present any medicinal activity, many 5,5-disubstituted hydantoins have shown interesting biological properties, and some of them are used in medicine. 5,5-Dimethyl-3-[4-nitro-3-(trifluoromethyl)phenyl]hydantoin [63612-50-0] (**23**) (105) is a pure antiandrogen that inhibits the testicular steroidogenic pathway, apparently acting by forming complexes with the cytoplasmic hormone receptors that are unable to undergo translocation into nuclei, thus blocking most of the biological response of the target cell to androgens (106). It has been proposed for a combined treatment with a potent LHRH agonist to block androgen formation, as the hormonal therapy of choice in prostatic carcinoma (107).

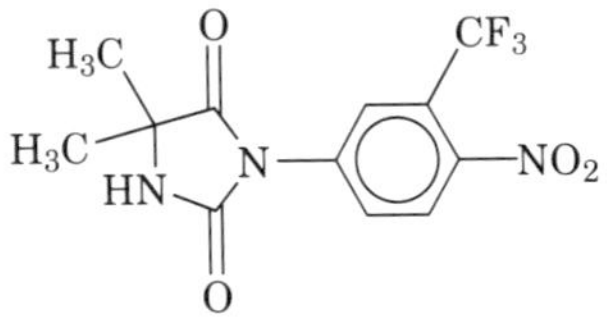

(**23**) 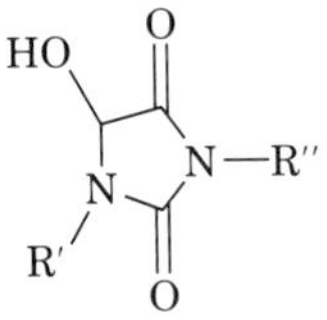   (**24**)

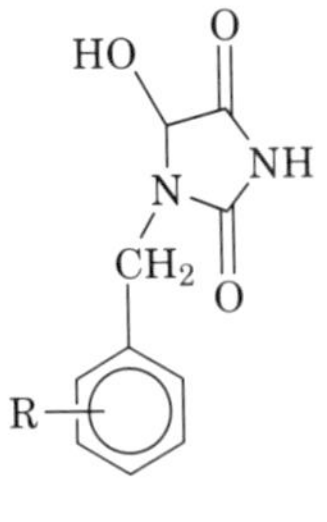

(**25**)

5-Hydroxyhydantoins. Some 5-hydroxyhydantoins such as (**24**) and (**25**) have shown antidiabetic activity (108). In some cases they are also diuretics and hypolipemics. These properties are probably related to the urea moiety found in these compounds.

5-Phenylhydantoins. In spite of the different examples of biological actions, 5,5-disubstituted hydantoins are mainly considered as anticonvulsant agents

(109), the most notable being 5,5-diphenylhydantoin (Dilantin, Phenytoin). This compound was introduced in 1938 (110), and despite significant toxic and teratological effects, is still a broadly used anticonvulsant for treatment of epilepsy. The mechanisms of the anticonvulsant action in this and related compounds have been well studied (111,112). A general model comprising two aromatic rings or their equivalents in a favored orientation, and a third region, usually a cyclic ureide, with a number of H-bond-forming groups has been proposed as pharmacophore (113). Its antiarrhythmic activity (114) as well as its plasma protein binding and metabolism, are relevant (115,116). In this context, tracers for phenytoin suitable for its determination in body fluids have been prepared by using *p*-aminophenytoin bound to carboxyfluorescein (117,118).

Another anticonvulsant, formerly used as a hypnotic, is 5-ethyl-5-phenylhydantoin [631-07-2] (Nirvanol, Mephenytoin). Its *S* isomer is stereoselectively eliminated in most subjects, a fact having clinical consequences with both desired and untoward effects (119).

5-Arylmethylenehydantoins. 5-Arylmethylenehydantoins such as (**26**) are cyclooxygenase and 5-lipoxygenase inhibitors and have been patented as antiinflammatory and antiallergy agents (120).

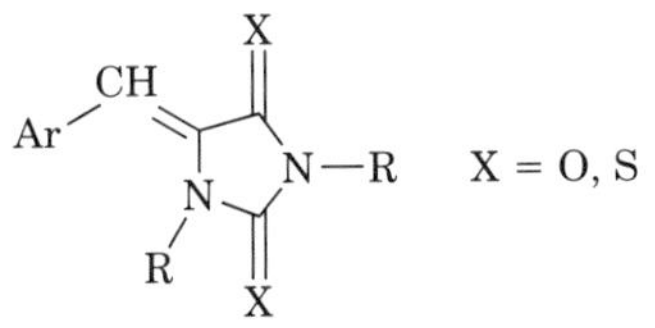

(**26**)

Other Derivatives. Among the different pharmacological activities described, the 5-nitro-2-furyl-methylimino derivatives such as nitrofurantoin [67-20-9] (**27**) show antimicrobial activity and have been specially proposed for urinary tract infections, but they may induce DNA damage and cytotoxicity (121,122). Some 3-arylhydantoins such as (**28**) are effective schistosomicides (123). Many others have been proposed as antifungals, insecticides, nematocides, and soil pesticides. Several types of hydantoin derivatives behave as herbicides, and a study of quantitative structure–activity relationships has been published (123) for compounds related to [60725-62-4] (**29**). Polyhalomethylsulfenylhydantoins exhibit fungicidal activity (124,125). Thus 1,3-bis(trihalomethylsulfenyl)-(**30**) and 3-aryl-1-[(1,1,2,2-tetrachloroethyl)thio]hydantoins (**31**), protect vines against *Plasmopara viticola*. Organophosphorous derivatives (phosphono- or phosphorothiolates) are useful as pesticides (126). As herbicides, hydantoins are included in the uracil group (127).

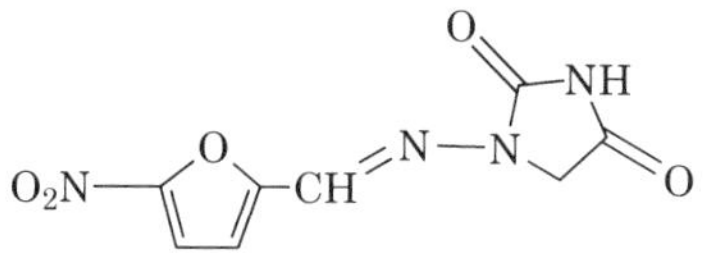

(27)

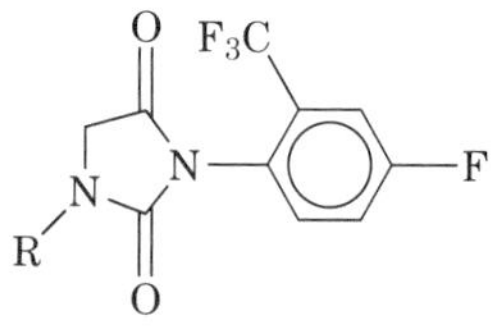

(28)

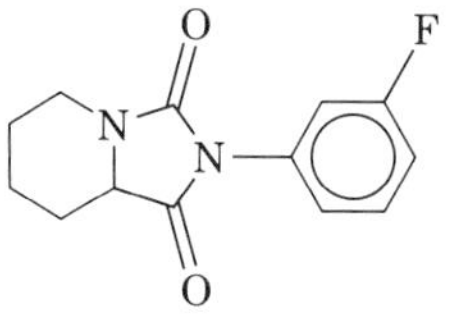

(29)

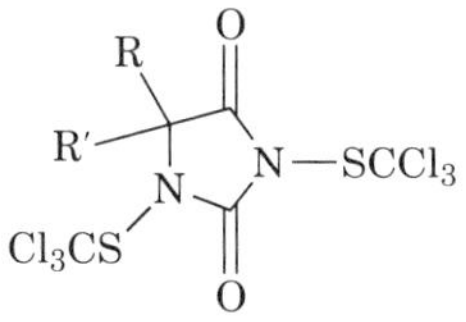

(30)

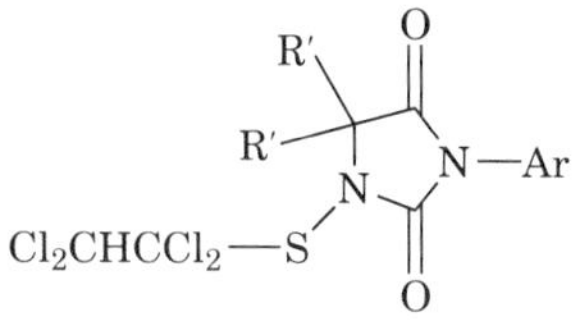

(31)

Some hydantoins are very useful carriers of the nitrogen mustard moiety bis (β-chloroethyl)amine, and are useful in several tumors and multiagent therapy regimens (128,129). Besides that, some 3-(2-chloro-ethyl)hydantoins (**32**) (130) and oxyranylmethylhydantoins, eg, 5-benzylidene-1,3-bis(oxyranylmethyl)hydantoin [*79413-02-8*] (**33**) (131), among other *N*-glycidylated oxo–nitrogen heterocycles, have shown antitumor activity.

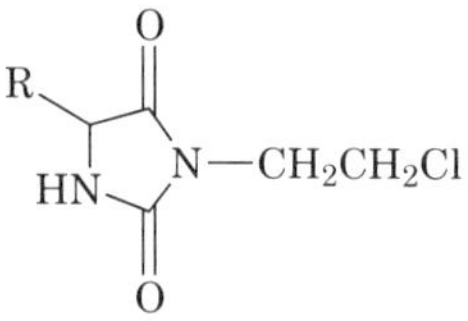

(32)

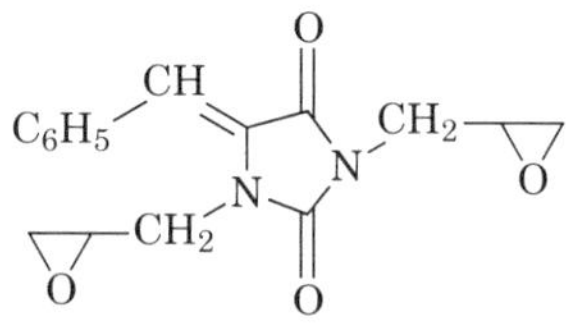

(33)

The 1-arylsulfonylhydantoins (**34**), especially 1-[(4-methylsulphenyl)]- and 1(4-bromophenyl)sulfonylhydantoins (132,133) and several 5,5-spirohydantoins, including sorbinil [*68367-52-2*] (**35**) and structures (**36**) and (**37**), are inhibitors of aldose reductase, and may be useful in the treatment of chronic diabetic complications such as cataract, because this disease results from the accumulation of polyols derived from sugars in the presence of aldose reductase (134). A computer-automated structure evaluation program has been used to study many relevant compounds and generate activating–inactivating fragments (135).

(34)   (35)   (36)   (37)

Structurally related to nitrofurantoins are Dantrolene [7261-97-4] (38), a peripherally acting muscle relaxant, and its analogues (39), which can be used as an antidote against succinylcholine-induced myopathy and in autoimmune myasthenia gravis therapy (136,137).

(38)   (39)

BIBLIOGRAPHY

"Hydantoin" in *ECT* 2nd ed., Vol. 2, pp. 141–164, by E. Smith, Olin Mathieson Chemical Corp.; "Hydantoin and Derivatives" in *ECT* 3rd ed., pp. 692–711, by J. H. Bateman, CIBA-GEIGY Corp.

1. G. R. Pettit and co-workers, *Can. J. Chem.* **68**, 1621 (1990).
2. E. Ware, *Chem. Rev.* **46**, 403 (1950).
3. E. S. Schipper and A. R. Day, in R. C. Elderfield, ed., *Heterocyclic Compounds*, Vol. 5, John Wiley & Sons, Inc., New York, 1957, p. 254.
4. C. Avendaño and G. G. Trigo, *Adv. Heterocycl. Chem.* **38**, 177 (1985).
5. S. A. Avetisyan, L. V. Azaryan, and S. L. Kockarov, *Arm. Khim. Zh.* **39**, 151 (1986) and **41**, 548 (1988).
6. J. T. Edward, *Chem. Org. Sulf. Comp.* **2**, 287 (1966).
7. R. E. Stuckey, *J. Chem. Soc.*, 5075 (1957).
8. M. J. Bausch, B. David, P. Dobrowolski, and V. J. Prassad, *J. Org. Chem.* **55**, 5806 (1990).
9. L. S. Rosenberg and J. L. Jackson, *Drug Der. Ind. Pharm.* **15**, 373 (1989).
10. H. Fujioka and T. Tan, *J. Pharm. Dyn.* **5**, 475 (1982).
11. S.-F. Tan, K.-P. Ang, Y.-F. Fong, and H. Jayachandran, *J. Chem. Soc. Perkin Trans. II*, 473 (1988).
12. M. J. Bausch and co-workers, *J. Org. Chem.* **56**, 5643 (1991).
13. E. Santos, I. Rosillo, B. del Castillo, and C. Avendaño, *J. Chem. Res., Synop.*, 131 (1982).
14. S. P. Agarwal and M. I. Blake, *J. Pharm. Sci.* **57**, 1434 (1968).
15. H. C. Carrington and W. S. Waring, *J. Chem. Soc.*, 354 (1950).

16. J. T. Edward and J. K. Liu, *Can. J. Chem.* **50**, 2423 (1972).
17. A. R. Katritzky and P. J. Taylor in A. R. Katritzky, ed., *Physical Methods in Heterocyclic Chemistry*, Vol. 4, Academic Press, New York, 1971, p. 265.
18. J. Bellanato, C. Avendaño, P. Ballesteros, and M. Martínez, *Spectrochim. Acta* **35A**, 807 (1979).
19. J. Derkosh, *Monatsch. Chem.* **92**, 361 (1961).
20. J. Bellanato, C. Avendaño, P. Ballesteros, E. Santos, and G. G. Trigo, *Spectrochim. Acta* **36A**, 879 (1980).
21. M. Willson, T. Boissou, R. Mathis, and F. Mathis, *Spectrochim. Acta* **40A**, 835 (1984).
22. R. A. Corral and O. O. Orazi, *Spectrochim. Acta* **21**, 2119 (1965).
23. S.-F. Tan, K.-P. Ang, H. Jayachandran, and Y.-F. Fong, *J. Chem. Soc. Perkin Trans. II*, 1043 (1987).
24. C. Pedregal, M. Espada, J. Albert, E. Cruz, and A. Virgili, *Magn. Res. Chem.* **29**, 1226 (1991).
25. M. Bausch, D. Selmarten, R. Gostowski, and P. Dobrowolski, *J. Phys. Org. Chem.* **4**, 67 (1991).
26. F. Cristiani, F. A. Devillanova, A. Díaz, F. Isaia, and G. Verani, *Spectrosc. Lett.* **21**, 767 (1988).
27. M. S. Fadeeva, R. S. Lebedev, T. I. Filaeva, and O. Y. Sdonova, *Deposited Doc.* VINITI 4969 (1982).
28. R. A. Corral, O. O. Orazi, A. M. Duffield, and C. Djerassi, *Org. Mass Spectrom.* **5**, 551 (1971).
29. R. A. Locock and R. T. Coutts, *Org. Mass Spectrom.* **3**, 735 (1972).
30. G. Ruecker, P. N. Natarajan, and A. F. Fell, *Arch. Pharm.* **304**, 833 (1971).
31. R. E. Ardrey and A. Darbre, *J. Chromatogr.* **87**, 499 (1973).
32. T. Sizuki, K. S. Song, and K. Tuzimura, *Org. Mass Spectrom.* **11**, 557 (1976).
33. B. M. Kown and S. Ch. Kim, *J. Chem. Soc. Perkin Trans. II*, 761 (1983).
34. H. M. Fales, Y. Nagai, G. W. A. Milne, H. B. Brewer, T. Brouzert, and J. J. Pisano, *Anal. Biochem.* **43**, 288 (1971).
35. O. O. Orazi, R. A. Corral, and H. Schuttenberg, *J. Chem. Soc. Perkin Trans. II*, 219 (1974).
36. O. O. Orazi and R. A. Corral, *Experientia* **21**, 508 (1965).
37. P. López-Alvarado, C. Avendaño, and J. C. Menéndez, *Tetrahedron Lett.* **33**, 6875 (1992).
38. O. O. Orazi and R. A. Corral, *Tetrahedron* **15**, 93 (1961).
39. J. W. Shaffer, R. Scheasley, and M. B. Winstead, *J. Med. Chem.* **10**, 739 (1967).
40. J. W. Shaffer, E. Steinberg, V. Krimsley, and M. B. Winstead, *J. Med. Chem.* **11**, 462 (1968).
41. S. Suri and R. D. Chapman, *Synthesis*, 743 (1988).
42. G. G. Trigo, C. Avendaño, E. Santos, J. T. Edward, and S. C. Wong, *Can. J. Chem.* **57**, 1456 (1979).
43. J. C. Menéndez and M. M. Söllhuber, *Heterocycles* **26**, 3203 (1987).
44. T. Moriya, K. Hagio, and H. Yoneda, *Chem. Pharm. Bull.* **28**, 1891 (1980).
45. S. Nagase, *Nippon Kagaku Zasshi* **81**, 938 (1960).
46. H. A. F. Daboun, S. E. Abdou, M. M. Hussein, and M. H. Elnagdi, *Synthesis*, 502 (1982).
47. S. Takahashi, *Kikan Kagaku Sosetu*, 111 (1989).
48. E. M. Meijer, W. H. J. Boesten, H. E. Schaemaker, and J. A. M. Van Balken, in J. Tramper, H. C. van der Plas, and P. Linko, eds., *Biocatalysis in Organic Synthesis*, Elsevier, Amsterdam, 1985, p. 135.
49. I. J. Wilk and W. Close, *J. Org. Chem.* **15**, 1020 (1950).
50. J. Cortes and H. Kohn, *J. Org. Chem.* **48**, 2247 (1983).

51. F. J. Marshall, *J. Am. Chem. Soc.* **78**, 3696 (1956).
52. E. de la Cuesta, P. Ballesteros, and G. G. Trigo, *Heterocycles* **16**, 1647 (1981).
53. J. Rubido, C. Pedregal, M. Espada, and J. Elguero, *Synthesis*, 307 (1985).
54. H. R. Divanfard, Y. A. Ibrahim, and M. M. Joullié, *J. Heterocycl. Chem.* **15**, 691 (1978).
55. J. C. Gramain and R. Remuson, *J. Chem. Soc. Perkin Trans. I*, 2341 (1982).
56. J. C. Gramain, J. P. Jendrau, J. Lemaire, and R. Remuson, *Rec. Trav. Chim. Pays-Bas* **109**, 325 (1990).
57. H. E. Zangg and W. B. Martin, *Org. React.* **14**, 60 (1965).
58. D. Ben-Ishai and E. Goldstein, *Tetrahedron* **27**, 3119 (1971).
59. S. M. Weinreb and co-workers, *J. Am. Chem. Soc.* **104**, 536 (1982).
60. W. T. Read, *J. Am. Chem. Soc.* **44**, 1766 (1922).
61. J. C. Menéndez, M. P. Díaz, C. Bellver and M. Söllhuber, *Eur. J. Med. Chem.* **27**, 61 (1992).
62. R. Sarges, H. R. Howard, and P. R. Kelbaugh, *J. Org. Chem.* **47**, 4081 (1982).
63. I. Ugi, *Angew. Chem.* **74**, 9 (1962).
64. M. Espada and co-workers, *Il Farmaco (Sci. Ed.)* **45**, 1237 (1990).
65. J. C. Menéndez and M. M. Söllhuber, *J. Heterocycl. Chem.* **28**, 923 (1991).
66. J. Knabe and W. Wutton, *Arch. Pharm.* **313**, 538 (1980).
67. S. Icli and L. D. Colebrock, *J. Pure Appl. Sci.* **9**, 39 (1976).
68. T. Ohashi, S. Takahashi, T. Nagamachi, K. Yoneda, and H. Yamada, *Agric. Biol. Chem.* **45**, 831 (1981).
69. J. T. Edward and C. Jitransgri, *Can. J. Chem.* **53**, 3339 (1975).
70. G. G. Trigo, C. Avendaño, E. Santos, J. T. Edward, and S. C. Wong, *Can. J. Chem.* **57**, 1456 (1979).
71. H. C. Carrington, *J. Chem. Soc.*, 681 (1947).
72. H. C. Carrington, C. H. Vasey, and W. S. Warings, *J. Chem. Soc.*, 396 (1959).
73. P. Macheras and A. Rosen, *Pharmazie* **39**, 322 (1984).
74. P. Messinger and H. Mayer, *Pharm. Ztg.* **122**, 2253 (1977).
75. G. Stajer, *Arch. Pharm.* **310**, 865 (1977).
76. H. Stamm, *Dtsch. Apoth. Ztg.* **110**, 1206 (1970).
77. W. Wiegrebe and L. Wehrhahn, *Arzneim-Forsch.* **25**, 517 (1975).
78. H. J. Roth, K. Eger, and R. Troschütz, *Pharmaceutical Chemistry*, Vol. 2, Ellis Horwood, Chichester, U.K., 1991, p. 302.
79. J. Philip, I. J. Holcomb, and S. A. Fusari, in K. Florey, ed., *Analytical Profiles of Drug Substances*, Vol. 13, Academic Press, Orlando, Fla., 1984, p. 417.
80. Y. Ishikawa, T. Kumazawana and T. Takahashi, *Z. Rechtsmed.* **99**, 253 (1988).
81. I. Junko and S. Takahashi, *Bunseki Kagaku* **38**, 659 (1989).
82. G. Jaouen, A. A. Ismail, and P. Brossier, PCI Int. Appl. WO 88 07,684 (Oct. 6, 1988).
83. G. L. Jones, G. H. Wimbish, W. E. McIntosh, *Med. Res. Rev.* **3**, 383 (1983).
84. H. Tatsumoto, R. Nakagawa, and S. Suzuki, *Kankyo Kagaku Kenkyu Hokoku (Chiba Daigaku)* **9**, 51 (1984).
85. R. A. Corral and O. O. Orazi, *J. Org. Chem.* **28**, 1100 (1963).
86. A. R. Suárez and O. A. Orio, *An. Asoc. Quím. Argent.* **65**, 163 (1977).
87. Ch. H. Cook, S. S. Jew, and Y. S. Chung, *Arch. Pharmacol. Res.* **5**, 103 (1982).
88. M. Rizk, M. I. Walash, A. A. Obou-Ouf, and F. Beeal, *Anal. Lett.* **14**, 1407 (1981).
89. I. Rizk, A. A. Obou-Ouf, and F. Belal, *Anal. Lett.* **16(A2)**, 129 (1983).
90. M. I. Walash, M. I. Rizk, A. M. Obou-Ouf, and F. Beeal, *Analyst* **108**, 626 (1983).
91. M. P. Radhamoma and P. Indrasenan, *Talanta* **30**, 49 (1983).
92. M. J. Blais, O. Enea, and G. Berthon, *Termochim. Acta* **30**, 45 (1979).
93. F. Barragán, M. T. Montaña, and J. L. Gómez-Ariza, *Microchem. J.* **25**, 524 (1980).
94. M. T. Montaña and J. L. Gómez-Ariza, *Microchem. J.* **25**, 360 (1980).

95. U.S. Pat. 4,297,224 (Oct. 27, 1981), N. T. Machiarolo, B. McGuire, and J. M. Scalise (to Great Lakes Chemical Corp.).
96. Eur. J. Pat. Appl. EP 176,163 (Apr. 2, 1986), R. A. Robinson, W. J. Boan, and G. D. Evans (to Purex Corp.).
97. S. D. Worley, W. B. Wheatley, and H. H. Kohl, *Ind. Eng. Chem. Prod. Res. Der.* **22**, 716 (1983).
98. J. Weiss, *Sci. Tech. Aerosp. Rep.* **21**, N83–22323 (1983).
99. D. Porret, *Makromol. Chem.* **108**, 73 (1967).
100. E. H. Catsiff, R. E. Coulehan, J. F. DiPrima, D. A. Gordon, and R. Seltzer, *Org. Coat. Plast. Chem.* **39**, 139 (1978).
101. Ger. Offen, DE 3,242,162 (May 17, 1984). K. Sirinyan, F. Jonas, and R. Merten (to Bayer A.G.).
102. Jpn. Kokai Tokkyo Koho JP 58,153,934 (Sept. 13, 1983), (to Konishiroku Photo Industry Co., Ltd.).
103. Jpn. Kokai Tokkyo Koho JP 58,198,570 (Nov. 18, 1983), (to Dainippon Ink and Chemicals Inc.).
104. Jpn. Kokai Tokkyo Koho JP 60,249,150 (Dec. 9, 1985), A. Ogawa, M. Tsuda, and K. Nakajo (to Fuji Photo Film Co., Ltd.).
105. T. Ojasoo and J. P. Raynaud, *Prog. Cancer Res. Ther.* **25**, 11 (1983).
106. J. G. Tezon, M. H. Vázquez, and J. A. Blaquier, *Endocrinology* **111**, 2039 (1982).
107. F. A. Lefebvre, C. Seguin, A. Belanger, S. Caron, M. R. Sariam, and F. Labrie, *Prostate* **3**, 569 (1982).
108. Jpn. Kokai Tokkyo Koho JP 60,188,373 (Sept. 25, 1985), K. Ienaga and K. Nakamura, (to Nippon Zoki Pharmaceutical Co., Ltd.).
109. A. Spinks and W. S. Waring, *Progr. Med. Chem.* **3**, 313 (1963).
110. H. H. Merrit and T. J. Putman, *J. Am. Med. Assoc.* **111**, 1068 (1938).
111. W. E. Stone and M. J. Javid, *Neurol. Res.* **7**, 202 (1985).
112. Y. Yaari, M. E. Selzer, and J. H. Pincus, *Ann. Neurol.* **20**, 171 (1986).
113. M. G. Wong, J. A. Defina, and P. R. Andrews, *J. Med. Chem.* **29**, 562 (1986).
114. W. Spinelly and M. R. Roseu, *J. Pharmacol. Exp. Ther.* **238**, 794 (1986).
115. E. Perucca, *Ther. Drug Monit.* **2**, 331 (1980).
116. D. Kadar, T. D. Fecycz, and W. Kalow, *J. Physiol. Pharmacol.* **61**, 403 (1983).
117. Ger. Offen. DE 3,205,506 (Sept. 16, 1982), C. H. J. Wang, S. D. Stroupe, and M. E. Jolley (to Abbott Laboratories).
118. K. Tasaka, R. Terao, Ch. Kamei, K. Hashigaki, and M. Yamato, *Int. J. Immunopharmacol.* **9**, 391 (1987).
119. W. S. Aslanian, E. Jacqz, C. B. McAllister, R. A. Branch, and G. R. Wilkinson, *J. Pharmacol. Exp. Ther.* **234**, 662 (1985).
120. Eur. Pat. Appl. EP 343,643 (Nov. 29, 1989), W. A. Cetenko, D. T. Connor, R. J. Sorenson, P. C. Unangst, and S. R. Stabler (to Warner-Lambert Co.).
121. N. E. McCarroll, B. H. Keech, and C. E. Piper, *Environ. Mutagen.* **3**, 607 (1981).
122. M. Korbelik, *Arch. Hig. Rada Toksikol.* **31**, 227 (1980).
123. H. Ohta, T. Jikihara, K. Wakabayashi, and T. Fujita, *Pestic. Biochem. Physiol.* **14**, 153 (1980).
124. U.S. Pat. 4,198,423 (Apr. 15, 1980), C. J. Mappes, E.-H Pommer, and B. Zeeh (to BASF A.G.).
125. Belg. Pat. 631,731 (May 2, 1962), E. Klauke, E. Kuehle, and F. Graven (to Bayer A.G.).
126. Eur. Pat. Appl. EP 186,124 (July 2, 1986), T. Haga, T. Koyanagi, K. Yoshida, O. Imai, and H. Okada (to Ishihara Sangyo Kaisha, Ltd.).
127. S. Saltzman, A. J. Acher, N. Brates, M. Horowitz, and A. Gevelberg, *Pestic. Sci.* **13**, 211 (1982).

128. G. Peng, V. Márquez, and J. S. Driscoll, *J. Med. Chem.* **18**, 846 (1975).

129. F. D. Deen, T. Hoshino, M. E. Williams, K. Nomura, and P. M. Bartle, *Cancer Res.* **39**, 4336 (1979).

130. J. C. Kim and co-workers, *Yakhak Hoechi* **27**, 309 (1983).

131. H. Fischer, H. Moeller, M. Budnowski, G. Atassi, P. Dumont, J. Venditti, and O. C. Yoder, *Arzneim.-Forsch.* **34**, 663 (1984).

132. Jpn. Kokai Tokkyo Koho JP 58,109,418 (June 29, 1983), H. Okuda.

133. K. Inagaki, I. Miwa, T. Yashiro, and J. Okuda, *Chem. Pharm. Bull.* **30**, 3244 (1982).

134. L. G. Humber, *Progr. Med. Chem.* **24**, 299 (1987).

135. G. Klopman and E. Buyukbingol, *Mol. Pharmacol.* **34**, 852 (1988).

136. M. Endo and S. Yagi, *Electroencephalogr. Clin. Neurophysiol. Suppl.* **36**, 261 (1982).

137. R. L. White, F. L. Wessels, T. J. Schwan, and K. O. Ellis, *J. Med. Chem.* **30**, 263 (1987).

CARMEN AVENDAÑO
J. CARLOS MENÉNDEZ
Universidad Complutense

HYDRAULIC FLUIDS

The moving parts of many industrial machines are actuated by fluid that is under pressure. A system used to apply the fluid can consist of a reservoir, a motor-driven pump, control valves, a fluid motor, and piping to connect these units, eg, a hydraulic system. Generally, petroleum (qv) lubricating oils, and sometimes water are used as the pressure-transmitting or hydraulic fluids. Lubricating oil is not only suitable for pressure transmission and controlled flow, but it also minimizes friction and wear of moving parts (see LUBRICATION AND LUBRICANTS) and protects ferrous surfaces from rusting (see CORROSION AND CORROSION CONTROL).

Hydraulic actuation is based on Pascal's discovery that pressure which has developed in a fluid acts equally and in all directions throughout the fluid and behaves as a hydraulic lever or force multiplier (see PRESSURE MEASUREMENT) (1). As shown in Figure 1, a 5-kg weight acting on a 10-cm^2 piston develops a 49-kPa (7.1-psi) pressure which, when transmitted to a 100-cm^2 piston, enables that piston to support a 50-kg weight. Pressure is transmitted easily around corners, and the two cylinders can be any reasonable distance apart (2). When motion occurs, the small (10 cm^2) piston must move 10 cm in order to move the large (100 cm^2) piston 1 cm. This is necessary because in this closed system the volume of liquid leaving one cylinder must equal the volume of liquid entering the other cylinder.

Following the invention of the hydraulic press in 1795 (3), the use of hydraulics expanded rapidly during the nineteenth century. The weight-loaded

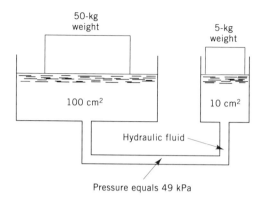

Fig. 1. Basis for hydraulic operation.

accumulator, invented ca 1850, was used to store energy in hydraulic systems. The elementary press circuit has several parts that are common to all hydraulic systems; a reservoir, a pump, piping, control valves, a motor, which in this case is a hydraulic cylinder or ram, and the hydraulic fluid. By ca 1860 hydraulic presses were used for forging, and an adjustable-speed hydraulic transmission was perfected in 1906 (2). The manufacture of hydraulically actuated machines attained industrial importance after 1920.

In 1840 a hydraulic power network, which involved large reciprocating pumps that were driven by steam engines, supplied fluid power to London. However, concurrent technology in steam (qv) turbines and the electric generators outmoded such networks until hydraulic systems were improved with the use of rotary pumps and oil. The rotary piston pump marked the transition from use of water to oil as the hydraulic fluid (4). The use of vacuum-distilled, refined mineral oils were instrumental in the success of rotary axial piston pumps and motors such as the Waterbury variable speed gear (5).

Hydraulic circuits have been used in numerous combinations in many industrial machines. Speed can be readily controlled by controlling the volume of fluid flow, and can be adjusted during operation, eg, rapid approach, slow cut or press, and rapid retraction are obtained easily. Force can be applied in any direction, transmitted around corners and to remote parts of a machine, and can be controlled easily by controlling fluid pressure. Great force is available with or without motion. Direction of movement is controlled by regulation of the direction of fluid flow. Smooth operation using inherent cushioning effect and protection against overload through oil-pressure relief is characteristic. Energy can be stored to meet sudden demands, and equipment is highly adaptable to remote and automatic control.

For proper operation under anticipated use, recommended lubricants are designated by the equipment designer, ie, the designer specifies both the type of fluid and the fluid's viscosity.

Viscosity Classification

The viscosity classification for hydraulic fluids and other industrial liquid lubricants is defined by ASTM D2422 (ISO STD 3448), which establishes 18 viscosity grades in the range of 2–1500 mm^2/s (=cS) covering approximately the range from kerosene to cylinder oils. Classification is based on the principle that the midpoint kinematic viscosity of each grade should be about 50% higher than that of the preceding one. Using this numbering system, viscosities are quoted as ISO viscosity grade (ISO VG) as shown in Table 1 (see also RHEOLOGICAL MEASUREMENTS).

Table 1. ISO Viscosity Classification

	Kinematic viscosity,[a] mm^2/s(=cS)		
ISO VG	Minimum	Midpoint	Maximum
2	1.98	2.2	2.42
3	2.88	3.2	3.52
5	4.14	4.6	5.06
7	6.12	6.8	7.48
10	9.00	10	11.0
15	13.5	15	16.5
22	19.8	22	24.2
32	28.8	32	35.2
46	41.4	46	50.6
68	61.2	68	74.8
100	90.0	100	110
150	135	150	165
220	198	220	242
320	288	320	352
460	414	460	506
680	612	680	748
1000	900	1000	1100
1500	1350	1500	1650

[a]At 40°C.

Different fluids have different rates of change in viscosity with temperature. The viscosity index (VI), a method of applying a numerical value to this rate of change, is based on a comparison with the relative rates of change of two arbitrarily selected types of oils that differ widely in this characteristic. A high VI indicates a relatively low rate of change of viscosity with temperature; a low VI indicates a relatively high rate of change of viscosity with temperature. A standard method for calculating the viscosity index is described in ASTM D2270.

Types

Antiwear premium hydraulic fluids represent the largest volume of hydraulic fluids used. Shortly after their introduction in 1960, a second product group was formulated, characterized by the same antiwear characteristics but having lower pour points and higher viscosity indexes. These were formulated for use in mobile and marine applications subject to temperature extremes.

The largest volume of hydraulic fluids are mineral oils containing additives to meet specific requirements. These fluids comprise over 80% of the world demand (ca 3.6×10^9 L (944×10^6 gal)). In contrast world demand for fire-resistant fluids is only about 5% of the total industrial fluid market. Fire-resistant fluids are classified as high water-base fluids, water-in-oil emulsions, glycols, and phosphate esters. Polyolesters having shear-stable mist suppressant also meet some fire-resistant tests.

Mineral Oil-Based Fluids.   Premium mineral oils are ideally suited for use in most hydraulic systems and are, by themselves, excellent hydraulic fluids. They are high viscosity index (VI) oils available in a wide range of viscosity grades. Unusually high VI products are especially suitable for use under low temperature conditions. All of the oils contain additives, eg, rust and oxidation inhibitors and antiwear materials. In the event that the additives are consumed or removed in service, these oils would continue to serve effectively for long periods. The oils are carefully processed to have good water-separating ability and resistance to foaming. Because of high oxidation resistance, these qualities are maintained over long service periods.

Lubricating Oils.   Lubricating oils generally include all classes of lubricating materials that are applied as fluids. Nearly all of the world's lubricating oils are made from the more viscous portion of crude oil which remains after removal of gas oil and lighter fractions by distillation. However, the crude oils from various parts of the world differ widely in properties and appearance, although there is relatively little difference in their elemental analysis. Much of the variation in physical characteristics and performance qualities of lubricating oils prepared from different crude sources can be accounted for by the variations that can exist in a single large hydrocarbon molecule. In order to minimize these variations to yield products that provide consistent performance in specific applications, four steps are followed in the manufacture of finished lubricating oils from the various available crudes: (*1*) selection and segregation of crudes according to the principal types of hydrocarbons present; (*2*) distillation of the crude to separate it into fractions containing hydrocarbons in the same boiling point range; (*3*) processing to remove undesirable constituents from the various fractions or to convert some of these materials to more desirable materials; and (*4*) blending to attain the physical characteristics that are required in the finished products and incorporating chemical agents to improve performance.

Factors in lube crude selection are supply (available quantities and constancy of composition), refining, production, and marketing. Two base stocks that are similar in viscosity are listed in Table 2. One base stock is made from a cycloparaffinic, eg, naphthenic, crude oil which contains no wax and has a low ($-46°C$) pour point. In contrast, the paraffinic stock requires dewaxing to reduce its pour point from about $+27$ to $-18°C$. Although both oils have identical viscosities at $38°C$, the viscosity of the cycloparaffinic oil is affected much more by temperature change than that of the paraffinic stock. This is reflected in the lower viscosity index of the cycloparaffinic oil. For products that operate over a wide temperature range, eg, automotive engine oils, the cycloparaffinic stock

Table 2. Lube-Base Stocks[a]

Property	Cycloparaffinic	Paraffinic
viscosity at 38°C, mm²/s (=cS)	20.5	20.5
pour point, °C	−46	−18
viscosity index	15	100
flash point, °C	171	199
specific gravity	0.9075	0.8615
color, ASTM	1.5	0.5

[a]Ref. 6.

is less desirable. The long-term supply of cycloparaffinic crudes is limited, and alternatives are being sought to replace them.

Fire-Resistant Hydraulic Fluids.   The four classifications of fire-resistant hydraulic fluids are listed below (7). Three of the four groups are fire resistant because they contain a significant amount of water which provides cooling and blanketing of the combustible materials.

Classification	Description
HF-A	high water content fluids (95/5 fluids); contain a maximum of 20% combustible material; range from milky to transparent in appearance
HF-B	water-in-oil (invert) emulsions: contain a maximum of 60% combustible material; water content normally is 40 or 45% with the continuous phase being the oil component; white, milky fluid
HF-C	water–glycol solutions; usually contain at least 35% water; transparent and are usually dyed
HF-D	water-free, pure chemical fluids; most common are phosphate esters; other types exist such as mist-suppressed polyol esters

The compressibility and thermal conductivity of mineral oils is compared to fire-resistant fluids in Table 3. All good hydraulic fluids must resist compression, and many fire-resistant fluids can operate at a lower temperature than mineral oils because of improved thermal conductivity.

Some of the tests and criterion used to define fire resistance may be found in the literature (9). Additionally, the compression–ignition and hot manifold tests as defined in MIL-H-19457 and MIL-H-5606, respectively; the Wick test as defined by Federal Standards 791, Method 352; flash point and fire point as defined in ASTM D92; autoignition temperature as defined in ASTM D2155; and linear flame propagation rate are defined in ASTM D5306 are used.

High Water-Base Fluids.   These water-base fluids have very high fire resistance because as little as 5% of the fluid is combustible. Water alone, however, lacks several important qualities as a hydraulic fluid. The viscosity is so low that it has little value as a sealing fluid; water has little or no ability to prevent wear or reduce friction under boundary-lubrication conditions; and water

Table 3. Compressibility and Thermal Conductivity Characteristics[a]

Fluid type	Compression[b] at 70 MPa,[c] %	Nominal thermal conductivity, $J/(cm \cdot s \cdot °C) \times 10^{-3}$
mineral oil	3.3	0.77
phosphate ester	2.5	1.27
water glycol	2.6	2.29
water-in-oil emulsions	3.5	
water	3.3	3.68[d]

[a]Ref. 8.
[b]Expressed as a percent reduction in volume.
[c]To convert MPa to psi, multiply by 145.
[d]At 50°C.

cannot prevent rust. These shortcomings can be alleviated in part by use of suitable additives. Several types of high water-based fluids commercially available are soluble oils, ie, oil-in-water emulsions; microemulsions; true water solutions, called synthetics; and thickened microemulsions. These last have viscosity and performance characteristics similar to other types of hydraulic fluids.

Water-in-Oil Emulsions. A water-in-oil or invert emulsion consists of a continuous oil phase which surrounds finely divided water droplets that are uniformly dispersed throughout the mixture. The invert emulsion ensures that the oil is in constant contact with the hydraulic system's moving parts, so as to minimize wear.

The fluid is formulated from a premium mineral oil-base stock that is blended with the required additive to provide antiwear, rust and corrosion resistance, oxidation stability, and resistance to bacteria or fungus. The formulated base stock is then emulsified with ca 40% water by volume to the desired viscosity. Unlike oil-in-water emulsions the viscosity of this type of fluid is dependent on both the water content, the viscosity of the oil, and the type of emulsifier utilized. If the water content of the invert emulsion decreases as a result of evaporation, the viscosity decreases; likewise, an increase in water content causes an increase in the apparent viscosity of the invert emulsion; at water contents near 50% by volume the fluid may become a viscous gel. A hydraulic system using a water-in-oil emulsion should be kept above the freezing point of water if the water phase does not contain an antifreeze. Even if freezing does not occur at low temperatures, the emulsion may thicken, or break apart with subsequent dysfunction of the hydraulic system.

Water–Glycol Solutions. These materials are transparent solutions of water and glycol having good low temperature properties. They frequently contain water-soluble additives to improve performance in corrosion resistance, antiwear, etc. A water-soluble polymer is commonly utilized to boost viscosity. As solutions their advantage over emulsions is their inherent stability.

Other Fire-Resistant Hydraulic Fluids. Phosphate and more recently polyol esters are marketed as fire-resistant compounds. They are formulated with additives to control wear, oxidation, corrosion, and misting. Seal compatibility and solvency characteristics of these fluids may be quite different from those of mineral oils.

Synthetic Fluids

The starting materials for synthetic lubricants are synthetic base stocks, often manufactured from petroleum, made by synthesizing compounds which have adequate viscosity for use as lubricants. The process of combining individual units can be controlled so that a large proportion of the finished base fluid is comprised of one or only a few compounds. Depending on the starting materials and the combining process that is used, the compound (or compounds) can have the properties of the most effective compounds in a mineral-base oil. It can also have unique properties, eg, miscibility with water or refrigerant, or complete nonflammability, that are not found in any mineral oil.

The primary performance features of synthetic lubricants are outstanding flow characteristics at extremely low temperatures and stability at extremely high temperatures. The comparative operating temperature limits of mineral oil and synthetic lubricants are shown in Figure 2; other advantages, as well as limiting properties, are outlined in Table 4. Synthesized hydrocarbons, organic esters, polyglycols, and phosphate esters account for over 90% of the volume of synthetic lubricant bases in use. Other synthetic lubricating fluids include a number of materials that generally are used in low volumes.

Hydrocarbons. Synthesized hydrocarbons are the most popular of the synthetic basestocks. These are pure hydrocarbons (qv) and are manufactured from raw materials derived from crude oil. Three types are used: olefin oligomers, alkylated aromatics, and polybutenes. Other types, such as cycloaliphatics, are also used in small volumes in specialized applications.

Olefin Oligomers. Olefin oligomers (poly-α-olefins) are formed by combining a low molecular weight material, usually ethylene (qv), into a specific olefin

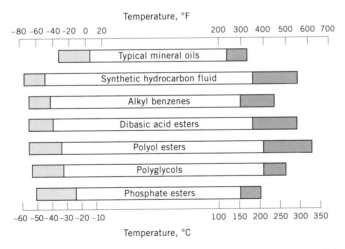

Fig. 2. Comparative temperature limits of mineral oil and synthetic lubricant, where ☐ represents continuous service; ▨ service dependent on starting torque; and ☐ intermittent service (6).

Table 4. Advantages and Limiting Properties of Synthetic Base Stocks[a]

Stock	Potential advantages	Possible limiting properties
synthetic hydrocarbon fluids	high temperature stability; long life; low temperature fluidity; high viscosity index; low volatility oil economy; compatibility with mineral oils and paints; no wax; hydrolytic stability	solvency–detergency;[b] seal compatibility[b]
alkylated benzenes	low temperature fluidity; low volatility; high temperature stability; hydrolytic stability	lubricity; solvency–detergency; low viscosity index
organic esters	high temperature stability; long life; low temperature fluidity; solvency–detergency	seal compatibility;[b] mineral oil compatibility;[b] antirust;[b] antiwear and extreme pressure;[b] hydrolytic stability; paint compatibility
phosphate esters	fire resistant; lubricating ability	seal compatibility; low viscosity index; paint compatibility; metal corrosion;[b] hydrolytic stability
polyglycols	water versatility; high viscosity index; low temperature fluidity; antirust; no wax	mineral oil compatibility; paint compatibility; oxidation stability[b]

[a]Comparisons are made on basestocks without additive packages.
[b]Limiting properties of synthetic base fluids which can be overcome by formulation chemistry.

which is oligomerized into a lubricating oil-type material and then is hydrogen-stabilized. In the oligomerization, a few (usually 3–10) of the basic building block molecules are combined to form the finished material. Therefore, the product may be formed having varying molecular weights and attendant viscosities to meet a broad range of requirements. A typical olefin oligomer-base oil molecule is the oligomer of 1-decene:

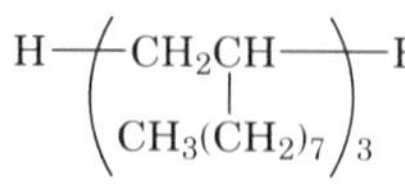

$$H \left[\begin{array}{c} CH_2CH \\ | \\ CH_3(CH_2)_7 \end{array} \right]_3 H$$

Olefin oligomers are a special type of paraffinic mineral oil comparable in properties to the most effective components found in petroleum-derived base oils (see OLEFIN POLYMERS). These oligomers have high (usually >135) viscosity indexes, excellent low temperature fluidity, very low pour points, and excellent shear and hydrolytic stability. Because of the saturated nature of the hydrocarbons, both the oxidation and thermal stability are good. Volatility is lower than that of comparably viscous mineral oils and evaporation loss at elevated temperatures is lower. In many applications, it is important that olefin oligomers are similar in composition to and compatible with mineral oils as well as with additive systems developed and machines designed to operate on mineral oils. The olefin oligomers do not cause any softening or swelling of typical seal materials; however, formulation of finished lubricants can promote a softening or seal swell effect, if desired (see OLEFINS, HIGHER).

Alkylated Aromatics. The alkylation (qv) process involves joining linear or branched alkyl groups to an aromatic molecule, usually benzene. Generally, the alkyl groups that are used contain from 10–14 carbon atoms and have normal paraffinic configurations. The properties of the product can be altered by changing the structure and position of the alkyl groups. Dialkylated benzene is a typical alkylated aromatic used as a lubricating oil.

Alkylated aromatics have excellent low temperature fluidity and low pour points. The viscosity indexes are lower than most mineral oils. These materials are less volatile than comparably viscous mineral oils, and more stable to high temperatures, hydrolysis, and nuclear radiation. Oxidation stability depends strongly on the structure of the alkyl groups (10). However it is difficult to incorporate inhibitors and the lubrication properties of specific structures may be poor. The alkylated aromatics also are compatible with mineral oils and systems designed for mineral oils (see BENZENE; TOLUENE; XYLENES AND ETHYLBENZENE).

Polybutenes. Polybutenes are produced by controlled polymerization of butenes and isobutene (isobutylene) (see BUTYLENES). A typical polyisobutylene structure is

$$\text{H}\!-\!\!\left(\text{CH}_2\overset{\overset{\displaystyle \text{CH}_3}{|}}{\underset{\underset{\displaystyle \text{CH}_3}{|}}{\text{C}}}\right)_{\!\!n}\!\!-\!\text{CH}_2\overset{}{\underset{\underset{\displaystyle \text{CH}_3}{|}}{\text{C}}}\!\!=\!\!\text{CH}_2$$

The low molecular weight materials produced by this process are used as lubricants, whereas the high molecular weight materials, the polyisobutylenes, are used as VI improvers and thickeners. Polybutenes that are used as lubricating oils have viscosity indexes of 70–110, fair lubricating properties, and can be manufactured to have excellent dielectric properties. Above their decomposition temperature (ca 288°C) the products decompose completely to gaseous materials.

Cycloaliphatics. Synthesized cycloaliphatics are generally not utilized as hydraulic fluids. Cycloaliphatics are synthesized for use as traction lubricants because, under high stress, they have high traction coefficients and excellent stability. A typical cycloaliphatic used as a synthetic traction fluid is 2,3-dicyclohexyl-2,3-dimethylbutane [5171-88-0], $C_{18}H_{34}$.

Substituted cyclopentane lubricants have been commercialized using cyclopentadiene as starting material. These specialty aerospace lubricants have low volatility and desirable optical properties.

Organic Esters. *Dibasic Acid Esters.* Dibasic acid esters (diesters) are prepared by the reaction of a dibasic acid with an alcohol that contains one reactive hydroxyl group (see ESTERS, ORGANIC). The backbone of the structure is formed by the acid. The alcohol radicals are joined to the ends of the acid. The physical properties of the final product can be varied by using different alcohols or acids. Compounds that are typically used are adipic, azelaic, and sebacic acids and 2-ethylhexyl, 3,5,5-trimethylhexyl, isodecyl, and tridecyl alcohols.

Dibasic acid esters have excellent low temperature fluidity and very low pour points. Viscosity indexes usually are high, some above 140, and the products are shear stable. Generally, however, the hydrolytic stability is not as good as mineral oils. Diesters have good lubricating properties, good thermal and oxidation stability, and lower volatility than comparably viscous mineral oils. These compounds also have the ability to suspend potential deposit-forming materials so that hot metal surfaces in contact with them remain clean. However, diesters cause more seal swelling than mineral oils. Also, they may not have good solubility for additives that have been developed for use in mineral oils, although good results are obtained with additives that are developed especially for these fluids. Additionally, these may affect paints and finishes more than mineral oils (see COATINGS; PAINT).

Polyol Esters. Polyol esters are formed by the reaction of an alcohol having two or more hydroxyl groups, eg, a polyhydric alcohol and a monobasic acid. In contrast to the diesters, the polyol in the polyol esters forms the backbone of the structure and the acid radicals are attached to it. The physical properties may be varied by using different polyols or acids. Trimethylolpropane [77-99-6], $C_6H_{14}O_3$, and pentaerythritol [115-77-5], $C_5H_{12}O_4$, are two commonly used polyols (see ALCOHOLS, POLYHYDRIC). Usually the acids that are used are obtained from animal or vegetable oils (qv) and contain 5–10 carbon atoms. Polyol esters have better high temperature stability than the diesters. The former's low temperature properties and hydrolytic stability are about the same as the latter's, but the viscosity indexes of the former may be lower. Volatility of polyol esters is equal to or lower than the diesters. The polyol esters affect paints and finishes more than mineral oils and often have different seal swelling characteristics (see also POLYESTERS).

Polyglycols. Polyglycols are the largest single class of synthetic lubricant bases and are most accurately described as polyalkylene glycol ethers (see GLYCOLS; POLYETHERS). Small quantities of simple glycols, eg, ethylene glycol [107-21-1], $C_2H_6O_2$, and poly(ethylene glycol), are used as hydraulic brake fluids. Polyglycols are used in both water-soluble and water-insoluble forms. Polyglycols decompose completely to form volatile compounds under high temperature oxidizing conditions, which results in low sludge buildup under moderate-to-high operating temperatures, or complete decomposition without deposits in certain extremely hot applications.

Polyglycols have low pour points and good viscosity–temperature characteristics, although at low temperatures these materials tend to become more viscous than some of the other synthesized bases. High temperature stability is fair to good and can be improved with additives. Thermal conductivity is high. Polyglycols are not compatible with mineral oils or additives that were developed for use in mineral oils, and may have considerable affect on paints and finishes. They have low solubility for hydrocarbon gases and for some refrigerants. Seal swelling is low, but care must be exercised in seal selection with the water-soluble types to be sure that the seals are compatible with water. The glycol fluid does have a tendency to adsorb moisture from the atmosphere.

Certain polyglycols apparently have compatibility with nonozone depleting refrigerants (see REFRIGERATION AND REFRIGERANTS).

Phosphate Esters. Phosphate esters are one of the larger volume classes of synthetic base fluids. A typical phosphate ester structure where R can be either an aryl or alkyl group is

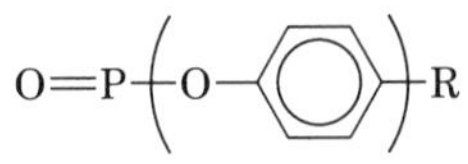

$$O{=}P{-}(O{-}\bigcirc{-})R$$

The phosphate esters have better fire resistance than mineral oils (see FLAME RETARDANTS, PHOSPHORUS COMPOUNDS). The lubricating properties are generally good; however, the high temperature stability is fair; and decomposition products can be corrosive. Generally, phosphate esters have poor viscosity–temperature characteristics, although pour points and volatility are low. Phosphate esters have considerable effect on paints and finishes and may cause swelling of many seal materials. Compatibility with mineral oils ranges from poor to good, depending on which ester is used; hydrolytic stability is fair. Phosphate esters have specific gravities greater than one which implies that water contamination tends to float rather than settle to the bottom, resulting in high pumping losses (see PHOSPHORUS COMPOUNDS).

Other Synthetic Lubricating Fluids. *Silicones.* Silicone fluids have a polymer-type structure except that the carbons in the backbone are replaced by silicon (see SILICON COMPOUNDS, SILICONES). Dimethylpolysiloxane [9016-00-6] one of the widely used silicone fluids, has the structure

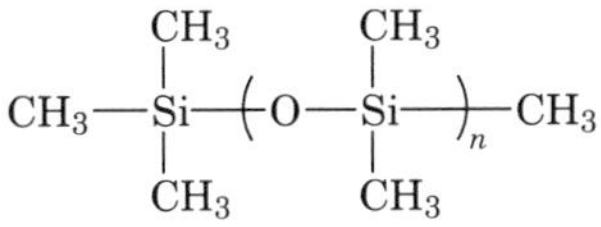

$$CH_3{-}\underset{\underset{CH_3}{|}}{\overset{\overset{CH_3}{|}}{Si}}{-}(O{-}\underset{\underset{CH_3}{|}}{\overset{\overset{CH_3}{|}}{Si}}{-})_n CH_3$$

Silicones have high viscosity indexes, some ≥ 300. Pour points are low and low temperature fluidity is good. These materials are chemically inert, nontoxic, fire resistant, and water repellent, and have low volatility. Seal swelling is low. Compressibility is considerably higher than for mineral oils. The thermal and oxidation stabilities of silicones are good up to high temperatures. However, if oxidation does occur, the oxidation products, which include silicon oxides, can be abrasive. A principal disadvantage of the common silicones is that these compounds have low surface tensions which permit extensive spreading on metal surfaces, especially on steel; consequently, effective adherent lubricating films do not form. The silicones that exhibit this characteristic also show poor response to wear- and friction-reducing additives.

Silicate Esters. Silicate esters, $Si(OR)_4$ where R is an aryl or alkyl group, have excellent thermal stability, and using proper inhibitors, show good oxidation stability. These have excellent viscosity–temperature characteristics, and the pour points and volatilities are low. Silicate esters have only fair lubricating properties, however, because resistance to hydrolysis is poor (see SILICON COMPOUNDS, ETHERS AND ESTERS).

Halogenated Fluids. Chlorocarbons, fluorocarbons, or combinations of the two are used to form lubricating fluids (see CHLOROCARBONS; FLUORINE COMPOUNDS, ORGANIC). Generally, these fluids are chemically inert, essentially nonflammable, and often show excellent resistance to solvents. Some have outstanding thermal and oxidation stability, because they are completely unreactive even in liquid oxygen, and extremely low volatility.

Additives

Practically all lubricating oils contain at least one additive; some oils contain several. The amount of additive that is used varies from <0.01 to 30% or more. Additives can have detrimental side effects, especially if the dosage is excessive or if interactions with other additives occur. Some additives are multifunctional, eg, certain VI improvers also function as pour-point depressants or dispersants. The additives most commonly used in hydraulic fluids include pour-point depressants, viscosity index improvers, defoamers, oxidation inhibitors, rust and corrosion inhibitors, and antiwear compounds.

Pour-Point Depressants. Pour-point depressants are high molecular weight polymers that inhibit formation of wax crystals which prevent oil flow at low temperatures. Two types which are used are alkylaromatic polymers, which are adsorbed by the wax crystals as they form, thereby preventing the crystals from growing and adhering to each other; and polymethacrylates, which cocrystallize with the wax thereby preventing crystal growth (see METHACRYLIC POLYMERS). The additives function by lowering the temperature at which a rigid structure forms. Depending on the type of oil being treated, a pour-point depression of up to 28°C can be achieved, although a lowering of ca 11–17°C is more common.

Viscosity Index Improvers. VI improvers are long-chain, high molecular weight polymers that increase the relative viscosity of an oil at high temperatures more than at low temperatures. In cold oil the molecules of the polymer adopt a compressed coiled form so that the affect on viscosity is minimized. In hot oil the molecules swell, and interaction with the oil produces a proportionally greater thickening effect. Although the viscosity of the oil–polymer mixture decreases as the temperature increases, viscosity does not decrease as much as the oil alone would decrease.

The VI improvers are subject to degradation as a result of mechanical shearing in service. Temporary shear breakdown occurs under certain conditions of moderate shear stress and results in a temporary loss in viscosity. Under these conditions the long molecules of the VI improver align in the direction of the stress with a consequential decrease in resistance to flow. When the stress is removed, the molecules return to their usual random arrangement and the temporary viscosity loss is recovered. This effect can temporarily reduce oil friction to facilitate hydraulic startup at low temperatures. Permanent shear breakdown occurs when the shear stress ruptures the backbone of the polymer, converting the polymer into low molecular weight materials which are less effective VI improvers. This results in a permanent viscosity loss. Permanent shear breakdown generally is the limiting factor controlling the maximum amount of VI improver that can be used in a particular oil blend.

The most common VI improvers are methacrylate polymers and copolymers, acrylate polymers (see ACRYLIC ESTER POLYMERS), olefin polymers and copolymers, and styrene–butadiene copolymers. The degree of VI improvement from these materials is a function of the molecular weight distribution of the polymer. VI improvers are used in engine oils, automatic transmission fluids, multipurpose tractor fluids, hydraulic fluids, and gear lubricants. Their use permits the formulation of products that provide satisfactory lubrication over a much wider temperature range than is possible using mineral oils alone.

Defoamers. The ability of oils to release entrained air and resist foaming varies considerably depending on the type of crude oil, type and degree of refining applied to it, and its viscosity. Silicone polymers used at a few ppm are the most widely used defoamers (qv). These materials are marginally soluble in oil and the correct choice of polymer size is critical if settling during long-term storage is to be avoided. Defoamers also may increase air entrainment in the oil. Organic polymers are sometimes used to overcome these drawbacks, although much higher concentrations are required.

Oxidation Inhibitors. When oil is heated in the presence of air, oxidation occurs. As a result of this oxidation, the oil viscosity and the concentration of organic acids in the oil increase, and varnish and lacquer deposits may form on hot metal surfaces that are exposed to the oil. In extreme cases, these deposits may be further oxidized to hard, carbonaceous materials. As the temperature increases, the rate of oxidation increases exponentially. Exposure to air, or more intimate mixing with it, also increases the rate of oxidation. Many materials, such as metals, particularly copper, and organic and mineral acids, may act as catalysts or oxidation promoters (see HEAT STABILIZATION; UV STABILIZERS).

The mechanism of oil oxidation is thought to proceed by a free-radical chain reaction. Reaction-chain initiators are formed from unstable oil molecules, and these react with oxygen to form peroxy radicals which in turn attack the unoxidized oil and form new initiators and hydroperoxides. The hydroperoxides are unstable and divide, thereby forming new initiators and continuing the reaction. Oxidation inhibitors may not entirely prevent oil oxidation when conditions of exposure are severe, and only some types of oils are inhibited to a great degree. Two general types of oxidation inhibitors are those that react with the initiators, peroxy radicals and hydroperoxides, to form inactive compounds, and those that decompose these materials to form less reactive compounds. At temperatures below 93°C, oxidation proceeds slowly and inhibitors of the first type are effective. Examples are hindered (alkylated) phenols, eg, 2,6-di(tert-butyl)-4-methylphenol [128-37-0], $C_{15}H_{24}O$, also known as 2,6-di(tert-butyl)-p-cresol (DBPC), and aromatic amines, eg, N-phenyl-α-naphthylamine [90-30-2],

$C_{16}H_{13}N$. These are used in turbines, circulation, and hydraulic oils that are intended for extended service at moderate temperatures (see ANTIOXIDANTS; ANTIOZONANTS; HYDROCARBON OXIDATION).

When the operating temperature exceeds ca 93°C, the catalytic effects of metals become an important factor in promoting oil oxidation. Inhibitors that reduce this catalytic effect usually react with the surfaces of the metals to form protective coatings (see METAL SURFACE TREATMENTS). Typical metal deactivators are the zinc dithiophosphates which also decompose hydroperoxides at temperatures above 93°C. Other metal deactivators include triazole and thiodiazole derivatives. Some copper salts intentionally put into lubricants counteract or reduce the catalytic effect of metals.

Corrosion and Rust Inhibitors. The two most troublesome types of corrosion caused by hydraulic fluids are corrosion by organic acids that develop in the oil and corrosion by contaminants that are picked up and carried by the oil. Corrosion by organic acids can occur in the high strength-bearing inserts used in internal combustion engines. Some of the metals used in these inserts, eg, the lead in copper–lead or lead–bronze, are readily attacked by organic acids in oil. The corrosion inhibitors form a protective film on the bearing surfaces and either may be adsorbed on the metal or chemically bonded to it. The most common additive used for this purpose is zinc dithiophosphate, but other sulfur- and phosphorus-containing materials also are used. Inclusion on highly alkaline materials in engine oil also helps to neutralize strong acids as they form, and thereby greatly reduce corrosion and corrosive wear (see CORROSION AND CORROSION CONTROL).

Rust inhibitors usually are corrosion inhibitors that have a high polar attraction toward metal surfaces and that form a tenacious, continuous film which prevents water from reaching the metal surface. Typical rust inhibitors are amine succinates and alkaline-earth sulfonates. Rust inhibitors can be used in most types of lubricating oils, but factors of selection include possible corrosion of nonferrous metals or formation of emulsions with water. Because rust inhibitors are adsorbed on metal surfaces, an oil can be depleted of its rust inhibitor. In certain cases, it is possible to correct the depletion by adding more inhibitor.

Antiwear Compounds. Additives are used in many lubricating oils to reduce friction, wear, and scuffing and scoring under boundary lubrication conditions, ie, when full lubricating films cannot be maintained. Two general classes of materials are used to prevent metallic contact.

Mild Wear and Friction-Reducing Compounds. Mild wear and friction-reducing compounds are polar materials, eg, fatty oils, acids, and esters. These compounds, which function under light to moderate loads, are long-chain molecules that form an adsorbed film on metal surfaces where the polar ends of the molecules are attached to the metal. The molecules are projected normal to the surface. Contact is between the projecting ends of the layers of molecules on the opposing surfaces. Friction is reduced and the surfaces move more freely relative to each other. Wear is reduced under mild sliding conditions, but under severe sliding conditions the layers of molecules can be rubbed off. Zinc dialkyl dithiophosphates are a family of friction-reducing compounds for antiwear hydraulic oils. The friction- and wear-reducing mechanism is quite complex.

Extreme Pressure Compounds.   At high temperatures or under heavy loads where severe sliding conditions exist, extreme pressure (EP) additives are required to reduce friction, control wear, and prevent severe surface damage. These materials function by reacting with the sliding metal surfaces to form oil-insoluble surface films (11). The sliding process can lead to some film removal, but replacement by further chemical reaction is rapid so that the loss of metal is extremely low. This process gradually depletes the amount of EP additive available in the oil. The severity of the sliding conditions and the additive-metal reactivity dictates which EP additives are required for maximum effectiveness. The optimum reactivity occurs when the additives minimize the adhesive or metallic wear but prevent appreciable corrosive or chemical wear. Additives that are too reactive form thick surface films which have less resistance to attrition, and thus some metal is lost by the sliding action. Because the chemical reaction is greatest on the asperities where contact is made and localized temperatures are highest, EP additives lead to polishing of the surfaces. Consequently, the load is distributed uniformly over a greater contact area which allows for a reduction in sliding severity, more effective lubrication, and reduced wear.

Extreme pressure agents usually contain sulfur, chlorine, or phosphorus, either alone or in combination. Sulfur compounds (qv), sometimes with chlorine or phosphorus compounds (qv), are used in many metal-cutting fluids whereas sulfur–phosphorus combinations are used in industrial gear lubricants. In some cases, borates are used in automotive gear lubricants (see BORON COMPOUNDS). These materials provide excellent protection against gear-tooth scuffing and are characterized by good oxidation stability, low corrosivity, seal compatibility, and low friction.

Properties

Hydraulic fluid functions include transmitting pressure and energy; sealing close-clearance parts against leakage; minimizing wear and friction in bearings and between sliding surfaces in pumps (qv), valves, cylinders, etc; removing heat; flushing away dirt, wear particles, etc; and protecting surfaces against rusting. The hydraulic-fluid properties that are used to characterize a suitable product and ASTM test designations are

Property	ASTM test designation
specific gravity	D1298
pour point	D97
flash point	D92
kinematic viscosity	D445
viscosity index	D2270
color, ASTM	D1500
acid number	D664 or 974
rust inhibition	D665
foaming characteristics	D892
oxidation stability	D943

hydrolytic stability	D2619
lubricity testing	
four-ball method	D2266
vane pump wear test	D2882
FZG method	D5182
emulsion characteristics	D1401
water content	D1744

Environmental Aspects

Developments in hydraulic fluids are driven by environmental concerns including disposal of waste, waste minimization, biotoxicity, effects on human health, and the ecology.

Used oil disposal trends include waste minimization such as by reclaiming used fluid on site, as well as recycling of mineral oil lubricants instead of disposing by incineration. The recycling effort involves a system where spent mineral oils are collected then shipped to specialty refineries where the materials are distilled, hydrofinished, and re-refined into fresh basestocks. These re-refined materials are virtually identical to virgin feedstocks.

Human and environmental welfare for lubricants and their use is addressed in Material Safety Data Sheets (MSDS). These MSDS address toxicology and health concerns based on the components in the lubricant as well as indicating the proper response in case of a spill. Environmental hazards of the lubricant are covered on European and Japanese MSDS as shown in Table 5.

Changes in fluid compositions include the reduction and removal of zinc from hydraulic fluids. Zinc-free antiwear hydraulic fluids, which may be ashless and free of phenol, were developed to meet wastewater treatment regulations for industrial sites by reducing the discharge of heavy metals and phenol into waterways.

Vegetable and seed oils as well as some synthetic basestocks present a new class of biodegradable basestocks. These fluids (10) have excellent biodegradation properties as measured by criteria developed by the Environmental Protection Agency (EPA) or Organization of Economic Cooperation and Development (OECD). OECD 301 and EPA 560/6-82-003 measure the biodegradation of lubricants. These tests were developed to measure the degradation of oil, especially

Table 5. MSDS Environmental Hazard Risk Phrases[a]

Risk code	Risk phrase	LC_{50},[b] mg/L
R50	very toxic to aquatic organisms	<1
R51	toxic to aquatic organisms	1–10
R52	harmful to aquatic organisms	10–100
R53	may cause long-term adverse effects in the aquatic environment	c

[a]Ref. 12.
[b]LC_{50} is the concentration in water that kills 50% of the organisms.
[c]Nonbiodegradable, potential bioaccumulator.

two-cycle oil, on waterways. Aquatic toxicity criteria toward fish is also found to be acceptable for this class of fluids as measured by EPA 560/6-82-002 and OECD 203:1-12.

Biodegradable hydraulic fluids are typically made from canola oil (rapeseed oil) or sunflower oil and contain performance additives for antiwear, demulsibility, etc (see SOYBEANS AND OTHER OILSEEDS). For this class of lubricants care must be taken that the fluid be kept sterile in use, otherwise the fluid may biodegrade in service. The degradation results in rancid fluids and inoperable hydraulic systems which requires extensive cleaning measures.

Economic Aspects

Hydraulic fluids are the second largest use of lubricants for automotive and industrial markets. Estimates for 1992 are that 1.089×10^9 L (81×10^6 gal) of hydraulic fluids were sold out of 8.9×10^9 L (2.3×10^9 gal) of total industrial lubricating fluids. The world market is shown in Table 6. Most hydraulic fluids were mineral oil-based products. The remainder represented principally fire-resistant hydraulic fluids and synthetic-based lubricants.

Table 6. Geographical Marketing of Hydraulic Fluids[a]

Geographical area	Sales, %	
	Hydraulic fluids	Industrial lubes[b]
North America	29	26
Western Europe	22	27
Central and Eastern Europe	26	20
Far and Middle East	23	27

[a]Ref. 3.
[b]South America and Africa also have about 4% of the industrial lubricant market.

Petroleum-Based Fluids. The usage and pricing of mineral oil-based hydraulic fluids formulated for use as petroleum-based hydraulic fluids are given in Table 7. The main suppliers of petroleum-based hydraulic fluids in the United

Table 7. Mineral Oil-Based Hydraulic Fluids

Type	1992	
	Usage, %	Price,[a] $/L
antiwear fluids		
premium	58	0.90
premium high VI	9	1.12
rust and oxidation inhibited fluids	15	0.93
all other, including synthetic fluids	18	
Total	100.0	

[a]Commercial posted price approximated for drum lots in the United States.

States are Amoco, Chevron, Citgo, Exxon, Mobil, Shell, Texaco, and Unocal (14). These eight companies supply about 62% of the hydraulic fluids in the United States. Over 80 other companies which supply general purpose hydraulic oils are listed in the Oil Daily's 1992 *Annual Lubricant Buyer's Directory*. Outside the United States, the main suppliers are British Petroleum, Exxon, Mobil, Shell, and Texaco.

The consumer industries involved and the market share of hydraulic fluids used therein include the following (13):

Industry	Hydraulic fluid market, %
manufacturing (machining)	13
mining	22
construction (transportation)	21
chemicals	7
basic metals	14
miscellaneous	23
Total	*100*

In 1992 U.S. lubricant sales exceeded 8.5×10^6 m^3. The total 1992 U.S. automotive lubricant sales and general industrial sales are given in Tables 8 and 9, respectively (15). The largest industrial segments using hydraulics are mining and construction.

Fire-Resistant Fluids. The total 1992 usage of fire-resistant fluids amounted to over 151,000 m^3 (4×10^7 gal) worldwide and includes the four principal categories shown in Table 10. The principal suppliers of fire-resistant fluids are listed in Table 11.

Except for fire-resistant fluids, synthetic lubricants have not captured a significant portion of the general lubricant or hydraulic markets, primarily because the cost is two to four times that of other premium lubricants. However, development of satisfactorily formulated products continues.

Specifications and Standards

The bulk of hydraulic fluids is specified and purchased on bid. Specifications and approval lists are issued by some manufacturers of hydraulic pumps and system components that require lubrication as well as power for control signal transmission. U.S. government military specifications for hydraulic fluids are listed in Table 12, and ASTM tests that are applicable to hydraulic fluids include the following:

Type and title of test	ASTM number
antiwear properties	
preliminary examination of hydraulic fluids	D2271
vane-pump testing of petroleum hydraulic fluids	D2882
evaluation of scuffing load capacity of oils by FZG	D5182
hydraulic fluid stability	

Table 8. Total U.S. Automotive Lubricant Sales, 10^3 m^3 [a,b]

Type	1978	1992
SAE J-183a, engine oils		
monograded	2408	936
multigraded	1720	3106
subtotal	*4128*	*4042*
non SAE J-183a, engine oils		
aircraft	65	34
gasoline-fueled two stroke	55	58
subtotal	*121*	*92*
transmission and hydraulic fluids		
automatic transmission	575	515
universal tractor	155	212
energy/shock absorber, power-steering	51	45
other (manual transmission, etc)	41	11
subtotal	*822*	*783*
gear lubricants		
GL-4 or less	24	28
GL-5 and 6	147	144
subtotal	*171*	*172*
Total	*5243*[c]	*5089*[c]

[a]Ref. 15.
[b]To convert m^3 to U.S. gal, multiply by 264.
[c]Automotive grease is not included.

thermal stability of hydraulic fluids	D2160
hydrolytic stability of hydraulic fluids	D2619
deposition tendencies of liquids in thin film and vapors	D3711
corrosiveness/oxidation stability of hydraulic oils/aircraft lubricants	D4636
fire-resistant tendencies	
flash and fire points by Cleveland open cup	D92
linear flame propagation rate	D5306

Some U.S. governmental lubricant requirements for nontactical equipment is now acquired as Commercial Item Descriptions (CID), rather than against specific military numbers. A new classification system for shear-stable, high VI hydraulic fluids was balloted by ASTM in 1994.

Uses

Hydraulic actuation is applied to machine tools, presses, draw benches, jacks, and elevators as well as to die-casting, plastic-molding, welding, coal-mining, and tube-reducing machines. Hydraulic loading is used for pressure, sugar-mill, and paper-machine press rolls, and calender stacks. The hydraulic press shown in Figure 3 is used for a wide variety of metalworking operations, including drawing, forging, straightening, cupping, embossing, and coining. The lifting

Table 9. Total U.S. General Industrial Sales, 10^3 m^3 [a,b]

Type	1978	1992
General industrial lubricants		
hydraulic oils	888	759
fire-resistant fluids		50
gear oils	191	124
turbine/circulation oils	317	186
refrigeration oils	38	14
way oils	29	31
compressor oils	23	19
rock-drill air tools	19	11
all others	142	89
other unspecified	13	16
subtotal	*1661*	*1299*
Industrial engine oils		
railroad diesel	194	108
marine	134	218
natural gas	157	174
subtotal	*486*	*500*
Metalworking oils		
metal-removing	242	123
metal-forming	67	55
metal-treating	28	20
metal-protecting	32	11
all other specified	8	13
unspecified	40	88
subtotal	*418*	*310*
Process oils		
electrical oils	390	250
rubber	304	334
white oils[c]	157	144
paraffinic	218	154
cycloparaffinic	257	116
other, specified	0	252
subtotal	*1326*	*1250*
Total general industrial lubes	*3890*	*3445*

[a]Ref 15.
[b]To convert m^3 to U.S. gal, multiply by 264.
[c]Does not include all production of white oils.

and tilting mechanism of forklift trucks also are hydraulically operated (2). Load capacities are 0.45–45 t, and operating fluid pressures are from 10.3–17.2 MPa (1500–2500 psi). In plants where forklift trucks must pass near molten metal, open flames, or other sources of ignition, there is a trend toward the use of fire-resistant fluids in the hydraulic systems. Hydraulic actuation also provides the required force as well as ease of control and adjustment of speed that is involved in broaching (2). However, the cost of hydrobroaches and work-holding fixtures

Table 10. Type of Fire-Resistant Fluid

	Fluid	Usage, %	Price, $/L
Classification	Type		
HF-A	high water-base fluids	22	1.06
HF-B	water-in-oil emulsions	32	1.00
HF-C	water–glycol solutions	26	1.85
HF-D	water-free chemical fluids	20	4.00
	Total	*100*	

Table 11. Fire-Resistant Fluid Suppliers

	Suppliers	
Fluid type	United States	Other countries
HF-A	DuraChem; Houghton; Mobil; Quaker; Sun; Texaco; Unocal	Aral; British Petroleum; Century; Exxon, Houghton; Mobil; Shell
HF-B	Conoco; Houghton; Hulbert; Mobil; Shell; Sun; Unocal	Century; Houghton; Mobil; Quaker; Shell
HF-C	Citgo; DuraChem; Houghton; Mobil; Nalco; Union Carbide; Unocal	British Petroleum; Houghton; Mobil; Union Carbide
HF-D	Akzo; Chevron; FMC; Houghton; Mobil Monsanto; Quaker	British Petroleum; Fina; Houghton; Mobil; Monsanto

Table 12. Hydraulic Fluids for Military Usage

Type of fluid or use	Viscosity	Specification number
arctic low pour hydraulic fluid	ISO VG 15	MIL-H-5606-F
hydraulic, steam turbine	ca ISO VG 68/100	MIL-H-17331-H
hydraulic and light turbine lubricating oil	ISO VG 32, 46, and 68 specified	MIL-H-17672-D
fire-resistant hydraulic fluid	ca ISO VG 46	MIL-H-19457-D
catapult hydraulic fluid	ca ISO VG 46	MIL-H-22072-C
high quality rust and oxidation inhibited oil	ISO VG 32, 46, 68, and 150 specified	MIL-H-46001-C
synthetic hydrocarbon, fire-resistant aircraft fluid	ISO VG 15	MIL-H-83282-C

limits hydrobroaching to mass-production where a large number of identical parts are machined.

Positive, adjustable-speed hydraulic transmissions are used for driving paper (qv) mills, wire-rope machines, and printing presses (see PRINTING PROCESSES). These transmissions are used on ships for steering gears, hoisting and mooring equipment, and, in the case of naval vessels, to elevate and train guns. Numerous other applications of hydraulics include mechanisms for tilting

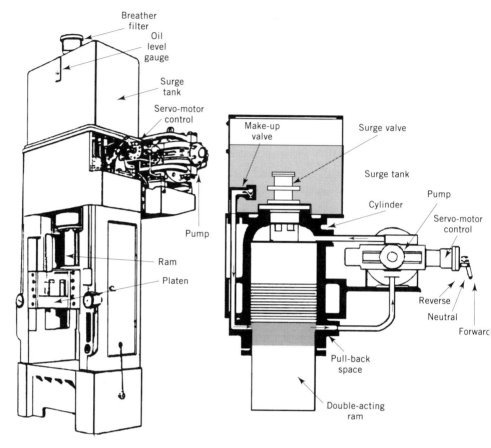

Fig. 3. Hydraulic press.

ladles and operating clamps, brakes, valves, furnace doors, and loading plat-
forms. There are also many hydraulic applications in aircraft, automobiles,
trucks, contractor, and farm equipment.

Hydrostatic Transmissions. The most recent use of hydraulic power has
been in hydrostatic transmissions which are used in many self-propelled harvesting machines and garden tractors and in large tractors and construction
machines. Applications in trucks for highway operation also are being developed.
No clutch is used and no gear shifting is involved, thus this type of transmission
could be called automatic, but in all other respects the hydrostatic transmission
has no similarity to the hydrokinetic automatic transmission (16).

The hydrokinetic transmission transfers power from the engine to the gear
box by first converting it into kinetic energy of a fluid in the pump. The kinetic
energy in the fluid is converted back to mechanical energy in the turbine. In
the hydrostatic system, engine power is converted into static pressure of a fluid
in the pump, and the static pressure acts on a hydraulic motor to produce the
output. Although the fluid moves through the closed circuit between the pump
and motor, energy is transferred primarily by the static pressure rather than by

the kinetic energy of the moving fluid. The relatively incompressible fluid acts like a solid link between the pump and motor.

The motor in a hydrostatic system can be any type of positive displacement hydraulic motor. Axial piston motors usually are used for large drives and, in some cases, for small drives. Gear and radial piston motors are used for low power drives; the motor usually is a fixed-displacement type. The direction of rotation is dependent on the direction of flow to the pump in the closed loop circuit. In addition to the pump and motor, connecting lines, relief valves, and a charge pump are required. The connecting lines may be passages where the pump and motor are in the same housing, or may be hoses where the motor is mounted away from the pump. The charge pump provides initial pressurization of the motor and replaces any fluid lost due to internal leakage. On small tractors it may also be used to supply fluid for remote hydraulic cylinders (16). A typical small tractor schematic diagram is shown in Figure 4. The low pressure make-up pump also is used to supply auxiliary hydraulic units. The drive system consists of a variable volume pump with a fixed displacement motor. Fluid is drawn through a strainer from the reservoir and excess fluid that is not required to charge the main pump flows through the filter back to the reservoir.

Hydrostatic drives allow for selection of any travel speed up to the maximum without a concurrent variance in engine speed. The engine can be operated at the governed speed to provide proper operating speeds for auxiliary elements, eg, the threshing section of a combine. A full range of travel speeds is available to adjust to terrain or crop conditions. Industrial applications for hydraulic systems and hydrostatic transmissions include the following (16):

Aircraft	Farm	Construction	Industrial
aircraft controls	tractors	road rollers and compactors	machine tools
constant speed alternator drives	combines, corn pickers	asphalt spreading and paving machines	mining, locomotives and power cranes
ground support equipment	cotton pickers, baler, miscellaneous fruit and vegetable harvesters	road graders, scrapers, front-end loaders, back hoes, trencher and ditching equipment, concrete mixer, truck cranes, aggregate plants, drilling rigs	lift and shop trucks, loggers chain

Electrorheological Fluids. Electrorheological fluids are a newer category of hydraulic fluids being actively pursued for use in shock absorbers. An electric field causes the fluid to thicken.

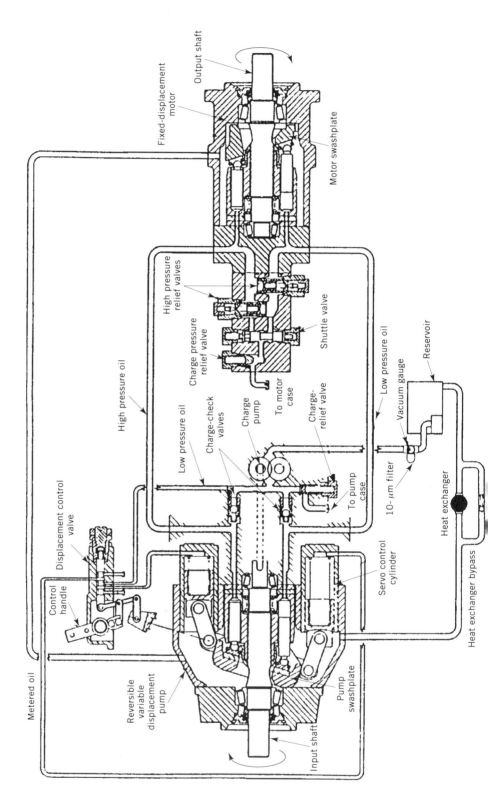

Output shaft

Fixed-displacement motor

Motor swashplate

High pressure relief valves

Charge pressure relief valve

Shuttle valve

To motor case

Charge-relief valve

To pump case

Vacuum gauge

Low pressure oil

Reservoir

10- μm filter

Heat exchanger

Heat exchanger bypass

Servo control cylinder

Pump swashplate

Input shaft

High pressure oil

Low pressure oil

Charge-check valves

Charge pump

Displacement control valve

Control handle

Reversible variable displacement pump

Metered oil

Fire-Resistant Hydraulic Fluids. Fire-resistant hydraulic fluids are used where the fluid could spray or drip from a break or leak onto a source of ignition, eg, a pot of molten metal or a gas flame (17). Conditions such as these exist in die-casting machines or in presses located near furnaces. Specific tests for fire resistance are conducted by Factory Mutual in the United States.

High water-content fluids are used in some hydraulic systems where work-stroke speeds are very low, eg, large freight elevators and large forging and extrusion presses. Pressures in these systems may be from 13.8–20.7 MPa (2000–3000 psi). Vertical in-line pumps with packed plungers and special axial–piston pumps are used with these fluids.

Water-in-oil emulsions are used as fire-resistant hydraulic fluids to replace petroleum hydraulic fluids in general industry, coal mines, and rolling mills where a fire hazard exists (1).

Synthetic Lubricants. Some of the primary applications for synthetic lubricants include the following (6):

Field of service	Synthetic fluids used
industrial	
circulating oils	polyglycols, synthetic hydrocarbon fluid (SHF), organic esters
gear lubricants	polyglycols, SHF
hydraulic fluids (fire-resistant)	phosphate esters, polyglycols
compressor oils	polyglycols, organic esters, SHF
gas turbine oils	SHF, organic esters
greases	SHF
automotive	
passenger car engine oils	SHF, organic esters
commercial engine oils	SHF, organic esters
gear lubricants	SHF
brake fluids	polyglycols
aviation	
gas turbines	organic esters
hydraulic fluids	SHF, phosphate esters, silicones
greases	silicones, organic esters, SHF

Olefin oligomers are used widely as automotive lubricants. They often are combined with some of the organic esters as base fluids in engine oils, gear oils, and hydraulic fluids, eg, for equipment intended for operation in extremely cold climates, and for premium oils, eg, for the service station market in temperate climates.

Alkylated aromatics are used as the base fluid in engine oils, gear oils, hydraulic fluids, and greases in subzero applications. They also are used as the base fluid in power transmission fluids and gas turbine, air compressor, and refrigeration compressor lubricants.

Polybutenes are used as electrical insulating oils, eg, as cable oils in high voltage underground cables, as impregnants to insulate paper for cables, as liquid dielectrics, and as impregnants for capacitors. Significant volumes are used as lubricants for rolling, drawing, and extrusion of aluminum before the aluminum is to be annealed. Other applications include gas compressor lubrication, open gear oils, food-grade lubricants, and as carriers for solid lubricants (such as chain lubricants).

Cycloaliphatics are used in stepless, variable-speed drives in which the torque is transmitted from the driving member to the driven member by the resistance to shear of the lubricating fluid. The high traction coefficients of the cycloaliphatics permit higher power ratings than conventional lubricants. Cycloaliphatics also are used in rolling element bearings to prevent skidding of the rolling elements.

Dibasic acid esters and polyol esters are used as the bases in all aircraft jet-engine lubricants. They also are employed in aircraft greases that are subjected to wide temperature ranges.

Polyglycol application depend on whether the water-soluble or water-insoluble types are used. The largest volume application of the water-soluble polyglycols is as hydraulic brake fluids. Other applications are in metalworking lubricants, where they can be removed by water flushing or burning, and in fire-resistant hydraulic fluids. In the latter application, the polyglycol is mixed with water, which provides the fire resistance. Some water-soluble glycols are used in quenching fluids because they become insoluble in water upon heating. Water-soluble polyglycols are also used in the preparation of water-diluted lubricants for rubber bearings and joints. Water-insoluble polyglycols are used as heat-transfer fluids and as the base fluid in certain industrial hydraulic fluids and in high temperature and bearing oils (see HEAT-EXCHANGE TECHNOLOGY, HEAT-TRANSFER MEDIA OTHER THAN WATER).

Phosphate esters are used predominantly in fire-resistant fluids. Hydraulic fluids for commercial aircraft are based on phosphate esters, as are many industrial fire-resistant hydraulic fluids. The latter are used in electrohydraulic control systems of steam turbines and industrial hydraulic systems where hydraulic fluid leakage might contact a source of ignition. In some cases, they are used in turbine bearing lubrication systems. Considerable quantities of phosphate esters are used as lubricants for compressors (where discharge temperatures are high) to prevent receiver fires which might occur with conventional lubricants. Some quantities are used in greases and miner oil blends are wear and friction reducing additives.

Silicones are used as compressor lubricants and as the base fluids in wide temperature range applications and in high temperature greases. They also are used in specialty greases designed to lubricate elastomeric materials that would be adversely affected by other types of lubricants. Silicones are used in specialty hydraulic fluids for liquid springs and torsion dampers where their high compressibility and minimal change in viscosity with temperature are beneficial. They also are being developed for use as hydraulic brake fluids.

Halogenated hydrocarbons that are inexpensive sometimes are used alone or in blends with phosphate esters as fire-resistant hydraulic fluids. Other halogenated fluids are used for oxygen-compressor lubricants, lubricants for vacuum pumps that are in contact with corrosive materials, solvent-resistant lubricants, and other lubricant applications where highly corrosive or reactive materials are being handled.

BIBLIOGRAPHY

"Hydraulic Fluids" in *ECT* 3rd ed., Vol. 12, pp. 712–733, by J. G. Wills, Mobil Oil Corp.

1. *Les Transmissions Hydrauliques*, Mobil, Revue Industrielle, Paris, 1963.
2. *Hydraulic Systems For Industrial Machines*, Mobil Oil Corp., New York, 1970.
3. R. E. Hatton, *Introduction to Hydraulic Fluids*, Reinhold Publishing Corp., New York, 1962.
4. U.S. Pat. 511,044 (1893), Cooper and Hamilton.
5. D. Fluty and co-workers, *Hydraulic Fluids for Marine Use*, International Marine and Shipping Conference, 1973.
6. J. G. Wills, *Lubrication Fundamentals*, Marcel Dekker, Inc., New York, 1980.
7. *Fluids for Hydraulic Transmissions-Fire Resistant Fluids—Classifications, Provisional Recommendation RP7.7H*, European Oil Hydraulic and Pneumatic Committee, 1976.
8. R. H. Warring, *Hydraulic Handbook*, 8th ed., Surrey, U.K., 1993.
9. U.S. Bureau of Mines, *Fire Resistant Hydraulic Fluids*, Schedule 30, Dec. 11, 1959; Factory Mutual Research Corp., *Approved Standard: Less Hazardous Hydraulic Fluids*, Aug. 26, 1969.
10. *Lubrication* **78**(4) (1992).
11. *Additives for Petroleum Oil*, Mobil Oil Corp., New York, 1969.
12. *Lubrizol NewsLine* **11**(5)(Sept. 1993).
13. Lubrizol Corp. estimates, Wickliffe, Ohio, 1993.
14. Ethyl Petroleum Additives estimates, Saint Louis, Mo., 1993.
15. *1992 Report on U.S. Lubricating Oil Sales*, National Petroleum Refiners Association, Washington, D.C., 1993.
16. C. L. Middleton, R. R. McCoy, and J. M. Stanck, *Modern Hydraulic and Hydrostatic Transmission Fluids*, Automotive Engineering Congress, SAE, Detroit, Mich., 1970.
17. K. G. Henrikson, *Fire-Resistant Fluids and Mobile Equipment*, Farm Construction and Industrial Machinery Meeting, SE, Milwaukee, Wisc., 1965.

General References

Unocal Product Data Sheets, Los Angeles, Calif., 1993.
E. R. Booser, *CRC Handbook of Lubrication*, Vol. 1, Boca Raton, Fla., 1983.

ALAN D. DENNISTON
Unocal Corporation

HYDRAULIC SEPARATION.   See MINERAL RECOVERY AND PROCESSING.

HYDRAZINE AND ITS DERIVATIVES

Hydrazine [302-01-2] (diamide), N_2H_4, a colorless liquid having an ammoniacal odor, is the simplest diamine and unique in its class because of the N—N bond. It was first prepared in 1887 by Curtius as the sulfate salt from diazoacetic ester. Thiele (1893) suggested that the oxidation of ammonia (qv) with hypochlorite should yield hydrazine and in 1906 Raschig demonstrated this process, variations of which constitute the chief commercial methods of manufacture in the 1990s.

The first large-scale use of hydrazine was as fuel for the rocket-powered German ME-163 fighter plane during World War II. Production in the United States began in 1953 at the Lake Charles, Louisiana plant of the Olin Corp., a facility then having a capacity of 2040 metric tons. In 1992 world capacity was about 44,100 metric tons N_2H_4.

Hydrazine and its simple methyl and dimethyl derivatives have endothermic heats of formation and high heats of combustion. Hence these compounds are used as rocket fuels. Other derivatives are used as gas generators and explosives (see EXPLOSIVES AND PROPELLANTS). Hydrazine, a base slightly weaker than ammonia, forms a series of useful salts. As a strong reducing agent, hydrazine is used for corrosion control in boilers and hot-water heating systems (see CORROSION AND CORROSION CONTROL); also for metal plating, reduction of noble-metal catalysts, and hydrogenation of unsaturated bonds in organic compounds. Hydrazine is also an oxidizing agent under suitable conditions. Having two active nucleophilic nitrogens and four replaceable hydrogens, hydrazine is the starting material for many derivatives, among them foaming agents for plastics, antioxidants (qv), polymers, polymer cross-linkers and chain-extenders, as well as fungicides (see FUNGICIDES, AGRICULTURAL), herbicides (qv), plant-growth regulators (see GROWTH REGULATORS, PLANT), and pharmaceuticals (qv). Hydrazine is also a good ligand; numerous complexes have been studied (see COORDINATION COMPOUNDS). Many heterocyclics are based on hydrazine, where the rings contain from one to four nitrogen atoms as well as other heteroatoms.

The many advantageous properties of hydrazine assure continued commercial utility. Hydrazine is available in anhydrous form as well as aqueous solutions, typically 35, 51.2, 54.4, and 64 wt % N_2H_4 (54.7, 80, 85, and 100% hydrazine hydrate).

Physical Properties

Anhydrous hydrazine is a colorless, hygroscopic liquid having a musty ammoniacal odor. It fumes in air owing to the absorption of water and perhaps also of carbon dioxide, forming carbazic acid [471-31-8], $CH_4N_2O_2$. Hydrazine is miscible with water, alcohol, amines, and liquid ammonia, but has only limited solubility in other solvents. Its physical properties are more like the isoelectronic hydroxylamine or hydrogen peroxide (qv) rather than ethane, owing to hydrogen bonding, which is exemplified in relatively high melting (2°C) and boiling (113.5°C) points, as well as an abnormally high (39.079 kJ/mol

(9.340 kcal/mol)) heat of vaporization as compared to 14.64 kJ/mol (3.50 kcal/mol) for the isoelectronic ethane.

The freezing point diagram for the hydrazine–water system (Fig. 1) shows two low melting eutectics and a compound at 64 wt % hydrazine having a melting point of −51.6°C. The latter corresponds to hydrazine hydrate [7803-57-8] which has a 1:1 molar ratio of hydrazine to water. The anomalous behavior of certain physical properties such as viscosity and density at the hydrate composition indicates that the hydrate exists both in the liquid as well as in the solid phase. In the vapor phase, hydrazine hydrate partially dissociates.

$$N_2H_5OH \rightleftharpoons N_2H_4 + H_2O \qquad K = \frac{p_{H_2O}\,p_{N_2H_4}}{p_{N_2H_5OH}} \qquad (1)$$

where p_x represents the partial pressure of substance x.

Equation 2 gives the temperature dependence of the equilibrium constant K when p is in kPa and T in Kelvin (1). Dissociation of the hydrate is extensive at elevated temperatures.

$$\log K = 10.56 - 3054/T \qquad (2)$$

Hydrazine forms a high (120.5°C) boiling azeotrope with water that has a composition of 58.5 mol % (71.48 wt %) N_2H_4 at 102.6 kPa (1.02 atm) pressure. This complicates the separation of hydrazine from water in the manufacturing process because it necessitates the removal of a large amount of water in order to approach the azeotropic composition.

Figure 2 shows the vapor pressures of anhydrous hydrazine (AH), monomethylhydrazine [60-34-4] (MMH), and unsymmetrical dimethylhydrazine

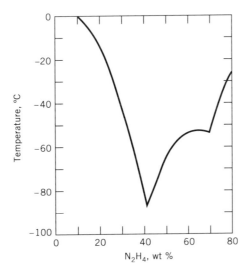

Fig. 1. Freezing-point diagram for hydrazine–water mixtures.

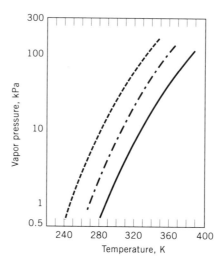

Fig. 2. Vapor pressure of (—) anhydrous hydrazine (AH), (—·—) monomethylhydrazine (MMH), and (---) unsymmetrical dimethylhydrazine (UDMH).

[57-14-7] (UDMH) as a function of temperature (2). The partial pressures of N$_2$H$_4$ over aqueous solutions of various concentrations are plotted in Figure 3.

Hydrazine has a gauche conformation, ie, one NH$_2$ group is rotated from the cis- or trans-positions. Maximum rotational energy barriers are 46.23 and 25.98 kJ/mol (11.05 and 6.21 kcal/mol), respectively (3,4). The high dipole moment, 6.17×10^{-30} C·m (1.85 D), supports this conformational assignment. The N—N bond length is 0.145 nm; the N—N—H bond angle is 112° (5), indicating sp^3 hybridization. The vacant tetrahedral positions are occupied by lone electron pairs, which are the source of the basic and nucleophilic character of the hydrazine molecule.

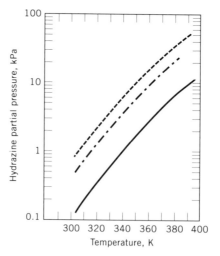

Fig. 3. Hydrazine partial pressure over aqueous solutions containing (---) 64, (—·—) 54.4, and (—) 35 wt % hydrazine.

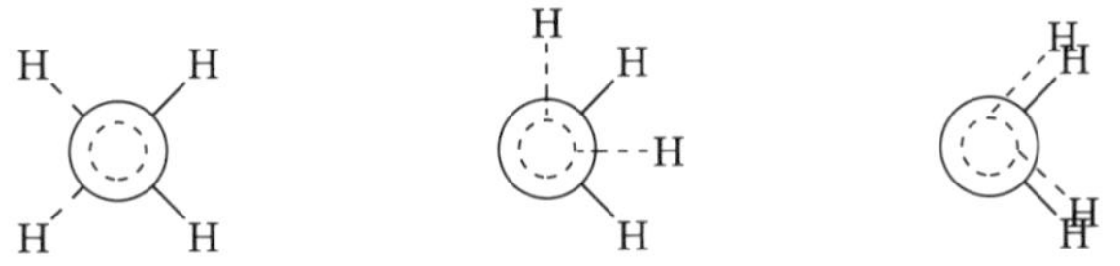

trans- or chair form　　　　　　gauche　　　　　cis- or boat form

Table 1 summarizes some of the physical properties of anhydrous hydrazine, hydrazine hydrate, monomethylhydrazine, and unsymmetrical dimethylhydrazine (6–8). A comprehensive review of the physical and thermodynamic properties of the hydrazines is available (9).

The explosive limits of hydrazine in air are 4.7–100 vol %, the upper limit (100 vol %) indicating that hydrazine vapor is self-explosive. Decomposition can be touched off by catalytic surfaces. The presence of inert gases significantly raises the lower explosive limit (10) (Table 2).

The hydrazine molecule represents an intermediate valence state for nitrogen, suggesting that hydrazine can function both as an oxidizing and reducing agent. Table 3 lists the standard electrode potentials for acidic and basic solutions (11). In basic solution, hydrazine is a powerful reducing agent (eq. 3). It is much less so, however, in an acidic medium (eq. 4). In fact, hydrazine solutions may be stabilized against air oxidation by acidification. The reduction of hydrazine to ammonia is thermodynamically favored in acid solution (eq. 5), but ordinarily the rate of this reaction is slow.

Chemical Properties

Thermal Decomposition. Hydrazine is a high energy compound having a high positive heat of formation; however, elevated ($> 200°C$) temperatures are needed before appreciable decomposition occurs. The decomposition temperature is lowered significantly by many catalysts, particularly copper, cobalt, molybdenum, ruthenium, iridium, and their oxides. Iron oxides (rust) also catalyze decomposition. Hydrazine, especially in high concentrations, should be handled with care using scrupulously clean systems. Reviews are available covering the kinetics and mechanisms of the thermal decomposition of hydrazine liquid and vapor (12,13). A detailed discussion of literature on this topic is in Reference 9. Simple gas-phase decomposition to nitrogen and hydrogen is seldom observed alone. Alternative, more energetic routes lead to the formation of some ammonia.

Acid–Base Reactions. Anhydrous hydrazine undergoes self-ionization to a slight extent, forming the hydrazinium, $N_2H_5^+$, and the hydrazide, $N_2H_3^-$, ions:

$$2\,N_2H_4 \rightleftharpoons N_2H_5^+ + N_2H_3^- \qquad K_i = 10^{-25} \tag{7}$$

Hydrazinium salts, $N_2H_5^+X^-$, are acids in anhydrous hydrazine, metallic hydrazides, $M^+N_2H_3^-$, are bases. Neutralization in this solvent system involves

Table 1. Physical Properties of Commercial Hydrazines

Property	Hydrazine	Hydrazine hydrate	MMH	UDMH
formula	N_2H_4	$N_2H_4 \cdot H_2O$	CH_3NHNH_2	$(CH_3)_2NNH_2$
molecular weight	32.0453	50.0607	46.0724	60.0986
CAS Registry Number	[302-01-2]	[7803-57-8]	[60-34-4]	[57-14-7]
freezing point, °C	2.0	−51.6	−52.4	−57.2
boiling point, °C	113.5	119.4	87.6	62.3
vapor pressure, 25°C, kPa[a]	1.92	1.2	6.62	22.3
critical constants				
P_c, MPa[b]	14.69		8.24	5.42
T_c, °C	380		312	250
d_c, g/mL	0.231		0.290	0.23
liquid density, 25°C, g/mL	1.004	1.032	0.874	0.786
surface tension, 25°C, mN/m(=dyn/cm)	66.45	74.3	33.83	24.09
liquid viscosity, 25°C, mPa·s(=cP)	0.913	1.5	0.775	0.492
refractive index, n_D^{25}	1.4683	1.4644	1.4284	1.4053
heat of vaporization, kJ/mol[c]	39.079	47.7	37.212	32.623
heat of fusion, kJ/mol[c]	12.66		10.42	10.07
heat capacity, 25°C, J/(g·K)[c]	3.0778		2.9296	2.945
heat of combustion, kJ/mol[c]	−622.1		−1304.2	−1979
heat of formation, kJ/mol[c]	50.434	−242.71	54.83	51.63
free energy of formation, kJ/mol[c]	149.24		179.9	206.69
entropy of formation, J/(mol·K)[c]	121.21		165.9	197.99
flash point, COC[d] °C	52	72	1.0	−15

[a]To convert kPa to mm Hg, multiply by 7.5.
[b]To convert MPa to atm, divide by 0.101.
[c]To convert J to cal, divide by 4.184.
[d]COC = Cleveland open cup.

Table 2. Lower Explosive Limits of Hydrazine in Other Gases

Mixture	N_2H_4 composition, vol %	Pressure, kPa^a	Temperature, °C
hydrazine–air	4.7	100.9	92–102
hydrazine–nitrogen	38.0	100.5	109–112
hydrazine–helium	37.0	100.9	105–118
hydrazine–water	30.9	91.8–118.5	130–135
hydrazine–water	37.4	28.5–35.3	98–100
hydrazine–heptane	86.8	53.9–110.2	104–133

^{a}To convert kPa to mm Hg, multiply by 7.5.

Table 3. Hydrazine Standard Electrode Potentials, 25°C

Half-reaction	$E°$, V	Equation
As reductant		
$N_2H_4 + 4\ OH^- \rightarrow N_2 + 4\ H_2O + 4\ e^-$	+1.16	(3)
$N_2H_5^+ \rightarrow N_2 + 5\ H^+ + 4\ e^-$	+0.23	(4)
As oxidant		
$2\ NH_4^+ \rightarrow N_2H_5^+ + 3\ H^+ + 2\ e^-$	−1.275	(5)
$2\ OH^- + 2\ NH_3 \rightarrow N_2H_4 + 2\ H_2O + 2\ e^-$	−0.1	(6)

the hydrazinium and hydrazide ions and is the reverse of equation 7. Metal hydrazides, formally analogous to the metal amides, are prepared from anhydrous hydrazine and the metals as well as from metal amides, alkyls, or hydrides. (The term hydrazide is also used for organic compounds where the carboxylic acid OH is substituted with a N_2H_3.) Sodium hydrazide [13598-47-5] is made from sodium or, more safely, from sodium amide (14):

$$2\ Na + 2\ N_2H_4 \longrightarrow 2\ NaN_2H_3 + H_2 \qquad (8)$$

The ionic hydrazides are extremely sensitive and explode on contact with air or upon heating. The alkaline-earth hydrazides, which have the general formula $M(N_2H_2)$, appear to be less sensitive (15). Hydrazides such as aluminum hydrazide [25546-96-7], $Al(N_2H_3)_3$, have also been made (16). The hydrazide anion is more nucleophilic than hydrazine and undergoes reactions not possible using hydrazine itself (17). Thus, styrene in ethyl ether solution at 0°C is converted in good yield to phenethylhydrazine [51-71-8] (**1**) and nitriles give amidrazones (**2**) which are useful in the synthesis of heterocycles. Such reactions probably proceed more safely using barium hydrazide [28330-62-3], BaN_2H_2, (18); however, little work has been carried out using the alkaline earth hydrazides.

$$\text{C}_6\text{H}_5-\text{CH}_2\text{CH}_2\text{NHNH}_2 \qquad \qquad R-C\overset{NH_2}{\underset{NNH_2}{<}}$$

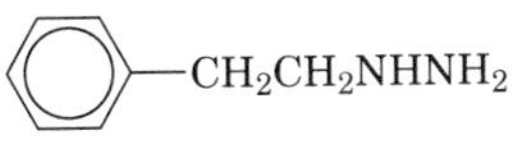

(**1**) 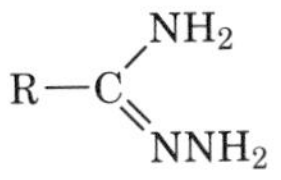   (**2**)

In water, hydrazine is a slightly weaker base than ammonia and ionizes in two steps:

$$N_2H_4 + H_2O \longrightarrow N_2H_5^+ + OH^- \qquad K_b \ (25°C) = 8.4 \times 10^{-7} \qquad (9)$$

$$N_2H_5^+ + H_2O \longrightarrow N_2H_6^{2+} + OH^- \qquad K_b \ (25°C) = 8.9 \times 10^{-16} \qquad (10)$$

Hydrazinium salts form by simple neutralization using the corresponding acid. Two series of salts are possible, containing the hydrazinium(I) ion, $N_2H_5^+$, and the hydrazinium(II) ion, $N_2H_6^{2+}$. The second ionization constant is so small, however, that salts containing the $N_2H_6^{2+}$ cation are extensively hydrolyzed in water. The $N_2H_6^{2+}$ cation is formed only with strong acids and exists only in strongly acidic solutions or in the solid state (19). Table 4 lists a few of the many known hydrazinium salts. Thermodynamic properties for most of these are available (20).

Hydrazine as Nucleophile. Reaction of hydrazine and carbon dioxide or carbon disulfide gives, respectively, hydrazinecarboxylic acid [471-31-8], $NH_2NHCOOH$, and hydrazinecarbodithioic acid [471-32-9], $NH_2NHCSSH$, in the form of the hydrazinium salts. These compounds are useful starting materials for further synthesis. For example, if carbon disulfide reacts with hydrazine in basic medium with an alkyl halide, an alkyl dithiocarbazate ester is obtained in a one-step reaction:

$$N_2H_4 + CS_2 + KOH + R'X \longrightarrow NH_2NHCSSR' + HOH + KX \qquad (11)$$

Such esters can be acylated on the amine nitrogen using acyl halides, $R\overset{\overset{\displaystyle O}{\|}}{C}X$, to give (**3**) which can be cyclized under appropriate conditions to yield 1,3,4-oxadiazoles (**4**) or thiadiazoles (**5**):

Table 4. Properties of Hydrazinium Salts

Compound	CAS Registry Number	Mp, °C	Solubility, 25°C, g/100 g water	Comments
$N_2H_4 \cdot HN_3$	[14662-04-5]	75.4 dec	very soluble	explosive
$N_2H_4 \cdot HBr$	[13775-80-9]	86.5	282.0	
$N_2H_4 \cdot 2HBr$	[23268-00-0]	195	86.4	
$N_2H_4 \cdot HCl$	[14011-37-1]	92.6	179	
$N_2H_4 \cdot 2HCl$	[5341-61-7]	198 dec	104.8	
$N_2H_4 \cdot HI$	[10039-55-1]	127	soluble	
$N_2H_4 \cdot HNO_3$	[37836-27-4]	70.71[a]	327.5	explosive
$N_2H_4 \cdot HClO_4$	[13762-80-6]	137		explosive
$N_2H_4 \cdot 2HClO_4$	[13812-39-0]			explosive
$N_2H_4 \cdot H_2SO_4$	[10034-93-2]	254	3.41	low solubility
$2N_2H_4 \cdot H_2SO_4$	[13464-80-7]	85	202	

[a]Value is for α form. A β form also exists having mp = 62.09°C.

RCNHNHCSR'
‖   ‖
O   S

(3)

N——N

R   O   R'

(4)

N——N

R   S   R'

(5)

Using thionyl chloride, and calcium oxide as HCl acceptor, hydrazine gives the calcium salt of hydrazinemonosulfinic acid, $Ca(NH_2NHSO_2)_2$, an extremely strong reducing agent (21). The monosulfonic acid (6) is formed from hydrazine and sulfur trioxide. Sulfuryl chloride gives sulfuryl dihydrazide (7); phosphorus oxychloride yields the corresponding trihydrazide (8). The phosphoric acid hydrazides have been reviewed (22).

NH_2NHSO_3H $O_2S(NHNH_2)_2$ $O{=}P(NHNH_2)_3$

(6)   (7)   (8)

Hydrazine–borane compounds are made by the reaction of sodium borohydride and a hydrazine salt in THF (23,24). The mono-$(N_2H_4{\cdot}BH_3)$ and di-$(N_2H_4{\cdot}2BH_3)$ adducts are obtained, depending on the reaction conditions. These compounds have been suggested as rocket fuels (25) and for chemical deposition of nickel–boron alloys on nonmetallic surfaces (see METALLIC COATINGS) (26).

Reductions. Hydrazine is a very strong reducing agent. In the presence of oxygen and peroxides, it yields primarily nitrogen and water with more or less ammonia and hydrazoic acid [7782-79-8]. Based on standard electrode potentials, hydrazine in alkaline solution is a stronger reductant than sulfite but weaker than hypophosphite; in acid solution, it falls between Sn^{2+} and Ti^{2+} (27).

Metal Reductions. Essentially all the metals of the transition, lanthanide, and actinide series have been reduced using hydrazine to either a lower valence state or to the metal. Such reductions can be used to make silver mirrors (28); for electroless plating (qv) of nickel (29,30), gold (31), cobalt, iron, chromium on metal or plastic (32); and for the preparation of noble-metal catalysts. It is possible to obtain exceedingly small and uniform metal particles by this technique, eg, gold particles having an average diameter of 6.8 nm (33), silver–palladium alloy having a specific surface area of 19 m^2/g (34), and ultrafine copper particles (35). The lanthanides and actinides are generally reduced to a lower valence state, not to the metals themselves (36). Chromate is readily reduced by hydrazine in basic solution to the insoluble chromium(III) hydroxide, which can then be removed by filtration (37). The final chromate concentration is below the limit detectable by standard procedures, ie, 2 ppb.

Hydrogenations. Reduction of organic compounds with hydrazine and its derivatives is covered in several reviews (38,39). These procedures have some advantages over conventional pressure hydrogenations in being more selective in their attack, sometimes stereospecific (cis addition), and in not requiring the use of hydrogen gas and high pressure equipment (see HYDROGENATION).

Carbonyl Reductions. The classical Wolff-Kishner reduction of ketones (qv) and aldehydes (qv) involves the intermediate formation of a hydrazone, which is then decomposed at high temperatures under basic conditions to give the methylene group, although sometimes alcohols may form (40).

$$R_2C{=}O + N_2H_4 \longrightarrow R_2C{=}NNH_2 \xrightarrow[\text{heat}]{\text{base}} R_2CH_2 + N_2 \qquad (12)$$

This reaction is discussed extensively in an earlier review (41). In the Huang-Minlon modification of the Wolff-Kishner reaction, a high boiling solvent such as diethylene glycol is used to achieve the required high temperatures (180–210°C) without the use of pressure equipment (42). The base is generally alkali hydroxide or an alkoxide (43).

Catalytic Hydrogenations. Nitro compounds, primarily aromatics, are reduced using hydrazine in the presence of standard hydrogenation catalysts such as Raney nickel or ruthenium on carbon. The products are generally the corresponding amines (qv), but in some cases partially hydrogenated intermediates, such as azo-, azoxy-, hydrazo-, and hydroxylamino aromatics, have been isolated (see AMINES BY REDUCTION). (Some aromatic hydroxylamines may be sensitive and one explosion has been reported when 2-chloro-5-methylphenylhydroxylamine formed in an attempt to reduce the corresponding nitro compound to 2-chloro-5-methylaniline (44)). This method of hydrogenation has also been applied to nitroso compounds, nitriles, oximes, nitrites, nitrates, and carbon–halogen compounds. In some cases, no catalyst is required (45), and yields may be better than in conventional hydrogenations. In the presence of several reducible bonds, the procedure is often selective. For example, dinitroarenes may be reduced in good yields to nitroanilines in the presence of Raney nickel (46). This catalyst is also effective for the selective reduction of nitro groups to amines in the presence of carbonyls and other reducible groups (47) and also for the reduction of nitriles to hydrazones (48). Dehalogenation of aromatics with hydrazine is favored by palladium on carbon.

Diazene Reductions. Olefins, acetylenes, and azo-compounds are reduced by hydrazine in the presence of an oxidizing agent. Stereochemical studies of alkene and alkyne reductions suggest that hydrazine is partially oxidized to the transient diazene [3618-05-1] (diimide, diimine) (**9**) and that the cis-isomer of diazene is the actual hydrogenating agent, acting by a concerted attack on the unsaturated bond:

$$N_2H_4 \xrightarrow{[O]} HN{=}NH + H_2O \qquad (13)$$

$$(\mathbf{9})$$

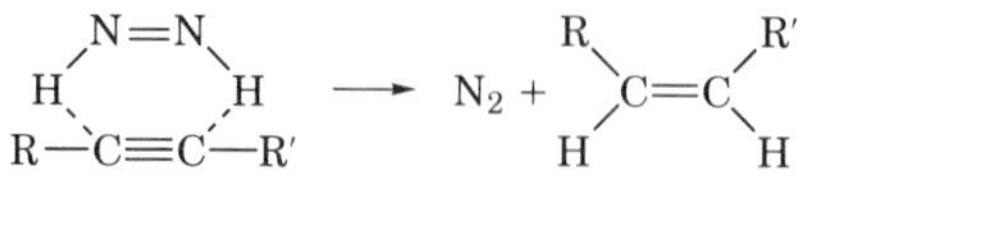

$$(14)$$

Suitable oxidants for the generation of diazene are oxygen, hydrogen peroxide, iodine, iodate, ferricyanide, and the cupric ion. Diazene is also formed by acidification of dipotassium azodicarboxylate [4910-62-7], KOOCN=NCOOK, or by thermolysis of p-toluenesulfonylhydrazide [1576-35-8]. Reductions with diazene are both selective and stereospecific: olefins (qv), acetylenes, and azo compounds are reduced easily; unsymmetrical bonds such as $C=N$, $C\equiv N$, $C=O$, NO_2, and NO are not attacked. Disulfides are not cleaved, nor are haloaromatics dehalogenated. This technique has been applied extensively to unsaturated fatty acids (49). Polybutadiene and its copolymers in xylene solution are hydrogenated to >99% using diazene generated thermally from p-toluenesulfonylhydrazide (50). Reduction of the double bond in acrylonitrile–butadiene rubber latices is efficiently achieved with hydrazine and hydrogen peroxide in the presence of Cu(II) as catalyst (51). Reductions using diazene have been reviewed (52).

Aldehyde Syntheses. Aromatic (Ar) carboxylic acids are reduced to the corresponding aldehydes by a sequence of steps known as the McFadyen-Stevens reaction. The acid is converted to the hydrazide, derivatized with benzenesulfonylchloride, then decomposed to the aldehyde in hot glycol in the presence of a base:

$$ArCOOH \xrightarrow{N_2H_4} ArCONHNH_2 \xrightarrow{C_6H_5SO_2Cl} ArCONHNHSO_2C_6H_5 \xrightarrow{base} ArCHO \qquad (15)$$

Aliphatic aldehydes can also be prepared in moderate yields by distilling the aldehyde from the mixture rapidly to avoid aldol addition or a Cannizzaro reaction (53).

Olefin Syntheses. Conversion of aldehydes and ketones to olefins by the base-catalyzed decomposition of p-toluenesulfonic (Ts) acid hydrazones (**10**) is known as the Bamford-Stevens reaction (54,55).

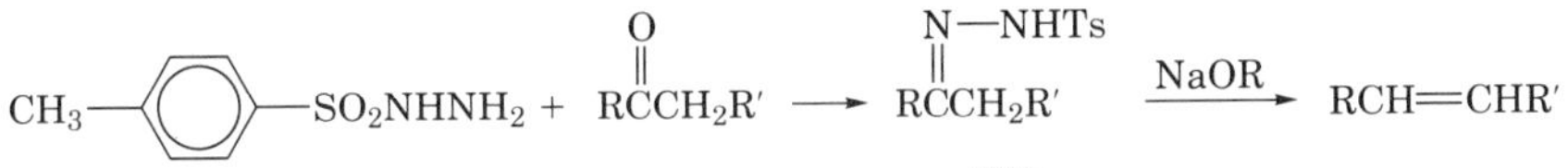

$$CH_3-\!\!\!\bigcirc\!\!\!-SO_2NHNH_2 + \underset{\overset{\|}{O}}{R}CCH_2R' \longrightarrow \underset{(\mathbf{10})}{\overset{N-NHTs}{\underset{\|}{R}CCH_2R'}} \xrightarrow{NaOR} RCH=CHR'$$

$$(16)$$

Alkylhydrazines. Mono- and higher substituted alkyl hydrazines can be made by alkylation of hydrazine using alkyl halides. For monoalkylhydrazines, however, the procedure is often unsatisfactory, producing a mixture of higher substituted hydrazines, including trialkylhydrazinium salts:

$$N_2H_4 \xrightarrow{RX} RNHNH_2 \xrightarrow{RX} R_2NNH_2 \xrightarrow{RX} R_3NNH_2^+X^- \qquad (17)$$

A large excess of hydrazine or bulky alkyl groups favor monosubstitution. For example, a 60–70% yield of monoisopropylhydrazine [2257-52-7] is achieved by reaction of isopropyl bromide and a fivefold excess of hydrazine (56).

In a variation of the Raschig process for making hydrazine, amines rather than ammonia are reacted with chloramine to give the corresponding alkyl hydrazine:

$$\begin{array}{ccc} RNH_2 & \nearrow & RNHNH_2 \\ R_2NH + NH_2Cl & \rightleftarrows & R_2NNH_2 \qquad (18) \\ R_3N & \searrow & R_3NNH_2{}^+Cl^- \end{array}$$

Monomethylhydrazine (MMH) and unsymmetrical dimethylhydrazine (UDMH) are produced commercially by this route for use as fuels in missiles and in the U.S. space program as well as for several commercial applications. Hydroxylamine-O-sulfonic acid [2950-43-8] can serve as a convenient aminating agent in place of chloramine but it has no cost advantage. Monoalkylhydrazines have also been made by the reaction of hydrazine and alcohols in the presence of phosphoric and hydrochloric acid at elevated temperatures in an autoclave (57). Unsymmetrical dialkylhydrazines, RR′NNH$_2$, can be prepared by nitrosation of dialkylamines followed by catalytic hydrogenation of the resulting nitrosamine, RR′NNO. UDMH was made commercially by this route (58), but this process has since been abandoned in the United States because the nitrosamine is such a potent carcinogen. UDMH can also be obtained by the catalytic reductive alkylation of acethydrazide [1068-57-1] using formaldehyde and hydrogen over Pd or Pt, followed by basic hydrolysis of the acetic acid dimethylhydrazide [6233-041-1] to remove the acetyl group (59,60) (eq. 19):

$$CH_3CONHNH_2 \xrightarrow[\text{catalyst}]{CH_2O/H_2} CH_3CONHN(CH_3)_2 \xrightarrow[H_2O]{\text{base}} CH_3COO^- + (CH_3)_2NNH_2 \qquad (19)$$

Reduction of hydrazones, R$_2$C=NNH$_2$, and azines, R$_2$C=NN=CR$_2$, with hydrogen (qv) or hydrides (qv) as hydrogenating agents provides another approach to mono- and disubstituted hydrazines (61,62). Complete hydrogenation leads to rupture of the N—N bond to yield primary amines (63). Alkylation of azines is also possible on the nitrogen, using alkylating agents such as alkyl sulfates.

Branched-chain alkenes react with hydrazine under acidic conditions to give the corresponding alkyl hydrazine. For example, isobutylene bubbled through an aqueous solution of hydrazine and HCl gives t-butylhydrazine in reasonably good yields (64).

Substituted Alkylhydrazines.   Substituted alkyl hydrazines are prepared from suitable alkylating agents. Epoxides yield hydroxyalkylhydrazines such as [109-84-2] (**11**) (65); aziridines, β-aminoalkylhydrazines, eg, β-amino-ethylhydrazine [14478-61-6] (**12**) (66,67); sultones, ω-sulfoalkylhydrazines, eg, ω-sulfopropylhydrazine [6482-66-2] (**13**) (68–70); and acrylonitrile, β-cyano-ethylhydrazine [353-07-1] (**14**) (71). These compounds are all potentially useful in further syntheses as each contains an active substituent on the alkyl group yet has an intact hydrazine moiety that can undergo many of the reactions of hydrazine itself.

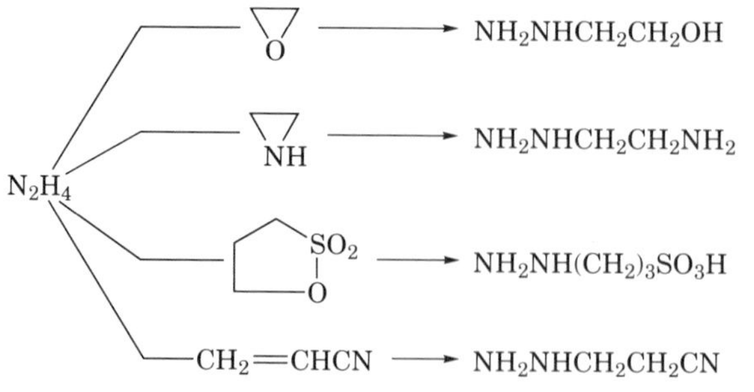

Hydroxyethylhydrazine (**11**) is a plant growth regulator. It is also used to make a coccidiostat, furazolidone, and has been proposed, as has (**14**), as a stabilizer in the polymerization of acrylonitrile (72,73). With excess epoxide, polysubstitution occurs and polyol chains can form to give poly(hydroxyalkyl) hydrazines which have been patented for the preparation of cellular polyurethanes (74) and as corrosion inhibitors for hydraulic fluids (qv) (75). Dialkylhydrazines, R_2NNH_2, and alkylene oxides form the very reactive amineimines (**15**) which react further with esters to yield aminimides (**16**):

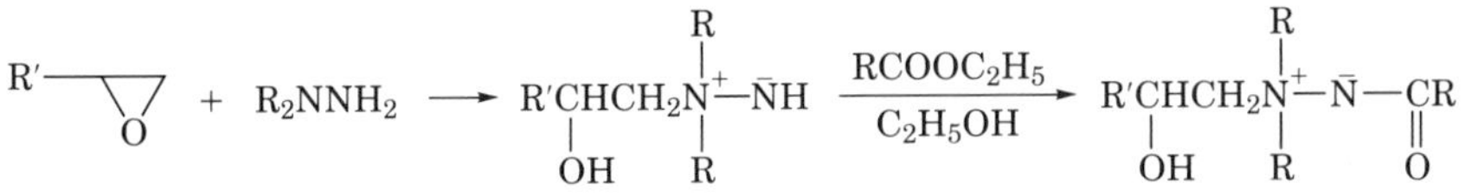

$$R'-\!\!\overset{\displaystyle\diagup\!\!\diagdown}{\underset{O}{}} + R_2NNH_2 \longrightarrow R'CHCH_2\overset{\overset{\displaystyle R}{\vert}}{\underset{\underset{\displaystyle R}{\vert}}{\underset{\displaystyle OH}{N}}}\!\!\overset{+}{\underset{}{}}\!\!-\bar{N}H \xrightarrow[\text{C}_2\text{H}_5\text{OH}]{\text{RCOOC}_2\text{H}_5} R'CHCH_2\overset{\overset{\displaystyle R}{\vert}}{\underset{\underset{\displaystyle OH}{\vert}}{N}}\!\!\overset{+}{\underset{\underset{\displaystyle R}{\vert}}{}}\!\!-\bar{N}\!-\!\overset{\overset{}{}}{\underset{\underset{\displaystyle O}{\vert\vert}}{C}}R \qquad (20)$$

$$(\mathbf{15}) \qquad\qquad\qquad\qquad (\mathbf{16})$$

The preparation and properties of these tertiary aminimides, as well as suggested uses as adhesives (qv), antistatic agents (qv), photographic products, surface coatings, and pharmaceuticals, have been reviewed (76). Thermolysis of aminimides causes N—N bond rupture followed by a Curtius rearrangement of the transient nitrene (**17**) intermediate to the corresponding isocyanate:

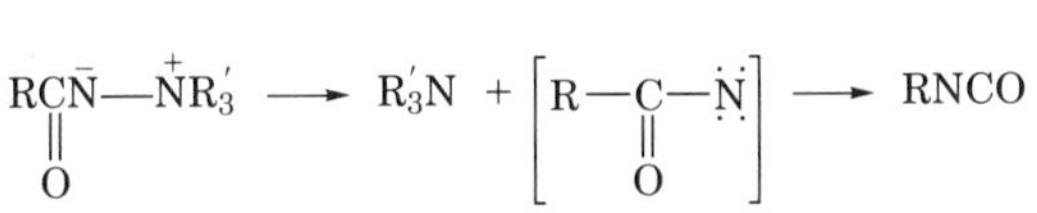

$$R\bar{C}\overset{}{\underset{\underset{\displaystyle O}{\vert\vert}}{N}}\!\!-\!\overset{+}{N}R_3' \longrightarrow R_3'N + \left[R\!-\!\overset{}{\underset{\underset{\displaystyle O}{\vert\vert}}{C}}\!-\!\ddot{N}\right] \longrightarrow RNCO \qquad (21)$$

$$(\mathbf{17})$$

Bisaminimides form diisocyanates. This reaction and its application in polymer chemistry have been reviewed (77). A review of the general subject of aminimide monomers and polymers is available (78).

Aromatic Hydrazines. A general synthesis for arylhydrazines is via diazotization of aromatic amines, followed by reduction of the resulting diazonium salt (**18**):

$$ArNH_2 \xrightarrow[\text{HCl}]{\text{NaNO}_2} (Ar\!-\!N\!\equiv\!N)^+Cl^- \xrightarrow[\text{NaOH}]{[H]} ArNHNH_2 \qquad (22)$$

$$(\mathbf{18})$$

In the industrial synthesis of phenylhydrazine [*100-63-0*], the reducing agent is sodium bisulfite. It is also possible to react aniline with chloramine as in the Raschig process (79):

$$C_6H_5NH_2 + NH_2Cl + NaOH \longrightarrow C_6H_5NHNH_2 + NaCl + H_2O \qquad (23)$$

Generally, aromatic hydrazines are not made from haloaromatics and hydrazine, unless the halo substituent is activated by neighboring electronegative groups. For example, 2,4-dinitrochlorobenzene reacts with hydrazine to give 2,4-dinitrophenylhydrazine [119-26-6] (**19**), which is used as an analytical reagent for the identification of aldehydes and ketones. Other haloaromatics that exchange halogen for hydrazine are 2-chloropyridine to give the 2-hydrazinopyridine [4930-98-7] (**20**) and 1-chlorophthalazine to form hydrazinophthalazine [86-54-4] (**21**), an antihypertensive drug (see CARDIOVASCULAR AGENTS). The preparation and properties of arylhydrazines are covered in some earlier reviews (80,81).

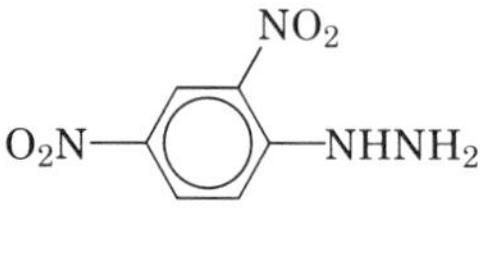

(**19**) 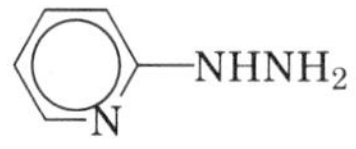   (**20**) 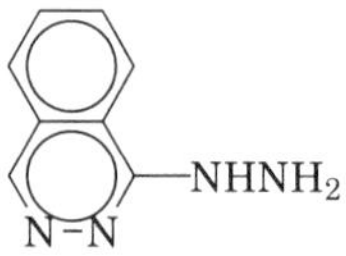   (**21**)

Hydrazides and Related Compounds.   Substitution of the hydroxyl group in carboxylic acids with a hydrazino moiety gives carboxylic acid hydrazides. In this formal sense, a number of related compounds fall within this product class although they are not necessarily prepared this way. Table 5 lists some of the more common of these compounds (82).

Carboxylic acid hydrazides are prepared from aqueous hydrazine and the carboxylic acid, ester, amide, anhydride, or halide. The reaction usually goes poorly with the free acid. Esters are generally satisfactory. Acyl halides are particularly reactive, even at room temperature, and form the diacyl derivatives (**22**), which easily undergo thermal dehydration to 1,3,4-oxadiazoles (**23**). Diesters give dihydrazides (**24**) and polyesters such as polyacrylates yield a polyhydrazide (**25**). The chemistry of carboxylic hydrazides has been reviewed (83,84).

(**22**) 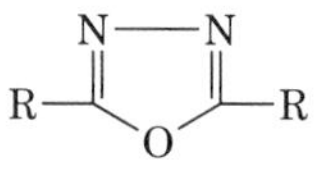   (**23**) 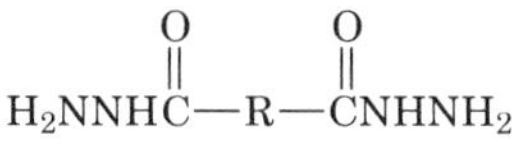   (**24**) 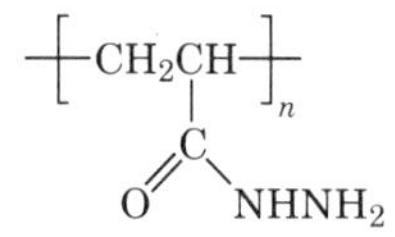   (**25**)

Thiohydrazides $\overset{\text{S}}{\overset{\|}{\text{RC}}}$—$NHNH_2$ are best prepared from hydrazine and dithioesters, $RCSSR'$. These are useful for the formation of heterocyclics and have been studied for their pharmacological and pesticidal properties. Several somewhat older reviews of the chemistry of thiohydrazides are available (85,86).

Sulfonic acid hydrazides, RSO_2NHNH_2, are prepared by the reaction of hydrazine and sulfonyl halides, generally the chloride RSO_2Cl. Some of these have commercial applications as blowing agents. As is typical of hydrazides generally, these compounds react with nitrous acid to form azides (**26**), which

Table 5. Hydrazides and Related Compounds

Type	CAS Registry Number	Structure
carboxylic hydrazides		RCNHNH_2 with $\|$ O below C
thiohydrazides		RCNHNH_2 with $\|$ S below C
sulfonylhydrazides		$\text{RSO}_2\text{NHNH}_2$
semicarbazide	[57-56-7]	$\text{H}_2\text{NCNHNH}_2$ with $\|$ O below C
thiosemicarbazide	[79-19-6]	$\text{H}_2\text{NCNHNH}_2$ with $\|$ S below C
carbohydrazide	[479-18-7]	$\text{H}_2\text{NNHCNHNH}_2$ with $\|$ O below C
thiocarbohydrazide	[2231-57-4]	$\text{H}_2\text{NNHCNHNH}_2$ with $\|$ S below C
amidrazones		$\text{RC}\overset{\displaystyle\nearrow\text{NH}}{\underset{\searrow\text{NHNH}_2}{}}$
hydrazidines		$\text{RC}\overset{\displaystyle\nearrow\text{NNH}_2}{\underset{\searrow\text{NHNH}_2}{}}$
aminoguanidine	[79-17-4]	$\text{H}_2\text{NC}\overset{\displaystyle\nearrow\text{NH}}{\underset{\searrow\text{NHNH}_2}{}}$
diaminoguanidine	[4363-78-7]	$\text{H}_2\text{NNH}-\text{C}\overset{\displaystyle\nearrow\text{NH}}{\underset{\searrow\text{NHNH}_2}{}}$
triaminoguanidine	[2203-24-9]	$\text{H}_2\text{NNH}-\text{C}\overset{\displaystyle\nearrow\text{NNH}_2}{\underset{\searrow\text{NHNH}_2}{}}$

decompose thermally to the very reactive, electron-deficient nitrenes (**27**). The chemistry of sulfonic acid hydrazides and their azides has been reviewed (87).

$$\text{RSO}_2\text{NHNH}_2 \xrightarrow{\text{HONO}} \underset{(\mathbf{26})}{\text{RSO}_2\text{N}_3} \xrightarrow{-\text{N}_2} \underset{(\mathbf{27})}{\text{RSO}_2\ddot{\text{N}}} \tag{24}$$

Hydrazinecarboxamides (semicarbazides) (**28**) and hydrazinecarbothioamides (thiosemicarbazides) (**29**) are starting materials for many useful products. They are generally prepared from hydrazine and either isocyanates or isothiocyanates. The unsubstituted parent semicarbazide (**28**, R = H) is best made from urea and hydrazine; using excess urea, hydrazodicarbonamide [110-21-4] is formed. Methods have been reported for the preparation of all positional isomers of thiosemicarbazide (85), with substituents in the 1, 2, and/or 4 positions (**30**):

$$
\begin{array}{ccc}
\underset{\text{(28)}}{\overset{\displaystyle\overset{O}{\overset{\|}{}}}{\text{RNHCNHNH}_2}} &
\underset{\text{(29)}}{\overset{\displaystyle\overset{S}{\overset{\|}{}}}{\text{RNHCNHNH}_2}} &
\underset{\text{(30)}}{\overset{\displaystyle\overset{S}{\overset{\|}{}}}{\underset{4\ \ 2\ \ 1}{\text{H}_2\text{NCNHNH}_2}}}
\end{array}
$$

A process for heating alkyldithiocarbamate salts (**31**) in the presence of hydrazine hydrate to produce 4-alkylthiosemicarbazides (**32**) has been described (88).

$$
\text{CH}_3\text{NH}_2 + 2\,\text{NH}_3 + \text{CS}_2 \longrightarrow \underset{\text{(31)}}{\text{CH}_3\text{NH}-\text{CS}-\text{S}^-\text{NH}_4^+} \xrightarrow{\text{N}_2\text{H}_4} \underset{\text{(32)}}{\text{CH}_3\text{NH}-\text{CS}-\text{NHNH}_2} \qquad (25)
$$

Thiosemicarbazides are useful in forming 1,3,4-thiadiazoles, a class of compounds having herbicidal activity.

Hydrazones and Azines.   Depending on reaction conditions, hydrazines react with aldehydes and ketones to give hydrazones (**33**), azines (**34**), and diaziridines (**35**), the latter formerly known as isohydrazones.

$$
\begin{array}{ccc}
\underset{\text{(33)}}{\overset{\displaystyle{\overset{R}{\underset{R}{}}}}{\text{C}=\text{N}-\text{NH}_2}} &
\underset{\text{(34)}}{\overset{\displaystyle{\overset{R}{\underset{R}{}}}}{\text{C}=\text{N}-\text{N}=\text{C}{\overset{R}{\underset{R}{}}}}} &
\underset{\text{(35)}}{\overset{\text{HN}}{\underset{\text{HN}}{}}\times{\overset{R}{\underset{R}{}}}}
\end{array}
$$

Hydrazones are formed from mono- and N,N-disubstituted hydrazines. Hydrazine itself can give either hydrazones or azines, depending mainly on the ratio of carbonyl component to hydrazine. The ease of formation of these compounds depends on the nature of the carbonyl constituent:

aldehydes > dialkyl ketones > alkylaryl ketones > diaryl ketones

Many of these compounds are highly colored and have found use as dyes and photographic chemicals. Several pharmaceuticals and pesticides are members of this class. An extremely sensitive analytical method for low hydrazine concentrations is based on the formation of a colored azine. They are also useful in heterocycle formation. Several reviews are available covering the chemistry of hydrazones (80,89) and azines (90).

Heterocyclics.   One of the most characteristic and useful properties of hydrazine and its derivatives is the ability to form heterocyclic compounds. Numerous pharmaceuticals, pesticides, explosives, and dyes are based on these rings. A review of the application of hydrazine in the synthesis of heterocyclics is available (91). For further information in the field of heterocyclic chemistry, see the *General References*.

Manufacture

The direct fixation of nitrogen and hydrogen to hydrazine is thermodynamically unfavorable. Ammonia is the preferred product. The commercially feasible processes involve partial oxidation of ammonia (or urea) using hypochlorite or hydrogen peroxide. Most hydrazine is produced by some variation of the Raschig process, which is based on the oxidation of ammonia using alkaline hypochlorite. Ketazine processes are modifications in which the oxidation is carried out in the presence of a ketone such as acetone or butanone. Bayer first commercialized such a process, using acetone, and hypochlorite as oxidant. A process developed by Produits Chimiques Ugine Kuhlmann (PCUK) and practiced by Elf Atochem (France) and Mitsubishi Gas (Japan) involves the oxidation of ammonia by hydrogen peroxide in the presence of butanone and another component that apparently functions as an oxygen-transfer agent. The oxidation of benzophenoneimine has received much attention but is not commercial.

Raschig Process. The Raschig process (92) is based on the oxidation of ammonia with hypochlorite according to the following overall reaction:

$$2\ NH_3 + NaOCl \longrightarrow N_2H_4 + NaCl + H_2O \tag{26}$$

This is a stepwise process in which chloramine [10599-90-3] is first formed from ammonia and hypochlorite in a rapid reaction at low temperature:

$$NH_3 + NaOCl \longrightarrow NH_2Cl + NaOH \tag{27}$$

The chloramine then reacts with excess ammonia to form hydrazine:

$$NH_2Cl + NH_3 + NaOH \longrightarrow N_2H_4 + NaCl + H_2O \tag{28}$$

This reaction is slow and requires elevated temperatures of 120–150°C under pressure. The kinetics (93,94) and mechanism (95,96) of these reactions have been studied. An undesirable competing reaction is the further oxidation of hydrazine by chloramine:

$$N_2H_4 + 2\ NH_2Cl \longrightarrow 2\ NH_4Cl + N_2 \tag{29}$$

This reaction is also slow but catalyzed by metal-ion impurities, especially copper(II). It is minimized by avoiding metal contamination or by adding chelating agents (qv) to complex the metal ions (Raschig used glue). Reaction 28 is favored over reaction 29 by using a large molar excess of ammonia over chloramine (30:1) and by keeping the hydrazine concentration low (ca 1–2%). Under these conditions, yields up to 80% (based on hypochlorite) can be achieved; the penalty for this favorable yield is an energy-intensive process involving large ammonia recycle and evaporative load.

Olin Raschig Process. One process used by the Olin Corp. in its Lake Charles, Louisiana, facility is an adaptation of the basic Raschig process (Fig. 4). Liquid chlorine is continuously absorbed in dilute sodium hydroxide to form sodium hypochlorite, cooled by an external brine recirculating loop to prevent hypochlorite decomposition and chlorate formation. The chlorination is automatically monitored by measuring the oxidation potential of the recirculating stream. Excess sodium hydroxide in the sodium hypochlorite is kept low (less than 1 g/L) to avoid caking in the salt crystallizer and to improve yields somewhat (97). The sodium hypochlorite is mixed with about a threefold molar excess of aqueous ammonia in the chloramine reactor; chloramine formation is almost instantaneous at about 5°C. To this chloramine-containing stream is added immediately a 20–30 molar excess of anhydrous ammonia under sufficient pressure to keep ammonia in the liquid phase. The heat of dilution raises the temperature rapidly to about 40°C. Further heating with steam to 120–130°C favors the hydrazine-forming reaction 28 and minimizes the competing destructive reaction 29. The nitrogen that is formed (eq. 29) is scrubbed with water to remove ammonia and is then used as an inert pad to prevent decomposition of hydrazine during the concentration steps.

The reactor effluent, containing 1–2% hydrazine, ammonia, sodium chloride, and water, is preheated and sent to the ammonia recovery system, which consists of two columns. In the first column, ammonia goes overhead under pressure and recycles to the anhydrous ammonia storage tank. In the second column, some water and final traces of ammonia are removed overhead. The bottoms from this column, consisting of water, sodium chloride, and hydrazine, are sent to an evaporating crystallizer where sodium chloride (and the slight excess of sodium hydroxide) is removed from the system as a solid. Vapors from the crystallizer flow to the hydrate column where water is removed overhead. The bottom stream from this column is close to the hydrazine–water azeotrope composition. Standard materials of construction may be used for handling chlorine, caustic, and sodium hypochlorite. For all surfaces in contact with hydrazine, however, the preferred material of construction is 304 L stainless steel.

Anhydrous hydrazine, required for propellant applications and some chemical syntheses, is made by breaking the hydrazine–water azeotrope with aniline. The bottom stream from the hydrate column (Fig. 4) is fed along with aniline to the azeotrope column. The overhead aniline–water vapor condenses and phase separates. The lower aniline layer returns to the column as reflux. The water layer, contaminated with a small amount of aniline and hydrazine, flows to a biological treatment pond. The bottoms from the azeotrope column consist of aniline and hydrazine. These are separated in the final hydrazine column to give an anhydrous overhead; the aniline from the bottom is recycled to the azeotrope column.

The small amount of aniline remaining in the anhydrous hydrazine can be objectionable in some monopropellant applications, such as in thruster engines used for the control of satellites and spacecraft; carbon containing impurities may foul the decomposition catalyst or cause metal embrittlement. These and similar applications require an even purer grade of anhydrous hydrazine. At one time Martin-Marietta Corp. practiced a process of fractional crystallization to remove aniline. A variation, zone freezing, has also been described (98). Another

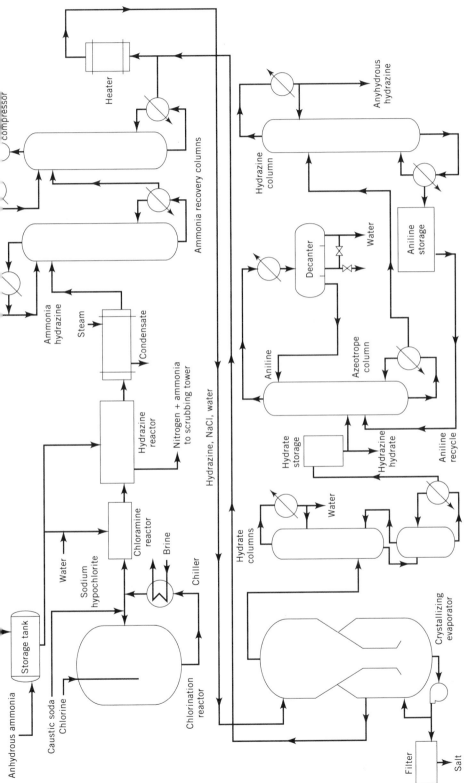

Fig. 4. Olin Raschig process flow sheet.

577

approach is to break the hydrazine–water azeotrope without the addition of an organic azeotroping agent. The ternary system, N_2H_4-NaOH-H_2O has a two-phase liquid region above 60°C (99). Addition of sufficient NaOH to hydrazine hydrate causes phase separation into a hydrazine-rich upper phase with a N_2H_4 content above that of the azeotrope. The lower phase contains the bulk of the water and caustic. The upper phase can then be distilled to obtain anhydrous hydrazine. Process variations based on this ternary system are described for the preparation of anhydrous hydrazine of ultrahigh purity by vacuum distillation (100). Olin Corp. produces such a product under the name Ultra Pure.

MMH and UDMH. Monomethylhydrazine and *unsym*-dimethylhydrazine are manufactured by Olin Corp. using the same Raschig process and equipment employed for anhydrous hydrazine. Chloramine, prepared as described above, reacts with methylamine or dimethylamine instead of with ammonia. These products are used primarily as propellants for rockets and missiles.

Ketazine Processes. The oxidation of ammonia by chlorine or chloramine in the presence of aliphatic ketones yields hydrazones (**36**), ketazines (**37**), or diaziridines (**38**), depending on the pH, ketone ratios, and reaction conditions (101).

$$R_2C{=}N{-}NH_2 \qquad\qquad R_2C{=}N{-}N{=}CR_2 \qquad\qquad \begin{array}{c} R \\ {>}C{<} \\ R \end{array}\begin{array}{c} NH \\ | \\ NH \end{array}$$

$$(\mathbf{36}) \qquad\qquad\qquad (\mathbf{37}) \qquad\qquad\qquad (\mathbf{38})$$

The hydrazones and diaziridines (isohydrazones) form ketazines with excess ketone. The advantage of these intermediate hydrazine derivatives lies in resistance to further oxidation compared to the behavior of hydrazine itself (eq. 29). Higher chemical yields are possible than in the Raschig process. After complete consumption of the oxidizing agents (Cl_2, NaOCl, NH_2Cl), intermediates can be concentrated and then hydrolyzed to hydrazine or its salts. Many patents describe embodiments of this reaction scheme (102–109).

Bayer Ketazine Process. Chlorine reacts continuously with dilute sodium hydroxide to form sodium hypochlorite and sodium chloride (Fig. 5). This hypochlorite solution reacts at 30–40°C under slight pressure with aqua ammonia and acetone (qv). With good mixing and a molar ratio of 1 hypochlorite:2 acetone:20 ammonia, a high yield based on chlorine is obtained. This dilute synthesis stream goes to the pressure ammonia stripper to remove excess ammonia, which is then absorbed in water and returned to the ketazine reactor. The bottoms from the ammonia recovery system, consisting of aqueous hydrazone, ketazine, and salt, go to the ketazine column where acetoneazine (**37**, R = CH_3) is taken overhead as a low boiling azeotrope with water (95°C, 55.5 wt % acetoneazine at 101.3 kPa) (110). In order to favor the ketazine equilibrium, additional acetone is added to this column so that all the hydrazine is present as ketazine (111).

$$(CH_3)_2C{=}N{-}NH_2 + (CH_3)_2C{=}O \longrightarrow (CH_3)_2C{=}N{-}N{=}C(CH_3)_2 + H_2O \qquad (30)$$

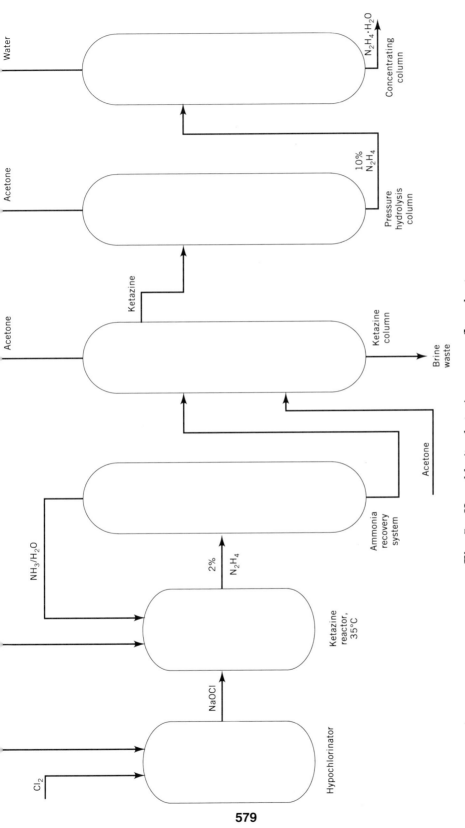

Fig. 5. Hypochlorite–ketazine process flow sheet.

579

The overhead acetoneazine from the ketazine column hydrolyzes in the pressure hydrolysis column at about 1 MPa (10 atm) to form acetone and hydrazine:

$$(CH_3)_2C = N - N = C(CH_3)_2 + 2\ H_2O \longrightarrow 2\ (CH_3)_2C = O + N_2H_4 \tag{31}$$

It is largely the pressure hydrolysis step that makes this ketazine process economic. Previous methods involved acid hydrolysis of ketazine to give the corresponding hydrazine salt. Because hydrazine, rather than a salt, is usually the desired product, an additional equivalent of base is needed in these processes to liberate the free hydrazine. Such processes require both one equivalent of acid and one of base to produce free hydrazine and this makes them uneconomical. When a salt such as hydrazine sulfate is the desired product, acid hydrolysis of the ketazine could become an option.

Acetone is withdrawn as overhead product from the pressure hydrolysis step and returned to the ketazine reactor. Excess acetone is also taken from the top of the ketazine column for recycle. The bottom stream from the ketazine column is essentially a sodium chloride solution with minor amounts of hydrazine and organic by-products. Before discharge this stream can be subjected to a clean-up chlorination treatment (112). The bottoms product from the pressure hydrolysis column is a 10–12% hydrazine solution. This is further fortified in a concentrating column where water is taken overhead, leaving essentially 64 wt % hydrazine, ie, 100% hydrazine hydrate, as a bottom product.

Comparison of the Raschig and Hypochlorite–Ketazine Processes. The formation of ketazine in the reaction step avoids destruction of hydrazine by chloramine, because the ketazine is not attacked by chloramine. The result is a synthesis yield (based on hypochlorite) that is substantially higher than that realized in the Raschig synthesis. Furthermore, the evaporative load is less because the hydrazine is removed from the bulk of the water as a low boiling ketazine–water azeotrope in the ketazine column. In the Raschig process, a much larger amount of water must be vaporized because hydrazine forms a high boiling azeotrope with water. This makes the ketazine process less energy intensive, an important consideration as energy costs escalate. Disadvantages of the ketazine process are the need to dispose of organic by-products, the loss of acetone, and the problem of handling acetone vapors. Also, it is difficult to remove all the organics from the final product so the total organic carbon (TOC) may be 500 to 1500 ppm. For some applications, this level of TOC may be objectionable and require further treatment, such as an additional distillation or adsorption on active carbon or other adsorptive surface (113). This is not a problem with hydrazine hydrate made by the standard Raschig process, and for most applications is not a problem at all.

Urea Process. In a further modification of the fundamental Raschig process, urea (qv) can be used in place of ammonia as the nitrogen source (114–116). This process has been operated commercially. Its principal advantage is low investment because the equipment is relatively simple. For low production levels, this process could be the most economical one. With the rapid growth in hydrazine production and increasing plant size, the urea process has lost importance, although it is reportedly being used, for example, in the People's Republic of China (PRC).

Peroxide–Ketazine Process. Elf Atochem in France operates a process patented by Produits Chimiques Ugine Kuhlmann (PCUK). Hydrogen peroxide (qv), rather than chlorine or hypochlorite, is used to oxidize ammonia. The reaction is carried out in the presence of methyl ethyl ketone (MEK) at atmospheric pressure and 50°C. The ratio of H_2O_2:MEK:NH_3 used is 1:2:4. Hydrogen peroxide is activated by acetamide and disodium hydrogen phosphate (117). Figure 6 is a simplified flow sheet of this process. The overall reaction results in the formation of methyl ethyl ketazine [5921-54-0] (**39**) and water:

$$H_2O_2 + 2\ NH_3 + 2\ \underset{CH_3}{\overset{C_2H_5}{>}}C{=}O \longrightarrow \underset{CH_3}{\overset{C_2H_5}{>}}C{=}N{-}N{=}C\underset{CH_3}{\overset{C_2H_5}{<}} + 4\ H_2O \qquad (32)$$

$$(\mathbf{39})$$

The mechanism of this reaction involves an activation of the ammonia and hydrogen peroxide because these compounds do not themselves react (118–121). It appears that acetamide functions as an oxygen transfer agent, possibly as the iminoperacetic acid (**41**) which then oxidizes the transient Schiff base formed between MEK and ammonia (**40**) to give the oxaziridine (**42**), with regeneration of acetamide:

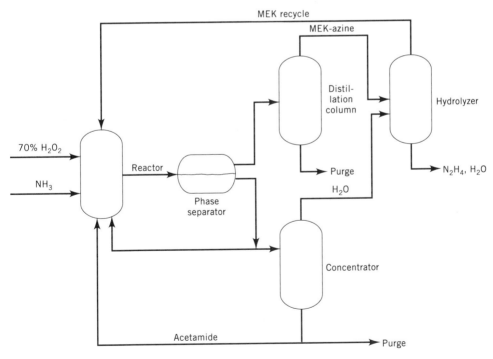

Fig. 6. Peroxide–ketazine process flow sheet. MEK = methyl ethyl ketone.

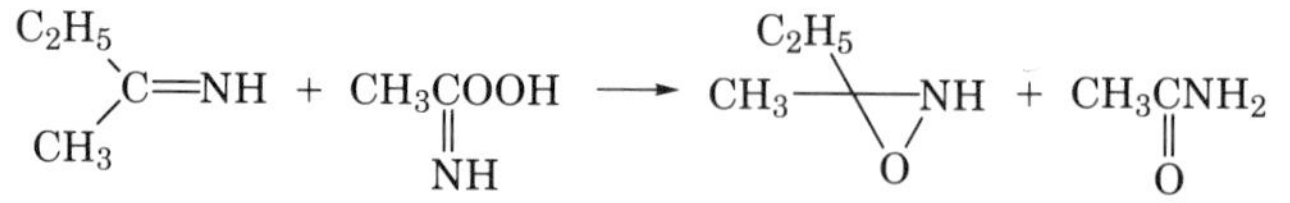

$$
\underset{(\mathbf{40})}{\underset{CH_3}{\overset{C_2H_5}{\diagdown}}C{=}NH} + \underset{(\mathbf{41})}{\underset{\overset{\|}{NH}}{CH_3\overset{}{C}OOH}} \longrightarrow \underset{(\mathbf{42})}{\underset{\overset{}{O}}{CH_3{-}\overset{C_2H_5}{\diagup}{-}NH}} + \underset{\overset{\|}{O}}{CH_3\overset{}{C}NH_2} \tag{33}
$$

The oxaziridine oxidizes a second molecule of ammonia to form a hydrazone (**43**) which, with excess ketone, forms the azine (**39**).

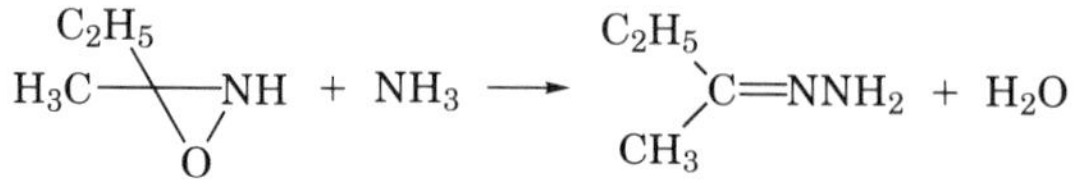

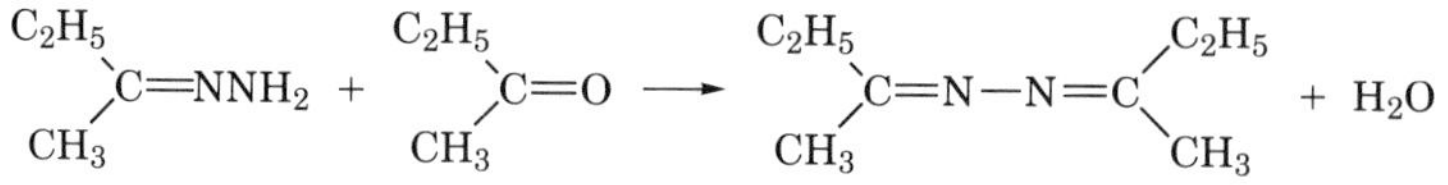

$$
\underset{\overset{}{O}}{H_3C{-}\overset{C_2H_5}{\diagup}{-}NH} + NH_3 \longrightarrow \underset{(\mathbf{43})}{\underset{CH_3}{\overset{C_2H_5}{\diagdown}}C{=}NNH_2} + H_2O \tag{34}
$$

$$
\underset{CH_3}{\overset{C_2H_5}{\diagdown}}C{=}NNH_2 + \underset{CH_3}{\overset{C_2H_5}{\diagdown}}C{=}O \longrightarrow \underset{(\mathbf{39})}{\underset{CH_3}{\overset{C_2H_5}{\diagdown}}C{=}N{-}N{=}C\overset{C_2H_5}{\underset{CH_3}{\diagup}}} + H_2O \tag{35}
$$

The methyl ethyl ketazine forms an immiscible upper organic layer easily removed by decantation. The lower, aqueous phase, containing acetamide and sodium phosphate, is concentrated to remove water formed in the reaction and is then recycled to the reactor after a purge of water-soluble impurities. Organic by-products are separated from the ketazine layer by distillation. The purified ketazine is then hydrolyzed under pressure ($0.2-1.5$ MPa ($2-15$ atm)) to give aqueous hydrazine and methyl ethyl ketone overhead, which is recycled (122). The aqueous hydrazine is concentrated in a final distillation column.

Comparison to the Raschig Process.   The economics of this peroxide process in comparison to the Raschig or hypochlorite–ketazine processes depend on the relative costs of chlorine, caustic, and hydrogen peroxide. An inexpensive source of peroxide would make this process attractive. Its energy consumption could be somewhat less, because the ketazine in the peroxide process is recovered by decantation rather than by distillation as in the hypcochlorite process. A big advantage of the peroxide process is the elimination of sodium chloride as a by-product; this is important where salt discharge is an environmental concern. In addition to Elf Atochem, Mitsubishi Gas (Japan) uses a peroxide process.

Benzophenone Process.   Benzophenone, $(C_6H_5)_2C{=}O$, reacts with ammonia to form diphenylmethanimine, $(C_6H_5)_2C{=}NH$. In the presence of copper catalysts, this is oxidized with oxygen to benzophenone azine, $(C_6H_5)_2C{=}N{-}N{=}C(C_6H_5)_2$. The formation of the imine and its further oxidation can be accomplished in a single step by passing ammonia and air through benzophenone at elevated temperature and pressure in the presence of zinc chloride and cuprous chloride. Acid hydrolysis of the azine gives hydrazine and regenerates the benzophenone. This process has been thoroughly reviewed (123); however, it is not commercial.

Economic Aspects

The estimated world production capacity for hydrazine solutions is 44,100 t on a N_2H_4 basis (Table 6). About 60% is made by the hypochlorite–ketazine process, 25% by the peroxide–ketazine route, and the remainder by the Raschig and urea processes. In addition there is anhydrous hydrazine capacity for propellant applications. In the United States, one plant dedicated to fuels production (Olin Corp., Raschig process), has a nominal capacity of 3200 t. This facility also produces the two other hydrazine fuels, monomethylhydrazine and unsymmetrical dimethylhydrazine. Other hydrazine fuels capacity includes AH in the PRC, Japan, and Russia; MMH in France and Japan; and UDMH in France, Russia, and the PRC.

Table 6. 1992 Hydrazine Solutions Production Capacity, N_2H_4, t

Country	Producer	Process	Capacity
United States	Miles	hypochlorite–ketazine	7,700
	Olin	hypochlorite–ketazine	8,800
Germany	Bayer	hypochlorite–ketazine	6,400
France	Atochem	peroxide–ketazine	6,100
Japan	Mitsubishi Gas	peroxide–ketazine	5,100
	Nippon Carbide	Raschig	1,500
Korea	Otsuka	hypochlorite–ketazine	3,600
PRC		Raschig/urea (?)	1,400
Russia		Raschig	3,500
Total			*44,100*

World demand for hydrazine solutions is about 31,000 t N_2H_4, excluding Eastern Europe, Russia, and mainland China. The demand is nearly equally divided between captive use and merchant business. Table 7 shows an estimated breakdown by use and by region for hydrazine solutions. Synthesis of chemical blowing agents (CBAs) for the foaming of plastics accounts for about 40%

Table 7. Hydrazine Solutions Applications, N_2H_4, t

Country	CBAs	Agriculture	Water treatment	Others[a]	Total
United States/Canada	1,815	2,495	910	2,040	7,260
Western Europe	1,590	3,860	1,905	1,315	8,670
Japan	2,680	1,135	1,950	365	6,130
Korea	3,400				3,400
Indonesia	1,500				1,500
Taiwan	680		90		770
Singapore	270		270	320	860
Australia/New Zealand			45		45
Africa/Middle East		180	410	365	955
Central and South America	545		500	180	1,225
Total	*12,480*	*7,670*	*6,080*	*4,585*	*30,815*

[a]Pharmaceuticals, photography, polymers, metal reduction, etc.

of demand; another 25% goes into agricultural chemicals (pesticides). Water treatment represents nearly 20% of the market. Pharmaceutical applications are significant but difficult to quantify; the remaining applications are fragmented among many industries.

Hydrazine is a mature product likely to grow worldwide in step with the gross national product. However, annual growth in chemical blowing agents in southeast Asia may be higher, perhaps $>5\%$. In water treatment, the growth might be somewhat less, perhaps 2–3%. A significant new application for hydrazine, in the manufacture of sodium azide for automobile air bags, could require an additional 3000 t/yr of N_2H_4.

Shipment and Specifications

Shipment of hydrazine solutions is regulated in the United States by the Department of Transportation (DOT) which classifies all aqueous solutions between 64.4 and 37% N_2H_4 as "Corrosive" materials with a subsidiary risk of "Poison". Hydrazine has been identified by both the Environmental Protection Agency and the DOT as a hazardous material and has been assigned a reportable quantity (RQ) of 0.450 kg (1 lb) if spilled. Drums for the shipment of these solutions must bear both the DOT specification "Corrosive" and "Poison" labels in association with the markings "RQ Hydrazine Aqueous Solution UN 2030." Aqueous solutions of 37% concentration or less are a hazard Class 6.1, UN 3293, Packing Group III and require "Keep Away From Food" placards and labels.

The aqueous grades of hydrazine are packaged in 208-L (55-gal) plastic or lined steel drums, 1304-L (345-gal) stainless steel returnable tote tanks and tank trucks. In addition, 35% Scav-Ox can be had in 114-L (30-gal) plastic drums and in 22.7-L (6-gal) or 25-kg pails. Anhydrous hydrazine (AH) and MMH are available in 114- and 208-L returnable drums made of 304 low Mo stainless steel. UDMH ships in 208-L mild steel drums.

Table 8 summarizes specifications for anhydrous grades of hydrazine. Propellant grades meet all requirements of the most recent Military Specification MIL-P-26536 D. Military specifications covering the other propellant hydrazines are MMH (Mil-P-27404B, May 22, 1979); UDMH (Mil-P-25604D, Amendment 1, Apr. 10, 1978, Supplemental Data Sheet, Jan. 24, 1984); Aerozine-50 (Mil-P-27402B, May 27, 1969); H-70 (Mil-P-87930, USAF, Oct. 25, 1977).

Table 9 lists specifications for the common aqueous hydrazine products available in the United States (124).

Whereas there is sometimes confusion in how concentrations of aqueous solutions of hydrazine are expressed, concentrations of wt % N_2H_4 are used herein. In many parts of the world, however, concentrations are often expressed in terms of wt % hydrazine hydrate, $N_2H_4 \cdot H_2O$. Hydrazine hydrate is 64 wt % N_2H_4, 36% H_2O. The correlation between the two systems is therefore:

wt % N_2H_4	64	54.4	51.2	35
wt % $N_2H_4 \cdot H_2O$	100	85	80	54.7

Table 8. Specifications for Propellant Hydrazine Grades[a]

Parameter	Regular[b]	Monopropellant[c]	High purity[c]	Ultra Pure
assay, wt %[d]	98	98.5	99.0	99.5
water, wt %	1.5	1.0	1.0	0.5
particulates, mg/L	10	1.0	1.0	1.0
ammonia, wt %			0.3	0.3
aniline, wt %	0.5	0.50	0.003	[e]
carbon dioxide, wt %		0.003	0.003	0.003
chloride, wt %	0.005	0.0005	0.0005	0.0005
iron, wt %	0.003	0.002	0.0004	0.0004
nonvolatile residue, wt %		0.005	0.001	0.001
other volatile carbonaceous matter, wt %		0.02	0.005	0.005

[a]All values are maximum allowed unless otherwise specified.
[b]Military Specification MIL-P-26536C, Amendment 2, Feb. 1, 1982.
[c]Military Specification MIL-P-26536D, July 27, 1987.
[d]Values are minimum allowed.
[e]Not detectable.

Table 9. Aqueous Hydrazine Specifications[a]

Specification	Hydrazine, wt %			
	64	54.4	51.2	35
assay,[b] wt %	64.0–64.8	54.4–55.1	51.2–51.8	35.0–35.7
water,[c] wt %	36.0	45.6	48.8	65.0
iron, ppm	3	3	3	2
sodium, ppm	4	4	4	3
chloride, ppm	3	3	3	2
fluoride, ppm	1	1	1	1
sulfate, ppm	1	1	1	1

[a]All values are maximum unless otherwise noted.
[b]Values are minimum to maximum.
[c]By difference.

Handling and Storage

Hydrazine is a base, a reducing agent, and a high energy compound; it is also volatile and toxic. These properties determine its proper handling, storage, use, and disposal. Inadvertent contact with acids or oxidizing agents must be avoided because extremely exothermic reactions and evolution of gases may result. Flash points, however, of aqueous and anhydrous hydrazine are relatively high compared to common combustibles such as alcohol or gasoline. It is much easier and safer to handle aqueous solutions than the anhydrous product; aqueous solutions containing less than 40% hydrazine have no flash point, ie, they cannot be ignited. Hydrazine is not sensitive to shock or friction. In the absence of decomposition catalysts, liquid anhydrous hydrazine has been heated to over 200°C without appreciable decomposition. Hydrazine fires are effectively combated using water because hydrazine is miscible with water in all proportions.

The broad explosive range of hydrazine vapor is a concern. An inert gas blanket helps avoid formation of explosive mixtures. Even for dilute aqueous solutions, a nitrogen pad prevents gradual loss of hydrazine by air oxidation. Because acid solutions of hydrazine are not readily oxidized, air oxidation can be minimized or prevented by acidification. This also serves to reduce the vapor pressure. Hydrazine may ignite wood, rags, paper, or other common organic materials. Thus these should not be used near hydrazine. Use protective clothing to avoid body contact and provide adequate ventilation to reduce inhalation danger.

Materials of Construction.   In choosing the proper materials of construction for storing and using hydrazine, it is necessary to consider both the effects of the material on the stability and quality of the hydrazine as well as the effect of the hydrazine on the material of construction. Hydrazine is thermally stable, storable for years without adverse effects either to the product or the storage container provided the recommended materials are used, all systems are clean, and an inert gas, ie, nitrogen, is maintained over the system at all times. Table 10 is a brief listing of materials compatibility (125).

As shown, materials generally considered satisfactory for all N_2H_4 concentrations, including anhydrous, are 304L and 347 stainless steels having less than 1.0 wt % molybdenum, a catalyst for the decomposition of hydrazine. For concentrations less than 10%, cold-rolled steel is satisfactory. Among the nonmetallic materials, poly(tetrafluoroethylene), polyethylene, and polypropylene are suitable; PVC is not recommended. Ethylene–propylene–diene monomer (EPDM) rubber (126) and polyketones and polyphenylene sulfides (127) are reportedly suitable for use with anhydrous hydrazine. Many factors are involved in materials compatibility; a final choice may require some testing. For example, low (< 250 ppm) concentrations of CO_2 in anhydrous hydrazine accelerate the decomposition of hydrazine in stainless steel (128). Long-term storage tests of hydrazine propellants in 17–7 PH stainless steel [12742-98-2] and AM350

Table 10. Materials Compatibility for Aqueous Hydrazine Solutions[a]

Material	Hydrazine concentration, wt % N_2H_4			
	<10	35	54.4	64
stainless steel				
304L	S	S	S	S
347	S	S	S	S
316[b]	S	S	S	S
cold-rolled steel	S	NR	NR	NR
copper	NS	NS	NS	NS
brass	NS	NS	NS	NS
aluminum	NS	NS	NS	NS
PTFE[c]	S	S	S	S
polyethylene	S	S	S	S
polypropylene	S	S	S	S

[a]S = generally satisfactory; NR = not recommended; and NS = not suitable owing either to decomposition or to adverse effects of the solution on the material of construction.
[b]Only up to 65°C.
[c]PTFE = poly(tetrafluoroethylene).

precipitation-hardened stainless steel [12731-98-5] at 50°C for about three years showed no pressure rise or N_2H_4 decomposition (129). Extensive information and bibliographies are available on suitable materials of construction for AH, MMH, and UDMH (7,130).

Disposal. Spills and wastewater containing hydrazines must be contained and treated. Proper disposal methods for the hydrazines make use of their reductive properties. Fuel-grade hydrazines may be burned, but aqueous solutions less than 50% may require supplementary fuel (131). Chemical destruction of dilute (preferably 5% or less) solutions can be achieved with various oxidants such as NaOCl, Ca(OCl)$_2$, H_2O_2, and acidified permanganate; however MMH and UDMH, may form mutagenic nitrosamines (132). A method is described for treating contaminated wastewater in which N_2H_4, MMH, UDMH, and N-nitrosodimethylamine [62-75-9] are effectively decomposed at a controlled pH of ~5 by a uv-induced chlorination (133) (see also PHOTOCHEMICAL TECHNOLOGY, PHOTOCATALYSIS). Ozonation of these three hydrazines yields a variety of products, including methanol, formaldehyde monomethylhydazone [36214-48-9], and formaldehyde dimethylhydrazone [2035-89-4]; and tetramethyl tetrazene [6130-87-6] from the oxidative coupling of two molecules of UDMH (134). Methyldiazene [26981-93-1] was also found as an oxidation product of MMH (135). The rates of decomposition from aqueous solutions are greatest at pH 9.1 in the presence of uv light (136). Blowdown from boilers treated with hydrazine for corrosion control is effectively treated by neutralization with lime and chlorination (137). A vapor suppressant foam system (ASE95 polyacrylic/MSAR combination) has been evaluated for covering hydrazine fuel spills to minimize the release of toxic fumes (138). An excellent review covering disposal methods for hydrazine, MMH, and UDMH is available (139). Although laboratory methods of decontamination are emphasized, these could easily be adapted for large-scale disposal.

Analytical Methods

As a strong reducing agent, hydrazine can be determined quantitatively using a variety of oxidants. Most convenient, perhaps, is an iodometric titration (140):

$$N_2H_4 + 2 I_2 \longrightarrow N_2 + 4 HI \qquad (36)$$

The pH must be kept at 7.0–7.2 for this method to be quantitative and to give a stable end point. This condition is easily met by addition of solid sodium bicarbonate to neutralize the HI formed. With starch as indicator and an appropriate standardized iodine solution, this method is applicable to both concentrated and dilute (to ca 50 ppm) hydrazine solutions. The iodine solution is best standardized using monohydrazinium sulfate or sodium thiosulfate. Using an iodide-selective electrode, low levels down to the ppb range are detectable (see ELECTROANALYTICAL TECHNIQUES) (141,142). Potassium iodate (143,144), bromate (145), and permanganate (146) have also been employed as oxidants.

Dilute hydrazine solutions may be determined colorimetrically by the formation of highly colored azines with salicylaldehyde or, preferably, p-dimethylaminobenzaldehyde. This is the basis for a qualitative spot test for hydrazine and some of its derivatives (147). Ammonia and hydroxylamine do not interfere. Using *p*-dimethylaminobenzaldehyde, the method is suitable for the quantitative determination of ppm levels or less of hydrazine (ASTM D1385-88). It is used for the analysis of hydrazine residuals in boiler water or wastewater, and in air by absorption of hydrazine from a known volume of air in an acidified solution of the aldehyde reagent. Reagents, color chips, color comparators, and spectrophotometers are available for these applications, eg, from Hach Chemical Co. (Ames, Iowa); LaMotte Co. (Chestertown, Maryland); CHEMetrics, Inc. (Calverton, Virginia); and others. Using a spectrophotometer set at 458 nm, it is possible to determine levels as low as about 6 ppb. A high speed continuous-flow monitor based on this method has been developed (148). An amperometric sensor is the basis of a continuous hydrazine-in-water analyzer offered by Ametek (Pittsburgh, Pennsylvania), especially for use in boiler water. The detection of hydrazine has been described down to 20 ppm in aqueous and alcoholic solutions via thin-layer chromatography (qv) of the *p*-dimethylaminobenzalazine [41463-93-8] (149). The basic ASTM D1385-88 procedure is adaptable to the determination of MMH and UDMH as these also form colored hydrazones. For mixtures of the three hydrazines mentioned, the sample can be derivatized using salicylaldehyde and separated by high pressure liquid chromatography, using uv for detection; for hydrazine, the uv detector should be set at 209 nm (150). Using an electrochemical detector, it is not necessary to derivatize first (151). Polarographic procedures sensitive to levels of 0.25–1.0 ppm (152) and 0–20 ppb (153) have been described. Gas chromatographic techniques for the propellant hydrazines, ie, hydrazine, MMH, and UDMH, have been developed for high concentrations as well as for dilute aqueous solutions (154).

Air monitoring of the workplace is important because of the inherent inhalation toxicity of hydrazine and its volatile derivatives. Hydrazine levels can be determined by passing a known volume of air through dilute sulfuric acid to absorb the hydrazine and measuring the concentration by the colorimetric ASTM D1385-88 method. An alternative is to trap the hydrazines on H_2SO_4-coated silica gel, eluting with water and derivatizing with 2-furaldehyde; the derivatives can then be measured by gas chromatography (155). A continuous air monitor is commercially available from MDA Scientific Inc. (Park Ridge, Illinois). It is based on the reduction by hydrazine of a phosphomolybdate-impregnated tape; the resulting blue spot is read photometrically. Models having a 0–0.5 ppm full scale are available. Methyl- and dimethylhydrazines can also be determined with this instrument, but not quantitatively in the presence of each other. An electrochemical gas sensor has been patented (156) and an electrochemical monitor has been developed (157). The feasibility of laser remote sensing of hydrazine, MMH, and UDMH in the gas phase has been studied (158).

Toxicology

The acute and chronic toxicity of hydrazine, the methylhydrazines, phenylhydrazine, and other hydrazine derivatives has received extensive study, and is comprehensively covered in several reviews (159–163).

Hydrazine is toxic and readily absorbed by oral, dermal, or inhalation routes of exposure. Contact with hydrazine irritates the skin, eyes, and respiratory tract. Liquid splashed into the eyes may cause permanent damage to the cornea. At high doses it can cause convulsions, but even low doses may result in central nervous system depression. Death from acute exposure results from convulsions, respiratory arrest, and cardiovascular collapse. Repeated exposure may affect the lungs, liver, and kidneys. Of the hydrazine derivatives studied, 1,1-dimethylhydrazine (UDMH) appears to be the least hepatotoxic; monomethylhydrazine (MMH) seems to be more toxic to the kidneys. Evidence is limited as to the effect of hydrazine on reproduction and/or development; however, animal studies demonstrate that only doses that produce toxicity in pregnant rats result in embryotoxicity (164).

The TLV is set at 0.1 ppm (hydrazine); 0.2 ppm (MMH); and 0.5 ppm (UDMH). The TLV is well below the olfactory limit of 3–5 ppm (hydrazine). The latter does not provide adequate warning when exposure exceeds the TLV; therefore, monitoring the working environment by suitable means and providing adequate ventilation is necessary.

Table 11 summarizes values for the median lethal dose (LD_{50}) for several species. In case of massive exposure, convulsions must be controlled, and glucose, fluid balance, and urinary output must be maintained. Medical surveillance requires checking for damage to the liver, the organ that apparently sustains initial damage, and monitoring for changes in the blood profile.

The International Agency for Research on Cancer (IARC) classifies hydrazine as a 2B or possible human carcinogen. The American Conference of Governmental Industrial Hygienists (ACGIH) classifies hydrazine as an A2 or suspect human carcinogen. Several animal studies have demonstrated that chronic oral and inhalation exposure to hydrazine can produce tumors. Nasal tumors were observed in long-term inhalation studies in rats, mice, hamsters, and dogs exposed to hydrazine vapors (165). These tumors were often associated with repeated tissue insult (chronic irritation). Long-term exposure of mice to hydrazine via oral administration resulted in the observation of tumors in multiple tissues, including lung, liver, and lymphatic system (166). In studies where hydrazine was administered to rats in their drinking water until the spontaneous death of all the rats (167), only under the toxicity of the highest concentrations

Table 11. Acute Toxicity of Hydrazine, MMH, and UDMH

Species	Administration route	LD_{50}, mg/kg		
		Hydrazine	MMH	UDMH
mouse	intraperitoneal	62	15	113
	intravenous	57	33	250
	oral	59	29	265
rat	intraperitoneal	59	21	102
	intravenous	55	17	119
	oral	60	32	122
rabbit	percutaneous	91	95	1060
	intravenous	20	12	70
dog	intravenous	25	12	60
monkey	intraperitoneal	20		60–100

could a weak carcinogenic effect of the hydrazine be detected. In a similar study, mice were exposed for their lifetime and no tumors were reported (168).

Hydrazine has not been shown to have carcinogenic potential in humans. An epidemiological study of 427 men occupationally exposed to hydrazine for varying times showed no obvious hazards associated with hydrazine exposure (169). In another, ongoing epidemiological study of a smaller group of occupationally exposed workers, no statistical differences in the incidence of cancers have been observed when compared to the control population (170). These results must be viewed with caution owing to the small number of individuals in both studies; also, the second study is yet in progress as of this writing.

Assays have shown that hydrazine can interact with DNA to elicit mutagenic effects. The mutagenicity of hydrazine has been reviewed (171). Among toxic pollutants that may enter the environment, hydrazine is one of the less persistent because it reacts with oxygen and ozone, particularly in the presence of catalytic surfaces such as metals, oxides, etc. The final products of these reactions are innocuous: nitrogen and water.

Uses

The principal applications of hydrazine solutions include chemical blowing agents, 40%; agricultural pesticides, 25%; and water treatment, 20%. The remaining 15% finds use in a variety of fields including pharmaceuticals, explosives, polymers and polymer additives, antioxidants, metal reductants, hydrogenation of organic groups, photography, xerography, and dyes.

Chemical Blowing Agents. Chemical blowing agents are gas-generating compositions used to produce flexible or rigid cellular polymeric or elastomeric foams by a controlled release of gases, generally by thermal decomposition or by reaction with a component of the polymer formulation. To accommodate a wide range of plastics, a series of products is needed, varying in decomposition temperatures, polymer compatibility, etc. The most important commercial products are hydrazine derivatives, which release nitrogen as the principal gaseous product. Some of the more important are listed in Table 12.

By far the largest (ca 85% of the total) volume chemical blowing agent is azodicarbonamide (**44**), made by the oxidation of hydrazodicarboxamide [*110-21-4*] (**51**) using chlorine or sodium chlorate. The hydrazo precursor is made by refluxing an aqueous solution of urea and hydrazine (172):

$$2\ \text{H}_2\text{N}\overset{\text{O}}{\overset{\|}{\text{C}}}\text{NH}_2 + \text{N}_2\text{H}_4 \xrightarrow{-2\ \text{NH}_3} \text{H}_2\text{N}\overset{\text{O}}{\overset{\|}{\text{C}}}\text{NHNH}\overset{\text{O}}{\overset{\|}{\text{C}}}\text{NH}_2 \xrightarrow{\text{Cl}_2} \text{H}_2\text{N}\overset{\text{O}}{\overset{\|}{\text{C}}}\text{N}{=}\text{N}\overset{\text{O}}{\overset{\|}{\text{C}}}\text{NH}_2 \qquad (37)$$

$$\text{(51)} \qquad\qquad\qquad \text{(44)}$$

All the sulfonic acid hydrazides are made from hydrazine or semicarbazide and the appropriate sulfonyl chloride in the presence of an HCl acceptor such as ammonia.

p-Toluenesulfonylhydrazide (**46**) has a higher decomposition temperature than the benzene derivative (**45**). Decomposing still higher is (**47**), which is also

Table 12. Common Hydrazine-Based Blowing Agents

Compound	CAS Registry Number	Molecular formula	Structure	Structure number
azodicarbonamide	[123-77-3]	$C_2H_4N_4O_2$	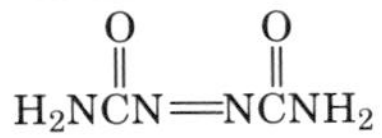 $H_2NCN{=}NCNH_2$ (with two C=O groups)	(44)
benzenesulfonylhydrazide	[80-17-1]	$C_6H_8N_2O_2S$	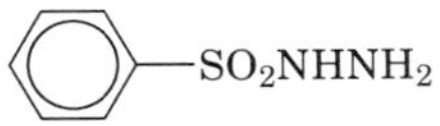 phenyl–SO_2NHNH_2	(45)
p-toluenesulfonylhydrazide	[1576-35-8]	$C_7H_{10}N_2O_2S$	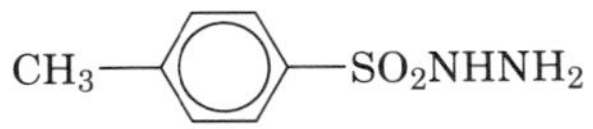 CH_3–phenyl–SO_2NHNH_2	(46)
4,4'-oxybis(benzenesulfonylhydrazide)	[80-51-3]	$C_{12}H_{14}N_4O_5S_2$	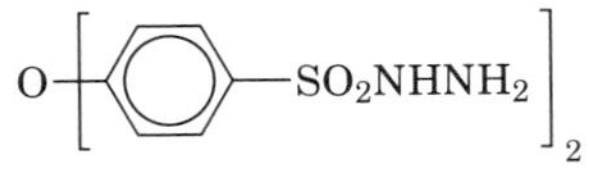 O[phenyl–SO_2NHNH_2]$_2$	(47)
p-toluenesulfonylsemicarbazide	[10396-10-8]	$C_8H_{11}N_3O_3S$	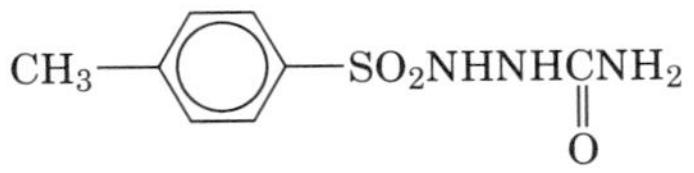 CH_3–phenyl–$SO_2NHNHCNH_2$ (C=O)	(48)
4,4'-oxybis(benzenesulfonylsemicarbazide)	[10195-67-2]	$C_{14}H_{16}N_6O_7S_2$	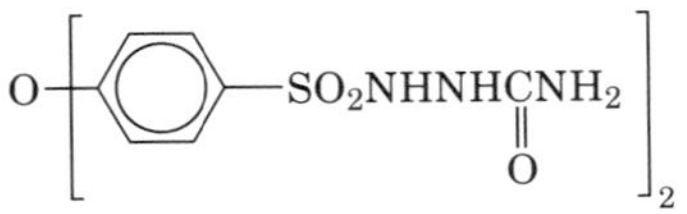 O[phenyl–$SO_2NHNHCNH_2$]$_2$ (C=O)	(49)
5-phenyl-3,6-dihydro-1,3,4-oxadiazin-2-one	[62501-39-7]	$C_9H_8N_2O_2$	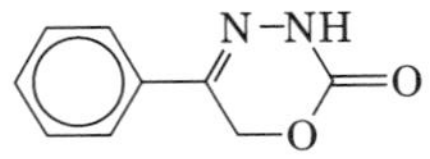 phenyl–N–NH ring with C=O	(50)

prepared from the corresponding bis(sulfonylchloride) (173). The sulfonylsemi-
carbazides (**48**) and (**49**) have relatively high gas yields. Decomposition temperatures are also high but if necessary can be lowered using activators, making these
compounds useful in the foaming of rubber (see RUBBER CHEMICALS). These are
made from the sulfonyl chlorides and semicarbazide or from the sulfonyl hydrazides and isocyanic acid (sodium cyanate/acid) (174).

5-Phenyl-3,6-dihydro-1,3,4-oxadiazin-2-one (**50**) is a high temperature blowing agent used primarily for polycarbonates (qv). It is prepared by the reaction
of α-hydroxyacetophenone and methyl carbazate (**52**), made from hydrazine and
dimethyl carbonate (175):

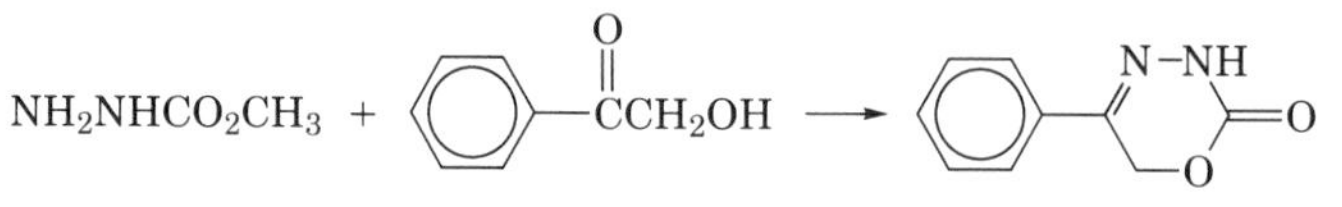

$$NH_2NHCO_2CH_3 \ + \qquad\qquad\qquad\qquad \longrightarrow \qquad\qquad\qquad\qquad \tag{38}$$

(52)   (50)

The primary domestic producers of these chemical blowing agents are Uniroyal
and Fairmount. European producers include Bayer (Germany), Atochem
(France), and FBC, Ltd. (U.K.). In Japan and the Pacific Rim the producers
include Dong Jin, Otsuka Chemical, Eiwa Chemical Ind., Kum Yang, Sankyo
Kasei, and Toyo Hydrazine. There are several excellent reviews covering the
technology of chemical blowing agents (176–178). A tabulation of producers,
trade names, and application data can be found in Reference 179.

Agricultural Uses.   Pesticides represent the second largest commercial
market for hydrazine. Hundreds of hydrazine derivatives have been patented
for a wide range of agricultural applications. Table 13 presents a sampling of
the 50–60 that are commercially available or developmental products. These
compounds are made from hydrazine, MMH, and UDMH and are for the most
part heterocyclic nitrogen compounds (see INSECT CONTROL TECHNOLOGY).

The earliest hydrazine derivative (1948) having extensive agricultural application was maleic hydrazide (**53**), a plant growth regulator that retards the
sprouting of onions and potatoes in storage and prevents sucker growth on tobacco plants. It is prepared from hydrazine and either maleic acid or maleic
anhydride (180,181). Activity in this area seems to have centered on triazole-
based fungicides such as typified by diclobutrazol (**54**) and flusilazol (**55**).

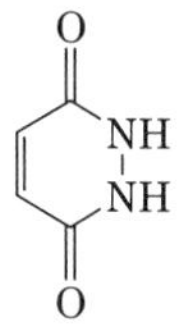

(53) 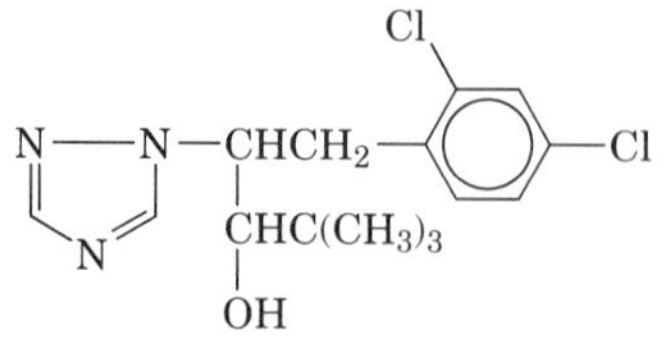   (54) 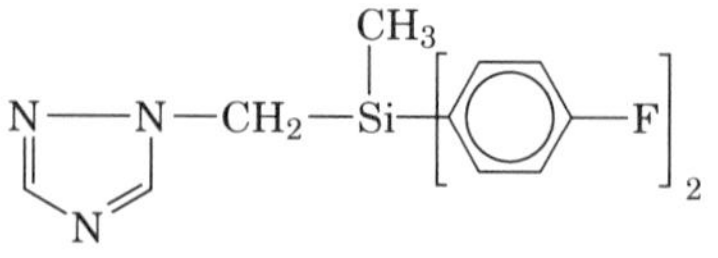   (55)

Table 13. Hydrazine-Based Pesticides

Generic name	CAS Registry Number	Compound class	Molecular formula	Trade name	Producer	Use[a]
amitrole	[61-82-5]	triazole	$C_2H_4N_4$	Ustilan	Bayer	H
bitertanol	[55179-31-2]	triazole	$C_{20}H_{23}N_3O_2$	Baycor	Miles	F
clofentezine	[74115-24-5]	tetrazine	$C_{14}H_8Cl_2N_4$	Apollo	Schering AG	I
cyproconazole	[94361-06-5]	triazole	$C_{15}H_{18}ClN_3O$	Alto	Sandoz	F
daminozide	[1596-84-5]	UDMH-hydrazide	$C_6H_{12}N_2O_3$	Dazide, Alar	Fine Agrochemical, Uniroyal	PGR
diclobutrazol	[75736-33-3]	triazole	$C_{15}H_{19}Cl_2N_3O$	Vigil	ICI	F
difenzoquat	[49866-87-7]	pyrazole	$C_{17}H_{17}N_2$	Avenge	American Cyanamid	H
flutriafol	[76674-21-0]	triazole	$C_{16}H_{13}F_2N_3O$	Impact	ICI	F
flusilazol	[85509-19-9]	triazole	$C_{16}H_{15}F_2N_3Si$	Punch	Du Pont	F
hexaconazole	[79983-71-4]	triazole	$C_{14}H_{17}Cl_2N_3O$	Anvil	ICI Americas	F
hydramethylnon	[67485-29-4]	azine	$C_{25}H_{24}F_6N_4$	Amdro	American Cyanamid	I
maleic hydrazide	[123-33-1]	dihydroxypyridazine	$C_4H_4N_2O_2$	Drexel MH, Royal MH-30	Drexel Chemical, Uniroyal	PGR
metamitron	[41394-05-2]	triazine	$C_{10}H_{10}N_4O$	Goltix	Bayer AG	H
metribuzin	[21087-64-9]	triazine	$C_8H_{14}N_4OS$	Sencor	Bayer AG	H
myclobutanil	[88671-89-0]	triazole	$C_{15}H_{17}ClN_4$	Rally	Rohm & Haas	F
paclobutrazol	[76738-62-0]	triazole	$C_{15}H_{20}ClN_3O$	Clipper	ICI Americas	PGR
propiconazole	[60207-90-1]	triazole	$C_{15}H_{17}Cl_2N_3O_2$	Tilt	Ciba-Geigy	F
pyridate	[55512-33-9]	pyridazine	$C_{14}H_{14}AsNO_2$	Tough	Agrolinz	H
tebuthiuron	[34014-18-1]	thiadiazole	$C_9H_{16}N_4OS$	Spike	Eli Lilly	H
triadimefon	[43121-43-3]	triazole	$C_{14}H_{16}ClN_3O_2$	Bayleton	Bayer	F
triadimenol	[55219-65-3]	triazole	$C_{14}H_{18}ClN_3O_2$	Baytan	Miles	F
tricyclazole	[41814-78-2]	triazole	$C_9H_7N_3S$	Beam	Eli Lilly	F
uniconazole	[76714-83-5]	triazole	$C_{15}H_{18}ClN_3O$	Prunit	Chevron	PGR

[a]F = fungicide; H = herbicide; I = insecticide; PGR = plant growth regulator.

The triazole [288-88-0] (**56**) required for these products can be made from for-mamide and hydrazine (182,183):

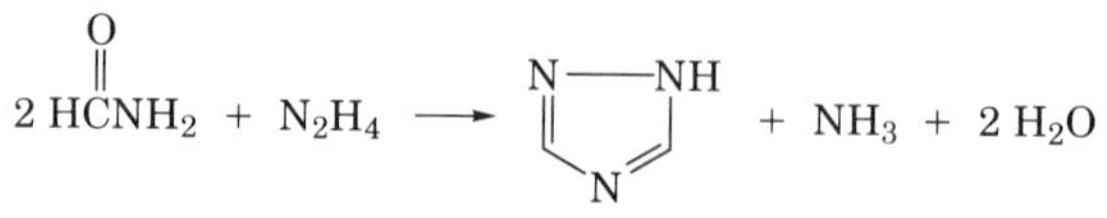

$$2 \ HCNH_2 + N_2H_4 \longrightarrow \underset{(56)}{\text{[structure]}} + NH_3 + 2 \ H_2O \tag{39}$$

This class of fungicides has been reviewed (184). Standard compilations of pesticides provide information on names, structures, and uses of these compounds (185,186).

Water Treatment.   Water treatment for corrosion protection represents the third largest market for hydrazine. It is used in boilers and hot-water heating systems to scavenge dissolved oxygen, a primary cause of corrosion. The application of hydrazine in boilers also includes wet lay-up for protection during idle periods (187). Hydrazine has been patented as a corrosion inhibitor for closed circulation cooling systems of internal combustion engines, especially for diesel engines in ships (188). The reaction removes oxygen down to essentially nondetectable levels (Indigo–Carmine Analytical Method, ASTM D888-87). The products of this reaction, nitrogen and water, are innocuous. Minor amounts of ammonia may form by thermal decomposition and this serves to raise the pH of the condensate and feedwater. For water treating, hydrazine is generally sold as a dilute aqueous solution, 35 wt % or less, because such solutions have no flash point and a relatively low partial pressure of hydrazine, minimizing the danger of inhalation by operating personnel.

Oxygen scavengers other than hydrazine are also used, especially catalyzed sodium sulfite, which reacts rapidly with oxygen even at room temperatures to form sodium sulfate. Catalyzed hydrazine formulations are now commercially available that react with oxygen at ambient temperatures at rates comparable to catalyzed sulfite (189). At elevated temperatures, the reaction rates are all similar. Table 14 lists the standard hydrazine solution products offered by Olin Corp. for sale to the water-treatment market. Other concentrations are available and other companies offer similar products.

Hydrazine has the advantage over sodium sulfite of not contributing dissolved solids to the boiler system, thereby reducing the need for purging. At the high temperatures encountered in utility boilers, hydrazine must be used because sulfite decomposes to corrosive sulfur dioxide and sodium sulfide (190,191).

Table 14. Composition of Scav-Ox Solutions

Parameter	Scav-Ox	Scav-Ox II	Scav-Ox Plus
N_2H_4 conc, wt %	35.0	35.0	35.0
catalyst			
type		organic	organometallic
wt %		0.2	0.8
color	colorless	light pink/orange	pink to red

Hydrazine also reacts with iron, probably in a sequence of steps involving oxidation of iron by water (192), to form an adherent magnetite coating on the metal surface that protects against further attack (eq. 40). In addition, hydrazine reduces red iron oxide, Fe_2O_3 (rust) to magnetite (eq. 41); loose magnetite is dense and settles to the bottom of the system for easy removal. This helps avoid deposits on the boiler heat-transfer surfaces that would otherwise impede heat transfer and cause boiler tubes to rupture. Boilers treated for longer periods of time with hydrazine do not develop much rust, thus requiring less clean out. A very low hydrazine residual concentration in the boiler system, generally less than 0.1 ppm, provides full corrosion protection.

$$3 \, Fe + 4 \, N_2H_4 + 4 \, H_2O \longrightarrow Fe_3O_4 + 8 \, NH_3 \tag{40}$$

$$N_2H_4 + 6 \, Fe_2O_3 \longrightarrow 4 \, Fe_3O_4 + 2 \, H_2O + N_2 \tag{41}$$

Concern for personnel exposure to hydrazine has led to several innovations in packaging to minimize direct contact with hydrazine, eg, Olin's E-Z drum systems. Carbohydrazide was introduced into this market for the same reason: it is a solid derivative of hydrazine, considered safer to handle because of its low vapor pressure. It hydrolyzes to release free hydrazine at elevated temperatures in the boiler. It is, however, fairly expensive and contributes to dissolved solids (carbonates) in the water (193). In field tests, catalyzed hydrazine outperformed both hydrazine and carbohydrazide when the feedwater oxygen and iron levels were critical (194). A published comparison is available (195) of these and other proposed oxygen scavengers, eg, diethylhydroxylamine, hydroquinone, methyethylketoxime, and isoascorbic acid.

Propellants and Explosives. Hydrazine fuels include anhydrous hydrazine (AH), monomethylhydrazine (MMH), and unsymmetrical dimethylhydrazine (UDMH) for military and space programs. These compounds are used mainly as bipropellant fuels, ie, with oxidizers, in rockets such as the Titan, MX missle, and the Ariane (UDMH/N_2O_4). Using oxygen or fluorine as the oxidizer, hydrazine is exceeded only by hydrogen in specific impulse, ie, kilograms of thrust developed for each kilogram of fuel consumed per second (196).

Hydrazine is also used as a monopropellant, no oxidant, by decomposing it over a catalyst to form propulsive gases (197) used for auxiliary power units and in thruster engines for attitude and in-orbit control of satellites and spacecraft. A significant advance was the development of catalysts that function at low temperatures without prior heating, such as the Shell 405 iridium on alumina (198). Iridium on alumina appears to be the preferred catalyst (199,200). The kinetics of this system have been studied (201). Various mixtures of iridium, ruthenium, cobalt, and molybdenum on Al_2O_3 as carrier are also reported as effective catalysts (202). Noncatalytic monopropellant thrusters having integral electric heaters have also been developed (203).

Aerozine-50, an equal mixture of AH and UDMH, was used to power the Titan II launch vehicle for the Gemini spacecraft. The Delta vehicles likewise use Aerozine-50/N_2O_4 in the third stage. UDMH and fuming nitric acid were used in the second stage of the Vanguard. The U.S. Space Shuttle is maneuvered in orbit with MMH/N_2O_4. Space vehicles propelled by hydrazine include the

Viking 1, ~er on Mars, the Pioneer and Voyager interplanetary probes, and the Giotto space, robe to Halley's comet.

Solid hydrazine derivatives such as the nitrate, monoperchlorate, and diperchlorate salts as well as aluminum perchlorate trihydrazinate [85962-45-4], $Al(ClO_4)_3 \cdot 3N_2H_4$, have also been studied as fuels (204). A review of solid hydrazines as hybrid rocket fuels has been published (205). An excellent reference source for declassified documents covering rocket, missile, space, and gun technology is available from the Chemical Propulsion Information Agency (CPIA) (206). The most complete bibliography of references in the field of hydrazine propellants is Reference 9.

Numerous explosives are based on hydrazine and its derivatives, including the simple azide, nitrate, perchlorate, and diperchlorate salts. These are sometimes dissolved in anhydrous hydrazine for propellant applications or in mixtures with other explosives (207). Hydrazine transition-metal complexes of nitrates, azides, and perchlorates are primary explosives (208).

Automobile safety air bags use sodium azide [26628-22-8], NaN_3, for gas generation. It can be made from hydrazine by refluxing ethyl or t-butyl nitrite with hydrazine hydrate and sodium hydroxide in alcohol (209,210):

$$N_2H_4 + t\text{-}C_4H_9ONO + NaOH \longrightarrow NaN_3 + t\text{-}C_4H_9OH + 2\,H_2O \qquad (42)$$

An alternative for air bags is 5-aminotetrazole [4418-61-5] (**57**), prepared by nitrosation of aminoguanidine (211). Other compounds under study as gas generators for air bags and fire extinguishers are the guanidinium (212) and triaminoguanidinium (213) salts of 5,5′-azobistetrazole [28623-02-1]. Tetrazene [31330-63-9] (**58**), an explosive primer (214), is made from aminoguanidine. Aminoguanidine, as well as di- and triaminoguanidine are prepared by the reaction of hydrazine with guanidine or cyanamide (215,216). Triaminoguanidine nitrate is a useful ingredient in gun propellants (217).

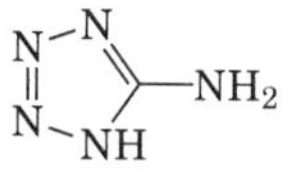

(57) 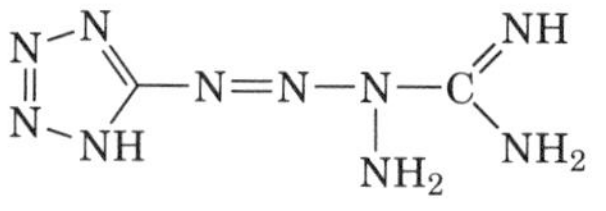   (58) 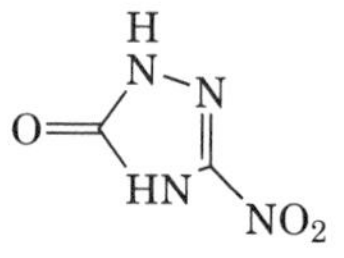   (59)

There is a drive to develop insensitive or less sensitive munitions, ie, those less likely to accidental or sympathetic detonation. A leading candidate is 3-nitro-1,2,4-triazolin-5-one [930-33-6] (**59**), made by the reaction of semicarbazide and formic acid to give 1,2,4-triazolin-5-one [932-64-9] followed by nitration of the triazolone (218).

Pharmaceuticals. The tuberculostatic drug isoniazid [54-85-3] (**60**), the hydrazide of 4-pyridinecarboxylic acid, was one of the earliest hydrazine-based drugs introduced in the United States. It is made by the reaction of hydrazine and an isonicotinic acid ester (219).

$$N\langle\bigcirc\rangle\text{—COOC}_2\text{H}_5 + \text{N}_2\text{H}_4 \longrightarrow N\langle\bigcirc\rangle\text{—CONHNH}_2 + \text{C}_2\text{H}_5\text{OH} \qquad (43)$$

(60)

Table 15 gives a sampling of other pharmaceuticals derived from hydrazine. Cefazolin, a thiadiazole tetrazole derivative, is one of the most widely used antibacterial drugs in U.S. hospitals (see ANTIBIOTICS, β-LACTAMS). Procarbazine, an antineoplastic, is a monomethylhydrazine derivative (220). Fluconazole has shown some promise in the treatment of AIDS-related fungal infections. Carbidopa is employed in the treatment of Parkinson's disease. Furazolidone is a verterinarian antibacterial.

Hydrazine sulfate [10034-93-2], $\text{N}_2\text{H}_4\cdot\text{H}_2\text{SO}_4$, originally advanced by the Syracuse Cancer Research Institute for treatment of cancerous cachexia and tumor inhibition (221), now has Investigational New Drug (IND) status in the United States. Clinical evaluations are underway at various institutions such as Harbor-UCLA Medical Center (222) and the Mayo Clinic. After extensive trials, hydrazine sulfate has been approved as an anticancer drug in Russia (223). Chemical structures for established drugs in the United States may be found in Reference 224.

Polymers. Polymers based on hydrazine include polyureylenes, polyacylsemicarbazides, polyhydrazides, polyoxadiazoles, and polyaminotriazoles. Aromatic polyhydrazides show excellent dyeability. As films, these materials exhibit a high degree of semipermeability to water, suggesting application as reverse osmosis (qv) membranes for the desalinization of seawater (225,226). Polyoxadiazoles fibers have been made from hydrazine–terephthalic acid copolymer [27027-96-9] and hydrazine–isophthalic acid–terephthalic acid copolymer [27924-75-0]. Dihydrazides have been patented as epoxy curing agents and chain extenders for polyurea polymers (227). Diacyl hydrazines are stabilizers for polyesters against thermal degradation (228) and similar compounds act as uv absorbers and heat stabilizers (qv) for plastics (229).

Other Uses. Fuel cells (qv) based on the reaction of hydrazine and oxygen or hydrogen peroxide have been studied for several decades (231). Reactions in this cell involve the anodic oxidation of hydrazine and cathodic reduction of the oxidizing agent in a potassium hydroxide electrolyte, generally on proprietary electrodes. The oxygen-based cell has a theoretical standard potential of 1.56 V. These cells operate quietly, at moderate (40–55°C) temperatures, and contain fairly dilute aqueous solutions of hydrazine. The products of reaction, nitrogen, and water, are innocuous. All of these factors have made such cells interesting for military applications but civilian use has been hampered by the high cost of hydrazine.

Many hydrazine derivatives find application in photography and other reprographic techniques. Heterocyclics such as triazoles, tetrazoles, tetrazolium salts, and 2,5-dimercapto-1,3,4-thiadiazole are used as stabilizers, fog-inhibitors, and spectral sensitizers. Pyrazoles and pyrazolones are coupling agents. Metal complexes based on formazans are used for textile dyeing, color photography (qv), and inks (qv) (231). Judging from the patent literature, a great deal of

Table 15. Pharmaceutical Chemicals Based on Hydrazine

Generic name	CAS Registry Number	Chemical type	Trade name	Producers	Use
carbidopa	[28860-95-9]	alkylhydrazine	Lodosyn	Merck	decarboxylase inhibitor
cefazolin	[25953-19-9]	tetrazole	Ancef, Kefzol	Smith Kline, Eli Lilly	antibacterial
fluconazole	[86386-73-4]	triazole	Diflucan	Pfizer	meningitis, antifungal
furazolidone	[67-45-8]	hydrazone		Teva (Israel)	antibacterial (vet)
hydralazine	[304-20-1]	phthalazine	Apresoline	Ciba-Geigy	antihypertensive
isoniazid	[54-85-3]	hydrazide	Cotinazin, Nydrazid	Pfizer, Squibb	tuberculostat
procarbazine	[366-70-1]	alkylhydrazine	Matulane	Hoffmann-LaRoche	antineoplastic
terconazole	[67915-31-5]	triazole		Janssen	antifungal
triazolam	[28911-01-5]	triazole	Halcion	Upjohn	sedative

work is being done with hydrazine derivatives in these areas. The vast majority of this effort is in Japan.

Many hydrazones and azines are colored and useful as dyestuffs. Examples are 2-hydroxynaphthazine, a yellow fluorescent dye (Lumogen LT Bright Yellow), and the pyridon–azino–quinone class of red-violet dyes. Numerous hydrazine derivatives are antioxidants and stabilizers by virtue of their reducing and chelating powers.

BIBLIOGRAPHY

"Hydrazine" in *ECT* 1st ed., Vol. 7, pp. 570–591, by L. F. Audrieth, P. H. Mohr, and E. B. Mohr, University of Illinois, and G. A. Michalek, Olin Mathieson Chemical Corp.; "Hydrazine and its Derivatives" in *ECT* 2nd ed., Vol. 11, pp. 164–196, by L. A. Raphealian, Olin Mathieson Chemical Corp.; in *ECT* 3rd ed., Vol. 12, pp. 734–771, by H. W. Schiessl, Olin Corp.

1. D. Altman and B. Adelman, *J. Am. Chem. Soc.* **74**, 3742–3744 (1952).
2. E. Schmidt, *Hydrazine and its Derivatives*, John Wiley & Sons, Inc., New York, 1984, pp. 141, 149, and 150.
3. L. Pederson and K. Morokuma, *J. Chem. Phys.* **46**, 3941 (1967).
4. K. Jones, in J. C. Bailer, ed., *Comprehensive Inorganic Chemistry*, Vol. 2, Pergamon Press, Ltd., Oxford, U.K., 1973, p. 255.
5. Y. Morino and co-workers, *Bull. Chem. Soc. Jpn.* **33**, 46 (1960).
6. D. R. Stull, ed., *JANAF Thermochemical Tables*, Dow Chemical Co., Midland, Mich., 1965.
7. *Liquid Propellants Manual*, contract NOw 62-0604-c, Chemical Propulsion Information Agency, Johns Hopkins University, Baltimore, Md., 1961, Unit 2 (AH), Unit 5 (UDMH) and Unit 11 (MMH); *Hazards of Chemical Rockets and Propellants Handbook*, Vol. 3, CPIA/194, publ. no. AD 870259, National Technical Information Service, U.S. Dept. of Commerce, Washington, D.C., May 1972, Unit 9 (AH and MMH), Unit 10 (UDMH).
8. G. H. Hudson, R. C. H. Spencer and J. T. P. Stern, in *Mellor's Comprehensive Treatise on Inorganic and Theoretical Chemistry*, Vol. 8, Suppl. 2, John Wiley & Sons, Inc., New York, 1967, pp. 69–113.
9. Ref. 2, exhaustive bibliography.
10. F. E. Scott, J. J. Burns, and B. Lewis, *Explosive Properties of Hydrazine, Report of Investigations 4460*, U.S. Dept. of the Interior, Bureau of Mines, Pittsburgh, Pa., May 1949.
11. W. M. Latimer, *The Oxidation States of the Elements and Their Potentials in Aqueous Solutions*, 2nd ed., Prentice Hall, Inc., Englewood Cliffs, N.J., 1952, p. 99.
12. L. F. Audrieth and B. A. Ogg, *The Chemistry of Hydrazine*, John Wiley & Sons, Inc., New York, 1951, pp. 86–94.
13. M. L. Davis, V. V. Vesselovsky, and H. L. Johnston, *Kinetics, Thermodynamics, Physico-Chemical Properties and Manufacture of Hydrazine*, Ohio State University, Wright-Patterson Air Force Base, Dayton, Ohio, Mar. 15, 1952.
14. L. De Bruyn, *Recl. Trav. Chim. Pays Bas* **14**, 87 (1895).
15. K. H. Linke and R. Taubert, *Z. Anorg. u. Allg. Chem.* **376**, 289 (1970).
16. H. Bock, *Z. Naturforschg.* **17b**, 429 (1962).
17. Th. Kaufmann, *Angew. Chem. Int. Ed. Eng.* **3**, 342–353 (1964).
18. K. H. Linke and co-workers, *Z. Naturforschg.* **26b**, 296 (1971).
19. Ref. 16, p. 426.

20. D. D. Wagman and co-workers, *U.S. Technical Note 270-3*, Natl. Bur. Stand. Washington, D.C., 1968, pp. 63, 70, 74, 75, 80.
21. M. Goehring and H. Kaspar, *Z. Anorg. u. Allg. Chem.* **278**, 255–259 (1955).
22. V. M. Ovrutskii and L. D. Protsenko, *Usp. Khim.* **55**(4), 652–678 (1986).
23. F. C. Gunderloy, *Inorganic Syntheses*, Vol. 9, McGraw-Hill Book Co., Inc., New York, 1967, pp. 13–16.
24. U.S. Pat. 3,323,878 (June 6, 1967), F. C. Gunderloy (to Esso Research and Engineering).
25. U.S. Pat. 3,367,115 (Feb. 6, 1968), L. Spenadel and W. J. Sparks (to Esso Research and Engineering).
26. USSR Pat. 272,762 (1970), A. Prokopcikas (to Chemistry and Chemical TE).
27. Ref. 11, pp. 74, 99, 109, 148, 268.
28. S. Wein, *Glass Ind.*, 413 (Aug. 1955).
29. Fr. Demande FR 2,590,595 (May 29, 1987), P. Josso and co-workers (to Office National d'Etudes et de Recherches Aerospatiales).
30. Eur. Pat. Appl. EP 376,494 (July 4, 1990) I. S. Mahmoud (to IBM Corp.).
31. Jpn. Kokai 3,215,677 A2 (Sept. 20, 1991), K. Shiokawa, T. Kudo, and N. Asaoka (to N. E. Chemcat Corp.).
32. U.S. Pat. 3,597,267 (Aug. 3, 1971), G. O. Mallory and D. W. Baudrand (to Allied Research Products, Inc.).
33. K. Meguro and co-workers, *Colloids Surf.* **34**(4), 381–388 (1989).
34. Eur. Pat. Appl. EP 249,366 (Dec. 16, 1987), T. Hayashi, A. Ushijima, and Y. Nakamura (to Mitsui Mining and Smelting Co.).
35. H. Itoh and co-workers, *Bull. Chem. Soc. Jpn.* **64**, 333–335 (1991).
36. A. M. Fedoseev and A. B. Yusov, *Dokl. Akad. Nauk SSSR* **317**(4), 916–919 (1991).
37. U.S. Pat. 3,537,896 (Nov. 3, 1970), W. Nohse and G. Fischer (to Lancy Laboratories, Inc.).
38. A. Furst, R. C. Berlo, and S. Hooton, *Chem. Rev.* **65**, 51 (1965).
39. H. O. House, *Modern Synthetic Reactions*, 2nd ed., W. A. Benjamin, Inc., Menlo Park, Calif., 1972, pp. 228–256.
40. P. S. Wharton and D. H. Bohlen, *J. Org. Chem.* **26**, 3615 (1961).
41. D. Todd, in R. Adams, ed., *Organic Reactions*, Vol. 4, John Wiley & Sons, Inc., New York, 1948, Chapt. 8.
42. Huang-Minlon, *Sci. Sinica (Peking)* **10**, 711 (1961); *Chem. Abstr.* **56**, 11047h (1962).
43. D. J. Cram and co-workers, *J. Am. Chem. Soc.* **84**, 1734 (1962).
44. C. S. Rondestvedt and T. A. Johnson, *Synthesis*, 851–852 (1977).
45. S. Abdel-Baky, M. Zhuang, and R. W. Giese, *Synth. Comm.* **21**(2), 161–165 (1991).
46. N. R. Ayyangar and co-workers, *Bull. Chem. Soc. Jpn.* **56**, 3159–3164 (1983).
47. D. Balcon and A. Furst, *J. Am. Chem. Soc.* **75**, 4334 (1953).
48. S. Pietra and C. Trinchera, *Gazz. Chim. Ital.* **86**, 1045–1053 (1956).
49. F. Aylward and M. Sawistowska, *Chem. Ind.*, 484 (1962).
50. S. F. Hahn, *J. Polym. Sci. Part A: Polym. Chem.* **30**(3), 397–408 (1992).
51. D. K. Parker, R. F. Roberts, and H. W. Schiessl, *Rubber Chem. Technol.* **65**(1), 245–258 (1992).
52. D. F. Pasto, in L. A. Paquette, ed., *Organic Reactions*, Vol. 40, John Wiley & Sons, Inc., New York, 1991, pp. 91–155.
53. H. Babad, W. Herbert, and A. W. Stiles, *Tetrahedron Lett.* **25**, 2927 (1966).
54. W. R. Bamford and T. S. Stevens, *J. Chem. Soc.*, 4735 (1952).
55. J. Cassanova and B. Waegell, *Bull. Soc. Chim. Fr.* **3–4**, Pt. 2, 922 (1975).
56. A. N. Kost and co-workers, *Zh. Obsch. Khim.* **33**, 867 (1963).
57. Jpn. Kokai 60 237,059 (Nov. 25, 1985), M. Sasaki and co-workers (to Nippon Carbide Ind. Co.).

58. U.S. Pat. 3,154,538 (Oct. 27, 1964), D. A. Lima (to FMC Corp.).
59. Belg. Pat. 839,664 (July 16, 1976), R. A. Grimm, N. A. Randen, and R. J. Small (to Ashland Oil Co.).
60. U.S. Pat. 4,127,611 (Nov. 28, 1978), R. J. Small (to Ashland Oil, Inc.).
61. J. Biel and co-workers, *J. Am. Chem. Soc.* **80**, 1519 (1958).
62. J. Biel and co-workers, *J. Am. Chem. Soc.* **82**, 2204 (1960).
63. Eur. Pat. Appl. EP 482,347 A2 (Apr. 29, 1992), K. Huthmacher and H. Schmitt (to Degussa AG).
64. U.S. Pat. 4,954,655 (Sept. 4, 1990), M. J. Kelly (to Rohm and Haas Co.).
65. U.S. Pat. 2,660,607 (Nov. 24, 1953), G. Gever and C. J. O'Keefe (to Eaton Laboratories, Inc.).
66. Ger. Pat. 1,108,233 (June 8, 1961), K. Eiter and E. Truscheit (to Farbenfabriken Bayer).
67. K. H. Mayer and S. Petersen, *Synthesis*, 370 (1971).
68. H. Dorn and K. Walter, *Z. Chem.* **7**, 151 (1967).
69. Ger. Pat. 1,287,589 (Jan. 23, 1969), W. Schindler (to VEB Filmfabrik).
70. I. Zeid and I. Ismail, *Chem. Ind.*, 380 (1973).
71. S. I. Suminov and A. N. Kost, *Zh. Obshch. Khim.* **33**, 2208 (1963).
72. Jpn. Pat. 70 37,553 (Nov. 28, 1970), H. Saki and co-workers (to Toray Ind.).
73. Jpn. Pat. 74 02,178 (Jan. 18, 1974), H. Sakai and co-workers (to Toray Ind.).
74. Brit. Pat. 987,354 (Mar. 24, 1965), (Farbenfabriken Bayer, AG).
75. U.S. Pat. 4,317,741 (Mar. 2, 1982), H. F. Lederle and F. J. Milnes (to Olin Corp.).
76. W. J. McKillip and co-workers, *Chem. Rev.* **73**, 255–281 (1973).
77. W. J. McKillip, in K. C. Frisch, ed., *Advances in Urethane Science and Technology*, Vol. 3, Technomic Publishing Co., Inc., Westport, Conn., 1974, pp. 81–107.
78. B. M. Culbertson, in J. I. Kroschwitz, ed., *Encyclopedia of Polymer Science and Engineering*, Vol. 1, John Wiley & Sons, Inc., New York, 1985, pp. 740–752.
79. K. Kaczorowski, M. Koperska, and M. Kwiatkowski, *Chem. Techn.* **39**(4), 172–175 (1987).
80. E. V. Brown and co-workers, *Method. Chim.* **6**, 73 (1975).
81. E. Enders, in Houben-Weyl, *Methoden der Organischen Chemie*, Vol. X/2, 4th ed., Georg Thieme Verlag, Stuttgart, 1967, pp. 169–692, 750–756.
82. P. A. S. Smith, *Derivatives of Hydrazine and Other Hydronitrogens Having N-N Bonds*, The Benjamin/Cummings Publishing Co., Reading, Mass., 1983, pp. 80–141.
83. H. Paulsen and D. Stoye, in J. Zabicky, ed., *The Chemistry of Functional Groups*, John Wiley & Sons, Inc., New York, 1970, pp. 515–600.
84. E. Müller, in Ref. 81, pp. 121–168.
85. K. A. Jensen and co-workers, *Acta. Chem. Scand.* **15**, 1067, 1087, 1109 (1961); **22**, 1 (1968); **23**, 1916 (1969).
86. W. Walter and K. J. Reubke, in Ref. 83, pp. 477–514.
87. R. Cremlyn, *Int. J. Sulfur Chem.* **8**(1), 133 (1973).
88. U.S. Pat. 4,132,736 (Jan. 2, 1979), C. Cramm, E. Kranz, and G. Hellrung (to Bayer, AG).
89. Ref. 82, pp. 43–79.
90. A. N. Kost and I. I. Grandberg, *Usp. Khim.* **28**(8), 921–947 (1959).
91. E. Hafez and co-workers, *Heterocycles* **22**(8), 1821–1877 (1984).
92. F. Raschig, *Ber.* **40**, 4587 (1907); Ger. Pat. 198,307 (June 7, 1907), F. Raschig; U.S. Pat. 910,858 (Jan. 26, 1909), F. Raschig; F. Raschig, *Schwefel und Stickstoffstudien*, Verlag Chemie, Berlin-Leipzig, 1924.
93. S. R. M. Ellis and co-workers, *IEC Proc. Des. Dev.* **3**(1), 18 (1964).
94. G. V. Jeffreys and J. T. Wharton, *IEC Proc. Des. Dev.* **4**(1), 71 (1965).
95. J. W. Cahn and R. E. Powell, *J. Am. Chem. Soc.* **76**, 2565 (1954).

96. G. Yagil and M. Anbar, *J. Am. Chem. Soc.* **84**, 1797 (1962).

97. U.S. Pat. 2,863,728 (Dec. 9, 1958), H. H. Sisler and R. S. Drago (to Ohio State University).

98. L. V. Litvinova and co-workers, *J. Appl. Chem. USSR* **51**, 2491–2492 (1978).

99. R. A. Penneman and L. F. Audrieth, *J. Am. Chem. Soc.* **71**, 1644–1647 (1949).

100. U.S. Pat. 4,804,442 (July 31, 1989), J. G. Rigsby (to Olin Corp.); U.S. Pat. 5,035,775 (July 30, 1991), A. B. Quackenbush and J. G. Rigsby (to Olin Corp.); Jpn. Kokai 03 109,209 (May 9, 1991), T. Ayabe (to Ishikawajima-Harima Heavy Industries Co.).

101. E. Schmitz, *Organische Chemie*, Vol. 9, Springer-Verlag, New York, 1967, pp. 67, 112.

102. U.S. Pat. 3,077,383 (Feb. 12, 1963), R. Mundil (to Farbenfabriken Bayer).

103. U.S. Pat. 3,773,831 (Nov. 20, 1973), G. V. Jeffreys and C. E. Marks (to Fisons Ltd.).

104. U.S. Pat. 3,622,566 (Nov. 23, 1971), G. Huck and S. R. Paulsen (to Bergwerksverband GmbH).

105. Ger. Pat. 1,126,395 (Mar. 29, 1962), S. R. Paulsen (to Bergwerksverband GmbH).

106. U.S. Pat. 3,382,281 (May 7, 1968), A. Jankowski and co-workers (to Bergwerksverband GmbH).

107. U.S. Pat. 3,189,411 (June 15, 1965), H. Kohnen, R. Mundil, and J. Steinbrecht (to Farbenfabriken Bayer).

108. U.S. Pat. 4,724,133 (Feb. 9, 1988), J. P. Schirmann and co-workers (to Atochem).

109. U.S. Pat. 3,481,701 (Dec. 2, 1969), M. Otsuka and co-workers (to Otsuka Chemical Co.).

110. K. W. Eichenhofer and E. Roos, in *Ullmanns Encyklopädie der Technischen Chemie*, Vol. 13, 4th ed., Verlag Chemie, Weinheim, Germany, 1977, p. 98.

111. U.S. Pat. 3,494,737 (Feb. 10, 1970), R. Mundil (to Farbenfabriken Bayer).

112. U.S. Pat. 4,056,469 (Nov. 1, 1977), K. W. Eichenhofer and R. Schliebs (to Bayer, AG).

113. U.S. Pat. 4,657,751 (Apr. 14, 1987), M. Alicot and J. Pierre (to Atochem).

114. E. Colton, M. M. Jones, and L. F. Audrieth, *J. Am. Chem. Soc.* **76**, 2572 (1954).

115. U.S. Pat. 3,442,612 (May 6, 1969), L. K. Huber and L. R. Ocone (to Pennsalt Chemicals Corp.).

116. R. Schönbeck and E. Kloimstein, *Chem. Ing. Tech.* **46**, 391 (1974).

117. U.S. Pat. 3,972,878 (Aug. 3, 1976), J. P. Schirmann, J. Combroux, and S. Y. Delavarenne (to Produits Chimiques Ugine Kuhlmann).

118. J. P. Schirmann and S. Y. Delavarenne, *Tetrahedron Lett.* 635 (1972).

119. E. G. E. Hawkins, *J. Chem. Soc. C*, 2663 (1969).

120. E. Schmitz and co-workers, *Chem. Ber.* **97**, 2521 (1964).

121. Can. Pat. 2,017,358 (Nov. 24, 1990), J. P. Schirmann, J. P. Pleuvry, and P. Tellier (to Atochem).

122. U.S. Pat. 4,725,421 (Feb. 16, 1988), J. P. Schirmann and co-workers (to Atochem).

123. H. Hayashi, *Catal. Rev. Sci. Eng.* **32**(3), 229–277 (1990).

124. Olin Corp. Product Bulletin 731-045, *Hydrazine*, Stamford, Conn., 1992.

125. Olin Corp. Product Bulletin 731-009R3, *Hydrazine Storage and Handling of Aqueous Solutions*, 1991. For propellant hydrazines (AH, MMH, UDMH), see *Air Force Systems Command Design Handbook, 1–6, System Safety*, 5th ed., Rev. 2, Wright Patterson Air Force Base, Dayton, Ohio, Apr. 13, 1990.

126. G. N. Kumar and co-workers, *Curr. Sci.* **57**(9), 1069–1071 (1988).

127. C. O. Arah and co-workers, *Int. SAMPE Symp. Exhib.* **36**(2), 1545–1560 (1991).

128. J. M. Bellerby, *J. Hazard. Mater.* **7**(3), 187–197 (1983).

129. E. J. King and H. G. Kammerer, *Gov. Rep. Announce. Index (U.S.)* **78**(2), 166 (1978).

130. Ref. 2, pp. 454–474.

131. M. G. MacNaughton, T. B. Stauffer, and D. A. Stone, *Aviation, Space, Env. Med.* **52**, 149–153 (1981).

132. M. Castegnaro and co-workers, *Am. Ind. Hyg. Assoc. J.* **47**(6), 360–364 (1986).
133. U.S. Pat. 4,402,836 (Sept. 6, 1983), E. G. Fochtman, R. L. Koch, and F. S. Forrest (to United States Dept. of the Air Force).
134. W. F. Cowen, R. A. Sierka, and J. A. Zirroli, in W. J. Cooper, ed., *Chemistry in Water Reuse*, Vol. 2, Ann Arbor Science, Mich., 1981, pp. 101–118.
135. D. A. Stone, *Proc. SPIE-Int. Soc. Opt. Eng.* **289**, 45–47 (1981).
136. R. A. Sierka and W. F. Cowen, *Proc. Ind. Waste Conf.* **35**, 406–415 (1980).
137. K. Morgenstern, *Wasserwirtsch.-Wassertech.* **31**(11), 388–389 (1981).
138. H. H. Takimoto, S. Lewis, and R. Hiltz, in J. Ludwigson, ed., *Hazardous Materials Spills Conference Proceedings*, Gov. Inst. Rockville, Md., 1984, pp. 125–128.
139. M. Castegnaro and co-workers, eds, *IARC Scientific Publication No. 54*, International Agency for Research on Cancer, Lyon, France, 1983.
140. I. M. Kolthoff and R. Belcher, *Volumetric Analysis*, Vol. 3, Wiley-Interscience, New York, 1957.
141. R. Christova, M. Ivanova, and M. Novkirishka, *Anal. Chim. Acta* **85**, 301–307 (1976).
142. C. Zhang, J. Chen, and Y. Zhao, *Fenxi Huaxue [FHHHDT]* **20**(3), 294–296 (1992); *Chem. Abstr.* **116**, 268136e (1992).
143. R. A. Penneman and L. F. Audrieth, *Anal. Chem.* **20**, 1058–1061 (1948).
144. W. R. McBride, R. A. Henry, and S. Skolnik, *Anal. Chem.* **23**, 890 (1951).
145. B. R. Sant and A. K. Mukherji, *Anal. Chim. Acta* **20**, 476 (1959).
146. I. M. Issa and co-workers, *J. Indian Chem. Soc.* **53**, 698 (1976).
147. F. Feigl, *Spot Tests in Organic Analyses*, 7th ed., Elsevier Scientific Publishing Co., New York, 1966, p. 277.
148. W. D. Basson and J. F. Van Staden, *Analyst* **103**, 998–1001 (1978).
149. M. Bordun and co-workers, *Anal. Chem.* **49**(1), 161–162 (1977).
150. G. D. George and J. T. Stewart, *Anal. Lett.* **23**(8), 1417–1429 (1990).
151. E. S. Fiala and C. Kulakis, *J. Chrom.* **214**, 229–233 (1981).
152. J. B. Gisclard, *Report AFFDL-TR-75-116*, Air Force Flight Dynamics Laboratory, Wright-Patterson Air Force Base, Dayton, Ohio, June 1975.
153. Jpn. Kokai JP 03 41,354 (Feb. 21, 1991), K. Yamada and co-workers (to Toa Electronics, Ltd.).
154. *NASA Tech Briefs 66-10586 and 67-10290*, Clearinghouse for Federal Scientific and Technical Information, Springfield, Va., 1966 and 1967.
155. L. R. Cook, R. E. Glenn, and G. E. Podolak, *Am. Ind. Hyg. Assoc. J.* **40**(1), 69–74 (1979).
156. U.S. Pat. 4,595,486 (June 17, 1986), J. C. Schmidt, D. N. Campbell, and S. B. Clay, (to Allied Corp.).
157. R. A. Saunders and co-workers, *Naval Research Laboratory (NRL) Report 8199*, U.S. Navy, Washington, D.C., Apr. 13, 1978.
158. D. K. Killinger, N. Menyuk, and A. Mooradian, *Gov. Rep. Announce. Index (U.S.)* **82**(20), 4082 (1982).
159. C. C. Haun and E. R. Kinkead, *Chronic Inhalation Toxicity of Hydrazine*, University of California, Irvine, Toxic Hazards Research Unit, Dayton, Ohio, Jan. 1975.
160. *Guide for Short Term Exposures of the Public to Air Pollutants, Guide for Hydrazine, Monomethylhydrazine and 1,1-Dimethylhydrazine*, Committee on Toxicology of the National Academy of Science, National Research Council, Washington, D.C., Jan., 1975.
161. W. W. Melvin and W. S. Johnson, *A Survey of Information Relevant to Occupational Health Standards for Hydrazines*, Environmental Health Labs., National Technical Information Service, U.S. Dept. of Commerce, Springfield, Va., Mar. 1976.
162. *Occupational Exposure to Hydrazines, NIOSH Criteria for a Recommended Standard*, U.S. Dept. of Health, Education and Welfare, National Institute of Occupational Safety and Health, Washington, D.C., June 1978.

163. B. Toth, *In vivo*, **2**, 209–242 (1988).
164. W. C. Keller, C. R. Olsen, and K. C. Back, *AFAMRL-TR-82-29 (AD/A119706)*, Air Force Aerospace Medicine Research Laboratory, Wright-Patterson Air Force Base, Dayton, Ohio, 1980.
165. E. H. Vernot and co-workers, *Fund. and App. Tox.* **5**, 1050–1064 (1985).
166. C. Biancifiori, *Lav. Inst. Anat. Istol. Patol. Perugia* **30**, 89 (1970); J. Juhaz, J. Balo, and B. Szende, *Nature* **210**, 1377 (1966).
167. D. Steinhoff and U. Mohr, *Exp. Pathol.* **33**(3), 133–143 (1988).
168. D. Steinhoff, U. Mohr, and W. M. Schmidt, *Exp. Pathol.*, **39**(1), 1–9 (1990).
169. N. Wald and co-workers, *Brit. J. Ind. Med.* **41**, 31–34 (1984).
170. J. C. Contassot and co-workers, "Epidemiological Study of Cancer: Morbidity Among Workers Exposed to Hydrazine," poster presented at the *XXII International Congress on Occupational Health*, in Sydney, Australia, Sept.–Oct., 1987.
171. R. F. Kimball, *Mutat. Res.* **39**(2), 111–126 (1977).
172. Ger. Pat. 2,452,016 (Aug. 19, 1976), G. Gollmer and D. Kashelikar (to Bayer AG).
173. U.S. Pat. 2,640,853 (June 2, 1953), N. K. Sundholm (to U.S. Rubber Co.).
174. U.S. Pat. 3,152,176 (Oct. 6, 1964), B. A. Hunter (to U.S. Rubber Co.).
175. U.S. Pat. 4,163,037 (July 31, 1979), G. E. Niznik (to General Electric Co.).
176. R. L. Heck and W. J. Peascoe, in J. I. Kroschwitz, ed., *Encyclopedia of Polymer Science and Engineering*, Vol. 2, John Wiley & Sons, Inc., New York, 1985, pp. 434–446.
177. B. A. Hunter and B. J. Geelan, *Plastics Engineering Handbook of the Society of the Plastics Industry*, 5th ed., Van Nostrand Reinhold, New York, 1991, pp. 502–510.
178. H. Hurnik, *Plastics Additives Handbook*, Carl Hanser Verlag, Munich, 1987, pp. 619–638.
179. "Blowing and Foaming Agents," in *Additives for Plastics*, 1st ed., D.A.T.A., Inc., San Diego, Calif., 1987, pp. 3–9.
180. U.S. Pat. 2,575,954 (Nov. 20, 1951), W. D. Harris and D. L. Schoene (to U.S. Rubber Co.).
181. U.S. Pat. 2,614,916 (Oct. 21, 1952), O. L. Hoffmann and D. L. Schoene (to U.S. Rubber Co.).
182. U.S. Pat. 4,390,704 (June 28, 1983), H. Beer (to Chemie Linz AG).
183. Ger. Offen. 3,328,835 (Feb. 21, 1985), C. Theis (to Dynamit Nobel AG).
184. P. A. Worthington, *Pestic. Sci.* **31**(4), 457–498 (1991).
185. *Farm Chemicals Handbook*, Meister Publishing Co., Willoughby, Ohio, annual.
186. H. Kidd and D. Hartley, eds., *Pesticide Index*, Royal Society of Chemistry/Crown Copyright, Cambridge, U.K., 1988.
187. *Hydrazine for Wet Lay-Up of Boilers*, Olin Corp., Stamford, Conn., 1990.
188. Jpn. Kokai 59 219,481 (Dec. 10, 1984), (to Mitsubishi Heavy Industries, Ltd.).
189. *SCAV-OX PLUS 35% Catalyzed Hydrazine Solution for Corrosion Protection in Industrial Boilers*, Olin Corp., Stamford, Conn., 1987.
190. *SCAV-OX Hydrazine Solutions to Control Oxygen Corrosion in Boiler Systems*, Olin Corp., Stamford, Conn., 1988.
191. *SCAV-OX 35% Hydrazine Solutions for Corrosion Protection in High and Medium Pressure Boilers*, Olin Corp., Stamford, Conn., 1989.
192. G. Bohnsack, *Vom Wasser* **53**, 147–161 (1979).
193. U.S. Pat. 4,269,717 (May 26, 1981), M. Slovinsky (to Nalco Chemical Co.).
194. T. B. Allgood, *Corros. Rev.* **8**(1–2), 1–9 (1988).
195. J. T. M. van der Wissel, *Polytech. Tijdschr.: Procestech.* **46**(7–8), 40–44 (1991).
196. R. L. Noland, *Chem. Eng.* **65**, 156 (May 19, 1958).
197. U.S. Pat. 4,938,932 (July 3, 1990), W. K. Burke (to Olin Corp.).

198. H. H. Voge and co-workers, *Development of Catalysts for Monopropellant Decomposition of Hydrazine, Final Report 3-13947, Contract NAS 7-97*, Shell Development Co., Emeryville, Calif., Apr.–Dec. 1964.
199. G. Schulz-Ekloff and R. Hoppe, *Catal. Lett.* **6**(3–6), 383–387 (1990).
200. U.S. Pat. 4,124,538 (Nov. 7, 1978), W. E. Armstrong, L. B. Ryland, and H. H. Voge (to Shell Oil Co.).
201. O. I. Smith and W. C. Solomon, *Ind. Eng. Chem. Fundam.* **21**(4), 374–378 (1982).
202. U.S. Pat. 4,324,819 (Apr. 13, 1982), P. J. Birbara (to United Aircraft Corp.).
203. U.S. Pat. Appl. 300,764 AO (Apr. 9, 1982), W. S. Davis (to U.S. Dept. of the Army).
204. India Pat. 150,422 (Oct. 2, 1982), (to Indian Institute of Science).
205. S. R. Jain, *J. Indian Inst. Sci.* **69**(3), 175–191 (1989).
206. *Selected Bibliographies, Handbooks, Manuals and Reviews: Chemical Propulsion*, Chemical Propulsion Information Agency, The Johns Hopkins University, G. W. C. Whiting School of Engineering, Columbia, Md.
207. H. H. Licht and A. Baumann, *Gov. Rep. Announce. Index (U.S.)* **86**(7), (1986).
208. K. C. Patil, C. Nesamani, and V. R. Pai Verneker, *Synth. React. Inorg. Met.-Org. Chem.* **12**(4), 383–395 (1982).
209. R. K. Salyaev, *Akad. Nauk SSSR*, 43–44 (1982).
210. U.S. Pat. 5,098,597 (Mar. 24, 1992), E. F. Rothgery, D. F. Gavin, and K. A. Thomas (to Olin Corp.).
211. Czech. Pat. 190,055 (Sept. 15, 1981), J. Arient and I. Voboril.
212. Ger. Offen. 4,034,645 A1 (May 7, 1992), K. M. Bucerius (to Fraunhofer-Gesellschaft zur Förderung der Angewandten Forschung).
213. U.S. Pat. 4,601,344 (July 22, 1986), R. Reed, M. L. Chan, and K. L. Moore (to United States of America).
214. U.S. Pat. 4,963,201 (Oct. 16, 1990), R. K. Bjerke and co-workers (to Blount, Inc.).
215. Fr. Demande 2,293,424 (July 2, 1976), to Rockwell International.
216. Eur. Pat. EP 167,251 (May 24, 1984), D. F. Doonan (to Olin Corp.).
217. U.S. Pat. 5,041,661 (Aug. 20, 1991), K. L. Wagaman and C. F. Clark (to United States of America).
218. U.S. Pat. 4,733,610 (Mar. 29, 1988), K. Y. Lee and M. D. Coburn (to U.S. Dept. of Energy).
219. H. Meyer and J. Mally, *Monatsh. Chem.* **33**, 393 (1912).
220. U.S. Pat. 3,520,926 (July 21, 1970), W. Bollag and co-workers (to Hoffmann-LaRoche).
221. U.S. Pats. 4,110,437 (Aug. 29, 1978); 4,328,246 (May 4, 1982); 4,867,978 (Sept. 19, 1989), J. Gold.
222. R. T. Chlebowski, *J. Clin. Oncol.* **8**(1), 9–15 (1990).
223. V. A. Filov and co-workers, *Vopr. Onkol.* **36**(6), 721–726 (1990).
224. *USAN and USP Dictionary of Drug Names*, United States Pharmacopeial Convention, Inc., Rockville, Md., 1990.
225. U.S. Pat. 3,954,607 (May 4, 1976), R. A. Halling (to Du Pont).
226. J. S. Shukla and S. K. Dixit, *Eur. Polym. J.* **28**(2), 199–202 (1992).
227. U.S. Pat. 5,075,503 (Dec. 24, 1991), J. Lin and G. P. Speranza (to Texaco Chemical Co.).
228. U.S. Pat. 5,114,997 (May 19, 1992), M. D. Golder and D. K. Walker (to Hoechst Celanese Corp.).
229. U.S. Pat. 5,041,545 (Aug. 20, 1991), T. N. Meyers (to Atochem North America).
230. M. A. Gutjahr, in G. Sandstede, ed., *From Electrocatalysis to Fuel Cells*, Univ. of Washington Press, Seattle, 1972, pp. 143–156.
231. W. Mennicke, in *Ullmann's Encyclopedia of Industrial Chemistry*, Vol. A16, 5th ed., VCH, Weinheim, 1990, pp. 321–333.

General References

E. Schmidt, *Hydrazine and Its Derivatives*, John Wiley & Sons, Inc., New York, 1984. Contains an exhaustive bibliography.

P. A. S. Smith, *Derivatives of Hydrazine and Other Hydronitrogens Having N-N Bonds*, The Benjamin/Cummings Publishing Co., Inc., Reading, Mass., 1983.

P. A. S. Smith, *The Chemistry of Open-Chain Nitrogen Compounds*, Vols. 1 and 2, W. A. Benjamin, Inc., Menlo Park, Calif., 1965–1966.

A. Weissberger and E. C. Taylor, eds., *The Chemistry of Heterocyclic Compounds*, John Wiley & Sons, Inc., New York.

A. R. Katritzky and A. J. Boulton, eds., *Advances in Heterocyclic Chemistry*, Academic Press, Inc., New York.

K. Jones, in J. C. Bailar, ed., *Comprehensive Inorganic Chemistry*, Vol. 2, Pergamon Press, Ltd., Oxford, 1973, pp. 250–265.

Houben-Weyl, *Methoden der Organischen Chemie*, 4th ed., Band X/2, Stickstoff Verbindungen 1, Teil 2, Georg Thieme Verlag, Stuttgart, Germany, 1967.

J. I. Kroschwitz, ed., *Encyclopedia of Polymer Science and Engineering*, John Wiley & Sons, Inc., New York.

Comprehensive Polymer Science, The Synthesis, Characterization, Reactions and Applications of Polymers, Pergamon Press, Ltd., Oxford, U.K.

HENRY W. SCHIESSL
Olin Corporation

HYDRIDES

Hydrides are compounds that contain hydrogen (qv) in a reduced or electron-rich state. Hydrides may be either simple binary compounds or complex ones. In the former, the negative hydrogen is bonded ionically or covalently to a metal, or is present as a solid solution in the metal lattice. In the latter, which comprise a large group of chemical compounds, complex hydridic anions such as BH_4^-, AlH_4^-, and derivatives of these, exist.

Although a few simple hydrides were known before the twentieth century, the field of hydride chemistry did not become active until around the time of World War II. Commerce in hydrides began in 1937 when Metal Hydrides Inc. used calcium hydride [7789-78-8], CaH_2, to produce transition-metal powders. After World War II, lithium aluminum hydride [16853-85-3], $LiAlH_4$, and sodium borohydride [16940-66-2], $NaBH_4$, gained rapid acceptance in organic synthesis. Commercial applications of hydrides have continued to grow, such that hydrides have become important industrial chemicals manufactured and used on a large scale.

Simple (Binary) Hydrides

Ionic Hydrides. The ionic or saline hydrides contain metal cations and negatively charged hydrogen ions. These crystallize in the cubic lattice similar to that of the corresponding metal halide, and when pure, are white solids. When dissolved in molten salts or hydroxides and electrolyzed, hydrogen gas is liberated at the anode. The densities are greater than those of the parent metal, and formation is exothermic. All are strong bases.

Alkali Metal Hydrides. Physical properties of the alkali metal hydrides are given in Table 1.

Lithium Hydride. Lithium hydride [7580-67-8] is very stable thermally and melts without decomposition. In the temperature range 600–800°C, the dissociation pressure for hydrogen, P_{H_2}, in units of kPa is expressed by

$$\log P_{H_2} = -9750/T + 10.632 \tag{1}$$

allowing quantitative preparation by the addition of hydrogen to molten lithium at 680–900°C at about 100 kPa (ca 1 atm) of hydrogen pressure.

Lithium hydride reacts vigorously with silicates above 180°C. Therefore, glass, quartz, and porcelain containers cannot be used in preparative processes. That only traces dissolve in polar solvents such as ether reflects its significant (60–75%) covalent bond character. It is completely soluble in, and forms eutectic melting compositions with, a number of fused salts.

Normally, lithium hydride ignites in air only at high temperatures. When heated it reacts vigorously with CO_2 and nitrogen. With the former, lithium formate is obtained. Reaction at high temperature with nitrogen produces lithium nitride. Therefore, dry limestone or NaCl powders are used to extinguish LiH fires. Lithium hydride reacts exothermically with moist air and violently with water.

$$\text{LiH} + \text{H}_2\text{O} \longrightarrow \text{LiOH} + \text{H}_2(g) \quad \Delta H_{298} = -132 \text{ kJ/mol } (-31.5 \text{ kcal/mol}) \tag{2}$$

Reaction with $AlCl_3$ gives lithium aluminum hydride, which is the main application of lithium hydride. Reaction with ammonia yields lithium amide (eq. 3).

$$\text{LiH} + \text{NH}_3 \xrightarrow{430°C} \text{LiNH}_2 + \text{H}_2(g) \tag{3}$$

Table 1. Physical Properties of Alkali Metal Hydrides

Hydride	CAS Registry Number	Mp, °C	$\Delta H_{(298)}$, kJ/mol[a]	$\Delta F_{(298)}$, kJ/mol[a]	S, J/(mol·K)[a]	Lattice energy, kJ/mol[a]	Density, g/cm³	Ref.
LiH	[7580-67-8]	688	−90.7	−70	25	916	0.77	1
NaH	[7646-69-7]	420 dec	−56.5	−37.7	48	791	1.36	2
KH	[7693-26-7]	dec	−57.9	−37.3	61	720	1.43	
RbH	[13446-75-8]	300 dec					2.60	
CsH	[13772-47-9]	dec					3.4	

[a]To convert J to cal, divide by 4.184.

An industrially important use of lithium and sodium hydride is in the generation of silane for semiconductor uses (see SILICON COMPOUNDS, SILANES; SEMICONDUCTORS). The exceptionally large amount of hydrogen available from the hydrolysis of lithium hydride (1 g LiH yields 2.82 L H_2) has led to use as an easily portable source of hydrogen for inflation of weather balloons and the like. However, calcium hydride is cheaper and therefore preferred. Lithium hydride is employed as a lightweight nuclear shielding material because of its high hydrogen content. Castings of up to 1 metric ton have been made (3).

In organic chemistry, LiH serves as a condensation agent. In the presence of trialkyl boranes very powerful reducing agents, $LiBHR_3$, which are soluble in THF, are obtained. These materials reduce aliphatic halides and in some cases highly stereospecific reductions can be accomplished.

Annual U.S. production is less than 100 t; the 1992 price was ca $72/kg.

Sodium Hydride.   Sodium hydride [7646-69-7] decomposes to its elements without melting, starting at ca 300°C. Decomposition is rapid at 420°C. The dissociation pressure in kPa between 100 and 600°C for the decomposition range 15–90% NaH can be found from

$$\log P_{H_2} = -6100/T + 10.78 \tag{4}$$

Sodium hydride is insoluble in organic solvents but soluble in fused salt mixtures and fused hydroxides such as NaOH. It oxidizes in dry air and hydrolyzes rapidly in moist air. The pure material reacts violently with water:

$$NaH + H_2O \longrightarrow NaOH + H_2(g) \tag{5}$$

and with carbon dioxide under mild conditions to form sodium formate:

$$NaH + CO_2 \longrightarrow HCOONa \tag{6}$$

Sodium hydride is manufactured by the reaction of hydrogen and molten sodium metal dispersed by vigorous agitation in mineral oil (4).

$$2\,Na + H_2 \xrightarrow[\text{1 MPa (10 atm)}]{250-265°C} 2\,NaH \tag{7}$$

A 25% dispersion of NaH crystals in oil is obtained. The commercial product, after filtration, is a 60% dispersion of NaH crystals (5–50 μm). The oil dispersions can be handled quite safely because the oil phase provides a barrier to air and moisture, whereas the unprotected crystals react vigorously. Traces of unreacted sodium metal give the product a gray color.

An important industrial use of NaH involves its *in situ* formation in molten NaOH or in fused eutectic salt baths. At concentrations of 1–2% NaH, these compositions are powerful reducing systems for metal salts and oxides (5). They have been used industrially for descaling metals such as high alloy steels, titanium, zirconium, etc.

In organic synthesis, NaH acts as a powerful base without substantial reducing activity. Because it is insoluble in organic solvents, its reactions are heterogeneous. Hence finely divided, high surface area NaH in oil is the preferred reagent, unless the mineral oil interferes with the isolation of the product. Sodium hydride has significant advantages over other bases such as sodium alcoholates, sodium amide, and metallic sodium. Sodium hydride is a stronger base and its high reactivity leads to clean reactions without by-product formation and side reactions, usually at lower temperatures than are required with competitive bases. Reactions are easily monitored by measuring the hydrogen evolution which accompanies the reaction of NaH and organics. Important industrial inorganic reactions of NaH include addition reactions using Lewis acids and reductions of inorganic halides. Industrial organic reactions include alkylations and acylations (see ALKYLATION), reactions with acidic organic compounds (metallations), a wide variety (aldol, Claisen, Dieckmann, Darzens, and Stobbe) of addition/condensation reactions, and polymerizations, especially of caprolactam (qv).

Annual U.S. production of the 60% NaH dispersion is less than 200 t; the 1992 price was ca $15/kg.

Potassium, Rubidium, and Cesium Hydrides. Although all the other alkali metal hydrides have been synthesized and some of the properties measured, only potassium hydride [7693-26-7] is commercially available. KH is manufactured in small amounts and sold as a mineral oil dispersion. It is a stronger base than NaH and is used to make the strong reducing agent $KBH(C_2H_5)_3$ and the super bases RNHK and ROK (6).

Alkaline-Earth Metal Hydrides. Table 2 gives thermochemical data of alkaline-earth metal hydrides. All form orthorhombic crystals.

Calcium Hydride. Calcium hydride [7789-78-8] dissociates into calcium and hydrogen ($P_{H_2} = 101$ kPa or 1 atm) at 990°C without melting. The dissociation equation between 25 and 96% CaH_2 when P_{H_2} is in kPa is

$$\log P_{H_2} = -7782/T + 8.195 \tag{8}$$

Calcium hydride is highly ionic and is insoluble in all common inert solvents. It can be handled in dry air at low temperatures without difficulty. When heated to about 500°C, it reacts with air to form both calcium oxide and nitride. Calcium hydride reacts vigorously with water in either liquid or vapor states at room

Table 2. Physical Properties of Alkaline-Earth Metal Hydrides

Hydride	CAS Registry Number	$\Delta H_{(298)}$, kJ/mol[a]	$\Delta F_{(298)}$, kJ/mol[a]	S, J/(mol·K)[a]	Density, g/cm^3
CaH_2	[7789-78-8]	−186.3	−147.4	42	1.90
SrH_2	[13598-33-9]	−180.5	−138.6	54	3.27
BaH_2	[13477-09-3]	−171.2	−132.3	67	4.16

[a]To convert J to cal, divide by 4.184.

temperature. The reaction with water provides 1.06 liters of hydrogen per gram CaH_2.

$$CaH_2 + 2 H_2O \longrightarrow Ca(OH)_2 + 2 H_2(g) \tag{9}$$

At elevated temperatures, CaH_2 reacts with halogens, sulfur, phosphorus, alcohols, and ammonia. At high temperatures, it reacts with refractory metal oxides and halides. Calcium hydride is substantially inert to organic compounds that do not contain acidic hydrogens.

Calcium hydride is prepared on a commercial scale by heating calcium metal to about 300°C in a high alloy steel, covered crucible under 101 kPa (1 atm) of hydrogen gas. Hydrogen is rapidly absorbed at this temperature and the reaction is exothermic.

The principal industrial applications for calcium hydride include (*1*) reduction at 600–1000°C of refractory oxides of metals such as titanium, zirconium, vanadium, niobium, uranium, thorium, chromium, and mixtures of these to produce metal and alloy powders (7); (*2*) use as a portable source of hydrogen. The military has used hydrogen generators based on the reaction of water and calcium hydride to inflate weather balloons for many years. A variety of generator sizes are available commercially; and (*3*) drying of liquids and gases (see DESICCANTS). Because of its low reactivity with most organic liquids and its high reactivity with water, calcium hydride is widely used as a superdrying agent for esters, ketones, halides, electrical insulator oils, silicones, solvents for Ziegler-Natta polymerization systems, monomers, and air and other gases. It is also used as an analytical reagent to determine water in organic liquids.

Annual U.S. production is less than 25 t; the 1992 price was ca $55/kg.

Other Alkaline-Earth Hydrides.   Strontium and barium hydrides resemble calcium hydride in properties and reactivity. They have no significant commercial applications.

Covalent Hydrides.   In all hydrides, hydrogen is bound to an atom of lower electronegativity ($X_H = 2.1$) than itself. In covalent hydrides, the hydrogen–metal bond is effected through a common electron pair. Beryllium and magnesium hydrides are included in this group and are polymeric materials, as is aluminum hydride. The simple hydrides of silicon, germanium, tin, and arsenic are gaseous or easily volatile compounds. Table 3 gives some properties of these compounds.

Main Group Metal Hydrides. Beryllium Hydride.   Beryllium hydride [*13597-97-2*] is an amorphous, colorless, highly toxic polymeric solid (H = 18.3%) that is stable to water but hydrolyzed by acid (8). It is insoluble in organic solvents but reacts with tertiary amines at 160°C to form stable adducts, eg, $(R_3N \cdot BeH_2)_2$ (9). It is prepared by continuous thermal decomposition of a di-*t*-butylberyllium-ethyl ether complex in a boiling hydrocarbon (10).

$$[(CH_3)_3C]_2Be \cdot O(C_2H_5)_2 \longrightarrow BeH_2 + 2(CH_3)_2C{=}CH_2 + (C_2H_5)_2O \tag{10}$$

Beryllium hydride was formerly of interest as a rocket fuel and as a moderator for nuclear reactors.

Table 3. Properties of Covalent Hydrides

Hydride	CAS Registry Number	Formula	Mp, °C	Bp, °C	Density[a] g/cm³	g/L
beryllium hydride[b]	[13597-97-2]	BeH_2	125 dec	220[c]		
magnesium hydride	[7693-27-8]	MgH_2	280 dec		1.45	
aluminum hydride	[7784-21-6]	AlH_3				
silane[d]	[7803-62-5]	SiH_4	−185[e]	−119.9	0.68[f] (−185)	1.44[g] (20)
germane	[7782-65-2]	GeH_4	−165	−90	1.523[f] (−142)	3.43[g] (0)
stannane	[2406-52-2]	SnH_4	−150	52		
arsine	[7784-42-1]	AsH_3	−116.9[e]	−62	1.604[f] (64)	2.695[g]

[a] Temperature in °C is given in parentheses.
[b] $\Delta H_{298} = 19.3$ kJ/mol (4.6 kcal/mol).
[c] Begins to dissociate.
[d] $\Delta H_{298} = 30.55$ kJ/mol (7.30 kcal/mol).
[e] Freezing point.
[f] Liquid.
[g] Gas at atmospheric pressure.

Magnesium Hydride. Magnesium hydride is a gray powder of about 97% purity which is insoluble in inert organic solvents. It is easily oxidized, and when heated to about 280°C, dissociates without melting; when P is in kPa

$$\log P_{H_2} = -3857/T + 8.91 \tag{11}$$

When prepared by direct reaction of the elements, magnesium hydride is stable in air and only mildly reactive with water.

$$Mg + H_2 \xrightarrow[\text{10–15 MPa (100–150 atm)}]{300-400°C} MgH_2 \tag{12}$$

However, when it is obtained by pyrolysis of diethylmagnesium or by reaction of diethylmagnesium and $LiAlH_4$ (11), it is very reactive with both air and water. This difference in reactivity mainly results from the much finer particle size of the product obtained by the pyrolysis route.

Aluminum Hydride. Aluminum hydride is a relatively unstable polymeric covalent hydride that received considerable attention in the mid-1960s because of its potential as a high energy additive to solid rocket propellants. The projected uses, including aluminum plating, never materialized, and in spite of intense research and development, commercial manufacture has not been undertaken. The synthetic methods developed were costly, eg,

$$3\ LiAlH_4 + AlCl_3 \xrightarrow{(C_2H_5)_2O} 4\ AlH_3 \cdot x(C_2H_5)_2O + 3\ LiCl \tag{13}$$

Silane. Silane is a colorless gas that is spontaneously flammable in air and slowly decomposed by water; in the presence of aqueous alkali it is completely hydrolyzed to form hydrogen and silicates. It is manufactured on a commercial scale and sold as a compressed gas in cylinders. It is prepared by the reaction

$$4 \text{ LiH} + \text{SiCl}_4 \longrightarrow \text{SiH}_4(g) + 4 \text{ LiCl} \tag{14}$$

Silane, pure or doped, is used to prepare semiconducting silicon by thermal decomposition at >600°C. Gaseous dopants such as germane, arsine, or diborane may be added to the silane at very low concentrations in the epitaxial growing of semiconducting silicon for the electronics industry. Higher silanes, eg, Si_2H_6 and Si_3H_8, are known but are less stable than SiH_4. These are analogues of lower saturated hydrocarbons.

Germane. Germane is a colorless gas, spontaneously flammable in air. It is manufactured in small amounts and is available as a compressed gas in cylinders.

$$\text{GeCl}_4 + 4 \text{ NaBH}_4 + 12 \text{ H}_2\text{O} \longrightarrow \text{GeH}_4 + 4 \text{ NaCl} + 4 \text{ B(OH)}_3 + 12 \text{ H}_2(g) \tag{15}$$

Germane is used primarily to produce high purity germanium metal or epitaxial deposits of germanium on substrates for electronics by thermal decomposition at about 350°C (see GERMANIUM AND GERMANIUM COMPOUNDS).

Stannane. Stannane is a colorless poisonous gas that decomposes rapidly at room temperature. A large number of organostannanes, eg, R_3SnH and R_2SnH_2, are known, and their properties as organic reducing agents have been extensively investigated. Tributyltin hydride [688-73-3] is used frequently to dehalogenate organic compounds (12).

Arsine. Arsine is a highly toxic colorless gas, made in small amounts as a dopant for silicon in the electronics industry by the reaction

$$4 \text{ H}_3\text{AsO}_3 + 3\text{BH}_4^- + 3 \text{ H}^+ \longrightarrow 4 \text{ AsH}_3 + 3 \text{ H}_3\text{BO}_3 + 3 \text{ H}_2\text{O} \tag{16}$$

Transition-Metal Hydrides. Transition-metal hydrides, ie, interstitial metal hydrides, have metallic properties, conduct electricity, and are less dense than the parent metal. Metal valence electrons are involved in both the hydrogen and metal bonds. Compositions can vary within limits and stoichiometry may not always be a simple numerical proportion. These hydrides are much harder and more brittle than the parent metal, and most have catalytic activity.

Titanium Hydride. Titanium hydride [7704-98-5], TiH_2, is a brittle, metallic-gray solid, density 3.8 g/cm^3, which produces 448 mL H_2 at STP per gram TiH_2. Titanium hydride powder is stable at room temperature and inert to water and most chemical reagents. At elevated temperatures TiH_2 is attacked by oxidizing agents and acids and by cold acid fluoride solutions. TiH_2 powder burns quietly when ignited but reacts violently when burned in the presence of oxidizers. Dissociation of TiH_2 begins at 300°C in vacuum and is nearly complete at 600°C at 1 mPa (7.5×10^{-3} μm Hg).

TiH$_2$ is prepared on an industrial scale by direct combination of hydrogen and the metal (sponge, ingot, scrap, etc) at 200–650°C, followed by cooling in an H$_2$ atmosphere. An alternative method is the reduction of the oxide using calcium hydride under hydrogen:

$$TiO_2 + 2\,CaH_2 \xrightarrow[H_2]{950-1000°C} TiH_2 + 2\,CaO + H_2(g) \tag{17}$$

The calcium oxide is leached out with acid (HCl), leaving a finely divided TiH$_2$ powder.

Titanium hydride is used as a source for Ti powder, alloys, and coatings; as a getter in vacuum systems and electronic tubes; as a sealer of metals; and as a hydrogen source.

Zirconium Hydride. Zirconium hydride [7704-99-6], ZrH$_2$, is a brittle, metallic-gray solid that is stable in air and water, and has a density of 5.6 g/cm^3. The chemical properties of ZrH$_2$ closely resemble those of titanium hydride. Thermal decomposition in vacuum (1 mPa (7.5 × 10^{-3} μm Hg)) begins at 300°C and is nearly complete at 500–700°C. It is prepared in the same manner as TiH$_2$.

Commercial uses are as a getter in the manufacture of vacuum tubes and other systems; as a hydrogen source for foaming metals; as a hydrogen reservoir; for the introduction of zirconium into powdered alloys; for metal–ceramic and metal–metal bonding; as a moderator in nuclear reactors; and as a source of Zr metal powder and alloys. Several organometallic compounds containing a zirconium hydrogen bond are known. Both TiH$_2$ and ZrH$_2$ have moderate fibrogenic and toxic action when inhaled (13).

Rare-Earth Hydrides. Activated rare-earth metals react directly with hydrogen even at room temperature. These metals are activated by heating to 300°C in H$_2$, followed by cooling under H$_2$. Lanthanum dihydride [13823-36-4], and lathanum trihydride [13864-01-2] are both known compositions (see LANTHANIDES). Cerium hydride [13569-50-1], CeH$_2$, and hydrides of higher hydrogen content (CeH$_{<3}$) have been made and studied (see CERIUM AND CERIUM COMPOUNDS) (14). Both cerium and lanthanum hydrides are black pyrophoric solids. Hydrogen content can be varied with hydrogen pressure and temperature over the metal hydride. Cerium and lanthanum hydrides are very reactive chemically and ignite spontaneously in air. They react with N$_2$ at 20°C and with water at 0°C. The most important uses of cerium and lanthanum hydrides are in the hydrogen-storage alloys LaNi$_5$ and CeMg$_2$. There is some interest in these materials as hydrogenation catalysts. The other rare earth metals (praseodymium, samarium, europium, gadolinium, and yttrium) react with hydrogen to form hydrides of varying composition.

Group 5 (VB) Hydrides. These hydrides (Table 4) are formed from the metals, preferably by heating the metals in powder form in a hydrogen atmosphere up to 1000°C. Trace impurities (oxides, nitrides) in the metal prevent complete hydriding. The hydrides are brittle powders that can be handled in air. Heating to above 400°C initiates hydrogen evolution; complete hydrogen removal is usually obtained at 700°C under vacuum. Tantalum and niobium hydrides are superconductors at <10 K. These hydrides are manufactured in small amounts mainly for research and development work in powder metallurgy (see METALLURGY, POWDER).

Table 4. Group 5 (VB) Hydrides

Compound	CAS Registry Number	Formula	Density, g/cm^3
vanadium hydride	[13761-67-6]	VH	5.4
niobium hydride	[13981-86-7]	NbH	6.6
niobium dihydride[a]	[13981-96-7]	NbH$_2$	
tantalum hydride	[13981-95-8]	TaH	15.1

[a]Decomposes slowly.

Hydrogen-Storage Alloys.   A number of metal alloys are very useful for safely storing large volumes of hydrogen because these easily dissolve hydrogen at relatively low temperatures and pressures, forming interstitial hydrides. The hydrogen is subsequently released by applying heat and lowering the pressure. Many metals and binary and ternary alloys have been thoroughly studied for this application. The most important appear to be

$$\text{FeTi} \underset{-\text{H}_2}{\overset{\text{H}_2}{\rightleftharpoons}} \text{FeTiH}_{1-2} \tag{18}$$

FeTi can also be modified with rare-earth metals, Ni, or Mn (15). AB$_5$ alloys where A is a rare-earth metal or mischmetal, Ca, or Th, and B is Co or Ni have also been used. LaNi$_5$ has shown special promise (16).

$$\text{LaNi}_5 \overset{\text{H}_2}{\longrightarrow} \text{LaNi}_5\text{H}_{6.7} \tag{19}$$

Magnesium titanium alloys form the hydrides Mg$_2$TiH$_6$ [74811-18-0] and MgTi$_2$H$_6$ [58244-88-5] (17). Traces of a third metal are often added to adjust dissociation pressures and/or temperatures to convenient ranges.

Hydrogen-storage alloys (18,19) are commercially available from several companies in the United States, Japan, and Europe. A commercial use has been developed in rechargeable nickel–metal hydride batteries which are superior to nickel–cadmium batteries by virtue of improved capacity and elimination of the toxic metal cadmium (see BATTERIES, SECONDARY CELLS–ALKALINE). Other uses are expected to develop in nonpolluting internal combustion engines and fuel cells (qv), heat pumps and refrigerators, and electric utility peak-load shaving.

Complex Hydrides

The complex hydrides are a large group of compounds in which hydrogen is combined in fixed proportions with two other constituents, generally metallic elements. These compounds have the general formula M(M'H$_4$)$_n$, where n is the valence of M, and M' is a trivalent Group 3 (IIIA) element such as boron, aluminum, or gallium. In the BH$_4^-$ and AlH$_4^-$ anions, the hydrogen atoms are arranged tetrahedrally around the boron or aluminum and retain significant hydride or electron-rich character. For this reason, the complex hydrides have achieved significant and broad use as reducing agents in many different areas of chemistry. Lithium, sodium, and potassium borohydrides, lithium aluminum hydride, and sodium dihydrobis(2-methoxyethoxy)aluminate are commercially available, but many others are known, such as the corresponding alkaline-earth and other metal, as well as quarternary ammonium and phosphonium complex hydrides. Sodium borohydride in particular and lithium aluminum hydride are the most important commercially. In addition, compounds have been prepared in which from one to three hydrogen atoms have been replaced by other groups. The most important complex hydrides are listed in Table 5.

Although the IUPAC has recommended the names tetrahydroborate, tetrahydroaluminate, etc, this nomenclature is not yet in general use.

Borohydrides. The alkali metal borohydrides are the most important complex hydrides. They are ionic, white, crystalline, high melting solids that are sensitive to moisture but not to oxygen. Group 13 (IIIA) and transition-metal borohydrides, on the other hand, are covalently bonded and are either liquids

Table 5. Complex Hydrides

Formula	CAS Registry Number	Density, g/cm^3	Mp,°C
$LiBH_4$	[16949-15-8]	0.66	278
$NaBH_4$	[16940-66-2]	1.074	505
KBH_4	[13762-51-1]	1.177	585
$Be(BH_4)_2$	[17440-85-6]	0.702	123 dec
$Mg(BH_4)_2$	[16903-37-0]		320 dec
$Ca(BH_4)_2$	[17068-95-0]		260 dec
$Zn(BH_4)_2$	[17611-70-0]		>50 dec
$Al(BH_4)_3$	[16962-07-5]	0.549	-64.5^a
$Zr(BH_4)_4$	[23840-95-1]	1.13	28.7
$Th(BH_4)_4$	[33725-13-2]	2.59	204 dec
$U(BH_4)_4$	[33725-14-3]	2.67	100 dec
$(CH_3)_4NBH_4$	[16883-45-7]	0.84	>310
$(C_2H_5)_4NBH_4$	[17083-85-1]	0.926	225 dec
$(C_4H_9)_4NBH_4$	[33725-74-5]		>300
$(C_8H_{17})_3CH_3NBH_4$	[17083-38-4]	0.9	ca 30
$C_{16}H_{33}(CH_3)_3NBH_4$		0.9	ca 160
$NaBH_3CN$	[25895-60-7]	1.20	240 dec
$NaBH(OCH_3)_3$	[16940-17-3]	1.24	230 dec
$LiAlH_4$	[16853-85-3]	0.917	190 dec
$NaAlH_4$	[13770-96-2]	1.28	178
$Mg(AlH_4)_2$	[17300-62-8]		140 dec
$Ca(AlH_4)_2$	[16941-10-9]		>230 dec
$LiAlH(OCH_3)_3$	[12706-93-6]		
$LiAlH(OC_2H_5)_3$	[17250-30-5]		
$LiAlH(OC_4H_9)_3$	[38884-28-5]	1.03	>400
$NaAlH_2(OC_2H_4OCH_3)_2$	[22722-98-1]	1.122	205 dec
$NaAlH_2(C_2H_5)_2$	[17836-88-3]		85

^{a}Bp, 44.5°C.

or sublimable solids. The alkaline-earth borohydrides are intermediate between these two extremes, and display some covalent character.

Lithium Borohydride.   Lithium borohydride [16949-15-8], $LiBH_4$, is made by metathesis between sodium borohydride and lithium chloride (20) in isopropylamine.

$$LiCl + NaBH_4 \longrightarrow LiBH_4 + NaCl \tag{20}$$

After the amine is removed, a single extraction with ethyl ether is sufficient to provide a 98% pure product in ca 75% yield. Lithium borohydride is a hygroscopic, white powder that decomposes slowly at its melting point, evolving hydrogen. The heat of formation is -190 kJ/mol (-45.4 kcal/mol).

Unlike many other borohydrides, lithium borohydride is highly soluble in ethers including aliphatic ethers, THF, and polyglycol ethers. It is also very soluble in amines and ammonia. Dissolution in water and lower aliphatic alcohols leads to extensive decomposition and hydrogen evolution.

Lithium borohydride contains 18.5% hydrogen by weight and, on complete hydrolysis, liberates 4.1 L (STP) hydrogen per gram.

$$LiBH_4 + 4 H_2O \longrightarrow LiOH + H_3BO_3 + 4 H_2 \tag{21}$$

Lithium borohydride is a more powerful reducing agent than sodium borohydride, but not as powerful as lithium aluminum hydride (Table 6). In contrast to sodium borohydride, the lithium salt, in general, reduces esters to the corresponding primary alcohol in refluxing ethers. An equimolar mixture of sodium or potassium borohydride and a lithium halide can also be used for this purpose (21,22).

Sodium Borohydride.   Sodium borohydride [16940-66-2] is a thermally stable, white crystalline solid that decomposes *in vacuo* above 400°C. The heat of formation is -192 kJ/mol (-45.9 kcal/mol). $NaBH_4$ is hygroscopic and absorbs water rapidly from moist air to form a dihydrate that decomposes slowly to sodium metaborate and hydrogen. It is soluble in many solvents including water, alcohols, liquid ammonia and amines, glycol ethers, and dimethyl sulfoxide.

Sodium borohydride and potassium borohydride [13762-51-1] are unique among the complex hydrides because they are stable in alkaline solution. Decomposition by hydrolysis is slow in water, but is accelerated by increasing acidity or temperature.

$$BH_4^- + 2 H_2O \longrightarrow BO_2^- + 4 H_2 \tag{22}$$

The decomposition rate of $NaBH_4$ solutions in water is conveniently estimated from equation 23 which expresses half-life in terms of the two most important variables, pH and temperature when $t_{1/2}$ is in minutes and T is in K (23).

$$\log_{10} t_{1/2} = pH - (0.034\,T - 1.92) \tag{23}$$

Table 6. Behavior of Various Functional Groups Toward Hydride Reagents[a]

Hydride reagents	Aldehyde	Ketone	Acid chloride	Ester	Carboxylic acid	Carboxylic salt	Amide
$NaBH_4$	+	+	+	±[b]	−	−	−
$LiBH_4$	+	+	+	+	−	−	−
$Zn(BH_4)_2$	+	+	+	−			
$NaBH_4/AlCl_3$	+	+	+	+	+	−	+
$NaBH_3CN$	+[c]	+[c]		−			−
$NaBH_2S_3$	+	+	−	−			+
$LiBH(R)_3$	+	+	+	+			
$NaBH(OOCCH_3)_3$	+	±[d]					
$NaBH_3(OOCCH_3)$				−			+
$NaBH(OOCCF_3)_3$							
$NaBH_3(OOCCF_3)$				−			+
$NaBH(OR)_3$	+	+	+	+[f]			
$NaBH_3(OH)$		+		+			
$NaBH_3$ (anilide)	+	+	+	+			−
$NaBH_2$ (ethanedithiolate)							+
$THF \cdot BH_3$	+	+	−	+[b]	+	−	+
$(3\text{-}CH_3\text{-}2\text{-}C_4H_9)_2BH$	+	+	−	−	−		
$(CF_3COO)_2BH$							
$LiAlH_4$	+	+	+	+	+	+	+
$LiAlH(OCH_3)_3$	+	+	+	+	+		
$LiAlH(O\text{-}t\text{-}C_4H_9)_3$	+	+	+[i]	−	−	−	−
$NaAlH_2(OC_2H_4\text{-}OCH_3)_2$	+	+	+	+	+	+	
AlH_3	+	+	+	+	+	+	+

[a] +, Reduction; −, no reduction; ±, reduction of some, but not all members of this class.
[b] Moderate reaction over long time period.
[c] At pH 3–4.
[d] <10% reduction in refluxing benzene.
[e] In enamines, quinolines, and indoles.
[f] At elevated temperature.
[g] Reaction in 1:1 ratio.
[h] Indole to indoline.
[i] Reverse addition yields aldehyde.

Solutions of $NaBH_4$ in methanol, and to a lesser degree ethanol, are subject to a similar decomposition reaction that evolves hydrogen; these solutions can be stabilized by alkali. The solubility of $NaBH_4$ in lower aliphatic alcohols decreases as the carbon chain length increases, but the stability increases. Solutions in 2-propanol and t-butanol are stable without alkali (22,24).

Reactions. Complete hydrolysis of $NaBH_4$ produces 2.37 L hydrogen (STP) per gram of borohydride; similarly, addition of acid to a cold aqueous solution liberates the theoretical amount of hydrogen. Rapid, controlled complete generation of hydrogen can be accomplished by adding an acidic compound or an appropriate metal salt such as nickel or cobalt chloride. These salts are effective catalytic accelerators. In the absence of acid, the hydrogen evolution slows down after a

Table 6.   *(Continued)*

Imide	Epoxide	Lactone	Carbinol	Imine	Nitrile	Nitro	Unsat. quat.	Halide	Tosyl-hydrazone	Olefin
−	±	±	+	−	−		+	−		−
	+	+		−	−			−		−
	−	−								−
	+	+		+	−					+
+				+	−		+	+	+	+
		+	+		+	+				
	+	+					+		+	+[e]
			+		+	−				
			+							+[e]
					+	−				
					+[f]					
					+	+				
					−	+				
+					+					
	+	+			+	−				+
		+[g]			+	+				+
										+[h]
+	+	+	+		+	+	+	+	+	−
		+			+					−
	−	−			−					−
+	+	+			−			+		−
	+	+			+	−			−	−

short time owing to the increase in pH caused by formation of the basic meta-borate ion. Conversely, dissolving the borohydride in a basic solution prevents initial hydrolysis and permits the reagent to be used in an aqueous solution.

The inorganic reductions of $NaBH_4$ are numerous and varied (Table 7). Comparatively few anions are reduced, yet the reduction of bisulfite to dithionite (hydrosulfite) (25), which is used in the pulp (qv) and paper (qv), clay (see CLAYS), and vat dyeing industries, is an important inorganic application of $NaBH_4$.

$$8\ HSO_3^- + BH_4^- \longrightarrow 4\ S_2O_4^{2-} + BO_2^- + 6\ H_2O \tag{24}$$

Iodate, chlorate, and hypochlorite are reduced quantitatively; permanganate and perrhenate are also reduced. Cation reductions in aqueous solution occur by one of four possible paths: reduction to a stable and soluble lower valence state, eg, Ce ($+4$ to $+3$), V ($+5$ to $+4$ to $+3$), Fe ($+3$ to $+2$), Cr ($+6$ to $+3$); reduction to the free metal, eg, Ag, As, Sb, Bi, Pb, Hg, and the noble metals; precipitation of a metal boride, eg, Ni, Co, and Cu; or formation of a volatile hydride, eg, B, Ge, Sn, As, Sb, and Bi. The borides and noble metals have been used as hydrogenation catalysts and to generate hydrogen quantitatively from aqueous $NaBH_4$.

Table 7. Summary of Inorganic Reductions by Sodium Borohydride[a]

1 (IA)	2 (IIA)	3 (IIIB)	4 (IVB)	5 (VB)	6 (VIB)	7 (VIIB)	8	9 (VIIIB)
H $1+ \rightarrow$ 0								
Li	Be							
Na	Mg							
K	Ca	Sc	Ti	V $5+ \rightarrow 4+$ $5+ \rightarrow 3+$	Cr $6+ \rightarrow 3+$ $3+ \rightarrow CrB_2$	Mn $7+ \rightarrow 2+$ $\rightarrow 4+$ $\rightarrow 6+$	Fe $3+ \rightarrow 2+$	Co $2+ \rightarrow Co_2B$
Rb	Sr	Y	Zr	Nb	Mo $6+ \rightarrow 3+$ $6+ \rightarrow 5+$	Tc	Ru $3+ \rightarrow 0$	Rh $3+ \rightarrow 0$
Cs	Ba	La-Lu	Hf	Ta	W $6+ \rightarrow$ W Blue	Re	Os $8+ \rightarrow 0$	Ir $4+ \rightarrow 0$
Fr	Ra	Ac						

La	Ce $4+ \rightarrow 3+$	Pr	Nd	Pm	Sm	Eu	Gd	Tb	Dy	Ho	Er	Tm	Yb	Lu
Ac	Th	Pa	U $6+ \rightarrow 4+$	Np	Pu	Am	Cm	Bk	Cf	Es	Fm	Md	No	

10	11 (IB)	12 (IIB)	13 (IIIA)	14 (IVA)	15 (VA)	16 (VIA)	17 (VIIA)	18 (VIIIA)
								He Ne
			B	C	N $3+ \rightarrow 3-$	O	F	
			Al	Si	P $3+ \rightarrow 3-$	S $4+ \rightarrow 3+$ $\rightarrow 0$ $\rightarrow 2-$	Cl $3+ \rightarrow 1-$ $1+ \rightarrow 1-$	Ar
Ni $2+ \rightarrow Ni_2B$	Cu $2+ \rightarrow 0$ $\rightarrow Cu_mB_n$	Zn	Ga	Ge $4+ \rightarrow GeH_4$ $\rightarrow Ge_2H_6$	As $3+ \rightarrow 0$ $\rightarrow AsH_3$ $\rightarrow As_2H_4$	Se $4+ \rightarrow 0$ $\rightarrow 2-$	Br	Kr
Pd $2+ \rightarrow 0$	Ag $1+ \rightarrow 0$	Cd $2+ \rightarrow 0$	In	Sn $4+ \rightarrow 2+$ $\rightarrow 0$ $\rightarrow SnH_4$	Sb $3+ \rightarrow 0$ $\rightarrow SbH_3$ $\rightarrow Sb_2H_4$	Te $4+ \rightarrow 0$ $\rightarrow 2-$	I $5+ \rightarrow 0$ $\rightarrow 1-$	Xe
Pt $4+ \rightarrow 0$	Au $3+ \rightarrow 0$	Hg $2+ \rightarrow 1+$ $\rightarrow 0$	Tl $3+ \rightarrow 1+$ $\rightarrow 0$	Pb $2+ \rightarrow 0$	Bi $3+ \rightarrow 0$ $\rightarrow BiH_3$	Po	At	Rn

[a] $\boxed{\rightarrow}$ Reducible by NaBH$_4$; $\boxed{\Rightarrow}$ reduction known to be quantitative.

619

Sodium borohydride reacts with boron halides to form diborane [*19287-45-7*], B_2H_6, which is more conveniently handled as the monomer BH_3 complexed with an ether, sulfide, or amine (see BORON COMPOUNDS). Sodium borohydride is used extensively for the reduction of organic compounds. Its broad synthetic utility is based not only on its ability to reduce aldehydes (qv) and ketones (qv) selectively and efficiently in the presence of other functional groups, but also to reduce other functional groups, eg, esters, di- and polysulfides, imines and quaternary iminium compounds, under special conditions or with added catalysts or coreagents.

Preparation.   Sodium borohydride is manufactured from sodium hydride and trimethyl borate in a mineral oil medium at about 275°C (26),

$$4\,NaH + (CH_3O)_3B \longrightarrow NaBH_4 + 3\,NaOCH_3 \qquad (25)$$

The process is shown schematically in Figure 1, which also shows the conversion to potassium borohydride. Commercial $NaBH_4$ is 97+% pure; yields are better than 90%. Other processes for manufacturing $NaBH_4$ have been described (27,28), but are not commercially important.

Storage and Handling.   Sodium borohydride is classified as a flammable solid. It is available as powder, caplets, and granules and as a 12% solution in caustic soda. The dry powder, caplets, and granules should be stored and handled in the same manner as other flammable hygroscopic material. In dry air or in sealed containers they are stable indefinitely. Fire extinguishers should be of the dry chemical type; carbon dioxide or halocarbon extinguishers must not be used. Accepted storage and handling procedures for the 12% solution are the same as for 50% liquid caustic soda. $NaBH_4$ solutions may decompose with hydrogen evolution when heated, subjected to acid conditions, or in the presence of the metal salts or finely divided metallic precipitates of nickel, cobalt, copper, or iron. Under normal storage conditions, the 12% solution loses less than 0.1% $NaBH_4$ per year by decomposition. Storage and reaction vessels should be provided with adequate venting. $NaBH_4$ has been widely used in chemical processes for many years and is considered safe to handle and use, with normal industrial safety precautions.

Economic Aspects.   Sodium borohydride is produced in large quantities mainly as powder and stabilized water solution. Potassium borohydride powder is produced in lesser amounts. Commercial quantities of sodium borohydride powder sell for ca $55/kg (1992 price); the 12% solution in caustic soda is priced at ca $47/kg of contained $NaBH_4$.

Uses.   The principal uses of $NaBH_4$ are in synthesis of pharmaceuticals (qv) and fine organic chemicals; removal of trace impurities from bulk organic chemicals; wood-pulp bleaching, clay leaching, and vat-dye reductions; and removal and recovery of trace metals from plant effluents.

In pharmaceutical applications, the selectivity of sodium borohydride is ideally suited for conversion of high value intermediates, such as steroids (qv), in multistep syntheses. It is used in the manufacture of a broad spectrum of products such as analgesics, antiarthritics, antibiotics (qv), prostaglandins (qv), and central nervous system suppressants. Typical examples of commercial aldehyde

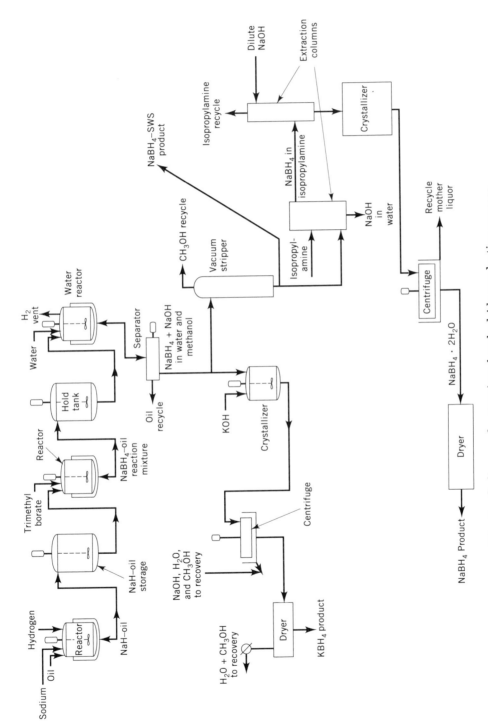

Fig. 1. Sodium and potassium borohydride production process.

reductions are found in the manufacture of Vitamin A (29) (see VITAMINS) and dihydrostreptomycin (30). An acyl azide is reduced in the synthesis of the antibiotic chloramphenicol (31) and a carbon–carbon double bond is reduced in an intermediate in the manufacture of the analgesic Talwin (32).

A particularly useful reaction has been the selective 1,2-reduction of α,β-unsaturated carbonyl compounds to allylic alcohols, accomplished by $NaBH_4$ in the presence of lanthanide halides, especially cerium chloride. Initially applied to ketones (33), it has been broadened to aldehydes (34) and acid chlorides (35). $NaBH_4$ by itself gives mixtures of the saturated and unsaturated alcohols.

Sodium borohydride is used for the removal of trace impurities, such as carbonyls or peroxides, from bulk organic chemicals (alcohols or glycols). This is called process-stream purification, and greatly improves stability in products sensitive to deterioration and development of undesirable colors and odors in, for example, plasticizer alcohols. Sodium borohydride is used widely to generate sodium hydrosulfite (eq. 24) for reductive bleaching of mechanical pulps for manufacture of several grades of paper, especially newsprint. This technology is also being applied in clay leaching and in vat dye reductions. Sodium borohydride provides a simple, efficient means of removing dissolved trace metals from manufacturing-plant effluents, eg, by precipitating mercury from chlor-alkali plant effluents, lead from wastewater streams of tetraethyllead producers, and copper and other metals from printed circuit board manufacturing operations. This precipitation method allows recovery and recycle of valuable metals, such as silver, gold, and noble metals from spent photographic fixer solutions, film manufacturing effluents, and used x-ray and photographic films, plating solutions, etc. There is also extensive use of $NaBH_4$ in the analytical determination of metals which form volatile hydrides (As, Sb, Bi, Sn, Pb, etc).

Other Borohydrides. Potassium borohydride was formerly used in color reversal development of photographic film and was preferred over sodium borohydride because of its much lower hygroscopicity. Because other borohydrides are made from sodium borohydride, they are correspondingly more expensive. Generally their reducing properties are not sufficiently different to warrant the added cost. Zinc borohydride [17611-70-0], $Zn(BH_4)_2$, however, has found many applications in stereoselective reductions. It is less basic than $NaBH_4$, but is not commercially available owing to poor thermal stability. It is usually prepared on site in an ether solvent. Zinc borohydride was initially applied to stereoselective ketone reductions, especially in prostaglandin syntheses (36), and later to aldehydes, acid halides, and esters (37).

Borohydride Derivatives. Modification of the BH_4^- anion has provided derivatives of widely differing reducing properties. Alkoxyborohydrides, such as sodium trimethyoxyborohydride [16940-17-3], $NaBH(OCH_3)_3$, exhibit enhanced reducing power but are less selective and more sensitive to decomposition by water. Sodium cyanoborohydride [25895-60-7], $NaBH_3CN$, on the other hand, shows weakened reducing properties and is unique among the complex hydrides because it is stable in acidic aqueous solutions to a pH of about 3. Typical behavior of a number of borohydride derivatives toward organic functional groups are summarized in Table 6.

Unusual reducing properties can be obtained with borohydride derivatives formed *in situ*. A variety of reductions have been reported, including hydrogenolysis of carbonyls and alkylation of amines with sodium borohydride in carboxylic acids such as acetic and trifluoroacetic (38), in which the acyloxy-borohydride is the reducing agent.

Sodium or tetramethylammonium triacetoxyborohydride has become the reagent of choice for diastereoselective reduction of β-hydroxyketones to anti-diols. Trialkylborohydrides, eg, alkali metal tri-*sec*-butylborohydrides, show outstanding stereoselectivity in ketone reductions (39).

Sodium Cyanoborohydride. $NaBH_3CN$ is a hygroscopic white solid that is remarkably stable, both thermally and hydrolytically. It is prepared by reaction of $NaBH_4$ and HCN in THF (40),

$$NaBH_4 + HCN \xrightarrow[THF]{} NaBH_3CN + H_2 \qquad (26)$$

or by the reaction of borane and sodium cyanide in THF (41),

$$THF \cdot BH_3 + NaCN \longrightarrow NaBH_3CN + THF \qquad (27)$$

which avoids handling toxic, hazardous hydrogen cyanide. Sodium cyanoborohydride and its solutions in THF are very toxic and must be handled with great care. Ingestion and skin contact must be avoided.

Sodium cyanoborohydride is remarkably chemoselective. Reduction of aldehydes and ketones are, unlike those with $NaBH_4$, pH-dependent, and practical reduction rates are achieved at pH 3 to 4. At pH 5–7, imines ($>C=N-$) are reduced more rapidly than carbonyls. This reactivity permits reductive amination of aldehydes and ketones under very mild conditions (42).

$$R_2C=O + HNR_2' \xrightarrow{H^+} R_2C=\overset{+}{N}R_2' \xrightarrow{NaBH_3CN} R_2CHNR_2' \qquad (28)$$

In addition, sodium cyanoborohydride reduces a wide variety of substrates (43) (see Table 6).

Sodium cyanoborohydride has become important in biochemical applications that require hydrolytic stability of the reducing agent and chemoselectivity, in sensitive molecules. It is also a preferred reagent for oxime reductions.

Aluminohydrides. Complex hydrides of aluminum have been known since the late 1940s, when lithium aluminum hydride was first made (44). In the ensuing decades, it was commercialized as a reducing agent in the synthesis of pharmaceuticals and other specialty organics. Sodium aluminum hydride [13770-96-2] (Table 5) was marketed in the 1960s and early 1970s, and has become available again in quantity in the 1990s, but as of this writing this less costly aluminohydride has not become widely used. A dialkoxy derivative of $NaAlH_4$, sodium dihydrobis(2-methoxyethoxy)aluminate [22722-98-1], $NaAlH_2(OC_2H_4OCH_3)_2$, is another aluminohydride which is commercially available. Other aluminohydrides have been prepared only on a laboratory scale.

In general, the aluminohydrides are more active and powerful reducing agents than the corresponding borohydrides. They decompose vigorously with water. Reaction also occurs with alcohols, although more moderately, providing a route to substituted derivatives.

Lithium Aluminum Hydride.   Freshly prepared lithium aluminum hydride [*16853-85-3*] is a white crystalline solid that tends to become gray during storage, although very little loss in purity occurs. The solid hydride is stable in dry air at room temperature, but decomposes above 125°C. Its heat of formation is -117 kJ/mol (-28.0 kcal/mol).

$$2 \text{ LiAlH}_4 \longrightarrow 2 \text{ LiH} + 2 \text{ Al} + 3 \text{ H}_2 \qquad (29)$$

Lithium aluminum hydride is hygroscopic. In moist air it decomposes slowly, but it reacts vigorously with water. Because of this potential for simultaneous evolution of significant amounts of heat and hydrogen gas (2.36 L/g at STP), careful storage and handling of LiAlH$_4$ is required. It is classified as a flammable solid. Complete protection from moisture, such as excessive humidity or condensate drip from cold water lines, is mandatory. Equipment in which the hydride is to be used must also be dry and purged with dry nitrogen, adequately vented, and fully grounded to eliminate static electricity hazards. Suitable explosion-resistant electrical services must also be provided.

LiAlH$_4$ is soluble in ethers, 35–40 g/100 g diethyl ether at 25°C. Solubility in THF, the other common solvent for LiAlH$_4$, is 13 g/100 g at 25°C. Polyethylene glycol dialkyl ethers are also good solvents.

Reactions.   Although lithium aluminum hydride is best known as a nucleophilic reagent for organic reductions, it converts many metal halides to the corresponding hydride, eg, Ge, As, Sn, Sb, and Si (45).

$$\text{SiCl}_4 + \text{LiAlH}_4 \longrightarrow \text{SiH}_4 + \text{LiCl} + \text{AlCl}_3 \qquad (30)$$

Organohalides of silicon can also be used to make organosilanes. Group 1 and 2 halides generally undergo metathesis, but the products are unstable, with few exceptions.

$$\text{MX}_n + n \text{ LiAlH}_4 \longrightarrow \text{M(AlH}_4)_n + n \text{ LiX} \qquad (31)$$

Aluminum hydride, formed from AlCl$_3$ (44), is a polymeric solid that is difficult to obtain completely free of the ether solvent used.

$$3 \text{ LiAlH}_4 + \text{AlCl}_3 \longrightarrow 4 \text{ AlH}_3 + 3 \text{ LiCl} \qquad (32)$$

Lithium aluminum hydride is a powerful reducing agent for organic functional groups, a property that has led to its extensive use especially in pharmaceutical synthesis. The voluminous literature describing its uses has resulted in several excellent reviews on the subject (see *General References*). Nearly all of the usual functional groups are reduced by LiAlH$_4$, olefinic bonds being the notable exception. These reductions are generally rapid and take place under mild conditions. Although isolated carbon–carbon multiple bonds are not reducible by LiAlH$_4$, reduction is possible when conjugation or suitable substituents are present.

Preparation. Commercial manufacture of LiAlH₄ uses the original synthetic method (44), ie, addition of a diethyl ether solution of aluminum chloride to a slurry of lithium hydride (Fig. 2).

$$4 \text{ LiH} + \text{AlCl}_3 \longrightarrow \text{LiAlH}_4 + 3 \text{ LiCl} \qquad (33)$$

The stoichiometry (4 mol lithium hydride to 1 mol LiAlH₄) makes this an inherently expensive process, even though high yields of pure product are obtained. For large-scale production, metathesis from NaAlH₄ is economically preferred.

$$\text{NaAlH}_4 + \text{LiCl} \longrightarrow \text{LiAlH}_4 + \text{NaCl} \qquad (34)$$

Sodium Aluminum Hydride. Sodium aluminum hydride can be prepared from NaH,

$$4 \text{ NaH} + \text{AlCl}_3 \longrightarrow \text{NaAlH}_4 + 3 \text{ NaCl} \qquad (35)$$

but direct synthesis from the elements is more economical (46,47),

$$\text{Na} + \text{Al} + 2 \text{ H}_2 \xrightarrow[\text{13.8 MPa (138 atm)}]{150°\text{C}} \text{NaAlH}_4 \qquad (36)$$

The direct synthesis in an aromatic hydrocarbon medium is patented, using a triethylaluminum catalyst (48); in this case, crystallization of the product from a solvent is not needed.

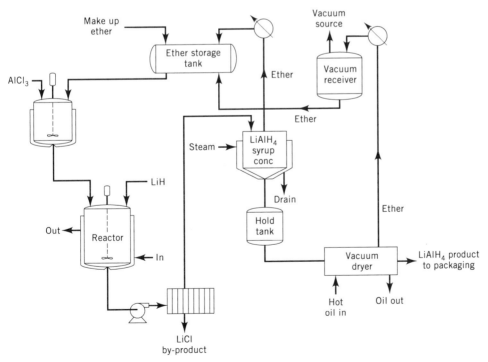

Fig. 2. Commercial manufacture of lithium aluminum hydride.

The reactions of $NaAlH_4$ are the same as those of $LiAlH_4$. However, it is much less soluble; THF is the only good solvent. Heterogeneous reductions in good yield with $NaAlH_4$ in hydrocarbon media have been reported (49).

Aluminohydride Derivatives. The few known derivatives of the aluminohydrides are principally alkoxy substitutions, including the trimethoxy, $LiAlH(OCH_3)_3$, triethoxy, $LiAlH(OC_2H_5)_3$, and tri-t-butoxy aluminohydrides, $LiAlH(O-t-C_4H_9)_3$. The alkoxyaluminohydrides are milder reducing agents than $LiAlH_4$. Steric and electronic effects in lithium tri-t-butoxyaluminohydride [38884-28-5] make this compound very chemoselective, resembling $NaBH_4$ more closely than $LiAlH_4$ in this regard. A key reaction of $LiAlH(O-t-C_4H_9)_3$ is the partial reduction of acid chlorides to the aldehyde, rather than to the alcohol, which can be carried out in the presence of a variety of other functional groups (50).

$$RCCl \xrightarrow[-78°C]{LiAlH(O-t-C_4H_9)_3} RCH \tag{37}$$

Sodium dihydrobis(2-methoxyethoxy)aluminate [22722-98-1], $NaAlH_2$-$(OC_2H_4OCH_3)_2$, has the largest commercial use among the aluminohydride derivatives. The pure compound is a slightly yellow, glassy solid, soluble in ethers and aromatic hydrocarbons. It is supplied as a 70% solution in benzene or toluene. Like $LiAlH_4$, it reduces a wide range of organic compounds, including aldehydes, ketones, acid chlorides, carboxylic acids and anhydrides, esters, lactones and lactams, amides, oximes, nitriles, and nitro compounds (51). Both aliphatic and aromatic halides are dehalogenated by the reagent. Convenience in handling and solubility in aromatic hydrocarbons are the principal advantages of this compound.

Health and Safety Factors

Generalizations on the toxicity of hydrides as a class of compounds are difficult to make because of variations in their chemical reactivity. Furthermore, there is little published toxicological information available.

Qualitative insight can, however, be obtained by focusing on the reactions with water, the extent and vigor of which can vary widely. In general, however, hydrides react exothermically with water, resulting in the generation of hydrogen.

$$MH_n + n H_2O \longrightarrow M(OH)_n + n H_2 \tag{38}$$

This hydrolysis reaction is accelerated by acids or heat and, in some instances, by catalysts. Because the flammable gas hydrogen is formed, a potential fire hazard may result unless adequate ventilation is provided. Ingestion of hydrides must be avoided because hydrolysis to form hydrogen could result in gas embolism.

Another aspect of the hydrolysis of hydrides is the alkalinity that results, especially from alkali metal and alkaline-earth hydrides. This alkalinity can cause chemical burns in skin and other tissues. Affected skin areas should be flooded with copious amounts of water.

The importance of hydrolysis potential, ie, whether moisture or water is present, is illustrated by the following example. In the normal dermal toxicity test, namely dry product on dry animal skin, sodium borohydride was found to be nontoxic under the classification of the Federal Hazardous Substances Act. Furthermore, it was not a skin sensitizer. But on moist skin, severe irritation and burns resulted.

Hydrolysis considerations obviously demand that hydrides be kept away from contact with acids.

Although there is little toxicity information published on hydrides, a threshold limit value (TLV) for lithium hydride in air of 25 $\mu g/m^3$ has been established (52). More extensive data are available (53) for sodium borohydride in the powder and solution forms. The acute oral LD_{50} of $NaBH_4$ is 50–100 mg/kg for $NaBH_4$ and 500–1000 mg/kg for the solution. The acute dermal LD_{50} (on dry skin) is 4–8 g/kg for $NaBH_4$ and 100–500 mg/kg for the solution. The reaction or decomposition by-product sodium metaborate is slightly toxic orally (LD_{50} is 2000–4000 mg/kg) and nontoxic dermally.

BIBLIOGRAPHY

"Hydrides" in *ECT* 1st ed., Vol. 7, pp. 595–598, by C. R. Hough, Polytechnic Institute of Brooklyn; in *ECT* 2nd ed., Vol. 11, pp. 200–224, by A. A. Hinckley, Ventron Corp.; in *ECT* 2nd ed., Suppl., pp. 499–509, by E. A. Sullivan, Ventron Corp.; in *ECT* 3rd ed., Vol. 12, pp. 772–792, by E. A. Sullivan and R. C. Wade, Thiokol/Ventron Div.

1. C. E. Messer, *A Survey Report on Lithium Hydride*, Tufts University, Medford, Mass., 1960.
2. J. Plesek and S. Hermanek, *Sodium Hydride: Its Use in the Laboratory and in Technology*, trans. by G. Jones, Academic, Prague, 1968.
3. F. H. Welch, *Nucl. Eng. Des.* **26**, 444 (1974).
4. M. D. Banus and A. A. Hinckley, *Adv. Chem. Ser.* **19**, 106 (1957).
5. H. L. Alexander, *Iron Steel Eng.* **24**(5), 44 (1947); M. Sittig, *Iron Age* **172**(25), 137 (1953); U.S. Pat. 3,721,626 (Mar. 20, 1973), V. Lhoty and J. Mostecky (to Valcovny Plechu, Narodni Podnik); Jpn. Pat. Appl. 83 84,984 (May 21, 1983) (to Mitsubishi Heavy Industries, Ltd).
6. C. A. Brown and S. Krishnamurthy, *J. Organomet. Chem.* **156**, 111 (1978); C. A. Brown, *J. Am. Chem. Soc.* **95**, 982 (1973).
7. Jpn. Pat. Appl. 87,262,407 (Nov. 14, 1987), K. Uchida and M. Tokunaga (to Hitachi Metals, Ltd.); Jpn. Pat. Appl. 89,110,705 (Apr. 27, 1989), H. Shibusawa (to Hitachi Metals, Ltd.).
8. G. E. Coats and F. Glocking, *J. Chem. Soc.*, 2526 (1954).
9. L. H. Shepherd, G. L. Ter Haar, and E. M. Martlett, *Inorg. Chem.* **8**, 976 (1969); U.S. Pat. 3,577,414 (May 4, 1971), L. H. Shepherd (to Ethyl Corp.); U.S. Pat. 3,806,520 (Apr. 23, 1974), G. Ter Haar (to Ethyl Corp.).
10. U.S. Pat. 3,816,608 (June 11, 1974), R. W. Baker, C. R. Bergeron, and A. Nugent (to Ethyl Corp.); U.S. Pats. 3,830,906 (Aug. 20, 1974) and 3,832,456 (Aug. 27, 1974), R. J. Larans, P. Kobetz, and R. W. Johnson (to Ethyl Corp.); U.S. Pat. 3,883,645 (May 13, 1975), C. R. Bergeron and R. W. Baker (to Ethyl Corp.).
11. E. C. Ashby and R. G. Beach, *Inorg. Chem.* **9**, 2300 (1970).
12. J. T. Groves and S. Kittisopikul, *Tetrahedron Lett.*, 4291 (1977); B. A. San Miguel, B. Maillard, and B. Delmond, *Tetrahedron Lett.* **28**, 2127 (1987); H. Oelschlager and co-workers, *Sci. Pharm.* **57**, 367 (1989).
13. G. A. Shkurko, *Gig. Tr.* **9**, 74 (1973).
14. A. C. Switendick, in R. Bau, *Adv. Chem. Ser.* **167**, 265 (1978).

15. J. J. Reilly and J. R. Johnson, *Report 1976, BNL-22062*; J. J. Reilly, *Report 1977, BNL-23047*, Brookhaven National Laboratory, Upton, NY.
16. J. H. N. van Vucht, F. A. Kuijpers, and H. C. A. M. Bruning, *Philips Res. Rep.* **25**, 133 (1970).
17. Jpn. Pat. 75,116,308 (Sept. 11, 1975), M. Kitada (to Hitachi Ltd.).
18. "Hydrogen Energy," *Proceedings of Hydrogen Economy, Miami Energy Conference, 1974*, Plenum Press, New York, 1975.
19. D. M. Cavagnaro, *Hydrogen Storage, Pt. 2: Hydrogen as a Hydride*, NTIS, Springfield, Va., 1978. A bibliography with abstracts.
20. H. I. Schlesinger and H. C. Brown, *J. Am. Chem. Soc.* **75**, 186 (1953).
21. R. Paul and N. Joseph, *Bull. Soc. Chim. Fr.* **19**, 550 (1952).
22. H. C. Brown, E. J. Mead, and B. C. Subba Rao, *J. Am. Chem. Soc.* **77**, 6209 (1955). H. C. Brown, S. Narasimhan, and Y. M. Choi, *J. Org. Chem.* **47**, 4702 (1982); K. Higashiura and K. Ienaga, *J. Org. Chem.* **57**, 764 (1992).
23. K. N. Mochalov, V. S. Khain, and G. G. Gil'manshin, *Kinet. Katal.* **6**, 541 (1965).
24. H. C. Brown and K. Ichekawa, *J. Am. Chem. Soc.* **83**, 4372 (1961).
25. G. S. Panson and C. E. Weill, *J. Inorg. Nucl. Chem.* **15**, 184 (1960).
26. H. I. Schlesinger, H. C. Brown, and A. E. Finholt, *J. Am. Chem. Soc.* **75**, 205 (1953).
27. T. P. Forbath, *Chem. Eng.* **68**, 92 (June 12, 1961).
28. U.S. Pat. 3,515,522 (June 2, 1970), V. Pecak, J. Vit, and V. Prochazka (to Czechoslovakia Academy of Sciences); Jpn. Pat. Appl. 90,217,304 (Aug. 30, 1990), T. Iwao and co-workers, (to Mitsui Toatsu Chemicals, Inc.); Jpn. Pat. Appl. 91,275,502 (Dec. 6, 1991), T. Iwao and co-workers (to Nippon Aluminum Alkyls, Ltd.).
29. Brit. Pat. 778,753 (July 10, 1957), O. D. Hawks (to Eastman Kodak Co.).
30. U.S. Pat. 2,790,792 (Apr. 30, 1957), M. A. Kaplan (to Bristol Labs, Inc.); U.S. Pat. 3,397,197 (Aug. 13, 1968), E. H. Makepeace (to Chas. Pfizer and Co., Inc.).
31. G. Ehrhart, W. Siedel, and H. Nahm, *Chem. Ber.* **90**, 2088 (1957).
32. U.S. Pat. 3,250,678 (May 10, 1966), S. Archer (to Sterling Drug Inc.).
33. J. L. Luche, *J. Am. Chem. Soc.* **100**, 2226 (1978); A. L. Gemal and J. L. Luche, *J. Am. Chem. Soc.* **103**, 5459 (1981); A. S. Lee, A. W. Norman, and W. H. Okamura, *J. Org. Chem.* **57**, 3846 (1992); L. A. Sandareses and J. L. Luche, *J. Org. Chem.* **57**, 2757 (1992).
34. J. D. White and co-workers, *J. Org. Chem.* **57**, 2270 (1992); C. J. Moody and co-workers, *J. Org. Chem.* **57**, 2105 (1992); G. Mehta, N. Krishnamurthy, and S. P. Karra, *J. Am. Chem. Soc.* **113**, 5765 (1991).
35. K. V. Lakshmy and co-workers, *Org. Prep. Proced. Int.* **17**, 251 (1985).
36. K. C. Nicolaou, W. E. Barnette, and R. L. Magolda, *J. Am. Chem. Soc.* **103**, 3472 (1981); A. E. Greene, A. Padilla, and P. Crabbe, *J. Org. Chem.* **43**, 4377 (1978); T. Nakata and T. Oishi, *Tetrahedron Lett.*, 1641 (1980).
37. B. C. Ranu and R. Chakraborty, *Tetrahedron Lett.* **31**, 7663 (1990); J. A. Marshall and R. Sedrani, *J. Org. Chem.* **56**, 5496 (1991); H. Kotsuki and co-workers, *Bull. Chem. Soc. Jpn.* **61**, 2684 (1988); S. Narasimhan, P. Palmer, and K. G. Prasad, *Indian J. Chem. Sect. B.* **30B**, 1150 (1991); T. Yamakawa, M. Masaki, and H. Nohira, *Bull. Chem. Soc. Jpn.* **64**, 2730 (1991).
38. G. W. Gribble and co-workers, *J. Am. Chem. Soc.* **96**, 7812 (1974); M. Soucek, J. Urban, and D. Saman, *Collect. Czech. Chem. Commun.* **55**, 761 (1990); A. F. Abdel-Magid, C. A. Maryannoff, and K. G. Carson, *Tetrahedron Lett.* **31**, 5595 (1990); G. W. Gribble and C. F. Nutaitis, *Org. Prep. Proced. Int.* **17**, 317 (1985); C. F. Nutaitis, *J. Chem. Ed.* **66**, 673 (1978); C. K. McClure and K-Y. Jung, *J. Org. Chem.* **56**, 2326 (1991); D. A. Evans and W. C. Black, *J. Am. Chem. Soc.* **114**, 2260 (1992).
39. H. C. Brown and S. Krishnamurthy, *J. Am. Chem. Soc.* **94**, 7159 (1972); H. C. Brown, S. C. Kim, and S. Krishnamurthy, *J. Org. Chem.* **45**, 1 (1980).

40. R. C. Wade and co-workers, *Inorg. Chem.* **9**, 2146 (1970).
41. U.S. Pat. 4,301,129 (Nov. 17, 1981), R. C. Wade and B. C. Hui (to Thiokol Corp.).
42. R. F. Borch, M. D. Bernstein, and H. D. Durst, *J. Am. Chem. Soc.* **93**, 2897 (1971).
43. R. O. Hutchins and N. R. Natale, *Org. Prep. Proced. Int.* **11**, 201 (1979).
44. A. E. Finholt, A. C. Bond and H. I. Schlesinger, *J. Am. Chem. Soc.* **69**, 1199 (1947).
45. A. E. Finholt and co-workers, *J. Am. Chem. Soc.* **69**, 2692 (1947).
46. E. C. Ashby, G. J. Brendel, and H. E. Redman, *Inorg. Chem.* **2**, 499 (1963).
47. U.S. Pat. 3,556,740 (Jan. 19, 1971), J. H. Murib (to National Distillers and Chemicals Corp.).
48. U.S. Pat. 4,045,545 (Aug. 30, 1977), E. C. Ashby (to Ethyl Corp.).
49. L. I. Zakharkin, D. N. Maslin, and V. V. Gavrilenko, *Izv. Akad. Nauk SSSR Ser. Khim.*, 561 (1964).
50. H. C. Brown and B. C. Subba Rao, *J. Am. Chem. Soc.* **80**, 5377 (1958).
51. V. Bazant and co-workers, *Tetrahedron Lett.*, 3303 (1968).
52. E. J. Fairchild, ed., *Registry of Toxic Effects of Chemical Substances*, U.S. Dept. of Health, Education and Welfare, Washington, D.C., 1977.

General References

T. E. Dergazarian and co-workers, *JANAF Thermochemical Data*, Dow Chemical Co., Midland, Mich., 1961. Thermochemical data from these tables were used whenever possible.
R. Bau, "Transition Metal Hydrides," *Advances in Chemistry Series*, Vol. 167, American Chemical Society, Washington, D.C., 1978.
G. D. James, *Rec. Chem. Prog.* **31**, 199 (1970). Inorganic chemistry of complex hydrides.
E. R. H. Walker, *Chem. Soc. Rev.* **5**, 23 (1976). Functional group selectivity of complex hydrides.
A. P. Scanzillo, *Encyclopedia Industrial Chemical Analysis*, Vol. 2, John Wiley & Sons, Inc., New York, 1971 pp. 242–260. Analysis of hydrides.
R. M. Adams and A. R. Siedle, *Boron, Metallo-Boron Compounds and Boranes*, Wiley-Interscience, New York, 1964, Chapt. 6, pp. 373–506. Borohydrides.
W. M. Mueller, J. P. Blackledge, and G. G. Libovitz, *Metal Hydrides*, Academic Press, Inc., New York, 1968, especially Chapt. 12, pp. 546–674.
B. D. James and M. G. H. Wallbridge, *Progr. Inorg. Chem.*, Vol. 11, Wiley-Interscience, New York, 1970, pp. 99–231. Borohydrides.
J. S. Pizey, *Synthetic Reagents*, Vol. 1, Ellis Harwood, Ltd., Chichester, U.K., 1974, Chapt. 2, pp. 101–294. Lithium aluminum hydride.
E. C. Ashby, *Adv. Inorg. Chem. Radiochem.* **8**, 283 (1966). Aluminohydrides.
N. G. Gaylord, *Reduction with Complex Metal Hydrides*, Interscience Publishers, New York, 1956.
M. Hudlicky, *Reductions in Organic Chemistry*, Halstead Press (div. of John Wiley & Sons, Inc.), New York, 1984.
A. Pelter, K. Smith, and H. C. Brown, *Borane Reagents*, Academic Press, Inc., London, 1988.
A. Hajos, *Complex Hydrides*, Elsevier/North-Holland, Inc., New York, 1979.

EDWARD A. SULLIVAN
Morton International

HYDROBORATION

Hydroboration is the addition of a boron–hydrogen bond across a double or triple carbon–carbon bond to give an organoborane:

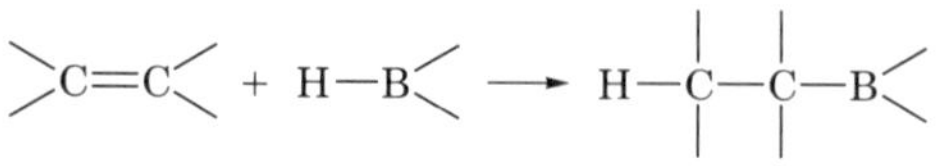

$$\text{>C=C<} \;+\; \text{H—B<} \;\longrightarrow\; \text{H—}\overset{|}{\underset{|}{C}}\text{—}\overset{|}{\underset{|}{C}}\text{—B<}$$

Other multiple bonds, eg, >C=O, —C≡N, —N=N—, also undergo the addition. However, those reactions not involving boron–carbon bond formation belong to reductions and are not described here. The boron atom in organoboranes can be replaced with other elements, usually with high stereoselectivity; many functional groups are tolerated. Consequently, organoboranes are among the most versatile synthetic intermediates, and their role in organic synthesis is constantly increasing.

One of the newer and more fruitful developments in this area is asymmetric hydroboration giving chiral organoboranes, which can be transformed into chiral carbon compounds of high optical purity. Other new directions focus on catalytic hydroboration, asymmetric allylboration, cross-coupling reactions, and applications in biomedical research. This article gives an account of the most important aspects of the hydroboration reaction and transformations of its products. For more detail, monographs and reviews are available (1–13).

The Hydroboration Reaction

Diborane [*19287-45-7*], the first hydroborating agent studied, reacts sluggishly with olefins in the gas phase (14,15). In the presence of weak Lewis bases, eg, ethers and sulfides, it undergoes rapid reaction at room temperature or even below 0°C (16–18). The catalytic effect of these compounds on the hydroboration reaction is attributed to the formation of monomeric borane complexes from the borane dimer, eg, borane–tetrahydrofuran [*14044-65-6*] (**1**) or borane–dimethyl sulfide [*13292-87-0*] (**2**) (19–21). Stronger complexes formed by amines react with olefins at elevated temperatures (22–24).

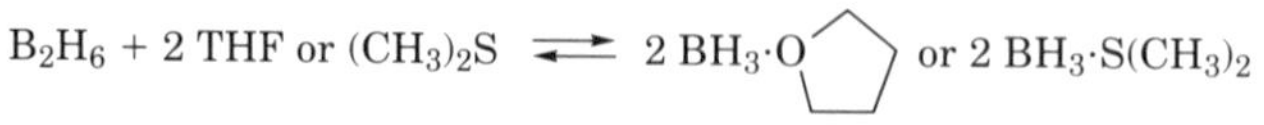

$$B_2H_6 + 2\text{ THF or }(CH_3)_2S \;\rightleftharpoons\; 2\text{ }BH_3\cdot O\!\!\bigcirc \text{ or } 2\text{ }BH_3\cdot S(CH_3)_2$$

(**1**)   (**2**)

Mono-, di-, and trialkylboranes may be obtained from olefins and the trifunctional borane molecule. Simple unhindered alkenes yield trialkylboranes and it is not possible to halt the reaction at the mono- or dialkylborane stage. With more hindered and trisubstituted alkenes the reaction can be controlled to stop at the dialkylborane stage.

$$BH_3 \xrightarrow{\text{olefin}} RBH_2 \xrightarrow{\text{olefin}} R_2BH \xrightarrow{\text{olefin}} R_3B$$

Tetrasubstituted and some hindered trisubstituted alkenes react rapidly only to the monoalkylborane stage. Rarely, when the tetrasubstituted double bond is incorporated in a cyclic structure, does hydroboration under normal conditions fail (25–27). However, such double bonds may react under conditions of greater force (25,28–31). Generally, trialkylboranes are stable at normal temperatures, undergoing thermal dissociation at temperatures above 100°C (32–34). In the presence of B—H bonds, trialkylboranes undergo a redistribution reaction (35–38).

$$R_3B + RBH_2 \rightleftharpoons 2\,R_2BH$$

Mono- and dialkylboranes usually exist as bridged dimers, *sym*-dialkyldiboranes and *sym*-tetraalkyldiboranes. Only very hindered alkylboranes, eg, bis(2,3-dimethyl-2-butyl) borane (39), are monomeric. However, for convenience of presentation monomers are shown in the equations.

Mechanism. The characteristic features of hydroboration were originally accounted for in terms of a simple four-center transition-state model (Fig. 1), serving as a useful working hypothesis (1,40). Participation of the p orbital of boron removes symmetry restrictions for the $2\pi_s + 2\sigma_s$ addition (41). The electron donation from the π orbital of an olefin to the p orbital of boron and back-donation from the boron–hydrogen bond to the π^* orbital would account for a concerted reaction involving the transition state shown in Figure 1. The gas-phase reaction of ethylene with the borane monomer proceeds with small, ca 8.4 kJ/mol activation energy (42). Calculations suggest the formation of a loose π-complex reorganizing to a four-center transition state having little or no activation energy. The experimental results obtained for solution reactions establish the dissociation to the free monomer and its reaction with an olefin as a general mechanism for all hydroborations with borane-electron donor base complexes as well as dialkylborane dimers (21). Thus the borane–dimethyl sulfide complex reacts with 2,3-dimethyl-2-butene in toluene by a two-step dissociation mechanism. The complex first liberates borane in an equilibrium and then the free borane reacts with the alkene.

$$BH_3{\cdot}S(CH_3)_2 \longrightarrow BH_3 + S(CH_3)_2$$

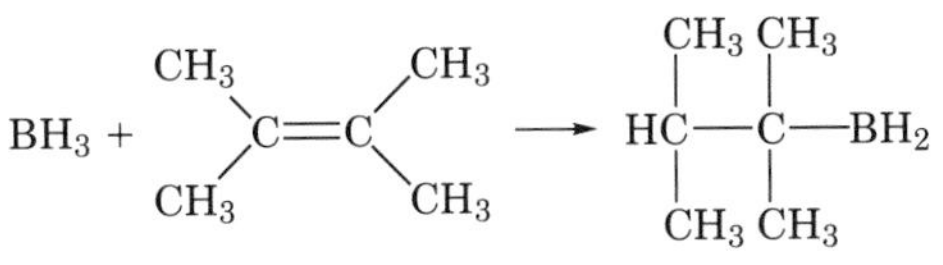

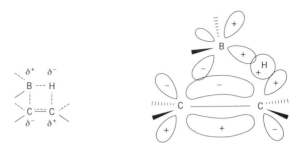

Fig. 1. Simple four-center transition-state model of hydroboration.

Retardation of the reaction rate by the addition of dimethyl sulfide is in accord with this mechanism. Borane–amine complexes and the dibromo-borane–dimethyl sulfide complex react similarly (43). Dimeric dialkylboranes initially dissociate (at rate k_1) to the monomers subsequently reacting with an olefin at rate k_2 (44). For highly reactive olefins $k_2 > k_{-1}$ (recombination) and the reaction is first-order in the dimer. For slowly reacting olefins $k_{-1} > k_2$ and the reaction shows 0.5 order in the dimer.

Hydroborating Agents

Mono- and dialkylboranes obtained by controlled hydroboration of hindered olefins and by other methods can serve as valuable hydroborating agents for more reactive olefins. Heterosubstituted boranes are also available and used for this purpose. These borane derivatives show differences in reactivity and selectivity.

Borane Complexes. Borane solutions in tetrahydrofuran are commercially available or can be prepared by absorbing gaseous diborane in tetrahydrofuran. Diborane can be conveniently generated from the reaction of sodium borohydride with boron trifluoride etherate (3) (see BORON COMPOUNDS, BORON HYDRIDES, HETEROBORANES, AND THEIR METALLA DERIVATIVES). Several other methods for its preparation or generation and use *in situ* are known (45–55). Although borane–THF (**1**) is a useful reagent, it must be stabilized with small amounts of sodium borohydride for longer storage at 0°C. The choice of solvent is limited to tetrahydrofuran, and concentration does not exceed 2 M in BH_3.

Borane–dimethyl sulfide complex (BMS) (**2**) is free of these inconveniences. The complex is a pure 1:1 adduct, ca 10 M in BH_3, stable indefinitely at room temperature and soluble in ethers, dichloromethane, benzene, and other solvents (56,57). Its disadvantage is the unpleasant smell of dimethyl sulfide, which is volatile and water insoluble. Borane–1,4-thioxane complex (**3**), which is also a pure 1:1 adduct, ca 8 M in BH_3, shows solubility characteristics similar to BMS (58). 1,4-Thioxane [15980-15-1] is slightly soluble in water and can be separated from the hydroboration products by extraction into water.

$$BH_3 \cdot S \underset{}{\bigcirc} O \qquad\qquad BH_3 \cdot N(CH_2CH_3)_3$$

$$\textbf{(3)} \qquad\qquad\qquad \textbf{(4)}$$

Borane–triethylamine complex (**4**) is used when slow liberation of borane at elevated temperatures is advantageous, eg, in the cyclic hydroboration of trienes to avoid the formation of polymers (59).

Monosubstituted Boranes. Only a few monoalkylboranes are directly available by hydroboration. Tertiary hexylborane, 2,3-dimethyl-2-butylborane [*3688-24-2*] (thexylborane, Thx) BH$_2$ (**5**), easily prepared from 2,3-dimethyl-2-butene, is the best studied (3,60–62). It should be used at 0°C, freshly prepared, to avoid isomerization at ambient temperatures. Its complex with N,N-diethylaniline is significantly more stable, fortunately retaining the reactivity (63).

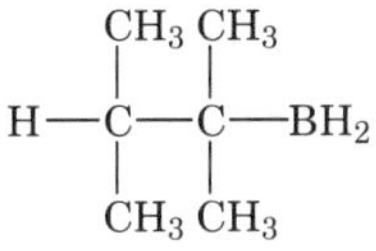

(5)    (6) 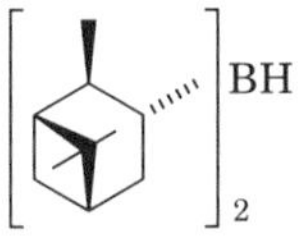   (8)

Monoisopinocampheylborane [*64234-27-1*], IpcBH$_2$ (**6**) is an important asymmetric hydroborating agent. It is prepared from α-pinene (**7**) either directly or better by indirect methods. The product obtained by hydroboration of α-pinene [*80-56-8*] is an equilibrium mixture. Optically pure monoisopinocampheylborane is best prepared from α-pinene via diisopinocampheylborane [*1091-56-1*], Ipc$_2$BH (**8**) (64,65). Both enantiomers are readily available.

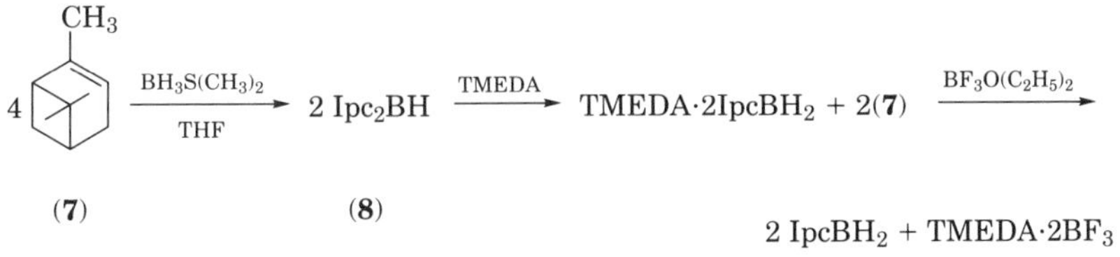

(7)   (8)

2 IpcBH$_2$ + TMEDA·2BF$_3$

(6)

The method does not require optically pure α-pinene because 100% enantiomeric excess (ee) is achieved by crystallization of the intermediate TMEDA·2IpcBH$_2$ adduct, where TMEDA = (CH$_3$)$_2$NCH$_2$CH$_2$N(CH$_3$)$_2$ (tetramethylethylenediamine). Other chiral monoalkylboranes derived from 2-alkyl- and 2-phenylapopinene are slightly more selective reagents as compared to monoisopinocampheylborane (66–68).

A number of less hindered monoalkylboranes is available by indirect methods, eg, by treatment of a thexylborane–amine complex with an olefin (69), the reduction of monohalogenoboranes or esters of boronic acids with metal hydrides (70–72), the redistribution of dialkylboranes with borane

(64) or the displacement of an alkene from a dialkylborane by the addition of a tertiary amine (73). To avoid redistribution, monoalkylboranes are best used *in situ* or freshly prepared. However, they can be stored as monoalkylborohydrides or complexes with tertiary amines. The free monoalkylboranes can be liberated from these derivatives when required (69,74–76). Methylborane, a remarkably unhindered monoalkylborane, exhibits extraordinary hydroboration characteristics. It hydroborates hindered and even unhindered olefins to give sequentially alkylmethyl- and dialkylmethylboranes (77–80).

Monohalogenoboranes are conveniently prepared from borane–dimethyl sulfide and boron trihalides (BX$_3$ where X = Cl, Br, I) by redistribution reaction, eg, for monochloroborane–dimethyl sulfide [*63348-81-2*] (**9**) (81–83). Other methods are also known (84–87).

$$2\ BH_3 \cdot S(CH_3)_2 + BCl_3 \cdot S(CH_3)_2 \longrightarrow 3\ H_2BCl \cdot S(CH_3)_2$$

$$(\mathbf{9})$$

The products are liquids, soluble in various solvents and stable over prolonged periods. Monochloroborane is an equilibrium mixture containing small amounts of borane and dichloroborane complexes with dimethyl sulfide (81). Monobromoborane–dimethyl sulfide complex shows high purity (82,83). Solutions of monochloroborane in tetrahydrofuran and diethyl ether can also be prepared. Strong complexation renders hydroboration with monochloroborane in tetrahydrofuran sluggish and inconvenient. Monochloroborane solutions in less complexing diethyl ether, an equilibrium with small amounts of borane and dichloroborane, show excellent reactivity (88,89). Monochloroborane–diethyl etherate [*36594-41-9*] (**10**) may be represented as H$_2$BCl·O(C$_2$H$_5$)$_2$.

Disubstituted Boranes.   Even slight differences in steric or electronic effects of substituents may have an effect on the hydroboration reaction course. These effects are well demonstrated in disubstituted boranes, and consequently a range of synthetically useful reagents has been developed.

Primary dialkylboranes react readily with most alkenes at ambient temperatures and dihydroborate terminal acetylenes. However, these unhindered dialkylboranes exist in equilibrium with mono- and trialkylboranes and cannot be prepared in a state of high purity by the reaction of two equivalents of an alkene with borane (35–38). Nevertheless, such mixtures can be used for hydroboration if the products are acceptable for further transformations or can be separated (90). When pure primary dialkylboranes are required they are best prepared by the reduction of dialkylhalogenoboranes with metal hydrides (91–93). To avoid redistribution they must be used immediately or be stabilized as amine complexes or converted into dialkylborohydrides.

In contrast to simple unhindered dialkylboranes, borinanes and borepanes do not redistribute readily. These boraheterocyclic reagents can be prepared by hydroboration of the corresponding dienes with borane, 9-BBN, or monochloroborane, followed by thermal isomerization or reduction, respectively (89,94–99). The hydroboration of a simple hindered alkene, 2-methyl-2-butene, with borane-THF stops at the dialkylborane stage to give bis(3-methyl-2-butyl)borane [*1069-54-1*] (disiamylborane), Sia$_2$BH (**11**) (100–102). The reagent is not stable over longer periods and it is recommended that it be used freshly prepared or stored at 0°C for only a few hours. It has high steric requirements, discriminates well between terminal and internal double bonds, and is often used in selective organic transformations (103–105).

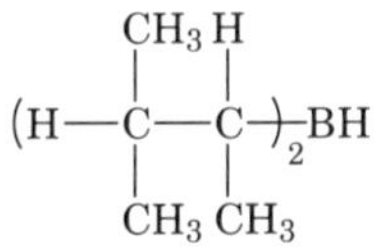

(11) 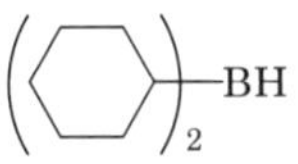   (12)

Dicyclohexylborane [*1568-65-6*], Chx$_2$BH (12) is prepared in quantitative yield by the same method. It is a white solid, sparingly soluble in ether or tetrahydrofuran. For most purposes isolation is not necessary and it can be used as a slurry in these solvents. Its steric requirements are slightly lower as compared to Sia$_2$BH (57,106–109). Much better thermal stability allows its use at higher temperatures.

9-Borabicyclo [3.3.1] nonane [*280-64-8*], 9-BBN (13) is the most versatile hydroborating agent among dialkylboranes. It is commercially available or can be conveniently prepared by the hydroboration of 1,5-cyclooctadiene with borane, followed by thermal isomerization of the mixture of isomeric bicyclic boranes initially formed (57,109).

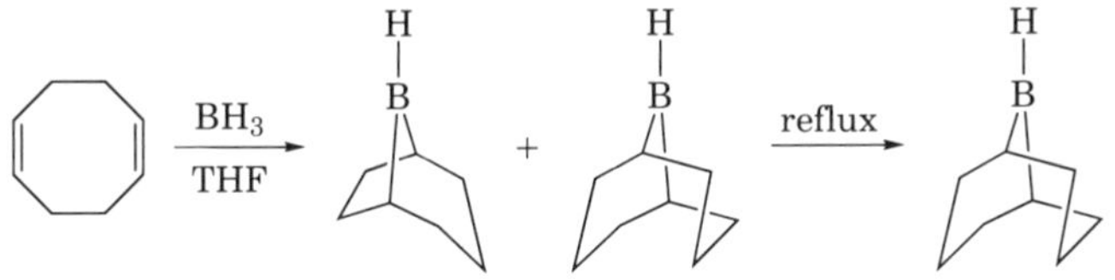

(13)

Other procedures have also been reported (38,110,111). The properties and chemistry of 9-BBN have been reviewed (112). The reagent is a white crystalline solid, stable indefinitely at room temperature, soluble in hexane, carbon tetrachloride, benzene, tetrahydrofuran, and diethyl ether. It exists as a dimer [*21205-91-4*] in the solid state and in non- or slightly complexing solvents, eg, hexane or diethyl ether. In tetrahydrofuran and dimethyl sulfide the dimer partially dissociates to the solvent complexed monomer (44). The thermal stability of 9-BBN is remarkable. It distills at 195°C at 1.6 kPa (12 mm Hg) without dissociation to the monomer and can be heated for several hours to 200°C without loss of the hydride activity. The reactivity of 9-BBN toward alkenes of different structures varies over the range of 10^5 and is somewhat lower compared to Sia$_2$BH (113). The thermal stability of 9-BBN enables hydroborations at 60–80°C. At this temperature most alkenes react in 1 h. Even tetrasubstituted double bonds, which fail to react with Sia_2BH, are hydroborated with 9-BBN. It is less sterically demanding than Sia_2BH and discriminates little between E and Z isomers (114). On the other hand, 9-BBN is quite sensitive to electronic factors. Thus, unique among dialkylboranes, it is more reactive toward alkenes than alkynes (115). The B-alkyl-9-BBN derivatives are exceptionally resistant to thermal isomerization. Consequently, isomerization sometimes encountered in the hydroboration of certain labile systems with borane can be circumvented by using 9-BBN (116).

The most hindered of all presently known hydroborating agents is possibly dimesitylborane, an air-stable white solid, slightly soluble in tetrahydrofuran, the best etheral solvent. It is commercially available or can be prepared according to the following reaction (117):

$$4 \text{ Mes}_2\text{BF} + \text{LiAlH}_4 \longrightarrow 4 \text{ Mes}_2\text{BH} + \text{AlF}_3 + \text{LiF}$$

Its reactions with olefins, governed by steric rather than electronic factors, are very sluggish. Even simple 1-alkenes require 8 h at 25°C for complete reaction. In contrast, alkynes are hydroborated with great ease to alkenylboranes, high steric requirements of the reagent preventing dihydroboration (117).

Dihalogenoboranes are conveniently prepared by the redistribution of borane–dimethyl sulfide with boron trihalide–dimethyl sulfide complexes (82,83), eg, for dibromoborane–dimethyl sulfide [55671-55-1] (14).

$$BH_3 \cdot S(CH_3)_2 + 2 \text{ BBr}_3 \cdot S(CH_3)_2 \longrightarrow 3 \text{ BHBr}_2 \cdot S(CH_3)_2$$

(14)

Although dichloroborane reacts directly with alkenes in the gas phase (118), its complexes with diethyl ether and dimethyl sulfide are so strong that direct hydroboration does not proceed (119,120). The addition of a decomplexing agent, eg, boron trichloride, is necessary for hydroboration. Dibromoborane–dimethyl sulfide is a more convenient reagent. It reacts directly with alkenes and alkynes to give the corresponding alkyl- and alkenyldibromoboranes (120–123). Dibromoborane differentiates between alkenes and alkynes hydroborating internal alkynes preferentially to terminal double and triple bonds (123). Unlike other substituted boranes it is more reactive toward 1,1-disubstituted than monosubstituted alkenes (124).

Several oxygen- and sulfur-substituted boranes have been reported (125–130). 1,3,2-Benzodioxaborole [274-07-7] (catecholborane, CB) (15) is the one best studied. It is commercially available or can be prepared by the reaction of catechol with borane·THF (57,131), or by other procedures (132). The product is a liquid existing as a monomer, remarkably stable to disproportionation. No

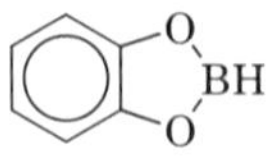

(15)

change in properties was observed after one year at 0°C. For hydroboration, it can be used neat or in various solvents, eg, carbon tetrachloride, chloroform, benzene, diethyl ether, or tetrahydrofuran (133,134). The electron donation from oxygen to boron reduces its reactivity. Consequently, hydroborations must be conducted at elevated temperatures. Alkenes require 100°C and alkynes 70°C with the neat reagent. Higher reactivity of alkynes allows selective hydroboration of triple bonds in the presence of double bonds (135). Catecholborane [274-07-7] and dialkoxyboranes hydroborate alkenes under mild conditions in the presence of lithium borohydride (130,136).

The hydroboration of alkenes with catecholborane under the influence of Wilkinson's catalyst proceeds at room temperature, whereas heating is necessary in the absence of catalyst (137). This result provided the impetus for studies of catalytic hydroboration. The studies focused on catecholborane, rhodium, and iridium catalysts, other catalytic systems being less explored (138–148). This rapidly evolving area of research has been reviewed (138). The rate of catalyzed reaction is sensitive to the substitution of the double bond, the reactivity of alkenes decreasing in the following order: monosubstituted > disubstituted terminal > disubstituted internal > trisubstituted (139). The regioselectivity of the addition to terminal alkenes is comparable to the uncatalyzed reaction. However, the mechanisms of the two processes are quite different and in allylic and homoallylic systems the regio- and diastereoselectivities of the catalyzed and uncatalyzed reactions are often opposite (138–140). The diastereoselectivity of the catalyzed hydroboration of the allylic alcohol derivative has been used advantageously in the synthesis of the C-1 to C-11 polypropionate fragment of the polyester antibiotic Ionomycin A (141). Directed hydroboration of phosphinite derivatives of allylic alcohols (stoichiometric ratios) and homoallylic amides (catalytic reaction) has been achieved (139).

Alkylchloro- and alkylbromoboranes are valuable reagents for the synthesis of di- and trialkylboranes having different alkyl groups. Thexylchloroborane [75030-54-5], ThxBHCl (16) is a very useful reagent.

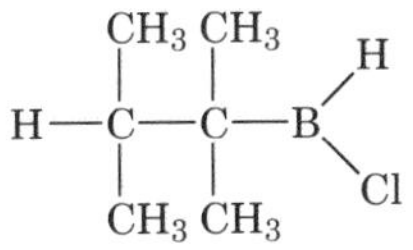

$$H-\underset{\underset{CH_3}{|}}{\overset{\overset{CH_3}{|}}{C}}-\underset{\underset{CH_3}{|}}{\overset{\overset{CH_3}{|}}{C}}-B\underset{Cl}{\overset{H}{<}}$$

(16)

It can be prepared by the reaction of monochloroborane with 2,3-dimethyl-2-butene (149), or from thexylborane (5) and hydrogen chloride (150). It undergoes rapid redistribution in diethyl ether, but its complex with dimethyl sulfide is remarkably stable at room temperature. Thexylchloroborane reacts with unhindered alkenes, producing the corresponding alkylthexylboranes cleanly. With more hindered alkenes the reaction is slow and accompanied by redistribution, leading to side products (149). Generally, the stepwise hydroboration of simple alkenes with monohaloboranes cannot be controlled precisely, although in certain cases monoalkylchloroboranes can be obtained (88). The opposite approach, stepwise reduction, works better, providing access to di- and trialkylboranes with different alkyl groups (91).

$$BHBr_2 \cdot S(CH_3)_2 \xrightarrow{alkene\ 1} BR^1Br_2 \cdot S(CH_3)_2 \xrightarrow{LiAlH_4} BHR^1Br \xrightarrow{alkene\ 2} BR^1R^2Br \cdot S(CH_3)_2 \xrightarrow[CH_3OH]{CH_3ONa}$$

$$BR^1R^2OCH_3 \xrightarrow{LiAlH_4} BHR^1R^2 \xrightarrow{alkene\ 3} BR^1R^2R^3$$

Among chiral dialkylboranes, diisopinocampheylborane (**8**) is the most important and best-studied asymmetric hydroborating agent. It is obtained in both enantiomeric forms from naturally occurring α-pinene. Several procedures for its synthesis have been developed (151–153). The most convenient one, providing product of essentially 100% ee, involves the hydroboration of α-pinene with borane–dimethyl sulfide in tetrahydrofuran (154). Other chiral dialkylboranes derived from terpenes, eg, 2- and 3-carene (155), limonene (156), and longifolene (157,158), can also be prepared by controlled hydroboration. A more tedious approach to chiral dialkylboranes is based on the resolution of racemates. *trans*-2,5-Dimethylborolane, which shows excellent enantioselectivity in the hydroboration of all principal classes of prochiral alkenes except 1,1-disubstituted terminal double bonds, has been prepared by this method (159).

Selectivity

Chemoselectivity.   Double and triple carbon–carbon bonds are more reactive than most other functionalities toward borane and substituted boranes. Consequently, many functional groups are tolerated in the hydroboration reaction (Table 1). Using a suitably chosen hydroborating agent, only aldehydes, ketones, and carboxylic acids must be protected (166,167). However, it may not be necessary to protect ketones in the catalytic hydroboration for isolated functionalities (137). Hydroboration of allylic and vinylic derivatives leads to α-, β-, or γ-substituted organoboranes prone to eliminations or rearrangements (168–175). In some cases such transformations may be synthetically useful. For example, the hydroboration of enamines enables a simple conversion of aldehydes and ketones into (Z)- or (E)-alkenes (176).

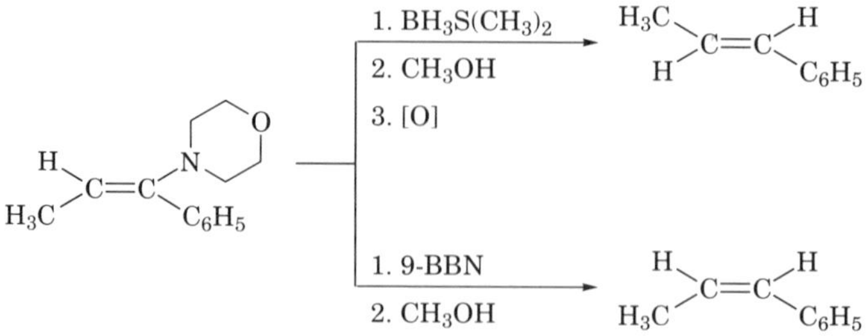

Table 1. Reactivities of Representative Functional Groups Toward Selected Boranes[a]

Functional groups	H₃B·THF (1)	ThxBH₂ (5)	ThxBHCl (16)	Sia₂BH (11)	9-BBN (13)	CB (15)
alkyne	f	f	f	f	f	m
alkene	f	f	f	f	f	s
aldehyde	f	f	f	f	f	f
ketone	f	f	f	f	f	m
carboxylic acid	f	vs	m[b]	n	vs	m
nitrile	f	vs	f	vs	vs	s
lactone	f	s	s	f	m	
carboxylic acid anhydride	f	f	f	vs	vs	m
epoxide	s	s	vs	s	vs	m
ester	s	vs	n	n	m	s
amide	s	s	vs	s	s	m
acid chloride	n	n	s	n	f	s
nitro	n	n	n	n	n	n
alkyl halide	n	n		n	n	n
Reference	160	161	162,163	164	165	133

[a]f, fast; m, moderate; s, slow; vs, very slow; n, no reaction.
[b]Reduced to aldehyde.

Conjugated ketones are either reduced to allylic alcohols (177) or undergo 1,4-addition to give enolboranes (178–181). Cisoid vinylic epoxides and aziridines give 1,4-addition products stereoselectively, probably via a six-membered transition state (182,183). However, sterically demanding hydroborating agents may react selectively with the double bond (184). The organoboranes initially formed from conjugated esters rearrange to the more stable borinates (185).

Many other examples in the literature illustrate the possibilities of chemoselective hydroborations (124,186–189). For example, selectivity between double and triple bonds has been shown (124).

Different reagents react with different bonds in the same molecule. PCC represents pyridinium chlorochromate.

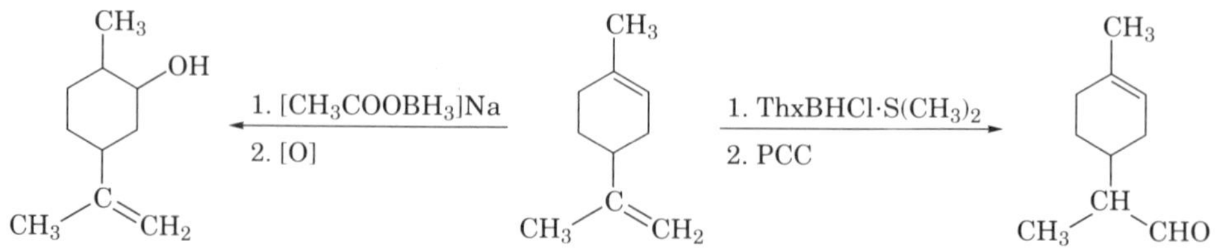

Among other examples are unsaturated amines (190,191), carbohydrates (192,193), heterocyclic olefins (194), phosphorus and sulfur compounds (195,196), organometallic compounds (148,197,198), functionalized intermediates in natural product syntheses (98–105,199,200), and many other compounds described in reviews (5,6,8,9,13).

Directive Effects. Hydroboration of olefins involves predominant placement of the boron atom at the less hindered site of the double bond. The direction of addition is governed by polarization of the boron–hydrogen bond and by combination of steric and electronic effects of substituents at the double bond. Simple 1-alkenes react with borane to place 94% of the boron at the terminal position. For most purposes this purity is sufficient. However, 6% of addition to the secondary position means that the organoborane may contain up to 18% of di-*n*-alkyl-*s*-alkylborane. This may complicate transformations involving selective migrations of alkyl groups. The selectivity of borane addition to disubstituted 1-alkenes and trisubstituted alkenes is higher, 98–99% of the boron going to the less substituted site. Internal disubstituted alkenes and alkynes are hydroborated with low selectivity even when the substituents differ in size, indicating low sensitivity of borane to steric effects. Most of these difficulties are circumvented by the application of substituted boranes showing high sensitivity to steric factors. Directive effects in the hydroboration of various alkenes and functional derivatives with representative hydroborating agents are shown in Table 2. 9-Borabicyclo [3.3.1] nonane is probably the most often used substituted borane because of its excellent chemo- and regioselectivity and a broad range of reactivity toward structurally different olefins. Dimesitylborane shows remarkable selectivity in the addition to internal alkynes (117).

Functional groups influence the regioselectivity of hydroboration by inductive, mesomeric, and steric effects, their magnitude depending on the proximity of the double bond and the functional group (168–175). An increased addition of boron to the secondary position (11–18%) is already observed for 3-butenyl derivatives (168). In allylic systems the electron-withdrawing substituents direct the boron atom to the β-position, the effect increasing with the increasing electronegativity of the substituent. Vinylic substituents with a strong +mesomeric effect, eg, —OR and —NR$_2$, direct boron to the β-position. The reverse is observed for −mesomeric substituents such as silicon and boron. For example, *gem*-diboryl compounds are obtained by dihydroboration of terminal alkynes even with bulky dialkylboranes such as (**12**)(201). Directive effects in the hydroboration of functionalized alkynes are similar to the corresponding alkenes.

$$RC\equiv CH + 2\ Chx_2BH \longrightarrow RCH_2CH(BChx)_2$$

(**12**)

Table 2. Directive Effects in the Hydroboration of Alkenes[a]

Hydroborating agent	Reference	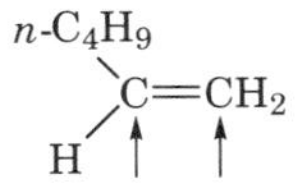 $n\text{-}C_4H_9\overset{\textstyle}{\underset{H}{}}C=CH_2$	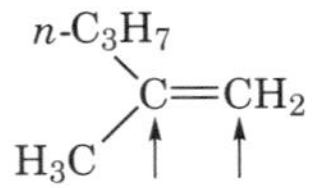 $n\text{-}C_3H_7\overset{\textstyle}{\underset{H_3C}{}}C=CH_2$	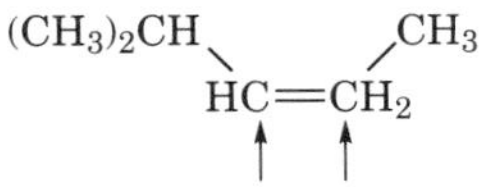 $(CH_3)_2CH\overset{\textstyle CH_3}{}HC=CH_2$	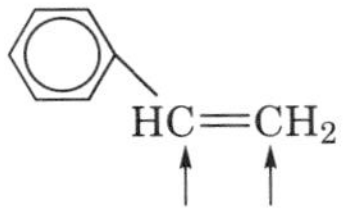 $\text{(phenyl)}HC=CH_2$
$BH_3\cdot THF$	40	6 94	1 99	43 57	19 81
Sia_2BH	101	1 99	3 97	3 97	2 98
9-BBN	109	0.1 99.9	0.2 99.8	0.2 99.8	2 98
$BH_2Cl\cdot O(C_2H_5)_2$	88	0.5 99.5	0.1 99.9	40 60	4 96
$BHBr_2\cdot S(CH_3)_2$	120,122	0.4 99.6	0.8 99.2		1 99
$ThxBHCl\cdot S(CH_3)_2$	149	0.6 99.4	>99.9	3 97	1 99

Hydroborating agent	Reference	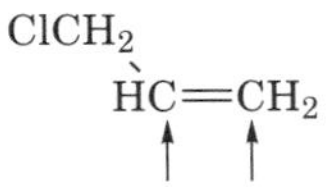 $ClCH_2\overset{\textstyle}{}HC=CH_2$	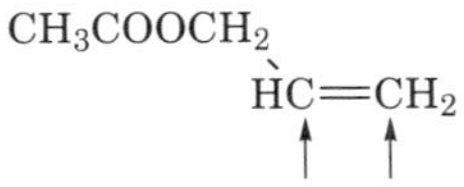 $CH_3COOCH_2\overset{\textstyle}{}HC=CH_2$	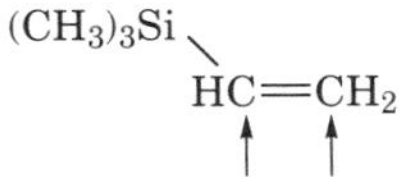 $(CH_3)_3Si\overset{\textstyle}{}HC=CH_2$	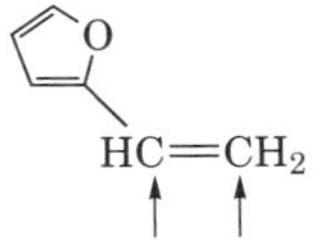 $\text{(2-furyl)}HC=CH_2$
$BH_3\cdot THF$	170	40 60	35 65	60 40[b]	16 84[c]
Sia_2BH	171	5 95	2 98	5 95[d]	0 100[c]
9-BBN	185	1.1 98.9	2.4 97.6	100[b]	3 97[c]

[a]Numbers are percentages.
[b]Ref. 194.
[c]Ref. 195.
[d]Ref. 196.

641

$$R-C\equiv C-CH_2Cl \xrightarrow{Chx_2BH} \begin{array}{c} R \\ H \end{array} C=C \begin{array}{c} CH_2Cl \\ BChx_2 \end{array}$$

Generally, borane shows low regioselectivity in the addition to allylic and vinylic derivatives. Fortunately, sterically demanding substituted boranes can overcome the electronic effects of functional groups and react with high selectivity. The possibility of controlling the direction of addition is important for the preparation of isomerically pure substituted organoboranes (202,203) (Table 2). The regioselectivities of catalyzed and uncatalyzed hydroborations may be different (139) (Fig. 2).

Stereoselectivity. The addition of a boron–hydrogen bond across the double bond proceeds cleanly in a cis fashion leading to simple diastereoselection for suitably substituted double bonds. Double bonds are approached by the hydroborating agent from the less sterically hindered face. The thermodynamically less stable addition products may result, as has been demonstrated for β-pinene and camphene (204,205). Borane discriminates well between faces differing significantly in steric hindrance. When the difference is small low selectivity results. Bulky, sterically demanding hydroborating agents show higher stereoselectivity. Functional groups may influence the stereochemistry of addition. For example, unsaturated bicyclic amines by strong complexation of borane may have one face of the double bond more hindered, and the addition is directed to the opposite side. In contrast, weakly complexing groups like ethers may facilitate the addition from the same side.

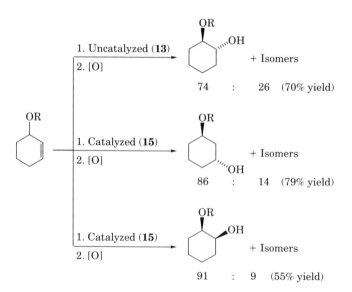

Fig. 2. 3-Alkoxycyclohexene yields different alcohols of varying isomeric purities and yields, as shown under the various conditions noted. The catalyst is $Rh[P(C_6H_5)_3]_3Cl$. R in the first two reactions is $Si(CH_3)_2(t\text{-}C_4H_9)$. In the last case $R = P(C_6H_5)_2$.

Reactions of Organoboranes

Organoboranes available by hydroboration and also by other methods are versatile synthetic intermediates. The organic groups attached to boron can be transferred, usually with high stereoselectivity, to hydrogen, oxygen, nitrogen, halogens, sulfur, selenium, metal atoms, and carbon. Consequently, carbon–hetero atom and carbon–carbon bonds can be constructed. The combination of hydroboration and functionalization corresponds overall to anti-Markovnikov addition of the elements of HX to a double bond.

$$RCH{=}CH_2 \xrightarrow{\quad HB{<}\quad} RCH_2CH_2B{<} \xrightarrow{\quad\quad} RCH_2CH_2X$$

REPLACEMENT OF BORON BY HYDROGEN OR A HETERO ATOM

Protonolysis. Simple trialkylboranes are resistant to protonolysis by alcohols, water, aqueous bases, and mineral acids. In contrast, carboxylic acids react readily with trialkylboranes, removing the first alkyl group at room temperature and the third one at elevated temperatures. Acetic and propionic acids are most often used. The reaction proceeds with retention of configuration of the alkyl group via a cyclic, six-membered transition state (206).

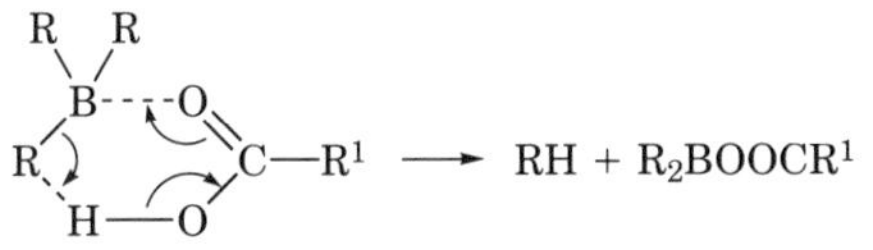

$$\longrightarrow RH + R_2BOOCR^1$$

Primary alkyl groups are more reactive than secondary and tertiary. Pivalic acid accelerates the rate of protonolysis of trialkylboranes with water and alcohols (207,208). The reaction can be controlled to give excellent yields of dialkylborinic acids and esters.

$$R_3B + HOR^1 \xrightarrow{(CH_3)_3CCOOH} R_2BOR^1 + RH$$

This reaction may be applied for the protection of hydroxyl groups in sugars (209). Allylic, benzylic, aryl, alkenyl, and alkynyl groups are more readily protonolyzed than alkyl groups. It is often sufficient to use water or methanol. For example, allylic organoboranes are hydrolyzed with water at room temperature, the reaction proceeding with the allylic rearrangement (14,210,211).

In general, hydroboration–protonolysis is a stereoselective noncatalytic method of cis-hydrogenation providing access to alkanes, alkenes, dienes, and enynes from olefinic and acetylenic precursors (108,212). Procedures for the protonolysis of alkenylboranes containing acid-sensitive functional groups under neutral or basic conditions have been developed (213,214).

Oxidation. The oxidation reactions of organoboranes have been reviewed (5,7,215). Hydroboration–oxidation is an anti-Markovnikov cis-hydration of carbon–carbon multiple bonds. The standard oxidation procedure employs 30% hydrogen peroxide and 3 M sodium hydroxide. The reaction proceeds with retention of configuration (216).

Sodium perborate can safely substitute for hydrogen peroxide (217). A variety of common organic functional groups such as double or triple bonds, halogens, ethers, esters, ketones, nitriles, and others are unaffected under the reaction conditions. However, certain functionally substituted organoboranes and functional groups present in the target molecules, eg, vicinal halo- and alkoxyorganoboranes or aldehydes, are sensitive to the alkaline medium. In such cases, buffering (218), simultaneous addition of base and peroxide (171), or reagents other than hydrogen peroxide, eg, trimethylamine N-oxide (219–224) or peracids (174,225,226), are used. The reactivity order of alkyl- and alkenylboranes toward trimethylamine N-oxide decreases in the following order: tertiary > secondary > primary alkyl > alkenyl groups (221,224). Alkynyl groups are not oxidized by the reagent (227). Organoboranes containing different groups can be selectively oxidized (222).

Usually, organoboranes are sensitive to oxygen. Simple trialkylboranes are spontaneously flammable in contact with air. Nevertheless, under carefully controlled conditions the reaction of organoboranes with oxygen can be used for the preparation of alcohols or alkyl hydroperoxides (228,229). Aldehydes are produced by oxidation of primary alkylboranes with pyridinium chlorochromate (188). Chromic acid at pH < 3 transforms secondary alkyl and cycloalkylboranes into ketones; pyridinium chlorochromate can also be used (230,231). A convenient procedure for the direct conversion of terminal alkenes into carboxylic acids employs hydroboration with dibromoborane–dimethyl sulfide and oxidation of the intermediate alkyldibromoborane with chromium trioxide in 90% aqueous acetic acid (232,233).

Halogenolysis. Alkylboranes are readily converted into the corresponding alkyl chlorides by a free-radical reaction with nitrogen trichloride (234). The reactivity order of alkyl groups is tertiary > secondary > primary, reflecting the stability of free radicals. The stereochemical integrity of alkyl groups is not retained, eg, tri(exo-2-norbornyl) borane gives 70% exo- and 30% endo-2-nobornyl chloride. Only two groups of the trialkylborane are readily accommodated in the reaction. Other reagents such as iron(III) chloride and dichloramine T can also be used (235,236). Although only one group is utilized by the latter reagent, its stereoselectivity is excellent.

The reactions of trialkylboranes with bromine and iodine are greatly accelerated by bases. The use of sodium methoxide in methanol gives good yields of the corresponding alkyl bromides or iodides. All three primary alkyl groups are utilized in the bromination reaction and only two in the iodination reaction.

Secondary groups are less reactive and the yields are lower. Both Br and I re-actions proceed with predominant inversion of configuration; thus, for example, tri(exo-2-norbornyl)borane yields >75% endo product (237,238). In contrast, the dark reaction of bromine with tri(exo-2-norbornyl)borane yields cleanly exo-2-norbornyl bromide (239). Consequently, the dark bromination complements the base-induced bromination.

The iodination reaction can also be conducted with iodine monochloride in the presence of sodium acetate (240) or iodine in the presence of water or methanolic sodium acetate (241). Under these mild conditions functionalized alkenes can be transformed into the corresponding iodides. Application of B-alkyl-9-BBN derivatives in the chlorination and dark bromination reactions allows better utilization of alkyl groups (235,242). An indirect stereoselective procedure for the conversion of alkynes into (E)-1-halo-1-alkenes is based on the mercuration reaction of boronic acids followed by in situ bromination or iodination of the intermediate mercuric salts (243).

Both (E)- and (Z)-1-halo-1-alkenes can be prepared by hydroboration of 1-alkynes or 1-halo-1-alkynes followed by halogenation of the intermediate boronic esters (244,245). Differences in the addition–elimination mechanisms operating in these reactions lead to the opposite configurations of iodides as compared to bromides and chlorides.

An alternative synthesis of (Z)-1-halo-1-alkenes involves hydroboration of 1-halo-1-alkynes, followed by protonolysis (246,247). Disubstituted (E)- and (Z)-alkenyl bromides can be prepared from (E)- and (Z)-alkenyl boronic esters, respectively, by treatment with bromine followed by base (248).

Replacement of Boron by Nitrogen. Organoboranes react with amino compounds containing good leaving groups, eg, chloramine, hydroxylamine-O-sulfonic acid, or mesitylsulfonylhydroxylamine to give primary amines (249–251). The replacement proceeds with retention of configuration (252). The yields are moderate because only one or two alkyl groups are utilized. How-ever, this inconvenience can be circumvented by the use of alkyldimethylboranes (249). Diamines can also be prepared (253).

Secondary amines are obtained in excellent yields by the reaction of alkyldichloroboranes with alkyl-, cycloalkyl-, or aryl azides (254). Reagents con-taining two leaving groups on nitrogen, eg, N-chloro-O-(dinitrophenyl)hydroxyl-amine, ClNHODNP, can also be used for the synthesis of secondary amines (255). Tertiary amines and other nitrogen compounds such as N-substituted aziridines (254), alkyldimethylamines (256), alkyl azides (257), and N-alkylsulfonylamides (258) can be prepared by the reaction of alkylboranes with β-iodoalkyl azides, N-chlorodimethylamine in the presence of galvinoxyl, sodium azide, and chlo-ramine T, respectively.

Replacement of Boron by Sulfur and Selenium. Trialkylboranes are cleaved by dialkyl- and diaryldisulfides in an air-catalyzed radical reaction producing mixed thioethers (259).

$$\text{B—R} + \text{R}^1\text{SSR}^1 \longrightarrow \text{RSR}^1 + \text{B—SR}^1$$

Alkylthiocyanates and alkylselenocyanates are obtained by treatment of trialkylboranes with potassium thiocycanate (260) and sodium selenoisocyanate (261), in
the presence of iron(III) compounds, respectively. Unsymmetrical trialkylboranes
react preferentially at the more highly branched alkyl group. Alkenylphenyl
selenides are obtained by the reaction of alkenylboronic acids with phenylselenyl
bromide (262).

Mercuration.   Mercury(II) salts react with alkyl-, alkenyl-, and arylboranes
to yield organomercurials, which are useful synthetic intermediates (263). For
example, dialkylmercury and alkylmercury acetates can be prepared from primary trialkylboranes by treatment with mercury(II) chloride in the presence of
sodium hydroxide or with mercury(II) acetate in tetrahydrofuran (3,264). Mercuration of s-alkylboranes is sluggish and requires prolonged heating. Alkenyl
groups are transferred from boron to mercury with retention of configuration
(243,265).

CARBON–CARBON BOND FORMATION

Coupling of Organic Groups Attached to Boron.   Treatment of alkylboranes with alkaline silver nitrate solution results in coupling of the alkyl groups.
Yields obtained for primary and secondary groups are in the range of 60–80%
and 35–50%, respectively. Unsymmetrical products can be obtained by the use of
an excess of one substrate. The reaction probably proceeds through the formation
of unstable silver alkyl intermediates breaking down into silver and free radicals
undergoing dimerization (3). Methylcopper couples dialkenylalkylboranes (266).
Alkenyldialkylboranes are also coupled with complete retention of configuration
by copper(I) compounds (267).

Cross-Coupling of Organoboranes With Organic Halides.   The   palladium-catalyzed cross-coupling reaction of alkenyl-, allyl-, aryl-, and alkylboranes
with all types of organic halides is widely used for the synthesis of alkenes;
(E,E)-, (E,Z)-, and (Z,Z)-dienes; and enynes (268–270). Even the most difficult alkyl–alkyl cross-coupling can be achieved with B-alkyl-9-BBN derivatives

and alkyl iodides (270). The reaction is tolerant of many functional groups on either coupling partner and stereochemically pure functionalized products can be obtained. An example follows (271). Many other applications have been reported (272–278).

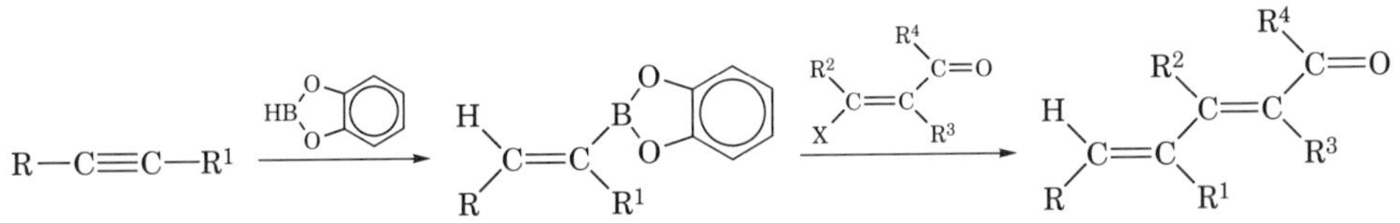

Coupling of alkylboranes, B-alkenyl-9-BBN derivatives and lithium trialkylmethylborates with organic halides is also catalyzed by copper(I) compounds (279–283). In general, palladium- and copper-catalyzed cross-couplings proceed with retention of configuration of the organic groups.

Organoborate Rearrangements. A variety of carbon–carbon bond forming reactions involving alkyl-, alkenyl-, and alkynylborates are anionotropic rearrangements in which organic moieties migrate from electron-rich boron to electron-deficient carbon. The electron deficiency can be created either by a polar bond, eg, to a halogen, or by the action of electrophiles on unsaturated borates. If the electrophile is iodine, the intermediate iodoorganoborane obtained from alkenyl- or alkynylborates breaks down to give an alkene or alkyne, respectively. For most other electrophiles the intermediate is stable and can be oxidized, hydrolyzed, or used for other transformations (Fig. 3). Retention of configuration of the migrating group is usual for this type of mechanism. Tertiary alkyl and methyl groups show low migratory aptitude. The general Zweifel alkene synthesis starts (Fig. 4) with the hydroboration of 1-alkynes or 1-haloalkynes with dialkylboranes (107,246,284). Treatment of the intermediate 1-alkenyldialkylborane with iodine and base yields (Z)-alkene. Base is sufficient to induce the rearrangement to (E)-alkenes since the electron-deficient center is already present in the haloalkenyl group (107,246). The scope of the original approach, limited by the availability of dialkylboranes, is considerably extended

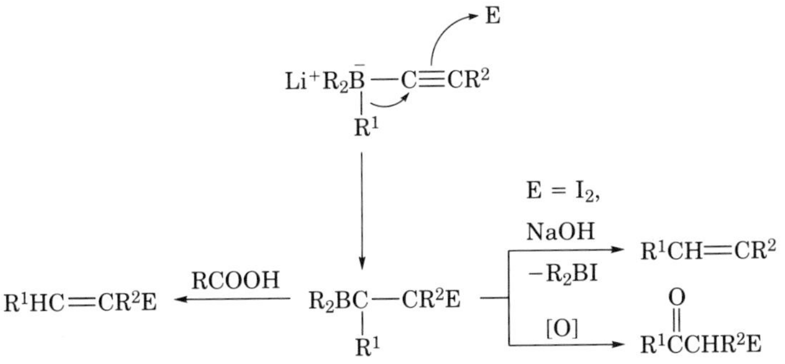

Fig. 3. Migration of an alkyl group from boron to carbon.

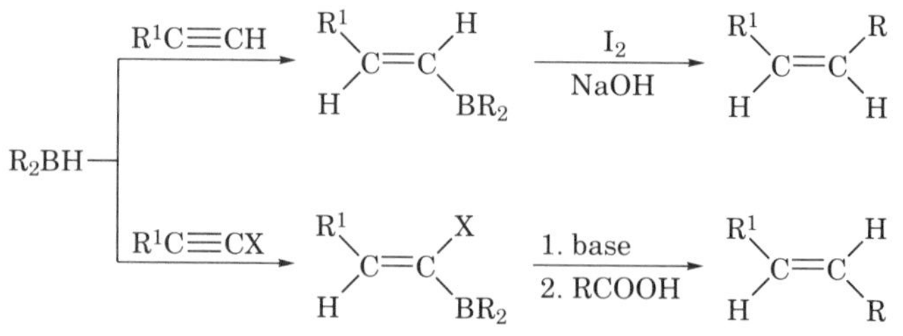

Fig. 4. Zweifel alkene synthesis.

by the application of dialkylhalo- and alkyldibromoboranes as their precursors and by the use of thexylborane derivatives (284–287).

A general synthesis of trisubstituted alkenes containing three different alkyl substituents is achieved by the following methodology (288):

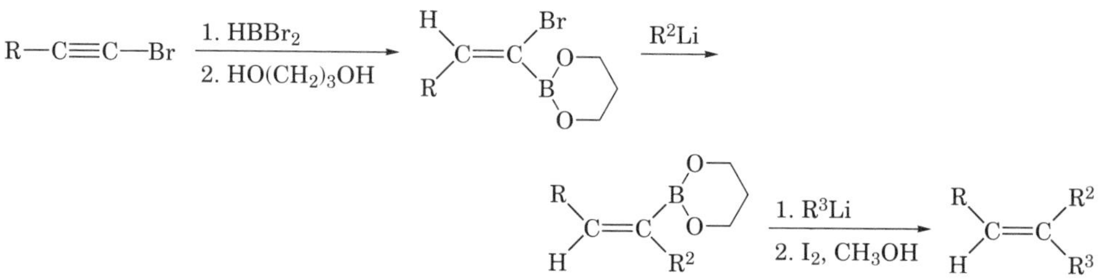

The organoborate intermediates can also be generated from alkenylboronic esters and alkyllithium or Grignard reagents, or from trialkylboranes and alkenyllithium compounds. Conjugated symmetrical and unsymmetrical diynes (289–291), stereochemically pure 1,3-dienes (292,293), and 1,3-enynes (294) including functionalized systems can be prepared (289,295).

Various electrophiles other than iodine have been used to induce alkenyl coupling (9). Alkyl halides and protic acids react with alkynylborates to yield mixtures of stereoisomeric alkenylboranes. Nevertheless, oxidation of these products is synthetically useful, providing single ketones (296–298). Alcohols are obtained from the corresponding alkenylborates.

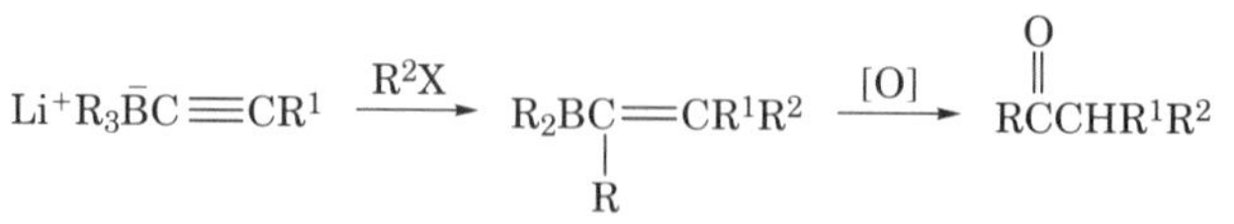

Markovnikov boranes not available by direct hydroboration can be prepared by protonolysis of alkylethenyl- and alkylalkynylborates (298,299).

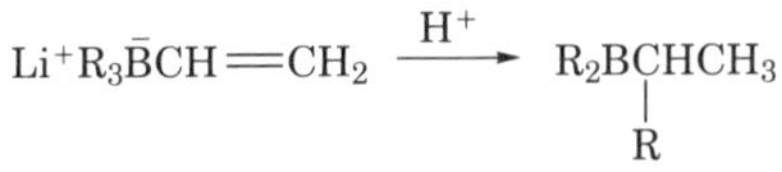

$$\text{Li}^+\text{R}_3\bar{\text{B}}\text{CH}=\text{CH}_2 \xrightarrow{\text{H}^+} \text{R}_2\text{BCHCH}_3 \atop \qquad\qquad\qquad | \atop \qquad\qquad\qquad \text{R}$$

Aldehydes react with alkenylborates to give 1,3-diols upon oxidation of the intermediate (300). Alkynylborates are transformed by epoxides into homoallylic alcohols and alkenylborates into 1,4-diols (300,301). Carbon dioxide reacts with alkenylborates to yield carboxylic acids (302). The scope of these transformations is further extended by the use of functionalized electrophiles and borates, often reacting with high stereoselectivity. For example, in the reactions of alkynylborates with α-bromocarbonyl compounds and related derivatives or trimethylsilyl chloride (303), the incoming group is placed cis to the migrating one (296).

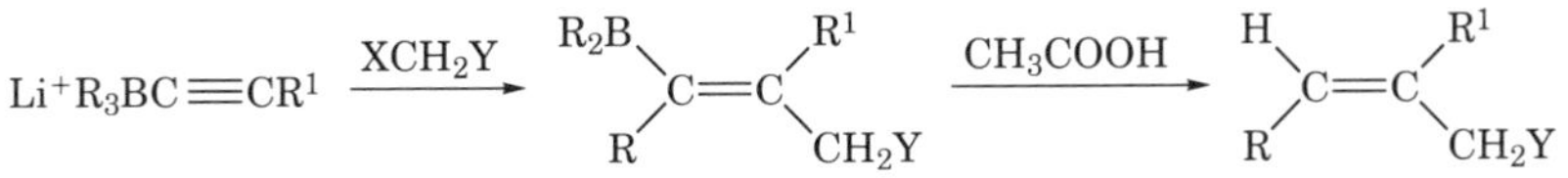

$$\text{Li}^+\text{R}_3\text{BC}\equiv\text{CR}^1 \xrightarrow{\text{XCH}_2\text{Y}} \underset{\text{R}}{\overset{\text{R}_2\text{B}}{>}}\text{C}=\text{C}\underset{\text{CH}_2\text{Y}}{\overset{\text{R}^1}{<}} \xrightarrow{\text{CH}_3\text{COOH}} \underset{\text{R}}{\overset{\text{H}}{>}}\text{C}=\text{C}\underset{\text{CH}_2\text{Y}}{\overset{\text{R}^1}{<}}$$

where X = Br; Y = COR^2, COOR^2, and C≡CH

Similar selectivity is observed in the synthesis of allylsilanes where X = CF_3SO_3^- and Y = $\text{Si(CH}_3)_3$ (304). Alkenyl- and alkynylborates containing a leaving group in the γ-position rearrange to allylic and allenic boranes, respectively (305).

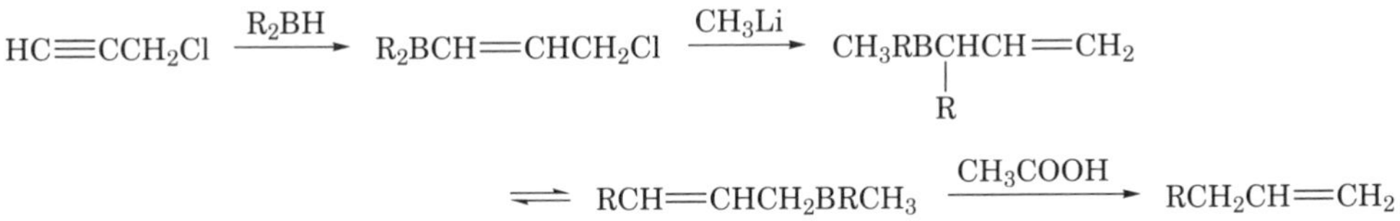

$$\text{HC}\equiv\text{CCH}_2\text{Cl} \xrightarrow{\text{R}_2\text{BH}} \text{R}_2\text{BCH}=\text{CHCH}_2\text{Cl} \xrightarrow{\text{CH}_3\text{Li}} \underset{\qquad\qquad |}{\text{CH}_3\text{RBCHCH}=\text{CH}_2} \atop \qquad\qquad\qquad\qquad\qquad\qquad\qquad\quad \text{R}$$

$$\rightleftharpoons \text{RCH}=\text{CHCH}_2\text{BRCH}_3 \xrightarrow{\text{CH}_3\text{COOH}} \text{RCH}_2\text{CH}=\text{CH}_2$$

Properly substituted organoborates can undergo elimination with ring formation. Thus cyclopropanes or cyclobutanes are formed when the organoboranes prepared by the hydroboration of allylic, propargylic, or homopropargylic chlorides and tosylates with 9-BBN are treated with base (306). The cyclopropane ring closure is stereospecific involving antiparallel alignment of the boron atom and the leaving group (307).

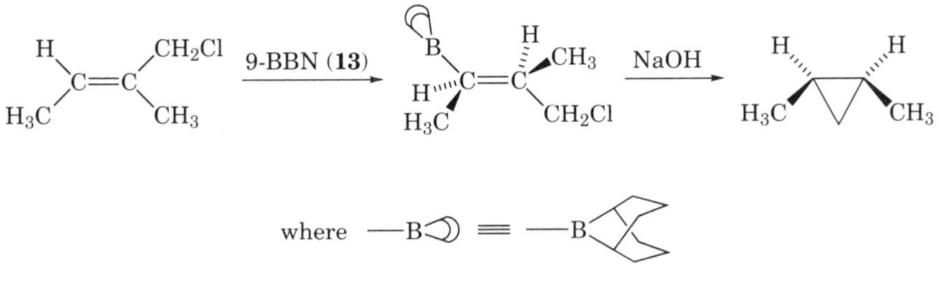

$$\underset{\text{H}_3\text{C}}{\overset{\text{H}}{>}}\text{C}=\text{C}\underset{\text{CH}_3}{\overset{\text{CH}_2\text{Cl}}{<}} \xrightarrow{\text{9-BBN (13)}} \underset{\text{H}_3\text{C}}{\overset{\text{B}}{\underset{\text{H}^{\cdots}}{>}}}\text{C}=\text{C}\underset{\text{CH}_2\text{Cl}}{\overset{\text{H}\quad\text{CH}_3}{<}} \xrightarrow{\text{NaOH}} \underset{\text{H}_3\text{C}}{\overset{\text{H}_{\cdots}}{\triangle}}\underset{\text{CH}_3}{\overset{\text{H}}{}}$$

$$\text{where} \quad -\text{B}\hspace{-0.3em}\bigcirc \equiv -\text{B}\hspace{-0.3em}\langle\rangle$$

Cyclopropanes can also be obtained by the reaction of vinyltrialkylborates with aldehydes followed by treatment with phosphorus pentachloride and base (300), and by the rearrangement of δ-substituted alkynyltrialkylborates (308). It is also possible to utilize this approach for the synthesis of five- and six-membered rings (3). Trans-1,4-elimination in cyclic systems leads to the formation of stereodefined acyclic 1,5-dienes or medium-ring dienes, depending on the starting compound (309).

Single-Carbon Insertion Reactions. Carbonylation, cyanidation, and "DCME," and related reactions are convenient general processes developed to bring about the transfer of organic groups from boron to a single-carbon atom.

Carbonylation. Trialkylboranes react with carbon monoxide at elevated temperatures and pressures (310,311), or more conveniently at $100-125\,^\circ\mathrm{C}$ in diglyme solution (312,313). Oxidation of the organoborane product yields a tertiary alcohol. The reaction proceeds stepwise and the migration of alkyl groups can be controlled to give the products of one, two, or three groups transferring. Primary, secondary, and tertiary alcohols, aldehydes, and ketones can be obtained, depending on the reaction conditions. Configuration of the migrating group is retained. Many functional groups, eg, ethers, esters, nitriles, and chloro, are tolerated.

The rate of carbon monoxide absorption is greatly increased in the presence of hydrides and the reaction proceeds at much lower temperatures (314,315). The intermediate formed after the first group migration (step 1) is reduced by the hydride (MH is potassium triisopropoxyborohydride or lithium trialkoxyaluminum hydride). Hydrolysis of the reduction product yields the methylol derivative RCH_2OH (314), oxidation yields the aldehyde RCHO (316), treatment with acid, inducing migration of the second alkyl group, followed by oxidation provides the secondary alcohol, R_2CHOH (317). For better utilization of alkyl groups it is advantageous to use the B-alkyl-9-BBN derivatives in these reactions (315,318).

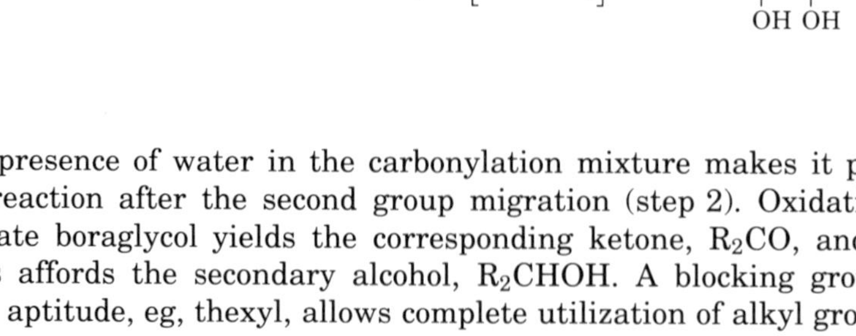

$$R_3B \xrightarrow{\ \ CO\ \ } [R_3\overset{-}{B}\overset{+}{C}O] \xrightarrow[\text{step 1}]{} [R_2B\overset{O}{\overset{\|}{C}}R] \xrightarrow{\ \ MH\ \ } R_2B\overset{OM}{\overset{|}{C}}HR$$

$$\text{step 2}$$

$$HOCR_3 \xleftarrow{\ [O]\ } O{=}BCR_3 \xleftarrow[\text{step 3}]{} \left[RB\overset{O}{\underset{CR_2}{\diagdown\diagup}} \right] \xrightarrow{\ H_2O\ } \underset{OH\ OH}{RB\!-\!CR_2}$$

The presence of water in the carbonylation mixture makes it possible to halt the reaction after the second group migration (step 2). Oxidation of the intermediate boraglycol yields the corresponding ketone, R_2CO, and alkaline hydrolysis affords the secondary alcohol, R_2CHOH. A blocking group of low migratory aptitude, eg, thexyl, allows complete utilization of alkyl groups (319).

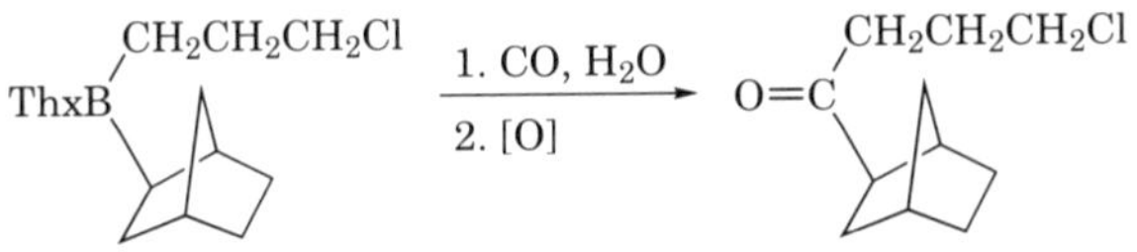

A convenient annulation procedure based on cyclic hydroboration–carbonylation of 1-vinyl- and 1-allylcycloalkenes with thexylborane provides trans-fused bicyclic ketones (312).

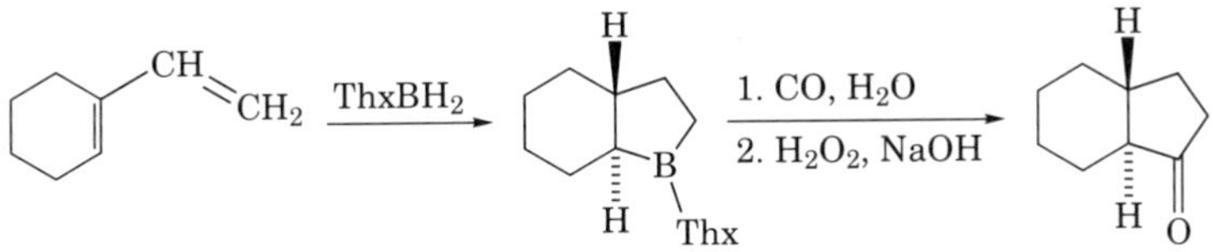

Moderate yields of acids and ketones can be obtained by palladium-catalyzed carbonylation of boronic acids and by carbonylation cross-coupling reactions (272,320,321). In an alternative procedure for the carbonylation reaction, potassium trialkylborohydride in the presence of a catalytic amount of the free borane is utilized (322). Finally, various tertiary alcohols including hindered and polycyclic structures become readily available by oxidation of the organoborane intermediate produced after migration of three alkyl groups (312,313,323).

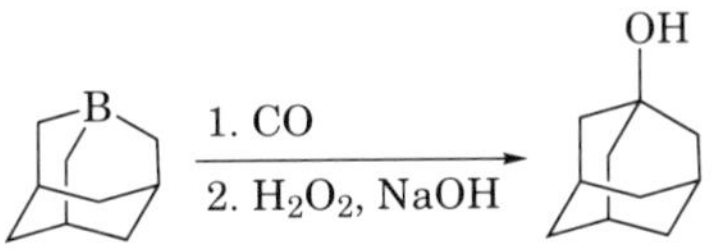

Cyanidation. The cyanide ion is isoelectronic with carbon monoxide. It coordinates with trialkylboranes, giving stable cyanoborate salts that must be treated with a suitable electrophile, eg, trifluoroacetic anhydride, to bring about the migration of groups from boron to carbon (324). Two groups migrate readily at room temperature to give a cyclic intermediate which can be oxidized with H_2O_2 to the ketone R_2CO. The third group migrates upon warming the intermediate with an excess of trifluoroacetic anhydride to give R_3COH (5,324). Conjugated ketones are obtained from alkenylboranes (325).

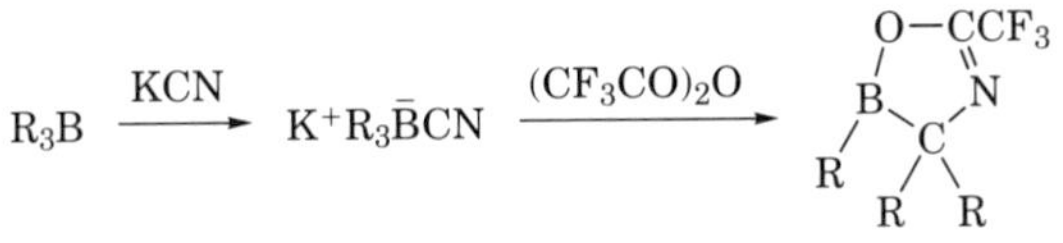

The cyanidation reaction proceeds under mild conditions and no special equipment is required. Stereochemistry of the product usually is the same as in the carbonylation reaction. However, in hindered systems stereoisomeric products may be formed (59,326). Annulation by cyclic hydroboration–cyanidation finds application in the synthesis of natural products (327,328).

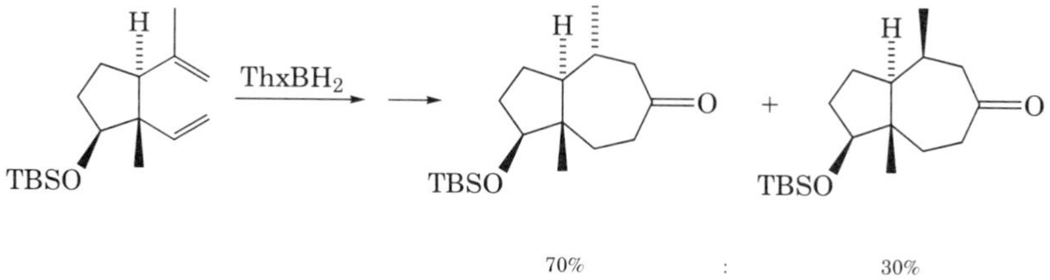

70%   :   30%

where TBS = $Si(CH_3)_2C(CH_3)_3$

Carbanion Coordination. The coordination of alkylboranes with carbanions generated from various halogeno- and thiomethane compounds, eg, chloroform, chlorodifluoromethane, dichloromethane, trimethylsilylchloromethane, tris(phenylthio)methane, 1,1-dichloromethyl methyl ether [4885-02-3] (DCME), and others, results in the migration of alkyl groups from boron to carbon (5,9,329–333). Generally, these reactions proceed under mild conditions, often at low temperatures. Very hindered trialkylboranes undergo transformation into the corresponding tertiary alcohols by the DCME reaction, which is also useful in the synthesis of ketones (89,334). Even the thexyl group migrates under mild conditions (334).

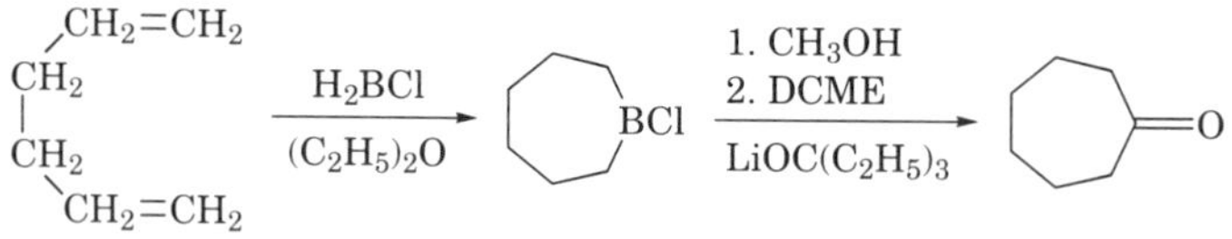

Carbanions derived from the other above mentioned methane derivatives react with alkylboranes, bornic, and boronic esters, providing rich possibilities for the preparation of single-carbon insertion products. Thus 2-alkyl-1,3,2-dithiaborolanes are converted into acids or thioacetals by trichloromethyllithium (335).

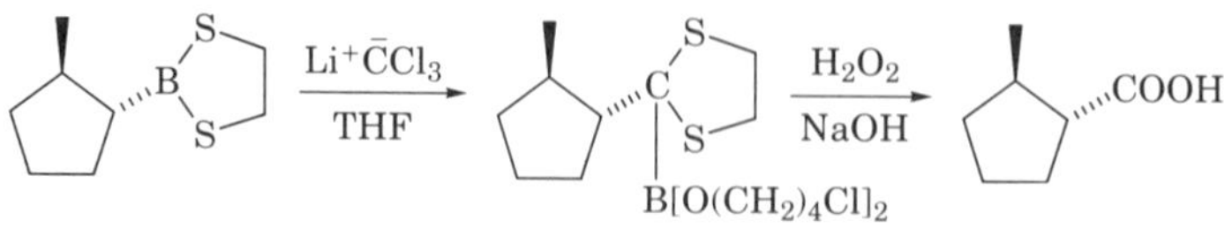

Aldehydes are conveniently synthetized by the reaction of boronic esters with dichloromethyllithium or (phenylthio)methoxymethyllithium (336,337). The synthesis of medium-ring boracyclane structures by stepwise ring enlargement is based on the reaction of B-methoxyboracycles with chloromethyllithium (338).

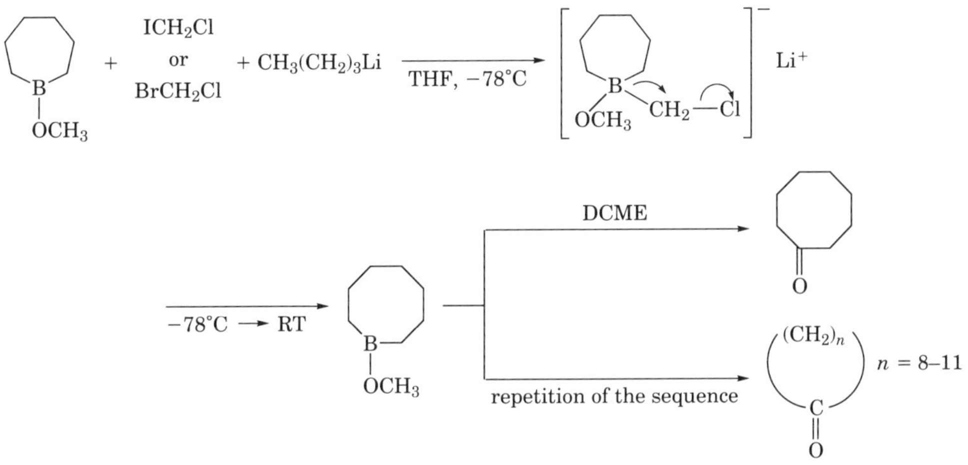

Dichloromethyllithium generated at $-100°C$ reacts with boronic esters to give dichloromethylborates, rearranging upon warming to room temperature to the homologated α-chloroboronic esters. Displacement of the chlorine atom with an alkyl group yields the homologated boronic ester and the sequence can be repeated (339,340). Primary, secondary, and tertiary alkyl, alkenyl, and aryl groups undergo migration. Functionalities, eg, benzyloxy, remote ketal, and ether groups are tolerated. The process has been developed into a useful methodology for the synthesis of optically active boronic acids and carbon compounds (341).

Reactions with Acyl Carbanion Equivalents. Alkyl substituted carbanions $\bar{C}RXY$ with potential leaving groups X, Y, and acyl carbanion equivalents or $\bar{C}HRX$ (342) react with alkylboranes, providing products with mixed alkyl groups derived from both reagents.

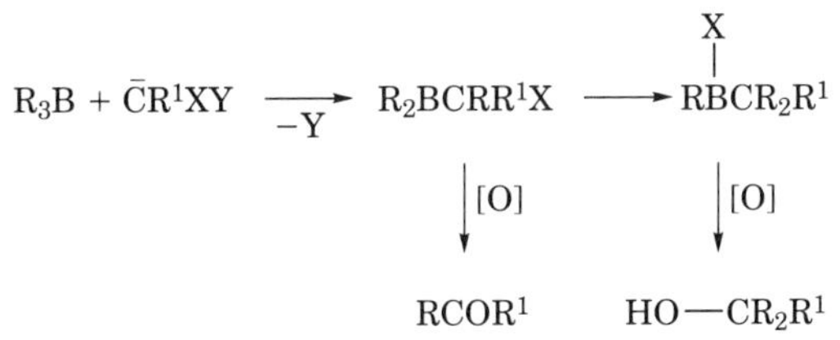

$$R_3B + \bar{C}R^1XY \xrightarrow[-Y]{} R_2BCRR^1X \longrightarrow \overset{X}{\underset{|}{RBCR_2R^1}}$$

$$\downarrow [O] \qquad\qquad \downarrow [O]$$

$$RCOR^1 \qquad HO-CR_2R^1$$

Lithiated thioacetals and sulfur ylides are convenient reagents for these transformations. Thus the anions derived from 2-substituted 1,3-benzodithioles are less sterically demanding than 1-lithio-1,1-bis(phenylthio)alkanes developed earlier and are advantageous for the synthesis of hindered ketones and tertiary alcohols (343–345). The first migration is spontaneous, the second requires the addition of an electrophile (HgCl$_2$ or hazardously toxic FSO$_2$CH$_3$). The preferential migration of primary to tertiary and 9-BBN groups has been observed (345). Sulfur ylides react with alkyl- and alkenylboranes to give the homologated organoboranes (346–348). The leaving ability of the thioalkoxy group is increased by the addition of an electrophile, eg, methyl iodide (347), copper(I) halides (342), or mercury(II) halides (347,349). Organoboranes react also with α-heterosubstituted carbanions containing a hetero atom-leaving group other than halogen or sulfur. Examples include lithiated enol ethers (349) and aldimines (350,351), nitrogen and phosphorus ylides (352,357), and ortholithiated furans and pyridines (354,355).

α-Alkylation of Carbonyl Compounds and Derivatives.   The organoborate intermediates generated by the reaction of alkylboranes with carbanions derived from α-halocarbonyl compounds and α-halonitriles rearrange to give α-alkylated products.

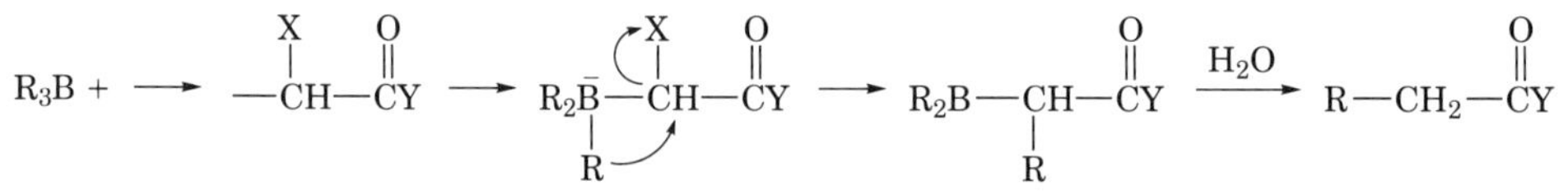

$$R_3B + \longrightarrow \underset{R}{\overset{X\quad O}{\underset{|}{\overset{||}{-CH-CY}}}} \longrightarrow R_2\bar{B}\underset{R}{\overset{X\quad O}{\underset{|}{\overset{||}{-CH-CY}}}} \longrightarrow R_2B\underset{R}{\overset{O}{\underset{|}{\overset{||}{-CH-CY}}}} \xrightarrow{H_2O} R-CH_2\overset{O}{\overset{||}{-CY}}$$

where X = halogen; Y = −H, −R, −OR

Two different groups can be introduced consecutively starting with α,α-dihalonitriles. The anions are generated by sterically hindered bases, eg, *tert*-butoxide, or better, 2,6-di-*tert*-butylphenoxide for sensitive compounds. Groups unreactive in S$_N$2 displacements such as alkenyl, aryl, or norbornyl can be introduced. Double alkylation with two secondary alkyl groups is possible. The use of B-alkyl-9-BBN derivatives instead of trialkylboranes allows better utilization of alkyl groups in the synthesis of ketones (356), esters (357), and nitriles (358). Malononitrile and 4-bromocrotonic acid can also be used (359,360).

α-Diazocarbonyl compounds react with trialkylboranes directly, in the absence of bases at 0–25°C (361–363). Mild conditions are advantageous when functionalities labile to bases are present (364,365). Di- and monoalkylchloroboranes are used for better utilization of alkyl groups (366,367). β,γ-Unsaturated carboxylic esters can be prepared in excellent isomeric purity by the reaction of 1-alkenyldichloroboranes with ethyl diazoacetate (368).

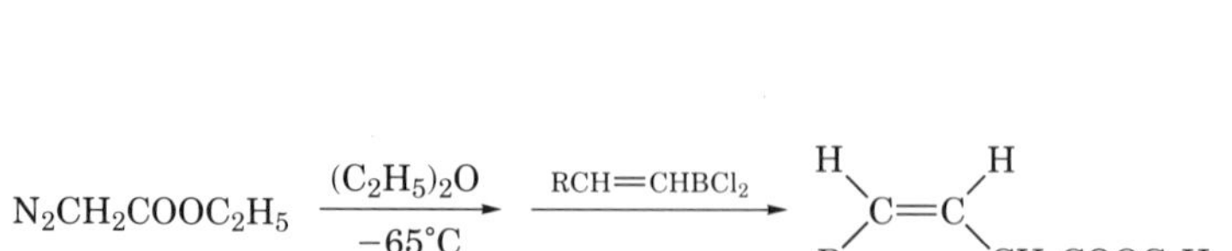

$$N_2CH_2COOC_2H_5 \xrightarrow[-65°C]{\underset{}{(C_2H_5)_2O} \quad RCH=CHBCl_2} \underset{R}{\overset{H}{\diagup}} C = C \underset{CH_2COOC_2H_5}{\overset{H}{\diagdown}}$$

α-Bromination Transfer.   The photochemical α-bromination of organoboranes with bromine proceeds readily (3). Weak bases such as water or tetrahydrofuran are sufficient to induce the migration of alkyl groups from boron to the α-brominated carbon. Consequently, bromination of an organoborane in the presence of water results in facile carbon–carbon bond formation. All three alkyl groups are utilized (369). It is also possible to halt the reaction after the first group migration.

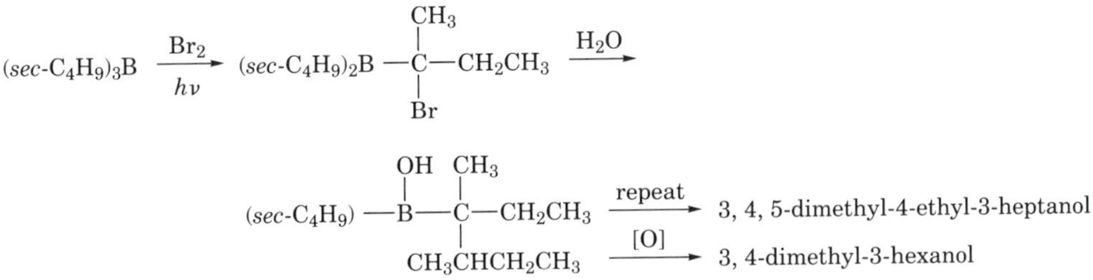

$$(sec\text{-}C_4H_9)_3B \xrightarrow[hv]{Br_2} (sec\text{-}C_4H_9)_2B\!-\!\underset{\underset{Br}{|}}{\overset{\overset{CH_3}{|}}{C}}\!-\!CH_2CH_3 \xrightarrow{H_2O}$$

$$(sec\text{-}C_4H_9)\!-\!\underset{\underset{CH_3CHCH_2CH_3}{|}}{\overset{\overset{OH\;\;CH_3}{|\quad|}}{B\!-\!C}}\!-\!CH_2CH_3 \;\;\begin{array}{l}\xrightarrow{repeat}\;\;\text{3, 4, 5-dimethyl-4-ethyl-3-heptanol}\\[4pt]\xrightarrow{[O]}\;\;\text{3, 4-dimethyl-3-hexanol}\end{array}$$

Selective bromination at the more substituted position makes possible connection of two different alkyl groups. The use of dialkylthexylboranes and dialkylhaloboranes allows full utilization of alkyl groups (370,371). Bicyclic and more complex boracycles can be transformed into carbon structures not readily available by other methods (372).

Addition to Carbonyl Compounds.   Unlike Grignard and alkyllithium compounds, trialkylboranes are inert to carbonyl compounds. The air-catalyzed addition to formaldehyde is exceptional (373). Alkylborates are more reactive and can transfer alkyl groups to acyl halides. The reaction provides a highly chemoselective method for the synthesis of ketones (374).

$$R_3B + R^1Li \longrightarrow LiR_3BR^1 \xrightarrow{R^2COX} R^1COR^2 + R_3B + LiX$$

The migration aptitude increases in the following order: secondary alkyl < primary alkyl < allyl, benzyl.

In contrast to alkylboranes, allylic organoboranes add readily to carbonyl compounds (14,375–378). Acid chlorides, esters, and amides undergo double addition (379). The reaction with aldehydes and ketones proceeds via a cyclic transition state with complete allylic rearrangement to give the homoallylic alcohols. Allylic organoboranes are prone to the allylic rearrangement leading to the loss of stereochemical integrity of the allylic group. In an equilibrium, the boron atom favors the least substituted position (211,380,381).

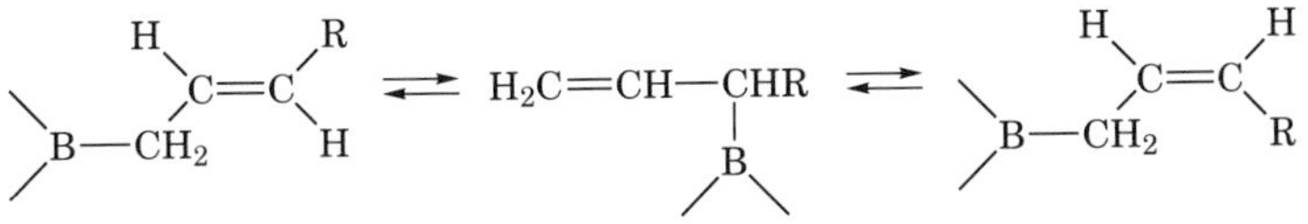

The sensitivity to this process depends on the substituents at the boron atom. Oxygen and nitrogen atoms attached to boron prevent the rearrangement at ambient temperatures but the reactivity of allylic boronic esters is lower compared to allylic dialkylboranes (382). The latter compounds must be used at low temperatures to avoid the rearrangement (383). The addition of stereodefined allylic organoboranes to aldehydes and ketones is highly stereospecific. The postulated chair-like transition state permits correlation of the product and substrate configurations (9,378). Typical diastereoselectivities are in the range of 80–95%. These additions are of importance finding application in the synthesis of natural products (383–386). A number of other synthetically useful additions of allylic organoboranes to carbonyl compounds, including 3-borolenes and dienylboranes, has been described (387–390), eg,

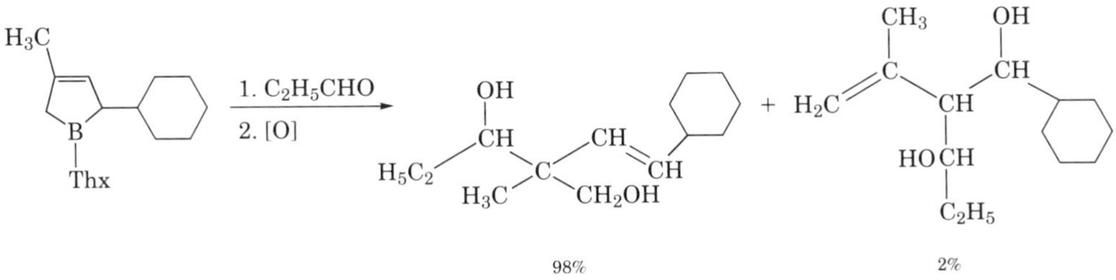

98%   2%

Allenyl and propargylic organoboranes react with aldehydes to give homopropargylic and homoallenyl alcohols, respectively (391,392). It is possible to prepare highly unsaturated enynols and trienols (392). B-Alkenyl-9-BBN derivatives transfer the alkenyl group to aldehydes providing (E)- or (Z)-allylic alcohols. Methyl vinyl ketone gives the 1,4-addition product (393). B-Alkynyl-9-BBN compounds are very mild, nonbasic reagents showing no reactivity toward various functional groups, eg, acid chlorides, anhydrides, esters, amides, nitriles, sulfoxides, and diethyl malonate. They can differentiate the sterically less hindered among aldehydes and ketones, for example, addition of B-alkynyl-9-BBN to a mixture of cyclopentanone and cyclohexanone results in a 71% yield of 1-alkylcyclohexanol and <1% 1-alkynylcyclopentanol (394). The simplest representative, B-ethynyl-9-BBN, decomposes above −78°C but B-2-trimethylsilyl(ethynyl)-9-BBN is a stable reagent (395).

β-Alkylation of Carbonyl Compounds. Trialkylboranes undergo air-catalyzed conjugate addition to various α,β-unsaturated carbonyl compounds (2). In these free-radical reactions the stereochemical integrity of the migrating group is lost. In certain cases, however, the addition may be selective. For example, mainly the trans product is obtained in the reaction of 2-methylcyclopentylborane with acrolein (396). Apart from the stereochemical aspects, several synthetically useful transformations have been described. Thus conjugated aldehydes and ketones such as acrolein (397,398), 2-bromoacrolein (399), methyl vinyl ketone (400), 2-cyclohexenone, and α-methylenecycloalkanones (401) react readily.

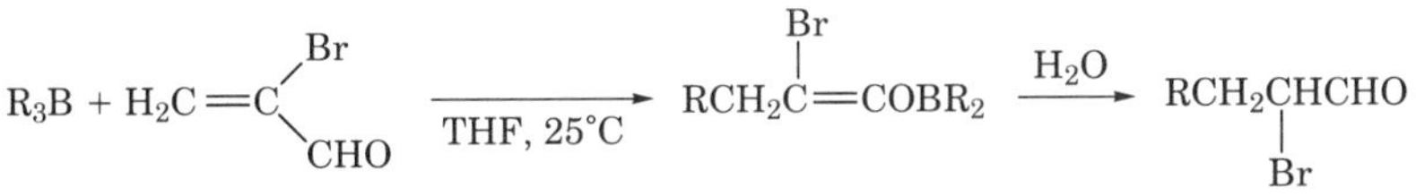

$$R_3B + H_2C=C\overset{Br}{\underset{CHO}{\big<}} \xrightarrow{\text{THF, 25°C}} RCH_2\overset{Br}{\underset{}{C}}=COBR_2 \xrightarrow{H_2O} RCH_2\underset{Br}{\underset{|}{CH}}CHO$$

The reaction with substituted derivatives, eg, crotyl aldehyde, is slower (402). Like other free-radical reactions, the reactivity order of alkyl groups is tertiary > secondary > primary. B-Alkyl-9-BBN derivatives react with ring opening except for the tertiary alkyl groups. For full utilization of the alkyl groups B-alkylborinane derivatives or diphenylalkylboranes are used (403–405). Simple trialkylboranes undergo related conjugate additions with enynones (406), quinones (404,407), and crotonaldimines (408). B-Alkenyl-9-BBN derivatives transfer the alkenyl group to cisoid enones with retention of configuration (393,409). B-Alkynyl-9-BBN derivatives add to cisoid enones (410,411) and ethenyl and ethynyl oxiranes (412,413), with ready formation of quaternary centers (410).

Boron-Stabilized Carbanions. Carbanions α to boron are stabilized by the mesomeric effect. The difficulty in deprotonation of alkylboranes lies in a strong tendency of bases to coordinate with boron rather than abstract a proton. Consequently, either very hindered bases or hindered groups attached to boron must be employed to suppress the rate of complex formation and enhance deprotonation. A highly hindered base, lithium 2,2,6,6-tetramethylpiperidide, deprotonates B-methyl-9-BBN, but it fails to generate carbanions from B-alkyl-9-BBN derivatives (414). Primary alkyldimesitylboranes and allyldimesitylborane are deprotonated by moderately hindered bases such as lithium dicyclohexylamide or mesityllithium (415). *gem*-Diboronic esters and more conveniently heterosubstituted boronic esters, eg, phenylthio- or trimethylsilylmethylboronates can also be deprotonated (416–420). Alternatively, boron-stabilized carbanions are generated by nucleophilic addition to hindered vinylic boranes (421) and by deboronation of *gem*-trialkylboranes (422,423) or methanetetraboronic esters (424). The anions add readily to aldehydes and ketones to give olefins, the so-called boron Wittig reaction (423,425–428). Stereoselective addition can be achieved. The reaction is sensitive to temperature and the amount of trifluoroacetic anhydride (TFAA) must be carefully controlled.

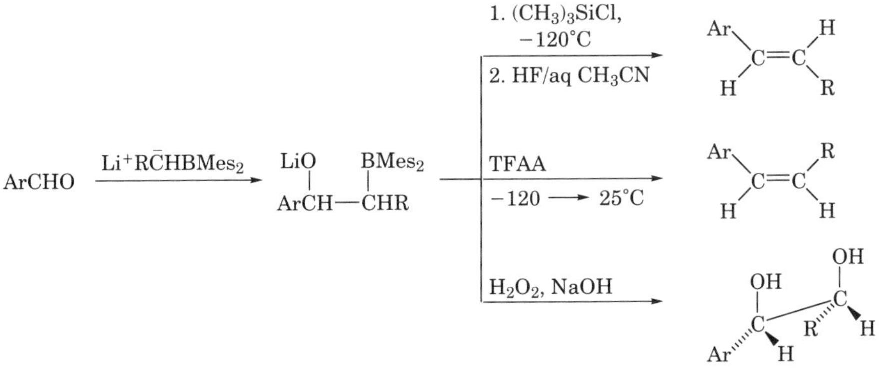

$$ArCHO \xrightarrow{Li^+R\bar{C}HBMes_2} \underset{ArCH-CHR}{\overset{LiO \quad BMes_2}{\underset{|}{\big|}\qquad\underset{|}{\big|}}}$$

1. $(CH_3)_3SiCl$, $-120°C$
2. $HF/aq\ CH_3CN$ → $\underset{H}{\overset{Ar}{\big>}}C=C\underset{R}{\overset{H}{\big<}}$

TFAA $-120 \longrightarrow 25°C$ → $\underset{H}{\overset{Ar}{\big>}}C=C\underset{H}{\overset{R}{\big<}}$

H_2O_2, NaOH → (diol product)

where Mes = mesityl

Other reactions of boron-stabilized carbanions involve alkylation by primary alkyl halides (418,429), addition to esters (418,430), and transformation into silyl-, tin-, and sulfur-substituted products and *gem*-dibora derivatives (431).

THERMAL ISOMERIZATION OF ORGANOBORANES

Trialkylboranes undergo isomerization under the action of heat, generally at temperatures above 100°C, the boron atom moving to the least hindered site of the alkyl group (3,432). It migrates past a tertiary but not a quaternary center. The rate of isomerization depends on the organoborane structure. Thus the relatively unhindered dichloroboryl, dibromoboryl, and 9-BBN groups migrate very slowly, whereas bulky hindered groups such as bis(2,5-dimethylcyclohexyl) boryl migrate readily (33,433). The isomerization appears to involve a dehydroboration–hydroboration mechanism. The addition of an excess of another olefinof equal or greater reactivity results in the displacement of the olefin from the organoborane (434). Consequently, the combination of hydroboration, isomerization, and displacement makes possible contrathermodynamic isomerization of olefins (435,436). In certain cases the displacement with an aldehyde is more convenient (154).

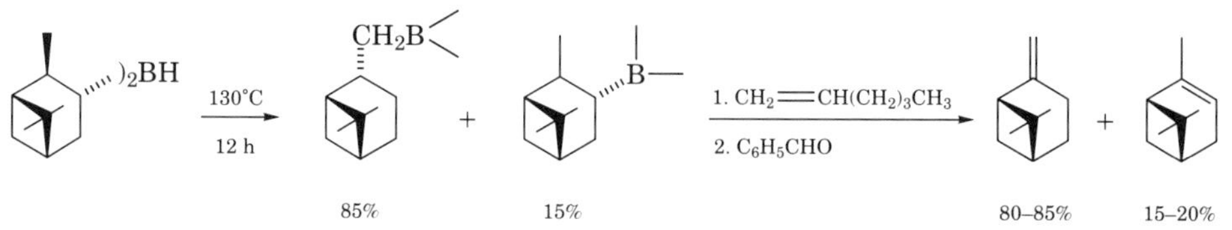

Long-chain primary alcohols, eg, triacontanol, can be prepared by the hydroboration, isomerization, and oxidation of the corresponding internal alkenes (437). The less thermodynamically stable stereoisomer can be transformed into the more stable one by heating, eg, *cis*- into *trans*-myrtanylborane (204).

CONCERTED REACTIONS OF ORGANOBORANES

Allylic organoboranes react via cyclic transition states not only with aldehydes and ketones, but also with alkynes, allenes, and electron-rich or strained alkenes. Bicyclic structures, which can be further transformed into boraadamantanes, are obtained from triallyl- or tricrotylborane and alkynes (323,438,439).

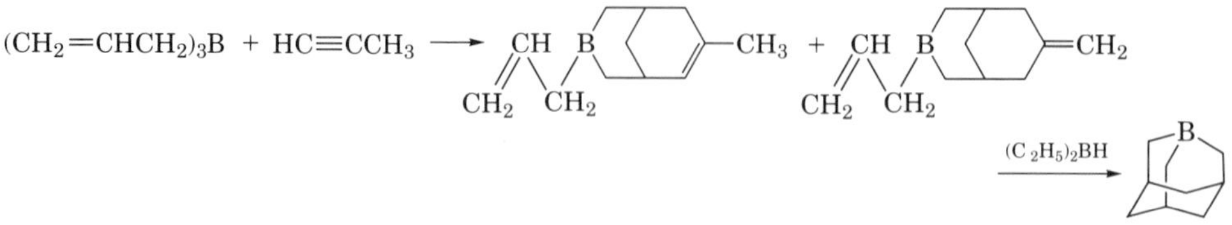

The addition proceeds in three discrete steps and the intermediates can be isolated. Simple alkenes are less reactive than alkynes and do not undergo the addition to allylic boranes, but electron-rich alkyl vinyl ethers react at moderate temperatures to give 1,4-dienes or dienyl alcohols (440).

Vinylboranes are interesting dienophiles in the Diels-Alder reaction. Alkenylboronic esters show moderate reactivity and give mixtures of exo and endo adducts with cyclopentadiene and 1,3-cyclohexadiene (441). Dichloroalkenylboranes are more reactive and dialkylalkenylboranes react even at room temperature (442–444). Dialkylalkenylboranes are omniphilic dienophiles insensitive to diene substitution (444). In situ formation of vinylboranes by transmetallation of bromodialkylboranes with vinyl trialkyltin compounds makes possible a one-pot reaction, avoiding isolation of the intermediate vinylboranes (443). Other cycloadditions of alkenyl- and alkynylboranes are known (445).

POLYMERIZATION

Hydroboration of α,ω-dienes with monoalkylboranes gives reactive organoboron polymers which can be transformed into polymeric alcohols or polyketones by carbonylation, cyanidation, or the DCME reaction followed by oxidation (446–448).

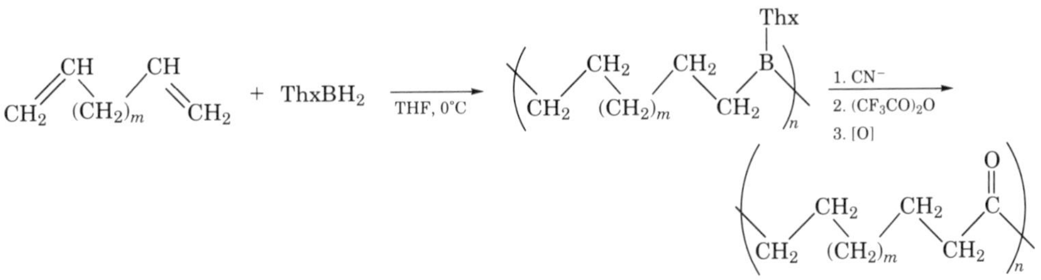

Air-stable boron-containing polymers can be prepared by the reaction of dicyano compounds with the tert-butylborane–trimethylamine complex (449). The properties of these new materials are being studied. Hydroboration is also applied for the conversion of double bonds in polymers into hydroxyl groups (450–454). Well-defined copolymers of ethylene–vinyl alcohol can be prepared (455).

SYNTHESIS OF ISOTOPICALLY LABELED COMPOUNDS

Organoborane reactions have been applied for the synthesis of isotopically labeled compounds important in chemical and biological research and in modern medical imaging techniques, such as positron emission tomography (pet) and magnetic resonance imaging (mri). Organoboranes tolerate a wide range of physiologically active functionalities and hence are well-suited intermediates in radiopharmaceutical pathways. Such use has been reviewed (456–458). Various isotopes both stable and short-lived can be incorporated into organic compounds via organoboranes. In $RCH_2CH_2B{<}$ → RCH_2CH_2X, R can contain NO_2, CN, COOR, halogen, OR, aryl, metal, etc, and X = D, T, ^{11}C, ^{13}C, ^{15}O, ^{13}N, ^{15}N, ^{123}I, ^{82}Br, and others. The stereoselective deuteration by deuterioboration or hydroboration–deuteriolysis is straightforward (207,459,460). Olefins monodeuterated in the allylic position are readily prepared by deuteriolysis of allylic organoboranes (461).

Special methodologies are often required for the incorporation of short-lived isotopes or due to the most convenient source of the isotope. For example, molecular oxygen is readily available labeled with essentially all known oxygen isotopes. Consequently, oxidation of organoboranes with molecular oxygen is preferred for the synthesis of alcohols (462). Bromination and iodination are carried out using halides instead of the free halogens due to the dangers and cost associated with radiohalogens (463,464). *In situ* generation of chloramine from ammonia has been developed for the amination reaction of organoboranes (465). Carbonylation and cyanidation with isotopically labeled carbon monoxide or cyanide ion, respectively, is used for the incorporation of carbon isotopes (466). Isotopically labeled steroids, carbohydrates, amino acids, and other compounds have been prepared by these methods (457).

Asymmetric Synthesis Via Chiral Organoboranes

There have been several reviews of asymmetric synthesis via chiral organoboranes (6,8,378,382,467–472). Asymmetric induction in the hydroboration reaction may result from the chirality present in the olefin (asymmetric substrate), in the reagent (asymmetric hydroboration), or in the catalyst (catalytic asymmetric hydroboration).

ORIGINS OF ASYMMETRY

Asymmetric Substrate. Excellent stereoselectivity is achieved in borane addition to rigid cyclic systems having one face of the double bond significantly more hindered than the other (473), eg, α-pinene (204). Less rigid monocyclic systems usually give mixtures of isomeric products (138,474–479). Electronegative substituents in allylic cycloalkenyl derivatives direct the boron atom to the trans-2-position (175,480,481). However, addition in the cis-2-position may be preferred for substituents complexing the hydroborating agent, eg, phosphinites (138). Acyclic diastereoselection is controlled by allylic, homoallylic, and more remote chiral centers (482–494) as in the following (493), in which R = $Si(CH_3)_2C(CH_3)_3$ and the product ratio of (**17**) to (**18**) is 13:1.

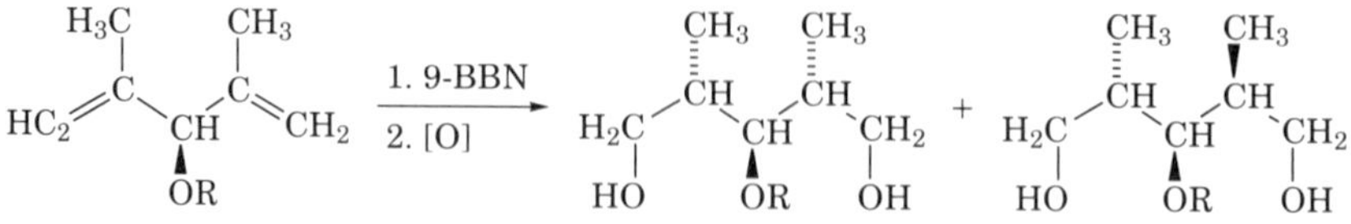

The stereochemical outcome of these reactions can be explained by considering the transition-state geometry. For example, applying the Houk model (495) to allylic alcohols and their derivatives, the smallest substituent at the preexisting chiral center is oriented "inside" over the face of the transition-state ring and the oxygen atom "outside" (483).

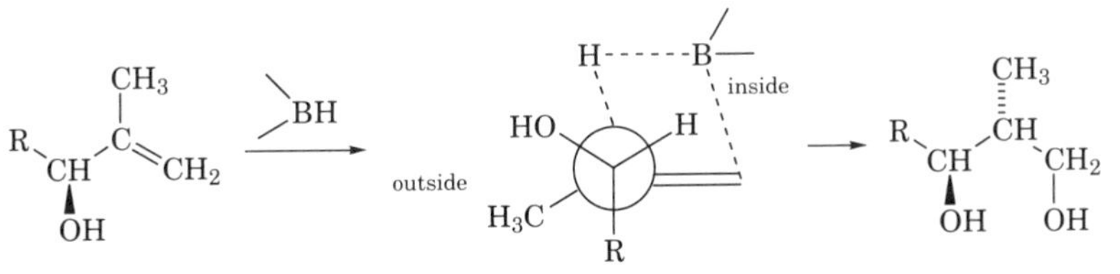

High levels of asymmetric induction have been achieved in the hydroboration of 1,3-, 1,4-, and 1,5-dienes with thexylborane (482,483,489,490). The first chiral center is formed by an intermolecular reaction. In the second step, the organoborane intermediate undergoes an intramolecular hydroboration, creating the second chiral center with high diastereoselectivity.

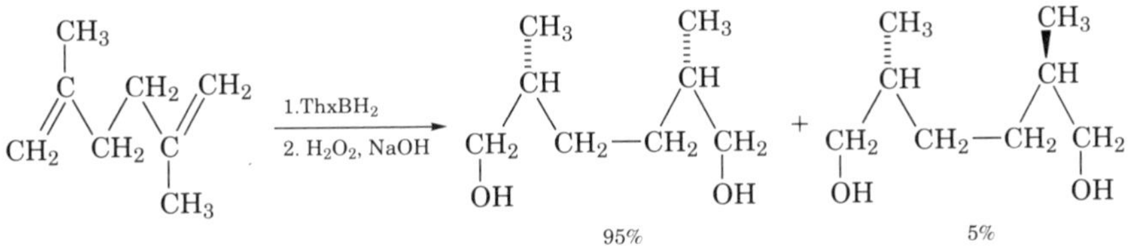

Asymmetric Hydroboration. Hydroboration–oxidation of (Z)-2-butene with diisopinocampheylborane was the first highly enantioselective asymmetric synthesis (496); the product was R(−)2-butanol in 87% ee. Since then several asymmetric hydroborating agents have been developed. Enantioselectivity in the hydroboration of significant classes of prochiral alkenes with representative asymmetric hydroborating agents is shown in Table 3.

2-Icr$_2$BH is bis(2-isocaranyl)borane [114533-27-6] (19), and Lgf$_2$BH is dilongifolylborane (20).

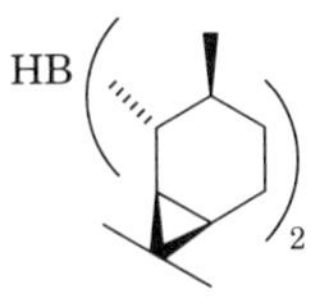

(19) 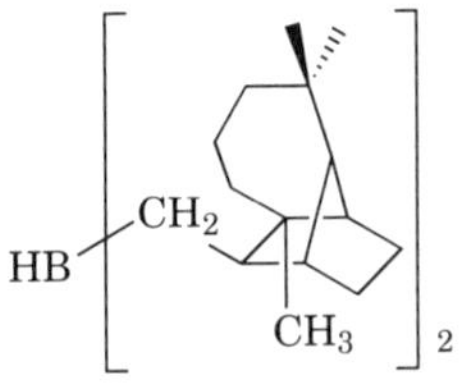   (20)

Table 3. Enantioselectivity in the Hydroboration of Prochiral Alkenes With Various Hydroborating Agents[a]

Class	Alkene	Ipc₂BH (8)	2-Icr₂BH (19)	Lgf₂BH (20)	TDMB (22)	IpcBH₂ (6)	EapBH₂ (21)
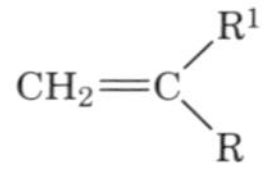 $CH_2=C(R^1)(R)$	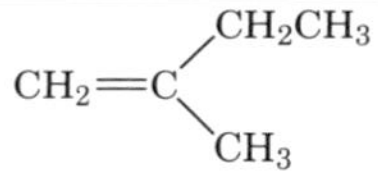 $CH_2=C(CH_2CH_3)(CH_3)$	21	15	1	1	1.5	2
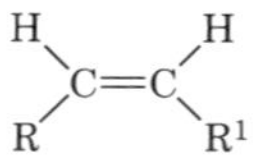 $(H)(R)C=C(H)(R^1)$	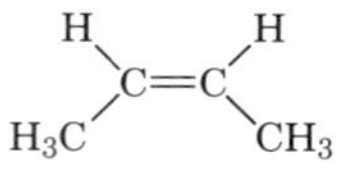 $(H)(H_3C)C=C(H)(CH_3)$	98.1	93	78	95.2	24	30
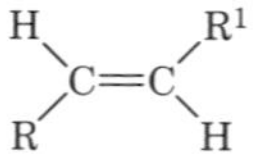 $(H)(R)C=C(R^1)(H)$	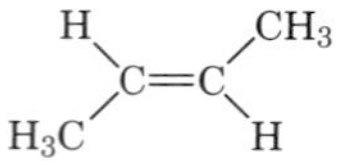 $(H)(H_3C)C=C(CH_3)(H)$		30	25	97	73	76
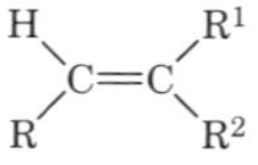 $(H)(R)C=C(R^1)(R^2)$	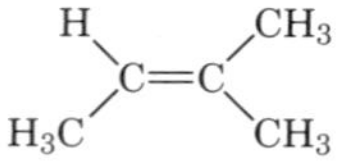 $(H)(H_3C)C=C(CH_3)(CH_3)$	13	37	70	94.2	53	68
Reference		152	155	471	159	497	66

[a]Numbers are % ee of the product alcohol.

Simple terminal prochiral alkenes are hydroborated with low selectivity by all asymmetric hydroborating agents known. The chiral centers of the reagent are probably too far from the incipient chiral center to exert good asymmetric induction. However, such alkenes may react with higher selectivity in catalytic asymmetric hydroboration. Chiral terminal alkenes may also react with high selectivity (225). Disubstituted (Z)-alkenes react readily with all asymmetric dialkylboranes. Diisopinocampheylborane (8) is the reagent of choice for this class, producing alcohols of very high optical purity with the same absolute configuration for similar structures. The relatively more hindered (E)-isomers and trisubstituted alkenes react sluggishly with diisopinocampheylborane. The reaction proceeds with the displacement of α-pinene and the organoborane product shows low optical purity. Less sterically demanding monoisopinocampheylborane (6) reacts readily with all classes of alkenes. Enantioselectivity of the addition to disubstituted (Z)-alkenes is low. In contrast, the (E) isomers and trisubstituted alkenes are hydroborated with much higher selectivity. Mono- and diisopinocampheylboranes are complementary hydroborating agents to each other. Mono(2-ethylapoisopinocampheyl)borane, EapBH$_2$ (21), shows slightly higher selectivity than monoisopinocampheylborane. However, its precursor, 2-ethylapopinene, is not commercially available. Similarly, trans-2,5-dimethylborolane (TDMB) (22), hydroborating three primary classes of alkenes with excellent selectivity, is not readily available. Models predicting the product configuration in the hydroboration of alkenes with mono- and diisopinocampheylboranes, based on the minimum energy calculations of the transition states, have been proposed (495).

(21)   (22)

As follows from the data presented in Table 3, the range of organoboranes produced in high optical purity directly by hydroboration is rather limited. Fortunately, the following features of the organoboranes obtained by hydroboration with mono- and diisopinocampheylborane considerably increase their potential as intermediates in asymmetric synthesis. First, the optical purity of IpcR*BH and Ipc$_2$R*B, where R* is a chiral group, can often be upgraded to essentially 100% ee by crystallization (65). Second, the chiral auxiliary isopinocampheyl group is readily removed from the organoborane by treatment with an aldehyde. Controlled treatment of Ipc$_2$R*B with aldehydes produces chiral boronates with higher enantiomeric purity than the substrate (498). Third, the boronates can be reduced by lithium aluminium hydride or lithium monoethoxyaluminium hydride to the corresponding mono- and dialkylborohydrides, which are stable over long periods (499). Mono- and dialkylboranes can be conveniently generated from the borohydrides by treatment with hydrogen chloride, methyl iodide, or chlorotrimethylsilane (76). In this way, optically pure mono- and dialkylboranes become available for further transformations into enantiomerically pure carbon compounds.

Finally, kinetic resolution of racemic olefins and allenes can be achieved by hydroboration. The reaction of an olefin or allene racemate with a deficient amount of an asymmetric hydroborating agent results in the preferential conversion of the more reactive enantiomer into the organoborane. The remaining unreacted substrate is enriched in the less reactive enantiomer. Optical purities in the range of 1–65% have been reported (471).

Catalytic Asymmetric Hydroboration. The hydroboration of olefins with catecholborane (an achiral hydroborating agent) is catalyzed by cationic rhodium complexes with enantiomerically pure phosphines, eg, [Rh(cod)$_2$]BF$_4$BINAP, where cod is 1,5-cyclooctadiene and BINAP is 2,2'-bis(diphenylphosphino)-1,1'-binaphthyl (138,147). The enantioselectivities are moderate; only styrenes give product alcohols exceeding 90% ee (147). Nevertheless, these are still preliminary results and improvement is possible. Noteworthy is higher enantioselectivity (69% ee) of the catalyzed hydroboration of 2,3,3-trimethyl-1-butene as compared to that achieved by the best asymmetric hydroborating agents (145). Advantages of the use of achiral hydroborating agents and only small amounts of chiral auxiliaries make catalytic asymmetric hydroboration very attractive (148,500–503).

SYNTHESIS OF CHIRAL MOLECULES

Synthesis of Chiral Alcohols, Ketones, Halides, Deuterated Hydrocarbons, and Amines. The chiral organoboranes produced by asymmetric hydroboration can be transformed into the heterosubstituted chiral products applying methodologies developed for achiral organoboron compounds. Thus, in most cases described above, the organoboranes obtained from alkenes are oxidized to the corresponding chiral alcohols (Table 3). Among other examples are the alcohols derived from heterocyclic olefins (504–506), dienes (383,507), functionalized olefins (225,508–512), and deuterium- or tritium-labeled chiral alcohols (513,514). Chiral ketones, eg, norcamphor, can be obtained by oxidation of chiral secondary alkylboranes with chromic acid (515,516). The hydroboration of (Z)-2-butene with diisopinocampheylborane followed by iodination with iodine in the presence of base or with iodine monochloride in the presence of sodium acetate affords 2-iodobutane of 87% optical purity (238,240). Chiral deuterated alkanes can be prepared either by asymmetric hydroboration–deuteriolysis or by asymmetric deuterioboration–protonolysis (514). Optically pure boronic esters available either by asymmetric hydroboration or homologation are particularly important intermediates, leading to optically pure products. Thus primary amines of high enantiomeric purity are produced through the intermediate formation of alkylmethylborinic esters (517).

Secondary amines having one or two chiral groups attached to the nitrogen atom are prepared from boronic esters by their conversion into alkyldichloroboranes, followed by treatment with organic azides (518). The second chiral group can be derived from an optically active azide.

$$\text{LiR}^*\text{BH}_3 \xrightarrow{\text{HCl}} \text{R}^*\text{BCl}_2 \xrightarrow{\text{RN}_3} \text{Cl}_2\text{BNRR}^* + \text{N}_2 \xrightarrow[\text{2. KOH}]{\text{1. H}_2\text{O}} \text{R}^*\text{NHR}$$

Synthesis of Chiral Alkanes, Alkenes, and Alkynes. An efficient general synthesis of α-chiral (Z)- and (E)-alkenes in high enantiomeric purity is based on the hydroboration of alkynes and 1-bromoalkynes, respectively, with enantiomerically pure IpcR*BH readily available by the hydroboration of prochiral alkenes with monoisopinocampheylborane, followed by crystallization (519).

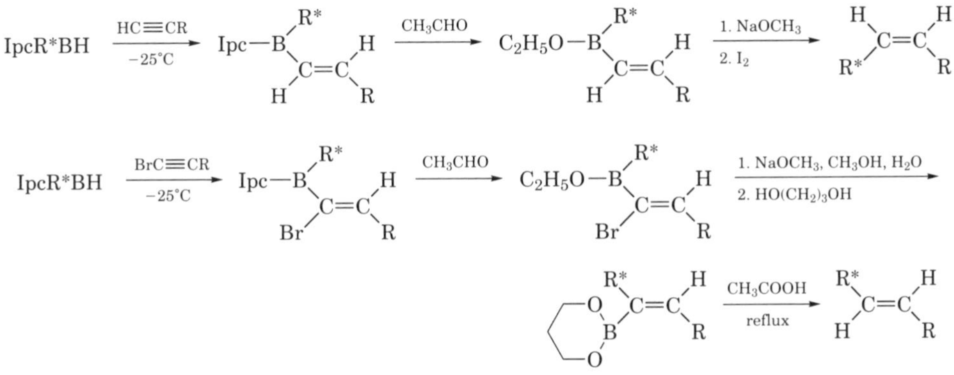

In the earlier, longer approach to (Z)- and (E)-alkenes, ThxR*BH was used instead of IpcR*BH. It is also possible to prepare α-chiral acetylenes and alkanes by this method (76,520). In a shorter synthesis of α-chiral alkynes, a prochiral disubstituted (Z)-alkene is hydroborated with diisopinocampheylborane and the trialkylborane produced is treated with alkynyllithium followed by iodine (521). Optical purity of the product is the same as the intermediate trialkylborane. The scope of this method is limited to alkenes handled by diisopinocampheylborane.

Synthesis of α-Chiral and Homologated Aldehydes, Acids, and β-Chiral Alcohols. A general approach to these compounds is based on the reaction of dichloromethyllithium with boronic esters. Rearrangement of the complex followed by reduction with potassium triisopropoxyborohydride provides the homologated boronic ester, which can be oxidized to the corresponding alcohol or transformed into the homologated aldehyde by reaction with methoxy(phenylthio) methyllithium (522,523). β-Chiral alcohols not available in high optical purity by asymmetric hydroboration of terminal alkenes are readily prepared by this method.

Synthesis of α- and β-Chiral Ketones, Esters, and Nitriles.   Chiral boronic esters are convenient precursors of α-chiral ketones (R*COR′), which can be prepared via the dialkylborinic ester or dialkylthexyl route (524,525).

The conversion of chiral boronic esters into optically pure B-alkyl-9-BBN derivatives followed by reaction with α-bromoketones, α-bromoesters, or α-bromonitriles leads to the homologated β-chiral ketones, esters, and nitriles, respectively (526).

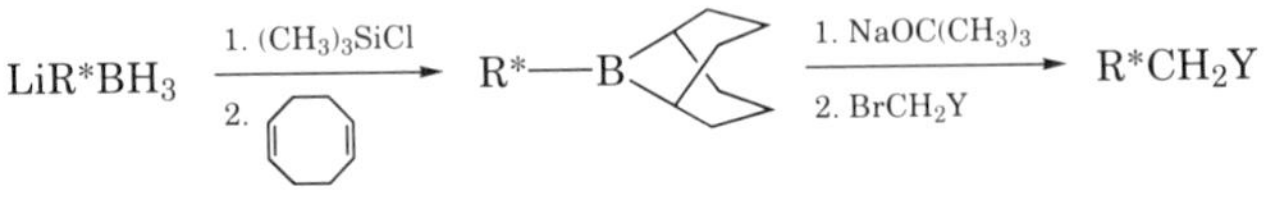

$$\text{LiR*BH}_3 \xrightarrow[\text{2.}]{\text{1. (CH}_3)_3\text{SiCl}} \text{R*}-\text{B} \xrightarrow[\text{2. BrCH}_2\text{Y}]{\text{1. NaOC(CH}_3)_3} \text{R*CH}_2\text{Y}$$

where Y = COR, COOR, CN

Asymmetric Allylboration.   Optically active allylic boronates and dialkyl-boranes transfer the allylic group to aldehydes enantioselectively. Optical purities of the product alcohols in a simple asymmetric synthesis using acetaldehyde and an allyl derivative of the ester (**23**) are in the range of 24–86% (384). In the following, (+)(**23**) yields 76% (+)(**24**) and 24% (+)(**25**), and (−)(**23**) yields 8% (−)(**24**) and 92% (−)(**25**). Correct pairing of a chiral ester and a chiral aldehyde results in high diastereoselectivity (527).

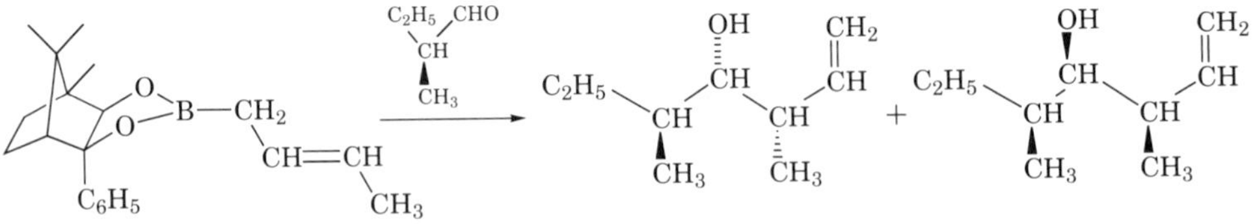

(23)   **(24)**   **(25)**

The asymmetric allyl- and crotylboration of aldehydes has emerged as an effective alternative to the aldol methodology in reactions involving acyclic stereoselection. Several reagents have been developed (Fig. 5), and their selectivity has been compared (535,536). Asymmetric allylboration has been applied to the synthesis of 2-deoxyhexoses (537); the AB disaccharide unit of olivomycin A (538); the C-17 to C-27 segment of rifamycin S (532); benzoyl–pedamide, the key building block of pederin (534); ipsdienol (539); and to the α- and γ-hydroxyallylation and γ-chloroallylation of aldehydes (529,534).

Homologation of Boronic Esters.   A convenient general method of enantioselective carbon–carbon bond formation, not involving hydroboration, is based on the homologation reaction of boronic esters derived from optically active 1,2-diols, eg, 2,3-pinanediol (341). The -ate complex formed by treating the boronate with dichloromethyllithium undergoes 1,2-migration of the alkyl group. The second alkyl group is introduced using an alkyllithium or Grignard reagent (qv). Oxidation provides a chiral alcohol of high optical purity. Consecutive chiral centers can be introduced by repeating the homologation sequence. Ultrahigh

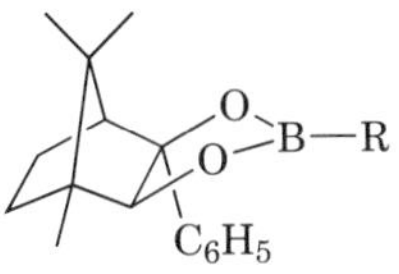

(**26**)

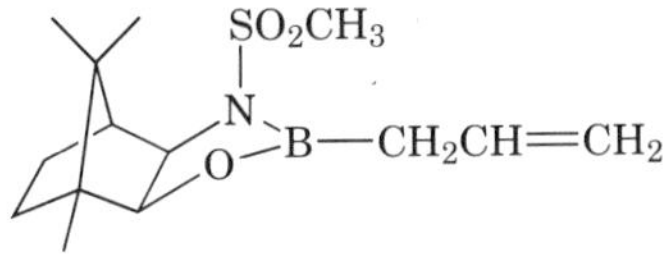

(**27**)

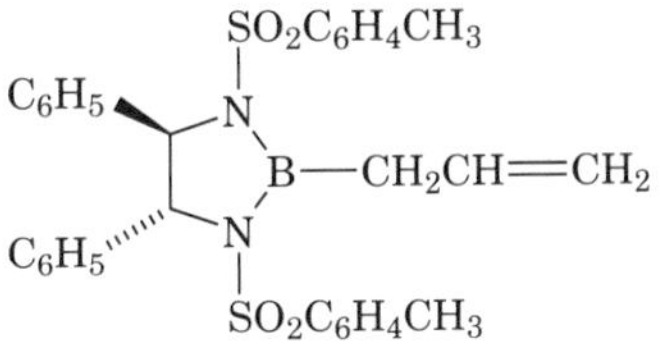

(**28**)

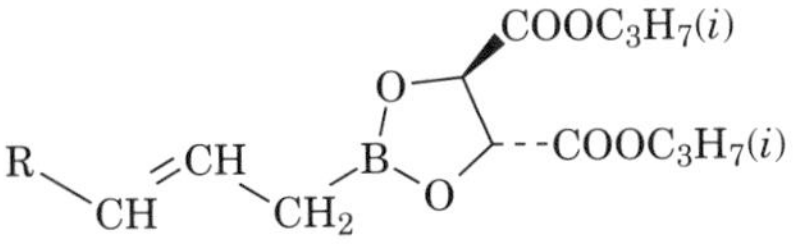

(**29**)

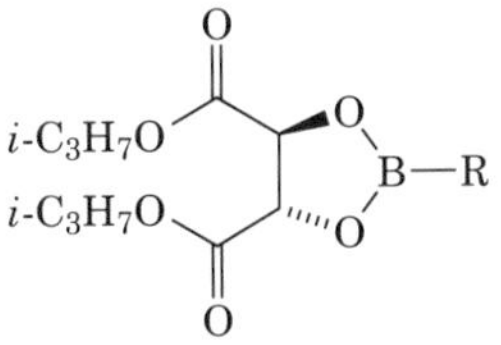

(**30**)

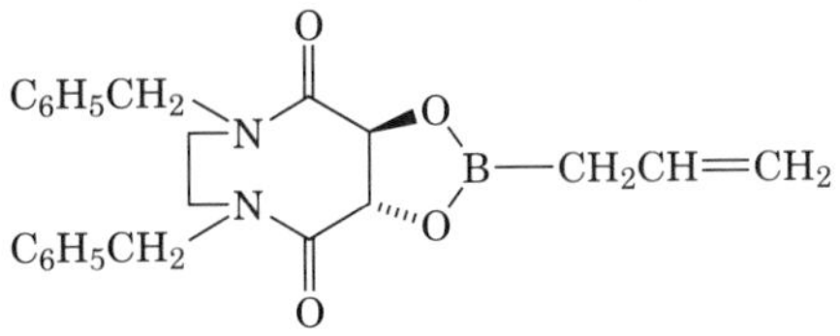

(**31**)

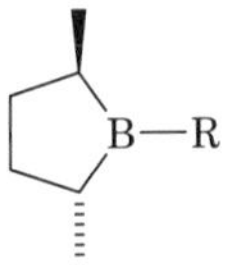

(**32**)

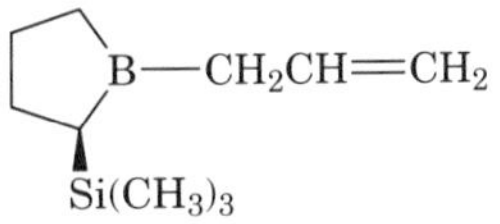

(**33**)

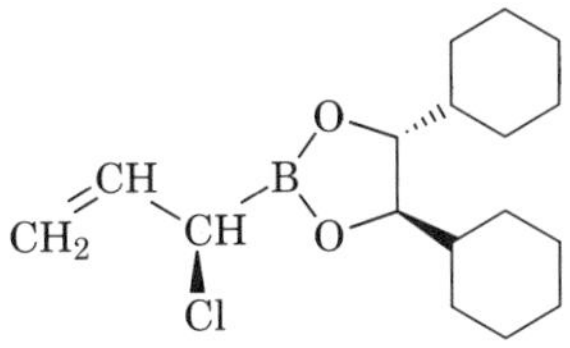

(**34**)

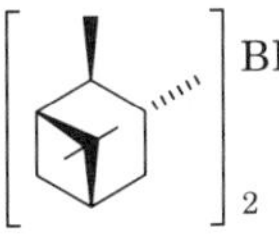

(**35**)

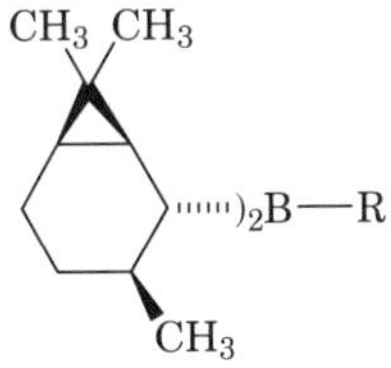

(**36**)

Fig. 5. Allylboration reagents. Fuller descriptions can be found in the references noted: (**26**), R = allyl or crotyl (384,472,527); (**27**) (528); (**28**) (386); (**29**) (R = H) is 2-(2-propenyl)-1,3,2-dioxaborolane-4,5-dicarboxylic acid diisopropyl ester [*99417-55-7*]. Other related reagents are (**29**), R = $(C_6H_{11}O)(CH_3)_2Si-$ and (**29**), R = $C_6H_5(CH_3)_2Si-$ (529); (**30**), R = allyl or crotyl (530,531); (**31**) (531,532); (**32**), R = allyl or crotyl (385,533); (**33**) (385); (**34**) (534); (**35**) (R = allyl) is B-allyldiisopinocampheylborane [*124821-92-7*], or R may be crotyl (382,535); (**36**) (R = allyl) is β-allylbis(2-isocaranyl)borane [*85116-38-7*], or R may be crotyl (382,535,536).

diastereomeric ratios (1000:1) for each new chiral center introduced have been achieved using boronic esters derived from 1,2-diols with C-2 symmetry (540). This chemistry has been applied to the synthesis of labile molecules, eg, (2*S*,3*S*)-2-methyl-3-benzyloxypentanal, the Anobiid beetle pheromones stegobiol and stegobione and carbohydrates (341,541).

Enolboration. The aldol reaction is one of the most powerful methodologies for the formation of carbon–carbon bonds in a stereodefined manner. Boron enolates are important intermediates for this transformation, since transition states of boron-mediated aldol reactions appear tightly organized, transmitting well the spatial arrangement to the aldol product (542). The stereochemical outcome of the addition to aldehydes depends on the configuration of the enolate, eg, (Z)-isomers of enol borinates produce the syn aldols, and E isomers produce the anti-aldols (543). Enolboration of aldehydes and ketones is best achieved with dialkylborane derivatives R_2BX in the presence of amines, where X is a good leaving group, eg, triflate or halogen (544–547). The stereoselective formation of (E)- or (Z)-enolates is influenced by substitutents on boron and the ketone, and also by the amine (547–551). Molecular modeling has been applied to design optimal chiral groups on boron for the anti-aldol products (552). Enolborinates can also be prepared by the addition of boranes to conjugated ketones (178–181). Boron enolates have been extensively applied to the synthesis of macrolides, ionophore antibiotics, and other compounds (542,553–556).

BIBLIOGRAPHY

"Boron Compounds" in *ECT* 1st ed., Suppl. 2, pp. 109–126, by J. W. Shepherd and E. B. Ayres, Callery Chemical Co.; "Boron Compounds (Organic)" in *ECT* 2nd ed., Vol. 3, pp. 707–723, by W. G. Woods, U.S. Borax Research Corp.; "Hydroboration" in *ECT* 3rd ed., Vol. 12, by H. C. Brown, Purdue University, and K. Smith, University College of Swansea, Wales, pp. 793–826.

1. H. C. Brown, *Hydroboration*, W. A. Benjamin, Inc., New York, 1962; 2nd printing, Benjamin/Cummings Publishing Co., Reading, Mass., 1980.
2. H. C. Brown, *Boranes in Organic Chemistry*, Cornell University Press, Ithaca, N.Y., 1972.
3. H. C. Brown, *Organic Syntheses via Boranes*, Wiley-Interscience, New York, 1975.
4. A. Pelter and K. Smith, in D. H. R. Barton and W. D. Ollis, eds., *Comprehensive Organic Chemistry*, Vol. 3, Pergamon Press, Oxford, U.K., 1979, p. 689.
5. A. Pelter, K. Smith, and H. C. Brown, *Borane Reagents*, Academic Press, London, 1988.
6. K. Smith and A. Pelter, in B. Trost, ed., *Comprehensive Organic Synthesis*, Vol. 8, Pergamon Press, Oxford, U.K., 1991, p. 703.
7. R. Köster, *Houben-Weyl Methoden der Organischen Chemie*, Vol. XIII/3a, 3b, 3c, G. Thieme, Stuttgart, Germany, 1982–1984.
8. D. S. Matteson, in F. R. Hartley, ed., *The Chemistry of the Metal–Carbon Bond*, Vol. 4, John Wiley & Sons, Inc., Chichester, U.K., 1987, p. 307.
9. M. Zaidlewicz, in G. Wilkinson, F. G. A. Stone, and A. W. Abel, eds., *Comprehensive Organometallic Chemistry*, Vol. 7, Pergamon Press, Oxford, U.K., 1982, p. 143.
10. E. Negishi, in ref. 9, p. 253.
11. H. C. Brown and M. Zaidlewicz, in J. J. Zuckermann and A. P. Hagen, eds., *Inorganic Reactions and Methods*, Vol. 10, VCH Publishers, New York, 1989, p. 35.
12. E. Negishi and M. J. Idacavage, *Organic Reactions* **33**, 1 (1985).
13. B. M. Mikhailov and Yu. N. Bubnov, *Organoboron Compounds in Organic Synthesis*, Harwood Academic Publishers, London, 1984.
14. D. T. Hurd, *J. Am. Chem. Soc.* **70**, 2053 (1948).
15. F. G. A. Stone and H. J. Emeleus, *J. Am. Chem. Soc.* **72**, 2755 (1950).

16. H. C. Brown and B. C. Subba Rao, *J. Am. Chem. Soc.* **78**, 2582 (1956).
17. H. C. Brown and B. C. Subba Rao, *J. Am. Chem. Soc.* **81**, 6423 (1959).
18. H. C. Brown and A. K. Mandal, *Synthesis*, 153 (1980).
19. D. J. Pasto, in E. L. Muetterties, ed., *Boron Hydride Chemistry*, Academic Press, New York, 1975, p. 197.
20. D. J. Pasto, B. Lepeska, T. C. Cheng, *J. Am. Chem. Soc.* **94**, 6083 (1972).
21. H. C. Brown and J. Chandrasekharan, *Gazz. Chim. Ital.* **117**, 517 (1987).
22. C. F. Lane, *Aldrichim. Acta* **6**, 51 (1973).
23. M. F. Hawthorne, *J. Am. Chem. Soc.* **83**, 2541 (1961).
24. E. C. Ashby, *J. Am. Chem. Soc.* **81**, 4791 (1959).
25. E. Mincione, *Ann. Chim. (Rome)* **67**, 119 (1977).
26. M. Rabinovitz, G. Salemnik, and E. D. Bergman, *Tetrahedron Lett.*, 3271 (1967).
27. M. Nussim, Y. Mazur, and F. Sondheimer, *J. Org. Chem.* **29**, 1120 (1964).
28. G. J. Abruscato and T. T. Tidwell, *J. Org. Chem.* **37**, 4151 (1972).
29. A. R. Hochstetler, *J. Org. Chem.* **40**, 1536 (1975).
30. H. C. Brown and U. S. Racherla, *Tetrahedron Lett.* **26**, 2187 (1985).
31. J. E. Rice and Y. Okamoto, *J. Org. Chem.* **47**, 4189 (1982).
32. H. C. Brown and U. S. Racherla, *J. Am. Chem. Soc.* **105**, 65 (1983).
33. H. C. Brown and U. S. Racherla, *J. Org. Chem.* **48**, 1389 (1983).
34. S. E. Wood and B. Rickborn, *J. Org. Chem.* **48**, 555 (1983).
35. M. E. Kuimova and B. M. Mikhailov, *Bull. Acad. Sci. USSR, Div. Chem. Sci.*, 288 (1979).
36. B. Wrackmeyer, *J. Organometal. Chem.* **117**, 313 (1976).
37. R. Köster and P. Binger, *Inorg. Synth.* **15**, 141 (1974).
38. H. C. Brown and M. Zaidlewicz, in Ref. 11, p. 88.
39. J. A. Soderquist and H. C. Brown, *J. Org. Chem.* **45**, 3571 (1980).
40. H. C. Brown and G. Zweifel, *J. Am. Chem. Soc.* **82**, 4708 (1960).
41. P. R. Jones, *J. Org. Chem.* **37**, 1886 (1972).
42. T. P. Fehlner, *J. Am. Chem. Soc.* **93**, 6366 (1971).
43. H. C. Brown and J. Chandrasekharan, *J. Org. Chem.* **53**, 4811 (1988).
44. K. K. Wang and H. C. Brown, *J. Am. Chem. Soc.* **104**, 7148 (1982).
45. R. W. Parry and M. K. Walter, *Prep. Inorg. React.* **5**, 45 (1968).
46. L. H. Long, *Progr. Inorg. Chem.* **15**, 1 (1972).
47. K. Becker, H. Keller-Rudek, and H. List, in K. Niedenzu and K. C. Buschbeck, eds., *Gmelin Handbuch der Anorganischen Chemie*, Vol. 52, Pt. 18, Springer-Verlag, Berlin, 1978, p. 67.
48. U.S. Pat. 4,388,284 (Jan. 14, 1983), S. G. Shore and M. A. Toft (to Ohio State University Research Foundation).
49. D. Männig and H. Nöth, *J. Chem. Soc., Dalton Trans.*, 1689 (1985).
50. Y. Aoyama, Y. Taneka, T. Fujisawa, T. Watanabe, H. Toi, and H. Ogoshi, *J. Org. Chem.* **52**, 2555 (1987).
51. A. Domb and Y. Arny, *J. Appl. Polym. Sci.* **30**, 3589 (1985).
52. R. Shundo, I. Nishiguschi, Y. Matsubara, and T. Hirashima, *Chem. Lett.*, 2033 (1989).
53. R. Shundo, Y. Matsubara, I. Nishiguchi, and T. Hirashima, *Bull. Chem. Soc. Jpn.* **65**, 530 (1992).
54. R. Soundararajan and D. S. Matteson, *J. Org. Chem.* **55**, 2274 (1990).
55. K. Smith and A. Pelter, Ref. 6, p. 709.
56. L. M. Braun, R. A. Braun, H. R. Crissman, M. Opperman, and R. M. Adams, *J. Org. Chem.* **30**, 1238 (1971).
57. H. C. Brown, A. K. Mandal, and S. U. Kulkarni, *J. Org. Chem.* **42**, 1392 (1977).
58. H. C. Brown and A. K. Mandal, *J. Org. Chem.* **57**, 4970 (1992).
59. H. C. Brown, E. Negishi, and W. C. Dickason, *J. Org. Chem.* **50**, 520 (1985).

60. E. Negishi and H. C. Brown, *Synthesis*, 77 (1974).
61. H. C. Brown, E. Negishi and J.-J. Katz, *J. Am. Chem. Soc.* **97**, 2791 (1975).
62. H. C. Brown, J.-J. Katz, C. F. Lane, and E. Negishi, *J. Am. Chem. Soc.* **97**, 2799 (1975).
63. A. Pelter, D. J. Ryder, and J. H. Sheppard, *Tetrahedron Lett.*, 4715 (1978).
64. H. C. Brown, A. K. Mandal, N. M. Yoon, B. Singaram, J. R. Schwier, and P. K. Jadhav, *J. Org. Chem.* **47**, 5069 (1982).
65. H. C. Brown and B. Singaram, *J. Am. Chem. Soc.* **106**, 1797 (1984).
66. H. C. Brown, R. S. Randad, K. S. Bhat, M. Zaidlewicz, S. A. Waissman, P. K. Jadhav, and P. T. Perumal, *J. Org. Chem.* **53**, 5513 (1988).
67. H. C. Brown, S. A. Waissman, P. T. Perumal, and U. P. Dhokte, *J. Org. Chem.* **55**, 1217 (1990).
68. K. K. Richter, M. Bonato, M. Follet, and J. M. Kamenka, *J. Org. Chem.* **55**, 2855 (1990).
69. H. C. Brown, J. R. Schwier, and B. Singaram, *J. Org. Chem.* **44**, 465 (1979).
70. H. C. Brown and S. K. Gupta, *J. Am. Chem. Soc.* **93**, 4062 (1971).
71. H. C. Brown, B. Singaram, and T. E. Cole, *J. Am. Chem. Soc.* **107**, 460 (1983).
72. H. C. Brown, N. G. Bhat, and V. Somayaji, *Organometallics* **2**, 1311 (1983).
73. H. C. Brown, J. R. Schwier, and B. Singaram, *J. Org. Chem.* **43**, 4395 (1978).
74. H. C. Brown and P. K. Jadhav, *J. Org. Chem.* **46**, 5047 (1981).
75. B. Singaram, T. E. Cole, and H. C. Brown, *Organometallics* **3**, 1520 (1984).
76. H. C. Brown, R. K. Bakshi, and B. Singaram, *J. Am. Chem. Soc.* **110**, 1529 (1988).
77. M. Srebnik, T. E. Cole, and H. C. Brown, *Tetrahedron Lett.* **28**, 3771 (1987).
78. M. Srebnik, T. E. Cole, and H. C. Brown, *J. Org. Chem.* **55**, 5051 (1990).
79. M. Srebnik, T. E. Cole, P. V. Ramachandran, and H. C. Brown, *J. Org. Chem.* **54**, 6085 (1989).
80. H. C. Brown, T. E. Cole, M. Srebnik, and K.-W. Kim, *J. Org. Chem.* **51**, 4925 (1986).
81. S. C. Shiner, C. M. Garner, and R. C. Hattiwanger, *J. Am. Chem. Soc.* **107**, 7167 (1985).
82. H. C. Brown and N. Ravindran, *Inorg. Chem.* **16**, 2938 (1077).
83. K. Kingberger and W. Siebert, *Z. Naturforsch. Teil B*, **30**, 55 (1975).
84. W. E. Paget and K. Smith, *J. Chem. Soc., Chem. Commun.*, 1169 (1980).
85. R. Bolton, P. N. Gates, and S. A. W. Jones, *Aust. J. Chem.* **40**, 987 (1987).
86. H. C. Brown and M. Zaidlewicz, *Chem. Stosow.* **26**, 155 (1982).
87. H. C. Brown and S. K. Kulkarni, *J. Organometal. Chem.* **239**, 23 (1982).
88. H. C. Brown and N. Ravindran, *J. Am. Chem. Soc.* **98**, 1785 (1976).
89. H. C. Brown and M. Zaidlewicz, J. Am. Chem. Soc. **98**, 4917 (1976).
90. R. Köster, G. Griaznow, W. Larbig, and P. Binger, *Justus Liebigs Ann. Chem.*, **672**, 1 (1964).
91. S. K. Kulkarni, D. Basavaiah, M. Zaidlewicz, and H. C. Brown, *Organometallics* **1**, 212 (1982).
92. P. J. Maddocks, A. Pelter, K. Rowe, K. Smith, and C. Subrahmaniam, *J. Chem. Soc., Perkin Trans. 1*, 653 (1981).
93. J. A. Soderquist and I. Rivera, *Tetrahedron Lett.* **29**, 3195 (1988).
94. E. Negishi, P. L. Burke, and H. C. Brown, *J. Am. Chem. Soc.* **94**, 7431 (1972).
95. E. Negishi and H. C. Brown, *J. Am. Chem. Soc.* **95**, 6757 (1973).
96. H. C. Brown and G. G. Pai, *Heterocycles* **17**, 77 (1982).
97. H. C. Brown and G. G. Pai, *J. Organometal. Chem.* **250**, 13 (1983).
98. H. C. Brown, G. G. Pai, and R. G. Naik, *J. Org. Chem.* **49**, 1072 (1984).
99. H. C. Brown and E. Negishi, *Tetrahedron* **33**, 2331 (1977).
100. H. C. Brown and A. W. Moerikofer, *J. Am. Chem. Soc.* **83**, 3417 (1961).
101. H. C. Brown and G. Zweifel, *J. Am. Chem. Soc.* **83**, 1241 (1961).

102. H. C. Brown and G. J. Klender, *Inorg. Chem.* **1**, 204 (1962).
103. D. E. Evans and W. C. Black, *J. Am. Chem. Soc.* **114**, 2260 (1992).
104. C. H. Heathcock, J. A. Stafford, and D. L. Clark, *J. Org. Chem.* **57**, 2575 (1992).
105. D. S. Dodd and A. C. Ochlschlager, *J. Org. Chem.* **57**, 2794 (1992).
106. G. Zweifel, N. R. Ayyangar, and H. C. Brown, *J. Am. Chem. Soc.* **85**, 2072 (1963).
107. G. Zweifel, H. Arzoumanian, and C. C. Whitney, *J. Am. Chem. Soc.* **89**, 3652 (1967).
108. G. Zweifel and N. L. Polston, *J. Am. Chem. Soc.* **92**, 4068 (1970).
109. H. C. Brown, E. F. Knights, and C. G. Scouten, *J. Am. Chem. Soc.* **96**, 7765 (1974).
110. Ref. 7, Vol. 13/3a, p. 321.
111. J. A. Soderquist and A. Negron, *J. Org. Chem.* **52**, 3441 (1987).
112. R. Köster and M. Yalpani, *Pure Appl. Chem.* **63**, 387 (1991).
113. H. C. Brown, R. Liotta, and C. G. Scouten, *J. Am. Chem. Soc.* **98**, 5297 (1976).
114. H. C. Brown, D. J. Nelson, and C. G. Scouten, *J. Org. Chem.* **48**, 641 (1983).
115. C. A. Brown and R. A. Coleman, *J. Org. Chem.* **44**, 2328 (1979).
116. H. C. Brown, R. Liotta, and L. Brener, *J. Am. Chem. Soc.* **99**, 3427 (1977).
117. A. Pelter, S. Singaram, and H. C. Brown, *Tetrahedron Lett.* **24**, 1433 (1983).
118. L. Lynds and D. R. Stern, *J. Am. Chem. Soc.* **81**, 5006 (1959).
119. H. C. Brown and N. Ravindran, *J. Am. Chem. Soc.* **98**, 1798 (1976).
120. H. C. Brown, N. Ravindran, and S. H. Kulkarni, *J. Org. Chem.* **45**, 384 (1980).
121. H. C. Brown and N. Ravindran, *J. Am. Chem. Soc.* **99**, 7097 (1977).
122. H. C. Brown and U. S. Racherla, *J. Org. Chem.* **51**, 895 (1986).
123. H. C. Brown and J. B. Campbell, Jr., *J. Org. Chem.* **45**, 389 (1980).
124. H. C. Brown and J. Chandrasekharan, *J. Org. Chem.* **48**, 644 (1983).
125. W. G. Woods and P. L. Strong, *J. Am. Chem. Soc.* **88**, 4667 (1966).
126. S. H. Rose and S. G. Shore, *Inorg. Chem.* **1**, 744 (1961).
127. S. Thaisrivongs and J. D. Wuest, *J. Org. Chem.* **42**, 3243 (1977).
128. S. Cabiddu, M. Maccioni, L. Mura, and M. Secci, *J. Heterocycl. Chem.* **12**, 169 (1975).
129. J. S. Cha and co-workers, *Heterocycles* **32**, 425 (1991).
130. A. Arase, Y. Nunokawa, Y. Masuda, and M. Hoshi, *J. Chem. Soc., Chem. Commun.*, 51 (1992).
131. H. C. Brown and S. K. Gupta, *J. Am. Chem. Soc.* **93**, 1816 (1971).
132. Ger. Offen. DE 3,528,321 (Feb. 12, 1987), D. Männig and H. Nöth (to Metallgesellschaft A.-G.).
133. C. F. Lane and G. N. Kabalka, *Tetrahedron* **32**, 981 (1976).
134. G. W. Kabalka, *Org. Prep. Proc. Int.* **9**, 133 (1977).
135. H. J. Bestmann, J. Suss, and O. Vostrowsky, *Tetrahedron Lett.* **21**, 2467 (1979).
136. A. Arase, Y. Nunokawa, Y. Masuda, and M. Hoshi, *J. Chem. Soc., Chem. Commun.*, 205 (1992).
137. D. Männig and H. Nöth, *Angew. Chem. Int. Ed. Engl.* **24**, 878 (1985).
138. K. Burgess and M. J. Ohlmeyer, *Chem. Rev.* **91**, 1179 (1991).
139. D. A. Evans, G. C. Fu, and A. H. Hoveyda, *J. Am. Chem. Soc.* **114**, 6671 (1992).
140. D. A. Evans, G. C. Fu, and B. A. Anderson, *J. Am. Chem. Soc.* **114**, 6679 (1992).
141. D. A. Evans and G. Sheppard, *J. Org. Chem.* **55**, 5192 (1990).
142. J. R. Knorr and J. S. Merola, *Organometallics* **9**, 3008 (1990).
143. P. Kocienski, K. Jarowicki, and S. Marczak, *Synthesis*, 1191 (1991).
144. K. Burgess, W. van der Donk, M. B. Jarstfer, and M. J. Ohlmeyer, *J. Am. Chem. Soc.* **113**, 6139 (1991).
145. K. Burgess and M. J. Ohlmeyer, *J. Org. Chem.* **53**, 5178 (1988).
146. K. Burgess and M. J. Ohlmeyer, *Tetrahedron Lett.* **30**, 5857 (1989).
147. T. Hayashi, Y. Matsumoto, and Y. Ito, *J. Am. Chem. Soc.* **111**, 3426 (1989).
148. J. F. Zhang, B. L. Lou, G. H. Guo, and L. X. Dai, *J. Org. Chem.* **56**, 1670 (1991).
149. H. C. Brown, J. A. Sikorski, S. U. Kulkarni, and D. H. Lee, *J. Org. Chem.* **47**, 863 (1982).

150. G. Zweifel and N. R. Pearson, *J. Am. Chem. Soc.* **102**, 5919 (1980).
151. H. C. Brown and N. M. Yoon, *Isr. J. Chem.* **15**, 12 (1977).
152. H. C. Brown, M. C. Desai, and P. K. Jadhav, *J. Org. Chem.* **47**, 5065 (1982).
153. H. C. Brown and B. Singaram, *J. Org. Chem.* **49**, 945 (1984).
154. H. C. Brown and N. N. Joshi, *J. Org. Chem.* **53**, 4059 (1988).
155. H. C. Brown, J. V. N. Vara Prasad, and M. Zaidlewicz, *J. Org. Chem.* **53**, 2911 (1988).
156. P. K. Jadhav and S. U. Kulkarni, *Heterocycles* **18**, 233 (1982).
157. P. K. Jadhav and H. C. Brown, *J. Org. Chem.* **46**, 2688 (1981).
158. P. K. Jadhav, J. V. N. Vara Prasad, and H. C. Brown, *J. Org. Chem.* **50**, 3203 (1985).
159. S. Masamune, B. M. Kim, J. S. Peterson, T. Sata, S. J. Veenstra, and T. Imai, *J. Am. Chem. Soc.* **107**, 4549 (1985).
160. H. C. Brown, P. Heim, N. M. Yoon, *J. Am. Chem. Soc.* **92**, 1637 (1970).
161. H. C. Brown, P. Heim, N. M. Yoon, *J. Org. Chem.* **37**, 2942 (1972).
162. H. C. Brown, B. Nazer, J. S. Cha, and J. A. Sikorski, *J. Org. Chem.* **51**, 5264 (1986).
163. H. C. Brown, J. S. Cha, N. M. Yoon, and B. Nazer, *J. Org. Chem.* **52**, 5400 (1987).
164. H. C. Brown, D. B. Bigley, S. K. Arora, and N. M. Yoon, *J. Am. Chem. Soc.* **92**, 7161 (1970).
165. H. C. Brown, S. Krishnamurthy, and N. M. Yoon, *J. Org. Chem.* **41**, 1778 (1976).
166. G. W. Kabalka and D. E. Bierer, *Synth. Commun.* **19**, 2783 (1989).
167. G. W. Kabalka and D. E. Bierer, *Organometallics* **8**, 655 (1989).
168. H. C. Brown and M. K. Unni, *J. Am. Chem. Soc.* **90**, 2902 (1968).
169. H. C. Brown and R. M. Gallivan, Jr., *J. Am. Chem. Soc.* **90**, 2906 (1968).
170. H. C. Brown and D. J. Cope, *J. Am. Chem. Soc.* **86**, 1801 (1964).
171. H. C. Brown and K. A. Kablys, *J. Am. Chem. Soc.* **86**, 1791, 1795 (1964).
172. H. C. Brown and R. L. Sharp, *J. Am. Chem. Soc.* **90**, 2915 (1968).
173. D. J. Pasto and S. R. Snyder, *J. Org. Chem.* **31**, 2773 (1966).
174. D. J. Pasto and J. Hickman, *J. Am. Chem. Soc.* **89**, 5608 (1967).
175. E. Dunkelblum, R. Levene, and J. Klein, *Tetrahedron* **28**, 1009 (1972).
176. B. Singaram, M. V. Rangaishenvi, H. C. Brown, C. T. Goralski, and D. L. Hasha, *J. Org. Chem.* **56**, 1543 (1991).
177. S. Krishnamurthy and H. C. Brown, *J. Org. Chem.* **41**, 1864 (1975).
178. Y. Matsumoto and T. Hayashi, *Synlett.*, 349 (1991).
179. D. A. Evans and G. C. Fu, *J. Org. Chem.* **55**, 5678 (1990).
180. G. P. Boldrini, F. Mancini, E. Tagliavini, C. Trombini, and A. Umani-Ronchi, *J. Chem. Soc., Chem. Commun.*, 1680 (1990).
181. G. P. Boldrini, M. Bortolotti, E. Tagliavini, C. Trombini, and A. Umani-Ronchi, *Tetrahedron Lett.* **32**, 1229 (1991).
182. M. Zaidlewicz, A. Uzarewicz and R. Sarnowski, *Synthesis*, 62 (1979).
183. R. Chaabouni, A. Laurent, and B. Marquet, *Tetrahedron* **36**, 877 (1980).
184. H. C. Brown and P. V. N. Vara Prasad, *J. Org. Chem.* **50**, 3002 (1985).
185. H. C. Brown and J. C. Chen, *J. Org. Chem.* **46**, 3978 (1981).
186. M. P. Cooke, Jr., *J. Org. Chem.* **57**, 1495 (1992).
187. Y. A. Andemichael and K. K. Wang, *J. Org. Chem.* **57**, 796 (1992).
188. H. C. Brown, S. U. Kulkarni, C. G. Rao, and V. D. Patil, *Tetrahedron* **42**, 5515 (1986).
189. V. K. Gatam, J. Singh, and R. S. Dihillon, *J. Org. Chem.* **53**, 187 (1988).
190. S. Kafka, P. Traska, J. Kytner, P. Taufman, and M. Ferles, *Coll. Czech. Chem. Commun.* **52**, 2047 (1987).
191. M. Baboullne, A. Lattes, and Z. Benmaarouf-Khalaayoun, *Synth. Commun.* **20**, 2091 (1990).
192. B. F. Molino, J. Cusmano, D. R. Mootoo, R. Faglih, and B. Fraser-Reid, *J. Carbohydr. Chem.* **6**, 479 (1987).
193. W. R. Friesen and A. K. Daljeet, *Tetrahedron Lett.* **31**, 6133 (1990).

194. H. C. Brown and M. V. Rangaishenvi, *J. Heterocycl. Chem.* **27**, 13 (1990).
195. A. S. Ionkin, S. N. Ignateva, Y. M. Nekhoroskov, J. J. Efremov, and B. A. Arbuzov, *Phosphorus, Sulfur, Silicon Rel. Elem.* **53**, 1 (1990); M. Hoshi, Y. Masuda, and A. Arase, *Bull. Chem. Soc. Jpn.* **63**, 447 (1990).
196. J. A. Sonderquist and A. Hassner, *J. Organomet. Chem.* **156**, C12 (1978).
197. D. Barrat, S. J. Davies, G. P. Elliot, J. A. K. Howard, D. B. Lewis, and F. G. A. Stone, *J. Organometal. Chem.* **325**, 185 (1987).
198. G. Erker and R. Aul, *Chem. Ber.* **124**, 1301 (1991).
199. M. Mori, N. Ueska, and M. Shibasaki, *J. Org. Chem.* **57**, 3519 (1992).
200. D. E. Moher, J. L. Collins, and P. Grieco, *J. Am. Chem. Soc.* **114**, 2764 (1992).
201. G. Zweifel and S. J. Backlund, *J. Organometal. Chem.* **156**, 159 (1978).
202. G. Zweifel, A. Horng, and J. T. Snow, *J. Am. Chem. Soc.* **92**, 1427 (1970).
203. H. C. Brown, K. P. Singh, and B. J. Garner, *J. Organometal. Chem.* **1**, 2 (1963).
204. G. Zweifel and H. C. Brown, *J. Am. Chem. Soc.* **86**, 393 (1964).
205. J. L'Homme and G. Ourisson, *Tetrahedron* **24**, 3201 (1968).
206. H. C. Brown and K. J. Murray, *Tetrahedron* **42**, 5497 (1986).
207. R. Köster, K.-L. Amen, and W. V. Dahlhoff, *Justus Liebigs Ann. Chem.*, 752 (1975).
208. R. Köster, W. Fenzl, and G. Seidel, *Justus Liebigs Ann. Chem.*, 352 (1975).
209. K. M. Taba, R. Köster, and W. V. Dahlhoff, *Justus Liebigs Ann. Chem.*, 463 (1987).
210. B. M. Mikhailov, *Organometal. Chem. Rev. Sect. A* **8**, 1 (1972).
211. M. Zaidlewicz, *J. Organometal. Chem.* **293**, 139 (1985).
212. A. J. Hubert, *J. Chem. Soc.*, 6669 (1965).
213. H. C. Brown and G. A. Molander, *J. Org. Chem.* **51**, 4512 (1986).
214. E. Negishi and K. W. Chiu, *J. Org. Chem.* **41**, 3484 (1976).
215. A. Pelter and K. Smith, in B. Trost, ed., *Comprehensive Organic Synthesis*, Vol. 7, Pergamon Press, Oxford, U.K., 1991, p. 593.
216. H. C. Brown, C. Snyder, B. C. Subba Rao and G. Zweifel, *Tetrahedron* **42**, 5505 (1986).
217. G. W. Kabalka, T. M. Shoup, and N. M. Geudgaon, *J. Org. Chem.* **54**, 5930 (1989).
218. H. C. Brown and R. A. Coleman, *J. Am. Chem. Soc.* **91**, 4606 (1969).
219. R. Köster and Y. Morita, *Angew. Chem., Int. Ed. Engl.* **5**, 580 (1966).
220. G. W. Kabalka and H. C. Hedgecock, Jr., *J. Org. Chem.* **40**, 1776 (1975).
221. G. W. Kabalka and S. W. Slayden, *J. Organometal. Chem.* **125**, 273 (1977).
222. R. W. Hoffmann and S. Dresley, *Synthesis*, 103 (1988).
223. C. T. Goralski, B. Singaram, and H. C. Brown, *J. Org. Chem.* **52**, 4014 (1987).
224. J. A. Soderquist and M. R. Najafi, *J. Org. Chem.* **51**, 1330 (1986).
225. S. Masamune, L. D. L. Lu, W. P. Jackson, T. Kaiho, and T. Toyoda, *J. Am. Chem. Soc.* **104**, 5523 (1982).
226. A. Pelter and M. Rowlands, *Tetrahedron Lett.* **28**, 1913 (1987).
227. R. Köster and Y. Morita, *Justus Leibigs Ann. Chem.* **704**, 70 (1967).
228. H. C. Brown, M. M. Midland, and G. W. Kabalka, *Tetrahedron* **42**, 5523 (1986).
229. H. C. Brown and M. M. Midland, *Tetrahedron* **43**, 4059 (1987).
230. V. V. Ramana Rao, D. Devaprabhkara and S. Chandrasekharan, *J. Organometal. Chem.* **162**, C9 (1978).
231. V. V. Brown and C. P. Garg, *Tetrahedron* **42**, 5511 (1986).
232. U. S. Racherla, V. V. Khanna, and H. C. Brown, *Tetrahedron Lett.* **34**, 1037 (1992).
233. C. H. Nguen, M. V. Mavrov, and E. P. Serebryakov, *Izv. AN SSSR Ser. Khim.*, 251 (1986).
234. H. C. Brown and N. R. De Lue, *Tetrahedron* **44**, 2785 (1988).
235. V. B. Jigajinni, W. E. Paget, and K. Smith, *J. Chem. Res. (S)*, 376 (1981).
236. D. J. Nelson and R. Soundararajan, *J. Org. Chem.* **53**, 5664 (1988).
237. H. C. Brown and C. F. Lane, *Tetrahedron* **44**, 2763 (1988).

238. H. C. Brown, M. W. Rathke, M. M. Rogic, and N. R. De Lue, *Tetrahedron* **44**, 2751 (1988).
239. H. C. Brown, C. F. Lane, and N. R. De Lue, *Tetrahedron* **44**, 2773 (1988).
240. G. W. Kabalka and E. E. Gooch III, *J. Org. Chem.* **45**, 3578 (1980).
241. G. W. Kabalka, K. A. R. Sastry, and K. U. Sastry, *Synth. Commun.* **12**, 101 (1982).
242. C. F. Lane and H. C. Brown, *J. Organometal. Chem.* **26**, C51 (1971).
243. H. C. Brown, R. C. Larock, S. K. Gupta, S. Rajagopalan, and N. G. Bhat, *J. Org. Chem.* **54**, 6079 (1989).
244. H. C. Brown, T. Hamaoka, N. Ravindran, C. Subrahmanyam, V. Somayaji, and N. G. Bhat, *J. Org. Chem.* **54**, 6075 (1989).
245. H. C. Brown and co-workers, *J. Org. Chem.* **54**, 6068 (1989).
246. G. Zweifel and H. Arzoumanian, *J. Am. Chem. Soc.* **89**, 5086 (1967).
247. H. C. Brown, C. D. Blue, D. J. Nelson, and N. G. Bhat, *J. Org. Chem.* **54**, 6064 (1989).
248. H. C. Brown and N. G. Bhat, *Tetrahedron Lett.* **22**, 21 (1988).
249. H. C. Brown, K-W. Kim, M. Srebnik, and B. Singaram, *Tetrahedron* 43, 4071 (1987).
250. G. W. Kabalka, K. A. R. Sastry, G. W. McCollum, and H. Yoshioka, *J. Org. Chem.* **46**, 5296 (1981).
251. Y. Tamura, J. Minamikawa, S. Fuji, and M. Ikeda, *Synthesis*, 196 (1973).
252. H. C. Brown, M. M. Midland, and A. B. Levy, *J. Am. Chem. Soc.* **95**, 2394 (1973).
253. G. W. Kabalka and Z. Wang, *Synth. Commun.* **20**, 2113 (1990).
254. H. C. Brown, M. M. Midland, A. B. Levy, A. Suzuki, S. Sono, and M. Itoh, *Tetrahedron* **43**, 4079 (1987).
255. R. H. Mueller and M. E. Thompson, *Tetrahedron Lett.* **22**, 1093 (1980).
256. A. G. Davies, S. C. W. Hook, and B. P. Roberts, *J. Organomet. Chem.* **23**, C11 (1970).
257. A. Suzuki, M. Ishidoya, and M. Tabata, *Synthesis*, 687 (1976).
258. V. B. Jigajinni, A. Pelter, and K. Smith, *Tetrahedron Lett.*, 181 (1978).
259. H. C. Brown and M. M. Midland, *J. Am. Chem. Soc.* **93**, 3291 (1971).
260. A. Arase, Y. Masuda, and A. Suzuki, *Bull. Chem. Soc. Jpn.* **47**, 2511 (1974).
261. A. Arase and Y. Masuda, *Chem. Lett.*, 1115 (1976).
262. S. Raucher, M. R. Hansen, and M. A. Colter, *J. Org. Chem.* **43**, 4885 (1978).
263. R. C. Larock, *Tetrahedron* **38**, 1713 (1982).
264. R. C. Larock and H. C. Brown, *J. Organometal. Chem.* **36**, 1 (1972).
265. R. C. Larock, S. K. Gupta, and H. C. Brown, *J. Am. Chem. Soc.* **94**, 4371 (1972).
266. Y. Yamamoto, H. Yatagi, K. Maruyama, A. Sonoda, and M. Murahashi, *Bull. Chem. Soc. Jpn.* **50**, 3427 (1977).
267. J. B. Campbell, Jr., and H. C. Brown, *J. Org. Chem.* **45**, 549 (1980).
268. E. Negishi, *Acc. Chem. Res.* **15**, 340 (1982).
269. A. Suzuki, *Pure Appl. Chem.* **57**, 1749 (1985).
270. A. Suzuki, *Pure Appl. Chem.* **63**, 419 (1991).
271. N. Satoh, T. Ishiyama, N. Miyaura, and A. Suzuki, *Bull. Chem. Soc. Jpn.* **62**, 3471 (1987).
272. N. Miyuara, T. Ishiyama, H. Sasaki, M. Ishikawa, M. Satoh, and A. Suzuki, *J. Am Chem. Soc.* **111**, 314 (1989).
273. N. Miyaura, M. Satoh, and A. Suzuki, *Tetrahedron Lett.* **27**, 3745 (1986).
274. T. Alves, A. B. De Oliveira, and V. Snieckus, *Tetrahedron Lett.* **29**, 2135 (1988).
275. J. M. Fu, M. J. Sharp, and V. Snieckus, *Tetrahedron Lett.* **29**, 5459 (1988).
276. M. A. Siddigni and V. Snieckus, *Tetrahedron Lett.* **29**, 5463 (1988).
277. A. Carpita, D. Neri, and R. Rossi, *Gazz. Chim. Ital.* **117**, 503 (1987).
278. I. Uenishi, J. M. Bean, R. W. Armstrong, and Y. Kishi, *J. Am. Chem. Soc.* **109**, 4756 (1987).
279. N. Miyaura, N. Sasaki, M. Itoh, and A. Suzuki, *Tetrahedron Lett.*, 3369 (1977).
280. N. Miyaura, M. Itoh, and A. Suzuki, *Bull. Chem. Soc. Jpn.* **50**, 2199 (1977).

281. H. C. Brown and J. B. Campbell, Jr., *J. Org. Chem.* **45**, 551 (1980).
282. H. C. Brown and G. A. Molander, *J. Org. Chem.* **46**, 645 (1981).
283. S. Sharma and A. C. Ochlschlager, *Tetrahedron Lett.* **29**, 261 (1988).
284. G. Zweifel, R. P. Fisher, J. T. Snow, and C. C. Whitney, *J. Am. Chem. Soc.* **93**, 6309 (1971).
285. H. C. Brown, D. Basavaiah, S. K. Kulkarni, N. G. Bhat, and J. V. N. Vara Prasad, *J. Org. Chem.* **53**, 239 (1988).
286. H. C. Brown, D. S. Basavaiah, S. U. Kulkarni, H. D. Lee, E. Negishi, and J. J. Katz, *J. Org. Chem.* **51**, 5270 (1991).
287. E. J. Corey and T. Ravindranathan, *J. Am. Chem. Soc.* **94**, 4013 (1972).
288. H. C. Brown and N. G. Bhat, *J. Org. Chem.* **53**, 6009 (1988).
289. A. Pelter, K. Smith, and M. Tabata, *J. Chem. Soc. Chem. Commun.*, 857 (1975).
290. J. A. Sinclar and H. C. Brown, *J. Org. Chem.* **41**, 1078 (1976).
291. A. Pelter, R. J. Hughes, K. Smith, and M. Tabata, *Tetrahedron Lett.*, 4385 (1976).
292. G. Zweifel, N. L. Polston, and C. C. Whitney, *J. Am. Chem. Soc.* **90**, 6243 (1968).
293. E. Negishi and T. Yoshida, *J. Chem. Soc. Chem. Commun.*, 606 (1973).
294. E. Negishi, R. M. Williams, G. Lew, and T. Yoshida, *J. Organomet. Chem.* **92**, C4 (1975).
295. H. C. Brown, N. G. Bhat, and D. Basavaiah, *Synthesis*, 674 (1986).
296. A. Pelter, K. J. Gould, and C. R. Harrison, *J. Chem. Soc., Perkin I*, 2428 (1976).
297. A. Pelter, C. R. Harrison, C. Shubrahmanyan, and D. Kirkpatrik, *J. Chem. Soc., Perkin I*, 2435 (1976).
298. N. Miyaura, Y. Yoshinari, M. Itoh, and R. Suzuki, *Tetrahedron Lett.*, 2961 (1974).
299. H. C. Brown, A. B. Levy, and M. M. Midland, *J. Am. Chem. Soc.* **97**, 5017 (1975).
300. K. Utimoto, K. Uchida, M. Yamada, and H. Nozaki, *Tetrahedron* **33**, 1945, 1949 (1977).
301. J. M. Mas, M. Malaeria, and J. Gore, *J. Chem. Soc. Chem. Commun.*, 1161 (1985).
302. M. Z. Deng, D. A. Lu, and W. H. Xu, *J. Chem. Soc. Chem. Commun.*, 1478 (1985).
303. B. Wrackmeyer and R. Zentgraf, *J. Chem. Soc. Chem. Commun.*, 402 (1978).
304. K. K. Wang and S. Dhumrongvaraporn, *Tetrahedron Lett.*, **28**, 1007 (1987).
305. G. Zweifel and A. Horng, *Synthesis*, 672 (1973).
306. H. C. Brown and S. P. Rhodes, *J. Am. Chem. Soc.* **91**, 4306 (1969).
307. H. L. Goering and S. L. Trenbeath, *J. Am. Chem. Soc.* **98**, 5016 (1976).
308. R. E. Merill, J. L. Allen, A. Abramovitch and E. Negishi, *Tetrahedron Lett.* 1019 (1977).
309. J. A. Marshall, *Synthesis*, 229 (1971).
310. M. E. D. Hillman, *J. Am. Chem. Soc.* **84**, 4715 (1962).
311. M. E. D. Hillman, *J. Am. Chem. Soc.* **85**, 982, 1626 (1963).
312. H. C. Brown, *Acc. Chem. Res.* **2**, 65 (1969).
313. H. C. Brown, *Science* **210**, 485 (1980).
314. M. W. Rathke and H. C. Brown, *J. Am. Chem. Soc.* **89**, 2740 (1967).
315. H. C. Brown, J. L. Hubbard, and K. Smith, *Synthesis*, 701 (1979).
316. H. C. Brown, R. A. Coleman, and M. W. Rathke, *J. Am. Chem. Soc.* **90**, 499 (1968).
317. J. L. Hubbard and H. C. Brown, *Synthesis*, 676 (1978).
318. H. C. Brown, T. M. Ford, and J. L. Hubbard, *J. Org. Chem.* **45**, 4067 (1980).
319. H. C. Brown and E. Negishi, *J. Am. Chem. Soc.* **89**, 5285 (1967).
320. T. Ohe, K. Ohe, S. Uemura, and N. Sugita, *J. Organometal. Chem.* **344**, C5 (1988).
321. Y. Wakita, T. Yasunaga, M. Akita, and M. Kojima, *J. Organomet. Chem.* **301**, C17 (1986).
322. H. C. Brown and J. L. Hubbard, *J. Org. Chem.* **44**, 467 (1979).
323. B. M. Mikhailov, V. N. Smirnov, and W. A. Kasparov, *Izv. AN SSSR, Ser. Khim.*, 2302 (1976).

324. A. Pelter, M. G. Huthings, and K. Smith, *J. Chem. Soc. Perkin Trans. I*, 142 (1975).
325. A. Pelter, A. Arase, and M. G. Hutchings, *J. Chem. Soc. Chem. Commun.*, 346 (1974).
326. A. Pelter, P. J. Maddocks and K. Smith, *J. Chem. Soc. Chem. Commun.*, 805 (1978).
327. T. A. Bryson and C. J. Reichel, *Tetrahedron Lett.* **21**, 2381 (1980).
328. M. C. Welch and T. A. Bryson, *Tetrahedron Lett.* **29**, 521 (1988).
329. C. Narayama and M. Periasamy, *Tetrahedron Lett.* **26**, 6361 (1985).
330. A. Pelter and J. M. Rao, *J. Organometal. Chem.*, **285**, 65 (1985).
331. H. C. Brown and S. M. Singh, *Organometallics* **5**, 994 (1986).
332. H. C. Brown and S. M. Singh, *Organometallics* **5**, 998 (1986).
333. D. S. Matteson and G. Hurst, *Organometallics* **5**, 1465 (1986).
334. H. C. Brown, J.-J. Katz, and A. B. Carlson, *J. Org. Chem.* **38**, 3968 (1973).
335. K. Brown and T. Imai, *J. Org. Chem.* **49**, 892 (1984).
336. M. W. Rathke, E. Chao, and G. Wu. *J. Organometal. Chem.* **127**, 145 (1976).
337. H. C. Brown and T. Imai, *J. Am. Chem. Soc.* **105**, 6285 (1983).
338. H. C. Brown, A. S. Phadke, and M. V. Rangaishenvi, *Heteroatom. Chem.* **1**, 83 (1990).
339. D. S. Matteson and D. Majumdar, *Organometallics* **2**, 1529 (1983).
340. D. S. Matteson and D. Majumdar, *J. Am. Chem. Soc.* **102**, 7588 (1980).
341. D. S. Matteson, *Pure Appl. Chem.* **63**, 339 (1991).
342. T. Mukaiyama, S. Yamamoto, and M. Shiono, *Bull. Chem. Soc. Jpn.* **45**, 2244 (1972).
343. S. Yamamoto, M. Shono, and T. Mukaiyama, *Chem. Lett.*, 961 (1973).
344. S. Ncube, A. Pelter, and K. Smith, *Tetrahedron Lett.* **21**, 1893 (1979).
345. *Ibid.*, p. 1895.
346. J. J. Tufariello, L. T. C. Lee, and P. Wojtkowski, *J. Am. Chem. Soc.* **89**, 6804 (1967).
347. E. Negishi, T. Yoshida, A. Silveira, Jr., and B. L. Chiou, *J. Org. Chem.* **40**, 814 (1975).
348. R. J. Hughes, S. Ncube, A. Pelter, K. Smith, E. Negishi, and Y. Yoshida, *J. Chem. Soc., Perkin Trans. I*, 1172 (1977).
349. A. B. Levy, S. J. Schwartz, N. Wilson, and B. Christie, *J. Organomet. Chem.* **156**, 123 (1978).
350. Y. Yamamoto, K. Kondo, and I. Moritani, *Bull. Chem. Soc. Jpn.* **48**, 3682 (1975).
351. Y. Yamamoto, K. Kondo, and I. Maritani, *J. Org. Chem.* **40**, 3644 (1975).
352. W. K. Musker and R. R. Stevens, *Tetrahedron Lett.*, 995 (1967).
353. R. Köster and B. Rickborn, *J. Am. Chem. Soc.* **89**, 2782 (1967).
354. A. Suzuki, N. Miyaura, and M. Itoh, *Tetrahedron* **27**, 2775 (1971).
355. K. Utimoto, N. Sakai, and H. Nozaki, *J. Am. Chem. Soc.* **96**, 5601 (1974).
356. H. C. Brown, H. Nambu, and M. M. Rogic, *J. Am. Chem. Soc.* **91**, 6852 (1969).
357. H. C. Brown, H. Nambu, and M. M. Rogic, *J. Am. Chem. Soc.* **91**, 6855 (1969).
358. H. C. Brown, H. Nambu, and M. M. Rogic, *J. Am. Chem. Soc.* **91**, 6854 (1969).
359. H. Nambu and H. C. Brown, *Organomet. Chem. Synth.* **1**, 95 (1970).
360. H. C. Brown and H. Nambu, *J. Am. Chem. Soc.* **92**, 1761 (1970).
361. J. Hooz and S. Linke, *J. Am. Chem. Soc.* **90**, 5936 (1968).
362. J. Hooz and S. Linke, *J. Am. Chem. Soc.* **90**, 6891 (1968).
363. J. Hooz, D. M. Gunn, and H. Kono, *Can. J. Chem.* **49**, 2371 (1971).
364. J. Hooz and F. G. Morrison, *Can. J. Chem.* **48**, 868 (1970).
365. H. C. Brown, N. G. Bhat, and J. B. Campbell, Jr., *J. Org. Chem.* **51**, 3398 (1986).
366. H. C. Brown, M. M. Midland, and A. B. Levy, *J. Am. Chem. Soc.* **94**, 3662 (1972).
367. J. Hooz, J. N. Bridson, J. G. Calzada, H. C. Brown, M. M. Midland, and A. B. Levy, *J. Org. Chem.* **38**, 2574 (1973).
368. H. C. Brown and A. M. Salunkhe, *Synlett.*, 684 (1991).
369. H. C. Brown and N. R. De Lue, *J. Am. Chem. Soc.* **96**, 311 (1974).
370. H. C. Brown, Y. Yamamoto, and C. F. Lane, *Synthesis*, 30 (1972).
371. H. C. Brown, Y. Yamamoto, and C. F. Lane, *Synthesis*, 304 (1972).

372. Y. Yamamoto and H. C. Brown, *J. Org. Chem.* **39**, 861 (1974).
373. H. C. Brown and Y. Yamamoto, *Synthesis*, 699 (1972).
374. E. Negishi, K. W. Chiu, and T. Yoshida, *J. Org. Chem.* **40**, 1676 (1975).
375. B. M. Mikhailov, Yu. N. Bubnov, A. V. Tsyban, and M. Sh. Grigoryan, *J. Organomet. Chem.* **154**, 131 (1978).
376. B. M. Mikhailov, Yu. N. Bubnov, and A. V. Tsyban, *J. Organomet. Chem.* **154**, 113 (1978).
377. J. Blais, A. L'Honore, J. Soulie, and P. Cadiot, *J. Organomet. Chem.* **78**, 323 (1974).
378. R. W. Hoffmann, U. Weidmann, R. Metternich, and J. Lanz, in St. Hermanek, ed., *Boron Chemistry*, World Scientific, Singapore, 1987, p. 337.
379. G. W. Kramer and H. C. Brown, *J. Org. Chem.* **42**, 2292 (1977).
380. B. M. Mikhailov, *Pure Appl. Chem.* **39**, 505 (1974).
381. G. W. Kramer and H. C. Brown, *J. Organomet. Chem.* **132**, 9 (1977).
382. H. C. Brown and P. V. Ramachandran, *Pure Appl. Chem.* **63**, 307 (1991).
383. H. C. Brown, K. S. Bhat, and P. K. Jadhav, *J. Chem. Soc. Perkin Trans. I*, 2633 (1991).
384. R. W. Hoffman and T. Herold, *Chem. Ber.* **114**, 375 (1981).
385. R. P. Short and S. Masamune, *J. Am. Chem. Soc.* **111**, 1892 (1989).
386. E. J. Corey, C.-M. Yu, and S. S. Kim, *J. Am. Chem. Soc.* **111**, 5495 (1989).
387. G. Zweifel and T. M. Shoup, in St. Hermanek, ed., *Boron Chemistry*, World Scientific, Singapore, 1987, p. 354.
388. Yu. N. Bubnov, *Pure Appl. Chem.* **59**, 895 (1987).
389. Yu. N. Bubnov, *Pure Appl. Chem.* **63**, 361 (1991).
390. M. Zaidlewicz, *Tetrahedron Lett.* **27**, 5135 (1986).
391. G. Zweifel, S. J. Backlund, and T. Leung, *J. Am. Chem. Soc.* **100**, 5561 (1978).
392. G. Zweifel and W. R. Pearson, *J. Org. Chem.* **46**, 829 (1981).
393. H. C. Brown, N. G. Bhat, and S. Rajagopalan, *Organometallics* **5**, 816 (1986).
394. H. C. Brown, G. Molander, S. Singh, and U. S. Racherla, *J. Org. Chem.* **50**, 1577 (1985).
395. J. C. Evans, C. T. Goralski, and D. L. Hasha, *J. Org. Chem.* **57**, 2941 (1992).
396. H. C. Brown, M. M. Rogic, M. W. Rathke, and G. W. Kabalka, *J. Am. Chem. Soc.* **91**, 2150 (1969).
397. H. C. Brown, M. M. Rogic, M. W. Rathke, and G. W. Kabalka, *J. Am. Chem. Soc.* **89**, 5709 (1967).
398. G. W. Kabalka, H. C. Brown, A. Suzuki, S. Honma, A. Arase, and M. Itoh, *J. Am. Chem. Soc.* **92**, 710 (1970).
399. H. C. Brown, G. W. Kabalka, M. W. Rathke, and M. M. Rogic, *J. Am. Chem. Soc.* **90**, 4165 (1968).
400. A. Suzuki, A. Arase, K. Matsumoto, M. Itoh, H. C. Brown, M. M. Rogic, and M. W. Rathke, *J. Am. Chem. Soc.* **89**, 5708 (1967).
401. H. C. Brown, G. W. Kabalka, M. W. Rathke, and M. M. Rogic, *J. Am. Chem. Soc.* **90**, 4166 (1968).
402. H. C. Brown and G. W. Kabalka, *J. Am. Chem. Soc.* **92**, 712 (1970).
403. H. C. Brown and E. Negishi, *J. Am. Chem. Soc.* **93**, 3777 (1971).
404. G. W. Kabalka, *Tetrahedron* **29**, 1159 (1973).
405. P. Jacob III, *J. Organomet. Chem.* **156**, 101 (1978).
406. A. Suzuki, S. Nozawa, M. Itoh, H. C. Brown, G. W. Kabalka, and G. W. Holland, *J. Am. Chem. Soc.* **92**, 3503 (1970).
407. K. Maruyama, K. Saimoto, and Y. Yamamoto, *J. Org. Chem.* **43**, 4895 (1978).
408. N. Miyaura, B. Kashiwagi, M. Itoh, and A. Suzuki, *Chem. Lett.*, 395 (1974).
409. P. Jacob III and H. C. Brown, *J. Am. Chem. Soc.* **98**, 7832 (1976).

410. J. A. Sinclar, G. A. Molander, and H. C. Brown, *J. Am. Chem. Soc.* **99**, 954 (1977).
411. G. A. Molander and H. C. Brown, *J. Org. Chem.* **42**, 3106 (1977).
412. A. Suzuki, N. Miyaura, M. Itoh, G. W. Holland, and E. Negishi, *J. Am. Chem. Soc.* **93**, 2792 (1971).
413. A. Suzuki, N. Miyaura, M. Itoh, H. C. Brown, and P. Jacob III, *Synthesis*, 305 (1973).
414. M. W. Rathke and R. Kow, *J. Am. Chem. Soc.* **94**, 6854 (1972).
415. A. Pelter, B. Singaram, and J. W. Wilson, *Tetrahedron Lett.* **24**, 631 (1983).
416. D. S. Matteson and R. J. Moody, *Organometallics* **1**, 20 (1982).
417. D. S. Matteson and R. J. Moody, *J. Am. Chem. Soc.* **99**, 3196 (1977).
418. D. S. Matteson and K. H. Arne, *Organometallics* **1**, 280 (1982).
419. D. S. Matteson and K. H. Arne, *J. Am. Chem. Soc.* **100**, 1325 (1978).
420. D. S. Matteson and D. Majumdar, *Organometallics*, **2**, 230 (1983).
421. G. M. Cooke, Jr., and R. K. Widener, *J. Am. Chem. Soc.* **109**, 931 (1987).
422. G. Cainelli, G. Dal Bello, and G. Zubrini, *Tetrahedron Lett.*, 4329 (1965).
423. G. Zweifel and H. Arzoumanian, *Tetrahedron Lett.*, 2535 (1966).
424. D. S. Matteson, *Synthesis*, 147 (1975).
425. A. Pelter and K. Smith, in B. Trost and I. Fleming, eds., *Comprehensive Organic Synthesis*, Vol. 1, Pergamon Press, Oxford, 1991, p. 487.
426. G. Cainelli, G. Dal Bello, and G. Zubiani, *Tetrahedron Lett.*, 315 (1966).
427. A. Pelter, D. Buss, and D. Colelough, *J. Chem. Soc. Chem. Commun.*, 297 (1987).
428. A. Mendoza and D. S. Matteson, *J. Org. Chem.* **44**, 1352 (1979).
429. A. Pelter, L. Williams, and J. W. Wilson, *Tetrahedron Lett.* **24**, 627 (1983).
430. R. Ray and D. S. Matteson, *J. Org. Chem.* **47**, 2479 (1982).
431. M. K. Garad, A. Pelter, B. Singaram, and J. W. Wilson, *Tetrahedron Lett.* **24**, 637 (1983).
432. H. C. Brown and G. Zweifel, *J. Am. Chem. Soc.* **88**, 1433 (1966).
433. H. C. Brown, U. S. Racherla, and H. Taniguchi, *J. Org. Chem.* **46**, 4313 (1981).
434. R. Köster, *Justus Liebigs Ann. Chem.* **618**, 31 (1958).
435. H. C. Brown and M. V. Bhatt, *J. Am. Chem. Soc.* **88**, 1440 (1966).
436. H. C. Brown, M. V. Bhatt, T. Munekata, and G. Zweifel, *J. Am. Chem. Soc.* **89**, 567 (1967).
437. K. Maruyama, K. Tarada, and Y. Yamamoto, *J. Org. Chem.* **45**, 737 (1980).
438. B. M. Mikhailov and V. N. Smirnov, *Izv. AN SSSR Ser. Khim.*, 1137 (1974).
439. B. M. Mikhailov, *Pure Appl. Chem.* **55**, 1439 (1983).
440. B. M. Mikhailov and Yn. W. Bubnov, *Tetrahedron Lett.*, 2127 (1967).
441. D. S. Matteson and J. D. Waldbilling, *J. Org. Chem.* **28**, 366 (1963).
442. N. Noriet, A. Youssofi, B. Carboni, and M. Vaultier, *J. Chem. Soc., Chem. Commun.*, 1105 (1992).
443. D. A. Singleton, J. P. Martinez, and G. M. Ndip, *J. Org. Chem.* **57**, 5768 (1992).
444. D. A. Singleton, J. P. Martinez, and J. V. Watson, *Tetrahedron Lett.* **33**, 1017 (1992).
445. E. Negishi, ref. 10, p. 318.
446. Y. Chujo, I. Tomita, Y. Hashiguchi, H. Tanigawa, E. Ihara, and T. Saegusa, *Macromolecules* **24**, 345 (1991); Y. Chujo, I. Tomita, Y. Hashiguchi, and T. Saegusa, *Polym. Bull.* **25**, 1 (1991).
447. Y. Chujo, I. Tomita, and T. Saegusa, *Polym. Bull.* **26**, 165 (1991).
448. Y. Chujo, I. Tomita, Y. Hashiguchi, and T. Saegusa, *Macromolecules* **25**, 33 (1992).
449. Y. Chujo, I. Tomita, N. Murata, H. Mauermann, and T. Saegusa, *Macromolecules* **25**, 27 (1992).
450. C. Pinazzi, J. C. Broose, A. Plandeam, and D. Reyo, *Appl. Polym. Symp.* **26**, 73 (1975).
451. T. C. Cheung, *Macromolecules* **21**, 865 (1988).
452. T. C. Cheung, *J. Polym. Sci. Pt. A, Polym. Chem.* **27**, 3251 (1989).
453. M. C. S. Perera, J. M. Elix, and J. H. Bradburg, *J. Appl. Polym. Sci.* **33**, 2731 (1987).

454. S. Stinson, *Chem. Eng. News,* **67**, 20 (1989).
455. S. Ramakrishnan, *Macromolecules* **24**, 3753 (1991).
456. G. W. Kabalka, Acc. Chem. Res. **17**, 215 (1984).
457. G. W. Kabalka, Pure Appl. Chem. **63**, 379 (1991).
458. G. W. Kabalka, in St. Hermanek, ed., *Boron Chemistry,* World Scientific, Singapore, 1987, p. 372.
459. H. C. Brown and G. Zweifel, *J. Am. Chem. Soc.* **83**, 3834 (1961).
460. G. W. Kabalka, R. J. Newton and J. Jacobus, *J. Org. Chem.* **44**, 4185 (1979).
461. M. Zaidlewicz and C. S. Panda, *Synthesis,* 645 (1987).
462. G. W. Kabalka and co-workers, *Int. J. Appl. Radiat. Isot.* **35**, 853 (1985).
463. G. W. Kabalka, R. S. Varma, Y. Z. Gai, and R. M. Baldwin, *Tetrahedron Lett.* **27**, 3843 (1986).
464. P. C. Srivastava and co-workers, *J. Med. Chem.* **28**, 408 (1985).
465. G. W. Kabalka and L. Wang, *J. Labelled Comp. Radiopharm.* **24**, 90 (1989).
466. G. W. Kabalka, M. C. Delgado, U. S. Kunda, and A. K. Sastry, *J. Org. Chem.* **49**, 174 (1984).
467. H. C. Brown and B. Singaram, *Acc. Chem. Res.* **21**, 287 (1988).
468. D. S. Matteson, *Acc. Chem. Res.* **21**, 294 (1988).
469. D. S. Matteson, *Chem. Rev.* **89**, 1535 (1989).
470. H. C. Brown, P. K. Jadhav, and B. Singaram, in R. Scheffold, ed., *Modern Synthetic Methods,* Vol. 4, Springer-Verlag, Berlin, 1986, p. 307.
471. H. C. Brown and P. K. Jadhav, in J. D. Morrison, ed., *Asymmetric Synthesis,* Vol. 2, Academic Press, New York, 1983, p. 1.
472. R. W. Hoffmann, *Angew. Chem. Int. Ed. Engl.* **21**, 555 (1982).
473. M. Zaidlewicz, Ref. 9, p. 201.
474. J. Klein and D. Lichtenberg, *J. Org. Chem.* **35**, 2654 (1970).
475. D. J. Pasto and F. M. Klein, J. Org. Chem. **33**, 1468 (1968).
476. M. H. Gordon and M. J. T. Robinson, *Tetrahedron Lett.,* 3867 (1975).
477. T. W. Bell, G. A. Crispino, and J. R. Vargas, J. Org. Chem. **54**, 1978 (1989).
478. D. H. Brithwistte, J. M. Brown, and M. W. Foxton, *Tetrahedron Lett.* **27**, 4367 (1986).
479. T. W. Bell, *Tetrahedron Lett.* **22**, 3443 (1980).
480. D. J. Pasto and J. Hickman, *J. Am. Chem. Soc.* **90**, 4445 (1968).
481. M. Zaidlewicz, Ref. 9, p. 229.
482. W. C. Still and J. C. Barrish, *J. Am. Chem. Soc.* **105**, 2487 (1983).
483. T. Harada, Y. Matsuda, J. Uchimura, and A. Oku, *J. Chem. Soc. Chem. Commun.,* 1429 (1989).
484. M. Midland and Y. C. Kwon, *J. Am. Chem. Soc.* **105**, 3723 (1983).
485. G. Schmid, T. Fukuyama, K. Akasaka, and Y. Kishi, *J. Am. Chem. Soc.* **101**, 259 (1979).
486. H. Nagaoka and Y. Kishi, *Tetrahedron* **37**, 3873 (1981).
487. Y. Kishi, *Aldrichimica Acta* **13**, 23 (1980).
488. C. H. Heathcock, E. T. Jarvi, and T. Rosen, *Tetrahedron Lett.,* 243 (1984).
489. Y. Yokoyama, H. Masaki, and H. Kawashima, *Chem. Lett.,* 453 (1989).
490. T. Harada, Y. Matsuda, I. Wada, J. Uchimura, and A. Oku, *J. Chem. Soc. Chem. Commun.,* 21 (1990).
491. Y. Yokoyama, H. Kawashima, M. Kohno, Y. Ogawa, and S. Uchida, *Tetrahedron Lett.* **32**, 1479 (1991).
492. Y. Yokoyama, Y. Tarada, and H. Kawashima, *Bull. Chem. Soc. Jpn.* **64**, 2563 (1991).
493. T. Harada, Y. Kagamihara, S. Tanaka, K. Sakamoto, and A. Oku, *J. Org. Chem.* **57**, 1637 (1992).
494. R. W. Hoffmann, *Chem. Rev.* **89**, 1841 (1989).
495. N. N. Houk, N. G. Rondon, Y. D. Wu, J. T. Metz, and M. N. Paddon-Row, *Tetrahedron* **40**, 2257 (1984).

496. H. C. Brown and G. Zweifel, *J. Am. Chem. Soc.* **83**, 486 (1961).

497. H. C. Brown, P. K. Jadhav, and A. K. Mandal, *J. Org. Chem.* **47**, 5074 (1982).

498. N. N. Joshi, C. Pyun, V. K. Mahindroo, B. Singaram, and H. C. Brown, *J. Org. Chem.* **57**, 504 (1992).

499. H. C. Brown, B. Singaram, and T. E. Cole, *J. Am. Chem. Soc.* **107**, 460 (1985).

500. M. Sato, N. Miyaura, and A. Suzuki, *Tetrahedron Lett.* **31**, 231 (1990).

501. T. Hayashi, Y. Matsumoto, and Y. Ito, *Tetrahedron Asym.* **2**, 601 (1991).

502. J. M. Brown and G. C. Lloyd-Jones, *Tetrahedron Asym.* **1**, 869 (1990).

503. Y. Matsumoto and T. Hayashi, *Tetrahedron Lett.* **32**, 3387 (1991).

504. H. C. Brown and J. V. N. Vara Prasad, *Heterocycles* **25**, 641 (1987).

505. H. C. Brown, A. K. Gupta, and J. V. N. Vara Prasad, *Bull. Chem. Soc. Jpn.* **61**, 93 (1988).

506. H. C. Brown and M. K. Rangaishenvi, *J. Heterocl. Chem.* **27**, 13 (1990).

507. J. J. Partridge, N. K. Chadha, and M. R. Uskokovic, *J. Am. Chem. Soc.* **95**, 532 (1973).

508. J. J. Partridge, N. K. Chadha, and M. R. Uskokovic, *J. Am. Chem. Soc.* **95**, 7171 (1973).

509. J. A. Soderquist and S. J. H. Lee, *Tetrahedron* **44**, 4033 (1988).

510. A. Rüttiman, G. Englert, H. Mayer, G. P. Moss, and B. C. L. Weddon, *Helv. Chim. Acta,* **66**, 1939 (1983).

511. M. A. Hobolski and E. Koft, *J. Org. Chem.* **57**, 965 (1992).

512. A. Rüttimann and H. Mayer, *Helv. Chim. Acta* **63**, 1456 (1980).

513. A. Burgos, J. M. Kamenka, A. M. Moustier, and B. Rousseau, *J. Labelled Comp. Radiopharm.* **29**, 1061 (1991).

514. A. Streitweiser, Jr., I. Schwagar, L. Verbit, and H. Rabitz, *J. Org. Chem.* **32**, 1532 (1976).

515. R. N. McDonald and R. N. Steppel, *J. Am. Chem. Soc.* **92**, 5664 (1970).

516. R. K. Hill and R. G. Edwards, *Tetrahedron* **21**, 1501 (1965).

517. H. C. Brown, K. W. Kim, T. E. Cole, and B. Singaram, *J. Am. Chem. Soc.* **108**, 6761 (1986).

518. H. C. Brown, A. M. Salunke, and B. Singaram, *J. Org. Chem.* **56**, 1170 (1991).

519. H. C. Brown, R. R. Iyer, V. K. Mahindroo, and N. G. Bhat, *Tetrahedron Asym.* **2**, 277 (1991).

520. H. C. Brown, V. K. Mahindroo, N. G. Bhat, and B. Singaram, *J. Org. Chem.* **56**, 1500 (1991).

521. C. A. Brown, M. C. Desai, and P. K. Jadhav, *J. Org. Chem.* **51**, 162 (1986).

522. H. C. Brown, R. G. Naik, R. K. Bakshi, C. Pyun, and B. Singaram, *J. Org. Chem.* **50**, 5586 (1985).

523. H. C. Brown, T. Imai, M. C. Desai, and B. Singaram, *J. Am. Chem. Soc.* **107**, 4980 (1985).

524. H. C. Brown, M. Srebnik, R. K. Bakshi, and T. C. Cole, *J. Am. Chem. Soc.* **109**, 5420 (1987).

525. H. C. Brown, A. K. Gupta, J. V. N. Vara Prasad, and M. Srebnik, *J. Org. Chem.* **53**, 1391 (1988).

526. H. C. Brown, N. N. Joshi, C. Pyun, and B. Singaram, *J. Am. Chem. Soc.* **111**, 1754 (1989).

527. R. W. Hoffmann, H.-J. Zeiss, W. Ladner, and S. Tabche, *Chem. Ber.* **115**, 2357 (1982).

528. M. T. Reetz and T. Zierke, *Chem. Ind.*, 663 (1988).

529. W. R. Roush and P. T. Grover, *Tetrahedron* **48**, 1981 (1992).

530. W. R. Roush, A. E. Walts, and L. K. Hoong, *J. Am. Chem. Soc.* **107**, 8186 (1985).

531. W. R. Roush, K. Ando, D. B. Powers, A. D. Palkowitz, and R. L. Haltermann, *J. Am. Chem. Soc.* **112**, 6339 (1990); **113**, 5133 (1991).

532. W. R. Roush, A. D. Palkowitz, and K. Ando, *J. Am. Chem. Soc.* **112**, 6348 (1990).

533. J. Garcia, B. M. Kim, and S. Masamune, *J. Org. Chem.* **52**, 4831 (1987).

534. R. W. Hoffmann and A. Schlapbach, *Tetrahedron* **48**, 1959 (1992).

535. H. C. Brown and R. S. Randad, *Tetrahedron* **46**, 4457 (1990).

536. H. C. Brown, R. S. Randad, K. S. Bhat, M. Zaidlewicz, and U. S. Racherla, *J. Am. Chem. Soc.*, **112**, 2389 (1990).

537. W. R. Roush, J. A. Straub, and M. S. Van Nieuwenzhe, *J. Org. Chem.* **56**, 1636 (1991).

538. W. R. Roush, X. Lin, and J. A. Straub, *J. Org. Chem.* **56**, 1649 (1991).

539. H. C. Brown and R. S. Randad, *Tetrahedron* **46**, 4463 (1990).

540. P. B. Tripathy and D. S. Matteson, *Synthesis*, 200 (1990).

541. D. S. Matteson, abstracts of papers, IMEBORON VIII, Knoxville, Tenn., 1993.

542. B. M. Kim, S. F. Williams, and S. Masamune, in B. M. Trost and I. Fleming, eds., *Comprehensive Organic Synthesis*, Vol. 2, Pergamon Press, Oxford, U.K., 1991, p. 239.

543. D. A. Evans, J. V. Nelson, E. Vogel, and T. R. Taber, *J. Am. Chem. Soc.* **103**, 3099 (1981).

544. T. Inoue and T. Mukaiyama, *Bull. Chem. Soc. Jpn.* **53**, 174 (1980).

545. T. Mukaiyama and T. Inoue, *Chem. Lett.*, 559 (1976).

546. T. Inoue, T. Uchimaru, and T. Mukaiyama, *Chem. Lett.*, 1531 (1977).

547. H. C. Brown, R. K. Dhar, R. K. Bakshi, P. K. Pandiarajan, and B. Singaram, *J. Am. Chem. Soc.* **111**, 3441 (1989).

548. H. C. Brown, K. Ganesan, and R. K. Dhar, *J. Org. Chem.* **58**, 147 (1993).

549. H. C. Brown, R. K. Dhar, K. Ganesan, and B. Singaram, *J. Org. Chem.* **57**, 499 (1992).

550. H. C. Brown, R. K. Dhar, K. Ganesan, and B. Singaram, *J. Org. Chem.* **57**, 2716 (1992).

551. H. C. Brown, K. Ganesan, and R. K. Dhar, *J. Org. Chem.* **57**, 3767 (1992).

552. C. Genari, I. Paterson, and co-workers, *J. Org. Chem.*, **57**, 5173 (1992).

553. I. Paterson, C. K. McClure, and R. C. Schumann, *Tetrahedron Lett.* **30**, 1293 (1989).

554. I. Paterson, M. A. Lister, and G. R. Ryan, *Tetrahedron Lett.* **32**, 1749 (1991).

555. I. Paterson and S. Osborne, *Tetrahedron Lett.* **31**, 2213 (1990).

556. D. A. Evans and J. Bartroli, *Tetrahedron Lett.* **23**, 807 (1982).

MAREK ZAIDLEWICZ
Nicolaus Copernicus University

HYDROBROMIC ACID.   See BROMINE COMPOUNDS.

HYDROCARBON OXIDATION

Oxidation has been of great interest to humanity since the discovery of fire (1). Some understanding of the subject began at the time of Lavoisier (2). Hydrocarbon oxidation has been studied for well over 100 years, but modern understanding of the mechanisms of free-radical chain reactions did not begin until the 1930s (3–12). Since that time, a vast body of multilingual literature has accumulated. There is significant agreement on underlying principles, but sharp controversies on various important points remain and, indeed, are probably growing, undoubtedly because the oxidation of even simple materials is complex. One simplification, however, is that radical reactions in the gas phase and in not strongly polar solvents do not differ appreciably in kinetic parameters. Differences can usually be traced to well-understood kinetic phenomena (13).

The scope of oxidation chemistry is enormous and embraces a wide range of reactions and processes. This article provides a brief introduction to the homogeneous free-radical oxidations of paraffinic and alkylaromatic hydrocarbons. Heterogeneous catalysis, biochemical and biomimetic oxidations, oxidations of unsaturates, anodic oxidations, etc, even if used to illustrate specific points, are arbitrarily outside the purview of this article. There are, even so, many unifying features among these areas.

Kinetics of Chain Reactions

One characteristic of chain reactions is that frequently some initiating process is required. In hydrocarbon oxidations radicals must be introduced and to be self-sustained, some source of radicals must be produced in a chain-branching step. Moreover, new radicals must be supplied at a rate sufficient to replace those lost by chain termination. In hydrocarbon oxidation, this usually involves the hydroperoxide cycle (eqs. 1–5).

$$RH + In\cdot \longrightarrow R\cdot + HIn \tag{1}$$

$$R\cdot + O_2 \longrightarrow ROO\cdot \tag{2}$$

$$ROO\cdot + RH \xrightarrow{k_p} ROOH + R\cdot \tag{3}$$

$$ROOH \xrightarrow{k_d} RO\cdot + OH\cdot \tag{4}$$

$$ROO\cdot + ROO\cdot \xrightarrow{k_t} \text{nonradical products} \tag{5}$$

In reaction 1, some initiating fragment (In·) reacts with the substrate hydrocarbon by abstracting a hydrogen atom, generating a carbon-centered radical with the structure of the original hydrocarbon. The origin of the initiating fragment has remained largely unresolved. A direct reaction of the hydrocarbon with oxygen (eq. 6) is frequently proposed (1,14). This reaction has such a high energy of activation that a termolecular alternative (eq. 7) is sometimes suggested (15,16).

$$RH + O_2 \longrightarrow R\cdot + HOO\cdot \tag{6}$$

$$2\,RH + O_2 \longrightarrow 2\,R\cdot + HOOH \tag{7}$$

However, reaction 7 suffers other shortcomings, eg, entropy problems. Other proposals range from trace peroxidic contaminants to ionic mechanisms for generating peroxides (1) to cosmic rays (17). In any event, the initiating reactions are significant only during the induction period (18).

Carbon-centered radicals generally react very rapidly with oxygen to generate peroxy radicals (eq. 2). The peroxy radicals can abstract hydrogen from a hydrocarbon molecule to yield a hydroperoxide and a new radical (eq. 3). This new radical can participate in reaction 2 and continue the chain. Reactions 2 and 3 are the propagation steps. Except under oxygen starved conditions, reaction 3 is rate limiting.

New radicals are introduced by thermolysis of the hydroperoxide by chain-branching decomposition (eq. 4). Radicals are removed from the system by chain-termination reaction(s) (eq. 5). Under steady-state conditions, the production of new radicals is in balance with the rate of radical removal by termination reactions and equation 8 applies for the scheme of equations 1–5 where r_i = rate of new radical introduction (eq. 4).

$$-d[RH]/dt = k_p[RH]\sqrt{r_i/2k_t} \tag{8}$$

New radicals come exclusively from the decomposition of the intermediate hydroperoxide (eq. 4), provided no other radical sources, eg, peroxidic impurities, are present. Hydroperoxides have varying degrees of stability, depending on their structure. They decompose by a variety of mechanisms and are not necessarily efficient generators of new radicals via thermolysis (19,20).

Deliberate use of a different radical source with a known radical generation rate, under conditions where the product hydroperoxide does not contribute appreciably to new radical generation, permits evaluation of $k_p/\sqrt{2k_t}$ by using equation 8. This factor, usually called the oxidizability of the hydrocarbon, is a measure of the relative susceptibilities of substrates to oxidation. It is independent of rates of initiation. The term oxidizability refers to oxidations of single substrates. A material with low oxidizability may actually have a high propagation rate constant but oxidize slowly because it also has a high termination rate constant. When cooxidized with a material of higher oxidizability, it may very well cooxidize faster than the more highly oxidizable substrate, depending on the ratio of propagation rate constants. Such ratios may be estimated in a number of ways (21–23) and can be useful in predicting product distributions. In cooxidations the high termination rate constant for the peroxy radicals produced from the low oxidizability component generally retards the rate of the cooxidation (22–24).

If a self-sustained oxidation is carried out under limiting rate conditions, the hydroperoxide provides the new radicals to the system (by reaction 4 or analogues) and is maintained at a low concentration (decomposition rate = generation rate). For these circumstances, the rate equation 9 holds, where n = average number of initiating radicals produced (by any means) per molecule of ROOH decomposed and f = fraction of RH consumed which disappears by ROO· attack (25).

$$-d[RH]/dt = nk_p^2[RH]^2/2fk_t \tag{9}$$

Under these conditions, a component with a low rate constant for propagation for peroxy radicals may be cooxidized at a higher relative rate because a larger fraction of the propagation steps is carried out by the more reactive (less selective) alkoxy and hydroxy radicals produced in reaction 4.

The simple hydroperoxide mechanism so far discussed is incomplete for representing reactions with significant products other than hydroperoxides. It can be adequate for oxidations of certain unsaturates, aldehydes, and alkylaromatics where the yield of the corresponding hydroperoxide can exceed 90%.

An important descriptor of a chain reaction is the kinetic chain length, ie, the number of cycles of the propagation steps (eqs. 2 and 3) for each new radical introduced into the system. The chain length for a hydroperoxide reaction is given by equation (10) where HPE = efficiency to hydroperoxide, %, and $2f$ = number of effective radicals generated per mol of hydroperoxide decomposed. For 100% radical generation efficiency, $f = 1$. For 90% efficiency to hydroperoxide, the minimum chain length ($f = 1$) is 14.

$$\text{chain length} = (HPE + 100/2f)/(100 - HPE) \tag{10}$$

Chain lengths of some oxidations can be quite long (>100), especially for substrates with easily abstractable hydrogens when they are oxidized under mild conditions at low conversions. Aldehydes are good examples of such substrates (26). Many other oxidations have chain lengths estimated from 3 to 10. At limiting rates, the chain length is near 1 (25).

Products other than hydroperoxides are formed in oxidations by reactions such as those of equations 11 and 12. Hydroxyl radicals (from eq. 4) are very energetic hydrogen abstractors; the product is water (eq. 11).

$$OH\cdot + RH \longrightarrow R\cdot + H_2O \tag{11}$$

Alkoxy radicals, such as those produced in reaction 4, can be vigorous hydrogen abstractors and may produce alcohols (eq. 12), but they can undergo other reactions as well.

$$RO\cdot + RH \longrightarrow R\cdot + ROH \tag{12}$$

The predominant radicals in hydrocarbon oxidation are usually alkylperoxy radicals, particularly when sufficient oxygen is present to scavenge alkyl radicals by reaction 2. These radicals are weak hydrogen abstractors and they build up to relatively high concentrations (27), becoming the dominant species in the radical flux. Depending on their structures, they can participate in various reactions other than chain propagation. They are usually the main participants in bimolecular radical reactions. Other radicals are less involved in bimolecular radical reactions because their high reactivities cause them to be present at very low concentrations.

There are two especially important radical–radical reactions of alkylperoxy radicals (28) both believed to proceed via formation of a transient tetroxide (eqs. 13 and 15).

$$ROO\cdot + \cdot OOR \longrightarrow ROOOOR \tag{13}$$

One decomposition of the tetroxide is not terminating, producing alkoxy radicals and oxygen (eq. 14).

$$ROOOOR \longrightarrow RO\cdot + O_2 + \cdot OR \tag{14}$$

An alternative chain-terminating decomposition of the tetroxide, known as the Russell mechanism (29), can occur when there is at least one hydrogen atom in an alpha position; the products are a ketone, an alcohol and oxygen (eq. 15). This mechanism is troubling on theoretical grounds (1). Questions about its validity remain (30), but it has received some recent support (31).

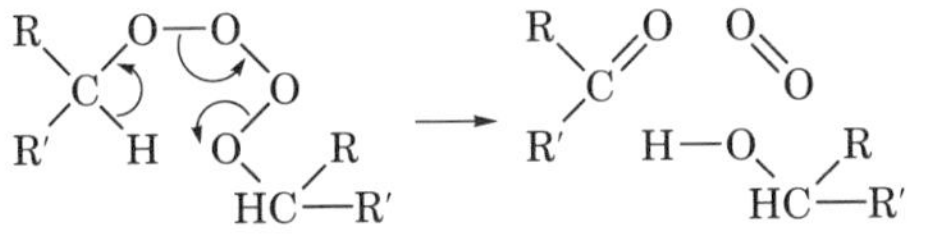

$$\tag{15}$$

Bimolecular reactions of peroxy radicals are not restricted to identical radicals. When both peroxy radicals are tertiary, reaction 15 is not possible. When an α-hydrogen is present, reaction 15 is generally the more effective competitor and predominates.

Alkoxy radicals can abstract hydrogen (eq. 12), but they may undergo a β-scission reaction (eq. 16).

$$\tag{16}$$

Under moderate conditions, primary alkoxy radicals tend to undergo reaction 12 whereas secondary and tertiary alkoxys tend to undergo β-scission. In general, the alkyl group that can form the lowest energy radical tends to become the departing radical. The β-scission of secondary alkoxy radicals yields aldehydes as the nonradical products; tertiary alkoxy radicals yield ketones.

Product Sequences in Hydrocarbon Oxidation

The products discussed so far (alcohols, aldehydes, and ketones) are produced by radical reactions involving the hydrocarbon substrate or from further reactions of the intermediate (and frequently transitory) hydroperoxides. Under conditions of low rates and high chain lengths, hydroperoxides may be the principal precursors for all other products. At high rates, however, significant quantities of chain-termination products (alcohols, ketones, and aldehydes) can be produced directly from peroxy radicals (eqs. 14 and 15) without going through hydroperoxides (25). In any event, there is a sense in which these products may be thought of as primary (22).

All components of the reaction mixture, whatever their source, are subject to the same kind of radical attacks as the starting substrate(s). Any free-radical oxidation is inevitably a cooxidation of substrate(s) and products. The yields of final products are determined by two factors: (1) how much is produced in the reaction sequence, and (2) how much product survives the reaction environment. By kinetic correlations and radiotracer techniques, it is possible to estimate these relationships and develop a mathematical model of the system (22).

Primary and secondary alcohols oxidize rapidly (frequently ca 6–10 times as fast as the parent hydrocarbon (22,32)), and probably efficiently, to the corresponding carbonyl compounds (22,33–35) (Fig. 1). The initial attack is highly selective for the hydrogen in the alpha position relative to the hydroxyl group. The addition of oxygen to the resulting radical is reversible (36). If the peroxy radical abstracts hydrogen (Path I), the hydroxyalkylhydroperoxide decomposes to hydrogen peroxide and the corresponding carbonyl compound (37,38). If, however, hydrogen is abstracted from the hydroxyalkyl radical by oxygen (Path II), the production of the carbonyl component is accompanied by the formation of a hydroperoxy radical. This radical is notable in that it has an exceptionally high chain-termination rate constant (1); hydroperoxy radicals may react readily with almost any other radical in the system to give oxygen and a nonradical product (eg, eq. 17).

$$HOO\cdot + R\cdot \longrightarrow RH + O_2 \tag{17}$$

Carbonyl compounds can be primary (from radicals or hydroperoxides) or secondary (from alcohols). Thus the picture emerges of hydrocarbon oxidations

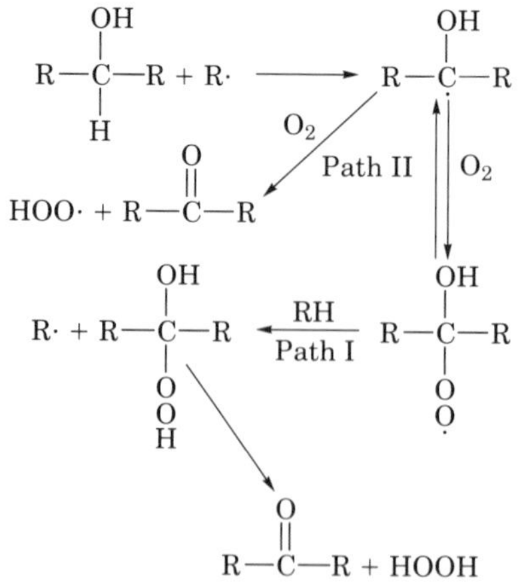

Fig. 1. Production of carbonyl compounds from alcohols by various oxidation routes.

occurring through complicated series-sequential pathways as in Figure 1, where clearly other reactions could be going on as well. All possible pathways are pursued to some extent; traffic along any pathway is a function of energy requirements and relative concentrations.

Carbonyl intermediates are also susceptible to further oxidation. Aldehydes can oxidize very rapidly to acids (39–42); peracids are likely intermediates.

$$
R-C\overset{O}{\underset{H}{\Big<}} + R\cdot \longrightarrow RH + R-C\overset{O}{\Big<}\cdot \xrightarrow{O_2} R-C\overset{O}{\underset{OO\cdot}{\Big<}} \tag{18}
$$

$$
R-C\overset{O}{\underset{OO\cdot}{\Big<}} + R-C\overset{O}{\underset{H}{\Big<}} \longrightarrow R-C\overset{O}{\underset{OOH}{\Big<}} + R-C\overset{O}{\Big<}\cdot \tag{19}
$$

$$
R-C\overset{O}{\underset{OOH}{\Big<}} + R-C\overset{O}{\underset{H}{\Big<}} \longrightarrow \longrightarrow 2\ R-C\overset{O}{\underset{OH}{\Big<}} \tag{20}
$$

Two important inefficient routes for aldehyde oxidation are (eqs. 21 and 22).

$$
R-C\overset{O}{\Big<}\cdot \longrightarrow R\cdot + CO \tag{21}
$$

$$
R-C\overset{O}{\parallel}OO\cdot + \cdot OO\overset{O}{\parallel}C-R \longrightarrow R-C\overset{O}{\diagup}\underset{OOOO}{\diagdown}\overset{O}{\diagdown}C-R \longrightarrow 2\ R\cdot + 2\ CO_2 + O_2 \tag{22}
$$

Reaction 21 is the decarbonylation of the intermediate acyl radical and is especially important at higher temperatures; it is the source of much of the carbon monoxide produced in hydrocarbon oxidations. Reaction 22 is a bimolecular radical reaction analogous to reaction 13. In this case, acyloxy radicals are generated; they are unstable and decarboxylate readily, providing much of the carbon dioxide produced in hydrocarbon oxidations. An in-depth article on aldehyde oxidation has been published (43).

Ketones oxidize about as readily as the parent hydrocarbons or even a bit faster (32). Although the reactivities of hydrogens on carbons adjacent to carbonyl groups are perhaps doubled, the effect is small because one methylene group is missing in comparison to the parent hydrocarbon. Ketones oxidize less readily than similar primary or secondary alcohols (35).

Acids are usually the end products of ketone oxidations (41,42,44) but vicinal diketones and hydroperoxyketones are apparent intermediates (45). Acids are readily produced from vicinal diketones, perhaps through anhydrides (via, eg, a Bayer-Villiger reaction) (46,47). The hydroperoxyketones reportedly decompose to diketones as well as to aldehydes and acids (45). Similar products are expected from radical–radical reactions of the corresponding peroxy radical precursors.

The carboxyl group of acids appears to deactivate the hydrogens on the alpha carbon atom toward attack by the free-radical flux in oxidation reactions. Acetic acid, therefore, is particularly inert toward further oxidation (hydrogens are both primary and deactivated) (48). For this reason, it is feasible to produce acetic acid by the oxidation of butane (in the liquid phase), even under rather severe oxidation conditions under which most other products are further oxidized to a significant extent (22).

Beginning with propionic acid, all the higher saturated acids are significantly less resistant to oxidation than acetic. For longer chain acids, all carbons in the gamma or higher positions are essentially equivalent to the corresponding carbons in the parent hydrocarbon with regard to susceptibility to oxidation (49–51). For the β-position, however, the picture is not so clear. Some investigators indicate deactivation of the β-position for hydrogen abstraction in autoxidation (49,50), but others report activation (by a factor of about 2) of this position for abstraction by hydroxyl radicals (51).

Esters are also formed in hydrocarbon oxidations. Although their source is still a matter of debate, there is rather convincing evidence that esters generally do not form in the early stages of an oxidation before the appearance of significant quantities of acids (41,42,52,53). The clear implication is that most esters come from esterification reactions of free acids and alcohols produced by the oxidation. There is widespread agreement that, once formed, most esters are exceptionally resistant to further oxidation, at least in the region of the ester group. Indeed, the consumption of lower molecular weight esters probably proceeds largely through prior hydrolysis and rapid oxidation of the liberated alcohol (54,55). Since primary and secondary alcohols oxidize readily (secondary > primary) and also esterify (primary > secondary), the role of alcohols in hydrocarbon oxidation can be even more complex than the role of many other intermediates.

Earlier reports have indicated that esters can form before significant amounts of acids accumulate (16). The Bayer-Villiger oxidations of ketones with intermediate hydroperoxides and/or peracids have been suggested as ester forming mechanisms (34,56). However, the reactions of simple aliphatic ketones with peracetic acid are probably too slow to support this mechanism (57,58). Very early proposals for ester formation, although imaginative, appear improbable (59).

In addition to production of simple monofunctional products in hydrocarbon oxidation there are many complex, multifunctional products that are produced by less well-understood mechanisms. There are also important influences of reactor and reaction types (plug-flow or batch, back-mixed, vapor-phase, liquid-phase, catalysts, etc).

Efficiency of Intermediate Formation.   The variation of the efficiency of a primary intermediate with conversion of the feed hydrocarbon can be calculated (22). Ratios of the propagation rate constants (k_2/k_1) and reactor type (batch or plug-flow vs back-mixed) are important parameters. Figure 2 shows that even materials which are rather resistant to oxidation ($k_2/k_1 = {\sim}0.1$) are consumed to a noticeable degree at high conversions. Also the use of plug-flow or batch reactors can offer a measurable improvement in efficiencies in comparison with back-mixed reactors. Intermediates that cooxidize about as readily as the feed

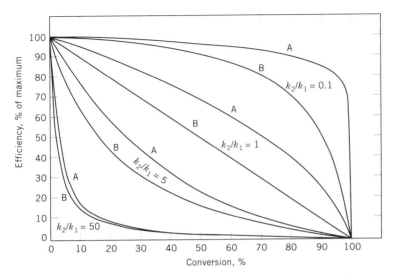

Fig. 2. Efficiency to a primary intermediate as % of maximum (zero conversion) efficiency; x axis is feed conversion. Parameters are oxidation rate-constant ratios (k_2/k_1) for primary intermediate vs feed and reactor type: A, plug-flow or batch; B, back-mixed.

hydrocarbon (eg, ketones with similar structure) can be produced in perhaps reasonable efficiencies but, except at very low conversions, are subject to considerable loss through oxidation. They may be suitable coproducts if they are also precursors to more oxidation-resistant desirable materials. Intermediates which oxidize relatively rapidly ($k_2/k_1 = 5$–50; eg, alcohols and aldehydes) are difficult to produce in appreciable amounts, even in batch or plug-flow reactors. Indeed, for $k_2/k_1 = 50$, to isolate 90% or more of the intermediate made, the conversion must not be greater than about 0.4%, even for a batch or plug-flow reactor.

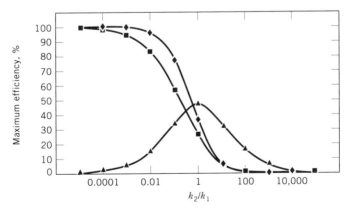

Fig. 3. Plot of maximum yield as a % of maximum (zero conversion) efficiency to a primary intermediate; x axis is ratio of oxidation rate constants (k_2/k_1) for primary intermediate vs feed: (◆) plug-flow or batch reactor; (■) back-mixed reactor; (▲) plug-flow advantage, %.

The maximum yield of a primary intermediate, as well as the efficiency (and conversion) at maximum yield, can also be calculated (22). Maximum yield plots are shown in Figure 3.

Vapor-Phase Oxidation

Above about 250°C, the vapor-phase oxidation (VPO) of many organic substances becomes self-sustaining. Such oxidations are characterized by a lengthy induction period. During this period, peroxides accumulate until they can provide a source of new radicals to sustain a chain reaction. Once a critical threshold peroxide concentration is reached, the reaction accelerates very rapidly.

THE NTC PHENOMENON

VPO reactions of typical alkanes may be considered conveniently in three temperature regions. Under some circumstances, particularly at pressures not greatly exceeding atmospheric, a curious and fundamentally important phenomenon known as the negative temperature coefficient (NTC) region is observed between the low and intermediate temperature ranges. In the NTC zone, increasing temperature actually results in lower reaction rates. A typical plot for propane VPO is presented in Figure 4. In the lower temperature regions, oxygenated materials are usually the principal products. As one progresses through the NTC zone, unsaturated products, usually with the same carbon skeleton as the feed (conjugate olefins), begin to predominate. Further increases in temperature are usually accompanied by production of lower molecular weight olefins as well as products derived from methyl radicals. Figure 5 illustrates these relationships for the VPO of butane.

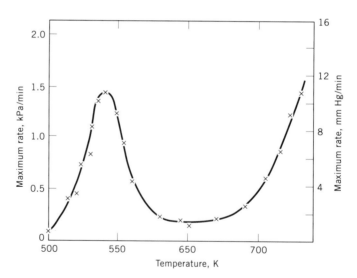

Fig. 4. Plot of maximum rate of pressure change for propane VPO showing NTC region; 5.33 kPa propane, 13.33 kPa O_2. Courtesy of Blackwell Scientific Publications, Ltd., Oxford (60). To convert kPa to mm Hg, multiply by 7.5.

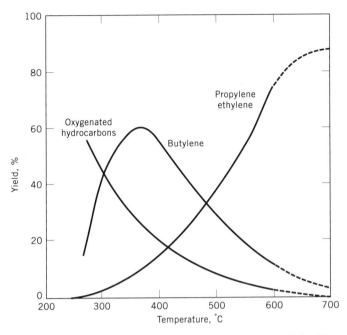

Fig. 5. The effect of temperature on product distribution in VPO of butane with air.

At still higher temperatures, when sufficient oxygen is present, combustion and "hot" flames are observed; the principal products are carbon oxides and water. Key variables that determine the reaction characteristics are fuel-to-oxidant ratio, pressure, reactor configuration and residence time, and the nature of the surface exposed to the reaction zone. The chemistry of hot flames, which occur in the high temperature region, has been extensively discussed (60–62) (see COMBUSTION SCIENCE AND TECHNOLOGY).

Modern real time instrumental methods permit analyses of unstable transient species and the free-radical intermediates as well. These methods have greatly expanded the scope and power of VPO studies, but important basic questions remain unresolved. Another complication is the role of surface. Peroxide decompositions and radical termination reactions can occur on a surface so that, depending on circumstances, surfaces can have either an inhibiting or accelerating effect. Each surface has varying amounts of adventitious contaminants and also accumulates deposits during reaction. Thus no two surfaces are exactly alike and each changes with time.

Reversibility of Equation 2. Notwithstanding the problems and conflicts, there is widespread agreement that the NTC phenomenon may well be related to the reversibility of equation 2 (13,60,63–67): $R\cdot + O_2 \leftrightarrows ROO\cdot$. In the low temperature regime, the equilibrium lies to the right and alkylperoxy radicals are the dominant radical species. They form hydroperoxides, the chain-branching agent, by reaction 3.

The principal isolated products are alcohols, aldehydes, and ketones, formed by reactions shown in equations 12, 15, 16, and Figure 1. Although aldehydes are very susceptible to further oxidation, acids are not usually significant products in low pressure VPO (below ~1.4 MPa (200 psig) in many cases). This appears to be because the decarbonylation of acyl radicals (eq. 21) competes very effectively with the addition of oxygen (eq. 18) under these conditions. As a result, carbon monoxide is a prominent product, along with lower molecular weight alcohols and ketones produced from the alkyl radical made in reaction 21. At higher pressures, reaction 18 becomes more competitive and acid yields rise (Table 1).

As the temperature approaches the NTC zone, the reversibility of reaction 2 comes into play and the steady-state concentration of alkyl radicals rises. There is a competing irreversible reaction of oxygen with radicals containing an alpha hydrogen which produces a conjugate olefin (eq. 23).

$$\underset{\underset{R}{|}}{\overset{\overset{H}{|}}{R-C}}-\underset{\underset{R}{|}}{\overset{\overset{R}{|}}{C}}\cdot + O_2 \longrightarrow \underset{\underset{R}{|}}{\overset{\overset{R}{|}}{C}}=\underset{\underset{R}{|}}{\overset{\overset{R}{|}}{C}} + HOO\cdot \tag{23}$$

Alternatively, a number of investigators (69–73) have proposed, on the basis of plausible kinetic arguments, that the conjugate olefin is produced by a rearrangement of alkylperoxy radicals (eq. 24).

$$RCH_2CH_2OO\cdot \longrightarrow [R\dot{C}HCH_2OOH] \longrightarrow RCH{=}CH_2 + HOO\cdot \tag{24}$$

This proposal, however, has been criticized on the basis of transition state theory (74). Hydroperoxy radicals produced in reaction 23 or 24 readily participate in chain-terminating reactions (eq. 17) and are only weak hydrogen abstractors. When they succeed in abstracting hydrogen, they generate hydrogen peroxide:

$$HOO\cdot + RH \longrightarrow HOOH + R\cdot \tag{25}$$

$$HOO\cdot + HOO\cdot \longrightarrow HOOH + O_2 \tag{26}$$

The sequence that makes alkylhydroperoxide has thus been diverted to make hydrogen peroxide, and with lower efficiency. Hydrogen peroxide generates radi-

Table 1. Products from the Reaction of a Propane:Air Mixture[a] at Various Pressures,%[b]

Compounds	At 0.101 MPa,[c] 378°C	At 2.03 MPa,[c] 281°C	At 6.08 MPa,[c] 251°C	At 10.1 MPa[c] 250°C
total aldehydes	20.5	21.8	13.5	13.7
n-alcohols	19.7	21.0	17.5	15.2
isopropyl alcohol	1.3	2.8	6.2	16.0
acetone	0.5	4.3	12.5	7.9
acids	4.3	17.0	19.0	18.9
carbon dioxide	7.3	17.1	21.4	20.6
carbon monoxide	21.3	16.0	9.9	7.7
propylene	25.1	0	0	0

[a] Ref. 68. Propane:air ratio = 1:3.6.
[b] Values are % carbon of propane burned.
[c] To convert MPa to atm, divide by 0.101.

cals thermally but its decomposition temperature is higher than that of alkyl-hydroperoxides. The overall effect is that the rate of generation of new radicals drops, the reaction rate is reduced, and the oxygenated products are displaced by olefins (mostly conjugate).

$$HOOH \longrightarrow 2\ OH\cdot \tag{27}$$

As the temperature is increased through the NTC zone, the contribution of alkylperoxy radicals falls. Little alkyl hydroperoxide is made and hydrogen peroxide decomposition makes a greater contribution to radical generation. Eventually the rate goes through a minimum. At this point, reaction 2 is highly displaced to the left and alkyl radicals are the dominant radical species.

At the higher temperatures a decomposition of alkyl radicals, which is an olefin-producing variation of the β-scission reaction, becomes competitive with reaction 23 (or sequence 2, 24):

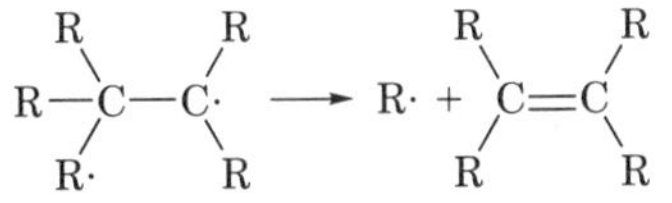

$$R-\underset{\underset{R\cdot}{|}}{\overset{\overset{R}{|}}{C}}-\underset{\underset{R}{|}}{\overset{\overset{R}{|}}{C}}\cdot \longrightarrow R\cdot + \underset{\underset{R}{|}}{\overset{\overset{R}{|}}{C}}=\underset{\underset{R}{|}}{\overset{\overset{R}{|}}{C} \tag{28}$$

In principle, this degradation can continue until the residual radical contains only hydrogen or methyl groups attached to the carbon with the odd electron. Those radicals which still contain a carbon–carbon bond can form an olefin via reaction 23 (or sequence 2, 24). Methyl radicals are a special case with limited options.

Cool Flames. An intriguing phenomenon known as "cool" flames or oscillations appears to be intimately associated with NTC relationships. A cool flame occurs in static systems at certain compositions of hydrocarbon and oxygen mixtures over certain ranges of temperature and pressure. After an induction period of a few minutes, a pale blue flame may propagate slowly outward from the center of the reaction vessel. Depending on conditions, several such flames may be seen in succession. As many as five have been reported for propane (75) and for methyl ethyl ketone (76); six have been reported for butane (77). As many as 10 cool flames have been reported for some alkanes (60). The relationships of cool flames to other VPO domains are depicted in Figure 6.

Below a certain critical temperature, which varies with pressure and stoichiometry, cool flames for several hydrocarbons propagate from the wall inward; above this temperature, they propagate from the center of the vessel (78). This transition is interpreted as evidence for a changeover from a predominantly heterogeneous preflame mechanism to a homogeneous one.

When the necessary conditions are met, a cool flame seems to arise when heat generation during the low temperature oxidation exceeds heat losses. This leads to increasing temperature and increasing rates because of a higher radical generation rate by the low temperature chain-branching agent, ROOH. At a critical temperature, a cool flame appears. This causes the temperature to rise into the NTC region. Provided the temperature does not rise enough to permit

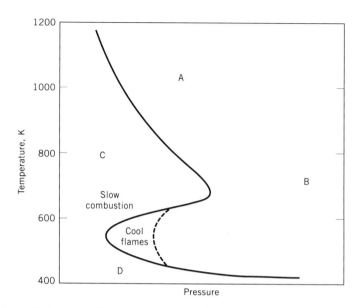

Fig. 6.   Schematic ignition diagram for a hydrocarbon + O_2 mixture, with applications. Region A, very rapid combustion, eg, a jet engine; region B, low temperature ignition, eg, internal combustion engine, safety hazards; regions C and D, slow oxidation to useful chemicals, eg, O-heterocyclic compounds in C and alcohols and peroxides in D. Courtesy of Blackwell Scientific Publications, Ltd., Oxford (60).

the intermediate temperature chain-branching agent, HOOH, to become effective in providing new radicals, the flame can be quenched and the temperature drops. When the temperature returns to the low temperature reaction region, if sufficient amounts of the reactants remain, the whole process can proceed through another cycle (79). If, on the other hand, the temperature rise is great enough to bring about significant reaction of the intermediate temperature chain-branching agent, two-stage ignition occurs, ie, the cool flame is followed, after ca 1 s, by a hot flame (explosion or rapid combustion) (80).

The blue luminescence observed during cool flames is said to arise from electronically excited formaldehyde (60,69). The high energy required indicates radical–radical reactions are producing hot molecules. Quantum yields appear to be very low (10^{-6} to 10^{-16}) (81). Cool flames never deposit carbon, in contrast to hot flames which emit much more intense, yellowish light and may deposit carbon (82).

The composition of an oxidizing mixture is altered extensively by the passage of a cool flame (66,83,84). Before passage of the flame, oxygenated materials are present. In the case of hexane oxidation, ROO· radicals are reportedly displaced by HOO· radicals above 563 K (85), in concordance with previous work (86,87). After the passage of a cool flame, olefins, some conjugate and others of lower molecular weight, are observed.

The relationships that give rise to cool flames in static reactors are also associated with oscillations in flow reactors (88–91). Flow reactors offer significant advantages in studying these phenomena because, during stable operation, greater quantities of intermediates are available for sampling and sampling probes can be moved to points of interest. Moreover, stationary temperatures can be much more accurately determined. Such systems are characterized by sudden shifts from one stable state to another or oscillations between them. Hysteresis is frequently encountered in approaching a particular steady state.

Using the flow technique, one group of investigators report evidence that ROO· radicals are not unstable at higher temperatures (ie, reaction 2 is not readily reversible) (88). The concentration of ROO· radicals is drastically reduced by the passage of a cool flame, but can be quite high, even at elevated temperatures, prior to the actual flame passage. The ROO· radicals may decrease because they are destroyed by RO· and HO· radicals generated in the cool flame and not because of the reversibility of reaction 2 (88).

Even though there is some challenge, the reversibility of reaction 2 as the root cause of the NTC and cool flame phenomena is widely accepted. It is also generally accepted that, as a consequence of reversibility, there is a temperature-dependent switchover from a low temperature chain-branching agent, ROOH, to an intermediate temperature chain-branching agent, HOOH. However, there are some instances where NTCs and cool flames are reported, but the generation of HOO· by either reaction 23 or 24 is not possible. The oxidations of acetone and methane are notable examples (76,92–94). These cases apparently require the presence of some other intermediate that can trigger the switchover to HOO· radicals. This role is often assigned to formaldehyde.

Additional evidence indirectly supports the reversibility of reaction 2. The addition of oxygen to a highly conjugated radical is readily reversible even at 40°C according to a study of the isomerization of methyl linolenate hydroperoxides in the presence of $^{18}O_2$ (95).

Effect of Pressure.   The effect of pressure in VPO has not been extensively studied but is informative. The NTC region and cool flame phenomena are associated with low pressures, usually not far from atmospheric. As pressure is increased, the production of olefins is suppressed and the NTC region disappears (96,97). The reaction rate also increases significantly and, therefore, essentially complete oxygen conversion can be attained at lower temperatures. The product distribution shifts toward oxygenated materials that retain the carbon skeleton of the parent hydrocarbon.

MOLECULAR STRUCTURE

Molecular structure strongly influences the reaction rate in low temperature VPO but has less effect in the intermediate and high temperature regions (96,98). At low temperature and low pressure, relative rates vary from 1.0 for pentane to 1380 for decane (96,99). This variation is much greater than the variation in the ease of abstractability of the hydrogen atoms. The relative efficiency of production of chain-branching agents is also involved (96). This is further borne out by the observation that branched-chain isomers oxidize more slowly than their straight-chain counterparts (96,98) even though the ease of hydrogen abstractability increases in the series primary < secondary < tertiary. Reaction 2 is more readily reversible for 3° radicals (63). Furthermore, they produce ketones (through reactions 2, 13, 14, and 16) rather than aldehydes.

Aldehydes oxidize very rapidly and are more effective in subsequent chain branching.

The similarity of oxidation rates of different hydrocarbons in the higher temperature regions is probably related to the predominance of alkyl radical cracking reactions under these conditions (reaction 28). The products of such reactions would be similar for most common hydrocarbons (96).

Methane. As our most abundant hydrocarbon, methane offers an attractive source of raw material for organic chemicals (see HYDROCARBONS). Successful commercial processes of the 1990s are all based on the intermediate conversion to synthesis gas. An alternative one-step oxidation is potentially very attractive on the basis of simplicity and greater energy efficiency. However, such processes are not yet commercially viable (100).

At ordinary pressures, the rate of homogeneous oxidation of methane [74-82-8] is not very high (<500°C) (101). Reaction 2 is highly reversed, ie, the ratio of $CH_3 \cdot$ to $CH_3OO \cdot$ is very high. The radicals present in highest concentration tend to be the principal participants in radical–radical reactions. The generation of intermediate ethane from methyl radicals has been found to be a significant path in the oxidation of methane (102). In oxidations near 1100 K, roughly half of the methyl radicals are directly oxidized; the remainder dimerize before oxidation. Moreover, the methyl radicals that are directly oxidized under these conditions tend not to react with oxygen molecules but with $HOO \cdot$ radicals or atomic oxygen. The principal reactions producing methyl radicals involve $HO \cdot$ or atomic oxygen.

A previous proposal for the initiation mechanism at temperatures up to 2000 K is an analogue of equation 6 (103):

$$CH_4 + O_2 \longrightarrow CH_3 \cdot + HOO \cdot$$

An alternative suggestion, based on a mathematical model fitted to experimental data, is that initiation occurs by thermolysis of a carbon–hydrogen bond:

$$CH_4 + M^* \longrightarrow CH_3 \cdot + H \cdot + M \tag{29}$$

where M* represents an energy-rich molecule (102).

Methane oxidations occur only by intermediate and high temperature mechanisms and have been reported not to support cool flames (104,105). However, others have reported that cool flames do occur in methane oxidation, even at temperatures >400°C (93,94,106,107). Since methyl radicals cannot participate in reactions 23 or 24, some other mechanism must be operative to achieve the quenching observed in methane cool flames. It has been proposed that the interaction of formaldehyde and its products with radicals decreases their concentrations and inhibits the whole oxidation process (93).

The reported characteristics of methane oxidation at high pressures are interesting. As expected, the reaction can be conducted at lower temperatures; eg, 262°C at 334 MPa (3300 atm) (100). However, the cool flame phenomenon is observed even under these conditions. At high pressures, in excess of about 5 MPa (50 atm), the product distribution shifts toward methanol (100,108–110).

Very extensive investigations have been conducted on the oxidation of methane over a number of heterogeneous catalysts, particularly Li^+–MgO (111,112). A typical oxidation is conducted at 700°C (113). Methyl radicals generated on the surface are effectively injected into the vapor space before further reaction occurs (114). Under these conditions, methyl radicals are not very reactive with oxygen and tend to dimerize. Ethane and its oxidation product ethylene can be produced in good efficiencies but maximum yield is limited to ca 20%. This limitation is imposed by the susceptibility of the intermediates to further oxidation (see Figs. 2 and 3). A conservative estimate of the lower limit of the oxidation rate constant ratio for ethane and ethylene with respect to methane is one, and the ratio for methanol may be at least 20 (115).

Some control over the split between methyl radical oxidation (to HCHO) and dimerization in heterogeneous oxidation can be achieved by varying conditions (116). For homogeneous oxidation, an efficiency of 70–80% to methanol has been claimed at 8–10% conversions (110). This is the high end of the reported range and is controversial. Even so, such technology appears unlikely to be competitive for regular commercial use until further advances are made (117). The critical need is to protect the products from further oxidation (118).

Although there are no new methane VPO competitive processes, current technology may be useful for the production of impure methanol in remote areas for use as a hydrate inhibitor in natural gas pipelines (119,120).

Ethane. Ethane VPO occurs at lower temperatures than methane oxidation but requires higher temperatures than the higher hydrocarbons (121). This is a transition case with mixed characteristics. Low temperature VPO, cool flames, oscillations, and a NTC region do occur. At low temperatures and pressures, the main products are formaldehyde, acetaldehyde ($HCHO:CH_3CHO$ = ca 5) (121–123), and carbon monoxide. These products arise mainly through ethylperoxy and ethoxy radicals (see eqs. 2 and 12–16 and Fig. 1).

As the temperature is raised, ethylene and hydrogen peroxide become important products. In the 600–630°C region, hydrogen peroxide is a significant product but it is isolated in much smaller quantities than ethylene (124). There is strong evidence that hydrogen peroxide must be made primarily by reaction 26 rather than reaction 25. In this mode, it does not serve as a chain-branching agent but rather as a delayed chain carrier through the following reaction scheme.

$$2\ C_2H_5\cdot + 2\ O_2 \longrightarrow 2\ HOO\cdot \longrightarrow HOOH \longrightarrow 2\ HO\cdot \xrightarrow{2\ C_2H_6} 2\ C_2H_5\cdot + 2\ H_2O$$
$$+ \qquad\qquad +$$
$$2\ C_2H_4 \qquad O_2$$

Chain-branching requirements are met by HOOH formed in reaction 25 and by HCHO oxidation (eqs. 30 and 31).

$$HCHO + O_2 \longrightarrow HOO\cdot + H\dot{C}O \qquad\qquad (30)$$
$$HOO\cdot + HCHO \longrightarrow HOOH + H\dot{C}O \qquad\qquad (31)$$

Ethylene oxide is a coproduct, probably formed by the reaction of ethylene and HOO· (124–126). Chain branching also occurs through further oxidation of ethylene; hydroxyl radicals are the main chain centers of propagation (127).

Most HCHO (>75%) is formed directly from ethane [74-84-0] (perhaps via reactions 2, 13, 14, 16, etc) without going through ethylene (126,128), although this is controversial (129).

Increasing pressure increases yields of methanol and ethanol, increases the $C_2H_5OH:CH_3OH$ ratio, and reduces yields of acetaldehyde and formaldehyde (96,97,130). Ethylene is insignificant at 790 kPa (7.8 atm) even at 460°C (96).

Propane. The VPO of propane [74-98-6] is the classic case (66,89,131–137). The low temperature oxidation (beginning at ca 300°C) readily produces oxygenated products. A prominent NTC region is encountered on raising the temperature (see Fig. 4) and cool flames and oscillations are extensively reported as complicated functions of composition, pressure, and temperature (see Fig. 6) (96,128,138–140). There can be a marked induction period. Product distributions for propane oxidation are given in Table 1.

Aldehydes are important products at all pressures, but at low pressures, acids are not. Carbon monoxide is an important low pressure product and declines with increasing pressure as acids increase. This is evidence for competition between reaction sequence 18–20 and reaction 21. Increasing pressure favors retention of the parent carbon skeleton, in concordance with the reversibility of reaction 2. Propylene becomes an insignificant product as the pressure is increased and the temperature is lowered. Both acetone and isopropyl alcohol initially increase as pressure is raised, but acetone passes through a maximum. This increase in the alcohol:carbonyl ratio is similar to the response of the methanol:formaldehyde ratio when pressure is increased in methane oxidation.

The changeover from ROO· radicals to HOO· radicals and the switch from organic peroxides to HOOH has been shown as temperature is increased in propane VPO (87,141). Tracer experiments have been used to explore product sequences in propane VPO (142–145). Propylene oxide comes exclusively from propylene. Ethylene, acetaldehyde, formaldehyde, methanol, carbon monoxide, and carbon dioxide come from both propane and propylene. Ethanol comes exclusively from propane.

Commercial VPO of propane–butane mixtures was in operation at Celanese Chemical Co. plants in Texas and/or Canada from the 1940s to the 1970s. The principal primary products were acetaldehyde, formaldehyde, methanol, and acetone. The process was run at low hydrocarbon conversion (3–10%) and a pressure in excess of 790 kPa (7.8 atm). These operations were discontinued because of various economic factors, mainly the energy-intensive purification system required to separate the complex product streams.

A process to produce propylene by VPO of propane was patented in the former USSR in 1987 (146). Similar processes have the potential to coproduce hydrogen peroxide. Yields of hydrogen peroxide as high as 1 mol/mol propylene produced have been reported with 60–70% propylene selectivity (147).

Butane. The VPO of butane (148–152) is, in most respects, quite similar to the VPO of propane. However, at this carbon chain length an important reaction known as back-biting first becomes significant. There is evidence that a β-dicarbonyl intermediate is generated, probably by intramolecular hydrogen abstraction (eq. 32). A postulated subsequent difunctional peroxide may very well be the precursor of the acetone formed.

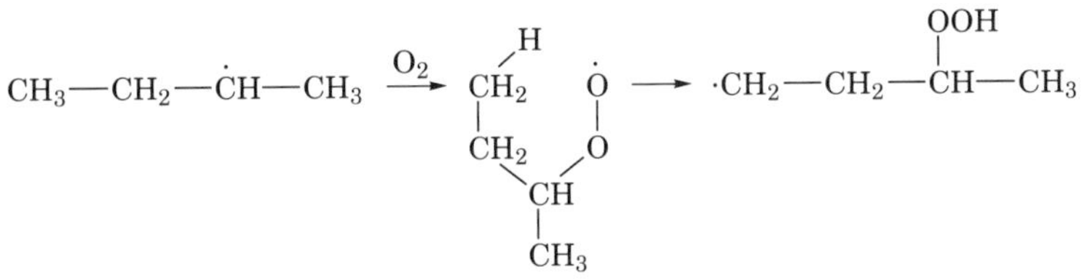

$$CH_3-CH_2-\overset{\cdot}{C}H-CH_3 \xrightarrow{O_2} \underset{\underset{\underset{CH_3}{|}}{\underset{\overset{|}{CH}}{CH_2}}}{\overset{\overset{H}{\diagup}}{CH_2}} \overset{\cdot}{O} \xrightarrow{\hspace{1cm}} \cdot CH_2-CH_2-\overset{\overset{\displaystyle OOH}{|}}{CH}-CH_3 \qquad (32)$$

At combustion temperatures, the oxidation of butane [106-97-8] is similar to that of propane (153). This is because most butyl radicals are consumed by carbon–carbon bond scission (reaction 28).

Isobutane. The VPO of isobutane [75-28-5] is similar to the VPO of other low molecular weight paraffins; however, among the significant differences, conjugate olefin can be a main product even in the low temperature region (78). The selectivity of isobutylene is ca 50% at 1% conversion at ca 290°C. The selectivity decreases with increasing temperature, passing through a minimum of ca 20% at ca 350°C. In the NTC range, the selectivity to isobutylene increases as expected. The low temperature mechanism for isobutylene is postulated as heterogeneous (78,154), whereas the higher temperature mechanism is the usual homogeneous one. In contrast, 80% selectivities of isobutylene have been reported in the initial products (<1% conversion) (155) over the range 270–330°C by a mechanism that was postulated to be mainly homogeneous. Increasing surface-to-volume ratio has a significant impact on product distribution (156).

Isobutane shows the usual NTC and cool flame phenomena (78,154,157, 158). As the pressure is increased, the expected increase in oxygenated products retaining the parent carbon skeleton is observed (96). Under similar conditions, isobutane oxidizes more slowly than n-butane (159). There are still important unresolved questions concerning isobutane VPO (160).

Higher Hydrocarbons. The VPO of higher hydrocarbons is similar to that of the lower members of the series with two significant additional complications: (1) the back-biting reactions of alkylperoxy radicals (eq. 32), particularly at positions 2 or 3 carbons removed from the peroxy position, and (2) above the NTC region, radical fragmentation (eq. 28).

Below the NTC region, intramolecular abstraction appears to generate β-dicarbonyl intermediates that are consumed during cool flames (161–164). Secondary attack on nonradical monofunctionals does not appear to be a significant source for these difunctional intermediates.

In the NTC region, back-biting reactions appear to be responsible for the formation of cyclic ethers (60,165–170). In addition to oxetanes and tetrahydrofurans, tetrahydropyrans, oxiranes, and others are also observed (60,96,169); the tetrahydrofurans are favored. O-Heterocycle yields of 25 to 30% have been reported for n-pentane (165,171). Conjugate and other olefins are also prominent products in this region (60,169–172).

More complete understanding of VPO of hydrocarbons, especially in the fuel range, is vital for continued development of modern technology (60). The quenching phenomena associated with NTC regions and cool flames have been recognized as producing significant emissions of unburned and partially oxidized fuel (102). A significant correlation between cool flame phenomena and octane number has been reported. The use of a flow reactor under cool flame conditions has been suggested as a rapid method for octane number determination (173).

Liquid-Phase Oxidation

Although there are many similarities between VPO and liquid-phase oxidation (LPO), there are sharp distinctions. First, of course, in LPO with a gaseous oxidant (such as air), it is necessary that the oxidant (oxygen) be transported from the vapor phase into the liquid phase before LPO can occur. Moreover, it is possible to have a homogeneous catalyst dissolved in the liquid phase. If the solvent and/or products are volatile, they may be vaporized in significant amounts into any vent gas issuing from the reactor. Thus each individual component may have its own residence time in the system. Under these conditions, the hydrocarbon conversion no longer has a simple relationship with the exposure of individual components to oxidation conditions and alternative relationships are needed for kinetic studies (22). LPOs generally occur at lower temperatures than VPOs. LPOs do not exhibit NTC phenomena; oscillations are known to occur (174,175), but these seem to be related to catalyst valence shift cycles and involve oxygen mass-transfer rate limitations. Another rate reduction effect is encountered if the temperature increases enough so that the vapor pressure approaches the imposed system pressure. In this case, oxygen is swept out of solution and the reaction stops; a boiling liquid does not support LPO. Generally, olefins are not significant products of the LPO of saturated materials. The high heat capacity and high heat of vaporization of the solvent, combined with limited oxygen supply, tend to damp temperature excursions in the liquid. Of course, runaway VPO reactions are possible if the LPO is quenched and unconverted oxygen collects in the vapor space.

The lower temperatures and reduced degree of oxygen starvation in LPO (vs VPO) generally reduce carbon monoxide production markedly by promoting reaction 18 and suppressing reaction 21. As a consequence, acids, from further oxidation of aldehydes, are usually the main products.

Mass Transfer.   The transfer of oxygen from the vapor to the liquid phase is a critical part of any LPO involving a gaseous oxidant. A simplified view of this process can provide a rationalization of the observed system behavior when air is the oxidant (176,177). The typical air-sparged LPO system can be considered to consist of two zones. The zone near the sparger is chemically rate limited since the liquid contains enough dissolved oxygen to scavenge alkyl radicals by reaction 2. Under these conditions, the reaction is zero order with respect to oxygen (178); the kinetics are thus determined by the initiating and chain-branching, chain-terminating, and chain-propagating reactions.

As the concentration of oxygen in the ascending bubbles falls, a point is reached where the mass transfer of oxygen from the bubble becomes rate limiting. Above this transition region (assumed to be negligible), if the partial pressure of oxygen in equilibrium with the dissolved oxygen is negligible compared to the partial pressure of oxygen in the bubble and if the bubbles are of constant size and rise at a uniform rate, the delivery of oxygen to the solution is approximately first order, in both time and distance, with respect to the oxygen in the bubble. The system therefore has a characteristic transfer height. As the bubbles rise through this distance, they deliver ca 63.2% ($100 \times (1 - 1/e)$) of the oxygen they contained at the start. The transfer characteristics of a mixed zone system can be approximated with a relatively simple equation (177).

In the chemically rate-limited zone, the energy of activation is on the order of 105 kJ/mol (25 kcal/mol) (25,176,177). Mass transfer, since it is a physical process, is likely to have an energy of activation <21 kJ/mol (5 kcal/mol) (179). The effective overall energy of activation for the oxygen-transport process is even lower because increasing temperature results in larger bubbles that rise faster (less surface per unit volume, lower contact time). The two zones can interact in such a way that the activation energy for the oxidation process can be very low as long as both zones are present. This lends stability to the operation. If the chemically rate-limited zone is so large that the energy of activation is significant, the operation becomes unstable because a small temperature fluctuation (reduction) can result in a significantly reduced reaction rate. This causes a reduction in the heat evolution rate and a further reduction of temperature ensues. The minimum stable operating temperature for a given system and set of conditions may be sharply defined; a reduction of as little as a few tenths of a degree in temperature may mean the difference between stable operations and loss of the reaction. The lowest minimum stable operating temperature can be approached when the temperature distribution in the reactor approaches uniformity.

Stable operation of LPO reactors thus requires the presence of a mass-transfer rate-limited zone. It is usually desirable to limit this zone to minimize oxygen starvation problems.

Pressure. Within limits, pressure may have little effect in air-sparged LPO reactors. Consider the case where the pressure is high enough to supply oxygen to the liquid at a reasonable rate and to maintain the gas holdup relatively low. If pressure is doubled, the concentration of oxygen in the bubbles is approximately doubled and the rate of oxygen delivery from each bubble is also approximately doubled in the mass-transfer rate-limited zone. The total number of bubbles, however, is approximately halved. The overall effect, therefore, can be small. The optimum pressure is likely to be determined by the permissible maximum gas holdup and/or the desirable maximum vapor load in the vent gas.

Reactor Configuration. The horizontal cross-sectional area of a reactor is a critical parameter with respect to oxygen mass-transfer effects in LPO since it influences the degree of interaction of the two types of zones. Reactions with high intrinsic rates, such as aldehyde oxidations, are largely mass-transfer rate-limited under common operating conditions. Such reactions can be conducted effectively in reactors with small horizontal cross sections. Slower reactions, however, may require larger horizontal cross sections for stable operation.

Catalysts and Promoters. The function of catalysts in LPO is not well understood. Perhaps they are not really catalysts in the classical sense because they do not necessarily speed up the reaction (25). They do seem to be able to alter relative rates and thereby affect product distributions, and they can shorten induction periods. The basic function in shortening induction periods appears to be the decomposition of peroxides to generate radicals (eq. 33).

$$ROOH + M^{n+} \longrightarrow RO\cdot + OH^- + M^{(n+1)+} \tag{33}$$

Although the decomposition rate of peroxide is thus increased, the consequent lowering of steady-state peroxide concentration leaves the effective rate unchanged in the simple peroxide cycle kinetic scheme (25). In real systems, at certain critical levels, a catalyst can become an inhibitor (2,180).

An important function of manganous ions is the ready reduction of alkylperoxy radicals (181) (eq. 34).

$$ROO\cdot + Mn^{2+} \longrightarrow ROO^- + Mn^{3+} \tag{34}$$

Higher valence-state metal ions can abstract hydrogen from a hydroperoxide (25) (eq. 35) or from a substrate (eq. 36).

$$ROOH + M^{(n+1)+} \longrightarrow ROO\cdot + H^+ + M^{n+} \tag{35}$$

$$RH + M^{(n+1)+} \longrightarrow R\cdot + H^+ + M^{n+} \tag{36}$$

Reaction 36 may occur through a peroxy radical complex with the metal ion (2,25,182). In any event, reaction 34 followed by reaction 36 is the equivalent of a metal ion-catalyzed hydrogen abstraction by a peroxy radical.

Reactions 33 and 35 constitute the two principal reactions of alkyl hydroperoxides with metal complexes and are the most common pathway for catalysis of LPOs (2). Both manganese and cobalt are especially effective in these reactions. There is extensive evidence that the oxidation of intermediate ketones is enhanced by a manganese catalyst, probably through an enol mechanism (34,96,183–185).

Mn(II) is readily oxidized to Mn(III) by just bubbling air through a solution in, eg, nonanoic acid at 95°C, even in the absence of added peroxide (186). Apparently traces of peroxide in the solvent produce some initial Mn(III) and alkoxy radicals. Alkoxy radicals can abstract hydrogen to produce R· radicals and Mn(III) can react with acid to produce radicals. The R· radicals can produce additional alkylperoxy radicals and hydroperoxides (reactions 2 and 3) which can produce more Mn(III). If the oxygen feed is replaced by nitrogen, the Mn(III) is rapidly reduced to Mn(II).

Chromium compounds decompose primary and secondary hydroperoxides to the corresponding carbonyl compounds, both homogeneously and heterogeneously (187–191). The mechanism of chromium catalyst interaction with hydroperoxides may involve generation of hexavalent chromium in the form of an alkyl chromate, which decomposes heterolytically to give ketone (192). The oxidation of alcohol intermediates may also proceed through chromate ester intermediates (193). Therefore, chromium catalysis tends to increase the ketone:alcohol ratio in the product (194,195).

Cupric ion has a unique ability to compete with oxygen for a carbon-centered free radical (compare reaction 2):

$$R\cdot + CuX_2 \longrightarrow [complex] \longrightarrow RX + CuX \text{ or } R'CH{=}CH_2 + HX + CuX \qquad (37)$$

Depending on the nature of the radical and the copper ligands, the complex formed in reaction 37 may form an olefin or a ligand-substituted product (2,196,197) where X can be acetate, halide, etc. When the radical is acyl, reaction 37 can compete with reaction 21 and improve efficiencies, especially for alpha-branched aldehydes (43,198). Since the acylium ion intermediate reacts with carboxyl groups to produce anhydride this approach has been used to make anhydrides (199,200). Copper catalysts generally retard LPOs, but soluble copper ions from corrosion have been identified as a primary cause of oxidative degradation of lubricating oils (201–203).

Zirconium and other nonmultivalent metal ions can act as promoters in some manner not currently clearly defined. For zirconium, the effect may be related to changes in monomer–dimer equilibria of Co(II) complexes (2,204,205).

Increasing efforts to heterogenize homogeneous catalysts for LPO are apparent (2,206–209). Significant advantages in product recovery, catalyst use, and catalyst recovery are recognized. In some instances, however, the active catalyst is reported to be material dissolved from the solid catalyst (210).

The selectivity to alcohol in LPO may be significantly increased when boric acid, *meta*-boric acid, or boric anhydride is present in stoichiometric amounts (2). The boron compounds appear to convert alkylhydroperoxides to alkyl borates and may also intercept alkylperoxy radicals, converting them to alkylperoxyboron compounds; these are later converted to alkyl borates. The alkyl borates are resistant to further oxidation; they are hydrolyzed to recover alcohols.

Propane.   Propane is difficult to oxidize in LPO because of its volatility and lack of reactivity. It can, however, be oxidized with a suitable solvent and sufficiently high pressures and temperatures (211). The principal products are acetone and isopropyl alcohol.

Butane.   Butane LPO has been a significant source for the commercial production of acetic acid and acetic anhydride for many years. At various times, plants have operated in the former USSR, Germany, Holland, the United States, and Canada. Only the Hoechst-Celanese Chemical Group, Inc. plants in Pampa, Texas, and Edmonton, Alberta, Canada, continue to operate. The Pampa plant, with a reported annual production of 250,000 t/yr, represents about 15% of the 1994 installed U.S. capacity (212). Methanol carbonylation is now the dominant process for acetic acid production, but butane LPO in established plants remains competitive.

The production of acetic acid from butane is a complex process. Nonetheless, sufficient information on product sequences and rates has been obtained to permit development of a mathematical model of the system. The relationships of the intermediates throw significant light on LPO mechanisms in general (22). Surprisingly, ca 25% of the carbon in the consumed butane is converted to ethanol in the first reaction step. Most of the ethanol is consumed by subsequent reaction.

The proposed mechanism for producing ethanol [64-17-5] from butane involves β-scission of a sec-butoxy radical (eq. 38). The sec-butoxy radicals are derived from sec-butylperoxy radicals (reaction 14 (213)) and/or through some sequence involving reaction 33. If 25% of the carbon forms ethanol, over 50% must pass through the sec-butoxy radical. Furthermore, the principal fate of sec-butoxy radicals must be the β-scission reaction; the ethoxy radical, on the other hand, must be converted to ethanol efficiently.

$$\underset{\displaystyle \overset{O \cdot}{\mid}}{CH_3CHCH_2CH_3} \longrightarrow CH_3CHO + CH_3CH_2 \cdot \xrightarrow{O_2} CH_3CH_2OO \cdot \longrightarrow$$

$$CH_3CH_2O \cdot \xrightarrow{RH} CH_3CH_2OH + R \cdot \quad (38)$$

Methyl ethyl ketone, a significant coproduct, seems likely to arise in large part from the termination reactions of sec-butylperoxy radicals by the Russell mechanism (eq. 15, where $R = CH_3$ and $R' = CH_2CH_3$). Since alcohols oxidize rapidly vs paraffins, the sec-butyl alcohol produced (eq. 15) is rapidly oxidized to methyl ethyl ketone. Some of the sec-butyl alcohol probably arises from hydrogen abstraction by sec-butoxy radicals, but the high efficiency to ethanol indicates this is a minor source.

Although it appears that methyl ethyl ketone [78-93-3] cannot be the principal product in butane LPO, it has been reported that the ratio of methyl ethyl ketone to acetic acid [64-19-7] can be as high as 3:1 in a plug-flow-type reactor (214). However, this requires a very unusual reactor (length:dia = 16,640:1). The reaction is very unstable and wall reactions may influence mechanisms.

Water has a strong inhibiting effect above ca 3% concentration in the feed (214); in contrast, others report that 14–16 wt % water increases the rate (215).

Acetone is a coproduct of butane LPO. Some of this is produced from isobutane, an impurity present in all commercial butane (by reactions 2, 13, 14, and 16). However, it is likely that much of it is produced through the backbiting mechanisms responsible for methyl ketone formation in the LPO of higher hydrocarbons (216).

Propionic acid made in butane LPO probably comes by a minor variation of reaction 38 that produces methyl radicals and propionaldehyde. It is estimated that up to 18% of the sec-butoxy radicals may decompose in this manner (213); this may be high since propionic acid is a minor product.

Between 6 and 30% of the radical attack on butane may occur at the primary hydrogen atoms (213). Since ca 6% of the butane goes to or through butyric acid (22), the middle of this range does not seem unreasonable. Because it is much more resistant to oxidation than its precursors or coproducts, acetic acid (qv) is the main product of butane LPO.

Butane LPO conducted in the presence of very high concentrations of cobalt catalyst has been reported to have special character (2,205,217–219). It occurs under mild conditions with reportedly high efficiency to acetic acid. It is postulated to involve the direct attack of Co(III) on the substrate. Various additives, including methyl ethyl ketone, p-xylene, or water, are claimed to be useful.

Isobutane. Isobutane can be oxidized noncatalytically to give predominantly t-butyl hydroperoxide [75-91-2] (TBHP) (reactions 2 and 3). The temperature is kept within the range 100 to 140°C. TBHP can serve as the initiator. At low rates of initiation and low conversions (which favor propagation reactions 2 and 3 over termination reaction 5), efficiencies greater than 90% can be achieved (2). Under practical conditions, the reaction is conducted at 125°C using di-*tert*-butyl peroxide, a more efficient initiator. TBHP is obtained in 75% efficiency at 8% isobutane conversion. *tert*-Butyl alcohol (21%), acetone (2%), and isobutyl derivatives (1%) are formed as side products. Using cobalt catalyst at 135°C gives predominantly *tert*-butyl alcohol (77%). Other branched-chain paraffins give mainly cleavage products under these conditions, indicating the *tert*-butoxy radical is more resistant to the β-scission reaction (reaction 16) than other *tert*-alkoxy radicals (205). With no catalyst at 150–180°C and 5 MPa (50 atm) in acetic acid solvent, the principal products are acetone and methyl acetate (220). Addition of a manganese catalyst lowers acetone and increases formic acid.

A significant outlet for TBHP is the molybdenum-complex catalyzed production of propylene oxide, a process developed by Oxirane (221–224). The reported U.S. capacity in 1991 was 0.55×10^6 t/yr (225). The *tert*-butyl alcohol coproduct is used mostly to make methyl *tert*-butyl ether, a gasoline additive.

Higher Paraffins. The LPO of paraffins to produce synthetic fatty acids (SFAs) has been practiced extensively in the former USSR and Eastern bloc countries. Production capacity reportedly exceeded 0.45×10^6 t/yr in 1975 (226). Elsewhere, varying degrees of interest in the process have been evident in Germany, Japan, China (227), and a number of other countries. Although the basic mechanisms are similar, several difficulties are encountered in higher paraffin LPO that are not so evident in the LPO of, eg, butane. For one thing, the desired product SFAs have virtually the same susceptibility to further oxidation, on a per carbon atom basis, as the feed hydrocarbon, contrasting sharply with acetic acid from butane LPO. The minimum impact of this lack of oxidation resistance can be viewed in Figure 2. As a limiting optimistic case, the SFAs can be treated as if they were primary products, made straightforwardly by a single oxidative attack (neither of these assumptions is true); the k_2/k_1 ratio would be about 1. Efficiency would be limited to ca 80% at a conversion of ca 35%, even with a plug-flow or batch reactor. The real situation would, of course, be worse.

An even more significant limitation is that a higher paraffin, once attacked, is highly susceptible to the back-biting reaction (eq. 32) before being converted to nonradical products (42,216). Keeping the conversion low reduces secondary attack on nonradical products by keeping the concentration of such products low with respect to the feed; however, the site of a back-biting attack cannot be diluted with respect to the attacking radical. Some of the products of back-biting attack are inevitably multifunctional products. Some are monofunctionals, such as methyl ketones, which can be subsequently converted to acids of lower molecular weight. Many of the multifunctionals, however, are essentially lost.

The typical SFA process uses a manganese catalyst with a potassium promoter (for solubilization) in a batch reactor. A manganese catalyst increases the relative rate of attack on carbonyl intermediates. Low conversions are followed by recovery and recycle of complex intermediate streams. Acid recovery and purification involve extraction with caustic and heat treatment to further decrease small amounts of impurities (particularly carbonyls). The fatty acids are recovered by freeing with sulfuric acid and, hence, sodium sulfate is a by-product.

Cyclohexane. The LPO of cyclohexane [110-82-7] supplies much of the raw materials needed for nylon-6 and nylon-6,6 production. Cyclohexanol (A) and cyclohexanone (K) may be produced selectively by using a low conversion process with multiple stages (228–232). The reasons for low conversion and multiple stages (an approach to plug-flow operation) are apparent from Figure 2. Several catalysts have been reported. The selectivity to A as well as the overall process efficiency can be improved by using boric acid (2,232,233). K/A mixtures are usually oxidized by nitric acid in a second step to adipic acid (233) (see CYCLOHEXANOL AND CYCLOHEXANONE).

A one-step LPO of cyclohexane directly to adipic acid (qv) has received a lot of attention (233–238) but has not been implemented on a large scale. The various versions of this process use a high concentration cobalt catalyst in acetic acid solvent and a promoter (acetaldehyde, methyl ethyl ketone, water).

Until the 1960s, adipic acid [124-04-9] was virtually the sole intermediate for nylon-6,6. However, much hexamethylenediamine is now made by hydrodimerization of acrylonitrile (qv) or via hydrocyanation of butadiene (qv). Cyclohexane remains the basis for practically the entire world output of adipic acid. The U.S. capacity for adipic acid for 1993 was 0.97×10^6 t/yr (233).

Alkylaromatics. The aromatic ring is fairly inert toward attack by oxygen-centered radicals. Aromatic acids consisting of carboxyl groups substituted on aromatic rings are good candidates for production by LPO of alkylaromatics since their k_2/k_1 ratios are low. Terephthalic acid [100-21-0] (TPA) and dimethyl terephthalate [120-61-6] (DMT) are the outstanding examples of high volume chemicals made in this way; efficiencies are ca 95%. The 1989 U.S. capacity for TPA was reported to be 4.0×10^6 t/yr (206). TPA is produced mostly by the LPO of p-xylene:

$$CH_3 - \bigcirc - CH_3 \xrightarrow{[O]} CH_3 - \bigcirc - COOH \xrightarrow{[O]} HOOC - \bigcirc - COOH \quad (39)$$

The intermediate p-toluic acid [99-94-5] (PTA) does not oxidize readily; its concentration builds up and the oxidation tends to stop before reaching high conversions to TPA (239,240). A common explanation is deactivation of the methyl group by the electron-withdrawing carboxyl group making the methyl particularly resistant to oxidation (2). This conclusion is questionable, however, since methyl p-toluate is readily oxidized even though the electron-withdrawing capability of simple carboalkoxy groups is equivalent to that of the carboxyl group (241,242). An appealing alternative explanation is that some phenolic material is produced during the oxidation of PTA. It is known that the oxidation of substituted aromatics may produce aryl radicals, probably through decarbonylation of aroyl radicals and decarboxylation of aroyloxy radicals (analogues of reactions 21 and 22); that some phenols are produced (perhaps via aryl radicals through analogues of reactions 2 and 15); and that phenols are powerful inhibitors of this type of reaction (54,240,243–247).

Various ways of overcoming the PTA oxidation problem have been incorporated into commercial processes. The predominant solution is the use of high concentrations of manganese and cobalt ions (2,248–254), optionally with various cocatalysts (204,255,256), in the presence of an organic or inorganic bromide promoter in acetic acid solvent. Operational temperatures are rather high (ca 200 °C). A lesser but significant alternative involves isolation of intermediate PTA, conversion to methyl p-toluate, and recycle to the reactor. The ester is oxidized to monomethyl terephthalate, which is subsequently converted to DMT and purified by distillation (248,257–264).

A third method utilizes cooxidation of an organic promoter with manganese or cobalt-ion catalysis. A process using methyl ethyl ketone (248,252,265–270) was commercialized by Mobil but discontinued in 1973 (263,264). Other promoters include acetaldehyde (248,271–273), paraldehyde (248,274), various hydrocarbons such as butane (270,275), and others. Other types of reported activators include peracetic acid (276) and ozone (277), and very high concentrations of cobalt catalyst (2,248,278).

The aromatic core or framework of many aromatic compounds is relatively resistant to alkylperoxy radicals and inert under the usual autoxidation conditions (2). Consequently, even somewhat exotic aromatic acids are resistant to further oxidation; this makes it possible to consider alkylaromatic LPO as a selective means of producing fine chemicals (206). Such products may include multifunctional aromatic acids, acids with fused rings, acids with rings linked by carbon–carbon bonds, or through ether, carbonyl, or other linkages (279–287). The products may even be phenolic if the phenolic hydroxyl is first esterified (288,289).

Using high concentrations of cobalt catalysts under mild conditions, it is even possible, when oxidizing a methylaromatic, to isolate relatively high yields of the aldehyde intermediate with consequently less production of the corresponding acid. This is possible because the initial attack on the aromatic substrate may occur through an electron-transfer mechanism (eq. 40) (2,205) rather than through direct hydrogen abstraction by a radical.

$$\text{C}_6\text{H}_5\text{—CH}_3 + \text{Co(III)} \longrightarrow [\text{C}_6\text{H}_5]^{+\cdot}\text{—CH}_3 \longrightarrow \text{C}_6\text{H}_5\text{—CH}_2\cdot + \text{H}^+ \qquad (40)$$
$$+\text{Co(II)}$$

The cation–radical intermediate loses a proton to become, in this case, a benzyl radical. The relative rate of attack (via electron transfer) on an aromatic aldehyde with respect to a corresponding methylarene is a function of the ionization potentials (8.8 eV for toluene, 9.5 eV for benzaldehyde); it is much lower than one would expect if the competition involved hydrogen abstraction by, eg, alkylperoxy radicals. For example, it has been reported that a 90% efficiency to benzaldehyde can be obtained at 10% conversion in the oxidation of toluene using a $\text{Co(OOCCH}_3)_2$ catalyst and a NaBr or paraldehyde promoter (290). From Figure 2, this implies a k_2/k_1 ratio near 1, which is orders of magnitude less than the ratio of reported propagation rate constants (291). By varying catalysts and conditions to take advantage of this effect, a wide variety of simple to complex aromatic aldehydes can be synthesized directly from the corresponding methylarenes with surprising efficiency (2,290–296).

An additional curious feature of alkylaromatic oxidation is that, under conditions where the initial attack involves electron transfer, the relative rate of attack on different alkyl groups attached to the same aromatic ring is quite different from that observed in alkane oxidation. For example, the oxidation of p-cymene can lead to high yields of p-isopropylbenzoic acid (2,205,297,298).

Synthetic phenol capacity in the United States was reported to be ca 1.6 × 10⁶ t/yr in 1989 (206), almost completely based on the cumene process (see CUMENE; PHENOL). Some synthetic phenol [108-95-2] is made from toluene by a process developed by The Dow Chemical Co. (2,299–301). Toluene [108-88-3] is oxidized to benzoic acid in a conventional LPO process. Liquid-phase oxidative decarboxylation with a copper-containing catalyst gives phenol in high yield (2,299–304). The phenolic hydroxyl group is located ortho to the position previously occupied by the carboxyl group of benzoic acid (2,299,301,305). This provides a means to produce meta-substituted phenols otherwise difficult to make (2,306). VPOs for the oxidative decarboxylation of benzoic acid have also been reported (2,307–309). Although the mechanism appears to be similar to the LPO scheme (309), the VPO reaction is reported not to work for toluic acids (310).

Dihydroxyarenes can be produced from the corresponding diisopropylarenes in a manner similar to the production of phenol from cumene (206,311–315).

An oxirane process utilizes ethylbenzene to make the hydroperoxide, which then is used to make propylene oxide [75-56-9]. The hydroperoxide-producing reaction is similar to the first step of cumene LPO except that it is slower (2,224,316–318). In the epoxidation step, α-phenylethyl alcohol [98-85-1] is the coproduct. It is dehydrated to styrene [100-42-5]. The reported 1992 capacity for styrene by this route was 0.59 × 10⁶ t/yr (319). The corresponding propylene oxide capacity is ca 0.33 × 10⁶ t/yr. The total propylene oxide capacity based on hydroperoxide oxidation of propylene [115-07-1] (coproducts are t-butyl alcohol and styrene) is 1.05 × 10⁶ t/yr (225).

BIBLIOGRAPHY

"Hydrocarbon Oxidation" in *ECT* 1st ed., Suppl. 1, pp. 401–418, by F. B. Marcotte and R. L. Mitchell, Celanese Corp. of America; in *ECT* 2nd ed., Vol. 11, pp. 224–241, by H. F. Hamil, Celanese Chemical Co.; in *ECT* 3rd ed., Vol. 12, pp. 826–851, by C. C. Hobbs, Celanese Chemical Co.

1. S. W. Benson and P. S. Nangia, *Acc. Chem. Res* **12**(7), 223–228 (1979).
2. R. A. Sheldon and J. K. Kochi, *Metal-Catalyzed Oxidations of Organic Compounds*, Academic Press, Inc., New York, 1981.
3. N. Semenov, *Chemical Kinetics and Chain Reactions*, Oxford University Press, London, 1935, p. 68.
4. J. L. Bolland, *Q. Rev. (London)* **3**, 1 (1949).
5. J. L. Bolland, *Trans. Faraday Soc.* **46**, 358 (1950).
6. J. L. Bolland and G. Gee, *Trans. Faraday Soc.* **42**, 236 (1946).
7. J. L. Bolland and H. P. Koch, *J. Chem. Soc.*, 445 (1945).
8. L. Bateman, *Q. Rev. (London)* **8**, 147 (1954).
9. L. Bateman and G. Gee, *Proc. R. Soc. (London)* A **195**, 376 (1948).

10. A. Robertson and W. A. Waters, *Trans. Faraday Soc.* **42**, 201 (1946).
11. F. R. Mayo, *J. Chem. Educ.* **63**(2), 97–99 (1986).
12. C. Walling, *Ibid.*, pp. 99–102.
13. S. W. Benson, in R. R. Gould and F. R. Mayo, eds., *Advances in Chemistry Series 76*, American Chemical Society, Washington, D.C., 1968, pp. 143–153.
14. Ref. 2, p. 20.
15. A. T. Betts and H. Uri, in Ref. 13, pp. 160–181.
16. B. D. Boss and R. N. Hazlett, *Can. J. Chem.* **47**, 4175–4182 (1969).
17. M. G. Simic, *J. Chem. Educ.* **58**(2), 125–131 (1981).
18. L. Sieg, in W. Jost, ed., *Low Temperature Oxidation*, Gordon and Breach, New York, 1965; p. 222.
19. C. Walling, *Free Radicals in Solution*, John Wiley & Sons, Inc., New York, 1957, pp. 503ff.
20. D. Swern, ed., *Organic Peroxides*, John Wiley & Sons, Inc., New York, 1970, p. 214.
21. G. A. Russell, in S. L. Friess, E. S. Lewis, and A. Weissberger, eds., *Technique of Organic Chemistry*, Vol. 8, Part 1, Interscience Publishers, New York, 1961, p. 344.
22. C. C. Hobbs and co-workers, *Ind. Eng. Chem. Proc. Des. Dev.* **11**, 59 (1972).
23. F. R. Mayo, *Acc. Chem. Res.* **1**(7), 193–201 (1968).
24. G. A. Russell, *J. Am. Chem. Soc.* **77**, 4583 (1955).
25. C. Walling, *J. Am. Chem. Soc.* **91**, 7590 (1969).
26. N. M. Emanuel, E. T. Denisov, and Z. K. Maizus, *Liquid Phase Oxidation of Hydrocarbons*, Plenum Press, New York, 1967.
27. Ref. 2, p. 22.
28. Ref. 2, p. 24.
29. G. A. Russell, *J. Am. Chem. Soc.* **79**, 3871 (1957).
30. S. W. Benson, *Oxid. Commun.* **2**(3–4), 169–188 (1982).
31. Q. J. Niu and G. D. Mendenhall, *J. Am. Chem. Soc.* **114**(1), 165–172 (1992).
32. B. D. Boss and R. N. Hazlett, *Ind. Eng. Chem., Prod. Res. Dev.* **14**(2), 135–138 (1975).
33. Ref. 26, p. 145.
34. F. R. Mayo, *Prepr. Div. Pet. Chem. Am. Chem. Soc.* **19**(4), 627 (1974).
35. A. Krzysztoforski and co-workers, *Ind. Eng. Chem. Process Des. Dev.* **25**(4), 894–898 (1986).
36. F. F. Rust, personal communication, 1970.
37. Ref. 2, p. 351.
38. Ref. 26, p. 147.
39. Ref. 26, p. 165.
40. Ref. 19, p. 411.
41. S. Blaine and P. E. Savage, *Ind. Eng. Chem. Res.* **30**(9), 2185–2191 (1991).
42. R. K. Jensen and co-workers, *J. Am. Chem. Soc.* **103**(7), 1742–1749 (1981).
43. D. R. Larkin, *J. Org. Chem.* **55**(5), 1563–1568 (1990).
44. F. Garcia-Ochoa, A. Romero, and J. Querol, *Ind. Eng. Chem. Res.* **28**(1), 43–48 (1989).
45. Ref. 26, pp. 154–155.
46. R. Stewart, *Oxidation Mechanisms*, W. A. Benjamin, New York, 1964, p. 30.
47. B. Plesnicar, in W. S. Trahanovsky, ed., *Oxidation in Organic Chemistry*, Academic Press, Inc., New York, 1978, p. 264.
48. C. Walling, *Pure Appl. Chem.* **15**(1), 69 (1967).
49. W. Pritzkow and V. Voerckel, *Oxid. Commun.* **4**(1–4), 223–228 (1983).
50. W. Pritzkow and V. Voerckel, *J. Prakt. Chem.* **326**(4), 572–578 (1984).
51. F. R. Hewgill and G. M. Proudfoot, *Aust. J. Chem.* **34**(2), 335–342 (1981).
52. S. Blaine and P. E. Savage, *Ind. Eng. Chem. Res.* **31**(1), 69–75 (1992).
53. S. Blaine and P. E. Savage, *Prepr. Div. Pet. Chem., Am. Chem. Soc.* **35**, 239–244 (1990).

54. P. Roffia, P. Calini, and L. Motta, *Ind. Eng. Chem. Prod. Res. Dev.* **23**(4), 629–634 (1984).
55. P. Roffia, P. Calini, and S. Tonti, *Ind. Eng. Chem. Res.* **27**, 765–770 (1988).
56. F. Broich and co-workers, *Erdöl Kohle* **16**, 284 (1963).
57. T. Horlenko, unpublished data, Celanese Chemical Co., 1964.
58. Ref. 47, p. 262.
59. A. W. Dawkins, *Eur. Chem. News, Norm. Paraffins Suppl.*, 49–58 (1966).
60. R. W. Walker, *Sci. Prog.* **74**(294, Pt. 2), 163–187 (1990).
61. C. K. Westbrook and F. L. Dryer, *Prog. Energy Combust. Sci.* **10**(1), pp. 1–57 (1984).
62. T. S. Norton and F. L. Dryer, *Symp. (Int.) Combust. [Proc.]*, **23**, 179–185 (1991).
63. S. W. Benson, *J. Am. Chem. Soc.* **87**(5), 972–979 (1965).
64. D. N. Koert and N. P. Ceransky, *Chem. Phys. Processes Combust.*, 17/1–17/4 (1991).
65. E. I. Finkel'shtein and G. N. Gerasimov, *Zh. Fiz. Khim.* **58**(4), 942–946 (1984).
66. R. D. Wilk, N. P. Ceransky, and R. S. Cohen, *Combust. Sci. Technol.* **49**(1–2), 41–78 (1986).
67. R. D. Wilk, N. P. Ceransky, and R. S. Cohen, *Combust. Sci. Technol.* **52**(1–3), 39–58 (1987).
68. D. M. Newitt and W. G. Schmidt, *J. Chem. Soc.*, 1665 (1937).
69. R. R. Baker and D. A. Yorke, *J. Chem. Ed.* **49**(5), 351 (1972); R. A. Geisbrecht and T. E. Daubert, *Ind. Eng. Chem. Prod. Res. Dev.* **15**(2), 115 (1976).
70. J. C. Dechaux and L. Delfosse, *Combust. Flame* **34**(2), 169–185 (1979).
71. R. I. Moshkina and co-workers, *Kinet. Katal.* **21**(6), 1379–1384 (1980).
72. R. D. Wilk, *Chem. Phys. Processes Combust.*, 11-1–11-4 (1991).
73. I. R. Slagle, J. Y. Park, and D. Gutman, *Symp. (Int.) Combust. [Proc.]*, 733–741 (1984).
74. S. W. Benson, in Ref. 13, pp. 143–153.
75. D. M. Newitt and L. S. Thornes, *J. Chem. Soc.*, 1656–1665 (1937).
76. J. A. Barnard, in Ref. 13, pp. 98–110.
77. T. Berry and co-workers, in Ref. 13, pp. 86–97.
78. J. G. Atherton and co-workers, *Symp. (Int.) Combust. [Proc.]* **14**, 513 (1973).
79. Ref. 18, pp. 197–198.
80. Ref. 18, p. 226.
81. S. W. Benson, *The Foundation of Chemical Kinetics*, McGraw Hill Book Co., Inc., New York, 1960, p. 485.
82. A. D. Walsh, in Ref. 18, pp. 285–327.
83. Ref. 18, p. 224.
84. J. W. Falconer, D. E. Hoare, and Z. F. Savaya, *Oxid. Commun.* **4**(1–4), 299–315 (1983).
85. Z. A. Mansurov and D. U. Bodykov, *Khim. Fiz.* **7**(10), 1430–1431 (1988).
86. M. Carlier and L. R. Sochet, *J. Chim. Phys. Phys.-Chim. Biol.* **72**, 623–630 (1975).
87. G. A. Sachyan, G. Sh. Alaverdyan, and A. B. Nalbandyan, *Dokl. Akad. Nauk SSSR* **204**, 883 (1972).
88. K. A. Sahetchian and co-workers, *Combust. Flame* **34**(2), 153–159 (1979).
89. A. A. Mantashyan, P. S. Gukasyan, and R. H. Sayadyan, *Arch. Termodyn. Spalania* **9**(2), 273–278 (1978).
90. A. A. Mantashyan, S. G. Bernatosyan, and T. R. Simonyan, *Oxid. Commun.* **5**(1–2), 207–223 (1983).
91. V. Caprio, A. Insola, and P. G. Lignola, *Fiamme Reaz. Flusso, Simp. Int. Din. Reaz. Chim.* **2**, 47–51 (1978).
92. J. A. Barnard and M. A. Sheikh, *Pak. J. Sci. Ind. Res.* **16**(3–4), p. 93 (1973).
93. V. Ya. Basevich and S. M. Kogarko, *Oxid. Commun.* **3**(3–4), 199–209 (1983).
94. G. A. Foulds and co-workers, *Prepr. Am. Chem. Soc., Div. Pet. Chem.* **37**(1), 51–60 (1992).

95. H. W.-S. Chan, G. Levett, and J. A. Matthew, *J. Chem. Soc. Chem. Comm.*, 756 (1978).
96. H. D. Medley and S. D. Cooley, *Adv. Pet. Chem. Ref.* **3**, 309 (1960).
97. U.S. Pat. 5,017,731 (May 21, 1991), H. D. Gesser and co-workers.
98. Ref. 82, p. 296.
99. C. F. Cullis and co-workers, *Disc. Faraday Soc.* **2**, 111, 114, 117, 128 (1947).
100. H. D. Gesser, N. R. Hunter, and C. B. Prakash, *Chem. Rev.* **85**(4), 235–244 (1985).
101. D. E. Hoare, in Ref. 18, pp. 125–167.
102. C. K. Westbrook and co-workers, *J. Phys. Chem.* **81**(25), 2542–2554 (1977).
103. W. M. Heffington and co-workers, *Symp. (Int.) Combust. [Proc.]* **16**, 997 (1977).
104. Ref. 82, p. 481.
105. A. D. Walsh, in Ref. 18, pp. 290, 305, 313.
106. G. G. Torchyan, A. A. Mantashyan, and A. B. Nalbandyan, *Vses. Konf. Kinet. Mekh. Gazofazn. Reakts.* **2**, 17 (1971).
107. L. A. Khachatryan and co-workers, *Dokl. Akad. Nauk SSSR Ser. Khim.* **224**, 1363 (1975).
108. R. I. Moshkina and co-workers, *Arm. Khim. Zh.* **41**(1–2), 32–36 (1988).
109. Can. Pat. 1,267,423 (Oct. 17, 1986), H. D. Gesser, N. R. Hunter, and L. A. Morton (to Canada, Minister of Energy, Mines, and Resources).
110. P. S. Yarlagadda and co-workers, *Ind. Eng. Chem. Res.* **27**(2), 252–256 (1988).
111. *Catal. Today* **6**(4) (1990).
112. *Catal. Today* **13**(4) (1992).
113. C. Shi, M. Hatano, and J. H. Lunsford, *Catal. Today* **13**(2–3), 191–199 (1992).
114. P. F. Nelson, C. A. Lukey, and N. W. Cant, *J. Phys. Chem.* **92**(22), 6176–6179 (1988).
115. J. A. Labinger, *Prepr. Am. Chem. Soc., Div. Pet. Chem.* **37**(1), 289–292 (1992).
116. J. S. J. Hargreaves, G. J. Hutchings, and R. W. Joyner, *Nature* **348**(6300), 428–429 (1990).
117. J. W. M. H. Geerts, J. H. B. J. Hoebink and K. Van der Wiele, *Catal. Today* **6**(4), 613–620 (1990).
118. J. H. Edwards and N. R. Foster, *Fuel Sci. Technol. Int.* **4**(4), 365–390 (1986).
119. P. M. Shcherbakov, S. A. Egorov, and S. F. Bobrova, *Neft. Gazov. Prom-st.*, (1), 40–41 (1986).
120. V. F. Budymka and co-workers, *Khim. Prom-st. (Moscow)*, (6), 330–331 (1987).
121. L. Sieg, in Ref. 18, p. 191.
122. A. B. Nalbandyan, *Dokl. Akad. Nauk SSSR* **66**, 413 (1949).
123. P. Gray, J. F. Griffiths, and S. M. Hasko, *Proc. R. Soc. London* **396**(1811), 227–255 (1984).
124. R. J. Sampson, *J. Chem. Soc.*, 5095 (1963).
125. R. I. Moshkina and co-workers, *Dokl. Akad. Nauk SSSR* **218**, 1147 (1974).
126. R. I. Moshkina and co-workers, *Kinet. Katal.* **19**(4), 830–839 (1978).
127. M. Cathonnet and H. James, *J. Chim. Phys. Phys.-Chim. Biol.* **74**(2), 156–167 (1977).
128. R. I. Moshkina and co-workers, *Kinet. Katal.* **17**, 1057 (1976).
129. J. H. Knox, in Ref. 13, pp. 1–21.
130. M. U. Fedurtsa and S. S. Abedzheva, *Khim. Prom-st., Ser.: Khlornaya Prom-st.*, (1), 5–6 (1980).
131. M. M. Aleksishvili, S. S. Polyak, and V. Ya. Shtern, *Kinet. Katal.* **15**, 290 (1974).
132. M. M. Aleksishvili, S. S. Polyak, and V. Ya. Shtern, *Kinet. Katal.* **17**, 1110 (1976).
133. E. A. Poladyan and A. A. Mantashyan, *Arm. Khim. Zh.* **28**, 949 (1975).
134. E. A. Polyadyan and G. L. Grigoryan, *Kinet. Katal.* **17**, 231 (1976).
135. E. A. Poladyan and A. A. Mantashyan, *Arm. Khim. Zh.* **29**, 131 (1976).
136. D. N. Koert, D. L. Miller, and N. P Ceransky, *Chem. Phys. Processes Combust.*, 8/1–8/4 (1991).

137. E. A. Poladyan and co-workers, *Kinet. Katal.* **17**, 304 (1976).
138. L. Sieg, in Ref. 18, p. 206.
139. A. A. Mantashyan and P. S. Gukasyan, *Dokl. Akad. Nauk SSSR* **234**, 379 (1977).
140. P. G. Lignola and co-workers, *Ber. Bunsenges. Phys. Chem.* **84**(4), 369–373 (1980).
141. I. K. Shakhnazaryan and co-workers, *Vses. Konf. Kinet. Mekh. Gazofaz. Reakts.* **2**, 4 (1971).
142. R. I. Moshkina and co-workers, *Dokl. Akad. Nauk SSSR* **249**(4), 908–912 (1979).
143. R. I. Moshkina and co-workers, *Oxid. Commun.* **3**(3–4), 241–257 (1983).
144. R. I. Moshkina and co-workers, *Khim. Fiz.*, (7), 976–987 (1982).
145. R. I. Moshkina, S. S. Polyak, and L. B. Romanovich, *Khim. Fiz.* **5**(9), 1249–1258 (1986).
146. Rus. Pat. SU 1,348,329 (Oct. 30, 1987), A. A. Mantashyan and M. D. Pogosyan (to Institute of Chemical Physics, Academy of Sciences, Armenian SSR).
147. T. Kunugi and co-workers, in Ref. 13, pp. 326–344.
148. J. C. Dechaux, J. L. Flament, and M. Lucquin, *Combust. Flame* **17**, 205 (1971).
149. R. R. Baldwin and R. W. Walker, *Combust. Flame* **21**, 55 (1973).
150. R. R. Baldwin, J. C. Plaistowe, and R. W. Walker, *Combust. Flame* **30**, 13 (1977).
151. M. Carlier and L. R. Sochet, *Combust. Flame* **25**, 309 (1975).
152. V. K. Proudler and co-workers, *Phi. Trans. R. Soc. Lond. A* **337**, 211–221 (1991).
153. W. J. Pitz and Charles K. Westbrook, *Symp. (Int.) Combust. [Proc.]* **20**, 831–843 (1985).
154. G. A. Luckett and R. T. Pollard, *Combust. Flame* **21**, 265 (1973).
155. J. Hay, J. H. Knox, and J. M. C. Turner, *Symp. (Int.) Combust. [Proc.]* **10**, 331 (1965).
156. G. McKay and J. A. Aga, *Combust. Flame* **40**(2), 221–224 (1981).
157. A. J. Brown and co-workers, *Mech. Hydrocarbon React. Symp.*, 751 (1975).
158. V. Caprio, A. Insola, and P. G. Lignola, *Symp. (Int.) Combust. [Proc.]* **16**, 1155 (1977).
159. S. D. Cooley and J. W. Walker, Celanese Chemical Co. Technical Center, Corpus Christi, Tex., unpublished data.
160. B. Vogin, F. Baronnet, and G. Scacchi, *Can. J. Chem.* **67**(5), 759–772 (1989).
161. J. R. Thomas and H. W. Crandall, *Ind. Eng. Chem.* **43**, 2761 (1951).
162. M. R. Barusch and co-workers, *Ind. Eng. Chem.* **43**, 2764 (1951).
163. *Ibid.*, p. 2766.
164. Ref. 18, pp. 222–223.
165. Ref. 77, p. 92.
166. J. H. Jones and D. A. Kurtz, *Am. Chem. Soc. Div. Pet. Chem. Prepr.* **17**, A98 (1972).
167. K. Richter, G. Oehlmann, and W. Schirmer, *Z. Phys. Chem. (Leipzig)* **253**(3–4), 207 (1973).
168. *Ibid.*, p. 217.
169. R. R. Baldwin, J. P. Bennett, and R. W. Walker, *Symp. (Int.) Combust.* **16**, 1041 (1977).
170. C. F. Cullis, M. M. Hirschler, and R. L. Rogers, *Proc. R. Soc. London, [Ser.] A* **382**(1783), 429–440 (1982).
171. R. R. Baldwin, J. P. Bennett, and R. W. Walker, *J. Chem. Soc., Faraday Trans. 1* **76**(5), 1075–1092 (1980).
172. M. M. Abdelkafi and A. Baklouti, *J. Soc. Chim. Tunis.* **2**(10), 7–12 (1991).
173. Ya. Yu. Stepanskii and co-workers, *Khim. Tekhnol. Topl. Masel*, (8), 54–56 (1981).
174. M. G. Roelofs and J. H. Jensen, *J. Phys. Chem.* **91**, 3380–3382 (1987).
175. M. G. Roelofs, E. Wasserman, and J. H. Jensen, *J. Am. Chem. Soc.* **109**, 4207–4217 (1987).
176. C. C. Hobbs and co-workers, *Ind. Eng. Chem. Prod. Res. Dev.* **11**, 220 (1972).
177. C. C. Hobbs and M. B. Lakin, in J. J. McKetta and W. A. Cunningham, eds., *Encyclopedia of Chemical Processing and Design*, Vol. 26, Marcel Dekker, Inc., 1987, pp. 351–373.

178. Ref. 26, p. 24.

179. K. J. Laidler, *J. Chem. Educ.* **49**, 343 (1972).

180. J. F. Black, *J. Am. Chem. Soc.* **100**, 527 (1978).

181. W. J. de Klein and E. C. Kooyman, *J. Catal.* **4**, 626 (1965).

182. I. I. Chuev and E. I. Zhilova, *Isv. Vyssh. Uchebn. Zaved., Khim. Kim. Technol.* **28**(1), 116–117 (1985).

183. J. Skriniarova, M. Hronec, and J. Ilavsky, *Oxid. Commun.* **10**(1–2), 51–68 (1987).

184. S. P. Prokopchuk, S. S. Abadzhev, and V. U. Shevchuk, *Ukr. Khim. Zh.* **49**(5), 505–508 (1983).

185. R. M. Dessau and E. I. Heiba, *Colloq. Int. C.N.R.S.* **278**, 271–274 (1978).

186. L. B. Levy, *J. Org. Chem.* **54**, 253–254 (1989).

187. N. Ikeda and K. Fukuzumi, *J. Am. Oil Chem. Soc.* **54**(3), 105 (1977).

188. U.S. Pat. 3,719,706 (Mar. 6, 1973), J. C. Brunie and co-workers (to Rhône-Poulenc SA).

189. Brit. Pat. 1,304,785 (Dec. 12, 1969), J. C. Brunie and co-workers (to Rhône-Poulenc SA).

190. U.S. Pat. 4,042,630 (Aug. 16, 1977), J. Wolters and J. L. J. P. Hennekens (to Stamicarbon B.V.).

191. Brit. Pat. 1,535,869 (Dec. 13, 1978), W. Voskuil and J. J. M. Van der Donck (to Stamicarbon BV).

192. S. V. Krylova and co-workers, *Oxid. Commun.* **10**(3–4), 243–254 (1987).

193. K. K. Sengupta, T. Samanta, and S. N. Basu, *Tetrahedron* **42**(2), 681–685 (1986).

194. O. A. Borislavskii and co-workers, *Vestn. L'vov. Politekh. Inst.* **211**, 120–122 (1988).

195. I. I. Korsak and co-workers, *Vestsi Akad. Navuk BSSR, Ser. Khim. Navuk*, (5), 37–40 (1981).

196. J. K. Kochi, *Pure Appl. Chem.* **4**, 377 (1971).

197. Ger. Offen. 2,704,077 (Aug. 11, 1977), D. W. Edwards and G. H. Jones (to Imperial Chemical Industries, Ltd.).

198. Ger. Offen. 3,029,700 (Feb. 26, 1981), C. C. Hobbs and H. H. Thigpen (to Celanese Corp.).

199. Ger. Offen. 2,757,222 (July 5, 1979), H. Erpenbach and co-workers, (to Hoechst AG).

200. Jpn. Kokai Tokkyo Koho 78 112,804 (Oct. 2, 1978), T. Maki (to Mitsubishi Chemical Industries Co., Ltd.).

201. N. V. Aleksandrov, *Kinet. Katal.* **19**(4), 1057–1060 (1978).

202. H. Yukawa and Y. Fujikawa, *Kenkyu Hokoku-Kanagawa-ken Kogyo Shikensho*, (52), 63–68 (1982).

203. D. A. Hutchison and J. L. Thompson, *Lubr. Eng.* **46**(7), 467–473 (1990).

204. A. W. Chester, E. J. Y. Scott, and P. S. Landis, *J. Catal.* **46**(3), 308–319 (1977).

205. J. E. Lyons, in B. E. Leach, ed., *Applied Industrial Catalysis*, Vol. 3, Academic Press, Inc., Orlando, Fla., 1984; Chapt. 6, pp. 131–214.

206. R. A. Sheldon, *Stud. Surf. Sci. Catal.* **66**, 573–594 (1991).

207. R. A. Sheldon, *Stud. Surf. Sci. Catal.* **59**, 33–54 (1991).

208. A. Ya. Yuffa and co-workers, *Homogeneous Heterog. Catal.* **5**, 727–739 (1988).

209. B. J. Hwang and T. C. Chou, *Ind. Eng. Chem. Res.* **26**(6), 1132–1140 (1987).

210. M. Hronec and Z. Hrabe, *Ind. Eng. Chem. Prod. Res. Dev.* **25**(2), 257–261 (1986).

211. Can. Pat. 1,058,637 (Oct. 16, 1973), B. W. Kiff and J. B. Saunby (to Union Carbide Corp.).

212. *World Petrochemicals Program, Ethylene and Derivatives 1992*, Vol. 5, private report from SRI International, Menlo Park, Calif., Jan. 1992, pp. UNIT-15 to UNIT-20.

213. Ref. 34, p. 631.

214. J. B. Saunby and B. W. Kiff, *Hydrocarbon Process* **55**(11), 247–252 (1976).

215. Jpn. Pat. 75 22,531 (July 31, 1975), K. Hori and co-workers (to Daicel Ltd.).

216. R. K. Jensen and co-workers, *J. Am. Chem. Soc.* **101**(25), 7574–7584 (1979).
217. U.S. Pat. 3,644,512 (Feb. 22, 1972), A. Onopchenko, J. G. D. Schulz, and R. Seekircher (to Gulf Research & Development Co.).
218. U.S. Pat. 4,032,570 (June 28, 1977), J. C. D. Schulz and R. Seekircher (to Gulf Research & Development Co.).
219. A. Onopchenko and J. G. D. Schulz, *J. Org. Chem.* **38**(5), 909–912 (1973).
220. M. U. Fedurtsa and S. S. Abadzhev, *Khim. Prom-st., Ser.: Khlornaya Prom-st.*, (5), 11–12 (1982).
221. Fr. Pat. 1,500,728 (Nov. 3, 1967), H. R. Grane (to Atlantic Richfield Co.).
222. Belg. Pat. 718,070 (July 12, 1968), H. R. Grane and L. S. Bitar (to Atlantic Richfield Co.).
223. U.S. Pat. 3,907,902 (Sept. 23, 1975), H. R. Grane (to Atlantic Richfield Co.).
224. *Propylene Oxide, Report 2b*, private report from SRI International's Process Economics Program, Menlo Park, Calif., Feb., 1971.
225. *World Petrochemicals Program, Propylene and Derivatives 1992*, Vol. 3, private report from SRI International, Menlo Park, Calif., Jan. 1992, pp. UNIT-164 to UNIT-170.
226. *Fatty Acids, Report 42*, private report from SRI International's Economic Program, Menlo Park, Calif., Oct. 1968.
227. W. Bao, *Riyong Huaxue Gongye*, (3), 135–137 (1991).
228. Ref. 213, p. 251.
229. Ger. Offen. 2,044,461 (Mar. 23, 1972), M. S. Furman.
230. U.S. Pat. 3,987,100 (Oct. 19, 1976), W. J. Barnette, D. L. Schmitt, and J. O. White (to E. I. du Pont de Nemours & Co., Inc.).
231. U.S. Pat. 3,530, 185 (Sept. 22, 1970), K. Pugi (to E. I. du Pont de Nemours & Co., Inc.).
232. K. Tanaka, *Chem. Technol.* **4**(9), 555 (1974).
233. A. Castellan, J. C. J. Bart, and S. Cavallaro, *Catal. Today* **9**(3), 237–254 (1991).
234. A. Onopchenko and J. G. D. Schulz, *J. Org. Chem.* **38**(21), 3729–3733 (1973).
235. U.S. Pat. 3,231,608 (Jan. 25, 1966), J. Kollar (to Gulf Research and Development Co.).
236. U.S. Pat. 4,032,569 (June 28, 1977), A. Onopchenko and J. G. D. Schulz (to Gulf Research and Development Co.).
237. A. Onopchenko and J. G. D. Schulz, *J. Org. Chem.* **38**, 3729 (1973).
238. U.S. Pat. 4,263,453 (Dec. 10, 1979), J. G. D. Schulz and A. Onopchenko (to Gulf Research and Development Co.).
239. H. S. Bryant and co-workers, *Chem. Eng. Prog.* **67**(9), 69 (1971).
240. W. H. Starnes, Jr., *J. Org. Chem.* **31**, 1436 (1966).
241. J. March, *Advanced Organic Chemistry*, 3rd ed., John Wiley & Sons, Inc., New York, 1985, p. 244.
242. G. Kohnstam and D. L. H. Williams, in Saul Patai, ed., *The Chemistry of Carboxylic Acids and Esters*, Wiley-Interscience, New York, 1969, Chapt. 16, p. 782.
243. I. M. Borisov, Yu. S. Zimin, and V. S. Martem'yanov, *Izv. Vyssh. Uchebn. Zaved., Khim. Khim. Tekhnol.* **34**(10), 46–49 (1992).
244. W. Partenheimer and J. A. Kaduk, *Stud. Surf. Sci. Catal.* **66**, 613–621 (1991).
245. Jpn. Kokai Tokkyo Koho JP 03 184,931 (Aug. 12, 1991), T. Maki, T. Masuyama, and T. Yokoyama (to Mitsubishi Kasei Corp.).
246. H. Buenger, *Compend.-Dtsch. Ges. Mineraloelwiss. Kohlechem.* **78–79**(1), 417–436 (1978).
247. D. E. Van Sickle, *Ind. Eng. Chem. Res.* **27**(3), 440–447 (1988).
248. P. Raghavendrachar and S. Ramachandran, *Ind. Eng. Chem. Res.* **31**(2), 453–462 (1992).

249. U.S. Pat. 2,833,816 (May 6, 1958), A. Saffer and R. S. Barker (to Mid-Century Corp.).
250. D. A. S. Ravens, *Trans. Fr. Soc.* **55**, 1768 (1959).
251. Brit. Pat. 1,063,451 (Mar. 30, 1967), W. J. Zimmerschied, D. E. Hanneman, and C. Serres, Jr. (to Standard Oil, Indiana).
252. Y. Ichikawa and Y. Takeuchi, *Hydrocarbon Process.* **51**(11), 103 (1972).
253. Jpn. Kokai 49-26240 (Mar. 8, 1974), M. Shigeyasu and co-workers (to Maruzen Oil).
254. M. Shigeyasu and T. Kitamura, *Shokubai* **20**(3), 155 (1978).
255. Jpn. Kokai 53 112,830 (Oct. 2, 1978), T. Yamaji and co-workers (to Teijin KK).
256. Jpn. Kokai 76 127,034 (Nov. 5, 1976), K. Teranishi and co-workers (to Mitsui Petrochemical Industries, Ltd.).
257. U.S. Pat. 2,653,165 (Sept. 22, 1953), I. E. Levine (to California Research Corp.).
258. U.S. Pat. 2,772,305 (Nov. 27, 1956), I. E. Levine and W. G. Toland, Jr. (to California Research Corp.).
259. Brit. Pat. 809,730 (Mar. 4, 1959), (to Imhausen Werke GmbH).
260. Ger. Offen. 2,310,824 (July 25, 1974), S. Takeda and co-workers (to Teijin Hercules Chemical).
261. Brit. Pat. 1,313,083 (Sept. 6, 1971), T. Harada and co-workers (to Teijin Hercules Chemical).
262. *Chemical Product Synopsis*, Mannsville Chemical Products, New York, Nov. 1978.
263. *Terephthalic Acid and Dimethyl Terephthalate*, report no. 75-1, Chem. Systems, Inc., New York, July 1975.
264. *Terephthalic Acid and Dimethyl Terephthalate, Report 9*, private report from SRI International's Process Economics Program, Menlo Park, Calif., Feb. 1966, Sept. 1970, and Aug. 1976.
265. W. F. Brill, *Ind. Eng. Chem.* **52**(10), 837–840 (1960).
266. H. S. Bryant and co-workers, *Chem. Eng. Prog.* **67**(9), 69 (1971).
267. U.S. Pat. 2,853,514 (Sept. 23, 1958), W. F. Brill (to Olin Mathieson Chemical Corp.).
268. U.S. Pat. 3,036,122 (May 22, 1962), A. E. Ardis and co-workers (to Olin Mathieson Chemical Corp.).
269. Brit. Pat. 1,129,398 (Oct. 2, 1968), (to Mobil Chem. Co.).
270. J. D. Behun, *Prepr. Div. Pet. Chem., Am. Chem. Soc.* **19**, 698 (1974).
271. U.S. Pat. 3,850,981 (Nov. 26, 1974), J. C. Trebellas, H. H. Thigpen, and W. M. Mays (to Celanese Corp. of America).
272. Fr. Pat. 1,348,386 (Aug. 10, 1962), B. Thompson and S. D. Neely (to Eastman Kodak Co.).
273. Jpn. Kokai 53 034,739 (Mar. 31, 1978), H. Torigata, M. Suematsu, and K. Nakaoka (to Toray Inds., Inc.).
274. K. Nakaoka and co-workers, *Ind. Eng. Chem. Prod. Res. Dev.* **12**, 150 (1973).
275. U.S. Pat. 3,215,733 (Nov. 2, 1975), A. F. MacLean, C. C. Hobbs, and A. L. Stautzenberger (to Celanese Corp. of America).
276. Fr. Pat. 1,549,026 (Dec. 6, 1968), E. R. Witt, K. Y. Zee-Cheng, and J. P. Cave (to Celanese Corp. of America).
277. G. A. Galstyan, V. A. Yakobi, and M. M. Dvortsevoi, *Neftekhimiya* **16**, 465 (1976).
278. O. Kobayashi and co-workers, *Nippon Kagaku Kaishi* **2**, 320 (1975).
279. J. S. Bawa and co-workers, *Proc. Natl. Symp. Catal.* **4**, 319–330 (1978).
280. U.S. Pat. 5,004,830 (Apr. 2, 1991), C. M. Park and W. P. Schammel (to Amoco Corp.).
281. G. N. Kogel and co-workers, *Zh. Org. Khim.* **24**(7), 1499–1504 (1988).
282. C. F. Hendriks, H. C. A. van Beek, and P. M. Heertjes, *Ind. Eng. Chem. Prod. Res. Dev.* **17**(3), 256–260 (1978).
283. U.S. Pat. 4,537,978 (Aug. 27, 1985), P. H. Kilner and co-workers (to Standard Oil, Indiana).
284. Jpn. Kokai Tokkyo Koho JP 61 27,942 (Feb. 7, 1986), M. Suematsu, S. Otoma, and K. Nakaoka (to Toray Industries, Inc.).

285. China Pat. CN 85,103,212 (Jan. 17, 1987), Z. Mi and co-workers (to Liming Chemical Institute).
286. Jpn. Kokai Tokkyo Koho JP 03 271,249 (Dec. 3, 1991), M. Yoshimizu, H. Takeuchi, and T. Kamei (to Nippon Steel Co., Ltd.).
287. Eur. Pat. Appl. EP 317,884 (May 31, 1989), F. Rohrscheid, G. Siegemund, and J. Lau (to Hoechst AG).
288. Jpn. Kokai 76 108,030 (Sept. 25, 1976), T. Aoyama and F. Ishihara (to Mitsubishi Gas Chemical Co., Inc.).
289. Jpn. Kokai Tokkyo Koho JP 61 24,541 (Feb. 3, 1986), S. Naito and K. Abe (to Mitsubishi Gas Chemical Co., Inc.).
290. H. V. Borgaonkar, S. R. Raverkar, and S. B. Chandalia, *Ind. Eng. Chem. Prod. Res. Dev.* **23**(3), 455–458 (1984).
291. R. A. Sheldon and N. De Heij, *Stud. Org. Chem. (Amsterdam)* **33**, 243–256 (1988).
292. K. Shimizu and co-workers, *Sekiyu Gakkaishi* **25**(1) 7–15 (1982).
293. Jpn. Kokai Tokkyo Koho JP 03,190,837 (Aug. 20, 1991), T. Yokoyama, N. Matsuyama, and T. Maki (to Mitsubishi Kasei Corp.).
294. N. Kitajima and co-workers, *Bull. Chem. Soc. Jpn.* **61**(3), 1035–1037 (1988).
295. Jpn. Kokai Tokkyo Koho 80 07,235 (Jan. 19, 1980), H. Imamura and K. Onisawa (to Agency of Industrial Sciences and Technology; Sanko Chemical Industry Co., Ltd.).
296. J. Imamura, *Yuki Gosei Kagaku Kyokaishi* **37**(8), 667–677 (1979).
297. N. N. Basaeva and co-workers, *Osnovn. Org. Sint. Neftekhim.* **6**, 11–16 (1976).
298. A. Onopchenko, J. G. D. Schultz and R. Seekircher, *J. Org. Chem.* **37**(9), 1414–1417 (1972).
299. W. W. Kaeding, *Hydrocarbon Process.* **43**(11), 173 (1964).
300. U.S. Pat. 2,727,926 (Dec. 20, 1955), W. W. Kaeding, R. O. Lindblom, and R. G. Temple (to The Dow Chemical Co.).
301. W. W. Kaeding and co-workers, *Ind. Eng. Chem. Proc. Des. Dev.* **4**(1), 97 (1965).
302. Belg. Pat 868,779 (July 5, 1978), (to Panclor Chemicals).
303. Ger. Offen 2,101,852 (July 29, 1971), L. L. Van Dierendonck, L. Laurentius, and J. P. H. Von den Hoff (to Stamicarbon NV).
304. L. Giuffre and co-workers, *Gazz. Chim. Ital.* **111**(5-6), 227–229 (1981).
305. Fr. Pat. 2,390,413 (Jan. 12, 1979), J. H. Atkinson (to ICI Chem. Ind. Ltd.).
306. Ger. Offen. 2,820,394 (Nov. 23, 1978), J. H. Atkinson (to Imperial Chemical Industries, Ltd.).
307. Ger. Offen. 2,844,195 (Oct. 11, 1978), A. P. Gelbein and A. M. Khonsari (to Lummus Co.).
308. A. P. Gelbein and A. S. Nislick, *Hydrocarbon Process.* **57**(11), 125–128 (1978).
309. M. Stolcova and co-workers, *J. Catal.* **101**(1), 153–161, (1986).
310. M. Hronec and co-workers, *Appl. Catal.* **69**(2), 201–204 (1991).
311. Jpn. Kokai Tokkyo Koho JP 63 27,473 (Feb. 5, 1988), M. Matsuno, K. Yoshiga, and K. Nakamura (to Mitsui Petrochemical Industries, Ltd.).
312. Jpn. Kokai Tokkyo Koho JP 60 84,235 (May 13, 1985), (to Mitsui Petrochemical Industries, Ltd.).
313. Jpn. Kokai Tokkyo Koho JP 60 89,440 (May 20, 1985), (to Mitsui Petrochemical Industries, Ltd.).
314. U.S. Pat. 4,503,262 (Mar. 5, 1985), B. F. Gupton and E. D. Little (to Virginia Chemicals, Inc.).
315. A. Tomita, *Numazu Kogyo Koto Semmon Gakko Kenkyu Hokoku* **13**, 49–59 (1978).
316. Kh. E. Kharlampidi and co-workers, *Khim. Tekhnol. Pererab. Nefti Gaza*, 10–13 (1985).
317. U.S. Pat. 3,459,810 (Aug. 5, 1969), C. Y. Choo and R. L. Golden (to Halcon International).

318. Neth. Pat. Appl. 75 13,860 (Mar. 31, 1976), (to Halcon International).

319. *World Petrochemicals Program, Aromatics and Derivatives 1992*, Vol. 5, private report from SRI International, Menlo Park, Calif., Jan. 1992, pp. UNIT-122 to UNIT-132.

General References

C. Walling, *Free Radicals in Solution*, John Wiley & Sons, Inc., New York, 1957.

F. R. Mayo, *Acc. Chem. Res.* **1**(7), 193–201 (1968).

S. W. Benson and P. S. Nangia, *Acc. Chem. Res.* **12**(7), 223–228 (1979).

S. W. Benson, *Thermochemical Kinetics*, 2nd ed., John Wiley & Sons, Inc., New York, 1976.

R. A. Sheldon and J. K. Kochi, *Metal-Catalyzed Oxidations of Organic Compounds*, Academic Press, Inc., New York, 1981.

R. W. Walker, *Sci. Prog. (Oxford)* **74**(294, Pt. 2), 163–187 (1990).

J. E. Lyons, in B. E. Leach, ed., *Applied Industrial Catalysis*, Vol. 3, Academic Press, Inc., Orlando, Fla., 1984, Chapt. 6, pp. 131–214.

CHARLES C. HOBBS
Consultant, sponsored by
Hoechst Celanese Corporation

HYDROCARBON RESINS

Hydrocarbon resin is a broad term that is usually used to describe a low molecular weight thermoplastic polymer synthesized via the thermal or catalytic polymerization of coal-tar fractions, cracked petroleum distillates, terpenes, or pure olefinic monomers. These resins are used extensively as modifiers in the hot melt and pressure sensitive adhesive industries. They are also used in numerous other applications such as sealants, printing inks, paints, plastics, road marking, carpet backing, flooring, and oil field applications. They are rarely used alone.

Typical hydrocarbon resins range in appearance from hard, brittle solids to viscous liquids. They may come in flakes, pellets, drums, or in molten form. Depending on application requirements, many resins are available as solutions in organic solvents or oils. Anionic, cationic, or nonionic emulsion forms are also manufactured. Hydrocarbon resins typically have a number average molecular weight (M_n) of less than 2000. The colors of these resins range from water-white to dark brown. Water-white resins usually are produced from the Lewis acid polymerization of pure olefinic monomers or by the hydrogenation of catalytically or thermally produced precursors. Colors are determined on the Gardner and Saybolt scales.

The first resins to be produced on a commercial scale were the coumarone–indene or coal-tar resins (1); production in the United States was started before 1920. These resins were dominant until the development of petroleum resins, which were established as important raw materials by the mid-1940s. Continued development of petroleum-based resins has led to a wide variety of aliphatic, cyclodiene, and aromatic hydrocarbon-based resins. The principal components of petroleum resins are based on piperylenes, dicyclopentadiene (DCPD), styrene, indene, and their respective alkylated derivatives.

With the improvement of refining and purification techniques, many pure olefinic monomers are available for polymerization. Under Lewis acid polymerization, such as with boron trifluoride, very light colored resins are routinely produced. These resins are based on monomers such as styrene, α-methylstryene, and vinyltoluene (mixed *meta*- and *para*-methylstyrene). More recently, purified *para*-methylstyrene has become commercially available and is used in resin synthesis. Low molecular weight thermoplastic resins produced from pure styrene have been available since the mid-1940s; resins obtained from substituted styrenes are more recent.

Terpene-based hydrocarbon resins are typically based on natural products such as α-pinene, β-pinene, and *d*-limonene [5989-27-5], which are obtained from the wood and citrus industries, respectively. These resins, which were originally the preferred tackifiers for natural rubber applications, possess similar properties to aliphatic petroleum resins, which were developed later. Terpene-based resins have been available since the mid-1930s and are primarily used in the adhesives industry.

Physical Properties

Most hydrocarbon resins are composed of a mixture of monomers and are rather difficult to fully characterize on a molecular level. The characteristics of resins are typically defined by physical properties such as softening point, color, molecular weight, melt viscosity, and solubility parameter. These properties predict performance characteristics and are essential in designing resins for specific applications. Actual characterization techniques used to define the broad molecular properties of hydrocarbon resins are Fourier transform infrared spectroscopy (ftir), nuclear magnetic resonance spectroscopy (nmr), and differential scanning calorimetry (dsc).

Softening Point. Softening point is the temperature at which a material shows a measurable flow or softening under a given weight. This property is usually measured by a standard ring and ball method and may be done manually (ASTM E28–E67) or automatically using an instrument such as the HERZOG Automatic Ring and Ball Tester MC-753 (ASTM D36–D2398). A typical procedure involves the pouring of a molten resin sample into a brass ring of defined diameter and thickness. The resin is then allowed to cool. A ball of a specific weight is then supported on the resin-filled ring and the apparatus is heated in a glycerol or water solvent until flow is detected.

Color. Colors of hydrocarbon resins are routinely determined on the Gardner scale (ASTM D154) or on the Saybolt scale (ASTM D156). Saybolt colors are determined from 2 g of resin in 18 g of toluene (10% solution) and are used for very light color to water-white resins. The Saybolt scale ranges from -30 to 30 with the latter being water-white. Gardner colors are determined as a 50% solution in toluene and are used for resins ranging in color from light yellow to dark brown. The Gardner scale ranges from 1 to 18 with the latter being dark brown. Molten resin colors are also reported in Gardner units. Aged color (or heat stability) is also an important property of hydrocarbon resins, which is determined by the high temperature oven aging of resins to predict thermal and oxidative stability.

Melt Viscosity.   Viscosities of resins at standard temperatures yield information about molecular weight and molecular weight distribution, as well as valuable information with respect to application logistics. Some customers prefer to receive resins in molten form. Melt viscosities help to determine the required temperature for a resin to be pumpable. Temperature–viscosity profiles are routinely supplied to customers by resin manufacturers. In general, a molten viscosity of 1–1.1 Pa·s (1000–1100 cP) or less at process temperatures is convenient for the pumping and handling of molten resin.

Solubility Parameter.   Compatibility between hydrocarbon resins and other components in an application can be estimated by the Hildebrand solubility parameter (2). In order for materials to be mutually soluble, the free energy of mixing must be negative (3). The solubility of a hydrocarbon resin with other polymers or components in a system can be approximated by the similarities in the solubility parameters of the resin and the other materials. True solubility parameters are only available for simple compounds and solvents. However, parameters for more complex materials can be approximated by relative solubility comparisons with substances of known solubility parameter.

Molecular Weight.   The determination of the molecular weight and molecular weight distribution of a hydrocarbon resin is very important for the prediction of performance parameters. Molecular weights are routinely determined using gel permeation (size exclusion) chromatography, which is commonly referred to as gpc or sec (4). Gpc quickly provides the relative weight average (M_w), number average (M_n), and Z-average (M_z) molecular weights, as well as the polydispersity (M_w/M_n) of a polymer or resin. In predicting solubility and compatibility with other polymers, the Z-average molecular weight has proven very useful. Other methods for determining the absolute molecular weight of polymers are light scattering (M_w) and vapor-phase osmometry (M_n). Solvent–nonsolvent separations of hydrocarbon resins have also been described (5–7).

Characterization.   In many cases, ftir is a timely and cost-effective method to identify and quantify certain functionalities in a resin molecule. Based on developed correlations, ftir is routinely used as an efficient method for the analysis of resin aromaticity, olefinic content, and other key functional properties. Near infrared spectroscopy is also quickly becoming a useful tool for on-line process and property control.

For most hydrocarbon resins where numerous structurally different monomers are polymerized, nmr is typically used as a general tool to quantify the aromatic and/or olefinic content of a resin. In conjunction with gpc and ftir, nmr measurements are used to identify and quantify particular functionalities or monomers present in hydrocarbon resins.

As applied to hydrocarbon resins, dsc is mainly used for the determination of glass-transition temperatures (T_g). Information can also be gained as to the physical state of a material, ie, amorphous vs crystalline. As a general rule of thumb, the T_g of a hydrocarbon resin is approximately 50°C below the softening point. Oxidative induction times, which are also determined by dsc, are used to predict the relative oxidative stability of a hydrocarbon resin.

Polymerization

Most commercial hydrocarbon resins produced from olefinic feedstocks are synthesized via carbocationic polymerization (8). Very similar catalyst systems are used in the synthesis of coumarone–indene, petroleum, terpene, and pure monomer-based resins (9). The first catalyst systems used were strong mineral acids, specifically sulfuric acid [8014-95-7] and phosphoric acid [766-38-2]. These acids typically afforded dark colored resins with very low softening points and molecular weights. More recently, resin synthesis technology has focused on the use of Lewis acids and Friedel-Crafts polymerization chemistry. Typical Lewis acids used in the synthesis of hydrocarbon resins are AlCl$_3$ [7446-70-7], BF$_3$ [7637-07-2], (CH$_3$CH$_2$)$_2$AlCl [96-10-6], (CH$_3$CH$_2$)AlCl$_2$ [563-43-9], and complexes of these acids with various electron-donating species (see FRIEDEL-CRAFTS REACTIONS).

Friedel-Crafts (Lewis) acids have been shown to be much more effective in the initiation of cationic polymerization when in the presence of a cocatalyst such as water, alkyl halides, and protic acids. Virtually all feedstocks used in the synthesis of hydrocarbon resins contain at least traces of water, which serves as a cocatalyst. The accepted mechanism for the activation of boron trifluoride in the presence of water is shown in equation 1 (10). Other Lewis acids are activated by similar mechanisms. In a more general sense, water may be replaced by any appropriate electron-donating species (eg, ether, alcohol, alkyl halide) to generate a cationic intermediate and a Lewis acid complex counterion.

$$BF_3 + H_2O \rightleftharpoons H^+ \, BF_3OH^- \tag{1}$$

The polymerization of monomers to form hydrocarbon resins is typically carried out by either the direct addition of catalyst to a hydrocarbon fraction or by the addition of feed to a solvent–catalyst slurry or solution. Most commercial manufacturers use a continuous polymerization process as opposed to a batch process. Reactor temperatures are typically in the range of 0–120°C.

In order to facilitate heat transfer of the exothermic polymerization reaction, and to control polymerizate viscosity, percent reactives are adjusted through the use of inert aromatic or aliphatic diluents, such as toluene or heptane, or higher boiling mixed aromatic or mixed aliphatic diluents. Process feed streams are typically adjusted to 30–50% polymerizable monomers.

Polymerizations are typically quenched with water, alcohol, or base. The resulting polymerizates are then distilled and steam and/or vacuum stripped to yield hard resin. Hydrocarbon resins may also be precipitated by the addition of the quenched reaction mixture to an excess of an appropriate poor solvent. As an example, aliphatic C-5 resins are readily precipitated in acetone, while a more polar solvent such as methanol is better suited for aromatic C-9 resins.

Coumarone–Indene or Coal-Tar Resins

Coumarone–indene or coal-tar resins, as the name denotes, are by-products of the coal carbonization process (coking). Although named after two particular components of these resins, coumarone (**1**) and indene (**2**), these resins are actually produced by the cationic polymerization of predominantly aromatic feedstreams. These feedstreams are typically composed of compounds such as indene, styrene, and their alkylated analogues. In actuality, there is very little coumarone in this type of feedstock. The fractions used for resin synthesis typically boil in the range of 150–250°C and are characterized by gas chromatography.

(**1**)   (**2**)

Cationic polymerization of coal-tar fractions has been commercially achieved through the use of strong protic acids, as well as various Lewis acids. Sulfuric acid was the first polymerization catalyst (11). More recent technology has focused on the Friedel-Crafts polymerization of coal fractions to yield resins with higher softening points and better color. Typical Lewis acid catalysts used in these processes are aluminum chloride, boron trifluoride, and various boron trifluoride complexes (12). Crude feedstocks typically contain 25–75% reactive components and may be refined prior to polymerization (eg, acid or alkali treatment) to remove sulfur and other undesired components. Table 1 illustrates the typical components found in coal-tar fractions and their corresponding properties.

As a result of incorporating rigid monomers such as indene and its alkylated derivatives, resin softening points of 150°C and greater are obtainable (13,14).

Table 1. Properties of Coal-Tar Monomers

Monomer	CAS Registry Number	Bp, °C	Density, g/mL at 20°C	n_D^{20}	Fp, °C
cyclopentadiene	[26912-33-4]	41–42	0.8041	1.4461	−85
dicyclopentadiene	[77-73-6]	170	0.988	1.512	−32.9
styrene	[100-42-5]	145.2	0.9059	1.5467	−30.6
o-methylstyrene	[611-15-4]	169.8	0.9036	1.5465	−68.6
m-methylstyrene	[100-80-1]	171.6	0.9113	1.5439	−86.3
p-methylstyrene	[622-97-9]	172.8	0.9106	1.5450	−34.2
coumarone	[271-89-6]	171.4	1.0948	1.5663	−28.9
2-methylcoumarone	[4265-25-2]	197	1.0534	1.5584	
indene	[95-13-6]	182.6	0.9960	1.5768	−1.6
2-methylindene	[2177-47-1]	187–199	0.9947	1.5723	

Colors typically range between 8–12 Gardner. Although much of the market has been replaced by hydrocarbon resins, coumarone–indene resins are employed in coatings, flooring, tiles, printing inks, and adhesives. They are also utilized as processing aids and pigment dispersing agents in the compounding of natural and synthetic rubber.

Petroleum Resins

Petroleum resins are low molecular weight thermoplastic hydrocarbon resins synthesized from steam cracked petroleum distillates. These resins are differentiated from higher molecular weight polymers such as polyethylene and polypropylene, which are produced from essentially pure monomers. Petroleum resin feedstocks are composed of various reactive and nonreactive aliphatic and aromatic components. The resins are usually classified as C-5 (aliphatic), C-9 (aromatic), C-5/C-9 (aromatic modified aliphatic), and cycloaliphatic diene-based resins. These range from viscous liquids to hard, friable solids, with colors ranging from water-white to dark brown. Petroleum resins are widely used in pressure sensitive and hot melt adhesives, as well as in sealants, inks, paints, coatings, plastics, carpet backing, road marking paints, and oil field applications.

Raw Materials. During the 1940s, while improved techniques for petroleum refining were being developed, the presence of reactive olefins was undesirable in aromatic solvents. In order to improve the purity of these aromatic fractions, methods were developed to remove reactive olefins and diolefins through the catalytic polymerization of these species using protic (15) and Lewis acids (16). During the same period of time, extensive surveys were being carried out to determine the best methods for cracking crude oil. These cracking methods were aimed at obtaining high yields of ethylene and propylene, as well as optimizing the formation of reactive monomers such as butadiene [106-99-0], isoprene [78-79-5], piperylene, styrene, and indene (17).

The feedstocks used in the production of petroleum resins are obtained mainly from the low pressure vapor-phase cracking (steam cracking) and subsequent fractionation of petroleum distillates ranging from light naphthas to gas oil fractions, which typically boil in the 20–450°C range (16). Obtained from this process are feedstreams composed of aliphatic, aromatic, and cycloaliphatic olefins and diolefins, which are subsequently polymerized to yield resins of various compositions and physical properties. Typically, feedstocks are divided into aliphatic, cycloaliphatic, and aromatic streams. Table 2 illustrates the predominant olefinic hydrocarbons obtained from steam cracking processes for petroleum resin synthesis (18).

Aliphatic: C-5–C-6. Aliphatic feedstreams are typically composed of C-5 and C-6 paraffins, olefins, and diolefins, the main reactive components being piperylenes (*cis*- [1574-41-0] and *trans*-1,3-pentadiene [2004-70-8]). Other main compounds include substituted C-5 and C-6 olefins such as cyclopentene [142-29-0], 2-methyl-2-butene [513-35-9], and 2-methyl-2-pentene [625-27-4]. Isoprene and cyclopentadiene may be present in small to moderate quantities (2–10%). Most steam cracking operations are designed to remove and purify isoprene from the C-5–C-6 fraction for applications in rubbers and thermoplastic elas-

Table 2. Unsaturated Monomers Obtained From Steam Cracking[a]

Component	Boiling range, °C
Olefins	
pentenes	20–40
hexenes	41–73
heptenes	72–98.5
Diolefins	
pentadienes	34–48
hexadienes	59–80
Cycloaliphatic olefins and diolefins	
cyclopentene	44
cyclopentadiene	41.5
cyclohexene	83
methylcyclopentadiene	73
Cycloaliphatic diolefin dimers	
dicyclopentadiene	170
methylcyclopentadiene dimer	200
Vinyl aromatic hydrocarbons	
styrene	145.2
α-methylstyrene	164
vinyltoluenes	166–170
indene	182.6
methylindenes	187–199

[a] Ref. 18.

tomers. Cyclopentadiene is typically dimerized to dicyclopentadiene (DCPD) and removed from C-5 olefin–diolefin feedstreams during fractionation (19).

Cycloaliphatic Diene: CPD–DCPD. Cycloaliphatic diene-based hydrocarbon resins are typically produced from the thermal or catalytic polymerization of cyclopentadiene (CPD) and dicyclopentadiene (DCPD). Upon controlled heating, CPD may be dimerized to DCPD or cracked back to the monomer. The heat of cracking for DCPD ($\Delta H_{cracking}$) is 24.6 kJ/mol (5.88 kcal/mol). In steam cracking processes, CPD is removed from C-5 and higher fractions by heat soaking to form the dimer, which is separable by controlled fractional distillation (18) (see CYCLOPENTADIENE AND DICYCLOPENTADIENE).

DCPD of varying purity is produced and used in the synthesis of hydrocarbon resins. Typically, cyclic feedstocks have DCPD concentrations ranging from 30–80%, although high purity (>95%) DCPD is available on a commercial scale and is sometimes used.

Aromatic. Aromatic feedstreams (C-8, C-9, C-10) derived from the steam cracking of petroleum distillates are composed of styrene, indene, vinyltoluenes (eg, *meta*- and *para*-methylstyrene), and their respective alkylated analogues. A typical aromatic feedstream might contain 50% reactive olefins with the remainder being alkylated benzenes and higher aromatics.

C-5 Aliphatic Petroleum Resins. Carbocationic polymerization of C-5 feedstreams has been accomplished with various Friedel-Crafts catalyst systems. Table 3 compares the efficiencies of selected Lewis acids in the polymerization of a typical C-5 stream containing 43 wt % C-5–C-6 diolefins and 47 wt %

Table 3. Comparative Efficiencies of Friedel-Crafts Catalysts[a,b]

Parameter	AlCl$_3$	AlBr$_3$	BF$_3$ etherate	BF$_3$ etherate/H$_2$O	TiCl$_4$	SnCl$_4$	AlCl$_3$/HCl/ o-xylene	DDB[c] sludge
concentration								
wt %	0.75	1.50	0.75[d]	0.38[d]	1.06	1.46	0.75[e]	0.75[e]
mmol/100 g	5.62	5.62	11.06[d]	5.62[d]	5.62	5.62		
resin								
yield, wt %	36.8	36.0	23.0	17.0	25.0	15.6	35.9	34.7
oligomer								
yield, wt %	1.9	1.5	17.4	16.0	3.4	3.4		
resin color (Gardner)	4−	4.5	2.5	4+	7	>18	6−	6+
after heating at 150°C								
3 h	8+	11+		8+	14+	>18		
16 h	14	15+	14+	14−	16+	>18		
resin softening point,								
°C	95	101	<20	<20	40	<20	107	113

[a] Ref. 20.
[b] At 50°C, 60 min.
[c] DDB = dodecylbenzene.
[d] As BF$_3$.
[e] As AlCl$_3$.

724

C-5–C-6 olefins (20). Based on weight percent yield of resin at equimolar concentrations of catalyst (5.62 mmol/100 g), efficiency follows $AlCl_3 \approx AlBr_3 > BF_3$ etherate–$H_2O > TiCl_4 > SnCl_4$. The most commonly used catalyst in petroleum resin synthesis is $AlCl_3$.

Although most referenced catalyst systems do not indicate the presence of water, typical feedstreams derived from steam cracking operations contain parts per million quantities of adventitious water unless they are scrupulously dried before polymerization. Water serves as a cocatalyst in Lewis acid catalyzed systems. In addition to powdered or gaseous single catalyst systems, liquid complexes containing $AlCl_3$ and anhydrous hydrochloric acid in aromatic solvents such as *ortho*-xylene have also been used and shown to exhibit similar activity based on weight percent $AlCl_3$ (20–24). $AlCl_3$ complexes with aqueous mineral acids, such as 86% H_3PO_4, have been used as well (25). Acidic liquid $AlCl_3$ sludges obtained from the alkylation of benzene and other aromatics have also been shown to have similar efficiency to $AlCl_3$. Table 3 contains an example using acidic sludge from a dodecylbenzene production facility (20).

$AlCl_3$ efficiency (based on g resin/g catalyst) can be markedly improved by polymerizing dry feeds (<10 ppm H_2O) with an $AlCl_3$/anhydrous HCl system. Proceeding from 250 ppm H_2O down to 10 ppm H_2O, catalyst efficiency improves from 30.6 to 83.0 (26). Low levels of tertiary hydrocarbyl chlorides have been shown to greatly enhance the activity of $AlCl_3$, while yielding resins with narrow molecular weight distributions relative to systems employing water or HCl (27).

The composition of feedstocks is very important for obtaining the desired properties of hydrocarbon resins such as softening point, color, and molecular weight profile. High C-5 diolefin content typically leads to higher softening points, as well as higher and broader molecular weights. These properties may be balanced by branched tertiary olefins, which typically lower softening point and lower/narrow molecular weight.

Under $AlCl_3$ catalysis, monomeric systems of greater than 85 wt % piperylenes tend to form unstable polymerization systems resulting in gel formation, which is undesirable in hydrocarbon resins (28). Softening point control in C-5 resins can be achieved by either the use of chain-transfer agents and/or through the use of alternative catalyst systems. Isobutylene [115-11-7] and branched tertiary alkenes function as chain-transfer agents due to the formation of a stable tertiary carbonium ion at the end of a growing polymer chain. This is illustrated in Figure 1 (counterions are not shown).

The $AlCl_3$ catalyzed polymerization of piperylenes in the presence of butadiene, isobutylene, and methylbutenes yields tackifying resins with softening points ranging from 25–80°C (29). When polymerized alone under $AlCl_3$ catalysis, isobutylene gives rubbery polymers, butadiene gives gummy materials, and methylbutenes form only low molecular weight residual oils (30). Due to their increased chain-transfer activity resulting in greater oligomer content, resin yields begin to decrease with greater concentrations of butadiene or isobutylene (31).

Blends of piperylenes and amylenes (mixed 2-methyl-1-butene and 2-methyl-2-butene) or UOP propylene dimers can be adjusted to produce softening points of 0–100°C and weight average molecular weights of <1200 (32,33). Careful control of the diolefin/branched olefin ratio is the key to consistent resin properties (34).

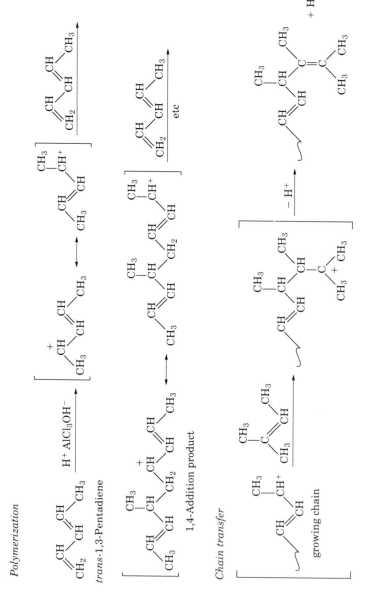

Fig. 1. Polymerization of piperylenes.

When used as chain-transfer agents in catalytic polymerizations, branched terminal olefins (eg, 2-methyl-1-butene) have been shown to suppress the softening point to a greater extent than branched internal olefins (ie, 2-methyl-2-butene) (35). Branched internal olefins, as opposed to the terminal olefins, can be used to decrease wax cloud point and molecular weight while maintaining the softening point (32). Wax cloud point is typically determined by the slow cooling of a molten blend of tackifier, wax, and polymer. The temperature at which the blend first begins to turn opaque is taken as the wax cloud point.

Due to the fact that BF_3 is a weaker Lewis acid than $AlCl_3$, structurally distinct resins are obtained upon the respective polymerization of a piperylenes–2-methyl-2-butene system with the two different Lewis acids. Much lower levels of branched olefin are required to achieve a softening point of $<40°C$ with the BF_3 catalyzed system (33,36). In fact, due to its weaker acidity, BF_3 is not useful for producing high softening point resins based on C-5 hydrocarbon feeds.

Levels of cyclopentadiene (CPD) and dicyclopentadiene (DCPD) in C-5 feedstreams have a great effect on the softening point, as well as the color and thermal stability of the resin. Typically, DCPD is added to C-5 feedblends to increase softening point. However, increased DCPD incorporation generally leads to higher color in the resin. Oligomers of CPD and alkylated analogues have been catalytically polymerized with piperylenes to yield resins with softening points greater than 200°C for printing ink applications (37).

In normal steam cracking operations, CPD levels in the C-5–C-6 hydrocarbon fractions are reduced to approximately 1–3 wt %. Studies have shown that these levels may be reduced to <0.5 wt % by heat soaking at 90–140°C followed by distillation. Heat soaking conditions of C-5 streams are usually optimized at 90–140°C to prevent dimerization of CPD with acyclic dienes. These studies have also shown that by increasing the piperylenes:CPD ratio in a C-5 feedblend from 42 to 247, the resulting resin color obtained from $AlCl_3$ improves three Gardner color units (38). Another less common method for reducing CPD levels in C-9 polymerization feedstocks is the prepolymerization of feedstocks with polar dienophiles. At initial CPD levels of <1 wt %, the resulting Diels-Alder adducts do not hinder the polymerization reaction (39).

C-5–C-9 Aromatic Modified Aliphatic Petroleum Resins. Compatibility with base polymers is an essential aspect of hydrocarbon resins in whatever application they are used. As an example, piperylene–2-methyl-2-butene based resins are substantially inadequate in enhancing the tack of 1,3-butadiene–styrene based random and block copolymers in pressure sensitive adhesive applications. The copolymerization of α-methylstyrene with piperylenes effectively enhances the tack properties of styrene–butadiene copolymers and styrene–isoprene copolymers in adhesive applications (40,41). Introduction of aromaticity into hydrocarbon resins serves to increase the solubility parameter of resins, resulting in improved compatibility with base polymers. However, the nature of the aromatic monomer also serves as a handle for molecular weight and softening point control.

Aromatic modified aliphatic resins are typically referred to as C-5–C-9 resins, although typical aromatic feedstreams contain C-8 to C-10 vinyl aromatic monomers. Aromatic monomers are typically introduced as either pure monomers or as C-8–C-10 aromatic fractions obtained from steam cracking operations. Styrene and its alkylated analogues are used to introduce aromaticity into hydrocarbon resins, although each has a different effect on softening point and molecular weight. Polymerization of C-5 diolefins and olefins in the presence of 20–40 wt % α-methylstyrene [98-83-9] produces resins with softening points in the range of 60–80°C when diolefin:olefin ratios are maintained from 0.6:1 to 1.4:1, respectively (40). Catalytic resins produced from 10–25 wt % styrene with C-5 diolefins and olefins possess similar properties, although softening points tend to be reduced to 40–60°C (42). This can be attributed to the structural differences and the differing chain-transfer mechanisms of styrene and α-methylstyrene. Higher softening point C-5–styrene resins may be obtained with vinyltoluenes or *para*-methylstyrene. Copolymerization of 5–30 wt % *p*-methylstyrene with a C-5 diolefin–olefin stream yields resins with softening points in the range of 75–120°C (43).

Although most aromatic modified C-5 resins are typically higher softening point resins, certain applications, such as adhesives, require lower softening points. Copolymerization of a C-8–C-10 vinyl aromatic fraction with piperylenes in the presence of a C-4–C-8 mono-olefin chain-transfer stream yields resins with softening points ranging from 0–40°C (44). A particular advantage of these liquid resins is the fact that they eliminate the need for plasticizers or oils in some pressure sensitive adhesive applications.

Catalysts used in the polymerization of C-5 diolefins and olefins, and monovinyl aromatic monomers, follow closely with the systems used in the synthesis of aliphatic resins. Typical catalyst systems are $AlCl_3$, $AlBr_3$, $AlCl_3$–HCl–*o*-xylene complexes and sludges obtained from the Friedel-Crafts alkylation of benzene. Boron trifluoride and its complexes, as well as $TiCl_4$ and $SnCl_4$, have been found to result in lower yields and higher oligomer content in C-5 and aromatic modified C-5 polymerizations.

The conversion of aromatic monomers relative to C-5–C-6 linear diolefins and olefins in cationic polymerizations may not be proportional to the feedblend composition, resulting in higher resin aromaticity as determined by nmr and ir measurements (43). This can be attributed to the differing reactivity ratios of aromatic and aliphatic monomers under specific Lewis acid catalysis. Intentional blocking of hydrocarbon resins into aromatic and aliphatic regions may be accomplished by sequential cationic polymerization employing multiple reactors and standard polymerization conditions (45).

C-9 Aromatic Petroleum Resins. Feedstocks typically used for aromatic petroleum resin synthesis boil in the approximate range of 100–300°C at atmospheric pressure, with most boiling in the 130–200°C range. The C-9 designation actually includes styrene (C-8) through C-10 hydrocarbons (eg, methylindene). Many of the polymerizable monomers identified in Table 1 for coumarone–indene type crudes from coal tar are also present in aromatic fractions from cracked petroleum distillates. Therefore, the technology developed for the polymerization of coal-tar crudes is also applicable to petroleum-derived aromatic feedstocks. In addition to availability, aromatic petroleum resins offer several advantages over coumarone–indene resins. These include improved color and odor, as well as uv and thermal stability (46).

Several methods have been developed for the further improvement of color and stability in aromatic resins. Thermal pretreatment of an aromatic fraction with a boiling range of 140–280°C with an α,β-unsaturated dicarboxylic acid anhydride has been found to be an effective method for color improvement. Treatment of an aromatic fraction with 8.1 wt % maleic anhydride (based on the total weight of polymerizable monomers) for 2 h at 100°C, followed by distillation and subsequent polymerization with a BF_3-phenol complex, yielded resins with color improvements of up to three Gardner units over resins produced without the prepolymerization feed treatment. Heat stability (3 h at 200°C) was also improved by four Gardner units (47). A related method involves the sequential treatment of a cracked petroleum fraction of boiling range 80–260°C with an acidic agent such as H_2SO_4 or acidic solid ion-exchange resins (eg, Amberlyst 15), followed by vacuum distillation and subsequent catalytic polymerization (48).

Catalyst systems used in the commercial synthesis of aromatic petroleum resins are very similar to those systems used in the manufacture of C-5 and C-5–C-9 type resins. The principal catalysts are $AlCl_3$ and BF_3 and their respective complexes. BF_3 catalysis usually yields resins with improved color over $AlCl_3$ catalyzed systems.

Cyclopentadiene/Dicyclopentadiene-Based Petroleum Resins. 1,3-Cyclopentadiene (CPD) is just one of the numerous compounds produced by the steam cracking of petroleum distillates. Due to the fact that DCPD is polymerized relatively easily under thermal conditions without added catalyst, resins produced from cycloaliphatic dienes have become a significant focus of the hydrocarbon resin industry.

Hydrocarbon resins based on CPD are used heavily in the adhesive and road marking industries; derivatives of these resins are used in the production of printing inks. These resins may be produced catalytically using typical carbocationic polymerization techniques, but the large majority of these resins are synthesized under thermal polymerization conditions. The rate constants for the Diels-Alder based dimerization of CPD to DCPD are well known (49). The ability to polymerize without Lewis acid catalysis reduces the amount of aluminous water or other catalyst effluents/emissions that must be addressed from an environmental standpoint. Both thermal and catalytically polymerized DCPD/CPD-based resins contain a high degree of unsaturation. Therefore, many of these resins are hydrogenated for certain applications.

Dicyclopentadiene-based feedstreams are routinely used in petroleum resin synthesis as a means of increasing softening point. High softening point resins (130–160°C) have been synthesized for adhesives and ink applications using a DCPD-based stream obtained from isoprene extraction operations. The $AlCl_3$ catalyzed polymerization of a hydrocarbon fraction containing 52% DCPD and 22.3% codimers of cyclopentadiene and C-5 diolefins, using a xylene solvent, has been shown to yield a resin having a softening point of 152°C (50). Xylene has been found to provide a lower polymerizate viscosity over toluene resulting in better heat-transfer properties and a higher softening point.

Gel formation and high levels of unsaturation are problems encountered with resins produced from the thermal polymerization (250–300°C) of high purity DCPD streams. Reduction of these undesirable properties has been accomplished by the use of acyclic conjugated dienes and codimers of CPD and acyclic dienes. Polymerization of a feedblend containing 432 g DCPD (93.5% purity), 48 g piperylenes, and 120 g xylene at 260°C under nitrogen for 3 h yielded a toluene-soluble resin with a softening point of 172°C and a color of 3–4 Gardner. Control experiments using 480 g DCPD (93.5% purity) and 120 g xylene, and the same polymerization conditions, produced a resin with a softening point of 218°C, which had a color of 5–6 Gardner and was only partially soluble in toluene (51). Monomer-to-solvent ratio has a significant effect on resin softening point and gel formation in thermal processes (52).

In addition to the use of C-4–C-5 acyclic conjugated dienes as a means of controlling resin properties, CPD codimers with conjugated acyclic dienes, specifically CPD–butadiene codimers, are also effective (53,54). Thermal polymerization of a blend of 30% DCPD (80% purity), 48% tetrahydroindene (THI) [7603-37-4], and 22% xylene, followed by removal of solvent and oligomers, produced a resin having a softening point of 86.5°C and a bromine value of 125. A comparative example using 30% DCPD and 70% xylene yielded a resin having a softening point of 129°C and a bromine value of 114 (53). Table 4 illustrates the effect of CPD–acyclic diene codimers on CPD-based polymerizations.

Distillation and stripping temperatures used for nonhydrogenated CPD-based resins are typically kept below 250°C and are usually conducted at reduced pressure (54). This is due to the high levels of unsaturation remaining in these resins and to the fact that thermal polymerization can be reinitiated at normal finishing temperatures of 250°C. Reactive, heat-reactive, or unsaturated resins react with drying oils in the range of 232–260°C to give products with improved performances in paints and varnishes.

In order to increase the solubility parameter of CPD-based resins, vinyl aromatic compounds, as well as other polar monomers, have been copolymerized with CPD. Indene and styrene are two common aromatic streams used to modify cyclodiene-based resins. They may be used as pure monomers or contained in aromatic steam cracked petroleum fractions. Addition of indene at the expense of DCPD in a thermal polymerization has been found to lower the yield and

Table 4. Effect of Cyclopentadiene–Acyclic Diene Codimers on DCPD-Based Thermal Polymerizations[a,b]

Example	1	2	3	4	5	6
DCPD, wt %	30	70	30	40	55	40
THI, wt %			48	41	30	
xylene, wt %	70	30	22	19	15	40[c]
yield, wt %	76.1	82.2	48.0	59.9	79.2	60.1
softening point,[d] °C	129.0	150.0	86.5	112.0	149.0	114.0
bromine value	114	95	125	118	101	112
color, Gardner	10	11	7	7	8	7
heat stability, °C	+15.0	+20.0	+3.0	+4.0	+4.0	+3.0

[a]Ref. 53.

[b]Polymerization conditions: 5 h at 260°C; finishing conditions: 210°C at 0.27 kPa (2 mm Hg).

[c]And 20 wt % vinylnorbornene.

[d]The increase in softening point measured after heating each resin at 220°C for 3 h. The smaller the increase, the better the heat stability.

softening point of the resin (55). Compatibility of a resin with ethylene–vinyl acetate (EVA) copolymers, which are used in hot melt adhesive applications, may be improved by the copolymerization of aromatic monomers with CPD. As with other thermally polymerized CPD-based resins, aromatic modified thermal resins may be hydrogenated.

Printing inks such as heat set or gravure inks typically employ polar resins such as rosin derivatives and alkyd resins, both of which have good affinity for pigments and solubility in drying oils. Petroleum resins obtained from the copolymerization of cyclopentadiene with polar monomers, such as unsaturated acid anhydrides or fatty acids followed by esterification, have also been shown to have comparable performance to rosin derivatives in these applications (56–61). As an example, CPD-based resins for printing ink applications have been produced by the reaction of 730 g DCPD (97% purity), 70 g maleic anhydride [108-31-6], and 190 g oleyl alcohol [143-28-2] in 300 g xylene for 5 h at 270–275°C. Distillation followed by steam stripping yielded a resin having a softening point of 135°C and an acid number of 21. In order to obtain gel-free resins when using unsaturated anhydrides such as maleic anhydride, stoichiometry should not exceed 0.5–1.0 moles of anhydride per mole of CPD (56).

The majority of thermal polymerizations are carried out as a batch process, which requires a heat-up and a cool down stage. Typical conditions are 250–300°C for 0.5–4 h in an oxygen-free atmosphere (typically nitrogen) at approximately 1.4 MPa (200 psi). A continuous thermal polymerization has been reported which utilizes a tubular flow reactor having three temperature zones and recycle capability (62). The advantages of this process are reduced residence time, increased production, and improved molecular weight control. Molecular weight may be controlled with temperature, residence time, feed composition, and polymerizate recycle.

Hydrogenation of Petroleum Resins.   Most petroleum resins produced by the catalytic or thermal polymerization of cracked petroleum fractions are colored and contain a certain degree of nonaromatic olefinic unsaturation. This unsaturation leads to reduced thermal and oxidative stability. High olefinic content may also lead to decreased resin compatibility with other components used in an application. Hydrogenation is used as a means of reducing resin color and odor, as well as improving stability and compatibility. The hydrogenation of petroleum resins may be achieved in solution or molten state and is usually accomplished by either a batchwise or, more commonly, a continuous process. The key parameters to be addressed with resin hydrogenation are catalyst activity, pressure, temperature, and feed rate.

Catalysts employed for the hydrogenation of petroleum resins are typically supported monometallic and bimetallic catalyst systems based on the group VI and VIII transition elements. Hydrogenations of thermally polymerized cyclopentadiene-based resins have been accomplished using excess hydrogen or a hydrogen-rich gas over a γ-Al$_2$O$_3$ supported sulfided nickel–tungsten (Ni–W) or nickel–molybdenum (Ni–Mo) catalyst system (63). Operating parameters for a 21% resin solution in Varsol were established at 250–330°C using a hydrogen pressure of 15–20 MPa (150–200 atm). Optimum resin decoloration and overall catalyst life were achieved using a catalyst with a specific fresh surface area of 181 m^2/g (64).

Processes for the hydrogenation of catalytic resins are usually carried out using the same catalysts and reaction conditions as those used for thermally polymerized precursors. Rainey nickel has been shown to be an effective catalyst for the hydrogenation of catalytic aromatic resins (65). By varying reaction times at a temperature of 280°C and a pressure of 19.6 MPa (200 kg/cm^2) in an autoclave, the degree of hydrogenation of the aromatic rings may be optimized to between 30 and 80%. Optimizing the reduction of the aromatic nuclei within a resin, as measured by the absorbance at 700 cm^{-1}, is conducted to optimize resin compatibility with ethylene–vinyl acetate copolymers.

Reduction of the aromatic nuclei contained in catalytic C-9 resins has also been accomplished in the molten state (66). Continuous downward concurrent feeding of molten resin (120°C softening point) and hydrogen to a fixed bed of an alumina supported platinum–ruthenium (1.75% Pt–0.25% Ru) catalyst has been shown to reduce approximately 100% of the aromatic nuclei present in the resin. The temperature and pressure required for this process are 295–300°C and 9.8 MPa (100 kg/cm^2), respectively. The extent of hydrogenation was monitored by the percent reduction in the uv absorbance at 274.5 nm.

Resins from Pure Monomers

Thermoplastic resins produced from pure monomers such as styrene, alkyl-substituted styrenes, and isobutylene are produced commercially. An advantage of these resins is the fact that they are typically lighter in color than Gardner 1 (water-white) without being hydrogenated. Among the earliest resins in this category were those made from styrene and sold as Piccolastic. Styrene and alkyl-substituted styrenes such as α-methylstyrene are very reactive toward Friedel-Crafts polymerization catalysts.

α-Methylstyrene resins possessing softening points of up to 165°C have been synthesized using BF$_3$, AlCl$_3$, (C$_2$H$_5$)AlCl$_2$, and (C$_2$H$_5$)$_2$AlCl at reaction temperatures of 0 to −50°C (67). Reaction temperature and monomer concentration must be carefully controlled to achieve the desired softening point and to minimize the dimer and trimer content of the finished polymerizate. Finishing techniques are comparable to those used for other petroleum-type resins.

Blends of vinyltoluene and α-methylstyrene (3:1, respectively) have been polymerized using BF$_3$ catalysis to yield resins with softening points ranging from 100–130°C and colors less than Gardner 1 (68). Polymerization of similar vinyltoluene:α-methylstyrene blends in ratios of 2.5:1 to 4.5:1, respectively, over an acidic clay catalyst (69) has been found to yield resins with softening points of up to 100°C. As compared to resins produced with boron trifluoride and having similar softening points, the maximum chain lengths of the acidic clay produced resins, as measured by gel permeation chromatography, were significantly shorter. This difference resulted in improved solubility in paraffinic waxes (70). BF$_3$ produced α-methylstyrene–styrene copolymers have also been synthesized for uses in the sizing and retexturing of fabrics (71), as well as in pressure sensitive adhesive applications (72–74).

Terpolymers from dimethyl-α-methylstyrene (3,4-isomer preferred)–α-methylstyrene–styrene blends in a 1:1:1 weight ratio have been shown to be useful in adhesive applications. The use of ring-alkylated styrenes aids in the solubility of the polymer in less polar solvents and polymeric systems (75). Monomer concentrations of no greater than 20% and temperatures of less than $-20°C$ are necessary to achieve the desired properties.

tert-Butyl styrene has been copolymerized with various vinyl aromatic and aliphatic monomers to yield resins with unique softening points and solubility profiles. *t*-Butylstyrene–α-methylstyrene copolymers having softening points of up to 140°C have been produced using BF_3 catalysis. Process control leads to aromatic resins soluble in low Kauri-butanol solvents at temperatures as low as 0°C (76). Increased amounts of either monomer beyond an approximate 3:2 ratio (α-methylstyrene:*t*-butylstyrene) may lead to linear blocks of respective monomer in the polymer, which is detrimental to performance properties. Copolymers of *t*-butylstyrene with piperylenes (77) and diisobutylene (78) have been prepared as tackifiers for pressure sensitive and hot melt adhesives. Polymerization of *t*-butylstyrene with piperylenes requires 1–2 wt % $AlCl_3$ or 0.25–1.0 wt % BF_3. Homopolymers of *t*-butylstyrene possessing softening points of 155°C have been produced by the use of 0.1–8.0 wt % zirconium tetrachloride as a polymerization catalyst (79).

In many cases, the softening points of vinyl aromatic resins are controlled by monomer concentration and polymerization temperature. Two general rules are that the lower the monomer concentration, the higher the softening point, and the lower the polymerization temperature, the higher the softening point. These techniques may also lead to high molecular weights, which may be detrimental to many applications. Recent processes have revealed the production of high softening point ($>140°C$) α-methylstyrene–*para*-methylstyrene resins which do not rely on low monomer concentration. Resin yields of greater than 96% (based on reactive monomers) have been obtained using 1.2 wt % BF_3 as the catalyst system (80).

Terpene-Based Resins

Terpenes, specifically monoterpenes, are naturally occurring monomers that are usually obtained as by-products of the paper and citrus industries. Monoterpenes that are typically employed in hydrocarbon resins are shown in Figure 2. Optically active *d*-limonene is obtained from various natural oils, particularly citrus oils (81). α and β-Pinenes are obtained from sulfate turpentine produced in the kraft (sulfate) pulping process. Southeastern U.S. sulfate turpentine contains approximately 60–70 wt % α-pinene and 20–25 wt % β-pinene (see TERPENOIDS). Dipentene, which is a complex mixture of *d,l*-limonene, α- and β-phellandrene, α- and γ-terpinene, and terpinolene, is also obtained from the processing of sulfate liquor (82).

Polymerization. Polymerization of monoterpenes is most commonly accomplished by carbocationic polymerization utilizing Friedel-Crafts-type catalyst systems, such as aluminum chloride. High energy radiation and Ziegler-type catalysts (83) have also been demonstrated to be effective. A number of Friedel-Crafts-type catalysts have been evaluated with respect to β-pinene (84). They are arranged here in a decreasing order of effectiveness: $AlCl_3 = AlBr_3$, $ZrCl_4$, $AlCl_3O(C_2H_5)_2$, $BF_3O(C_2H_5)_2$, $SnCl_4$, $BiCl_3$, $SbCl_3$, and $ZnCl_2$. Typical polymerization procedures entail the addition of monomer to a cooled solvent–catalyst

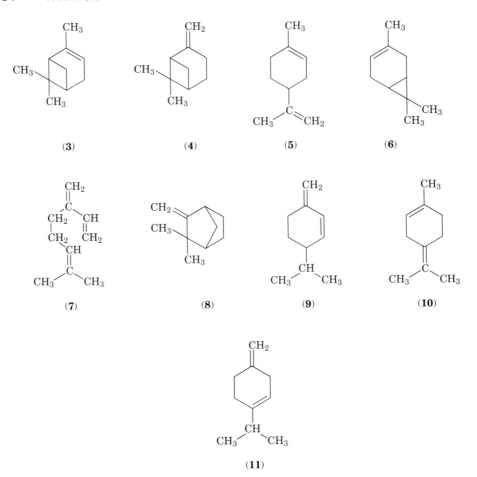

Fig. 2. Monoterpenes in hydrocarbon resins: α-pinene [2437-95-8] (**3**), β-pinene [18172-67-3] (**4**), limonene [7705-14-8] (**5**), 3-carene [13466-76-9] (**6**), myrcene [123-35-3] (**7**), camphene [5794-03-6] (**8**), β-phellandrene [555-10-2] (**9**), terpinolene [586-62-9] (**10**), and β-terpinene [99-86-5] (**11**).

slurry over a determined period of time. Routine reaction temperatures are 20–55°C, although some polymerizations may employ lower temperatures. Finishing procedures are identical to those used for catalytic petroleum resins.

Colors of commercial terpene-derived resins typically range from 2 to 5 Gardner. β-Pinene and limonene generally give better yields of hard resin than α-pinene in solution polymerization using AlCl₃. However, the addition of trialkylsilicon halides (85), silicon tetrahalides (86,87), or dibutyltin dichloride (88) as cocatalysts have been shown to significantly improve the yields of α-pinene polymerizations. Typical cocatalyst ratios are 5:1, AlCl₃:R₃SiX (X = Cl, Br, I).

Particular drawbacks of using alkylsilicon and alkyltin halides with AlCl₃ for the cationic polymerization of terpenes are low yields and the fact that they require rigorously dried feeds (≤50 ppm H₂O) to be effective. Increased water content results in lower yields and lower softening points (85). Catalyst systems comprised of AlCl$_3$ with antimony halides in the presence or absence of a lower alkyl, alkenyl, or aralkyl halide are particularly effective in systems containing up to 300 ppm H$_2$O (89,90). Use of 2–12 wt % of a system composed of 2–3 parts AlCl$_3$, 0.7–0.9 parts SbCl$_3$, and 0–0.2 parts of an organic halide provides good resin yields (94% based on reactive monomers) and softening points greater than 110°C. Softening points of 150°C have been achieved for β-pinene resins produced with an AlCl$_3$–SbCl$_3$ catalyst system (90).

Organogermanium halide cocatalysts have also been shown to be effective in the cationic polymerization of less reactive terpenes, as well as other hydrocarbon streams (91). Slow addition of 300 g α-pinene to a slurry of 2 wt % AlCl$_3$/(C$_2$H$_5$)$_3$GeCl (5:1, respectively) in 210 g xylene at −15°C was followed by gradual warming to 20°C. Catalyst deactivation was accomplished with a volume of water equal to the value of α-pinene used. Atmospheric distillation followed by steam stripping yielded a 120°C softening point resin in a 96.7% yield based on reactive monomers (92). Addition of an organic halide to the AlCl$_3$–R$_3$GeX catalyst system reduces the amount of organogermanium halide needed without affecting yield or resin properties. This reduction results in a more economical catalyst system than the antimony, silicon, or tin-based systems (93,94).

To avoid high resin chloride content associated with the use of high concentrations of aluminum trichloride, a trialkylaluminum–water cocatalyst system in a 1.0:0.5 to 1.0 mole ratio has been used in conjunction with an organic chloride for the polymerization of β-pinene (95). Softening points up to 120°C were achieved with 1–3 Gardner unit improvement in color over AlCl$_3$ produced resins.

Boron trifluoride etherate–n-hexanol complexes have successfully been used to polymerize β-pinene, as well as dipentene, to yield resins with softening points ≥70°C (82). Limonene or dipentene sulfate has been polymerized with aluminum chloride in a mixed toluene/high boiling aliphatic naphtha to give high yields of light colored resins (96). For the polymerization of dipentene or limonene, 4–8 wt % of AlCl$_3$ has been used. Polymerization of β-pinene typically requires lower levels of catalyst relative to limonene or dipentene.

Terpene Copolymers. Terpenes are routinely polymerized with other terpenes or with nonterpene-type monomers (97–102). The AlCl$_3$ catalyzed polymerization of β-pinene, dipentene, and terpene oligomers (oily dimers and trimers) has been found to yield resins with softening points ranging from 0–40°C (103).

Copolymerization of 10–40 wt % Δ-3-carene with a heat soaked C-5–C-6 olefin and diolefin stream under AlCl$_3$ catalysis has been shown to yield resins having softening points in the range of 60–140°C (104). These resins exhibit superior peel strength and tack properties over low softening point C-5 petroleum resins in pressure sensitive adhesive applications. Δ-3-Carene provides a rigid cyclic structure in the polymer backbone, which contributes to high softening points. This monomer also provides resins with narrower molecular weight distributions and better wax compatibilities than DCPD-modified resins.

A process of polymerization of isomerized α-pinene or turpentine with vinylbenzenes has been disclosed (105). α-Pinene or turpentine is isomerized by flash pyrolysis at 518 ± 5°C in a hot tube reactor to yield a mixture of predominantly dipentene and *cis*-alloocimene (2,6-dimethyl-2,4,6-octatriene [*673-84-7*]). Catalytic copolymerization of the isomerized fraction with styrene yields resins with softening points ranging from 70–110°C and properties suitable for tackifiers in hot melt and pressure sensitive adhesive applications.

Equal portions of styrene and β-pinene have been copolymerized using a 1:1 $AlCl_3$–organic ketone complex to yield resins having narrower molecular weight properties, lower softening points, and lighter color than resins produced from $AlCl_3$ catalysis alone (106). This Lewis acid–ketone complex has been demonstrated to be effective for monomers of differing reactivities by producing unimodal molecular weight distributions. Although producing equivalent number average molecular weights (M_n) to polymerizations using the complex, $AlCl_3$-catalyzed polymerizations of β-pinene with styrene or C-5 dienes tend to yield resins with higher softening points and broad bimodal molecular weight distributions. Other Lewis acid complexes utilizing oxygenates such as alcohols, ethers, esters, or carboxylic acids tend to give high oligomer contents and lower yields.

Heteroatom functionalized terpene resins are also utilized in hot melt adhesive and ink applications. Diels-Alder reaction of terpenic dienes or trienes with acrylates, methacrylates, or other α,β-unsaturated esters of polyhydric alcohols has been shown to yield resins with superior pressure sensitive adhesive properties relative to petroleum and unmodified polyterpene resins (107). Limonene–phenol resins, produced by the BF_3 etherate-catalyzed condensation of 1.4–2.0 moles of limonene with 1.0 mole of phenol have been shown to impart improved tack, elongation, and tensile strength to ethylene–vinyl acetate and ethylene–methyl acrylate-based hot melt adhesive systems (108). Terpene polyol ethers have been shown to be particularly effective tackifiers in pressure sensitive adhesive applications (109).

Economic Aspects

The approximate regional demands for hydrocarbon resins in 1980 and 1990 are shown in Table 5. Based on these figures, the average worldwide total growth rate of petroleum resins from 1980 to 1990 was 22,600 t/yr. The average worldwide total growth for natural resins for the same time period was 17,000 t/yr. The projected annual growth rate for hydrocarbon resins through the early to mid-1990s has been approximated at 4–8% based on various sources. Production figures may be gained from the U.S. International Trade Commission reports or from various consulting firms, which specialize in monitoring hydrocarbon resin and related industries. Consumption and growth predictions with respect to specific end uses may also be gained from these sources.

Table 5. Worldwide Hydrocarbon Resin Demand, 10^3 t

Resin	United States		Europe		Asia		Total	
	1980	1990	1980	1990	1980	1990	1980	1990
petroleum	197	304	97	127	48	137	342	568
natural[a]	121	239	109	173	72	60	302	472
Total	*318*	*543*	*205*	*300*	*121*	*197*	*644*	*1040*

[a]Rosin esters and terpenes.

Applications

Hydrocarbon resins are used extensively as modifiers in adhesives, sealants, printing inks, paints and varnishes, plastics, road marking, flooring, and oil field applications. In most cases, they are compounded with elastomers, plastics, waxes, or oils. Selection of a resin for a particular application is dependent on composition, molecular weight, color, and oxidative and thermal stability, as well as cost. A listing of all hydrocarbon resin suppliers and the types of resins that they produce is impractical. A representative listing of commercially available hydrocarbon resins and their suppliers is included in Table 6.

Table 6. Hydrocarbon Resin Suppliers

Manufacturer	Resin type	Trade name
Arakawa	C-9 aromatic, hydrogenated	Arkon
Arizona Chemical Co.	C-5 aliphatic	Betaprene, Sta-Tac
	C-5–C-9 aromatic modified aliphatic	Sta-Tac
	polyterpene	Nirez, Zonarez
	modified polyterpene	Nirez, Zonatec
Eastman	DCPD, hydrogenated	Eastotac
Exxon	C-5 aliphatic	Escorez 1000 series
	C-5–C-9 aromatic modified aliphatic	Escorez 2000 series
	C-9 aromatic	Escorez 7000 series
	DCPD, nonhydrogenated	Escorez 8000 series
	DCPD, hydrogenated	Escorez 5000 series
Goodyear	C-5 aliphatic	Wingtack 95
	C-5–C-9 aromatic modified aliphatic	Wingtack Plus, Extra, 86
Hercules	C-5 aliphatic	Piccopale series
	C-5–C-9 aromatic modified aliphatic	Hercotac
	pure C-9 aromatic, nonhydrogenated	Kristalex
		Piccolastic
		Piccotex
	pure C-9 aromatic, hydrogenated	Regalrez
	DCPD, nonhydrogenated	Piccodiene
	polyterpene	Piccolyte
	modified polyterpene	Piccofyn
Interesin	C-9 aromatic	Norsolene
Lawter	C-9 aromatic	Petro Rez
Neville	C-5 aliphatic	Nevtac
	C-5–C-9 aromatic modified aliphatic	Super Nevtac
	C-9 aromatic	Nevchem

Adhesives. The largest use for hydrocarbon resins is in adhesives (qv). There are numerous classes of adhesives, but the largest classes utilizing hydrocarbon resins are hot melt adhesives (HMA) and pressure sensitive adhesives (PSA). Hot melt adhesives are typically composed of a higher molecular weight polymer, a tackifier, a wax, and an antioxidant. HMAs do not utilize a solvent. Typical polymers that are used in HMA applications are ethylene–vinyl acetate (EVA), ethylene–methyl acrylate (EMA), and low molecular weight polyethylene. Depending on the comonomer content (eg, vinyl acetate) of the polymers (polarity), useful tackifiers may range from C-5 aliphatic to C-9 aromatic resins and rosin esters.

Pressure sensitive adhesives typically employ a polymer, a tackifier, and an oil or solvent. Environmental concerns are moving the PSA industry toward aqueous systems. Polymers employed in PSA systems are butyl rubber, natural rubber (NR), random styrene–butadiene rubber (SBR), and block copolymers. Terpene and aliphatic resins are widely used in butyl rubber and NR-based systems, whereas PSAs based on SBR may require aromatic or aromatic modified aliphatic resins.

Styrenic block copolymers (SBCs) are also widely used in HMA and PSA applications. Most hot melt applied pressure sensitive adhesives are based on triblock copolymers consisting of SIS or SBS combinations (S = styrene; I = isoprene; B = butadiene). Pressure sensitive adhesives typically employ low styrene, high molecular weight SIS polymers, while hot melt adhesives usually use higher styrene, lower molecular weight SBCs. Resins compatible with the mid-block of an SBC improves tack properties; those compatible with the end blocks control melt viscosity and temperature performance.

In general, hydrocarbon resin tackifiers used in adhesive applications serve to enhance adhesion by increasing the wetting properties of a polymer. Increased wetting allows for extensive interfacial contact between the adhesive and a substrate. From a rheological standpoint, tackifiers serve to reduce the storage modulus of a particular polymer, while increasing the glass-transition temperature (T_g). This serves to enhance cohesive strength. Alternatively, plasticizers (qv) tend to decrease the T_g of the system and reduce cohesive strength.

Printing Inks. Hydrocarbon resins are used extensively in printing inks and have replaced many of the natural resins that are used as the binder portion of inks (qv). They perform well in heat-set letterpress, heat-set web offset, publication gravure, and lithographic printing inks. Hydrocarbon resins are also used in alcohol and water-based flexographic printing inks. Resins suitable for printing ink applications, specifically heat set and gravure inks, should have a reasonably high softening point (130–160°C), light color, good solubility in typical ink solvents, the ability to wet pigment, and a reasonably high solution viscosity.

For rotogravure printing, aromatic, functionally modified aromatic or aliphatic, and cycloaliphatic-based petroleum resins perform well in Type A and B rotogravure publication inks and Type T packaging printing inks. In heat-set inks, both web-offset and letterpress, the aromatic and functionally modified aromatic resins exhibit fast solvent release, good pigment wetting properties, solvency in ink oils, acceptable compatibility with alkyds and other film formers, and the tack required for good printing.

Rubber Compounding. Hydrocarbon resins are used in the production of various types of rubber-based products, including tires, shoe soles and heels, hoses, industrial belting, mats, electrical wire insulation, and roll coverings. Coumarone–indene and aromatic resins reinforce mineral-loaded SBR stock and increase tensile strength, elongation, and resistance to flex cracking. The aliphatic- and terpene-based resins are used as tackifiers for NR and NR/SBR combinations in the formulation of tires and molded goods. Aromatic modified aliphatic or functionally modified aromatic resins can act as both tackifying and reinforcing resins. Normally a range of 5–15 parts of resin per 100 parts of elastomer is used in rubber compounding.

Protective Coatings. Paint (qv) and varnishes for industrial and trade sale applications frequently contain hydrocarbon resins. Aromatic petroleum resins are incorporated into air-dry and low bake industrial primers containing medium and long-oil alkyds, as well as in gloss and semigloss industrial and trade sale enamels to speed up drying (see ALKYD RESINS). Aromatic resins are also employed in leafing aluminum paints to improve leafing properties and speed drying time, and in oil and varnish stains to increase penetrating characteristics and impart water resistance.

Nonhydrogenated dicyclopentadiene-based resins are oxidizable resins that develop a degree of solvent resistance in aged films. Their unsaturation causes copolymerization in drying oils (qv) used in cooked varnishes and alkyds, and thus reduces drying time. Pure monomer resins are used in aerosol paints where they retain pigments in excellent condition and promote high gloss and fast solvent release. The low solution viscosity of these resins is also helpful in formulating these low solids coatings.

To control the degree of moisture evaporation and setting time, freshly poured concrete is sprayed with solvent solutions of aromatic, dicyclopentadiene, or aliphatic resins (see CEMENT).

Other Uses. Large quantities of hydrocarbon resins are used in mastics, caulks, and sealants (qv). Polymers for these adhesive products include neoprene, butyl rubber, polyisoprene, NR, SBR, polyisobutylene, acrylics, polyesters, polyamides, amorphous polypropylene, and block copolymers. These adhesives may be solvent or water-borne and usually contain inorganic fillers.

Terpene resins, because of their low odor and acceptable FDA clearance, are used as tackifiers for the natural and synthetic gum bases used in chewing gum. Selected petroleum resins are also used as gum bases.

Coumarone–indene resins and styrene-modified aromatic resins, representing 20–30% of the binder portion, are used in the manufacture of light colored asphalt floor tile. Electrostatic toners used in copy machines employ resins made from styrene and styrene copolymer resins to disperse the pigment needed to produce the image. Comonomer resins composed of α-methylstyrene and vinyltoluene or styrene are used by dry-cleaning establishments as retexturizing agents to provide the proper hand, drape, and creasability to textiles after cleaning.

High softening point modified aliphatic resins are used as diverting agents for hydraulic fracturing, acid treating, and fluid loss operations in crude oil recovery. Key attributes of suitable resins are the solubility in crude oil and softening points high enough to accommodate application temperatures.

BIBLIOGRAPHY

"Coumarone–Indene Resins" in *ECT* 1st ed., Vol. 4, pp. 594–600, by L. M. Geiger, The Neville Co.; "Petroleum Resins" in *ECT* 1st ed., Suppl. 1, pp. 659–667, by P. O. Powers, Pennsylvania Industrial Chemical Corp.; "Hydrocarbon Resins" in *ECT* 2nd ed., Vol. 11, pp. 242–262, by P. O. Powers, Amoco Chemicals Corp.; in *ECT*, 3rd ed., Vol. 12, pp. 852–869, by J. F. Holohan, Jr., J. Y. Penn and W. A. Vredenburgh, Hercules Inc.

1. P. O. Powers, in N. M. Bikales ed., *Encyclopedia of Polymer Science and Technology*, Vol. 4, John Wiley & Sons, Inc., New York, 1966, p. 272.
2. J. H. Hildebrand, *The Solubility of Non-Electrolytes*, Van Nostrand Reinhold, New York, 1950, pp. 129, 424–439.
3. G. A. Davies and co-workers, in I. Skeist, ed., *Handbook of Adhesives*, 3rd ed., Van Nostrand Reinhold, New York, 1991, pp. 266–269.
4. C. W. DeWalt, *Adhes. Age* **13**, 38 (1970).
5. H. E. Adams and P. O. Powers, *Ind. Eng. Chem. Anal. Ed.* **15**, 711 (1943).
6. P. O. Powers, *Ind. Eng. Chem.* **44**, 380 (1952).
7. P. O. Powers, *Rubber Rev.* **36**, 1542 (1963).
8. J. P. Kennedy and E. Marechal, *Carbocationic Polymerization*, John Wiley & Sons, Inc., New York, 1982, pp. 15–156.
9. *Ibid.*, pp. 10–14.
10. A. G. Evans, G. W. Meadows, and M. Polanyi, *Nature* **158**, 94 (1946); P. H. Pleish, M. Polanyi, and H. A. Skinner, *J. Chem. Soc.*, 247 (1947); A. G. Evans and M. Polanyi, *J. Chem. Soc.*, 252 (1947).
11. J. A. Kinney, in J. J. Mattiello, ed., *Protective and Decorative Coatings*, Vol. 1, John Wiley & Sons, Inc., New York, 1941, p. 360.
12. U.S. Pat. 4,946,915 (Aug. 7, 1990), H. Sato and M. Makino (to Nippon Oil Co.).
13. Brit. Pat. 950,602 (Feb. 26, 1964), (to Centre Nationale de la Recherche Scientifique).
14. M. Epimakhov and co-workers, *Koks i Khim.* **3**, 36 (1964).
15. U.S. Pat. 2,733,285 (Mar. 19, 1952), G. P. Hamner (to Esso Research and Engineering Co.).
16. U.S. Pat. 2,849,512 (Aug. 26, 1958), F. W. Banes, J. F. Nelson, and R. F. Leary (to Esso Research and Engineering Co.).
17. Ref. 1, p. 273.
18. J. Findlay, in N. M. Bikales, ed., *Encyclopedia of Polymer Science and Technology*, Vol. 9, John Wiley & Sons, Inc., New York, 1968, p. 853.
19. U.S. Pat. 3,676,412 (July 11, 1972), C. P. Senyard and J. C. Winkler (to Esso Research and Development Co.); U.S. Pat. 3,855,187 (Dec. 17, 1974), C. P. Senyard and J. C. Winkler (to Esso Research and Engineering Co.).
20. U.S. Pat. 4,078,132 (Mar. 7, 1978), A. Lepert (to Exxon Research and Engineering Co.).
21. U.S. Pat. 3,987,123 (Oct. 19, 1976), A. Lepert (to Exxon Research and Engineering Co.).
22. U.S. Pat. 4,008,360 (Feb. 15, 1977) Y. Kitagawa, K. Kudo, and H. Kuribayashi (to Sumitomo Chemical Co.).
23. U.S. Pat. 4,068,062 (Jan. 10, 1978), A. Lepert (to Exxon Research and Engineering Co.).
24. U.S. Pat. 4,151,338 (Apr. 24, 1979), J. Disteldorf and co-workers (to Veba-Chemie Aktiengesellschaft).
25. U.S. Pat. 4,131,567 (Dec. 26, 1978), R. T. Wojcik (to Arizona Chemical Co.).
26. U.S. Pat. 4,419,503 (Dec. 6, 1983), F. M. Benitez and M. F. English (to Exxon Research and Engineering Co.).

27. U.S. Pat. 4,068,062 (Jan. 10, 1978), A. Lepert (to Exxon Research and Engineering Co.).
28. U.S. Pat. 3,893,986 (July 8, 1975), A. Ishikawa, H. Komai, and H. Tsubaki (to Nippon Zeon Co., Ltd.).
29. U.S. Pat. 3,661,870 (May 9, 1972), H. L. Bullard (to Goodyear Tire and Rubber Co.).
30. U.S. Pat. 3,813,357 (May 28, 1974), D. R. St. Cyr (to Goodyear Tire and Rubber Co.).
31. U.S. Pat. 3,960,823 (June 1, 1976), H. Komai and A. Ishikawa (to Nippon Zeon Co., Ltd.).
32. U.S. Pat. 4,403,080 (Sept. 6, 1983), V. L. Hughes (to Exxon Research and Engineering Co.).
33. U.S. Pat. 4,916,192 (Apr. 10, 1990), S. G. Hentges (to Exxon Chemical Patents, Inc.).
34. U.S. Pat. 4,514,554 (Apr. 30, 1985), F. M. Benitez and V. L. Hughes (to Exxon Research and Engineering Co.).
35. U.S. Pat. 4,098,983 (July 4, 1978), H. L. Bullard and R. A. Osborn (to Goodyear Tire and Rubber Co.).
36. U.S. Pat. 3,853,826 (Dec. 10, 1974), D. R. St. Cyr (to Goodyear Tire and Rubber Co.).
37. U.S. Pat. 3,987,123 (Oct. 19, 1976), A. Lepert (to Exxon Research and Engineering Co.).
38. U.S. Pat. 4,562,233 (Dec. 31, 1985), F. M. Benitez, V. L. Hughes, and A. B. Small (to Exxon Research and Engineering Co.).
39. U.S. Pat. 4,230,840 (Oct. 28, 1980), M. Aoki and S. Katayama (to Mitsui Petrochemical Industries, Ltd.).
40. U.S. Pat. 4,037,016 (July 19, 1977), G. W. Feeny and B. W. Habeck (to Goodyear Tire and Rubber Co.).
41. U.S. Pat. 4,230,842 (Oct. 28, 1980), H. L. Bullard and R. A. Osborne (to Goodyear Tire and Rubber Co.).
42. U.S. Pat. 4,623,698 (Nov. 18, 1986), M. L. Evans, L. E. Jacob, and A. Lepert (to Exxon Research and Engineering Co.).
43. U.S. Pat. 4,636,555 (Jan. 13, 1987), M. L. Evans and co-workers (to Exxon Research and Engineering Co.).
44. U.S. Pat. 4,933,409 (Jun. 12, 1990) M. L. Evans and S. G. Hentges (to Exxon Chemical Patents, Inc.).
45. U.S. Pat. 5,021,499 (June 4, 1991), S. Budo and co-workers (to Exxon Research and Engineering Co.).
46. U.S. Pat. 3,799,913 (Mar. 26, 1974), A. S. Andrews, J. A. Schlademan, and H. B. Wheeler (to Neville Chemical Co.).
47. U.S. Pat. 4,105,843 (Aug. 8, 1978), Y. Iwase, S. Katayama, and T. Nakano (to Mitsui Petrochemical Industries, Ltd.).
48. U.S. Pat. 4,684,707 (Aug. 4, 1987), M. L. Evans (to Exxon Chemical Patents, Inc.).
49. *Dicyclopentadiene*, Technical Data Bulletin No. 48100-1, Velsicol Chemical Corporation, Chicago, 1968.
50. U.S. Pat. 4,330,655 (May 18, 1982), H. L. Bullard (to Goodyear Tire and Rubber Co.).
51. U.S. Pat. 3,968,088 (July 6, 1976), H. Asai and A. Wada (to Nippon Zeon Co., Ltd.).
52. Jpn. Pat. 49-2344 (Jan. 19, 1974), H. Sato and co-workers (to Nippon Oil Co., Ltd.).
53. U.S. Pat. 4,413,067 (Nov. 1, 1983), H. Hayashi, A. Oshima, and S. Tsuchiya (to Nippon Oil Co., Ltd.).
54. U.S. Pat. 4,419,497 (Dec. 6, 1983), H. Hayashi, A. Oshima, and S. Tsuchiya (to Nippon Oil Co., Ltd.).
55. Brit. Pat. 2,032,439 (May 8, 1980), H. Hayashi, A. Oshima, and S. Tsuchiya (to Nippon Oil Co., Ltd.).
56. U.S. Pat. 3,957,736 (May 18, 1976), H. Hayashi, H. Sato, and S. Tsuchiya (to Nippon Oil Co., Ltd.).

57. U.S. Pat. 3,968,088 (July 6, 1976), H. Asai and A. Wada (to Nippon Zeon Co., Ltd.).
58. U.S. Pat. 4,028,291 (June 7, 1977), H. Hayashi and co-workers (to Nippon Oil Co., Ltd.).
59. U.S. Pat. 4,056,098 (Nov. 1, 1977), J. J. Laurito (to Neville Chemical Co.).
60. U.S. Pat. 4,189,410 (Feb. 19, 1980), J. J. Laurito (to Neville Chemical Co.).
61. U.S. Pat. 4,433,100 (Feb. 21, 1984), J. J. Laurito (to Neville Chemical Co.).
62. U.S. Pat. 5,109,081 (Apr. 28, 1992), R. B. Pannell and G. E. Wissler (to Exxon Chemical Patents, Inc.).
63. U.S. Pat. 4,328,090 (May 4, 1982), J. R. Shutt and A. N. Stuckey (to Exxon Research and Engineering).
64. U.S. Pat. 4,629,766 (Dec. 16, 1986), B. Bossaert, A. Malatesta, and J. Mourand (to Exxon Research and Engineering).
65. U.S. Pat. 3,926,878 (Dec. 16, 1975), N. Minami and K. Shimizu (to Arakawa Rinsan Kagaku Kogyo Kabushiki Kaishi).
66. U.S. Pat. 4,540,480 (Sept. 10, 1985), N. Azuma and S. Suetomo (to Arakawa Kagaku Kogyo Kaubshiki Kaisha).
67. U.S. Pat. 3,669,947 (June 13, 1972), T. Hokama and H. Kahn (to Velsicol Chemical Corp.).
68. U.S. Pat. 3,000,868 (Sept. 19, 1961), P. O. Powers (to Pennsylvania Industrial Chemical Corp.).
69. U.S. Pat. 3,630,981 (Dec. 28, 1971), C. C. Campbell and D. A. Finfinger (to Pennsylvania Industrial Chemical Corp.).
70. U.S. Pat. 3,956,250 (May 11, 1976), C. C. Campbell and D. A. Finfinger (to Hercules Inc.).
71. U.S. Pat. 3,879,334 (Apr. 22, 1975), P. S. Douglas and W. A. Vredenburgh (to Hercules Inc.).
72. U.S. Pat. 3,932,332 (Jan. 13, 1976), P. S. Douglas, A. P. Patellis, and W. A. Vredenburgh (to Hercules Inc.).
73. U.S. Pat. 4,075,404 (Feb. 21, 1978), P. S. Douglas, A. P. Patellis, and W. A. Vredenburgh (to Hercules Inc.).
74. U.S. Pat. 4,113,801 (Sept. 12, 1978), P. S. Douglas, A. P. Patellis, and W. A. Vredenburgh (to Hercules Inc.).
75. U.S. Pat. 3,429,843 (Feb. 25, 1969), V. A. Arnold and R. J. Lee (to Standard Oil Co.).
76. U.S. Pat. 3,654,250 (Apr. 4, 1972), B. J. Davis (to Reichold Chemicals Inc.).
77. U.S. Pat. 3,622,551 (Nov. 23, 1971), B. J. Davis (to Reichold Chemicals Inc.).
78. U.S. Pat. 3,702,842 (Nov. 14, 1972), B. J. Davis and W. J. Ramson (to Reichold Chemicals Inc.).
79. U.S. Pat. 3,919,181 (Nov. 11, 1975), R. A. Meader, Jr. and K. C. Peterson (to Schenectady Chemicals Inc.).
80. U.S. Pat. 4,732,936 (Mar. 22, 1988), J. F. Holohan, Jr. (to Hercules Inc.).
81. U.S. Pat. 4,973,787 (Nov. 27, 1990), H. A. Colvin (to Goodyear Tire and Rubber Co.).
82. U.S. Pat. 4,670,504 (Jun. 2, 1987), C. G. Cardenas and co-workers (to Sylvachem Corp.).
83. J. R. Hadley, C. S. Marvel, and L. T. Longone, *J. Polym. Sci.* **40**, 551 (1959).
84. W. J. Roberts, *J. Am. Chem. Soc.* **72**, 1226 (1950).
85. U.S. Pat. 3,478,007 (Nov. 11, 1969), L. B. Barkley and A. P. Patellis (to Pennsylvania Industrial Chemical Corp.).
86. U.S. Pat. 3,816,381 (Jun. 11, 1974), R. F. Phillips (to Arizona Chemical Co.).
87. U.S. Pat. 3,852,218 (Dec. 3, 1974), R. F. Phillips (to Arizona Chemical Co.).
88. U.S. Pat. 3,354,132 (Nov. 21, 1967), A. D. Sproat (to Pennsylvania Industrial Chemical Corp.).
89. U.S. Pat. 4,016,346 (Apr. 5, 1977), E. R. Ruckel and R. T. Wojcik (to Arizona Chemical Co.).

90. U.S. Pat. 4,048,095 (Sept. 13, 1977), E. R. Ruckel and R. T. Wojcik (to Arizona Chemical Co.).

91. U.S. Pat. 4,011,385 (Mar. 8, 1977), E. R. Ruckel and L. S. Wang (to Arizona Chemical Co.).

92. U.S. Pat. 4,077,905 (Mar. 7, 1978), E. R. Ruckel and L. S. Wang (to Arizona Chemical Co.).

93. U.S. Pat. 4,057,682 (Nov. 8, 1977), E. R. Ruckel and L. S. Wang (to Arizona Chemical Co.).

94. U.S. Pat. 4,113,653 (Sept. 12, 1977), E. R. Ruckel and L. S. Wang (to Arizona Chemical Co.).

95. U.S. Pat. 4,487,901 (Dec. 11, 1984), G. D. Malpass (to Hercules Inc.).

96. U.S. Pat. 3,297,673 (Jan. 10, 1967), H. G. Sellers and H. F. McLaughlin (to Tenneco Chemicals, Inc.).

97. U.S. Pat. 3,466,267 (Sept. 6, 1969), J. M. Derfer (to SCM Corp.).

98. U.S. Pat. 3,737,418 (Jun. 5, 1973), H. G. Arlt, R. F. Phillips and E. R. Ruckel (to Arizona Chemical Co.).

99. U.S. Pat. 3,761,457 (Sept. 25, 1973), H. G. Arlt and E. R. Ruckel (to Arizona Chemical Co.).

100. Neth. Pat. Appl. 6,611,789 (Feb. 27, 1967), (to Tenneco Chemicals, Inc.).

101. U.S. Pat. 3,413,246 (Nov. 26, 1968), Y. Yen and H. P. Weymann (to Tenneco Chemicals, Inc.).

102. U.S. Pat. 3,622,550 (Nov. 23, 1971), J. F. Holohan and A. P. Patellis (to Pennsylvania Industrial Chemical Corp.).

103. U.S. Pat. 4,052,549 (Oct. 4, 1977), J. W. Booth (to Arizona Chemical Co.).

104. U.S. Pat. 4,245,075 (Jan. 13, 1981), A. Lepert (to Exxon Research and Engineering Co.).

105. U.S. Pat. 4,797,460 (Jan. 10, 1989), C. B. Davis (to Arizona Chemical Co.).

106. U.S. Pat. 5,051,485 (Sept. 24, 1991), J. W. Booth and J. J. Schmid (to Arizona Chemical Co.).

107. U.S. Pat. 4,709,084 (Nov. 24, 1987), M. S. Pavlin and R. L. Veazey (to Union Camp Corp.).

108. U.S. Pat. 3,929,938 (Dec. 30, 1975), R. M. Hill and R. H. White (to Reichold Chemicals, Inc.).

109. U.S. Pat. 4,650,822 (Mar. 17, 1987), J. O. Bledsoe, Jr. and R. L. Veazey (to Union Camp Corp.).

R. Derric Lowery
Exxon Chemical Company

HYDROCARBONS

SURVEY

Hydrocarbons, compounds of carbon and hydrogen, are structurally classified as aromatic and aliphatic; the latter includes alkanes (paraffins), alkenes (olefins), alkynes (acetylenes), and cycloparaffins. An example of a low molecular weight paraffin is methane [74-82-8]; of an olefin, ethylene [74-85-1]; of a cycloparaffin, cyclopentane [287-92-3]; and of an aromatic, benzene [71-43-2]. Crude petroleum oils [8002-05-9], which span a range of molecular weights of these compounds, excluding the very reactive olefins, have been classified according to their content as paraffinic, cycloparaffinic (naphthenic), or aromatic. The hydrocarbon class of terpenes is not discussed here. Terpenes, such as turpentine [8006-64-2] are found widely distributed in plants, and consist of repeating isoprene [78-79-5] units (see ISOPRENE; TERPENOIDS).

In the paraffin series, methane, CH_4, to n-butane, C_4H_{10}, are gases at ambient conditions. Propane, C_3H_8, and butanes are sometimes considered in a special category because they can be liquefied at reasonable pressures. These compounds are commonly referred to as liquefied petroleum gases (qv) (LPG). The pentanes, C_5H_{12}, to pentadecane [629-62-9], $C_{15}H_{32}$, are liquids, commonly called distillates, which include gasoline [8006-61-9], kerosene [8008-20-6], and diesel fuels (see GASOLINE AND OTHER MOTOR FUELS). n-Hexadecane [544-76-3], $C_{16}H_{34}$, and higher molecular weight paraffins are solids at ambient conditions and are referred to as waxes (qv). All classes of hydrocarbons are used as energy sources and feedstocks (qv) for petrochemicals.

Hydrocarbons are important sources for energy and chemicals and are directly related to the gross national product. The United States has led the world in developing refining and petrochemical processes for hydrocarbons from crude oil and natural gas [8006-14-2]. In 1861 the United States produced over 99% of the world's output of crude. About 100 years later U.S. production amounted to 35% of the world's production (1). Hydrocarbons from crude oil have become the energy sources of the industrial world, largely replacing wood and even displacing coal. However, in the United States, crude oil production peaked at 1.3×10^6 t/d (9.6×10^6 bbl/d) (conversion factors vary depending on oil source) in 1970, causing increased reliance on foreign oil sources (2). Since the crude oil embargo in 1973, a number of alternative energy sources have been investigated to reduce the U.S. international trade deficit. The fossil-fuel era may turn out to have been a brief interlude between the wood-burning era of the nineteenth century and the renewable energy sources era of the twenty-first century (see FUEL RESOURCES).

Hydrocarbons were first used in the field of medicine by the Romans. Bitumen was used in ancient Mesopotamia as mortar for bricks, as a road construction material, and to waterproof boats. Arabia and Persia have a long history of producing oil.

With the beginning of the industrial revolution around 1800, oil became increasingly important for lubrication and better illumination. Expensive vegetable oils were replaced by sperm whale oil [8002-24-2], which soon became scarce and its price skyrocketed. In 1850 lubrication oil was extracted from coal and oil shale (qv) in England, and ultimately about 130 plants in Great Britain and 64 plants in Pennsylvania, West Virginia, and Kentucky employed this process.

The earliest oil marketed in the United States came from springs at Oil Creek, Pennsylvania, and near Cuba, New York. It was used for medicinal purposes and was an article of trade among the Seneca Indians. At that time, the term Seneca Oil applied to all oil obtained from the earth. The first oil well was drilled in 1859 in Pennsylvania to a depth of 21.2 m for Seneca Oil Co. It produced 280 t (2000 bbl) in that year. This was the beginning of crude oil production.

Because crude oil is a complex mixture of hydrocarbons, early products such as kerosene were not uniform, and with new refining processes a whole new technology was developed. By 1920 the demand for gasoline exceeded that for kerosene and lubricating oils. The development of thermal cracking, followed by catalytic cracking, provided more gasoline and petrochemicals (3) (see CATALYSIS). During World War II the need for higher octane gasoline increased the demand for aromatic hydrocarbons. This led to several refining developments to increase gasoline octane and catalytic hydroforming and reforming to produce aromatics.

Hydrocarbon resources can be classified as organic materials which are either mobile such as crude oil or natural gas, or immobile materials including coal, lignite, oil shales, and tar sands. Most hydrocarbon resources occur as immobile organic materials which have a low hydrogen-to-carbon ratio. However, most hydrocarbon products in demand have a H:C higher than 1.0.

Products	Molar H:C ratio
natural gas	4.0
LPG	2.5
gasoline	2.1
fuel oil	
light	1.8
heavy	1.3
coal	0.8

Immobile hydrocarbon sources require refining processes involving hydrogenation. Additional hydrogen is also required to eliminate sources of sulfur and nitrogen oxides that would be emitted to the environment. Resources can be classified as mostly consumed, proven but still in the ground, and yet to be discovered. A reasonable estimate for the proven reserves for crude oil is estimated at 140×10^9 t (1.0×10^{12} bbl) (4). In 1950 the United States proven reserves were 32% of the world's reserve. In 1975 this percentage had decreased to 5%, and by 1993 it was down to 2.5%. Since 1950 the dominance of reserves has been in the eastern hemisphere and in offshore fields. Proved world gas reserves are nearly 4×10^{12} trillion metric feet (5) with 31% in the Middle East.

Another factor to be considered is the fraction of crude that can be obtained from a reservoir. The average primary recovery from a reservoir was about 25–30% of the crude oil in place in 1993.

Hydrocarbons as Energy Sources

Hydrocarbons from petroleum (qv) are still the principal energy source for the United States as shown in Table 1. About 60% of the world's energy is supplied by gas and oil and about 27% from coal (6–8). The annual energy demand for oil in different world areas is given in Table 2.

The use of natural gas as a hydrocarbon source depends on transportation. Over long distances and waterways, liquefied natural gas (LNG) is delivered in cryogenic tankers or trucks (see GAS, NATURAL; PIPELINES). In the United States, about 22% of the fossil-fuel energy used in 1990 was gas, but in Japan this percentage was much less.

A significant obstacle to increased gas use is the lack of sufficient transportation and distribution systems. Environmental concerns have encouraged

Table 1. U.S. Energy Sources[a]

Source	Energy consumption, %	
	1973	1990
oil	46.9	41.2
gas	30.3	23.9
coal	17.5	23.5
hydroelectric	4.1	3.9
nuclear	1.2	7.6
Total	*100.0*	*100.0*

[a]Ref. 6–8.

Table 2. Worldwide Oil Consumption Trends,[a] 10^6 t/d[b]

Country/Area	1980	1983	1990
North America	2.62	2.32	2.55
Latin America	0.66	0.67	0.77
Western Europe	2.00	1.75	1.86
USSR and Central Europe	1.58	1.53	1.40
Middle East	0.29	0.35	0.43
Africa	0.21	0.23	0.28
Asia	1.42	1.35	1.83
Australasia[c]	0.10	0.096	0.11
Total World	*8.88*	*8.29*	*9.24*

[a]Ref. 6
[b]To convert t/d to bbl/d, multiply by 7.
[c]Australia and New Zealand.

reliance on natural gas as a cleaner burning fuel. Combustion of natural gas emits about half the CO_2 that coal generates at equivalent heat output. However, low oil prices have caused the number of operating drilling rigs in the United States to drop to well below the peak in the 1980s, cutting production of gas.

Table 3 lists the refinery product yields in North America and worldwide, illustrating patterns of consumption. The United States refines about 25% of the world's crude oil, and because of its declining oil reserves, must import additional crude oil.

Table 3. Petroleum Product Market Share,[a] %

Country/Area	1976	1982	1990
North America[b]			
gasoline	38.5	41.5	41.4
middle distillates	25.7	26.9	29.6
fuel oil	18.0	12.9	8.8
other	17.8	18.7	20.2
Total	*100.0*	*100.0*	*100.0*
World[c]			
gasoline	26.4	28.1	28.6
middle distillates	29.2	32.0	35.5
fuel oil	29.6	23.8	18.2
other	14.8	16.1	17.7
Total	*100.0*	*100.0*	*100.0*

[a] Courtesy of British Petroleum; Ref. 6.
[b] Includes the United States and Canada.
[c] Excludes Communist and ex-Communist areas.

Natural gas imports have grown more slowly because imports from overseas require governmental licenses and cryogenic liquefaction plants are very expensive. Natural gas imports are chiefly by pipeline from Canada (see GAS, NATURAL).

Gas and oil are the principal energy sources even though the United States has large reserves of coal. Although the use of coal and lignite is being encouraged as an energy source, economic and environmental considerations have kept petroleum consumption high (see also AVIATION AND OTHER GAS TURBINE FUELS). The use of compressed natural gas (CNG) is expected to grow in response to the Clean Air Act of 1990. Reliance on foreign imports has remained high, increasing since the collapse of oil prices in 1986. Imports reached 47.2% in 1990 (7).

In 1990, U.S. energy consumption by end user sector was 35.8% residential and commercial, 37% industrial, and 27.2% transportation (8). The breakdown of consumption by source was 41.2% petroleum, 23.8% natural gas, 23.5% coal, 7.6% nuclear, and 3.9% hydroelectric and other (6,8,9).

Hydrocarbons as Chemical Intermediates

Because of the time lag involved in collecting information on U.S. chemicals production and sales literature figures are always 2–3 years out of date. In 1991, total production of primary chemical products (ie, C_2–C_5 olefins and paraffins,

C_6-C_8 aromatics, plus miscellaneous other compounds used as intermediates in synthesis of other chemicals) was 5.41×10^7 kg. Only about half that amount was actually sold on the open market; the rest was used internally as feedstock for other chemicals. Total U.S. primary chemical sales in 1991 amounted to 2.76×10^7 kg for a total value of $\$9.63 \times 10^9$. Table 4 gives a breakdown of these figures by chemical type; because petroleum and gas account for more than 98% of the total, no other figures are given. All figures in this section do not consider use of hydrocarbons as fuel, with the possible exception that some benzene, toluene, and xylene may be used in aviation fuel.

Raw Materials. Petroleum and its lighter congener, natural gas, are the predominant sources of hydrocarbon raw materials, accounting for over 95% of all such materials. Assuring sources of petroleum and natural gas has become a primary goal of national policies all over the world, and undoubtedly was one of the principal justifications for the 1992 Gulf War.

The dominant role of petroleum in the chemical industry worldwide is reflected in the landscapes of, for example, the Ruhr Valley in Germany and the U.S. Texas/Louisiana Gulf Coast, where petrochemical plants connected by extensive and complex pipeline systems dot the countryside. Any movement to a different feedstock would require replacement not only of the chemical plants themselves, but of the expensive infrastructure which has been built over the last half of the twentieth century. Moreover, because petroleum is a liquid which can easily be pumped, change to any of the solid potential feedstocks (like coal and biomass) would require drastic changes in feedstock handling systems.

Coal is used mainly to produce synthesis gas, a mixture of CO and hydrogen. Much of the production of synthesis gas is unreported, as it typically is never isolated. For economic reasons, few chemicals are made from synthesis gas in the United States. The Fischer-Tropsch process, used by Germany in both world wars to supplement its hydrocarbon supplies, can produce a full range of hydrocarbons

Table 4. Primary Products from Petroleum and Natural Gas, 1991[a]

Product	Production, 10^6 kg	Sales, 10^6 kg	Value, 10^6 \$
	Aliphatic hydrocarbons		
ethylene/acetylene	18,260	6,994	2,832
propylene	9,774	5,588	2,026
C-4 hydrocarbons	6,340	4,132	1,072
C-5 hydrocarbons	1,697	905	242
all other aliphatics	5,558	2,583	1,099
Total aliphatics	*41,629*	*20,201*	*7,272*
	Aromatics and cyclics		
benzene	5,209	3,706	1,380
toluene	2,857	1,441	406
xylenes	2,866	1,235	326
others	1,537	1,057	250
Total aromatics	*12,469*	*7,440*	*2,362*
Total	*54,096*	*27,641*	*9,634*

[a] Ref. 10.

from methane to heavy wax, by passing synthesis gas over iron-based catalysts at high temperature and pressure. Fischer-Tropsch chemistry has been used most recently in South Africa at the Sasolburg complex. However, with the exception of such politically dictated operations, Fischer-Tropsch production of hydrocarbons is much more expensive than production and refining of petroleum. The cost of Fischer-Tropsch hydrocarbons from coal has historically been $10–$15 per barrel higher than petroleum.

There are some chemicals that can be made economically from coal or coal-derived substances. Methanol and CO are used to make acetic anhydride and acetic acid. Methanol itself can be made from synthesis gas over a copper–zinc catalyst (see FEEDSTOCKS, COAL CHEMICALS).

Though there has been much discussion about using biomass as a renewable resource for hydrocarbon production, few chemicals are currently made in significant quantities from biological feedstocks. The most important is ethanol (qv), used as an oxygenated additive to reformulated gasoline, which is meant to burn more cleanly than normal gasoline. That use is made economically feasible only by significant government subsidies in the form of tax exemptions. Furfural is made by fermentation from corncobs. The only other biomass-derived chemicals of any importance are glycerol (qv) and sugars such as sorbitol and mannitol.

Synthesis Gas Chemicals. Hydrocarbons are used to generate synthesis gas, a mixture of carbon monoxide and hydrogen, for conversion to other chemicals. The primary chemical made from synthesis gas is methanol, though acetic acid and acetic anhydride are also made by this route. Carbon monoxide (qv) is produced by partial oxidation of hydrocarbons or by the catalytic steam reforming of natural gas. About 96% of synthesis gas is made by steam reforming, followed by the water gas shift reaction to give the desired H_2/CO ratio.

steam reforming	$CH_4 + H_2O \longrightarrow CO + 3\ H_2$
water gas shift	$CO + H_2O \longrightarrow CO_2 + H_2$

Aliphatic Chemicals. The primary aliphatic hydrocarbons used in chemical manufacture are ethylene (qv), propylene (qv), butadiene (qv), acetylene, and *n*-paraffins (see HYDROCARBONS, ACETYLENE). In order to be useful as an intermediate, a hydrocarbon must have some reactivity. In practice, this means that those paraffins lighter than hexane have little use as intermediates. Table 5 gives 1991 production and sales from petroleum and natural gas. Information on uses of the C_1–C_6 saturated hydrocarbons are available in the literature (see HYDROCARBONS, C_1–C_6).

Cyclic Hydrocarbons. The cyclic hydrocarbon intermediates are derived principally from petroleum and natural gas, though small amounts are derived from coal. Most cyclic intermediates are used in the manufacture of more advanced synthetic organic chemicals and finished products such as dyes, medicinal chemicals, elastomers, pesticides, and plastics and resins. Table 6 details the production and sales of cyclic intermediates in 1991. Benzene (qv) is the largest volume aromatic compound used in the chemical industry. It is extracted from catalytic reformates in refineries, and is produced by the dealkylation of toluene (qv) (see also BTX PROCESSING).

Table 5. Aliphatic Hydrocarbons, 1991 Production and Sales[a]

Product	Production, 10^6 kg	Sales, 10^6 kg	Value, 10^6 \$
C-2 hydrocarbons			
ethylene	18,123	6,930	2,785
acetylene for chemicals	137	64	47
Total C-2	*18,260*	*6,994*	*22,832*
propylene			
butadiene/butylenes	1,047	693	139
1,3-butadiene for elastomers	1,385	1,388	433
Total propylene	*9,774*	*5,588*	*2,026*
C-4 hydrocarbons			
butene-1	425	212	92
isobutene	499	459	103
all other C-4s	2,542	1,179	222
Total C-4	*6,340*	*4,132*	*1,072*
C-5 hydrocarbons			
isoprene	214	164	58
mixed pentenes	189		
all other C-5s	1,294	741	184
Total C-5	*1,697*	*905*	*242*
all other aliphatics			
α-olefins			
C-6 to C-10	469	220	173
C-11+	392	215	168
dodecene	157	143	66
hexane		172	50
n-heptane	52	23	19
nonene	254	77	36
miscellaneous *n*-paraffins	673	467	141
all other	3,560	1,233	447
Total others	*5,558*	*2,583*	*1,099*
Total	*41,629*	*20,202*	*7,271*

[a] Ref. 10.

Hydrocarbons as End Use Chemicals

Lubricants. Petroleum lubricants continue to be the mainstay for automotive, industrial, and process lubricants. Synthetic oils are used extensively in industry and for jet engines; they, of course, are made from hydrocarbons. Since the viscosity index (a measure of the viscosity behavior of a lubricant with change in temperature) of lube oil fractions from different crudes may vary from +140 to as low as −300, additional refining steps are needed. To improve the viscosity index (VI), lube oil fractions are subjected to solvent extraction, solvent dewaxing, solvent deasphalting, and hydrogenation. Furthermore, automotive lube oils typically contain about 12–14% additives. These additives may be oxidation inhibitors to prevent formation of gum and varnish, corrosion inhibitors, or detergent dispersants, and viscosity index improvers. The United States consumption of lubricants is shown in Table 7.

Lubricating oils are also used in industrial and process applications such as hydraulic and turbine oils, machine oil and grease, marine and railroad diesel,

Table 6. Cyclic Hydrocarbon Intermediates, 1991[a]

Product	Production, 10^6 kg	Sales, 10^6 kg	Value, 10^6 \$
benzene, all grades	5,209	3,706	1,380
ethylbenzene	4,024	160	68
styrene	3,681	1,634	952
xylenes, mixed	2,866	1,235	326
toluene, all grades	2,857	1,441	406
p-xylene	2,427	1,203	526
cumene (isopropylbenzene)	1,890	1,510	682
cyclohexane	1,047	936	392
o-xylene	348	276	93
dicyclopentadiene	66	48	20
all other aromatics	1,537	1,057	250
all other cyclic intermediates	3,015	3,556	3,042
Total	*28,967*	*16,762*	*8,137*

[a]Ref. 10.

Table 7. United States Lubricant Consumption[a]

Year	Demand, 10^3 t[b]
1970	7099
1980	8342
1990	8551
1992	7870

[a]Ref. 11.
[b]To convert metric tons to gal, multiply by 294.

and metalworking oils. Process oils are used in the manufacture of rubber, textiles, leather, and electrical goods. The distribution of lube oils used in these applications in 1992 is as follows: automotive, 4571 t; industrial, 2229 t; and process, 1070 t (~315,000 gal) (11).

Synthetic lubricants are tailored molecules which have a higher viscosity index and a lower volatility for a given viscosity than lube oils from petroleum. Synthetic oils have the following advantages (12): energy conservation, extended drain periods, fuel economy, oil economy, high temperature performance, easier cold starting, cleaner engines, cleaner intake valves, and reduced wear.

Synthetic oils have been classified by ASTM into synthetic hydrocarbons, organic esters, others, and blends. Synthetic oils may contain the following compounds: dialkylbenzenes, poly(α-olefins); polyisobutylene, cycloaliphatics, dibasic acid esters, polyol esters, phosphate esters, silicate esters, polyglycols, polyphenyl ethers, silicones, chlorofluorocarbon polymers, and perfluoroalkyl polyethers.

Very high VI (120–145) lubestocks made by hydrocracking and wax isomerization are also becoming important at lower cost than the synthetics. These are primarily isoparaffins or mononaphthenes with long isoparaffin side chains. The demand for engine oils is expected to grow about 2% between 1990 and

2000 with higher growth in the industrial and process applications. New technology will be developed for additives and synthetic oils (see LUBRICATION AND LUBRICANTS; HYDRAULIC FLUIDS).

To conserve hydrocarbons, certain reclaiming technologies have been developed, involving re-refining used lubricating oils and reclaiming rubber. In 1988 about 2% of the used lubricating oil was re-refined (12). Refining oil had been curtailed because of environmental problems with acid sludge. New technology for re-refining oils without creating the acid sludge disposal problems is being marketed by KTI, UOP, Chemical Engineering Partners, and Enprotec. U.S. production of re-refined oils fell from 1.02×10^6 t (7×10^6 bbl) in 1960 to 170,000 t (1.2×10^6 bbl) in 1975 and in 1993 was 290,000 t (2×10^6 bbl) (12).

Agriculture/Food. *Traditional Uses.* Large quantities of hydrocarbons are used in agriculture, particularly as energy sources. Although solar energy is a cheap alternative, the convenience and reproducibility of drying with LPG has made hydrocarbon-derived energy the drying method of choice for such diverse applications as curing tobacco and drying peanuts, corn, and soybeans. In addition to these uses, hydrocarbons are used as the feedstock for a large variety of pesticides (Table 8) (see INSECT CONTROL TECHNOLOGY).

In addition to these uses related to crop production, hydrocarbons are used extensively in packaging, particularly in plastic films and to coat boxes with plastic and (to a much lesser extent) wax. Polymeric resins derived from hydrocarbons are also used to make trays and cases for delivery of packaged foodstuffs (see FILM AND SHEETING MATERIALS; PACKAGING; PAPER).

Highly pure *n*-hexane is used to extract oils from oilseeds such as soybeans, peanuts, sunflower seed, cottonseed, and rapeseed. There has been some use of hydrocarbons and hydrocarbon-derived solvents such as methylene chloride to extract caffein from coffee beans, though this use is rapidly being supplanted by supercritical water and/or carbon dioxide, which are natural and therefore more acceptable to the public.

Table 8. Pesticides Derived from Hydrocarbons, 1991[a]

Product	Production, 10^6 kg	Sales, 10^6 kg	Value, 10^6 $
Cyclic			
herbicides and growth regulators	227	167	1,815
fungicides	37	32	242
insecticides/rodenticides	33	40	762
all other cyclic pesticides	3	3	16
Total cyclic	*300*	*242*	*2,835*
Acyclic			
insecticides/rodenticides	86	105	318
herbicides and growth regulators	48	84	774
organophosphorus insecticides	13	11	121
metham [*144-54-7*]	10	39	17
fungicides	8	5	39
all other acyclic pesticides	72	64	234
Total acyclic	*151*	*203*	*1,184*
Total	*451*	*445*	*4,019*

[a]Ref. 10.

Feedstock for Protein.   Certain microorganisms, such as some bacteria, fungi, molds, and yeasts can metabolize hydrocarbons and hydrocarbon-derived materials. Because single-cell proteins (SCPs) are about 50% protein by weight, it was believed early in the development phase that the economics of SCP production would be favorable. That belief has proven essentially correct, but acceptance of SCPs as a primary source of food protein has been very slow. Except for limited uses as flavor enhancers and similar additive, SCPs have not made a significant impact on the markets for proteins. The future for hydrocarbons as a feedstock for SCPs is not bright, as the original *n*-paraffin feeds have been largely supplanted by alcohols derivable from nonpetroleum sources.

Surfactants.   Surfactants (qv) are chemicals, natural or synthetic, that reduce the surface tension of water or other solvents, and are used chiefly as soaps, detergents, dispersing agents, emulsifiers, foaming agents, and wetting agents. Surfactants may be produced from natural fats and oils, from silvichemicals such as lignin, rosin, and tall oil, and from chemical intermediates derived from coal and petroleum.

The greatest amount of surfactant consumption is in packaged soaps and detergents for household and industrial use. The remainder is used in processing textiles and leather, in ore flotation and oil-drilling operations, and in the manufacture of agricultural sprays, cosmetics, elastomers, food, lubricants, paint, pharmaceuticals, and a host of other products.

Table 9 gives U.S. production and sales of hydrocarbon-based surfactants by class for 1991. All quantities are reported in terms of 100%-active agent; diluents and other additives in the products as sold are omitted.

Table 9. Surfactants Derived from Hydrocarbons, 1991[a]

Product	Production, 10^6 kg	Sales, 10^6 kg	Value, 10^6 \$
amphoteric	10	9	23
anionic			
sulfonic acids and their salts	620	284	280
sulfuric acid esters	414	125	195
carboxylic acids and their salts	406	214	137
phosphoric acid esters	32	28	54
all other anionics	19	19	35
Total anionic	*1,491*	*670*	*701*
cationic			
amines and amine oxides	181	89	178
quaternary ammonium salts	72	59	157
all other cationics	4	1	3
Total cationic	*257*	*149*	*338*
nonionic			
ethers	623	595	736
carboxylic acid			
esters	156	119	226
amides	39	29	48
all other nonionics	9	6	13
Total nonionic	*827*	*749*	*1,023*
Total	*2,585*	*1,577*	*2,085*

[a] Ref. 10.

Coatings. Protective and decorative coatings (qv) for homes, vehicles, and a variety of industrial uses provide a large market for hydrocarbons. At one time, most paints, varnishes, and other coatings utilized organic chemical solvents. However, due to environmental concerns and solvent cost, approximately 40% of all coatings are waterborne, or even dispense with the solvent altogether (powder coating).

Vinyl, alkyd, and styrene–butadiene latexes are used as film formers in most architectural coatings. Because alkyd resins (qv) require organic solvents, their use has decreased substantially for architectural coatings, but is still holding up in industrial applications, where their greater durability justifies the added expense (see LATEX TECHNOLOGY).

Polymers. Hydrocarbons from petroleum and natural gas serve as the raw material for virtually all polymeric materials commonly found in commerce, with the notable exception of rayon which is derived from cellulose extracted from wood pulp. Even with rayon, however, the cellulose is treated with acetic acid (qv), much of which is manufactured from ethylene (see FIBERS, REGENERATED CELLULOSICS).

Synthetic Fibers. Virtually all synthetic fibers are produced from hydrocarbons, as follows:

Fiber	Hydrocarbon precursor
nylon	cyclohexane
cellulose acetate	ethylene, methane
acrylics	propylene
polyesters	p-xylene, ethylene
polyolefins	propylene, ethylene
carbon fibers	pitch

Worldwide synthetic fiber production for 1990 was ~17.5×10^6 t (see FIBERS, SURVEY).

Elastomers. Elastomers are polymers or copolymers of hydrocarbons (see ELASTOMERS, SYNTHETIC; RUBBER, NATURAL). Natural rubber is essentially polyisoprene, whereas the most common synthetic rubber is a styrene–butadiene copolymer. Moreover, nearly all synthetic rubber is reinforced with carbon black, itself produced by partial oxidation of heavy hydrocarbons. Table 10 gives U.S. elastomer production for 1991. The two most important elastomers, styrene–butadiene rubber (qv) and polybutadiene rubber, are used primarily in automobile tires.

Plastics and Resins. Plastics and resin materials are high molecular weight polymers which at some stage in their manufacture can be shaped or otherwise processed by application of heat and pressure. Some 40–50 basic types of plastics and resins are available commercially, but literally thousands of different mixtures (compounds) are made by the addition of plasticizers, fillers, extenders, stabilizers, coloring agents, etc.

The two primary types of plastics, thermosets and thermoplastics, are made almost exclusively from hydrocarbon feedstocks. Thermosetting materials are

Table 10. U.S. Elastomer Production, 1991[a]

Product	Production, 10^6 kg	Sales, 10^6 kg	Value, 10^6 $
styrene–butadiene (SBR)	902	603	593
polybutadiene (BR)	371	170	160
thermoplastic elastomers	215	176	535
ethylene–propylene (EP)	206	194	394
butadiene–acrylonitrile (nitrile)	73	75	164
silicone elastomers	60	38	358
acrylic elastomers	4	6	27
all other elastomers	335	268	748
Total	*2,166*	*1,530*	*2,979*

[a] Ref. 10.

those that harden during processing (usually during heating, as the name implies) such that in their final state they are substantially infusible and insoluble. Thermoplastics may be softened repeatedly by heat, and hardened again by cooling.

Table 11 shows U.S. production and sales of the principal types of plastics and resins. Some materials are used both as plastics, ie, bulk resin, and in other applications. For example, nylon is used in fibers, urethanes as elastomers. Only their use as plastics is given in Table 11; their uses in other applications are listed with those applications.

Toxicity

Most hydrocarbons are nontoxic or of low toxicity (13,14).

Paraffins.   Methane and ethane are simple asphyxiants, whereas the higher homologues are central nervous system depressants. Liquid paraffins can remove oil from exposed skin and cause dermatitis or pneumonia in lung tissue. Generally, paraffins are the least toxic class of hydrocarbons.

Olefins, Diolefins, and Acetylenes.   Members of this category having up to four carbon atoms are both asphyxiants and anesthetics, and potency for the latter effect increases with carbon chain length. Skin-contact effects are similar to those of paraffins.

Cycloparaffins.   Members of this class produce effects much like the paraffins, except that unsaturated cycloparaffins are more noxious than the saturated counterparts. Breathing high concentrations of cycloparaffin vapors can result in irritation and anesthesia.

Aromatic Hydrocarbons.   These are the most toxic of the hydrocarbons and inhalation of the vapor can cause acute intoxication. Benzene is particularly toxic and long-term exposure can cause anemia and leukopenia, even with concentrations too low for detection by odor or simple instruments. The currently acceptable average vapor concentration for benzene is no more than 1 ppm. Polycyclic aromatics are not sufficiently volatile to present a threat by inhalation (except from pyrolysis of tobacco), but it is known that certain industrial products, such as coal tar, are rich in polycyclic aromatics and continued exposure of human skin to these products results in cancer.

Table 11. U.S. Plastics and Resins, 1991[a]

Product	Production, 10^6 kg	Sales, 10^6 kg	Value, 10^6 $
Thermosets			
phenolics	1,201	552	593
urea–formaldehydes	1,147	756	268
polyether/polyester polyols	789	690	848
unsaturated polyesters	513	454	644
alkyd resins	356	280	335
epoxies	254	180	529
melamine–formaldehydes	116	95	224
polyurethanes	95	81	324
all other thermosets	73	55	163
Total thermosets	*4,544*	*3,143*	*3,928*
Thermoplastics			
polyethylene	9,429	9,387	7,285
vinyl resins	4,231	3,828	3,433
styrenics	3,310	2,978	3,795
polypropylene	2,664	2,403	1,793
polyesters	1,689	1,162	2,048
acrylics	743	622	1,553
engineering resins	488	351	1,295
polyamides (including nylon)	316	324	974
petroleum hydrocarbon resins	174	164	183
modified rosins	170	163	219
fluorocarbons	23		
all other thermoplastics	533	262	1,634
Total thermoplastics	*23,770*	*21,644*	*24,212*
Total	*28,314*	*24,787*	*28,140*

[a] Ref. 10.

Regulatory Issues

Regulatory issues are increasingly driving the petroleum industry, taking an ever growing share of capital. New environmental regulations govern every aspect of operation from drilling to refining (15–19). The number of regulations has increased dramatically since the 1970s.

Clean Air Act. The 1990 Clean Air Act Amendment (CAAA) mandated significant reductions in refinery air emissions of nitrogen and sulfur oxides, carbon monoxide, and particulates (see AIR POLLUTION). Fugitive emissions from leaks and hazardous air pollutants must also be controlled. Even more stringent air regulations have been issued by the California South Coast Air Quality Management District (SCAQMD). The total U.S. costs of refinery air control has been estimated (18) to amount to $7.5 billion by 2010.

The CAAA has also had a drastic impact on refiners by mandating the reformulation of gasoline and diesel fuel (15–17). The goal is to achieve specific reductions in emissions of volatile organic compounds, toxic compounds, and carbon monoxide without increasing emissions of nitrogen oxides. Significant operating changes and investments are required.

Gasoline. The CAAA establishes a permanent role for oxygenates in U.S. gasoline, requiring 2% oxygen content in the nine worst ozone nonattainment areas (15). In these areas, reformulated gasoline must have lower vapor pressure and benzene content (<1%). Toxics reductions also require lower total aromatics of about <25%, depending on the benzene content. It must also meet 1990 baselines for olefins, sulfur, and 90% distillation point. In California, even more severe changes are required by the California Air Resources Board (CARB).

The result is to require significant operating changes and refining investments (16). Benzene precursors are diverted around the catalytic reformers, whose severity is typically reduced significantly. This reduces hydrogen production, which may require adding a hydrogen plant. The butane content is reduced to lower vapor pressure. Isomerization of C_5-C_6 hydrocarbons is made essential. Production of methyl tertiary butyl ether (MTBE) or similar oxygenates is increased. Catalytic feed hydrotreating is often required to reduce sulfur contents of the FCC gasoline, as well as improve yields. The cost of these changes has been estimated (16) at 0.53–0.8¢/L (2–3¢/gal) for U.S. EPA gasoline and 3.2–4.5¢/L (12–17¢/gal) for gasoline meeting CARB regulations.

Diesel Fuel. Federal diesel specifications were changed to specify a maximum of 0.05% sulfur and a minimum cetane index of 40 or a maximum aromatics content of 35 vol % for on-road diesel. For off-road diesel, higher sulfur is allowed. CARB specifications require 0.05% sulfur on or off road and 10% aromatics maximum or passage of a qualification test. Process technologies chosen to meet these specifications include hydrotreating, hydrocracking, and aromatics saturation.

Clean Water Act. The Water Quality Act of 1987 and Clean Water Act of 1977 amended the Water Pollution Control Act of 1972, and are known collectively as the Clean Water Act (CWA). Their objective is to restore and maintain the integrity of U.S. waters. There are spill prevention, control, and containment requirements with which to comply. It requires replacement of older storage tanks or installation of double bottoms or seals.

The CWA covers refinery wastewater, including both process and storm water. It sets effluent limits and treatment standards for wastewater treating facilities. Much more capacity for storm water storage must be provided. It requires increasing process water reuse after tertiary treatment. This may involve two-stage activated sludge biological treatment using activated carbon, which would also minimize discharge of suspended solids.

Resource Conservation and Recovery Act. The Resource Conservation and Recovery Act (RCRA) of 1976, as amended in a more comprehensive form in 1984, governs the hazardous (lethal) and nonhazardous (benign) waste sectors and the gray area between. There are four classes of hazardous: ignitable, corrosive, reactive, and toxic. The regulations cover handling of hazardous waste from beginning to end, with the burden put on the generator to identify and track it. There are strict management standards overseen by the U.S. EPA for collecting, labeling, and accumulating it.

The generator is given 90 days to get the hazardous waste off-site to a treatment, storage, and disposal facility. The act covers design, permitting, operation, and closure standards for such facilities. All facilities must have at least an interim Part A permit, and eventually obtain a comprehensive Part B permit. There are strict financial requirements for funds during closure. The act mandates corrective action such as cleanup of spills and remediation of contaminated soil to protect groundwater. Finally, it requires that generation be minimized wherever possible.

Comprehensive Environmental Response, Compensation, and Liability Act.   The Comprehensive Environmental Response, Compensation, and Liability Act (CERCLA/Superfund) was amended by SARA in 1986. It provides for the liability, compensation, cleanup, and emergency response for hazardous substances released into the environment and the cleanup of inactive hazardous waste sites. As such it subjects the petroleum industry to involvement wherever wastes have been improperly disposed of in the past. CERCLA also regulates shipping under the Hazardous Material Transportation Act. It makes the carriers responsible for damages or remedial acts resulting from the release of a hazardous substance during shipping. Costs are increasing rapidly and criminal penalties are being imposed, hence this is an area of growing concern to the industry.

Toxic Substances Control Act.   The Toxic Substances Control Act (TSCA) of 1976, as amended various times, regulates exposure to toxic substances that present hazards to health or the environment. It authorizes the U.S. EPA to develop adequate data on the health effects of chemical substances and to regulate exposures to those found to pose unreasonable risks to health and environment. All chemicals must be listed on the TSCA Inventory of Commercial Chemicals. No one may manufacture a new chemical without first giving a Pre-Manufacturing Notice (PMA). TSCA requires reporting within 15 working days of releases, conditions, or instances posing risks of injury to human health and the environment to the U.S. EPA. Records must be kept of any allegations a company receives about such risks. TSCA noncompliance could result in fines of up to \$25,000/d per violation and/or imprisonment.

Occupational Safety and Health Act.   The Occupational Safety and Health Act (OSHA), as amended in 1990, encourages reduction of occupation safety and health hazards and promotes safe and healthful conditions.

Hazardous Waste Operations and Emergency Response.   In response to an EPA mandate in SARA, Hazardous Waste Operations and Emergency Response (HAZWOPER) regulations were issued. These address emergency responders, training of those working at Superfund sites, and cleanup operations.

Criminal Enforcement.   As the environmental regulations become more complex, there has been an increasing drive in the United States to criminal enforcement for significant environmental harm and culpable conduct. The key is a pattern of behavior that has a cumulative adverse effect. This behavior has become recognized as morally reprehensible, with increasingly severe penalties imposed. Concerns regarding such enforcement have provided a strong driving force to the industry to avoid pollution.

Pollution Prevention.   Waste minimization is an important consideration in meeting the environmental regulations. Companies are finding that this can save valuable materials as well as minimize pollution and the possibility of environmental harm. Processes are increasingly designed with pollution prevention in mind. In the Pollution Prevention Act of 1990, U.S. Congress declared pollution prevention to be the national policy. The bill called on the U.S. EPA to address this need by establishing a source reduction program to collect and disseminate information, provide financial assistance to individual states, and implement other activities such as removing regulatory barriers.

BIBLIOGRAPHY

"Hydrocarbons" in *ECT* 1st ed., Vol. 7, pp. 598–652 by G. Egloff and M. Alexander; "Survey" under "Hydrocarbons" in *ECT* 2nd ed., Vol. 11, pp. 262–287, by W. Nelson Axe, Phillips Petroleum Co.; in *ECT* 3rd ed., Vol. 12, pp. 870–892, by G. H. Dale and D. P. Montgomery, Phillips Petroleum Co.

1. M. W. Ball, D. Ball, and O. S. Turner, *The Fascinating Oil Business*, Bobbs Merrill Co., Inc., Indianapolis, Ind., 1965.
2. M. L. Mesnard, ed., *The Oil Producing Industry in Your State*, Independent Petroleum Association of America, Washington, D.C., 1978.
3. I. I. Nesterov and F. K. Salmanov, in R. F. Meyer, ed., *The Future Supply of Nature-Made Petroleum and Gas*, Pergamon Press, New York, 1977, p. 185.
4. L. F. Ivanhoe and G. G. Leckie, *Oil Gas J.*, 87 (Feb. 15, 1993).
5. *Oil Gas J. Spec.*, 39 (Dec. 28, 1992).
6. R. J. Beck, "Oil Industry Outlook, 1992–1996" *Oil Gas. J.* **8** (1992).
7. W. Leprowski, *Chem. Eng. News*, 20 (June 7, 1991).
8. Technical data, *World Energy Outlook*, Chevron Corp., Richmond, Calif., Apr. 1990.
9. G. J. Mascetti and H. M. White, *Utilization of Used Oil*, Vol I, *Aerospace Report ATR-77 (7384)-1*, El Segundo, Calif., Dec., 1977.
10. *Synthetic Organic Chemicals*, U.S. International Trade Commission.
11. R. E. Clark, "Utility of Annual Lubrication Sales," quarterly *Lubricants Meeting*, Houston, Tex., Nov. 4–5, 1993.
12. P. H. Voorhees, "Generation and Flow of Used Oil on the United States in 1988," *Petroleum Re-refiners/Project ROSE Conference*, Baltimore, Nov. 30, 1989.
13. G. D. Clayton and F. E. Clayton, eds., Patty's, *Industrial Hygiene and Toxicology*, 4th ed., Vol. 2, Pt. B, John Wiley & Sons, Inc., New York, 1993.
14. *Industrial Hygiene Monitoring Manual for Petroleum Refineries and Selected Petrochemical Operations*, American Petroleum Institute, Washington, D.C., 1979.
15. G. H. Unzelman, *Oil Gas J.*, 44 (Apr. 15, 1991).
16. R. Ragsdale, *Oil Gas J.*, 51 (Mar. 21, 1994).
17. J. G. Grant and R. A. Purciau, "Gasoline Reformulation", NPRA FL-91-115, *1991 NPRA Fuels and Lubricants Meeting*, Nov. 7–9, 1991, Houston, Tex.
18. A. K. Rhodes, *Oil Gas J.*, 39 (Nov. 29, 1993).
19. *Environmental Statutes, 1993 Edition*, Government Institutes, Inc., Rockville, Md., Feb. 1993.

DAVID E. MEARS
Unocal

ALAN D. EASTMAN
Phillips Petroleum

ACETYLENE

Acetylene, C_2H_2, is a highly reactive, commercially important hydrocarbon. It is used in metalworking (cutting and welding) and in chemical manufacture. Chemical usage has been shrinking due to the development of alternative routes to the same products based on cheaper raw materials. The reactivity of acetylene is related to its triple bond between carbon atoms and, as a consequence, its high positive free energy of formation. Because of its explosive nature, acetylene is generally used as it is produced without shipping or storage.

Physical Properties

The physical properties of acetylene [74-86-2] have been reviewed in detail (1). The triple point is at $-80.55°C$ and 128 kPa (1.26 atm). The temperature of the solid under its vapor at 101 kPa (1 atm) is $-83.8°C$. The vapor pressure of the liquid at 20°C is 4406 kPa (43.5 atm). The critical temperature and pressure are 35.2°C and 6190 kPa (61.1 atm). The density of the gas at 20°C and 101 kPa is 1.0896 g/L. The specific heats of the gas, C_p and C_v (at 20°C and 101 kPa) are 43.91 and 35.45 J/mol·°C (10.49 and 8.47 cal/mol, °C), respectively. The heat of formation ΔH_f at 0°C is 227.1 kJ/mol (54.3 kcal/mol). Tables and diagrams of thermodynamic properties have been given (1) and data on the solubility of acetylene in organic liquids in relation to temperature and pressure have been reviewed and correlated (1). The dissolving powers of some of the better solvents are compared in Table 1. The solubility in water at 20°C is 16.6 g/L at 1520 kPa (15.0 atm) and 1.23 g/L at 101 kPa. Acetylene forms a hydrate of approximate

Table 1. Solubility of Acetylene in Some Organic Liquids

Solvent	CAS Registry Number	Bp, °C	Acetylene solubility[a]
acetone	[67-64-1]	56.5	237
acetonitrile	[75-05-8]	81.6	238
N,N-dimethylformamide	[68-12-2]	153	278
dimethyl sulfoxide	[67-68-5]	189	269
N-methyl-2-pyrrolidinone	[872-50-4]	202	213
γ-butyrolactone[b]	[96-48-0]	206	203

[a]g/L of solution at 15°C and 1520 kPa (15.0 atm) total pressure.
[b]Butanoic acid, 4-hydroxy-, lactone.

stoichiometry $C_2H_2 \cdot 6H_2O$. The dissociation pressure of the hydrate is 582 kPa (5.75 atm) at 0°C and 3343 kPa (33 atm) at 15°C. Its heat of formation at 0°C is 64.4 kJ/mol (15.4 kcal/mol) (1).

Chemical Properties

Acetylene is highly reactive due to its triple bond and high positive free energy of formation. Extensive reviews of acetylene chemistry are available (2–8). Important reactions involving acetylene are hydrogen replacements, additions to the triple bond, and additions by acetylene to other unsaturated systems. Moreover, acetylene undergoes polymerization and cyclization reactions. The formation of a metal acetylide is an example of hydrogen replacement, and hydrogenation, halogenation, hydrohalogenation, hydration, and vinylation are important addition reactions. In the ethynylation reaction, acetylene adds to a carbonyl group (see ACETYLENE-DERIVED CHEMICALS).

Many of the reactions in which acetylene participates, as well as many properties of acetylene, can be understood in terms of the structure and bonding of acetylene. Acetylene is a linear molecule in which two of the atomic orbitals on the carbon are sp hybridized and two are involved in π bonds. The lengths and energies of the C—H σ bonds and C≡C $\sigma + 2\pi$ bonds are as follows:

Bond	Bond length, nm	Energy, kJ/mol
≡C—H	0.1059	506
—C≡C—	0.1205	837

The two filled π orbitals result in a greater concentration of electron density between the carbon atoms than exists in ethylene. The resulting diminution of electron density on the carbon atom makes acetylene more susceptible to nucleophilic attack than is ethylene. The electron withdrawing power of the triple bond polarizes the C—H bond and makes the proton more acidic than the protons of ethylene or ethane. The pK_a of acetylene is 25 (9), and the acidic nature of acetylene accounts for its strong interaction with basic solvents in which acetylene is highly soluble (10,11). Acetylene forms hydrogen bonds with basic solvents (12), and as a result, the vapor pressures of such solutions deviate greatly from Raoult's law (13). The high concentration of electrons in the triple bond enables acetylene to behave as a Lewis base toward strong acids; it forms an adduct with HCl (14).

Metal Acetylides. The replacement of a hydrogen atom on acetylene by a metal atom under basic conditions results in the formation of metal acetylides which react with water in a highly exothermic manner to yield acetylene and the corresponding metal hydroxide. Certain metal acetylides can be prepared by reaction of the finely divided metal with acetylene in inert solvents such as xylene, dioxane, or tetrahydrofuran at temperatures of 38–45°C (15).

Acetylides of the alkali and alkaline-earth metals are formed by reaction of acetylene with the metal amide in anhydrous liquid ammonia.

$$C_2H_2 + MNH_2 \longrightarrow MC\equiv CH + NH_3$$

Aluminum triacetylide [61204-16-8] is formed from $AlCl_3$ and sodium acetylide [1066-26-8] in a mixture of dioxane and ethylbenzene at 70–75°C (16).

$$3\ NaC\equiv CH + AlCl_3 \longrightarrow Al(C\equiv CH)_3 + 3\ NaCl$$

Copper acetylides form under a variety of conditions (17–19). Cuprous acetylides are generally explosive, but their explosiveness is a function of the formation conditions and increases with the acidity of the starting cuprous solution. They are prepared by the reaction of cuprous salts with acetylene in liquid ammonia or by the reaction of cupric salts with acetylene in basic solution in the presence of a reducing agent such as hydroxylamine. Acetylides also form from copper oxides and salts produced by exposing copper to air, moisture, and acidic or basic conditions. For this reason, copper or brasses containing over 66% copper or brazing materials containing silver or copper should not be used in an acetylene system. Silver and mercury form acetylides in a manner similar to copper.

Acetylene Grignard reagents, which are useful for further synthesis, are formed by the reaction of acetylene with an alkylmagnesium bromide.

$$C_2H_2 + 2\ RMgBr \longrightarrow BrMgC\equiv CMgBr + 2\ RH$$

With care, the monosubstituted Grignard reagent can be formed and it reacts with aldehydes and ketones to produce carbinols (see GRIGNARD REACTIONS).

Hydrogenation.   Acetylene can be hydrogenated to ethylene and ethane. The reduction of acetylene occurs in an ammoniacal solution of chromous chloride (20) or in a solution of chromous salts in H_2SO_4 (20). The selective catalytic hydrogenation of acetylene to ethylene, which proceeds over supported Group VIII metal catalysts, is of great industrial importance in the manufacture of ethylene by thermal pyrolysis of hydrocarbons (21–23). Nickel and palladium are the most commonly used catalysts. Partial hydrogenation to ethylene is possible because acetylene is adsorbed on the catalyst in preference to ethylene.

Halogenation and Hydrohalogenation.   Halogens add to the triple bond of acetylene. $FeCl_3$ catalyzes the addition of Cl_2 to acetylene to form 1,1,2,2-tetrachloroethane which is an intermediate in the production of the industrial solvents 1,2-dichloroethylene, trichloroethylene, and perchloroethylene (see CHLOROCARBONS AND CHLOROHYDROCARBONS). Acetylene can be chlorinated to 1,2-dichloroethylene directly using $FeCl_3$ as a catalyst and a large excess of acetylene. trans-$C_2H_2Cl_2$ is formed from acetylene in solutions of $CuCl_2$, $CuCl$, and HCl (24–26). Bromine in solution or as a liquid adds to acetylene to form first 1,2-dibromoethylene and finally tetrabromoethylene. Iodine adds less readily and the reaction stops at 1,2-diiodoethylene. Hydrogen halides react with acetylene to form the corresponding vinyl halides. An example is the formation of vinyl chloride which is catalyzed by mercuric salts.

Hydration. Water adds to the triple bond to yield acetaldehyde via the formation of the unstable enol (see ACETALDEHYDE). The reaction has been carried out on a commercial scale using a solution process with $HgSO_4/H_2SO_4$ catalyst (27,28). The vapor-phase reaction has been reported at 250–400°C using a wide variety of catalysts (28) and even with no catalyst (29). Vapor-phase catalysts capable of converting acetic acid to acetone directly convert the steam–acetylene mixture to acetone (28,30,31).

$$2\ C_2H_2 + 3\ H_2O \longrightarrow CH_3COCH_3 + CO_2 + 2\ H_2$$

Addition of Hydrogen Cyanide. At one time the predominant commercial route to acrylonitrile was the addition of hydrogen cyanide to acetylene. The reaction can be conducted in the liquid (CuCl catalyst) or gas phase (basic catalyst at 400 to 600°C). This route has been completely replaced by the ammoxidation of propylene (SOHIO process) (see ACRYLONITRILE).

Vinylation. Acetylene adds weak acids across the triple bond to give a wide variety of vinyl derivatives. Alcohols or phenols give vinyl ethers and carboxylic acids yield vinyl esters (see VINYL POLYMERS).

$$ROH + C_2H_2 \longrightarrow ROCH{=}CH_2$$
$$RCOOH + C_2H_2 \longrightarrow RCOOCH{=}CH_2$$

Vinyl ethers are prepared in a solution process at 150–200°C with alkali metal hydroxide catalysts (32–34), although a vapor-phase process has been reported (35). A wide variety of vinyl ethers are produced commercially. Vinyl acetate has been manufactured from acetic acid and acetylene in a vapor-phase process using zinc acetate catalyst (36,37), but ethylene is the currently preferred raw material. Vinyl derivatives of amines, amides, and mercaptans can be made similarly. N-Vinyl-2-pyrrolidinone is a commercially important monomer prepared by vinylation of 2-pyrrolidinone using a base catalyst.

Ethynylation. Base-catalyzed addition of acetylene to carbonyl compounds to form -yn-ols and -yn-glycols (see ACETYLENE-DERIVED CHEMICALS) is a general and versatile reaction for the production of many commercially useful products. Finely divided KOH can be used in organic solvents or liquid ammonia. The latter system is widely used for the production of pharmaceuticals and perfumes. The primary commercial application of ethynylation is in the production of 2-butyne-1,4-diol from acetylene and formaldehyde using supported copper acetylide as catalyst in an aqueous liquid-filled system.

Polymerization and Cyclization. Acetylene polymerizes at elevated temperatures and pressures which do not exceed the explosive decomposition point. Beyond this point, acetylene explosively decomposes to carbon and hydrogen. At 600–700°C and atmospheric pressure, benzene and other aromatics are formed from acetylene on heavy-metal catalysts.

Cuprous salts catalyze the oligomerization of acetylene to vinylacetylene and divinylacetylene (38). The former compound is the raw material for the production of chloroprene monomer and polymers derived from it. Nickel catalysts with the appropriate ligands smoothly convert acetylene to benzene

(39) or 1,3,5,7-cyclooctatetraene (40–42). Polymer formation accompanies these transition-metal catalyzed syntheses.

Explosive Behavior

Gaseous Acetylene. Commercially pure acetylene can decompose explosively (principally into carbon and hydrogen) under certain conditions of pressure and container size. It can be ignited, ie, a self-propagating decomposition flame can be established, by contact with a hot body, by an electrostatic spark, or by compression (shock) heating. Ignition is generally more likely the higher the pressure and the larger the cross-section of the container. The wire temperature required for ignition decreased from 1252 to 850°C with increasing pressure from 170 to 2000 kPa (~20 atm) (43) when a platinum or nickel resistance wire of 0.25 mm diameter and 25 or 75 mm length was heated gradually in pure acetylene; the pure acetylene was initially at room temperature in a tube 50 mm in diameter and 256 mm in length.

When the wall of the container is heated, ignition occurs at a temperature that depends on the material of the wall and the composition of any foreign particles that may be present. In clean steel pipe, acetylene at 235–2530 kPa (2.3–25 atm) ignites at 425–450°C (44,45). In rusted steel pipe, acetylene at 100–300 kPa ignites at 370°C (46). In steel pipe containing particles of rust, charcoal, alumina, or silica, acetylene at 200–2500 kPa ignites at 280–300°C (44). Copper oxide causes ignition at 250°C (47) and solid potassium hydroxide causes ignition at 170°C (44).

For local, short-duration heat sources, such as electrostatic sparks, the reported ignition energies for different pressures are on the orders of magnitude given in Table 2 (48–50).

Once a decomposition flame has formed, its propagation through acetylene in a pipe is favored by large diameter and high pressure. In a long pipe of given diameter, there is a pressure below which continued propagation of a flame, even though temporarily established, is very unlikely and may be impossible. In a pipe that is so short that heating by the ignition source raises the pressure, the required initial pressure is less than in a long pipe. Figure 1 shows pressures at which a flame travels through room temperature acetylene in long horizontal pipes of various diameters as the result of thermal (nonshock) ignition. The plotted points represent values reported in the literature (51–59) and the results of unpublished work (46). Many points representing detonation at higher pressures have been omitted. Propagation at pressures below atmospheric (101 kPa) require high energy ignition sources (100–1200 J or 24–287 cal). Pressure–diameter conditions near the curve, drawn at approximately the minimum pressure for propagation, tend to lead to deflagration rather than detonation, as

Table 2. Ignition Energy of Gaseous Acetylene at Various Pressures

pressure, kPa[a]	65	100	150	200	300	1000	2000
energy, J[b]	1200	100	10	2	0.3	0.002	0.0002

[a]To convert kPa to atm, divide by 101.3.
[b]To convert J to cal, divide by 4.184.

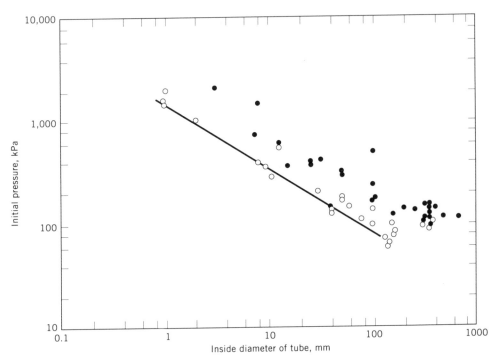

Fig. 1. Pressure required for propagation of decomposition flame through commercially pure acetylene free of solvent and water vapor in long horizontal pipes. Gas initially at room temperature; ignition by thermal nonshock sources. Curve shows approximate least pressure for propagation: (●), detonation; (○), deflagration (46,51–59). To convert kPa to atm, divide by 101.3.

indicated. However, the firing of a high explosive charge can cause a detonation wave to propagate at pressures even lower than those plotted as deflagrations, eg, 47 kPa in a 76 mm diameter tube and 80 kPa in a 13 mm diameter tube (60).

Deflagration flames in vessels and relatively short pipes usually propagate at an increasing velocity without becoming detonation waves, and develop pressures about ten times the initial pressure. In long pipes these flames usually become detonation waves. A slowly propagating deflagration flame in a long pipe occasionally neither accelerates nor dies out, but continues indefinitely. Decomposition flames in acetylene at pressures of 160–200 kPa in pipes of 50–150 mm diameter and 1500–6400 diameters in length have been observed to travel the full length at average velocities of 0.2–1 m/s (46). Generally, the velocity increases with increasing diameter. The pressure rises were less than 7 kPa (1 psi).

The calculated detonation velocity in room temperature acetylene at 810 kPa is 2053 m/s (61). Measured values are about 1000–2070 m/s, independent of initial pressure but generally increasing with increasing diameter (46,60–64). In a time estimated to be about 6 s (65), an accidental fire-initiated decomposition flame in acetylene at ca 200 kPa in an extensive piping system traveled successively through 1830 m of 76–203-mm pipe, 8850 m of 203-mm pipe, and 760 m of 152-mm pipe.

The predetonation distance (the distance the decomposition flame travels before it becomes a detonation) depends primarily on the pressure and pipe diameter when acetylene in a long pipe is ignited by a thermal, nonshock source. Figure 2 shows reported experimental data for quiescent, room temperature acetylene in closed, horizontal pipes substantially longer than the predetonation distance (44,46,52,56,58,64,66,67). The predetonation distance may be much less if the gas is in turbulent flow or if the ignition source is a high explosive charge.

The pressure developed by decomposition of acetylene in a closed container depends not only on the initial pressure (or more precisely, density), but also on whether the flame propagates as a deflagration or a detonation, and on the length of the container. For acetylene at room temperature and pressure, the calculated explosion pressure ratio, $P_{final}/P_{initial}$, is ca 12 for deflagration and ca 20 for detonation (at the Chapman-Jouguet plane). At 800 kPa (7.93 atm) initial pressure, the ratio is about the same for deflagration and 21.6 for detonation (61). The explosion pressure ratio for detonation refers to the pressure at the Chapman-Jouguet plane (of an ideal one-dimensional detonation wave) at which the heat-producing reaction has just come to equilibrium. This is the maximum effective pressure at the side wall of the pipe through which an established wave travels except at and near the end struck by the wave, where the pressure is increased by reflection. The calculated ratio, $P_{reflected}/P_{initial}$, for pressure devel-

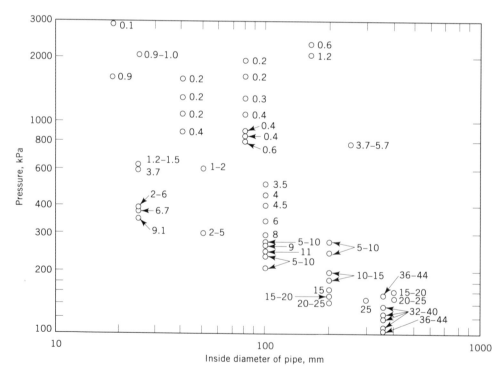

Fig. 2. Predetonation distances (in m) observed in acetylene at various pressures in horizontal pipes of various diameters. Gas quiescent, at room temperature, ignition by thermal nonshock sources (44,46,52,56,58,64,66,67). To convert kPa to atm, divide by 101.3.

oped by reflection of a detonation wave at the rigid end of a pipe is 48.5 for room temperature and pressure, and 52.5 for room temperature and 800 kPa initial pressure (46).

The measured explosion pressure ratio for deflagration in a container only a few diameters in length approaches the theoretical value; often it is about 10. However, in a pipe hundreds or thousands of diameters in length, deflagration may cause very little pressure rise because only a small fraction of the contents is hot at any time.

Explosion pressure ratios that have been measured in experiments involving detonation vary over a wide range, and depend not only on the density of the acetylene through which the detonation wave travels, but on the location of the pressure sensor and its dynamic response. Unpublished Union Carbide work, using bursting diaphragms and bonded-strain gauge sensors, records values ranging from the theoretical reflected pressure ratio of ca 50 to 238. Russian work using crusher gauges indicates ratios as high as 658 (56,58). The Russian workers do not state whether the doubling effect of rapid rise inherent in the crusher gauge (68) was taken into account. The higher ratios have been indicated when the acetylene pressure has been only slightly above the minimum required for development of detonation. Evidently in these cases the acetylene in the far end of the pipe is compressed substantially while the flame moves at subsonic velocity toward it. The flame moves through acetylene at a density that is higher than the initial density after transition to detonation, particularly when the predetonation distance is a large fraction of the pipe length.

Flame Arresters. Propagation of a decomposition flame through acetylene in a piping system (by either deflagration or detonation) can be stopped by a hydraulic back pressure valve in which the acetylene is bubbled through water (65,69). It can also be stopped by filling the pipe with parallel tubes of smaller diameter, or randomly oriented Raschig rings (54,70–72). The small tubes should have a diameter less than that indicated (for the pressure to be used) by the curve of Figure 1, and the packed section should be long enough so that any decomposition products that are pushed through will not ignite the gas downstream. The presence of water or oil (on the walls or as mist) increases the effectiveness of the arrangement. Beds of granular ceramic material are effective with acetylene at cylinder pressure.

Ignition of Gaseous Acetylene Mixtures. The initial pressure required for ignition by fused resistance wire is given for several acetylene–diluent mixtures in Figure 3. Air in concentrations of less than ca 13% inhibits ignition, but oxygen in any concentration promotes it. The data were obtained with relatively small containers and low ignition energies. With larger containers and higher ignition energies, the minimum pressure for ignition may be somewhat lower.

Acetylene–Air and Acetylene–Oxygen Mixtures. The flammability range for acetylene–air at atmospheric pressure is ca 2.5–80% acetylene in tubes wider than 50 mm. The range narrows to ca 8–10% as the diameter is reduced to 0.8 mm. Ignition temperatures as low as 300°C have been reported for 30–75% acetylene mixtures with air and for 70–90% mixtures with oxygen. Ignition energies are lower for the mixtures than for pure acetylene; a spark energy of 0.02 mJ (200 erg) has been found sufficient to ignite a 7.7% acetylene–air mixture at atmospheric pressure and room temperature (76,77).

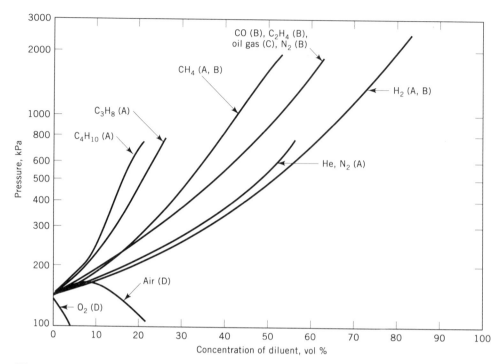

Fig. 3. Pressure required for ignition of mixtures of acetylene and a diluent gas (air, oxygen, butane, propane, methane, carbon monoxide, ethylene, oil gas, nitrogen, helium, or hydrogen) at room temperature. Initiation: fused resistance wire. Container: A, 50 mm dia × 305 mm length (73); B, 269 mm dia × 269 mm length (52); C, 102 mm dia × 254 mm length (74); and D, 120 mm dia sphere (75). To convert kPa to atm, divide by 101.3.

In acetylene–air mixtures, the normal mode of burning is deflagration in relatively short containers and detonation in pipes. In oxygen mixtures, detonation easily develops in both short and long containers. Measurements of predetonation distances of acetylene–oxygen at 100 kPa and 40°C in 25 and 50 mm diameter tubes 2.9 and 3.6 m long gave values of 25–35 mm for mixtures of 25–50% acetylene (78). For gas mixtures at atmospheric temperature and pressure, the maximum detonation velocity has been calculated to be 2020 m/s for acetylene–air (at 15% acetylene) (79) and 2944–2960 m/s for acetylene–oxygen (at 50% acetylene) (79,80). The corresponding explosion pressure ratios for detonations in mixtures of these compositions have been calculated to be 21.9–22.4 for acetylene–air (79,81) and 43.5–50.2 for acetylene–oxygen (79,80), both referring to the Chapman-Jouguet plane. The $P_{reflected}/P_{initial}$ ratios for the same mixtures are estimated to be ca 50 for acetylene–air and 110 for acetylene–oxygen (46). Except for the detonation pressures of acetylene–air mixtures, these quantities have been measured and the results agree approximately with the calculated values (46).

Several studies of spherical and cylindrical detonation in acetylene–oxygen and acetylene–air mixtures have been reported (82,83). The combustion and oxidation of acetylene are reviewed extensively in Reference 84. A study of the characteristics and destructive effects of detonations in mixtures of acetylene (and other hydrocarbons) with air and oxygen-enriched air in earthen tunnels and large steel pipe is reported in Reference 81.

Liquid and Solid Acetylene. Both the liquid and the solid have the properties of a high explosive when initiated by detonators or by detonation of adjoining gaseous acetylene (85). At temperatures near the freezing point neither form is easily made to explode by heat, impact, or friction, but initiation becomes easier as the temperature of the liquid is raised. Violent explosions result from exposure to mild thermal sources at temperatures approaching room temperature.

The minimum ignition energy of liquid acetylene under its vapor, when subjected to electrostatic sparks, has been found to depend on the temperature as indicated in Table 3 (86). Ignition appears to start in gas bubbles within the liquid.

Table 3. Minimum Ignition Energy of Liquid Acetylene

temperature, °C	-78	-50	-40	-35	-30	-27
vapor pressure, kPa^a	145	537	779	931	1103	1234
minimum ignition energy, J^b	>11	1.5	0.98–4.1	0.98	0.68	0.13

^{a}To convert kPa to atm, divide by 101.3.
^{b}To convert J to cal, divide by 4.184.

Analysis

General Methods. Traces of acetylene can be detected by passing the gas through Ilosvay's solution which contains a cuprous salt in ammoniacal solution. The presence of acetylene is indicated by a pink or red coloration caused by the formation of cuprous acetylide, Cu_2C_2. The same method can be used for the quantitative determination of acetylene in parts per billion concentrations; the copper acetylide is measured colorimetrically (87).

The preferred quantitative determination of traces of acetylene is gas chromatography, which permits an accurate analysis of quantities much less than 1 ppm. This procedure has been highly developed for air pollution studies (88) (see AIR POLLUTION CONTROL METHODS). Other physical methods, such as infrared and mass spectroscopy, have been widely used to determine acetylene in various mixtures.

Acetylene can be determined volumetrically by absorption in fuming sulfuric acid (or more conveniently in sulfuric acid activated with silver sulfate); or by reaction with silver nitrate in solution and titration of the nitric acid formed:

$$HC\equiv CH + 3\ AgNO_3 \longrightarrow (AgC\equiv CAg)AgNO_3 + 2\ HNO_3$$

The precipitated acetylide must be decomposed with hydrochloric acid after the titration as a safety measure. Concentrated solutions of silver nitrate or silver perchlorate form soluble complexes of silver acetylide (89). Ammonia and hydrogen sulfide interfere with the silver nitrate method which is less accurate than the sulfuric acid absorption method. Acetylene and monosubstituted acetylenes may also be determined by means of potassium mercuric iodide and potassium hydroxide in methanol solution by back-titration of the excess potassium hydroxide (90).

$$2\,RC\equiv CH + K_2HgI_4 + 2\,KOH \longrightarrow (RC\equiv C)_2Hg + 4\,KI + 2\,H_2O$$

Acetylene Derived from Calcium Carbide. The analysis of acetylene derived from calcium carbide includes the determination of phosphorus, sulfur, and nitrogen compounds which are always present in the crude gas. Gas chromatographic methods which are accurate and convenient have been developed for phosphine, arsine, hydrogen sulfide, and ammonia (91). Chemically, the quantitative determination of phosphorus and sulfur can be achieved by oxidation with calcium or sodium hypochlorite. Phosphine and hydrogen sulfide present in the gas are oxidized to phosphate and sulfate and are measured gravimetrically as phosphomolybdate and barium sulfate, respectively. Ammonia is determined by the Nessler method after absorption in dilute hydrochloric acid. Oxygen, if present, can be determined by gas chromatography or by paramagnetic oxygen analyzer. It can also be determined chemically by absorption in alkaline pyrogallol after the acetylene has been removed by fuming sulfuric acid (20%) or it can be determined electrometrically after removal of interfering impurities. Qualitatively, hydrogen sulfide and phosphine can be detected in concentrations as low as 10 ppm by the brown to black discoloration of moist silver nitrate paper.

Acetylene Derived from Hydrocarbons. The analysis of purified hydrocarbon-derived acetylene is primarily concerned with the determination of other unsaturated hydrocarbons and inert gases. Besides chemical analysis, physical analytical methods are employed such as gas chromatography, ir, uv, and mass spectroscopy. In industrial practice, gas chromatography is the most widely used tool for the analysis of acetylene. Satisfactory separation of acetylene from its impurities can be achieved using 50–80 mesh Porapak N programmed from 50–100°C at 4°C per minute.

Handling of Acetylene

The design of equipment for the handling and use of acetylene must take into consideration the possibility of acetylene decompositions. The design parameters must consider various factors, namely pressure, temperature, source of ignitions, and ultimate pressures which may result from a decomposition. Decompositions do not occur spontaneously but must have a source of ignition. Decompositions in small vessels and short piping systems used at moderate pressures of 103 kPa (15 psi) result in a maximum pressure not greater than approximately 12 times the initial pressure (1240 kPa or 180 psi), in this example. Theoretically, during constant volume deflagration at ~100–500 kPa (1–5 atm) without loss of heat, the pressure rises to 11.5–12 times the initial pressure. The minimum acetylene pressure at which a deflagration flame can propagate throughout a long tube of any diameter has been determined experimentally. As a typical example, acetylene deflagrates in a 2.54 cm inside diameter tube at pressures above 241 kPa (35 psi) and a detonation does not develop. In pipelines of considerable lengths and of diameter sufficient to permit a true detonation to develop, the maximum pressure developed is ~ 50 times the initial pressure. Thus, the maximum pressure expected in a 2.54 cm inside diameter tubing at 277 kPa (40 psi) is 13.8 MPa (277 kPa × 50). In long pipelines an effect called cascading may develop in which case the maximum pressure may be several hundred times the initial pressure.

Exact design criteria on equipment for handling acetylene is not readily available because of the great number of factors involved. However, recommendations have been made concerning the equipment, piping, compressors, flash arresters, and proper materials (69,92–94).

Acetylene in Cylinders. Acetylene cylinders are constructed to stabilize acetylene and, thereby, safely avoid the hazard of a detonation (95). Cylinders constructed for other gases do not have the same features and it is extremely important that such cylinders not be charged with acetylene. Likewise, acetylene cylinders should not be charged with other gases, even though they are capable of containing those gases up to the service pressure of the cylinder. The basic feature of an acetylene cylinder that is different from all other cylinders is that it is entirely filled with a monolithic porous mass. It is this monolithic mass that stabilizes the acetylene and permits its safe shipment. The monolithic mass is a unique technical development because it must have high porosity (up to 92%) to be economical and yet possess sufficient strength that it will not break down, crack, or crumble during many years of service. Thus whereas the porous mass completely fills the interior of the cylinder, it occupies only about 10% of the total volume. There is a slight clearance between the outside surface of the filler mass and the inside surface of the pressure shell, through which the acetylene flows to the valve for discharge. After the cylinder has been manufactured, a specified quantity of solvent is added, usually acetone. Acetone dissolves many times its own volume of acetylene and its purpose is to increase the amount of acetylene that may be safely charged and shipped. The acetylene is, therefore, not a free gas but is in solution. The solubility of acetylene in acetone increases with rising pressure and with diminishing temperature. Thus a pressure gauge attached to an acetylene cylinder reads the solution pressure and is not a direct measure of the amount of acetylene contained. The pressure is greatly affected by changes in cylinder temperature. For example, the gauge pressure of a cylinder may be 1590 kPa (230 psi) when the temperature is 21°C, and less than 692 kPa (100 psi) at −17.7°C without any acetylene having been withdrawn. It is therefore obvious that the contents of an acetylene cylinder, unlike oxygen or nitrogen cylinders, cannot be determined accurately by pressure gauge readings alone. Acetylene cylinder contents can, however, be accurately measured by weight, and it is on this basis that cylinder charging operations are conducted. Weight of acetylene can be converted into standard volume (atmospheric pressure, and 21°C) by the factor of 0.906 L/kg (14.7 ft^3/lb).

Acetylene cylinders are fitted with safety devices to release the acetylene in the event of fire. Cylinders manufactured in the United States are equipped with safety devices which contain a fusible metal that melts at 100°C. In large cylinders the safety devices are in the form of a replaceable, threaded steel plug with a core of fusible metal. Small cylinders (0.28 and 1.12 m^3; 10 and 40 ft^3, respectively) may have the fusible metal in passages in the cylinder valve.

The most common sizes of acetylene cylinders are those with nominal capacities of 0.28, 1.12, 2.80, 8.4, and 11.2 m^3 (10, 40, 100, 300, and 400 ft^3). The largest size produced in any quantity is about 28.0 m^3 (1000 ft^3) capacity and is commonly referred to as a lighthouse cylinder because of its use in lighting buoys, etc, for marine navigation purposes.

The manufacture and shipping of acetylene cylinders in the United States are in compliance with the specifications and regulations of the Department of Transportation (96). The specifications are verified by the Bureau of Explosives of the American Association of Railroads. DOT-8 and DOT-8AL specify the requirements for manufacture and testing of steel shell, porous mass, and quantity of acetone (therefore, acetylene) which may be charged. These regulations specify such things as the chemical analysis and physical properties of the shell material, certain critical fabrication limits, heat treatment and the tests which must be conducted on each cylinder, as well as destructive sampling tests. To be a legal article of commerce, each cylinder must bear the following markings: DOT Specification (for 8AL), serial number, registered symbol, inspector symbol, date of manufacture, and tare weight. The cylinder must be registered with the Bureau of Explosives through a manufacturing report which also includes the test results for that group of cylinders.

The tare weight (sometimes called stencil weight because it is cut into the cylinder metal) is the total weight of the cylinder and contents, but does not include a removable valve protection cap, if such is used. The saturation gas part of the tare weight is a calculated number which allows for the 11.4 g of acetylene required to saturate each 453.6 g of contained acetone at atmospheric pressure. The correct tare weight is an absolute necessity to the safe charging of acetylene cylinders.

The specifications set the maximum vol % of solvent that may be added to the cylinder shell (measured by its water capacity). The volume of solvent also varies with the capacity of the cylinder. Cylinders in the 90−92% porosity range with a capacity above 9.1 kg of water may contain a maximum acetone charge of 43.4%, whereas those with 9.1 kg or less water capacity may contain up to 41.8 vol %. The first category of cylinders are normally referred to as welding cylinders and the latter as small tanks (those with 0.28 and 1.12 m^3 acetylene capacity).

The volume of acetylene that may be charged is limited by the DOT regulations to the amount which would produce an equilibrium pressure of 1833 kPa (266 psi) at 21°C. This maximum ratio by weight is 0.58 units of acetylene per unit of acetone. Welding cylinders designed and charged to these limits have a liquid-full temperature of about 65.5°C and the small tanks have a liquid-full temperature of about 79.4°C. The liquid-full temperature represents the point at which the liquid (acetylene−acetone) has expanded to completely fill the container. The DOT regulations specify that an acetylene cylinder shall not be filled and shipped at a pressure which would exceed the equivalent pressure of 1833 kPa at 21°C. A cylinder designed to the limits indicated above and charged to 1833 kPa at 21°C would have a pressure of approximately 2142 kPa (311 psi) at 32°C.

The DOT regulations also stipulate that prior to manufacture of cylinders of new design, or with substantially modified design, prototype cylinders must pass a set of Designed Qualification Tests administered by the Bureau of Explosives. These tests include a drop test to show that the porous mass will not compact, break down, or disintegrate during normal use; a flash test to show that the porous mass–acetone–shell combination will extinguish a flash which enters the cylinder (this requirement is the heart of acetylene cylinder design); a fire test to show the adequacy of the safety devices; and an impact test to show that the full charge of acetylene is stable under conditions such as falling off a fast-moving truck.

After a design has proved to be satisfactory by passing these tests, the regulations require inspection during manufacture to make certain that the cylinders produced are of equal quality.

Filling Cylinders with Acetylene. The filling and shipping of acetylene cylinders are subject to the regulations of the U.S. Department of Transportation. To completely charge acetylene cylinders in a reasonable period of time requires compression of acetylene to pressures above 1833 kPa, usually in the range of 2074–2419 kPa (300–350 psi). Because acetylene at these pressures detonates, if a source of ignition is present, the cylinder charging plant must be carefully designed and constructed taking into account all of the safety hazards. The acetylene industry has prepared a basic set of guidelines (94).

Many factors must be taken into account in the charging operation, such as the rate of charging, cooling during charging, and interstage cooling of the gas to remove as much water as possible prior to the final drying. Water reduces the quantity of acetylene which may safely be carried, because the solubility of acetylene in water is less than in acetone. Other important factors include the mechanical reliability of the cylinder, valves and safety devices, the residual acetylene, and the presence of sufficient acetone to maintain the 0.58 ratio of acetylene:solvent. The charging of acetylene cylinders can be hazardous and should only be undertaken with the consent of the owner and by persons having full knowledge of the subject (97).

Manufacture

FROM CALCIUM CARBIDE

Acetylene is generated by the chemical reaction between calcium carbide [75-20-7] and water with the release of 134 kJ/mol (900 Btu/lb of pure calcium carbide).

$$CaC_2 + 2 H_2O \longrightarrow Ca(OH)_2 + C_2H_2$$

Because of the exothermic reaction and the evolution of gas, the most important safety considerations in the design of acetylene generators are the avoidance of excessively high temperatures and high pressures. The heat of reaction must be dissipated rapidly and efficiently in order to avoid local overheating of the calcium carbide which, in the absence of sufficient water, may become incandescent and cause progressive decomposition of the acetylene and the development of explosive pressures. Maintaining temperatures less than 150°C also minimizes polymerization of acetylene and other side reactions which may form undesirable contaminants. For protection against high pressures, industrial acetylene generators are equipped with pressure relief devices which do not allow the pressure to exceed 204.7 kPa (15 psig). This pressure is commonly accepted as a safe upper limit for operating the generator.

Most carbide acetylene processes are wet processes from which hydrated lime, $Ca(OH)_2$, is a by-product. The hydrated lime slurry is allowed to settle in a pond or tank after which the supernatant lime-water can be decanted and reused in the generator. Federal, state, and local legislation restrict the methods of storage and disposal of carbide lime hydrate and it has become increasingly important to find consumers for the by-product. The thickened hydrated lime is marketed for industrial wastewater treatment, neutralization of spent pickling acids, as a soil conditioner in road construction, and in the production of sand–lime bricks.

Carbide-to-Water Generation. This process is the one most widely used in the United States for generating acetylene from calcium carbide. Standards for the design and construction of acetylene-generating equipment using this technique have been developed over the years by the acetylene industry. Underwriters Laboratories, Inc. have generally accepted design criteria for acetylene generating equipment (98). A water capacity of 3.78 L (1 gal) per 0.454 kg of carbide and a gas-generating rate of 0.028 m^3 (1 ft^3) per hour per 0.454 kg of carbide hopper capacity is considered normal. These design criteria apply to gravity feed generators where it is possible to have an uncontrolled release of the entire carbide hopper contents into the generating chamber. Other high capacity generators (up to 283 m^3/h) are designed so that it is impossible to have uncontrolled feed of carbide (screw-feed type); therefore, the chamber water capacity can be reduced, the carbide hopper capacity can be increased, and the gas production capacity can be raised. These high capacity generators also must pass prescribed safety tests before sale.

There are two classes of acetylene generators: the low pressure generator which operates below 108.2 kPa (15.7 psi), and the medium pressure generator which operates between 108.2 and 204.7 kPa (29.7 psi). The latter is more prevalent in the United States.

There are numerous variations in the design of commercially available carbide-to-water acetylene generators. Basically, however, they are practically identical in that they consist of a water vessel or reaction chamber, a carbide feed mechanism, and a carbide storage container that empties into the feed mechanism. The water vessel is equipped with a means of filling with water and draining the lime slurry. Agitation, either hand or power driven, is provided for keeping the lime hydrate and reacting carbide in suspension. Pressure gauges and relief devices are also incorporated in the generating chamber. The water shell or generating chamber in the more common generator supports the feed mechanism and the carbide hopper, which is also fitted with pressure gauges and safety relief valves. The valved gas outlet from the generator leads directly to a flash arrester which protects the equipment against flashbacks originating from acetylene-consuming equipment downstream. The gas in the high capacity generator, owing to the higher operating temperatures, is first cooled in a water scrubber, which is an integral part of the generator, before passing through the flash arrester. The continuous supply of cooling water in the scrubber is also a source of water for the reaction with carbide. Most commercial acetylene generators are fitted with either mechanical, pneumatic, or electrical interference mechanisms which for safety reasons enforce a prescribed sequence of procedures for the operator using the equipment. Some of these safety mechanisms are high and low water level shutdown, high temperature cut out, high acetylene pressure shutdown, and shutdown on loss of either electric or pneumatic power to the generator controls. All safety conditions must be satisfied before start-up or during operation.

The gas demand dictates the rate of acetylene generation which is satisfied by the rate of carbide feed. One method of feed of properly sized carbide, which is used in medium pressure generators, employs gravity flow controlled by a valve activated by a spring-loaded rubber diaphragm. The motion of the diaphragm reflects the change of internal generator pressure and is transmitted to the carbide feed valve which, in turn, either opens or closes as the pressure decreases or increases; the generator operating pressure is set by the spring load applied to the rubber diaphragm. High capacity generators are equipped with screw conveyors for carbide feed with either constant speed on–off operation or of the variable-speed close-pressure control type. The on–off type is controlled by a simple electric pressure switch which pneumatically signals the required feed-screw operation to match the gas demand and can have pressure fluctuations up to 128.9 kPa (18.7 psi) or less.

Carbide-to-water generators exhibit a noticeable temperature rise in the course of normal operation. In properly designed units employing the recommended ratio of water to carbide, the temperature rises 21–27°C and thus attains a temperature on the order of 60–65°C. In high capacity continuous generators, where water addition is through sprays in the cooling tower or the generating chamber, temperatures in the slurry are allowed to rise to 82.2°C. With this type of generator the lime slurry must be dumped periodically to avoid flooding and there is a certain amount of gas lost each time. Since the solubility of acetylene in lime slurry is greatly diminished at 71.1–82.2°C, the gas loss through solubility is minimized by operation at these elevated temperatures.

Carbide of proper size, such as 14 ND or 6 × 2 mm, is important to the trouble-free operation of carbide-to-water generators. Larger size carbide interferes with the proper closing of the carbide valve causing uncontrolled feed. In screw feeds the carbide size is limited only to the clearance between the screw and its housing. Generators that do not have continuous paddle agitation or water sprays to submerge and wet the freshly introduced carbide must use oiled carbide to reduce the reaction rate. Otherwise, small size (6 × 2 mm) carbide, which reacts rapidly with water, can be carried to the surface by the evolved gas where, without sufficient dissipation of heat, it can become incandescent and may initiate explosive decomposition of acetylene. Low pressure generators use a large grade of carbide, 35 × 9 mm. The low pressure generator is no longer made in the United States; however, some are still in use after more than 40 years of service. Any generator equipped with effective water sprays or paddles can use mixtures of dust carbide. It is imperative that the dust be completely wetted; otherwise, islands of dust will float on the surface where they can become incandescent and initiate explosive acetylene decomposition.

Water-to-Carbide Generation. This method of acetylene production has found only limited acceptance in the United States and Canada but has been used frequently in Europe for small-scale generation. The rate of generation is regulated by the rate of water flow to the carbide. Hazardous hot spots may occur and overheating may lead to the formation of undesirable polymer by-products. This method is, therefore, used mainly in small acetylene generators such as portable lights or lamps where the generation rate is slow and the mass of carbide is small.

Dry Generator. This water-to-carbide acetylene generation method is used in certain large-scale operations. The dry process uses about a kg of water per kg of carbide and the heat of reaction is dissipated by the vaporization of the water. Absolute control of the addition of water is critical and the reacting mass of dry lime and unreacted carbide must be continuously mixed to prevent hazardous localized overheating and formation of undesirable polymer by-products. The gas stream is filtered to remove lime dust. The dry lime by-product is considered to be advantageous compared to the wet lime by-product. Lime from dry generators is very fine and requires storage in silos or protection from scattering by wind currents. Transport must be in closed containers, such as bags.

Purification of Carbide Acetylene. The purity of carbide acetylene depends largely on the quality of carbide employed and, to a much lesser degree, on the type of generator and its operation. Carbide quality in turn is affected by the impurities in the raw materials used in carbide production, specifically, the purity of the metallurgical coke and the limestone from which the lime is produced. The nature and amounts of impurities in carbide acetylene are shown in Table 4.

The maximum amount of impurities in U.S. Grade B acetylene (99) (Carbide Generated Acetylene) is 2% on a dry basis. This gas meets commercial requirements for acetylene used in cutting and welding (qv). Production of U.S. Grade A

Table 4. Impurities in Carbide Acetylene

Type	Amount, approx
PH_3	a few hundred ppm
$(CH_2{=}CH)_2S$	100 ppm (as H_2S)
NH_3	a few hundred ppm
O_2	250 ppm or less
N_2	few tenths of a percent (<1.0)
ArH_3	3 ppm or less
CH_4, CO_2, CO, H_2	a few hundred ppm
SiH_4	10 ppm or less
$CH_2{=}CH{-}C{\equiv}CH$	50 ppm
$CH_2{=}CH{-}C{\equiv}C{-}CH{=}CH_2$	50 ppm
$HC{\equiv}C{-}C{\equiv}H$	a few hundred ppm
$(CH_2{=}C{=}CH_2)^a$	traces (variable according to carbide quality)

^{a}And other dienes, eg, hexadiene and butadienylacetylene; also methylacetylene.

acetylene (99) used in sensitive chemical reactions requires further purification to reduce impurities to 0.5%. There are four main impurities: phosphine [7803-51-2], ammonia [7664-41-7], hydrogen sulfide [7783-06-4], and organic sulfides. The purification involves oxidation of phosphine to phosphoric acid, the neutralization and absorption of ammonia, and the oxidation of hydrogen sulfide and organic sulfur compounds. Many processes are employed depending on the type and amount of impurities and the end use of the gas. These wet or dry processes range from simply passing the gas over purifying media to multistep chemical treatments.

The most commonly used dry methods employ oxidizing agents such as chromic acid or chromates, hypochlorite, permanganate, and ferric salts deposited on solid carriers such as diatomaceous earth arranged in beds or layers through which the gas is passed at ambient temperature. Some of the purifying media can be regenerated several times with diminishing effectiveness until they eventually lose their activity. Because of the high material and labor requirements, dry purification of acetylene is not practiced where large volumes of gas have to be treated. Large-scale acetylene installations exclusively employ continuous, wet purification processes. Elaborate continuous purification methods have been developed in Europe, where at certain locations relatively low grade carbide is used to generate acetylene of low purity; these involve successive contact with water, dilute caustic solution, and chlorine–water or hypochlorite solution, followed in certain locations by a final treatment with activated carbon (100–102). Such an intensive purification of acetylene is beneficial in cases where the gas is to be used in processes employing sensitive catalytic systems.

BIBLIOGRAPHY

"Properties" under "Acetylene" in *ECT* 3rd ed., Vol. 1, pp. 192–203, by C. M. Dietz and H. B. Sargent, Union Carbide Corp.; "Handling," pp. 203–210, by R.O. Tribolet, Union Carbide Corp.; "Manufacturing," pp. 203–210, by R. P. Shaffer, Union Carbide Corp.

1. S. A. Miller, *Acetylene—Its Properties, Manufacture and Uses*, Vol. 1, Academic Press, Inc., New York, 1965.
2. J. W. Copenhaver and M. H. Bigelow, *Acetylene and Carbon Monoxide Chemistry*, Reinhold Publishing Co., New York, 1949.
3. S. A. Miller, *Acetylene—Its Properties, Manufacture and Uses*, Vol. 2, Academic Press Inc., New York, 1966.
4. R. A. Raphael, *Acetylene Compounds in Organic Synthesis*, Butterworths, London, 1955.
5. T. F. Rutledge, *Acetylenic Compounds—Preparation and Substitution Reactions*, Reinhold Book Corp., New York, 1968.
6. T. F. Rutledge, *Acetylenes and Allenes*, Reinhold Book Corp., New York, 1969.
7. H. G. Viehe, *Chemistry of Acetylenes*, Marcel Dekker, Inc., New York, 1969.
8. R. J. Tedeschi, *Acetylene-based Chemicals from Coal and Other Natural Resources*, Marcel Dekker, Inc., New York, 1982.
9. Ref. 6, p. 8.
10. A. C. McKinnis, *Ind. Eng. Chem.* **27**, 2928 (1962).
11. Ref. 1, p. 74.
12. R. C. West and C. S. Kraihanzel, *J. Am. Chem. Soc.* **83**, 765 (1961).
13. H. J. Copley and C. E. Holley, Jr., *J. Am. Chem. Soc.* **61**, 1599 (1939).

14. D. Cook, Y. Lupien, and W. G. Schneider, *Can. J. Chem.* **34**, 957 (1956).
15. T. Mole and J. R. Suertes, *Chem. Ind. London*, 1727 (1963).
16. U.S. Pat. 3,321,487 (May 23, 1967) P. Chini and A. Baradal.
17. V. F. Bramfield, M. T. Clark, and A. P. Seyfang, *J. Soc. Chem. Ind. London* **65**, 346 (1947).
18. W. Feitknecht and L. Hugi-Carmes, *Schweiz. Arch. Angew. Wiss. Tech.* **23**, 328 (1957).
19. W. Pakulat, *Schweisstechnik Berlin* **19**, 374 (1964).
20. Ref. 3, p. 1.
21. G. Bond and P. B. Wells, in D. D. Eley, H. Pines, and P. B. Weisz, eds., *Advances in Catalysis*, Vol. 15, Academic Press, Inc., New York, 1964, p. 155.
22. Ref. 3, pp. 1–22.
23. *Proceedings of the 4th Ethylene Producers Conference*, Vol. 1, American Institute of Chemical Engineers, New York, 1992, pp. 177–211.
24. U.S. Pat. 2,440,997 (May 4, 1948), E. Adler.
25. A. L. Klebanskii, A. S. Vol'kenshtein, and N. A. Orlova, *Zh. Obshch. Khim.* **5**, 1255 (1935); *J. Prakt. Chem.* **1**, 145 (1936).
26. Ger. Pat. 968,921 (Apr. 10, 1958); 969,191 (May 8, 1959); and 1,097,977 (Jan. 26, 1961), A. Jacabowsky and K. Sennewald.
27. J. A. Nieuwland and R. R. Vogt, *The Chemistry of Acetylene*, Reinhold, New York, 1945, pp. 115–123.
28. Ref. 3, pp. 134–145.
29. U.S. Pat. 3,291,839 (Dec. 13, 1966), R. W. Carney and G. A. Renberg.
30. Yu. A. Gorin, I. K. Gorin, and N. A. Rozenberg, *Zh. Prikl. Khim.* (*Leningrad*) **40**, 399 (1967).
31. Yu. A. Gorin, I. K. Gorin, and A. E. Kalaus, *Chem. Abstr.* **57**, 12305a (1962).
32. Ref. 2, pp. 37–40.
33. Ref. 3, pp. 198–207.
34. U.S. Pat. 3,370,095 (Feb. 20, 1968), J. F. Vitcha.
35. V. A. Sims and J. F. Vitcha, *I&EC Product Res. Dev.* **2**, 293 (1963).
36. Brit. Pat. 1,125,055 (Aug. 28, 1968), C. H. A. Borsboom, P. A. Gautier, and D. Medema.
37. Brit. Pat. 1,100,038 (Jan. 24, 1968), L. G. Smith.
38. J. A. Nieuwland and co-workers, *J. Am. Chem. Soc.* **53**, 4197 (1931).
39. U. Rosenthal and W. Schulz, *J. Organomet. Chem.* **321**, 103 (1987).
40. Ref. 2, p. 189.
41. W. Reppe and co-workers, **560**, 1 (1948).
42. G. N. Schrauzer, P. Glockner, and S. Eichler, *Angew. Chem. Internat. Edit.* **3**, 185 (1964).
43. C. M. Detz, *Combust. Flame* **26**, 45 (1976).
44. Ref. 1, p. 485.
45. W. Rimarski and M. Konschak, *Forschungsarb. Schweissens Schneidens* **4**, 43 (1929).
46. Technical data, Union Carbide Corp., South Charleston, W. Va., unpublished.
47. W. Rimarski and M. Konschak, *Forschungsarb. Schweissens Schneidens* **5**, 100 (1930).
48. B. A. Ivanov and S. M. Kogarko, *Zh. Prikl, Mekh. Tek. Fiz.* **59**(3) (1963).
49. B. A. Ivanov and S. M. Kogarko, *Nauch, Tekh. Probl. Goreniya Vzryva* **105**(2)(1965).
50. Y. Hashiguchi and T. Fujisaki, *Kogyo Kagaku Zasshi* **61**(6), 515 (1958).
51. P. Hölemann, R. Hasselmann, and G. Dix, *Forschungsber. Landes Nordrhein Westfalen* **102** (1954).
52. N. A. Copeland and M. A. Youker, "German Techniques for Handling Acetylene in Chemical Operations," *FIAT Final Report No. 720 (PB 20078)*, 1946; Includes translations of reports by Boesler, Rimarski, and Weissweiler.
53. M. Gugger, Chemische Werke Hüls, private communication, 1954.

54. H. Schmidt and K. Haberl, *Tech. Ueberwach.* **7**(12), 423 (1955).
55. W. Rimarski and M. Konschak, *Forschungsarb. Schweissens Schneidens* **9**, 105 (1934).
56. S. M. Kogarko, A. G. Lyamin, and V. A. Mikhailov, *Khim. Prom. Moscow* **41**(8), 621 (1965); *Dokl. Akad. Nauk SSSR* **162**(4), 857 (1965).
57. S. M. Kogarko and B. A. Ivanov, *Dokl. Phys. Chem.* **140**(1), 676 (1961).
58. S. M. Kogarko and co-workers, *Khim. Prom. Moscow* **7**, 496 (1962).
59. E. Barsalou, *Mem. Poudres* **43**, 63 (1961).
60. R. E. Duff, H. T. Knight, and H. R. Wright, *J. Chem. Phys.* **22**, 1618 (1954).
61. E. Penny, *Discuss. Faraday Soc.* **22**, 157 (1956).
62. M. Berthelot and H. LeChatelier, *C. R. Acad. Sci.* **129**, 427 (1899).
63. Ref. 1, p. 495.
64. P. Hölemann, R. Hasselmann, and G. Dix, *Forschungsber. Landes Nordrhein Westfalen*, **382** (1957).
65. M. E. Sutherland and H. W. Wegert, *Chem. Eng. Prog.* **69**(4), 48 (1973).
66. W. Rimarski and M. Konschak, *Forschungsarb. Schweissens Schneidens* **6**, 92 (1931).
67. Ref. 1, p. 496.
68. A. B. Arons, *Underwater Explosion Research*, Vol. 1, U.S. Office of Naval Research, 1950.
69. *Acetylene Transmission for Chemical Synthesis, Pamphlet G 1.3*, Compressed Gas Association, Arlington, Va., 1984.
70. H. Beller, *Weld. J. (Miami, Fla.)* **37**, 1090 (1958).
71. S. M. Kogarko, A. G. Lyamin, and V. A. Mikhailov, *Khim. Prom. Moscow* **4**, 275 (1964).
72. B. A. Ivanov and S. M. Kogarko, *Int. Chem. Eng. Process. Ind.* **4**(4), 670 (1964).
73. G. W. Jones, R. E. Kennedy, and I. Spolan, *U.S., Bur. Mines Rep. Invest.* **4196** (1948).
74. W. Rimarski and M. Konschak, *Forschungsarb. Schweissens Schneidens* **8**, 113 (1933).
75. Ya. M. Landesman, L. M. Savichkaya, and M. A. Glikin, *Khim. Prom. Moscow* **47**(5), 347 (1971).
76. H. F. Calcote and co-workers, *Ind. Eng. Chem.* **44**, 2656 (1952).
77. J. B. Fenn, *Ind. Eng. Chem.* **43**, 2865 (1951).
78. L. E. Bollinger, M. C. Fong, and R. Edse, *Amer. Rocket Soc. J.* **31**, 588 (1961).
79. N. Manson, *Propagation des Détonations et des Déflagrations dans les Mélanges Gazeux*, Office National Etudes Recherches Aéronautiques et Institut Français Pétroles, Paris, 1947.
80. G. B. Kistiakowsky, H. T. Knight, and M. E. Malin, *J. Chem. Phys.* **20**, 884 (1952).
81. D. S. Burgess and co-workers, *U.S. Bur. Mines Rep. Invest.* **7196**, (1968).
82. G. A. Carlson, *Combust. Flame* **21**(3), 383 (1973).
83. J. H. Lee, B. H. K. Lee, and I. Shanfield, *10th Symposium Combustion, University Cambridge, 1964*, pp. 805–813, discussion 813–815 (Pub. 1965).
84. A. Williams and D. B. Smith, *Chem. Rev.* **70**, 267 (1970).
85. H. A. Mayes and H. J. Yallop, *Chem. Eng. London* **1965** (185).
86. D. W. Breck, H. R. Gallisdorfer, and R. P. Hamlen, *J. Chem. Eng. Data* **7**, 281 (1962).
87. E. E. Hughes and R. Gorden, Jr., *Anal. Chem.* **31**, 94 (1959).
88. *Test for C_1 through C_5 Hydrocarbons in the Atmosphere by Gas Chromatography*, ASTM Analysis D2820-72, Vol. 26.
89. L. Barnes, Jr., and I. J. Molinini, *Anal. Chem.* **27**, 1025 (1955).
90. J. G. Hanna and S. Siggia, *Anal. Chem.* **21**, 1469 (1949).
91. L. Chelmu, *Chim. Anal. Bucharest* **2**, 212 (1972).
92. W. G. Schepman, paper presented at the *International Symposium Acetylene Association Meeting*, Philadelphia, Pa., 1958.
93. H. Schmidt and K. Haberl, *Tech. Ueberwach.* **423**(12), (1955); A. Ebert, *Explosivstoffe* **44**, 245 (1956).

94. *Acetylene Cylinder Charging Plants*, Pamphlet NFPA SIA, National Fire Prevention Association, Boston, Mass., 1974.

95. *Guidelines for Periodic Visual Inspection and Requalification of Acetylene Cylinders, Pamphlet C-13*, Compressed Gas Association, Arlington, Va., 1992; *Acetylene, Pamphlet G-1*, Compressed Gas Association, Arlington, Va., 1990.

96. Department of Transportation, *Code of Federal Regulations 49*, items 173.303; 173.306; 178.59; and 179.60, Washington, D.C., Oct. 1, 1973.

97. Department of Transportation, *CFR 49(10-1-73)* item 173.30(b), p. 207.

98. Underwriters Laboratories Inc., *Standards for Acetylene Generators No. 297 Portable Medium Pressure*, May 1973; *No. 408 Stationary Medium Pressure*, May 1973.

99. Federal Specification, *Acetylene Technical Dissolved BB-A-106B*, Washington, D.C.

100. Grimm, "Large-Scale Purification of Carbide Acetylene with Special Regard to the Production of Acetaldehyde and Ethylene," *OTS Report, PB 35209*, U.S. Department of Commerce, 1944.

101. Merkel, "Results of Acetylene Purification at Ludwigshafen," *OTS Report, PB 35211*, U.S. Department of Commerce, 1944.

102. W. E. Alexander, "Purification and Drying of Acetylene for Chemical Use," *OTS Report, PB 44943*, U.S. Department of Commerce, Washington, D.C.

ROBERT M. MANYIK
Union Carbide Corporation

C. M. DIETZ
H. B. SARGENT
R. O. THRIBOLET
R. P. SCHAFFER
Consultants

Manufacture

FROM HYDROCARBONS

Although acetylene production in Japan and Eastern Europe is still based on the calcium carbide process, the large producers in the United States and Western Europe now rely on hydrocarbons as the feedstock. Now more than 80% of the acetylene produced in the United States and Western Europe is derived from hydrocarbons, mainly natural gas or as a coproduct in the production of ethylene. In Russia about 40% of the acetylene produced is from natural gas.

Development of the modern processes for the manufacture of acetylene from hydrocarbons began in the 1920s when Badische Anilin- und Soda-Fabrik (BASF) initiated an intensive research program based on Berthelot's early (1860) laboratory investigations on the conversion of low molecular weight aliphatic hydrocarbons to acetylene by means of thermal cracking. BASF's development of the electric arc process led to the first commercial plant for the manufacture of acetylene from hydrocarbons. This plant was put into operation at Chemische Werke Hüls in Germany in 1940. In the United States, commercial manufacture of acetylene from hydrocarbons began in the early 1950s; expansion was rapid until the mid to late 1960s, when acetylene was gradually supplanted by cheaper ethylene as the main petrochemical intermediate.

Theory. The hydrocarbon to acetylene processes that have been developed to commercial or pilot-plant scale must recognize and take advantage of the unique thermodynamic properties of acetylene. As the free energy data shown in Figure 4 indicate, the common paraffinic and olefinic hydrocarbons are more stable than acetylene at ordinary temperatures. As the temperature is increased, the free energy of the paraffins and olefins become positive while that of the acetylene decreases, until at >1400 K acetylene is the most stable of the common hydrocarbons. However, it is also evident that, although it has the lowest free energy of the hydrocarbons at high temperature, it is still unstable in relation to its elements C and H_2. Thus it is necessary to heat the feedstock extremely fast to minimize its decomposition to its elements and, for a similar reason, the quench must be extremely rapid to avoid the decomposition of the acetylene product. Numerous acetylene production processes have been developed, each in its own way and with varying degrees of success, accommodating the unique thermodynamics and pyrolysis kinetics of acetylene.

Examination of the equilibrium composition of the product gas mixture under relevant reactor conditions indicates the restrictive process conditions required to optimize the production process. Figure 5 illustrates the equilibrium composition for the carbon–hydrogen system with a C/H ratio of 1 to 4 at 101.3 kPa (1 atm) and at temperatures to 7000 K. This diagram is relevant to the pyrolysis of methane at atmospheric pressure. It is immediately evident that the hydrocarbon feedstock, CH_4, decomposes into its thermodynamically preferred state of C and H_2 at well below 1000 K, whereas appreciable amounts of acetylene are not present until about 3000 K. Fortunately, the rate of the formation of acetylene is greater than the CH_4 decomposition rate. Thus it is important to heat the reactant as rapidly as possible to avoid decomposition of the feedstock to C and H_2 and to maximize the C_2H_2 formation. In a study to design an electric arc reactor (1) for producing acetylene from coal, it was found that the acetylene reaches equilibrium concentrations in less than 1 millisecond. Thus with rapid mixing and heating, it is possible to attain appreciable concentrations of acetylene with relatively little degradation of the feedstock to carbon.

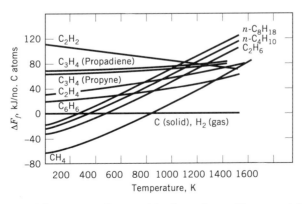

Fig. 4. Free energy of formation of several hydrocarbons. To convert kJ to kcal, divide by 4.184.

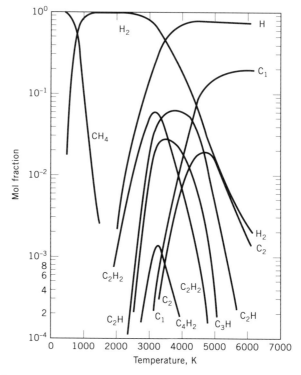

Fig. 5. Equilibrium diagram for carbon–hydrogen system at 101.3 kPa (1 atm). (C/H = 1/4).

Optimum reactor design and operating conditions can be further explored through equilibrium diagrams and computer models. The effects of feedstock composition pressures and temperature on the product composition have been explored in this manner in a study (1) in which the added constraint that the heating would be rapid enough to preempt the degradation of the feedstock to carbon was imposed. Thus it was shown (Fig. 6) that at a C/H of 1 to 6 and at 51 kPa (0.5 atm), the concentration of acetylene at equilibrium conditions was as high as 16% and the temperature of peak concentration was 2000 K. Thus the process conditions of methane pyrolysis in excess hydrogen at reduced pressure are more promising than pyrolysis at atmospheric conditions as depicted in Figure 5.

Addressing the second step of the reaction, ie, the quench step, it is most important to quench the equilibrium mixture as quickly as possible in order to preserve the high acetylene concentration. In a study of the quenching mechanism (2), the acetylene-forming step was separated from the acetylene-preserving step by injecting known amounts of acetylene into a carbon-free plasma stream. The effect of various gases injected into the stream not only indicated the effectiveness of the quenching medium, but also revealed a great deal about the dynamics of the high temperature equilibrium composition. Hydrogen is much more effective in preserving the acetylene than the inert gases argon, helium, or nitrogen. Thus hydrogen injection allows recovery of as much as 90% of the

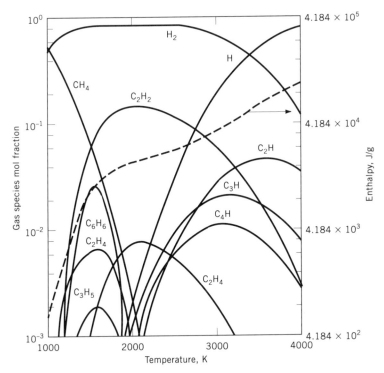

Fig. 6. Equilibrium diagram for carbon–hydrogen system at 51 kPa (0.5 atm). (C/H = 1/6). To convert J to cal, divide by 4.184.

acetylene, whereas with the other gases less than 50% of the acetylene was recovered. The observed effect of pressure was that in the 25–50 kPa (0.25–0.50 atm) range 85–90% of the acetylene is recovered, but at pressures between 50 and 100 kPa recovery decreases to 70%.

Because it was not possible to explain the differences in the effectiveness of hydrogen as compared to other gases on the basis of differences in their physical properties, ie, thermal conductivity, diffusivity, or heat capacity differences, their chemical properties were explored. To differentiate between the hydrogen atoms in the C_2H_2 molecules and those injected as the quench, deuterium gas was used as the quench. The data showed that although 90% of the acetylene was recovered, over 99% of the acetylene molecules had exchanged atoms with the deuterium quench to form C_2HD and C_2D_2.

To extend the study of the apparent decomposition recombination reaction, and specifically to determine if the carbon atoms exchange with other atoms in other acetylene molecules, tests using carbon isotopes were conducted. A mixture of 50% regular acetylene, $^{12}C_2H_2$, and 50% heavy acetylene, $^{13}C_2H_2$, was injected into the plasma stream. The results showed that, as before, 90% of the acetylene was recovered and that 97% of the acetylene molecules had exchanged carbon atoms.

These isotope exchange reactions not only provide an explanation for the effectiveness of hydrogen as a quench for the acetylene mixture, but also provide insight into the nature of the dynamic equilibrium present in hydrocarbon–hydrogen mixtures at high temperatures. The data indicate that essentially all of the acetylene molecules underwent total atom exchange, but because only a fraction of the energy required for decomposition of the C_2H_2 was supplied by the electric plasma, a chain or shuffle reaction is implied. The shuffle is initiated by the fragmentation of a relatively few C_2H_2 molecules into C_2H, C_2, CH, and H species. These fragments collide with C_2H_2 molecules, exchanging atoms and splitting off additional fragments. Allowing a residence time of 0.1 milliseconds at an average plasma temperature of 3000 K, it can be estimated that each molecule undergoes approximately 2×10^4 collisions. If an efficiency of 10% is assumed, each molecule experiences 2000 viable collisions in the first 2.5 cm of residency in the plasma stream. As the reaction mixture cools downstream, the number of collisions decrease and the chain reaction terminates as two CH fragments or a C_2H and H fragment collide to reform an acetylene molecule. Thus 90% of the acetylene is recovered but each C_2H_2 molecule has undergone some 2000 atom exchanges. As long as the exchanges occur between C and H species, high acetylene yield can be preserved. The introduction of inert species, A, He, or N_2, however, terminates the chain reaction and leads to acetylene degradation.

Process Technology. The processes designed to produce acetylene as the main product of a hydrocarbon feedstock are generally classified according to their energy source, ie, electricity or combustion. Using this classification, several processes that are now or have been operated commercially are listed in Table 5 and are described in the subsequent text. Two special cases, the production of acetylene by steam hydrocracking in oil refineries and the potentially commercial process of producing acetylene from coal, are also discussed.

Electric Discharge Processes. The synthetic rubber plant built by the I. G. Farbenindustrie during World War II at Hüls, contained the first successful

Table 5. Acetylene Process Technology

Energy source	Process designation	Feedstock	Typical cracked gas concentrations, mol %	
			Acetylene	Ethylene
Electricity				
electric arc	Hüls	natural gas	15	0.9
arc plasma	Hüls	crude oil	14	7
	Hoechst	naphtha	14	7
Combustion				
partial comb.	BASF, SBA, Montecatini	natural gas, naphtha	8	0.2
pyrolysis	Hoechst HTP, SBA	natural gas	11	15
	BASF Submerged Flame	naphtha, bunker C	6	6
	Wulff	range of hydrocarbons	14	8
	Kureha	crude oil	8^a	8^a

[a]Concentrations depend on severity of pyrolysis. At a high severity ($\sim$2000°C) acetylene/ethylene ratio is 1, but at lower severity acetylene concentration is reduced and ethylene is increased.

commercial installation for the electric arc cracking of lower hydrocarbons to acetylene. The plant, with a capacity of 200 t/d, was put into operation in August 1940.

The electric discharge processes can supply the necessary energy very rapidly and convert more of the hydrocarbons to acetylene than in regenerative or partial combustion processes. The electric arc provides energy at a very high flux density so that the reaction time can be kept to a minimum (see FURNACES, ELECTRIC–ARC FURNACES).

There have been many variations in the design of electric arc reactors but only three have been commercialized. The most important is the installation at Hüls. The other commercial arc processes were those of Du Pont (3) (a high speed rotating arc) and a Romanian process that produced both ethylene and acetylene. The Hüls process and the Romanian process (at Borzesti, Romania) are still operating, but the Du Pont process has been shut down since 1969.

Hydrocarbon, typically natural gas, is fed into the reactor to intersect with an electric arc struck between a graphite cathode and a metal (copper) anode. The arc temperatures are in the vicinity of 20,000 K inducing a net reaction temperature of about 1500°C. Residence time is a few milliseconds before the reaction temperature is drastically reduced by quenching with water. Just under 11 kW·h of energy is required per kg of acetylene produced. Low reactor pressure favors acetylene yield and the geometry of the anode tube affects the stability of the arc. The maximum theoretical concentration of acetylene in the cracked gas is 25% (75% hydrogen). The optimum obtained under laboratory conditions was 18.5 vol % with an energy expenditure of 13.5 kW·h/kg (4).

Hüls Arc Process. The design of the Hüls arc furnace is shown in Figure 7. The gaseous feedstock enters the furnace tangentially through a turbulence chamber, E, and passes with a rotary motion through pipe H (length approx 1.5 m, diameter 85–105 mm). The arc, G, burns between the bell-shaped cathode, C, and the anode pipe, H (grounded). Due to the rotary motion of the gas, the starting points of the arc rotate within the hollow electrodes. The cathodic or anodic starting point of the arc can move upward or downward freely. With the exception of the insulator, D, all parts of the furnace are made of iron. The wall thickness of the electrodes is 10–20 mm.

The arc is about 100 cm long and extends about 40–50 cm into the anode pipe. About 20 cm below the anodic starting point of the arc, cold hydrocarbons (C ≥ 2) are introduced into the tube through several nozzles to prequench the hot (about 1750 K) reaction gases. The quench feed becomes partly cracked (mainly to ethylene). Immediately below the anode pipe the hot reaction mixture is cooled to a temperature of about 450 K by means of a water spray, I. The electrodes are water-jacketed. The cathode is insulated from the other parts of the furnace which are grounded (insulator D). The arc is started by means of an ignition electrode. The arc is operated at 8000 kW, 7000 V, and a direct current of 1150 A. Off-peak power is generally used (5).

The feed to the arc consists of a mixture of fresh hydrocarbons and recycle gas. Table 6 indicates the composition of a typical feedstock as well as the composition of the gas leaving the arc furnace.

Taking into account the purification losses, the following operating requirements are necessary in order to obtain 100 kg of purified acetylene: 200 kg

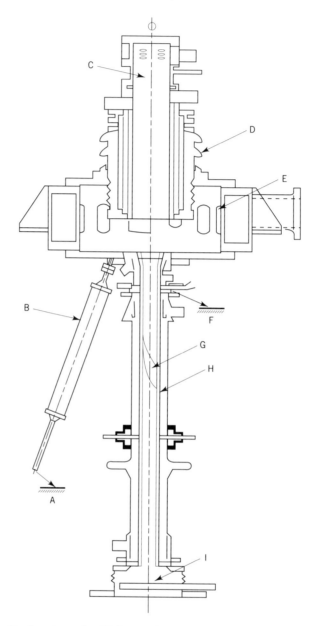

Fig. 7. Schematic drawing of a Hüls arc furnace. A, ground; B, ignition electrode; C, bell-shaped cathode; D, insulator; E, turbulence chamber; F, ground; G, arc; H, anode pipe; and I, water spray. Courtesy of Hüls AG.

hydrocarbons (feedstock plus quench), 1030 kW·h electric energy for the arc, 250 kW·h electric energy for the separation unit, and 150 kg steam.

The by-products amount to 49.5 kg ethylene, 29 kg carbon black, 15 kg residual oil, and 280 m³ hydrogen.

A considerable amount of carbon is formed in the reactor in an arc process, but this can be greatly reduced by using an auxiliary gas as a heat carrier.

Table 6. Composition of Feedstock and Reaction Product, Arc Process

Component	Feed gas, including recycle, vol %	Cracked gas,[a] vol %
C_2H_2	1.2	15.9
C_3H_4	1.0	1.0
C_4H_2	0.8	0.5
C_4H_4	0.7	0.5
C_2H_4	1.7	7.1
C_3H_6	2.3	0.9
C_4H_8	1.0	0.4
C_4H_6	0.4	0.3
CH_4	53.4	17.0
C_2H_6	10.2	1.2
C_3H_8	7.9	0.8
C_4H_{10}	12.5	2.1
C_5H_6	0.2	0.2
C_6H_6	0.4	0.4
C_7H_8		0.1
H_2	2.8	50.1
CO	0.8	0.7
N_2	2.7	0.8

[a]The cracked gas contains the products produced in the arc from the feedstock as well as the products obtained from the quench hydrocarbons. The liquid quench feed amounts to 120 kg/1000 kW·h and is composed of 25 kg C_3H_8, 60 kg n-C_4H_{10}, and 35 kg iso-C_4H_{10}.

Hydrogen is a most suitable vehicle because of its ability to dissociate into very mobile reactive atoms. This type of processing is referred to as a plasma process and it has been developed to industrial scale, eg, the Hoechst WLP process. A very important feature of a plasma process is its ability to produce acetylene from heavy feedstocks (even from crude oil), without the excessive carbon formation of a straight arc process. The speed of mixing plasma and feedstock is critical (6).

Farbwerke Hoechst AG and Hüls AG have cooperated in the development of industrial-scale plasma units up to 10,000 kW (7). Yields of acetylene of 40–50 wt % with naphtha feedstock, and about 27 wt % with crude oil feedstock, have been obtained. Acetylene concentration in the cracked gas is in the 10–15 vol % range.

Hoechst WLP Process. The Hoechst WLP process uses an electric arc-heated hydrogen plasma at 3500–4000 K; it was developed to industrial scale by Farbwerke Hoechst AG (8). Naphtha, or other liquid hydrocarbon, is injected axially into the hot plasma and 60% of the feedstock is converted to acetylene, ethylene, hydrogen, soot, and other by-products in a residence time of 2–3 milliseconds. Additional ethylene may be produced by a secondary injection of naphtha (Table 7, Case A), or by means of radial injection of the naphtha feed (Case B). The oil quenching also removes soot.

Hüls Plasma Process. In the Hüls plasma process, the hydrocarbon is injected tangentially into the hot hydrogen. In crude oil cracking, a residence time of 2–4 ms converts 20–30% of the crude (8). Crude oil data are given in Table 7, and data for naphtha and light hydrocarbon feeds are given in Table 8. In general, the arc processes achieve high temperatures easily, produce high

Table 7. Characteristic Data of Electric Plasma Processes[a]

Data	Hoechst WLP process[b]		Hüls[c]
	Case A	Case B	
output, kW	10,000	9,000	8,500
naphtha input/100 kg acetylene, kg	192	250	
crude oil/100 kg acetylene, kg			367
quenching oil, kg	53	63	
products/100 kg acetylene,			
ethylene, kg	50	95	48
C_1 to C_6 hydrocarbons, kg			82
hydrogen, m^3 (kg)	145 (13)	150 (13.5)	112 (10)
quenching oil, 20% carbon, kg	75	100	
crude oil residue with 20% carbon, kg			127
energy consumption/100 kg acetylene, kW·h	930	1,095	980
analysis of cracking gases, vol %			
C_2H_2	13.7	10.8	14.5
C_2H_4	6.4	9.8	6.5
yield (C_2H_2 + C_2H_4), wt %	78	78	56

[a]Ref. 7.
[b]Hydrogen plasma process using naphtha. Case A: secondary injection of naphtha; Case B: radial injection of the naphtha feed.
[c]Hydrogen plasma process using crude oil.

Table 8. Operational Results of the Hüls Plasma Process in the Cracking of Light Hydrocarbons[a,b]

Data	Propane	n-Butane	Benzene	Naphtha
acetylene in the cracking gas, vol %	13.7	14.6	18.1	14.8
energy consumption, kW·h/100 kg C_2H_2	960	960	900	990
acetylene to ethylene ratio	2.2	1.7	18.0	1.8
carbon (rust) formation, kg/100 kg C_2H_2	2.3	3.1	44.5	6.1
yield (C_2H_2 + C_2H_4), wt %	61	61	56	54

[a]Courtesy of Applied Science Publishers Ltd.
[b]Ref. 7.

yields of acetylene and few by-products, but can be handicapped by excessive carbon formation. On the strongly negative side are the high power consumption and the difficulty of controlling the arc geometry. Preheating the feed gas is one method to reduce cost in arc processes.

Electric arcs have been struck between grains of coal submerged in liquid hydrocarbons, such as kerosene and crude oil (9,10), to produce a gas with 30 vol % acetylene and 5–11 vol % ethylene (11). The energy consumption in those cases is about 9 kW·h/kg acetylene.

Flame or Partial Combustion Processes. In the combustion or flame processes, the necessary energy is imparted to the feedstock by the partial combustion of the hydrocarbon feed (one-stage process), or by the combustion of residual gas, or any other suitable fuel, and subsequent injection of the cracking stock into the hot combustion gases (two-stage process). A detailed discussion of the kinetics for the pyrolysis of methane for the production of acetylene by partial oxidation, and some conclusions as to reaction mechanism have been given (12).

There are several commercial versions of this partial combustion technique, including the widely used BASF process (formerly called the Sachsse process) and its various modifications with an overall similar design (13). Natural gas or other methane-rich feedstock is mixed with a limited amount of oxygen (insufficient for complete combustion), and fed through a specially designed distributor or burner to a single reaction zone in which ignition occurs. Preheating of the oxygen and methane, which is usually carried up to 500°C or above, supplies part of the energy and thus, by using less oxygen, reduces dilution of the acetylene by carbon oxides and hydrogen.

The design of the burner is of considerable importance (see COMBUSTION TECHNOLOGY). Combustion must be as brief and uniform as possible across the reaction chamber. Preignition, stability and blow-off of the flame, the possibility of backfiring through the ports of the burner head, and the deposition of carbon on the burner walls depend on the burner design and the velocities of the gas and the flame. The feasibility of partial combustion processes results from the high rates of reaction together with the relatively slow rate of decomposition of acetylene and hydrocarbon to carbon and hydrogen.

So-called tonnage oxygen, with a purity of 95–98%, is normally used as the oxidant. Although more expensive than air, its use gives several economic advantages, including a higher acetylene concentration in the cracked gas which results in lower purification costs. In addition, the plant off-gas obtained after separation of the acetylene contains high concentrations of hydrogen and carbon monoxide which, after further treatment, can be used for the synthesis of methanol or ammonia. The utilization of the off-gas is of considerable importance in establishing satisfactory economics for the partial combustion processes.

BASF Process.   The basic design of the BASF process converter is shown in Figure 8. The burner is made of mild steel and is water-cooled. The hydrocarbon feed can be methane, LPG, or naphtha, and these are separately preheated and mixed with oxygen. Self-ignition occurs if methane is preheated to 650°C and naphtha to 320°C. The oxygen and hydrocarbon feed are mixed in a venturi and passed to a burner block with more than 100 channels. The gas mixture speed in the channels is kept high enough to avoid backfiring but low enough to avoid blowout. The flame stability is enhanced by the addition of small amounts of oxygen flowing downward from the spaces between the channels. About one-third of the methane feed is cracked to acetylene and the remainder is burned.

The reaction gas is rapidly quenched with injected water at the point of optimum yield of acetylene, which happens to correspond with the point of maximum soot production. Coke will deposit on the walls of the burner and must be removed from time to time by a scraper.

The composition of the cracked gas with methane and naphtha and the plant feed and energy requirements are given in Table 9. The overall yield of acetylene based on methane is about 24% (14). A single burner with methane produces 25 t/d and with naphtha or LPG produces 30 t/d. The acetylene is purified by means of N-methylpyrrolidinone.

SBA Process.   Two partial combustion processes have been developed by the Société Belge de l'Azote et des Produits Chimiques de Marly (located near

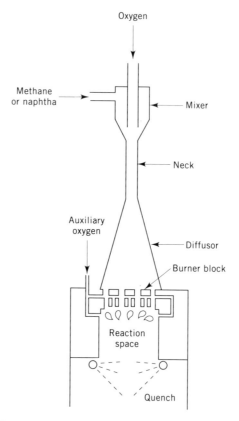

Fig. 8. BASF burner for the production of acetylene from methane or light naphtha (14). Courtesy of Verlag Chemie GmbH, Weinheim.

Brussels). The first is a single-stage process using an entirely metallic converter. It produces 20–25 t/d of acetylene from methane, with an oxygen consumption of 4.6 kg/kg of acetylene produced. Methane and oxygen are preheated separately to 700°C and mixed. The oxygen is mixed with the methane through a series of holes in the internal shell. The volume flow rate of the mixed gas is set, using an inverted cone-shaped device, to that required at the point of ignition.

The flame-space walls are stainless steel and are water cooled. No mechanical coke scraper is required. A water quench cools the cracked gas stream rapidly at the point of maximum acetylene and this is followed by a secondary water quench. The primary quench point can be adjusted for variation in throughput, to accommodate the dependence of acetylene yield on residence time in the flame space.

Purification of the cracked gas is accomplished by water scrubbing, an electrostatic precipitator, and liquid ammonia absorption.

The SBA two-stage converter (Fig. 9) consists of two superimposed chambers. In the first (combustion) chamber, the combustion in oxygen of a hydrogen-rich gas is effected in the presence of superheated steam. By means of a special design, the combustion takes place with the formation of a ring of short flames, surrounded by steam. The energy required for pyrolysis is highly concentrated

Table 9. BASF Process Consumptions and By-Product Yields and Cracked Gas Composition[a,b]

Component	Methane	Naphtha
Feed and energy requirements		
hydrocarbon, kg/100 kg C_2H_2	410	430
oxygen, kg	490	430
N-methylpyrrolidinone, kg	0.5	0.5
electric energy consumption, kW·h	230	210
steam requirement, kg	450	450
residual gas, m^3	850	760
carbon, kg	5	30
Cracked gas, vol %		
C_2H_2	8.0	9.3
C_2H_4	0.2	0.2
CH_4	4.2	5.0
CO_2	3.4	3.8
CO	25.9	36.9
H_2	56.8	43.2
N_2	0.8	0.7
O_2	0.2	0.2
other hydrocarbons	0.5	0.7

[a] Courtesy of Verlag Chemie GmbH, Weinheim.
[b] Ref. 14.

and thermal losses are reduced to a minimum (15). In the second (pyrolysis) chamber, the hydrocarbon feedstock is injected into the hot combustion gases. The reaction products are thoroughly quenched to avoid all parasitic reactions.

With this type of burner, a wide variety of raw materials, ranging from propane to naphtha, and heavier hydrocarbons containing 10–15 carbon atoms, can be used. In addition, the peculiar characteristics of the different raw materials that can be used enable the simultaneous production of acetylene and ethylene (and heavier olefins) in proportions which can be varied within wide limits without requiring basic modifications of the burner.

Montecatini Process. This partial combustion process operates at higher pressure, 405–608 kPa (4–6 atm), than the BASF and SBA processes. The burner dimensions are proportionately smaller. Because of the higher pressure, the danger of premature ignition of the methane–oxygen mixture is higher so that 2 vol % of steam is added to the gas mixture to alter the flammability limits.

The cracked gas composition is shown in Table 10 for the water quench operation (16). One thousand cubic meters of methane and 600 m^3 of oxygen produce 1800 m^3 of cracked gas. If a naphtha quench is used, additional yields are produced, consuming 130 kg of naphtha/1000 m^3 of methane (17). The volume and distribution of this additional yield is shown in Table 10. Purification of the acetylene is by methanol absorption.

Hoechst HTP Process. The two-stage HTP (high temperature pyrolysis) process was operated by Farbwerke Hoechst in Germany. The cracking stock for the HTP process can be any suitable hydrocarbon. With hydrocarbons higher than methane, the ratio of acetylene to ethylene can be varied over a range of

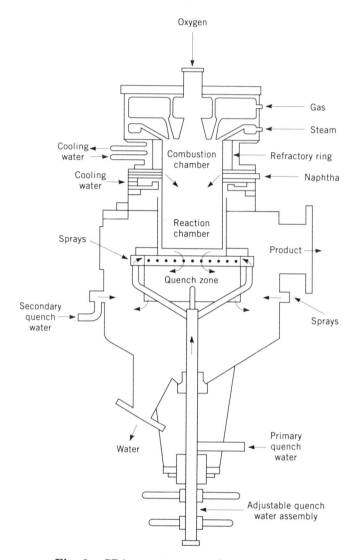

Fig. 9.   SBA two-stage acetylene converter.

70:30 to 30:70. Total acetylene and ethylene yields, as wt % of the feed, are noted in Table 11.

The Hoechst burner is a water-cooled unit of all-metal construction. Fuel gas, which may be hydrogen, hydrocarbons, or off-gas from the process, is burned with oxygen in near stoichiometric amount in the combustion chamber. The hot combustion gases (tempered with dilution steam), together with the hydrocarbon feedstock injected, preferably as a vapor, enter the reaction zone where cracking of the feedstock takes place. Residence time in the reaction zone is very short, of the order of 1 ms. A rapid quench in specially designed equipment is effected to reduce the gas temperature below cracking temperatures.

Table 10. Montecatini Process Cracked Gas Composition[a]

Component	Composition, vol %	STP,[b] m³
C_2H_2	8.5	19.2
CO_2	3.8	2.6
CH_4	6.5	19.7
H_2	54.3	57.8
CO	25.2	9.8
C_2H_4 and higher hydrocarbons	1.7	31.2
Total	*100.0*	*140.3*

[a]Ref. 16.
[b]Per 100 kg of naphtha added.

Table 11. High Temperature Pyrolysis Process Yields

Feed	Yield, wt %	Product
methane[a]	40.0	acetylene
butane	54.8	acetylene + ethylene (50:50)
light naphtha	54.0	acetylene + ethylene (30:70)
	50.0	acetylene + ethylene (70:30)

[a]Methane recycled.

BASF Submerged-Flame Process. This process can make acetylene from a wide range of feedstocks (naphtha to Bunker C oil) and, of course, crude oil itself. Oil is burned below the surface in an electrically ignited, oxygen-fed flame and quenching is immediate by the surrounding oil. The operating pressure is 900 kPa (9 bars) (14). The temperature of the oil is regulated at 200–250°C by circulation to a waste-heat boiler. The soot content of the oil is purged by burning it in the reactor. Crude oils with 12.4 wt % hydrogen can be cracked with a resulting soot level in the oil of 30%. Lower hydrogen content crudes can be handled by a separate purge of the oil to remove excess soot. An average composition of the cracked gas is shown in Table 12; it does not vary much with feedstock changes. The capacity of the commercial burner is 25 tons of acetylene and 30 tons of ethylene per day.

Table 12. BASF Submerged-Flame Process-Average Cracked Gas Composition[a,b]

Component	Vol %
CO	43
H_2	29
CO_2	7
CH_4	4
C_2H_4	6.7
C_2H_2	6.2
C_3–plus higher hydrocarbons	4.0
H_2S	0.03–0.3

[a]Courtesy of Verlag Chemie GmbH,   eim.
[b]Ref. 14.

In summary, the bad features of partial combustion processes are the cost of oxygen and the dilution of the cracked gases with combustion products. Flame stability is always a potential problem. These features are more than offset by the inherent simplicity of the operation, which is the reason that partial combustion is the predominant process for manufacturing acetylene from hydrocarbons.

Regenerative Furnace Processes. The regenerative furnace processes supply the necessary energy for the cracking reaction by heat exchange with a solid refractory material. An alternating cycle operation is employed whereby the hydrocarbon feed is heated by the hot refractory mass to produce acetylene. Following this period, during which carbon and tars are deposited on the refractories, the process employs a combustion step in which the refractory mass is heated in an oxidizing atmosphere and the carbon and tar deposits are removed by burning. The refractories must resist both reducing and oxidizing atmospheres at ca 1200°C. The refractories must also withstand the frequent and rapid heating and cooling cycles and abrasion in the case of moving refractory beds (pebbles).

Wulff Process. The regenerative technique is best exemplified by the Wulff process, licensed by Union Carbide Corp. The furnace consists basically of two masses of high purity alumina refractory tile having cylindrical channels for gas flow and separated by a central combustion space as shown in Figure 10. Its cyclic operation has four distinct steps, each of approximately one minute in duration, the sequence being pyrolysis and heat in one direction followed by pyrolysis and heat in the other direction. Continuity of output is achieved by paired installations.

The regenerative nature of the Wulff operation permits the recovery of most of the sensible heat in the cracked gas. The gases leave the furnace at temperatures below 425°C, thus obviating the need for special high temperature alloys in the switch valve and piping system.

This type of regenerative process runs at low pressure (just below atmospheric) and uses a considerable amount of dilution steam (two to three times

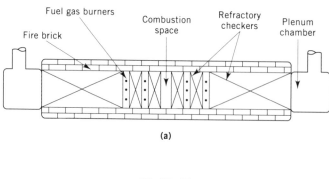

(a)

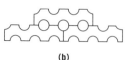

(b)

Fig. 10. (a) Wulff furnace design. (b) Checker detail of Wulff furnace refractory.

the hydrocarbon feed). To crack methane, a reaction temperature of 1500°C must be reached, but higher hydrocarbons can be pyrolyzed to acetylene at lower temperatures, eg, 1200°C. Up to 15 vol % acetylene can be obtained in the cracked gas, but ethylene can also be produced at lower average cracking temperatures and with lower acetylene yields. When cracking propane to acetylene and ethylene, the acetylene concentration in the cracked gases ranges from 14 to greater than 16 mol %, and the ethylene concentration ranges from 8 to 13 mol % (18). Typical yields for acetylene plus ethylene (once-through cracking) on propane feed range from 51 to 59 wt % for acetylene to ethylene ratios of 3.5:1 and 1:3.5, respectively. Dimethylformamide is the purification solvent used (19).

Regenerative pyrolysis processing is very versatile; it can handle varied feedstocks and produce a range of ethylene to acetylene. The acetylene content of the cracked gases is high and this assists purification. On the other hand, the plant is relatively expensive and requires considerable maintenance because of the wear and tear on the refractory of cyclic operation.

Pyrolysis by Direct Firing. Pyrolysis of hydrocarbon in direct-fired tubes with steam dilution is practiced extensively to make ethylene (qv). This technique is operated generally at the limits of metallurgy and at the maximum severity permissible (combination of time and temperature), while avoiding excessive coking rates inside the cracking tubes. The manufacture of acetylene requires even higher cracking temperatures. Such severe conditions normally induce an extreme rate of coking and an inoperable situation. If, however, the requisite high temperature can be reached without a high cracking severity and without excessively hot reactor walls, catastrophic coking rates can be avoided and useful operation is possible. Kureha Chemical Industries (Japan) developed a process based on this principle that operates at a level of pyrolysis severity and allows acceptable levels of acetylene production.

The unit Kureha operated at Nakoso to process 120,000 metric tons per year of naphtha produces a mix of acetylene and ethylene at a 1:1 ratio. Kureha's development work was directed toward producing ethylene from crude oil. Their work showed that at extreme operating conditions, 2000°C and short residence time, appreciable acetylene production was possible. In the process, crude oil or naphtha is sprayed with superheated steam into the specially designed reactor. The steam is superheated to 2000°C in refractory lined, pebble bed regenerative-type heaters. A pair of the heaters are used with countercurrent flows of combustion gas and steam to alternately heat the refractory and produce the superheated steam. In addition to the acetylene and ethylene products, the process produces a variety of by-products including pitch, tars, and oils rich in naphthalene. One of the important attributes of this type of reactor is its ability to produce variable quantities of ethylene as a coproduct by dropping the reaction temperature (20–22).

Separation and Purification of Hydrocarbon-Derived Acetylene. The pyrolysis of methane results in a cracked gas that is relatively low in acetylene content and that contains predominantly a mixture of hydrogen, nitrogen, carbon monoxide, carbon dioxide, unreacted hydrocarbons, acetylene, and higher homologues of acetylene. In cases where a higher hydrocarbon than methane is used as feedstock, the converter effluent also contains olefins (ethylene, propylene, propadiene, butadiene), aromatics (benzene, naphthalene), and miscellaneous higher hydrocarbons. Most acetylene processes produce significant amounts of carbon black and tars which have to be removed before the separation of acetylene from the gas mixture.

The isolation of the acetylene from the various converters presents a complicated problem. The unstable, explosive nature of acetylene imposes certain restrictions on the use of the efficient separation techniques developed for other hydrocarbon systems. The results of decomposition and detonation studies on acetylene and its mixtures with other gases indicate that operating conditions where the partial pressure of acetylene exceeds 103–207 kPa (15–30 psi) should be avoided. Similar limitations apply to the operating temperatures which should not exceed 95–105°C. Low temperatures may lead to the appearance of liquid or solid acetylene or its homologues with concomitant danger of unexpected decompositions. In view of these severe operating restrictions, it is not surprising that all commercial processes for the recovery of hydrocarbon-derived acetylene are based on absorption–desorption techniques using one or more selective solvents.

Of the many solvents proposed, only a few have found commercial application, including water (Hüls), anhydrous ammonia (SBA), chilled methanol (Montecatini), N-methylpyrrolidinone (BASF), butyrolactone, acetone, dimethylformamide, and hydrocarbon fractions.

The separation and purification of acetylene is further complicated by the presence in the pyrolysis gas of higher acetylenes which polymerize rather easily in solution. The removal of these constituents is a necessity, particularly in view of the utilization of the acetylene in chemical synthesis. This can be accomplished by scrubbing with small amounts of a suitable mineral oil or other organic solvent (SBA, Wulff) or by low temperature fractionation (Hüls). In the latter case, the concentrated, dry acetylene is cooled close to the freezing point (195 K), whereby the higher acetylenes are liquefied and removed as a solution in methanol or benzene.

The carbon black (soot) produced in the partial combustion and electrical discharge processes is of rather small particle size and contains substantial amounts of higher (mostly aromatic) hydrocarbons which may render it hydrophobic, sticky, and difficult to remove by filtration. Electrostatic units, combined with water scrubbers, moving coke beds, and bag filters, are used for the removal of soot. The recovery is illustrated by the BASF separation and purification system (23). The bulk of the carbon in the reactor effluent is removed by a water scrubber (quencher). Residual carbon clean-up is by electrostatic filtering in the case of methane feedstock, and by coke particles if the feed is naphtha. Carbon in the quench water is concentrated by flotation, then burned.

The BASF process uses N-methylpyrrolidinone as the solvent to purify acetylene in the cracked gas effluent. A low pressure prescrubbing is used to remove naphthalenes and higher acetylenes. The cracked gas is then compressed to 1 MPa (10 atm) and fed to the main absorption tower for acetylene removal. Light gases are removed from the top of this tower.

Stripping of acetylene from the solvent takes place at atmospheric pressure. Pure acetylene is removed from the side of the stripper; light impurities are removed overhead and recycled to the compressor. Higher acetylenes are removed from the side of a vacuum stripper with the acetylene overheads being recycled to the bottom of the acetylene stripper.

The gases leaving the purification system are scrubbed with water to recover solvent and a continuous small purge of solvent gets rid of polymers. The acetylene purity resulting from this system is 99%. The main impurities in the acetylene are carbon dioxide, propadiene, and a very small amount of vinylacetylene.

COPRODUCT ACETYLENE FROM STEAM CRACKING

In the steam cracking of petroleum liquids to produce olefins, mainly ethylene, small concentrations of acetylene are produced. Although the concentrations are small, the large capacities of the olefin plants result in appreciable quantities of coproduct acetylene which, in many cases, are sufficient to satisfy the modest growth in acetylene demand due to specialty chemicals such as 1,4-butanediol (see ACETYLENE-DERIVED CHEMICALS). Because specifications for polymer-grade ethylene limit the acetylene contamination to below 5 ppm, the refinery operator must decide on whether to hydrogenate the contaminate acetylene to ethylene or to separate it as a by-product. The decision is influenced by the concentration of the acetylene, which is sensitive to the composition of the feedstock and the severity of the cracking conditions, and the availability of an over-the-fence use because acetylene by its nature cannot be economically transported any distance or stored. If a convenient use is available, it is generally cost-effective to recover the acetylene and sell it as a by-product, since it generally attracts a higher price than the ethylene.

The quantity of coproduct acetylene produced is sensitive to both the feedstock and the severity of the cracking process. Naphtha, for example, is cracked at the most severe conditions and thus produces appreciable acetylene; up to 2.5 wt % of the ethylene content. On the other hand, gas oil must be processed at lower temperature to limit coking and thus produces less acetylene. Two industry trends are resulting in increased acetylene output: (1) the ethylene plant capacity has more than doubled, and (2) furnace operating conditions of higher temperature and shorter residence times have increased the cracking severity.

Worldwide, approximately 180,000 t/yr acetylene product is recovered as a by-product within olefin plants. This source of acetylene is expected to increase as plant capacity and furnace temperature increase. The recovery may include compression and transfer of the acetylene product via pipelines directly to the downstream consumer.

Acetylene Recovery Process. A process to recover coproduct acetylene developed by Linde AG (Fig. 11), and reduced to practice in 11 commercial plants, comprises three sections: acetylene absorption, ethylene stripper, and acetylene stripper.

Acetylene Absorption. The gaseous feedstock containing the C_2 hydrocarbons is introduced into the acetylene absorption tower at a pressure range of 0.6–3 MPa (6–30 bar) depending on availability within the process design of the olefin plant. The absorption takes place in a countercurrent lean solvent flow, which is preferably dimethylformamide (DMF) and with less frequency N-methylpyrrolidinone (NMP). The overhead gas fraction is partially condensed against refrigerant to avoid any solvent losses. The acetylene absorption tower is designed from thermodynamic and hydraulic points of view to minimize the

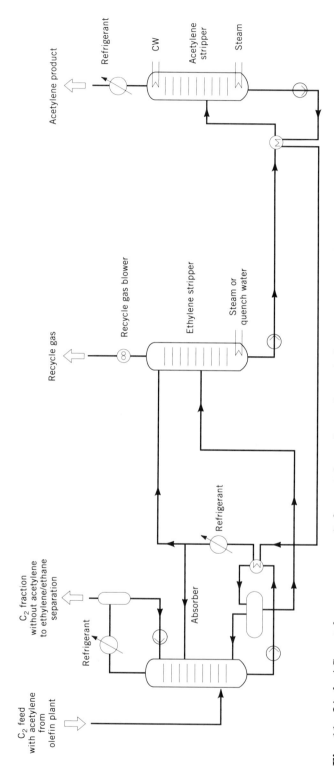

Fig. 11. Linde AG acetylene recovery unit for acetylene absorption, ethylene stripping, and acetylene stripping. CW = cooling water. Courtesy of P. Cl. Haehn, Linde AG.

acetylene content in the overhead product (less than 0.2 ppm) and the recycle flow rate to the cracked gas compressor in the olefin plant (less than 2%).

Ethylene Stripping. The acetylene absorber bottom product is routed to the ethylene stripper, which operates at low pressure. In the bottom part of this tower the loaded solvent is stripped by heat input according to the purity specifications of the acetylene product. A lean DMF fraction is routed to the top of the upper part for selective absorption of acetylene. This feature reduces the acetylene content in the recycle gas to its minimum (typically 1%). The overhead gas fraction is recycled to the cracked gas compression of the olefin plant for the recovery of the ethylene.

Acetylene Stripping. The loaded solvent with acetylene and traces of other basically olefinic components is pumped to the acetylene stripper tower for the delivery of the acetylene product in the overhead (typically 99.9% purity minimum). Solvent traces can be eliminated by chilling or water washing depending on downstream process requirements. The bottom product (lean solvent) is pumped back to the acetylene absorber and ethylene stripper towers after exchanging the maximum possible amount of its energy within the recovery process for economic reasons. The recovery process uses commercial solvents without the addition of an antifoaming agent. The applied solvents are not corrosive or fouling.

ACETYLENE FROM COAL

Coal, considered a solid hydrocarbon with a generic formula of $CH_{0.8}$, was explored by numerous workers (24–31) as a feedstock for the production of acetylene. Initially, the motivation for this work was to expand the market for the use of coal in the chemical process industry, and later when it was projected that the cost of ethylene would increase appreciably if pretroleum resources were depleted or constrained.

Acetylene traditionally has been made from coal (coke) via the calcium carbide process. However, laboratory and bench-scale experiments have demonstrated the technical feasibility of producing the acetylene by the direct pyrolysis of coal. Researchers in Great Britain (24,28), India (25), and Japan (27) reported appreciable yields of acetylene from the pyrolysis of coal in a hydrogen-enhanced argon plasma. In subsequent work (29), it was shown that the yields could be dramatically increased through the use of a pure hydrogen plasma.

Based on the bench-scale data, two coal-to-acetylene processes were taken to the pilot-plant level. These were the AVCO and Hüls arc-coal processes. The Avco process development centered on identifying fundamental process relationships (29). Preliminary data analysis was simplified by first combining two of three independent variables, power and gas flow, into a single enthalpy term. The variation of the important criteria, specific energy requirements (SER), concentration, and yield with enthalpy are indicated in Figure 12. As the plots show, minimum SER is achieved at an enthalpy of about 5300 $kW/(m^3/s)$ (2.5 kW/cfm), whereas maximum acetylene concentrations and yield are obtained at about 7400 $kW/(m^3/s)$ (3.5 kW/cfm). An operating enthalpy between these two values should, therefore, be optimum. Based on the results of this work and the need to demonstrate the process at sufficient size to judge industrial applicability,

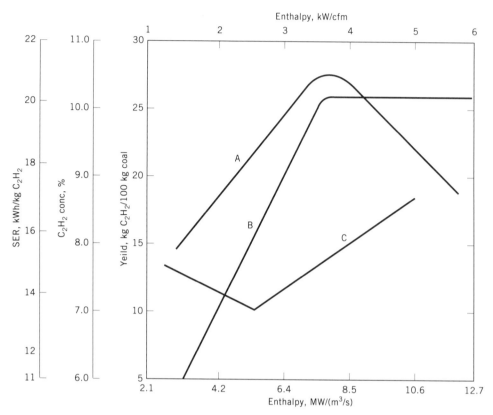

Fig. 12. Critical process parameters as a function of gas enthalpy where A is yield, B is concentration, and C is the specific energy requirement (SER).

AVCO built and operated a 1 MW reactor in 1979–1982. This project was jointly funded by the U.S. Department of Energy (30).

The AVCO reactor design is called a rotating arc reactor. In this design (Fig. 13), the arc is spread out radially from a center cathode to the walls which serve as the anode. In order to ensure temperature uniformity of the gas as well as the reactor wall, the arc is rotated using a magnetic field. The coal, which has been ground to conventional power plant grind, ie, 80% through a 74 μm (200 mesh) screen, is suspended in a hydrogen carrier gas and is fed through the top of the arc-coal reactor. The acetylene formed in the arc region is stabilized by rapidly quenching the gas stream to below 1400 K using a variety of quench media (hydrogen, methane, coal, hydrocarbons, water, etc).

The overall energy efficiency of the arc reactor is greatly enhanced by using a two-stage approach, ie, by using a chemically active quench in which further acetylene is produced. The active quench takes advantage of the latent heat of the gas below the arc to form additional acetylene. Two quench materials investigated were additional coal and a hydrocarbon (propane). In each case, the additional feed was injected below the arc reactor zone. Included in these experiments was a secondary water quench to freeze the acetylene yield.

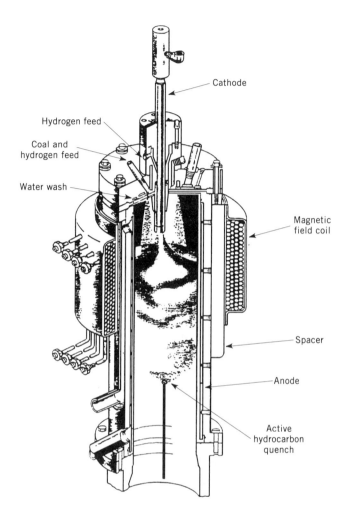

Fig. 13. AVCO rotating arc reactor.

An analytical model of the process has been developed to expedite process improvements and to aid in scaling the reactor to larger capacities. The theoretical results compare favorably with the experimental data, thereby lending validity to the application of the model to predicting directions for process improvement. The model can predict temperature and compositional changes within the reactor as functions of time, power, coal feed, gas flows, and reaction kinetics. It therefore can be used to project optimum residence time, reactor size, power level, gas and solid flow rates, and the nature, composition, and position of the reactor quench stream.

The economics of the arc-coal process is sensitive to the electric power consumed to produce a kilogram of acetylene. Early plant economic assessments indicated that the arc power consumption (SER = kW·h/kg C_2H_2) must be below 13.2. The coal feedcoal quench experiments yielded a 9.0 SER with data that indicated a further reduction to below 6.0 with certain process improvements. In the propane quench experiment, ethylene as well as acetylene is produced. The combined process SER was 6.2 with a C_2H_2/C_2H_4 production ratio of 3 to 2. Economic analysis was completed utilizing the achieved acetylene yields, and an acetylene price approximately 35% lower than the price of ethylene was projected.

In subsequent work at Hüls (31) similar results were obtained. That is, using German coals it was also found that the magnetically rotated arc was the preferred reactor design and that the product mixture could be enriched through the use of a hydrocarbon quench. In this two-stage reactor a SER of 11.5 kW·h/kg C_2H_2 was achieved, but it was projected that this could be reduced through further development work.

Uses

Acetylene is used primarily as a raw material for the synthesis of a variety of organic chemicals (see ACETYLENE-DERIVED CHEMICALS). In the United States, this accounts for about 80% of acetylene usage and most of the remainder is used for metal welding or cutting. The chemical markets for acetylene are shrinking as ways are found to substitute lower cost olefins and paraffins for the acetylene, with some products now completely derived from olefinic starting materials. Metalworking applications, however, have held up better than chemical uses.

Chemical Uses. In Europe, products such as ethylene, acetaldehyde, acetic acid, acetone, butadiene, and isoprene have been manufactured from acetylene at one time. Wartime shortages or raw material restrictions were the basis for the choice of process. Coking coal was readily available in Europe and acetylene was easily accessible via calcium carbide.

The principal chemical markets for acetylene at present are its uses in the preparation of vinyl chloride, vinyl acetate, and 1,4-butanediol. Polymers from these monomers reach the consumer in the form of surface coatings (paints, films, sheets, or textiles), containers, pipe, electrical wire insulation, adhesives, and many other products which total billions of kg. The acetylene routes to these monomers were once dominant but have been largely displaced by newer processes based on olefinic starting materials.

Vinyl chloride (chloroethene) is a significant market for acetylene (see VINYL POLYMERS). The reaction of acetylene and hydrogen chloride is carried out in the vapor phase at 150–250°C over a mercuric chloride catalyst. The acetylene route is usually coupled with an ethylene chlorination unit so that the hydrogen chloride derived from cracking dichloroethane can be consumed in the reaction with acetylene. Thus one mole each of ethylene, acetylene, and chlorine give two moles of vinyl chloride with a minimum of by-products. The oxychlorination of ethylene, however, eliminates by-product hydrogen chloride and thus much of the incentive for using the acetylene-based process. Hard cracking of hydrocarbons to a 1:1 molar mixture of ethylene and acetylene for use as feedstock for vinyl chloride production is done primarily outside the United States.

Vinyl acetate (ethenyl acetate) is produced in the vapor-phase reaction at 180–200°C of acetylene and acetic acid over a cadmium, zinc, or mercury acetate catalyst. However, the palladium-catalyzed reaction of ethylene and acetic acid has displaced most of the commercial acetylene-based units (see ACETYLENE-DERIVED CHEMICALS; VINYL POLYMERS). Current production is dependent on the use of low cost by-product acetylene from ethylene plants or from low cost hydrocarbon feeds.

Minor amounts of acetylene are used to produce chlorinated ethylenes. Trichloroethylene (trichloroethene) and perchloroethylene (tetrachloroethene) are prepared by successive chlorinations and dehydrochlorinations (see CHLOROCARBONS AND CHLOROHYDROCARBONS). The chlorinations take place in the liquid phase using uv radiation and the dehydrochlorinations use calcium hydroxide in an aqueous medium at 70–100°C. Dehydrochlorination can also be carried out thermally (330–700°C) or catalytically (300–500°C).

Tetrachloroethylene can be prepared directly from tetrachloroethane by a high temperature chlorination or, more simply, by passing acetylene and chlorine over a catalyst at 250–400°C or by controlled combustion of the mixture without a catalyst at 600–950°C (32). Oxychlorination of ethylene and ethane has displaced most of this use of acetylene.

Acetylene is condensed with carbonyl compounds to give a wide variety of products, some of which are the substrates for the preparation of families of derivatives. The most commercially significant reaction is the condensation of acetylene with formaldehyde. The reaction does not proceed well with base catalysis which works well with other carbonyl compounds and it was discovered by Reppe (33) that acetylene under pressure (304 kPa (3 atm), or above) reacts smoothly with formaldehyde at 100°C in the presence of a copper acetylide complex catalyst. The reaction can be controlled to give either propargyl alcohol or butynediol (see ACETYLENE-DERIVED CHEMICALS). 2-Butyne-1,4-diol, its hydroxyethyl ethers, and propargyl alcohol are used as corrosion inhibitors. 2,3-Dibromo-2-butene-1,4-diol is used as a flame retardant in polyurethane and other polymer systems (see BROMINE COMPOUNDS; FLAME RETARDANTS).

Much more important is the hydrogenation product of butynediol, 1,4-butanediol [110-63-4]. The intermediate 2-butene-1,4-diol is also commercially available but has found few uses. 1,4-Butanediol, however, is used widely in polyurethanes and is of increasing interest for the preparation of thermoplastic polyesters, especially the terephthalate. Butanediol is also used as the starting material for a further series of chemicals including tetrahydrofuran, γ-butyrolactone, 2-pyrrolidinone, N-methylpyrrolidinone, and N-vinylpyrrolidinone (see ACETYLENE-DERIVED CHEMICALS). The 1,4-butanediol market essentially represents the only growing demand for acetylene as a feedstock. This demand is reported (34) as growing from 54,000 metric tons of acetylene in 1989 to a projected level of 88,000 metric tons in 1994.

A small amount of acetylene is used in condensations with carbonyl compounds other than formaldehyde. The principal uses for the resulting acetylenic alcohols are as intermediates in the synthesis of vitamins (qv).

Another small-scale use for acetylene is in the preparation of vinyl ethers from alcohols, including polyols and phenols. A base such as sodium or potassium hydroxide is used as catalyst in a liquid-phase high pressure reaction at 120–180°C. This general reaction is also a product of the acetylene research done at I. G. Farbenindustries by J. W. Reppe. A wide variety of alcohols can be vinylated, but only a few have achieved any commercial use. The most important is methyl vinyl ether (methoxyethene), which is used as a monomer and comonomer with maleic anhydride (Gantrez resins, GAF) for the preparation of adhesives, coatings, and detergents, as well as starting materials for further synthesis.

Acetylene black is prepared by the partial combustion of acetylene and has specialty uses in batteries. Only about 3500 t/yr are produced in the United States.

Vinyl fluoride (fluoroethene), is manufactured from the catalyzed addition of hydrogen fluoride to acetylene. It is used to prepare poly(vinyl fluoride) which has found use in highly weather-resistant films (Tedlar film, Du Pont). Poly(vinylidene fluoride) also is used in weather-resistant coatings (see FLUORINE COMPOUNDS, ORGANIC). The monomer can be prepared from acetylene, hydrogen fluoride, and chlorine but other nonacetylenic routes are available.

At one time, the only commercial route to 2-chloro-1,3-butadiene (chloroprene), the monomer for neoprene, was from acetylene (see ELASTOMERS, SYNTHETIC). In the United States, Du Pont operated two plants in which acetylene was dimerized to vinylacetylene with a cuprous chloride catalyst and the vinylacetylene reacted with hydrogen chloride to give 2-chloro-1,3-butadiene. This process was replaced in 1970 with a butadiene-based process in which butadiene is chlorinated and dehydrochlorinated to yield the desired product (see CHLOROCARBONS AND CHLOROHYDROCARBONS).

Fuel Uses.   At one time acetylene was widely used for home, street, and industrial lighting. These applications disappeared with the advent of electrical lighting during the 1920s. However, one of the first fuel uses for acetylene, metalworking with the oxyacetylene flame, continues to consume a significant amount of acetylene (35).

Fusion welding is the process of uniting metallic parts by heating the surfaces of the portions to be joined until the metal flows. Many electrical and chemical means are used to provide the heat for various welding processes, but the oxyacetylene flame remains the preferred choice of the gas welding processes. Cheaper fuels are available, such as propane and butane, but they do not reach the high flame temperature (3200°C) or achieve acetylene's combustion intensity (product of the burning velocity and the heating value of the fuel). The cheaper fuels are reserved for use in specialized applications where their properties are applicable. The oxyacetylene flame can be used in joining most metals and thus has a versatility advantage (see WELDING).

Large quantities of acetylene are used in metal cutting which involves the combustion and melting of the metal; the oxyacetylene flame supplies the heat to initiate the process. Acetylene seems to have the advantage over other fuels because of the need for less oxygen and a shorter preheat time. Although oxyacetylene cutting is used in the field for construction and demolition, most cutting operations are performed in steel mills or fabricating shops.

Other uses of oxyacetylene flames in mill operations are in building up or hardfacing metal, lancing (piercing a hole in a metal mass), and a variety of metal cleaning procedures. A minor but interesting fuel use of acetylene is in flame spectrophotometry where oxygen and nitrous oxide are used as oxidants in procedures for a wide variety of the elements.

Economic Aspects

The relative economics of acetylene for chemical uses from calcium carbide and from hydrocarbon partial combustion or arc processes have swung rather clearly in favor of the hydrocarbon-based processes. Even more economically attractive is the acetylene produced as an unavoidable by-product in the manufacture of ethylene (qv). The economics apply to chemical uses, not industrial gases where calcium carbide does have advantages of scale which overcome its higher production cost. However, the key economic factor in the use of acetylene is the lower price of alternative materials which have decreased or eliminated some of the largest outlets for acetylene. Acetylene's triple bond inherently consumes more energy of formation than olefins; thus acetylene is more expensive. There seems no likelihood of reversing the decline in acetylene usage unless there is a change in raw material costs or more by-product acetylene is recovered.

Most by-product acetylene from ethylene production is hydrogenated to ethylene in the course of separation and purification of ethylene. In this process, however, acetylene can be recovered economically by solvent absorption instead of hydrogenation. Commercial recovery processes based on acetone, dimethylformamide, or N-methylpyrrolidinone have a long history of successful operation. The difficulty in using this relatively low cost acetylene is that each 450,000 t/yr world-scale ethylene plant only produces from 7000–9000 t/yr of acetylene. This is a small volume for an economically scaled derivatives unit.

The average price of pipeline shipments of acetylene in the United States in 1990 was $0.86/kg. In comparison, the average price of ethylene was about $0.53/kg. The range of prices due to the different sources of the acetylene can only be roughly estimated since each process has by-products and coproducts which may be credited or debited in more than one way. However (36), the relative prices of acetylene by the three primary categories of process may be calculated to be 2:1.5:1 g from calcium carbide, from hydrocarbon partial combustion, and for by-product acetylene, respectively. Pricing of the by-product acetylene from ethylene production at a value equivalent to ethylene plus recovery costs could reverse the trend away from acetylene-based processes. There is a commercial swing toward this objective as high cost calcium carbide sources are shut down.

Acetylene from calcium carbide can be advantageous in that calcium carbide may be shipped to the point of acetylene usage and acetylene generated on the spot. This avoids the necessity for low pressure, low pressure-drop gaseous acetylene pipelines, or high pressure cylinders for shipping acetylene. The carbide route is the preferred method of operation for most industrial gas operations. It is well suited to small-scale consumers. The high cost of acetylene in industrial gas applications reflects these scale, handling, and shipping factors.

SUPPLY AND DEMAND

United States.   The demand for acetylene generally peaked between 1965 and 1970, then declined dramatically until the early 1980s, and has been slowly increasing at between 2 and 4% per year since. The dramatic decline was related to increased availability of low cost ethylene, an alternative feedstock for many chemicals, and the recent increase is due to the modest growth of acetylenic chemicals, particularly 1,4-butanediol.

In the United States, the acetylene production exceeded 450,000 t/yr between 1963 and 1970, but then declined until it hit a minimum production level below 150,000 t/yr in 1982. Of this production, about 40,000 t were dedicated to industrial use, ie, welding, etc. Thus only slightly more than 100,000 t were produced for the chemical process industry. Figure 14 illustrates the 17-year decline in acetylene production and indicates the reduced derivative demand to which the accumulated decline is attributed (37).

As Figure 14 also shows, the only acetylene derivatives to sustain growth during this period were the so-called acetylenic chemicals. These include 1,4-butanediol, vinyl ethers, N-vinyl-2-pyrrolidinone, and butanediol. Of these, 1,4-butanediol, a principal feed for tetrahydrofuran, accounts for over 90% of the acetylenic chemicals demand (38).

In 1984, acetylene production received a significant influx with the increase of capacity at the Borden Co. plant in Geismar, Louisiana. This influx provided an additional 33,000 t/yr, which were absorbed by the vinyl chloride monomer (VCM) and acetylenic chemicals market. Acetylene demand has exhibited a modest growth rate of about 2% per year since 1984. In fact, demand exceeded production by about 10,000 t in 1987. This imbalance has generally been corrected by added capacity, particularly by the Quantum Chemical Co. plant at Deer Park, Texas, which came on line in 1991.

The U.S. Department of Commerce estimates total production of about 163,000 t in 1990. Other estimates based on demand data indicate that it was as high as 175,000 t. With demand and supply in balance, it is estimated

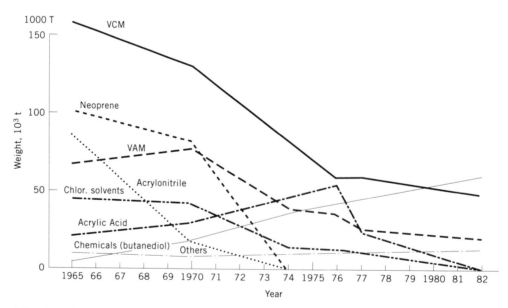

Fig. 14. Acetylene for chemical use in the United States (37). Courtesy of Chem Systems. Total consumption for specified years in 10^3 t as follows: 1965, 478; 1970, 385; 1974, 195; 1976, 174; 1977, 142; and 1982, 109–116.

that in 1997 the demand will be 185,000 t. The distribution in product demand is projected to be the following: 1,4-butanediol and other acetylenic chemicals (45%), vinyl chloride monomer (45%), acetylene black[a] (5%), and industrial use (5%).

In 1991, U.S. plant capacity for producing acetylene was estimated at 176,000 t/yr. Of this capacity, 66% was based on natural gas, 19% on calcium carbide, and 15% on ethylene coproduct processing. Plants currently producing acetylene in the United States are listed in Table 13.

It is difficult to indicate a representative price for acetylene because it is generally produced either for captive use or on contract. The price seems to be dictated mainly by the price movement of ethylene, often a coproduct as well as an alternative feedstock competing with acetylene. That is, in 1981 when ethylene was $0.55 per kg, acetylene was $1.12 per kg; and when in 1987 the price of ethylene dropped to $0.31 per kg, acetylene dropped to $0.68 per kg.

Western Europe. Acetylene demand in Western Europe exceeds by far that of any other geographical region. Prior to the unification of Germany in 1990, acetylene consumption in Western Europe was estimated to be 430,000 t/yr; with the addition of 280,000 t of consumption attributed to the former German Democratic Republic, total consumption increased to 710,000 t.

Table 14 lists the acetylene-producing plants in Western Europe as of 1991. Of the 782,000 t of annual capacity, 48% is produced from natural gas, 46% from calcium carbide, 4% from naphtha, and 2% as ethylene coproduct.

Of the estimated 710,000 t consumed in 1990, 25% was used to produce vinyl chloride [75-01-4] monomer (VCM), 14% for vinyl acetate [108-05-4] monomer (VAM), 23% for butanediol, 14% for industrial use, and the balance to produce other products such as acrylic acid, synthetic rubber, chlorinated

Table 13. Plants Producing Acetylene in the United States, 1991

Operator/location	Process	Feedstock	Capacity, 10^3 t/yr	Market
Borden-BASF Geismar, La.	BASF part. ox.	natural gas	90	VCM and 1,4-butanediol
Carbon-Graphite Group Calvert City, Ky.		calcium carbide	34	acetylenic chemicals
Chevron Cedar Bayou, Tex.	ethylene coproduct	petroleum	8	acetylene black
Rohm and Haas Deer Park, Tex.	BASF part. ox.	natural gas	25	1,4-butanediol
Quantum Chemical Deer Park, Tex.	ethylene coproduct	petroleum	14[a]	1,4-butanediol
Union Carbide Seadrift, Tex.	ethylene coproduct	petroleum	6	acetylenic chemicals
Taft, La.	ethylene coproduct	petroleum	5	industrial gas
Texas City, Tex.	ethylene coproduct	petroleum	8	acetylenic chemicals

[a]Estimated.

Table 14. Plants Producing Acetylene in Western Europe, 1991

Location/operator	Process	Feedstock	Capacity, 10^3 t/yr	Market
Germany				
BASF	part. ox.	natural gas	85	acrylics and
	coproduct	petroleum	6	acetylenic
				chemicals
Hoechst AG	calcium		6	acetylene
	carbide			black
Hüls	electric arc	natural gas	130	VCM and
				acetylenic
				chemicals
former GDR	calcium		350	synthetic rubber
plants	carbide			VAM, VCM
Italy				
EniChem Anic Sol	part. ox.	natural gas	70	VCM
Montedipe	part. ox.	natural gas	41	VAM and tetra-
				chloroethane
Netherlands				
DSM NV	coproduct	petroleum	2	captive
Shell	coproduct	petroleum	9	captive
Switzerland				
Lonze AG	Montecatini	naphtha	30	acetaldehyde
				and acetylenic
				chemicals
Austria				
Donan Chemic AG	calcium		3	captive
	carbide			
France				
Rhône-Poulenc	BASF	natural gas	50	VAM

solvents, and acetylene black. The demand for PVC is expected to decrease as legislation limiting its use in packaging is pending. Consequently, VCM consumption will also suffer.

Growth in the use of acetylene for the production of 1,4-butanediol is projected to continue at the rate of about 5% per year. However, competition from a new technology based on maleic anhydride may impact the use of acetylene in this market.

Eastern Europe. Production of acetylene in Eastern Europe is dominated by the capacity of states of the former Soviet Union. In 1990, production from Eastern Europe amounted to between 530,000 and 535,000 t/yr, with 51% provided by the states of the former Soviet Union. Of the balance, Poland with 23% and Rumania with 17% are the main producers. Table 15 indicates the capacity levels of each of the countries of Eastern Europe.

Japan and China. During the 1980s, acetylene demand in Japan suffered a significant decline. Chemical use declined from over 100,000 to 42,000 t, acetylene black production declined from 20,000 to less than 10,000 t, and industrial use went from 42,000 to 30,000 t. Thus, based on 1990 estimates, Japan has an

Table 15. Production of Acetylene in Eastern Europe

Country	Feedstock	Capacity, 10^3 t/yr	Market
Bulgaria	CaC_2	11.6	VCM
Czechoslovakia	CaC_2	20	VAM, VCM acetic acid
Hungary	CaC_2	5.4	VCM
Poland	CaC_2	123	VCM, acetaldehyde
Rumania	natural gas	155	VCM, VAM acetic acid and acetylenic chemicals
former Soviet Union states	natural gas and CaC_2	270	VAM, VCM, acetic acid, acetaldehyde
Yugoslavia	Ca_2C	12.6	VCM

excess capacity for acetylene production with capabilities for 247,000 t/yr and a demand of only 82,000 t.

Almost all (98%) of the acetylene produced in Japan is based on the calcium carbide process, with only about 2% being produced as a coproduct of ethylene production. The reliance on calcium carbide and its attractions for small users can be partially related to market demand. Thus, only a single plant with a maximum annual capacity of 89,000 t, operated by Danki Kagaku Koggs KK, is dedicated to the chemical process industry. Two smaller plants are dedicated to the production of acetylene black, but 64 plants with annual capacities generally less than 5000 t serve the industrial acetylene needs, ie, metals cutting and welding.

Very little data are readily available on China's supply and demand situation. It is known, however, that they rely almost entirely on calcium carbide for acetylene production and that there are numerous low capacity plants, a situation that is probably not too different from that of Japan.

The contribution of the section on coproduct acetylene by P.Cl. Haehn of Linde AG is gratefully acknowledged.

BIBLIOGRAPHY

"Acetylene" in *ECT* 1st ed., Vol. 1, pp. 101–123, by G. R. Webster, Carbide and Carbon Chemicals Corp., R. L. Hasche, Tennessee Eastman Co., and K. Kaufman, Shawinigan Chemicals Ltd.; in *ECT* 1st ed., Suppl. 2, pp. 1–35, by H. B. Sargent, Linde Co., Div. of Union Carbide Co., and W. G. Schepman, Union Carbide Olefins Co.; in *ECT* 2nd ed., Vol. 1, pp. 171–211, by H. Beller and J. M. Wilkinson, Jr., GAF Corp.; "From Hydrocarbons," pp. 211–237, by D. A. Duncan, Institute of Gas Technology; "Economic Aspects," pp. 237–243, by R. M. Manyik, Union Carbide Corp.

1. C. S. Kim, R. E. Gannon, and S. Ubhayakar, *Proceedings of 87th Annual Meeting*, American Institute of Chemical Engineering, Boston, Mass., Aug. 1979.
2. V. J. Krukonis and R. E. Gannon, *Advances in Chemistry Series*, V 131, American Chemical Society, Washington, D.C., 1974, pp. 29–41.
3. *Chem. Week* **94**, 64 (Jan. 18, 1964).

4. S. A. Miller, *Acetylene, Its Properties, Manufacture and Uses*, Vol. 1, Academic Press, Inc., New York, 1965, p. 402.

5. Ref. 4, p. 395.

6. J. E. Anderson and L. K. Case, *Ind. Eng. Chem. Process Des. Dev.* **1**(3), 161 (1962).

7. K. Gehrmann and H. Schmidt, *World Pet. Congr. Proc. 8th* **4**, 379 (1971).

8. H. Hoefermann and co-workers, *Chem. Ind.* **21**, 863 (1969).

9. U.S. Pat. 2,632,731 (Mar. 24, 1953), W. von Ediger (to Technical Assets Inc.).

10. L. Andrussow, *Erdoel Kohle* **12**, 24 (1959).

11. H. Kroepelin and co-workers, *Chem. Ing. Technol.* **28**, 703 (1956).

12. P. J. Leroux and P. M. Mathieu, *Chem. Eng. Prog.* **57**(11), 54 (1961).

13. E. Bartholome, *Chem. Ing. Technol.* **26**, 245 (1954).

14. H. Friz. *Chem. Ing. Technol.* **40**, 999 (1968).

15. U.S. Pat. 3,019,271 (May 18, 1959), F. F. A. Braconier (to Société Belge de L'Azote).

16. G. Fauser, *Chim. Ind. (Milan)* **42**(2), 150 (1960).

17. Brit. Pat. 932,429 (July 24, 1963), (to Montecatini).

18. U.S. Pat. 2,796,951 (June 25, 1957), M. S. P. Bogart (to The Lummus Co.).

19. G. H. Bixler and C. W. Coberly, *Ind. Eng. Chem.* **45**(12), 2596 (1953).

20. J. Happel and L. Kramer, *Ind. Eng. Chem.* **59**(1), 39 (1967).

21. A. Holman, O. A. Rokstad, and A. Solbakken, *Ind. Eng. Chem. Process Des. Dev.* **15**(3), 439 (1976).

22. J. M. Reid and H. R. Linden, *Chem. Eng. Prog.* **56**(1), 47 (1960).

23. *Hydrocarbon Process.* (Nov. 1971).

24. R. L. Bond and co-workers, *Nautre (London)* **200**(4913), 1313 (Dec. 28, 1963).

25. S. C. Chakravartty, D. Dutta, and A. Lahiri, *Fuel* **55**(1), 43 (1976).

26. R. L. Coates, C. L. Chen, and B. J. Pope, *Adv. Chem. Ser.* **131**, 92 (1974).

27. Y. Kawana, *Chem. Econ. Eng. Rev.* **4**(1), (45), 13 (1972).

28. W. R. Ladner and R. Wheatley, *Fuel* **50**(4), 443 (Oct. 1971).

29. R. E. Gannon and V. Krukonis, "Arc-Coal Process Development," *R&D Report No. 34—Final Report, Contract No. 14-01-0001-493*, prepared for Office of Coal Research by AVCO Corp., 1972.

30. *Avco Arc-Coal Acetylene Process Development Program*, final report, contract DE-ACO2-79-C-S40214, prepared for U.S. Dept. of Energy by Avco Systems Division, Apr. 1981.

31. R. Muller and co-workers, *Proceedings of 8th International Symposium on Plasma Chemistry International Union of Pure and Applied Chemistry*, Vol. 2, Tokyo, 1987.

32. U.S. Pat. 2,538,723 (Jan. 16, 1951), O. Fruhwirth and co-workers (to Donau Chemie A. G.).

33. W. Reppe and co-workers, *Justus Liebigs Ann. Chem.* **596**, 1 (1955).

34. A. M. Brownstein, *Chem. Tech.* (Aug. 1991).

35. *Oxyacetylene Handbook*, 3rd ed., Union Carbide Corp., Linde Division, New York.

36. O. Horn, *Erdoel Kohle Erdgas Petrochem. Brennst. Chem.* **26**(3), 129 (1973).

37. H. Wittcoff, Chem Systems, Inc., private communication, Apr. 1994.

38. K. L. Ring and co-workers, *Chemical Economics Handbook*, SRI International, Menlo Park, Calif., 1994.

General References

H. K. Kamptner, W. R. Krause, and H. P. Schilken, "Acetylene from Naphtha Pyrolysis," *Chem. Eng. N.Y.*, 80 (Feb. 28, 1966).

V. J. Krukonis and R. E. Gannon, "Deuterium and Carbon-13 Tagging Studies of the Plasma Pyrolysis of Coal," *Adv. Chem. Ser.* **131**, 29 (1974).

R. F. Goldstein and A. L. Waddams, *The Petroleum Chemicals Industry*, 3rd ed., E. & F. N. Spon Ltd., London, 1967, pp. 303–316.

"Acetylene Production Using Hydrogen Plasma," *Oil Gas J.*, 82 (Mar. 12, 1973).

"Acetylene: Winning With Wulff?" *Chem. Week*, 89 (Apr. 16, 1966).

"Acetylene and Ethylene Processes—Conference Report," *Chem. Process Eng. (London)*, 101 (May 1968).

R. B. Stobaugh, W. C. Allen, Jr., and Van R. H. Sterberg, "Vinyl Acetate: How, Where, Who—Future," *Hydrocarbon Process*, **51**(5), 153 (1972).

R. J. Parsons, "Progress Review No. 61: The Use of Plasmas in Chemical Synthesis," *J. Inst. Fuel* **43**(359), 524 (Dec. 1970).

G. Duembgen and co-workers, "Untersuchungen zur Acetylen-Herstellung durch Methan- und Leicht-benzin-Spaltung," *Chem. Ing. Technol.* **40**, 1004 (1968).

L. S. Lobo and D. L. Trimm, "Carbon Formation from Light Hydrocarbons on Nickel," *J. Catal.* **29**(1), 15 (Apr. 1973).

D. T. Illin and co-workers, "Production of Acetylene by Electrocracking of Natural Gas in a Coaxial Reactor," translated from *J. Appl. Chem. USSR* **42**(3), 648 (1969).

H. K. Kamptner, W. R. Krause, and H. P. Schilken, "HTP: After Five Years," *Hydrocarbon Process Pet. Refiner* **45**(4), 187 (1966).

H. Bockhorn and co-workers, "Production of Acetylene in Premixed Flames and of Acetylene–Ethylene Mixtures," *Chem. Ing. Technol.* **44**(14), 869 (1972).

"Thermal Decomposition of Ethane in a Plasma Jet," *Kogyo Kagaku Zasshi* **74**(9), 83 (1971).

"Production of Acetylene and Ethylene by Submerged Combustion," *Khim. Promst. (Moscow)* **49**(5), 330 (1973).

"Wulff Furnaces Make Acetylene, Ethylene," *Oil Gas J.*, 81 (Mar. 12, 1973).

K. Gerhard Baur, "Acetylene From Crude Oil Makes Debut in Italy," *Chem. Eng.*, 82 (Feb. 10, 1969).

"Acetylene–Badische Anilin- & Soda-Fabrik AG," *Hydrocarbon Process.*, 118 (Nov. 1971).

"Acetylene–Wulff Process," *Hydrocarbon Process.* **46**(11), 139 (1967).

"Process Costs, Wulff Actylene," *Chem. Process. Eng.* **47**(2), 71 (1966).

"Procedeul de obtinere a acetilenei si etenei cu flacara imersata," *Rev. Chim. (Bucharest)* **22**(12), 715 (Dec. 1971).

H. K. Kamptner, W. R. Krause, and H. P. Schilken, "High-Temperature Cracking," *Chem. Eng.*, 93 (Feb. 28, 1966).

K. L. Ring and co-workers, *Chem. Ec. Handbook*, SRI International, Menlo Park, Calif., 1994.

Chem. Week, 41 (June 21, 1993).

Chem. Marketing Reporter (June 1993).

Acetylene, report no. 76-2, Chem Systems, Inc., 1976.

H. Witcoff, Chem. Systems, Inc., private communication, Apr. 1994.

T. Wett, "Marathon Tames the Wulffs at Burghausen," *Oil Gas J.* **70**, 101 (Sept. 4, 1972).

H. Gladisch, "Acetylen-Herstellung im elektrishen Lichtbogen," *Chem. Ing. Technol.* **41**(4), 204 (1969).

E. A. Schultz, "Das Marathon-Werk Burghausen," *Erdoel Kohle Erdgas Petrochem.* **21**, 481 (1968).

R. B. Stobaugh, *Petrochemical Manufacturing & Marketing Guide*, Gulf Publishing Co., Houston, Tex., 1966, pp. 1–17.

RICHARD E. GANNON
Textron Defense Systems

C$_1$–C$_6$

METHANE, ETHANE, AND PROPANE

Physical Properties

Methane, ethane, and propane are the first three members of the alkane hydrocarbon series having the composition, C_nH_{2n+2}. Selected properties of these alkanes are summarized in Table 1 (1).

Manufacturing and Processing

The main commercial source of methane, ethane, and propane is natural gas, which is found in many areas of the world in porous reservoirs; they are associated either with crude oil (associated gas) or in gas reservoirs in which no oil is present (nonassociated gas). These gases are basic raw materials for the organic chemical industry as well as sources of energy. The composition of natural gas varies widely but the principal hydrocarbon usually is methane (see GAS, NATURAL). Compositions of typical natural gases are listed in Table 2.

Gas is classified as dry or wet depending on the amount of condensable hydrocarbons contained in it. A dry gas that is low in hydrogen sulfide and carbon dioxide needs little or no treatment prior to its use as a fuel or chemical feedstock; however, if these acidic contaminants are present, they must be removed (see FEEDSTOCKS). Processes using regenerable bases frequently are used for such purposes, ie, the weak acidity of carbon dioxide and hydrogen sulfide allows them to be separated from natural gas or other gaseous feeds by adsorption in aqueous bases. There are numerous proprietary processes which isolate the acid gases, typically by absorption in aqueous solution of mono- or diethanolamines (MEA and DEA) or their heavier analogues. The processes differ in the amines used, the plant configuration, and the presence and identity of cosolvents (3).

Condensable hydrocarbons are removed from natural gas by cooling the gas to a low temperature and then by washing it with a cold hydrocarbon liquid to absorb the condensables. The uncondensed gas (mainly methane with a small amount of ethane) is classified as natural gas. The condensable hydrocarbons (ethane and heavier hydrocarbons) are stripped from the solvent and are separated into two streams. The heavier stream, which largely contains propane with some ethane and butane, can be liquefied and is marketed as liquefied petroleum gas (LPG) (qv). The heavier fractions, which consist of C$_5$ and heavier hydrocarbons, are added to gasoline to control volatility (see GASOLINE AND OTHER MOTOR FUELS).

Relatively small amounts of methane, ethane, and propane also are produced as by-products from petroleum processes, but these usually are consumed

Table 1. Selected Properties of Methane, Ethane, and Propane[a]

Property	Methane	Ethane	Propane
CAS Registry Number	[74-82-8]	[74-84-0]	[74-98-6]
molecular formula	CH_4	C_2H_6	C_3H_8
molecular weight	16.04	30.07	44.09
mp, K	90.7	90.4	85.5
bp, K	111	185	231
explosivity limits, vol %	5.3–14.0	3.0–12.5	2.3–9.5
autoignition temperature, K	811	788	741
flash point, K	85	138	169
heat of combustion, kJ/mol[b]	882.0	1541.4	2202.0
heat of formation, kJ/mol[b]	84.9	106.7	127.2
heat of vaporization, kJ/mol[b]	8.22	14.68	18.83
vapor pressure at 273 K, MPa[c]		2.379	0.475
specific heat, J/(mol·K)[b]			
at 293 K	37.53	54.13	73.63
at 373 K	40.26	62.85	84.65
density, kg/m³[d]			
at 293 K	0.722	1.353	1.984
at 373 K	0.513	0.992	1.455
critical point			
pressure, MPa[c]	4.60	4.87	4.24
temperature, K	190.6	305.3	369.8
density, kg/m³[d]	160.4	204.5	220.5
triple point			
pressure, MPa[c]	0.012	1.1×10^{-6}	3.0×10^{-10}
temperature, K	90.7	90.3	85.5
liquid density, kg/m³[d]	450.7	652.5	731.9
vapor density, kg/m³[d]	0.257	4.51×10^{-5}	1.85×10^{-8}
dipole moment	0	0	0
hazards	fire, explosion, asphyxiation[e]	fire, explosion, asphyxiation[e]	fire, explosion, asphyxiation[e]

[a]Ref. 1.
[b]To convert J to cal, divide by 4.184.
[c]To convert MPa to atm, divide by 0.101.
[d]To convert kg/m³ to lb/ft³, divide by 16.0.
[e]No significant toxic effects.

as process or chemical feedstock fuel within the refineries. Some propane is recovered and marketed as LPG.

There are, however, a variety of other sources of methane that have been considered for fuel supply. For example, methane present in coal (qv) deposits and formed during mining operations can form explosive mixtures known as fire damp. In Western Europe, some methane has been recovered by suction from bore holes drilled in coal beds and the U.S. Bureau of Mines has tested the economic practicality of such a system. Removal of methane prior to mining the coal would reduce explosion hazards associated with coal removal. As much as 11.3×10^9 m³ (400 trillion (10^{12}) cubic feet or 400 TCF) of methane might be recoverable from U.S. coal beds.

Table 2. Composition of Typical Natural Gases[a]

Component, vol %	United States			Location	
	Salt Lake, Utah	Webb, Texas	Klifside, Texas	Sussex, England	Lacq, France
methane	95.0	89.4	65.8	93.2	70.0
ethane	0.8	6.0	3.8	2.9	3.0
propane	0.2	2.2	1.7		1.4
butanes		1.0	0.8		0.6
pentanes and heavier hydrocarbons		0.7	0.5		
hydrogen sulfide				1.0	15.0
carbon dioxide	3.6	0.6			
helium, nitrogen	0.4	0.1	25.6		
helium			1.8		
Total	100	100	100	97.1[b]	90[b]

[a]Ref. 2.
[b]Components present in trace quantities are not included.

Methane also is commonly produced by the decomposition of organic matter by a variety of bacterial processes, and the gas is used as a fuel in sewage plants (see WATER, SEWAGE). Methane also is called marsh gas because it is produced during the decay of vegetation in stagnant water.

There has been considerable research into the production of substitute natural gas (SNG) from fractions of crude oil, coal, or biomass (see FUELS, SYNTHETIC; FUELS FROM BIOMASS; FUELS FROM WASTE). The process involves partial oxidation of the feedstock to produce a synthesis gas containing carbon monoxide and hydrogen. After removal of acid gases (CO_2 and H_2S) and water gas shift ($CO + H_2O \rightarrow CO_2 + H_2$) to produce the necessary ratio of H_2 to CO, methane is produced by the following reaction:

$$CO + 3\,H_2 \longrightarrow CH_4 + H_2O$$

A commercial-scale SNG plant, the Great Plains Plant in North Dakota, was actually built and operated for several years using a scheme based on coal. However, upon termination of the government subsidy, the plant's owner, ANR Inc., found it uneconomical to continue plant operation and shut down the plant in the late 1980s.

Production and Shipment

World natural gas reserves and production are shown in Table 3 (see also FUELS, SYNTHETIC). The deposits of natural gas are extensive and provide sources of feedstock and fuel.

The large-scale use of natural gas requires a sophisticated and extensive pipeline system (see PIPELINES). In many underdeveloped areas, large quantities of natural gas are being flared because they must be produced with crude

Table 3. World Natural Gas Production,[a] 1993

Country	Production, 10^9 m^{3b}	Share, %
CIS (former USSR)	761.0	35.0
United States	544.5	25.0
Canada	156.5	7.2
the Netherlands	85.5	3.9
United Kingdom	63.2	2.9
Indonesia	53.3	2.4
Algeria	50.6	2.3
Mexico	37.1	1.7
Saudi Arabia	32.1	1.5
Iran	31.4	1.3
Norway	24.8	1.1
United Arab Emirates	24.3	1.1
Australia	24.0	1.1
next seven	135.2	6.1
all others	156.3	7.2
Total	*2179.8*	*100.0*

[a]Ref. 4.
[b]To convert m^3 to ft^3, multiply by 35.3.

oil. However, the opportunity for utilizing the streams or for bringing the gas to industrial markets is being developed. Several large-scale ammonia plants have been built in developing countries (Pakistan, Saudi Arabia, Iran, etc). In some cases, pipeline delivery is feasible, namely from Algeria to France and from Libya to Italy. A third possibility is liquefaction of the methane and shipment in specially designed refrigerated tanker ships (see CRYOGENICS). The process for liquefying, transporting, and revaporizing the natural gas requires advanced technology and particular attention to safety. Liquefaction of natural gas requires removal of its sensible and latent heats either by an adiabatic expansion process or by multistage mechanical refrigeration (qv). In the expanding cycle, the pressure of the gas is the energy source. The efficiency of the expansion process is low but capital and operating costs are lower than those of the multistage refrigeration process.

As indicated in Table 4, large-scale recovery of natural gas liquid (NGL) occurs in relatively few countries. This recovery is almost always associated with the production of ethylene (qv) by thermal cracking. Some propane also is used for cracking, but most of it is used as LPG, which usually contains butanes as well. Propane and ethane also are produced in significant amounts as by-products, along with methane, in various refinery processes, eg, catalytic cracking, crude distillation, etc (see PETROLEUM). They either are burned as refinery fuel or are processed to produce LPG and/or cracking feedstock for ethylene production.

Uses

Methane.   The largest use of methane is for synthesis gas, a mixture of hydrogen and carbon monoxide. Synthesis gas, in turn, is the primary feed for

Table 4. World NGL[a] Production,[b] 1993

Country	Production, 10^3 m^3/d	Share, %
United States	278.4	36.68
Canada	154.9	20.41
Mexico	98.7	13.00
CIS (former USSR)[c]	51.7	6.82
Algeria	23.0	3.03
Indonesia	20.7	2.73
Venezuela	15.9	2.10
Kuwait	13.3	1.75
Australia	13.2	1.74
Egypt	11.1	1.46
all others	78.1	10.29
Total	*759.0*	*100.00*

[a]NGL = natural gas liquid.
[b]Ref. 5.
[c]1991 data (latest available).

the production of ammonia (qv) and methanol (qv). Synthesis gas is produced by steam reforming of methane over a nickel catalyst.

Methane is also used for the production of several halogenated products, principally the chloromethanes. Due to environmental pressures, this outlet for methane is decreasing rapidly.

At one time, methane was widely used to produce acetylene (qv), by processes involving either electric arcs or partial oxidation. The so-called Reppe chemicals (ie, 1,4-butanediol and derivatives), once made solely from acetylene, can now be made from butane; the outlook for continued acetylene demand from methane is poor. In 1993, in fact, acetylene production for chemicals was only about a third of that in 1970 (see ACETYLENE-DERIVED CHEMICALS). Much interest has been shown in direct conversion of methane to higher hydrocarbons, notably ethylene. Development of such a process would allow utilization of natural gas from remote wells. Much gas is currently flared (burned) from such wells because the pipeline gathering systems needed for such gas tend to be prohibitively expensive. If the gas could be converted on-site to a condensable gas or pumpable liquid, bringing those hydrocarbons to market would be facilitated. In the early 1990s, partial oxidative coupling of methane to higher hydrocarbons (chiefly C₂s) achieved by passing methane and an oxygen-containing gas over a basic oxide catalyst at high temperatures (600–700°C) and low pressures (<1 atm) has been the method of choice. However, despite enormous efforts, C₂ yields higher than about 30% have not yet been realized. Direct methane conversion to other materials, such as methanol, has similarly not yielded commercially interesting results, mainly due to the extreme temperatures and very low throughput required for high selectivity to the desired products (7).

Ethane and Propane. The most important commercial use of ethane and propane is in the production of ethylene (qv) by way of high temperature (ca 1000 K) thermal cracking. In the United States, ca 60% of the ethylene is produced by thermal cracking of ethane or ethane/propane mixtures. Large ethylene plants have been built in Saudi Arabia, Iran, and England based on ethane recovery from natural gas in these locations. Ethane cracking units have been installed in Australia, Qatar, Romania, and France, among others.

Ethane has been investigated as a feedstock for production of vinyl chloride, at scales up to a large pilot plant, but nearly all vinyl chloride is still produced from ethylene.

Propane's largest use outside of steam cracking is as fuel, since propane is the chief constituent of NGL. Historically, NGLs have been used for homes and businesses located away from natural gas systems. Recently, environmental concerns coupled with the clean-burning nature of NGL (since virtually all of the sulfur and other pollutants are removed during processing of the natural gas) have stimulated research on and field trials of propane as a fuel source for internal combustion engines in cars, buses, and so on. Several oil companies have even established fueling stations on interstate highways. Propane's main competition in the replacement fuel market is compressed natural gas (CNG). Compared to CNG, NGLs have better driveability, longer range, and more simple conversion from gasoline.

BUTANES

Butanes are naturally occurring alkane hydrocarbons that are produced primarily in association with natural gas processing and certain refinery operations such as catalytic cracking and catalytic reforming. The term butanes includes the two structural isomers, n-butane [106-97-8], $CH_3CH_2CH_2CH_3$, and isobutane [75-28-5], $(CH_3)_2CHCH_3$ (2-methylpropane).

Properties

The properties of butane and isobutane have been summarized in Table 5 and include physical, chemical, and thermodynamic constants, and temperature-dependent parameters. Graphs of several physical properties as functions of temperature have been published (17) and thermodynamic properties have been tabulated as functions of temperature (12).

The alkanes have low reactivities as compared to other hydrocarbons. Much alkane chemistry involves free-radical chain reactions that occur under vigorous conditions, eg, combustion and pyrolysis. Isobutane exhibits a different chemical behavior than n-butane, owing in part to the presence of a tertiary carbon atom and to the stability of the associated free radical.

Reactions of n-Butane. The most important industrial reactions of n-butane are vapor-phase oxidation to form maleic anhydride (qv), thermal cracking to produce ethylene (qv), liquid-phase oxidation to produce acetic acid (qv) and oxygenated by-products, and isomerization to form isobutane.

Maleic Anhydride. The largest chemical use for n-butane is as feedstock for maleic anhydride. A dilute air–butane mixture is passed over a vanadium–

Table 5. Properties of Butane

Property	n-Butane	Isobutane	Reference
molecular weight	58.124	58.124	
normal fp in air at 101.3 kPa,[a] K	134.79	113.55	8
normal bp at 101.3 kPa,[a] K	272.65	261.43	8,9
flammability limits at 293.15 K and 101.3 kPa,[a] vol %			
in air			
lower	1.8	1.8	10
upper	8.4	8.4	10
in oxygen			
lower	1.8	1.8	10
upper	ca 40	ca 40	10
autoignition temperature at 101.3 kPa,[a] K			
in air	693	693	10
in oxygen	558	558	10
flash point, K	199	190	11
heat of combustion, kJ/mol[b]			
gross[c]			
gas	2880	2866	11
liquid	2853	2847	11
net[c]			
gas	2653	2645	11
liquid	2634	2627	11
ΔH_f°, kJ/mol[b]			
ideal gas at 298.15 K	−126.15	−134.51	9,12–14
liquid at 298.15 K	−147.7	−158.4	14
ΔG_f° at 101.3 kPa[a] and 298.15 K, kJ/mol[b]	−17.15	−20.88	8,12,13
$\log_{10} K_f$ at 298.15 K	3.0035	3.6	8
heat of fusion, kJ/mol[b]	4.660	4.540	10
heat of vaporization at normal bp, kJ/mol[b]	22.39	21.30	10,13
vapor pressure at 310.93 K, kPa[a]	356	498	9
Antoine vapor pressure equation[d,e]			
A	5.9340	5.8731	
B	935.86	882.80	
C	−34.42	−33.15	
T max	290	280	
T min	195	187	
thermal conductivity at 101.3 kPa,[a] W/(m·K)			10
at 273.15 K	0.0136	0.0140	
at 323.15 K	0.0182	0.0185	
at 373.15 K	0.0234	0.0242	
coefficient of thermal expansion for air saturated liquid at 288.7 K, saturation pressure, $(1/\text{vol})\,(d\,\text{vol}/dT)_p$, K^{-1}	0.00211	0.00214	11
density, kg/m^3			
gas, at 101.3 kPa[a]			11
at 288.7 K	2.5379	2.5285	
at 298.15 K	2.4553		8
liquid, at saturation pressure and 298.15 K	572.87	551.0	8

Table 5. (Continued)

Property	n-Butane	Isobutane	Reference
critical point			
pressure, MPa^g	3.797	3.648	8
temperature, K	425.16	408.13	8
density, kg/m^3	228.0	221.0	8,10
volume, cm^3/mol	225	263	8
compressibility factor	0.274	0.283	8
$S°$, ideal gas at 298.15 K, J/mol·K^b	309.9	295.4	8
$C_p°$, ideal gas, J/mol·K^b			
at 288.7 K	95.04	94.16	11
at 298.15 K	98.49	96.65	8
$C_v°$, ideal gas at 288.7, J/mol·K^b	86.72	85.85	11
$C_p°/C_v°$, ideal gas at 288.7 K	1.096	1.097	11
$C_p°$, liquid at 288.7 K and 101.3 kPa,^{a} J/mol·K^a	137.1	138.5	9,11
dT/dp at 101.3 kPa,^{a} K/kPa^a	0.260	0.253	8
dipole moment, C·m^h	0.0	3.34 × 10^{-31}	13
surface tension at saturation pressure, mN/m(=dyn/cm)			
at 223.15 K	20.88	18.7	8,11
at 273.15 K	14.84	13.0	8,11
refractive index, n_D^{25}			
liquid at saturation pressure	1.32592	1.3503	10
gas at 101.3 kPa^a	1.001286		8
stoichiometric combustion flame temperature, K			
in air	2243	2246	10
in oxygen	3118	3118	10
maximum flame speed, m/s			
in air	0.37	0.36	10
in oxygen	3.31	3.3	10

^{a}To convert kPa to atm, divide by 101.3.
^{b}To convert J to cal, divide by 4.184.
^{c}Real gas at 101.3 kPa^a and 288.7 K; liquid at saturation pressure and 298.15 K.
^{d}Equations for vapor pressure, liquid volume, saturated liquid density, liquid viscosity, heat capacity, and saturated liquid surface tension are described in Refs. 13, 15, and 16.
^{e}Log$_{10}$ $P = A - B/(T + C)$ where P = vapor pressure, kPa^f; T = temperature, K; A, B, and C = constants.
^{f}To convert kPa to mm Hg, multiply by 7.5.
^{g}To convert MPa to atm, divide by 0.1013.
^{h}To convert C·m to debye, divide by 3.336 × 10^{-30}.

phosphorus catalyst 400–500°C to produce maleic anhydride [108-31-6] in good yield. Formerly benzene was used as feedstock, but in the last few years nearly all maleic anhydride in the United States, and an increasing proportion worldwide, is made from butane.

　　Thermal Cracking.　 n-Butane is used in steam crackers as a part of the mainly ethane–propane feedstream. Roughly 0.333–0.4 kg ethylene is produced per kilogram n-butane. Primary by-products include propylene (50–57 kg/100 kg ethylene), butadiene (7–8.5 kg/100 kg), butylenes (5–20 kg/ 100 kg), and aromatics (6 kg/100 kg).

Dehydrogenation. Dehydrogenation of *n*-butane was once used to make 1,3-butadiene, a precursor for synthetic rubber. There are currently no on-purpose butadiene plants operating in the United States; butadiene is usually obtained as a by-product from catalytic cracking units.

Liquid-Phase Oxidation. Liquid-phase catalytic oxidation of *n*-butane is a minor production route for acetic acid manufacture. Formic acid (qv) also is produced commercially by liquid-phase oxidation of *n*-butane (18) (see HYDROCARBON OXIDATION).

Isomerization. Structural isomerization of *n*-butane to isobutane is commercially useful when additional isobutane feedstock is needed for alkylation (qv). The catalysts permit low reaction temperatures which favor high proportions of isobutane in the product. The Butamer process also is well known for isomerization of *n*-butane.

Reactions of Isobutane. *Alkylation.* The addition of isobutane to various C$_3$–C$_4$ alkenes is used in the production of high quality gasoline blending stock (19). The highly branched C$_7$–C$_8$ hydrocarbons that are produced have high octane ratings; eg, the significant quantities of trimethylpentanes that are produced are 100–109.6 research octane number-clear (see GASOLINE AND OTHER MOTOR FUELS). The alkylation reaction is promoted by acidic systems such as AlCl$_3$–HCl, H$_2$SO$_4$, and HF. A complicated series of reactions occurs by a carbonium ion mechanism. As many as 40 products are produced in concentrations of 100 ppm or more (20). Licensed commercial processes have been described (21) (see ALKYLATION; FRIEDEL-CRAFTS REACTIONS).

Other Reactions. *n*-Butane or mixtures of *n*-butane and isobutane may be catalytically converted to propane (22) in order to overcome propane shortage, absorb excess butane, reduce worldwide LPG consumption, and satisfy seasonal variations in demand for propane. Dehydrogenation of isobutane to isobutylene has been suggested (23) as a method to increase the quantity of isobutylene feedstock available for methyl *tert*-butyl ether (MTBE) production; MTBE is a high octane gasoline-blending stock (see GASOLINE AND OTHER MOTOR FUELS; ETHERS). Aromatics such as xylenes, ethylbenzenes, toluene, and benzene may be made by dehydrocyclodimerization of butanes (24) (see XYLENES AND ETHYLBENZENE; TOLUENE; BENZENE; BTX PROCESSING). Other commercial reactions of butanes include nitration (qv) and halogenation.

Manufacture and Processing

Butanes are recovered from raw natural gas and from petroleum refinery streams that result from catalytic cracking, catalytic reforming, and other refinery operations. The most common separation techniques are based on a vapor–liquid, two-phase system by which liquid butane is recovered from the feed gas.

Although raw natural gas is an important source of butanes, the concentrations therein are relatively small, eg, 0.30 mol % *n*-butane and 0.35 mol % isobutane (25). The primary method of recovery is absorption (qv) of the propane, butanes, and heavier hydrocarbons in a refrigerated absorption oil that is composed of natural gasoline components which have been recovered from the gas. The natural-gas feed is contacted with the absorption oil at 5.52–8.27 MPa (54.5–81.6 atm) and 233–255 K. The oil from the absorber contains propane, butanes, natural gasoline, and dissolved ethane and methane, and flows to a de-ethanizing absorber that operates at lower pressure. Methane and ethane are taken overhead and the bottom product, which contains propane and butane, enters a debutanizer where propane and butane are separated from the gasoline absorption oil. Some of the gasoline is removed as a product and the rest is cooled and returned to the absorbers. The propane–butane stream undergoes final product separation in a depropanizer column (26).

Refrigeration (qv), adsorption, expansion, compression, fractionation, and cryogenic processes are used to recover products from natural gas. Advantages of the turbo expanders in cryogenic processes are high thermodynamic efficiencies and simplicity. Chilled feed gas, containing butanes, is fed to the expander side of an expander–compressor. The pressure is lowered by near isentropic expansion to the demethanizer operating pressure resulting in efficient further cooling of the feed gas to the demethanizer temperature. The expanding gas delivers shaft work to the compressor side of the expander–compressor which partially recompresses the residue gas. Details of a recovery process based on a turbo expander have been described (27); different processing schemes for butane recovery from natural gas have been reviewed (25).

The butane-containing streams in petroleum refineries come from a variety of different process units; consequently, varying amounts of butanes in mixtures containing other light alkanes and alkenes are obtained. The most common recovery techniques for these streams are lean oil absorption and fractionation. A typical scheme involves feeding the light hydrocarbon stream to an absorber-stripper where methane is separated from the other hydrocarbons. The heavier fraction is then debutanized, depropanized, and de-ethanized by distillation to produce C$_4$, C$_3$, and C$_2$ streams, respectively. Most often the C$_4$ stream contains butylenes and other unsaturates which must be removed by additional separation techniques if pure butanes are desired.

Shipment

Butanes are shipped by pipeline, rail car, sea tanker, barge, tank truck, and metal bottle throughout the world. All U.S. container shipments must meet Department of Transportation regulations. Domestic water shipments are regulated by the U.S. Coast Guard.

Economic Aspects

The principal sources of butanes are petroleum refining and natural gas liquids. Refinery processes of catalytic reforming and catalytic cracking produce the majority of butanes and hydrocracking, coking, and visbreaking contribute less. Because catalytic cracking and reforming are the most widespread gasoline-producing processes, the supply of butanes is closely associated with gasoline production. Extensive use of butanes by the refineries complicates accurate, quantitative estimates of butane production at the refineries.

Specifications, Standards, Quality Control, and Storage

Large quantities of butane are shipped under contract standards rather than under national or worldwide specifications. Most of the petrochemical feedstock materials are sold at purity specifications of 95–99.5 mol %. Butane and butane–petroleum mixtures intended for fuel use are sold worldwide under specifications defined by the Gas Processors Association, and the specifications and test methods have been published (28). Butanes may be readily detected by gas chromatography. Butanes commonly are stored in caverns (29) or refrigerated tanks.

Health and Safety

n-Butane and isobutane are colorless, flammable, and nontoxic gases (30). They are simple asphyxiants, irritants, and anesthetics at high concentrations. Isobutane causes drowsiness in a short time in concentrations of 1 vol %; however, there are no apparent injuries from either hydrocarbon after 2 h exposures at concentrations of up to 5%. A threshold limit value (TLV), of 600 ppm for *n*-butane has been recommended (31). The extreme flammability of these hydrocarbons necessitates handling and storage precautions. Storage in well-ventilated areas away from heat and ignition sources is recommended. Because they are heavier than air, they should not be used near sparking motors or other nonexplosion-proof equipment. Contact of the liquid form of the hydrocarbons with the skin can cause frostbite. Both butane and isobutane form solid hydrates with water at low temperatures. Hydrate formation in liquefied light petroleum product pipelines and certain processing equipment can lead to pluggage and associated safety problems. Isobutane hydrate forms more readily than *n*-butane hydrates.

Uses

Butanes are used as gasoline blending components, liquefied gas fuel, and in the manufacture of chemicals. *n*-Butane and small amounts of isobutane are blended directly into motor fuel to control the fuel's volatility. Larger amounts of butanes are used in the winter, particularly in cold climates, to make engine starting easier. Recent environmental regulations in the United States have limited the permitted vapor pressure of gasolines, particularly in the summer months. Strictly speaking, the regulations apply only to areas which cannot meet air quality standards (nonattainment areas) but gasolines are usually blended to meet the strictest requirements in their sales areas. As recently as the late 1970s butanes accounted for 6–8 vol % of the gasoline sold, and nearly 85% of the total butane consumption. In the early 1990s, the volume of butane in gasoline is falling rapidly, and it is still unclear where the bottom will be.

Although *n*-butane is used directly in motor gasoline, isobutane is alkylated with C$_3$–C$_4$ olefins to produce highly branched C$_7$–C$_8$ hydrocarbons having high octane ratings. High octane blend stocks, eg, alkylates and reformates, are useful for raising the octane rating of the total refinery gasoline pool. In 1987 ca 89% of the isobutane consumption in the United States was for motor fuel alkylate (32). In addition to its use as a motor fuel alkylate, isobutane is a reactant in the production of propylene oxide (qv) by peroxidation of propylene (qv) (33).

In 1987 nonmotor fuel uses of butanes represented ca 16% of the total consumption. Liquid petroleum gas (LPG) is a mixture of butane and propane, typically in a ratio of 60:40 butane–propane; however, the butane content can vary from 100 to 50% and less (see LIQUEFIED PETROLEUM GAS). LPG is consumed as fuel in engines and in home, commercial, and industrial applications. Increasing amounts of LPG and butanes are used as feedstocks for substitute natural gas (SNG) plants (see FUELS, SYNTHETIC). n-Butane, propane, and isobutane are used alone or in mixture as hydrocarbon propellents in aerosols (qv).

Production of maleic anhydride by oxidation of n-butane represents one of butane's largest markets. Butane and LPG are also used as feedstocks for ethylene production by thermal cracking. A relatively new use for butane of growing importance is isomerization to isobutane, followed by dehydrogenation to isobutylene for use in MTBE synthesis. Smaller chemical uses include production of acetic acid and by-products. Methyl ethyl ketone (MEK) is the principal by-product, though small amounts of formic, propionic, and butyric acid are also produced. n-Butane is also used as a solvent in liquid–liquid extraction of heavy oils in a deasphalting process.

PENTANES

There are three isomeric pentanes, ie, saturated aliphatic hydrocarbons of molecular formula C$_5$H$_{12}$. They are commonly called n-pentane [109-66-0], isopentane [78-78-4] (2-methylbutane), and neopentane [463-82-1] (2,2-dimethylpropane).

Properties

Each isomer has its individual set of physical and chemical properties; however, these properties are similar (Table 6). The fundamental chemical reactions for pentanes are sulfonation to form sulfonic acids, chlorination to form chlorides, nitration to form nitropentanes, oxidation to form various compounds, and cracking to form free radicals. Many of these reactions are used to produce intermediates for the manufacture of industrial chemicals. Generally the reactivity increases from a primary to a secondary to a tertiary hydrogen (37). Other properties available but not listed are given in equations for heat capacity and viscosity (34), and saturated liquid density (36).

Occurrence and Recovery

Pentanes occur chiefly in straight-run gasoline, natural gasoline, and in certain refinery streams. Straight-run gasoline is the gasoline boiling range material recovered from crude oil by distillation (ie, before any other processing). Natural gasoline is the C$_5$+ fraction of the liquids recovered from natural gas.

Table 6. Properties of Pentanes

Property	n-Pentane	Isopentane	Neopentane	Ref.
molecular weight	72.151	72.151	72.151	34
normal freezing point, K	143.429	113.250	256.57	8
normal bp, K	309.224	301.002	282.653	8
water solubility at 25°C, g C$_5$H$_{12}$/100 kg H$_2$O	9.9	13.2		11
spontaneous ignition temperature in air, K	557.0	700.0	729.0	11
flash point, K	233.0	213.0	198.0	35
critical point				
pressure, MPa^a	3.369	3.381	3.199	8
temperature, K	469.7	460.39	433.75	8
density, kg/m^3	231.9	234.0	237.7	36
volume, m^3/mol	304 × 10^{-6}	306 × 10^{-6}	303 × 10^{-6}	8
compressibility factor	0.262	0.273	0.269	8
heat of combustion, kJ/mol^b at 298 K				
liquid	3245	3239	3230	8
gas	3272	3264	3253	8
heat of fusion, kJ/mol^b	8.39	5.15	3.15	8
heat of vaporization, kJ/mol^b	25.77	24.69	22.75	8
entropy of fusion, kJ/(mol·K)^b	5.852 × 10^{-2}	4.548 × 10^{-2}	1.226 × 10^{-2}	8
entropy of vaporization, kJ/(mol·K)^b	8.335 × 10^{-2}	8.203 × 10^{-2}	8.050 × 10^{-2}	8
Antoine vapor pressure equation^{c,d}				
A	6.00122	5.91458	5.72918	
B	1075.78	1020.01	883.420	34
C	−39.94	−40.05	−45.37	
T, max	330	322	305	
T, min	220	216	260	
dielectric constant	1.843	1.843	1.801	35
dipole moment, C·m^f	0.0	3.336 × 10^{-31}	0.0	34
surface tension, mN/m (=dyn/cm)				
at 20°C	16.00	15.00	12.05	35
at 30°C	14.95	13.93	10.98	35
refractive index, n_D^{25}				
liquid	1.35472	1.35088	1.339	8
gas	1.001585			8
ASTM octane number				
research	61.8	93.0	85.5	11
motor	63.2	89.7	80.2	11

^{a}To convert MPa to atm, divide by 0.101.
^{b}To convert J to cal, divide by 4.184.
^{c}Equations for four-parameter vapor pressure, Harlacher vapor pressure, and liquid volume can be found in Refs. 36, 34, and 36, respectively.
^{d}Log$_{10}$ $P = A - B/(T + C)$ where P = vapor pressure, kPa^e; T = temperature, K; A, B, and C = constants. To convert Antoine equation to mm Hg, add 0.8751 to A.
^{e}To convert kPa to atm, divide by 101.3.
^{f}To convert C·m to debye, divide by 3.336 × 10^{-30}.

Appreciable quantities of pentanes are produced in catalytic cracking, while smaller amounts come from hydrocracking and catalytic reforming. Table 7 shows typical pentane concentration in these streams.

Most pentanes are still blended into motor fuel, though increasingly strict vapor pressure regulations may end this practice in the United States by the year 2000. Most of the gasoline range material from refinery units is added to the gasoline pool. Natural and straight-run gasolines, however, are often de-pentanized by distillation, and the resulting pentane fraction processed through a splitting column or molecular sieve unit to separate the normal and isopentane. The isopentane (research octane = 93) is added to the gasoline pool, while the *n*-pentane (research octane = 61.8) is isomerized over fixed-bed platinum catalysts.

Table 7. Estimated Concentration of Pentanes, %

Stream	Pentane	Isopentane
straight-run gasoline	6.8	6.3
natural gasoline	6.8	6.3
catalytic cracker naphtha	1	5

Health and Safety

Pentanes are only slightly toxic. Because of their high volatilities and, consequently, their low flash points, they are highly flammable. Pentanes are classified as nonreactive, ie, they do not react with fire-fighting agents. The fire hazard properties for pentanes are listed in Table 8 (38).

The threshold limit value for the time-weighted average (8-h) exposure to pentanes is 600 ppm or 1800 mg/m^3 (51 mg/SCF); the short-term exposure limit (15 min) is 750 ppm or 2250 mg/m^3 (64 mg/SCF) (39). Pentanes are classified as simple asphyxiants and anesthetics (qv).

The ICC classifies all three pentanes as flammable liquids and requires that they be affixed with a red label for shipping. Because of their high vapor pressures, *n*- and isopentane are transported in heavy-walled drums and neopentanes are transported in cylinders (see PACKAGING MATERIALS, INDUSTRIAL PRODUCTS).

Table 8. Pentane Fire Hazard Properties

Pentane	Bp, °C	Flash point, °C	Ignition temperature, °C	Flammability limits, vol %	
				Lower	Upper
isopentane	27.8	<−51	420	1.4	7.6
n-pentane	36.1	<−40	260	1.5	7.8
neopentane	9.5	gas	450	1.4	7.5

Uses

The main use for pentanes has been in motor fuel, though regulations limiting fuel vapor pressure are decreasing the amount of pentanes, particularly isopentane, present in gasoline during warm parts of the year. At one time, significant quantities of pentane were used as feedstock for ethylene units. However, most U.S. ethylene capacity is now based on ethane–propane feedstock; only limited amounts of heavier material can be tolerated without overloading the downstream purification train.

Isopentane can be alkylated with light olefins (qv) to give gasoline material; however, the resulting alkylate is lower quality (research octane = 74–80) than that produced from isobutane (research octane clear = 90–98). The demand for higher gasoline-pool octane discourages isopentane alkylation. Nevertheless, some outlet has to be found for the increasing amount of pentane displaced from gasoline by vapor pressure regulation, and it is likely that much of that pentane will find its way into alkylation streams.

Some isopentane is dehydrogenated to isoamylene and converted, by processes analogous to those which produce methyl *t*-butyl ether [*1634-04-4*] (MTBE) to *t*-amyl methyl ether [*994-05-8*] (TAME), which is used as a fuel octane enhancer like MTBE. The amount of TAME which the market can absorb depends mostly on its price relative to MTBE, ethyl *t*-butyl ether [*637-92-3*] (ETBE), and ethanol, the other important oxygenated fuel additives.

HEXANES

Hexane refers to the straight-chain hydrocarbon, C_6H_{14}; branched hydrocarbons of the same formula are isohexanes. Hexanes include the branched compounds, 2-methylpentane, 3-methylpentane, 2,2-dimethylbutane, 2,3-dimethylbutane, and the straight-chain compound, *n*-hexane. Commercial hexane is a narrow-boiling mixture of these compounds with methylcyclopentane, cyclohexane, and benzene (qv); minor amounts of C_5 and C_7 hydrocarbons also may be present. Hydrocarbons in commercial hexane are found chiefly in straight-run gasoline which is produced from crude oil and natural gas liquids (see GASOLINE AND OTHER MOTOR FUELS; GAS, NATURAL). Smaller volumes occur in certain petroleum refinery streams.

Properties

Properties of the principal hydrocarbons found in commercial hexane are shown in Table 9. The flash point of *n*-hexane is −21.7°C and the autoignition temperature is 225°C. The explosive limits of hexane vapor in air are 1.1–7.5%. Above 2°C the equilibrium mixture of hexane and air above the liquid is too rich to fall within these limits (42).

Manufacture

Commercial hexanes are manufactured by two-tower distillation of a suitable charge stock, eg, straight-run gasolines that have been distilled from crude oil

Table 9. Properties of Hydrocarbons Found in Commercial Hexanes[a]

Hydrocarbon	CAS Registry Number	Freezing point, °C	Normal bp, °C	Liquid density, kg/m³ at 20°C	Liquid refractive index, n_D^{20}	Antoine vapor pressure equation[b–d]			
						A	B	C	range, K
2-methylbutane	[78-78-4]	−159.900	27.852	619.67	1.35373				
n-pentane	[109-66-0]	−129.730	36.065	626.20	1.35748				
cyclopentane	[287-92-3]	−93.866	49.262	745.38	1.40645				
2,2-dimethylbutane	[75-83-2]	−99.870	49.741	649.16	1.36876	5.8797	1081.2	−43.81	230–350
2,3-dimethylbutane	[79-29-8]	−128.538	57.988	661.64	1.37495	5.9347	1127.2	−44.25	235–354
2-methylpentane	[107-83-5]	−153.660	60.271	653.15	1.37145	5.9640	1135.4	−46.58	240–370
3-methylpentane	[96-14-0]		63.282	664.31	1.37652	5.9738	1152.4	−46.02	240–365
n-hexane	[110-54-3]	−95.322	68.736	659.33	1.37486	6.0027	1171.6	−48.78	245–370
methylcyclopentane	[96-37-7]	−142.455	71.812	748.64	1.40970	5.9878	1186.1	−47.11	250–375
benzene	[71-43-2]	5.533	80.100	879.01	1.50112	6.0305	1211.0	−52.36	280–377
cyclohexane	[110-82-7]	6.554	80.738	778.55	1.42623	5.9662	1201.5	−50.50	280–380
2,2-dimethylpentane	[590-35-2]	−123.811	79.197	673.85	1.38215				
2,4-dimethylpentane	[108-08-7]	−119.242	80.500	672.70	1.38145				
1,1-dimethylcyclopentane	[1638-26-2]	−69.795	87.846	754.48	1.41356				

[a] Ref. 40.
[b] Ref. 34.
[c] Equations for Harlacher vapor pressure, vapor heat capacity, saturated liquid volume, and liquid viscosity can be found in Refs. 34 and 41.
[d] $\text{Log}_{10} P = A - B/(T + C)$ where P = vapor pressure, kPa[e]; T = temperature, K; A, B, and C = constants.
[e] To convert P (kPa) to P (mm Hg), add 0.875 to A; to convert kPa to mm Hg, multiply by 7.5.

or natural gas liquids that have been stripped from natural gas. Product composition is a function of the charge stock used and the degree of separation achieved in the fractionators. Because benzene forms minimum boiling azeotropes with n-hexane, methylcyclopentane, and cyclohexane, it cannot be eliminated by fractionation. Another source of hexanes is the BTX raffinate which remains after the removal of aromatics from catalytic reformates (see BTX PROCESSING). Catalytic reformates are the products obtained from reforming naphthas for the conversion of cycloparaffins to aromatics. Raffinates are low in cycloparaffins and aromatics, and hexanes that have been obtained from them are highly paraffinic.

Highly pure n-hexane can be produced by adsorption on molecular sieves (qv) (see ADSORPTION, LIQUID SEPARATION) (43). The pores admit normal paraffins but exclude isoparaffins, cycloparaffins, and aromatics. The normal paraffins are recovered by changing the temperature and/or pressure of the system or by elution with a liquid that can be easily separated from n-hexane by distillation. Other than benzene, commercial hexanes also may contain small concentrations of olefins (qv) and compounds of sulfur, oxygen, and chlorine. These compounds cannot be tolerated in some chemical and solvent applications. In such cases, the commercial hexanes must be purified by hydrogenation.

The composition and properties of and the range of hydrocarbon distributions and impurities encountered in three commercial hexanes are listed in Table 10. Hexane A is derived from fractionation of natural gas liquids. Because there is not an available effective treatment to remove impurities, the benzene and sulfur contents in hexane A are high. Hexane B is highly pure relative to A and is recovered in a refinery operation that involves hydrogenation. Hence its benzene, sulfur, and olefin contents are low. Hexane C is typical of a stream that meets specifications for polymerization-grade material.

Health and Safety

Hexane is classified as a flammable liquid by the ICC, and normal handling precautions for this type of material should be observed. According to the ACGIH, the maximum concentration of hexane vapor in air to which a worker may be exposed without danger of adverse health effects is 125 ppm; benzene is rated at 10 ppm.

n-Hexane can be grouped with the general anesthetics (qv) in the class of central nervous system depressants. Hexane vapors are mildly irritating to mucous membranes. Exposure to concentrations in excess of 1% hexane may cause dizziness, unconsciousness, prostration, and death. Prolonged skin contact with hexane results in irritation and dermatitis. Direct contact with lung tissue can result in chemical pneumonitis, pulmonary edema, and hemorrhage (44).

Uses

Other than fuel, the largest volume appliction for hexane is in extraction of oil from seeds, eg, soybeans, cottonseed, safflower seed, peanuts, rapeseed, etc. Hexane has been found ideal for these applications because of its high solvency for oil, low boiling point, and low cost. Its narrow boiling range minimizes losses, and its low benzene content minimizes toxicity. These same properties also

Table 10. Compositions of Typical Commercial Hexanes

Property	Hexane A	Hexane B	Hexane C
Hydrocarbon analysis, liquid vol %			
2,3-dimethylbutane	0.05	0.16	
2-methylpentane	3.48	1.49	0.30
3-methylpentane	9.38	5.40	3.27
n-hexane	63.91	81.23	88.19
methylcyclopentane	19.43	11.71	8.23
cyclohexane	0.78		
benzene	2.81	0.004	0.01
dimethylpentanes	0.16		
Physical and chemical properties			
distillation (ASTM D1078) initial bp, °C	68.3	68.2	67.1
dry point	71.0	69.0	67.3
flash point, °C		−23	
aniline point, °C (ASTM D611)		65.3	66.1
Reid vapor pressure, kPa,^{a} at 37.8°C	33.8	35.9	35.2
specific gravity (°API)	0.689 (73.9)	0.674 (78.4)	0.672 (79.0)
bromine number		0.0001	0.0016
peroxides (as H$_2$O$_2$), ppm		<1	<1
carbonyls (as acetone), ppm		<1	3.8
acidity		nil	
sulfur, ppm (ASTM D1266 app. I)	25	<1	0.4
phenols, ppm (ASTM D52-R)			<1.0
water, ppm			54
color (ASTM D156)		+30	+30

^{a}To convert kPa to psi, multiply by 0.145.

make hexane a desirable solvent and reaction medium in the manufacture of polyolefins, synthetic rubbers, and some pharmaceuticals. The solvent serves as catalyst carrier and, in some systems, assists in molecular weight regulation by precipitation of the polymer as it reaches a certain molecular size. However, most solution polymerization processes are fairly old; it is likely that those processes will be replaced by more efficient nonsolvent processes in time.

CYCLOHEXANE

Cyclohexane [*110-82-7*], C$_6$H$_{12}$, is a clear, essentially water-insoluble, noncorrosive liquid that has a pungent odor. It is easily vaporized, readily flammable, and less toxic than benzene. Structurally, it is a cycloparaffin. Cyclohexane was synthesized by Baeyer in 1893 and it was discovered by Markovnikov in Caucasian petroleum fractions shortly thereafter. Its presence in United States crude oils was established in 1931 (45). Cyclohexane was produced first by hydrogenation of benzene in 1898 (46).

Properties

Properties of cyclohexane are given in Table 11, and a number of binary azeotropes that are formed with cyclohexane are listed in Table 12.

Stereochemistry. Cyclohexane can exist in two molecular conformations: the chair and boat forms. Conversion from one conformation to the other involves rotations about carbon–carbon single bonds. Energy barriers associated with this type of rotation are low and transition from one form to the other is rapid. The predominant stereochemistry of cyclohexane has no influence in its use as a raw material for nylon manufacture or as a solvent.

Reactions. The most important commercial reaction of cyclohexane is its oxidation (in liquid phase) with air in the presence of soluble cobalt catalyst or boric acid to produce cyclohexanol and cyclohexanone (see HYDROCARBON OXI-DATION; CYCLOHEXANOL AND CYCLOHEXANONE). Cyclohexanol is dehydrogenated with zinc or copper catalysts to cyclohexanone which is used to manufacture caprolactam (qv).

Cyclohexane is dehydrogenated easily to benzene over platinum or palla-dium catalysts on charcoal or alumina at 300–320°C (50,51). Dehydrogenation of cyclohexane over palladium or platinum begins at 170°C and is a reversible reac-tion; below 200°C, the equilibrium favors cyclohexane formation. Dehydrogena-tion occurs smoothly without ring scission (52). In contrast, nickel on charcoal gives appreciable quantities of methane. The activity of nickel is moderated by using alumina as a support and, at 300–310°C, dehydrogenation occurs without ring scission (53).

Isomerization of cyclohexane in the presence of aluminum trichloride cat-alyst with continuous removal of the lower boiling methylcyclopentane by dis-tillation results in a 96% yield of the latter (54). The activity of $AlCl_3$–HCl catalyst has been determined at several temperatures. At 100°C, the molar ratio of methylcyclopentane to cyclohexane is 0.51 (55).

Occurrence

Cyclohexane is present in all crude oils in concentrations of 0.1–1.0%. The cy-cloparaffinic crude oils, such as those from Nigeria and Venezuela, have high cy-clohexane concentrations, and the highly paraffinic crude oils, such as those from Indonesia, Saudi Arabia, and Pennsylvania, have low concentrations; and con-centrations of cycloparaffins in crude oils from Texas, Oklahoma, and Louisiana tend to fall in between (see PETROLEUM).

Manufacture and Shipment

Essentially all high purity cyclohexane is made by hydrogenation of benzene (qv). A small amount of cyclohexane of lower purity is produced by fractional distillation from crude oil and from catalytic reformer effluent (56). Hydrogena-tion of benzene to cyclohexane can be carried out in either the liquid or vapor phase in the presence of hydrogen. Various processes have been developed for producing cyclohexane and all are catalytic and involve nickel, palladium, or platinum as the catalysts. Generally, the metal is put on a support although

Table 11. Properties of Cyclohexane

Property	Value
mol wt	84.156
fp, °C	6.554
molal fp lowering, °C	20.3
bp, °C	80.738
flammability limits (in air), vol %	1.3–8.4
flash point (closed up), °C	−17
heat of transition, kJ/kg[a]	80.08
heat of fusion, kJ/kg[a]	31.807
heat of vaporization, kJ/kg[a]	
at 25°C	392.50
at 80.7°C	357.44
vapor pressure, kPa[b]	
at 30°C	16.212[c]
at 40°C	24.613[c]
at 50°C	36.237[c]
at 60°C	51.901[c]
at 70°C	72.521[c]
at 80°C	99.095[c]

Antoine equation[d]

  at 1.333–199.95 kPa[b,e]

$$\log_{10} P = 5.965 - 2766.63/(T - 50.50)$$

  at 1.333 kPa[b] to P_c^f

$$\log_{10} P = 22.373 + 5562.12/T - 2.303/\log_{10} T + 4.22P/T^2$$

Property	Value
transition point, °C	−87.05
critical pressure, kPa[b]	4110.00[c]
critical temperature, °C	281.0
critical density, g/cm³	0.2718
surface tension at 20°C, N/m[g]	0.0253 ± 0.3
$n_D^{20\,h}$	1.4623
$d_4^{20\,h}$	0.77855
dielectric constant at 10⁵ Hz and 25°C	2.023
kinematic viscosity at 20°C, mm²/s (=cSt)	1.259
dynamic viscosity at 20°C, mPa·s (=cP)	0.980
specific heat relative to water at 25.9°C	0.440
cryoscopic constant $(A)^i$, mole fraction, °C	0.00411

[a] To convert J to cal, divide by 4.184.
[b] To convert kPa to atm, divide by 101.3.
[c] Ref. 47.
[d] Equations for saturated liquid density, liquid heat capacity, vapor heat capacity, and liquid viscosity can be found in Ref. 34.
[e] To convert log $P_{(kPa)}$ to log $P_{(mm\ Hg)}$, add 0.875 to 5.965.
[f] To convert log $P_{(kPa)}$ to log $P_{(mm\ Hg)}$, add 0.875 to 22.373.
[g] To convert N/m to dyn/cm, divide by 0.001.
[h] For air-saturated liquid at 101.3 kPa.
[i] For use in calculating mol % purity, p, by using the equation $\log p = 2.00000 - (A/2.30259)(T' - T)$ where T' is the freezing point of a given sample (48,49).

Table 12. Binary Azeotropes by Cyclohexane

Second component	Cyclohexane, wt %	Bp, °C
water	91.6	69.0
benzene	45	77.5
methanol	39	54.2
ethyl alcohol	70	64.9
n-propyl alcohol	80	74.3
isopropyl alcohol	67	68.6
n-butyl alcohol	96	79.8
isobutyl alcohol	86	78.1

Raney nickel is used in one process. Because of the equilibrium relationship between cyclohexane and benzene, temperature control of the reaction is critical; however, this is complicated because hydrogenation is exothermic. Consequently, most commercial cyclohexane processes involve multistage reactors in which recycling of cyclohexane, staged injection of benzene feed, and inter-reactor cooling to absorb the heat of hydrogenation occur. A generalized flow scheme for a vapor-phase multistage process is shown in Figure 1.

Some processes use only one reactor (57) or a combination of liquid- and vapor-phase reactors (58). The goal of these schemes is to reduce energy consumption and capital cost. Hydrogenation normally is carried out at 2–3 MPa (20–30 atm). Temperature is maintained at 300–350°C to meet a typical specification of less than 500 ppm benzene in the product; at higher temperatures, thermodynamic equilibrium shifts to favor benzene and the benzene specification is impossible to attain. Also, at higher temperatures, isomerization of cyclohexane to methylcyclopentane occurs; typically there is a 200 ppm specification limit on methylcyclopentane content.

Carbon monoxide and sulfur compounds are catalyst deactivators. Small amounts of carbon monoxide can be present in the hydrogen from catalytic reformers and ethylene units (which are the typical sources) even after cryogenic concentration of the hydrogen. Therefore the hydrogen that is fed to the unit usually is passed through a methanator before being introduced into the hydrogenation reactors. The methanator converts the carbon monoxide to methane and water. In most instances, there is no sulfur in the hydrogen: the content of sulfur compounds in benzene usually is specified to be less than 1 ppm as sulfur. At this level, a catalyst life of several years can be achieved. The majority of cyclohexane that is produced is shipped in bulk by tank car, tank truck, barge, or tanker.

Economic Aspects

In a well-designed multistage hydrogenation unit, operating costs are small as a result of recovery of the heat of hydrogenation between reactor stages by steam generation or integration with other process units, or by more efficient one- and two-stage processes (57,58). Consequently, the principal costs in cyclohexane manufacture are maintenance expenses, interest and return charges on the plant

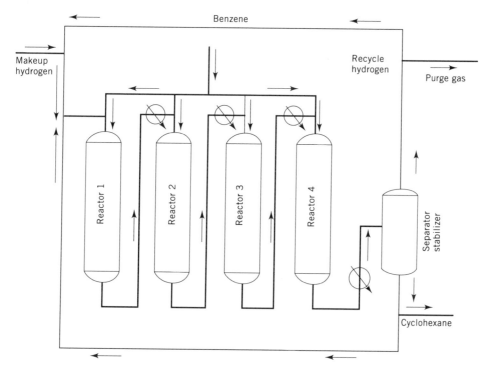

Fig. 1. Hydrogenation of benzene to cyclohexane.

and working capital, and the cost of benzene and high purity hydrogen. The cost of hydrogen recovery from the gases produced from catalytic reforming of naphtha or from ethylene manufacture is included in the manufacturing cost. The price of cyclohexane is dependent on the price of benzene (qv). Virtually all cyclohexane goes to the production of nylon (see POLYAMIDES).

The United States accounts for about a third of the world's consumption of cyclohexane, or 3.785×10^6 m³/yr (about 1 billion gallons per year). U.S. producers and their 1990 capacities are listed in Table 13. Texaco has announced that it is leaving the cyclohexane business, but the timing is not yet certain. Over 90% of all cyclohexane goes to the production of nylon through either adipic acid (qv) or caprolactam (qv). The balance is used to produce 1,6-hexamethylenediamine [124-09-4] (HMDA) and for various solvent uses (see DIAMINES AND HIGHER AMINES, ALIPHATIC; POLYAMIDES).

Table 13. U.S. Cyclohexane Producers,[a] 1993

Producer	Capacity, 10^3 m³/yr[b]
Phillips Petroleum	852
Texaco	284
Chevron	144
CITGO	114
Total	1394

[a]Ref. 59
[b]To convert m³/yr to gal/yr, multiply by 264.

Specifications

For nylon manufacture, a typical purity specification of cyclohexane is 99.8%. The principal contaminants are unconverted benzene, aliphatic hydrocarbons that have boiling points near benzene and that are in the benzene that is fed to the hydrogenation unit, and methylcyclohexane which is formed by hydrogenation of the small amount of toluene in the benzene. Freezing point and specific gravity are used as determinants of purity (see Table 11). Also, a distillation range for the product of 1°C with 80.7°C included in the range usually is specified. However, gas chromatography generally is used for production and quality control as it is a more exact analytical method. A sulfur level of less than 1 ppm also is specified for high purity cyclohexane and is measured by ASTM method D3120.

Analysis

Hydrocarbon mixtures containing cyclohexane can be analyzed using gas chromatography. Specific retention volumes for several stationary phases have been determined (60). Cyclohexane may be distinguished from other cycloparaffins and most aliphatic hydrocarbons, such as methylcyclopentane and hexane, respectively, by boiling point difference. Cyclohexane is differentiated from benzene (bp, 80.1°C) and 2,4-dimethylpentane (bp, 80.5°C) by chemical means, eg, by dehydrogenation over platinum. Cyclohexane also is distinguished from benzene by determining its inertness to fuming sulfuric acid unless exposed for prolonged periods of time.

Quantitative estimation of cyclohexane in the presence of benzene and aliphatic hydrocarbons may be accomplished by a nitration–dehydrogenation method described in Reference 61. The mixture is nitrated with mixed acid and under conditions that induce formation of the soluble mononitroaromatic derivative. The original mixture of hydrocarbons then is dehydrogenated over a platinum catalyst and is nitrated again. The mononitro compounds of the original benzene and the benzene formed by dehydrogenation of the cyclohexane dissolve in the mixed acid. The aliphatic compound remains unattacked and undissolved. This reaction may be carried out on a micro scale.

Health and Safety

The threshold limit value (TLV) for cyclohexane is 300 ppm (1050 mg/m^3). With prolonged exposure at 300 ppm and greater, cyclohexane may cause irritation to eyes, mucous membranes, and skin. At high concentrations, it is an anesthetic and narcosis may occur. Because of its relatively low chemical reactivity, toxicological research has not been concentrated on cyclohexane.

Uses

Almost all of the cyclohexane that is produced in concentrated form is used as a raw material in the first step of nylon-6 and nylon-6,6 manufacture. Cyclohexane also is an excellent solvent for cellulose ethers, resins, waxes (qv), fats, oils, bitumen, and rubber (see CELLULOSE DERIVATIVES, ETHERS; RESINS, NATURAL; FATS AND FATTY OILS; RUBBER, NATURAL). When used as a solvent, it usually is in admixture with other hydrocarbons. However, a small amount is used as a reaction diluent in polymer processes.

The cyclohexane in crude oil has three primary dispositions. Some of it is included in a light fraction (35–75°C) that is distilled from crude oil and is blended with other materials into motor gasoline (see GASOLINE AND OTHER MOTOR FUELS). Alternatively, this fraction is used as a feed to ethylene manufacture, particularly in Europe. The third, and most important disposition, is as a feed to a catalytic reformer where the naturally occurring cyclohexane is converted to benzene by dehydrogenation (see BTX PROCESSING). The cyclohexane used in this fashion often is reconverted by hydrogenation for use in nylon manufacture after recovery of high purity benzene by solvent extraction of the liquid product from the catalytic reformer. Direct recovery of cyclohexane from crude oil for chemical applications is practiced to a limited extent because of the small volumes of cyclohexane needed and the difficulty in fractionally distilling it from the many hydrocarbons in crude oil that boil at similar temperatures.

BIBLIOGRAPHY

"Hydrocarbons" in *ECT* 1st ed., Vol. 7, pp. 598–652, by G. Egloff and M. Alexander, Universal Oil Products Co.; in *ECT* 2nd ed., Vol. 11, pp. 262–293, by W. Nelson Axe, Phillips Petroleum (Survey) and N. W. Gerade, Esso Research and Engineering Co. (Toxicity); "Butanes" in *ECT* 2nd ed., Vol. 3, pp. 815–821, by C. E. Morrell, Esso Research and Engineering Co.; "Methane" in *ECT* 2nd ed., Vol. 13, pp. 364–369, by H. R. Heichelheim, Texas Technological College; "Ethane" in *ECT* 2nd ed., Vol. 8, pp. 383–385, by A. P. Lurie, Eastman Kodak Co.; "Propane" in *ECT* 2nd ed., Vol. 16, pp. 546–548, by C. E. Morrell, Enjay Chemical Intermediates Laboratory; "Pentanes" in *ECT* 2nd ed., Vol. 14, pp. 707–716, by J. L. Chadwick and R. L. Greene, Purvin & Gertz, Inc.; "Hexanes" in *ECT* 2nd ed., Vol. 11, pp. 1–15, by C. A. Porter, American Mineral Spirits Co., a Division of Union Oil Co. of California; "Cyclohexane" in *ECT* 1st ed., Vol. 4, pp. 760–769, by M. D. Barnes, Lion Oil Co.; in *ECT* 2nd ed., Vol. 6, pp. 675–682, by J. C. Kirk, Continental Oil Co.; "Methane, Ethane, and Propane" under "Hydrocarbons C₁–C₆" in *ECT* 3rd ed., Vol. 12, pp. 901–909, by D. S. Maisel, Exxon Chemical Co.; "Butanes" under "Hydrocarbons C₁–C₆" in *ECT* 3rd ed., Vol. 12, pp. 910–919, by D. D. Boesiger, R. H. Nielsen, and M. A. Albright, Phillips Petroleum Co.; "Pentanes" under "Hydrocarbons C₁–C₆" in *ECT* 3rd ed., Vol. 12, pp. 919–925, by T. W. Schmidt and D. M. Haskell, Phillips Petroleum Co.; "Hexanes" under "Hydrocarbons C₁–C₆" in *ECT* 3rd ed., Vol. 12, pp. 926–930, by G. H. Dale and L. E. Drehman, Phillips Petroleum Co.; "Cyclohexane" under "Hydrocarbons C₁–C₆" in *ECT* 3rd ed., Vol. 12, pp. 931–937, by M. L. Campbell, Exxon Chemical Co.

1. R. D. Goodwin, *NBS Technical Note 653*, U.S. Department of Commerce, NBS, Cryogenics Div., Boulder, Col., Apr. 1974; R. D. Goodwin, R. M. Roder, and G. C. Straty, *NBS Technical Note 684*, Aug. 1976; R. D. Goodwin, *NBS PB272-355*, July 1977.
2. *Hydrocarbon Process*, 192 (May 1977).
3. *Synthetic Organic Chemicals, United States Production and Sales, 1991*, U.S. International Trade Commission Publication 2607, Washington, D.C., Feb. 1993.
4. *Oil Gas J.*, 87 (Mar. 14, 1994).
5. *Oil Gas J.*, 52 (June 13, 1994).

6. L. F. Albright, B. L. Crynes, and S. Nowak, eds., *Novel Production Methods for Ethylene, Light Hydrocarbons, and Aromatics*, Marcel Dekker, Inc., New York, 1992; E. E. Wolf, ed., *Methane Conversion by Oxidative Processes*, Van Nostrand Reinhold, New York, 1992; J. S. Lee and S. T. Oyama, *Catal. Rev.-Sci. Eng.* **30**(2), 249–280 (1988).

7. J. M. Fox III, T-P Chen, and B. D. Degen, *Chem. Eng. Prog.*, 42–50 (Apr. 1990).

8. *Selected Values of Properties of Hydrocarbons and Related Compounds, Research Project 44*, American Petroleum Institute, Thermodynamics Research Center, Texas A&M University, College Station, Texas.

9. *Gas Processors Suppliers Association Engineering Data Book*, 9th ed., Gas Processors Suppliers Association, Tulsa, Okla., 1972, 1977.

10. *L'Air Liquide Division Scientifique, Encyclopedie Des Gaz*, Engl. transl. by N. Marshall, Elsevier Scientific Publishing Co., Amsterdam, the Netherlands, 1976.

11. *Phillips 66 Engineering Standards*, Phillips Petroleum Co., Bartlesville, Okla., 1976.

12. D. R. Stull, E. F. Westrum, Jr., and G. C. Sinke, *The Chemical Thermodynamics of Organic Compounds*, John Wiley & Sons, Inc., New York, 1969.

13. R. C. Reid, J. M. Prausnitz, and T. K. Sherwood, *The Properties of Gases and Liquids*, 3rd ed., McGraw-Hill, New York, 1977.

14. D. M. Himmelblau, *Basic Principles and Calculations in Chemical Engineering*, Prentice Hall, Inc., Englewood Cliffs, N.J., 1962.

15. *Tables of Stored Parameters, from Computer Stored Thermophysical Data System*, Management Information & Control, Phillips Petroleum Co., Bartlesville, Okla., 1976.

16. J. J. Jasper, *J. Phys. Chem. Ref. Data* **1**, 841 (1972).

17. R. W. Gallant, *Hydrocarbon Process.* **44**(7), 95 (1965); **45**(12), 113 (1966).

18. J. B. Saunby and B. W. Kiff, *Hydrocarbon Process*, **55**(11), 247 (1976).

19. *Prepr. Am. Chem. Soc. Div. Pet. Chem.* **22**(3), 1020 (1977).

20. L. J. McGovern, *Prepr. Am. Chem. Soc. Div. Pet. Chem.* **17**(3), B19 (1972).

21. L. F. Albright, *Oil Gas J.*, 79 (Nov. 12, 1990) and 70 (Nov. 26, 1990).

22. N. H. Fromager, J. R. Bernard, and M. Grand, *Chem. Eng. Prog.* **73**(11), 89 (1977).

23. J. L. Menfils and co-workers, *Hydrocarbon Process.* (2), 47 (1992).

24. S. M. Csicsery, *Ind. Eng. Chem. Proc. Des. Dev.* **18**(2), 191 (1979).

25. J. F. Wolfe, *Chem. Eng. Prog. Symp. Ser. 103* **66**, 7 (1970).

26. *Hydrocarbon Process*, **58**(4), 123 (1979).

27. Ref. 26, p. 126.

28. *Gas Processors Association Publication 2140-77*, Gas Processors Association, Tulsa, Okla., 1977.

29. *North American Storage Capacity for Light Hydrocarbons 1977*, Gas Processors Association, Tulsa, Okla.

30. W. Braker and A. L. Mossman, *Matheson Gas Data Book*, 5th ed., Matheson Gas Products, East Rutherford, N.J., 1971.

31. W. Braker, A. L. Mossman, and D. Siegel, *Effects of Exposure To Toxic Gases—First Aid and Medical Treatment*, 2nd ed., Matheson Gas Products, East Rutherford, N.J., 1977.

32. *Pace Petroleum Service Annual Issue*, Cameron Engineers, Inc., Denver, Colo., 1987.

33. E. Seaton, *Oil Gas J.* **76**(28), 66 (1978).

34. R. C. Reid, J. M. Prausnitz, and T. K. Sherwood, *The Properties of Gases and Liquids*, 3rd ed., McGraw-Hill, New York, 1977.

35. R. R. Dreisbach, "Physical Properties of Chemical Compounds, II," *Advances in Chemistry Series 22*, American Chemical Society, Washington, D.C., 1959.

36. *Computer Stored Thermophysical Data System*, Management Information and Control, Phillips Petroleum Co., Bartlesville, Okla., 1976.

37. H. H. Szmant, *Organic Chemistry*, Prentice-Hall, Englewood Cliffs, N.J., 1957, p. 69.

38. *Publication No. 325 M*, National Fire Protection Association, Boston, Mass., 1969.

39. *Threshold Limit Valves for Chemical Substances in Workroom Air Adopted by ACGIH for 1976*, paper presented at the *American Conference of Governmental Industrial Hygienists*, Atlanta, Ga., May 1976.
40. C. E. Miller and F. D. Rossini, eds., *Physical Constants of Hydrocarbons C_1 to C_{10}*, American Petroleum Institute, New York, 1961.
41. R. W. Hankinson and G. H. Thomson, *Hydrocarbon Process.* **58**, 277 (1979).
42. J. W. Smalling, *J. Am. Oil Chem. Soc.* **42**, 102A (1965).
43. E. Guccione, *Chem. Eng.* **72**, 104 (1965).
44. *Industrial Hygiene Monitoring Manual for Petroleum Refineries and Selected Petrochemical Operations*, prepared for American Petroleum Institute, contract no. LER-40-73, by Clayton Environmental Consultants, Inc., API, Washington, D.C., p. 69.
45. J. H. Bruun and M. M. Hicks-Bruun, *J. Res. Natl. Bur. Std. A* **7**, 607 (1931).
46. P. Sabatier, *Ind. Eng. Chem.* **18**, 1005 (1926).
47. G. Scatchard, S. E. Wood, and J. M. Mochel, *J. Am. Chem. Soc.* **61**, 3206 (1930).
48. *Selected Values of Hydrocarbons and Related Compounds, American Petroleum Institute Research Project 44*, Carnegie Press, Pittsburgh, Pa., 1960, Table 23-2 (3.1110)-m, p. 1.
49. P. R. Dreisbach, *Adv. Chem. Ser.* (15), 441 (1955).
50. R. P. Linstead and S. L. S. Thomas, *J. Chem. Soc.*, 1127 (1940).
51. N. Zelinski, *J. Russ. Phys. Chem. Soc.* **43**, 1220 (1911).
52. N. Zelinski, *Ber. Dtsch. Chem. Ges.* **44**, 3121 (1911).
53. N. Zelinski and W. Kommaresky, *Ber. Dtsch. Chem. Ges.* **57**, 667 (1924).
54. N. I. Shuikin and co-workers, *Neftekhimiya* **1**, 756 (1961).
55. A. P. Lein and co-workers, *Ind. Eng. Chem.* **44**, 351 (1952).
56. U.S. Pat. 3,009,002 (Nov. 14, 1961), C. M. Kron (to Phillips Petroleum Co.).
57. U.S. Pat. 3,767,719 (Oct. 23, 1973), J. H. Colvert and co-workers (to Texaco, Inc.).
58. U.S. Pat. 3,597,489 (Aug. 3, 1971), O. D. Vu and R. Odello (to Institut Francais de Petrole).
59. *1993 Directory of Chemical Producers*, SRI International, Menlo Park, Calif., 1993.
60. A. B. Littlewood, *J. Gas Chromatog.* **1**(5), 5 (1963).
61. E. Berl and W. Koerber, *Ind. Eng. Chem., Anal. Ed.* **12**, 175 (1940).

ALAN D. EASTMAN
Phillips Petroleum

DAVID MEARS
Unocal Corporation

HYDROCARDANOL, *m*-PENTADECYLPHEN. See ALKYLPHENOLS.

HYDROCHLORIC ACID. See HYDROGEN CHLORIDE.

HYDROCOLLOIDS. See GUMS.

HYDROGEL. See CONTACT LENSES.

HYDROGEN

Hydrogen, the lightest element, has three isotopes: hydrogen [12385-13-6], H, at wt 1.0078; deuterium [16873-17-9], D, at wt 2.0141; and tritium [15086-10-9], T, at wt 3.0161 (1). Hydrogen is very abundant, being one of the atoms composing water; deuterium and tritium occur naturally on Earth, but at very low levels. Tritium, a radioactive low energy beta-emitter with a half-life of 12.26 yr (2), is useful as a tracer in hydrogen reactions (see DEUTERIUM AND TRITIUM; RADIOACTIVE TRACERS).

Whereas hydrogen atoms exist under certain conditions, the normal state of pure hydrogen is the hydrogen molecule [1333-74-0], H_2, which is the lightest of all gases. Molecular hydrogen is a product of many reactions, but is present at only low levels (0.1 ppm) in the Earth's atmosphere. The hydrogen molecule exists in two forms, designated ortho-hydrogen and para-hydrogen, depending on the nuclear spins of the atoms. Many physical and thermodynamic properties of H_2 depend on the nuclear spin orientation, but the chemical properties of the two forms are the same.

Hydrogen is a very stable molecule having a bond strength of 436 kJ/mol (104 kcal/mol), and is not particularly reactive under normal conditions. However, at elevated temperatures and with the aid of catalysts, H_2 undergoes many reactions. Hydrogen forms compounds with almost every other element, often by direct reaction of the elements. The explanation for its ability to form compounds with such chemically dissimilar elements as alkali metals, halogens, transition metals, and carbon lies in the intermediate electronegativity of the hydrogen atom.

Hydrogen is one of the most important industrial commodities. It is used in the production of ammonia (qv), urea (qv), methanol (qv) and higher alcohols, and hydrochloric acid (see HYDROGEN CHLORIDE); as a reducing agent; and to desulfurize or hydrogenate various petroleum (qv) and edible oils (see HYDROGENATION). Hydrogen, produced as a by-product, is used in a multitude of industrial processes as a fuel, and liquid hydrogen is an important cryogenic fluid (see CRYOGENICS). Almost all commercial hydrogen is produced by reaction of water and hydrocarbons (qv), the steam reforming reaction, or by partial oxidation of hydrocarbons.

Hydrogen is seen by many as having a central role in the future energy equation (see HYDROGEN ENERGY). This role is envisioned as one of an energy carrier. Hydrogen would be produced from primary renewable energy sources (qv), eg, solar energy (qv), by various water (qv) splitting techniques. The hydrogen produced would be shipped to various locations and used as a fuel or chemical commodity. Upon combustion, hydrogen returns to water, accompanied by virtually no pollution and no greenhouse gas production, in contrast to what occurs when hydrocarbons are burned (see AIR POLLUTION; ATMOSPHERIC MODELING).

Physical and Thermodynamic Properties

The spins of the atomic nuclei in a hydrogen molecule can be coupled in two distinct ways: with nuclear spins parallel (ortho-hydrogen) or nuclear spins antiparallel (para-hydrogen). Because molecular spins are quantized, ortho- and para-hydrogen exist in different quantum states. As a result, there are differences in many properties of the two forms of hydrogen. In particular, those properties that involve heat, such as enthalpy, entropy, and thermal conductivity, can show definite differences for *ortho-* vs *para*-hydrogen. Other thermodynamic properties show little difference.

The equilibrium between *ortho-* and *para*-hydrogen is a function of temperature. Equilibrium compositions for various temperatures ranging from absolute zero to 500 K are shown in Table 1. For any given hydrogen sample, equilibrium conditions of *ortho-* and *para*-hydrogen are often not realized, however, because the uncatalyzed interconversion of the two forms is relatively slow at low temperatures. At high temperatures, where molecular dissociation occurs, self-conversion rates are more rapid. Hence, when hydrogen is prepared, the equilibrium ortho–para ratio characteristic of the temperature of preparation can persist for relatively long periods of time at other temperatures.

The equilibrium 3:1 ratio of *ortho-* to *para*-hydrogen that occurs at about room temperature is called normal hydrogen. Physical properties are typically given for normal hydrogen and for *para*-hydrogen (20.4 K equilibrium hydrogen). *para*-Hydrogen is the lower energy form of hydrogen. The equilibrium mixture at very low temperatures is almost pure *para*-hydrogen, and the conversion of *ortho*-hydrogen to *para*-hydrogen is an exothermic process. The energy released on conversion of normal hydrogen to *para*-hydrogen is given in Table 1 for several temperatures; additional data are also available (5). The energy released on converting liquid normal hydrogen to 90% *para*-hydrogen is sufficient to vaporize 64% of the original liquid (6). For this reason, catalysts have been developed that rapidly convert normal to *para*-hydrogen, greatly facilitating the storage of liquid hydrogen. Many materials have been found to catalyze this conversion, including hydrous iron(III) oxide, rare earths, uranium compounds, and carbon (7–11). Strong magnetic fields also enhance the ortho–para conversion rate.

Table 1. Temperature Dependence of the Equilibrium *Ortho-*, *Para*-Hydrogen Composition[a,b]

Temperature, K	*para*-Hydrogen at equilibrium, %	ΔH^c, kJ/mol[d]
0	100.0	
10	100.0	-1.0627
20	99.82	
25	99.01	
30	97.02	
35	93.45	
40	88.73	
50	77.05	-1.062
75	51.86	
100	38.62	-0.9710
150	28.60	
200	25.97	-0.3302
300	25.07	-0.0556
500 and higher	25.00	

[a] Refs. 3–5.
[b] Normal hydrogen contains ca 3:1 ratio of *ortho-* to *para*-hydrogen. See text.
[c] Heat of reaction of the conversion of normal to *para*-hydrogen.
[d] To convert kJ to kcal, divide by 4.184.

For the deuterium molecule [7782-39-0], D_2, the ortho–para relationship is the opposite of that in H_2, ie, the ortho form is the more prevalent one at low temperatures. For instance, at 20 K, the equilibrium concentration of D_2 is 97.97% ortho, and at 220 K it is 66.66% ortho-deuterium (12).

Tables 2, 3, and 4 outline many of the physical and thermodynamic properties of para- and normal hydrogen in the solid, liquid, and gaseous states, respectively. Extensive tabulations of all the thermodynamic and transport properties listed in these tables from the triple point to 3000 K and at 0.01–100 MPa (1–14,500 psi) are available (5,39). Additional properties, including accommodation coefficients, thermal diffusivity, virial coefficients, index of refraction, Joule-Thomson coefficients, Prandtl numbers, vapor pressures, infrared absorption, and heat transfer and thermal transpiration parameters are also available (5,40). Thermodynamic properties for hydrogen at 300–20,000 K and 10 Pa to 10.4 MPa (10^{-4}–103 atm) (41) and transport properties at 1,000–30,000 K and 0.1–3.0 MPa (1–30 atm) (42) have been compiled. Enthalpy–entropy tabulations for hydrogen over the range 3–100,000 K and 0.001–101.3 MPa (0.01–1000 atm) have been made (43). Many physical properties for the other isotopes of hydrogen (deuterium and tritium) have also been compiled (44).

As can be seen from Tables 2–4, many of the corresponding physical properties of normal and para-hydrogen are significantly different from each other. These differences have often been used to advantage in analysis. For instance, at 120–190 K the thermal conductivity for para-hydrogen is more than 50% greater than that of ortho-hydrogen. Hence, thermal conductivity offers a means of determining the ortho–para ratio of a given hydrogen sample. The thermal conductivity of hydrogen gas is the highest of all common gases, about seven times that of air. Thus thermal conductivity is also used to detect hydrogen in the presence of other gases. Second and third virial coefficients for para-hydrogen have been

Table 2. Physical and Thermodynamic Properties of Solid Hydrogen

Property	Hydrogen		Refs.
	para-	Normal	
mp, K (triple point)	13.803	13.947	13–15
vapor pressure at mp, kPa^a	7.04	7.20	15, 16
vapor pressure at 10 K, kPa^a	0.257	0.231	16
density at mp, $(mol/cm^3) \times 10^3$	42.91	43.01	17, 18
heat of fusion at mp, J/mol^b	117.5	117.1	15
heat of sublimation at mp, J/mol^b	1023.0	1028.4	19, 20
C_p at 10 K, $J/(mol \cdot K)^b$	20.79	20.79	21
enthalpy at mp, $J/mol^{b,c}$	−740.2	321.6	14, 15, 18, 19
internal energy at mp, $J/mol^{b,c}$	−740.4	317.9	15, 22
entropy at mp, $J/(mol \cdot K)^{b,c}$	1.49	20.3	14, 15, 18, 19
thermal conductivity at mp, $mW/(cm \cdot K)$	9.0	9.0	23
dielectric constant at mp	1.286	1.287	24
heat of dissociation at 0 K, kJ/mol^b	431.952	430.889	16

^{a}To convert kPa to mm Hg, multiply by 7.5.
^{b}To convert J to cal, divide by 4.184.
^{c}Base point (zero values) for enthalpy, internal energy, and entropy are 0 K for the ideal gas at 101.3 kPa (1 atm) pressure.

Table 3. Physical and Thermodynamic Properties of Liquid Hydrogen

Property	Hydrogen para-	Normal	Refs.
mp, K (triple point)	13.803	13.947	13–15
normal bp, K	20.268	20.380	13–15
critical temperature, K	32.976	33.18	13–15
critical pressure, kPa[a]	1292.8	1315	13–15
critical volume, cm³/mol	64.144	66.934	13–15, 17, 18
density at bp, mol/cm³	0.03511	0.03520	13–15, 17, 18
density at mp, mol/cm³	0.038207	0.03830	13–15, 17, 18
compressibility factor, $Z = PV/RT$			
at mp	0.001606	0.001621	15, 22
bp	0.01712	0.01698	15, 22
critical point	0.3025	0.3191	15, 22
adiabatic compressibility, $(-\partial V/V\partial P)_s$, MPa^{-1b}			
at triple point	0.00813	0.00813	25
bp	0.0119	0.0119	25
coefficient of volume expansion, $(-\partial V/V\partial T)_p$, K^{-1}			
at triple point	0.0102	0.0102	26
bp	0.0164	0.0164	26
heat of vaporization, J/mol[c]			
at triple point	905.5	911.3	15, 22
bp	898.3	899.1	15, 22
C_p, J/(mol·K)[c]			
at triple point	13.13	13.23	13–15, 17, 18
bp	19.53	19.70	13–15, 17, 18
C_v, J/(mol·K)[c]			
at triple point	9.50	9.53	13–15, 17, 18
bp	11.57	11.60	13–15, 17, 18
enthalpy, J/mol[c,d]			
at triple point	−622.7	438.7	13–15, 17, 18
bp	−516.6	548.3	13–15, 17, 18
internal energy, J/mol[c,d]			
at triple point	−622.9	435.0	13–15, 17, 18
bp	−519.5	545.7	13–15, 17, 18
entropy, J/(mol·K)[c,d]			
at triple point	10.00	28.7	13–15, 17, 18
bp	16.08	34.92	13–15, 17, 18
velocity of sound, m/s			
at triple point	1273	1282	13, 15, 27–29
bp	1093	1101	13, 15, 27–29
viscosity, mPa·s(=cP)			
at triple point	0.026	0.0256	15, 29–32
bp	0.0133	0.0133	15, 29–32
thermal conductivity, mW/(cm·K)			
at triple point	0.73	0.73	15, 28, 30, 31, 33
bp	0.99	0.99	15, 28, 30, 31, 33

Table 3. *(Continued)*

| Property | Hydrogen | | Refs. |
	para-	Normal	
dielectric constant			
at triple point	1.252	1.253	15, 24
bp	1.230	1.231	15, 24
surface tension,			
mN/m(=dyn/cm)			
at triple point	2.99	3.00	15, 34
bp	1.93	1.94	15, 34
isothermal compressibility,			
$1/V(\partial V/V\partial P)_T$, MPa$^{-1 b}$			
at triple point	−0.0110	−0.0110	35
bp	−0.0199	−0.0199	35

[a] To convert kPa to mm Hg, multiply by 7.5.
[b] To convert MPa to atm, divide by 0.101.
[c] To convert J to cal, divide by 4.184.
[d] Base point (zero values) for enthalpy, internal energy, and entropy are 0 K for the ideal gas at 101.3 kPa (1 atm) pressure.

tabulated from 14 to 500 K (45,46). Equations are given for calculating the virial coefficients over this temperature range. These values of the virial coefficients for *para*-hydrogen agree with values for normal hydrogen at 100 K (47).

Expansion from high to low pressures at room temperature cools most gases. Hydrogen is an exception in that it heats upon expansion at room temperature. Only below the inversion temperature, which is a function of pressure, does hydrogen cool upon expansion. Values of the Joule-Thomson expansion coefficients for hydrogen have been tabulated up to 253 MPa (36,700 psi) (48), and the Joule-Thomson inversion curve for *para*-hydrogen has been determined (49,50).

The vapor pressure of liquid *para*-hydrogen as a function of temperature can be calculated from the following equation (51):

$$\ln(P/P_t) = ax + bx^2 + cx^3 + d(1 - x)^e$$

where $a = 3.05300134164$; $b = 2.80810925813$; $c = -0.655461216567$; $d = 1.59514439374$; $e = 1.5814454428$; $x = (1 - T_t/T)/(1 - T_t/T_c)$; $T_t = 13.8$ K; $T_c = 32.938$ K; $P_t = 0.007042$ MPa; T, the temperature, is in K; and P, the vapor pressure, is in MPa. The vapor pressure of liquid normal hydrogen as a function of temperature can be calculated from the following equation (16):

$$\log_{10} P = -A/T + B + CT$$

where $A = 44.9569$; $B = 6.79177$; $C = 0.020537$; T is temperature in K; and P is vapor pressure in Pa. Tables of vapor pressure for liquid and solid normal hydrogen (52), sublimation pressures of *para*-hydrogen from 1 K to the triple point (53), and equations for estimating sublimation pressures of normal and *para*-hydrogen (16) are all available in the literature.

Table 4. Physical and Thermodynamic Properties of Gaseous Hydrogen[a]

Property	Hydrogen		Refs.
	para-	Normal	
density at 0°C, (mol/cm^3) $\times$ 10^3	0.05459	0.04460	13, 14, 17, 18
compressibility factor, $Z = PV/RT$, at 0°C	1.0005	1.00042	15, 22
adiabatic compressibility, $(-\partial V/V\partial P)_s$, at 300 K, MPa^{-1}[b]	7.12	7.03	25
coefficient of volume expansion, $(\partial V/V\partial P)_p$, at 300 K, K^{-1}	0.00333	0.00333	26
C_p at 0°C, J/(mol·K)[c]	30.35	28.59	13–15, 17, 18
C_v at 0°C, J/(mol·K)[c]	21.87	20.30	13–15, 17, 18
enthalpy at 0°C, J/mol[c,d]	7656.6	7749.2	13–15, 17, 18
internal energy at 0°C, J/mol[c,d]	5384.5	5477.1	13–15, 17, 18
entropy at 0°C, J/(mol·K)[c,d]	127.77	139.59	13–15, 17, 18
velocity of sound at 0°C, m/s	1246	1246	13, 15, 27–29
viscosity at 0°C, mPa·s (=cP)	0.00839	0.00839	15, 29–32
thermal conductivity at 0°C, mW/(cm·K)	1.841	1.740	15, 28, 30, 31, 33
dielectric constant at 0°C	1.00027	1.000271	15, 24
isothermal compressibility $1/V(\partial V/\partial P)_T$, at 300 K, MPa^{-1}[b]	−9.86	−9.86	35
self-diffusion coefficient at 0°C, cm^2/s		1.285	36
gas diffusivity in water at 25°C, cm^2/s		4.8 $\times$ 10^{-5}	37
Lennard-Jones parameters			
collision diameter, σ, m $\times$ 10^{10}		2.928	38
interaction parameter, ϵ/k, K		37.00	38
heat of dissociation at 298.16 K, kJ/mol[c]	435.935	435.881	16

[a]All values at 101.3 kPa (1 atm).
[b]To convert MPa to atm, divide by 0.101.
[c]To convert J to cal, divide by 4.184.
[d]Base point (zero values) for enthalpy, internal energy, and entropy are 0 K for the ideal gas at 101.3 kPa (1 atm) pressure.

Hydrogen gas diffuses rapidly through many materials, including metals. This property is used in separating hydrogen from other gases and in purifying hydrogen on an industrial scale. Hydrogen diffusion through metals is also used as an analytical technique for hydrogen determination in gas chromatography (54). Hydrogen is only slightly soluble in water but is somewhat more soluble in organic compounds. For instance, at 0°C and 0.1 MPa (1 atm) pressure, the solubility of H_2 in water (STP) is 0.0214 cm^3/g; in benzene the solubility is 0.0585 cm^3/g (55). A method has been outlined for estimating the solubility of hydrogen gas in various solvents as a function of temperature and pressure (56). The solubility of hydrogen in water as a function of temperature and pressure in the range 0–100°C and 0.1–10 MPa (1–100 atm) is available (55), as is the solubility of hydrogen in a number of other aqueous solutions as well as organic solvents (55).

Solid hydrogen usually exists in the hexagonal close-packed form. The unit cell dimensions are $a_0 = 378$ pm and $c_0 = 616$ pm. Solid deuterium also exists in the hexagonal close-packed configuration, and $a_0 = 354$ pm, $c_0 = 591$ pm (57–59).

In addition to H_2, D_2, and molecular tritium [100028-17-8], T_2, the following isotopic mixtures exist: HD [13983-20-5], HT [14885-60-0], and DT [14885-61-1]. Table 5 lists the vapor pressures of normal H_2, D_2, and T_2 at the respective boiling points and triple points. As the molecular weight of the isotope increases, the triple point and boiling point temperatures also increase. Other physical constants also differ for the heavy isotopes. A 98% *ortho*–2% *para*-deuterium mixture (the low temperature form) has the following critical properties: $P_c = 1.650$ MPa (16.28 atm), $T_c = 38.26$ K, $V_c = 60.3$ cm^3/mol (61). The thermal conductivity of gas-phase deuterium is about 0.73 times that of gas-phase hydrogen. This thermal conductivity difference offers a convenient method for analysis of H_2–D_2 mixtures. Other physical properties of D_2, T_2, HD, DT, and HT are listed in the literature (60).

A mixture of solid and liquid *para*-hydrogen, termed slush hydrogen, is thought to be better for fuel purposes than liquid normal hydrogen because of the greater density and higher heat capacity of the solid–liquid mixture. Some thermodynamic properties of slush hydrogen and oxygen are given in the literature (62). As of this writing the National Aeronautics and Space Administration (NASA) is researching the possibility of using slush hydrogen as a fuel for the space shuttle and for hypersonic planes. Slush hydrogen is a highly energetic hydrogen slurry that takes up 15% less volume than conventional liquid hydrogen. Use of slush hydrogen could cut a plane's gross lift-off weight by as much as 30% (63). The slush is made by passing liquid hydrogen through a vessel cooled by helium and collecting solid hydrogen that forms on the vessel walls.

Solid hydrogen is known to undergo phase transitions as the pressure is increased. One phase of solid hydrogen that is postulated to exist under conditions of extreme pressure is metallic hydrogen (64–66). Metallization of hydrogen at extremely high pressures was first predicted from theory in 1935. Metallic hydrogen, predicted to have unusual properties, including very high

Table 5. Vapor Pressures of Hydrogen Isotopes, Normal Species[a,b]

Parameter	Temperature, K	Vapor pressure, kPa[c]		
		H_2	D_2	T_2
bp of T_2	25.04	323.0	147.9	101.3
bp of D_2	23.67	237.6	101.3	66.72
triple point of T_2	20.62	108.0	37.24	21.60
bp of H_2	20.38	101.3	34.01	19.24[d]
triple point of D_2	18.73	59.63	17.13	8.012[d]
triple point of H_2	13.95	7.199	0.674[d]	0.197[d]

[a]Refs. 16 and 60.
[b]Liquid unless at triple point or otherwise indicated.
[c]To convert kPa to mm Hg, multiple by 7.5.
[d]Solid.

temperature superconductivity, could store 100 times more energy compared with the same mass of liquid hydrogen.

Hydrogen gas chemisorbs on the surface of many metals in an important step for many catalytic reactions. A method for estimating the heat of hydrogen chemisorption on transition metals has been developed (67). These values and metal–hydrogen bond energies for 21 transition metals are available (67).

Chemical Properties

Hydrogen-Producing Reactions. *Industrial.* The main means of producing hydrogen industrially are steam reforming of hydrocarbons

$$CH_4 + H_2O \longrightarrow CO + 3\,H_2$$

partial oxidation of hydrocarbons

$$C_nH_{2n} + {}^n\!/_2\,O_2 \longrightarrow n\,CO + n\,H_2$$

and water electrolysis

$$2\,H_2O \longrightarrow 2\,H_2 + O_2$$

Mixtures of $CO-H_2$ produced from hydrocarbons, as shown in the first two of these reactions, are called synthesis gas. Synthesis gas is a commercial intermediate from which a wide variety of chemicals are produced. A principal, and frequently the only source of hydrogen used in refineries is a by-product of the catalytic reforming process for making octane-contributing components for gasoline (see GASOLINE AND OTHER MOTOR FUELS), eg,

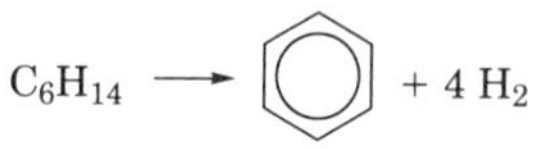

$$C_6H_{14} \longrightarrow \bigcirc + 4\,H_2$$

Hydrogen is also a significant by-product of other industrial processes, such as steam pyrolysis (68) of hydrocarbons to produce ethylene (qv), eg,

$$C_2H_6 \longrightarrow C_2H_4 + H_2$$

This by-product hydrogen can be used as fuel or purified and used in other chemical or refinery operations.

Laboratory. Hydrogen is produced on a laboratory scale from the action of an aqueous acid on a metal or from the reaction of an alkali metal in water:

$$Zn + 2\,HCl \longrightarrow H_2 + ZnCl_2$$
$$2\,Na + 2\,H_2O \longrightarrow H_2 + 2\,NaOH$$

These reactions can be carried out at room temperature. Hydrogen gas can also be produced on a laboratory scale by the electrolysis of an aqueous solution. Production of hydrogen through electrolysis is also used industrially. This involves the following reaction at the cathode of the electrochemical cell:

$$H^+ \text{ (aq)} + e^- \longrightarrow 1/2\ H_2$$

Hydrogen atoms can be produced in significant quantities in the gas phase by the action of radiation on or by extreme heating of H_2 (3000 K). Although hydrogen atoms are very reactive, these atoms can persist in the pure state for significant periods of time because of the inability to recombine without a third body to absorb the energy of bond formation.

Bonding of Hydrogen to Other Atoms.   The hydrogen atom can either lose the $1s$ valence electron when bonding to other atoms, to form the H^+ ion, or conversely, it can gain an electron in the valence shell to form the hydride ion, H^- (see HYDRIDES). The formation of the H^+ ion is a very endothermic process:

$$1/2\ H_2 \text{ (g)} \longrightarrow H^+ \text{ (g)} + e^- \qquad \Delta H = 1310\ \text{kJ/mol (313.1 kcal/mol)}$$

Hence, H^+ exists only when hydrogen is bonded to the most electronegative atoms. In aqueous solutions, H^+ hydrates to form H_3O^+ ion.

The formation of the hydride ion is also endothermic:

$$1/2\ H_2 \text{ (g)} + e^- \longrightarrow H^- \text{ (g)} \qquad \Delta H = 151\ \text{kJ/mol (36.1 kcal/mol)}$$

Hydride ions only form when hydrogen reacts with very electropositive materials.

Most hydrogen compounds are formed through covalent bonding of hydrogen to the other atoms. Hydrogen can bond with itself to form the hydrogen molecule. Because the hydrogen molecule has a high bond strength (436 kJ/mol or 104 kcal/mol), it is not particularly reactive under normal conditions. For this reason, high temperatures and catalysts are often used in reactions involving hydrogen.

Reactions of Synthesis Gas.   The main hydrogen manufacturing processes produce synthesis gas, a mixture of H_2 and CO. Synthesis gas can have a variety of H_2-to-CO ratios, and the water gas shift reaction is used to reduce the CO level and produce additional hydrogen, or to adjust the H_2-to-CO ratio to one more beneficial to subsequent processing (69):

$$CO + H_2O \longrightarrow CO_2 + H_2$$

Synthesis gas is used mainly to produce ammonia (qv) and methanol (qv) (70).

$$CO + 2\ H_2 \longrightarrow CH_3OH$$
$$3\ H_2 + N_2 \longrightarrow 2\ NH_3$$

In methanol production, zinc copper–chromium catalysts are used with reaction conditions of 200–400°C and pressures ≥ 10 MPa (100 atm) (71,72). Methanol is an industrial solvent and chemical intermediate from which formaldehyde (qv), acetic acid, methyl chloride, and methylamines are made (70). The gasoline additive methyl *tert*-butyl ether (MTBE), used for octane boosting, is produced from methanol and isobutene. Methanol production has received increased attention owing to the possibility of methanol-based fuels (see ALCOHOL FUELS). Ammonia production is an exothermic reaction thermodynamically favored by low temperatures, but high temperatures are needed to get reasonable rates of reaction. The approximate equilibrium constant K_p for the equation as written at 1 MPa (10 atm) is 7.08×10^{-4} at 350°C and 1.45×10^{-5} at 500°C (73). Industrially, the reaction is catalyzed with promoted iron oxides (74). Other materials such as ruthenium also catalyze this reaction (75). Operating conditions in an industrial plant cover a wide range of pressures (14–101 MPa (140–1000 atm)) and temperatures (ca 400–520°C). Ammonia is used to produce a wide variety of other chemicals, including nitrogen-based fertilizers (qv).

Synthesis gas is used in the production of substitute natural gas (SNG), ie, methane, and higher hydrocarbons,

$$CO + 3\,H_2 \longrightarrow CH_4 + H_2O$$

$$m\,CO + (2m + 1)\,H_2 \longrightarrow C_m H_{2m+2} + m\,H_2O$$

as well as in the production of olefins (qv), glycols (qv), and higher alcohols (see ALCOHOLS, HIGHER ALIPHATIC). Several reviews are available summarizing these reactions (76,77). The first reaction, called methanation, is used not only to produce methane (SNG) but also to remove small quantities of carbon monoxide from a gas stream (78) (see FUELS, SYNTHETIC–GASEOUS FUELS). Nickel catalysts are used with temperatures of about 315°C. Other catalysts have also been used for methanation (79–82).

The second reaction is called the Fischer-Tropsch synthesis of hydrocarbons. Depending on the conditions and catalysts, a wide range of hydrocarbons from very light materials up to heavy waxes can be produced. Catalysts for the Fischer-Tropsch reaction include iron, cobalt, nickel, and ruthenium. Reaction temperatures range from about 150 to 350°C; reaction pressures range from 0.1 to tens of MPa (1 to several hundred atm) (77). The Fischer-Tropsch process was developed industrially under the designation of the Synthol process by the M. W. Kellogg Co. from 1940 to 1960 (83).

Ethylene glycol, propylene glycol, and glycerol (qv) as well as higher alcohols can be prepared from synthesis gas. A series of patents describes this reaction, catalyzed by a rhodium homogeneous catalyst (84–86). Hydroformulation, also called the oxo process (qv), is a well-established industrial reaction involving synthesis gas. An aldehyde such as *n*-butyraldehyde, an important chemical intermediate (87), is produced from an olefin using homogeneous catalysis:

$$CH_3CH{=}CH_2 + CO + H_2 \longrightarrow CH_3CH_2CH_2CHO$$

A second principal application for this reaction is the production of higher alcohols from the aldehydes made from the olefins (88). These alcohols are used to manufacture detergents (see DETERGENCY; SURFACTANTS):

$$RCH_2CH_2CHO + H_2 \longrightarrow R(CH_2)_3OH$$

Other reactions that involve synthesis gas are various hydrogenation reactions, for example,

$$H_2 + CH_3C\equiv CH \longrightarrow CH_3CH=CH_2$$

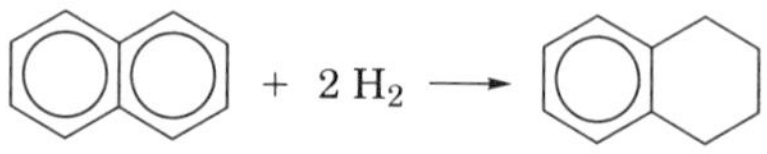

$+ \ 2 \ H_2 \longrightarrow$

In hydrogenation, it is often desirable to hydrogenate selectively, leaving some unsaturated bonds intact. A review of hydrogenation reactions is available (89).

Other Reactions of Hydrogen. *Heteroatom Removal from Fuels.* Sulfur, nitrogen, and oxygen are heteroatoms, which are abundant in many fuel sources such as petroleum (qv), coal (qv), and oil shale (qv). These elements are considered pollutants and detriments to the refining process. Hydrogen is used to reduce the levels of these contaminants. Coal contains both inorganic sulfur, ie, pyrite [1309-36-0], FeS_2, and organic sulfur, which undergo the following reactions when subjected to hydrogen at high temperatures (90–93):

$$FeS_2 + H_2 \longrightarrow FeS + H_2S$$
$$FeS + H_2 \longrightarrow Fe + H_2S$$

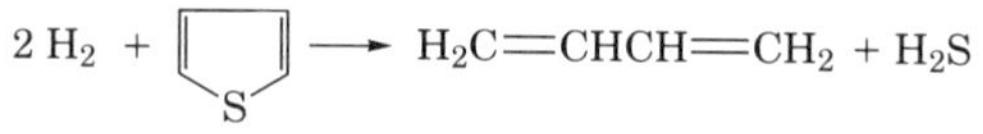

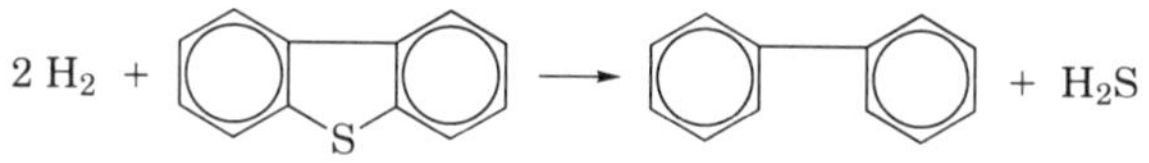

$2 \ H_2 +$ $\longrightarrow H_2C=CHCH=CH_2 + H_2S$

$2 \ H_2 +$ $\longrightarrow$ $+ \ H_2S$

Thiophene [110-02-1], C_4H_4S, and dibenzothiophene [132-65-0], $C_{12}H_8S$, are models for the organic sulfur compounds found in coal, as well as in petroleum and oil shale. Cobalt–molybdenum and nickel–molybdenum catalysts are used to promote the removal of organic sulfur (see COAL CONVERSION PROCESSES, CLEANING AND DESULFURIZATION). Hydrogen also reacts with other sulfur compounds:

$$RSH + H_2 \longrightarrow RH + H_2S$$
$$COS + H_2 \longrightarrow H_2S + CO$$

Petroleum, particularly shale oil, also contains organic oxygen and nitrogen compounds. Model reactions for the removal of these materials with hydrogen include

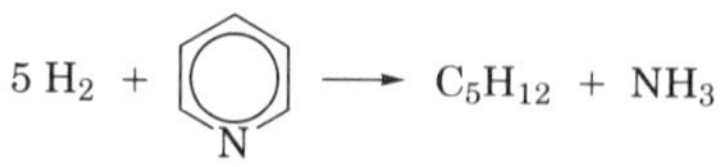

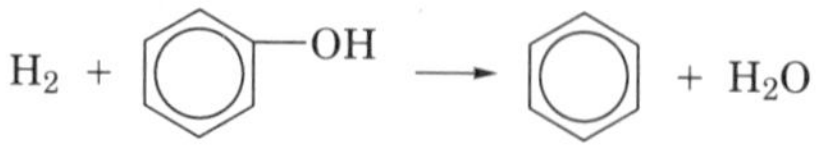

$5 \ H_2 +$ $\longrightarrow C_5H_{12} + NH_3$

$H_2 +$ $\longrightarrow$ $+ \ H_2O$

As a Reducing Agent. Hydrogen reacts with a number of metal oxides at elevated temperatures to produce the metal and water. Examples of these reactions are

$$FeO + H_2 \longrightarrow Fe + H_2O$$
$$Cr_2O_3 + 3 H_2 \longrightarrow 2 Cr + 3 H_2O$$
$$NiO + H_2 \longrightarrow Ni + H_2O$$
$$Bi_2O_3 + 3 H_2 \longrightarrow 2 Bi + 3 H_2O$$

Reduction of metal oxides with hydrogen is of interest in the metals refining industry (94,95) (see METALLURGY). Hydrogen is also used to reduce sulfites to sulfides in one step in the removal of SO_2 pollutants (see AIR POLLUTANTS) (96). Hydrogen reacts directly with SO_2 under catalytic conditions to produce elemental sulfur and H_2S (97–98). Under certain conditions, hydrogen reacts with nitric oxide, an atmospheric pollutant and contributor to photochemical smog, to produce N_2:

$$2 NO + 2 H_2 \longrightarrow N_2 + 2 H_2O$$

A ruthenium catalyst is particularly active for promoting this reaction.
Organic compounds can also be reduced with hydrogen:

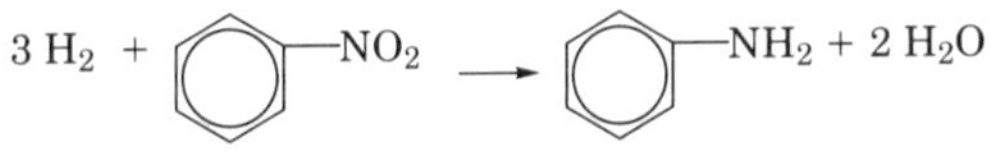

$$3 H_2 + \underset{}{\bigcirc}\!-\!NO_2 \longrightarrow \underset{}{\bigcirc}\!-\!NH_2 + 2 H_2O$$

$$2 H_2 + RCOOH \longrightarrow RCH_2OH + H_2O$$

$$H_2 + RCHO \longrightarrow RCH_2OH$$

Reactions of Hydrogen and Other Elements. Hydrogen forms compounds with almost every other element. Direct reaction of the elements is possible in many cases. Hydrogen combines directly with the halogens, X_2, to form the corresponding hydrogen halide.

$$H_2 + X_2 \longrightarrow 2 HX$$

The reaction with fluorine occurs spontaneously and explosively, even in the dark at low temperatures. This hydrogen–fluorine reaction is of interest in rocket propellant systems (99–102) (see EXPLOSIVES AND PROPELLANTS, PROPELLANTS). The reactions with chlorine and bromine are radical-chain reactions initiated by heat or radiation (103–105). The hydrogen–iodine reaction can be carried out thermally or catalytically (106).
Hydrogen combines directly with oxygen, either thermally or with the aid of a catalyst.

$$2 H_2 + O_2 \longrightarrow 2 H_2O$$

Many materials catalyze this reaction, among them metals and metal oxides such as Pt, Pd, NiO, and Co_3O_4 (107–109). One application for this reaction is in the removal of trace impurities of oxygen in a nitrogen stream using a palladium catalyst (110). Oxygen–hydrogen mixtures present a hazard because the mixture ignites explosively under certain conditions. Many studies have been made of the explosion and detonation limits of a hydrogen–oxygen system (111–112). Industrially one of the most important reactions of hydrogen is in the production of ammonia where synthesis gas is most often employed. Hydrogen reacts with graphite to form methane:

$$2\ H_2 + C_{(graphite)} \longrightarrow CH_4$$

Thermodynamically, the formation of methane is favored at low temperatures. The equilibrium constant K_p is $10^{8.82}$ at 300 K and is $10^{-1.0}$ at 1000 K (113). High temperatures and catalysts are needed to achieve appreciable rates of carbon gasification, however. This reaction was studied in the range 820–1020 K, and it was found that nickel catalysts speed the reaction by three to four orders of magnitude (114). The literature for the carbon–hydrogen reaction has been surveyed (115).

Hydrogen reacts directly with a number of metallic elements to form hydrides (qv). The ionic or saline hydrides are formed from the reaction of hydrogen with the alkali metals and with some of the alkaline-earth metals. The saline hydrides are salt-like in character and contain the hydride, ie, H^-, ion. Saline hydrides form when pure metals and H_2 react at elevated temperatures (300–700°C). Examples of these reactions are

$$Li + 1/2\ H_2 \longrightarrow LiH$$
$$Ca + H_2 \longrightarrow CaH_2$$

The saline hydrides are very reactive and are strong reducing agents. All saline hydrides decompose in water, often violently, to form hydrogen:

$$NaH + H_2O \longrightarrow H_2 + NaOH$$

Catalysts can be beneficial in the preparation of some saline hydrides (116).

Other metals also form compounds with hydrogen, either through direct heating of the elements, or during electrolysis with the metal as an electrode. These metallic hydrogen compounds, also called hydrides, are in fact covalently bonded and do not contain H^-. Many metallic hydrides are nonstoichiometric in nature and appear to be metal alloys, having properties typical of metals, such as high electrical conductivity. Some compounds, such as MgH_2, are intermediate in properties between the saline hydrides and the metallic hydrides. A review of hydrides is available (117).

Reactions of Atomic Hydrogen. Atomic hydrogen is a very strong reducing agent and a highly reactive radical that can be produced by various means.

Subjecting H_2 at 0.1 MPa (1 atm) pressure to a temperature of 4000 K produces about 62% hydrogen atoms (118).

$$H_2 \longrightarrow 2\,H\cdot$$

Hydrogen atoms can also be formed on catalytic surfaces, during electrolysis and upon decomposition of hydrocarbon radicals.

$$CH_3CH_2\cdot \longrightarrow CH_2{=}CH_2 + H\cdot$$

Hydrogen atoms are thought to play a principal role in the mechanistic steps of many reactions, including hydrocarbon thermolysis (119). Some reactions of atomic hydrogen with olefins and paraffins are the following (120–122):

$$H\cdot + C_4H_8 \longrightarrow CH_3\cdot + C_3H_6$$
$$H\cdot + C_3H_6 \longrightarrow C_3H_7\cdot$$
$$H\cdot + C_2H_6 \longrightarrow H_2 + C_2H_5\cdot$$

Other reactions (118,123) of atomic hydrogen include

$$H\cdot + Cl_2 \longrightarrow HCl + Cl\cdot$$
$$H\cdot + O_2 \longrightarrow O\cdot + OH\cdot$$
$$H\cdot + O_3 \longrightarrow HO\cdot + O_2$$
$$H\cdot + NO_2 \longrightarrow HO\cdot + NO$$

Hydrogen atoms also react with a graphite surface at elevated temperatures to produce methane and acetylene (124,125).

Absorption of Hydrogen in Metals. Many metals and alloys absorb hydrogen in large amounts. A striking example is a palladium electrode which, during electrolysis, can absorb several hundred times its volume of hydrogen. The absorption is largely reversible for palladium and for some other metals and alloys. Many metals can store more hydrogen per unit volume than a liquid-hydrogen Dewar vessel (126). Thus the hydrogen is compressed in the metal to a density greater than in the liquid state. Some metal systems that have been studied are TiFe, $LaNi_5$, $SmCo_5$, Mg_2Ni, Mg_2Ca, YCo_5, and $ThCo_5$ (127–131). These systems are being developed for energy storage.

Hydrogen diffuses and absorbs in many metals, with detrimental effects. Hydrogen exposure, under certain conditions, can seriously weaken and embrittle steel (qv) and other metals. In one study, iron (qv) in a hydrogen atmosphere at high pressures and 400–450°C degenerated in all mechanical properties (139). It is thought that atomic hydrogen on the surface of the metal diffuses to voids in the metal, forming hydrogen gas at very great pressures. Eventually, the metal may yield to the high hydrogen pressures (133). The limits to the use of steel in hydrogen service and additives to improve resistance to hydrogen attack have been discussed (134,135).

Hydrogen at elevated temperatures can also attack the carbon in steel, forming methane bubbles that can link to form cracks. Alloying materials such as molybdenum and chromium combine with the carbon in steel to prevent decarburization by hydrogen (132).

Manufacture

The principal commercial processes specific for the manufacture of hydrogen are steam reforming, partial oxidation, coal gasification, and water electrolysis. However, these are not of equal economic importance. In the United States, the bulk of the industrial hydrogen is manufactured by steam reforming of natural gas (see GAS, NATURAL). Relatively small quantities of hydrogen are produced by steam reforming of naphtha, partial oxidation of oil, coal gasification (see COAL CONVERSION PROCESSES, GASIFICATION), or water electrolysis. Worldwide, hydrogen as a raw material for the chemical industry is derived as follows: 77% from natural gas/petroleum, 18% from coal, 4% by water electrolysis, and 1% by other means (136). Significant quantities of hydrogen, especially to satisfy refinery H_2 demand, are produced as by-product $^-H_2$.

These processes all produce hydrogen from hydrocarbons and water.

steam reforming $CH_4 + 2 H_2O \longrightarrow CO_2 + 4 H_2$

naphtha reforming $C_n H_{2n} + 2 + n H_2O \longrightarrow n CO + (2n + 1) H_2$

resid partial oxidation $CH_{1.8} + 0.98 H_2O + 0.51 O_2 \longrightarrow CO_2 + 1.88 H_2$

coal gasification $CH_{0.8} + 0.6 H_2O + 0.7 O_2 \longrightarrow CO_2 + H_2$

water electrolysis $2 H_2O \longrightarrow 2 H_2 + O_2$

Process selection criteria focuses on a number of factors: hydrogen content of feedstock; hydrogen yield from the process; economics, including cost of feedstocks; capital and operating costs; energy requirements; environmental considerations; and intended use of the hydrogen. Proceeding from natural gas to liquid hydrocarbons and then to solid feedstocks, the processing difficulty and manufacturing costs increase. The partial oxidation and coal gasification processes require more capital investment than the steam reforming plants because an air-separation plant (see NITROGEN; OXYGEN), larger water gas shift and CO_2-removal facilities, and gas cleanup (see AIR POLLUTION CONTROL METHODS) are needed. The capital cost of water electrolysis plants is comparable to those of steam reforming in small-capacity plants, but electric power costs are almost prohibitive. In large-capacity plants, the capital cost of the electrolysis process significantly exceeds that of other processes. Table 6 gives the relative capital, thermal efficiency, and feedstock requirements for the principal commercial processes for the manufacture of hydrogen (137).

Steam Reforming. In steam reforming, light hydrocarbon feeds ranging from natural gas to straight run naphthas are converted to synthesis gas (H_2, CO, CO_2) by reaction with steam (qv) over a catalyst in a primary reformer furnace. This process is usually operated at 800–870°C and 2.17–2.86 MPa (300–400 psig), using a Ni-based catalyst. Temperatures up to 1000°C and

Table 6. Characteristics and Efficiencies for Producing Hydrogen Processes[a−c]

Parameter	Steam reforming (SR)	Partial oxidation (POX)	Texaco gasification (TG)	Water electrolysis
feedstock	natural gas	residual oil	bituminous coal	water and electricity
requirement per day	1.1×10^6 m³	1020 m³	2320 t	507 MW
thermal efficiency, %	78.5[d]	76.8	63.2	27.2
by-product	steam	sulfur	sulfur	oxygen
by-product capacity, t/d	1.7	30	70	695
capital cost, $ $\times 10^6$	83.2	205	316	132
production costs, $/(100 m³)				
feedstock	4.46	3.86	3.93	19.21
capital	2.14	5.39	8.32	3.32
o and m[e]	0.75	1.93	3.29	0.93
Total	7.35	11.18	15.54	23.46
by-product credit	−0.16	−0.03	−0.08	−0.83
net H₂ production cost				
$/(100 m³)	7.19	11.15	15.46	22.63
$/GJ[f]	5.60	9.10	12.58	16.91
net production cost ranking	1	2	3	4

[a]Ref. 3.
[b]Mid-1987 U.S. dollars.
[c]A H₂ production capacity of 2.8×10^6 m³/d (100×10^6 SCF/d) gas at 21−42 kg/cm² is assumed. To convert m³/d to SCF/d, multiply by 35.71.
[d]By-products account for an additional <0.1% of the thermal efficiency.
[e]Operation and maintenance.
[f]High heating value basis. To convert $/GJ to $/(10⁶ Btu), multiply by 0.054.

853

pressures up to 3.79 MPa (550 psia) are used in an autothermal-type reformer, or secondary reformer, when the hydrogen is used for ammonia, or in some cases methanol, production.

Nickel catalysts are also used for steam methane reforming. Moreover, nickel catalysts containing potassium to inhibit coke formation from feedstocks such as LPG and naphtha have received wide application.

Because hydrocarbon feeds for steam reforming should be free of sulfur, feed desulfurization is required ahead of the steam reformer (see SULFUR REMOVAL AND RECOVERY). As seen in Figure 1, the first desulfurization step usually consists of passing the sulfur-containing hydrocarbon feed at about 300–400°C over a Co–Mo catalyst in the presence of 2–5% H_2 to convert organic sulfur compounds to H_2S. As much as 25% H_2 may be used if olefins are present in the feed. This is then followed by adsorption (qv) of H_2S over ZnO catalyst to reduce the sulfur level to less than 0.1 ppmwt. When the hydrocarbon feed contains large amounts of sulfur, for example several hundred ppm or higher, bulk removal of H_2S is usually employed by such solvents as monoethanolamine (MEA) [141-43-5] prior to the ZnO desulfurization step. In this case, the effluent from the Co–Mo reactor must be cooled for bulk H_2S removal and reheated for final ZnO purification.

The gas and process steam mixture can then be introduced into the primary reformer. This reformer is a direct-fired chamber containing single or multiple rows of high nickel-alloy tubes: HK-40, HP-Modified, Incoloy 800, or other alloys are selected according to operating pressures and temperatures. The tubes are normally 72–110 mm ID and 10–13 m long. The catalyst contains 5 to 25% Ni (lower contents also include other metal promoters) as NiO supported on calcium aluminate, alumina, calcium aluminate titanate, or magnesium aluminate. Space velocities (SV) are usually on the order of 5000–8000 h^{-1} based on wet feed. Steam-to-carbon ratios are usually in the range of 3.0–5.0, outlet gas temperatures, 800–870°C, and pressure, 2.16–2.51 MPa (300–350 psig). The outlet gas composition corresponds to a 0–25°C temperature approach to steam-reforming equilibrium. That is, an equilibrium temperature is lower than actual at start and end of run catalyst activity. The flue gas temperatures are 980–1040°C exiting the fired section of the furnace. In the convection section, the flue gases are cooled by superheating the steam for drivers, generating steam, preheating the hydrocarbon feed for desulfurization, and preheating the feed-plus-steam mixture before entering the radiant section of the furnace. In order to obtain high overall furnace efficiency, the stack temperature can be lowered to 150–170°C by preheating combustion air for the radiant section burners.

The gas leaving the primary reformer with a high outlet temperature is about 76.7% H_2, 12% CO, 10% CO_2, and 1.3% CH_4 on a dry gas basis. Up to 95% conversion of CH_4 can be achieved in the primary reformer.

In the next step, the CO is converted to CO_2 and hydrogen by the water gas shift reaction step:

$$CO + H_2O \longrightarrow H_2 + CO_2 + 38.4 \text{ MJ/mol (9200 kcal/mol) of CO at } 371°C$$

This reaction is first conducted on a chromium-promoted iron oxide catalyst in the high temperature shift (HTS) reactor at about 370°C at the inlet. This

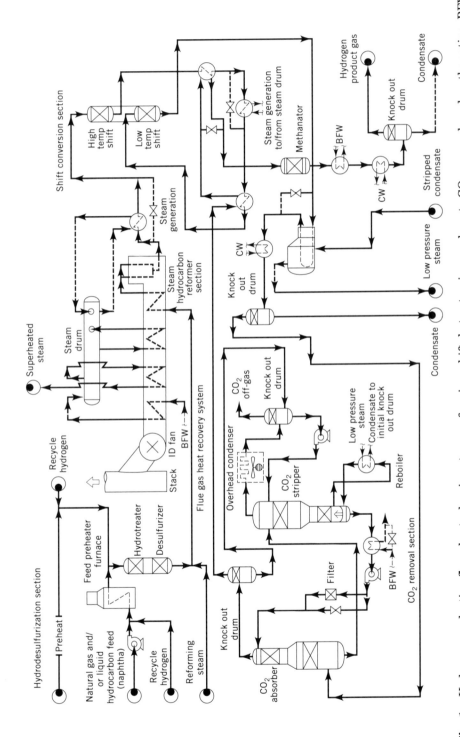

Fig. 1. Hydrogen production flow sheet, showing steam reforming, shift, hot potassium carbonate CO_2 removal, and methanation. BFW = boiling feed water, CW = cooling water, and ID = inner diameter.

catalyst is usually in the form of 6×6-mm or 9.5×9.5-mm tablets, SV about $4000\ h^{-1}$. Converted gases are cooled outside of the HTS by producing steam or heating boiler feed water and are sent to the low temperature shift (LTS) converter at about $200-215°C$ to complete the water gas shift reaction. The LTS catalyst is a copper–zinc oxide catalyst supported on alumina. CO content of the effluent gas is usually $0.1-0.25\%$ on a dry gas basis and has a $14°C$ approach to equilibrium, ie, an equilibrium temperature $14°C$ higher than actual, and SV about $4000\ h^{-1}$. Operating at as low a temperature as possible is advantageous because of the more favorable equilibrium constants. The product gas from this section contains about 77% H_2, 18% CO_2, 0.30% CO, and 4.7% CH_4.

The gas is then cooled with as much heat recovery as possible. CO_2 is scrubbed out by hot potassium carbonate or other processes such as MEA, inhibited MEA, methyldiethanolamine (MDEA), Selexol, Sulfinol, and Rectisol solutions (see CARBON DIOXIDE). The scrubbed gas contains about 98.2% H_2, 0.3% CO, 0.01% CO_2, and 1.5% CH_4. Remaining carbon oxides are converted to methane by passing the gases reheated to about $315°C$ over a methanation catalyst at SV $6000\ h^{-1}$. This catalyst has about 35% Ni supported on silica or other refractory supporting materials. The Ni content can vary from $15-48\%$. On this catalyst, CO and CO_2 are hydrogenated to CH_4. The outlet gases are cooled using heat recovery, and entrained water is removed in a droplet separator. A typical hydrogen product is 98.2% H_2 and 1.8% CH_4, but may be as low as $92-95\%$ H_2 if acceptable.

As an alternative to scrubbing out the CO_2 followed by methanation, the shifted gas can be purified by pressure-swing adsorption (PSA) when high purity hydrogen is desirable.

Pressure-Swing Adsorption Purification.   In nearly all cases where high purity ($>99\%$) hydrogen is needed, PSA is used in preference to cryogenic separation in the newer steam-reforming hydrogen plants and other hydrogen purification applications. Pressure-swing adsorption utilizes the fact that larger molecules such as CO, CO_2, CH_4, and H_2O, and also N_2, C_2H_6, and other light hydrocarbons, can be effectively separated from the smaller hydrogen gas by selective adsorption on high surface area materials such as molecular sieves (qv). Hydrogen has a very weak affinity for adsorption. The process of pressure-swing adsorption is capable of producing very pure ($>99.9\%$) hydrogen at recoveries of $70-90\%$, depending on the number of adsorption stages (138,139). The PSA system, shown in Figure 2, is operated under a pressurization–depressurization cycle at ambient temperatures. The adsorption is exothermic, whereas the desorption process is endothermic. Regeneration is accomplished by depressurization of the adsorbent bed, followed by a purge of hydrogen obtained from one of the adsorbent beds undergoing depressurization. Some hydrogen discharges into the purge-gas stream as a result of the countercurrent depressurization and the purging of impurities from the adsorbent.

This purge-gas hydrogen, along with other desorbed impurities such as methane and carbon monoxide, accounts for up to 90% of the fuel requirement in the reformer burners, thereby reducing the external fuel requirement. To make up for the hydrogen discharged to the purge stream, more hydrogen must be produced in the reforming train, which increases the feed requirement and the size of the reformer furnace. Offsetting this disadvantage is efficient heat

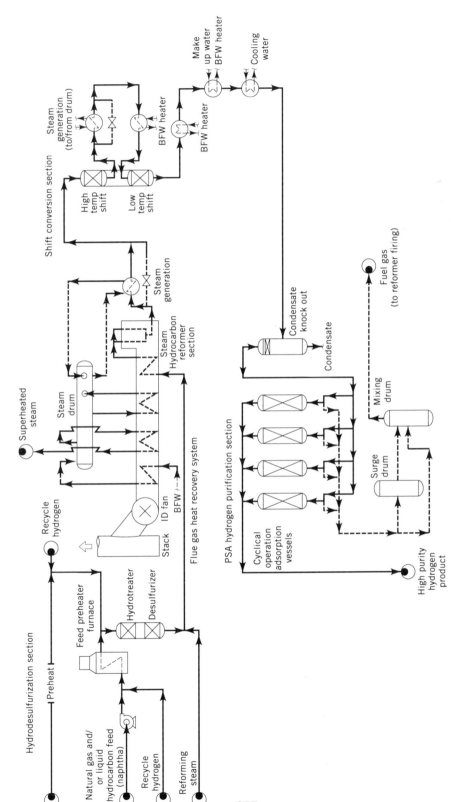

Fig. 2. Hydrogen production flow sheet showing steam reforming, shift, and pressure swing adsorption (PSA). BFW = boiling feed water.

recovery. The effect is that the overall hydrogen production cost, ie, operating expense and capital, is frequently less than for a conventional plant. For high purity hydrogen plants, PSA has become the preferred route, whereas the classical or conventional route is used mainly when CO_2 is the main product or when high purity H_2 is not required, as in ammonia production.

Use of a low temperature shift converter in a PSA hydrogen plant is not needed; it does, however, reduce the feed and fuel requirements for the same amount of hydrogen production. For large plants, the inclusion of a low temperature shift converter should be considered, as it increases the thermal efficiency by approximately 1% and reduces the unit cost of hydrogen production by approximately $0.70/1000 m^3 (2¢/1000 ft^3) (140,141).

The PSA system requires larger feed desulfurizers, reformer furnaces, and shift converters. The design of the reforming section of a hydrogen plant based on absorption and methanation depends on the degree of CO conversion in the shift converters, and the extent of CO_2 removal from the hydrogen stream. The less the CO conversion and CO_2 removal, the greater the hydrogen chemical consumption in the methanator, and the greater the steam-to-carbon ratio or temperature required in the reformer to maintain a given product purity. Because the PSA unit removes all of the CH_4, CO, and CO_2 from the hydrogen stream, the design of the reformer is independent of hydrogen product purity and can, therefore, be improved to the extent desired. Usually, a lower methane conversion is acceptable for a PSA-based hydrogen plant.

Some advantages of using the PSA system are that fewer processing units are required at lower maintenance cost and increased reliability; the risk of dangerous reaction runaways in case of CO_2 breakthrough is removed through elimination of the methanator unit; and heat can be recovered from the process more efficiently because no thermal energy is required for the regeneration of a CO_2-removal system, or for preheat of the methanator feed.

Naphtha at one time was a more popular feed, and alkali-promoted catalysts were developed specifically for use with it. As of 1994 the price of naphtha in most Western countries is too high for a reformer feed, and natural gas represents the best economical feedstock. However, where natural gas is not available, propane, butane, or naphtha is preferentially selected over fuel oil or coal.

Naphtha desulfurization is conducted in the vapor phase as described for natural gas. Raw naphtha is preheated and vaporized in a separate furnace. If the sulfur content of the naphtha is very high, after Co–Mo hydrotreating, the naphtha is condensed, H_2S is stripped out, and the residual H_2S is adsorbed on ZnO. The primary reformer operates at conditions similar to those used with natural gas feed. The nickel catalyst, however, requires a promoter such as potassium in order to avoid carbon deposition at the practical levels of steam-to-carbon ratios of 3.5–5.0. Deposition of carbon from hydrocarbons cracking on the particles of the catalyst reduces the activity of the catalyst for the reforming and results in local uneven heating of the reformer tubes because the firing heat is not removed by the reforming reaction.

The first naphtha steam-reforming furnaces were operated by ICI in the U.K. There are now many similar plants in operation all over the world. The amount of CO_2 that must be removed from the gas stream is higher than for a natural gas feedstock. Process and utility requirements per 30,000 m^3 (1.06 × 10^6 ft^3) of 97% H$_2$ gas are listed in Table 7.

Advances in Steam Reforming.   The direct-fired atmospheric tubular reformer is the industry workhorse for steam reforming of natural gas for the production of hydrogen. However, depending on process and product requirements, a number of alternative reforming methods have been proposed and commercialized. These include partial oxidation or autothermal or secondary reforming, heat-exchange-based reforming, and adiabatic pre-reforming. Additionally, a number of ideas are emerging to improve reforming operations heat management. These include pressurized fireboxes, helium-heated reformers, electrically heated reformers, molten bath reforming and integrated hydrogen separation reforming (to facilitate a shift in the equilibrium toward higher feedstock conversion).

Partial Oxidation of Natural Gas.   Partial oxidation of natural gas may be economically competitive with steam methane reforming (SMR) if high CO:H$_2$ ratios and high synthesis gas pressures are desired or if plant feed flexibility is required (138). Texaco has developed a process, called HyTex, that uses a noncatalytic partial oxidation gasifier to generate high purity H$_2$ from natural gas and off-gases. The gasification step is similar to partial oxidation of heavy oils. The feedstocks are converted to H$_2$ and carbon monoxide, ie, synthesis gas, at 1200–1350°C in an oxygen-lean atmosphere. The product gas is cooled and water scrubbed and fed to a shift reactor to convert CO into CO$_2$ for hydrogen plants. Final purification of the hydrogen is by pressure swing adsorption. Texaco and Union Carbide have formed a company, HydroGEN, to market the new technology in competition with conventional steam-reforming processes. The process may be competitive if oxygen is available over-the-fence at an attractive price (140). Shell Oil Co. also has partial oxidation technology in commercial operation for generating high purity H$_2$ from natural gas and other hydrocarbons.

Reforming Exchanger.   An advance in reforming technology is the commercialization of complementary and supplementary reforming technologies. Waste heat reforming, which refers to the direct use of high temperature process heat

Table 7. Process and Utility Requirements of Producing Hydrogen by PSA[a]

	Feedstock	
Requirement	Naphtha[b]	Natural gas
process feed and fuel, m^{3}[c]	9.67	9712
thermal equivalent, GJ[d]	303.1	297
energy, kWh	1080	1080
cooling water, m^3	598	393.6
demineralized water, m^3	60.6	56.4
export steam at 4.24 MPa (600 psig), t	41.8	41.7

[a]Requirements per 30,000 m^3 (1.6 × 10^6 ft^3) of 97% H$_2$ gas.
[b]Mol wt, 100; sp gr, 1.695 (48°API).
[c]To convert m^3 to ft^3, multiply by 35.3.
[d]To convert GJ to 10^3 Btu, divide by 1.055.

generated in an autothermal or primary reformer, can be used to provide part of the reforming energy input. Typically, the most advantageous scheme is to use a heat-exchange type of reformer, which has reforming catalyst in the tubes and the heat of reaction supplied by exchange with process gas on the shell-side. A number of reforming exchanger designs have been commercialized: ICI's gas-heated reformer (GHR) for the production of ammonia and Air Product's enhanced heat-transfer reforming (EHTR) for the production of hydrogen. The Kellogg Reforming Exchanger System (KRES) allows for elimination of the capital- and maintenance-intensive direct-fired primary reformer. Not only is hydrogen generation made more economical, but the amount of nitrous oxides that are released to the atmosphere are significantly reduced compared to conventional reforming. Some technologies, eg, Uhde's combined autothermal reformer (CAR), physically integrate autothermal and heat-exchange-based primary reforming duties in a single vessel.

Pre-Reformer. A pre-reformer is based on the concept of shifting reforming duty away from the direct-fired reformer, thereby reducing the duty of the latter. The pre-reformer usually occurs at about 500°C inlet over an adiabatic fixed bed of special reforming catalyst, such as sulfated nickel, and uses heat recovered from the convection section of the reformer. The process may be attractive in case of plant retrofits to increase reforming capacity or in cases where the feedsock contains heavier components.

Autothermal Reformer. Sometimes referred to as secondary reforming, autothermal reforming combines partial oxidation and steam reforming in a single vessel. It is most applicable when used for synthesis gas generation in the manufacture of ammonia, because air can be used for combustion and to provide the 3-to-1 $H_2:N_2$ stoichiometric ratio for ammonia synthesis. The oxygen in the air provides the partial oxidation medium. The nitrogen introduced is consumed in the ammonia synthesis. Air (oxygen) is mixed with steam and injected into the process gas where specially designed burners initiate combustion. The combustion heat provides the endothermic heat required to drive the reforming over a bed of nickel-based catalyst.

Secondary reforming is conducted at elevated temperatures; therefore the vessel is typically a refractory-lined, water-jacketed vessel. Exit temperatures are in the range of 925–1000°C. The high level heat is recovered downstream of the autothermal reformer by generating high pressure steam. Autothermal reforming is typically used in tandem with primary reforming in ammonia plants. Stand-alone autothermal reforming has been suggested for the production of synthesis gas for manufacture of ammonia, methanol, or oxosynthesis. A wide range of feedstocks can be employed, ranging from natural gas to LPG, naphtha, olefinic refinery off-gases, acetylene off-gases, and coke-oven gas. This process would be most economical at a site where low cost oxygen is available.

Electro-Reforming. The concept of using electricity to provide the endothermic heat of reforming has been proposed. Nuclear waste heat can be contained in high temperature helium gas which is brought into heat exchange with a natural gas feedstock (142).

Membrane Reforming. Membrane reforming involves increasing conversion beyond what is possible by equilibrium limitations by removing the hydrogen product from the reformed gas through special ceramic or metal membranes. High temperature-resistant membranes such as palladium foils on ceramic cylindrical inserts, have been used successfully on a laboratory scale. As of this writing, mechanical integrity and membrane times life under cyclic operations using actual feedstocks have not been determined (see MEMBRANE TECHNOLOGY).

By-Product Hydrogen. *Off-Gas Technologies.* The ratio of directly produced hydrogen to by-product H_2 is about 1:1 in the United States. Table 8 (143) gives typical compositions of hydrogen-rich off-gases. Refinery operations are both generators and consumers of hydrogen. Hydrogen is mainly produced by steam reforming and by recovery from various H_2-rich gases from dehydrogenation processes. Hydrogen is typically recovered from catalytic cracking, hydrotreating, and catalytic, ie, platinum–rhenium, reforming processes.

There has been an increasing interest in utilizing off-gas technology to produce ammonia. A number of ammonia plants have been built that use methanol plant purge gas, which consists typically of 80% hydrogen. A 1250-t/d methanol plant can supply a sufficient amount of purge gas to produce 544-t/d of ammonia. The purge gas is first subjected to a number of purification steps prior to the ammonia synthesis.

Pressure Partial Oxidation of Hydrocarbons. There are two commercial processes for producing hydrogen and hydrogen-containing synthesis gases by the noncatalytic partial oxidation of hydrocarbons under pressure. These are the Texaco process, which began commercial operation in 1954, and the Shell gasification process, in commercial operation since 1956. Both processes carry out the partial oxidation by burning hydrocarbons with oxygen or oxygen-rich gas mixtures to produce a gas that contains hydrogen and carbon monoxide and small quantities of carbon dioxide, water vapor, and methane. Typical synthesis gas from a heavy oil partial oxidation process might contain about, 46% H_2, 46% CO, 6% CO_2, 1% CH_4, and 1% N_2 plus Ar.

The principal advantage of the pressure noncatalytic partial oxidation processes over steam reforming is that it can operate on any hydrocarbon feedstocks that can be compressed or pumped, from natural gas to crude oil, residual oil,

Table 8. Typical Compositions of By-Product Hydrogen-Rich Stream[a]

Source	Composition, mol %[b]				
	H_2	CO	CH_4	N_2	Other
methanol purge gas	80.0	2.0	14.0	1.0	3.0[c]
ethylene plant tail gas	84.4	0.2	15.4		
CO plant H_2 gas	97.0	1.95	0.40	0.09	0.56
chloralkali plant H_2	99.86			0.08	0.06[d]
aromatics formation	96.6		3.4		
coke-oven gas	60.0	6.5	22.5	6.5	4.5[e]

[a]Ref. 143.
[b]Dry basis.
[c]2.6 mol % CO_2 and 0.4 mol % CH_3OH.
[d]Oxygen.
[e]2.5 mol % CO_2, 0.5 mol % O_2, and 1.5 mol % unidentified other.

or asphalts (qv), with heavy fuel oils and residual oils being the predominate feedstocks. Feedstock must have a sufficiently low viscosity at preheat temperatures to atomize effectively. No desulfurization is required prior to the partial oxidation step. Consequently, by 1965 partial oxidation processes were being installed for producing hydrogen, primarily in locations where natural gas or lighter hydrocarbons, including naphtha, were unavailable or were uneconomical as compared to residual fuel oil or crude oil. The principal disadvantage is the necessity for providing a supply of 95–99% pure oxygen (qv), ordinarily obtained in an air-separation plant, which adds appreciably both to the plant investment and the operating cost.

The Texaco process was first utilized for the production of ammonia synthesis gas from natural gas and oxygen. It was later (1957) applied to the partial oxidation of heavy fuel oils. This application has had the widest use because it has made possible the production of ammonia and methanol synthesis gases, as well as pure hydrogen, at locations where the lighter hydrocarbons have been unavailable or expensive such as in Maine, Puerto Rico, Brazil, Norway, and Japan.

Chemistry of Partial Oxidation. The process is carried out by injecting preheated hydrocarbon, preheated oxygen, and steam through a specially designed burner into a closed combustion vessel, where partial oxidation occurs at 1250–1500°C, using substoichiometric oxygen for complete combustion. Pressure is typically set by downstream product requirements, but is usually in the range of 3–8 MPa (435–1160 psi). The overall reaction is represented by

$$C_n H_m + n/2\ O_2 \longrightarrow n\ CO + m/2\ H_2$$

The overall process can be divided into three phases, the heating and cracking phase, the reaction phase, and the soaking phase.

In the heating and cracking phase, preheated hydrocarbons leaving the atomizer are intimately contacted with the steam-preheated oxygen mixture. The atomized hydrocarbon is heated and vaporized by back radiation from the flame front and the reactor walls. Some cracking to carbon, methane, and hydrocarbon radicals occurs during this brief phase.

In the reaction phase, hydrocarbons react with oxygen according to the highly exothermic combustion reaction. Practically all of the available oxygen is consumed in this phase.

$$C_n H_m + (n + m/4)\ O_2 \longrightarrow n\ CO_2 + m/2\ H_2O$$

The remaining unoxidized hydrocarbons react endothermically with steam and the combustion products from the primary reaction. The main endothermic reaction is the reforming of hydrocarbon by water vapor:

$$C_n H_m + n\ H_2O \rightleftharpoons n\ CO + (n + m/2)\ H_2$$

The complex of reactions results in a thermal equilibrium at 1300–1400°C.

The soaking phase takes place in the rest of the reactor where the gas is at high temperatures. A portion of the carbon disappears by reactions with CO_2 and steam. Some carbon, about 1–3 wt % of the oil feed, is present in the product gas. Natural gas feedstock produces only about 0.02 wt % of carbon.

The final composition of the reactor product gas is established by the water gas shift equilibrium at the reactor outlet waste-heat exchanger inlet where rapid cooling begins. Some units quench instead of going directly to heat exchanger.

$$CO + H_2O \rightleftharpoons CO_2 + H_2$$

In the reducing atmosphere of the reactor, sulfur compounds form hydrogen sulfide and small amounts of carbonyl sulfide [463-58-1], COS, in a molar ratio of approximately 24:1.

Figure 3 gives a typical configuration of the gasification section of the Shell partial oxidation process (144). The downstream purification sequence is not shown. The synthesis gas product from the gasification process unit is treated for sulfur removal. The CO-shift unit includes high temperature shift catalyst followed by low temperature shift catalyst. The heat released during the exothermic CO-shift reaction is recovered by raising high pressure and low pressure steam and by supplying makeup heat for several intermediate gas streams. Carbon dioxide removal is carried out as in steam-reforming processes. The Texaco partial oxidation process (145) is basically the same as the Shell process but has been operated at higher pressures of up to 8.0 MPa (1160 psi).

Equipment. Partial-oxidation gasification section equipment in many plants consists essentially of (*1*) the gasification reactor; (*2*) the waste-heat

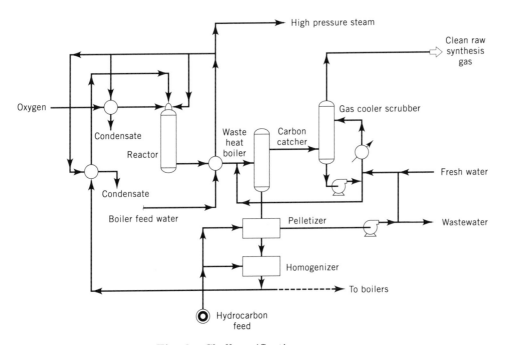

Fig. 3. Shell gasification process.

exchanger for heat recovery from the hot reactor gas or direct quench system; (3) the economizer heat exchanger for further heat recovery; (4) the carbon removal system for separating carbon from the reactor product gas; and (5) the carbon recovery system for recycle of carbon.

The gasification reactor is a vertical, empty, steel pressure vessel with a refractory lining into which preheated feedstock and steam are introduced premixed with oxygen. Steam-to-oil weight ratio is 0.35:1; the oxygen-to-oil ratio is 1.05:1.

Heat is recovered from the reactor product gas by generating high pressure steam in a waste-heat exchanger of special design using helical coils mounted in the exchanger shell. Sensible heat recovered from the reactor effluent gas plus the potential heat of combustion of the product gas equals about 95% of the hydrocarbon feedstock heating value, which is higher heating value (HHV). Problems of soot in conventional exchanger tubes require the use of helical tubes and proper gas velocity. The steam is generated at a pressure of at least 1 MPa (10 atm) greater than the reactor pressure so that the steam can be used directly as moderating steam. Waste-heat exchangers are designed for steam pressures of about 10 MPa (100 atm).

The gas leaving the heat recovery equipment contains soot and ash; some ash is deposited in the bottom of the reactor for removal during periodic inspection shutdowns. The gas passes to a quench vessel containing multiple water-sprays which scrub most of the soot from the gas. Additional heat recovery can be accomplished downstream of the quench vessel by heat exchange of the gas with cold feed water. Product gas contains less than 5 ppm soot.

The water–carbon slurry formed in the quench vessel is separated from the gas stream and flows to the carbon recovery system needed for environmental reasons and for better thermal efficiency. The recovered carbon is recycled to the reactor dispersed in the feedstock. If the fresh feed does not have too high an ash content, 100% of the carbon formed can be recycled to extinction.

In pelletizing, the water–carbon slurry is contacted with a low viscosity oil which preferentially wets the soot particles and forms pellets that are screened from the water and homogenized into the oil feed to the gasification reactor (see SIZE ENLARGEMENT).

When the recycle soot in the feedstock is too viscous to be pumped at temperatures below 93°C, the water–carbon slurry is first contacted with naphtha; carbon–naphtha agglomerates are removed from the water slurry and mixed with additional naphtha. The resultant carbon–naphtha mixture is combined with the hot gasification feedstock which may be as viscous as deasphalter pitch. The feedstock carbon–naphtha mixture is heated and flashed, and then fed to a naphtha stripper where naphtha is recovered for recycle to the carbon–water separation step. The carbon remains dispersed in the hot feedstock leaving the bottom of the naphtha stripper column and is recycled to the gasification reactor.

Desulfurization of Synthesis Gas. Removal of acidic constituents such as H_2S, CO_2, and COS from a gas stream is achieved by solvent scrubbing. These solvents provide simultaneous physical absorption (qv) and chemical adsorption (qv) under feed-gas conditions. Regeneration is accomplished by release of the acidic constituents at near atmospheric pressure and a somewhat elevated temperature. The gas is contacted by the solvent countercurrently in an absorber column. The rich solvent is regenerated by pressure reduction, heating, and stripping in the regenerator column where the acid gases are liberated. The lean solvent is cooled by heat exchange with the rich solvent before returning via a cooler to the absorber for reuse.

The conversion of CO to CO_2 can be conducted in two different ways. In the first, gases leaving the gas scrubber are heated to 260°C and passed over a cobalt–molybdenum catalyst. These catalysts typically contain 3–4% cobalt(II) oxide [1307-96-6], CoO; 13–15% molybdenum oxide [1313-27-5], MoO_3; and 76–80% alumina, Al_2O_3, and are offered as 3-mm extrusions, SV about 1000 h^{-1}. On these catalysts any COS and CS_2 are converted to H_2S. Operating temperatures are 260–450°C. The gases leaving this shift converter are then scrubbed with a solvent as in the desulfurization step. After the first removal of the acid gases, a second shift step reduces the CO content in the gas to 0.25–0.4%, on a dry gas basis. The catalyst for this step is usually Cu–Zn, which may be protected by a layer of ZnO.

The second CO_2 removal is conducted using the same solvent employed in the first step. This allows a common regeneration stripper to be used for the two absorbers. The gases leaving the second absorption step still contain some 0.25–0.4% CO and 0.01–0.1% CO_2 and so must be methanated as discussed earlier. The CO, CO_2, and possibly small amounts of CH_4, N_2, and Ar can also be removed by pressure-swing adsorption if desired.

Hydrogen From Coal. The production of hydrogen containing synthesis gas by the gasification of coal generated extensive research in the 1980s, driven by the prospect of rising gas and oil prices. The gasification of coal is well-established technology, but is not yet economically competitive with steam reforming of natural gas, LPG, or naphtha for production of hydrogen. However, a number of cost-effective coal gasification installations have been commissioned in areas where natural gas and oil are not readily available and where coal is abundant. For example, coal is used to produce synthesis gas for the Fischer-Tropsch synthesis of gasoline at SASOL in South Africa and for ammonia synthesis in China.

There are a number of well-established coal gasification technologies (see COAL CONVERSION PROCESSES, GASIFICATION; FUELS, SYNTHETIC–GASEOUS FUELS). In general, all types of coal can be processed. Gasification technologies are differentiated by the type of gasifier and the operating conditions employed. There are three principle gasifier types: fixed bed (Lurgi and British Gas Co. (BGC)), fluidized bed (Lurgi, Winkler), and entrained-flow (Koppers-Totzek, Shell, Texaco). Coal gasification processes are also classified according to the gasification operating temperature as low, medium, or high.

The Texaco coal gasification system is an example of a high (>1300°C) temperature–(>2 MPa) pressure system used for raw-gas generation. Figure 4 gives a typical simplified processing sequence for coal gasification. Pulverized coal, shown to be most efficient, is used as the feedstock. Chemical equilibrium at elevated temperatures favors the formation of H_2 and CO. Under high temperature conditions, methane formation is minimized, and no tars and oils are produced. Although hydrogen yield is slightly reduced, high pressure gasification

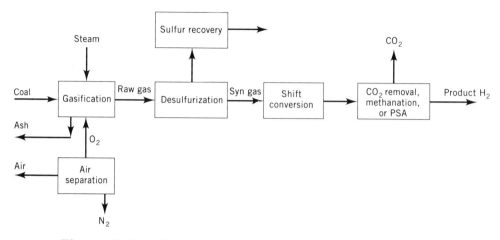

Fig. 4. Coal gasification process. PSA = pressure-swing adsorption.

results in significant power savings from elimination of raw gas compression. Low (700–800°C) temperature gasification processes, such as the Lurgi gasifier or BGC–Lurgi slagging gasifier, require a more complex processing sequence. Considerable amounts of methane, tars, and oils are formed. Recovered methane must be sold or steam-reformed for more hydrogen production. Tar and oil can be used as boiler fuel or in other ways dictated by economics. The energy requirement of the Texaco pressurized gasification process is 108,000 kJ/mol (25,812 kcal/mol) H_2, lower heating value (LHV). This corresponds to a thermal efficiency of 63.2% on a higher heating value (HHV) basis. This conversion efficiency is nearly 20% higher than that reported for atmospheric gasification systems (145). Typical synthesis gas composition from the Texaco gasification process using bituminous coal is 34% H_2, 48% CO, 17% CO_2, and 1% N_2 + Ar.

Coal gasification plants include substantial coal receiving and storage facilities; the process units; coal preparation, solid waste disposal, and water-treatment facilities; a cooling water system, a steam system, and electric power-generating facilities (147). Coal gasification may eventually replace natural gas and oil as the primary feedstock route for hydrogen manufacture because of ample coal supplies and the ability to maintain chemical industry infrastructure. Therefore there is continued incentive to develop and refine coal-based technologies.

Other Processes Using Hydrocarbon Feeds. Several other processes use hydrocarbon feeds. Coke-oven gas, having a typical analysis 60% H_2, 22.5% CH_4, 6.5% CO, 1.5% higher hydrocarbons, 0.5% O_2, 2.5% CO_2, and 6.5% N_2, has been used in Europe and to some extent in the United States for the production of hydrogen. During the first half of the twentieth century, coke-oven gas (COG) was an important source of H_2 for ammonia production. The COG process includes raw gas treatment and H_2 separation. In the pretreatment steps, benzene and higher hydrocarbons are scrubbed out using oil, and sulfur compounds and CO_2 are removed. After drying, H_2 is purified by cryogenic separation or pressure-swing adsorption. The system operates at about 2.03 MPa (20 atm). Linde AG and Montecatini offer such a process.

Thermal decomposition of hydrocarbons or natural gas on a brick check-erwork heated to 1100°C by hot combustion gases has been studied in the United States. A patent for the process was issued based on the decomposition of methane at 650–980°C, 101–203 kPa (1–2 atm) in a fluidized-catalyst-bed reactor (148). Carbon collected on the catalyst is burned to provide heat. The catalysts are oxides of Al, Li, Mg, Zn, and Ti, activated with compounds of Ni, Fe, and Co. Using a reactor temperature of 840°C, a regenerator temperature of 870°C, catalyst recirculation of 300 kg/kg CH_4 feed, and with a deposition of 0.25 wt % C on the catalyst, a product gas containing 93.3% H_2, 6.5% CH_4, 0.1% CO, and 0.1% N_2 was reported (148–150).

A process for the oxidation of hydrocarbons with steam and oxygen or with steam only at 425–730°C has been patented (151). Oxidation of hydrocarbons by steam which gives 95% H_2 at 980–1200°C on a catalyst based on NiO (FeO, CoO) on a zirconia–silica carrier has been described (152). The British Gas Council patented a reforming catalyst based on Ni—NiO—U_3O_8—UO_3 for reforming of liquids at 600°C (153). Reforming of hydrocarbons in molten iron oxide at high pressures using a CaO + ZnO flux was also proposed.

Water-Splitting Techniques.   Only one water-splitting method, electroly-sis, is practiced industrially for the production of hydrogen, and that only to a limited extent.

Direct, One-Step Thermal Water Splitting.   The water decomposition reaction has a very positive free energy change, and therefore the equilibrium for the reaction is highly unfavorable for hydrogen production.

$$H_2O\ (g) \longrightarrow H_2\ (g) + \tfrac{1}{2}\ O_2\ (g)$$

This situation does not improve greatly with increasing temperature, because the entropy change is small. At 2000 K (1727°C) and 101.3 kPa (1 atm), the hydrogen mole fraction at equilibrium is 0.036; at 3000 K (2727°C) the hydrogen mole fraction is about 0.2 (154).

There are significant problems for one-step thermal water splitting. In fu-ture nuclear and solar facilities, about 927°C is considered the upper tempera-ture range, which is not sufficient for this reaction. Even if high temperature heat sources were available, materials of construction would present difficulties. There would also be separation problems (155).

Electrochemical Water Splitting.   Electrochemical water splitting, ie, elec-trolysis, is an old and proven process to convert water to hydrogen, and is used industrially on a limited scale. The main problem is that the electricity used to drive the process is 3–5 times more expensive than fossil fuel-derived energy. Hence, electrolysis is of limited use. It has many desirable features, however. Electrolysis is a very clean, reliable process, and the hydrogen is very pure. As of this writing, electrolysis is the only proven water-splitting technique for hydrogen production. Electrolysis linked to renewable electricity-producing tech-nology could become more important in the future (156) (see ELECTROCHEMICAL PROCESSING, INORGANIC).

Electrochemical water splitting occurs when two electrodes are placed in water and a direct current is passed between the electrodes.

cathode $2 H_2O + 2 e^- \longrightarrow H_2 + 2 OH^-$

anode $2 OH^- \longrightarrow {}^{1/2} O_2 + H_2O + 2 e^-$

The standard free energy, enthalpy, and entropy are, respectively, $\Delta G = 1.23$ V or 237.19 kJ/mol (56.69 kcal/mol); $\Delta H = 285.85$ kJ/mol (68.32 kcal/mol); and $\Delta S = 70.08$ J/(mol·K) (16.72 cal/(mol·K)). The ΔG of 1.23 V is the reversible voltage, ie, the absolute minimum voltage needed to get the reaction to proceed. The total energy required for the reaction to proceed (ΔH) can be supplied by a combination of electricity and heat. Because $\Delta G = \Delta H - T\Delta S$ and ΔS is positive, the electrical work needed (ΔG) can be lowered by operating at higher temperatures, as shown in Figure 5. That is, more and more of the total energy needed can be supplied by heat, with increasing operating temperature. This is usually desirable, because heat is generally less expensive than electricity.

Three types of electrochemical water-splitting processes have been employed: (*1*) an aqueous alkaline system; (*2*) a solid polymer electrolyte (SPE); and (*3*) high (700–1000°C) temperature steam electrolysis. The first two systems are used commercially; the last is under development.

Aqueous Alkaline Electrolysis. The traditional process employs potassium hydroxide, KOH, added to the water to improve the conductivity through the cell. Table 9 shows operating parameters for industrial electrolyzers. All of these systems use a diaphragm to separate the cathode and anode, and keep the product oxygen and hydrogen from mixing. There are basically two types of units offered: tank type and filter press. In the tank type, many individual cells are connected in parallel and fed from one low voltage source. This requires large current flows at low voltage, as well as large transformers and rectifiers. Most commercial electrolyzers are of the filter press type, where cells are stacked and connected in series. The back side of the cathode for one cell is the anode for the next. This is called a bipolar arrangement. The voltage required to run the whole module is the sum of the voltages for each individual cell, so low voltages are not needed. However, a series arrangement means that if one cell fails, the module fails. Some units operate at high pressures. This is considered an efficient way to compress hydrogen. Much work is being directed toward improving traditional alkaline electrolysis (157,158). New cell geometries that lower resistances, better electrodes to reduce overvoltages, and better diaphragm materials, so that higher temperatures can be used, are all being considered. Higher temperatures enable the electrodes to function more efficiently. Improvements in design and materials are manifested in higher cell current densities.

Solid Polymer Electrolyte. The electrolyte in solid polymer electrolyte (SPE) units is Nafion, a solid polymer developed by Du Pont, which has sulfonic acid groups attached to the polymer backbone. Electrodes are deposited on each side of the polymer sheet. H^+ ions produced at the anode move across the polymer to the cathode, and produce hydrogen. The OH^- ions at the anode produce oxygen. These units have relatively low internal resistances and can operate at higher temperatures than conventional alkaline electrolysis units. SPE units are now offered commercially.

High Temperature Steam Electrolysis. Steam electrolysis occurs at very high temperatures, so that more of the energy to drive the reaction is in the

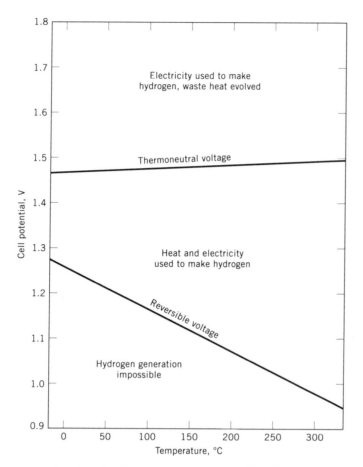

Fig. 5. Idealized operating conditions for electrolyzer.

form of heat rather than electricity. At 1000°C, 46% of the energy could be in the form of heat (159). Here, a ceramic is used as the electrolyte, and O^{2-} ions are transported through the ceramic material. As of this writing, steam electrolysis is in the development stage.

Comparison of Technologies. Figure 6 compares the electrochemical water splitting technologies in terms of current density and voltage. The SPE methodology is better than the conventional alkaline, but is no better than advanced alkaline using the modern zero-gap cell geometry. Design constraints may make SPE more suitable for small markets, rather than large ones (160).

Multistep Thermochemical Water Splitting. Multistep thermochemical hydrogen production methods are designed to avoid the problems of one-step water splitting, ie, the high temperatures needed to achieve appreciable ΔG reduction, and the low efficiencies of water electrolysis. Although water electrolysis itself is quite efficient, the production of electricity is inefficient (30–40%). This results in an overall efficiency of 24–35% for water electrolysis.

Table 9. Operating Conditions for Hydrogen Production Electrolyzers[a]

				Manufacturer		
Parameter	Electrolyzer Corp. Ltd.[b]	BBC	Norsk Hydro	de Nora	Lurgi[c]
operating temperature, °C	70	80	80	80	90
electrolyte, % KOH	28	25	25	29	25
current density, j, kA/m^2	1.34	2.00	1.75	1.50	2.00
cell voltage, V	1.9	2.04	1.75	1.85	1.86
current yield, %	>99.9	>99.9	>98	>98.5	98.75
O$_2$ purity, %	99.7	≥99.6	99.3 ⋯ 99.7	99.6	99.3 ⋯ 99.5
H$_2$ purity, %	99.9	≥99.8	98.8 ⋯ 99.9	99.9	99.8 ⋯ 99.9
energy requirements,[d] kWh	4.9	4.9	4.3	4.6	4.5

[a]All cells are bipolar operating at normal, atmospheric pressure unless otherwise noted.
[b]Monopolar cells are used.
[c]Operating pressure is 3 MPa (30 bar).
[d]On the basis of 1 m^3 of H$_2$ produced.

870

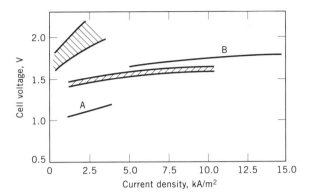

Fig. 6. Comparison of current density and cell voltage characteristics of the electrolysis systems where lines A and B represent steam electrolysis and the use of SPE, respectively, ▨, the conventional KOH water electrolysis, and ▨, zero-gap cell geometry employing 40% KOH, at 120–140°C.

In multistep thermochemical water splitting, two or more reactions are used to produce hydrogen from water. A hypothetical example is

$$H_2O + 2\,A \longrightarrow 2\,AH + {}^{1}/_{2}\,O_2$$

$$2\,AH \longrightarrow 2\,A + H_2$$

where A represents some chemical entity. These two reactions sum to give water decomposition to hydrogen and oxygen. In the ideal process, no other material would have a net consumption or production, so water is the only material input. The goal of this type of process is to be purely thermal and avoid the inefficiencies of electricity production needed in electrolysis, yet operate at a relatively low temperature.

Many reaction schemes have been proposed (161,162). All reaction schemes are designed such that reaction steps having positive ΔS values are operated at high (625–725°C) temperatures, whereas reaction steps having negative ΔS values are operated at low (about 225°C) temperatures. The purpose is to lower the free energy change, ie, the work requirement, and increase the thermal requirement, for improved efficiency. Other considerations, such as reaction kinetics, corrosion, cost of materials, and side reactions must also be taken into account.

Only a few of the proposed processes have received serious attention. One such is the chloride cycle, called the Mark 9, developed at the European Joint Research Center, (JRC Euratom) at Ispra, Italy:

$$6\,FeCl_2 + 8\,H_2O \longrightarrow 2\,Fe_3O_4 + 12\,HCl + 2\,H_2 \qquad 650°C$$

$$2\,Fe_3O_4 + 3\,Cl_2 + 12\,HCl \longrightarrow 6\,FeCl_3 + 6\,H_2O + O_2 \qquad >200°C$$

$$6\,FeCl_3 \longrightarrow 6\,FeCl_2 + 3\,Cl_2 \qquad 350°C$$

Many problems have been reported (163), and the process has been abandoned because of the difficulty in handling solids. Processes which are thought to have the best likelihood of success are based on sulfuric acid decomposition. Three prominent cycles are based on this reaction:

the General Atomics iodine–sulfur cycle

$$I_2 + SO_2 + 2\,H_2O \longrightarrow 2\,HI + H_2SO_4 \qquad\qquad 25°C$$

$$2\,HI \longrightarrow H_2 + I_2 \qquad\qquad 300°C$$

$$H_2SO_4 \longrightarrow H_2O + SO_2 + \tfrac{1}{2}\,O_2 \qquad 871°C$$

the Westinghouse S cycle

$$SO_2 + 2\,H_2O \longrightarrow H_2SO_4 + H_2 \qquad\qquad 25°C,\ 0.17\ V$$

$$H_2SO_4 \longrightarrow H_2O + \tfrac{1}{2}\,O_2 + SO_2 \qquad 871°C$$

the JRC Euratom Mark 13

$$2\,HBr \longrightarrow H_2 + Br_2 \qquad\qquad 80–250°C,\ 1.066\ V$$

$$Br_2 + SO_2 + 2\,H_2O \longrightarrow 2\,HBr + H_2SO_4 \qquad 20–120°C$$

$$H_2SO_4 \longrightarrow H_2O + SO_2 + \tfrac{1}{2}\,O_2 \qquad 650–850°C$$

Two of these cycles have an electrolysis step. Although one of the purposes of the thermochemical cycles is to avoid electrolysis and the associated inefficiencies of electricity production, the electrolysis steps proposed use much less electrical energy than water electrolysis. The Mark 13 is regarded as the most advanced thermochemical cycle, with overall efficiency of about 40%, including the electrolysis step (164).

A detailed discussion of thermochemical water splitting is available (155,165–167). Whereas many problems remain to be solved before commercialization is considered, this method has the potential of being a more efficient, and hence more cost-effective way to produce hydrogen than is water electrolysis.

Solar Processes. The use of solar energy (qv) for the production of hydrogen has been described (168). There are three groups of technical processes. The first group of processes for hydrogen production from sunlight and water uses biophotolysis, photocatalysis, photoelectrolysis, thermochemical reactions, and direct thermal water splitting. The second group includes photovoltaics, solar thermionics, and the solar thermal process (see PHOTOVOLTAIC CELLS; PHOTOCHEMICAL TECHNOLOGY). The third group includes processes for hydrogen production using indirect forms of solar energy: ocean thermal energy conversion, wind energy conversion, ocean-wave energy conversion, hydropower, and energy from biomass and wastes (see FUELS FROM BIOMASS; FUELS FROM WASTE).

Large-scale demonstrations of solar- and wind-power for use in water splitting for hydrogen production are underway. In Germany a windmill-powered electrical generator is used to split water using standard water electrolysis technology to produce 1 m^3 of 99.999% H$_2$ from 4 L of feedwater (169). In Pakistan a demonstration plant uses solar energy to generate 140 kW for electrolysis of water to produce hydrogen.

Among uses of solar energy for hydrogen production is a scheme for conversion of solar energy in photovoltaic cells that use an HI electrolyte to produce hydrogen. Direct decomposition of water at temperatures exceeding 2200°C, reached by the use of parabolic mirrors, has also been studied. In this process, H_2 and O_2 were separated by diffusion through high temperature membranes.

Production From Bacteria.   The process of biochemical hydrogen formation has been known since the 1940s. Extensive research has demonstrated the hydrogen-producing ability of algae and photobiosynthetic bacteria and microorganisms. However, studies using cultures of *Rhodopseudomonas capsulata* have indicated the need for a significant investment in solar generators to supply the required energy. Alternatively, nonphotosynthetic bacteria, which can obtain energy from chemical substrates such as starch (qv), cellulose (qv), and glucose, are also capable of producing hydrogen. Researchers have identified two enzymes, hydrogenase and formate dehydrogenase, present in hydrogen-producing nonphotosynthetic bacteria. Hydrogen is also known to be produced in the human intestine, and a process for collection of gases containing hydrogen evolved from human and animal wastes has been suggested (170). Future possibilities include large-scale fermentation (qv) of genetically engineered microorganisms to produce hydrogen (see GENETIC ENGINEERING, MICROBES). The production of hydrogen from landfill and sewage-sludge-derived incola and pure cultures of coliform bacteria during the digestion of pure sugars and sugar-rich substrates has been described (171). In some cases, the effluent gases from the digestion process contained up to 90% hydrogen with carbon dioxide as a by-product.

Hydrogen production by filamentous, heterocystous cyanobacteria (blue-green algae) could be the basis of a biophotolysis system. Biophotolysis is the light-driven splitting of water, in reactions which involve the enzymes nitrogenase and hydrogenase. Hydrogen production by nitrogen-starved cultures of *Anabaena cylindrica* has been described (172,173). The rate of H_2 production is 30 mL/(h·L) of culture; NH_4Cl increases production rates. The H_2 to O_2 ratio is 4:1 in complete nitrogen starvation, about 1.7:1 with addition of NH_4Cl. Thermodynamic efficiency of the conversion of light energy to free energy of hydrogen via algae photosynthesis is 0.4%. Hydrogen production by filamentous nonheterocystous *Cyanobacterium plectonema boryanum* has been reported (174).

Nuclear-Based Process.   Hydrogen production based on the use of dedicated nuclear facilities has been described (175). Efficiency of 43%, 50% for the nuclear heat-to-hydrogen energy conversion, is claimed. The process uses an acyclic a-c generator and SPE electrolyzers.

The thermochemical heat requirement can be met by fission, fusion, or solar energy (see FUSION ENERGY). High temperature nuclear reactors are capable of generating potential thermochemical heat in the heat-carrier temperature range from 900–1000°C. The development of these reactors is underway in Germany, France, and the former Soviet Union (176). The high temperature nuclear heat is also suggested for electro, thermal, plasma, and chemical methods for hydrogen-producing reactions.

In Germany, a process that combines the EVA-ADAM process of steam reforming with a high temperature, gas-cooled reactor (HTGR) for hydrogen production has been developed (177). A HTGR can provide heat at about 950°C, suitable for steam reforming. Chemical energy is transported in the form of hydrogen and carbon oxides, with subsequent release of energy for consumption via a methanation step (178). A single-stage, internally cooled methanation reactor (IRMA) had to be developed because traditional fixed-bed methanation catalysts would operate at temperature ranges too high for commercially available methanation catalysts.

Coal can be processed to H_2 by heat from a high temperature, gas-cooled reactor at a process efficiency of 60–70%. Process steps are coal liquefaction, hydrogasification of the liquid, and steam reforming of gaseous products (179).

Metal–Water Processes.   The steam-iron process, one of the oldest methods to produce hydrogen, involves the reaction of steam and spongy iron at 870°C. Hydrogen and iron oxide are formed. These then react further with water gas to recover iron. Water gas is produced by reaction of coal with steam and air.

$$3\ Fe + 4\ H_2O \longrightarrow Fe_3O_4 + 4\ H_2$$

$$Fe_3O_4 + 2\ H_2 + 2\ CO \longrightarrow 3\ Fe + 2\ H_2O + 2\ CO_2$$

A more efficient route is based on use of zinc vapor. Zinc vapor is absorbed in molten lead to form a 20–30% solution, which is contacted with steam at 300–500°C and 10–20 MPa (100–200 atm). The H_2 product is withdrawn, and the ZnO removed from the lead is reduced and recycled (180).

Miscellaneous H_2 Processes.   Production of hydrogen by irradiation of coal using laser beams has been described (181). Operation is at 1200°C and atmospheric pressures (see LASERS). Pyrolysis of coal using a ruby laser produces H_2 and acetylene (see HYDROCARBONS, ACETYLENE). A low energy continuous CO_2 laser yields H_2 and CO and a low ash solid having a lowered H–C ratio (182). Cracking of hydrocarbons by passing them through molten iron or iron alloys at 1200°C has been described (183). Using natural gas feed, 99.8% H_2 is obtained. The carbon formed is dissolved in the metal. In a separate zone, this carbon is oxidized by air-blowing to CO and CO_2, and the regenerated metal is returned.

Hydrogen can also be obtained by decomposition of ammonia or methanol. The electronics and metal industries make use of a hydrogen generator that decomposes ammonia to obtain a very high purity hydrogen. Hydrogen production from the gasification of biomass is being studied (184).

Hydrogen from Hydrogen Sulfide.   Attention has been given to the prospect of splitting H_2S as a source of hydrogen. The impetus is derived largely from the extensive sources of hydrogen sulfide found in fossil fuels. Gas oil and residuum desulfurization is practiced extensively in the petroleum (qv) industry. Burning of H_2S-containing fossil fuels is not environmentally desirable. A number of reaction processes have been proposed for splitting hydrogen sulfide, using energy derived from thermochemical, electrical, or microwave means, along with high temperature catalysis. In the widely used Claus process, H_2S is partially oxidized to water and elemental sulfur in an exothermic reaction. Thus the hydrogen forms water and is invariably lost.

If the hydrogen could be reduced, the coproduction of hydrogen and valuable side products, eg, sulfur, sulfuric acid, and calcium sulfate, from H_2S could become economically competitive.

The direct splitting of H_2S, analogous to the splitting of water, is not economically feasible because of the high energy input requirement for the endothermic reaction.

$$2\ H_2S \longrightarrow 2\ H_2 + 2\ S$$

Direct splitting requires temperatures above 977°C. Yields of around 30% at 1127°C are possible by equilibrium. The use of catalysts to promote the reaction can lower the temperature to around the 327–727°C range. A number of transition metal sulfides and disulfides are being studied as potential catalysts (185). Thermal decomposition of H_2S at 1130°C over a Pt–Co catalyst with about 25% H_2 recovery has been studied.

A more economical route appears to be the indirect route, using a two-step reaction sequence via a sulfurization and desulfurization of a metal sulfide. Decomposition using the mixed oxidation-state Ni_3S_2 has been proposed:

$$Ni_3S_2 + H_2S+ \longrightarrow 3\ NiS + H_2$$
$$3\ NiS \longrightarrow Ni_3S_2 + S$$

The first reaction proceeds around 502–602°C; the desulfurization reaction goes at 827°C.

Desulfurization processes using halogens have been extensively studied. A typical approach (186) uses the reaction scheme (187):

$$H_2S + I_2 \longrightarrow S + 2\ HI$$
$$2\ HI \longrightarrow H_2 + I_2$$

Additionally, there are a number of useful electrochemical reactions for desulfurization processes (185). Solar–thermal effusional separation of hydrogen from H_2S has been proposed (188). The use of microporous Vicor membranes has been proposed to effect the separation of H_2 from H_2S at 1000°C. These membrane systems function on the principle of upsetting equilibrium, resulting in a twofold increase in H_2 yield over equilibrium amounts.

The direct reaction of methane and hydrogen sulfide to yield hydrogen and carbon disulfide is being studied (189).

$$CH_4 + 2\ H_2S \rightleftharpoons 4\ H_2 + CS_2$$

The thermal catalytic route proposed involves heating the fresh reactant feed plus recycle up to 790°C and feeding this material into a MoS_2 catalyst fixed-bed reactor operating at 0.1 MPa (1 atm). The route yields a production of H_2 almost 50% higher than the decomposition of H_2S route.

Hydrogen Purification

A wide range and a number of purification steps are required to make available hydrogen/synthesis gas having the desired purity that depends on use. Technology is available in many forms and combinations for specific hydrogen purification requirements. Methods include physical and chemical treatments (solvent scrubbing); low temperature (cryogenic) systems; adsorption on solids, such as active carbon, metal oxides, and molecular sieves, and various membrane systems. Composition of the raw gas and the amount of impurities that can be tolerated in the product determine the selection of the most suitable process.

The impurities usually found in raw hydrogen are CO_2, CO, N_2, H_2O, CH_4, and higher hydrocarbons. Removal of these impurities by shift catalysis, H_2S and CO_2 removal, and the pressure-swing adsorption (PSA) process have been described (vide supra). Traces of oxygen in electrolytic hydrogen are usually removed on a palladium or platinum catalyst at room temperature.

Cryogenically produced hydrogen usually has a 90–99% purity. Petrochemical and refinery off-gases are the most frequent feeds for cryogenic purification systems. Ethylene, methane, and carbon monoxide are easily recovered as by-products. Raw H_2 suitable for the cryogenic purification process is a 30–80% H_2 stream. A prepurification step is frequently needed to remove such impurities as higher hydrocarbons, CO_2, water, and H_2S. In the system, the feed is cooled cryogenically by indirect heat exchange with the product. This is achieved by one or more stages of partial condensation. The hydrocarbon-rich feed liquid is expanded to provide the Joule-Thompson refrigeration that drives the process: the product is vaporized by the feed. Hydrocarbon separation is obtained by cooling and partial condensation of the feed against the entering streams at cryogenic temperatures. Purity is improved by distillation (qv). The capacity of cryogenic systems is usually above 70,000 m^3/d (2.5×10^6 ft^3/d). When large volumes of raw hydrogen are handled and the removal of some of the impurities is not critical, cryogenic separation is more economical than PSA. Upgrading of various refinery waste gases is nearly always more economical than H_2 production by steam reforming.

Cryogenic separation of hydrogen from ammonia-plant purge gases as a cost-effective means of improving plant efficiency has been the subject of great interest (190). A cryogenic purge gas recovery unit provides energy savings of up to 0.7 GJ/t of ammonia. Typical systems can provide 90% hydrogen purity at 90–95% recoveries.

Membrane modules have found extensive commercial application in areas where medium purity hydrogen is required, as in ammonia purge streams (191). The first polymer membrane system was developed by Du Pont in the early 1970s. The membranes are typically made of aromatic polyaramide, polyimide, polysulfone, and cellulose acetate supported as spiral-wound hollow-fiber modules (see HOLLOW-FIBER MEMBRANES).

The capacity of palladium to absorb large quantities of H_2 at high temperatures was known for many years. Union Carbide developed a palladium-based H_2 process and in 1965 had in operation overall capacities of about 1.13×10^6 m^3/d (40×10^6 ft^3/d), producing >99.9% H_2.

At temperatures of 300–400°C, molecular hydrogen dissociates to atomic hydrogen on a palladium surface; the hydrogen atoms dissolve in the palladium, diffuse through it, and recombine to molecular hydrogen on the opposite surface. No other molecules exhibit this property. Thus, absolutely pure hydrogen can be produced if the palladium film is free from mechanical defects that permit leakage of other gases. Feed gases containing 50% or more hydrogen are treated to remove heavy hydrocarbons, hydrogen sulfide, or olefins, which poison the palladium surface. The gases are heated to operating temperature and fed into the diffusion cell at pressures of 1.4–3.45 MPa (14–34 atm). The purified hydrogen is obtained at pressures of 448–690 kPa (4.4–6.8 atm). H_2 purification using precious metal membranes is proposed for small-scale uses, where very high (99.999%) purity H_2 is required. Optimum membrane composition appears to be 77% Pd and 23% Ag (192).

Other methods suggested for H_2 purification include adsorption and desorption on active carbon and fractional crystallization (qv) of impurities from liquid hydrogen. Centrifuging at 1–5.1 MPa (10–50 atm) and 64–110 K while keeping the temperature below the dew point of impurities that collect as a liquid film in the centrifuge has also been suggested. Purities over 99% can be obtained by this method (192).

Environmental Considerations

Methods for the large-scale production of hydrogen must be evaluated in the context of environmental impact and cost. Synthesis gas generation is the principal area requiring environmental controls common to all syngas-based processes. The nature of the controls depends on the feedstock and method of processing.

Short-term environmental concerns associated with hydrogen production are minimized by the use of steam reforming of natural gas and by the recovery of hydrogen as a by-product. The methane steam reforming process is one of the most environmentally acceptable. The environmental concerns become greater for the partial oxidation of heavier fuels and coal gasification technologies. Although the technology exists for effective environmental controls for these latter processes, the controls place a heavy economic burden on the overall system.

Coal feedstocks present the most serious environmental problems because of potential particulate emissions from coal-handling and processing facilities. Additionally, disposal of ash and slag solids removed from the gasification step must be in an environmentally safe manner. Some gasification produces significant amounts of liquid by-products such as tars, phenols, and naphthas which must be either recovered or incinerated. Some coals contain significant amounts of sulfur which must be stripped out of the raw syngas as hydrogen and carbonyl sulfides, necessitating further processing in a sulfur recovery unit. Condensate streams from gasifiers may also contain hydrogen cyanide and soluble metals in addition to ammonia which further complicates disposal.

Partial oxidation of heavy liquid hydrocarbons requires somewhat simpler environmental controls. The principal source of particulates is carbon, or soot, formed by the high temperature of the oxidation step. The soot is scrubbed from the raw synthesis gas and either recycled back to the gasifier, or recovered as solid pelletized fuel. Sulfur and condensate treatment is similar in principle to that required for coal gasification, although the amounts of potential pollutants generated are usually less.

Reforming of natural gas or naphtha is the cleanest of synthesis gas generations. Most natural gases contain sufficiently low levels of sulfur that it is possible to remove sulfur in a simple fixed-bed adsorption system. Higher levels of sulfur are amenable to treatment in conventional solvent absorption–stripping systems for acid gas removal, such as MEA. Organic sulfur compounds in naphtha are usually hydrotreated over a cobalt–molybdenum catalyst and stripped as hydrogen sulfide. Residual sulfur is removed in a fixed-bed system similar to that used for natural gas before reforming.

Process condensate from reforming operations is commonly treated by steam stripping. More recently the stripper has been designed to operate at a sufficiently high pressure to allow the overhead stripping steam to be used as part of the reformer steam requirement. In older plants, the stripper overhead was vented to the atmosphere. Contaminants removed from the process condensate are reformed to extinction, so disposal to the environment is thereby avoided. This system not only reduces atmospheric emissions, but contributes to the overall efficiency of the process by recovering condensate suitable for boiler feedwater makeup, because the process is a net water consumer.

All fired equipment, whether process furnaces or utility boilers, is subject to federal, via the 1990 Amendment to the Clean Air Act (CCA), and often local regulation, usually in the form of sulfur and nitrogen oxide, NO_x, limitations (see EXHAUST CONTROL, INDUSTRIAL). Use of low NO_x burners reduces the nitrous oxide in the flue gas. Flue gases can be suitably brought into regulation compliance by a combination of conventional control techniques including selective catalytic reduction and flue gas scrubbing. Effluents from other utility systems, such as blowdowns from steam drums, cooling water systems and polishers, can normally be handled by neutralization.

Economic Aspects

The United States consumes about 1.2 EJ (1.1×10^{15} Btu (1.1 quad)) of hydrogen annually (49). Most U.S. hydrogen production, estimated at over 6.5×10^{10} m^3/yr (2.3×10^{12} ft^3/yr), is used captively (193) in the production of ammonia and methanol as well as in refinery operations. Sales or merchant use may total about 2.0×10^9 m^3, ca 3%, of production and is principally divided between two U.S. producers, Praxair and Air Products. Additional merchant hydrogen capacity is located in Canada. Merchant hydrogen uses are divided among chemicals (83%), electronics (5%), metals (5%), government (4%), float glass (1%), and foods (1%). The balance is miscellaneous uses. Table 10 gives prices for commonly delivered hydrogen (194).

Table 10. 1993 Delivered Hydrogen Prices[a]

Type	Cost, $/m³
Gaseous hydrogen	
pipeline[b]	0.01–0.06
bulk grade	
standard	1.79
electronics	8.93
Liquid hydrogen	
bulk[c]	0.27–0.71

[a] Ref. 194.
[b] Higher range based on a 2.8×10^6 m³/d steam reformer of natural gas at $1.90/kJ.
[c] Cost varies based on volume delivered.

Shipment and Storage

Whereas the safe storage of hydrogen has been practiced for many years, as of this writing almost all hydrogen is used near the production site.

Hydrogen Gas. Hydrogen is supplied by many vendors as a high pressure gas in steel cylinders. Pressures are typically 15–40 MPa (150–400 atm). Storage facilities provided on manufacturing sites use either low pressure gas holders, high pressure steel storage tanks, or cryogenic storage. Small amounts of hydrogen are shipped in steel gas cylinders which hold up to 7.45 m^3 (263 ft^3) of H$_2$ at 16.6 MPa (164 atm). High pressure tube trailers are sized at 708–5100 m^3 (25,000–180,000 ft^3). Air Products offers the following grades in % purity of hydrogen gas: VLSI, 99.9996; Electronic, 99.999; Research, 99.9995; Ultrapure Carrier, 99.9993; UHP/Zero, 99.999; and High Purity, 99.995.

Future storage methods may involve existing underground formations that previously held natural gas (195–197). Storage costs for hydrogen are expected to be higher than methane on an energy basis because of the lower energy per volume in hydrogen. Moreover, the existing pipeline system for natural gas could probably be used to ship hydrogen long distances (198). However, shipping costs would be expected to be greater than for natural gas, owing again to lower energy per unit volume for hydrogen (198). A 140-mile hydrogen pipeline is in use in the Houston/Port Arthur region of Texas.

Liquid Hydrogen. The use of liquid hydrogen is a well established technology because of its use in the space program. Liquid hydrogen is, however, more difficult to produce and maintain than liquid natural gas (199,200). Refrigeration costs are high owing to the low (bp = 20.4 K) liquefication temperature. There are a number of special problems associated with liquid hydrogen. Examples are the need to precool the gas to the inversion temperature before the hydrogen can cool on expansion to liquefy, and the exothermic ortho-to-para conversion after liquefaction.

Large-scale use of liquid hydrogen has led to the construction of large insulated storage tanks such as the 1893 m^3 (500,000 gal) liquid hydrogen storage sphere erected at the then Atomic Energy Commission (now DOE) test site in Nevada in 1963 (201). Liquid hydrogen has been transported by rail in tank cars of 36 and 107 m^3 (9,500 and 28,000 gal) capacity. The 107 m^3-capacity, jumbo tank cars are 23.7 m in length and are specially built for hauling liquid hydrogen. These latter have a special Linde cryogenic insulation and operate under less than 133 mPa (1 μm Hg) absolute pressure with a heat-transfer coefficient of 0.1163 W/(m^2·K). An insulation layer consisting of a vacuum jacket and multilayer radiation shielding keeps evaporation losses in a jumbo tank car down to 0.3%/d when the liquid hydrogen is stored at −253°C. Using such large tank cars is possible because of the low (70 kg/m^3) density of liquid hydrogen. These tank cars have proved to be a useful source of standby hydrogen gas for industrial operations (202): 107 m^3 (28,000 gal) of liquid hydrogen is equivalent to 83,000 m^3 (2.93 × 10^6 ft^3) of hydrogen gas.

The ICC classifies hydrogen as a flammable gas and requires that it carry a red label. Data on storage is available (203). The production and handling of flammable gases and liquefied flammable gases is regulated by OSHA (204).

Storage as Hydrides. The discovery of metal compounds that reversibly absorb hydrogen is relatively recent. In the 1970s, the AB$_5$ and AB family of alloys, which reversibly absorb hydrogen at room temperature and low pressure, were identified (205). Both A and B are metals. As of this writing many such compounds are known; $LaNi_5$ and TiFe are examples.

The use of hydrides as a means of storing hydrogen is not yet (ca 1994) of commercial importance. Hydride storage has been used in demonstrations, eg, to power automobiles (206). Hydride formulations and properties are available (131,207–209).

Storing hydrogen in hydrides has the following advantages: (*1*) the storage capacity is very high. Hydrogen can be stored in the alloys at a density greater than in its liquid form without the need for cryogenic technology. And (*2*), hydrides are safer than other means of storing hydrogen. Because hydrogen is in a low pressure form, the rupture of a fuel tank in an accident would not be as dangerous as the rupture of a high pressure gas cylinder or liquid hydrogen cylinder.

Storing hydrogen as hydrides has the following disadvantages: (*1*) hydrides as a class are quite expensive; (*2*) the containers for the hydrides must have heat exchangers to remove heat during charging; (*3*) hydrides can be unstable and affected by poisons, ie, they can degenerate on use; and (*4*) hydrides only hold about 2–4% hydrogen on a weight basis. One hydride which has received attention is iron titanium hydride [39433-92-6], which absorbs hydrogen in the following manner:

$$FeTi + H_2 \longrightarrow FeTiH_2$$

Heat is produced during hydrogen absorption, and must be supplied during its release. The weight of hydrogen that can be released from the hydride, about 1.5 wt %, requires approximately the same volume as liquid hydrogen, but the hydride is much heavier. The hydrogen released from a hydride has a very high purity, but removal of solid fines may be needed in some cases.

Before a metal can begin absorbing hydrogen to form the hydride, it must be heated to 300°C under vacuum to expel all other absorbed gases. Absorption of hydrogen is conducted at 3.4 MPa (34 atm) at room temperature. Formation of hydride causes a volume increase and embrittlement of alloy particles (210–213). An alloy containing Mn, such as 85–90 mol % FeTi and 10–15 mol % Mn, is more easily activated and cycled. Such an alloy releases H_2 within the molar range of 1.6–0.1 with a change of hydrogen of 1.5 wt % under a thermal load of 29 kJ/mol H_2 (6500 Btu/lb). Hydrogen reservoirs usually consist of an assembly of 16-mm diameter stainless steel tubes containing the hydride. A typical overall weight of a storage tank is about 560 kg, holding 6.35 kg of H_2. Hydrogen-release temperatures vary between 30°C and 80°C. Typical H_2 equilibrium pressures over an Fe–Ti alloy are shown in Figure 7. H_2 is produced at 103.4 kPa (776 mm Hg). Alloys other than Fe–Ti are undergoing development. Mg_2NiH_4 has a storage capacity of 5–6 wt % H_2 (210–218). An alloy, $Ca_xM_{1-x}Ni$, where M is misch metal, ie, a cerium alloy (see CERIUM AND CERIUM COMPOUNDS), is suggested for operation at 50.7 kPa to 30.4 MPa (0.5–300 atm). A LaNi metal fused at 1400°C can store hydrogen at a ratio of 170 L/kg metal. An alloy of 26% Mg, 7.4% misch metal, and 66.6% Ni has an enthalpy change of 50 kJ/mol

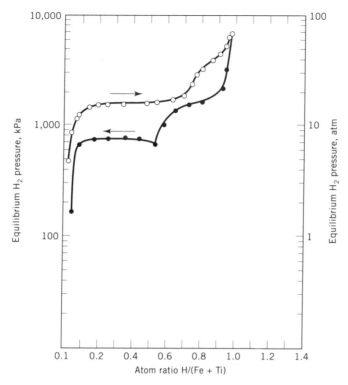

Fig. 7. Pressure–composition relationships for the FeTi–H system at 40°C during formation of the hydride (○) and release of hydrogen (●).

(12 kcal/mol) of H_2; an alloy of 26% Mg, 74% misch metal has an enthalpy change of 76.6 kJ/mol (18.3 kcal/mol) H_2.

Storage in Microcapsules.   Storage in microcapsules is under development and has no commercial significance as of this writing. Small glass spheres of about 0.1 mm diameter are heated to about 300–400°C and subjected to high (80 MPa (800 atm)) hydrogen pressures. At this temperature, hydrogen passes through the glass walls. Upon cooling the glass spheres contain about 5–10% by weight hydrogen, which can be released with heating (219,220). Glass microcapsules behave very much like metal hydrides.

Analysis

The determination of hydrogen content of an organic compound consists of complete combustion of a known quantity of the material to produce water and carbon dioxide, and determination of the amount of water. The amount of hydrogen present in the initial material is calculated from the amount of water produced. This technique can be performed on macro (0.1–0.2 g), micro (2–10 mg), or submicro (0.02–0.2 mg) scale. Micro determinations are the most common. There are many variations of the method of combustion and determination of water (221,222). The oldest and probably most reliable technique for water determination is a gravimetric one where the water is absorbed onto a desiccant, such as magnesium perchlorate. In the macro technique, which is the most accurate, hydrogen content of a compound can be routinely determined to within ±0.02%. Instrumental methods, such as gas chromatography (qv) (223) and mass spectrometry (qv) (224), can also be used to determine water of combustion.

Methods are available for determining the hydrogen content of metals (225–231). A combustion technique, similar to that for hydrocarbons, has been used. The metal is placed in a strong oxidizing agent and ignited in oxygen. The amount of water resulting indicates the amount of hydrogen present in the metal. Other methods of analyzing for hydrogen in metals are vacuum fusion and hot extraction techniques in which the metal is heated in a vacuum and the hydrogen content of the metal is released as H_2 (225). Analysis for the hydrogen gas is most often by gas chromatography (227–230) or mass spectrometry (226).

To determine quantitatively the amount of hydrogen present in a gas mixture, gas chromatography using a thermal conductivity detector is often employed. This technique can determine hydrogen in the presence of oxygen and nitrogen (232), and in the very complex gas streams that might be found in a refinery environment (233–237) involving H_2, He, O_2, H_2S, N_2, CO, CO_2, and various hydrocarbons. In one technique, hydrogen's unique ability to diffuse through metals is used in a gas chromatography: hydrogen is separated from a gas stream by diffusion through a metal and is quantitatively determined (54). Gas chromatography has also been used to determine the amounts of trace impurities in hydrogen (238). The isotopes of hydrogen gas (H_2, HD, D_2) can be separated using gas chromatography (239).

Thermal conductivity is used as an analytical tool in the determination of hydrogen. Because the thermal conductivities of *ortho*- and *para*-hydrogen are different, thermal conductivity detectors are used to determine the ortho:para ratio of a hydrogen sample (240,241). In one method (242), an analyzer is described which splits a hydrogen sample of unknown ortho:para ratio into two separate streams, one of which is converted to normal hydrogen with a catalyst. The measured difference in thermal conductivity between the two streams is proportional to the ortho:para ratio of the sample.

Hydrogen has a very high thermal conductivity compared to that of most other gases. This difference is used to detect hydrogen in the presence of other gases. In one method (243), hydrogen is oxidized selectively over a catalyst in a gas mixture. The amount of hydrogen initially present is determined by the thermal conductivity of the gas mixture before and after the removal of hydrogen. The thermal conductivity difference between H_2 and D_2 has led to a method for determining the D_2 content of an H_2 sample (244–246). The deuterium content of water can be determined in a similar way after decomposing the water into hydrogen and oxygen.

Mass spectrometry has been used to determine the amount of H_2 in complex gas mixtures (247), including those resulting from hydrocarbon pyrolysis (68). Mass spectrometry can also be used to measure hydrogen as water from hydrocarbon combustion (224,248). Moreover, this technique is also excellent for determining the deuterium:hydrogen ratio in a sample (249,250).

Nuclear magnetic resonance (nmr) is a nondestructive means of measuring the amount of hydrogen in various materials; for example, nmr has been used to determine the hydrogen content of coals (251).

Several methods have been developed for detecting traces of hydrogen in the atmosphere (252–255). In one method (252), hydrogen in the atmosphere reacts with HgO to produce mercury vapor. The amount of Hg vapor produced, which is proportional to the amount of hydrogen present, is monitored by the Hg absorption of 253.7 nm light. This method is reported to detect H_2 at 0.5 ppm levels to within $\pm 3\%$; the limit of detection is 0.01 ppm H_2. Another technique for detecting small quantities of hydrogen in a gas uses sound velocity as a measure of the amount of hydrogen present (256). It is reported that very small quantities of hydrogen cause detectable increases in sound velocity.

The detection of trace impurities in hydrogen typically involves an enrichment process to concentrate the impurities. Large volumes of hydrogen are passed over an absorbent material such as SiO_2 at very low temperatures. The impurities in the hydrogen are concentrated on the SiO_2, then desorbed and analyzed with gas chromatography (238), by mass spectrometry (257,258) or by other means (259). It is claimed that traces of methane in H_2, He, and Ne in the ppb range can be detected using this method (238). Detection of carbon monoxide in hydrogen at 2 ppm levels is claimed (259). Determination of traces of impurities in hydrogen by other techniques is also possible (244,247,260,261). Although hydrogen gas is not directly detectable by infrared spectroscopy, other compounds present are often detectable. The determination of methane in H_2 at levels from 0.005% to 2% is reported using infrared spectroscopy (260).

Health and Safety Factors

Hydrogen gas is not considered toxic but it can cause suffocation by the exclusion of air. The main danger in the use of liquid and gaseous hydrogen lies in its extreme flammability in oxygen or air. Some of the properties of hydrogen important in safety considerations are shown in Table 11. Hydrogen is odorless and colorless, and is, therefore, not easily detectable. Also, hydrogen burns with a nearly invisible flame. The detonation and flammability limits for hydrogen–air mixtures are much wider than those of gasoline–air or methane–air mixtures. Hydrogen is also explosive in mixtures with other materials such as fluorine. Tritium is a radioactive material. A comparison of the properties of hydrogen, methane, and gasoline, as related to safety may be found in the literature (266).

Mandatory regulations governing the distribution of liquid or gaseous hydrogen are available (267,268) as are guidelines on the safe use of liquid hydrogen (269) and gaseous hydrogen (270). Other reports concerning hydrogen safety may be found (271–273).

Uses

Hydrogen is an important industrial chemical in petroleum refining and in the synthesis of ammonia and methanol. These three areas account for more than 94% of U.S. industrial hydrogen consumption. The balance is consumed in the manufacture of various chemicals, eg, cyclohexane, benzene by toluene dealkylation, oxo-alcohols, and aniline; metallurgical processing; reducing gas blanketing; vegetable-oil hydrogenation; government space applications and transportation

Table 11. Properties of Hydrogen of Interest in Safety Considerations

Property	Value	Ref.
diffusion coefficient in air, cm^2/s	0.61	262
limits in air, vol %		
flammability	4.0–75.0	263
detonation	18.3–59.0	262
limits in oxygen, vol %		
flammability	4.5–94.0	264
detonation	15.0–90.0	264
ignition temperature, °C		
in oxygen	560	264
in air	585	264
flame temperature, °C	2045	263
heat of combustion, kJ/g[a]	119.9–141.9	266
% thermal energy radiated from flame	17–25	266
burning velocity in air, cm/s	265–325	266
detonation velocity in air, km/s	1.48–2.15	266

[a]To convert J to cal, divide by 4184.

fuel; float glass manufacture; and in the electronics industry. Table 12 shows estimates of U.S. hydrogen consumption in 1988 (193). Totals include intentionally produced hydrogen as well as by-product hydrogen. The captive market represents 97% of the hydrogen consumed. The merchant market covers the balance.

Refinery hydrogen requirement is met either by direct manufacture or indirect by-product recovery. Manufacture is typically by steam methane reforming. In addition significant quantities of captive hydrogen are generated as a by-product within the refinery itself, by purification of catalytic reforming streams, or from other dehydrogenation processes. The hydrogen requirement for refinery operations is expected to increase as the demand for upgrading liquid products and reformulated gasolines materializes.

High temperature steam reforming of natural gas accounts for 97% of the hydrogen used for ammonia synthesis in the United States. Hydrogen requirement for ammonia synthesis is about 336 m^3/t of ammonia produced for a typical 1000-t/d ammonia plant. The near-term demand for ammonia remains stagnant. Methanol production requires 560 m^3 of hydrogen for each ton produced, based

Table 12. U.S. Consumption of Hydrogen, 1988[a]

Industry	Consumption	
	$10^9 \ m^3$	%
ammonia	30.3	49
petroleum refining	22.7	37
methanol	4.6	8
small		
captive users	1.9	3
merchant users	2.0	3
Total	*61.5*	*100*

[a]Ref. 193.

on a 2500-t/d methanol plant. Methanol demand is expected to increase in response to an increased use of the fuel−oxygenate methyl t-butyl ether (MTBE) (see OCTANE IMPROVERS).

Hydrogen Economy

As of 1990 nearly 80% of the world's energy demand was met by fossil fuels (274). But the inevitable exhaustion of fossil fuels has led to concern and developments surrounding alternative renewable energy sources. The concept of a hydrogen economy was introduced in the 1960s as a vision for future energy requirements for the planet earth. In this economy hydrogen is envisioned along with electricity to be a dominant energy carrier. Solar power would supply the primary energy.

Hydrogen is preferred because it is storable and transportable. The concept of the hydrogen economy is predicated not only on the wide use of hydrogen as an energy carrier, but also on its exploitation as a clean, renewable, and nonpolluting fuel. Use of hydrogen as a fuel for surface vehicular propulsion or air transportation is attractive for a number of reasons. Hydrogen burns with increased efficiency (15−22% higher thermal efficiency than gasoline). From an environmental standpoint, hydrogen combustion results in no emissions of the greenhouse gases, CO, CO_2, SO_2, etc, although the formation of NO_x is greater than for gasoline-based engines.

Serious hydrogen-powered vehicle development programs are underway in Germany and Japan (275). BMW, Mercedes, and Mazda have developed state-of-the-art hydrogen fueled vehicles. BMW utilizes liquid hydrogen storage, Mercedes relies on compressed H_2 gas, and Mazda makes use of a metal hydride storage system. Requirements for cost-effective implementation of hydrogen-powered vehicles need to solve costly hydrogen generation problems and develop safe on-board hydrogen storage systems.

Hydrogen use as a fuel in fuel cell applications is expected to increase. Fuel cells (qv) are devices which convert the chemical energy of a fuel and oxidant directly into d-c electrical energy on a continuous basis, potentially approaching 100% efficiency. Large-scale (11 MW) phosphoric acid fuel cells have been commercially available since 1985 (276). Molten carbonate fuel cells (MCFCs) are expected to be commercially available in the mid-1990s (277).

Whereas most of the technology for hydrogen production, transportation, and usage is viable as of 1994, research efforts are needed to make them more economically attractive.

BIBLIOGRAPHY

"Hydrogen" in *ECT* 1st ed., Vol. 7, pp. 675−692, by R. M. Reed and N. C. Updegraff, The Girdler Corp.; in *ECT* 2nd ed., Vol. 11, pp. 338−379, by R. M. Reed, Girdler Corp., Subsidiary of the Chemical & Industrial Corp.; in *ECT* 3rd ed., Vol. 12, pp. 938−982, by B. G. Mandelik and D. S. Newsome, Pullman Kellogg.

1. A. G. Sharpe, *Inorganic Chemistry*, 3rd ed., Longman Scientific and Technical, Burnt Hill, Essex, U.K., (co-published in U.S. by John Wiley & Sons, Inc. New York), 1992, p. 211.

2. K. M. Mackay and M. F. A. Dove in J. C. Bailar, H. J. Emeleus, R. Nyholm, and A. F. Trotman-Dickenson, eds., *Comprehensive Inorganic Chemistry*, Vol. 1, Pergamon Press, New York, 1973, p. 93.

3. K. M. Mackay in Ref. 2, p. 11.

4. R. D. McCarty, *Hydrogen Technological Survey—Thermophysical Properties*, NASA SP-3089, U.S. Government Printing Office, Washington, D.C., 1975, pp. 518–519.

5. R. D. McCarty, J. Hord, H. M. Roder, *Selected Properties of Hydrogen (Engineering Design Data)*. U.S. Dept. of Commerce, National Bureau of Standards, Washington, D.C., 1981, pp. 6-291.

6. R. B. Heslop and P. L. Robinson, *Inorganic Chemistry*, Elsevier, Science Publishing Co., Inc., New York, 1967, p. 256.

7. W. K. Hall, *Accounts Chem. Res.* **8**(8), 257 (1975).

8. T. Tanabe, H. Adachi, and S. Imoto, *Technol. Rep. (Osaka Univ.)* **23**(1121–1154), 721 (1973).

9. K. N. Zhavoronkova and L. M. Korabel'nikova, *Kinet. Katal.* **14**, 966 (1973).

10. Y. Ishikawa, L. G. Austin, D. E. Brown, P. L. Walker, Jr., *Chem. Phys. Carbon* **12**, 39 (1975).

11. D. H. Weitzel, W. V. Loebenstein, J. W. Draper, and D. E. Park, *J. Res. Nat. Bur. Stand.* **60**, 221 (1958).

12. Ref. 2, p. 86.

13. H. M. Roder, L. A. Weber, and R. D. Goodwin, *National Bureau of Standards Monograph*, No. 94, Washington, D.C., Aug. 1965.

14. H. W. Woolley, R. B. Scott, and F. G. Brickwedde, *J. Res. Nat. Bur. Stand.* **41**, 379 (1948).

15. Ref. 5, pp. 6-127, 274.

16. Ref. 3, p. 14.

17. H. M. Roder and R. D. Goodwin, *National Bureau of Standards Technical Note*, No. 130, Washington, D.C., Dec. 1961.

18. J. W. Dean, *National Bureau of Standards Technical Note*, No. 120, Washington, D.C., Nov. 1961.

19. J. C. Mullins, W. T. Ziegler, and B. S. Kirk, *Technical Report No. 1*, Georgia Institute of Technology, Atlanta, Ga., Nov. 1961.

20. Ref. 3, p. 5.

21. Ref. 3, p. 15.

22. Ref. 4, p. 512.

23. C. Y. Ho, R. W. Powell, and P. E. Liley, *Standard Reference Data on the Thermal Conductivity of Selected Materials*, part 3, final report NBS-NSR05 Contr. CST-1346, Purdue University, Lafayette, Ind., Sept. 1968.

24. R. J. Corruccini, *National Bureau of Standards Technical Note*, No. 144, Washington, D.C., Apr. 1962.

25. Ref. 4, p. 169.

26. Ref. 4, p. 184.

27. R. D. McCarty and L. A. Weber, *National Bureau of Standards Technical Note*, No. 617, Washington, D.C., Apr. 1972.

28. V. J. Johnson, ed., *A Compendium of the Properties of Materials at Low Temperatures-Phase 1: Part 1, Properties of Fluids*, WADD Tech. Rep. 60-56, Contract No. AF33 (616) 58-4, National Bureau of Standards Cryogenic Engineering Laboratory, New York, July 1960.

29. J. G. Hust and R. B. Stewart, *National Bureau of Standards*, Rep. No. 8812, Washington, D.C., May 1965.

30. D. E. Diller, *J. Chem. Phys.* **42**, 2089 (1965).

31. P. E. Angerhofer and H. J. M. Hanley, *National Bureau of Standards Report*, No. 10700, Boulder, Colo., Aug. 1971.

32. J. Hilsenrath, ed., *National Bureau of Standards Circular*, No. 564, Washington, D.C., 1955.
33. H. M. Roder and D. E. Diller, *J. Chem. Phys.* **52**, 5928 (1970).
34. R. J. Corruccini, *National Bureau of Standards Technical Note*, No. 322, Washington, D.C., Aug. 1965.
35. Ref. 4, p. 98.
36. J. O. Hirschfelder, C. F. Curtiss, and R. B. Bird, *Molecular Theory of Gases and Liquids*, John Wiley & Sons, Inc., New York, 1954, p. 581.
37. J. E. Vivian and C. J. King, *AIChE J.* **10**, 220 (1964).
38. Ref. 36, p. 1110.
39. Ref. 4, pp. 255–507.
40. Ref. 4, p. 93.
41. R. F. Kubin and L. Presley, *Thermodynamic Properties and Mollier Chart for Hydrogen from 300 K to 20,000 K*, NASA SP-3002, The Office of Technical Services, U.S. Dept. of Commerce, Washington, D.C., 1964, 69 pp.
42. J. M. Yos, *Transport Properties of Nitrogen, Hydrogen, Oxygen, and Air to 30,000 K*, NASA Doc. N63-16525, The Office of Technical Services, U.S. Dept. of Commerce, Washington, D.C., 1963, 70 pp.
43. F. Bosnjukovic, W. Springe, and K. F. Knoche, *Pyrodynamics* **1**, 283 (1964).
44. Ref. 2, pp. 77–116.
45. Ref. 4, p. 115.
46. Ref. 5, p. 6-126.
47. R. D. Goodwin, D. E. Diller, H. M. Roder, and L. A. Weber, *J. Res. Nat. Bur. Stand.* **A68**(1), 121 (1964).
48. A. Michels, W. DeGraaf, and G. J. Wolkers, *Appl. Sci. Res.* **A12**(1), 9 (1963).
49. R. B. Stewart and H. M. Roder, in R. B. Scott, ed., *Technology and Uses of Liquid Hydrogen*, Macmillan Company, New York, 1964, pp. 379–404.
50. Ref. 5, p. 6-128.
51. Ref. 4, p. 163.
52. Ref. 5, pp. 6-275, 6-288.
53. J. C. Mullins, W. T. Ziegler, and B. S. Kirk, *Adv. Cryog. Eng.* **8**, 116 (1963).
54. U.S. Pat. 4,067,227 (Jan. 10, 1978), T. Johns and E. A. Berry (to Carle Instruments, Inc.).
55. H. F. Beeghly, in I. M. Kolthoff and P. J. Elving, eds., *Treatise on Analytical Chemistry*, Vol. 1, Part II, Wiley-Interscience, New York, 1961, pp. 45–68.
56. Ref. 4, p. 171.
57. V. S. Kogan, A. S. Bulatov, and L. F. Yakimenko, *Zh. Eksp. Teor. Fiz.* **46**(1), 148 (1964).
58. C. S. Barrett, L. Meyer, and J. Wasserman, *J. Chem. Phys.* **45**, 834 (1966).
59. Ref. 5, p. 6-281.
60. Ref. 2, pp. 77–116.
61. R. D. Arnold and H. J. Hoge, *J. Chem. Phys.* **18**, 1295 (1950).
62. H. M. Roder, *The Thermodynamic Properties of Slush Hydrogen and Oxygen*, PB Rep. No. PB-274186, National Technical Information Service, Springfield, Va., 1977, 44 pp.
63. *Chemical Week*, 80, (Dec. 20–27, 1989).
64. M. Ross and C. Shishkevish, *Molecular and Metallic Hydrogen*, R-2056-ARPA, Rand Corp., Santa Monica, Calif., 1977.
65. F. E. Harris and J. Delhalle, *Phys. Rev. Lett.* **39**, 1340 (1977).
66. A. K. McMahan, *Metallic Hydrogen: Recent Theoretical Progress*, report 1977, Lawrence Livermore Lab., University of California, Livermore, UCRL-79910, 1977.
67. E. Miyazaki, *Surf. Sci.* **71**, 741 (1978).

68. H. P. Leftin, D. S. Newsome, T. J. Wolff, and J. C. Yarze, *Industrial and Laboratory Pyrolysis*, ACS Symposium Series Vol. 32, American Chemical Society, Washington, D.C., 1976, p. 363.

69. David S. Newsome, *Catal. Rev.* **21**(2), 275–318 (1980).

70. M. T. Gillies, ed., *Chemical Technology Review, No. 209: C1 Based Chemicals From H$_2$ and Carbon Monoxide*, Noyes Data Corp., Park Ridge, N.J., 1982.

71. U.S. Pat. 3,888,896 (June 10, 1975), R. L. Espino and T. S. Pletzke (to Chemical Systems, Inc.).

72. Ger. Offen. 1,965,007 (Oct. 15, 1970), (to Catalysts and Chemicals, Inc.).

73. A. T. Larson and R. L Dodge, *J. Amer. Chem. Soc.* **45**, 2918 (1923).

74. A. Nielsen, *An Investigation on Promoted Iron Catalysts for the Synthesis of Ammonia*, Jul. Gjellerups Forlag, Copenhagen, Denmark, 1968, p. 12.

75. K. Aika and A. Ozaki, *J. Catal.* **35**(1), 61 (1974).

76. I. Wender, *Catal. Rev. Sci. Eng.* **14**(1), 97 (1976).

77. Y. T. Shaw and A. J. Perrotta, *Ind. Eng. Chem. Prod. Res. Dev.* **15**, 123 (1976).

78. A. Rehmat and S. S. Randhava, *Ind. Eng. Chem. Prod. Res. Dev.* **9**, 512 (1970).

79. Can. Pat. 979,914 (Dec. 16, 1975), J. M. Lalancette (to Ventron Corp.).

80. U.S. Pat. 3,958,957 (May 25, 1976), K. K. Koh, R. E. Pennington, L. W. Vernon, and N. C. Nahas (to Exxon Research and Engineering Co.).

81. V. T. Coon, T. Takeshita, W. E. Wallace, and R. S. Craig, *J. Phys. Chem.* **80**, 1878 (1976).

82. U.S. Pat. 3,947,483 (Mar. 30, 1976), T. P. Kobylinski and H. E. Swift (to Gulf Research and Development Co.).

83. L. W. Garrett, Jr., *Chem. Eng. Prog.* **56**(4), 39 (1960).

84. Ger. Offen. 2,559,057 (July 8, 1976), L. Kaplan (to Union Carbide Corp.).

85. U.S. Pat. 3,944,588 (Mar. 16, 1976), L. Kaplan (to Union Carbide Corp.).

86. Ger. Offen. 2,531,070 (Jan. 29, 1976), J. N. Cawse (to Union Carbide Corp.).

87. R. A. Sheldon, *Chemicals From Synthesis Gas: Catalytic Reactions of CO and H$_2$*, D. Reidel Publishing Company, 1983, p. 86.

88. *Ibid.*, pp. 86, 114.

89. P. N. Rylander, *Hydrogenation Methods*, Academic Press, Inc., Orlando, Fla., 1985.

90. L. Robinson, *Hydrocarbon Process.* **57**(11), 213 (1978).

91. N. Sotani and M. Hasegawa, *Bull. Chem. Soc. Jpn.* **46**(1), 25 (1973).

92. S. S. Block, J. B. Sharp, and L. J. Darlage, *Fuel* **54**(2), 113 (1975).

93. PB Rep. PB-185882, Clearinghouse for Federal Scientific Technical Information, Washington, D.C., 1969; *U.S. Gov. Res. Develop. Rep.* **69**(22), 48 (1969).

94. M. Onoda, M. Ohtani, and K. Sanbonji, *Tohoku Daigaka Senko Seiren Kenkyusho* **21**, 159 (1965).

95. U.S. Pat. 3,303,017 (Feb. 7, 1967), F. X. Mayer and R. G. Tripp (to Esso Research and Engineering Co.).

96. U.S. Pat. 3,551,108 (Dec. 19, 1970), L. F. Grantham (to North American Rockwell Corp.).

97. D. L. Murdock and G. A. Atwood, *Ind. Eng. Chem. Process Des. Develop.* **13**, 254 (1974).

98. Fr. Pat. 2,112,925 (July 28, 1972), W. E. Watson (to Allied Chemical Corp.).

99. J. B. Levy, *U.S. Gov. Res. Develop. Rep.* **41**(8), 34 (1966).

100. H. A. Arbit and S. D. Clapp, NASA Accession No. N66-32923, Rep. No. NASA CR-54978, Clearinghouse for Federal Scientific and Technical Information, Washington, D.C., 1966, 224 pp.

101. D. A. Bittker, NASA Accession No. N66-34652, Rep. No. NASA TN-D-3607, Clearinghouse for Federal Scientific and Technical Information, Washington, D.C., 1966, 19 pp.

102. D. J. MacLean, *Gov. Rep. Announce. (U.S.)* **73**(14), 272 (1973); U.S. National Technical Information Service, AD Rep. (760770), 1972, 31 pp.
103. C. Vidal, *J. Chim. Phys. Physicochim. Biol.* **68**, 1360 (1971).
104. *Ibid.*, 854 (1971).
105. L. S. Bernstein and L. F. Albright, *AIChE J.* **18**(1), 141 (1972).
106. B. R. Puri and K. C. Kalra, *Chem. Ind. (London)* **50**, 1810 (1969).
107. V. Ponec, *J. Catal.* **6**(3), 362 (1966).
108. G. K. Boreskov, V. V. Popovskii, and V. A. Sazonov, *Proc. Int. Congr. Catal., 4th 1968* **1**, 439 (1971).
109. C. Borgianni, F. Cramarossa, F. Paniccia, and E. Molinari, *Proc. Int. Congr. Catal., 4th 1968* **1**, 102 (1971).
110. M. I. Silich, L. L. Klinova, G. N. Ivanova, N. M. Malygina, and A. I. Patsukova, *Khim. Prom.* (Moscow) **49**, 447 (1973).
111. R. L. Mathews, *Explosion and Detonation Limits for an Oxygen–Hydrogen–Water Vapor System*, U.S. Atomic Energy Commission KAPL-M-6564, Clearinghouse for Federal Scientific and Technical Information, Washington, D.C., 1966, 54 pp.
112. S. Kaye and R. T. Murray, *Adv. Cryog. Eng.* **13**, 545 (1967).
113. P. L. Walker, Jr., F. Rusinko, Jr., and L. G. Austin in D. D. Eley, P. W. Selwood, and P. B. Weisz, eds., *Advances in Catalysis*, Vol. XI, Academic Press, Inc., New York, 1959, pp. 134–221.
114. J. L. Figueiredo and D. L. Trimm, *J. Catal.* **40**, 154 (1975).
115. R. A. Krakowski and D. R. Orlander, U.S. Atomic Energy Comm. UCRL-19149, Clearinghouse for Federal Scientific and Technical Information, Washington, D.C., 1970.
116. V. Prochazka and J. Subrt, *Collect. Czech. Chem. Commun.* **41**, 522 (1975).
117. K. M. Mackay in J. C. Bailar, H. J. Emeleus, R. Nyholm, and R. F. Trotman-Dickenson, eds., *Comprehensive Inorganic Chemistry*, Vol. 1, Pergamon Press, New York, 1973, pp. 23–76.
118. Ref. 3, p. 16.
119. H. P. Leftin and A. Cortes, *Ind. Eng. Chem. Process Des. Develop.* **11**, 613 (1972).
120. R. A. Kalinenko, S. I. Korochuk, K. P. Lavrovskii, Y. V. Maksimov, and Y. P. Yarnpol'skii, *Dokl. Akad. Nauk.* (SSSR) **204**, 1125 (1972).
121. W. E. Faleoner and W. A. Sunder, *Int. J. Chem. Kinet.* **3**(5), 395 (1971).
122. R. D. Kelley, R. Klein, and M. D. Seheer, *J. Phys. Chem.* **74**, 4301 (1970).
123. T. A. Brabbs, F. E. Belles, and R. S. Brokaw, NASA Special Publ. NASA SP-239, Clearinghouse for Federal Scientific and Technical Information, Washington, D.C., 1970, pp. 105–117.
124. R. K. Gould, *J. Chem. Phys.* **63**, 1825 (1975).
125. M. Balooeh and D. R. Olander, *J. Chem. Phys.* **63**, 4772 (1975).
126. R. W. Cahn, *Nature* **276**, 665 (Dec. 14, 1978).
127. J. H. N. Van Vueht, F. A. Kuijpers, and H. C. A. M. Bruning, *Philips Res. Rep.* **25**(2), 133 (1970).
128. F. A. Kuijpers and H. H. Van Mal, *J. Less-Common Met.* **23**, 395 (1971).
129. H. H. Van Mal, K. H. J. Busehow, and A. R. Miedema, *J. Less-Common Met.* **35**, 65 (1974).
130. T. Takeshita, W. E. Wallaee, and R. S. Craig, *Inorg. Chem.* **13**, 2282 (1974).
131. F. L. Schlapback, ed., *Hydrogen in Intermetallic Compounds*, Springer-Verlag, Berlin, 1988, Chapt. 5, pp. 197–237.
132. V. P. Teodorovieh, N. N. Kolgatin, and V. I. Deryadina, *Khim. Neft. Mashinostr.* **12**, 21 (1966).
133. C. A. Zapffe and C. E. Sims, *Trans. Amer. Inst. Mech. Eng.* **145**, 225 (1941).
134. G. A. Nelson, *Hydrocarbon Process.* **44**(5), 185 (1965).

135. G. A. Nelson, *Werkst. Korros.* **14**, 65 (1963).
136. M. Fischer, G. Kreysa, and G. Sandstede, *Int. J. Hydrogen Energy*, **12**(1), 39–46 (1987).
137. M. Steinberg and H. C. Cheng, *Int. J. Hydrogen Energy*, **14**(11), 797–820 (1989).
138. S. C. Nirula, "Syngas by the Partial Oxidation of Natural Gas", PEP Review No. 90-3-3, SRI International, Nov. 1991.
139. R. Dupont and P. R. Degand, *Hydrocarbon Process.* (July 1986).
140. W. F. Fong and M. E. Quintana, *HyTex—A Novel Process for Hydrogen Production*, NPRA 89th Annual Meeting, March 17–19, 1991, San Antonio, Tex.
141. J. L. Heek and T. Johansen, *Hydrocarbon Process.*, 175 (Jan. 1978).
142. E. K. Nazarov, A. T. Nikitin, N. N. Ponomarev-Stepnoy, A. N. Protsenko, A. Y. Stolyarevskii, and N. A. Doroshenko, *Int. J. Hydrogen Energy*, **15**(1), 45–54 (1990).
143. T. A. Czuppon and J. M. Lee, *Oil Gas J.*, 42–50 (Sept. 7, 1987).
144. C. L. Reed and C. J. Kuhre, ACS/CSJ Chemical Congress at Honolulu, Hawaii, Apr. 1–6, 1979.
145. *Hydrocarbon Process.* **71**(4), 140 (Apr. 1992).
146. Coal gasification technical data, Texaco Development Corp., Jan. 1978.
147. D. O. Moore, T. A. Czuppon, and B. G. Mandelik, paper presented at 14th Intersociety Energy Conversion Conference, Boston, Mass., Aug. 1979.
148. U.S. Pat. 3,129,060 (Apr. 14, 1964), J. B. Pohlenz (to Universal Oil Products Co.).
149. Brit. Pat. 979,720 (Jan. 6, 1965), (to Universal Oil Prod. Co.).
150. *Hydrocarbon Process. Pet. Ref.* **43**(9), 232 (1964).
151. U.S. Pat. 3,615,299 (Oct. 26, 1971), P. E. Fischer and M. M. Holm (to Chevron Research).
152. Ger. Pat. 2,126,664 (Dec. 16, 1971), J. R. Kiovosky (to Norton Co.).
153. Fr. Pat. 1,529,097 (June 14, 1968) (to British Gas Council).
154. C.-J. Winter and J. Nitsch, eds., *Hydrogen as an Energy Carrier: Technology, Systems, Economy*, Springer-Verlag, Berlin, 1988, pp. 166–205.
155. J. O'M. Bockris, *Energy Options: Real Economics and the Solar Hydrogen System*, John Wiley & Sons, Inc., New York, 1980, p. 314.
156. Ref. 154, p. 209.
157. Ref. 154, p. 185.
158. L. O. Williams, *Hydrogen Power: An Introduction to Hydrogen Energy and its Applications*, Pergamon Press, Oxford, U.K., 1980, p. 61.
159. Ref. 155, p. 329.
160. Ref. 154, p. 189.
161. Ref. 158, p. 77.
162. K. D. Williams, Jr. and F. J. Edeskuty, eds., *Recent Developments in Hydrogen Technology*, Vol. 1, CRC Press, Boca Raton, Fla., 1986, Chapter 1, p. 13.
163. Ref. 154, p. 193.
164. Ref. 154, p. 197.
165. Ref. 154, p. 190.
166. Ref. 162, p. 2.
167. Ref. 158, p. 74.
168. J. A. Hanson, paper presented at *IGT Symposium Hydrogen for Energy Distribution*, July 1978.
169. *Chemical Week*, 20 (July 13, 1988).
170. S. D. Huang, C. K. Secor, R. Ascione, and R. M. Zweig, *Int. J. Hydrogen Energy*, **10**(4), 227–231 (1985).
171. S. Roychowdhury, D. Cox, and M. Levandowsky, *Int. J. Hydrogen Energy*, **13**(7), 407–410 (1988).

172. J. C. Weisman and J. R. Benemann, *Appl. Environ. Microbiol.* **33**(1), 123 (1977).

173. G. D. Smith, G. D. Ewart, and W. Tucker, *Int. J. Hydrogen Energy*, **17**(9), 695–698 (1992).

174. S. Sarkar, K. D. Pandley, and A. K. Kashyap, *Int. J. Hydrogen Energy*, **17**(9), 689–694 (1992).

175. S. E. Foh, W. J. D. Esher, and T. D. Donakowski, paper presented at IGT Symposium Hydrogen for Energy Distribution, July 1978.

176. S. Yalcin, *Int. J. Hydrogen Energy*, **14**(8), 551–561 (1989).

177. Harth-Hoehzein, *World Hydrogen Energy Conference*, Coral Gables, Fla., 1974.

178. M. V. Twigg, ed., *Catalyst Handbook*, 2nd ed., Imperial Chemical Industries, Wolfe Publishing, Frome, UK, pp. 378–383.

179. R. N. Quade, *World Hydrogen Energy Conference*, Coral Gables, Fla., 1974.

180. U.S. Pat. 3,928,550 (Dec. 23, 1975) W. H. Seitzer (to Sun Ventures Inc.).

181. F. S. Karn, R. A. Friedel, and H. G. Sharkey, *Fuel* **48**, 297 (1969).

182. F. S. Karn, R. A. Friedel, and H. G. Sharkey, *ACS Div. Fuel Chem.* **14**(4), 1970.

183. Brit. Pat. 1,187,782 (Apr. 15, 1970) I. G. Nizon.

184. M. A. DeLuchi, E. D. Larson, and R. H. Williams, *Hydrogen and Methanol: Production and Use in Fuel Cell and Internal Combustion Engine Vehicles—A preliminary Assessment*, Vol. 12, *Solid Fuel Conversion for the Transportation Sector*, ASME, Fuels and Combustion Technologies Division, New York, 1991, pp. 55–70.

185. G. H. Schuetz, *Int. J. Hydrogen Energy*, **10**(7/8), 439–446 (1985).

186. U.S. Pat. 4,066,739 (Jan. 3, 1978), W. C. Chen.

187. M. Dokiya, H. Fukuda, and T. Kameyama, *Bull. Chem. Soc. Jpn.* **51**(1), 150 (1978).

188. E. A. Fletcher, J. E. Noring, and J. P. Murray, *Int. J. Hydrogen Energy* **9**, 587–593 (1984).

189. S. K. Megalofonos and N. G. Papayannakos, *Int. J. Hydrogen Energy*, **16**(5), 319–327 (1991).

190. A. J. Finn, *Nitrogen* **175**, 25–32 (Sept.-Oct. 1988).

191. *Nitrogen*, **173**, 25–29 (May-June 1988).

192. G. Meunier, J. P. Manaud, *Int. J. Hydrogen Energy* **17**(8), 599–602 (1992).

193. B. Heydron, H. Schwendener, and M. Tashiro, *Hydrogen*, CEH Product Review, No. 743.5, SRI International, Menlo Park, Calif., 1990.

194. Technical data, Airco Gases, March 18, 1993.

195. Kenneth E. Cox and K. D. Williamson, Jr., eds., *Hydrogen: Its Technology and Implications: Implications of Hydrogen Energy*, Vol. 5, CRC Press Inc., Boca Raton, Fla., 1979, p. 35.

196. Ref. 155, p. 244.

197. Ref. 162, Vol. 2, Chapt. 2, p. 19.

198. Ref. 158, p. 93.

199. Ref. 158, pp. 92–102.

200. Ref. 155, pp. 247–253.

201. D. Gidaspow and Y. Liu, *Proceedings Intersoc. Energy Convers. Eng. Conf.*, Vol. 1, 1976, pp. 920–925.

202. *Chem. Eng.* **68**(18), 66 (1961).

203. M. T. Hodge, C. H. Meyers, and R. E. McCoskey, Miscellaneous Publication M191 National Bureau of Standards, Washington, D.C., 1948.

204. *Occupational Safety and Health Act of 1970*, PL 91-596 (12-29-70) in U.S. Statutes at Large, 91st Congress, Second Session, Vol. 84, U.S. Government Printing Office, Washington, D.C., 1971, Part 2, pp. 1590–1620.

205. Ref. 131, p. 197.

206. Ref. 162, Vol. 2, Chapt. 3.

207. Ref. 155, p. 239.

208. Ref. 162, Vol. 2, Chapt. 2, p. 21.

209. R. G. Barnes, ed., *Hydrogen Storage Materials*, Trans Tech Publications, Aedermannsdorf, Switzerland, 1988, pp. 1–15.

210. G. Strickland in Ref. 175.

211. D. Frankel and Y. Shabtai, *Miami International Conference, Alternative Energy Sources*, Hemisphere Pub. Corp., Washington, D.C., 1977, pp. 557–559.

212. G. D. Sandrock, *Intersoc. Energy Conv. Eng. Conf.* **12**(1), 951 (1977).

213. Jpn. Kokai 77 60,211 (May 18, 1977), M. Suwa and Y. Kita (to Hitachi Ltd.); Jpn. Kokai 77 70,916 (June 13, 1977), M. Suwa and Y. Kita (to Hitachi Ltd.); Jpn. Kokai 77 20,911 (Feb. 17, 1977), N. Yanagihara and T. Yamashita (to Mutsushita Electric Industrial Co. Ltd.).

214. L. W. Jones, *World Hydrogen Energy Conference*, 1C-61, University of Miami, Coral Gables, Fla., 1976.

215. E. M. Dickson, *World Hydrogen Energy Conference*, 2C-5, University of Miami, Coral Gables, Fla., 1976.

216. J. Finegold and W. Van Vorst, *Energy Environ. Proc. Natl. Conf.*, New York, 15–20 (1974).

217. R. G. Lundberg in Ref. 175.

218. J. M. Kelley and R. Manvi in Ref. 175.

219. Ref. 158, p. 102.

220. Ref. 162, Vol. 2, Chapt. 2, p. 23.

221. G. Ingram and M. Lonsdale, in I. M. Kolthoff and P. J. Elving, eds., *Treatise on Analytical Chemistry, Part II, Analytical Chemistry of Inorganic and Organic Compounds*, Vol. 11, Wiley-Interscience, New York, 1965, pp. 297–403.

222. J. Mitchell, Jr., in I. M. Kolthoff and P. J. Elving, eds., *Treatise on Analytical Chemistry, Part II, Analytical Chemistry of the Elements*, Vol. 1, Wiley-Interscience, New York, 1961, pp. 69–206.

223. G. Dugan, *Anal. Lett.* **10**, 639 (1977).

224. H. C. E. Van Leuven, *Fresenius Z. Anal. Chem.* **264**, 220 (1973).

225. H. F. Beeghly in Ref. 188, pp. 45–68.

226. V. l. Yavoiskii, L. B. Kosterev, V. L. Safonov, and M. I. Afanas'ev, *Sb. Mosk. Inst. Stali Splavov* **62**, 57 (1970).

227. M. Hosoya, *Sci. Rep. Res. Inst. Tohoku Univ. Ser. A* **22**(5–6), 183 (1971).

228. Swed. Pat. 382,353 (Jan. 26, 1976), K. F. Alm, L. H. Andersson, and J. Ruokolahti.

229. P. Escoffier, *Chim. Anal.* **49**(4), 208 (1967).

230. H. Goto and M. Hosoya, *Nippon Kinzoju Gakkaishi* **35**(1), 16 (1971).

231. Louis Raymond, ed., *Hydrogen Embrittlement: Prevention and Control*, American Society for Testing and Materials, Philadelphia, Pa., 1988.

232. R. C. Orth and H. B. Land, *J. Chromatogr. Sci.* **9**(6), 359 (1971); Y. S. Su, *Anal. Chim. Acta* **36**, 406 (1966).

233. M. Shykles, *Anal. Chem.* **47**, 949 (1975).

234. D. R. Deans, M. T. Huckle, and R. M. Peterson, *Chromatographia* **4**(7), 279 (1971).

235. E. W. Cook, *Chromatographia* **4**(4), 176 (1971).

236. R. I. Jerman and L. R. Carpenter, *J. Gas Chromatogr.* **6**(5), 298 (1968).

237. Czech. Pat. 153,678 (June 15, 1974), M. Krejci and K. Tesarik.

238. F. Zocchi, *J. Gas Chromatogr.* **6**(4), 251 (1968).

239. H. A. Smith and D. P. Hunt, *J. Phys. Chem.* **64**, 383 (1960); I. Yasumori and S. Ohno, *Bull. Chem. Soc. Jpn.* **39**, 1302 (1966).

240. U.S. Pat. 3,352,644 (Nov. 14, 1967), I. Lysyj (to North American Aviation, Inc.).

241. J. Dericbourg, *J. Chromatogr.* **123**, 405 (1976).

242. K. Kikuchi and M. Takahashi, *Bull. Inst. Int. Froid Annexe* **2**, 237 (1970).
243. U.S. Pat. 3,549,327 (Dec. 22, 1970), G. J. Fergusson (to Scientific Research Instruments Corp.).
244. G. Ciuhandu and A. Chicu, *Lucr. Conf. Nat. Chim. Anal., 3rd* **3**, 239 (1971).
245. A. Farkas and L. Farkas, *Proc. R. Soc. Ser. A* **144**, 467 (1934).
246. A. Farkas and L. Farkas, *Nature* **132**, 894 (1933).
247. M. S. Chupakhin and L. T. Duev, *Zh. Anal. Khim.* **22**, 1072 (1967).
248. H. C. E. Van Leuven, *Anal. Chim. Acta* **49**, 364 (1970).
249. I. V. Abashidza, V. G. Artemchuk, V. E. Vetshtein, and I. V. Gol'denfel'd, *Prib. Tekh. Eksp.* **2**, 182 (1971).
250. S. Gaona and P. Morales, *Rev. Mex. Fis.* **20**(Suppl.), 91 (1971).
251. B. C. Gerstein and R. G. Pembleton, *Anal. Chem.* **49**(1), 75 (1977).
252. U. Schmidt and W. Seiler, *J. Geophys. Res.* **75**, 1713 (1970).
253. U.S. Pat. 3,325,378 (June 13, 1967), M. W. Greene and R. I. Wilson (to Beckman Instruments, Inc.).
254. S. A. Hoenig, C. W. Carlson, and J. Abramowitz, *Rev. Sci. Instrum.* **38**(1), 92 (1967).
255. U.S. Pat. 4,030,340 (June 21, 1977), S. Chang (to General Monitors, Inc.).
256. U.S. Pat. 3,429,177 (Feb. 25, 1969), A. C. Krupnick and D. P. Lucero (to NASA).
257. E. K. Vasil'eva and A. G. Zakomornyi, *Zavod. Lab.* **33**, 471 (1967).
258. S. V. Starodubtsev and co-workers, *Fiz. Svoistva Osobo Chist. Metal. Poluprov. Akad. Nauk Uzb. SSR Fiz-Tekh Inst.*, 18 (1966).
259. E. Zielinski and K. Mayer, *Chem. Anal.* **18**(4), 745 (1973).
260. G. A. Salamatina and M. G. Sarina, *Fiz-Khim. Metody Anal.* **1**, 89 (1970).
261. Fr. Demande 2,115,080 (Aug. 11, 1972), (to Kombinat Mess and Regelungstechnik).
262. J. Hord in *Symposium Papers: Hydrogen for Energy Distribution, Institute of Gas Technology,* Institute of Gas Technology, Chicago, July 24–28, 1978, p. 613.
263. B. Lewis and G. von Elbe, *Combustion, Flames, and Explosions of Gases*, 2nd ed., Academic Press, Inc., New York, 1961.
264. Ref. 4, p. 194.
265. *Fire Hazard Properties of Flammable Liquids, Gases and Volatile Solids*, Rep. No. 325, National Fire Protection Assoc., Washington, D.C., May 1960.
266. Ref. 5, p. 6-289.
267. *Code of Federal Regulations, Title 49*, Transportation, Materials Transportation Bureau, Department of Transportation, U.S. Government Printing Office, Washington, D.C., 1976, Chapt. 1, Parts 100–199.
268. *Ibid.*, Chapt. III, Subchapt. B, Parts 390–397.
269. *Standard for Liquified Hydrogen Systems at Consumer Sites*, Natl. Fire Protect. Assoc. pamphlet no. 50B (ANSI Z292.3), 1973.
270. *Standard of Gaseous Hydrogen Systems at Consumer Sites*, Natl. Fire Protect. Assoc. pamphlet no. 50A (ANSI Z292.2), 1973.
271. *Hydrogen Safety Manual*, Report TM-X-52454, National Aeronautics and Space Administration, Washington, D.C., 1968.
272. B. Rosen, V. H. Dayan, and R. L. Proffit, *Hydrogen Leak and Fire Detection: A Survey*, Rep. SP-5092, National Aeronautics and Space Administration, Washington, D.C., 1979.
273. W. Balthasar and J. P. Schoedel, *Hydrogen Saftey Manual*, Comm. European Communities, Luxembourg, 1983.
274. G. R. Davis, *Scientific Amer.* **263**, 55–62 (1990).
275. M. A. DeLuchi, *Int. J. Hydrogen Energy*, **14**(2), 81–130 (1989).
276. J. E. Sinor, *Technical Economics, Synfuels, and Coal Energy 1989*, ASME, pp. 103–109.

277. D. S. Cameron, *Int. J. Hydrogen Energy*, **15**(9), 669–675 (1990); *Business Week*, 40–41 (Dec. 24, 1990).

T. A. Czuppon
S. A. Knez
D. S. Newsome
The M. W. Kellogg Company

HYDROGEN BROMIDE.   See Bromine compounds.

HYDROGEN CHLORIDE

History and Occurrence

Hydrogen chloride [7647-01-0], HCl, exists in solid, liquid, and gaseous states and is very soluble in water. The aqueous solution, hydrochloric acid, owes its name to Davy whose studies in 1810 of chlorine and its compounds proved that the gas consists of only hydrogen and chlorine. The term muriatic acid, proposed by Lavoisier in 1789, to indicate the presence of chlorine in an inorganic compound, is used in the U.S. industry to refer to the commercial forms of hydrochloric acid. Similarly, the arbitrary specific gravity scale devised in the eighteenth century by the French chemist Baumé (°Bé) is still used to characterize the commercial grades of the acid, whereas the Twaddell density scale (°Tw) is less commonly used. The relationship between these density scales and the concentration of three standard grades of hydrochloric acid is summarized in Table 1.

Hydrogen chloride was discovered in the fifteenth century by the German alchemist Basilius Valentinus who treated the green vitriol, $FeSO_4 \cdot 7H_2O$, and common salt and obtained what was then called spirit of salt (1). In the seventeenth century, Glauber prepared hydrochloric acid from common salt [7647-14-5] and sulfuric acid [7664-93-9]. Commercial production of hydrochloric acid began in England in 1823. However in 1863, legislation was passed prohibiting

Table 1. Density and Concentration of Commercial Grades of Hydrochloric Acid

Specific gravity	Density unit[a]		HCl,%
	°Bé	°Tw	
1.1417	18	28.34	27.92
1.1600	20	32.00	31.45
1.1789	22	35.78	35.21

[a]$[145/(145 - °Be)] = (0.005°Tw + 1) =$ specific gravity.

the indiscriminate discharge of waste hydrogen chloride from the crude scrubbers into the atmosphere. This legislation forced manufacturers to develop uses for the HCl generated in the Leblanc process for soda ash which involved reaction between NaCl and H_2SO_4. The industrial synthesis of hydrogen chloride followed the development of the chlor-alkali electrolytic process early in the twentieth century (see ALKALI AND CHLORINE PRODUCTS). The route involving direct combination of H_2 and Cl_2, produced electrolytically, and the chemical routes based on reaction between alkali metal and/or alkaline-earth chlorides and H_2SO_4 or $NaHSO_4$, were used to produce high purity hydrochloric acid. However, as of the mid-1990s, these processes are becoming obsolete because of the large amounts of hydrogen chloride generated as by-product from several manufacturing processes, a typical example being the industrial production of chlorinated hydrocarbons (see CHLOROCARBONS AND CHLOROHYDROCARBONS).

Hydrochloric acid is found naturally in the gases evolved from volcanoes, particularly those in Mexico and South America. Its formation is attributed to the high temperature reaction of water with the salts found in seawater. The original atmosphere of the earth is considered to have contained water (qv), carbon dioxide (qv), and hydrogen chloride in the ratio of 20:3:1, giving an early ocean consisting of about $1N$ HCl, which dissolved the crustal minerals, leading to the ocean salinity. Hydrogen chloride was also detected in the atmosphere of the planet Venus. The dissociation of HCl is considered the source of chlorine detected in the spectra of distant stars.

Hydrochloric acid is also present in the digestive system of most mammals (2). The gastric mucosa lining the human stomach produces about 1.5 L/d of gastric juices, containing an acid concentration in the range of 0.05 to 0.1 N. A deficiency of hydrochloric acid impairs the digestive process, particularly of carbohydrates and proteins, and excess acid causes gastric ulcers.

Physical and Thermodynamic Properties

Anhydrous Hydrogen Chloride. Anhydrous hydrogen chloride is a colorless gas that condenses to a colorless liquid and freezes to a white crystalline solid. The physical and thermodynamic properties of HCl are summarized in Table 2 for selected temperatures and pressures. Figure 1 shows the temperature dependence of some of these properties.

The high thermal stability of hydrogen chloride is a consequence of the large enthalpy of its formation. The calculated percent dissociation at different temperatures is given in Table 3. The dissociation of gaseous hydrogen chloride into its elements is important for selecting proper materials of construction, because of the formation of free chlorine which is highly corrosive in the presence of moisture.

The dielectric constant of liquid hydrogen chloride is low relative to other solvents having similar ionization characteristics, and many organic compounds form conducting solutions in liquid hydrogen chloride. The narrow temperature range of the liquid phase, −94 to −85°C at atmospheric pressure, and the low temperatures required for liquefaction (triple point = −114.25°C), however, severely limit studies in solutions of liquid hydrogen chloride. The liquid vapor

Table 2. Physical and Thermodynamic Properties of Anhydrous Hydrogen Chloride

Property	Value
melting point, °C	−114.22
boiling point, °C	−85.05
heat of fusion at −114.22°C, kJ/mol[a]	1.9924
heat of vaporization at −85.05°C, kJ/mol[a]	16.1421
entropy of vaporization, J/(mol·K)[a]	85.85
triple point, °C	−114.25
critical temperature, T_c, °C	51.54
critical pressure, P_c, MPa[b]	8.316
critical volume, V_c, L/mol	0.069
critical density, g/L	424
critical compressibility factor, Z_c	0.117
ΔH_f° at 198 K, kJ/mol[a]	−92.312[c]
	−100.4[d]
ΔG_f° at 298 K, kJ/mol[a]	−95.303
S° at 298 K, J/(mol·K)[a]	186.786
dissociation energy at 298 K, kJ[a]	431.62[c]
	427.19[d]
compressibility coefficient	0.00787
internuclear separation, nm	0.12510
dipole moment, C·m[e]	3.716[c]
	3.74[d]
ionization potential, J[a]	20.51[c]
	20.45[d]
heat capacity, C_p, J/(mol·K)[a]	
vapor (constant pressure)	
at 273.16 K	29.162
at 973.2 K	30.554
liquid at 163.16 K	60.378
solid at 147.16 K	48.98
surface tension at 118.16 K, mN/cm(=dyn/cm)	23
viscosity, mPa·s(=cP)	
liquid at 118.16 K	0.405
vapor at 273.06 K	0.0131
vapor at 523.2 K	0.0253
thermal conductivity, mW/(m·K)	
liquid at 118.16 K	335
vapor at 273.16 K	13.4
density, g/cm³	
liquid	
at 118.16 K	1.045
at 319.15 K	0.630
solid	
rhombic at 81 K	1.507
cubic at 98.36 K	1.48
cubic at 107 K	1.469

Table 2. (Continued)

Property	Value
refractive index	
liquid at 283.16 K	1.254
gas at 273.16 K	1.0004456
dielectric constant	
liquid at 158.94 K	14.2
gas at 298.16 K	1.0046
electrical conductivity, $(\Omega \cdot m)^{-1}$	
at 158.94 K	1.7×10^{-7}
at 185.56 K	3.5×10^{-7}

[a] To convert J to cal, divide by 4.184.
[b] To convert MPa to atm, divide by 0.101.
[c] Measured value.
[d] Calculated value. See Refs. 3 and 4.
[e] To convert C·m to debye, divide by 3.34×10^{-30}.

pressure, P, from 160 to 260 K is given by

$$\log P = -905.53 \, T^{-1} + 1.75 \log T - 0.0050077 \, T + 3.78229$$

where P is in kPa and T is in Kelvin.

The entropy value of gaseous HCl is a sum of contributions from the various transitions summarized in Table 4. Independent calculations based on the spectroscopic data of $H^{35}Cl$ and $H^{37}Cl$ separately, show the entropy of HCl at 298 K to be 186.686 and 187.372 J/(mol·K) (44.619 and 44.783 cal/(mol·K)), respectively. The low temperature (rhombic) phase is ferroelectric (6). Solid hydrogen chloride consists of hydrogen-bonded molecular crystals consisting of zigzag chains having an angle of 93.5° (6). Proton nmr studies at low temperatures have also shown the existence of a dimer $(HCl)_2$ (7).

Temperature dependence of viscosity of the gas over a wide range of temperatures is given by equation 1 where T is in Kelvin and η_0 is the value of η at 273 K.

$$\eta = \eta_0 (T/273.1)^{1.03} \tag{1}$$

Gaseous diffusion and thermal diffusion data may be found in References 8 and 9.

The vapor pressure of solid and liquid hydrogen chloride is described by equations 2 and 3, respectively,

$$\log p = -911.31 \, T^{-1} + 2.1875 + 4.313 \times 10^{-3} \, T \tag{2}$$

$$\log p = -8.555 \times 10^{-2} \, T^{-1} + 2.553 \tag{3}$$

where p is the vapor pressure in mPa, and T the temperature in Kelvin. These equations are accurate over a wide temperature range (10). Temperature dependence of specific heat, C_p, for hydrogen chloride expressed in J/(mol·K) is given by equation 4 (see Fig. 1a) (11).

$$C_p = 28.1663 + 1.8096 \times 10^{-3} \, T + 15.4692 \times 10^{-7} \, T^2 \tag{4}$$

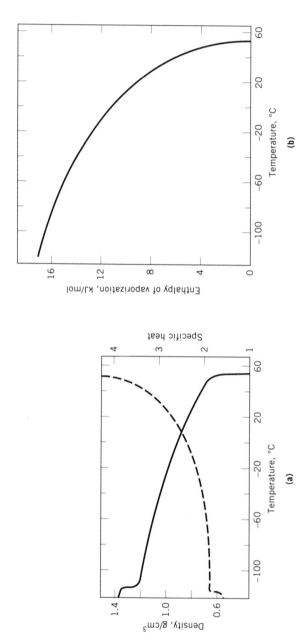

Fig. 1. (a) Variation of the density (——) and specific heat (– – –) of liquid HCl with temperature; (b) enthalpy of vaporization of HCl (5). To convert J to cal, divide by 4.184.

898

Table 3. Thermal Stability of Gaseous Hydrogen Chloride

Temperature, K	Log K_p	Dissociation, %
298	16.70	4.5×10^{-7}
473	10.70	4.5×10^{-4}
673	7.648	1.5×10^{-2}
873	5.981	0.10
1073	4.929	0.342
1273	4.205	0.73
1773	3.099	2.74
2273	2.461	5.55
2773	2.039	8.73
4000	1.59	14.7

Table 4. Components of the Entropy of Hydrogen Chloride

Entropy change	Temp, K	Value, J/(mol·K)[a]
0 K to transition point	0–98.36	30.79
solid transition: rhombic to cubic	98.36	12.09
transition to melting point	96.36–158.94	21.13
entropy of fusion	158.94	12.55
melting point to boiling point	158.94–188.11	9.87
entropy of vaporization	188.11	85.86
vapor	188.11–298	14.48
Total	*298*	*186.77*

[a]To convert J to cal, divide by 4.184.

The value of the specific heat at constant pressure and constant volume is 1.404 at 0°C.

Hydrogen Chloride–Water System. Hydrogen chloride is highly soluble in water and this aqueous solution does not obey Henry's law at all concentrations. Solubility data are summarized in Table 5. The relationship between the pressure and vapor composition of unsaturated aqueous hydrochloric acid solutions is given in Reference 12. The vapor–liquid equilibria for the water–hydrogen chloride system at pressures up to 1632 kPa and at temperatures ranging from −10 to +70°C are documented in Reference 13.

Hydrogen chloride and water form four hydrates. The dihydrate is formed when a saturated solution is cooled at atmospheric pressure. It dissociates at −18.3°C in open vessels and has a melting point of −17.7°C in a sealed tube. The structure has been shown by x-ray analysis to be $(H_2O)_2H^+Cl^-$ (14). The monohydrate has a melting point of −15.35°C; the trihydrate has a melting point of −24.9°C; the hexahydrate is very unstable and has a melting point of −70°C (14). Addition of hydrogen chloride to pure water lowers the freezing point until a eutectic temperature of about −85°C is reached at 25% HCl, a concentration that closely corresponds to the composition of the hexahydrate. Continued addition of HCl raises the freezing point first to that of the trihydrate and subsequently to that of the dihydrate (see Fig. 2).

Hydrogen chloride and water form constant boiling mixtures. The properties of these mixtures, determined with great accuracy, and often used as

Table 5. Solubility of Hydrogen Chloride in Water[a]

Temperature, °C	Solubility, g HCl/100 g solution	H₂O in vapor, mol %
−18.3	48.98	
−15	48.27	
−10	47.31	0.0070
0	45.15	0.0178
10	44.04	0.0460
20	42.02	0.1230
30	40.22	0.2850
40	38.68	0.6350
50	37.34	
60	35.94	

[a]At 101.3 kPa (1 atm).

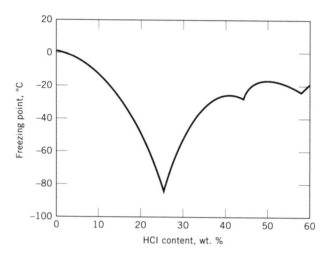

Fig. 2.   Freezing point of aqueous solutions of HCl (15).

analytical standards (16), are summarized in Table 6 and graphically depicted in Figure 3.

An equation of state of the form $PV = RT$ was developed (17) for the vapor of concentrated and dilute hydrochloric acid which is valid up to a HCl mole fraction, x, of 0.23, at temperatures up to 780 K and pressures up to 15.0 MPa. A simplified Redlich-Kwong equation,

$$PV = RT + BP \tag{5}$$

where $B = b - aT^{1.5}R^{-1}$, and $a/(\text{cm}^6 \cdot \text{Pa·K}^{0.5} \cdot \text{mol}^2) = 3283 - 4035x_{\text{HCl}}$; $b/(\text{mol} \cdot \text{cm}^3) = 139.8 - 218.4x_{\text{HCl}}$; and $R = 8.3143$ MPa·cm³/(mol·K), was found to describe this data. The saturation line is defined by equation 6 where P is in kPa and T is in Kelvin.

$$\log_{10} P = 3.515 - 2056\,T^{-1} + x_{\text{HCl}}(2.064 - 988\,T^{-1}) \tag{6}$$

Table 6. Properties of Constant Boiling Hydrochloric Acid

Pressure, kPa[a]	Boiling point, °C	Density at 25°C	HCl, wt %
6.7	48.724	1.1118	23.42
33	81.205	1.1042	21.883
66	97.578	1.0993	20.916
93	106.424	1.0966	20.360
97		1.0963	20.293
101	108.584	1.0959	20.222
104			20.173
106	110.007	1.0955	20.155
133	116.185	1.0933	19.734

[a]To convert kPa to mm Hg, multiply by 7.5.

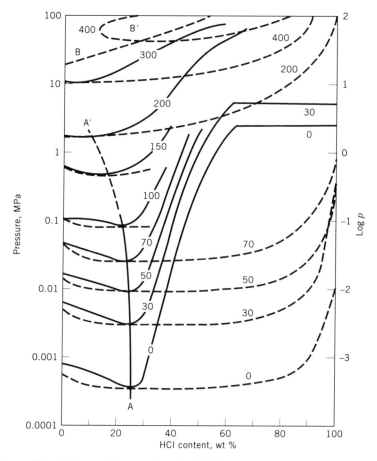

Fig. 3. Vapor–liquid-phase diagram for the HCl–H$_2$O system (5) where (–––) represents the demarcation between the two-phase region and the gas phase; (—) denotes the demarcation of the two-phase region from the liquid phase; and the numbers associated with the curves correspond to temperatures in °C. A–A′ connects the azeotropic points and B–B′ represents the critical segregation curve above the critical point of water, ie, point B occurs at 341.1°C at 21.43 MPa.

901

Hydrogen chloride is completely ionized in aqueous solutions at all but the highest concentrations. Thermodynamic functions have been determined electrochemically for equations 7 and 8. Values are given in Table 7.

$$HCl \text{ (g)} \rightleftharpoons H^+ \text{ (aq)} + Cl^- \text{ (aq)} \tag{7}$$

$$HCl \text{ (aq)} \rightleftharpoons H^+ \text{ (aq)} + Cl^- \text{ (aq)} \tag{8}$$

The viscosity of hydrochloric acid solutions, η, increases slightly with increasing concentration and is related to the molar concentration c by

$$(\eta - \eta_0)\eta_0 = 0.0030 + 0.0625\, c^{0.5} + 0.0008\, c \tag{9}$$

where η_0 is the viscosity of the water, 0.8904 mPa·s(=cP) at 25°C. The surface tension of dilute hydrochloric acid solutions is very close to that of water (71.97 mN/m·(=dyn/cm) at 25°C) and decreases slightly as the concentration of HCl increases.

The variation of the dielectric constant of the HCl + H_2O mixtures is not appreciably different from that of pure water (78.30) at 25°C until the hydrogen chloride concentration reaches a minimum of 0.2%. It increases slightly over the dielectric constant of water as the concentration increases.

The specific heat of aqueous solutions of hydrogen chloride decreases with acid concentration (Fig. 4). The electrical conductivity of aqueous hydrogen chloride increases with temperature. Equivalent conductivity of these solutions are summarized in Table 8. Other physicochemical data related to HCl may be found in the literature (5).

Hydrogen Chloride–Water–Inorganic Compound Systems. *Salts.* Salting out metal chlorides from aqueous solutions by the common ion effect upon addition of HCl is utilized in many practical applications. Typical data for ferrous chloride [13478-10-9], $FeCl_2$, potassium chloride [7447-40-7], KCl, and NaCl are shown in Table 9. The properties of the $FeCl_2 \cdot HCl \cdot H_2O$ system are important to the steel-pickling industry (see METAL SURFACE TREATMENTS; STEEL). Other metal chlorides that are salted out by the addition of hydrogen chloride to aqueous solutions include those of magnesium, strontium, and barium.

Metal chlorides which are not readily salted out by hydrochloric acid can require high concentrations of HCl for precipitation. This property is used to recover hydrogen chloride from azeotropic mixtures. A typical example is the calcium chloride [10043-52-4] addition used to break up the HCl–H_2O azeotrope

Table 7. Thermodynamic Functions of Aqueous Hydrochloric Acid

Property[a]	Value	
	Equation 7	Equation 8
$\Delta H°$, kJ/mol	−74.852	−57.32
$\Delta G°$, kJ/mol	−35.961	−39.7
$\Delta S°$, J/(mol·K)	−130.33	−56.53

[a]To convert J to kcal, divide by 4.184.

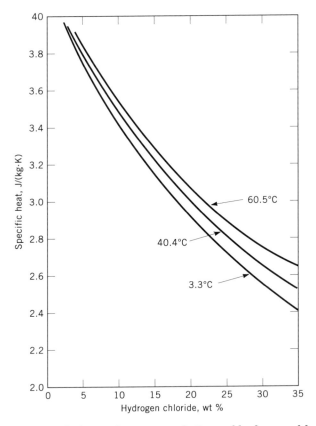

Fig. 4. Specific heat of aqueous solutions of hydrogen chloride.

Table 8. Equivalent Conductivity of Hydrochloric Acid

	Equivalent conductivity, $(\Omega \cdot cm)^{-1}$	
Concentration, wt %	25°C	65°C
0	426.2	666.8
0.91	377.4	578.2
3.58	332.3	509.2
7.90	270.0	416.3
13.60	200.1	310.8
20.66	134.7	209.5
28.63	83.5	130.0

and permit recovery of HCl gas by distillation (see DISTILLATION, AZEOTROPIC AND EXTRACTIVE).

Salts of acids other than hydrochloric acid commonly show increased solubility in hydrochloric acid. This phenomenon has been explained by the Debye-Hückel theory for strong electrolytes (17–19).

Chlorine. The solubility of chlorine [7782-50-5] in hydrochloric acid is an important factor in the purification of by-product hydrochloric acid. The concentration of chlorine in solution, S, is proportional to the partial pressure of

Table 9. Dependence of Solubility of Metal Chlorides on Hydrogen Chloride Concentration in Aqueous Solutions

Concentration of HCl, wt %	Temperature, °C	Solubility of M_xCl_y, wt %
Ferrous chloride		
3	21.5	34.0
12	22.7	20.6
20	22.1	10.6
Potassium chloride		
0	25.0	26.8
11.5	25.0	9.59
21.8	25.0	1.92
Sodium chloride		
0	25.0	26.46
11.1	25.0	10.63
17.1	25.0	4.73

chlorine, ρ, in the gas phase and follows Henry's law, $S = Hp$, in the temperature range of 30–90°C, and at partial pressures of chlorine of <101.3 kPa (1 atm) for HCl concentrations varying in the range of 2–10 N. Henry's coefficient, H, is a function of temperature and HCl concentration, C, as expressed by equation 10:

$$H = \alpha C + B \tag{10}$$

where

$$\log \alpha = -1.21 \times 10^{-2}\, T - 1.603 \tag{11}$$

and

$$B = 2.14 \times 10^2\, T^{-1.21} \tag{12}$$

The units for B are g/L Cl_2 at 101 kPa (1 atm) and the unitless α represents the ratio of the concentration in g/L of Cl_2 to HCl at atmospheric pressure. The temperature is expressed in degrees Celsius. The empirical equation for the Gibbs free energy change was found to be linear with temperature for $\Delta G°$ in kJ/mol, T in Kelvin.

$$\Delta G° = 8.983 + 0.121\, T \tag{13}$$

The entropy change of 121 J/(mol·K) (28.9 cal/(mol·K)) is temperature independent.

The effect of pressure on the solubility of chlorine in hydrochloric acid has been reported for pressures varying from about 100 to 6500 kPa (1–6.5 atm) (20). At pressures above 200 kPa, there is a linear dependence of pressure on the solubility in the acid concentration range of 0.1–5.0 N.

Hydrogen Chloride–Organic Compound Systems. The solubility of hydrogen chloride in many solvents follows Henry's law. Notable exceptions are HCl in polyhydroxy compounds such as ethylene glycol (see GLYCOLS), which have characteristics similar to those of water. Solubility data of hydrogen chloride in various organic solvents are listed in Table 10.

Table 10. Solubilities of Hydrogen Chloride in Common Solvents[a]

	Solubility, mol HCl/mol solvent		
Compound	0°C	20°C	Other[b]
water	0.409	0.3578	
dimethylformamide	2.41		
diethyl ether	1.123	0.67	
methanol	0.92	0.74	
ethanol	0.97	0.82	
2-propanol	1.00	0.83	
n-octanol	1.00		
ethyl acetate	0.73	0.49	
propyl acetate	0.76	0.53	
acetic acid		0.14	
benzene	0.065	0.039	
octane		0.029	
dodecane		0.031	
1,1,1-trichloroethane		0.031	
1,1,2,2-tetrachloroethane		0.027	
hexane		0.017	
n-hexadecane			0.024_{29}
			0.014_{102}
			0.010_{177}
toluene		0.054	
diisopropyl ether			0.978_{10}
dibutyl ether			0.893_{10}
dioxane			1.046_{10}
tetrahydrofurane			1.284_{10}
chloroform			0.028_{15}
			0.023_{25}
carbon tetrachloride		0.025	
			0.016_{50}

[a]Ref. 5.   [b]Temperatures in °C given as subscript.

Chemical Properties

Reactions of Anhydrous Hydrogen Chloride. *Inorganic Compounds.* Hydrogen chloride reacts with inorganic compounds by either heterolytic or homolytic fission of the H–Cl bond. However, anhydrous HCl has high kinetic barriers to either type of fission and hence, this material is relatively inert.

Reactions with Salts of Main Group Anions. Anhydrous HCl protonates the Group 15 (V) hydrides, MH_3, where M = N, P, and As (see HYDRIDES).

$$MH_3 + HCl \longrightarrow MH_4^+ + Cl^- \qquad (14)$$

Reactions of HCl and nitrides, borides, silicides, germanides, carbides, and sulfides take place at significant rates only at elevated ($\geq 650°C$) temperatures. The products are the metal chlorides and the corresponding hydrides. The reactions most studied are those involving nitrides of aluminum, magnesium, calcium, and titanium, where ammonia (qv) is formed along with the corresponding metal chloride.

The reaction of HCl and silicon, germanium, and boron hydrides is catalyzed by aluminum chloride and is useful for preparing chloro-substituted silanes and germanes.

$$MH_4 + HCl \longrightarrow MH_3Cl + H_2 \tag{15}$$

$$MH_3Cl + HCl \longrightarrow MH_2Cl_2 + H_2 \tag{16}$$

Reaction with Metal Oxides. The reaction of hydrogen chloride with the transition-metal oxides at elevated temperatures has been studied extensively. Fe_2O_3 reacts readily at temperatures as low as $300°C$ to produce $FeCl_3$ and water. The heavier transition-metal oxides require a higher reaction temperature, and the primary reaction product is usually the corresponding oxychlorides. Similar reactions are reported for many other metal oxides, such as Sb_2O_3, BeO, Al_2O_3, and TiO_2, which lead to the formation of relatively volatile chlorides or oxychlorides.

Reaction with Metals. Thermodynamic considerations for the reaction

$$M + n\ HCl \longrightarrow MCl_n + n/2\ H_2 \tag{17}$$

indicate that most metals should react with HCl. However, this reaction is kinetically slow at all but elevated temperatures. The reaction of the vapors of sodium and potassium metals has been reported (21). Hydrogen chloride reacts with powdered silicon at $250°C$ to give $SiHCl_3$ and $SiCl_4$ (22).

Reaction with Oxidizing Agents. Hydrogen chloride and oxygen react in the gaseous state to liberate chlorine:

$$4\ HCl + O_2 \longrightarrow 2\ Cl_2 + 2\ H_2O \tag{18}$$

The rate of this reaction is significantly enhanced over catalysts such as copper chloride which is the basis for the Deacon process for producing Cl_2 from HCl. The relationship between the equilibrium constant K_p and the temperature in Kelvin for the reaction is expressed by equation 19.

$$\log\ K_p = 5500/T - 4.31\ \log\ T + 0.0015\ T + 5.18 \tag{19}$$

The Weldon-Pechiney process for manufacturing Cl_2 from HCl involves the use of MnO_2 as the oxidizing agent instead of the O_2 employed in the Deacon reaction.

Reaction with Other Inorganic Halogen Compounds. Anhydrous HCl forms addition compounds at lower temperatures with halogen acids such as HBr and HI, and also with HCN. These compounds are stable at room temperature.

Reaction with Oxyacids and Salts.   Hydrogen chloride reacts with sulfur trioxide yielding liquid chlorosulfuric acid [7790-94-5] (qv).

$$HCl + SO_3 \longrightarrow ClSO_3H \tag{20}$$

The reaction can also be carried out with oleum, distilling the chlorosulfuric acid as it forms. Reaction with oxidizing oxyacids such as HNO_3 liberates chlorine. Anhydrous sulfates of the heavy metals form addition compounds with HCl that can be released by heating the complex to elevated temperatures. The complex $CuSO_4 \cdot 2HCl$ has been used for storage and transport of HCl (23).

Reaction with Organic Compounds.   Addition to Olefins and Acetylenes. Hydrogen chloride adds to carbon–carbon double and triple bonds in a variety of organic compounds. Both 1,2- and 1,4-additions are possible with conjugated dienes. The specific orientation follows the Markonikov rule, which states that halogen attaches itself to the site having lower electron density. This reaction was initially used to make vinyl chloride from acetylene and hydrogen chloride but has since been replaced by oxychlorination (or oxyhydrochlorination) technology, using ethylene (qv) and HCl (or chlorine) as the raw materials. Oxychlorination accounts for more consumption of HCl than any other process (21–26).

Acetylene and hydrogen chloride historically were used to make chloroprene [126-99-8]. The olefin reaction is used to make ethyl chloride from ethylene and to make 1,1-dichloroethane from vinyl chloride. 1,1-Dichloroethane is an intermediate to produce 1,1,1-trichloroethane by thermal (26) or photochemical chlorination (27) routes.

Replacement of Aliphatic Hydroxyl with Chloride.   Lower alcohols such as methanol (qv) can be converted to the corresponding alkyl chlorides by carrying out the reaction

$$ROH + HCl \longrightarrow RCl + H_2O \tag{21}$$

using either a liquid (28) or a solid catalyst (29). The generally used liquid catalyst is zinc chloride although ferric chloride has been used in the past. Solid catalysts include metal chlorides deposited on a carrier such as carbon, silica gel, pumice, alumina, polyphosphoric acids, and activated alumina. In the case of higher alcohols, catalysts such as zinc chloride are used to promote the reaction in the liquid phase.

Chloromethylation Reactions.   The introduction of the chloromethyl group to both aliphatic and aromatic compounds is carried out by reaction of paraformaldehyde [30525-89-4] and hydrogen chloride. This method is used for synthesizing methyl chloromethyl ether [107-30-2], benzyl chloride [100-44-7], and chloromethyl acetate.

Hydrochloric Acid.   *Reaction with Inorganic Compounds.*   Most metals and alloys react with aqueous hydrochloric acid via

$$M + n\, H_3O^+ \longrightarrow M^{n+} + n\, H_2O + n/2 H_2 \tag{22}$$

This is essentially a corrosion reaction involving anodic metal dissolution where the conjugate reaction is the hydrogen (qv) evolution process. Hence, the rate depends on temperature, concentration of acid, inhibiting agents, nature of the surface oxide film, etc. Unless the metal chloride is insoluble in aqueous solution eg, Ag or Hg^{2+}, the reaction products are removed from the metal or alloy surface by dissolution. The extent of removal is controlled by the local hydrodynamic conditions.

Oxides and hydroxides react with HCl to form a salt and water as in a simple acid–base reaction. However, reactions with low solubility or insoluble oxides and hydroxides is complex and the rate is dependent on many factors similar to those for reactions with metals. Oxidizing agents such as H_2O_2, H_2SeO_4, and V_2O_5 react with aqueous hydrochloric acid, forming water and chlorine.

Electrolytic Processes. HCl can be electrolyzed to produce H_2 and chlorine.

$$2\ HCl \longrightarrow H_2 + Cl_2 \tag{23}$$

Electrolytic cells for this reaction are manufactured by Hoechst-Uhde (see ELECTROCHEMICAL PROCESSING, INORGANIC (30,31).

Reaction with Organic Compounds. Many organic reactions are catalyzed by acids such as HCl. Typical examples of the use of HCl in these processes include conversion of lignocellulose to hexose and pentose, sucrose to inverted sugar, esterification of aromatic acids, transformation of acetaminochloroben-zene to chloroanilides, and inversion of methone [1074-95-9].

Manufacturing and Processing

Hydrogen chloride is produced by the direct reaction of hydrogen and chlorine, by reaction of metal chlorides and acids, and as a by-product from many chemical manufacturing processes such as chlorinated hydrocarbons.

Synthesis from Hydrogen and Chlorine. Less than 10% of the U.S. production capacity of HCl is made by the direct reaction of the elements.

$$H_2 + Cl_2 \longrightarrow 2\ HCl + 184\ kJ\ (44\ kcal) \tag{24}$$

Because this reaction is highly exothermic, the equilibrium flame temperature for the adiabatic reaction with stoichiometric proportions of hydrogen and chlorine can reach temperatures up to 2490°C where the equilibrium mixture contains 4.2% free chlorine by volume. This free hydrogen and chlorine is completely converted by rapidly cooling the reaction mixture to 200°C. Thus, by properly controlling the feed gas mixture, a burner gas containing over 99% HCl can be produced. The gas formed in the combustion chamber then flows through an absorber/cooler to produce 30–32% acid. The HCl produced by this process is known as burner acid.

Commercially, the burner chamber and the absorber cooler sections are combined as a single unit for small-scale production. However, in large capacity plants, these units are separated. A typical commercial unit is schematically described in Figure 5 (32).

In general, silica has proved to be a good material of construction for the burner. Cast iron, steel, or graphite was sometimes used. Gaseous HCl produced

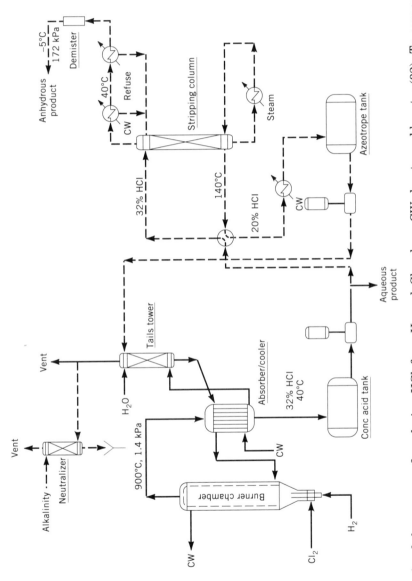

Fig. 5. Schematic of the process of producing HCl from H_2 and Cl_2 where CW denotes cold water (32). To convert kPa to psig, multiply by 0.145.

by this method is very pure and can be used to manufacture pure hydrochloric acid by the adiabatic absorption process (33) or falling film absorption process (34).

Decomposition of Metal Chlorides by Acids.   Two commercial processes employing the acidic decomposition of metal chlorides are the salt–sulfuric acid process and the Hargreaves process. Although these processes are declining in importance, they are used mainly because of the industrial demand for salt cake [7757-82-6] by the paper (qv) and glass (qv) industries. In the United States, however, little HCl is produced this way.

The Salt–Sulfuric Acid Process.   The reaction between NaCl and sulfuric acid occurs in two stages, both endothermic.

$$NaCl + H_2SO_4 \longrightarrow NaHSO_4 + HCl \tag{25}$$

$$NaCl + NaHSO_4 \longrightarrow Na_2SO_4 + HCl \tag{26}$$

The first of these reactions takes place at temperatures of about 150°C, the second reaction proceeds at about 550–660°C. Typical furnaces used to carry out the reaction include cast-iron retorts; the Mannheim mechanical furnace, which consists of an enclosed stationary circular muffle having a concave bottom pan and a domed cover; and the Laury furnace, which employs a horizontal two-chambered rotating cylinder for the reaction vessel. The most recent design is the Cannon fluid-bed reactor in which the sulfuric acid vapor is injected with the combustion gases into a fluidized bed of salts. The Mannaheim furnace has also been used with potassium chloride as the feed.

The Hargreaves Process.   This process, which follows the scheme given by equation 27, is exothermic.

$$4\,NaCl + 2\,SO_2 + O_2 + 2\,H_2O \longrightarrow 2\,Na_2SO_4 + 4\,HCl \tag{27}$$

This reaction is self-sustaining without the need for additional heating once the reactants are heated to 430–450°C; 10–12% HCl is produced by this process as compared to 30–60% HCl from the Mannheim furnace process.

Hydrogen Chloride as By-Product from Chemical Processes.   Over 90% of the hydrogen chloride produced in the United States is a by-product from various chemical processes. The crude HCl generated in these processes is generally contaminated with impurities such as unreacted chlorine, organics, chlorinated organics, and entrained catalyst particles. A wide variety of techniques are employed to treat these HCl streams to obtain either anhydrous HCl or hydrochloric acid. Some of the processes in which HCl is produced as a by-product are the manufacture of chlorofluorohydrocarbons, manufacture of aliphatic and aromatic hydrocarbons, production of high surface area silica (qv), and the manufacture of phosphoric acid [7664-38-2] and esters of phosphoric acid (see PHOSPHORIC ACID AND THE PHOSPHATES).

Hydrogen Chloride Produced from Incineration of Waste Organics.   Environmental regulations regarding the disposal of chlorine-containing organic wastes has motivated the development of technologies for burning or pyrolyzing the waste organics and recovering the chlorine values as hydrogen chloride. Several catalytic and noncatalytic processes have been developed (34–37) to treat these wastes to produce hydrogen chloride (see INCINERATORS).

Hydrogen Chloride from Hydrochloric Acid Solutions. Gaseous hydrogen chloride is obtained by partially stripping concentrated hydrochloric acid using an absorber–desorber system. The stripper is operated at a pressure of 100–200 kPa (1–2 atm) for improved recovery of HCl. The overhead vapors consisting of 97% HCl and 3% H_2O are cooled to remove most of the water as concentrated HCl, and the residual water vapor is removed by drying the gas with sulfuric acid. The stripped acid which is close to the azeotropic composition, is recycled to the absorber.

Stripping is accomplished by dehydration using sulfuric acid (38), lithium chloride [7447-41-8] (39), and tertiary amines containing from 14–32 carbon atoms in an organic solvent immiscible with water followed by thermal treatment of the HCl–organic complex (40).

Purification. *Anhydrous Hydrogen Chloride.* Gaseous HCl from all the manufacturing processes described invariably contains moisture, and sometimes organic species. H_2SO_4 drying can be used to remove small amounts of water, reducing the residual water content to less than 0.02%. If the water content is high, it can be removed as concentrated hydrochloric acid by cooling the gas mixture before drying with sulfuric acid. Addition of chlorosulfuric acid to this stream reduces the water content to less than 10 ppm (41). This mixture also removes unsaturated organics such as ethylene and vinyl chloride (42) and certain organic compounds such as monochloroacetic acid (43).

Chlorine can be removed by either activated carbon adsorption or by reaction with olefins such as ethylene over-activated carbon at temperatures of 30–200°C (44). Addition of liquid high boiling paraffins can reduce the chlorine content in the HCl gas to less than 0.01% (45).

Solid absorbents generally remove the organics from HCl. Polystyrene foam is effective for removal of aromatic compounds (46) and synthetic zeolites have been used for removal of toluene and chlorosilane (47). Activated carbon is effective for removal of chloroacetylene (48). Low boiling organic compounds can be removed by scrubbing HCl using a high boiling organic liquid (49) and low boiling unsaturated organics such as ethylene and vinyl chloride can be converted to high boiling saturates by reaction with HCl in the presence of $AlCl_3$ (50). These reaction products can be removed by adsorption (qv) on granulated activated carbon. Because HCl has a lower boiling point than any of the organic impurities present, impurities such as $COCl_2$ and CCl_4 can be removed by compressing and cooling the gas to partially or completely liquefy the HCl (51). A two-stage solvent scrubbing process, having solvents resistant to chlorination such as CCl_4 and C_4Cl_6 (52), can also be used for purifying crude HCl gas containing both chlorine and chlorocarbons.

Crude HCl recovered from production of chlorofluorocarbons by hydrofluorination of chlorocarbons contains unique impurities which can be removed by processes described in References 53–62. $ClCN–Cl_2$ mixtures generated by reaction of hydrogen cyanide and Cl_2 during the synthesis of $(ClCN)_3$ can be removed from the by-product HCl, by fractional distillation and recycling (see CYANIDES) (59).

Hydrochloric Acid.   Use of air or purified HCl gas as stripper is practiced to remove volatile dissolved organics and chlorine from aqueous HCl (63). Coalescing and decanting is needed for complete removal of suspended organics (64). A countercurrent stripping system is recommended to minimize the loss of hydrogen chloride (65). Whereas residual organics and chlorine can be removed by adsorption on activated carbon, aromatic impurities have also been removed from hydrochloric acid by extraction with a liquid paraffin (66). By-product HCl from chloroflurocarbon processes contains residual hydrogen fluoride which can be removed from the acid by contacting with a $CaCl_2$ solution (67) or with activated alumina or silica gel (68) (see FLUORINE COMPOUNDS).

Materials of Construction, Storage, and Handling

Gaseous Hydrogen Chloride.   Cast iron (qv), mild steel, and steel alloys are resistant to attack by dry, pure HCl at ambient conditions and can be used at temperatures up to the dissociation temperature of HCl. The corrosion rate at 300°C is reported to be 0.25 cm/yr and no ignition point has been found for mild steel at 760°C, at which temperature HCl is dissociated to the extent of 0.2%.

When moisture films are formed, water vapor can accelerate the corrosion rate. Hence, it is necessary to maintain the temperature above the dew point of the gas mixture by at least 20°C, to prevent the formation of moisture films. A temperature of 130°C or above, at atmospheric pressure, can be used for all mixtures of HCl gas and water vapor because the azeotropic boiling point is 108.6°C. The boiling point of azeotropic mixtures can be used as a guide at other pressures (see Table 6).

HCl gas reacts with metal oxides to form chlorides, oxychlorides, and water. Therefore, all the steel equipment should be pickled to remove the oxide scales before it is put in service. Because oxidizing agents in the HCl gas such as oxygen or chlorine significantly affect the corrosion rate, it is essential that the operating temperature of the steel equipment be kept below the temperature (316°C) at which ferric chloride is vaporized from the metal surface.

Stainless steel alloys show excellent corrosion resistance to HCl gas up to a temperature of 400°C. However, these are normally not recommended for process equipment owing to stress corrosion cracking during periods of cooling and shut down. The corrosion rate of Monel is similar to that of mild steel. Pure (99.6%) nickel and high nickel alloys such as Inconel 600 can be used for operation at temperatures up to 525°C where the corrosion rate is reported to be about 0.08 cm/yr (see NICKEL AND NICKEL ALLOYS).

Anhydrous HCl is stored in liquid form under pressure at −25°C and these pressure containers are rated for 2–2.4 MPa (300–350 psig). The material of construction for this purpose is carbon steel which should be ductile at very low (−87°C) temperatures.

Aqueous Hydrochloric Acid.   *Metals.*   Most metals react with aqueous HCl following equation 22. The reaction rate is dependent on the concentration of the acid, oxidizing, reducing, or complexing agents, and corrosion inhibitors, in addition to the metallurgical characteristics of the material and the prevailing hydrodynamic conditions (see CORROSION AND CORROSION CONTROL).

Tantalum and zirconium exhibit the highest corrosion resistance to HCl. However, the corrosion resistance of zironium is severely impaired by the presence of ferric or cupric chlorides. Tantalum–molybdenum alloys containing more than 50% tantalum are reported to have excellent corrosion resistance (see MOLYBDENUM AND MOLYBDENUM ALLOYS) (69). Pure molybdenum and tungsten are corrosion resistant in hydrochloric acid at room temperature and also in 10% acid at 100°C but not in boiling 20% acid.

Corrosion resistance of iron can be improved by alloying with Ni, Cr, Mo, Cu, Mn, W, and Sb. A Japanese stainless steel containing, on a wt % basis, 16–18 Ni, 9–11 Cr, 6–8 Mo, 2 Mn, 1 Si, and 0.04 P is reported to exhibit good resistance to hot hydrochloric acid (70).

Nickel-based alloys have superior corrosion resistance to iron-based alloys. The only alloys recommended for hot hydrochloric acid use are Ni–Mo alloys containing 60–70% Ni and 25–33% Mo. Chlorimet (63 Ni, 32 Mo, 3 Fe) and Hastelloy (60 Ni, 28 Mo, 6 Fe) are found to be stable at all acid concentrations in the absence of air and iron chlorides. Electroless nickel, a Ni–P alloy containing 2–10% P, shows excellent resistance to hot hydrogen chloride (71). The corrosion resistance increases with phosphorus content. This coating can be deposited on cast iron, wrought iron, mild steel, stainless steels, brass, bronze, and aluminum (qv).

Although lead chloride is moderately soluble in the acid, lead is also used occasionally in hydrochloric acid service. Addition of 6–25% Sb increases the corrosion resistance. Air and ferric chloride accelerate the corrosion. Durichlor (14.5% Si, 3% Mo, 82% Fe), a silica-based alloy, shows excellent resistance to hot hydrochloric acid in the absence of ferric chloride.

Plastics and Elastomers. Common plastics and elastomers (qv) show excellent resistance to hydrochloric acid within the temperature limits of the materials. Soft natural rubber compounds have been used for many years as liners for concentrated hydrochloric acid storage tanks up to a temperature of 60°C (see RUBBER, NATURAL). Semihard rubber is used as linings in pipe and equipment at temperatures up to 70°C and hard rubber is used for pipes up to 50°C and pressures up to 345 kPa (50 psig). When contaminants are present, synthetic elastomers such as neoprene, nitrile, butyl, chlorobutyl, hypalon, and ethylene–propylene–diene monomer (EPDM) are preferred to natural rubber. Standard plastics such as polypropylene, poly(vinyl chloride), Saran, and acrylonitrile–butadiene–styrene (ABS) show good resistance to hydrochloric acid. The fluorocarbon plastics exhibit extremely high corrosion resistance and a high upper temperature limit of operation.

Carbon and Graphite. Carbon and graphite rendered impervious with 10–15% phenolic, epoxy, or furan resin are among the most important materials for hydrochloric acid service up to 170°C. The most important applications of these materials for hydrochloric acid service are heat exchangers and centrifugal pumps.

Glass and Ceramics. Glass and ceramic-coated equipment is widely used for handling hydrochloric acid. The glass lining is normally 0.5–1.0 mm (20–40 mil) thick and can be applied to various base metals. A variety of glasses is used by different equipment fabricators and the corrosion resistance varies with the type of glass (see REFRACTORY COATINGS). Fused-cast refractories (qv) such as alumina, silica, zirconia, and chrome alumina are also used for hydrochloric acid service for lining equipment such as brick-lined towers. Mono-lithic refractories have also been used for special applications such as stone-ware vessels.

Production and Economic Aspects

Over 90% of the HCl produced in the United States originates as a coproduct from various chlorination processes; direct generation of HCl from H_2 and Cl_2 accounts for only about 8% of the total production. Table 11 describes the production contribution of HCl from significant sources through the period 1980 to 1992 (72). Figure 6 illustrates the historical production growth of HCl in the United States (73). The growth rate, about 5–6% from 1955 to 1975, slowed to ~1% because of disparity between supply and demand (see Table 12). The production capacity in 1993 was about 2.92 million metric tons, down 9.6% from the 1992 production of 3.24 million metric tons (74).

As capital costs increase, operating costs decrease, as shown in Table 13; those costs involved in the production of HCl by various routes are presented in Table 13 for some nominal production capacities. These data indicate the direct route from H_2 and Cl_2 to be most economic among all the technologies considered here. Details, including the assumptions involved in estimating the various costs in Table 13, may be found in the literature (75,76).

The 1990 price of anhydrous HCl was about $330/t; the 1993 price of 20° Bé (31.4% HCl) was about $73/t (77). Prices depend on plant location, transporta-tion burden, and on-site demand. These factors all influence the selling price significantly, sometimes carrying zero or negative value.

Analysis

Solutions of Hydrogen Chloride.   The concentration of aqueous solutions of hydrogen chloride can be estimated from specific gravity tables for most indus-trial needs. HCl in aqueous solutions is determined quantitatively by volumetric analysis using standard alkali, whereas HCl in methanol or acetic acid media is estimated from conductometric titrations using standard solutions of lithium, sodium, or potassium acetate. Precise determination of chloride ion concentration in solutions containing mixtures of halides can be accomplished by differential potentiometry to determine the end point, using silver nitrate as the reagent.

Determination in the Atmosphere.   Trace amounts of HCl in the atmos-phere are detected using krypton homologues as detectors (78), Zn, Al, and Cu kryptonates respond to HCl concentrations in the range of 97–250 ppb; or from photometric determination of NH_4Cl (79) formed by reaction of ammonia and HCl in a helium reservoir (see HELIUM GROUP, COMPOUNDS). A congo red indicator for chlorimetric determination of HCl in air has been found effective for concen-trations as low as 5 ppm (80). Colorimetric procedures have also been developed for determining HCl in air by absorption in water containing a suitable indi-cator such as methyl red. Absorption of gaseous HCl on a filter treated with KOH–triethanoloamine is used to determine the HCl in ambient air at the

Table 11. U.S. Production of Hydrochloric Acid, 100% HCl, t × 10^6 [a]

Year	Vinyl chloride monomer[b]	Isocyanates[c]	Fluorocarbons	By-product Chlorinated hydrocarbons C$_1$[d]	C$_2$	Other[e]	Total	Chlorine and hydrogen	Salt and sulfuric acid	Total
1980	1.7	0.36	0.27	0.36	0.45	0.63	3.8	0.27	0.9	4.3
1985	2.1	0.45	0.27	0.27	0.27	0.63	4.0	0.18	0.9	4.4
1990	2.7	0.54	0.27	0.27	0.27	0.63	4.7	0.36	f	5.0
1993[g]	3.3	0.54	0.27	0.27	0.18	0.63	5.2	0.36	f	5.5

[a] Ref. 72.
[b] Essentially all HCl generated during vinyl chloride monomer production is recycled to make intermediate ethylene dichloride.
[c] Estimated HCl production during isocyanate manufacture represents net HCl. This value excludes HCl consumed in the reaction process to make methylene diphenylene diamine (MDA) and polymethylene polyamine (PMPPA) intermediates in the production of 4,4′-methylenebis(phenylisocyanate) (MDI) and polymethylene polyphenylisocyanate (PMPPI).
[d] Most HCl generated during chlorinated C$_1$ production is recycled to make methyl chloride.
[e] Other sources of by-product HCl include allyl chloride, chlorobenzenes, chlorinated paraffins, linear alkylbenzene, silicone fluids and elastomers, magnesium, fluoropolymers, chlorotoluenes, benzyl chloride, potassium sulfate, and agricultural chemicals.
[f] Included under other by-product column. The U.S. company producing HCl and coproduct sodium sulfate from sodium chloride and sulfuric acid shut down its plant in 1992.
[g] Preliminary.

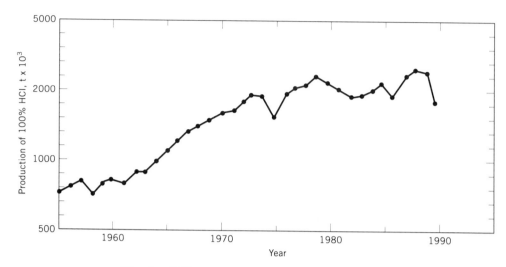

Fig. 6.   HCl production from 1955 to 1990 (73).

ppb level (81). To trap the HCl in the atmosphere, a nylon filter can also be used. The filter, after exposure to HCl, is sonicated in water to extract the HCl which is analyzed by conventional methods (82).

Determination of Impurities.   Impurities in hydrochloric acid primarily arise from the products of the manufacturing process used to produce it. Because most hydrogen chloride is a by-product of processes making chlorinated organic compounds, HCl contains a significant amount of organics. Organic compounds containing chlorine, at the ppm level, can be analyzed by the reduction of $KMnO_4$ to MnO_2 in alkaline solution (83). Aromatic compounds can be determined from the uv absorption spectrum of the aliphatic hydrocarbon extract of the HCl sample. Organic chloride impurities which are of particular concern in food-grade hydrochloric acid, can be determined by the FDA procedure employing two extractions with cold petroleum ether to measure impurities in the 5–10 ppm range (84). Reagent-grade hydrochloric acid is generally analyzed for metallic impurities by methods such as inductively coupled plasmas (icp). Other procedures can also be used to determine trace metals at levels of 10^{-6}–$10^{-7}\%$ (85,86) (see TRACE AND RESIDUE ANALYSIS).

Health and Safety Factors

Hydrogen chloride in air is an irritant, severely affecting the eye and the respiratory tract. The inflammation of the upper respiratory tract can cause edema and spasm of the larynx. The vapor in the air, normally absorbed by the upper respiratory mucous membranes, is lethal at concentrations of over 0.1% in air, when exposed for a few minutes. HCl is detectable by odor at 1–5 ppm level and becomes objectionable at 5–10 ppm. The maximum concentration that can be tolerated for an hour is about 0.01% which, even at these levels, causes severe throat irritation. The maximum allowable concentration under normal working conditions has been set at 5 ppm.

Table 12. U.S. Consumption of Hydrochloric Acid 100% HCl, t × 10³ [a,b]

Product	1989	1993	Average annual growth rate, %
Anhydrous hydrogen chloride			
ethylene dichloride (EDC) and vinyl chloride	2611	3345	6.4
methyl chloride[c]	336	349	1.0
chlorinated C_2	41	36	−2.9
ethyl chloride	43	24	−13.8
other	238	257	1.9
subtotal, anhydrous HCl	3269	4011	5.3
subtotal, less captive EDC and methyl chloride[d]	322	317	−0.4
Muriatic acid			
brine treating (chlorine/caustic)	222	213	−1.0
steel pickling	204	181	−2.9
food	145	181	5.7
oil-well acidizing	136	122	−2.6
calcium chloride	100	145	9.8
chlorine	63	63	0.0
other	625	662	1.4
subtotal, muratic acid	1496	1567[e]	1.2
Total	4765	5578	4.0
Total, less captive EDC and methyl chloride	*1818*	*1885*	*0.9*

[a]Ref. 73.
[b]Data include captive and merchant HCl demand.
[c]Chlorinated C_1.
[d]Most HCl consumed for the production of EDC and methyl chloride is recycled acid generated in an integrated process and therefore does not affect net supplies of HCl. The exclusion of HCl consumption for EDC and methyl chloride production gives a better indication of net HCl supply. This figure gives only an order of magnitude of net HCl available for captive consumption and sales.
[e]May include some HCl that is neutralized for disposal.

Hydrogen chloride in the lungs can cause pulmonary edema, a life threatening condition. In order for HCl in air to reach the lungs, it must be transported either as an aerosol or as a deposit on soot particles of less than 3 μm in diameter. A procedure for the removal of 99% of the HCl from municipal waste incinerators has been developed (87). Lime is employed as a dry adsorbent which is collected in a filter bag system.

Hydrogen chloride in air can also be a phytotoxicant (88). Tomatoes, sugar beets, and fruit trees of the Prunus family are sensitive to HCl in air.

Exposure of concentrated hydrochloric acid to the skin can cause chemical burns or dermatitis. Whereas the irritation is noticed readily, the acid can be water flushed from the exposed area. Copious use of running water is the only recommended safety procedure for any external exposure. Ingestion is seldom a problem because hydrochloric acid is a normal constituent of the stomach juices. If significant quantities are accidently swallowed, it can be neutralized by antacids.

Table 13. Capital and Operating Costs[a] for HCl Production by Process[b]

Process	Concentration, °Bé	Capacity, t × 10³/yr	Capital cost,[c] $ × 10⁶	Net production cost,[d] $/kg	Product cost,[e] $/kg
		Hydrochloric acid			
direct	22	9.9	2.8	0.23	0.30
		19.9	3.9	0.19	0.24
Mannheim	22	9.9	10.5	0.34	0.61
		19.9	17.3	0.26	0.48
salt by a	22	9.9	7.9	0.29	0.49
fluidized bed		19.9	11.6	0.21	0.36
Hargreaves	22	9.9	14.1	0.42	0.78
		19.9	23.2	0.30	0.59
incineration gas	18	7.6	4.7	0.21	0.36
		15.2	6.9	0.15	0.27
spent pickle	18[f]	6.6	6.2	0.32	0.55
liquor		13.2	9.3	0.20	0.37
		Anhydrous HCl			
direct	100[f]	19.9	4.3	0.20	0.25
		33.0	5.7	0.18	0.22
incineration gas	100[f]	6.9	6.0	0.37	0.59
		13.9	9.2	0.27	0.43

[a]Materials costs employed in these estimates in $/kg are, Cl_2, 0.10; H_2, 0.30; slaked lime, 0.044; salt, 0.02; H_2SO_4, 0.062; S, 0.064; $FeCl_3$, 1.08; ferric oxide, 0.088; Na_2SO_4, 0.044; utility costs are cooling water, 0.1¢/m³; steam at 1 MPa (150 psig), $7.52/t; electricity, 3¢/kWh; process water, 1.4¢/m³; fuel oil, $2.36/GJ; and natural gas, $2.23/GJ.
[b]Ref. 75.
[c]Includes off sites.
[d]This includes labor costs, variable costs, overhead, taxes, and depreciation.
[e]This is a sum of the production cost and return-on-investment (ROI) before taxes at 25%/yr of total capital costs.
[f]Units are wt %.

Storage and Handling

All Department of Transportation (DOT), Environmental Protection Agency (EPA), and Occupational Safety and Health Act (OSHA) rules and regulations should be reviewed prior to handling hydrochloric acid and all the regulations must be followed. All employees handling HCl must be trained to ensure that they are familiar with the appropriate materials safety data sheets and applicable regulations.

The Department of Transportation classifies HCl as a corrosive material and requires that it be transported in DOT-approved delivery vessels. Tank cars must conform to 103B, 103B-W, or DOT 111A60W5 specifications. Tank trailers must conform to DOT MC-310, MC-311, MC-312, or DOT-412 specifications with display of a corrosive placard on both sides, front, and rear of the tank. The United Nations identification number for muriatic acid is UN1789, which must appear on all shipping papers and placards.

Tables 14 and 15 list many of the typical materials of construction used in the industry and the corresponding relative resistance to hydrochloric acid (15).

Table 14. Corrosion Resistance of Metals to HCl[a,b]

	HCl concentration, wt %		
Material	1–20	>20	>2[c]
aluminum 3003	NR	NR	NR
carbon–graphite, resin impregnated	A[d]	A[d]	A[d]
carbon steel 1018	NR	NR	NR
copper nickel, 90/10 or 70/30	B-NR	NR[e]	NR
Hastelloy B	A	A[f]	B[f]
Hastelloy C	A	A	C[g]
Inconel	B-NR	NR	NR
Monel	NR	NR	NR
nickel	A-B	C-NR	NR
stainless steel 303, 304, 316	NR	NR	NR
tantalum	A	A[h]	A
titanium	B-C	C-NR[i]	NR
zirconium	A[j]	A[k]	A

[a]A is little to no attack; B, good resistance; C, limited resistance; and NR, not recommended (15).
[b]At a temperature of 21°C for metals unless otherwise indicated.
[c]At a temperature of 79.4°C unless otherwise indicated.
[d]Up to 249°C.
[e]At 37 wt % HCl.
[f]Up to 100 wt % HCl.
[g]Up to 10 wt % HCl.
[h]Up to 149°C.
[i]At 39 wt % HCl.
[j]Up to 100°C.
[k]Up to 37 wt % HCl at 100°C.

Uses

Hydrogen chloride and the aqueous solution, muriatic acid, find application in many industries. In general, anhydrous HCl is consumed for its chlorine value, whereas aqueous hydrochloric acid is often utilized as a nonoxidizing acid. The latter is used in metal cleaning operations, chemical manufacturing, petroleum well activation, and in the production of food and synthetic rubber.

Most of the HCl produced is consumed captively, ie, at the site of production, either in integrated operations such as ethylenedichloride–vinyl chloride monomer (EDC/VCM) plants and chlorinated methane plants or in separate HCl consuming operations at the same location. Captive use of anhydrous HCl accounted for 80–85% of the total demand in 1989. The combined merchant market for anhydrous and aqueous HCl in that same year was about 9.1×10^5 metric tons on the basis of 100% HCl (see Table 12) (73).

Anhydrous Hydrogen Chloride. *Ethylene Dichloride and Vinyl Chloride.* In the United States, all ethylene dichloride [*107-06-2*] (EDC) is produced from

Table 15. Corrosion Resistance of Nonmetals to HCl[a,b]

Material	HCl concentration, wt %		
	1–20	>20	>2[c]
acid-proof brick			
carbon		A[d] (499)	
fireclay		A[d] (1371)	
epoxy	A (to 21)	A[d] (21)	NR (93)
	B-C (50)	AB[e] (66)	
glass		A[d] (232)	
glass-reinforced epoxy or polyester		A (104)	
graphite		A[d] (399)	
Halar, E-CTFE	A (93)	A (to 540)	A (93)
KEL-F	A (21)	A[f] (149)	A (93)
Kynar	A (to 135)	A[f] (135)	A (to 135)
neoprene		A[d] (to 104)	
Noryl	A (93)	A[d] (93)	A[d] (93)
nylon	A[g] (21)	NR[h] (21)	NR (93)
polycarbonate	A (93)	A[i] (93)	A[i]
polyethylene			
UHMW	A (21)	A[j] (49)	A[j]
HMW	A[j] (66)	AB[k] (66)	
polypropylene	A-C (to 79)	B-NR[l] (to 62)	A-C[m]
polysulfone	A (93)	A[d] (93)	A[d] (93)
poly(vinyl chloride)	A (to 60)	A-B[j] (to 60)	NR (93)
rubber-lined carbon steel	A (to 66)	A[d] (to 66)	
Ryton	A (93)	A[d] (93)	A (93)
Teflon			
TFE	A (93)	A[f] (to 127)	A (93)
FEP	A (21)	A[j] (to 93)	A (93)
PFA	A (to 120)	A[d] (to 120)	A (93)
Tefzel, ETFE	A (21)	A[j] (106)	A (93)

[a]A is little to no attack; B, good resistance; C, limited resistance; and NR, not recommended (15).
[b]Temperatures in °C are given in parentheses.
[c]At a temperature of 79.4°C unless otherwise indicated.
[d]At 37 wt % HCl.
[e]At 50 wt % HCl.
[f]At 100 wt % HCl.
[g]At 10 wt % HCl.
[h]At >10 wt % HCl.
[i]At 20 wt % HCl.
[j]Concentrated HCl.
[k]At 40 wt % HCl.
[l]At 35 wt % HCl.
[m]To 40 wt % HCl.

ethylene, either by chlorination or oxychlorination (oxyhydrochlorination). The oxychlorination process is particularly attractive to manufacturers having a supply of by-product HCl, such as from pyrolysis of EDC to vinyl chloride [75-01-4] monomer (VCM), because this by-product HCl can be fed back to the oxychlorination reactor. EDC consumption follows demand for VCM which consumed about 87% of EDC production in 1989. VCM is, in turn, used in the manufacture of PVC resins. Essentially all HCl generated during VCM production is recycled to produce precursor EDC (see CHLOROCARBONS AND CHLOROHYDROCARBONS; VINYL POLYMERS).

   Methyl Chloride. Most of the HCl consumed in the manufacture of methyl chloride [74-87-3] from methanol (qv) is a recycled product. The further reaction of methyl chloride with chlorine to produce higher chlorinated methanes generates significant amounts of HCl which are fed back into methyl chloride production. Another source of recycled HCl is silicone production based on methyl chloride.

   Chlorine. Several methods are available for generating chlorine from HCl. These include electrolysis of metallic chloride solutions, electrolysis of hydrochloric acid, oxidation of hydrogen chloride to chlorine with nitric acid, and oxidation of hydrogen chloide to chlorine using oxygen in the presence of catalysts (Deacon process and the modified Deacon process). As of this writing, only about 60,000 to 65,000 metric tons of Cl_2 is produced via electrolysis of HCl in the United States annually. Details related to electrolytic and chemical routes for manufacturing Cl_2 from HCl are given in the literature (89) (see ALKALI AND CHLORINE PRODUCTS, CHLORINE AND SODIUM HYDROXIDE).

   Chlorinated C_2. Perchloroethylene (PCE) and trichloroethylene (TCE) can be produced either separately or as a mixture in varying proportions by reaction of C_2-chlorinated hydrocarbons, eg, C_2-chlorinated waste streams or ethylene dichloride, with a mixture of oxygen and chlorine or HCl.

   Ethyl Chloride. Most ethyl chloride [75-00-3] is produced by the hydrochlorination of ethylene (qv) using anhydrous HCl. Historically, the primary use of ethyl chloride was for the manufacture of tetraethyllead (TEL), a primary component of antiknock mixes in gasolines. Use has declined as a result of the environmental concern regarding lead and TEL is no longer produced in the United States. Other uses of ethyl chloride include the production of ethyl cellulose, which is formulated into lacquers, adhesives (qv), inks (qv), and varnishes.

   Miscellaneous. Other uses for anhydrous HCl include use in cottonseed delinting and disinfecting (see COTTON), as a catalyst promotor for petroleum (qv) isomerization, in the production of agricultural chemicals, and in the preparation of hydrochloride salts in the pharmaceutical industry (see PHARMACEUTICALS). It is also used in the production of ethylene chlorohydrin (see CHLOROHYDRINS), intermediate in the production of polysulfide elastomers (qv). In addition, anhydrous HCl is used in the electronic industry as an etching and cleaning agent during the production of silicon wafers and semiconductors and for the cleaning the circuit boards. Anhydrous HCl is also used in the production of trichlorosilane and chlorosulfonic acid.

   Aqueous Hydrochloric Acid. Muriatic acid consumption in 1993 was about 1.57 million metric tons (100% basis). The largest captive use of aqueous HCl is for brine acidification prior to electrolysis in chlorine/caustic cells and the largest merchant markets for HCl are steel pickling and oil-well acidizing, which accounted for 25 and 16% of merchant production, respectively, during 1989.

Brine Treatment.   The principal use of aqueous HCl is for the acidification of brine prior to feeding it to the electrolytic cells for producing chlorine and caustic soda. Almost all of this HCl comes from captive sources. An estimated 213 thousand metric tons of HCl (100% basis) was used for brine treatment in 1993 (74).

Metal Cleaning.   About 204 thousand metric tons of HCl (100% basis) was consumed in 1993 for steel pickling, wherein the hydrochloric acid readily dissolves all of the various oxides present in the scale formed during the hot rolling process. Using suitable inhibitors such as alkyl pyridines, HCl reacts very slowly with the base metal rendering the surface so clean that it must be passivated with a mild alkaline rinse.

Hydrochloric acid is also used in other metal cleaning processes which range from large-scale process equipment such as tube and shell heat exchangers to solder fluxing agents and household cleaners for plumbing fixtures. Use of HCl for metal cleaning purposes is projected to decline in view of the problems associated with the disposal of the spent acid. Some steel mills regenerate spent acid and hence require only makeup HCl.

Oil-Well Acidizing.   HCl is used both to clean out old oil wells and to encourage the flow of crude oil or gas to the well. Oilfield service companies use HCl concentrations in the range of 5–27% depending on geological and other factors. HCl consumption follows rig activity for oil production and is dependent on the price of crude oil. Rig activity is expected to grow by about 5–6% a year in the mid-1990s. However, many new wells are intended for gas production only and hence HCl use in well acidizing will probably grow by 2% per year.

Food.   The food industry uses hydrochloric acid for processing a variety of products such as high fructose corn syrup for sweetening soft drinks (see CARBONATED BEVERAGES), hydrolyzed vegetable protein and soy sauce, gelatin, acidification of vegetable juices and can foods, and artificial sweeteners (qv) (see FOOD PROCESSING).

Production of Calcium Chloride.   Calcium chloride [*10043-52-4*] is produced by reaction of limestone and hydrochloric acid (see CALCIUM COMPOUNDS). The HCl consumption varies depending on the quality of limestone used (see LIME AND LIMESTONE). The largest use of calcium chloride is for highway deicing and other uses including dust control, oil recovery, concrete treatment, and tire ballasting. Demand for HCl to produce calcium chloride for deicing depends on the weather conditions during the year, coupled with environmental constraints which include increasing regulatory restrictions (90,91).

Minerals and Metals.   HCl is consumed in many mining operations for ore treatment, extraction, separation, purification, and water treatment (see MINERAL RECOVERY AND PROCESSING). Significant quantities are also used in the recovery of molybdenum (see MOLYBDENUM AND MOLYBDENUM ALLOYS) and gold (see GOLD AND GOLD COMPOUNDS). This market consumed about 36 thousand metric tons in 1993.

Miscellaneous.   Hydrochloric acid is used for the recovery of semiprecious metals from used catalysts, as a catalyst in synthesis, for catalyst regeneration (see CATALYSTS, REGENERATION), and for pH control (see HYDROGEN-ION AC-

TIVITY), regeneration of ion-exchange (qv) resins used in wastewater treatment, electric utilities, and for neutralization of alkaline products or waste materials. In addition, hydrochloric acid is also utilized in many production processes for organic and inorganic chemicals.

BIBLIOGRAPHY

"Hydrochloric Acid" in ECT 1st ed., Vol. 7, pp. 652–674, by S. Gershon, Merritt-Chapman & Scott Corp.; in ECT 2nd ed., Vol. 11, pp. 307–337, by W. R. Kleckner and R. C. Sutter, Diamond Alkali Co.; "Hydrogen Chloride" in ECT 3rd ed., Vol. 12, pp. 983–1015, by D. S. Rosenberg, Hooker Chemical Co.

1. L. Reti, *Chymia* **10**, 11 (1965).
2. H. R. Koeltz, *Scand. J. Gastroenterol Supp.*, 27 (1993).
3. P. A. Straub and A. D. McLean, Theor. Chim. Acta **32**, 227 (1974).
4. A. Blumen and L. Merkel, J. Phys. B. **10**(15), L555 (1977).
5. S. Austin and A. Glowacki, *Ullmann's Encyclopedia of Industrial Chemistry*, 5th ed., Vol. A13, VCH, Weinheim, Germany, 1989, p. 283.
6. S. Hoshino, K. Shinaoka, and N. Nimuva, *Phys. Rev. Lett.* **19**, 1286 (1967).
7. J. E. Kohl, M. G. Semack, and D. White, *J. Chem. Phys.* **69**(12), 5378 (1978).
8. H. Braune and F. Zehle, *Z. Phys. Chem. Abt. B* **49**, 247 (1941).
9. A. J. E. Welch, *Ann. Rep. Chem. Soc.* **37**, 153 (1940).
10. H. Chihara and A. Inaba, *J. Chem. Thermodynam.* **8**, 915 (1976).
11. H. M. Spencer and J. L. Justice, *J. Am. Chem. Soc.* **56**, 2311 (1934).
12. D. F. Othmer and L. M. Naphtali, *Ind. Eng. Chem.* **1**(1), 6 (1956).
13. J. T. F. Kao, *J. Chem. Eng. Data* **15**(3), 362 (1970).
14. I. Taesler and J. O. Lundgren, *Acta Crystallogr. Sect. B* **34**, 2424 (1978).
15. *Muriatic Acid—Bulk Storage and Handling*, brochure, Occidental Chemical Corp., Niagara Falls, N.Y., 1991.
16. E. T. McGuinness and K. S. Clarke, *J. Chem. Educ.* **45**(11), 740 (1968).
17. W. Kindler and G. Wuester, *Ber. Bunsenges. Phys. Chem.* **82**, 543 (1978).
18. W. F. Linke, *Solubilities, Inorganic and Metal-Organic Compounds*, Vol. 1, 4th ed., American Chemical Society, Washington, D.C., 1958, p. 664.
19. R. K. Gupta, *J. Appl. Chem.* **18**, 49 (1968).
20. V. A. Smirnov, Z. M. Aliev, V. V. Karapysh, and I. I. Gurchin, *Russ. J. Phys. Chem.* **48**(5), 720 (1974).
21. C. E. H. Baron and A. G. Evans, *Trans. Faraday Soc.* **31**, 1932 (1935).
22. R. Schwarz and F. Heinrich, *Z. Anorg. Chem.* **221**, 227 (1935).
23. U.S. Pat. 2,356,334 (Aug. 22, 1944), A. H. Maude and D. S. Rosenberg (to Hooker Electrochemical Co.).
24. R. W. McPherson, C. M. Starks, and G. J. Fryar, *Hydrocarbon Process.* **59**, 75 (1979).
25. "Ethylene Dichloride," *Chemical Economics Handbook*, SRI International, Menlo Park, Calif., June 1992.
26. Ger. Pat. 2,050,745 (Apr. 22, 1971), A. Campbell and A. Thoburn; U.S. Pat. 3,825,609 (July 23, 1974), A. Campbell and A. Thoburn (to Imperial Chemical Industries, Ltd.).
27. U.S. Pat. 3,019,175 (Jan. 30, 1962), A. J. Haefner and F. Conrad (to Ethyl Corp.).
28. U.S. Pat. 4,922,043 (May 1, 1990), J. T. Petrosky (to Vulcan Materials Co.).
29. U.S. Pat. 5,041,406 (Aug. 20, 1991), D. A. Harley and M. T. Holbrook (to Dow Chemical Co.).
30. K. Kerger, *Chem. Eng. Tech.* **43**, 167 (1971).
31. *Chlorine and Hydrogen from HCl by Electrolysis*, Uhde, Dortmund, Germany, 1987.

32. J. E. Buice, R. L. Bowlin, K. W. Mall, and J. A. Wilkinson, *Encyclopedia of Chemical Processing and Design*, Vol. 26, Marcel Dekker, Inc., New York, 1987, p. 396.
33. J. B. Bingeman and L. B. Reynolds, *Chem. Eng. Prog.* **56**, 37 (1960).
34. G. Guerreri and C. J. King, *Hydrocarbon Process.* **53**, 131 (1974).
35. J. C. Zimmer and R. Guaitella, *Hydrocarbon Process.* **55**, 117 (1976).
36. U.S. Pat. 4,073,87 (Feb. 14, 1977), W. Opitz and H. Hennen (to Hoechst Aktiengesellschaft.).
37. Ger. Pat. 2,532,075 (Feb. 5, 1976), R. M. Kovach and H. J. Essig (to B. F. Goodrich Co.).
38. Ger. Pat. 1,914,579 (Sept. 24, 1970), F. Bandmaier and co-workers (to Sigri Electrographit GmbH).
39. U.S. Pat. 3,763,019 (Oct. 2, 1973), A. Yodis (to Allied Chemical Corp.).
40. U.S. Pat. 4,115,530 (Sept. 19, 1978), A. Coenen, K. Kosswig, and G. Prominski (to Chemische Werker Hüls AG).
41. Jpn. Pat. 73 63,991 (Sept. 5, 1973), N. Nakajima, T. Matsumoto, and M. Matsuda (to Seitetsu Chem. Industries Co., Ltd.).
42. U.S. Pat. 3,681,014 (Aug. 1, 1972), H. L. Hackett and D. A. Cullison (to Continental Oil Co.).
43. Ger. Pat. 2,522,286 (Nov. 25, 1976), K. Gehrmann and A. Mainski (to Hoechst AG).
44. Fr. Pat. 2,252,288 (June 20, 1975), Y. Correia and J. Lesparre (to Rhône-Progil).
45. U.S. Pat. 3,131,027 (Apr. 28, 1964), W. L. Borkowski and J. J. Van Venvooy (to Sunoil Co.).
46. Jpn. Pat. 72 40,630 (Oct. 14, 1972), M. Aumi and S. Kawase (to Osaka Soda Co., Ltd.).
47. V. P. Timoshenko, *Zh. Prikl. Khim. Leningrad* **47**(2), 261 (1974).
48. U.S. Pat 3,979,502 (Nov. 25, 1975), Y. Correia and F. Muller (to Rhône Progil).
49. USSR Pat. 395,320 (Aug. 28, 1973), E. R. Berlin and co-workers.
50. U.S. Pat. 3,446,586 (May 27, 1969), D. M. Young (to Dow Chemical).
51. Ger. Pat. 2,143,994 (Mar. 15, 1973), P. Fischer and co-workers (to Bayer AG).
52. U.S. Pat. 2,841,243 (July 1, 1958), T. Hooker, E. J. Geering, and A. H. Maude (to Hooker Electrochemical Co.).
53. Ger. Pat. 2,229,571 (Dec. 28, 1972), D. R. Merchant (to Pennwalt Corp.).
54. Ger. Pat. 2,617,689 (Apr. 25, 1975), M. C. Sze and J. E. Paustian (to Lumnus Co.).
55. Jpn. Pat. 72 16,405 (May 15, 1972), H. Aiso and T. Takeuchi (to Daikin Kogyo Co., Ltd.).
56. Jpn. Pat. 75 95,196 (July 29, 1975), H. Oshio, T. Toyoda, and K. Matsuoka (to Central Glass Co., Ltd.).
57. U.S. Pat. 3,253,029 (May 24, 1966), F. S. Fawcett (to E. I. du pont de Nemours & Co., Ltd.).
58. U.S. Pat. 4,092,403 (May 30, 1978), C. E. Rectenwald and H. B. Hinckley (to Union Carbide Corp.).
59. Ger. Pat. 1,900,972 (Sept. 4, 1969), J. A. Riethmann and L. Scheck (to Agripat SA).
60. Ger. Pat. 1,809,607 (June 18, 1970), F. Geiger, W. Weigert, and T. Luessling (to Deutsche Gold and Silver-Scheideanstalt).
61. U.S. Pat. 3,807,139 (Apr. 30, 1974), L. DeFlore and S. Quarta (to Societa Italiana Resine).
62. U.S Pat. 3,425,188 (Feb. 4, 1969), C. E Kircher and R. J. Jones (to Dextrex Chemical Corp.).
63. U.S. Pat. 3,597,167 (Aug. 3, 1971), D. R. Marks and C. R. Hanson (to Velsicol Chemical Corp.).
64. U.S. Pat. 3,855,400 (Dec. 17, 1974), R. Paolieri, T. A. Pitts, and J. W. McCloskey (to Hooker Chemicals & Plastics Corp.).
65. Ger. Pat. 2,413,043 (Sept. 25, 1975), H. G. Ufermann, H. Wiechers, and E. Zirngiebl (to Bayer AG).

66. U.S. Pat. 3,445,197 (May 20, 1969), J. N. Yarbrough (to Continental Oil Co.).
67. U.S. Pat. 3,140,916 (July 16, 1964), F. R. Lowdermilk (to Pennsalt Chemical Corp.).
68. U.S. Pat. 3,411,879 (Nov. 19, 1968), R. E. Whitfield (to Dow Chemical Co.).
69. W. C. Schumb, S. F. Radtke, and M. B. Bever, *Ind. Eng. Chem.* **42**, 876 (1950).
70. F. Tsukamoka, *Met. Prog.* **85**(1), 107 (1964).
71. W. J. Suski, *Chem. Eng. News*, 148 (Nov. 21, 1966).
72. "Hydrochloric Acid," *Chemical Economics Handbook*, SRI International, Menlo Park, Calif., June 1991.
73. *Current Industrial Reports*, MA28A, U.S. Department of Commerce, Bureau of Census, Washington, D.C., 1993.
74. *Chem. Eng. News*, 13 (Apr. 11, 1994).
75. Y.-R. Chin, SRI International, Menlo Park, Calif., private communication, 1994.
76. *Process Economics Program #134*, SRI International, Menlo Park, Calif., 1979.
77. *Hydrochloric Acid*, Chemical Products Synopsis, Mansville Chemical Products Corp., Asbury, N.J., Jan. 1993.
78. C. Levy, O. Cuccira, and D. Cheleck, *NASA accession #N65-15250*, report #AD609721, Waltham, Mass., 1964.
79. *Environ. Sci. Tech.*, 1078 (Nov. 1976).
80. Brit. Pat. 1,057,984 (Feb. 8, 1967) (to Mine Safety Appliances Co.).
81. K. R. Williams and S. H. Jacobi, *Atmos. Environ.*, **12**, 2509 (1978).
82. D. Grosjean, *Environ. Sci. Technol.* **24**(1), 77 (1990).
83. O. Steinhauser, *Mitt. Ver. Grosskesselbesitzer* **48**(2), 143 (1968).
84. J. G. Cummings and K. T. Zee, *J. Assoc. Anal. Chem.* **50**, 1262 (1967).
85. J. Ditz, J. Dvorak, and J. Marecek, *Chem. Prum.* **15**(11), 677 (1965).
86. M. Zygmunt, *Chem. Anal. Warsaw* **8**(6), 849 (1963).
87. *Environ. Sci. Tech.*, 138 (Feb. 1979).
88. O. C. Taylor, *Technical Report*, TR-76-125, Aerospace Medical Research Lab, Wright-Patterson AFB, Ohio, 1976.
89. *A Survey of Potential Chlorine Production Process*, contract 31-109-38-4211, Rept. # ANL-OEPM-79-1, Versar, Inc. Argonne National Laboratory, Argonne, Ill., Apr. 1979.
90. G. Busch, *Chem. Mark. Rep.*, 19 (Jan. 18, 1993).
91. G. Busch, *Chem. Mark. Rep.*, 3 (Mar. 9, 1992).

MOHAMED W. M. HISHAM
TILAK V. BOMMARAJU
Occidental Chemical Corporation

HYDROGEN ENERGY

As fossil fuel reserves diminish and the negative environmental impact of the fossil fuel-based economy becomes increasingly apparent, more and more worldwide attention is being turned toward clean energy sources such as that derivable from hydrogen (qv). Fossil fuels provided about 80% of total world energy demand in 1990 (1), and if oil and natural gas usage continues at the mid-1990s rate, known reserves are not expected to last into the year 2030 (2) (see FUEL

RESERVES; GAS, NATURAL; PETROLEUM). Coal (qv) reserves are much larger, but of all the fossil fuels coal is considered to be the least environmentally benign (3).

The burning of fossil fuels emits various pollutants into the atmosphere, ie, carbon dioxide (qv); nitric and nitrous oxides, NO_x; as well as sulfur dioxide from sulfur-containing fuels such as coal; and products of incomplete combustion including carbon monoxide (qv), unburned hydrocarbons (qv), and particulates (see AIR POLLUTION). Hydrogen, however, when combusted, produces mainly water. Although in early hydrogen internal combustion engine conversions significant quantities of NO_x were measured in vehicle emissions owing to the high combustion chamber temperature, specialized processes, such as water induction, eliminate virtually all of the nitrogen-containing by-products. Therefore, whether hydrogen is used to provide energy for combustion or chemical reaction, as in the case of the hydrogen fuel cell, the process is essentially pollution-free (see FUEL CELLS).

Hydrogen, the most abundant element in the universe, is not normally found in its unreacted state in nature; it is almost always necessary to synthetically produce the hydrogen to be utilized. Hydrogen is made by splitting water into hydrogen and oxygen, a process that utilizes energy; therefore hydrogen is not usually considered a source of energy, but rather a form of stored energy. The energy to produce hydrogen from water can come from a wide variety of sources including solar energy (qv), wind, hydroelectric, nuclear, natural gas reformation, or coal gasification. The hydrogen produced can then be stored, transported, and utilized in conventional energy consumption devices such as automobiles and gas appliances.

When hydrogen is oxidized, the formation of water and the release of nearly as much energy as was required to synthesize the hydrogen results. The fraction of the energy liberated relative to that initially required represents the utilization efficiency or efficiency of the conversion cycle. In many cases, hydrogen's chemical properties make much higher utilization efficiencies possible than for conventional fuels. Thus it is often possible to accomplish more useful work by converting energy to hydrogen rather than utilizing the primary fuel source.

One of the advantages of hydrogen energy over conventional energy sources is that hydrogen offers a convenient method for capturing and storing energy supplied by renewable sources (see RENEWABLE ENERGY RESOURCES). Solar, wind, and geothermal power (see GEOTHERMAL ENERGY) can be harnessed and collected for times of demand by using the renewable energy source and water to produce hydrogen, which can then be combusted or chemically combined with oxygen, producing the pollution-free required energy.

Hydrogen Fuel

Properties. When compared to other common fuels on a per unit mass basis, hydrogen is found to have a considerably higher heating value than many of the hydrocarbon fuels in common usage. For example, hydrogen's heating value is 120 kJ/g (28.7 kcal/g) compared to 50 kJ/g (12 kcal/g) for natural gas and 45 kJ/g (11 kcal/g) for gasoline (see GASOLINE AND OTHER MOTOR FUELS) (4). This makes hydrogen especially attractive for use where fuel weight is of considerable importance, such as in aviation or space applications. Although hydrogen offers a much higher heating value per unit weight, it is also substantially more voluminous than hydrocarbon fuels; 3.4 m³ of hydrogen contains the same amount of energy as 1 m³ of methane (Table 1). The energy required to ignite hydrogen is an order of magnitude less than that required to ignite hydrocarbon fuels, but that energy must be supplied at a higher temperature than that required for most hydrocarbon fuels.

Production. Although hydrogen has been estimated to make up more than 75% of the mass of the universe, less than 1 ppm by volume exists in its free state in the atmosphere. On earth, hydrogen occurs primarily combined with oxygen in water, but is also found in all organic matter, including living plants, petroleum, natural gas, coal, etc. Proposed methods for hydrogen production range from natural gas reformation to photosynthetic production utilizing blue-green algae.

Natural Gas Reformation. In the United States, most hydrogen is presently produced by natural gas reformation or methane–steam reforming. In this process, methane mixed with steam is typically passed over a nickel oxide catalyst at an elevated temperature. The reforming reaction is highly endothermic and is usually performed in a fire tube reformer where the catalyst is loaded in the tubes. The methane–steam reforming reaction is

$$CH_4 + H_2O \longrightarrow CO + 3\,H_2 \tag{1}$$

This reaction is affected by the steam-to-carbon ratio, temperature, and pressure, as well as catalyst activity.

Table 1. Combustion Properties for Common Hydrocarbon Fuels[a]

Fuel	Quantity of fuel[b] per GJ[c]	Flammability limit in air, vol % gas		Maximum flame speed, cm/s	Spontaneous ignition temperature, °C	Ignition energy,[d] mJ[c]
		Lower	Higher			
hydrogen, m³	98[e]	4	75	265	571	0.02
methane, m³	29[e]	4.	16	34	632	0.47
propane, m³	11[e]	2	11	40	504	0.31
butane, m³	9[e]	2	10	37	431	0.76
methanol, kg	47[e]	6	50	49	470	
gasoline, kg	22	0.8	6	34	447	1.35
diesel (heptane), kg	22	1	8	40	247	0.70
coal, kg[f]						
anthracite	30					
bituminous	36					
lignite	58					

[a]Ref. 4.
[b]Lower heat value fuel.
[c]To convert J to cal, divide by 4.184.
[d]Ref. 5.
[e]Ref. 6.
[f]Ref. 7.

After the steam reforming step, the product gas stream contains a considerable quantity of carbon monoxide which may undergo reaction with additional steam, thereby increasing the hydrogen yield:

$$CO + H_2O \longrightarrow CO_2 + H_2 \qquad (2)$$

This reaction is commonly known as the CO or water gas shift reaction. The conversion of CO by this reaction is slightly exothermic and favored by lower temperatures. The lower practical operating temperature range for this reaction is between 180 and 200°C (8).

Coal Gasification. Coal gasification is the second most common method used for producing hydrogen and its chemistry is similar to natural gas reformation. Steam reacts with carbon in coal, thereby forming carbon monoxide and hydrogen. This low energy–density gas is subsequently sent through a CO shift reaction as used in the methane–steam reforming process (9). The resulting products are hydrogen, carbon dioxide, and ash. The basic reactions for coal gasification are

$$C + H_2O \longrightarrow CO + H_2 \qquad (3)$$
$$CO + H_2 + H_2O \longrightarrow 2\,H_2 + CO_2 \qquad (4)$$

In this process, any sulfur present in the coal exits the gasifier as hydrogen sulfide which is removed by various processes such as a Holmes-Stretford unit where the sulfide is absorbed and regenerated. The resulting sulfur is filtered out as a cake (39 wt %) which is sold as a valuable feedstock (see COAL CONVERSION PROCESSES, GASIFICATION; SULFUR REMOVAL AND RECOVERY).

Electrolysis. By utilizing water and electrical energy, hydrogen can be produced with no negative environmental effects by electrolysis. When two electrodes are immersed in an aqueous ion-conducting electrolyte in the presence of a catalyst and electricity is applied across the cell, water is decomposed to produce hydrogen from the negatively charged cathode and oxygen from the positively charged anode (see ELECTROCHEMICAL PROCESSING).

Various techniques and schemes are used to increase the efficiency of this conversion process. The higher the voltage, the faster the hydrogen and oxygen evolution, but the greater the quantity of wasted heat. The theoretical voltage for the decomposition of water at 25°C is 1.23 V. Typical commercial electrolysis equipment has an efficiency ranging between 40 and 80%. Early electrolysis systems typically used potassium hydroxide as the electrolyte. The equipment was heavy, bulky, and required extensive maintenance. Additional water had to be continually added and electrolyte replacement was frequently required as impurities began to accumulate.

Since the early 1970s, the solid polymer electrolyte (SPE) has begun replacing potassium hydroxide in electrolysis systems. Electrodes coming into contact with the water-saturated membrane encounter an acidic environment equivalent to that of a sulfuric acid solution (10 wt %), although only pure water is actually circulated through the system. The membrane is proton-selective, allowing only hydrogen ions to pass through. On the anode side of the cell, the electric current pulls a hydrogen ion away from a water molecule and then transports it through the membrane; on the cathode side, the hydrogen ion receives an electron, thereby evolving hydrogen.

Electrolysis systems based on solid polymer electrolyte technology offer many advantages over the older technology. These systems are compact and lightweight, and because the electrolyte is a solid membrane material and not a liquid, it does not wash away. In an SPE electrolyzer, a small amount of water is routinely drained from the cell, removing any impurities. Because of the high (60–80%) efficiency, SPE electrolyzers are ideal for converting renewable energy resources into hydrogen.

Biological Production of Hydrogen.   Hydrogen production by photosynthetic microorganisms is another method for capturing and accumulating energy. Blue-green algae, also known as cyanobacteria, as well as other anaerobic bacteria are capable of splitting water into hydrogen and oxygen by a light-driven process known as biophotolysis (10). Cyanobacteria typically possess three hydrogen-metabolizing enzymes: nitrogenase, membrane-bound uptake hydrogenase, and soluble hydrogenase. In general, the capacity to photoproduce hydrogen varies widely from one species to another depending on the levels of nitrogenase and hydrogenase in the organism. When incubated under anaerobic conditions in the presence of 1% carbon monoxide, as much H_2 as 1400 nmol/(mg·h) have been reported for the *Anabaena* strain CA (11).

Transporting.   Small volumes of gaseous hydrogen are typically transported in high pressure containers via truck, train, or barge. Transmission of large quantities of gas using these methods, however, is impractical. The preferred method for transporting large volumes of hydrogen efficiently is through an underground pipeline system similar to existing natural gas networks (12) (see PIPELINES).

Economic Aspects

Although hydrogen can be produced utilizing water and renewable energy, as of this writing the majority of hydrogen is synthesized from fossil fuels, eg, by methane reformation and coal gasification processes. As natural gas and petroleum reserves become depleted, a shift from a liquid fossil fuel base to cleaner burning fuels is expected. At such time the cost of remaining liquid fossil fuels would be expected to increase, and fuel would have to be acquired from renewable energy sources or from coal gasification (13).

The use and effective costs of various energy alternatives are shown in Table 2. Use or internal costs include production, transportation, and distribution. Effective costs take into account the use costs; estimated external costs, which include costs associated with damage to the environment caused by utilization of various fossil fuels; and fuel utilization efficiencies, ie, the efficiency of converting fuels into mechanical, electrical, or thermal energy. The effective costs are expressed as $/GJ of fossil fuel equivalent (15). The overall equation for the effective cost is

$$C_{\text{eff}} = (C_{\text{in}} + C_{\text{ex}})(\eta_F/\eta_S) \qquad (5)$$

Table 2. Projected Synthetic and Fossil Fuel Prices, $/GJ[a,b]

Fuel	Use cost	Effective cost[c]
	Gaseous fuel	
hydrogen		
coal	12.08	20.86
		10.17[d]
hydro	15.01	12.31
		6.00[d]
solar	18.94	15.53
		7.58[d]
synthetic natural gas	11.45	24.81
natural gas	7.04	
	Liquid fuel	
hydrogen		
coal	14.23	25.36
hydro	17.89	14.67
solar	22.80	18.70
synthetic gasoline	19.85	34.97
synthetic jet fuel	15.65	30.77
fuel oil	9.14	
gasoline	11.23	21.40
jet fuel	7.42	17.59

[a]To convert J to cal, divide by 4.184.
[b]Estimates for the year 2000 in 1990 $U.S. (14).
[c]In internal combustion engine, unless otherwise noted.
[d]Fuel cells.

where C_{eff} = effective cost, C_{in} = internal cost, C_{ex} = external cost, and η_F/η_S = utilization efficiency.

In the fuel cell hydrogen is used two to three times as efficiently as in an internal combustion engine. Hence, when utilized in a fuel cell, hydrogen can cost two to three times that of more conventional fossil fuels and still be competitively priced, ie, as of this writing the market price for hydrogen when used in a fuel cell and produced by electrolysis is competitively priced with gasoline.

Hydrogen Storage

Numerous methods of storing hydrogen are used including compressed gaseous hydrogen, liquid hydrogen, and metal hydrides. Most commercially viable hydrogen energy applications employ one or more of these storage technologies. Other hydrogen storage methods have been proposed, including solid hydrogen (slush) and microspheres, but as yet that technology has not proven feasible (16,17).

Compressed Gaseous Hydrogen.   Gaseous hydrogen is the most economical of the practical hydrogen storage technologies; it is also the heaviest and most voluminous. Compressed gas storage is a good method for stationary hydrogen storage installations, but it is not a practical choice for vehicle storage systems.

Compressed hydrogen (640.8 kg/GJ (1490 lb/10^6 Btu)) weighs almost 20 times more than gasoline (30.7 kg/GJ), and takes up nearly 15 times the space at 0.407 and 0.026 m^3/GJ (15 and 0.96 lb/10^6 Btu), respectively. The very high (13.8 MPa (2000 psi)) pressures of compressed hydrogen could be dangerous in the event of a malfunction or a collision.

Liquid Hydrogen. The preferred method of storing hydrogen for space and aircraft applications where light weight is the primary consideration is as a liquid. Hydrogen gas is transformed into a liquid by cooling it to 20.3 K. This process is energy intensive and requires specialized equipment. Consequently, liquid hydrogen is considerably more expensive than gaseous hydrogen.

Liquid hydrogen is stored in specially designed dewar flasks. Any heat leaking into the liquid hydrogen causes a proportionate amount of the liquid to flash or vaporize into the gaseous form, thus super insulation, employed in cryogenic storage containers, is used (see CRYOGENICS). The common method of making superinsulated containers involves a lightweight, thin-walled inner vessel, which actually holds the liquid, surrounded by a jacket consisting of alternating layers of reflective Mylar and an insulating layer of foam (see INSULATION, THERMAL). The Mylar and foam layers are encapsulated in a vacuum to further minimize heat conduction (see VACUUM TECHNOLOGY). The entire vessel is then enclosed in an outer vessel which makes the maintenance of the vacuum possible.

A problem associated with the design of a cryogenic hydrogen vessel is the method of supporting the inner vessel without substantial vaporization. Various technologies have been developed and are commercially in use. In a well-designed cryogenic storage container, a boil-off rate of less than one-half of one percent is achieved; ie, the inner vessel is so well insulated that the amount of heat penetrating the insulation is only sufficient to vaporize one-half of one percent of the hydrogen every 24 hours. Such a tank would be able to store liquid hydrogen for a maximum of 200 days. After that time, all of the liquid would have been transformed back into the gaseous state. Meanwhile, as the hydrogen changes back into the gaseous form it becomes voluminous and must be vented, or the pressure inside the liquid hydrogen tank begins to increase. For use in a liquid hydrogen vehicle, some hydrogen would need to be used almost every day. Alternatively, the vaporized hydrogen could be run through a reliquefaction machine and put back into the tank; this approach, however, is more practical for larger-scale applications.

When a liquid hydrogen tank is filled initially or is refilled after sitting for some time, the inner vessel must be cooled to liquid hydrogen temperature by pumping liquid hydrogen into the inner vessel and allowing it to flash-off. The hydrogen that is vaporized into gaseous form by the latent heat of the inner containment vessel (flash-off) must be considered when refueling liquid hydrogen containers. The amount of flash-off is usually equal to only 1 or 2% of the tank capacity and flash-off only occurs when a liquid hydrogen tank is completely empty and has been empty long enough for the inner vessel to warm.

To prevent hydrogen from vaporizing during the refueling process, it is transported in special vacuum-jacketed transfer lines. If liquid hydrogen is transported in fuel lines that are not vacuum-jacketed, oxygen and nitrogen from the air that comes in contact with the lines are cooled to below their liquefaction temperatures, and as a result the liquid nitrogen and liquid oxygen drip off the line. Liquefied oxygen can be dangerous, causing severe cold burns when brought into contact with the skin. Furthermore, the density of oxygen in liquid air can cause combustible materials to oxidize at more rapid rates than normal.

Metal Hydrides. Certain metal alloys form reversible metal hydrides in the presence of hydrogen (18) (see HYDRIDES). It has been demonstrated that metal hydrides are capable of safely storing hydrogen in a powdered form (19). These hydrides have an improved weight density as compared to compressed hydrogen gas and a volume density similar to liquid hydrogen. Metal hydride technology is considered the safest method for storing hydrogen on board a vehicle (19,20).

The properties of metal hydrides vary greatly from metal to metal. The hydride reaction becomes especially complex when two or more metals, each having different hydriding properties, are mixed together into an alloy of varying percentages. Over 200 alloys have been characterized and studied to determine applicability to an automotive storage system (21–23). Some alloys were found to be extremely heavy, storing only small quantities of hydrogen; others were extremely expensive and therefore impractical; and a third category of metal hydride alloys was classified as dangerous. For example, mischmetal hydride is pyrophoric, spontaneously igniting if brought into contact with air. In 1974 the first successful vehicle hydride storage vessel was installed by Billings Research Corp. (24). This vessel used the iron–titanium family of metal hydrides. Refinements to the system continue to be made.

One of the problems with early hydride systems was decrepitation of the alloy. Each time the metal hydride storage tank was recharged the particles would break down and eventually the particles became so small that they began to pass through the 5-μm sintered metal filter which kept the hydride inside the tank. Addition of one-half of one percent manganese, which caused the decrepitation process to cease once the particles reached a size of about 10 μm, solved this problem.

One of the principal advantages of hydrides for hydrogen storage is safety (25). As part of a study to determine the safety of the iron–titanium–manganese metal hydride storage system, tests were conducted in conjunction with the U.S. Army (26). These tests simulated the worst possible conditions resulting from a serious collision and demonstrated that the metal hydride vessels do not explode.

Another important attribute of metal hydride storage systems is longevity. Metal hydride storage containers have been tested by charging and discharging in thousands of cycles, equivalent to millions of kilometers of driving (27,28). During the first 100 cycles, the hydride actually improves with each cycle. After the first 100 cycles, the improvement phenomenon is no longer observed, but there is no degradation in the performance of the tank even after thousands of cycles. For this reason, used metal hydride storage vessels could be salvaged from old vehicles, reactivated, and used again.

Although metal hydride storage systems do not degrade, hydrides can be poisoned by impurities in the hydrogen, including carbon monoxide, oxygen, or even water vapor. However it is easy to remove these impurities from a hydrogen gas stream even on a small scale, such as in electrolysis. Electrolysis units, which have been built since the early 1980s, produce hydrogen directly out of the cell at pressures adequate for recharging hydride vessels, and although the electrolyzers work at pressures of up to 4.1 MPa (40 atm), only 1.7 MPa (17 atm) is necessary to recharge a metal hydride storage vessel. Simple yet reliable methods of purifying hydrogen to the hydride grade have been successfully employed in the electrolysis process. These methods include de-oxo catalysts and pressure swing adsorption technologies employing molecular sieves (qv) (29,30).

Chemical properties of metal hydride are presented in Figure 1 (31–33). In a hydride storage container, the higher the temperature, the higher the pressure. Typical hydriding alloys have a pressure plateau, or, in some cases, several pressure plateaus. As the tank is charged, the pressure rises rapidly until it reaches the plateau pressure for that particular temperature. Then, as additional hydrogen is added to the container, the hydrogen begins to be absorbed by the alloy in the vessel, and the pressure does not increase. This phenomenon continues until the reactive sites within the metal alloy are completely occupied. At this point addition of more hydrogen causes the pressure to increase again, indicating that the tank's storage capacity is saturated.

The iron–titanium family of hydrides store hydrogen having weight percents of hydrogen that range between 1.6 and 2.3. Other alloys have much higher weight percents. For example, magnesium hydrides have hydrogen weight percents as high as 7.6. These latter alloys require very high temperatures to release the hydrogen, making them impractical for most automotive storage applications. The energy required to dissociate various metal hydride alloys is shown in Table 3.

Metal Hydride Storage in Mass Transit Vehicles. The first metal hydride transit vehicle to go into service was a passenger bus powered by a Dodge engine put into public transit service in 1975 in Utah (20). To provide the bus with sufficient range, two Fe–Ti hydride vessels were fabricated, each having a total mass of 714.4 kg for a total hydride system mass of 1429 kg. The actual amount of hydride alloy contained inside the stainless steel vessels was 1016 kg, which could store 12.7 kg hydrogen, representing a 1.25 wt % alloy and a 0.9 wt % storage system. The metal hydride performance data is presented in Figure 2.

In a 1978 test, an Argosy bus was retrofitted for hydrogen operation in Riverside, California. In this bus, radiator water was used to supply the heat of dissociation for the metal hydride storage system. The overall weight of the hydride storage system was substantially reduced by substituting aluminum vessels for the stainless steel used in the earlier prototype. The aluminum pressure vessels were enclosed in an exterior water jacket, thereby allowing radiator water to flow over the outer aluminum vessel, providing heat to the metal hydride bed. The cooling water from the engine provided the heat for metal hydride dissociation; the metal hydride dissociation provided the cooling for the engine. To recharge a metal hydride vessel it is necessary to supply hydrogen to the vessel while dissipating the heat given off by the exothermic reaction. In the Riverside Bus Project, the recharge waste heat was dissipated by connecting an external heat exchanger with two quick connects (Fig. 3). After the lines were connected to the bus, a coolant circulation pump caused the fluid to circulate in a loop through the metal hydride vessel and then through an external finned-tubing-to-air heat exchanger.

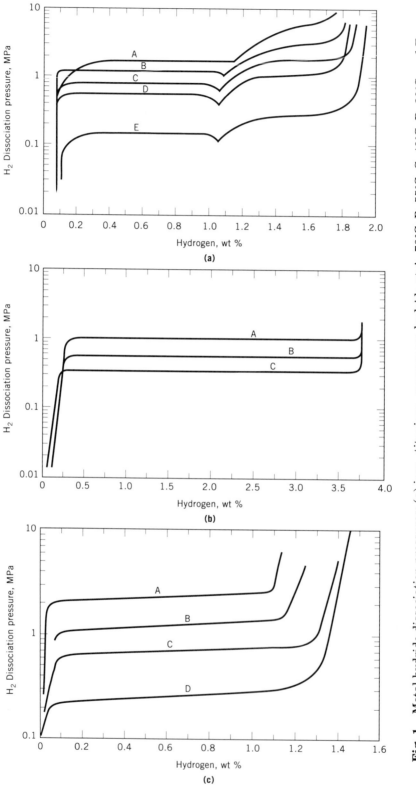

Fig. 1. Metal hydride dissociation pressures: (**a**) iron–titanium–manganese hydride at A, 70°C; B, 55°C; C, 40°C; D, 30°C; and E, 0°C; (**b**) magnesium–nickel hydride at A, 349°C; B, 322°C; and C, 298°C; and (**c**) lanthanum–nickel hydride at A, 81°C; B, 66°C; C, 52°C; and D, 21°C. To convert MPa to psi, multiply by 145 (31–33).

Table 3. Heat of Formation of Metal Hydrides[a]

Material	Hydrogen, wt %	Heat of formation, kJ/g[b,c]
FeTi[d]	1.9	−14.2
Mg$_2$Ni	3.6	−32.2
Mg	7.6	−38.9
LaNi$_5$	1.5	−15.1

[a]Ref. 28.
[b]To convert J to cal, divide by 4.184.
[c]Based on grams of hydrogen gas.
[d]Contains a trace amount of Mn.

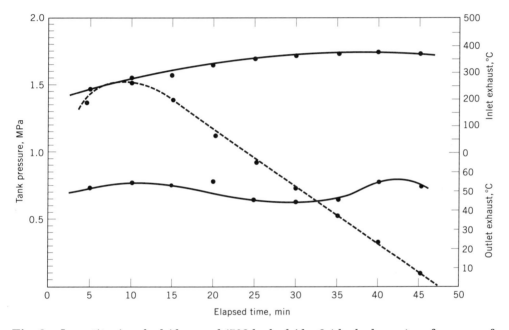

Fig. 2. Iron–titanium hydride vessel (508 kg hydride, 6.4 kg hydrogen) performance of the Provo-Orem (Utah) bus at 3.4 MPa (493 psi) charge pressure, having 22 cylinders, each with a 75 mm dia and 1750 mm length, where (---) indicates tank pressure. To convert MPa to psi, multiply by 145. See text.

Hydrogen Energy Applications

Hydrogen Internal Combustion Engine. Standard gasoline and diesel-powered internal combustion engine vehicles can be converted to run on hydrogen. These vehicles have often been found considerably less polluting, safer, and more efficient than their fossil fuel-burning counterparts.

Backfire. Early hydrogen engine conversions were typically affected by an unexpected occurrence known as backfire or backflash. Backfire, the occurrence of one or the other (or sometimes both) of two distinguishable phenomena, occurs when an explosive mixture of fuel and air builds up in the exhaust system of an engine and is then ignited. The explosion comes out the tailpipe of the vehicle. Backfire is common in poorly tuned engines that run on gasoline and is fairly

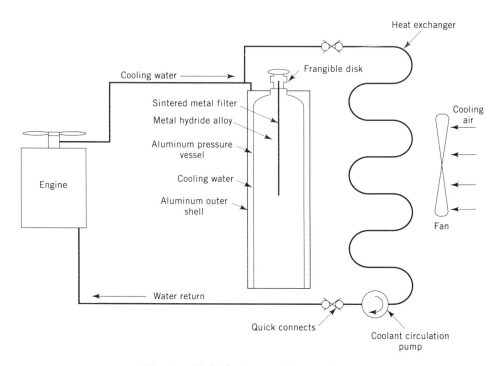

Fig. 3. Hydride heat-exchanger design.

rare in hydrogen engines. A second type of backfire, more properly referred to as backflash, occurs when a combustible mixture of hydrogen and air in the intake system of the engine is ignited. In this case the explosion comes out of the carburetor of the vehicle, often damaging the equipment. Backflash is a common problem in hydrogen engines.

In a high performance, gasoline-burning engine, fuel is ignited at the appropriate time in four, six, or eight alternative cylinders, usually by means of a high voltage ignition system directed to the cylinder ready to be fired by a distributor cap and a rotor. When the ignition system fires a spark in the proper cylinder, simultaneously a much smaller induced spark is created in other cylinders. These induced sparks are usually caused by electromagnetic interference between the cables running from the distributor cap to each individual spark plug. The induced sparks are sometimes present in a cylinder during the intake stroke and come in direct contact with a flammable mixture of fuel and air. In a conventional gasoline engine, induced sparks are too weak to ignite the gasoline/air mixture. In the hydrogen engine, however, because less than one-tenth as much energy is required for ignition, the hydrogen/air mixture ignites readily. If the cylinder is in the intake cycle, the flame burns up through the intake valve and into the intake manifold, causing backflash (Fig. 4).

Another significant cause of backflash in the hydrogen engine is from pre-ignition sources. When the intake filter is removed and a small quantity of very fine dust particles is introduced into the engine intake, backflash occurs. This procedure creates no extraordinary results if the engine is operating on gasoline, but when the engine is operating on hydrogen, the result is an almost immediate

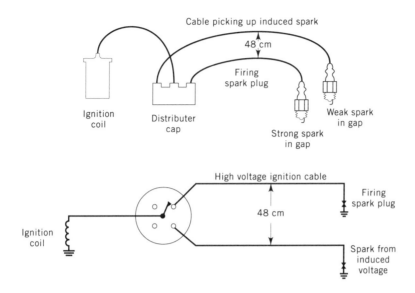

Fig. 4. Induced ignition in hydrogen engines.

backflash. The particles are drawn into the combustion chamber during the intake stroke. The fuel is then compressed, and the charge is ignited, resulting in the power stroke. During the exhaust stroke, the exhaust valve opens and the burnt charge, along with most of the particles, is pushed out into the exhaust system. The cycle repeats itself with the intake valve opening and a new intake stroke beginning. At this point, however, some of the dust particles, which are still hot from the last cycle, remain in the combustion chamber. As the hydrogen fuel comes into contact with these hot particles, the incoming fuel charge is ignited and burns quickly back into the intake manifold, resulting in detonation. In the case of the hydrocarbon engine, these small particles do not have sufficient thermal mass to ignite the hydrocarbon fuel.

Other sources of pre-ignition backflash are carbon buildup within the combustion chamber, caused by previous operation of the engine on hydrocarbon fuels, and rusting spark plugs. If carbon buildup comes to a sufficiently high temperature during engine operation, it can ignite the incoming charge of hydrogen and air during the intake stroke, causing backflash. In the same way, iron oxide on the spark plugs is an excellent catalyst for the ignition of hydrogen at high temperatures. In this case, the incoming hydrogen–air charge comes into contact with the hot spark plug which provides a source of catalytic ignition and the resulting backflash. To prevent these problems, several things can be done. In the case of carbon buildup, operating an engine on hydrogen quickly burns out any carbon deposits left over from gasoline operation, and the problem essentially takes care of itself. For the iron oxide source of backflash, the best solution is to exchange conventional spark plugs for plugs made of noncorrosive material such as stainless steel.

Water Induction.   One method of controlling backflash, and a host of other problems associated with hydrogen engine conversions, is water induction (34).

The technology employed consists of inducting tiny droplets of liquid water into the combustion chamber with the air–fuel charge. When the charge is compressed and ignited, the droplets of water become involved in the ensuing chemical reaction. For example, a charge inside a combustion chamber is ignited by the spark plug and begins to propagate through the combustion chamber. Outside the arc of flame, the air–fuel mixture is filled by a mist of water. As the flame front encounters these droplets, it instantly transforms them from liquid water into water vapor. In the process, a significant amount of heat energy is consumed, and because the peak combustion temperature is substantially reduced, autoignition sources are not heated to a high enough temperature to cause the autoignition phenomenon to take place. For this reason, water induction virtually eliminates backflash from this source.

Water induction is also effective in reducing the formation of nitric oxide, NO (35), which results when the nitrogen and oxygen in the air are heated to high temperatures. The formation of nitric oxide in an uncontrolled hydrogen engine is presented in Figure 5a as a function of hydrogen equivalence ratio. If a water induction system is installed on the engine and adjusted to provide sufficient droplets of water so as to maintain the peak combustion temperature below the nitric oxide formation threshold of 700°C, this reaction is virtually eliminated, as shown in Figure 5b. As a result, the nitric oxide pollution coming from a hydrogen engine with water induction properly installed is one part per million or 0.01 g/km.

The water induction technique of hydrogen engine conversion has other benefits, such as an increase in engine brake thermal efficiency as water induction is applied. There is also a corresponding increase in power output (see Fig. 5) that occurs as the propagating flame encounters the water droplets, causing them to vaporize into steam and expand dramatically. Although the vaporizing of the water droplets reduces the peak combustion temperature, the amount of energy consumed as the water undergoes a phase change generates a greater increase in pressure than would result if that same energy were utilized to heat and expand the air and combustion products normally encountered in the combustion chamber. It is the pressure of the gases that forces the piston down and not the temperature. Therefore, if a high pressure can be achieved inside the combustion chamber without high temperature, the hydrogen engine has increased power and no backflash.

There are some problems with water induction, however. In order to be effective, the water droplets must be uniformly dispersed throughout the combustion chamber. Any pockets of combustible product within the chamber devoid of the water droplets generate high combustion temperatures, nitric oxide, and possibly even cause backflash. Figure 6 shows a typical intake manifold system having a water-atomizing spray nozzle on the right side of the common air inlet. The stream of incoming air sweeps the water droplets into the intake manifold. In this configuration the majority of them end up in the fourth cylinder. A similar nozzle installed over the inlet to each cylinder results in droplets evenly distributed between each of the cylinders, but test results indicate that there are pockets within the cylinders devoid of the water droplets. Nitric oxide concentrations average 100 to 500 times higher than in better designed systems where water droplets are evenly distributed within each of the cylinders.

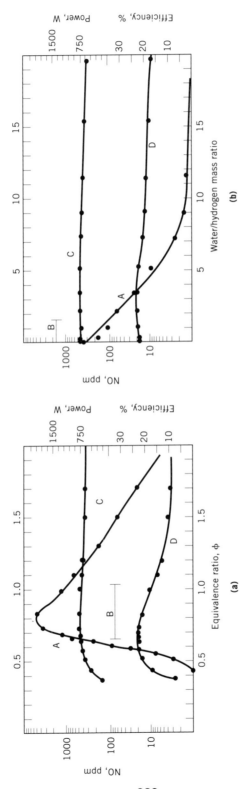

Fig. 5. NO formation in a hydrogen engine having spark at 17° before top-dead center (BTC); rpm, 2900; and compression ratio, 5.5:1, where A is nitric oxide; B, backfire; C, power; and D, brake thermal efficiency. (**a**) Effect of equivalence ratio, ϕ; and (**b**), effect of water induction at $\phi = 0.625$.

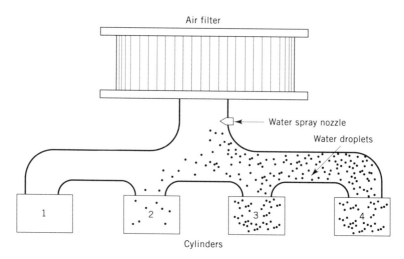

Fig. 6. Intake manifold system with a water-atomizing spray nozzle distributing to four individual cylinders. See text.

A specially designed water induction system was used in the Provo-Orem bus to increase the water induction mass ratio when operating at or near full power setting. Engine performance data as a function of the equivalence ratio and water injection mass ratio are shown in Figure 7.

Direct Cylinder Induction. One of the disadvantages of converting an internal combustion (IC) engine to hydrogen is the resulting loss of 25–35% power from the engine. This loss of power is not an intrinsic characteristic of hydrogen engines. To the contrary, hydrogen, because of its higher flame speed, actually has the potential of producing more power than conventional gasoline fuels. This loss of power phenomenon may be understood by examining the operation of the internal combustion engine. During the intake stroke, a mixture of fuel and air is drawn into the combustion chamber. The carburetion system on the engine adjusts the ratio of fuel-to-air for proper and complete combustion. In the case of gasoline, the droplets of fuel occupy such a small area that during the intake stroke the droplets constitute <1% of the total volume of fuel and air drawn into the combustion chamber.

The case with hydrogen is very different. If the carburetor is set to mix hydrogen and air at 0.8 stoichiometry, then 25% of the volume of gas in the intake manifold is hydrogen and 75% is air. In a cylinder having 1000 mL displacement, 990 mL of air would be drawn into the chamber during the intake stroke if the engine were operated on gasoline. Conversion of the same engine to hydrogen would mean 750 mL of air drawn into the cylinder during intake. The rest of the combustion chamber would already be filled with the hydrogen. Because one-fourth of the air found in an IC engine is displaced by the hydrogen, there is not as much oxygen available to react with the fuel, and the amount of power is proportionately reduced by one-fourth. Thus many hydrogen engines are less powerful than the gasoline counterparts because not as much air fits in the engine during the intake cycle. The remedy to this problem is a technique known as direct cylinder injection (DCI).

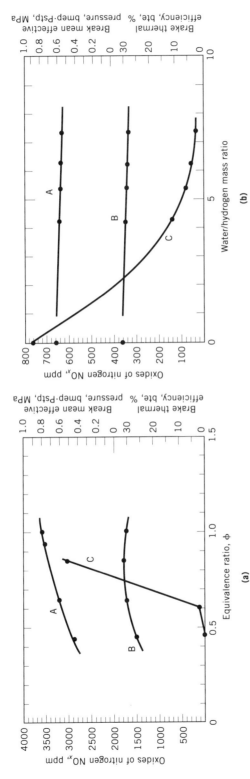

Fig. 7. NO$_x$ formation for the Provo-Orem bus run at a compression ratio of 12:1 at 30°C, 3000 rpm, where A is brake mean effective pressure; B, brake thermal efficiency; and C, oxides of nitrogen. (**a**) Effect of equivalence ratio, ϕ, at a water/H$_2$ mass ratio of 6.0 and spark = 17° before top-dead center (BTC) and (**b**), effect of water injection where ϕ = 0.60 and spark = 14°BTC. To convert MPa to psi, multiply by 14.

In a DCI system there is no throttle limiting the amount of gases going into the intake port of the engine and no fuel premixed with air. During the intake stroke, the engine draws in a full charge (1000 mL) of air. After the chamber is filled with air, while the intake valve is closing and at the beginning of and during the compression stroke, hydrogen is injected into the combustion chamber under pressure. The pressurized hydrogen entering the cylinder increases the pressure over that drawn in naturally by the engine, which results not only in equivalent, but even enhanced power output as compared to operation using gasoline.

There are many benefits of a DCI system. There is no throttle on a DCI engine, thus the pumping loss at part throttle settings is eliminated. In fact, this engine has the main efficiency advantage of a diesel engine. Furthermore, because the hydrogen is injected into the combustion chamber during the intake stroke, or in some cases at the end of the compression stroke, the gases in the chamber do not have an opportunity to mix completely. As a result, the charge is stratified, resulting in a much slower flame speed and consequently a lower peak combustion temperature. In this way the tendency of the engine to detonate is greatly diminished and nitric oxide formation is reduced. Another advantage of DCI is its effect on backflash. When the intake valve opens in a DCI system, a cool charge of pure air is drawn into the combustion chamber. This cool air quenches any residual flame, hot catalytic particles, or any other source of ignition within the combustion chamber, all before the hydrogen fuel is introduced into the system.

Problems associated with the implementation of a direct cylinder injection system are (1) the injection of the hydrogen gas must be carefully synchronized or controlled to be in perfect phase with the internal combustion engine, which is accomplished by placing a position sensor on the crankshaft. The sensor is used to electronically trigger a computer control system, or, alternatively, to mechanically actuate an injector system; and (2) the injection of the large volume of hydrogen fuel required for each combustion cycle. An injector having a substantial orifice size and operating fast enough to keep up with the engine at high speeds is needed. Newer engines have two intake valves: one for hydrogen, the other for air.

Improved Vehicle Performance.   One of the advantages of operating engines on hydrogen, as compared to hydrocarbon fuels, is the higher utilization efficiency that can be achieved. In a testing program sponsored by the United States Postal Service, two identical vehicles were operated on a rural mail route (36). One vehicle had a conventional gasoline engine, the other was retrofitted for hydrogen operation. The engine conversion method employed was water injection and hydrogen was stored in an iron–titanium–manganese hydride. Miles traveled and fuel consumed was monitored for each vehicle daily to determine the efficiency of a hydrogen engine in a mail delivery application. The hydrogen-fueled vehicle proved more efficient than the gasoline counterpart as can be seen in Figure 8.

Hydrogen in Aviation.   Hydrogen is of considerable interest in aviation because of its extremely high energy content per unit mass; it is the lightest fuel available and has the potential to increase aircraft payload greatly (37). Another advantage of using liquid hydrogen in aircraft is the airframe cooling which such

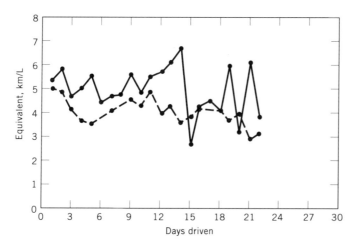

Fig. 8. Vehicle fuel efficiency (average) where (—) is hydrogen (4.77 km/L) and (---) is gasoline (3.93 km/L).

a system offers. Titanium and other heavy alloys are required for supersonic aircraft frames because these are the only materials having the strength needed at the high temperature to which the exterior surface of supersonic aircraft are exposed (see HIGH TEMPERATURE ALLOYS). If cooling from liquid hydrogen were available, lightweight alloys such as aluminum and boron could be used, further increasing range and payload (38).

The Hydrogen Homestead. Hydrogen is an ideal fuel for a wide variety of residential applications. This versatility was demonstrated in the Hydrogen Homestead Project in the mid–1970s (39). The fuel for Hydrogen Homestead, a home in which the appliances were converted to operate on hydrogen (Fig. 9), was produced by electrolysis based on solid polymer electrolyte technology and was stored in an iron–titanium–manganese metal hydride vessel (Table 4). The vessel was instrumented with temperature probes, micro-expansion measurement transducers, and an internal vessel viewing window. The stainless outer vessel was encapsulated in a 10-cm layer of foam to insulate it from the external environment. Heating and cooling of the vessel was accomplished with an internal heat exchanger. The vessel stored 31 kg hydrogen, a quantity adequate to maintain operation of the Hydrogen Homestead for about 10 days.

Hydrogen Range. A natural gas range was modified for operation on hydrogen which contains approximately one-third the energy per unit volume of natural gas, yet its low viscosity allows the hydrogen to flow through an orifice approximately three times as fast as natural gas. Thus the natural gas orifice sizes on the range were suitable for hydrogen conversion without modification. Hydrogen has such a high flame speed that it is better if it is not premixed with air as is done in the case of natural gas. Therefore the primary air intake is eliminated. The secondary air of the natural gas appliance, which mixes with the fuel and supports combustion after the fuel has passed through the burner head, is more than adequate for hydrogen combustion where there is no problem with incomplete combustion. Even without primary air, however, the hydrogen burns too quickly, resulting in regions of high temperature in which nitrogen

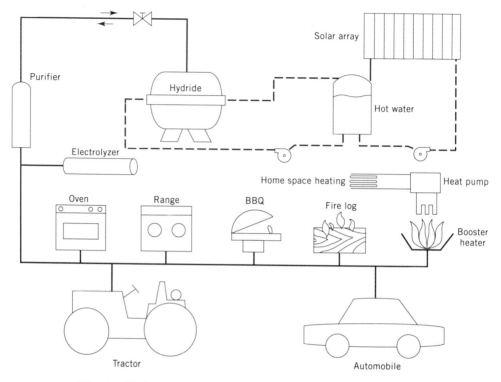

Fig. 9. Hydrogen Homestead energy system (39). See Table 4.

Table 4. Homestead Metal Hydride Vessel[a]

Property	Value
height, cm	123.2
diameter, cm	97.3
wall thickness, cm	2.38
internal volume, m^3	0.5969
service pressure, kPa[b]	3447
test pressure, kPa[b]	6895
hydride composition	51% Ti, 44% Fe, 5% Mn
hydride mass, kg	1791
service temperature, °C	55
pressure excursion, kPa[b]	7–3447
useable quantity, wt %	1.72
stored hydrogen, kg	30.81
stored energy, higher heating value, GJ[c]	4.37

[a]Vessel material is mild steel containing, on a wt % basis, 0.15 C, 1.1 Mn, 0.09 Cr, and 0.005 N.
[b]To convert kPa to psi, multiply by 0.145.
[c]To convert J to cal, divide by 4.184.

and oxygen from the air combine to form nitric oxide. In a conventional range burner, a nitric oxide concentration of 35 ppm was measured when the burner was operated on natural gas. Operation of the burner on hydrogen without primary air resulted in 250 ppm nitric oxide. These concentrations are too high for indoor operation.

A simple solution is available to eliminate this problem. At high temperature, stainless steel is an excellent catalyst for hydrogen combustion. Utilizing this property, a stainless steel catalyst can be carefully fitted around the outside of a burner head to perform three functions: (1) inhibition of the mixing of hydrogen and air, thereby causing a zone immediately around the burner head where the concentration of hydrogen is very high and the concentration of air is very low; (2) catalytic support of the reaction between hydrogen and air at concentrations which are not flammable. Because there is no high temperature zone, nitric oxide formation is virtually eliminated; and (3) production of a visible flame. Normally when hydrogen burns, the flame is invisible. The stainless steel catalyst glows a bright orange, making it apparent that the flame is burning, and at the same time giving some indication of how high the burner is set. Catalytic burners were utilized in virtually all of the natural gas-to-hydrogen conversions in the Hydrogen Homestead. Only 1–5 ppm of nitric oxide were generated by such a burner.

Hydrogen Water Heater and Furnace. A significant advantage of using hydrogen instead of conventional fuels was the increase in utilization efficiency when appliances were converted to hydrogen. One such case was the hydrogen water heater. The stainless steel catalyst was not necessary for this appliance because it was possible to control the mixing of hydrogen and air in such a way that the actual combustion took place on the surface of the water heater's internal heat exchanger. The burner consisted of a piece of stainless steel tubing which sprayed the hydrogen against the metal wall of the heat exchanger (Fig. 10). Because the metal was a good conductor of heat and was constantly being cooled by the water on the other side of the wall, the flame was effectively quenched with virtually no (3 ppm) nitric oxide formation. In the hydrogen water

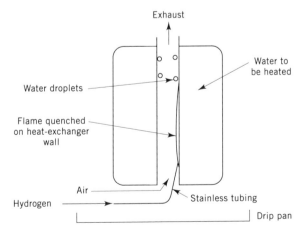

Fig. 10. Hydrogen water heater conversion.

heater, it was possible to cool the steam in the exhaust gases until it condensed on the sides of the internal heat exchanger, draining into a pan underneath. Thus the hydrogen water heater was able to take advantage of the higher heating value of the hydrogen fuel. This advantage was also realized in the conversion of the furnace. When using a conventional furnace during the cold months of the year the humidity inside a home becomes uncomfortably low. Expensive heating systems employ a power humidifier to put additional moisture into the air, thereby increasing the comfort level. Because the only by-product of burning hydrogen is water vapor, it was possible to modify the furnace so that all of the exhaust gases went directly into the house, eliminating the waste of heat going up the chimney. In addition, all of the steam was captured within the living environment of the home, increasing the humidity and lowering the fuel bill.

    The Hydrogen Fuel Cell.   A hydrogen fuel cell is an electrolytic cell in which hydrogen and oxygen combine to form water (40). The resulting energy is liberated as electricity. Unless a blower is used to supply air to the fuel cell, it has no moving parts, makes no noise, has no air pollution, and efficiently utilizes hydrogen fuel. In the fuel cell (Fig. 11a), a solid polymer membrane (41) is located between two electrodes. On the side of the anode, hydrogen gas is supplied to the cell; on the cathode side, oxygen is supplied. The anode reaction is

$$H_2 \longrightarrow 2\,H^+ + 2\,e^- \tag{6}$$

That is, hydrogen dissociates in the presence of the catalyst, forming hydrogen ions and giving up electrons to the anode. The hydrogen ions are transported across the membrane to the cathode. At the cathode, hydrogen ions react with oxygen to form H_2O.

$$2\,H^+ + 2\,e^- + {}^1\!/_2\,O_2 \longrightarrow H_2O \tag{7}$$

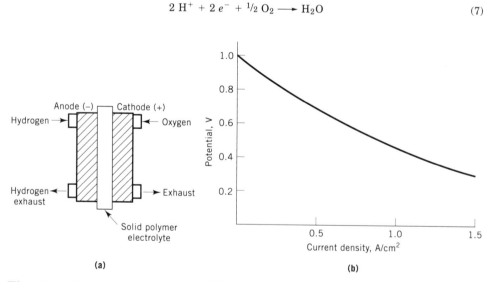

(a)   (b)

Fig. 11. Solid polymer electrolyte (SPE) fuel cell: (**a**) cell design and (**b**) power curve at 25°C.

The electrons, liberated at the anode, travel by electrical cable through the external load, such as an electric motor, to the cathode. If the external circuit is open the reaction is stopped, no fuel is consumed, and no power is generated. The electrolytic reaction, then, is controlled by the load connected to the cell. The overall fuel cell reaction is

$$H_2 + \tfrac{1}{2} O_2 \longrightarrow H_2O \tag{8}$$

Each cell generates a maximum potential of just over one volt. The cells are stacked in series to achieve higher voltages as required.

As can be seen from Figure 11**b**, the output voltage of a fuel cell decreases as the electrical load is increased. The theoretical polarization voltage of 1.23 V/cell (at no load) is not actually realized owing to various losses. Typically, solid polymer electrolyte fuel cells operate at 0.75 V/cell under peak load conditions or at about a 60% efficiency. The efficiency of a fuel cell is a function of such variables as catalyst material, operating temperature, reactant pressure, and current density. At low current densities efficiencies as high as 75% are achievable.

Some fuel cells have the capability of being operated in reverse as an electrolyzer. The dual-role capability of these devices makes them contenders for off-peak electric power storage applications. In the case of the fuel cell car, it is possible to operate the fuel cell in reverse at night by connecting a source of electricity and water to the unit. In this electrolysis mode, hydrogen is generated which is used to recharge the hydrogen storage vessel. The next day, the electric power and water are disconnected, and the unit switches over to the fuel cell mode, generating the electricity to power the vehicle.

Hydrogen Fuel Cell Vehicle Demonstrations. The International Academy of Science (Independence, Missouri) is testing a prototype automobile powered by an SPE fuel cell (42). Hydrogen is stored on board the prototype in an iron–titanium hydride vessel. The pressure of hydrogen gas, evolved from the hydride, is utilized in a diaphragm compressor to compress the air to the cell. Thermoelectric heat pumps were included in the heat-exchanger design to boost heat transfer in either direction should it become necessary during the operation of the vehicle. Using a metal hydride mass of 136 kg, the vehicle has a total capacity of 2.3 kg of hydrogen and a range of 300 km.

Ballard Power Systems, in conjunction with the province of British Columbia and the government of Canada, have converted a diesel bus for Vancouver, B.C. Transit (43). This 9.1-m vehicle is powered by a 105-kW fuel cell. Gaseous hydrogen, stored on board the bus in DOT-approved glass-wound composite cylinders operating at 20.7 MPa (3000 psi), provides the necessary fuel required for the 150-km projected vehicle range.

BIBLIOGRAPHY

"Hydrogen Energy" in *ECT* 3rd ed., Vol. 12, pp. 1015–1037, by J. O'M. Bockris, Texas A&M University.

1. G. R. Davis, *Sci. Am.* **263**, 55–62 (1990).
2. W. Fulkerson, R. J. Judkins, and M. K. Sanghvi, *Sci. Am.* **263**, 129–135 (1990).

3. F. Barbir and T. N. Veziroglu, *Int. J. Hydrogen Energy*, **17**, 301 (1992).
4. R. H. Perry and C. H. Chilton, *Chemical Engineers' Handbook*, 5th ed., McGraw-Hill Book Co., New York, 1973, p. 9-19.
5. E. Obert, *Internal Combustion Engines and Air Pollution*, Intext Educational Publishers, New York, 1973, p. 96.
6. Ref. 4, p. 9-16.
7. Ref. 4, p. 9-4.
8. *Feedstock Purification & Manufacture of Synthesis Gases: Hydrogen, Ammonia, Methanol*, product bulletin, United Catalysts, Inc., Louisville, Ky.
9. R. E. Billings, *Hydrogen From Coal: A Cost Estimation Guidebook*, PennWell Publishing Co., Tulsa, Okla., 1983.
10. G. D. Smith, G. D. Ewart, and W. Tucker, *Int. J. Hydrogen Energy* **17**, 695–698 (1992).
11. D. Kumar and H. D. Kumar, *Int. J. Hydrogen Energy* **17**, 847–852 (1992).
12. D. A. Mathis, *Hydrogen Technology for Energy*, Noyes Data Corp., Park Ridge, N.J., 1976, pp. 24–25.
13. Ref. 3, pp. 299–300.
14. Ref. 3, pp. 299–308.
15. Ref. 3, p. 303.
16. F. V. Grivorev, S. B. Kormer, O. L. Mihailova, A. P. Tolochko, and V. D. Urlin, *ZhETF Science* **294** (1974).
17. R. J. Teitel, "Experimental Studies on Microcavity Hydrogen Storage," *3rd Miami International Conference on Alternative Energy Sources*, Miami Beach, Fla., 1980.
18. K. Sapru, "An Elementary Approach to Designing Metal Hydrides for Practical Applications," *Project Hydrogen '91, Conference*, International Academy of Science, Independence, Mo., 1991.
19. R. L. Woolley and H. M. Simmons, "Hydrogen Storage in Vehicles—An Operational Comparison of Alternative Prototypes," *Society of Automotive Engineers Fuel and Lubricants Meeting*, St. Louis, Mo., 1976.
20. R. E. Billings, *A Hydrogen-Powered Mass Transit System*, Technical Library, International Academy of Science, 1976.
21. G. D. Sandrock, J. J. Reilly, and J. R. Johnson, *Metallurgical Considerations in the Production and Use of FeTi Alloys for Hydrogen Storage*, The International Nickel Company, Inc., Suffern, N.Y., 1976.
22. H. W. Newkirk, *Hydrogen Storage by Binary and Ternary Intermetallics for Energy Applications—A Review*, Lawrence Livermore Laboratory, University of California, Livermore, 1976.
23. D. A. Rohy, J. F. Nachman, A. N. Hammer, and T. E. Duffy, *Automotive Storage of Hydrogen Using Modified Magnesium Hydrides*, final report, U.S. Department of Energy, USDOE Publication Number SAN-1167-1, Washington, D.C., 1979.
24. D. L. Hendriksen and co-workers, "Prototype Hydrogen Automobile Using Metal Hydride," *World Hydrogen Energy Conference*, Miami Beach, Fla., 1976.
25. R. L. Woolley and G. J. Germaine, "Dynamic Tests of Hydrogen-Powered IC Engines," *World Hydrogen Energy Conference*, Miami Beach, Fla., 1976.
26. R. L. Woolley, *Performance of a Hydrogen-Powered Transit Vehicle*, Technical Library, International Academy of Science, Independence, Mo., 1976.
27. R. M. Hartley, ed., Hydride Pioneers, *Hydrogen Progress*, Billings Energy Corp., Provo, Utah, 1976, pp. 7–9.
28. R. E. Billings, *The Hydrogen World View*, International Academy of Science, Independence, Mo., 1991.
29. *G74C/D-0379 Catalysts*, product bulletin, United Catalysts Inc., Louisville, Ky.
30. H. A. Stewart and J. L. Heck, *Hydrogen Purification By Pressure Swing Adsorption*, Union Carbide Corp., Linde Division, New York.

31. J. J. Reilly and R. H. Wiswall, Jr., *Inorg. Chem.* **13**, 218 (1974).
32. K. C. Hoffman and co-workers, "Metal Hydride Storage for Mobile and Stationary Applications," *SAE Fuels and Lubricants Meeting*, St. Louis, Mo., 1976.
33. J. H. N. van Vucht, F. A. Kuijpers, and H. C. A. M. Bruning, *Philips Res. Rep.* **25**, 133 (1978).
34. R. L. Woolley and D. L. Henriksen, "Water Induction in Hydrogen-Powered IC Engines," Vol. 6C, *World Hydrogen Energy Conference*, Miami Beach, Fla., 1976, pp. 3–24.
35. R. L. Woolley, "Hydrogen Engine NO$_x$ Control by Water Induction," *NATO/CCMS Fourth International Symposium on Automotive Propulsion Systems*, Washington, D.C., 1977.
36. V. R. Anderson, "Hydrogen Energy in United States Post Office Delivery Systems," *World Hydrogen Energy Conference*, Vol. 2, Zurich, Switzerland, 1978.
37. J. O.'M. Bockris, *Energy: The Solar-Hydrogen Alternative*, John Wiley & Sons, Inc., New York, 1975, p. 292.
38. Ref. 27, p. 293.
39. R. E. Billings, "Hydrogen Homestead," Technical Library, International Academy of Science, Independence, Mo., 1978.
40. J. O.'M. Bockris and S. Srinivasan, *Fuel Cells: Their Electrochemistry*, McGraw-Hill, New York, 1970.
41. W. G. F. Grot and co-workers, *Perfluorinated Ion Exchange Membranes*, E. I. du Pont de Nemours and Co., Inc., Wilmington, Del., 1972.
42. R. E. Billings, M. Sanchez, P. A. Cherry, and D. B. Eyre, "LaserCell Prototype Vehicle," *Project Hydrogen '91 Conference*, International Academy of Science, Independence, Mo., 1991.
43. B. N. Gorbell and K. B. Prater, "The Ballard/BC Bus Demonstration Program", *Project Hydrogen '91 Conference*, American Academy of Science, Independence, Mo., 1991.

General References

J. O.'M. Bockris, N. Bonciocat, and F. Gutmann, *An Introduction to Electrochemical Science*, Wydeham Press, London, 1974.
J. O.'M. Bockris and S. Srinivasan, *Fuel Cells: Their Electrochemistry*, McGraw-Hill, New York, 1970.
J. O.'M. Bockris, T. N. Veziroglu, and D. Smith, *Solar Hydrogen Energy: The Power to Save the Earth*, MacDonald Optima, London, 1991.
L. W. Skelton, *The Solar-Hydrogen Energy Economy: Beyond the Age of Fire*, Van Nostrand Reinhold Co., Inc., New York, 1984.
T. N. Veziroglu and R. E. Billings, eds., *Project Hydrogen '91 Conference*, American Academy of Science, Independence, Mo., 1991.
R. E. Billings, ed., *Project Energy '93 Conference*, International Academy of Science, Independence, Mo., 1993.

Roger E. Billings
International Academy of Science

HYDROGEN FLUORIDE. See Fluorine compounds, inorganic.

HYDROGEN IODIDE. See Iodine compounds.

HYDROGEN-ION ACTIVITY

In a world in which most chemical and biological reactions occur in aqueous media, it is well recognized that the hydrogen ion is of fundamental importance. Hydrogen ions are involved in a wide variety of natural and industrial reactions, and the equilibrium positions as well as the rates of these reactions are therefore dependent on hydrogen-ion concentration. The hydrogen ion is more correctly termed hydronium ion. The unhydrated proton does not exist in aqueous solution but rather is bound to several molecules of water. This ion, sometimes represented as $H(H_2O)_n^+$, is usually written simply as H^+. More important is the distinction between the hydrogen ion concentration and its activity. The hydrogen ion concentration, or total acidity, is obtained by titration and corresponds to the total concentration of hydrogen ions available in a solution, ie, free, unbound hydrogen ions as well as hydrogen ions associated with weak acids. The hydrogen ion activity refers to the effective concentration of unbound hydrogen ions, ie, the form which affects physicochemical reaction rates and equilibria. The effective concentration of hydrogen ion in solution is expressed in terms of pH, which is the negative logarithm of the hydrogen-ion activity, a_{H^+}

$$pH = -\log_{10} a_{H^+} \tag{1}$$

The relationship between activity, a, and concentration, c, is

$$a = \gamma c \tag{2}$$

where the activity coefficient γ is a function of the ionic strength of the solution and approaches unity as the ionic strength decreases; ie, the difference between the activity and the concentration of hydrogen ion diminishes as the solution becomes more dilute. The pH of a solution may have little relationship to the titratable acidity of a solution that contains weak acids or buffering substances; the pH of a solution indicates only the free hydrogen-ion activity. If total acid concentration is to be determined, an acid–base titration must be performed.

Thermodynamically, the activity of a single ionic species is an inexact quantity, and a conventional pH scale has been adopted that is defined by reference to specific solutions with assigned pH(S) values. These reference solutions, in conjunction with equation 3, define the pH(X) of the sample solution.

$$pH(X) = pH(S) - \frac{(E_X - E_S)F}{2.303RT} \tag{3}$$

E_S is the electromotive force (emf) of the cell:

reference electrode$|KCl(\geq 3.5M)||$solution $S|H_2(g)$, Pt

and E_X is the emf of the same cell when the reference buffer solution S is replaced by the sample solution X. The quantities R, T, and F are the gas constant, the thermodynamic temperature, and the Faraday constant, respectively.

For routine pH measurements, the hydrogen gas electrode [$H_2(g)$,Pt] usually is replaced by a glass membrane electrode.

The availability of multiple pH(S) reference solutions makes possible an alternative definition of pH(X):

$$pH(X) = pH(S_1) + [pH(S_2) - pH(S_1)]\frac{(E_X - E_{S1})}{(E_{S2} - E_{S1})} \qquad (4)$$

where E_{S1} and E_{S2} are the measured cell potentials when the sample solution X is replaced in the cell by the two reference solutions S_1 and S_2 such that the values E_{S1} and E_{S2} are on either side of, and as near as possible to, E_X. Equation 4 assumes linearity of the pH vs E response between the two reference solutions, whereas equation 3 assumes both linearity and ideal Nernstian response of the pH electrode. The two-point calibration procedure is recommended if a pH electrode other than the hydrogen gas electrode is used for the measurements.

pH Determination

Two methods are used to measure pH: electrometric and chemical indicator (1–7). The most common is electrometric and uses the commercial pH meter with a glass electrode. This procedure is based on the measurement of the difference between the pH of an unknown or test solution and that of a standard solution. The instrument measures the emf developed between the glass electrode and a reference electrode of constant potential. The difference in emf when the electrodes are removed from the standard solution and placed in the test solution is converted to a difference in pH. Electrodes based on metal–metal oxides, eg, antimony–antimony oxide (see ANTIMONY AND ANTIMONY ALLOYS; ANTIMONY COMPOUNDS), have also found use as pH sensors (8), especially for industrial applications where superior mechanical stability is needed (see SENSORS). However, because of the presence of the metallic element, these electrodes suffer from interferences by oxidation–reduction systems in the test solution.

More recently, two different types of nonglass pH electrodes have been described which have shown excellent pH-response behavior. In the neutral-carrier, ion-selective electrode type of potentiometric sensor, synthetic organic ionophores, selective for hydrogen ions, are immobilized in polymeric membranes (see MEMBRANE TECHNOLOGY) (9). These membranes are then used in more-or-less classical glass pH electrode configurations, although they have the advantages of being more easily fabricated into unique designs and are less prone to protein fouling in biological fluids. Another type of pH sensor is based on an integrated ion-selective electrode and insulated-gate field-effect transistor (10). These sensors (qv), usually termed ion-selective field-effect transistors (ISFETS), are based on the modulation of the transistor source-drain current by a potential (or charge) applied to the transistor gate region. In the case of a pH ISFET, this potential is produced by a pH-responsive material applied to the gate region and in contact with the test solution. The most successful materials for this purpose have been crystalline solids, such as silicon nitride or tantalum oxide, and polymeric membranes containing the ionophores selective for hydrogen ions. Sensors based on the ISFET principle have the potential for extreme miniaturization and multisensor arrays on a single transistor chip (see also ELECTROANALYTICAL TECHNIQUES).

The second method for measuring pH, the indicator method, has more limited applications. The success of this procedure depends on matching the color that is produced by the addition of a suitable indicator dye to a portion of the unknown solution with the color produced by adding the same quantity of the same dye to a series of standard solutions of known pH. Alternatively, the color is matched against a color comparison chart for the particular dye. The results obtained by the indicator method are less accurate relative to those obtained using a pH meter. The indicator method, however, is inexpensive and simple to apply. In addition to being used by direct addition to the test solution, indicator dyes can be immobilized onto paper strips, eg, litmus paper or, more recently, onto the distal end of fiber-optic probes which, when combined with spectrophotometric readout, provide more quantitative indicator-dye pH determinations.

Reference Buffer Solutions. To define the pH scale and permit the calibration of pH measurement systems, a series of reference buffer solutions have been certified by the U.S. National Institute of Standards and Technology (NIST). The pH values at 25°C of these primary and secondary reference buffer solutions are listed in Table 1 (2). The pH 6.86 and 7.41 phosphate buffers have long been accepted as the primary reference standards for blood pH measurements. It is well known that the ionic strength of these buffers is significantly different from that of blood, thus biasing (however reproducibly) the pH measurement in blood owing to the residual liquid-junction potential. Two concentrations of two zwitterionic buffer systems, HEPES/HEPESate and MOPSO/MOPSOate, have been certified as pH buffers at ionic strengths comparable to that of blood

Table 1. Solution pH Standards, Molality Scale[a]

Composition[b]	Solution pH at 25°C
Primary standards	
potassium hydrogen tartrate (saturated at 25°C)	3.557
0.05 m potassium dihydrogen citrate	3.776
0.05 m potassium hydrogen phthalate	4.006
0.025 m KH_2PO_4 + 0.025 m Na_2HPO_4	6.863
0.008695 m KH_2PO_4 + 0.03043 m Na_2HPO_4	7.410
0.01 m $Na_2B_4O_7 \cdot 10H_2O$	9.180
0.025 m $NaHCO_3$ + 0.025 m Na_2CO_3	10.011
Secondary standards	
0.05 m potassium tetraoxalate dihydrate	1.681
0.05 m HEPES + 0.05 m NaHEPESate	7.503
0.08 m HEPES + 0.08 m NaHEPESate	7.516
0.05 m MOPSO + 0.05 m NaMOPSOate	6.867
0.08 m MOPSO + 0.08 m NaMOPSOate	6.865
0.01667 m TRIS + 0.05 m TRIS·HCl	7.699
$Ca(OH)_2$ (saturated at 25°C)	12.454

[a]Ref. 2.
[b]HEPES = N-2-hydroxyethylpiperazine-N'-2-ethanesulfonic acid [7365-45-9]; MOPSO = 3-(N-morpholino)-2-hydroxypropane sulfonic acid [68399-77-9]; and TRIS = tris(hydroxymethyl)aminomethane [77-86-1].

and consequently should minimize the residual liquid-junction potential problem. The International Union of Pure and Applied Chemistry (IUPAC) recommends the NIST primary standards plus a series of operational standards, measured vs the phthalate reference value standard in a cell having liquid junction, for the definition of the pH scale (11).

Accuracy and Interpretation of Measured pH Values. The acidity function which is the experimental basis for the assignment of pH, is reproducible within about 0.003 pH unit from 10 to 40°C. If the ionic strength is known, the assignment of numerical values to the activity coefficient of chloride ion does not add to the uncertainty. However, errors in the standard potential of the cell, in the composition of the buffer materials, and in the preparation of the solutions may raise the uncertainty to 0.005 pH unit.

The reproducibility of the practical scale that has been defined using the seven primary standards includes the possible inconsistencies introduced in the standardization of the instrument using seven different standards of different composition and concentration. These inconsistencies are the result of variations in the liquid-junction potential when one solution is replaced by another and are unavoidable. The accuracy of the practical scale from 10 to 40°C therefore appears to be from 0.008 to 0.01 pH unit.

Variations in the liquid-junction potential may be increased when the standard solutions are replaced by test solutions that do not closely match the standards with respect to the types and concentrations of solutes, or to the composition of the solvent. Under these circumstances, the pH remains a reproducible number, but it may have little or no meaning in terms of the conventional hydrogen-ion activity of the medium. The use of experimental pH numbers as a measure of the extent of acid–base reactions or to obtain thermodynamic equilibrium constants is justified only when the pH of the medium is between 2.5 and 11.5 and when the mixture is an aqueous solution of simple solutes in total concentration of ca $\leq 0.2\ M$.

Sources of Error. pH electrodes are subject to fewer interferences and other types of error than most potentiometric ionic-activity sensors, ie, ion-selective electrodes (see ELECTROANALYTICAL TECHNIQUES). However, pH electrodes must be used with an awareness of their particular response characteristics, as well as the potential sources of error that may affect other components of the measurement system, especially the reference electrode. Several common causes of measurement problems are electrode interferences and/or fouling of the pH sensor, sample matrix effects, reference electrode instability, and improper calibration of the measurement system (12).

In general, the potential of an electrochemical cell, E_{cell}, is the sum of three potential terms:

$$E_{cell} = E_{pH} - E_{ref} + E_{lj} \tag{9}$$

where E_{pH} and E_{ref} are the potentials of the pH and reference electrodes, respectively, and E_{lj} is the ubiquitous liquid-junction potential. After substitution of the Nernst equation for the pH electrode potential term in equation 9,

$$E_{cell} = E_{pH}^o - \frac{RT}{F} \ln a_H - E_{ref} + E_{lj} \tag{10}$$

it can be calculated that a 1 mV error in any of the potential terms corresponds to an error of ca 4% in the hydrogen-ion activity. Under carefully controlled experimental conditions, the potential of a pH cell can be measured with an uncertainty as small as 0.3 mV, which corresponds to a ± 0.005 pH unit uncertainty.

The measurement of pH using the operational cell assumes that no residual liquid-junction potential is present when a standard buffer is compared to a solution of unknown pH. Although this may never be strictly true, especially for complex matrices, the residual liquid-junction potential can be minimized by the appropriate choice of a salt-bridge solution and calibration buffer solutions.

Other problems occur in the measurement of pH in unbuffered, low ionic strength media such as wet deposition (acid rain) and natural freshwaters (see AIR POLLUTION; GROUNDWATER MONITORING) (13). In these cases, studies have demonstrated that the principal sources of the measurement errors are associated with the performance of the reference electrode liquid junction, changes in the sample pH during storage, and the nature of the standards used in calibration. Considerable care must be exercised in all aspects of the measurement process to assure the quality of the pH values on these types of samples.

pH Measurement Systems

Glass Electrodes. The glass electrode is the hydrogen-ion sensor in most pH-measurement systems. The pH-responsive surface of the glass electrode consists of a thin membrane formed from a special glass that, after suitable conditioning, develops a surface potential that is an accurate index of the acidity of the solution in which the electrode is immersed. To permit changes in the potential of the active surface of the glass membrane to be measured, an inner reference electrode of constant potential is placed in the internal compartment of the glass membrane. The inner reference compartment contains a solution that has a stable hydrogen-ion concentration and counterions to which the inner electrode is reversible. The choice of the inner cell components has a bearing on the temperature coefficient of the emf of the pH assembly. The inner cell commonly consists of a silver–silver chloride electrode or calomel electrode in a buffered chloride solution.

Immersion electrodes are the most common glass electrodes. These are roughly cylindrical and consist of a barrel or stem of inert glass that is sealed at the lower end to a tip, which is often hemispherical, of special pH-responsive glass. The tip is completely immersed in the solution during measurements. Miniature and microelectrodes are also used widely, particularly in physiological studies. Capillary electrodes permit the use of small samples and provide protection from exposure to air during the measurements, eg, for the determination of blood pH. This type of electrode may be provided with a water jacket for temperature control.

The membrane of pH-responsive glass usually is made as thin as is consistent with adequate mechanical strength. Nevertheless, its electrical resistance is high, eg, $10-250$ MΩ. Therefore, an electronic amplifier must be used to obtain adequate accuracy in the measurement of the surface potential of a glass electrode. The versatility of the glass electrode results from its mechanism of operation, which is one of proton exchange rather than electron transfer; hence, oxidizing and reducing agents in the solution do not affect the pH response.

Most modern electrode glasses contain mixtures of silicon dioxide [7631-86-9], either sodium oxide [1312-59-3] or lithium oxide [12057-24-8], and either calcium oxide [1305-78-8], barium oxide [1304-28-5], cesium oxide [12018-61-0], or lanthanum oxide [1312-81-8]. The latter oxides are added to reduce spurious response to alkali metal ions in high pH solutions. The composition of the glass has a profound effect on the electrical resistance, the chemical durability of the pH-sensitive surface, and the accuracy of the pH response in alkaline solutions (see GLASS). Both the electrical and the chemical resistance of the electrode glasses decrease rapidly with a rise in temperature. Therefore, it is difficult to design an electrode that is sufficiently durable for extended use at high temperatures and yet, when used at room temperature, free from the sluggish response often characteristic of pH cells of excessively high resistance. Most manufacturers use different glass compositions for electrodes, depending on their intended use.

The mechanism of the glass electrode response is not entirely understood. It is clear, however, that when a freshly blown membrane of pH-responsive glass is first conditioned in water, the sodium or lithium ions that occupy the interstices of the silicon–oxygen network in the glass surface are exchanged for protons from the water. The protons find stable sites in the conditioned gel layer of the glass surface. Exchange of the labile protons between these sites and the solution phase appears to be the mechanism by which the surface potential reflects changes in the acidity of the external solution. When the glass electrode and the hydrogen gas electrode are immersed in the same solution, the potentials usually differ by a constant amount, even though the pH of the medium is raised from 1 to 10 or greater. In this range, the potential of a glass electrode, E_g, may be written

$$E_g = E_g^o + \frac{RT}{F} \ln a_{H^+} \tag{11}$$

where E_g^o is the standard (formal) potential of that particular glass electrode on the hydrogen scale.

Departures from the ideal behavior expressed by equation 11 usually are found in alkaline solutions containing alkali metal ions in appreciable concentration, and often in solutions of strong acids. The supposition that the alkaline error is associated with the development of an imperfect response to alkali metal ions is substantiated by the successful design of cation-sensitive electrodes that are used to determine sodium, silver, and other monovalent cations (3).

The advantage of the lithium glasses over the sodium glasses in the reduction of alkaline error is attributed to the smaller size of the proton sites remaining after elution of the lithium ions from the glass surface. This view is consistent with the relative magnitudes of the alkaline errors for various cations. These errors decrease rapidly as the diameter of the cation becomes larger. The error observed in concentrated solutions of the strong acids is characterized by a marked drift of potential with time, which is thought to result from the penetration of acid anions, as well as protons, into the glass surface (14).

The immersion of glass electrodes in strongly dehydrating media should be avoided. If the electrode is used in solvents of low water activity, frequent conditioning in water is advisable, as dehydration of the gel layer of the surface causes a progressive alteration in the electrode potential with a consequent drift of the measured pH. Slow dissolution of the pH-sensitive membrane is unavoidable, and it eventually leads to mechanical failure. Standardization of the electrode with two buffer solutions is the best means of early detection of incipient electrode failure.

Fouling of the pH sensor may occur in solutions containing surface-active constituents that coat the electrode surface and may result in sluggish response and drift of the pH reading. Prolonged measurements in blood, sludges, and various industrial process materials and wastes can cause such drift. Therefore, it is necessary to clean the membrane mechanically or chemically at intervals that are consistent with the magnitude of the effect and the precision of the results required.

Reference Electrodes and Liquid Junctions. The electrical circuit of the pH cell is completed through a salt bridge that usually consists of a concentrated solution of potassium chloride [7447-40-7]. The solution makes contact at one end with the test solution and at the other with a reference electrode of constant potential. The liquid junction is formed at the area of contact between the salt bridge and the test solution. The mercury–mercurous chloride electrode, the calomel electrode, provides a highly reproducible potential in the potassium chloride bridge solution and is the most widely used reference electrode. However, mercurous chloride is converted readily into mercuric ion and mercury when in contact with concentrated potassium chloride solutions above 80°C. This disproportionation reaction causes an unstable potential with calomel electrodes. Therefore, the silver–silver chloride electrode and the thallium amalgam–thallous chloride electrode often are preferred for measurements above 80°C. However, because silver chloride is relatively soluble in concentrated solutions of potassium chloride, the solution in the electrode chamber must be saturated with silver chloride.

The commercially used reference electrode–salt bridge combination usually is of the immersion type. The salt-bridge chamber usually surrounds the electrode element. Some provision is made to allow a slow leakage of the bridge solution out of the tip of the electrode to establish the liquid junction with the standard solution or test solution in the pH cell. An opening is usually provided through which the electrode chamber may be refilled with the salt-bridge solution. Various devices are used to impede the outflow of bridge solution, eg, fibers, porous ceramics, capillaries, ground-glass joints, and controlled cracks. Such commercial electrodes normally give very satisfactory results, but there is some evidence that the type and structure of the junction may affect the reference potential when measurements are made at very low pH and, possibly, at high alkalinities.

Combination electrodes have increased in use and are a consolidation of the glass and reference electrodes in a single probe, usually in a concentric arrangement, with the reference electrode compartment surrounding the pH sensor. The advantages of combination electrodes include the convenience of using a single probe and the ability to measure small volumes of sample solution or in restricted-access containers, eg, test tubes and narrow-neck flasks. In addition, the surrounding electrolyte solution in the reference electrode compartment provides excellent electrical shielding of the pH sensor, which reduces noise and susceptibility to polarization.

Theoretical considerations favor liquid junctions by which cylindrical symmetry and a steady state of ionic diffusion are achieved. Special cells in which a stable junction can be achieved are not difficult to construct and are available commercially. A solution of potassium chloride that is saturated at room temperature usually is used for the salt bridge. It has been shown that the higher the concentration of the solution of potassium chloride, the more effective the bridge solution is in reducing the liquid-junction potential (15). Also, the saturated calomel reference electrode is stable, reproducible, and easy to prepare. However, the saturated electrode is not without its disadvantages. For example, it shows a marked hysteresis with changes of temperature. After long periods and on temperature lowering, the salt-bridge chamber may become filled with large crystals of potassium chloride that block the flow of bridge solution and thereby impair the reproducibility of the junction potential and raise the resistance of the cell. A slightly undersaturated (eg, 3.5 M) solution of potassium chloride is preferred.

Samples that contain suspended matter are among the most difficult types from which to obtain accurate pH readings because of the so-called suspension effect, ie, the suspended particles produce abnormal liquid-junction potentials at the reference electrode (16). This effect is especially noticeable with soil slurries, pastes, and other types of colloidal suspensions. In the case of a slurry that separates into two layers, pH differences of several units may result, depending on the placement of the electrodes in the layers. Internal consistency is achieved by pH measurement using carefully prescribed measurement protocols, as has been used in the determination of soil pH (17).

Another effect that may result in spurious pH readings is caused by streaming potentials. Presumably, these are attributable to changes in the reference electrode liquid junction that are caused by variations in the flow rate of the sample solution. Factors that affect the observed pH include the magnitude of the flow-rate changes, the geometry of the electrode system, and the concentration of the salt-bridge electrolyte; therefore, this problem may be avoided by maintaining constant flow and geometry characteristics and calibrating the system under operating conditions that are identical to those of the sample measurement.

pH Instrumentation.   The pH meter is an electronic voltmeter that provides a direct conversion of voltage differences to differences of pH at the measurement temperature. One class of instruments is the direct-reading analogue, a deflection meter having a large scale calibrated in mV and pH units. Most modern direct-reading meters have digital displays of the emf or pH. The types range from very inexpensive meters that read to the nearest 0.1 pH unit to research models capable of measuring pH with a precision of 0.001 pH unit and drifting less than 0.003 pH unit over 24 h; however, the fundamental meaning of these measured values is considerably less certain than the precision of the measurement.

Because of the very large resistance of the glass membrane in a conventional pH electrode, an input amplifier of high impedance (usually $10^{12}-10^{14}$ Ω)

is required to avoid errors in the pH (or mV) readings. Most pH meters have field-effect transistor amplifiers that typically exhibit bias currents of only a pico-ampere (10^{-12} ampere), which, for an electrode resistance of 100 MΩ, results in an emf error of only 0.1 mV (0.002 pH unit).

In addition, most devices provide operator control of settings for temperature and/or response slope, isopotential point, zero or standardization, and function (pH, mV, or monovalent–bivalent cation–anion). Microprocessors are incorporated in advanced-design meters to facilitate calibration, calculation of measurement parameters, and automatic temperature compensation. Furthermore, pH meters are provided with output connectors for continuous readout via a strip-chart recorder and often with binary-coded decimal output for computer interconnections or connection to a printer. Although the accuracy of the measurement is not increased by the use of a recorder, the readability of the displayed pH (on analogue models) can be expanded, and recording provides a permanent record and also information on response and equilibrium times during measurement (5).

Temperature Effects. The emf, E, of a pH cell may be written

$$E = E_g^{o'} - k\,\mathrm{pH} \tag{12}$$

where k is the Nernst factor $(2.303\ RT)/F$, and $E_g^{o'}$ includes the liquid-junction potential and the half-cell emf on the reference side of the glass membrane. Changes of temperature alter the scale slope because k is proportional to T. The scale position also is changed because the standard potential is temperature dependent: $E_g^{o'}$ is usually a quadratic function of the temperature.

The objective of temperature compensation in a pH meter is to nullify changes in emf from any source except changes in the true pH of the test solution. Nearly all pH meters provide automatic or manual adjustment for the change of k with T. If correction is not made for the change of standard potential, however, the instrument must always be standardized at the temperature at which the pH is to be determined. In industrial pH control, standardization of the assembly at the temperature of the measurements is not always possible, and compensation for shift of the scale position, though imperfect, is useful. If the value of $E_g^{o'}$ were a linear function of T, it would be easy to show that the straight lines representing the variation of E and pH at different temperatures would intersect at a point, the isopotential point or pH_i. Even though $E_g^{o'}$ does not usually vary linearly with T, these plots intersect at about pH_i when the range of temperatures is narrow. By providing a temperature-dependent bias potential of $k\,\mathrm{pH}_i$, an approximate correction for the change of the standard potential with temperature can be applied automatically (1,5).

Nonaqueous Solvents

The activity of the hydrogen ion is affected by the properties of the solvent in which it is measured. Scales of pH only apply to the medium, ie, the solvent or mixed solvents, eg, water–alcohol, for which the scales are developed. The comparison of the pH values of a buffer in aqueous solution to one in a nonaqueous solvent has neither direct quantitative nor thermodynamic significance. Consequently, operational pH scales must be developed for the individual solvent systems. In certain cases, correlation to the aqueous pH scale can be made, but in others, pH values are used only as relative indicators of the hydrogen-ion activity.

Other difficulties of measuring pH in nonaqueous solvents are the complications that result from dehydration of the glass pH membrane, increased sample resistance, and large liquid-junction potentials. These effects are complex and highly dependent on the type of solvent or mixture used (1,5).

Indicator pH Measurements

The indicator method is especially convenient when the pH of a well-buffered colorless solution must be measured at room temperature with an accuracy no greater than 0.5 pH unit. Under optimum conditions an accuracy of 0.2 pH unit is obtainable. A list of representative acid–base indicators is given in Table 2 with the corresponding transformation ranges. A more complete listing, including the theory of the indicator color change and of the salt effect, is also available (1).

Because they are weak acids or bases, the indicators may affect the pH of the sample, especially in the case of a poorly buffered solution. Variations in the ionic strength or solvent composition, or both, also can produce large uncertainties in pH measurements, presumably caused by changes in the equilibria of the indicator species. Specific chemical reactions also may occur between solutes in the sample and the indicator species to produce appreciable pH errors. Examples of such interferences include binding of the indicator forms by proteins and colloidal substances and direct reaction with sample components, eg, oxidizing agents and heavy-metal ions.

Table 2. Acid–Base Indicators

Indicator	pH Range	Color	
		Acid	Base
acid cresol red [1733-12-6]	0.2–1.8	red	yellow
methyl violet [8004-87-3]	0.5–1.5	yellow	blue
acid thymol blue [76-61-9]	1.2–2.8	red	yellow
bromophenol blue [115-39-9]	3.0–4.6	yellow	blue
methyl orange [547-58-0]	3.2–4.4	red	yellow
bromocresol green [76-60-8]	3.8–5.4	yellow	blue
methyl red [493-52-7]	4.4–6.2	red	yellow
bromocresol purple [115-40-2]	5.2–6.8	yellow	purple
bromothymol blue [76-59-5]	6.0–7.6	yellow	blue
phenol red [143-74-8]	6.6–8.2	yellow	red
cresol red [1733-12-6]	7.2–8.8	yellow	red
thymol blue [76-61-9]	8.0–9.6	yellow	blue
phenolphthalein [77-09-8]	8.2–9.8	colorless	red
tolyl red [6410-10-2]	10.0–11.6	red	yellow
parazo orange [547-57-9]	11.0–12.6	yellow	orange
trinitrobenzoic acid [129-66-8]	12.0–13.4	colorless	orange

Industrial Process Control

Specialized equipment for industrial measurements and automatic control have been developed (18) (see PROCESS CONTROL). In general, the pH of an industrial process need not be controlled with great accuracy. Consequently, frequent standardization of the cell assembly may be unnecessary. On the other hand, the ambient conditions, eg, temperature and humidity, under which the industrial control measurements are made, may be such that the pH meter must be much more robust than those intended for laboratory use. To avoid costly downtime for repairs, pH instruments may be constructed of modular units, permitting rapid removal and replacement of a defective subassembly.

The pH meter usually is coupled to a data recording device and often to a pneumatic or electric controller. The controller governs the addition of reagent so that the pH of the process stream is maintained at the desired level.

Immersion-cell assemblies are designed for continuous pH measurement in tanks, troughs, or other vessels containing process solutions at different levels under various conditions of agitation and pressure. The electrodes are protected from mechanical damage and are sometimes provided with devices to remove surface deposits as they accumulate. Process flow chambers are designed to introduce the pH electrodes directly into piped sample streams or bypass sample loops that may be pressurized. Electrode chambers of both types usually contain a temperature-sensing element that controls the temperature-compensating circuits of the measuring instrument.

Glass electrodes for process control do not differ materially from those used for pH measurements in the laboratory, but the emphasis in industrial application is on rugged construction to withstand both mechanical stresses and high pressures. Pressurized salt bridges, which ensure slow leakage of bridge solution into the process stream even under very high pressures, have been developed. For less severe process monitoring conditions, reference electrodes are available with no-flow polymeric or gel-filled junctions that can be used without external pressurization.

BIBLIOGRAPHY

"Hydrogen-Ion Concentration" in *ECT* 1st ed., Vol. 7, pp. 711–726, by R. G. Bates and E. R. Smith, National Bureau of Standards; in *ECT* 2nd ed., Vol. 11, pp. 380–390, by R. G. Bates, National Bureau of Standards; "Hydrogen-Ion Activity" in *ECT* 3rd ed., Vol. 13, pp. 1–11, by R. A. Durst, National Bureau of Standards, and R. G. Bates, University of Florida.

1. R. G. Bates, *Determination of pH, Theory and Practice*, 2nd ed., Wiley-Interscience, New York, 1973.

2. Y. C. Wu, W. F. Koch, and R. A. Durst, *Standardization of pH Measurements*, National Bureau of Standards Special Publication 260-53, U.S. Government Printing Office, Washington, D.C., 1988; National Institute of Standards and Technology, Gaithersburg, Md., private communication, 1993.

3. G. Eisenman, ed., *Glass Electrodes for Hydrogen and Other Cations*, Marcel Dekker, New York, 1967.

4. G. Mattock, *pH Measurement and Titration*, Macmillan, New York, 1961.

5. C. C. Westcott, *pH Measurements*, Academic Press, Inc., New York, 1978.

6. *pH of Aqueous Solutions with the Glass Electrode*, ASTM method E70-77, American Society for Testing and Materials, Philadelphia, Pa., 1977.

7. H. Galster, *pH Measurement: Fundamentals, Methods, Applications, Instrumentation*, VCH, New York, 1991.

8. S. Glab and co-workers, *Crit. Rev. Anal. Chem.* **21**, 29 (1989).

9. D. Ammann and co-workers, *Ion-Selective Electrode Rev.* **5**, 3 (1983).

10. J. Janata, *Chem. Rev.* **90**, 691 (1990); J. Janata and R. J. Huber, eds., *Solid State Chemical Sensors*, Academic Press, Inc., New York, 1985; G. F. Blackburn, in A. P. F. Turner, I. Karube, and G. S. Wilson, eds., *Biosensors*, Oxford University Press, Oxford, 1987.

11. A. K. Covington, R. G. Bates, and R. A. Durst, *Pure Applied Chem.* **57**, 531 (1985).

12. R. A. Durst, in H. Freiser, ed., *Ion-Selective Electrodes in Analytical Chemistry*, Plenum Publishing, New York, 1978, Chapt. 5.

13. R. A. Durst, W. Davison, and W. F. Koch, *Pure Appl. Chem.* **66**, 649 (1994).

14. K. Schwabe, in H. W. Nürnberg, ed., *Electroanalytical Chemistry*, John Wiley & Sons, Inc., New York, 1974, Chapt. 7.

15. E. A. Guggenheim, *J. Am. Chem. Soc.* **52**, 1315 (1930).

16. H. Jenny and co-workers, *Science* **112**, 164 (1950).

17. A. M. Pommer, in Ref. 3, Chapt. 14.

18. F. G. Shinskey, *pH and pIon: Control in Process and Waste Streams*, John Wiley & Sons, Inc., New York, 1973.

General References

Refs. 1–7 are also general references.

RICHARD A. DURST
Cornell University

ROGER G. BATES
University of Florida

HYDROGEN PEROXIDE

Hydrogen peroxide [7722-84-1], H_2O_2, mol wt 34.016, is a strong oxidizing agent commercially available in aqueous solution over a wide range of concentrations. It is a weakly acidic, nearly colorless clear liquid that is miscible with water in all proportions. The atoms are covalently bound in a nonpolar H—O—O—H structure having association (hydrogen bonding) somewhat less than that found in water. Thenard discovered hydrogen peroxide in 1818, producing it by the action of dilute acids on barium peroxide [1304-29-6], BaO_2. Hydrogen peroxide has been an article of commerce since the mid-nineteenth century, initially as weak 3–7% solutions. Its scale of manufacture and use have increased markedly since 1925 when electrolytic processes were introduced to the United States and industrial bleach applications were developed. Now manufactured primarily in large, strategically located anthrahydroquinone autoxidation processes, its many uses include bleaching wood pulp and textiles, preparing other peroxygen compounds, and serving as a nonpolluting oxidizing agent (see also PEROXIDES AND PEROXY COMPOUNDS, INORGANIC; PEROXIDES AND PEROXY COMPOUNDS, ORGANIC).

Physical Properties

Properties of pure hydrogen peroxide are listed in Table 1. In aqueous solution the hydrogen bonds (association) between water and H_2O_2 molecules are appreciably more stable than those between molecules of the individual species. This increase in attraction forces is evidenced from many properties such as heat of mixing, vapor pressure, viscosity, dielectric constant, etc. Physical constants have been determined or calculated for aqueous H_2O_2 solutions, the only form in which hydrogen peroxide is commercially available (Table 2). Numerous other physical property data appear in the literature, including approximation coefficients for free energy function calculations, coefficients of diffusion, partition coefficients, spectroscopic studies, thermodynamic properties, and third law entropy (1,2). Mathematical correlations for vapor pressure, surface tension, heat of vaporization, heat capacity, liquid density, thermal conductivity, and viscosity values have been established for the 0–450°C range, as well as heat and free energy of formation, vapor heat capacity, thermal conductivity, and viscosity values for the 0–1200°C range (3).

Table 1. Properties of Hydrogen Peroxide

Property	Value
mp,[a] °C	−0.41
bp, °C	150.2
density at 25°C, g/mL	1.4425
viscosity at 20°C, mPa·s(=cP)	1.245
surface tension at 20°C, mN/m(=dyn/cm)	80.4
specific conductance at 25°C, $(\Omega \cdot cm)^{-1}$	4×10^{-7}
heat of fusion, J/g[b]	367.52
specific heat at 25°C, J/(g·K)[b]	2.628
heat of vaporization at 25°C, kJ/g[b]	1.517
dissociation constant[c] at 20°C	1.78×10^{-12}
heat of dissociation, kJ/mol[b]	34.3

[a] Tends to supercool.
[b] To convert J to cal, divide by 4.184.
[c] At zero ionic strength.

Chemical Properties

Hydrogen peroxide is a weak acid, having a $pK_a = 11.75$.

$$H_2O_2 + H_2O \longrightarrow HO_2^- + H_3O^+ \qquad (1)$$

Table 2. Physical Properties of Aqueous Hydrogen Peroxide

Liquid, wt % H_2O_2	Freezing point, °C	Boiling point,[a] °C	Vapor,[a] wt % H_2O_2	Density at 25°C, g/mL	ΔH_{vap} at 25°C, kJ/g[b]
10	−6.4	101.7	0.9	1.0324	2.357
20	−14.6	103.6	2.1	1.0694	2.274
30	−25.7	106.2	4.2	1.1081	2.192
40	−41.4	109.6	7.6	1.1487	2.105
50	−52.2	113.8	13.0	1.1914	2.017
60	−55.5	119.0	20.8	1.2364	1.926
70	−40.3	125.5	33.4	1.2839	1.832
80	−24.8	132.9	51.5	1.3339	1.733
90	−11.5	141.3	75.0	1.3867	1.627

[a]At 101.3 kPa (1 atm).
[b]To convert J to cal, divide by 4.184.

Dissociation of the second proton is insignificant. The pH of its aqueous solutions can be measured reproducibly with a glass electrode, but a correction dependent on the concentration must be added to obtain the true pH value. Correction values for the most common commercial solutions are listed in Table 3. The apparent pH of commercial product solutions can be affected by the type and amount of stabilizers added, and many times the pH is purposely adjusted to a grade specification range.

The reactions of hydrogen peroxide include

Decomposition

$$H_2O_2 \longrightarrow 2\ OH\cdot \tag{2}$$

$$H_2O_2 + \cdot OH \longrightarrow \cdot OOH + H_2O \tag{3}$$

$$\cdot OOH + \cdot OH \longrightarrow H_2O + O_2 \tag{4}$$

or

$$2\ H_2O_2 \longrightarrow 2\ H_2O + O_2 \tag{5}$$

Molecular additions

$$H_2O_2 + Y \longrightarrow Y \cdot H_2O_2 \tag{6}$$

Substitutions

$$H_2O_2 + RX \longrightarrow ROOH + HX \tag{7}$$

$$H_2O_2 + 2\ RX \longrightarrow ROOR + 2\ HX \tag{8}$$

Oxidations

$$H_2O_2 + W \longrightarrow WO + H_2O \tag{9}$$

Reductions

$$H_2O_2 + Z \longrightarrow ZH_2 + O_2 \tag{10}$$

Table 3. Apparent and True pH of Aqueous Hydrogen Peroxide

H_2O_2 conc, wt %	Equivalence point[a]	True pH	Correction factor
35	3.9	4.6	+0.7
50	2.8	4.3	+1.5
70	1.6	4.4	+2.8
90[b]	0.2	5.1	+4.9

[a]Measured using glass electrode.
[b]Routine manufacture has been discontinued.

Hydrogen peroxide may react directly or after it has first ionized or dissociated into free radicals. Often, the reaction mechanism is extremely complex and may involve catalysis or be dependent on the environment. Enhancement of the relatively mild oxidizing action of hydrogen peroxide is accomplished in the presence of certain metal catalysts (4). The redox system Fe(II)—Fe(III) is the most widely used catalyst, which, in combination with hydrogen peroxide, is known as Fenton's reagent (5).

Free-Radical Formation. Hydrogen peroxide can form free radicals by homolytic cleavage of either an O—H or the O—O bond.

$$\text{HOOH} \longrightarrow \text{H·} + \text{·OOH} \qquad \Delta H = 380 \text{ kJ/mol (90 kcal/mol)} \qquad (11)$$

$$\text{HOOH} \longrightarrow 2 \text{ ·OH} \qquad \Delta H = 210 \text{ kJ/mol (50 kcal/mol)} \qquad (12)$$

Equation 11 predominates in uncatalyzed vapor-phase decomposition and photochemically initiated reactions. In catalytic reactions, and especially in solution, the nature of the reactants determines which reaction is predominant.

Fenton chemistry is dependent on the formation of free radicals.

$$\text{Fe(II)} + \text{H}_2\text{O}_2 \longrightarrow \text{Fe(III)} + {}^-\text{OH} + \text{·OH}$$

$$\text{Fe(III)} + \text{H}_2\text{O}_2 \longrightarrow \text{·OOH} + \text{Fe(II)} + \text{H}^+ \qquad (13)$$

The radicals are then involved in oxidations such as formation of ketones (qv) from alcohols. Similar reactions are finding value in treatment of waste streams to reduce total oxidizable carbon and thus its chemical oxygen demand. These reactions normally are conducted in aqueous acid medium at pH 1–4 to minimize the catalytic decomposition of the hydrogen peroxide. More information on metal and metal oxide-catalyzed oxidation reactions (Milas' oxidations) is available (4–7) (see also PHOTOCHEMICAL TECHNOLOGY, PHOTOCATALYSIS).

Decomposition. The decomposition of hydrogen peroxide may be homogeneous or heterogeneous and can occur in the vapor or the condensed phase. Although there is considerable evidence that the decomposition occurs as a chain reaction involving free radicals, the products of the decomposition are water and oxygen gas. Decomposition of hydrogen peroxide must be controlled at all times, in part because of the economic impact, but more importantly because the resultant simultaneous generation of oxygen and heat may cause serious safety problems. For the decomposition of pure hydrogen peroxide at 298.16 K (eq. 5), $\Delta H = -105.26$ kJ/mol (-25.26 kcal/mol) of H_2O_2 when H_2O_2 is in the vapor state, and $\Delta H = -98.20$ kJ/mol (-23.47 kcal/mol) when H_2O_2 is a liquid.

The mechanism and rate of hydrogen peroxide decomposition depend on many factors, including temperature, pH, presence or absence of a catalyst (7–10), such as metal ions, oxides, and hydroxides, etc. Some common metal ions that actively support homogeneous catalysis of the decomposition include ferrous, ferric, cuprous, cupric, chromate, dichromate, molybdate, tungstate, and vanadate. For combinations, such as iron and copper together, the decomposition rate is greater than for the individual components. A key factor is that copper reduces iron and thus assures a greater concentration of the more active ferrous ions. Other combinations yield similar results. Catalase enzyme and the halide ions except for fluoride also can be active. Active common catalytic surfaces include copper, mild steel, iron, silver, palladium, platinum, and oxides of iron, lead, nickel, manganese, and mercury.

The stability of pure hydrogen peroxide solutions increases with increasing concentration and is maximum between pH 3.5–4.5. The decomposition rate of ultrapure hydrogen peroxide increases 2.2–2.3-fold for each 10°C rise in temperature from ambient to about 100°C. This approximates an Arrhenius-type response with activation energy of about 58 kJ/mol (13.9 kcal/mol). However, decomposition increases as low as 1.6-fold for each 10°C rise have been noted for impure, unstabilized solutions.

The decomposition of aqueous hydrogen peroxide is minimized by various purification steps during manufacture, use of clean passive equipment, control of contaminants, and the addition of stabilizers. The decomposition is zero-order with respect to hydrogen peroxide concentration.

Stabilization. Pure hydrogen peroxide solutions are relatively stable and can be stored for extended periods in clean passive containers. Commercial solutions, however, invariably contain or may be exposed to varying amounts of catalytic impurities and must therefore contain reagents which deactivate these impurities, either by adsorption (qv) or through formation of complexes (see CHELATING AGENTS). For example, sodium pyrophosphate [7758-16-9], $Na_2H_2P_2O_7$, added to acidic hydrogen peroxide solutions, acts as a complexing agent, whereas sodium stannate trihydrate [12058-66-1], $Na_2SnO_3 \cdot 3H_2O$, forms protective colloids (qv). Alkaline solutions of hydrogen peroxide are inherently less stable than acidic solutions, and often magnesium or silicate ions are added to alkaline solutions to form soluble or colloidal compounds which deactivate tramp metal ions. Additionally, the use of amine-substituted organophosphonic acids or their alkali metal salts for stabilizing weak hydrogen peroxide solutions (0.01–5.0%) in the pH range of 7–12.5 has been patented (11).

Many stabilizer systems have been tailored to a particular industry need or for particular areas where dilution water quality is poor. These grades are heavily stabilized and may contain organic sequestering agents, ie, stannate, phosphates, and nitrate ions, so that the weak solutions produced by dilution from hard water retain acceptable stability. The nitrate is not a stabilizer, but it inhibits corrosion of aluminum storage tanks by chloride ion.

Several patents (12–14) claim that adding small amounts of organic sequestering agents to aqueous hydrogen peroxide solutions, that contain normal stabilizers such as stannates and phosphates, improves the resistance against polyvalent metal cation-induced decomposition. Other patented stabilizer packages include combinations of organic compounds and organometallic salts with or without stannates and phosphates (15–21).

Special hydrogen peroxide stabilizer packages have been tailored for uses such as cleaning high capacity semiconductor chips and other critical electronic components for the computer industry (see SEMICONDUCTORS). These grades require extremely low metal ion content and, in some cases, the use of all-organic stabilizers (22–28) is cited. Some very pure, unstabilized 30% hydrogen peroxide containing <1 ppb of total metal ions has reportedly been used in this market. Because container corrosion would be a source of metal ion contamination, the containers are either special glass or are lined.

Molecular Addition.   Oxyacid salts, metal peroxides, nitrogen compounds, and others form crystalline peroxyhydrates in the presence of hydrogen peroxide. When dissolved in water, the peroxyhydrates react as solutions of their components. The peroxyhydrates formed from sodium carbonate and urea are commercially available. Examples of peroxyhydrates can be found in the literature (29,30).

Sodium perborate tetrahydrate [10486-00-7] is formed from hydrogen peroxide and sodium borate [1303-96-4] and, although cited often as an example of a peroxyhydrate, has been shown to be a true peroxy compound, sodium peroxyborate trihydrate (31). The peroxyborate is a key bleaching ingredient in laundry detergent formulations (see BLEACHING AGENTS). Most recently, sodium perborate monohydrate [10338-33-9] has been used. This pure compound has no water of hydration and properly should merely be called sodium peroxyborate.

Substitution.   A variety of peroxygen compounds can be formed through substitution reactions of hydrogen peroxide with organic reagents. These compounds are commercially useful as catalysts for polymerizations and oxidizing agents for a number of specialized reactions. The reactant and principal product of such reactions include alkylating agents, ie, alkylhydroperoxides; carboxylic and peroxy acids; acid anhydrides or chlorides, ie, diacyl peroxides; and ketones and ketone peroxides. These derivatives are considerably more hazardous than hydrogen peroxide and should be handled with extreme care, per manufacturer instructions.

Inorganic peroxygen compounds can be prepared through similar reactions with inorganic reagents. Alkaline-earth metal peroxides are prepared from hydrogen peroxide and the corresponding hydroxide, and monoperoxysulfuric acid by reacting hydrogen peroxide and sulfur trioxide or sulfuric acid.

Oxidation.   Hydrogen peroxide is a strong oxidant. Most of its uses and those of its derivatives depend on this property. Hydrogen peroxide oxidizes a wide variety of organic and inorganic compounds, ranging from iodide ions to the various color bodies of unknown structure in cellulosic fibers. The rate of these reactions may be quite slow or so fast that the reaction occurs on a reactive shock wave. The mechanisms of these reactions are varied and dependent on the reductive substrate, the reaction environment, and catalysis. Specific reactions are discussed in a number of general and other references (4,5,32–35).

Reduction.   Hydrogen peroxide reduces stronger oxidizing agents such as chlorine, sodium hypochlorite, potassium permanganate, and ceric sulfate. The last two are used for the volumetric determination of hydrogen peroxide. The ability of hydrogen peroxide to reduce chlorine and hypochlorite leads to the use of H_2O_2 as a pollution abatement treatment for industrial waste streams (see WASTES, INDUSTRIAL).

Manufacture

Hydrogen peroxide is composed of equal molar amounts of hydrogen and oxygen and can be formed directly by catalytically combining the gaseous elements. It can also be formed from compounds that contain the peroxy group; from water and oxygen by thermal, photochemical, electrochemical or similar processes; and by the uncatalyzed reaction of molecular oxygen with appropriate hydrogen-containing species. It has been manufactured commercially by processes based on the reaction of barium peroxide or sodium peroxide with an acid, the electrolysis of sulfuric acid and related compounds, the autoxidation of 2-alkylanthrahydroquinones, isopropyl alcohol, and hydrazobenzene, and more recently by the Huron-Dow (36–38) process through the cathodic reduction of oxygen in an electrolytic cell using dilute sodium hydroxide as the electrolyte. By far, the majority of hydrogen peroxide produced since 1957 has been based on the autoxidation of 2-alkylanthrahydroquinones.

AUTOXIDATION METHODS

Anthrahydroquinone Autoxidation. It was discovered in 1901 that hydroquinone [123-31-9] and hydrazobenzene [122-66-7] would each react with O_2, forming H_2O_2 quantitatively. Then in 1935 it was found that 2-alkylanthraquinones (2-alkyl-9,10-anthracenediones) were well suited for use as the reaction hydrogen carrier in a cyclic process (39). The first semiworks-scale anthrahydroquinone autoxidation process plant, producing H_2O_2 at a rate of 30 t/mo, was operated as a production unit by I. G. Farbenindustrie in Germany during World War II (40). All subsequent anthrahydroquinone autoxidation processes retain the basic features of this, the Riedl-Pfleiderer process, shown in Figure 1.

A 2-alkylanthraquinone (RAQ) (**1**) dissolved in a suitable solvent or solvent mixture is catalytically hydrogenated to the corresponding 2-alkylanthrahydroquinone (RAHQ) (**2**), ie, 2-alkyl-9,10-anthracenediol or 2-alkylanthraquinol (eq. 15). The 2-alkylanthraquinone is commonly called the reaction carrier, hydrogen carrier, or working material; the 2-alkylanthraquinone–solvent mixture is called the working solution. Carriers in industrial use include 2-*tert*-amylan-

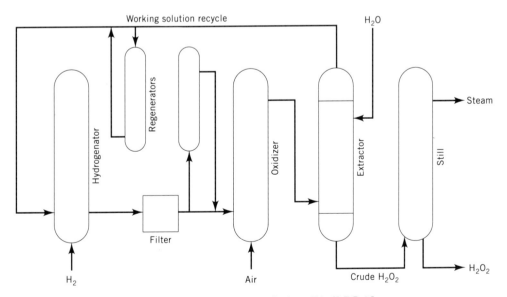

Fig. 1. Anthrahydroquinone autoxidation, Riedl-Pfleiderer process.

thraquinone [32588-54-8], 2-iso-sec-amylanthraquinone [68892-24-4], 2-tert-butylanthraquinone [84-47-9], and 2-ethylanthraquinone [84-51-5].

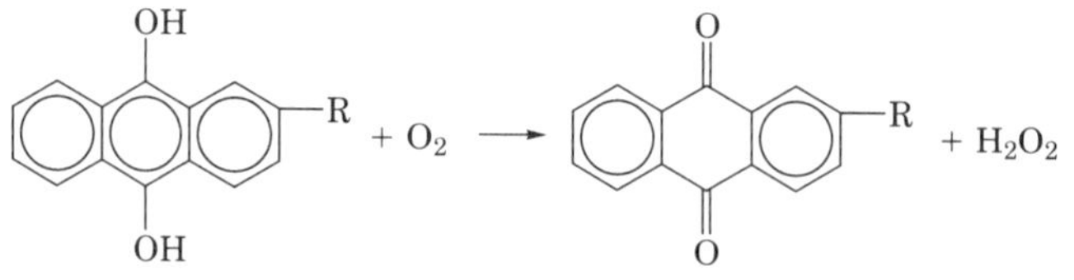

$$(15)$$

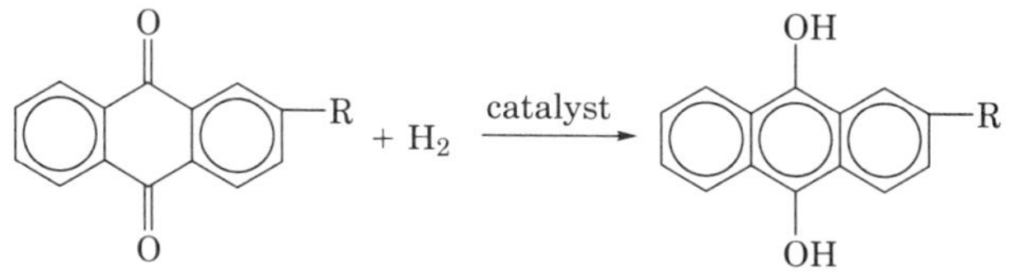

$$(16)$$

(1)   (2)

All have the alkyl group in the 2-position; thus the terms alkylanthrahydroquinone and alkylanthraquinone are taken to mean C-2 substitution herein. The working solution containing the carrier product alkylanthrahydroquinone is separated from the hydrogenation catalyst and aerated with an oxygen-containing gas, usually compressed air, to reform the alkylanthraquinone carrier while simultaneously producing hydrogen peroxide (eq. 16). The hydrogen peroxide is then extracted from the oxidized working solution using demineralized water, and the aqueous extract is purified and concentrated by fractionation to the desired strength. The extracted working solution is dried and recycled back to hydrogenation.

When the process is first started, the working solution contains only the alkylanthraquinone specie. The alkylanthraquinone (RAQ) acts in close concert with the catalyst, which is usually palladium metal, and can be envisioned as Pd:RAQ. The complex reacts with hydrogen, yielding Pd:RAHQ. The alkylanthrahydroquinone is subject to many other secondary reactions that occur continuously during each process cycle. During the short time that the alkylanthrahydroquinone is in contact with the catalyst, a minor amount of catalytic reduction of the unsubstituted aromatic ring occurs at the 5, 6, 7, and 8 positions, yielding the tetrahydroalkylanthrahydroquinone (3) shown in Figure 2. Further ring reduction may occur in the 1, 2, 3, and 4 positions of the substituted ring yielding octahydroalkylanthrahydroquinone (4). Once this compound is formed, it remains until purged owing to its essentially nonexistent oxidation rate. Transannular tautomerization of the alkylanthrahydroquinone also yields the hydroxyanthrones (oxanthrones) (5) and (6) which can be reduced to the anthrones (7) and (8). Another possible chemical route to the oxanthrone is via the formation and subsequent breakup of a quinhydrone-type charge-transfer complex. Although the formation of tetrahydroalkylanthrahydroquinones (3) is slow, these oxidize yielding H_2O_2 and the corresponding tetrahydroalkylanthraquinones which are the apparent precursors to the tetrahydroalkylanthraquinone epoxide (9). This series of reactions repeats with

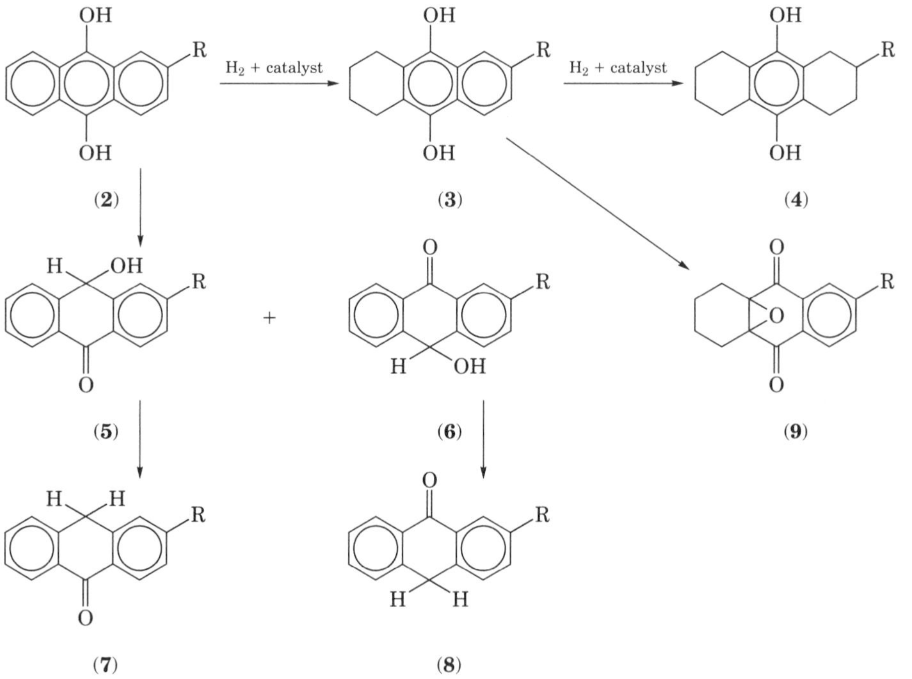

Fig. 2.   Possible secondary reactions of 2-alkylanthrahydroquinones.

each subsequent cycle of equations 15 and 16, resulting in a continuing buildup of the tetrahydroalkylanthraquinones, until these secondary products could become the dominant anthraquinone species if not reverted. These compounds react directly with alkylanthrahydroquinone species accepting its hydrogen to form the tetrahydroalkylanthrahydroquinone (THRAQ) (**10**) and thus freeing the alkylanthraquinone (eq. 17) to continue the hydrogenation cycle.

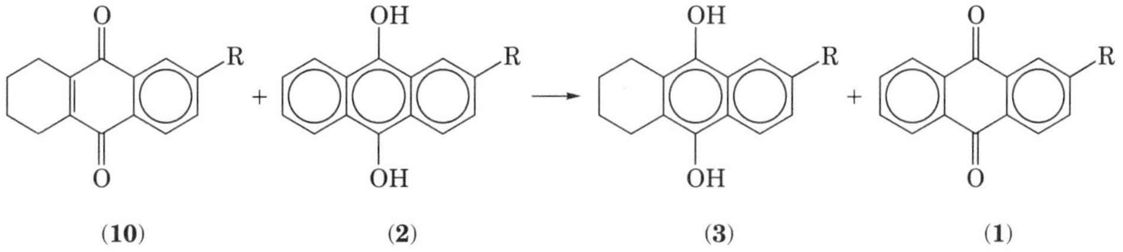

$$(17)$$

Although considered an active participant in the process cycle, the tetrahydroalkylanthraquinone (**10**) may not be a significant part of the catalytic hydrogenation because, dependent on the concentration in the working solution, these could all be converted to the hydroquinone by the labile shift per equation 17 and not be available to participate. None of the other first- or second-generation anthraquinone derivatives produce hydrogen peroxide, but most are susceptible to further reaction by oxidative or reductive mechanisms.

The chemical yield of hydrogen peroxide and the anthraquinone per process cycle is very high, but other secondary reactions necessitate regeneration of the working solution and hydrogenation catalyst, and the removal of organic material from the extracted hydrogen peroxide.

The first commercial-scale anthrahydroquinone autoxidation process in the United States was put into operation by E. I. du Pont de Nemours & Co., Inc. (Memphis, Tennessee) in 1953, followed by FMC Corporation (West Virginia), LaPorte Chemicals, Ltd. (U.K.), Degussa (Germany), Mitsubishi-Gas Chemical Co. (Japan), and others.

Working Solution Composition. The working solution in an anthraquinone process is composed of the anthraquinones, the by-products from the hydrogenation and oxidation steps, and solvents. The solvent fraction usually is a blend of polar and aromatic solvents which together provide the needed solubilities and physical properties. Once the solution has been defined, its composition and physical properties must be maintained within prescribed limits for achieving optimum operation.

Each working solution has an inherent maximum capacity which depends on the anthraquinone, anthrahydroquinone, water, and hydrogen peroxide solubilities. The capacity is defined as the maximum amount of hydrogen peroxide in g/L of working solution that can be produced per process cycle. The theoretical production capability of the process is the product of this capacity and the working solution flow rate. Usually the limiting factors are the solubilities of the anthrahydroquinones and hydrogen peroxide and the distribution coefficient of the working solution–H_2O_2–water system. The capacity can be altered somewhat by using various substituent groups on the anthraquinones or a combination of them with various solvents. The working solution and each of its individual ingredients should be chemically stable throughout the process cycle. Although the principal hydrogen peroxide producers have only used 2-ethylanthraquinone (EAQ), 2-*tert*-butylanthraquinone (BAQ), and a mix of branched isomers 2-*tert*- and 2-iso-*sec*-amylanthraquinones (AAQs), other compounds have been patented (41–44) for use in autoxidation processes. Whereas the anthraquinones are usually monosubstituted at the 2-position, the literature includes examples of C-1 substituents as well as 1,2-disubstituted alkylanthraquinones and naphthoquinones.

The solvents must be nontoxic and reasonably high boiling to minimize vapor emissions and the subsequent effect on the environment, possess a flash point and physical properties (density, viscosity, diffusivities) consistent with planned operating conditions, have reasonable solvency for hydrogen and oxygen, be relatively nonreactive and resistant to oxidative attack, and be nearly insoluble in the aqueous extract, yet favorably partition so that a reasonably high strength hydrogen peroxide solution can be safely extracted. Alcohols, ureas, amides, caprolactams, esters, and pyrrolidones have been used or cited in the literature as usable polar solvents. Table 4 lists the working solution components of the principal U.S. producers. Although a high distribution coefficient increases the extraction step efficiency, if too high it creates potential safety problems; especially if a free phase forms as a result of slightly exceeding the water solubility of the oxidized but unextracted working solution. The distribution coefficient is defined as weight of solute per unit weight of extract, divided

Table 4. U.S. Hydrogen Peroxide Producers' Working Solution Components

	Producer				
Parameter	Degussa	Du Pont	EKA	FMC	Interox
RAQ	EAQ	EAQ/BAQ/AAQ	EAQ	EAQ	AAQ
solvent system[a]					
AQ	A/B	A/B	A/B	A/B	A/B
HAQ	TBU/TOP	DIBC	TBU/TOP	TOP	DIBC
Reference	45	46	47	48	49

[a]A/B = alkylated benzenes; TBU = tetrabutyl urea; TOP = trioctyl phosphate; and DIBC = diisobutyl carbinol. See text.

by the weight of solute per unit weight of solvent. Table 5 shows the effect of the distribution coefficient (DC) on the maximum achievable aqueous hydrogen peroxide strength vs the working solution capacity (WSC).

In extraction (qv), the distribution coefficient value is the slope of the equilibrium line. In practice, the slope of the operating line is set at a value somewhat less than the distribution coefficient to provide driving force and fix the required theoretical extraction stages at some reasonable number.

Hydrogenation and Catalyst. The carrier alkylanthraquinone is catalytically reduced by hydrogen to the corresponding anthrahydroquinone (anthraquinol) in the hydrogenator. The reaction is carried out at slightly elevated (100–400 kPa/g (14.5–58 psi/g)) pressure, and below 75°C. Most patent examples cite temperatures in the 40–50°C range. The exothermic hydrogenation reaction accounts for nearly 55% of the 188.7 kJ/mol (45.1 kcal/mol) heat of formation of H_2O_2 from the elements. The excess heat is removed by conventional means such as precooling the working solution feed, cooling jackets or internal cooling, recirculating a cooled stream from the outlet to the inlet, or by combinations of these.

The extremes appear to be the 25–40°C operating temperatures proposed by Mitsubishi Gas-Chemical (MGC) (50,51) for a substantially tetrahydroalkyl-anthraquinone-free working solution (anthra system) and the 60–75°C indicated by LaPorte Chemicals (Interox-Solvay), where the working solutions contain more tetrahydroalkylanthraquinone than alkylanthraquinone (tetra system) (52,53). Degussa apparently uses the highest operating pressure, near 300 kPa (43.5 psi) overpressure (54). The conversion of quinone to hydroquinone

Table 5. Maximum Hydrogen Peroxide Achievable, %

	WSC[a]			
DC	7.7	10.0	12.5	15.0
50	28.9	35.2	40.5	45.0
75	37.9	44.9	50.5	55.1
100	44.8	52.1	57.7	62.1
150	54.9	62.0	67.1	71.1
200	61.9	68.5	73.2	76.6

[a]Density of 0.93 g/mL.

or degree of hydrogenation is normally 45–50% to minimize secondary reactions, but conversions above 80% have been cited.

The hydrogenation rate is maintained nearly constant by either the periodic addition or exchange of catalyst. The rate can also be varied somewhat by adjusting the hydrogen partial pressure and the temperature between certain limits. However, the rate apparently is not affected by pressures greater than 405 kPa (4 atm). Ammonia (qv) gas added to the hydrogenator, or water-soluble amines or ammonium salts added to the working solution, have been cited as effective ways to increase the hydrogenation rate (55).

Anthra System. Operating in the all anthraquinone–anthrahydroquinone system provides benefits in the oxidizer because the anthra species oxidize 5–10 times faster than the tetra species (56). To maintain a low tetra content, combinations of milder operating temperature and low hydrogen partial pressure, more selective catalysts and solvents, and special reaction carriers have been proposed. MGC proposed the continuous dehydrogenation of the tetra specie using a high temperature catalytic technique in which an olefin gas, eg, ethylene, is converted to the corresponding alkane, eg, ethane (57–58). Some ring dehydrogenation of the tetra specie to the anthra specie per equation 2 is one of several functions performed by the basic alumina in the working solution regenerators. The reaction is reasonably slow.

$$3 \text{ THRAQ} \xrightarrow{\text{basic Al}_2\text{O}_3} \text{RAQ} + 2 \text{ THRAHQ} \tag{18}$$

Increased temperature and low initial hydroquinone content favor reversion.

The main anthraquinone degradation route in the anthra system is the formation of anthrone, presumably by reduction of oxanthrone, the transannular tautomer of the alkylanthrahydroquinone. The oxanthrone is readily regenerated by passing through activated alumina, but the anthrone is only marginally affected. Anthrone is subject to other reaction such as formation of nonregenerable dianthrone.

Tetra System. Most principal producers operate within the tetra mode because no specific measures are taken to either suppress the formation of tetra during hydrogenation or to substantially revert it back to anthra as it forms. With time, the working solution contains more tetra than anthra species, and the hydrogenated anthraquinone occurs exclusively as the 2-alkyltetrahydroanthrahydroquinone (**3**) (see eq. 17). This mode of operation is known as the all-tetra system. If the concentration of tetra species increases much above 75% of the total quinone content, further hydrogenation of the substituted ring may occur and lead to octahydroanthrahydroquinone (**4**) formation. Regardless of the operating mode, the formation of the tautomer oxanthrone occurs.

Conventional continuous stirred-tank reactors (CSTRs), tubular, draft-tube agitated, and fixed-bed (59,60) hydrogenator designs have been patented, as has a special carbon candlefilter for use with fine palladium black catalyst (61). Reduction methods involving other than catalytic fixing of hydrogen have been proposed but have little value in large commercial plants.

The catalyst must be active, selective, and stable over a rather long life-span. Slurry-type catalysts are easy to remove and add without need for a process shutdown. These can therefore be rejuvenated by various external means and returned to maintain average catalytic activity at a reasonably high level. Alternatively the hydrogenation system can be started at relatively low catalyst concentration and more catalyst added as necessary to maintain the desired result. As long as the deactivated catalyst does not adversely impact the chemistry, this type of operation can continue until the concentration of catalyst reaches a level where maintaining suspension is limited by system energy input. The catalyst is then removed, rejuvenated, or processed for metal recovery and the cycle repeated (see CATALYSTS, REGENERATION). A fixed-bed-type hydrogenator is not burdened by suspension difficulties. If the palladium remains fixed and the support resists attrition, the preoxidation filtration needs should be less stringent. Maintaining favorable chemistry impact as the bed slowly deactivates is a prime consideration, given the high cost of the anthraquinone species. Periodically the hydrogenator must be shut down either to replace a portion of the bed or for *in situ* rejuvenation of the entire bed.

The Raney nickel catalyst used in the original Riedl-Pfleiderer process was active, caused excessive ring hydrogenation, was easily deactivated by oxygen or hydrogen peroxide, and was pyrophoric. The use of a palladium catalyst, used by all principal producers as of this writing, avoids the problems associated with Raney nickel. It can be used as palladium black, as wire screen or gauze, or supported on a carrier for use in slurry or fixed-bed applications. The activity and selectivity of supported catalyst are influenced by both chemical and physical properties of the support and perhaps even more by the preparation procedure. The supports generally have a high crush strength and are reasonably smooth and rounded to resist attrition, abrasion, and fracturing. A truly round support with minimum asparities and with an aspect ratio near 1.0 would be ideal to avoid attrition. Treating the palladium-supported catalysts with hydrogen at high temperature or adding other platinum group metals to the palladium reportedly improves selectivity.

The most recent and novel fixed-bed-type hydrogenator utilizes a honeycomb monolith having alumina- and silica-modified surfaces on which palladium has been deposited (62). The design is claimed to offer excellent long-term stability and mass-transfer characteristics. A hybrid tubular design incorporating static mixing to improve hydrogen mass transfer has also been claimed (63).

Oxidation. The hydrogenated working solution, which has been filtered or in some other manner freed of catalyst, is oxidized by the noncatalytic reaction with an oxygen-containing gas, which is usually filtered and compressed ambient air. Oxygen or oxygen-enriched air can be used but avoidance of the obvious hazards become more complicated. (None of the principal producers use oxygen or oxygen-enriched air.) Either co- or countercurrent flow of gas and liquid in a single- or multistage system with or without packing or trays can be employed. With coflow up, liquid holdup is maximized, but the design may permit high strength aqueous-phase hydrogen peroxide that separates from the solution to drain back down the vessel walls and collect in the base. Such a solution can approach the theoretical maximum concentration as controlled by the system operating titer and distribution coefficient and could become a serious safety concern. Using countercurrent flow liquid holdup is less, but because the working solution flows down the column, a separate phase, if formed, is flushed out with the working solution into the extractor where it is safely diluted.

Because the reaction takes place in the liquid, the amount of liquid held in the contacting vessel is important, as are the liquid physical properties such as viscosity, density, and surface tension. These properties affect gas bubble size and therefore phase boundary area and diffusion properties for rate considerations. Chemically, the oxidation rate is also dependent on the concentration of the anthrahydroquinone, the actual oxygen concentration in the liquid, and the system temperature (64). The oxidation reaction is also exothermic, releasing the remaining 45% of the heat of formation from the elements. Temperature can be controlled by the various options described under hydrogenation. Added heat release can result from decomposition of hydrogen peroxide or direct reaction of H_2O_2 and hydroquinone (HQ) at a catalytic site (eq. 19).

$$H_2O_2 \text{ (l)} + HQ \text{ (l)} \xrightarrow{\text{catalyst}} 2\,H_2O \text{ (l)} + Q \text{ (l)} \quad \Delta H = -279.9 \text{ kJ/mol} \quad (66.90 \text{ kcal/mol}) \quad (19)$$

The oxidation of alkylanthrahydroquinones by forming noncoalescing foams (65) between the working solution and oxygen or oxygen-containing gas and continuous gas–liquid dispersions (66) has been patented. Minimized investment and energy consumption are claimed and the short contact time reportedly minimizes detrimental effects on working solution components.

Hydrogen Peroxide Recovery. Hydrogen peroxide formed in the oxidation step is usually recovered by countercurrent extraction of the oxidized working solution, using demineralized water in liquid–liquid sieve tray columns. Working solutions used by the principal producers are less dense than water so these would enter near the base of the column and flow upward as the dispersed phase. Water enters the column at the top and increases in hydrogen peroxide content and density as it flows downward as the continuous phase. All known principal producers use sieve tray columns having these flow paths for extraction. Dependent on the type and composition of the working solution, concentrations of hydrogen peroxide up to 45 wt % are obtained by extraction. For safety reasons, 45 wt % aqueous hydrogen peroxide extract is a reasonable limit for nonmiscible organic systems (67). The columns and trays are usually constructed from 304 or 316 stainless steel or low carbon equivalents. Aluminum and high grade aluminum alloys are also adequate materials. The sieve tray extractor's particular advantages are high throughput, reasonably high tray efficiency, and, because they have no moving parts, they are economically maintained. Rate turndown is about 2:1, limited by the dispersed phase droplet size or tray stability. Other extract methods involving use of rotating mechanical devices, packed columns, spray columns, and unfilled columns have been claimed.

The tray-free area can be decreased at intervals from top to bottom as the density differential between the aqueous phase and the working solution widens. This adjustment maintains a nearly constant depth of coalesced working solution beneath each tray. For this type extractor the distance between trays (spacings) is constant from top to bottom. Alternatively, the tray area can be held constant and the height of the coalesced layer beneath the tray permitted to vary, thus providing the needed pressure drop for flow.

Liquid system properties including density, viscosity, and interfacial tension, together with design factors including orifice size and mass flow rate, influence droplet formation, transport surface area, and droplet coalescence. Considerable engineering research has been conducted in Europe on sieve tray column performance and design factors (68,69). For most working solutions the optimum individual tray hole diameter is 2.5–3.0 mm. Tray spacings are best determined experimentally. Hydrophobic netting made from polyethylene, polypropylene, poly(vinyl chloride), and polytetrafluoroethylene has been used in the working solution layer beneath the trays to aid aqueous droplet coalescence and separation.

Hydrogen peroxide can also be recovered directly from the working solution by vacuum distillation or by stripping with organic solvents. The organic solutions obtained by solvent stripping can be used to prepare peroxycarboxylic acids or they can be extracted with water to obtain aqueous hydrogen peroxide of higher strength. The use of extract water containing sodium metaborate has been claimed to give sodium perborate, or that containing aliphatic carboxylic acids to give peroxycarboxylic acid. Other patents describe transfer of hydrogen peroxide from one solvent to another.

Working Solution Regeneration and Purification. Economic operation of an anthraquinone autoxidation process mandates frugal use of the expensive anthraquinones. During each reduction and oxidation cycle some finite amount of anthraquinone and solvent is affected by the physical and chemical exposure. At some point, control of tetrahydroanthraquinones, tetrahydroanthraquinone epoxides, hydroxyanthrones, and acids is required to maintain the active anthraquinone concentration, catalytic activity, and favorable density and viscosity. This control can be by removal or regeneration.

Treating the working solution or isolated quinone mixtures with a dehydrogenation catalyst converts the tetrahydroanthraquinones to the corresponding anthraquinone. Treating with an olefin in the presence of hydrogenation catalysts converts oxanthrones and minor amounts of tetrahydroanthraquinones back to the corresponding anthraquinone (**1**) (70). Treating the working solution with activated alumina or sodium aluminum silicate at moderate temperature (60–90°C) in the presence of a proton source such as the anthrahydroquinone (**2**) converts some of the tetrahydroanthraquinone (**3**) to the corresponding anthraquinone, tetrahydroanthraquinone epoxide (**9**) to the corresponding tetrahydroanthraquinone epoxy alcohol, and oxanthrone to the corresponding anthraquinone. Total conversion of the tetrahydroanthraquinone epoxide also requires dehydration of the intermediate epoxy alcohol. Other procedures for working solution regeneration include treatment with alkali, acids, metal oxides, chlorides, silicates or aluminosilicates, and dithionates, with oxygen, ozone, or purifying by washing with water. *In situ* oxidative regeneration of anthrone upon addition of small amounts of N,N-di-n-butylaniline to the working solution is claimed (71).

Recovery of working solution components includes recrystallization of the anthraquinones, extraction of crude mixtures with lower alcohols, use of anion-exchange resin, or isolation of the hydroquinones (71). Dilution of a heavily degraded working solution with n-heptane is claimed to form two separate phases where most of the degraded products are in the heptane phase. The low boiling heptane is recovered by distillation (72). Another method used to separate the degradation products from the working solution involves extraction with a C-9-alkylated aromatic and a carbon dioxide fluid at 10–20 MPa (1450–2900 psi) and −4 to 10°C (73).

The alumina or sodium aluminosilicate catalysts used to regenerate degraded working solutions lose activity with time as active soda sites are neutralized, but these too can be regenerated.

Hydrogenation Catalyst Regeneration. Procedures for recovering palladium from spent catalyst as well as methods for *in situ* or external regeneration have been developed. These latter methods include treating with wet steam, strong oxidants, liquid ammonia or concentrated ammonium hydroxide, other highly alkaline solutions, carboxylic acids, hydrogen peroxide (74), organic solvents, or heating in air at temperatures ranging from 250 to 700°C. Many of the methods involve multiple steps having various sequential treatments with polar solvents, aqueous ammonium hydroxide, steam, and oxygen-containing gases to remove adsorbed organic matter prior to palladium dissolution and deposition. Periodic reduction of the hydrogen partial pressure and treatment with an inert gas tends to prolong catalyst life by stripping catalyst poisons. Only those treatments which essentially reconstitute the catalyst return it to near-original activity. Other methods improve economics or prolong catalyst life by stripping catalyst poisons.

Alcohol Autoxidation. The noncatalytic partial oxidation of primary and secondary alcohols using air or oxygen in liquid phase gives hydrogen peroxide and the corresponding aldehydes or ketones in good yield. If not operated closed loop, the economics of this process depends on the value of the by-product aldehyde or ketone. A liquid-phase process based on the autoxidation of isopropyl alcohol [67-63-0] was operated for about 20 years at Norco, Louisiana, by Shell Chemical Co. (see PROPYL ALCOHOLS, ISOPROPYL ALCOHOL). The by-product acetone (qv) was marketed and most of the hydrogen peroxide was used captively to produce glycerol (qv).

$$(CH_3)_2CHOH + O_2 \longrightarrow (CH_3)_2C{=}O + H_2O_2 \tag{20}$$

This process was shut down in the late 1970s for economic reasons.

Alcohol autoxidation is carried out in the range of 70–160°C and 1000–2000 kPa (10–20 atm). These conditions maintain the product and reactants as liquids and are near optimum for practical hydrogen peroxide production rates. Several additives including acids, nitriles, stabilizers, and sequestered transition-metal oxides reportedly improve process economics. The product mixture, containing hydrogen peroxide, water, acetone, and residual isopropyl alcohol, is separated in a wiped film evaporator. The organics and water are taken overhead and further refined to recover by-product acetone and the isopropyl alcohol for recycle. The hydrogen peroxide is concentrated at the bottom and must be continuously diluted with water to maintain its concentration near 45% to avoid safety problems that result from hot, concentrated hydrogen peroxide in intimate contact with organic materials. The recovered hydrogen peroxide is purified by scrubbing with solvent, passed through both anion- and cation-exchange resins, and stabilized. Several modifications, special materials of construction treatments, and use of other secondary reactions have been noted (75–77).

In the former USSR, there reportedly are two technologies in use; one is old anthrahydroquinone autoxidation technology and the other is closed-loop isopropyl alcohol oxidation technology. Production facilities include several smaller, 100–150-t/yr isopropyl alcohol oxidation plants and a larger, 15,000-t/yr plant, which reportedly is being expanded to 30,000-t/yr. Differences in this technology as compared to the Shell Chemical Co. process are the use of oxygen-enriched air in the oxidation step and, catalytic reduction of the coproduct acetone back to isopropyl alcohol per equation 21.

$$(CH_3)_2C{=}O + H_2 \longrightarrow (CH_3)_2CHOH \tag{21}$$

Alcohol oxidation patent activity in the 1980s and 1990s by ARCO (78–83) involved the oxidation of methylbenzyl alcohol [98-85-10] (sec-phenethylalcohol) to acetophenone [98-86-2] and hydrogen peroxide. Methylbenzyl alcohol is the main coproduct of propylene oxide [75-56-9] (qv) produced from propylene [115-07-1] (qv) and ethylbenzene hydroperoxide [3071-32-7]. The last is produced by the oxidation of ethylbenzene [100-41-7]. Methylbenzyl alcohol has no significant market value and as of this writing is dehydrated to styrene (qv). Acetophenone is the main by-product and is catalytically hydrogenated back to methylbenzyl alcohol. In the process, ethylbenzene hydroperoxide is catalytically decomposed to methylbenzyl alcohol and hydrogen peroxide, which is then separated by distillation and extraction. The methylbenzyl alcohol is oxidized by molecular oxygen in liquid phase, giving acetophenone and hydrogen peroxide. After separating the hydrogen peroxide, the acetophenone is hydrogenated back to methylbenzyl alcohol completing the cycle, as shown in equations 22 and 23.

$$C_6H_5CH(OH)CH_3 + O_2 \longrightarrow C_6H_5COCH_3 + H_2O_2$$
$$C_6H_5COCH_3 + H_2 \longrightarrow C_6H_5CH(OH)CH_3 \tag{22}$$

to give the net reaction

$$H_2 + O_2 \longrightarrow H_2O_2 \tag{23}$$

As of this writing, the process has not been commercialized, but apparently the alcohol can be separated from its propylene oxide coproduct process to maintain an economically competitive position. The formation of organic hydroperoxides is a concern, as it was in the Shell process.

Hydrocarbon Autoxidation. Hydrogen peroxide can be prepared directly by the vapor-phase partial oxidation of hydrocarbons and indirectly from several hydrocarbon hydroperoxides. Although many patents have been granted, non-processes are known to be practiced on a commercial scale. These vapor-phase oxidation processes generate a great number of by-products which complicate the recovery, purification, and concentration of the hydrogen peroxide. Several methods for recovery have been listed in the literature, including extractive distillation of the crude reaction mixture using acetic acid.

Organic hydroperoxides can be prepared by liquid-phase oxidation of selected hydrocarbons in relatively high yield. Several cyclic processes for hydrogen peroxide manufacture from hydroperoxides have been patented (84,85), and others (86–88) describe the reaction of *tert*-butyl hydroperoxide with sulfuric acid to obtain hydrogen peroxide and coproduct *tert*-butyl alcohol or *tert*-butyl peroxide.

Nitrogen Compound Autoxidation.   Cyclic processes based on the oxidation of hydrazobenzene and dihydrophenazine to give hydrogen peroxide and the corresponding azobenzene–phenazine were developed in the United States and Germany during World War II. However, these processes could not compete economically with the anthrahydroquinone autoxidation process.

Hydrolysis of Peroxycarboxylic Systems.   Peroxyacetic acid [79-21-0] is produced commercially by the controlled autoxidation of acetaldehyde (qv). Under hydrolytic conditions, it forms an equilibrium mixture with acetic acid and hydrogen peroxide. The hydrogen peroxide can be recovered from the mixture by extractive distillation (89) or by precipitating as the calcium salt followed by carbonating with carbon dioxide. These methods are not practiced on a commercial scale. Alternatively, the peroxycarboxylic acid and alcohols can be treated with an esterifying catalyst to form H_2O_2 and the corresponding ester (90,91) (see PEROXIDES AND PEROXY COMPOUNDS).

ELECTROLYTIC METHODS

The electrolytic processes for commercial production of hydrogen peroxide are based on (*1*) the oxidation of sulfuric acid or sulfates to peroxydisulfuric acid [13445-49-3] (peroxydisulfates) with the formation of hydrogen and (*2*) the double hydrolysis of the peroxydisulfuric acid (peroxydisulfates) to Caro's acid and then hydrogen peroxide. To avoid electrolysis of water, smooth platinum electrodes are used because of the high oxygen overvoltage. The overall reaction is

$$2\,HOH + \text{electrical energy} \longrightarrow HOOH + H_2 \qquad (24)$$

This electrolytic process technology is no longer used because of the extensive and continuous electrolyte purification needs, the high capital and power requirements, and economic inability to compete with large-scale anthrahydroquinone autoxidation processes.

The Huron-Dow Process.   The Huron-Dow (H-D) process is a refinement of the cathodic reduction of oxygen in an alkaline electrolyte yielding low strength hydrogen peroxide directly. Earlier attempts relied on neutralizing the excess caustic or forming insoluble metal peroxides (92). The two reactions involved are

Cathode $$O_2 + H_2O + 2\,e^- \longrightarrow OH^- + HOO^- \qquad (25)$$

where oxygen is reduced, giving hydroxyl and perhydroxyl ions, and

Anode $2 \, OH^- \longrightarrow \frac{1}{2} \, O_2 + H_2O + 2 \, e^-$ (26)

where hydroxyl ion is oxidized, giving oxygen and water. The net reaction, in absence of losses, requires passing $2 \, F$ of electricity (192,978 C) through the solution to produce 1 mol of hydrogen peroxide.

The H-D process development, begun in the early 1980s, is intended for on-site production of dilute alkaline hydrogen peroxide for direct use in the pulp and paper industry. Operated on a pilot scale for many years, it was commercialized in 1991 when improvements in the electrolytic cell design and its current efficiency were achieved (93–100). The newest cell design (94–96) consists of a porous cathode and a platinum-coated titanium anode separated by a diaphragm and an ion-exchange (qv) membrane (see MEMBRANE TECHNOLOGY; METAL ANODES). The diaphragm consists of multiple layers of a porous polypropylene composite, which assures uniform flow of the electrolyte. The electrolyte enters near the base of the anode, and the product solution exits near the base of the cathode. Oxygen gas enters near the top of the porous cathode, and oxygen made through the anodic oxidation of hydroxyl ion exits the top of the anode compartment. The ion-exchange membrane is claimed to control the migration of ions into and out of the cathode compartment, reducing peroxide losses and improving the current efficiency. This latter is claimed to be 95% for the electrolysis.

Because the electrolyte contains excess alkali which could cause spontaneous decomposition of the hydrogen peroxide, the H_2O_2 in the product solution is limited to 30–40 g/L. The chelating agent (ethylenedinitro)tetraacetic acid (EDTA) is added at up to 0.5 g/L to protect against tramp metal ion induced decomposition (96). For use in a pulp (qv) mill it may be necessary to blend with purchased hydrogen peroxide to achieve the proper ratio of caustic to hydrogen peroxide (see ELECTROCHEMICAL PROCESSING, INORGANIC).

OTHER METHODS

Direct Combination of Hydrogen and Oxygen. Hydrogen peroxide can be formed directly by thermal, electric-discharge, or metal-activated reaction between hydrogen and oxygen. Silent electric discharge processes have been patented but the power requirements are too high for commercial use. Since about 1910, the study of forming hydrogen peroxide by directly combining the elements has challenged researchers. Investigations during the 1960s and 1970s focused on platinum-group metal-catalyzed processes (101–106). More recent investigations have included use of homogeneous catalysts such as organic complexes of palladium, iron and platinum, iridium, and the use of promoters (107–109); heterogeneous catalysts using acidic aqueous solutions and some additives (110–118); and those processes using heterogeneous catalysts and organic solvents including acetone, methanol, fluorocarbons, and acetonitrile (119–121). Platinum-group metal carriers have included hydrophobic materials such as Teflon, polyethylene, polypropylene, and polystyrene (109); silicic acid (106); and carbon, charcoal, silica, and alumina (116).

In Du Pont patents (116) the catalyst is prepared by spray-drying a mixture of colloidal silica or other carriers and Pt/Pd salts. Aqueous hydrogen peroxide solutions up to 20 wt % are reported for reaction conditions of 10–17°C and

13.7 MPa (140 kg/cm²) with 60–70% of the hydrogen feed selectively forming hydrogen peroxide.

Interest has continued in on-site manufacture of hydrogen peroxide from the elements, particularly for remote sites located considerable distances from world-scale anthraquinone processes. However, no commercial-scale direct combination plants have been constructed as of this writing.

Inorganic Methods. Before the development of electrolytic processes, hydrogen peroxide was manufactured solely from metal peroxides. Early methods based on barium peroxide, obtained by air-roasting barium oxide, used dilute sulfuric or phosphoric acid to form hydrogen peroxide in 3–8% concentration and the corresponding insoluble barium salt. More recent patents propose acidification with carbon dioxide and calcination of the by-product barium carbonate to the oxide for recycle.

Acid ion-exchange resins have been proposed for sodium peroxide-based processes. Alternatively, use of an acid and a water-insoluble organic has been suggested to avoid the hydrogen peroxide recovery problems otherwise encountered in using the more soluble sodium salts. A cyclic synthesis of hydrogen peroxide and hydrazine has been proposed, using an alkali metal and alkali metal peroxide as the common electron donor-acceptor pair (122).

PURIFICATION AND CONCENTRATION

The crude product from any hydrogen peroxide process can be used as such, but commercial grades are further purified, concentrated, and stabilized. Crude products from organic-based processes contain organic impurities which affect color, odor, surface tension, and stability, and are normally pretreated to reduce the carbon content before final purification and concentration by various distillation methods.

Procedures include solvent extraction followed by optional air stripping to remove residual solvent and treatment with synthetic resins, polyethylene, waxes, carbon, and aluminum and magnesium hydroxides and alumina. Active ion-exchange resins have been used to remove both metallic and acidic impurities. More recent patented methods for purifying crude hydrogen peroxide include further contact with an aromatic gasoline in static mixers, followed by a series of coalescing steps to effect phase separation (123), passing the solution through columns packed with halogen-containing porous styrene-divinylbenzene copolymer resin (124), and passing the solution through an anion-exchange resin which has been pretreated with various chelating agents to remove metal ions and organics (125). The amount of nonvolatile organic impurities present in extracted crude hydrogen peroxide from anthrahydroquinone autoxidation processes may be reduced by washing the working solution either continuously or batchwise with water or dilute caustic and discarding the washings.

Concentration of hydrogen peroxide prepared by the autoxidation processes can be carried out safely and conveniently by distillation at reduced pressure. The distillations include liquid or vapor feed to columns consisting of stripping and rectification; vapor feed to columns consisting of rectification only, with or without reboil; or combinations involving both liquid and vapor feeds. General procedures analogous to those developed for concentrating electrolytic process product are used to alleviate yield problems and assure safe operations with heating and concentrating hydrogen peroxide with organic materials.

The hydrogen peroxide is concentrated during distillation and is usually marketed as a 50–70-wt % product. It can be concentrated to higher strengths by redistillation or through freeze–melt techniques. A procedure for concentrating very dilute solutions via formation of organic nitrogen-compound peroxyhydrates has been patented (126). Other patents describe the production of a nearly water-free organic solution of hydrogen peroxide by mixing aqueous hydrogen peroxide with alkyl or cycloalkyl esters of saturated aliphatic carboxylic acids and separating by azeotropic removal of water (127), and production of water-free hydrogen peroxide for nuclear hydroxylation of phenols by mixing the phenol with 70-wt % aqueous hydrogen peroxide and dichloromethane followed by removing water and the dichloromethane through azeotropic distillation (128).

Economic Aspects

U.S. production of hydrogen peroxide is shown in Table 6 (129). The driving influence on H_2O_2 capacity and production has come from the pulp and paper sector, as this industry converts from chlorine bleaching and the attendant dioxin and chloroform formation to the environmentally more benign peroxide chemistry. Since the 1970s, U.S. and North American production has increased significantly owing to start-up of nine new plants and expansions to existing plants. Table 7 lists all North American manufacturers and their estimated capacities.

Total U.S., North American, and both Americas capacities for 1993, as estimated from commissioning notices and public announcements by the manufacturers, are $355–360 \times 10^3$, $505–515 \times 10^3$, and $555–565 \times 10^3$ t/yr,

Table 6. U.S. Production of Hydrogen Peroxide[a,b]

Year	Production, t $\times 10^3$
1960	26.0
1970	55.7
1980	105.8
1982	98.5
1984	126.3
1986	138.0
1988	161.0
1990	216.6
1992	271.8
1994	360[c]
1996	470[c]
1998	580[c]
2000	700[c]

[a] Ref. 129. Courtesy U.S. Department of Commerce, Bureau of the Census.
[b] 100% basis.
[c] Value is estimated.

Table 7. U.S. and North American Manufacturers of Hydrogen Peroxide[a,b]

Manufacturer	Site	Capacity, t × 10³/yr U.S.	Capacity, t × 10³/yr N.A.
Chemprox[c]	Becancour, P.Q.		30
Degussa[d]	Theodore, Ala.	75	75
Du Pont	Memphis, Tenn.	64	64
	Maitland, Ont.		36
	Gibbons, Alta.		36
Eka Nobel	Columbus, Miss.	35	35
FMC Corp.	Spring Hill, W.Va.	38	38
	Bayport, Tex.	43	43
	Prince George, B.C.		41
	Santa Clara, Mexico		10
Fort Howard[e]	Muskogee, Okla.	3	3
Solvay Interox	Deer Park, Tex.	50	50
	Longview, Wash.	50	50
Total		*358*	*511[f]*

[a]100% basis.

[b]All producers use the alkylanthraquinone autoxidation process except where noted.

[c]Chemprox (Oxysynthese) is considering building a 20 × 10³ t/yr plant at Portland, Oreg. or Tacoma, Wash. Estimated start-up in 1995 or 1996.

[d]Degussa is considering adding a 50 × 10³ t/yr plant in North America in the 1997–1998 time frame.

[e]Uses the Huron-Dow process.

[f]Kemira has announced plans for a 34 kt/yr plant at a site in the U.S. or Canada, with estimated start-up to be in late 1995.

respectively. Worldwide capacity is estimated at 1.8–1.9 × 10⁶ t/yr, of which one-half is in Europe.

The January 1993 price for technical-grade H_2O_2 is shown in Table 8. Because of overcapacity and slower growth during the mid-1991 through 1993 recession period, prices, especially to large accounts, had declined significantly as the manufacturers bid for market share to maintain volume. Somewhat higher pricing reflected a more typical marketplace with increasing demand for the last half of 1994 when prices rose about $0.05/kg. Capacity increases are forecast to be required starting in the 1995–1996 period.

Table 8. 1993 Technical-Grade Hydrogen Peroxide Price Schedule[a]

Concentration, wt %	Price,[b] $/kg Commodity basis	Price,[b] $/kg 100% basis
35	0.541	1.543
50	0.755	1.510
	1.091[c]	2.182[c]
70	1.052	1.503

[a]Ref. 130.

[b]Fob nearest producer, material in tank cars or trailers unless otherwise noted.

[c]Material in 0.208-m³ drums.

Specifications and Standards

Aqueous hydrogen peroxide is sold in grades ranging from 3 to 86 wt %, most often containing 35, 50, and 70 wt % H_2O_2. In the United States, the only use for 86% H_2O_2 (ca 1994) is for attitude (flight path behavior) control of space craft, etc. This minor amount is imported.

The 3–6% H_2O_2 solutions for cosmetic and medicinal use are obtained by diluting a more concentrated grade, usually with the addition of extra stabilizer. The 3% grade has a USP specification (131) (see COSMETICS).

The 30% reagent-grade hydrogen peroxide is purer than the industrial grades, is covered by ACS reagent specification, and is used as a laboratory reagent and in some specialty uses (see FINE CHEMICALS). Several grades are also marketed for electronics use and thus have exceptionally low impurity levels. Some of these latter contain very little or no stabilizers (see ELECTRONIC MATERIALS).

The 35 and 50% H_2O_2 concentrations are used for most industrial applications. The standard grades contain sufficient stabilizers to ensure safety. Grades having lower stabilizer content and evaporative residue are available for specialty uses and for aseptic cleansing of containers in food processing and marketing.

The 70% H_2O_2 concentration is used in chemical processing and for certain organic oxidations. Dilution-grade 70% H_2O_2 is stabilized to allow for later dilution to 35–50% storage concentration.

Concentrations over 8 wt % are classified as corrosive liquids by the Department of Transportation (DOT). The Bureau of Explosives regulation (132) classifies all solutions containing 20 wt % and greater as oxidizers and corrosives. The product containers must have identifying labels (drums) or placards (tank cars, tank trailers) indicating that the contents are an oxidizer and corrosive material, UN 2014 or UN 2015. Bills of lading must also be so identified. Tank cars and tank trailers are constructed from high purity aluminum or 300 series stainless steel.

Up to 52 wt % is permitted to be shipped in uv light-protected polyethylene drums. Double headed aluminum drums are permitted for 70% shipments but have not been used for some years.

Analytical and Test Methods

Analytical methods for the qualitative or quantitative determination of hydrogen peroxide are based on its redox or physical properties. Aqueous hydrogen peroxide solutions are usually titrated with potassium permanganate, a strong oxidizer capable of oxidizing many organic materials. Therefore, titrations of hydrogen peroxide in the presence of organic matter may be biased on the high side. Colorimetric and specific procedures are available (133,134), as are other procedures (impurities and stability) (131,135–137).

High pressure liquid chromatography (qv) (138) and coulometry can be used to detect and quantify anthraquinones and their derivatives in a hydrogen peroxide process working solution.

Health and Safety Factors

Hydrogen peroxide, especially in high concentrations, is a high energy material and a strong oxidant. The Comprehensive Environmental Response, Compensation, and Liability Act (CERCLA) reportable spill quantity for greater than 52% hydrogen peroxide is 1 lb (0.45 kg). It is considered an acute, reactive, and pressure hazard under Superfund and Reauthorization Amendments (SARA) Title III. However, it can be handled safely if proper personal protective equipment is worn and the proper precautions are observed (136). Some generally applicable control measures and precautions for handling hydrogen peroxide include the use of adequate ventilation to keep airborne concentrations below exposure limits, 8 h TWA, 1.4 mg/m^3, use of coverall chemical splash goggles in combination with a full-length face shield if spraying is a potential occurrence, use of a NIOSH/MSHA-approved respirator if airborne concentration can exceed exposure limits, and use of neoprene or other impervious and compatible gloves. Other clothing items such as impervious aprons, pants, jackets, hoods, boots, and totally encapsulating chemical suits with breathable air supply should be available for use as necessary. Leather (qv) gloves and shoes (uppers or soles) should not be worn because these can ignite within three minutes following contact with 50% hydrogen peroxide or greater. Cotton (qv) items should be avoided for the same reason.

Health and Physiological Effects. Hydrogen peroxide is irritating to the skin, eyes, and mucous membranes. However, low concentrations (3–6%) are used in medicinal and cosmetic applications. Precautions should be taken with higher strengths to prevent inhalation of the vapors, ingestion, or splashing into the eyes. In case of accidental contact with hydrogen peroxide, the affected area should be washed immediately with excess water.

Tests using rabbits indicate that transition from an eye irritant to a corrodant occurs at about 10% H$_2$O$_2$. At strengths 10% and above, a splash in the eye can cause severe ulceration and irreversible corneal damage unless flushed immediately with water. Inhalation of concentrated vapors can cause irritation of the nose and throat with chest discomfort, cough, difficult breathing, and shortness of breath. Ingestion can cause irritation of the gastrointestinal tract with pain, bleeding, and, distension of the stomach and esophagus, resulting from the liberation of oxygen. Gross overexposure by ingestion may be fatal. No component in industrial hydrogen peroxide solutions is listed as a carcinogen.

Decomposition and Explosive Hazards. The principal hazards associated with hydrogen peroxide include (1) decomposition of H$_2$O$_2$ with unrelieved pressure buildup; (2) spontaneous combustion of mixtures of H$_2$O$_2$ and readily oxidizable material; (3) inadvertent admission of incompatible materials into a tank containing H$_2$O$_2$ or vice versa; (4) decomposition of H$_2$O$_2$ to form an oxygen-rich vapor phase; (5) deflagration, detonation of a condensed-phase mixture of H$_2$O$_2$ and organics initiated by shock or thermal effects; and (6) explosive reaction of H$_2$O$_2$ vapor.

Hydrogen peroxide decomposes with the generation of heat and oxygen. The decomposition is promoted by catalytic impurities and its rate increases by a factor of about 2.2–2.3 for each 10°C rise in temperature over the range of 20–100°C (136). If the system is contained, and decomposition starts, the temperature increases and the volume of oxygen released increases the pressure so that the rate becomes self-accelerating. Pressures reaching several thousand kPa are achievable and ruptures of tanks, valves, piping, etc, occur. The hazard from decomposition increases with increasing concentration because of relative gas generation. Aqueous solutions containing more than 65 wt % can totally vaporize by adiabatically absorbing the heat of decomposition, resulting in much higher system temperature. For example, under ideal adiabatic conditions 1 m^3 of 70% H_2O_2 can generate 2545 m^3 of gas and reach 233°C, but 1 m^3 of 50% H_2O_2 can generate 1280 m^3 of gas at close to 100°C. Thus it is imperative that decomposition be controlled for safe storage and handling. Containers and equipment must be constructed from compatible materials, be adequately vented, and maintained free of contaminants (136). Drums and other portable shipping containers should preferably be stored or shipped alone; so, in the event of an accident, the H_2O_2 cannot mix with common combustibles or flammable liquids and gases.

In organic processes, such as organic oxidations using high strength hydrogen peroxide, decomposition can result in an oxygen-enriched organic vapor space. The higher oxygen content has little effect on the lower explosive limit but significantly lowers the required ignition energy. Another potential is the formation of a condensed-phase organic–hydrogen peroxide mixture at or near the stoichiometric ratio. Such a stoichiometric mixture can deflagrate or detonate if the mixing is sufficient to approach emulsification and sufficient external energy in the form of heat or shock is added. The external energy requirement is usually quite large, so condensed-phase explosions would be rare. A vapor-phase explosion could possibly propagate to the condensed phase by this mechanism.

Many inorganic reagents are incompatible with hydrogen peroxide. Among these are nitric (139) and sulfuric acid (67) above certain concentrations, and mercurous oxide (140), which can react with explosive violence. Many organic compounds can form unstable peroxides during reaction with hydrogen peroxide. The most frequent explosive rupture incidents involve strong bases and hydrogen peroxide. Hydrogen peroxide is less stable at high pH; thus the inadvertent admission of caustic into a tank of hydrogen peroxide could result in runaway decomposition and tank destruction. A review of each installation should be made to assure that design features are in place so as to avoid such contamination. A thorough safety review is a must for all new use applications, and explosion testing may be required.

Hydrogen peroxide concentrations of 35% and above may, and 50% and above do, cause spontaneous combustion of dry grasses, wood, and leaves. Spills of concentrated hydrogen peroxide should be diluted with excess amounts of water immediately. Practically all solid combustible materials contain sufficient quantities of catalytic impurities to rapidly decompose hydrogen peroxide, especially at 70% concentration and above.

It is impossible to obtain a propagating detonation of commercial hydrogen peroxide at ambient temperature under normal conditions of storage (141). Concentrations of 86% and above can detonate, but only if subject to a high energy source. In the vapor phase, explosions occur under certain conditions. The lower explosive limit at atmospheric pressure is about 26 mol % H_2O_2, which is close to the equilibrium vapor concentration above boiling 75% hydrogen peroxide.

Explosive vapor concentration above 90% hydrogen peroxide is avoided by maintaining its temperature below ca 115°C. Under vacuum, the lower explosive concentration limit increases, which provides a safe route to concentrating by fractionation.

Uses

Hydrogen peroxide is used in many applications throughout a wide variety of industries. The principal use areas are shown in Table 9. Most are based on the oxidizing properties of hydrogen peroxide. Some are derived from substitution, decomposition, or the formation of perhydrates.

Bleaching. The largest single use for hydrogen peroxide in the United States and North America is wood pulp bleaching, but consumption for the manufacture of chemicals, environmental applications, and for bleaching cotton (qv), wool (qv), and other textiles (qv) is significant.

Environmental concerns have led the pulp and paper industry to turn to alkaline solutions of hydrogen peroxide as a replacement for chlorine and, in some cases, for hypochlorite and chlorine dioxide in bleaching applications (143–152). These include mechanical and kraft pulp bleaching and deinking (153) of recycled paper (see RECYCLING, PAPER). Kraft pulp bleaching is the largest application in the United States, where hydrogen peroxide has been used in the final bleaching sequence to obtain very white product having maximum color stability. A variety of bleaching sequences are in use and hydrogen peroxide is now being used in a multiple number of the steps. The annual growth rate of hydrogen peroxide in this industry through the early 1990s has been about 12–16%. Future growth rate is expected to depend on how fast and how far the pulp manufacturers move toward chlorine-free bleaching. Most of the hydrogen peroxide use in Canada is at more recently constructed large chemithermal mechanical pulp (CTMP) mills. These modern mills can consume 5000–10,000 t/yr of hydrogen peroxide.

The bleaching of cotton textiles was once the single largest use for hydrogen peroxide, with lesser quantities used to bleach wool, silk (qv), cotton–synthetic blends and some vegetable and animal fibers (qv). The stabilized alkaline hydrogen peroxide bleaching of cotton and cotton blend fabrics is done in continuous processes. A primary advantage for hydrogen peroxide in this industry is that

Table 9. Usage of Hydrogen Peroxide in North America

Use area	H_2O_2, t $\times$ 10^3/yr[a]					
	1978		1980		1991	
pulp and paper	10.2	(10.0)	31.4	(29.1)	144	(49.1)
textiles	25.5	(25.0)	17.1	(15.8)	24	(8.2)
chemicals[b]	28.6	(28.0)	30.0	(27.7)	44	(15.0)
environmental	10.2	(10.0)	18.1	(16.7)	43	(14.7)
all others	27.5	(27.0)	11.6	(10.7)	38	(13.0)
Total	*100.2*		*108.2*[c]		*293.0*[c]	

[a]Values in parentheses are percentages.
[b]Includes plasticizers.
[c]Ref. 142.

it has no effect on many modern dyes. It has been estimated that 85% of all cotton fabrics are bleached with hydrogen peroxide. Added to its other advantages, hydrogen peroxide is a nonpolluting oxidant which is of significant and increasing importance.

Hydrogen peroxide is also used to bleach solid surfaces such as wood (qv) or linoleum, and to improve the color of oils and waxes.

Environmental Applications.   Hydrogen peroxide is an ecologically desirable pollution-control agent because it yields only water or oxygen on decomposition. It has been used in increasingly greater amounts to convert domestic and industrial effluents to an environmentally compatible state (see WASTES, INDUSTRIAL; WATER).

Hydrogen peroxide is used to treat wastewaters (144–161) and sewage effluents, and to control hydrogen sulfide generated by the anaerobic reaction of raw sewage in sewer lines or collection points, thus minimizing or eliminating disagreeable odors. It has been proposed as a supplemental oxygen source for overloaded activated sludge plants, and reportedly controls denitrification in secondary clarifiers and improves bulking conditions. It also has been used as a flotation (qv) assistant (162–163). Activated sludge-treated wastewater chemical oxygen demand (COD) values have been decreased using hydrogen peroxide with an iron salt catalyst (162–163). It has also been generated *in situ* in wastewater reservoirs by the cathodic reduction of oxygen (164).

Industrial liquid and gaseous detoxification systems based on hydrogen peroxide have been developed (149,150,157–161,165–173). H_2O_2 is reported to be especially suitable for cyanide-containing wastes having high free cyanide concentration or those also containing organic impurities (173). Hydrogen peroxide systems with formaldehyde (169), copper salt, water glass, or iodide and silver ion added to increase reaction rates and efficiency have been described or patented. A method and apparatus for cyanide destruction by monoperoxysulfuric acid generated as needed from hydrogen peroxide and sulfuric acid has been claimed. Procedures have been described for removing nitrite ion from waste streams and for recovering arsenic acid or arsenates. Similar procedures using hydrogen peroxide have been developed for detoxifying organic pollutants (168), including formaldehyde (qv), phenol mixtures (169), acetic acid, lignin sugars, surfactants (qv), amines and glycol ethers (159–161), and sulfur derivatives (170–172). Most recently, there has been interest in using hydrogen peroxide as an *in situ* oxygen source for maintaining aerobic bacteria needed for the bioremediation of organics, eg, gasoline that has leaked into the soil from underground tanks.

Toxic or malodorous pollutants can be removed from industrial gas streams by reaction with hydrogen peroxide (174,175). Many liquid-phase methods have been patented for the removal of NO_x gases (138,142,174,176–178), sulfur dioxide, reduced sulfur compounds, amines (154,171,172), and phenols (169). Other effluent treatments include the reduction of biological oxygen demand (BOD) and COD, color, odor (142,179,180), and chlorine concentration.

Chemical Uses.   Hydrogen peroxide or a peroxycarboxylic acid made from H_2O_2 is used in the manufacture of a number of organic and inorganic chemicals (181). The electrophilic epoxidation of soybean oil, linseed oil, and related unsaturated esters with peroxyformic or peroxyacetic acid formed *in situ* from aqueous hydrogen peroxide has been used to prepare plasticizers (qv) and stabilizers. Processes have been developed for continuous epoxidation of low molecular weight olefins, such as propylene (182–185), using water-immiscible organic solutions of a peroxycarboxylic acid, and for the continuous preparation of peroxy acid solutions using hydrogen peroxide (186). The epoxidation of water-soluble olefins by hydrogen peroxide with metal or fluoroketone catalysts has been described (187) in the literature. The hydroxylation of olefins is an important use for hydrogen peroxide, especially in the synthesis of glycerol (qv) from propylene. The catalyzed hydroxylation of phenols yields catechol and hydroquinone (181,188–191).

Hydrogen peroxide or its derivatives have been used to prepare long-chain amine oxides for detergent use, thiourea dioxide for oxidizing sulfur and vat dyes in textile processing, di-*tert*-butyl-peroxy-cyclohexanes (192) for use in rubber vulcanization, magnesium silicate sols (193) for fiber coating and forming, hydroxyimidazoles (194) for use as antimycotics (see ANTIPARASITIC AGENTS) and herbicides (qv), and 5-hydroxy-hydantoin (194) for use in penicillin and cephalosporin synthesis (see ANTIBIOTICS; HYDANTOIN AND ITS DERIVATIVES).

Interest in hydrogen peroxide as an active chemical ingredient for preparing other large-volume chemicals is growing. Processes have been developed for the preparation of hydrazine (see HYDRAZINE AND ITS DERIVATIVES) (181) and caprolactam (qv) (181,195). Other chemical uses for hydrogen peroxide include the preparation of cyanogen, cyanogen chloride (181,195), bromine (197), and iodic acid (198).

Derivative Formation.   Hydrogen peroxide is an important reagent in the manufacture of organic peroxides, including *tert*-butyl hydroperoxide, benzoyl peroxide, peroxyacetic acid, esters such as *tert*-butyl peroxyacetate, and ketone derivatives such as methyl ethyl ketone peroxide. These are used as polymerization catalysts, cross-linking agents, and oxidants (see PEROXIDES AND PEROXY COMPOUNDS).

Zinc, calcium, and magnesium peroxides prepared from hydrogen peroxide are used as specialized oxidants where a slow release of hydrogen peroxide is desired. These find use in various medicinal treatments.

Hydrogen peroxide is used, particularly in Europe with the hotter wash temperatures, to manufacture sodium peroxyborate for household use and detergent bleach applications and for some chemical reactions which need a solid source of hydrogen peroxide. Sodium carbonate perhydrate and urea peroxyhydrate are used in similar applications. Potassium monoperoxysulfate is manufactured for uses requiring a highly effective bleaching agent.

Mining.   Hydrogen peroxide, in combination with various carbonates or bicarbonates, is used as an oxidant for the in-place solution mining of low grade uranium ores (199–202). It is also used to precipitate uranium from ion-exchange eluates and solvent-extraction strip liquors and has been proposed as a fracturing agent in the *in situ* mining of low grade copper (qv) ores and residues. Procedures for leaching copper, manganese, zinc (203), and gold and silver (204) from milled ore using hydrogen peroxide and for the recovery of ruthenium (205,206) and uranium (207,208) have been patented.

Propellant.   The catalytic decomposition of 70% hydrogen peroxide or greater proceeds rapidly and with sufficient heat release that the products are oxygen and steam (see eq. 5). The thrust developed from this reaction can be used to propel torpedoes and other small missiles (see EXPLOSIVES AND PROPELLANTS). An even greater amount of energy is developed if the hydrogen peroxide or its decomposition products are used as an oxidant with a variety of fuels.

Other Uses. There are numerous small speciality uses for hydrogen peroxide. These include oxidizing metal ions to a higher valence state to facilitate subsequent removal, chemical polishing metal surfaces, and other metal surface treatments (qv). By catalytically decomposing hydrogen peroxide, the oxygen can be used as an *in situ* blowing agent for preparing certain foam rubbers and plastics (see FOAMED PLASTICS). Hydrogen peroxide, peroxyhydrates, and organic peroxides are used as disinfectants (see DISINFECTANTS AND ANTISEP-TICS) and antimycotics to sterilize contact lenses (qv). Small amounts are used in cosmetic preparations which employ its bleaching and oxidizing properties. Minor amounts are also used to prepare aseptic packaging for foods (see FOOD PACKAGING). Patents have been granted which claim using hydrogen peroxide as an oxygen source for various purposes including the maintenance of breathable concentrations (see OXYGEN-GENERATING SYSTEMS).

BIBLIOGRAPHY

"Hydrogen Peroxide" in *ECT* 1st ed., Vol. 7, pp. 727–741, by N. A. Milas, Massachussetts Institute of Technology; Suppl. 1, pp. 418–429, by H. A. Bewick and J. K. Farrell, Solvay Process Division, Allied Chemical & Dye Corp.; in *ECT* 2nd ed., Vol. 11, pp. 391–417, by A. F. Chadwick and G. L. K. Hoh, E. I. du Pont de Nemours & Co., Inc.; in *ECT* 3rd ed., Vol. 13, pp. 12–38, by J. R. Kirchner, E. I. du Pont de Nemours & Co., Inc.

1. I. B. Rozhdestvenskii, V. N. Gutov, and N. A. Zhigulskaya, *Sb. Tr.-Glavniiprockt Energ. Inst.* **7**, 526 (1962); O. K. Borggaard, *Acta Chem. Scand.* **26**, 3393 (1973); H. Nakamura, T. Hatamoto, and I. Nakamori, *Kogaku Kogaku Rombunshu* **2**, 606 (1976); J. L. Arnau, P. A. Giguere, A. Motoko, and R. C. Taylor, *Spectrochim. Acta* **30A**, 777 (1974); P. A. Giguere and T. K. K. Srinivasan, *J. Raman Spectrosc.* **2**(2), 225 (1974); J. L. Arnau and P. A. Giguere, *Phys. Chem. Ice Pap. Symp.* 1972, 66 (1973).
2. M. G. Faulkner, *Diss. Abstr. Int.* **B35**, 4096 (1975); J. L. Arnau and P. A. Giguere, *Can. J. Spectrosc.* **17**(4), 121 (1972); H. Chen and P. A. Giguere, *Spectrochim. Acta* **29A**, 1611 (1973); K. Osafune and K. Kimura, *Chem. Phys. Lett.* **25**(1), 47 (1974); A. W. Ellenbroek and A. Dynamus, *Chem. Phys.* **31**(1), 107 (1978); L. I. Nekrasov, *Zh. Fiz. Khim.* **46**(11), 2143 (1972); P. A. Giguere, *J. Chem. Thermodyn.* **6**, 1013 (1974).
3. C. L. Yaws and H. S. N. Setty, *Chem. Eng.* **81**, 67 (1974).
4. J. G. Wallace, *Hydrogen Peroxide in Organic Chemistry*, 1st ed., E. I. du Pont de Nemours & Co. Inc., Wilmington, Del., 1960, pp. 60–62.
5. J. K. Kochi, *Organometallic Mechanisms and Catalysis*, Academic Press, Inc., New York, 1978, pp. 65–68.
6. *Ibid.*, Chapt. 4.
7. J. E. Lyons, *Aspects Homogeneous Catal.* **3**, 1 (1977).
8. I. L. Vasilenko, *Sb. Tr. Belgorod. Tekhnol. Inst. Materialov* **15**, 87 (1975).
9. A. McAuley, *Inorg. React. Mech.* **5**, 107 (1977).
10. J. Mackenzie, *Chem. Eng.*, 84–90 (June 1990).
11. U.S. Pat. 4,294,575 (1981) (to Monsanto Industrial Chemicals); U.S. Pat. 4,304,762 (1981) (to Unilever Inc.); Can. Pat. CA 1,152,292 (1983) (to Oxysynthese).

12. U.S. Pat. 4,034,064 (July 5, 1977), J. A. Cook (to PPG Industries, Inc.); U.S. Pat. 4,061,721 (Dec. 6, 1977), W. A. Strong (to PPG Industries, Inc.).
13. U.S. Pat. 4,362,706 (Dec. 7, 1982), P. E. Willard (to FMC Corp.).
14. Brit. Pat. GB008210 (Apr. 12, 1989), (to Interox Chemicals Ltd.).
15. Eur. Pat. EP49808 (1982), (to Air Products and Chemicals).
16. Jpn. Kokai 78-102297 (Sept. 6, 1978), S. Shiga and T. Inada (to Furukawa Electric Co.).
17. Jpn. Kokai 53-102895 (Sept. 7, 1978), S. Shiga and co-workers (to Furukawa Electric Co.).
18. U.S. Pat. 4,140,772 (Feb. 2, 1979), T. F. Korenowski (to Dart Industries).
19. U.S. Pat. 4,744,968 (May 17, 1988), M. J. Malin (to Technicon Instruments).
20. U.S. Pat. 4,879,043 (1989), (to Du Pont).
21. Eur. Pat. EP 483170-A (1990), (to Interox Chemicals, Ltd.).
22. U.S. Pat. 4,070,442 (Jan. 24, 1978), J. C. Watts (to Du Pont).
23. U.S. Pat. 4,132,762 (Jan. 2, 1979), L. Kim (to Shell Oil Co.).
24. U.S. Pat. 4,133,869 (Oct. 31, 1979), L. Kim (to Shell Oil Co.).
25. U.S. Pat. 4,770,808 (Sept. 13, 1988), C. F. McDonough and co-workers (to Interox Chemicals, Ltd.).
26. Belg. Pat. BE-000796 (1989), (to Interox Chemicals, Ltd.).
27. Brit. Pat. GB 025376 (1990), (to Interox Chemicals, Ltd.).
28. U.S. Pat. 5,078,672 (1992), (to FMC Corp.).
29. Ger. Offen. 2,742,907 (Oct. 26, 1978) (to USDA).
30. P. G. Cookson, A. G. Davies, and N. Fazal, *J. Organomet. Chem.* **99**(2), 1 (1975).
31. A. Hansson, *Acta. Chem. Scand.* **15**, 934 (1961).
32. A. G. Davies, *Organic Peroxides*, Butterworth & Co. Inc., London, 1961.
33. R. Criege, in K. Wiberg, ed., Academic Press, Inc., New York, 1965.
34. D. F. Sangster, in S. Patai, ed., Vol. 1, Wiley-Interscience, New York, 1971.
35. D. J. Hucknall, *Selective Oxidation of Hydrocarbons*, Academic Press, Inc., London, 1974.
36. U.S. Pats. 3,969,201 (July 13, 1976) and 4,118,305 (Oct. 3, 1978), C. W. Olomam and co-workers (to Canadian Patents and Development).
37. U.S. Pats. 4,224,129 (Sept. 23, 1980), 4,317,704 (Mar. 2, 1982), 4,406,758 (Sept. 27, 1983), and 4,431,494 (Feb. 14, 1984), J. A. McIntyre and co-workers (to The Dow Chemical Co.).
38. U.S. Pat. 5,106,464 (1992), (to H-D Tech. Inc.).
39. U.S. Pats. 2,158,525 (May 16, 1939) and 2,215,883 (Sept. 24, 1940), H. Reidl and G. Pfleiderer (to I. G. Farbenindustrie).
40. Ger. Offen. 649,234 (1934), G. Pfleiderer (to I. G. Farbenindustrie).
41. Jpn. Pats. 76 20,198, 76 20,199 (June 23, 1976), (to Mitsubishi Gas-Chemical Co., Ltd.).
42. Ger. Pat. 1,667,515 (Nov. 20, 1975) V. I. Franchuk and co-workers.
43. Brit. Pat. 1,425,202 (Feb. 18, 1976), L. G. Vaughn (to Du Pont).
44. U.S. Pat. 3,923,967 (Dec. 2, 1975), J. R. Kirchner and L. G. Vaughn (to Du Pont).
45. U.S. Pat. 3,676,878 (1973), (to Degussa).
46. U.S. Pats. 2,657,980 (1953) and 4,046,868 (1977); Ger. Pats. 888,849 (1949) and 1,0030,314 (1966) (to Du Pont).
47. U.S. Pats. 4,800,073 and 4,800,074 (1989), (to EKA Nobel AB).
48. U.S. Pats. 3,996,341 (1976) and 4,394,369 (1983), (to FMC Corp.).
49. Ger. Pats. 933,088 (1951), (to LaPorte Chemicals Ltd.) and 2,013,299 (1970), (to Solvay & Cie), and Brit. Pat. 1,390,408 (1975), (to Solvay & Cie).
50. Fr. Pat. 1,234,036 (1958), (to Mitsubishi Edogawa Chemical Co.).
51. Jpn. Pat. 70 71894 (1970), (to Mitsubishi Edogawa Chemical Co.).

52. Ger. Pat. 938,252 (1952), (to LaPorte Chemicals Ltd.).
53. Brit. Pat. 718,307 (Nov. 10, 1954), J. A. Williams and C. W. Lefeuvre (to LaPorte Chemicals Ltd.).
54. U.S. Pat. 3,423,176 (Jan. 21, 1969), G. Kabisch and H. Herzog (to Degussa).
55. Ger. Pat. 1,244,129 (July 13, 1967) H. Herzog and G. Kabisch (to Degussa); U.S. Pat. 3,307,909 (Mar. 7, 1967), V. J. Reilly (to Du Pont).
56. H. Pistor, in K. Winnacker and L. Kuchler, eds., *Chemische Technologie, I. Anorganische, Technologie 1*, Hanser, Munich, Germany, 1969, pp. 41–42.
57. Fr. Pat. 1,234,036 (1958), (to Mitsubishi Edogawa Chemical Co.).
58. Jpn. Pat. 70 71,894 (1970), (to Mitsubishi Edogawa Chemical Co.).
59. Ger. Offen. 1,925,034 (1969) and 2,151,104 (1972), (to FMC Corp.); U.S. Pat. 4,800,075 (Jan. 24, 1989), C. L. Jenkins (to Du Pont).
60. U.S. Pat. 4,552,748 (1985), T. Berglin (to EKA AB).
61. U.S. Pat. 3,433,358 (1969), (to Degussa).
62. U.S. Pat. 4,552,748 (Nov. 12, 1985) and EP 0 102,934 B1 (1986), (to Eka Nobel AB).
63. Brit. Pat. 2,236,746 A (Apr. 17, 1991), I. Turunen and E.-L. Mustonen (to Kemira Oy).
64. E. Santacesaria, R. Ferro, S. Ricci, and S. Cara, *Ind. Eng. Chem. Res.* **26**, 155–159, (1987).
65. Can. Pat. 1,283,273 (Apr. 4, 1991) and Austria Pat. A 1361/85 (July 5, 1985), J. Kemnade (Germany) and B. Maurer (Austria) (to Osterreichische Chemische Werke GmbH).
66. Brit. Pat. Appl. GB 2,236,746A (Apr. 17, 1991), I. Turunen and E.-L. Mustonen (to Kemira Oy).
67. *Concentrated Hydrogen Peroxide*, Shell Technical Bulletin No. SC59:44R, Shell Chemical Corp., 1959, p. 69.
68. H.-D. Muller and Th. Pilhofer, *Chem. Ing. Techn.* **48**(11), 1069 (1976).
69. D. Mewes and K. Kunkel, *Ger. Chem. Eng.* **1**(2), 111–115 (1978).
70. U.S. Pat. 3,965,251 (June 6, 1976), H. Shin and co-workers (to Mitsubishi Gas-Chemical Co., Ltd.).
71. U.S. Pat. 4,668,499 (1987), J. D. Rushnere (to Du Pont).
72. Ger. Offen. 2,012,988 (Sept. 24, 1979), W. R. Logan (to LaPorte Industries, Ltd.).
73. U.S. 4,668,436 (May 26, 1987), D. S. Sethi (to FMC Corp.).
74. U.S. 4,824,609 (Apr. 25, 1989), D. S. Sethi (to FMC Corp.).
75. Jpn. Pat. 76 14,116 (Feb. 25, 1976), E. Y. Pneva.
76. Can. Pat. 1,003,189 (Jan. 11, 1977), A. A. Shkurkina.
77. Hung. Pat. 12,766 (Jan. 28, 1977), E. Y. Pneva.
78. U.S. Pat. 4,393,038 (July 12, 1983), R.-H. Sun (to Atlantic Richfield Co.).
79. U.S. Pat. 4,547,354 (Oct. 15, 1985).
80. U.S. Pats. 4,897,085 and 4,897,252 (Jan. 30, 1990), R. N. Cochran and co-workers (to ARCO Chemical Technology).
81. U.S. Pat. 4,975,266 (Dec. 4, 1990), R. S. Albal, R. N. Cochran, and L. M. Candela (to Atlantic Richfield Co.).
82. U.S. Pats. 4,994,625 (Feb. 19, 1991), R. S. Albal and R. N. Cochran; 5,039,508 (Aug. 13, 1991), R. N. Cochran and L. M. Candela; 5,046,680 (Aug. 20, 1991), R. S. Albal, R. N. Cochran, and A. P. Woinsky (to Atlantic Richfield Co.).
83. U.S. Pat. 5,149,414-A (1991), (to Atlantic Richfield Co.).
84. U.S. Pat. 4,131,646 (Dec. 26, 1978), L. W. Gosser (to Du Pont).
85. U.S. Pat. 3,899,576 (Aut. 12, 1974), R. Rosenthal and J. A. Kieras (to Atlantic Richfield Co.).
86. Ger. Offen. 2,100,784 (July 22, 1971), J. O. Turner (to Sun Oil Co.).
87. U.S. Pat. 3,737,518 (June 5, 1973), G. Bonetti, R. Rosenthal, and J. A. Kieras (to Atlantic Richfield Co.).

88. Neth. Pat. Appl. 7,533/74 (Dec. 20, 1974), (to Atlantic Richfield Co.).
89. U.S. Pat. 3,341,297 (Sept. 12, 1967), A. F. MacLean and A. L. Stautzenberger (to Celanese Corp.).
90. U.S. Pat. 3,124,421 (Mar. 10, 1964), W. Lohringer and J. Sixt (to Wacker GmbH).
91. Jpn. Kokai 76 2,700 (Jan. 10, 1976), N. Isogai, T. Ikawo, and T. Tokeda (to Mitsubishi Gas-Chemical Co., Ltd.).
92. Ger. Offen. 2,331,296 (Jan. 16, 1975) and 2,501,342 (July 22, 1976), B. Kastening and W. Paul (to Kernforschungsanlage Juelich GmbH).
93. U.S. Pats. 4,317,704 (Mar. 2, 1982); 4,224,129 (Sept. 23, 1980); 4,260,469 (Apr. 7, 1981); 4,341,606 (July 27, 1982); 4,189,510 (Feb. 19, 1980); 4,345,429 (Sept. 7, 1982); 4,204,918 (May 27, 1980); 4,187,350 (Feb. 5, 1980), J. A. McIntyre and co-workers (to Dow Chemical).
94. U.S. Pat. 4,431,494 (Feb. 14, 1984), J. A. McIntyre and co-workers (to Dow Chemical).
95. U.S. Pat. 4,406,758 (Sept. 27, 1983), J. A. McIntyre and co-workers (to Dow Chemical).
96. U.S. Pat. 4,511,411 (April 16, 1985), J. A. McIntyre and co-workers (to Dow Chemical).
97. U.S. Pat. 4,872,957 (Oct. 10, 1989), D. F. Dong and co-workers (to H-D Technologies).
98. U.S. Pat. 4,891,107 (Jan. 2, 1990), D. F. Dong and co-workers (to H-D Technologies).
99. J. A. McIntyre and co-workers in *Proc. 1983 Electrochem. Soc. Symp., Electrochem. Proc. Plant Design (1982)*, pp. 79–97.
100. A. Clifford and co-workers *Electrosynthesis of Alkaline Hydrogen Peroxide*, from *The Electrochemical Society meeting, Kinsgton, Ontario, Canada*, May 19, 1990.
101. Brit. Pats. 1,041,045 and 1,041,045 (Sept. 1, 1966); Brit. Pats. 1,056,121–1,056,126 (Jan. 25, 1967); Brit. Pat, 1,094,804 (Dec. 13, 1967); Fr. Pat. 1,366,253 (June 1, 1964), (to Imperial Chemicals, Ltd.).
102. Fr. Pat. 1,214,015 (Apr. 5, 1960), (to Engelhard Industries, Inc.).
103. Jpn. Kokai 75 145,394, 75 145,395, and 75 145,396 (Nov. 21, 1975); Jpn. Kokai 75 146,596 (Nov. 25, 1975); Ger. Offen. 2,655,920 (Aug. 4, 1977); Jpn. Kokai 78 72,799 (June 28, 1978), (to Tokuyama Soda Co., Ltd.).
104. Ger. Offen. 2,615,625 (Oct. 21, 1976), L. Kim and G. Schoenthal (to Shell International Research Maatschappij BV).
105. U.S. Pat. 4,128,627 (Dec. 5, 1978), P. N. Dyer and F. Mosely (to Air Products and Chemicals, Inc.).
106. U.S. Pat. 4,009,252 (Feb. 22, 1977); Jpn. Pats. 54 39,836 (Nov. 30, 1979), 55 10,521 (Mar. 17, 1980), 54 39,837 (Nov. 30, 1979), and 57 246 (Jan. 6, 1982), Y. Izumi and co-workers (to Tokuyama Soda Co., Ltd.).
107. U.S. Pats. 4,336,240 (June 22, 1982) and 4,369,128 (Jan. 18, 1983), F. Mosely and co-workers (to Air Products and Chemicals, Inc.).
108. U.S. Pats. 4,347,231 (Aug. 31, 1982) and 4,347,232 (Aug. 31, 1982), R. C. Michaelson (to FMC Corp.).
109. Jpn. Kokai 1 133909 (May 26, 1989), T. Kyora (to Mitsui Toatsu Chemicals).
110. U.S. Pat. 5,104,635 (Aug. 31, 1990), T. Kanada, K. Nagai, and T. Nawata (to Mitsubishi Gas-Chemical Co., Ltd.).
111. U.S. Pat. 5,132,099 (Sept. 20, 1991), H. Nagashima, U. Nagashima, Y. Ishiuchi, and Y. Hiramatsi (to Mitsubishi Gas-Chemical Co., Ltd.).
112. E. P. Appl. 504741 (Sept. 23, 1992), H. Nagashima, Y. Ishiuchi, and Y. Hiramatsi (to Mitsubishi Gas-Chemical Co., Ltd.).
113. U.S. Pat. 5,169,618 (Jan. 13, 1992 filed), M. J. Maraschino (to Kerr-McGee).
114. Belg. Pat. 92 15520 (Sept. 17, 1992), J. V. Weynbergh, J. SchoeLooschts, and J. C. Coleny, (to Interox Int. SA).

115. Ger. Offen. 4,127,918 (Sept. 10, 1992), U. Lueckoff, G. Luft, and H. Pauckscht (to Interox International SA).
116. U.S. Pats. 4,681,751 (July 21, 1987), 4,832,938 (May 23, 1989), 4,889,705 (Dec. 26, 1989), and 4,772,488 (Sept. 20, 1988), L. W. Gosser and co-workers (to Du Pont); Eur. Pat. 342,047 (Nov. 15, 1989), L. W. Gosser (to Du Pont).
117. U.S. Pat. 5,135,731 (Aug. 4, 1992), L. W. Gosser and M. A. Paoli (to Du Pont) and U.S. Pat. 5,112,702 (Dec. 31, 1990), T. Berzins and L. W. Gosser (to Du Pont).
118. U.S. Pats. 4,335,092 (June 15, 1982), 4,336,238 (June 22, 1982), 4,336,239 (June 22, 1982), 4,379,778 (Apr. 12, 1983), and 4,389,390 (June 21, 1983), A. I. Dalton, Jr. and co-workers (to Air Products & Chemicals).
119. U.S. Pat. 4,007,256 (Feb. 8, 1977), L. Kim and co-workers (to Shell Oil) and Brit. Pat. 1,539,962 (Feb. 7, 1979), L. Kim and co-workers (to Shell Internationale Research Maatschappij).
120. Jpn. Kokai 1 192,710 (Aug. 2, 1989), (to Mitsui Toatsu Chemicals).
121. U.S. Pat. 4,009,252 (Feb. 22, 1977), Y. Izumi, H. Miyazaki, and S. Kawahara (to Tokuyama Soda Co., Ltd.).
122. Ger. Offen. 2,654,514 (June 1, 1978), R. Radebold and W. Seiler.
123. U.S. Pat. 4,759,921 (July 26, 1988), W. Kunkel and co-workers (to Degussa).
124. Jpn. Kokai 63 156,004 (June 29, 1988), S. Togo and co-workers (to Mitsubishi Gas-Chemical Co., Ltd.).
125. Jpn. Kokai 1 153,509 (June 15, 1989), (to Tokai Denka Kogyo).
126. Ger. Offen. 2,233,159 (Jan. 24, 1974), A. Becker and U. Schwenk (to Farbwerke Hoechst AG).
127. U.S. Pat. 4,564,514 (Jan. 14, 1986), K. Drauz and co-workers (to Degussa).
128. U.S. Pats. 4,686,010 (Aug. 11, 1987) and 4,760,199 (July 26, 1988), K. Drauz and co-workers (to Degussa).
129. *Inorganic Chemicals*, Current Industrial Reports MA28A, Economics and Statistics Administration, Bureau of the Census, U.S. Department of Commerce, Washington, D.C.
130. *Chem. Mktg. Rep.*, 1994.
131. *The United States Pharmacopeia XX (USP XX-NF XV)*, The United States Pharmacopeial Convention Inc., Rockville, Md., 1980, p. 318.
132. BOE 6000 K, Bureau of Explosives, Washington, D.C., 1993.
133. G. A. Parker, *Chem. Anal. (NY)* **8**, 253 (1978).
134. J. Gallus-Olender and B. Franc, *Chem. Anal. (Warsaw)* **19**(1), 203 (1974).
135. *Food Chemical Codex*, 3rd ed., National Academy of Sciences–National Research Council, Washington, D.C., 1980.
136. Material Safety Data Sheet 8250CR, Du Pont, Wilmington, Del., Dec. 13, 1991.
137. N. J. Stalter, *Soap. Chem. Spec.* **45**(6), 62 (1969).
138. Jpn. Kokai 78 33,975 (Mar. 30, 1978), H. Hayasaka, Y. Sekiguchi, and N. Okigami (to Hitachi Ship Building and Equipment Co., Ltd.).
139. F. J. Miner and P. G. Hagan, *Rate of Hydrogen Peroxide Decomposition in Nitric Acid Solutions*, Dow Chemical USA, Golden, Colo.
140. *Manual of Hazardous Chemical Reactions No. 491*, National Fire Protection Association, 1971.
141. *U.S. Bureau of Mines Information Circular 8387*, Washington, D.C., 1968.
142. Can. Pat. 960,437 (Jan. 7, 1975), P. B. Lonnes and co-workers (to Environmental Research Corp.).
143. M. Coeyman and N. Alperowicz, *Chem. Wk.*, 42, 43, (Feb. 17, 1993).
144. S. Rothenburg and co-workers, *Tappi* **58**, 182 (1975).
145. Ger. Pat. 2,327,900 (Aug. 3, 1978), (to Degussa).
146. Fr. Demande 2,367,859 (May 12, 1978), (to Degussa).

147. U.S. Pat. 5,169,495 (Apr. 30, 1991), D. Lachenal (to Elf Atochem).

148. USSR Pat. SU4,754,659 (Aug. 7, 1991), (to Paper Research Institute).

149. Ger. Pat. Appl. DE4,035,813A (Nov. 10, 1990), (to AKZO GmbH).

150. U.S. Pat. Appl. 91US-770628 (Nov. 12, 1990), (to Repap Technologies Inc.).

151. Eur. Pat. Appl. EP514,608-A1 (Apr. 30, 1990), (to Elf Atochem).

152. Eur. Pat. Appl. EP514,609-A1 (Apr. 30, 1990), (to Elf Atochem).

153. G. Papageorges and J. Deceuster, *Indian Chem. Age* **27**, 451 (1976).

154. W. H. Kibbel, Jr., *Ind. Water Eng.* **13**(4), 6 (1976).

155. W. G. Strunk, *Treat. Disposal Ind. Waste Waters Residues Proc. Nat'l. Conf.* 119 (1977).

156. J. Eley, *J. Mo. Water Sewerage Conf.* **47**, 25 (1976).

157. U.S. Pat. Appl. 89US-354039 (May 19, 1989), (to Degussa).

158. Eur. Pat. Appl. EP495707-A1 (Jul. 22, 1992), (to OTV (Omnium Traitement Valorisa)).

159. Eur. Pat. Appl. EP509382-A2 (Apr. 17, 1991) (to W. R. Grace & Co.).

160. Pat. Appl. WO9211208A-1 (Dec. 18, 1990), (to Eastman Kodak Co. Inc.).

161. Jpn. Pat. Appl. 04100594-A (Dec. 18, 1990), (to Fuji Photo Film Co.).

162. Ger. Offen. 2,446,511 (Apr. 15, 1976), N. Wolters and U. Loll.

163. Jpn. Kokai 78 128,145 (Nov. 8, 1978), T. Yoshida and T. Iwammoto (to Nippon Peroxide Co.).

164. Jpn. Kokai 76 32,057 (Mar. 18, 1976) and 78 123,556 (Oct. 28, 1978), (to Toa Gosei Chemical Industry Co., Ltd.).

165. *Ind. Miljoe* **6**(6), 27, 33 (1973). (1976).

166. W. H. Kibble Jr., *Ind. Wastes (Chicago)* **24**(3), 26 (1978).

167. J. Shapiro and Y. Thiffault, *Eau Que* **11**(1), 113 (1978).

168. Brit. Pat. 1,526,190 (Sept. 27, 1978), (to Erdochemie GmbH).

169. Belg. Pats. 863,321 and 863,322 (July 25, 1978), (to Degussa).

170. Jpn. Kokai 78 95,170 (Aug. 19, 1978), (to Wako Pure Chemical Ind.).

171. Jpn. Kokai 78 108,068 (Sept. 20, 1978), S. Ikuta and T. Shimomura (to Mitsubishi Gas-Chemical Co., Inc.).

172. Jpn. Pat. 78 40,591 (Oct. 27, 1978), (to Nippon Kogyo Senjo).

173. U.S. Pat. Appl. 5,120,453 (Dec. 24, 1990), (to UOP).

174. U.S. Pat. 5,112,587 (May 6, 1991), (to Degussa).

175. Ger. Offen. 2,754,932 (June 15, 1978), S. Azuhata and co-workers (to Hitachi, Ltd.).

176. Fr. Pat. 2,373,327 (Aug. 11, 1978), (to Hitachi KK).

177. Ger. Pat. 2,524,115 (Aug. 10, 1978), (to Mitsubishi Jukogyo).

178. Ger. Pat. 2,6654,324 (July 10, 1978), (to Ugine Kuhlmann).

179. Jpn. Kokai 78 110,961 (Sept. 28, 1978), (to C. Soda).

180. Belg. Pat. 867,389 (Nov. 23, 1978), (to Degussa).

181. W. M. Weigert and co-workers, *Chem. Ztg.* **99**(3), 106 (1975).

182. Fr. Pats. 2,379,519 and 2,379,520 (Oct. 6, 1978), (to Interox).

183. U.S. Pat. 4,137,242 (Jan. 30, 1978), G. Prescher and co-workers (to Degussa).

184. Belg. Pats. 841,208 (Oct. 28, 1976) and 847,664 (Apr. 27, 1977), (to Bayer AG).

185. Brit. Pat. 1,520,821 (Aug. 9, 1978), (to Olin Corp.).

186. Ger. Offen. 2,807,344 (Aug. 31, 1978), A. M. Hildon, T. D. Manly, and A. J. Jaggers (to Propylox).

187. Fr. Pats. 2,378,773 and 2,378,774 (Sept. 29, 1978) and 2,372,161 (July 28, 1978); Belg. Pats. 848,522 (May 20, 1977) and 863,237 (July 24, 1978); Ger. Pat. 2,446,830 (Aug. 24, 1978), (to Ugine Kuhlmann).

188. Fr. Pat. 2,318,851 (Mar. 25, 1977), (to Rhône-Poulenc SA).

189. Ger. Offen. 2,167,040 (Oct. 20, 1977), (to Brichma SpA).

190. Jpn. Kokai 75 130,727 (Oct. 16, 1975), (to Ube Industries, Inc.).

191. U.S. Pat. 4,053,523 (Oct. 11, 1977), H. Siefert and co-workers (to Bayer AG).

192. USSR Pat. 1,680,695 (Apr. 20, 1989), (to Fedorova EV).

193. U.S. Pat. 5,153,031 (Nov. 14, 1990), J. M. Burlitch (to Cornell Research Federation, Inc.).

194. U.S. Pat. 5,112,985 (Sept. 9, 1989), (to BASF).

195. Brit. Pat. 1,177,495 (Jan. 14, 1979), (to Farbenfabriken Bayer, AG).

196. U.S. Pat. 4,046,862 (Sept. 6, 1977), W. Heimberger and G. Schreyer (to Degussa).

197. Ger. Pat. 2,534,541 (Aug. 17, 1978), (to Ugine Kuhlmann); Ger. Offen. 2,713,345 (Sept. 28, 1978), (to Chemische Fabrik Kalk GmbH).

198. Jpn. Kokai 78 108,096 (Sept. 20, 1978), (to Ise Chemical Industries Co., Ltd.).

199. B. C. Lawes, *In Situ* **2**(2), 75 (1978).

200. Fr. Pat. 2,299,410 (Oct. 1, 1976), (to Wyoming Mineral Corp.).

201. Fr. Pat 2,376,215 (Sept. 1, 1978), (to Minatonic Corp.).

202. Fr. Pat. 2,380,410 (Oct. 13, 1978), (to Union Oil Co., CA).

203. Belg. Pat. 866,937 (Nov. 13, 1978), (to Interox Chemicals).

204. Brit. Pat. 1,534,485 (Dec. 6, 1978), (to Soc. Mines Fond Zinc SA).

205. Brit. Pat 1,527,758 (Oct. 11, 1978), (to Japan Carlit KK).

206. U.S. Pat. 4,132,569 (Jan. 2, 1979), R. S. DePablo, D. E. Harring, and D. E. Bramstedt (to Diamond Shramrock Corp.).

207. Ger. Pat. 2,623,977 (Aug. 17, 1978), (to Nukim Nuclear-Chem. GmbH).

208. Jpn. Pat. 78 41,120 (Oct. 31, 1978), (to Nippon Nuclear Fuel).

General References

W. C. Schumb, C. N. Satterfield, and R. L. Wentworth, *Hydrogen Peroxide*, Reinhold Publishing Corp., New York, 1955.

H. Pistor, in K. Winnacker and L. Kuchler, eds., *Chemische Technologie, Band 1 Anorganische Technolgie 1*, Hanser, Munich, 1969 (industrial processes).

J. G. Wallace, *Hydrogen Peroxide in Organic Chemistry*, E. I. du Pont de Nemours & Co., Inc., Wilmington, Del., 1962.

R. Powell, *Hydrogen Peroxide Manufacture*, Chemical Process Review No. 20, Noyes Development Corporation, Park Ridge, N.J., 1968.

W. M. Weigert, ed., *Wasserstoffperoxid ana Seine Derivate: Chemie and Anwendung*, Huethig, Heidelberg, Germany, 1978.

W. Machu, *Das Wasserstoff peroxyd und die Perverbindungen*, Springer-Verlag, Vienna, 1951.

G. Dusing, P. Kleinschmit, G. Knippschild, W. Kunkel, and S. Habersang, eds., in *Ullmans Encyclopedia of Technical Chemistry*, 4th ed., Vol. 17, Verlag Chemie GmbH, Weinheim, Germany, 1979.

O. Kausch, *Das Wasserstoffperoxyd*, Wilhelm Knapp, Halle, Germany, 1938 (photolithoprinted, Edwards Bros., Ann Arbor, Mich.).

C. Crampton, G. Faber, R. Jones, J. P. Leaver, and S. Schelle, *Chem. Soc. Spec. Publ.* **31**, 232 (1977).

W. S. Wood, *Hydrogen Peroxide*, Monograph No. 2., Royal Institute of Chemistry, London, 1954.

WAYNE T. HESS
E. I. du Pont de Nemours & Co., Inc.

HYDROGEN SULFIDE. See SULFUR COMPOUNDS.

HYDROMETALLURGY. See METALLURGY, EXTRACTIVE.

HYDROPROCESSES. See HYDROTHERMAL PROCESSING; PETROLEUM, RE-
FINERY PROCESSES.

HYDROQUINONE, RESORCINOL, AND CATECHOL

Hydroquinone [123-31-9], resorcinol [108-46-3], and catechol [120-80-9] (or
pyrocatechol) are represented by structures (**1**), (**2**), and (**3**), respectively. This
article reviews their syntheses and derivatives, production and use, and
toxicology.

Manufacture and Processing

Dihydroxybenzenes (dihydric phenols) are industrially prepared according to
four different routes: (1) oxidation of aniline (selective access to hydroquinone);
(2) alkali fusion of m-benzenedisulfonic acid (selective access to resorcinol);
(3) oxidation of p- or m-diisopropylbenzene (selective access to hydroquinone
or resorcinol); and (4) hydroxylation of phenol by hydrogen peroxide (simultaneous access to hydroquinone and catechol). Routes starting from o- or p-
dichlorobenzene are no longer used.

During the 1980s few innovations were disclosed in the literature. The
hydroxylation of phenol by hydrogen peroxide has been extensively studied
in order to improve the catalytic system as well as to master the ratio of
hydroquinone to catechol. Other routes, targeting a selective access to one of
the dihydroxybenzenes, have appeared. World production capacities according
to countries and process types are presented in Table 1.

Aniline Oxidation. Even though this is quite an old process, it still has
limited use to produce hydroquinone on a commercial scale. In the first step,

aniline is oxidized by manganese dioxide in aqueous sulfuric acid. The resulting benzoquinone, isolated by vapor stripping, is reduced in a second step by either an aqueous acidic suspension of iron metal or by catalytic hydrogenation.

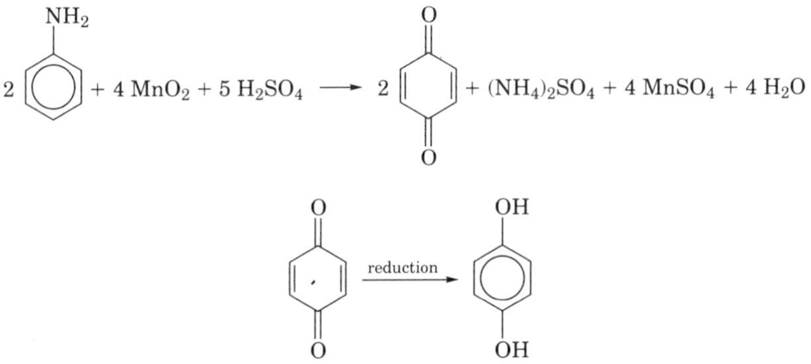

$$2\ \underset{}{C_6H_5NH_2} + 4\ MnO_2 + 5\ H_2SO_4 \longrightarrow 2\ \underset{}{C_6H_4O_2} + (NH_4)_2SO_4 + 4\ MnSO_4 + 4\ H_2O$$

Table 1. Manufacturing Processes and World Production Capacities for Dihydroxybenzenes

Dihydroxybenzene	World capacity,[a] t	Process	Location
hydroquinone	~45,000– 50,000	aniline oxidation	ex Comecon[b] People's Republic of China
		phenol hydroxylation	France United States Italy Japan
		p-diisopropylbenzene hydroperoxidation	United States Japan
resorcinol	30,000– 35,000	benzenedisulfonic acid alkali fusion	Japan United States Italy Germany United Kingdom Puerto Rico
		m-diisopropylbenzene hydroperoxidation	Japan
catechol	>25,000	phenol hydroxylation	France United States Italy Japan
		coal-tar distillation	United Kingdom ex Comecon[b]

[a] Estimated for 1994.
[b] Comecon = Council for Mutual Economic Assistance (Communist-bloc nations).

The yield of hydroquinone is 85 to 90% based on aniline. The process is mainly a batch process where significant amounts of solids must be handled (manganese dioxide as well as metal iron finely divided). However, the principal drawback of this process resides in the massive coproduction of mineral products such as manganese sulfate, ammonium sulfate, or iron oxides which are environmentally not friendly. Even though purified manganese sulfate is used in the agricultural field, few solutions have been developed to dispose of this unsuitable coproduct. Such methods include $MnSO_4$ reoxidation to MnO_2 (1), or $MnSO_4$ electrochemical reduction to metal manganese (2). None of these methods have found applications on an industrial scale. In addition, since 1980, few innovative studies have been published on this process (3).

Alkali Fusion of *m*-Benzenedisulfonic Acid. Even though this process like the previous one is a very ancient one, it is still the main route for the synthesis of resorcinol. It has been described in detail previously and does not seem to have drastically evolved since 1980. It involves the reaction of benzene with sulfuric acid to form *m*-benzenedisulfonic acid which is then converted to its disulfonate sodium salt by treatment with sodium sulfite. In a second step, this salt is heated to 350°C in the presence of sodium hydroxide yielding the sodium resorcinate and sodium sulfite.

After sulfuric acid work-up (accompanied by the formation of sodium sulfate), the resorcinol is extracted and isolated in a 94% yield based on *m*-benzenedisulfonic acid [98-48-6]. In addition to the technical complexity that goes along with the manipulation of solids at high temperature, this process produces large amounts of salts (sulfite and sulfate salts) which economically as well as environmentally are not always desired.

Hydroperoxidation of *m*- or *p*-Diisopropylbenzene. This is an important industrial route to resorcinol and hydroquinone. The process in principle is identical to the cumene process for the manufacturing of phenol (qv).

Diisopropylbenzenes (DIPB) are readily obtained via Friedel-Crafts alkylation of benzene or cumene by propylene. This reaction in liquid phase has not evolved drastically since 1980 with the exception of the large variety of heterogeneous acid catalysts that are now being used, mainly zeolites, type HZSM-12, giving a para/meta ratio = 0.7 (4). In fact, propylene can also be replaced by isopropyl alcohol coming from the hydrogenation of acetone that is produced at the end of the process and can therefore be recycled (5).

Another mode of preparation includes the dismutation of cumene (qv) in DIPB and benzene in presence of an HZSM-12 catalyst at 200°C (6). The ratio p/m increases with the temperature. Finally, but with purification difficulties, DIPB can be coproduced with cumene (7).

Resorcinol or hydroquinone production from m- or p-diisopropylbenzene [100-18-5] is realized in two steps, air oxidation and cleavage, as shown above. Air oxidation to obtain the dihydroperoxide (DHP) coproduces the corresponding hydroxyhydroperoxide (HHP) and dicarbinol (DC). This formation of alcohols is inherent to the autooxidation process itself and the amounts increase as DIPB conversion increases. Generally, this oxidation is carried out at 90–100°C in aqueous sodium hydroxide with eventually, in addition, organic bases (pyridine, imidazole, citrate, or oxalate) (8) as well as cobalt or copper salts (9).

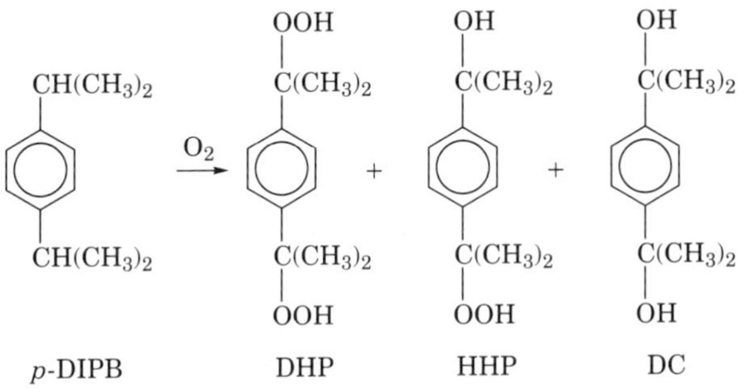

Prior to acid cleavage, the mixture is treated with H_2O_2 in a biphasic water–toluene system to convert DC and HHP to DHP. This improvement has been specifically described in the case of DC (10). Acidic cleavage of the organic phase can be directly realized in the presence of hydrogen peroxide (11). However, due to the presence of both the hydrogen peroxide and acetone in the acidic medium, formation of acetone peroxides occur. These are sensitive explosive compounds and their formation represents a safety hazard. A mixture of m- and p-DIPB can also be oxidized to obtain simultaneously resorcinol and hydroquinone (12), even though m-diisopropylbenzene [99-62-7] in such a mixture is less easily oxidized than p-DIPB (12).

Hydroxylation of Phenol with Hydrogen Peroxide. Phenol [106-44-5] can be hydroxylated to hydroquinone and catechol in the presence of homogeneous or heterogeneous catalysts.

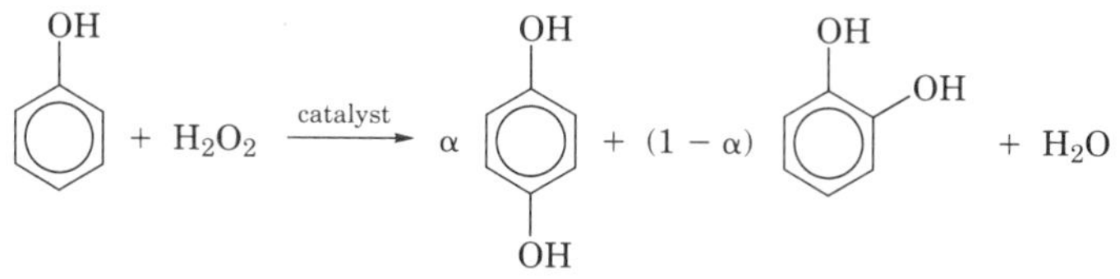

This process has been widely studied and led to the construction of new and original industrial units. Interest in the reaction stems from the simplicity of the process as well as the absence of undesirable by-products. However, in order to be economically reliable, such a process has to give high yield of dihydroxybenzenes (based on hydrogen peroxide as well as phenol) and a great flexibility for the isomeric ratio of hydroquinone to catechol. This last point generated more research and led to original and commercial processes.

Homogeneous Catalysis. The oldest system is known as the Fenton reaction in which the catalyst is a soluble cobalt, copper, and iron salt. The reaction is carried out in an aqueous acid phase (pH = 2–3). Yields in dihydroxybenzenes based on H_2O_2 are generally lower than 60%. Adding iron complexant does not change the isomeric ratio (13), catechol being the strongest iron complexant in the medium (catechol:hydroquinone ~2). Furthermore, this is an intrinsic radical mechanism, which generates small amounts of resorcinol, which is difficult to separate from hydroquinone. A few innovations have been made: catalysis with $FeSO_4/Al_2(SO_4)_3$ (14), catalysis with metallic phthalocyanides (15), catalysis by metal ions other than iron such as Hg^{2+}, WO_4^{2-}, Ce^{3+}, etc (16), and cocatalysis by β-cyclodextrins (17) which increases the yield in hydroquinone.

Catalysis by Protons. The discovery of hydrogen peroxide hydroxylation of phenol in the presence of strong acids such as perchloric, trifluoromethanesulfonic, or sulfuric acids allows suppression of all previous drawbacks of the process (18,19). This mode of hydroxylation gives high yields (85% based on H_2O_2 at phenol conversion of 5–6%). It can be run without solvents and does not generate resorcinol. Its main advantage relies on the fact that the ratio hydroquinone:catechol can be tuned (0.5–1) depending on the market need. Two industrial units based in France and in the United States are operating using this process. The reaction can be viewed as follows:

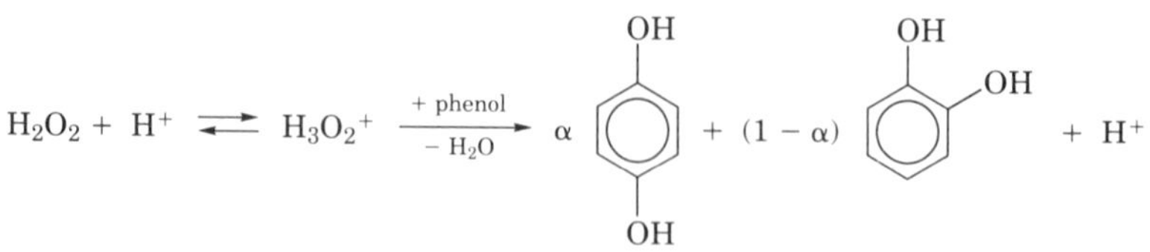

The ratio of hydroquinone to catechol increases with the acidity of the catalyst that is being used; a ratio of 1.8 can be obtained in trifluoromethanesulfonic acid (20). The rate of hydroxylation can be increased by adding to the acid catalyst a carbonyl compound (21–23). In the case of aromatic aldehydes and benzophenones with electron-donating groups, not only the rate of the reaction increases but also the para selectivity. By using a mixture of the acid and its corresponding alkali or alkaline-earth salt, the yield can also be increased (24). The catalyst can also be replaced by metalloids such as S, Se, P, Te, etc (25), by phosphorous compounds such as P_2O_5 (26) or by the use of a strong acid salt in conjunction with phosphoric acid (27). Many other catalytic systems have been described: catalysis by SO_2 or SeO_2 with anhydrous solutions of H_2O_2 (28); catalysis by sulfur derivatives (such as dithiolanemethione, sulfamic acid, dithiophosphoric acid) which probably generates sulfonic acid catalysts in the presence of H_2O_2, generally in the presence of ketones (29); and catalysis by arsenic or antimony derivatives where remarkable selectivities in catechol are observed, hydroquinone:catechol = 0.3 (even though yields do not exceed 60%) (30). Catalysis by heteropolyacids has been described leading selectively to hydroquinone or catechol depending on the heteropolyacid used. In this case, yields based on H_2O_2 are relatively low (31). Catalysis using cationic Pt(II) complexes has also been reported leading to high selectivity in catechol via intermediate electrophilic metalation (32).

Heterogeneous Catalysis. The main discovery of the 1980s was the use of titanium silicalite (TS-1) a synthetic zeolite from the ZSM family containing no aluminum and where some titanium atoms replace silicon atoms in the crystalline system (Ti/Si = 5%) (33). This zeolite can be obtained by the hydrolysis of a silicate and an alkyl titanate in the presence of quaternary ammonium hydroxide followed by heating to 170°C. Mainly studies have been devoted to the structure of TS-1 and its behavior toward H_2O_2 (34). The oxidation properties of the couple H_2O_2/TS-1 have been extensively developed in many areas: alkane oxidation to alcohols and ketones, olefin epoxidation, ether oxidation, caprolactam synthesis etc (35–38). Still, the main area of research on TS-1 is phenol hydroxylation by H_2O_2 where innovations have been outstanding; for example, at 80°C, in two hours, in acetone as a solvent (phenol:acetone = 1.3 by weight), 88% yield of dihydroxybenzenes based on H_2O_2 and 96% yield of dihydroxybenzenes based on phenol are obtained. In this case, a 20% conversion of phenol can be obtained and the isomeric ratio para/ortho is close to 1 (39). The ratio can be increased by carrying out the reaction in an alcoholic medium; for example, in *t*-butyl alcohol the ratio is 3.3 (40). For its industrial use, TS-1 can be converted to very resistant microspheres (41). A process based on this catalyst is operative in Italy (10,000 t/yr). Many efforts have been devoted to the optimization and the simplification of TS-1 synthesis (42), hence a more active TS-1 containing large pores was obtained (43). Moreover, a vanadium silicalite has been synthesized. Its catalytic properties, which appear to be less active than the regular TS-1, are now being studied (44).

From a mechanistic point of view, the fact that aluminum atoms are absent from the structure tends to suggest a nonacidic pathway. In fact, it is highly probable that titanyl species are the active sites to which H_2O_2 adds leading to electrophilic, very reactive entities (Ti–OOH) toward the nucleophilic substrate (45). Although TS-1 appears to be the heterogeneous catalyst most studied and the best able to carry out the hydroxylation of phenol with H_2O_2, other heterogeneous catalysts are capable of performing the same reaction. Such catalysts include germanozeosilicates (46), titanozeosilicates (47), stanno- and zirconozeosilicates (48), and titanium oxide (in this case, a ratio of hydroquinone to catechol <0.25 can be obtained) (49). In addition, another class of catalysts acting only by their intrinsic acidity can be cited: acidic clay (50), bridged clay (51), zirconium phosphates in acetic acid medium (52), and the nafion resin exchanged by vanadium(V) or Ti(IV), leading to a selective access to hydroquinone in low yield (53). Finally, HZSM-5 exchanged with metal cations might provide hydroquinone selectively and quantitatively (54). None of these catalysts have yet found industrial applications.

Miscellaneous Preparations.   Many laboratory innovations have been run.

Starting from Cyclohexene.   Industrial synthesis of cyclohexene is now well established by partial hydrogenation of benzene according to the ASAHI procedure (55) using $Ru°/Zn^{II}$, ZrO_2 catalysis; conversion in benzene = 70%; selectivity in cyclohexene = 80%. Numerous studies have appeared (56) dealing with the selective synthesis of catechol via the epoxide, the 1,2-diol, and the dehydrogenation of 1,2-cyclohexanediol. Even though the respective yield of each step is high, optimization studies are being conducted.

Cyclohexene can also be oxidized in cyclohexene-2-one which is hydrated into cyclohexan-1-ol-3-one. Dehydrogenation of this compound gives resorcinol selectively (57).

Starting from Phenol.   Phenol can be selectively oxidized into *p*-benzoquinone with oxygen. The reaction is catalyzed by cuprous chloride. At low catalyst concentration, the principal drawback of this method is the high pressure of oxygen that is required, leading to difficult safety procedures. It appears that a high concentration of the catalyst (50% of Cu(I)−phenol) allows the reaction to proceed at atmospheric pressure (58).

Starting from Chlorophenols.   Even though this process has almost disappeared due to the formation of sodium chloride during the sodium hydroxide hydrolysis, continuing studies have demonstrated that it is now possible to carry out the hydrolysis using water at high temperature (300−700°C) and high pressure in conjunction with a heterogeneous catalyst such as La_2O_3 (59) or silicoaluminates doped with Ni−Cu and Ni−Ag (60).

Starting from Bisphenol A.   This method allows a selective access to hydroquinone starting from bisphenol A [80-05-7] according to the following scheme:

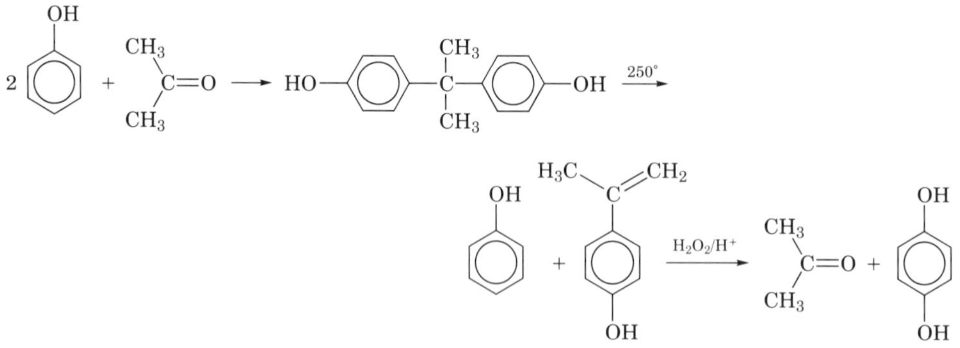

The yield of hydroquinone based on bisphenol A is close to 90%. The phenol and the acetone formed can easily be recycled. However, this process has not been industrialized.

Pyrolysis of Vegetals. Many publications concern the synthesis of dihydroxybenzenes by wood, lignites, and tree bark pyrolysis (61). The selective extraction of these compounds in low concentration from the crude mixture remains a significant problem. So far, the price of the extraction overcomes the advantage of starting from a cheap starting material.

Biochemical Routes. Enzymatic oxidation of benzene or phenol leading to dilute solution of dihydroxybenzenes is known (62). Glucose can be converted into quinic acid [77-95-2] by fermentation. The quinic acid is subsequently oxidized to hydroquinone and *p*-benzoquinone with manganese dioxide (63).

Starting from Benzene. In the direct oxidation of benzene [71-43-2] to phenol, formation of hydroquinone and catechol is observed (64). Ways to favor the formation of dihydroxybenzenes have been explored, hence CuCl in aqueous sulfuric acid medium catalyzes the hydroxylation of benzene to phenol (24%) and hydroquinone (8%) (65). The same effect can also be observed with Cu(II)–Cu(0) as a catalytic system (66). Efforts are now directed toward the use of Pd° on a support and Cu^{II} in aqueous acid and in the presence of a reducing agent such as CO, H$_2$, or ethylene (67). Aromatic hydroxylation can also be achieved with oxygen and hydrogen using Pd°–TS-1 as catalyst (68).

Other Methods. A variety of other methods have been studied, including phenol hydroxylation by N$_2$O with HZSM-5 as catalyst (69), selective access to resorcinol from 5-methyloxohexanoate in the presence of Pd/C (70), cyclotrimerization of carbon monoxide and ethylene to form hydroquinone in the presence of rhodium catalysts (71), the electrochemical oxidation of benzene to hydroquinone and *p*-benzoquinone (72), the air oxidation of phenol to catechol in the presence of a stoichiometric CuCl and Cu(0) catalyst (73), and the isomerization of dihydroxybenzenes on HZSM-5 catalysts (74).

Synthesis of Derivatives

Hydroquinone and catechol are important industrial intermediates, and there has been significant research and development of processes for manufacturing their derivatives.

Catechol Derivatives. An elegant synthesis of trimethoxybenzaldehyde [86-81-7] (**4**) starting from guaiacol [90-05-1] (**5**) and formaldehyde (75) has been developed. The reaction sequence is as follows:

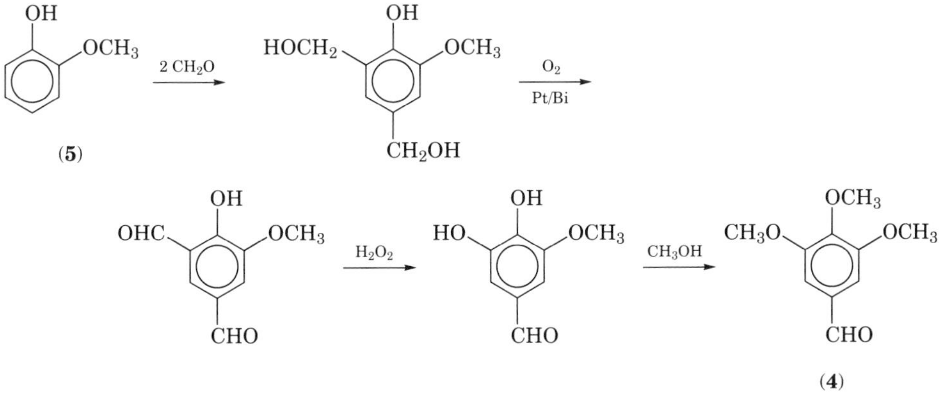

(**5**)

(**4**)

A new synthesis of carbofuran [*1563-66-2*] (**6**) has been described via 2,3-dihydro-2,2-dimethyl-7-hydroxybenzofuran [*1563-38-8*] (**7**), starting from *o*-methallyloxyphenol [*4790-71-0*] (**8**) in the presence of trivalent aluminium derivatives (76):

(**8**)   (**7**)   (**6**)

Hydroquinone Derivatives.   Thermotropic polymers are polymers in which the main chain contains a mesomorphic state between the disordered liquid and the crystalline state. Many important industrial developments have been devoted to this class of polymers. Melting point decrease of polyesters containing only rigid units in the principal chain can be obtained in many ways, but in particular by introducing substituents or functional groups allowing internal rotations of the rigid units to occur. Therefore, development of hydroquinone containing substituents such as halogens, eg, chlorohydroquinone [*615-67-8*] (**9**), alkyls (**10**), or phenyl (**11**) was initiated (77,78).

(**9**)   (**10**)   (**11**)

o-Phenylphenol [*90-43-7*] (**12**) can be oxidized by air to form the corresponding phenylbenzoquinone [*363-03-1*] (**13**) (79).

(**12**)   (**13**)

Original routes involving the direct oxidation of aromatic precursors (**14,15**) into quinols (**16,17**) followed by a thermal transformation of the latter have been patented for the synthesis of methylhydroquinone [*95-71-6*] (**10**) and phenylhydroquinone [*1079-21-6*] (**11**) (80,81).

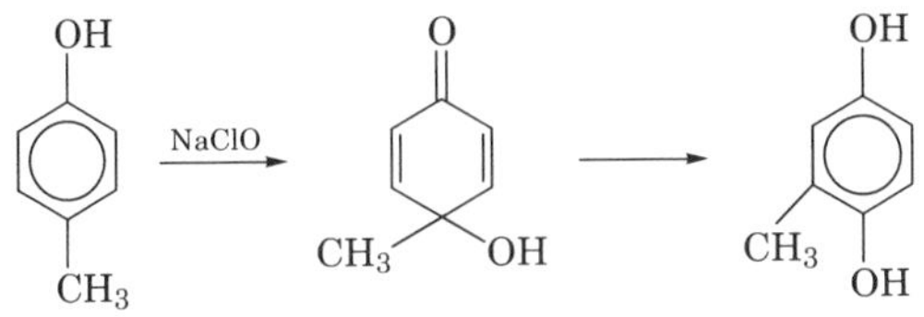

(14)   (16)   (10)

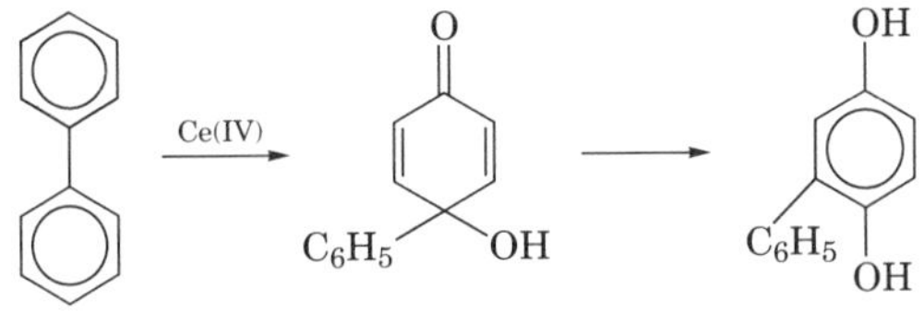

(15)   (17)   (11)

However, the vast majority of research has been devoted to synthesis involving electrophilic substitution on the aromatic ring of hydroquinone. Hence, phenylhydroquinone can be obtained by the reaction of phenyldiazonium salts (**18**) with hydroquinone (82).

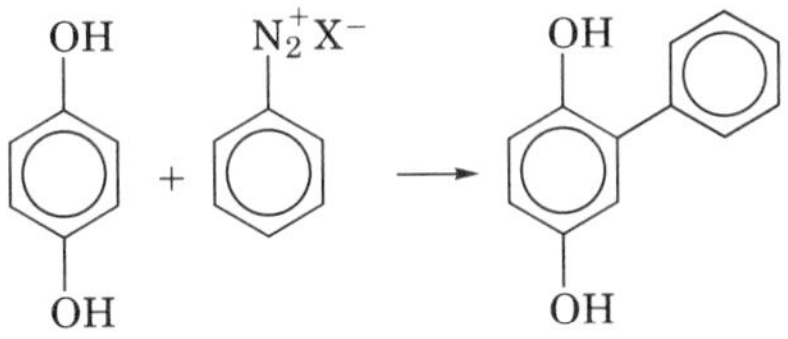

(18)   (11)

Access to phenylhydroquinone based on the following scheme has been extensively studied (83).

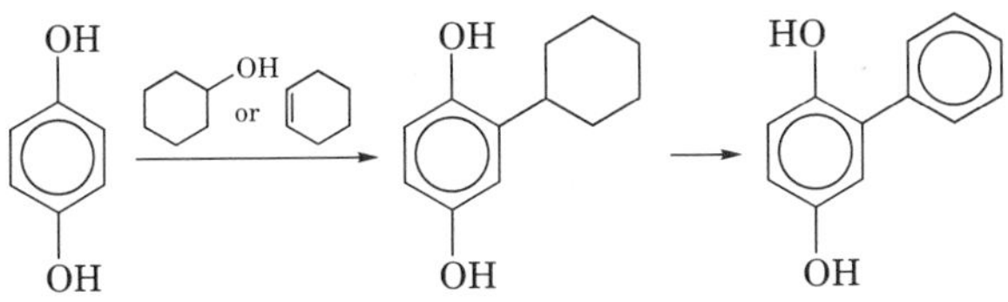

(19)

The condensation of cyclohexanol or cyclohexene is generally carried out in the presence of phosphoric acid, pyrophosphoric acid, or HY zeolites; the aromatization of intermediate cyclohexylhydroquinone [4197-75-5] (**19**) is realized in the presence of a dehydrogenation catalyst.

The selective monochlorination of hydroquinone by $SOCl_2$ (84) or a combination of HCl and H_2O_2 (85) has been studied in various solvents.

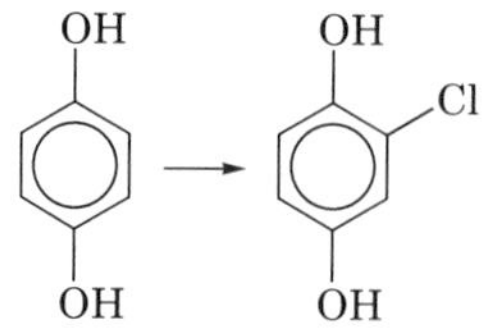

The synthesis of chloranil [118-75-2] (**20**) has been improved. The old processes starting from phenol or 2,4,6-trichlorophenol have been replaced by new ones involving hydroquinone chlorination. These processes allow the preparation of chloranil of higher purity, avoiding traces of pentachlorophenol. Different types of chlorination conditions have been disclosed. The reaction can be performed according to the following stoichiometry, operating with chlorine in aqueous acetic acid (86,87), biphasic medium (88), or in the presence of surfactants (89).

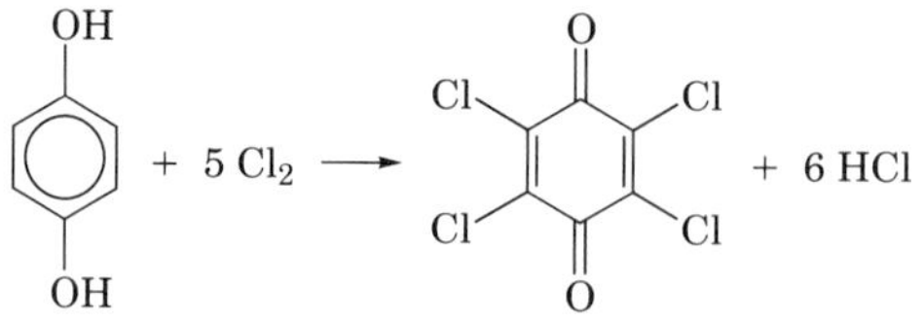

(**1**)   (**20**)

Other patents (86,90) describe the synthesis without the coproduction of HCl, using the mixture of HCl and H_2O_2 according to

$$(\mathbf{1}) + 5\,H_2O_2 + 4\,HCl \longrightarrow (\mathbf{20}) + 10\,H_2O$$

Resorcinol Derivatives. Aminophenols (qv) are important intermediates for the syntheses of dyes or active molecules for agrochemistry and pharmacy. Syntheses have been described involving resorcinol reacting with amines (91). For these reactions, a number of catalysts have been used: *p*-toluene sulfonic acid (92), zinc chloride (93), zeolites and clays (94), and oxides supported on silica (95). In particular, catalysts performing the condensation of ammonia with resorcinol have been described: gadolinium oxide on silica (96), nickel, or zinc phosphates (97), and iron phosphate (98).

2,4-Dihydroxybenzophenones are used for the syntheses of dyes, polymers, and medicines. They are prepared by the condensation of resorcinol with benzoic acids. Catalysts used for this transformation are sulfonic resins (99), boron trifluoride (100), or zinc chloride in the presence of $POCl_3$ (101). In a more classical manner, the condensation can be carried out using acid chlorides in the presence of aluminum chloride (102) or trichloromethylbenzene (103).

The synthesis of 2,4-dihydroxyacetophenone [89-84-9] (**21**) by acylation reactions of resorcinol has been extensively studied. The reaction is performed using acetic anhydride (104), acetyl chloride (105), or acetic acid (106). The esterification of resorcinol by acetic anhydride followed by the isomerization of the diacetate intermediate has also been described in the presence of zinc chloride (107). Alkylation of resorcinol can be carried out using ethers (108), olefins (109), or alcohols (110). The catalysts which are generally used include sulfuric acid, phosphoric and polyphosphoric acids, acidic resins, or aluminum and iron derivatives. 2-Chlororesorcinol [6201-65-1] (**22**) is obtained by a sulfonation–chloration–desulfonation technique (111). 1,2,4-Trihydroxybenzene [533-73-3] (**23**) is obtained by hydroxylation of resorcinol using hydrogen peroxide (112) or peracids (113).

(**21**)   (**22**)   (**23**)

The condensation of an aldehyde with resorcinol gives rise to calix (3) arene (114). Isoprene reacts with resorcinol under acidic conditions to give benzodipyran (**24**) and monochroman (**25**) (115).

(**24**)   (**25**)

Resorcinol carboxylation with carbon dioxide leads to a mixture of 2,4-dihydroxybenzoic acid [89-86-1] (**26**) and 2,6-dihydroxybenzoic acid [303-07-1] (**27**) (116). The condensation of resorcinol with chloroform under basic conditions, in the presence of cyclodextrins, leads exclusively to 2,4-dihydroxybenzaldehyde [95-01-2] (**28**) (117). Finally, the synthesis of 1,3-bis(2-hydroxyethoxy)benzene [102-40-9] (**29**) has been described with ethylene glycol carbonate in basic medium (118), in the presence of phosphines (119). Ethylene oxide, instead of ethyl glycol carbonate, can also be used (120).

(**26**)   (**27**)   (**28**)   (**29**)

Economic Aspects and Uses

The total hydroquinone capacity is estimated to be 45,000–50,000 t/yr; demand for 1992 was around 40,000 t/yr. Hydroquinone is used in a broad range of applications such as photographic developers, polymerization inhibitors, rubber antioxidants, food antioxidants, synthesis intermediates, and water treatment.

Catechol is produced by coproduction with hydroquinone starting from phenol. Other techniques such as coal extraction remain marginal. The installed capacities (~25,000 t/yr) are now sufficient to cover the demand. Catechol is mainly used for synthesis in food, pharmaceutical, or agrochemical ingredients. A specific application of *tert*-butylcatechol is as a polymerization inhibitor.

The main uses of resorcinol are in the manufacture of rubber and wood adhesives. Manufacturing capacity is about 30,000 t/yr worldwide. World average prices in November 1993 are reported for the main derivatives in Table 2.

Table 2. Prices of Dihydroxybenzenes and Their Derivatives[a]

Compound	CAS Registry Number	Grade	Price, $/kg
hydroquinone	[123-31-9]	photo	4.20
		technical	3.80
guaiacol	[90-05-1]	technical	8.00
catechol	[120-80-9]	standard	6.00
resorcinol	[108-46-3]	USP	14.85
		technical	5.39
butylated hydroxyanisole (BHA)	[25013-16-5]	food	19.67
eugenol	[97-53-0]	USP	8.00
glyceryl guaiacolate	[93-14-1]		15.00
heliotropine	[120-57-0]		25.50
isoeugenol	[97-54-1]		9.67
vanillin	[123-33-5]	USP	17.60

[a]Data courtesy of Rhône-Poulenc.

Specification

Hydroquinone is available in photographic, inhibitor, and technical grade. Manufacturer's specifications are given in Table 3.

Catechol is available in standard and extra pure grade. Manufacturer's specifications for catechol and resorcinol are given in Table 4.

Table 3. Manufacturer's Specifications for Hydroquinone[a]

Specification	Photographic	Inhibitor	Technical
appearance	crystals	crystals	crystals
color	white	white to light tan	light tan
melting point, °C	171.0–174.0	171.0–174.0	170.0 min
assay, wt % (min)	99.5	99	99
heavy metal content, eg, Pb % (max)	0.001		
ash content, wt % (max)	0.05	0.05	
iron content, wt % (max)	0.001		

[a]Data courtesy of Rhône-Poulenc.

Table 4. Manufacturer's Specifications for Catechol and Resorcinol[a]

Specification	Catechol		Resorcinol	
grade	standard	extra pure	technical	USP XX
appearance	flakes	white crystals	flakes[b]	crystals[c]
melting point, °C	102	102		109.1
assay, wt (min)	98	99.5	99	99
ash content, wt % (max)			0.005	0.05

[a]Data courtesy of Rhône-Poulenc.
[b]White or slightly colored.
[c]White or nearly white.

Health and Safety

Dihydroxybenzenes (DHBs) are slightly more acutely toxic than phenol (Table 5). Contact with dihydroxybenzene through oral, dermal, or respiratory routes can induce significant systemic exposure. Skin or eye effects have been demonstrated during chronic or accidental professional exposure. No systemic effect has been described in such circumstances.

DHBs are rapidly absorbed through the digestive tract. Their absorption through skin is low in aqueous solution but may be increased in suitable solvents or creams. As vapor pressure is very low, significant respiratory exposure occurs only when dust is present. DHBs are irritants for skin and mucous membranes.

DHBs after absorption distribute rapidly and widely among tissues but bioaccumulation is low (121). They are metabolized to their respective benzoquinone and then detoxified by conjugation and excreted in the urine mainly as conjugates. Some deconjugations may occur in the urine. Resorcinol is also excreted in the urine in a free and conjugated state, essentially glucuronide and sulfate.

In experimental animals and *in vitro*, DHBs show a variety of biological effects including binding of metabolites to various proteins. Clastogenic effects have been observed *in vitro* and in some *in vivo* studies with the three compounds. No reproductive effects have been shown by conventional studies with either hydroquinone, catechol, or resorcinol (122). Hydroquinone has been shown to induce nephrotoxicity and kidney tumors at very high doses in some strains of rat (123); catechol induces glandular stomach tumors at very high dose (124). Repeated dermal application of resorcinol did not induce cancer formation (125).

In humans, cases of dermatitis have been described after contact with DHBs. Combined exposure to hydroquinone and quinone airborne concentrations

Table 5. Toxicity Data for Dihydroxybenzenes

Route	Hydroquinone	Catechol	Resorcinol
LD$_{50}$ oral rat,[a] mg/kg	320	260	301
LD$_{50}$ dermal rabbit, mg/kg[b]	>3800	800	3360
TWA (OSHA), mg/m^3	2	20	45

[a]Ref. 91.
[b]Rat.

causes eye irritation, sensitivity to light, injury of the corneal epithelium, and visual disturbances (126). Cases with an appreciable loss of vision have occurred (127). Long-term exposure causes staining due to irritation or allergy of the conjunctiva and cornea and also opacities. Resorcinol and catechol are also irritants for eyes.

Acute intoxication with DHBs occurs mainly by the oral route; symptoms are close to those induced by phenol poisoning including nausea, vomiting, diarrhea, tachypnea, pulmonary edema, and CNS excitation with possibility of seizures followed by CNS depression. Convulsions are more frequent with catechol as well as hypotension due to peripheral vasoconstriction. Hypotension and hepatitis seem more frequent with hydroquinone and resorcinol. Methemoglobinemia and hepatic injury may be noted within a few days after intoxication by DHBs.

All operations producing dust require the usual measures to prevent dust in the atmosphere exceeding the allowable daily concentration. If this is not feasible, personal protection devices should be used. Especially when hydroquinone is present as a powder, adequate eye protection should be provided.

BIBLIOGRAPHY

"Hydroquinone" in *ECT* 1st ed., Vol. 7, pp. 755–762, by P. W. Vittum, Eastman Kodak Co.; "Resorcinol" in *ECT* 1st ed., Vol. 11, pp. 711–720, by R. A. V. Raff, Koppers Co., Inc.; "Pyrocatechol" in *ECT* 1st ed., Vol. 11, pp. 307–314, by R. A. V. Raff, Koppers Co., Inc.; "Hydroquinone, Resorcinol, and Pyrocatechol" in *ECT* 2nd ed., Vol. 11, pp. 462–492, by R. Raff and B. V. Ettling, Washington State University; "Hydroquinone, Resorcinol, and Catechol" in *ECT* 3rd ed., Vol. 13, pp. 39–69, by J. Varagnat, Rhône-Poulenc.

1. *Huaxue Shijie* **26**, 209 (1985).
2. China Pat. 87,105,905 (Sept. 23, 1987), (to Qinejiane Calcium Carbide).
3. *Huaxue Shijie* **27**, 442 (1986).
4. Eur. Pat. 148,584 (Jan. 4, 1984), W. W. Kaeding (to Mobil Oil).
5. Jpn. Pat. 61 221,135 (Mar. 27, 1985), K. Kasano and co-workers (to Keishitsu Ryubun).
6. Eur. Pat. 149,508 (Jan. 4, 1984), W. W. Kaeding (to Mobil Oil).
7. Eur. Pat. 271,623 (Dec. 19, 1986), S. Koskimies and T. Haimela (to Neste Oy).
8. Jpn. Pat. 63 27,473 (July 18, 1986), M. Masamitsu and co-workers (to Mitsu Petro).
9. Eur. Pat. 482,573 (Oct. 25, 1990), H. Iwane, K. Kaneko, and N. Suzuki (to Mitsubishi Petro.); Jpn. Pat. 04 059,755 (June 26, 1990) and 04 059,756 (June 26, 1990), H. Koitsu and co-workers (to Nippon Shokubai).
10. Eur. Pat. 368,292 (Nov. 9, 1988), M. Dorn, E. Hagel, and W. Zeiss (to Peroxid Chemie).
11. U.S. Pat. 4,463,198 (Aug. 23, 1982), E. Nowak and co-workers (to Goodyear).
12. Eur. Pat. 271,623 (Dec. 19, 1986), S. Koskimies and T. Haimela (to Neste Oy).
13. Fr. Pat. 2,121,000 (Apr. 7, 1971), K. Matsuzawa and co-workers (to Mitsubishi Chem.); Fr. Pat. 2,222,344 (Mar. 23, 1973), S. Tamara and co-workers (to Ube Ind.).
14. USSR 1,502,559 (Oct. 29, 1987), A. Mikhailyk and Y. Mitnik (to Moscow Mendeleev Chem. Inst.).
15. *Stud. Surf. Sci. Catal.* **66**, 455 (1991).
16. *J. Chem. Tech. Biotechnol.* **35A**, 365 (1985).
17. *React. Kinet. Catal. Lett.* **43**(2), 419 (1991).

18. Fr. Pat. 2,071,464 (Dec. 30, 1969), F. Bourdin and co-workers (to Rhône-Poulenc).
19. *Ind. Eng. Chem. Prod. Res. Dev.* **15**(3), 212 (1976).
20. Ger. Pat. 2,633,302 (July 25, 1975), M. Jouffret and co-workers (to Rhône-Poulenc).
21. Fr. Pat. 2,336,364 (Dec. 24, 1975), M. Costantini, M. Jouffret, and R. Nantermet (to Rhône-Poulenc).
22. Eur. Pat. 480,800 (Oct. 8, 1990), M. Costantini and M. Jouffret (to Rhône-Poulenc).
23. Fr. Pat. 2,266,683 (Apr. 4, 1974), S. Umemura and co-workers (to Ube).
24. Fr. Pat. 2,655,332 (Dec. 5, 1989), M. Costantini and D. Laucher (to Rhône-Poulenc).
25. Fr. Pat. 2,201,279 (Sept. 28, 1978), P-E. Bost and co-workers (to Rhône-Poulenc).
26. Fr. Pat. 2,182,668 (May 3, 1972), P-E. Bost (to Rhône-Poulenc).
27. Eur. Pat. 408,418 (July 11, 1989), M. Costantini and D. Laucher (to Rhône-Poulenc).
28. Eur. Pat. 139,194 (Sept. 27, 1983), K. Drauz and A. Kleemann (to Degussa AG).
29. Jpn. Pat. 03 236339 (Feb. 9, 1990), M. Uohama and T. Akiyama; Jpn. Pat. 02 311433 (May 23, 1989), M. Uohama and A. Akiyama; Jpn. Pat. 02 311434 (May 23, 1989), M. Uohama and co-workers; Jpn. Pat 03 34948 (June 29, 1989), M. Uohama and K. Takahashi, Eur. Pat. 468,477 (Sept. 26, 1990), M. Uohama and T. Katsuji (to Dainippon Ink Chem.).
30. Jpn. Pat. 58 039,633 (Sept. 3, 1981), I. Kuriyama and T. Minoru (to Mitsubishi Gas).
31. World Pat. 9,214,691 (Feb. 16, 1991), S. W. Brown and co-workers (to Interox Chem.).
32. *Organometallics* **11**, 3578 (1992).
33. Brit. Pat. 2,024,790 (June 22, 1978), M. Taramasso and co-workers (to Snamprogetti SpA).
34. *Chimica Industria* **72**, 610 (1990); *J. Catal.* **133**, 220 (1992); *J. Mol. Catal.* **71**, 129 (1992); *Catal. Lett.* **16**, 85 (1992) and **16**, 109 (1992); *Appl. Catal. A* **92**, 113 (1992) and **92**, 93 (1992).
35. Eur. Pat. 412,596 (Aug. 9, 1989), B. Anfossi and co-workers (to Eniricherche); Jpn. Pat. 3,220,142 (Jan. 25, 1990), T. Tatsumi and M. Tominaga (to Tosoh Corp.); *Catal. Lett.* **11**, 291 (1991); *Appl. Catal.* **68**, 249 (1991); *Stud. Surf. Catal.* **72**, 21 (1992).
36. Eur. Pat. 190,609 (Feb. 5, 1985), F. Maspero and U. Romaru (to Enichem Sintesi); Eur. Pat. 230,949 (Jan. 28, 1986), M. G. Clerici and U. Ramaru; Eur. Pat. 315,247 (Nov. 2, 1987), M. G. Clerici and G. Bellussi; Eur. Pat. 315,248 (Feb. 11, 1987) G. Bellussi and co-workers (to Eniricherche).
37. Eur. Pat. 301,486 (July 29, 1987), P. Roffia and co-workers; Eur. Pat. 208,311 (July 10, 1985), P. Roffia and co-workers (to Montedipe); Eur. Pat. 496,385 (Jan. 23, 1991), V. Gervasutti and co-workers (to Enichem. Anic.); *J. Catal.* **137**, 252 (1992).
38. BE 1,003,840 (May 25, 1990), F. Maspero and U. Romano (to Enichem Synthesis).
39. Fr. Pat. 2,523,575 (Mar. 19, 1982), A. Esposito, C. Neri, and F. Buonomo (to Anic).
40. Fr. Pat. 2,657,346 (Jan. 19, 1990), M. Marinelli and co-workers (to Enichem Synthesis); *J. Catal.* **131**, 294 (1991); *J. Mol. Catal.* **68**, 45 (1991).
41. Eur. Pat. 200,260 (Apr. 23, 1985), G. Bellussi and co-workers (to Enichem Sintesi).
42. *J. Catal.* **130**, 1 (1991); *Chem. Comm.*, 589 (1992); *J. Mol. Catal.* **71**, 373 (1992); Eur. Pat. 492,697 (Dec. 12, 1990), A. Carati and co-workers (to Eniricherche).
43. *Chem. Comm.*, 589 (1992).
44. *Chem. Comm.*, 1245 (1992); *J. Catal.* **137**, 225 (1992); *Appl. Catal. A.* **93**, 123 (1993).
45. *Catal. Lett.* **8**, 237 (1991).
46. Eur. Pat. 346,250 (June 8, 1988), M. Costantini and co-workers (to Rhône-Poulenc).
47. Eur. Pat. 314,582 (Oct. 29, 1987), J-M. Popa and co-workers (to Rhône-Poulenc).
48. Eur. Pat. 466,545 (Aug. 31, 1990), J-L. Guth and co-workers (to Rhône-Poulenc).
49. Eur. Pat. 385,882 (Feb. 28, 1989), E. Garcin and co-workers (to Rhône-Poulenc).
50. Eur. Pat. 314,583 (Oct. 29, 1987), M. Costantini and co-workers (to Rhône-Poulenc).
51. Eur. Pat. 299,893 (July 17, 1987), M. Costantini and co-workers (to Rhône-Poulenc).
52. World Pat. 9,218,449 (Apr. 11, 1991), A. Johnstone and co-workers (to Solvay Interox).

53. Eur. Pat. 132,783 (July 18, 1983), R. A. Bull (to FMC).

54. U.S. Pat. 4,578,521 (Feb. 4, 1985), C. D. Chang and S. D. Hellring (to Mobil Oil).

55. Eur. Pat. 220,525 (July 10, 1986), H. Nagahara and M. Konishi (to Asahi Kasei); Jpn. Pat. 3,227,946 (Nov. 16, 1989), K. Matsuoka (to Daicel Chem.).

56. Jpn. Pat. 4,041,449 (June 7, 1990), S. Yokota and K. Matsuoka; Jpn. Pat. 4,026,641 (May 18, 1990), K. Matsuoka and K. Tagawa; Jpn. Pat. 4,046,133 (June 12, 1990), S. Yokota; Jpn. Pat. 3,240,746 (Feb. 19, 1990), K. Matsuoka, (to Daicel Chem); Jpn. Pat. 58 055,439 (Sept. 25, 1981), T. Maki and K. Murayoma (to Mitsubishi Chem.).

57. U.S. Pat. 4,861,921 (Mar. 3, 1987), J. D. Fellmann, R. J. Saxton, and P. Tung (to Catalytica).

58. Czech. Pat. 232,629 (Apr. 21, 1983), J. Sykora and M. Pado.

59. Jpn. Pat. 63 267,740 (Apr. 27, 1987), H. Sato and Y. Kawashima (to Idemitsu Kosan).

60. Jpn. Pat. 04 117,338 and Jpn. Pat. 04 117,339 (Sept. 6, 1990), F. Masahiko and co-workers (to Keishitsu Ryubun).

61. *Holzforschung* **43**, 99 (1989); *J. Wood Tech.* **9**, 251 (1989); *ACS Symp. Ser.* **397**, 228 (1989); *Fund. Thermochem. Biomass Convers.*, 453 (1985); *Org. Geochem.* **6**, 417 (1984).

62. *Biotechnol. Prog.* **8**, 78 (1992); Eur. Pat. 158,424 (Apr. 3, 1984), A. Yoskikawa and H. Sato (to Agency of Ind. Sci. Tech.); Eur. Pat. 268,331 (Nov. 20, 1986), P. J. Geary and J. E. Hawes, Eur. Pat. 253,438 (July 8, 1986), P. R. Betteridge and co-workers, Eur. Pat. 252,567 (July 8, 1986), P. R. Betteridge and co-workers (to Shell); Eur. Pat. 315,949 (Nov. 9, 1987), S. Doi and K. Horikoshi (to Idemitsu Kosan); U.S. Pat. 4,894,337 (Jan. 17, 1989), P. J. Oriel and G. Gurujeyala (to Michigan State University); Eur. Pat. 73,134 (Aug. 19, 1981), M. C. Hall (to Pfizer Inc.); *Bio. Ind. Z.*, 12 (1990).

63. *J. Am. Chem. Soc.* **114**, 9725 (1992).

64. Jpn. Pat. 60 009,889 (June 27, 1983), K. Sasaki and S. Ito (to Mitsui Petro.); Jpn. Pat. 2,138,233 (Feb. 22, 1989), K. Sasaki and co-workers (to Tosoh Corp.).

65. *J. Org. Chem.* **51**, 3471 (1986) and **53**, 296 (1988).

66. Jpn. Pat. 04 273,836 (Feb. 28, 1991), F. Matsuda and K. Kato (to Mitsui Toatsu).

67. *J. Mol. Catal.* **73**, 237 (1992); *Catal. Lett.* **11**, 11 (1991); *Bull. Chem. Soc. Jpn.* **62**, 2613 (1989); Jpn. Pat. 04 178,341 (Nov. 14, 1990), K. Sasaki and A. Kuriai (to Sasaki K.); Jpn. Pat. 04 244,039 (Jan. 30, 1991), F. Matsuda and K. Kato; Eur. Pat 519,084 (Dec. 26, 1990), K. Inoue and co-workers (to Mitsui Toatsu).

68. *Chem. Comm.*, 1446 (1992); Eur. Pat. 469,662 (Aug. 1, 1990), M. G. Clerici and co-workers (to Eniricherche).

69. Eur. Pat. 341,164 (May 2, 1988), M. Gubelmann and P. J. Tirel (to Rhône-Poulenc); *Russ. Chem. Rev.* **61**, 1130 (1992).

70. U.S. Pat. 4,431,848 (Apr. 26, 1982), N. P. Greco (to Koppers).

71. *Organometallics* **6**, 2021 (1987).

72. *Tetrah. Lett.* **30**, 205 (1989).

73. *Tetrah. Lett.* **23**, 1577 (1982).

74. *J. Org. Chem.* **52**, 921 (1987).

75. Fr. Pat. 2,486,523 (July 11, 1980), K. Formanek, D. Michelet, and D. Petre (to Rhône-Poulenc).

76. U.S. Pat. 4,324,731 (Apr. 13, 1982), D. Michelet and S. Veralini (to Rhône-Poulenc).

77. P. W. Lenz, *Recent Advances in Liquid Crystalline Polymers*, Elsevier Applied Science Publishers, Denmark, 1983.

78. W. J. Jackson, in E. J. Vandenberg, ed., *Contemporary Topics in Polymer Science*, Vol. 5, Plenum Publishing Corp., New York, 1984, p. 177.

79. U.S. Pat. 4,909,965 (Aug. 3, 1981), P. M. Burke (to E. I. du Pont de Nemours & Co., Inc.).

80. Jpn. Pat. 62 167,740 (Jan. 21, 1986), I. Kimura (to CK Fine Chemicals).

81. U.S. Pat. 4,945,183 (Nov. 30, 1989), E. F. Moran, Jr. (to E. I. du Pont de Nemours & Co., Inc.).

82. U.S. Pat. 4,960,957 (May 30, 1989), P. W. Wojtkowski (to E. I. du Pont de Nemours & Co., Inc.).

83. Jpn. Pat. 63 297,335 (May 29, 1987), S. Ishizawa (to CK Fine Chemicals); World Pat. 8,900,987 (Aug. 3, 1987), (to Eastman Kodak); Jpn. Pat. 01 093,552 (Oct. 5, 1987), R. Kondo and T. Okubo (to Toray); Jpn. Pat. 2,115,136 (Oct. 21, 1988), M. Takara and co-workers (to Mitsui Petrochem.); Jpn. Pat. 3,206,059 (Jan. 9, 1990), M. Furuya and H. Nakajima (to Asahi).

84. U.S. Pats. 4,439,595, C. S. Chiang, and 4,439,596 (Mar. 21, 1983), R. S. Irwin (to E. I. du Pont de Nemours & Co., Inc.).

85. Fr. Pat. 2,574,785 (Dec. 19, 1984), S. Ratton (to Rhône-Poulenc).

86. Eur. Pat. 326,455 (Oct. 1, 1989), J. R. Desmurs and I. Jouve (to Rhône-Poulenc).

87. Jpn. Pat. 03 258745 (Mar. 8, 1990), K. Ashida and co-workers (to Hon Shu).

88. Jpn. Pat. 04 134045 (Sept. 25, 1990), H. Doi and K. Yoshitani (to Takuyama Soda).

89. Eur. Pat. 278,378 (Feb. 6, 1987), O. Arndt and T. Papenfuhs (to Hoechst).

90. Eur. Pat. 278,377 (Feb. 6, 1987), O. Arndt and T. Papenfuhs (to Hoechst).

91. Jpn. Pat. 05 085,993 (Sept. 3, 1991), M. Kawabara and co-workers (to Mitsui Petro.); Jpn. Pat. 03 020,251 (June 16, 1989), M. Kondo and co-workers (to Mitsui Petro.).

92. Ger. Pat. 2,850,391 (Nov. 21, 1978), F. Werner (to Bayer).

93. Jpn. Pat. 05 238,994 (Feb. 27, 1992), H. Fujii and co-workers (to Nippon Soda); Jpn. Pat. 03 072,447 (Aug. 10, 1989), A. Otsuji and co-workers (to Mitsui Toatsu).

94. Eur. Pat. 449,546 (Mar. 26, 1990), H. Dressler (to Indspec Chemical Corp.).

95. Jpn. Pat. 55 053,250 (Oct. 16, 1978), K. Iwata and H. Arita (to Mitsui Petro.).

96. Jpn. Pat. 55 053,246 (Oct. 16, 1978), K. Iwata and H. Arita (to Mitsui Petro.).

97. Jpn. Pat. 55 015,412 (July 18, 1978), T. Nagaoka and S. Shimizu (to Mitsui Petro.).

98. Jpn. Pat. 55 004,338 (June 27, 1978), T. Nagaoka and co-workers (to Mitsui Petro.).

99. Jpn. Pat. 61 282,335 (June 6, 1985), T. Abe and co-workers (to Honshu Paper).

100. Jpn. Pat. 59 190,943 (Apr. 12, 1983), S. Hanaoka and co-workers (to Shipuro Kasei).

101. Jpn. Pat. 03 167,151 (Nov. 27, 1989), Y. Abe and K. Okawa (to Shipuro Kasei).

102. Ger. Pat. 3,206,129 (Feb. 20, 1982), K. Eiglmeier and J. Schulz (to Riedel-De Haen).

103. Pol. Pat. 145,626 (Dec. 23, 1985), A. Olszanowski and co-workers (to Politechnika Pozanska); Eur. Pat. 154,092 (Mar. 2, 1984), K. C. Liu (to GAF Corp.); Eur. Pat. 32,275 (Dec. 31, 1981), A. Paucot, P. Malfroid, and J. Mulders (to Solvay).

104. Jpn. Pat. 02 011,536 (June 30, 1988), T. Kanechika and co-workers (to Sumitomo Chem.).

105. *Indian J. Chem.* **20B**(11), 989 (1981); Jpn. Pat. 61 180,738 (Feb. 6, 1985), K. Tanaka and co-workers (to Nipoon Zeon); Eur. Pat. 219,282 (Oct. 11, 1985), K. Tanaka (to Sumitomo Chem.).

106. *Shanghai Diyi Yixueyuan Xuebao* **11**(4) 312 (1984); Jpn. Pat. 59 065,039 (Oct. 6, 1982), (to Sumitomo Chem.).

107. Jpn. Pat. 02 011,536 (June 30, 1988), T. Kanechika and co-workers (to Sumitomo Chem.).

108. Czech. Pat. 265,262 (Mar. 24, 1987), J. Kroupa and co-workers.

109. Jpn. Pat. 05 201,903 (Oct. 16, 1992), M. Ito and S. Iimuro (to Mitsui Toatsu); *Zh. Org. Khim.* **24**, 827 (1988); Jpn. Pat. 57 142,935 (Feb. 27, 1981), T. Ogura (to Showa Denko).

110. Brit. Pat. 1,581,428 (Mar. 13, 1978), B. E. Leach and C. M. Starks (to Conoco Inc.); Jpn. Pat. 58 072,530 (Oct. 28, 1981), K. Taniguchi and co-workers (to Mitsui Petro.).

111. Eur. Pat. 146,923 (Dec. 21, 1983), F. Della Valle and A. Romeo (to Fidia SpA).

112. Ger. Pat. 3,607,924 (Mar. 11, 1986), G. Prescher, G. Ritter, and H. Saurstein (to Degussa).

113. Ger. Pat. 3,632,075 (Sept. 20, 1986), G. Prescher, G. Ritter, and H. Saurstein (to Degussa).
114. Ger. Pat. 290,412 (Dec. 12, 1989), Ehrhardt and co-workers (to Universitat Leipzig); Ger. Pat. 287,158 (Sept. 23, 1988), F. Weinelt and co-workers (to Universitat Leipzig).
115. *J. Chem. Soc.* **1**, 335 (1982).
116. Jpn. Pat. 62 010,043 (Sept. 4, 1985), I. Tomino and co-workers (to Mitsui Petro.); Ger. Pat. 3,832,076 (Sept. 21, 1988), U. Eichenauer and P. Neuman (to BASF); Eur. Pat. 552,912 (Oct. 23, 1992), T. Nakamatsu and co-workers (to Sumitomo).
117. *Makromol. Chem.* **2**, 707 (1981).
118. Jpn. Pat. 02 096,545 (Oct. 3, 1988), O. Kimura and M. Yamashita (to Takoa Chem.); World Pat. 9,116,292 (Apr. 16, 1990), C. E. Summer, B. J. Hitch, and B. L. Bernard (to Eastman Kodak).
119. U.S. Pat. 5,059,723 (July 13, 1990), (to Indspec. Chem. Corp.).
120. Jpn. Pat. 04 001,151 (Apr. 16, 1990), (to Mitsui Petro.).
121. Y. C. Kim and H. B. Matthews, *Fund. Appl. Toxicol.* **9**(3), 409–414 (1987).
122. J. C. DiNardo and co-workers, *Toxicol. Appl. Pharmacol.* **78**(1), 63–66 (1985).
123. National Toxicology Program, *Technical Report on the Toxicology and Carcinogenesis Studies of Hydroquinone in F344/N Rats and B6C3F1 Mice (Gavages Studies)*, NTP TR 366, NTP, Research Triangle Park, N.C., 1989.
124. M. Hirose and co-workers, *Jpn. J. Cancer Res.* **81**(9), 857–861 (1990).
125. F. Stenback and P. Shubik, *Toxicol. Appl. Pharmacol.* **30**, 7–13 (1974).
126. J. H. Sterner, F. L. Oglesby, and B. Anderson, *J. Ind. Hyg. Toxicol.* **26**, 60–73 (1947).
127. B. Anderson and F. L. Oglesby, *AMA Arch. Ophthalmol.* **59**, 495–501 (1958).

LÉON KRUMENACKER
MICHEL COSTANTINI
P. PONTAL
J. SENTENAC
Rhône-Poulenc Recherches

HYDROTHERMAL PROCESSING

Hydrothermal processing encompasses a broad set of technologies which share a range of operating temperatures and pressures and require the use of engineered pressure systems. Table ? compares the distinguishing features of these technologies, all of which utilize water heated to temperatures above its boiling point (nominally 100°C). These diverse technologies have common needs for equipment and plant design. Many of the designs can be translated from one application to another (see also HIGH PRESSURE TECHNOLOGY).

Pressure hydrometallurgy, which involves the treatment of ores using water-based chemical processes at elevated temperatures and pressures, typically results in the extraction of the desired metals into the aqueous phase by

Table 1. Features of Hydrothermal Processes

Process name	Temperature range, °C	Pressure range, MPa[a]	Conditions	Application	Commercial use
pressure leaching/ precipitation	100–200	<5	reducing or inert	treatment of ores to recover metals	alumina, cobalt, nickel
hydrothermal synthesis	100–350	<20	oxidizing or inert	preparation of fine particle oxides	zeolites, titanates, ferrites, silicates
hydrothermal crystallization	350–450	<200	inert	preparation of large crystals and gemstones	α-quartz
wet oxidation	150–350	<20	oxidizing	destruction of organic sludges	municipal water treatment
supercritical water oxidation	400–600	22–30	oxidizing	destruction of hazardous and toxic waste	paper pulp, chemical warfare agents, propellants, mixed radioactive waste, industrial hazardous waste[b]

[a]To convert MPa to psi, multiply by 145.
[b]In demonstration stage only.

leaching (see METALLURGY, EXTRACTIVE; MINERAL RECOVERY AND PROCESSING). The extracting solution can contain inorganic acids or bases, as well as other effective complexing reagents (see CHELATING AGENTS). Once the metals have been extracted, a number of processes can be used to separate and recover valuable metals and/or inorganic chemicals. The primary methods are precipitation, electrowinning, solvent extraction, and ion exchange (qv). In some cases, leaching or precipitation reactions are carried out at elevated temperatures and pressures. These hydrothermal processes are used to improve the yield and reduce reaction time or to effect a better separation in the case of complex ores or other raw materials.

A commercial process which uses hydrothermal leaching on a large scale is the Bayer process for production of aluminum oxide (see ALUMINUM COMPOUNDS). This process is used to extract and precipitate high grade aluminum hydroxide (gibbsite [14762-49-3]) from bauxite [1318-16-7] ore. The hydrothermal process step is the extraction step in which concentrated sodium hydroxide is used to form a soluble sodium aluminate complex:

$$Al(OH)_3 + NaOH \longrightarrow NaAl(OH)_4(aq)$$

or

$$AlOOH + NaOH + H_2O \longrightarrow NaAl(OH)_4(aq)$$

Reaction conditions depend on the composition of the bauxite ore, and particularly on whether it contains primarily gibbsite, $Al(OH)_3$, or boehmite [1318-23-6], AlOOH. The dissolution process is conducted in large, stirred vessels or alternatively in a tubular reactor. The process originated as a batch process, but has been converted to a continuous one, using a series of stirred tank reactors or a tubular reactor.

Many other commercial hydrometallurgical processes employ a hydrothermal leach step. In the titanium dioxide sulfate process, ilmenite [12168-52-4] ore, primarily consisting of $FeO \cdot TiO_2$, is completely dissolved in concentrated H_2SO_4 at 110 to 120°C. The liquor is purified by removal of SiO_2 and $FeSO_4 \cdot 7H_2O$, and concentrated. Hydrous TiO_2, precipitated by hydrolysis, is then calcined to produce TiO_2 which is widely used as a pigment. This process, the so-called sulfate route, is responsible for production of greater than 40% of TiO_2 production (see TITANIUM COMPOUNDS). Ilmenite may also be pressure leached in an HCl solution.

Nickel and cobalt are recovered by processes that employ both pressure leaching and precipitation steps. The raw materials for these processes can be sulfide concentrates, matte, arsenide concentrates, and precipitated sulfides. Typically, acidic conditions are used for leaching; however, ammonia is also effective in leach solutions because of the tendency for soluble cobalt and nickel ammines to form under the leach conditions.

Pressure precipitation involves the separation of metals, metal oxides, or other inorganic materials by taking advantage of the retrograde solubility of a number of common compounds. Products are often of a well-defined form, eg, particle size, crystal structure, habit, and of relatively high purity, despite origination from a relatively low grade source. A combination of pressure leaching and precipitation cycles can result in the conversion of a low grade ore or inorganic waste material to a commercially valuable product.

Two other hydrothermal processes related to hydrometallurgical precipitation are hydrothermal synthesis and hydrothermal crystallization. These terms, sometimes used interchangeably, refer to the preparation of a wide range of high purity oxides through a crystallization (qv) process. Herein hydrothermal synthesis refers to the precipitation of very fine crystallites, typically <10 μm in diameter. The feed materials for a hydrothermal synthesis process are often purified chemicals including hydroxides, oxides, and metal salts. However, liquors or solids generated from hydrometallurgical or other chemical processes can be used; an example is the use of titanium hydrolyzate, a by-product of the sulfate route to titanium dioxide, for preparation of barium titanate or titanium dioxide pigment. Commercial processes which utilize hydrothermal synthesis include those for manufacture of a wide range of zeolites used as catalysts and molecular sieves (qv). A number of processes have also been developed for a wide range of electronic, magnetic, and ceramic materials.

Most hydrothermal synthesis processes are carried out at moderate (in the range of 100 to 300°C) temperatures and at the corresponding solution vapor pressure. The basic process, including conventional unit operations, is depicted in Figure 1. A feed slurry or solution consisting of oxides, hydroxides, and salts of the corresponding metal oxide product is pumped into a batch reactor (autoclave) or continuous flow reactor which heats the slurry to the desired reaction temperature. At this temperature the feed materials react and/or transform, primarily through dissolution and precipitation to the stable oxide form. An

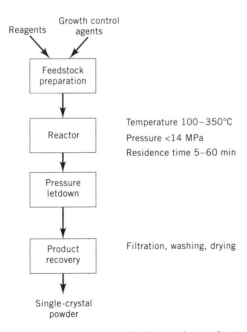

Fig. 1. General process operations for hydrothermal synthesis. Feedstocks may be oxides, hydroxides or salts, gels, organic materials, or acids or bases. The atmosphere within the reactor may be oxidizing or reducing. To convert MPa to psi, multiply by 145.

example is the formation of barium titanate from barium hydroxide and titanium oxide which occurs at temperatures between 150 and 250°C:

$$Ba(OH)_2(aq) + TiO_2 \longrightarrow BaTiO_3(s) + H_2O$$

Reaction times can be as short as 10 minutes in a continuous flow reactor (1). In a typical batch cycle, the slurry is heated to the reaction temperature and held for up to 24 hours, although hold times can be less than an hour for many processes. After reaction is complete, the material is cooled, either by batch cooling or by pumping the product slurry through a double-pipe heat exchanger. Once the temperature is reduced below approximately 100°C, the slurry can be released through a pressure letdown system to ambient pressure. The product is then recovered by filtration (qv). A series of wash steps may be required to remove any salts that are formed as by-products. The clean filter cake is then dried in a tray or tunnel dryer or reslurried with water and spray dried.

Hydrothermal crystallization processes occur widely in nature and are responsible for the formation of many crystalline minerals. The most widely used commercial application of hydrothermal crystallization is for the production of synthetic quartz (see SILICA, SYNTHETIC QUARTZ SILICA). Piezoelectric quartz crystals weighing up to several pounds can be produced for use in electronic equipment. Hydrothermal crystallization takes place in near- or supercritical water solutions (see SUPERCRITICAL FLUIDS). Near and above the critical point of water, the viscosity (300–1400 mPa·s (=cP) at 374°C) decreases significantly, allowing for relatively rapid diffusion and growth processes to occur.

Hydrothermal crystallization processes are batch in nature, and can take several days to complete crystal growth (2). A typical quartz heating cycle lasts between 15 and 60 days. In this process, sodium hydroxide or sodium carbonate solutions are used to dissolve pieces of relatively inexpensive quartz feedstock ~1.25 cm in diameter. The crude quartz is placed in the bottom of a pressure vessel with a relatively high length-to-diameter ratio (Fig. ?). The basic solution fills about 80% of the vessel volume at room temperature and expands to fill the vessel as temperature is increased. External heaters are used to create a temperature differential between the upper and bottom reactor zones. A typical production cycle would have a temperature of 425°C at the bottom of the vessel where nutrient is dissolved, and 375°C at the top where silica is deposited onto the surface of growing crystals. The solution phase fills the entire vessel under the reaction conditions, and generates a pressure of about 150 MPa (15 atm). Seeds of high quality quartz crystals are used in the upper region to control the crystallization of silica into the desired crystal orientation.

Two other chemical processes that rely on hydrothermal processing chemistry are wet oxidation and supercritical water oxidation (SCWO). The former process was developed in the late 1940s and early 1950s (3). The primary, initial application was spent pulp (qv) mill liquor. Shortly after its inception, the process was utilized for the treatment of industrial and municipal sludge. Wet oxidation is a term that is used to describe all hydrothermal oxidation processes carried out at temperatures below the critical temperature of water (374°C), whereas SCWO reactions take place above this temperature.

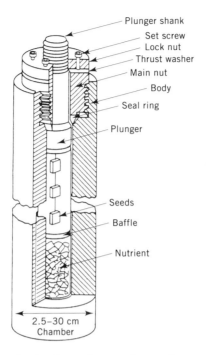

Fig. 2. Hydrothermal crystal growth apparatus.

Water Treatment Based on Wet Oxidation and SCWO

All wet oxidation processes involve the reaction of oxygen and suspended or dissolved organic substances. Wet oxidation processes are typically operated at temperatures between 150 and 325°C and at pressures of 2 to 20 MPa (290–2900 psi). The process is analogous to simple burning, although it is more difficult to initiate and proceeds at much lower temperatures. Wet oxidation processes can become thermally self-sustaining if the organic content is about 0.5 wt %, corresponding to a chemical oxygen demand (COD) of roughly 15,000 ppm. In order to recover energy, a COD of about 20,000 ppm is required. In general, wet oxidation processes work best at CODs of 40,000 to 100,000 ppm. The more efficient the process is for destruction of organic materials, the closer it can be operated to the 15,000 ppm COD lower limit without additional fuel. Wet oxidation is best suited for treatment of hazardous organic waste streams that are too dilute to be incinerated and too concentrated or toxic to be biologically treated (see INCINERATORS).

Studies have been made on the destruction of dozens of compounds by wet oxidation (4–13). The application of wet oxidation processes for industrial waste treatment was in practice as early as 1912 in Norway for partial oxidation of sulfite process liquor. Subsequently, the process was studied for the destruction of various organic substances in alkaline liquors and a process for recovering energy from various wastes was patented in 1954 (see FUELS FROM WASTE) (14). More recently, emphasis has turned toward destruction of a number of organic-containing sludges, including coke oven gas scrubber liquid, coal gas condensate, sulfitic refinery spent caustic, pesticides, cyanide plating baths, wood (qv) preservatives, mercaptobenzothiazole, diphenylamine, phenol, solvent still bottoms, acrylonitrile (qv), petrochemical waste, petroleum sludge, paper (qv) mill sludge, pink water, textile sludge, and a wide variety of chlorinated and nonchlorinated compounds (see WASTES, INDUSTRIAL).

SCWO processes take advantage of the increased solubility of oxygen and organic wastes as well as faster reaction rates which occur at temperatures above the critical point (374°C) of water. Because of the higher process temperatures, reactors must withstand higher pressures and increased tendency toward corrosion, but can be designed to be relatively compact owing to the reduced residence time necessary (see HIGH TEMPERATURE ALLOYS). SCWO processes are designed to provide complete (>99.99%) destruction of organics to the ultimate combustion products of carbon dioxide and water. Nitrogen-bearing wastes are converted to ammonia (qv) or nitrate salts, depending on feed composition. Sulfur (qv) and chlorine-containing wastes are converted to soluble salts or acids depending on reaction pH. The residence time for SCWO reactions can be shorter than five minutes. At higher temperatures, the residence time is decreased. The pressure for SCWO reactions is slightly above the critical pressure (21.8 MPa (3160 psi)), regardless of reaction temperature.

Work on SCWO process development was performed in the early 1970s and in the 1980s, many of the developments were performed under private contract to industrial or federal agencies. The technology for waste treatment and generation of energy from waste or low grade materials was commercialized by Modar (Natick, Massachusetts).

The performance of SCWO for waste treatment has been demonstrated (15,16). In these studies, a broad number of refractory materials such as chlorinated solvents, polychlorinated biphenyls (PCBs), and pesticides were studied as a function of process parameters (17). The success of these early studies led to pilot studies which showed that chlorinated hydrocarbons, including 1,1,1-trichloroethane [71-55-6], o-chlorotoluene [95-49-8], and hexachlorocyclohexane, could be destroyed to greater than 99.99997, 99.998, and 99.9993%, respectively. In addition, no traces of organic material could be detected in the gaseous phase, which consisted of carbon dioxide and unreacted oxygen. The pilot unit had a capacity of 3 L/min of liquid effluent and was operated for a maximum of 24 hours.

The demonstration unit was later transported to the CECOS facility at Niagara Falls, New York. In tests performed in 1985, approximately 3400 L of a mixed waste containing 2-chlorophenol [95-57-8], nitrobenzene [98-95-3], and 1,1,2-trichloroethane [79-00-5] were processed over 145 operating hours; 2-propanol was used as a supplemental fuel; the temperature was maintained at 615 to 635°C. Another 95-h test was conducted on a PCB containing transformer waste. Very high destruction efficiencies were achieved for all compounds studied (17). A later bench-scale study, conducted at Smith Kline and French Laboratories in conjunction with Modar (18), showed that simulated chemical and biological wastes, a fermentation broth, and extreme thermophilic bacteria were all completely destroyed within detection limits.

Although these tests showed high destruction efficiencies, two significant potential operating problems, corrosion and formation of sticky solids which rapidly plug the reactor, were exposed. A reactor design addressing these problems, by removing salts before plugging can occur, has been patented (19).

Synthesis of Advanced Inorganic Materials

Hydrothermal synthesis has proven a useful and relatively flexible tool for production of a wide variety of advanced materials. Advanced oxide products typically require high purity, tight control over particle size and distribution, and incorporation of special properties such as controlled inhomogeneity in the chemical and physical structure. Attention to these concerns in the process development and engineering stage can allow specifications to be met while keeping costs at or near those for conventional milling and calcination technology.

Zeolites. Manufacturing of zeolites is perhaps the largest commercial application of hydrothermal synthesis. The hydrothermal synthesis mechanisms, process conditions, and underlying chemistry are better understood for zeolites than for any other material synthesized at elevated temperatures and pressures (20). These materials are used as molecular sieves, adsorbents and drying agents, and ever increasingly as catalysts. The unique appeal of zeolites is a crystalline structure and composition which combines a high degree of catalytic activity with shape and size selectivity toward both reactants and products. Patented in 1972, ZSM-5 is one of a family of aluminosilicate catalysts (21). This catalyst was developed for the methanol-to-gasoline (MTG) process and generated a large body of research.

Synthesis processes for zeolites depend on the composition of the material, the desired particle size, morphology, etc. A principal distinguishing factor is the composition. ZSM-5 is a sodium aluminum silicate having a $SiO_2:Al_2O_3$ ratio of greater than five. High silica zeolites in general crystallize faster than those with high aluminum or substituted metal cation concentrations. The synthesis process is sensitive to numerous process variables. Among these are temperature, pH, origin of silica and alumina, identity of alkaline cation, reaction time, the use of a templating agent to control morphology and crystalline structure, and the use of seeding to control rate of nucleation and particle size. There is considerable trial and error in the design of a chemical process for a zeolite having specific properties, despite the wide body of literature on process effects. The general range of processing conditions encompasses the entire range of subcritical water temperatures and pressures. Synthesis is conducted almost exclusively by batch reaction; a typical cycle lasts several hours up to several days, depending on process and product composition.

Electronic Ceramics. Whereas as of this writing most ceramic oxide powders are made by a ball-milling and calcination process, interest is growing in the development of hydrothermal synthesis processes for production of titanate powders used in making electronic ceramic components including capacitors, actuators, sensors (qv), and transducers (see ADVANCED CERAMICS; CERAMICS AS ELECTRONIC MATERIALS). The value of the advanced oxide powder market is expected to reach $67 million by 1995 (22). Starting materials are typically oxides or carbonates of reasonable purity which are wet milled together to achieve a mixture of particles in the size range of 1–10 μm. After drying, the mixture is heated to several hundred degrees Celsius in a calciner where the reaction between oxides takes place. For example, barium carbonate and titanium dioxide combine to form barium titanate [12047-27-7] and gaseous carbon dioxide. This product must be remilled to reduce the particle size to the 1–5 μm size range suitable for most ceramic-forming processes. The material is often spray dried to improve the flow characteristics of the powder. The advantages of hydrothermal synthesis over conventional processes are primarily improved control over powder stoichiometry and uniformity of the crystalline material, which can lead to greater reliability and performance in the finished ceramic. In addition, powders are more reactive, resulting in lower temperature requirements for the firing process in which green molded ceramic bodies are densified into finished products. The particle size of hydrothermally synthesized powders is uniform and typically <1 μm in diameter. A typical microstructure for a hydrothermal barium titanate powder is shown in Figure **?a**. Proper dispersion of these particles followed by forming into ceramic shapes and firing can result in a uniformly fine-grained ceramic having improved mechanical strength and electrical properties.

The most significant commercial product is barium titanate, $BaTiO_3$, used to produce the ceramic capacitors found in almost all electronic products. As electronic circuitry has been miniaturized, demand has increased for capacitors that can store a high amount of charge in a relatively small volume. This demand led to the development of highly efficient multilayer ceramic capacitors. In these devices, several layers of ceramic, from 25–50 μm in thickness, are separated by even thinner layers of electrode metal. Each layer must be dense, free of pin-holes and flaws, and ideally consist of several uniform grains of fired ceramic. Manufacturers are trying to reduce the layer thickness to 10–12 μm.

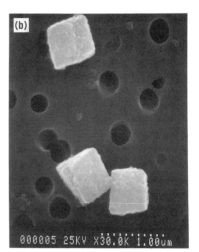

Fig. 3.   SEM micrographs of (**a**) hydrothermal $BaTiO_3$ and (**b**) hydrothermal $Pb(Zr,Ti)O_3$.

Conventionally prepared ceramic powders cannot meet the rigorous demands of these applications, therefore an emphasis has been placed on production of advanced powders by hydrothermal synthesis and other methods.

Several manufacturers of ceramic powders are involved in commercialization of hydrothermally derived powders. In the United States, Cabot (Boyertown, Pennsylvania) has built a small manufacturing plant and is supplying materials to capacitor manufacturers. Other manufacturers include Sakai Chemical and Fuji Titanium in Japan. Sakai Chemical is reportedly producing 1 t/d in its demonstration plant. A comparison of the characteristics of commercially available powders is given in Table ?.

A further advantage of hydrothermal synthesis of electronic titanates is the ability to chemically dope the material with commercially important oxides. For example, barium titanate is rarely used in its pure form for production of capacitors. Other oxides including ZrO_2, SnO_2, SrO, and CaO are added to improve electrical properties such as dielectric permittivity, dielectric loss, etc. In the conventional process, these must be dry measured, co-milled, then diffused into the crystal structure of the host oxide by high temperature calcination. The processes are energy intensive, difficult to control, and contribute undesired impurities to the final oxide powder. Hydrothermal synthesis, on the other hand, allows for adding dopants in a controlled manner and achieving mixing at a molecular scale. This is accomplished by using a coprecipitation step to prepare a well-mixed feed material, typically a hydrous oxide gel (24). The gel is prepared by codissolving salts of the insoluble oxide components, eg, titanium, zirconium, and tin. The solution is then precipitated using sodium hydroxide or ammonia. The resulting gel containing relatively soluble components, eg, barium hydroxide

Table 2. Commercial Hydrothermal Barium Titanate Powders[a]

| | | | Fuji Titanium | |
Parameter	Cabot BT-10	Sakai	BTWH	BTWS
Ba/Ti ratio	1.002 ± 0.003	0.999/1.000	0.996	0.998
purity, ppm				
Na_2O		<10	620	50
CaO	70	10	20	20
SrO	480	14/20	30	100
SiO_2	100	70/100	50	50
Al_2O_3	20	10/100	50	40
Fe_2O_3	15	20/40	40	30
Cl	30		530	50
particle size, μm	0.1–0.5	0.1/0.5	0.1/0.3	0.1/0.5
surface area, m^2/g	8.5–10.5	11/2.3	11.6/5.2	10.5/2.5
density, g/cm^3		5.65/5.88[b]	5.85/5.75[b]	
electrical[c]				
ϵ, 20°C		4090/3750		
ρ, $\Omega\cdot$cm		$3.5 \times 10^{11}/10^{12}$		

[a]Ref. 23
[b]At 1200°C.
[c]ϵ = dielectric constant; ρ = resistivity.

and strontium hydroxide, then reacts under hydrothermal conditions, typically 150–250°C, depending on composition, to form uniform crystallites of target composition. The crystallization process itself can occur through a number of mechanisms, including adsorption of metal ions on the hydroxide gel followed by transformation to the stable oxide form, and dissolution followed by classical supersaturation-driven crystallization.

Another important class of titanates that can be produced by hydrothermal synthesis processes are those in the lead zirconate–lead titanate (PZT) family. These piezoelectric materials are widely used in manufacture of ultrasonic transducers, sensors, and miniature actuators (see PIEZOELECTRICS). The electrical properties of these materials are derived from the formation of a homogeneous solid solution of the oxide end members. The process consists of preparing a coprecipitated titanium–zirconium hydroxide gel. The gel reacts with lead oxide in water to form crystalline PZT particles having an average size of about 1 μm (Fig. 3b). A process has been developed at Battelle (Columbus, Ohio) to the pilot-scale level (5-kg/h).

Pigments. A number of inorganic pigments can be produced by hydrothermal synthesis. Colored pigments include red and black iron oxides, Fe_2O_3 and Fe_3O_4; green chromium sesquioxide [1308-38-9], Cr_2O_3; blue-green chromium oxyhydroxide [20770-05-2], CrOOH; red and brown zinc ferrite [12063-19-3], $ZnFe_2O_4$; yellow iron oxide [20344-49-4], FeOOH; blue cobalt aluminate [1333-88-6], $CoAl_2O_4$; and many others (see COLORANTS FOR CERAMICS; PIGMENTS, INORGANIC). All processes involve the reaction of oxides, hydroxides, and salts in water at elevated temperatures according to the process steps depicted in Figure 1. Compound oxides such as ferrites (qv), cobaltites, and aluminates are typically formed in alkaline solutions. Reaction temperature depends on the reactivity of the feed materials and the composition of the product. The iron oxide and chromium oxide processes have been modified for production of magnetic pigments used for manufacture of recording media. Many of the process details are proprietary to the pigment manufacturers. One development is in lustrous iron oxide pigment preparation. Luster may be imparted to iron oxide by doping with alumina and silica and controlling the particle morphology by hydrothermal synthesis.

Titanium dioxide can be precipitated directly from a chloride or sulfate solution by heating to temperatures in the range of 110 to 150°C, and corresponding vapor pressure. This method is not used commercially, however, owing primarily to the environmental superiority of the chloride vapor process. It is also not possible to control particle size in the desired size of 0.2–0.3 μm. When precipitating TiO_2 directly from solution, very rapid homogeneous nucleation occurs because of supersaturation effects, resulting in the formation of extremely fine particles (see TITANIUM COMPOUNDS, INORGANIC).

A process has been developed by J. M. Huber Co. to treat kaolin clay pigments using a hydrothermal process (see CLAYS) (25). The products, called synthetic alkali metal aluminosilicates (SAMS), have superior pigmentary qualities for paper (qv) coating.

Silicates. In 1990, a continuous hydrothermal production plant was started up in Sulitjelma, Norway for manufacture of 9000 m^3/yr of calcium silicate [10101-39-0] having average density of 0.255 g/cm³. The calcium silicate is of the xonotlite [12141-77-4] form, $Ca_6Si_6O_{17}(OH)_2$, which can easily be transformed into β-wollastonite [14567-51-2] refractory by firing at 865°C. The material has many advantages, including little bound water compared to tobermorite, a lower grade form of calcium silicate. Refractory made from hydrothermal xonotlite has outstanding mechanical properties and thermal resistance. Lightweight refractory can be manufactured, having a maximum density of 0.5 g/cm³. In addition, xonotlite crystals (powders) can be used as fillers in toothpaste, as a thixotropic agent in paint (qv), and as a filler in rubber manufacture (see SILICON COMPOUNDS, SYNTHETIC INORGANIC SILICATES).

The technology for the plant (26) involves

$$6\ Ca(OH)_2 \longrightarrow 6\ Ca^{2+} + 12\ OH^-$$

$$6\ SiO_2 + 12\ H_2O \longrightarrow 6\ Si(OH)_4(aq)$$

$$6\ Ca^{2+} + 6\ Si(OH)_4(aq) + 12\ OH^- \longrightarrow Ca_6Si_6O_{17}(OH)_2 + 17\ H_2O$$

A small excess of hydrated lime is used to ensure complete reaction and amorphous or crystalline silica is used. The reaction to xonotlite takes place at temperatures above 225°C and pressure of 2.55 MPa (370 psi). Residence time is one hour when using amorphous silica and 3.25 hours when using less reactive silica flour (crystalline). Production is accomplished in a batch reactor operated in a semicontinuous mode. Some of the process heat is recovered as steam during pressure letdown and returned to heat the next batch. An important aspect of the design is the utilization of pressure systems which prevent the degradation of product when releasing it to ambient pressure, while avoiding a long batch cooling cycle.

Process Equipment

Hydrothermal Synthesis Systems. Of the unit operations depicted in Figure 1, the pressurized sections from reactor inlet to pressure letdown are key to hydrothermal process design. In consideration of scale-up of a hydrothermal process for high performance materials, several criteria must be considered. First, the mode of operation, which can be either continuous, semicontinuous, or batch, must be determined. Factors to consider are the operating conditions, the manufacturing demand, the composition of the product mix (single or multiple products), the amount of waste that can be tolerated, and the materials of construction requirements. Criteria for the selection of hydrothermal reactor design may be summarized as

single or multiple products	typically a single product plant benefits from continuous reactor design, either tubular or continuous stirred tank reactors (CSTRs) in series
capacity	very large production plants benefit from continuous design
flexibility	variable production schedules may dictate the use of multiple batch production lines

reaction conditions	low temperature and pressure reactions can be operated economically in either batch or continuous modes; long (>10 min) residence times may dictate batch reaction
space	in plant retrofits, available space may dictate reactor selection to some degree
physical and chemical characteristics of products	changeover from stirred batch to continuous tubular may affect product properties

These factors can all be shown to influence the operating and/or capital costs of the plant (1). In addition a key concern is to reproducibly manufacture products having the desired physical and chemical properties.

Conventional reactors used for conducting hydrothermal reactions, leaching or precipitation, are frequently called autoclaves. Vessels can be vertical, horizontal, or spherical. Vertical autoclaves are generally steam heated and agitated, whereas horizontally oriented vessels may be mechanically agitated. Spherical autoclaves are rotated to impart gentle mixing to the slurry. Steam agitated vessels are fabricated from welded stainless steel cylinders with spherical heads. Steam is generally fed from the bottom, and a liner of acid-proof brick may be incorporated. Horizontally oriented vessels are divided into several mixed sections. The body can be lined with rubber, lead, or alloy steel.

Tubular continuous autoclaves were invented in the early 1930s and have been used extensively in Germany for leaching bauxite. Most of the heat can be recovered by exchanging the heat from exiting slurry with the incoming material. Hence, tubular systems are more efficient and can provide very high throughput. Continuous systems require the design of pressure delivery and let-down systems which can withstand the operating conditions of the hydrothermal processing. As of this writing, the only publicly acknowledged use of a continuous reactor system in the United States for hydrothermal synthesis was for a pilot-scale system designed and operated by Battelle (Columbus, Ohio) at production rates up to about 5-kg/h. Both tubular (Fig. ?) and continuous stirred tank reactor designs were tested for hydrothermal synthesis of electronic ceramics. Flow at the reactor inlet was controlled using a diaphragm or piston (positive displacement) pump. The letdown system was based on the use of a heat exchanger to cool the exit flow to approximately ambient temperature. The cooled slurry was transported to a pressurized tank used to control flow and system pressure. The letdown tank was fitted with a continuous level probe which could be used to monitor fill volume. A signal from the level probe opened and shut a conventional needle valve ensuring that the flows in and out of the vessel were balanced. A separate pressure control loop maintained a constant pressure in the head space. This pressure was transmitted backward through the system via a liquid seal in the letdown tank, thus constant system pressure and flow could be assured. This system was operated at temperatures up to 320°C and pressures to 13 MPa (1880 psi).

A commercial design based on semicontinuous operation was developed for manufacture of silicate powders (27). A slurry, prepared containing the feed materials and water, is fed to the reactor tank and heated by circulating a

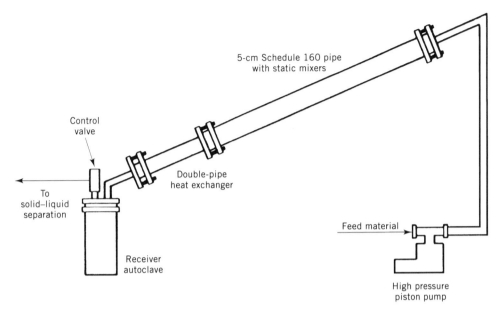

Fig. 4. Continuous tubular reactor design. Courtesy of Battelle.

heat-exchange fluid in channels located on the outside vessel wall. A six-bladed stirrer is operated at about 100 rpm in order to keep reagents well mixed. Once the slurry reaches the operating temperature, the vessel heat is maintained until reaction is complete. For most fine-particle products, this time is less than one hour.

In order to facilitate pressure letdown and heat recovery, an antipressure or receiving vessel is used. The vapor phase between vessels is allowed to come to equilibrium by opening a connecting valve in a line connecting the top of the reactor and receiving vessels. A liquid line connecting the bottom of the first vessel to the second vessel is then opened. No flow occurs at first because the two vessels are at equal pressure. However, by slowly venting steam from the second vessel, the liquid begins to flow. The flow rate is controlled by a pressure and flow control loop. By controlling the flow rate, laminar flow is ensured which protects the crystallized material from damage as it is passed through the valve. Also, heat can be recovered from the exiting process fluid and returned to heat the incoming fluid for the next batch. Once all material is transferred between the vessels, valves between vessels are shut, and the next charge is added to the reactor vessel. Product fluid is pumped from the receiving vessel to filtration and product collection equipment.

Wet Oxidation Reactor Design. Several types of reactor designs have been employed for wet oxidation processes. Zimpro, the largest manufacturer of wet oxidation systems, typically uses a tower reactor system. The reactor is a bubble tower where air is introduced at the bottom to achieve plug flow with controlled back-mixing. Residence time is typically under one hour. A horizontal, stirred tank reactor system, known as the Wetox process, was initially developed by Barber-Colman, and is also offered by Zimpro.

Wetox uses a single-reactor vessel that is baffled to simulate multiple stages. The design allows for higher destruction efficiency at lower power input and reduced temperature. Its commercial use has been limited to one facility in Canada for treatment of a complex industrial waste stream. Kenox Corp. (North York, Ontario, Canada) has developed a wet oxidation reactor design (28). The system operates at 4.1–4.7 MPa (600 to 680 psi) with air, using a static mixer to achieve good dispersion of liquid and air bubbles.

Another wet oxidation system is designed to be operated in a deep well, up to 1500 m below the ground surface. Called a vertical tube reactor, it may be suitable for destruction of municipal sludge and toxic wastes in some geological locations having large (>380 L/min) waste flows. Two companies, Vertech and Oxidyne (Dallas, Texas) independently developed similar designs which incorporate inexpensive low pressure pumps, because gravity generates enough pressure to maintain the liquid at the bottom of the well. A full-scale demonstration of the Vertech system was made in Longmont, Colorado (29).

Other above-ground continuous flow systems have been designed and operated for SCWO processes. A system developed by Modell Development Corp. (Modec) uses a tubular reactor and can be operated at temperatures above 500°C. It employs a pressure letdown system in which solid, liquids, and gases are separated prior to pressure release. This simplifies valve design and material selection on the liquid leg.

BIBLIOGRAPHY

1. W. J. Dawson and M. K. Han, "Development and Scale-Up of Hydrothermal Processes for Synthesis of High Performance Materials," *Proceedings of the Milton E. Wadsworth IV International Symposium on Hydrometallurgy*, Aug. 1–5, 1993.
2. R. A. Laudise, *Chem. Eng. News* (Sept. 28, 1987).
3. G. S. Seiler, *Sludge Manage. Ser.* **17**, 100–105 (1987).
4. T. L. Randall, in C. P. Huang, ed., *Proceedings of the 13th Mid-Atlantic Conference*, 1981, pp. 501–508.
5. C. R. Baillod and B. M. Faith, "Wet Oxidation and Ozonation of Specific Organic Pollutants," report to the Office of Research and Development, U.S. Environmental Protection Agency, Washington, D.C., Aug. 1983.
6. G. B. O. Biezugbe and B. G. Place, "Wet-Air Oxidation of TS-1S Wastewater," U.S. Department of Energy, Washington, D.C., Sept. 1985.
7. J. Kalman, G. Palmai, and I. Szebenyi, in S. T. Kolaczkowski and B. D. Crittenden, eds., *Management of Hazardous and Toxic Wastes for the Process Industry, (International Congress)*, Elsevier, New York, 1986, pp. 594–602.
8. G. Friedhofen, H. Kerres, J. Rosembaum, and R. Thiel, "Wet Air Oxidation of Waste Water," report to Bundesministerium fur Forshung und Technologie (BMFT), Dec. 1980.
9. *FEMS Sym.* **12**, 65–74 (1981).
10. "Characterization and Treatment of Wastes from Metal-Finishing Operations," PEI Associates, Inc., report to U.S. Environmental Protection Agency, Nov. 1990.
11. A. K. Chowdery and W. C. Copa, *Ind. Chem. Eng.* **28**(3), 3–10 (1986).
12. W. Copa, J. Heimbunch, and P. Schaefer, report to U.S. Environmental Protection Agency, 1984, pp. 60–64.

13. P. J. Canney and P. T. Schaefer, in M. D. LaGrega and L. K. Hendrian, eds., *Proceedings of the 15th Toxic Hazardous Waste Processing Mid-Atlantic Industrial Waste Conference*, Butterworth, U.K., 1983, pp. 277–284.
14. U.S. Pat. 2,665,249.
15. M. Modell, "Detoxification and Disposal of Hazardous Organic Chemicals by Processing with Supercritical Water," Final Report, U.S. Medical Research and Development Command, 1985.
16. S. H. Timberlake, G. T. Hong, M. Simson, and M. Modell, "Supercritical Water Oxidation for Wastewater Treatment: Preliminary Study of Urea Destruction," SAE Technical Paper Series Number 820872, 1982.
17. K. C. Swallow, W. R. Killilea, K. C. Malinkowski, and C. N. Staszak, *Waste Manage.* **9**, 19–26 (1989).
18. J. B. Johnston, and co-workers, *Bio/Technol.* **6**, 1423 (1988).
19. U.S. Pat. 5,200,093 (Apr. 6, 1993), H. E. Barner, C-Y. Huang, W. R. Killilea, and G. T. Hong (to ABB Lummus Crest Inc.).
20. R. M. Barrer, *Hydrothermal Chemistry of Zeolites*, Academic Press, Inc., New York, 1982.
21. U.S. Pat. 3,702,886 (1972) (to Mobil Oil Corp.).
22. *Ceramic Bull.* **69**(5) (1990).
23. C. Pommier, "La Synthese Hydrothermale," in J. C. Niepce, ed., *BaTiO$_3$: Materiau de base pour le dielectrique des condensateurs ceramiques*, Septima, Collection Forceram, 1993.
24. U.S. Pat. 5,112,433 (May 12, 1992), W. J. Dawson and S. L. Swartz (to Battelle Memorial Institute).
25. U.S. Pat. 5,223,235 (June 29, 1993), S. K. Wasson (to J. M. Huber Corp).
26. U.S. Pat. 4,238,240 (1980), 4,366,121 (1982), and 4,753,787 (1988), P. Krijgsman.
27. U.S. Pat. 5,026,527 (June 25, 1991), P. Krijgsman.
28. U.S. Pat. 4,604,215 (Aug. 5, 1986).
29. "Aqueous Phase Oxidation of Sludge using the Vertical Reaction Vessel System," report to U.S. Environmental Protection Agency, Washington, D.C., Mar. 1987.

WILLIAM J. DAWSON
Chemical Materials International

PIETER KRIJGSMAN
CEC Company

HYDROXYBENZALDEHYDES

Hydroxybenzaldehydes are organic compounds of the general formula:

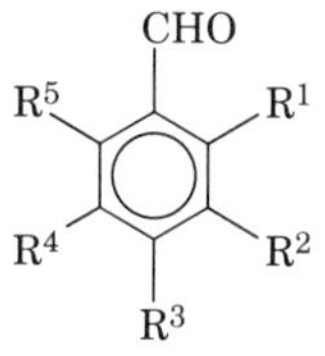

where R^1 through R^5 = H or OH but at least one R group is OH. All of the iso-meric mono-, di-, and trihydroxybenzaldehydes have been isolated (Table 1). The higher polyhydroxybenzaldehydes are unknown (12). This article deals primar-ily with p-hydroxybenzaldehyde [123-08-0] and salicylaldehyde [90-02-8] which together represent more than 99% of the hydroxybenzaldehydes market.

Of the two commercially important monohydroxybenzaldehydes the ortho isomer (salicylaldehyde) is the more important one. Salicylaldehyde (salicylic aldehyde, salicylal) is a colorless, oily liquid, the only hydroxybenzaldehyde liquid at room temperature with a pungent irritating odor. It occurs naturally in beer, oils of spirea, bird cherries and cassia, and in coffee, grape, tea, and tomato (13,14). (see BENZALDEHYDE). Salicylaldehyde and its derivatives are utilized as ingredients in agricultural chemicals, electroplating, perfumes, petroleum chemicals, polymers, and fibers (qv).

p-Hydroxybenzaldehyde (4-formylphenol) is a colorless to faint tan solid, with a slight, agreeable, aromatic odor. It occurs naturally in some plants in small amounts (15–17).

Physical Properties

Physical constants for salicylaldehyde and p-hydroxybenzaldehyde are listed in Table 2. Spectral data have been published (Table 2).

The location of the hydroxyl and aldehyde groups ortho to one another in salicylaldehyde permits intramolecular hydrogen bonding, and this results in the lower melting point and boiling point and the higher acid dissociation constant observed relative to p-hydroxybenzaldehyde.

Chemical Properties

The effect of the aldehyde group on the phenolic hydroxyl group is primarily an increase in its acidity: both 2-hydroxy- and 4-hydroxybenzaldehydes are stronger acids than phenol (pK_a (H_2O, 20°C) = 9.89; see Table 2 for comparison). The aldehyde group, however, has little effect on the reaction of the hydroxyl group. The deactivating effect of the phenolic hydroxyl on the aldehyde group is more pronounced, but the hydroxybenzaldehydes still undergo most of the normal aldehyde reactions.

Table 1. Physical Properties of Hydroxybenzaldehydes

Chemical Abstracts name	Mol wt	CAS Registry Number	Common name	Mp, °C	Reference	Bp, °C[a]
2-hydroxybenzaldehyde	122.12	[90-02-8]	salicylaldehyde	–7	1	197
3-hydroxybenzaldehyde	122.12	[100-83-4]	m-hydroxybenzaldehyde	108	1	240
4-hydroxybenzaldehyde	122.12	[123-08-0]	p-hydroxybenzaldehyde	117	1,2	310
2,3-dihydroxybenzaldehyde	138.12	[24677-78-9]	o-pyrocatechualdehyde	105–107	3	
2,4-dihydroxybenzaldehyde	138.12	[95-01-2]	β-resorcylaldehyde	201–202	1	220–228[b]
2,5-dihydroxybenzaldehyde	138.12	[1194-98-5]	gentisaldehyde	97–98	4	99[c]
2,6-dihydroxybenzaldehyde	138.12	[387-46-2]	γ-resorcylaldehyde	154–155	5	
3,4-dihydroxybenzaldehyde	138.12	[139-85-5]	protocatechualdehyde	154 (dec)	1	
3,5-dihydroxybenzaldehyde	138.12	[26153-38-8]	α-resorcylaldehyde	161–162	6	
2,3,4-trihydroxybenzaldehyde	154.12	[2144-08-3]		179–180	7	
2,3,5-trihydroxybenzaldehyde	154.12	[74186-01-9]		187	8	
2,3,6-trihydroxybenzaldehyde	154.12	[64168-39-4]		170–180 (dec)	9	
2,4,5-trihydroxybenzaldehyde	154.12	[35094-87-2]		223	10	
2,4,6-trihydroxybenzaldehyde	154.12	[487-70-7]	phloroglucinaldehyde	292–295 (dec)	7	
3,4,5-trihydroxybenzaldehyde	154.12	[13677-79-7]	gallaldehyde	210	11	

[a] At 100 kPa = 0.987 atm unless otherwise noted.
[b] At 2.9 kPa[d].
[c] At 1.6 kPa[d].
[d] To convert kPa to mm Hg, multiply by 7.5.

Table 2. Physical and Spectral Data for *o*- and *p*-Hydroxybenzaldehydes

	Salicylaldehyde		*p*-Hydroxybenzaldehyde	
Property	Value	Reference	Value	Reference
	Physical data			
bp, °C (kPa)[a]	93 (3.3)	18	170 (1.3)	23
	80 (1.6)	19		
vapor pressure		20		28
density, g/cm³	1.167 (20°C)	21	1.143 (117°C)	29
solubility, g/100 g water	1.7 (86°C)	22	1.3 (30.5°C)	22
acid dissociation constant pK$_a$, H$_2$O, 25°C	8.14	23	7.6	23
	Spectral data			
infrared		24		30
ultraviolet		25		25
proton nmr		26		31
mass spectra		27		32

[a]To convert kPa to mm Hg, multiply by 7.5.

Reactions of the Hydroxyl Group. The hydroxyl proton of hydroxybenz-aldehydes is acidic and reacts with alkalies to form salts. The lithium, sodium, potassium, and copper salts of salicylaldehyde exist as chelates. The cobalt salt is the most simple oxygen-carrying synthetic chelate compound (33). The stability constants of numerous salicylaldehyde–metal ion coordination compounds have been measured (34). Both salicylaldehyde and 4-hydroxybenzaldehyde are read-ily converted to the corresponding anisaldehyde by reaction with a methyl halide, methyl sulfate (35–37), or methyl carbonate (38). The reaction shown produces *p*-anisaldehyde [*123-11-5*] in 93.3% yield. Other ethers can also be made by the use of the appropriate reagent.

Reactions of the Aromatic Ring. The aromatic ring of hydroxybenzalde-hydes participates in several typical aromatic electrophilic reactions.

Halogenation. Chlorination and bromination yield mono- and dihalo derivatives, depending on reaction conditions. Bromination of *p*-hydroxy-benzaldehyde in chloroform yields 65–75% of the product shown (39).

Sulfonation and Diazonium Coupling. Like phenol, salicylaldehyde reacts easily in these reactions (40,41), with a high para-selectivity.

Reactions of the Aldehyde Group. *Oxidation.* Oxidation of hydroxy-benzaldehydes can result in the formation of a variety of compounds, depending on the reagents and conditions used. Replacement of the aldehyde function by a hydroxyl group results when 2- or 4-hydroxybenzaldehydes are treated with hydrogen peroxide in acidic (42) or basic (43) media: pyrocatechol or hydroquinone are obtained, respectively.

Salicylaldehyde is readily oxidized, however, to salicylic acid by reaction with solutions of potassium permanganate, or aqueous silver oxide suspension. 4-Hydroxybenzaldehyde can be oxidized to 4-hydroxybenzoic acid with aqueous silver nitrate (44). Organic peracids, in basic organic solvents, can also be used for these transformations into benzoic acids (45). Another type of oxidation is the reaction of salicylaldehyde with alkaline potassium persulfate, which yields 2,5-dihydroxybenzaldehyde (46).

Canizzaro Reaction. Both 2- and 4-hydroxybenzaldehydes undergo this self-oxidation–reduction reaction, but much less readily than benzaldehyde; the reaction requires metal catalysts such as nickel, cobalt, or silver to yield the corresponding hydroxybenzoic acids and hydroxybenzyl alcohols (47–48).

Reduction. These hydroxybenzaldehydes can be reduced by catalytic hydrogenation over palladium or platinum to yield the corresponding hydroxybenzyl alcohols, but the electrolytic reduction in an alkaline medium gives the coupling product 1,2-bis(4-hydroxyphenyl)ethane-1,2-diol in very good yield from 4-hydroxybenzaldehyde (49–51).

Reactions with Amines and Amides. Hydroxybenzaldehydes undergo the normal reactions with aliphatic and aromatic primary amines to form imines and Schiff bases; reaction with hydroxylamine gives an oxime, reaction with hydrazines gives hydrazones, and reactions with semicarbazide give semicarbazones. The reaction of 4-hydroxybenzaldehyde with hydroxylamine hydrochloride is a convenient method for the preparation of 4-cyanophenol (52,53).

Many excellent chelating agents are prepared by the reaction of salicyl-aldehyde with 1,2- or 1,3-diaminoalkanes and have wide use in chemistry, for example as oxidation catalysts (see CHELATING AGENTS) (54–56).

The reaction product of salicylaldehyde and a secondary aniline is the benzylic alcohol, with total para-selectivity (57). The yield is 93%.

Primary amides condense with hydroxybenzaldehydes in a manner similar to amines. This reaction is often conducted in the presence of sodium acetate or an organic base such as pyridine. For example, the reaction of salicylaldehyde and propionamide produces salicylidene propionamide (58).

Hydroxybenzaldehydes readily react with compounds containing methyl or methylene groups bonded to one or two carboxyl, carbonyl, nitro, or similar strong electron-withdrawing groups. The products are usually β-substituted styrenes. 4-Hydroxybenzaldehyde, for example, reacts with 2-methylquinazolines (where R = H, Cl) to give compounds which have anti-inflammatory activity (59).

Aldol Reactions. In the same way, hydroxybenzaldehydes react readily with aldehydes and ketones to form α,β-unsaturated carbonyl compounds in the Claisen-Schmidt or crossed-aldol condensation (60).

Perkin Reaction. A product of significant commercial importance, coumarin [*91-64-5*], is made by the reaction of salicylaldehyde with acetic anhydride and sodium acetate, a Perkin reaction (61).

Other Reactions. The reaction of 4-hydroxybenzaldehyde with sodium cyanide and ammonium chloride, Strecker synthesis, yields *p*-hydroxyphenylglycine [*938-97-6*], a key intermediate in the manufacture of semisynthetic penicillins and cephalosporins (see ANTIBIOTICS, β-LACTAMS).

Manufacture

The main processes for the manufacture of hydroxybenzaldehydes are based on phenol. The most widely used process is the saligenin process. Saligenin (2-hydroxybenzyl alcohol [*90-01-7*]) and 4-hydroxybenzyl alcohol [*623-05-2*] are produced from base-catalyzed reaction of formaldehyde with phenol (35). Air oxidation of saligenin over a suitable catalyst such as platinium or palladium produces salicylaldehyde (62).

Reaction of phenyl metaborate with formaldehyde, followed by catalytic oxidation, has been reported to give salicylaldehyde selectively and directly from phenol without isolation of any intermediate products (63).

Although 4-hydroxybenzaldehyde can be made by the saligenin route, it has been made historically by the Reimer-Tiemann process, which also produces salicylaldehyde (64). Treatment of phenol with chloroform and aqueous sodium hydroxide results in the formation of benzal chlorides, which are rapidly hydrolyzed by the alkaline medium into aldehydes. Acidification of the phenoxides results in the formation of the final products, salicylaldehyde and 4-hydroxybenzaldehyde. The ratio of ortho and para isomers is flexible and can be controlled within certain limits. The overall reaction scheme is shown in Figure 1. Product separation is accomplished by distillation, but this process leads to environmental problems because of the quantities of sodium chloride produced.

Other routes for hydroxybenzaldehydes are the electrolytic or catalytic reduction of hydroxybenzoic acids (65,66) and the electrolytic or catalytic oxidation of cresols (67,68). (see SALICYLIC ACID AND RELATED COMPOUNDS). Salicylaldehyde is available in drums and bulk quantities. The normal specification is a freezing point minimum of 1.4°C. 4-Hydroxybenzaldehyde is available in fiber drums, and has a normal specification requirement of a 114°C initial melting point. More refined analytical methods are used where the application requires more stringent specifications.

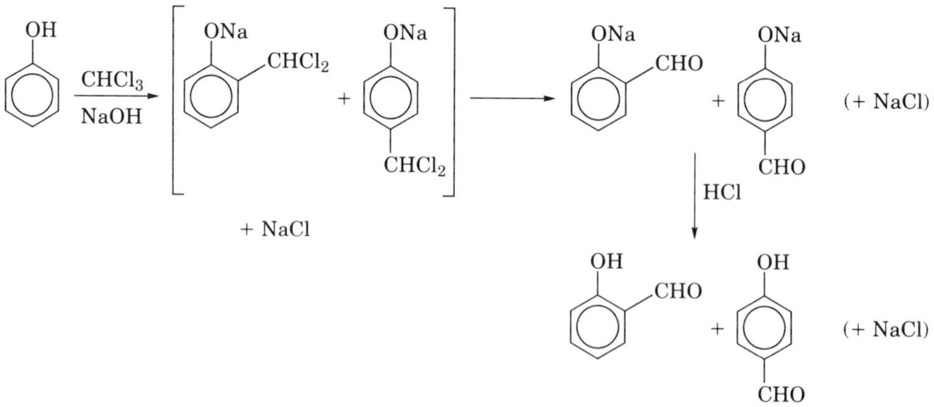

Fig. 1.   Production of hydroxybenzaldehydes by the Reimer-Tiemann process.

Economic Aspects

Rhône-Poulenc (RP), producing both in Europe and the United States, is the only producer of salicylaldehyde worldwide, for merchant sales. A large portion of it is used captively in the manufacture of coumarin. The remainder is available for the merchant market.

Worldwide capacity figures for salicylaldehyde are not published; however, the estimated capacity is approximately 4000–6000 t/y. The supply–demand picture for salicylaldehyde has been well balanced in the 1990s, as RP has expanded capacity to meet the growing market need. The price of salicylaldehyde was fairly stable at approximately $2.20/kg during the late 1970s and early 1980s; however, prices rose rapidly in the mid-1970s. The 1980 price of salicylaldehyde was in the $6.00/kg range; the 1994 price is now stable in the same range.

Chuo Kasein (Japan), various Chinese companies, and Hoechst (France) are producers of p-hydroxybenzaldehyde.

Health and Safety Factors

Salicylaldehyde has a moderate acute oral toxicity; the LD_{50} for rats is 0.3–2.0 g/kg of body weight. p-Hydroxybenzaldehyde has a low acute oral toxicity; the LD_{50} for rats is 4.0 g/kg of body weight. Neither material is likely to present a problem from ingestion incidental to its handling and industrial use. It should be recognized, however, that serious effects may result if substantial amounts are swallowed.

Tests performed on rabbits indicate that neither material is absorbed through the skin in toxic amounts. Skin contact with p-hydroxybenzaldehyde is essentially nonirritating; however, contact with salicylaldehyde is capable of causing a severe burn, especially in case of prolonged or repeated contact. Hence, such contact should be avoided. p-Hydroxybenzaldehyde is slightly irritating to the eyes and can cause slight transient irritation and slight transient corneal injury. Salicylaldehyde is appreciably irritating to the eyes and may cause pain, irritation, and some corneal injury.

Uses

The hydroxybenzaldehydes are used primarily as chemical intermediates for a variety of products. The largest single use of salicylaldehyde is in the manufacture of coumarin. Coumarin is an important commercial chemical used in soaps, flavors and fragrances, and electroplating (see COUMARIN). Other significant uses of both salicylaldehyde and p-hydroxybenzaldehyde are as follows.

Agricultural Chemicals. Salicylaldehyde is a valuable intermediate in the manufacture of herbicides (qv) and pesticides. The phenylhydrazones of salicylaldehyde are used to inhibit cereal rusts (69), and as herbicides for a variety of weed species (70), eg, *Amaranthus retroflexus*. In addition, both the hydrazone and phenylhydrazone of salicylaldehyde have been found to be effective antimicrobials (71) (see DISINFECTANTS; INDUSTRIAL ANTIMICROBIAL AGENTS). The N-methyl- and N,N-dimethylcarbamates of salicylaldehyde acetals and mercaptals are effective insecticides; the N-methylcarbamate of the cyclic acetal of salicylaldehyde (from salicylaldehyde and ethylene glycol) is an important commercial insecticide (72). This product, dioxacarb [6988-21-2], [2-(1,3-dioxolan-2-yl)phenyl-N-methylcarbamate] (1), is widely used in European and African countries for the protection of potatoes and cocoa (73) (see INSECT CONTROL TECHNOLOGY). Another important use of salicylaldehyde is in the synthesis of micronutrients. In particular, ferric ion chelates have been used on alkaline or calcerous soils which can occur in citrus or olive groves (see FERTILIZERS; MINERAL NUTRIENTS).

p-Hydroxybenzaldehyde has extensive use as an intermediate in the synthesis of a variety of agricultural chemicals. Halogenation of *p*-hydroxybenzaldehyde, followed by conversion to the oxime, and subsequent dehydration results in the formation of 3,5-dihalo-4-hydroxybenzonitrile (**2**). Both the dibromo- and diiodo-compounds are commercially important contact herbicides, bromoxynil [*1689-84-5*] (**2**) where X = Br, and ioxynil [*1689-83-4*] (**2**), where X = I respectively (74). Several hydrazone derivatives have also been shown to be active herbicides (70).

(1)   (2)

Electroplating. Salicylaldehyde is a starting material in the synthesis of coumarin (qv) which is widely used by the electroplating (qv) industry as a brightener and leveling agent in nickel plating. The imine resulting from the reaction of salicylaldehyde with an alkanolamine (containing a primary amine group) is an effective brightening agent in the electroplating of zinc on iron and steel (75). In another zinc electroplating (qv) process in a noncyanide alkaline bath containing a polyaminesulfone, salicylaldehyde itself was found to aid in producing a bright zinc electroplate on steel (76).

Both *p*-hydroxybenzaldehyde and its methyl ether, *p*-methoxybenzaldehyde [*123-11-5*] (*p*-anisaldehyde) have found extensive use in electroplating. The most widespread application has been in alkaline bright zinc plating, both in non-cyanide (77) and in cyanide-containing (78) baths. The aldehydes act as both brightening and leveling agents.

Flavors and Fragrances. Salicylaldehyde is a starting material in the synthesis of coumarin, which finds extensive use in the soap (qv) and perfume (qv) industries and salicylaldehyde can be used itself as a preservative in essential oils and perfumes (see OILS, ESSENTIAL). The antibacterial activity of salicylaldehyde is strong enough to allow its use at very low concentrations (79).

p-Hydroxybenzaldehyde has an agreeable aromatic odor, but is not itself a fragrance. It is, however, a useful intermediate in the synthesis of fragrances. The methyl ether of *p*-hydroxybenzaldehyde, ie, *p*-anisaldehyde, is a commercially important fragrance. *p*-Anisaldehyde can be made in a simple one-step synthesis from *p*-hydroxybenzaldehyde and methyl chloride. Another important fragrance, 4-(*p*-hydroxyphenyl)butanone, commonly referred to as raspberry ketone, can be prepared from the reaction of *p*-hydroxybenzaldehyde and acetone, followed by reduction (see FLAVORS AND SPICES).

Petroleum Products. Condensation products of salicylaldehyde and amines are used in various forms for the removal or neutralization of the metallic ions that cause oxidative degradation in petroleum products. The product formed from propylenediamine and salicylaldehyde, ie N,N'-disalicylidene-1,2-propanediamine, has proven itself to be a commercially important chelating agent, primarily because of its combination of outstanding chelation properties plus its solubility in oil (80). Other adducts are used as detergents in lubricating oils and gasoline (81), as sludge inhibitors in fuel oils and gasoline (82) and as antioxidants to improve the high temperature stability of polyester lubricants (83), gasoline (84), and petroleum oils (85).

Pharmaceuticals. p-Hydroxybenzaldehyde is often a convenient intermediate in the manufacture of pharmaceuticals (qv). For example, 2-(p-hydroxyphenyl)glycine can be prepared in a two-step synthesis starting with p-hydroxybenzaldehyde (86). This amino acid is an important commercial intermediate in the preparation of the semisynthetic penicillin, amoxicillin (see ANTIBIOTICS, β-LACTAMS). Many cephalosporin-type antibiotics can be made by this route as well (87). The antiemetic, trimethobenzamide [*138-56-7*], is conveniently prepared from p-hydroxybenzaldehyde (88) (see GASTROINTESTINAL AGENTS).

Polymer Applications. The reaction of salicylaldehyde with poly(vinyl alcohol) to form an acetal has been used to provide dye receptor sites on poly(vinyl alcohol) fibers (89) and to improve the light stability of blend fibers from vinyl chloride resin and poly(vinyl alcohol) (90) (see FIBERS, POLY(VINYL ALCOHOL).

The metal coordination complexes of both salicylaldehyde phenylhydrazone (91) and salicylaldoxime provide antioxidant (92) protection and uv stability to polyolefins (see ANTIOXIDANTS; UV ABSORBERS). In addition, the imines resulting from the reaction of salicylaldehyde and aromatic amines, eg, p-aminophenol or α-naphthylamine, can be used at very low levels as heat stabilizers (qv) in polyolefins (93).

The resiliency and dyeability of poly(vinyl alcohol) fibers is improved by a process incorporating p-hydroxybenzaldehyde to provide a site for the formation of a stable Mannich base. Hydroxyl groups on the fiber are converted to acetal groups by p-hydroxybenzaldehyde. Subsequent reaction with formaldehyde and ammonia or an alkylamine is rapid and forms a stable Mannich base that is attached to the polymer backbone (94).

Miscellaneous. The reaction products of salicylaldehyde with certain compounds containing active methylene groups, eg, acetylacetone, are excellent uv absorbers. Films containing these compounds can be used as uv filters to protect light-sensitive foods, wood products, paper, dyes, fibers, and plastics (95).

The reaction product of salicylaldehyde and hydroxylamine, salicylaldoxime, has been found to be effective in photography in the prevention of fogging of silver halide emulsions on copper supports (96). It also forms the basis for an electrolytic facsimile-recording paper (97) and in combination with a cationic polymer, is used in another electrolytic dry-recording process (98) (see ELECTROPHOTOGRAPHY).

The copper-chelating ability of salicylaldoxime has been used to remove copper from brine in a seawater desalination plant effluent. A carbon–sorbate bed produced by sorption of the oxime on carbon proved to be extremely effective in the continuous process (99). In another application, the chelating ability of salicylaldoxime with iron and copper was used to stabilize bleaching powders containing inorganic peroxide salts (100).

BIBLIOGRAPHY

"Phenolic Aldehydes" in *ECT* 1st ed., Vol. 10, pp. 320–325, by W. R. Brookes, General Electric Co.; in *ECT* 2nd ed., Vol. 15, pp. 160–165, by D. B. G. Jacquiss, General Electric Co.; "Hydroxybenzaldehydes" in *ECT* 3rd ed., Vol. 13, pp. 70–79, by R. M. Mullins, Dow Chemical Co.

1. D. R. Lide, ed., *Handbook of Chemistry and Physics, 1990–1991*, 71st ed., CRC Press, Boca Raton, Fla., 1992.
2. M. Suzuki, *Chem. Eng. Sci.* **33**, 271–273 (1978).
3. F. C. Hoyng, *Org. Prep. Proced.* **13**(2), 175–178 (1981).
4. R. T. Borchardt and co-workers, *Synthesis* **21**, 710–712 (1988).
5. J. R. Merchant and A. J. Mountvala, *J. Org. Chem.* **23**, 1774–1776 (1958).
6. E. Späth and K. Kromp, *Ber.* **74B**, 867–869 (1941).
7. A. KrentzBerger, *Angew. Chem. Int. Ed. Engl.* **6**(11), 940 (1967).
8. R. E. Corbett and co-workers, *J. Chem. Soc.*, 1–6 (1950).
9. M. Hirama and S. Ito, *Chem. Lett.* **6**, 627–630 (1977).
10. H. Gross, A. Rieche, and G. Matthey, *Ber.* **96**, 308–313 (1963).
11. K. Freudenberg and H. H. Hübner, *Chem. Ber.* **85**, 1181–1191 (1952).
12. F. Wesseley and F. Lechner, *Monatsh. Chem.* **60**, 159 (1932).
13. Flavor and Extract Manufacturers' Association of the United States (FEMAUS), *Gov. Rep. Annouce. (U.S.)* **85**(6), 48 (1985).
14. D. L. J. Opdyke, *Food Cosmet. Toxicol.* **17**, 903–905 (1979).
15. U. Schumarzmaier, *Chem. Ber.* **109**, 3379–3380 (1976).
16. P. J. Scheuer and T. Higa, *J. Chem. Soc. Perkin Trans.* **1**, 1350 (1974).
17. B. W. Staddon and J. Weatherston, *Tetrahedron Lett.* 4567 (1967).
18. T. S. Corswell and C. E. Pfeifer, *J. Am. Chem. Soc.* **50**, 1766 (1928).
19. L. Kahovec and K. W. F. Kohlrausch, *Z. Phys. Chem.* **38**, 119–134 (1938).
20. L. H. Thomas, *J. Chem. Soc.*, 4906–4908 (1960).
21. H. G. Wallmann, *Pharmazie* **29**, 708–709 (1974).
22. N. V. Sidgwick and E. N. Allott, *J. Chem. Soc.* **123**, 2819 (1923).
23. H. LeFranc, Rhône-Poulenc, internal data, 1980.
24. F. Fukushima, *Bull. Chem. Jpn.* **38**, 1694 (1965).
25. R. A. Morton and A. L. Stubb, *J. Chem. Soc.*, 1347–1359 (1940).
26. Reuben, *J. Am. Chem. Soc.* **98**, 3726–3727 (1976).
27. F. W. Lafferty and F. M. Bockhoff, *Anal. Chem.* **50**, 69–72 (1978).
28. G. H. Parsons, C. H. Rochester, and C. E. C. Wood, *J. Chem. Soc., Sect. B*, 533 (1971).
29. A. Buramoy and I. Markowitsch-Buramoy, *J. Chem. Soc.*, 36–39 (1936).
30. H. H. Freedman, *J. Am. Chem. Soc.* **83**, 2901 (1961).
31. R. J. Highet and P. F. Highet, *J. Org. Chem.* **30**, 902–905 (1965).
32. H. Scheuer, *J. Chem. Soc. Perkin Trans.* **I**, 1350 (1974).
33. R. H. Bailes and M. Calvin, *J. Am. Chem. Soc.* **69**, 1886 (1947).
34. D. P. Meller and L. Maley, *Nature* **159**, 370 (1947).
35. K. C. Eapen and L. M. Yeddanapalli, *Makromol. Chem.* **119**, 4 (1968).
36. Jpn. Pat. 142,098 (June 18, 1986), M. Osu and co-workers (to Sumitomo Chemical Co.).
37. N. Kitajima and co-workers, *Bull. Chem. Soc. Jpn.* **61**, 967 (1988).
38. Ger. Offen, 2,807,762 (Feb. 23, 1978), F. Merger and co-workers (to BASF AG).
39. Ger. Offen, 2,717,515 (Apr. 27, 1976), M. Pauly (to Laboratoires Serobiologiques).
40. U.S. Pat. 4,332,950 (Mar. 2, 1981), C. A. Kelly and F. A. Meneghini (to Polaroid Corp.).
41. U.S. Pat. 4,098,783 (June 17, 1971), R. F. W. Cieciuch and co-workers, (to Polaroid Corp.).

42. Eur. Pat. 44,260 (July 11, 1980), K. Formanek and co-workers (to Rhône-Poulenc Industries).
43. H. D. Dakin, *Organic Syntheses*, Vol. III, John Wiley & Sons, Inc., New York, 1923, p. 27.
44. U.S. Pat. 645,438 (Jan. 1, 1991), A. B. De Milo and R. N. Huettel (to National Institutes of Health).
45. Jpn. Pat. 03 15,7345 (Nov. 15, 1989), H. Matsuoka, Y. Asabe, and H. Kawaguchi (to Nippon Terpene Chemical Co.).
46. W. Baker and N. C. Brown, *J. Chem. Soc.* **151**, 2303 (1948).
47. I. A. Pearl, *J. Org. Chem.* **12**, 85 (1947).
48. G. I. Kudryaustiv and E. I. Shilov, *Dokl. Akad. Nauk. SSSR* **64**, 73 (1949).
49. D. F. Tomkins and J. H. Wagenknecht, *J. Electrochem. Soc.* **125**, 372 (1978).
50. U.S. Pat. 4,157,286 (Apr. 4, 1978), C. J. H. King (to Monsanto Co.).
51. U.S. Pat. 4,133,729 (Dec. 19, 1977), C. J. H. King (to Monsanto Co.).
52. T. Van Es, *J. Chem. Soc.*, 1564 (1965).
53. Pol. Pat. 115,568 (Dec. 21, 1979) W. Zamlynski and M. Jawdosiuk (to Politechnika Warszawska).
54. Fr. Pat. 233,1549 (Feb. 28, 1977), R. R. Gaudette and J. L. Ohlson (to W. R. Grace and Co.).
55. Jpn. Pat. 00 78,932 (Oct. 3, 1983), K. Tanaka and K. Shioda (to Sumitomo Chemical KK).
56. Ger. Pat. 3302,498 (Jan. 26, 1983) M. Strozel and co-workers (to BASF AG).
57. K. A. Petrov, L. V. Treshchalina, and V. M. Chizhov, *Zh. Obskch. Khim.* **47**, 2741 (1977).
58. K. C. Padya and T. S. Sohdi, *Proc. Ind. Acad. Sci.* **7A**, 361 (1938).
59. G. P. Zhikhareva and co-workers, *Khim. Farm. Zh.* **11**, 58 (1977).
60. J. E. Baldwin and co-workers, *J. Org. Chem.* **42**, 3846 (1977).
61. W. H. Perkin, *J. Chem. Soc.* **21**, 53 (1868); **31**, 388 (1877).
62. Ger. Offen. 2,612,844 (Oct. 7, 1976) J. Le Ludec (to Rhône-Poulenc SA).
63. U.S. Pat. 3,321,526 (May 23, 1967) P. A. R. Marchand, and J. B. Grenet (to Rhône-Poulenc SA).
64. H. Wynberg, *Chem. Rev.* **60**, 169 (1960).
65. K. S. Udupa, G. S. Subramanian, and H. V. K. Udupa, *Ind. Chemist* **39**, 238 (1963).
66. Jpn. Pat. 01 26,242-A (Dec. 14, 1983), T. Maki and T. Yokoyama (to Mitsubishi Chem. Ind. Ltd.).
67. U.S. Pat. 4,471,140 (Aug. 19, 1982), A. T. Au (to Dow Chemical Co.).
68. Jpn. Pat. 61 24,535 (July 12, 1984), H. Kaneda and co-workers (to Kansai Tek. KK).
69. U.S. Pat. 2,818,367 (Dec. 31, 1957), E. G. Jaworsko and V. R. Gaertner (to Monsanto Chemical Co.).
70. M. Mazza, L. Montanari, and F. Pavanetto, *Farmaco, Ed. Sci.* **31**(5), 334 (1976).
71. M. N. Rotmistrov and co-workers, *Mikrobiol. Zh. (Kiev)* **36**(2), 244 (1974).
72. E. F. Nikles, *J. Agr. Food Chem.* **17**(5), 939 (1969).
73. R. P. Ouellette and J. A. King, *Chemical Week Pesticides Register*, MacGraw-Hill Book Co., Inc., New York, 1976.
74. U.S. Pat. 3,397,054 (Aug. 13, 1968), R. D. Hart and H. E. Harris (to Schering Corp.).
75. Ger. Offen. 1,961,812 (Oct. 1, 1970), R. P. Cope and J. A. Von Pless (to Stauffer Chemical Co.).
76. Ger. Offen. 2,608,644 (Sept. 9, 1976), S. Fujita, K. Murayama, and T. Kaneda (to Japan Metal Finishing Co.).
77. U.S. Pat. 3,871,974 (Mar. 18, 1975), J. R. Duchene and P. J. DeChristopher (to Richardson Chemical Co.).
78. Ger. Offen. 1,919,665 (Oct. 15, 1970), S. Acimovic (to Riedel and Co.).

79. M. G. deNavarre, ed., *The Chemistry and Manufacture of Cosmetics*, Vol. 3, 2nd ed., 1975, pp. 85–100.
80. P. Polss, *Hydrocarbon Process.*, 61. (Feb. 1973).
81. U.S. Pat. 3,919,094 (Nov. 11, 1978), S. Schiff (to Phillips Petroleum Co.).
82. Brit. Pat. 1,077,760 (Aug. 2, 1967), H. J. Andress (to Mobil Oil Corp.).
83. U.S. Pat. 3,634,248 (Jan. 11, 1972), H. J. Andress (to Mobil Oil Corp.).
84. U.S. Pat. 3,399,041 (Aug. 27, 1968), L. J. McCabe (to Mobil Oil Corp.).
85. Ger. (DDR) Pat. 60,835 (Mar. 20, 1968), D. Hoerding.
86. Neth. Appl. 6,607,754 (Dec. 5, 1966), H. Fink and G. Schröder (to Rhohm & Haas GmbH).
87. U.S. Pat. 3,946,003 (Mar. 23, 1976), R. D. Cooper (to Eli Lilly and Co.).
88. U.S. Pat. 2,879,293 (Mar. 24, 1959), M. W. Goldberg (to Hoffmann-La Roche, Inc.).
89. Jpn. Pat. 5,561 (Aug. 9, 1955), T. Kenichi and co-workers (to Kurashiki Rayon Co.).
90. Jpn. Kokai 73 87,198 (Nov. 16, 1973), M. Furuno and S. Hoshino (to Asahi Dow Ltd.).
91. U.S. Pat. 3,208,968 (Sept. 28, 1965), H. A. Cyba and A. K. Sparks (to Universal Oil Products and Sun Oil Co.).
92. Neth. Appl. 6,614,765 (Apr. 24, 1967), P. J. Briggs and R. J. Hurlock (to Imperial Chemical Industries Ltd.).
93. USSR Pat. 253,349 (Sept. 30, 1969) E. N. Matveeva and co-workers (to State Scientific Research Institute of Polymerized Plastics).
94. K. Matsubayashi and K. Tanabe, *Kogyo Kagaku Zasshi* **62**, 1753 (1959).
95. Ger. Pat. 1,087,902 (Aug. 25, 1960), D. Lauerer and M. Pestemer (to Farbenfabriken Bayer Akt. Ges.).
96. Fr. Demande 2,003,606 (Nov. 7, 1969), T. I. Abbott (to Eastman Kodak Co.).
97. U.S. Pat. 2,864,748 (Dec. 16, 1958), A. H. Mones (to Faximile, Inc.).
98. Jpn. Kokai 73 45,342 (June 28, 1973) Y. Sekine and W. Shimotsuma (to Matsushita Electric Industrial Co. Ltd.).
99. R. H. Moore, *U.S. Office Saline Water, Research and Development Progress Report, 651*, 1971, 87 pp.
100. Ger. Offen. 2,420,009 (Nov. 7, 1974), T. Fujino, M. Yamanaka, and K. Deguchi (to Kao Soap Co., Ltd.).

CHRISTIAN MALIVERNEY
MICHEL MULHAUSER
Rhône-Poulenc Recherches

HYDROXYCARBOXYLIC ACIDS

Lactic Acid

Lactic acid [50-21-5] (2-hydroxypropanoic acid), $CH_3CHOHCOOH$, is the most widely occurring hydroxycarboxylic acid and thus is the principal topic of this article. It was first discovered in 1780 by the Swedish chemist Scheele. Lactic acid is a naturally occurring organic acid that can be produced by fermentation or chemical synthesis. It is present in many foods both naturally or as a product of *in situ* microbial fermentation, as in sauerkraut, yogurt, buttermilk, sourdough breads, and many other fermented foods. Lactic acid is also a principal metabolic intermediate in most living organisms, from anaerobic prokaryotes to humans.

Although lactic acid is ubiquitous in nature and has been produced as a fermentation by-product in many industries (for example, corn steep liquor, a principal by-product of the multimillion-ton per year corn wet-milling industry, contains approximately 25 wt % lactic acid), it has not been a large-volume chemical. By 1990 its worldwide production volume had grown to approximately 40,000 t/yr with two significant producers, CCA Biochem bv of the Netherlands, with subsidiaries in Brazil and Spain, and Sterling Chemicals, Inc. in Texas City, Tex., as the primary manufacturers (1). CCA uses carbohydrate feedstocks and fermentation technology, and Sterling uses a chemical technology. Thus lactic acid has been considered a relatively mature fine chemical in that only its use in new applications, eg, as a monomer in plastics or as an intermediate in the synthesis of high volume oxygenated chemicals, would cause a significant increase in its anticipated demand (1). Announcements of new lactic acid production plants by chemical and agriprocessing companies may usher in new technologies for efficient, low cost manufacture of lactic acid and its derivatives for new applications (2–5).

Physical Properties. Pure, anhydrous lactic acid is a white, crystalline solid with a low melting point. However, it is difficult to prepare the pure anhydrous form of lactic acid; generally, it is available as a dilute or concentrated aqueous solution. The properties of lactic acid and its derivatives have been reviewed (6). A few important physical and thermodynamic properties from this reference are summarized in Table 1.

Lactic acid is also the simplest hydroxy acid that is optically active. L(+)-Lactic acid [*79-33-4*] (**1**) occurs naturally in blood and in many fermentation products (7). The chemically produced lactic acid is a racemic mixture and some fermentations also produce the racemic mixture or an enantiomeric excess of D(−)-lactic acid [*10326-41-7*] (**2**) (8).

$$
\begin{array}{cc}
\begin{array}{c}
\mathrm{COOH} \\
| \\
\mathrm{HO-C-H} \\
| \\
\mathrm{CH_3}
\end{array}
&
\begin{array}{c}
\mathrm{COOH} \\
| \\
\mathrm{H-C-OH} \\
| \\
\mathrm{CH_3}
\end{array}
\\
(\mathbf{1}) & (\mathbf{2})
\end{array}
$$

Many of the physical properties are not affected by the optical composition, with the important exception of the melting point of the crystalline acid, which is estimated to be 52.7–52.8°C for either optically pure isomer, whereas the reported melting point of the racemic mixture ranges from 17 to 33°C (6). The boiling point of anhydrous lactic acid has been reported by several authors; it was primarily obtained during fractionation of lactic acid from its self-esterification product, the dimer lactoyllactic acid [*26811-96-1*]. The difference between the boiling points

Table 1. Physical and Thermodynamic Properties of Lactic Acid

Property	Value
density, g/mL at 20°C	1.2243
viscosity,[a] mPa·s(=cP)	36.9
dissociation constant, pK_a at 25°C	3.862
heat capacity, J/(g·K)[b]	
crystalline	1.41
liquid, 25°C	2.34
heat of dissociation at 25°C, J/mol[b]	−263
free energy of dissociation, kJ/mol[b]	20.9
heat of solution, L(+) at 25°C, kJ/mol[b]	7.79
heat of fusion, kJ/mol[b]	
racemic	11.33
L(+)	16.86
heat of combustion, MJ/mol[b]	
racemic	−1.355
L(+)	−1.343
heat of formation, MJ/mol[b]	
crystalline L(+)	−0.693
dilute solution	−0.686
free energy of formation, MJ/mol[b]	
crystalline L(+)	−0.522
liquid racemic	−0.529

[a]88.6 wt % solution at 25°C.
[b]To convert J to cal, divide by 4.184.

of racemic and optically active isomers of lactic acid is probably very small (6). The uv spectra of lactic acid and dilactide [95-96-5], which is the cyclic anhydride from two lactic acid molecules, as expected show no chromophores at wavelengths above 250 nm, and lactic acid and dilactide have extinction coefficients of 28 and 111 at 215 nm and 225 nm, respectively (9,10). The infrared spectra of lactic acid and its derivatives have been extensively studied and a summary is available (6).

Chemical Properties. Its two functional groups permit a wide variety of chemical reactions for lactic acid. The primary classes of these reactions are oxidation, reduction, condensation, and substitution at the alcohol group.

Oxidation. Lactic acid oxidation by strong oxidizing agents such as permanganate, chromate, hydrogen peroxide, or halogens under photochemical activation leads to the formation of a multitude of decomposition products, including varying amounts of pyruvate, acetaldehyde, acetate, carbon dioxide, etc (11–14). The yield and specificity of the products depend on the type of the oxidant and the reaction mechanism. The production of these oxidized products using lactic acid as a feedstock is not of potential commercial or synthetic interest.

Biochemical oxidation of lactate to pyruvate by lactate dehydrogenase is a well-known enzymatic reaction in metabolic pathways.

$$\underset{H_3C-CH-COOH}{\overset{OH}{|}} + NAD \xrightarrow{\text{lactate}\atop\text{dehydrogenase}} \underset{H_3C-C-COOH}{\overset{O}{\|}} + NADH_2$$

Reduction. Lactic acid has been reduced with hydrogen iodide to propionic acid (15). The carboxyl group can be reduced to an alcohol group by catalytic hydrogenolysis or by reduction with hydrogenating chemicals such as lithium aluminium hydride or sodium borohydride. Usually the esters of lactic acid have been converted to propylene glycol by reduction. When hydrogenating chemicals were used with optically active ester, the optical configuration was retained in the propylene glycol (16). Reportedly, in catalytic hydrogenolysis using copper chromite or Raney nickel, racemization occurred (17) and the propylene glycol was optically inactive.

Advances in catalysts and process engineering development have enabled the conversion of organic acids to corresponding alcohols with high selectivity and rates under moderate temperatures and pressure, thus improving the process economics (18,19). Using such processes for the conversion of lactate to propylene glycol [57-55-6] (3) could be an important industrial process in the future in the manufacture of large-volume industrial chemicals from fermentation of renewable carbohydrates (1,20).

$$
\underset{\underset{\displaystyle H_3C-CH-COOR}{|}}{OH} + 2\,H_2 \xrightarrow{\text{catalyst}} \underset{\underset{\displaystyle H_3C-CH-CH_2OH}{|}}{OH} + ROH
$$

$$(3)$$

Condensation. A variety of condensation reactions involving the hydroxyl or the carboxyl or both groups occur with lactic acid. The important reactions where products can be obtained in high yields are esterification (both intramolecular and with another alcohol or acid), dehydration, and aminolysis.

Because lactic acid has both hydroxyl and carboxyl functional groups, it undergoes intramolecular or self-esterification and forms linear polyesters, lactoyllactic acid (4) and higher poly(lactic acid)s, or the cyclic dimer 3,6-dimethyl-*p*-dioxane-2,5-dione [95-96-5] (dilactide) (5). Whereas the linear polyesters, lactoyllactic acid and poly(lactic acid)s, are produced under typical condensation conditions such as by removal of water in the presence of acidic catalysts, the formation of dilactide with high yield and selectivity requires the use of special catalysts which are primarily weakly basic. The use of tin and zinc oxides and organostannates and -titanates has been reported (6,21,22).

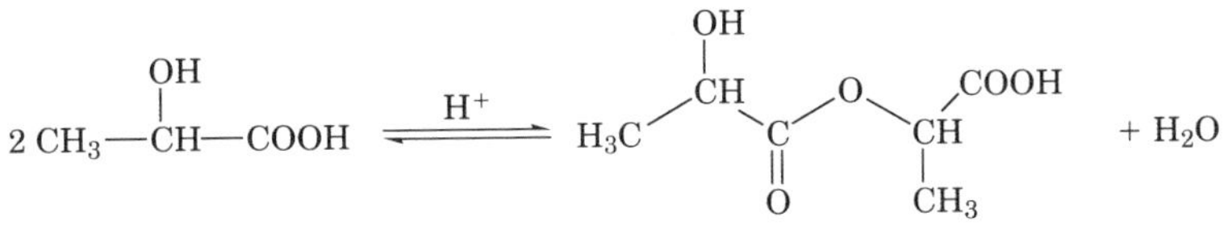

$$
2\,CH_3-\underset{\underset{\displaystyle}{|}}{\overset{\displaystyle OH}{CH}}-COOH \underset{\longleftarrow}{\overset{H^+}{\rightleftharpoons}} \quad (4) \quad + H_2O
$$

$$(4)$$

$$
2\,CH_3-\underset{\underset{\displaystyle}{|}}{\overset{\displaystyle OH}{CH}}-COOH \overset{\text{catalysts}}{\rightleftharpoons} \quad (5) \quad + H_2O
$$

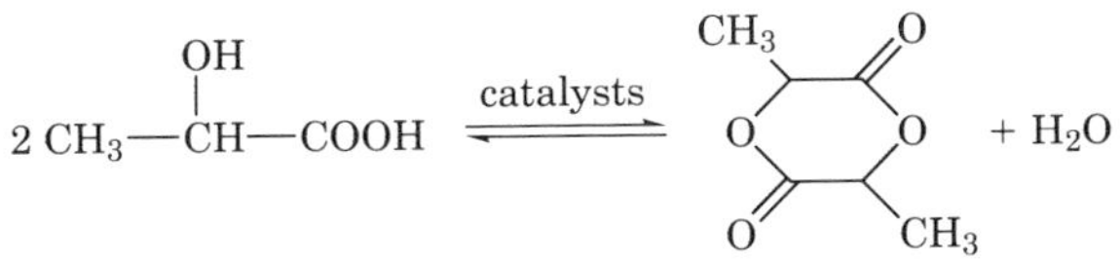

$$(5)$$

Dilactide (**5**) exists as three stereoisomers, depending on the configurations of the lactic acid monomer used. The enantiomeric forms wherein the methyl groups are cis are formed from two identical lactic acid molecules, D- or L-, whereas the dilactide formed from a racemic mixture of lactic acid is the optically inactive meso form, with methyl groups trans. The physical properties of the enantiomeric dilactide differ from those of the meso form (**6**), as do the properties of the polymers and copolymers produced from the respective dilactide (23,24).

Polylactide is the generally accepted term for highly polymeric poly(lactic acid)s. Such polymers are usually produced by polymerization of dilactide; the polymerization of lactic acid as such does not produce high molecular weight polymers. The polymers produced from the enantiomeric lactides are highly crystalline, whereas those from the meso lactide are generally amorphous. Usually dilactide from L-lactic acid is preferred as a polymerization feedstock because of the availability of L-lactic acid by fermentation and for the desirable properties of the polymers for various applications (1,25).

Esterification of the carboxyl group is another important reaction which is used either to recover and purify lactic acid from impure solutions or to produce the ester as the desired end product. Several methods can be used to prepare lactic acid esters; the important ones are direct reaction with an alcohol, transesterification of one ester into another with alcohol, alcoholysis of ammonium lactate solution, and reaction of a metal lactate with an alkyl halide. Direct esterification is usually catalyzed by mineral acids such as sulfuric, hydrochloric, or phosphoric acids or *p*-toluenesulfonic acid: lactic acid reacts with excess alcohol under reflux and the water formed in the reaction is removed. With lower alcohols this method does not give satisfactory yields because the alcohol volatilizes instead of reacting. A more effective method has been to pass the vapors of alcohol through lactic acid previously heated to a temperature above the boiling point of the alcohol. This method is more efficient than the reflux still method and has been used for purification of lactic acid from impure streams such as fermentation broths (26,27). Transesterification is often used to prepare higher alcohol esters from methyl or ethyl lactate. Generally good yields ranging from 65 to 90% are obtained (6,28).

Ammonium lactate [*34302-65-3*] in concentrated aqueous solutions has been converted to ammonia and the ester by alcoholysis at temperatures ranging from 100–200°C using a variety of alcohols and water entrainers, such as toluene. Ester yields ranging from 50–80% were obtained. This method has also been suggested as a recovery and purification method from impure solutions of lactate (29). However, a considerable amount of the lactate is not converted to the recoverable ester and is lost as lactamide (**6**).

$$CH_3-\overset{\overset{\displaystyle OH}{|}}{CH}-COO^-\ NH_4^+ + ROH \quad \begin{array}{c} \nearrow \\ \\ \searrow \end{array} \quad \begin{array}{l} CH_3-\overset{\overset{\displaystyle OH}{|}}{CH}-COOR + NH_3 + H_2O \\[2em] CH_3-\overset{\overset{\displaystyle OH}{|}}{CH}-\overset{\overset{\displaystyle O}{||}}{C}NH_2 + H_2O + ROH \end{array}$$

$$(\mathbf{6})$$

Reaction of a metal lactate (such as silver lactate) with an alkyl halide is a classic method of preparation of the ester, but it is too expensive to be of commercial relevance. Lactamide [2043-43-8] is another high yielding condensation product from lactic acid. It can be produced by aminolysis of dilactide or lactate ester such as methyl or ethyl lactate.

Substitution at the Alcohol Group. Acylation of the OH group by acylating agents such as acid chlorides or anhydrides is one of the important high yielding substitution reactions at the OH group of lactic acid and its functional derivatives. Aliphatic, aromatic, and other substituted derivatives can be produced.

$$\underset{\substack{|\\CH_3-CH-COOR}}{OH} + R'-\overset{O}{\overset{||}{C}}-O-\overset{O}{\overset{||}{C}}-R' \longrightarrow \underset{\substack{|\\CH_3-CH-COOR}}{OOCR'} + R'COOH$$

Several emulsifiers used in bakery and other food processing applications, such as stearoyl-2-lactylic acid and mono- and diglyceride derivatives are important commercial products produced by substitution reactions at the OH group. Upon pyrolysis of acylated esters, the corresponding acrylic acid ester (**7**) is produced with the simultaneous elimination of the acylating acid molecule. This reaction, when further developed and integrated with economical lactic acid production technology, could lead to economical process for acrylic acid manufacture (1,20).

$$\underset{\substack{|\\CH_3-CH-COOR}}{OOCR'} \xrightarrow[\text{catalyst}]{\Delta} CH_2{=}CH-COOR + R'COOH$$
$$(\mathbf{7})$$

Manufacture and Processing. Lactic acid can be manufactured either by chemical synthesis or by carbohydrate fermentation; both are used for commercial production. In the United States, lactic acid is manufactured synthetically by Sterling Chemicals, Inc. using the lactonitrile route. In Japan, Musashino Chemical Co. uses this technology for all of Japan's production. CCA Biochemical bv of the Netherlands uses carbohydrate fermentation technology in plants in Europe and Brazil and markets worldwide. Prior to 1991, the annual U.S. consumption of lactic acid was estimated at 18,500 t, with domestic production of approximately 8600 t by Sterling Chemical and the rest imported from Europe and Brazil. Worldwide consumption was estimated at approximately 40,000 t/yr. Several agriprocessing and chemical companies such as Archer Daniels Midland, Cargill, and Ecological Chemical Products Co. (a joint venture between Du Pont and ConAgra, Inc.) have announced plans to produce lactic acid from carbohydrate fermentation, targeting very large-volume opportunities in biodegradable polymer feedstocks. These developments are expected to boost the fermentation-derived lactic acid capacity and also introduce new technologies for efficient and economical production (2–5).

Chemical Synthesis.   The commercial process is based on lactonitrile [78-97-7] which used to be a by-product of acrylonitrile synthesis. It involves base-catalyzed addition of hydrogen cyanide to acetaldehyde to produce lactonitrile (8). This is a liquid-phase reaction and occurs at atmospheric pressures. The crude lactonitrile is then recovered and purified by distillation and is hydrolyzed to lactic acid using either concentrated hydrochloric or sulfuric acid, producing the corresponding ammonium salt as a by-product. This crude lactic acid is esterified with methanol, producing methyl lactate. The latter is recovered and purified by distillation and hydrolyzed by water under acid catalysts to produce lactic acid, which is further concentrated, purified, and shipped under different product classifications, and methanol, which is recycled.

$$CH_3CHO + HCN \xrightarrow{\text{catalyst}} CH_3 — CHOH — CN$$

$$(8)$$

$$CH_3CHOHCN + 2\,H_2O + \tfrac{1}{2}\,H_2SO_4 \longrightarrow CH_3CHOHCOOH + \tfrac{1}{2}\,(NH_4)_2SO_4$$

$$CH_3CHOHCOOH + CH_3OH \longrightarrow CH_3CHOHCOOCH_3 + H_2O$$

$$CH_3CHOHCOOCH_3 + H_2O \longrightarrow CH_3CHOHCOOH + CH_3OH$$

Other possible chemical synthesis routes for lactic acid include base-catalyzed degradation of sugars; oxidation of propylene glycol; reaction of acetaldehyde, carbon monoxide, and water at elevated temperatures and pressures; hydrolysis of chloropropionic acid (prepared by chlorination of propionic acid); nitric acid oxidation of propylene; etc. None of these routes has led to a technically and economically viable process (6).

Carbohydrate Fermentation.   The existing commercial production processes use homolactic organisms such as *Lactobacillus delbrueckii*, *L. bulgaricus*, and *L. leichmanii*. A wide variety of carbohydrate sources, eg, molasses, corn syrup, whey, dextrose, and cane or beet sugar, can be used. The use of a specific carbohydrate feedstock depends on the price, availability, and its purity. Proteinaceous and other complex nutrients required by the organisms are provided by corn steep liquor, yeast extract, soy hydrolysate, etc. Excess calcium carbonate is added to the fermenters to neutralize the acid produced and produce a calcium salt of the acid in the broth. The fermentation is conducted batchwise, taking 4–6 days to complete, and lactate yields of approximately 90 wt % from a dextrose equivalent of carbohydrate are obtained. It is usually desired to keep the calcium lactate in solution so that it can be easily separated from the cell biomass and other insolubles, and this limits the concentration of carbohydrates that can be fed in the fermentation and the concentration lactate in the fermentation broth, which is usually around 10 wt %. The calcium lactate-containing broth is filtered to remove cells, carbon-treated, evaporated, and acidified with sulfuric acid to convert the salt into lactic acid and insoluble calcium sulfate, which is removed by filtration. The filtrate is further purified by carbon columns and ion exchange and evaporated to produce technical- and food-grade lactic acid, but not a heat-stable product, which is required for the stearoyl lactylates, polymers, and other value-added applications. The technical-grade lactic acid can be esterified with methanol or ethanol and the ester is recovered by distillation, hydrolyzed by water, evaporated, and the alcohol recycled. This separation process produces a highly pure product which, like the synthetic product, is water-white and heat stable.

$$C_6H_{12}O_6 + Ca(OH)_2 \xrightarrow{\text{fermentation}} (CH_3CHOHCOO^-)_2Ca^{2+} + 2H_2O$$

$$(CH_3CHOHCOO^-)_2Ca^{2+} + H_2SO_4 \longrightarrow 2CH_3CHOHCOOH + CaSO_4$$

Some of the economic hurdles and process cost centers of this conventional carbohydrate fermentation process, schematically shown in Figure 1, are in the complex separation steps which are needed to recover and purify the product from the crude fermentation broths. Furthermore, approximately a ton of gypsum, $CaSO_4$, by-product is produced and needs to be disposed of for every ton of lactic acid produced by the conventional fermentation and recovery process (30). These factors have made large-scale production by this conventional route economically and ecologically unattractive.

Advances in membrane-based separation and purification technologies, particularly in micro- and ultrafiltration and electrodialysis, have led to the inception of new processes for lactic acid production. These processes should, when developed and commercialized, lead to low cost production of lactic acid with reduction of nutrient needs, and without creating the by-product gypsum problem (31–35). Desalting electrodialysis requires low amounts of energy to recover, purify, and concentrate lactate salts from crude fermentation broths (31,32). Advances in water-splitting electrodialysis membranes enable the efficient production of protons and hydroxyl ions from water and can thus produce acid and base from a salt solution (33). Patents (31,32) describe an efficient and potentially economical process for lactic acid production and purification. Using an osmotolerant strain of lactic acid bacteria and a configuration of desalting electrodialysis, water-splitting electrodialysis, and ion-exchange purification steps, a concentrated lactic acid product containing less than 0.1% of proteinaceous impurities could be produced from a carbohydrate fermentation. The electric power requirement for the electrodialysis steps was approximately 1 kW·h/kg lactic acid. The process produces no by-product gypsum and only a small amount of by-product salt from the ion-exchange regeneration. Such a process can be operated in a continuous manner, can be scaled up for large-volume production, and forms the basis for commercial development for one of the companies that has announced its intention to be a commercial producer of lactic acid and its derivative products (5).

Another entrant, Ecochem, a Du Pont–Conagra partnership, has developed a recovery and purification process that produces a by-product ammonium salt instead of insoluble gypsum cake, and the company intends to sell this as a low cost fertilizer. A 10,000 t/yr demonstration-scale plant has been completed to prove the process and develop products and markets for polymers and derivatives (4). These advances and activities show that efficient and economical production of purified lactic acid is feasible, and very large-volume production systems could be potentially built in the future. The utilization of the purified lactic acid to produce polymers and other chemical intermediates requires the development and integration of catalytic chemical conversion process steps with the lactic acid production processes. Examples of such process steps are dilactide production

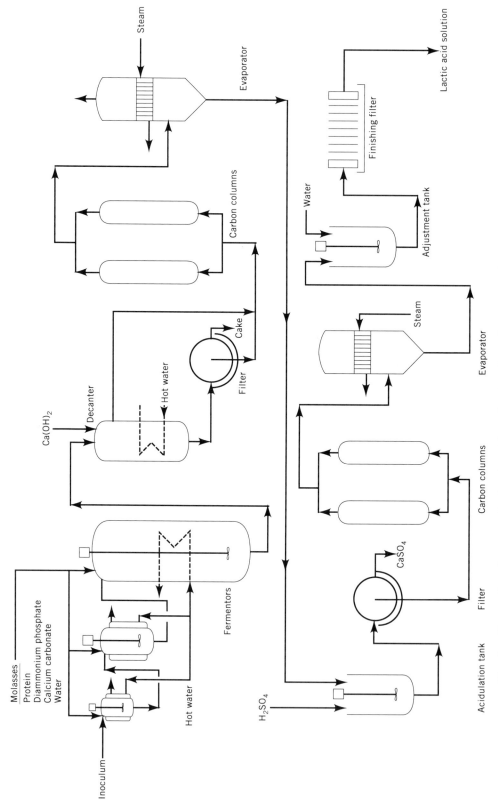

Fig. 1. Conventional process for lactic acid production from dextrose, molasses, or whey.

for polymerization to make high mol wt polymers or copolymers, hydrogenolysis to make propylene glycol (a large-volume intermediate chemical), etc. In the past, little effort was devoted to developing efficient and potentially economical processes for such integrations because only small-volume, high margin specialty polymers for biomedical applications or specialty chemicals were the target products.

Several advances in catalysts and process improvements have occurred, and proprietary technologies have been developed that may enable the commercialization of integrated processes for large-scale production in the future. In a patent issued to Cargill Inc. (36), the development of a continuous process for manufacture of lactide polymers with controlled optical purity from purified lactic acid is described. The process uses a configuration of multistage evaporation followed by polymerization to low molecular weight prepolymer which is then catalytically converted to dilactide and the purified dilactide is recovered in a distillation system with partial condensation and recycling. The dilactide can be used to make high mol wt polymers and copolymers. The process has been able to use fermentation-derived lactic acid, and the claimed ability to recycle and reuse the acid and prepolymers could make such a process very efficient and economical (36). In patents issued to Du Pont (37–44), processes to make cyclic esters, dilactide, and glycolide from their corresponding acid or prepolymer are described. This process uses an inert gas such as nitrogen to sweep away the cyclic esters from the reaction mass and then recovers and purifies the volatilized cyclic ester by scrubbing with an appropriate organic liquid, and separates the cyclic ester from the liquid by precipitation or crystallization and filtration of the solids. High purity lactide with minimal losses due to racemization has been claimed as a product in this process. Recycling and reuse of the lactic moiety in the various process streams have been claimed to be feasible (43). Both Cargill and Du Pont are developing their particular process to commercial scale, with the aim of large-scale production of biodegradable polymers (4,5).

Hydrogenolysis reaction technology designed to produce alcohol from organic acids or esters has also advanced recently; new catalysts and processes give high selectivity and rates and operate at moderate pressures (18,19,45). This technology has been commercialized to produce 1,4-butanediol, tetrahydrofuran, and other four-carbon chemical intermediates from maleic anhydride. In the future such technologies could be integrated with low cost lactic acid production processes to make propylene glycol and other intermediate chemicals (20).

Economic Aspects. The annual U.S. production of lactic acid in the decade spanning 1980–1990 ranged between 8000 and 9000 t, with only one producer, Sterling Chemicals Inc., manufacturing by chemical synthesis. The annual consumption increased from approximately 12,000 t to approximately 18,500 t during this period. Thus, imports, primarily of fermentation-derived product from Europe or Brazil, supplied approximately 50% of the demand. The worldwide demand and production for the year 1990 was estimated to be approximately 40,000 t. The principal manufacturers are Sterling (synthetic, the United States), Musashino (synthetic, Japan), and CCA (fermentation: CSM (Holland); IQSF (Brazil); Ayuso (Spain)). Croda (fermentation, U.K.) and Benkeiser (fermentation, Germany) have discontinued their production. The price, which depends on the product grade, has ranged from \$2200 to \$2500/t. Thus, since the early 1980s, lactic acid has been a mature fine chemical with growing but established markets and a few producers manufacturing in small plants with old technology.

The incentive for economical production of lactic acid is coming from the development of new, large-volume uses of lactic acid, particularly as feedstocks for biodegradable polymers and oxygenated chemicals (1,20). To meet these potential opportunities, efficiently integrated processes have to be developed and commercialized and integrated with large low cost production facilities for carbohydrate feedstocks. The advent and deployment of highly efficient, membrane-based separation processes (31–33) and chemical and catalytic conversion technologies, together with commercial interest by several agriprocessing and chemical companies (2–5), will lead to the production of low cost lactic acid, which in turn will result in new opportunities for large-scale use of lactic acid.

Specifications, Quality Control, and Analytical Methods. Lactic acid is generally sold under four general product categories: *synthetic*: a highly purified product from a chemical synthesis process. It is water-white, has excellent heat stability, and can be used in both food and industrial applications; *fermentation*: a food-grade product from carbohydrate fermentation refined by ion exchange and activated carbon. The product contains residual carbohydrate or protein impurities and is not heat stable; *heat-stable fermentation*: a highly refined, heat-stable product from esterification of fermentation-derived lactic acid, followed by hydrolysis of the recovered ester to produce the acid; and *technical*: a crude product from either a synthetic or fermentation process, used in industrial applications where high purity is not required.

The fermentation-derived food-grade product is sold in 50, 80, and 88% concentrations; the other grades are available in 50 and 88% concentrations. The food-grade product meets the *Food Chemicals Codex III* and the pharmaceutical grade meets the FCC and the *United States Pharmacopoeia XX* specifications (7). Other lactic acid derivatives such as salts and esters are also available in well-established product specifications. Standard analytical methods such as titration and liquid chromatography can be used to determine lactic acid, and other gravimetric and specific tests are used to detect impurities for the product specifications. A standard titration method neutralizes the acid with sodium hydroxide and then back-titrates the acid. An older standard quantitative method for determination of lactic acid was based on oxidation by potassium permanganate to acetaldehyde, which is absorbed in sodium bisulfite and titrated iodometrically.

Lactic acid is generally recognized as safe (GRAS) for multipurpose food use. Lactate salts such as calcium and sodium lactates and esters such as ethyl lactate used in pharmaceutical preparations are also considered safe and nontoxic (7). The U.S. Food and Drug Administration lists lactic acid (all isomers) as GRAS and sets no limitations on its use in food other than current good manufacturing practice (46).

Uses. Currently, the principal use of lactic acid is in food and food-related applications, which in the United States accounts for approximately 85% of the demand. The rest (~15%) of the uses are for nonfood industrial applications. The expected advent of the production of low cost lactic acid in high volume can open new applications for lactic acid and its derivatives, because it is a versatile molecule that can be converted to a wide range of industrial chemicals or polymer feedstocks (1,6,20).

As a food acidulant, lactic acid has a mild acidic taste, in contrast to other food acids. It is nonvolatile, odorless, and is classified GRAS for general-purpose food additives by the FDA in the United States and by other regulatory agencies elsewhere. It is a good preservative and pickling agent for sauerkraut, olives, and pickled vegetables. It is used as acidulant, flavoring, pH buffering agent, or inhibitor of bacterial spoilage in a wide variety of processed foods such as candy, breads and bakery products, soft drinks, soups, sherbets, dairy products, beer, jams and jellies, mayonnaise, processed eggs, and many other processed foods, often in conjunction with other acidulants (7). An emerging new use for lactic acid or its salts is in the disinfection and packaging of carcasses, particularly those of poultry and fish, where the addition of aqueous solutions of lactic acid and its salts during processing increase shelf life and reduce the growth of anaerobic spoilage organisms such as *Clostridium botulinum* (47,48).

A large fraction (>50%) of the lactic acid for food-related uses goes to produce emulsifying agents used in foods, particularly for bakery goods. The emulsifying agents are esters of lactate salts with longer-chain fatty acids. Four important products are calcium and sodium stearoyl-2-lactylate [25383-99-7], glyceryl lactostearate [1338-10-9], and glyceryl lactopalmitate [1335-49-5]. Of the stearoyl lactylates, the calcium salt [5793-94-2] is a good dough conditioner and the sodium salt is both a conditioner and an emulsifier for yeast-leavened bakery products. The glycerates and palmitates are used in prepared cake mixes and other bakery products and in liquid shortenings. In prepared cake mixes the palmitate improves cake texture, whereas the stearate increases cake volume and permits mixing tolerances (7). The manufacture of these emulsifiers requires heat-stable lactic acid; hence only the synthetic or the heat-stable fermentation grades are used for this application.

Technical-grade lactic acid has long been in use in the leather tanning industry as an acidulant for deliming hides and in vegetable tanning. In various textile-finishing operations and acid dying of wool, technical-grade lactic acid was used extensively. Cheaper inorganic acids are now more commonly used in these applications. The future availability of lower cost lactic acid and the increasing environmental restrictions on waste salt disposal may reopen these markets for lactic acid.

Lactic acid is currently used in a wide variety of small-scale, specialized industrial applications where the functional speciality of the molecule is desirable. Examples include pH adjustment of hardening baths for cellophane used in food packaging, terminating agent for phenol–formaldehyde resins, alkyd resin modifier, solder flux, lithographic and textile printing developers, adhesive formulations, electroplating and electropolishing baths, and detergent builders (with maleic anhydride to form carboxymethoxysuccinic acid-type compounds). Owing to the current high cost and low volume of production, these applications account for only 5–10% of the consumption of lactic acid (7).

Lactic acid and ethyl lactate [97-64-3] have long been used in pharmaceutical and cosmetic applications and formulations, particularly in topical ointments, lotions, parenteral solutions, and biodegradable polymers for medical applications such as surgical sutures (qv), controlled-release drugs, and prostheses. A substantial part of pharmaceutical lactic acid is used as the sodium salt [16595-31-6] for parenteral and dialysis applications. The calcium salt [5743-48-6] is widely used for calcium deficiency therapy and as an effective anticarries agent. As humectants in cosmetic applications, the lactates are often superior as natural products and more effective than polyols (7). Ethyl lactate is the active ingredient in many antiacne preparations. The use of the chirality of lactic acid for synthesis of drugs and agrochemicals is an opportunity for new applications for optically active lactic acid or its esters. The chiral synthesis routes to R-(+)-phenoxypropionic acid [1129-46-0] and its derivatives using S-(−)-lactate ester as a chiral synthon have been described (7). These compounds are used in herbicide production. Another use, as an optically active liquid crystal using lactic acid as a chiral synthon, has been described (49). These advances could open new small-volume specialty chemical opportunities for optically active lactic acid and its derivatives.

Polymers of lactic acids are biodegradable thermoplastics that can be made from a variety of renewable carbohydrate resources. A fairly wide range of properties is obtainable by copolymerization with other functional monomers such as glycolide, caprolactone, polyether polyols, etc. The polymers are transparent, which is important for packaging applications. They offer good shelf life because they degrade slowly by hydrolysis which can be controlled by adjusting the composition and molecular weight. The properties of lactic copolymers which approach that of large-volume petroleum-derived polymers such as polystyrene, flexible poly(vinyl chloride) (PVC), vinylidene chloride, etc, have been summarized (1). There are numerous patents and articles on lactic acid polymers and copolymers, their properties, potential uses, and processes, dating back to the early work by Carothers at Du Pont. Several reference articles and patents provide further information (1,50–53).

Large-volume oxygenated chemicals such as propylene glycol (1990 U.S. production, 400,000 t), propylene oxide (1990 U.S. production, 1.6×10^6 t), acrylic acid and acrylate esters (1990 U.S. production, 500,000 t), and other chemical intermediates such as lactate ester plasticizers can potentially be made from lactic acid (1,20). The advances made in hydrogenolysis technology can be further developed and integrated to make propylene glycol from lactic acid in the future. Advances in catalysis and process technologies are needed for efficient conversion of lactate to acrylate, or propylene glycol to propylene oxide (qv), even though some of the early patents show the technical feasibility of such conversions (54–57). Common derivatives of lactic acid, particularly the esters of low molecular weight alcohols, are food-safe, biodegradable products that could find wider applications as solvents and plasticizers, particularly as regulations and consumer preferences increase the demand for such "green" chemicals.

Low molecular weight polymers of L-lactic acid having a degree of polymerization (DP) 2–10 have plant growth promotion activity for a variety of crops and fruits when applied at a low level (58,59). These findings may lead to products and formulations that would incorporate L-poly(lactic acid) as or into controlled-release or degradable mulch films for large-scale agricultural applications. Thus in addition to its current applications lactic acid may also be used as a very large-volume polymer feedstock and chemical intermediate as new products and production technologies are deployed.

Hydroxyacetic Acid

Hydroxyacetic acid [79-14-1] (glycolic acid), $HOCH_2COOH$, is the first and simplest member of the family of hydroxycarboxylic acids. It occurs naturally as the chief acidic constituent of sugar-cane juice and also occurs in sugar beets and unripe grape juice. It was first synthesized in 1848 by reaction of glycine with nitrous acid, and it was characterized in 1851. It is widely used as a cleaning agent for a variety of industrial applications, and also as a specialty chemical and biodegradable copolymer feedstock (30,53).

Properties. Glycolic acid is a colorless, translucent solid; mp = 10°C; bp = 112°C; d at 25°C = 1.26 g/mL; K_a at 25°C = 1.5×10^{-4}; pH at 25°C = 0.5; heat of combustion = 697.1 kJ/mol (166.6 kcal/mol); heat of solution = −11.55 kJ/mol; and flash point >300°C (53). Crystalline glycolic acid has unit space cell dimensions of a = 0.8965 nm, b = 1.0563 nm, c = 0.7826 nm, and Z = 8 with unit cell of P2./c type. The crystal structure consists of a loose, three-dimensional H-bonded network of two closely similar but crystallographically distinct types of molecules.

Glycolic acid is soluble in water, methanol, ethanol, acetone, acetic acid, and ethyl acetate. It is slightly soluble in ethyl ether and sparingly soluble in hydrocarbon solvents.

Reactions. Because it contains both a carboxyl and a primary hydroxyl group, glycolic acid can react as an acid or an alcohol or both. Thus some of the important reactions it can undergo are esterification, amidation, salt formation, and complexation with metal ions, which lead to many of its uses. As a fairly strong acid it can liberate gases (often toxic) when it reacts with the corresponding salts, eg, carbon dioxide from carbonates; hydrogen cyanide from cyanide salts; hydrogen sulfide from metal sulfides; chlorine from some active chlorine compounds such as hypochlorites, chlorinated isocyanurates, and chlorinated hydantoins; and hydrogen from some metals.

Direct esterification of the free acid or poly(glycolic acid) or transesterification of the methyl ester leads to the facile formation of alkyl esters. Esters of the hydroxyl group are readily prepared by reaction with the corresponding acid chloride. Self-esterification or dehydration is another important reaction; it occurs upon heating in concentrated solution with loss of water to form a mixture of the linear dimer carboxymethyl hydroxyacetate, $HOCH_2COOCH_2COOH$; cyclic dimer diglycolide (glycolide [502-97-6]); and polymeric polyglycolide (poly(hydroxyacetic acid) [26124-68-5]). Di- and polyglycolides are useful, low molecular weight, biodegradable polymers with several applications, and they are also copolymers with lactic acid polymers for biomedical applications and have potential for large-volume usage as biodegradable packaging polymers (1,60). N-Hydroxyacetamide can be prepared by reaction of ammonia with the polyacid or by reaction of an ester of glycolic acid with aqueous ammonium hydroxide. Substituted amines are prepared by the reaction of an amine with the acid.

Glycolic acid also undergoes reduction or hydrogenation with certain metals to form acetic acid, and oxidation by hydrogen peroxide in the presence of ferrous salts to form glyoxylic acid [298-12-4], HCOCOOH, and in the presence of ferric salts in neutral solution to form oxalic acid, HOOCCOOH; formic acid, HCOOH;

and liberate CO_2 and H_2O. These reduction and oxidation reactions are not commercially significant.

Manufacture, Processing, and Economic Aspects. Hydroxyacetic acid is produced commercially in the United States as an intermediate by the reaction of formaldehyde with carbon monoxide and water.

$$HCHO + CO + H_2O \xrightarrow{catalyst} HOCH_2 \text{---} COOH$$

Du Pont Chemicals is the primary manufacturer in the United States and Hoechst in Europe. Annual production is estimated to be approximately 10,000 t and selling prices (1992) range between $1.15 and $1.50/kg, depending on the product grade. Formaldehyde reacts with carbon monoxide in the presence of water and an acid catalyst such as HF; other mineral acids such as sulfuric, phosphoric, or hydrochloric; or acidic ion-exchange resins. The pressures and temperatures are believed to range between 10–50 MPa (100–500 atm) and 50–100°C, respectively; the actual conditions used in the process are not disclosed. The reaction mixture is purified by ion exchange, carbon treatment, and steam stripping to remove ionic, color, and volatile impurities, respectively.

From 1940 to about 1968 Du Pont produced ethylene glycol by a multistep glycolic acid, methanol, and formaldehyde route. Glycolic acid as the intermediate was esterified by methanol, methyl glycolate was then hydrogenated to produce ethylene glycol, and the methanol was recycled. This process was thought to be run at a 75,000 t/yr scale but the advent of inexpensive ethylene and improved ethylene oxide processes made this process economically unattractive for ethylene glycol manufacture. Glycolic acid, however, is made by this route, and prior manufacturing experience suggests that large-scale production of glycolic acid by this route is feasible should the demand for glycolic acid rise.

Other methods of production include hydrolysis of glycolonitrile [107-16-4] with an acid (eg, H_3PO_3 or H_2SO_3) having a pK_a of about 1.5–2.5 at temperatures between 100–150°C; glycolonitrile produced by reaction of formaldehyde with hydrogen cyanide; recovery from sugar juices; and hydrolysis of monohalogenated acetic acid. None of these has been commercially and economically attractive.

Specifications. Hydroxyacetic acid is available in three grades: as a 70% technical aqueous solution; a 70% low sodium aqueous solution; or, in smaller quantities, in a 99+% pure crystalline form. The aqueous solutions are a clear, light amber liquid with a mild odor resembling burnt sugar. The specifications have been provided (60). The acid and the dimer (glycolide) are not regulated as hazardous materials, and the primary safety and handling precautions are based on their acidic nature. They are only slightly toxic by ingestion. The approximate oral LD_{50} is 4240 mg/kg for glycolic acid, and the approximate lethal dose for glycolide is 3400 mg/kg (oral, rats) (60).

Uses. The primary uses of hydroxyacetic acid are in cleaning and metal processing, with many other specialized applications in a variety of industries. It has many properties which make it an ideal hard-surface cleaner. It is nonvolatile and biodegradable, for safe, environmentally friendly use. Its hard water salts and iron oxide dissolution ability give it excellent cleaning characteristics. Its metal complexes are water soluble, providing superior rinsibility. Its use is compatible with most other acids and with cleansing compounds (eg, surfactants, glycol ethers, scents, and dyes), which leads to its use in a variety of cleaning formulations for superior performance.

In applications as hard surface cleaners of stainless steel boilers and process equipment, glycolic acid and formic acid mixtures are particularly advantageous because of effective removal of operational and preoperational deposits, absence of chlorides, low corrosion, freedom from organic iron precipitations, economy, and volatile decomposition products. Ammoniated glycolic acid in mixture with citric acid shows excellent dissolution of the oxides and salts and the corrosion rates are low.

Other cleaning applications are in cooling tower–heat exchanger scale build-up removal, well water light iron and carbonate scale removal, dairy equipment milkstone and other deposit removal, nonferrous metal cleaning formulations, food processing equipment, masonry, paper felt, and as a laundry sour in detergent formulations. Details of some of these cleaning applications are available (60). In metal processing, glycolic acid is used in a variety of applications in pickling, etching, electroplating, electropolishing, and brightening. In pickling operations it is used to replace volatile organic acids and reduces losses and emissions at elevated temperatures. In etching, the soluble nature of its metal complexes is useful for prevention of unwanted precipitates. In electroplating (qv), the sodium and potassium glycolates are excellent substitutes for Rochelle salts in bath additives, and because glycolic acid forms complexes with virtually all multivalent metals, its salts are used in many electroplating baths, eg, chrome, lead, copper, nickel, cobalt, and tin. Electropolishing of stainless steel utilizes glycolic acid in combination with either phosphoric or sulfuric acids. The mixture gives a bath with long life and reduced sludging characteristics. Brightening of copper and copper alloys utilizes glycolic acid baths to enhance and retain luster.

Other specialized applications include biomedical uses, printed wire board flux, adhesives, textiles, hydrogen sulfide abatement, tanning, oil well acidification, and biodegradable polymers and copolymers (60). For biomedical applications, highly pure polymers of glycolic acid and lactic acid derived from the condensation polymerization of glycolide and lactide are used to manufacture bioabsorbable surgical sutures (qv), staples, and clips, sustained-release drug delivery systems, and implantable prosthetic devices. Some of the future applications of these polymers may be in the manufacture of larger volume biodegradable polymers for commercial consumer use as the production processes and economics become favorable (1,21,40–48). The low sodium grade of the acid is used as an intermediate flux in the electronic industry, as well as for cleaning and deoxidizing circuit boards before soldering.

The ability of glycolic acid to react with casein to form cationic casein has been utilized in the manufacture of adhesives and textile sizing compounds. The FDA has given clearance for its use in food packaging adhesives (61). In the textile applications, glycolic acid has been used to acidify dye baths; complex metal ions present in bath water, rinse, etc; in dyeing of chrome colors; solvation of liquid cationic dye products; neutralization of fabric following alkaline treatments; and many other applications where the low volatility of the acid is useful. In the tanning process, glycolic acid is used to replace mineral acids where high quality leather is desired. It is used in the deliming operation to provide pH adjustment of the tanning liquor, where glycolic acid prevents staining of leather caused by the highly colored metal tanning salts.

A complex of glycolic acid with ferrous ion can catalyze the oxidation of hydrogen sulfide. This ability is used in the abatement of dissolved hydrogen sulfide in steam condensates from geothermal power generators. Glycolic acid and its low molecular weight polymers are used for oil well acidification, complexing with metals (especially iron) during cleaning operations, or water flooding. The easy degradability is advantageous for these applications.

Other Hydroxy Acids

Apart from lactic and hydroxyacetic acids, other α- and β- hydroxy acids have been small-volume specialty products produced in a variety of methods for specialized uses. γ-Butyrolactone [96-48-0], which is the monomeric inner ester of γ-hydroxybutyric acid [591-81-1], is a large-volume chemical derived from 1,4-butanediol (see ACETYLENE-DERIVED CHEMICALS).

Preparation. The general preparation of α-*hydroxy acids* is by the hydrolysis of an α-halo acid or by the acid hydrolysis of the cyanohydrins of an aldehyde or a ketone.

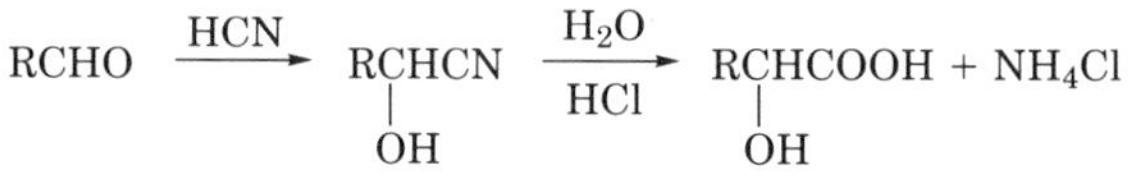

$$RCHO \xrightarrow{\text{HCN}} \underset{\underset{\text{OH}}{|}}{RCHCN} \xrightarrow[\text{HCl}]{\text{H}_2\text{O}} \underset{\underset{\text{OH}}{|}}{RCHCOOH} + NH_4Cl$$

Aliphatic α-hydroxy acids that do not have side chains can be prepared in good yields by the hydrolysis of α-nitrato acids with aqueous sulfite solutions. The α-nitrato acids are obtained by the reaction of olefins and N_2O_4 in the presence of oxygen. The α-hydroxy acids that are obtained can be esterified or acylated directly to yield anhydro ester acids which, in turn, give the α-hydroxy acid on saponification. For example, 2-nitrooxyoctanoic acid, $C_6H_{13}CH(ONO_2)COOH$ (from 1-octene), is added to a solution of sodium sulfite (2.5 mol/mol acid) at 60–90°C; it is then cooled and acidified. The organic layer yields 95% 2-hydroxyoctanoic acid [617-73-2], $C_6H_{13}CHOHCOOH$ (mp, 55–65°C, saponification no. 354 and acid no. 320) (62).

β-*Hydroxy acids* may be made by catalytic reduction of β-keto esters followed by hydrolysis. β-Hydroxy acids can also be prepared by the Reformatsky reaction. This reaction is analogous to the Grignard reaction (qv). An aldehyde or ketone reacts with an α-haloester in the presence of activated zinc in anhydrous ethyl ether (63). Though ether is the commonly used solvent, benzene, toluene, xylene, or mixtures of these with ether, depending on the temperature of the reaction, may also be used. The reaction product is treated with an aqueous acid to hydrolyze the organozinc complex to produce the β-hydroxy ester. The organozinc compounds are considerably less reactive toward the carbonyl group than Grignard reagents. Thus the carbonyl groups that may be used include saturated and unsaturated aliphatic aldehydes, aromatic aldehydes, and aliphatic, aromatic, alicyclic, saturated, and unsaturated ketones. Among the halo esters, the α-bromo esters generally give the best results; α-chloro esters react slowly or not at all and the α-iodo esters are not commonly available. β-Hydroxypropionic acid [503-66-2] can be formed by the reaction of formaldehyde and lead tetraacetate trihydrate with catalytic amounts of pyridine and hydroquinone under pressure at 175°C for two hours. After acidification with excess HCl, the reaction yields lead chloride and a red solution which, when extracted with a suitable solvent, produces the desired β-hydroxy acid (64).

γ-*Hydroxy acids* are seldom obtained in the free state because of the ease with which they form monomeric inner esters, which form stable five-membered rings. Thus the lactones of these acids are the common chemical forms and among these lactones γ-butyrolactone (**9**) is one of the larger volume specialty chemicals derived from dehydrogenation of 1,4-butanediol.

$$\text{HOCH}_2(\text{CH}_2)_2\text{CH}_2\text{OH} \xrightarrow[\Delta]{\text{catalyst}} \underset{\text{(9)}}{\text{[lactone]}} + 2\,\text{H}_2$$

γ-Butyrolactone production in 1992 in the United States was estimated to be approximately 45,000 t/yr; GAF Corp. and Du Pont are the primary manufacturers in their integrated acetylenic chemical manufacturing plants.

Reactions and Uses. The common reactions that α-hydroxy acids undergo such as self- or bimolecular esterification to oligomers or cyclic esters, hydrogenation, oxidation, etc, have been discussed in connection with lactic and hydroxyacetic acid. A reaction that is of value for the synthesis of higher aldehydes is decarbonylation under boiling sulfuric acid with loss of water. Since one carbon atom is lost in the process, the series of reactions may be used for stepwise degradation of a carbon chain.

$$\text{RCHOHCOOH} \xrightarrow{\text{H}^+} \text{RCHOHC(OH)}_2 \xrightarrow{-\text{H}_2\text{O}} \text{RCHOHCO}^+ \xrightarrow{-\text{CO}} \xrightarrow{-\text{H}^+} \text{RCHO}$$

β-Hydroxy acids lose water, especially in the presence of an acid catalyst, to give α,β-unsaturated acids, and frequently β,γ-unsaturated acids. β-Hydroxy acids do not form lactones readily because of the difficulty of four-membered ring formation. The simplest β-lactone, β-propiolactone, can be made from ketene and formaldehyde in the presence of methyl borate but not from β-hydroxypropionic acid. β-Propiolactone [57-57-8] is a useful intermediate for organic synthesis but caution should be exercised when handling this lactone because it is a known carcinogen.

γ-Hydroxybutyric acid and its derivatives, particularly its sodium salt, have been studied and used as anesthetics, tranquilizers, sedatives, and hypnotics in surgery and general obstetrics. Some of the physicochemical properties, metabolism, toxicity, pharmacological and clinical aspects, and uses of γ-hydroxybutyric acid have been reviewed (65,66).

Certain bacterial species produce polymers of γ-hydroxybutyric acid and other hydroxyalkanoic acids as storage polymers. These are biodegradable polymers with some desirable properties for manufacture of biodegradable packaging materials, and considerable effort is being devoted by ICI Ltd. and others to the development of bacterial fermentation processes to produce these polymers at a high molecular weight (66).

γ-Butyrolactone (**9**) undergoes amination reactions with methylamine or ammonia to produce N-methyl-2-pyrrolidinone [*872-50-4*] (NMP) (**10**, R=CH$_3$) or 2-pyrrolidinone [*616-45-5*] (PDO) (**10**, R=H), respectively, both of which are commercially important derivatives: NMP is a commercially important solvent used in lube oil and other solvent extractions and reaction solvents in the electronics industry; PDO reacts with acetylene to make N-vinylpyrrolidinone [*88-12-0*] (**11**), which is used to make poly(vinylpyrrolidinone) polymers (**12**) and copolymers for a number of commercial uses in cosmetics, hair spray, pharmaceutical formulations, germicides, and other consumer and industrial products (67).

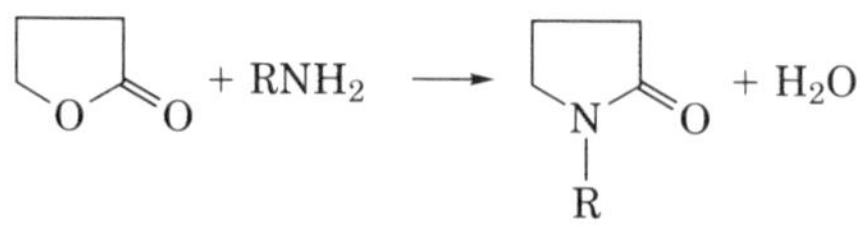

$$\text{(9)} + RNH_2 \longrightarrow \text{(10)} + H_2O$$

(**9**)   (**10**)

$$\text{(PDO)} + HC{\equiv}CH \longrightarrow \text{(11)} \longrightarrow \text{(12)}$$

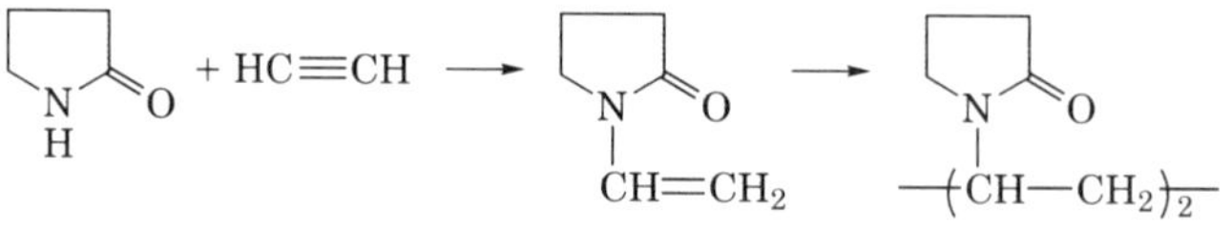

(**11**)   (**12**)

Other multifunctional hydroxycarboxylic acids are mevalonic and aldonic acids which can be prepared for specialized uses as aldol reaction products (mevalonic acid [*150-97-0*] (**13**)) and mild oxidation of aldoses (aldonic acids).

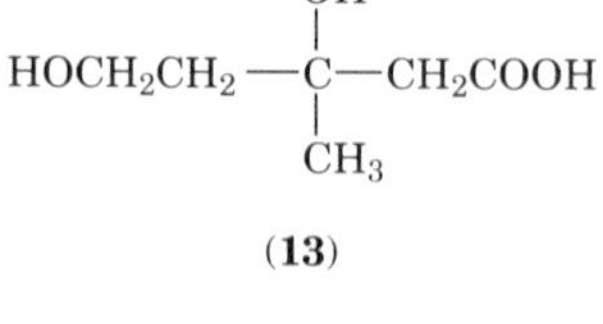

$$HOCH_2CH_2 \overset{\overset{\displaystyle OH}{|}}{\underset{\underset{\displaystyle CH_3}{|}}{C}} CH_2COOH$$

(**13**)

BIBLIOGRAPHY

"Lactic Acid" in *ECT* 1st ed., Vol. 8, pp. 167–180, by E. M. Filachione, Eastern Regional Research Laboratory, Bureau of Agriculture and Industrial Chemistry, U.S. Department of Agriculture; in *ECT* 2nd ed., Vol. 12, pp. 170–188, by G. T. Peckham, Jr., Clinton Corn Processing Co., and E. M. Filachione, U.S. Department of Agriculture, Wyndmoor,

Pa.; "Hydroxy Carboxylic Acids" in *ECT* 3rd ed., Vol. 13, pp. 80–103, by J. H. Van Ness, Monsanto Industrial Chemicals Co.

1. E. S. Lipinsky and R. G. Sinclair, *Chem. Eng. Progr.* **82**(8), 26–32 (1986).
2. *Chem. Mktg. Rep.*, 22 (Oct. 15, 1990).
3. *Chem. Mktg. Rep.*, 5 (Oct. 16, 1989).
4. *Chem. Week*, 32 (June 26, 1991); *Chem. Eng. News*, 6 (Aug. 6, 1992).
5. *Chem. Mktg. Rep.*, 9 (Oct. 21, 1991).
6. C. H. Holten, A. Muller, D. Rehbinder, *Lactic Acid*, International Research Association, Verlag Chemie, Copenhagen, Denmark, 1971.
7. *Lactic Acid and Lactates*, product bulletin, Purac Inc., Arlington Heights, Il., 1989.
8. S. C. Prescott and C. G. Dunn, *Industrial Microbiology*, 3rd ed., McGraw-Hill Book Co., Inc., New York, 1959.
9. P. A. Levine and A. Rothen, *J. Biol. Chem.* **107**, 533–553 (1934).
10. R. C. Schulz and J. Schwaab, *Makromol. Chem.* **87**, 90–102 (1965).
11. H. J. Fenton and H. O. Jones, *J. Chem. Soc.* **77**, 69–76 (1900).
12. E. J. Witzemann, J. Am. Chem. Soc. **48**, 211–222 (1926).
13. G. V. Bakore and S. Narain, *J. Am. Chem. Soc.* **85**, 3419–3424 (1963).
14. P. S. MacMahon and T. N. Srivastava, *J. Indian Chem. Soc.* **23**, 261–269 (1946).
15. P. A. Levene and H. L. Halla, *J. Biol. Chem.* **67**, 329–332 (1926).
16. H. Adkins and H. R. Billica, *J. Am. Chem. Soc.* **70**, 3121–3125 (1948).
17. E. Bowden and H. Adkins, *J. Am. Chem. Soc.* **56**, 689–691 (1934).
18. Ger. Pat. DE 3443277 A1 (June 5, 1985), J. Kipax, N. Harris, K. Turner, B. Anthony, and C. Rathmell (to Davy Mckee, Ltd.).
19. U.S. Pat. 4,777,303 (Oct. 11, 1988), M. Kitson and P. S. Williams (to BP Chemicals, Ltd.).
20. R. Datta and co-workers, *Proceedings of the International Congress on Chemicals from Biotechnology*, Hannover, Germany, 1993, in press.
21. Brit. Pat. 1,007,347 (Oct. 13, 1965) E. I. du Pont de Nemours & Co., Inc.
22. U.S. Pat. 4,797,468 (Jan. 10, 1989), K. S. De Vries (to Akzo NV Netherlands).
23. U.S. Pat. 4,766,182 (Aug. 23, 1988); J. R. Murdoch and G. L. Loomis (to E. I. du Pont de Nemours & Co., Inc.).
24. U.S. Pat. 4,800,219 (Jan. 24, 1989), J. R. Murdoch and G. L. Loomis (to E. I. du Pont de Nemours & Co., Inc.).
25. D. L. Wise and co-workers, in G. Gregoriadis, ed., *Drug Carriers in Biology and Medicine*, Academic Press, New York, 1979, pp. 237–270.
26. E. M. Filachione and C. H. Fisher, *Ind. Eng. Chem.* **38**, 228–232 (1946).
27. R. Ueda and co-workers, *Hakko Kogaku Zasshi* **40**, 555–563 (1962).
28. M. L. Fein and C. H. Fisher, *J. Org. Chem.* **15**, 530–534 (1950).
29. E. M. Filachione and E. J. Costello, Ind. Eng. Chem. **44**, 2189–2191 (1952).
30. *Biomass Process Handbook*, Technical Insights, Inc., Fort Lee, NJ., 1982, pp. 96–103.
31. U.S. Pat. 4,885,247 (Dec. 5, 1989), R. Datta (to Michigan Biotechnology Institute).
32. Eur. Pat. Appl. 0 393 818 A1 (Feb. 20, 1990), D. Glassner and R. Datta (to Michigan Biotechnology Institute).
33. K. N. Mani, *J. Membrane Sci.* **58**, 117–138 (1991); AQUATECH systems product brochure, AlliedSignal Corp., Morristown, N.J., 1985.
34. M. Hongo, Y. Nomura, M. Iwahara, *Appl. Envir. Microbiol.* **52**(2), 314–319 (1986).
35. P. X. Yao and K. Toda, *J. Gen. Appl. Microbiol.* **36**, 111–125 (1990).
36. U.S. Pat. 5,142,023 (Aug. 25, 1992), P. R. Gruber and co-workers (to Cargill Inc.).
37. U.S. Pat. 4,835,293 (May 30, 1989), K. K. Bhatia (to E. I. du Pont de Nemours & Co., Inc.).
38. U.S. Pat. 5,023,349 (June 11, 1991), K. K. Bhatia (to E. I. du Pont de Nemours & Co., Inc.).

39. U.S. Pat. 5,023,350 (June 11, 1991), K. K. Bhatia (to E. I. du Pont de Nemours & Co., Inc.).

40. U.S. Pat. 5,043,458 (Aug. 27, 1991), K. K. Bhatia (to E. I. du Pont de Nemours & Co., Inc.).

41. U.S. Pat. 5,091,544 (Feb. 25, 1992), K. K. Bhatia (to E. I. du Pont de Nemours & Co., Inc.).

42. U.S. Pat. 5,117,008 (May 26, 1992), K. K. Bhatia, N. E. Drysdale, and J. R. Kosak (to E. I. du Pont de Nemours & Co., Inc.).

43. U.S. Pat. 5,136,057 (Aug. 4, 1992), K. K. Bhatia (to E. I. du Pont de Nemours & Co., Inc.).

44. U.S. Pat. 5,138,074 (Aug. 11, 1992), K. K. Bhatia and H. E. Bellis (to E. I. du Pont de Nemours & Co., Inc.).

45. U.S. Pat. 5,037,996 (Aug. 6, 1991), H. Suzuki, H. Inagaki, and H. Ueno (to Tonen Corp.).

46. *Code of Federal Regulations*, U.S. Food and Drug Administration, Washington, D.C., April 1, 1992.

47. U.S. Pat. 4,888,191 (Dec. 19, 1989), R. J. Anders, J. G. Cerveny, and A. L. Milkowski (to Oscar Mayer Foods Corp.).

48. U.S. Pat. 5,017,391 (May 21, 1991), R. J. Anders, J. G. Cerveny, and A. L. Milkowski (to Oscar Mayer Foods Corp.).

49. U.S. Pat. 4,812,259 (Mar. 14, 1989), K. Yoshinaga, K. Katagiri, and K. Shinjo (to Canon Kabushiki Kaisha).

50. P. Dave and co-workers, *Polymer Prep.*, **31**(1), 442–443 (1990).

51. U.S. Pats. 4,045,418, 4,057,537 (1977), R. G. Sinclair (to Gulf Oil Corp.).

52. J. Zhu and co-workers, *Proc. of C-MRS Intl. Symp.* **3**, 387–390 (1990).

53. T. Nakamura and co-workers, *Adv. Biomater., Biomater. Clin. Appl.* **7**, 759–764, (1987).

54. U.S. Pat. 2,859,240 (Nov. 4, 1958), R. E. Holmen (to 3M Co.).

55. U.S. Pat. 4,729,978 (Mar. 8, 1988), R. A. Sawicki (to Texaco Inc.).

56. U.S. Pat. 4,158,008 (June 12, 1979), H. M. Weitz, J. Hartig, and R. Platz (to BASF AG).

57. U.S. Pat. 4,226,780 (Oct. 7, 1980), G. Fouquet, F. Merger, and K. Baer (to BASF AG).

58. U.S. Pat. 4,813,997 (Mar. 21, 1989), A. Kinnersley and co-workers (to CPC International).

59. U.S. Pat. 5,059,241 (Oct. 22, 1991), D. C. Young (to Union Oil Co.).

60. *Glycolic (Hydroxyacetic) Acid: Properties, Uses, Storage, and Handling; Glycolide S.G.: Properties, Uses, Storage, and Handling*, bulletins, Du Pont Chemicals, Wilmington, Del., 1992.

61. *Code of Federal Regulations*, 21-CFR 175.105, U.S. Food and Drug Administration, Washington, D.C.

62. Ger. Pat. 1,257,766 (Jan. 4, 1968), W. Mueller and co-workers (to Lentia GmbH).

63. C. R. Noller, *Chemistry of Organic Compounds*, 3rd ed., W. B. Saunders Co., Philadelphia, Pa., 1965.

64. U.S. Pat. 3,202,702 (Aug. 24, 1965), K. L. Oliver (to Union Oil Co.).

65. A. Pesce and G. Rugerio, *Minerva Anestesiol.* **42**(1), 88 (1976).

66. R. Keeler, R & D Magazine, 46–52, (Jan. 1991).

67. *BLO-γ-butyrolactone*, product brochure, GAF Corp., Wayne, N.J., 1987.

RATHIN DATTA
Consultant

HYDROXY DICARBOXYLIC ACIDS

Many natural and synthetic organic compounds are hydroxy dicarboxylic acids (see also HYDROXYCARBOXYLIC ACIDS). This article discusses mainly malic and tartaric acids; thiomalic acid is included because of its structural similarity to malic acid.

Malic Acid

Malic acid [6915-15-7] (hydroxysuccinic acid, hydroxybutanedioic acid, or 1-hydroxy-1,2-ethanedicarboxylic acid), $C_4H_6O_5$, is a white, crystalline material. The levorotatory isomer, $S(-)$-malic acid [97-67-6] (L-malic acid), is a natural constituent and common metabolite of plants and animals. The racemic compound, R,S-malic acid [617-48-1] (DL-malic acid), is a widely used food acidulant. This material is also used in some industrial applications as a sequestrant and as a buffer for pH control. $R(+)$-Malic acid [636-61-3] (D-malic acid) is available only as a laboratory chemical. In the United States, R,S-malic acid was first produced synthetically in 1923. Until the early 1960s, it was produced batchwise on a small scale and had limited industrial application. Following the introduction of a modern, continuous manufacturing process in the early 1960s, malic acid gradually became a large-volume industrial organic acid.

Physical Properties. Malic acid crystallizes from aqueous solutions as white, translucent, anhydrous crystals. The $S(-)$ isomer melts at 100–103°C (1) and the $R(+)$ isomer at 98–99°C (2). On heating, D,L-malic acid decomposes at ca 180°C, by forming fumaric acid and maleic anhydride. Under normal conditions, malic acid is stable; under conditions of high humidity, it is hygroscopic.

Malic acid is a relatively strong acid. Its dissociation constants are given in Table 1. The pH of a 0.001% aqueous solution is 3.80, that of 0.1% solution is 2.80, and that of a 1.0% solution is 2.34. Many of its physical properties are similar to those of citric acid (qv). Solubility characteristics are shown in Figure 1 and Table 1, densities of aqueous solutions are listed in Table 2, and pH values vs concentration are shown in Figure 2.

Chemical Properties. Because of its chiral center, malic acid is optically active. In 1896, when tartaric acid was first reduced to malic acid, the levorotatory enantiomer, $S(-)$, was confirmed as having the spatial configuration (1) (5,6). The other enantiomer (2) has the R configuration. A detailed discussion of configuration assignment by the sequence rule or the R and S system is available (7).

$$
\begin{array}{cc}
\begin{array}{c}
COOH \\
| \\
HO\!-\!C\!\leftarrow\!H \\
| \\
CH_2COOH
\end{array}
&
\begin{array}{c}
COOH \\
| \\
H\!\leftarrow\!C\!\leftarrow\!OH \\
| \\
CH_2COOH
\end{array}
\\
(\mathbf{1}) & (\mathbf{2})
\end{array}
$$

Table 1. Physical Properties of *R,S*-Malic Acid[a]

Property	Value
mol wt	134.09
appearance	white, crystalline
crystal system	triclinic
melting point, °C	ca 130
d_4^{20}	1.601
dissociation constant	
K_1	4×10^{-4}
K_2	9×10^{-6}
heat of combustion (at 20°C), MJ/mol[b]	1.340
heat of solution, kJ/mol[b] solute	−20.515
viscosity (50% aqueous solution at 25°C), mPa·s(=cP)	6.5
solubility[c] in nonaqueous solvents, % wt/wt	
ethanol	45.5
acetone	17.8
methanol	82.7

[a]Ref. 3.
[b]To convert J to cal, divide by 4.184.
[c]Ref. 4.

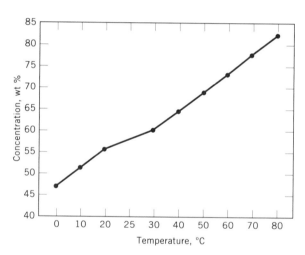

Fig. 1. *R,S*(±)-Malic acid solubility in water, showing maximum solubilities vs temperature (4).

The optical activity of malic acid changes with dilution (8). The naturally occurring, levorotatory acid shows a most peculiar behavior in this respect; a 34% solution at 20°C is optically inactive. Dilution results in increasing levo rotation, whereas more concentrated solutions show dextro rotation. The effects of dilution are explained by the postulation that an additional form, the epoxide (**3**), occurs in solution and that the direction of rotation of the normal (open-

Table 2. Density of Aqueous Malic Acid Solutions at 15°C[a]

Concentration, g/L	d_{15}^{15}, g/mL	(°Bé)	Concentration, g/L	d_{15}^{15}, g/mL	(°Bé)
30	1.115	(14.9)	46	1.169	(21.0)
32	1.121	(15.7)	48	1.179	(22.0)
34	1.129	(16.6)	50	1.186	(22.7)
36	1.138	(17.6)	52	1.192	(23.4)
38	1.146	(18.5)	54	1.199	(24.1)
40	1.151	(19.0)	56	1.208	(25.0)
42	1.158	(19.8)	58	1.212	(25.4)
44	1.165	(20.5)	60	1.220	(26.1)

[a]At 25°C specific gravity ranges from 1.04 at 10 wt % to 1.25 at 55 wt %.

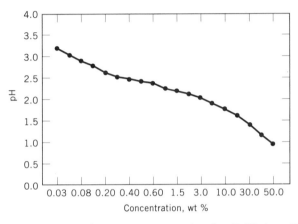

Fig. 2. Malic acid pH values vs concentration for $R,S(\pm)$-malic acid (4).

chain) and epoxide forms is reversed (8). Synthetic (racemic) R,S-malic acid can be resolved into the two enantiomers by crystallization of its cinchonine salts.

$$\text{HOOCCH}_2\text{CH}\overset{\displaystyle\diagdown\ \diagup}{\underset{\displaystyle\text{O}}{}}\text{C(OH)}_2$$

(3)

Reactions. Malic acid undergoes many of the characteristic reactions of dibasic acids, monohydric alcohols, and α-hydroxycarboxylic acids. When heated to 170–180°C, it decomposes to fumaric acid and maleic anhydride which sublimes on further heating (see MALEIC ANHYDRIDE, MALEIC ACID, AND FUMARIC ACID). Malic acid forms two types of condensation products: linear malomalic acids and the cyclic dilactone or malide; it does not form an anhydride.

As a dibasic acid, malic acid forms the usual salts, esters, amides, and acyl chlorides. Monoesters can be prepared easily by refluxing malic acid, an alcohol, and boron trifluoride as a catalyst (9). With polyhydric alcohols and polycarboxylic aromatic acids, malic acid yields alkyd polyester resins (10) (see ALCOHOLS, POLYHYDRIC; ALKYD RESINS). Complete esterification results from the reaction of the diester of malic acid with an acid chloride, eg, acetyl or stearoyl chloride (11).

Alkyl halides in the presence of silver oxide react with alkyl malates to yield alkoxy derivatives of succinic acid, eg, 2-ethoxysuccinic acid, $HOOCCH_2CH(OC_2H_5)COOH$ (12,13). A synthetic approach to produce ethers of malic acid is the reaction of malic esters and sodium alkoxides which affords 2-alkoxysuccinic esters (14).

Amides are obtained when alkyl esters of malic acid are treated with ammonia in alcoholic solution. Hydrazine reacts in a similar manner to yield malic dihydrazide (15) (see HYDRAZINE AND ITS DERIVATIVES). Depending on the proportions of water that are present in the reaction, aniline and malic acid form N,N'-diphenylmalamide or the cyclic compound, N-phenylmalanil (16). When monoanilinium malate is distilled under reduced pressure, a mixture of C-anilino-N-phenylsuccinimide, N-phenylmalanil, and N-phenylsuccinimide is obtained (17,18).

Malic acid yields coumalic acid when treated with fuming sulfuric acid (19). Similar treatment of malic acid in the presence of phenol and substituted phenols is a facile method of synthesizing coumarins that are substituted in the aromatic nucleus (20,21) (see COUMARIN). Similar reactions take place with thiophenol and substituted thiophenols, yielding, among other compounds, a red dye (22) (see DYES AND DYE INTERMEDIATES). Oxidation of an aqueous solution of malic acid with hydrogen peroxide (qv) catalyzed by ferrous ions yields oxalacetic acid (23). If this oxidation is performed in the presence of chromium, ferric, or titanium ions, or mixtures of these, the product is tartaric acid (24). Chlorals react with malic acid in the presence of sulfuric acid or other acidic catalysts to produce 4-ketodioxolones (25,26).

In aqueous solution, malic acid can be mildly corrosive toward aluminum and corrosive to carbon steel. Under normal conditions, it is not corrosive to stainless steels, which usually are the construction materials for processes involving malic acid. Malic acid is also virtually noncorrosive to tinplate and other materials used to package acidulated foods and beverages (Table 3) (27).

At proper pH in aqueous media, malic acid forms complexes or chelates with metal ions (see CHELATING AGENTS). These chelating reactions are useful in industrial processes requiring elimination or control of metal-ion catalysis (eg, of oxidation), removal of corrosion products (eg, rust), lowering of metal oxidation potentials (electroplating), etc. The chelating properties of malic acid vary with different metal cations, ionic strength, pH, etc, and in many cases approximate those of other hydroxy carboxylic acids (28). Formation constants for malic acid

Table 3. Corrosion by 25 wt % Solution of Malic Acid

Metal	Corrosion rate, mm/yr[a]	Temperature, °C
aluminum	<0.5	below 100
carbon steel	>1.3	[b]
stainless steel (304)	<0.05	below 50
stainless steel (316)	<0.05	below 120

[a]To convert mm/yr to mils penetration/yr (Mpy), divide by 2.54×10^{-2}.
[b]At all temperatures.

chelates with various metal ions are as follows: Ca, 1.8; Cu, 3.4; Mg, 2.2; and Zn, 2.8. Malic acid forms a weak buffer from approximately pH 3.0 to 6.0 (Fig. 3).

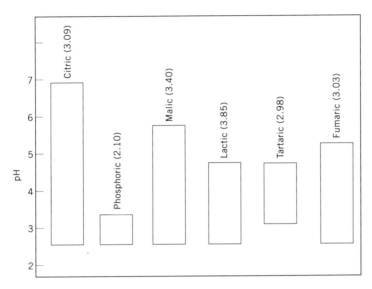

Fig. 3. Effective buffer ranges for food acidulants; pK_a values are given in parentheses.

Occurrence. $S(-)$-Malic acid occurs widely in biological systems. It is the predominant acid in many fruits (Table 4). However, malic acid occurs in relatively low concentrations, thus making its isolation from natural sources expensive and impractical.

In addition to its presence in fruits, $S(-)$-malic acid has been found in cultures of a variety of microorganisms including the aspergilli, yeasts, species of *Sclerotinia*, and *Penicillium brevicompactum*. Yields of levorotatory malic acid as high as 74% of theoretical have been reported. Iron, manganese, chromium,

Table 4. Malic Acid in Fruits[a]

Fruit	Total acid, %	Fruit	Total acid, %
apple	97.2	orange pulp	trace
apricot	23.7–69.8	peach	50.0–96.2
banana	53.7–92.3	pear	33.0–86.6
blueberry	6.0	persimmon	100.0
cherry	94.2	pineapple	12.5
cranberry	19.1–23.5	plum	98.5
gooseberry	46.2	quince	100.0
grape (Concord)	60.0	rhubarb	77.0
grapefruit	5.6	strawberry	9.9–11.0
lemon	4.5	watermelon	100.0
orange peel	59.6–80.0		

[a]Refs. 29–31.

or aluminum ions reportedly enhance malic acid production. $S(-)$-Malic acid is involved in two respiratory metabolic cycles: the Krebs tricarboxylic acid cycle (see CITRIC ACID) and the glyoxylic acid cycle. These metabolic cycles account for the terminal oxidation system which supplies energy and provides the carbon skeletons from which many of the amino acids of proteins are derived.

Manufacture. In the United States, Canada, and Europe, only the synthetic R,S-malic acid is produced commercially, whereas both the S and R,S forms are produced in Japan.

Biosynthesis of S(-)-Malic Acid. Aqueous fumaric acid is converted to levorotatory malic acid by the intracellular enzyme, fumarase, which is produced by various microorganisms. A Japanese process for continuous commercial production of $S(-)$-malic acid from fumaric acid is based on the use of immobilized *Brevibacterium flavum* cells in carrageenan (32). The yield of pyrogen-free $S(-)$-malic acid that is suitable for pharmaceutical use is ca 70% of the theoretical.

Commercial Synthesis of R,S-Malic Acid. The commercial synthesis of R,S-malic acid involves hydration of maleic acid [110-16-7] or fumaric acid [110-17-8] at elevated temperature and pressure. A Japanese patent (33) describing a manufacturing procedure for malic acid claims the direct hydration of maleic acid at 180°C and 1.03–1.21 MPa (150–175 psi).

The conventional commercial processes are commonly carried out on maleic anhydride in aqueous solution at elevated temperatures above 150°C, at pressures above 1.4 MPa (200 psi), and for a residence time of 3–5 h (34). The resulting mixture contains primarily malic acid and fumaric acid in equilibrium with a small percentage of maleic acid. The purification of the malic acid that is obtained can be accomplished by a two-stage crystallization process (35) involving the following steps: (*1*) adjusting the aqueous solution to a malic acid concentration of 40% by weight at 40°C; (*2*) cooling the solution to ca 15°C until equilibrium is reached; (*3*) separating solid fumaric acid from the slurry, preferably by filtration; (*4*) concentrating the mother liquor from the previous step to a malic acid concentration of at least 62% by weight at a temperature of at least 40°C to effect the crystallization of malic acid; (*5*) separating solid malic acid from the resulting slurry at about 40°C; (*6*) washing the malic acid with an aqueous solution that is substantially free from maleic and fumaric acids; (*7*) redissolving the washed malic acid crystals in water; (*8*) removing insolubles, principally fumaric acid, by filtration; (*9*) passing the mother liquor through a carbon column to remove colored contaminants; (*10*) adjusting the resulting solution to a malic acid concentration of 62% by weight at about 40°C and maintaining this temperature to effect the crystallization of malic acid; (*11*) separating solid malic acid at ca 40°C; and (*12*) washing the solid malic acid with an aqueous solution that is substantially free of maleic and fumaric acids. The purified malic acid contains less than 0.05% maleic acid and less than 1% fumaric acid. Additional purification can be achieved by use of cation- and anion-exchange columns after the carbon treatment and before the second-stage crystallization, ie, between steps (*9*) and (*10*) (36) (see ION EXCHANGE). The cation exchange removes heavy-metal ions that were introduced into the system and the anion exchange removes unsaturated organic acids, primarily maleic and fumaric acids. The purified malic acid crystals that are obtained after the final washing step are dried and classified before packaging.

The initial step of production is carried out in a titanium reactor (34) because of the high corrosivity of maleic acid to most metals under the drastic reaction conditions used. The other steps are performed in stainless steel equipment. Improved purification processes for malic acid have been patented (37,38).

Energy and Environmental Considerations. The energy requirements to produce malic acid via conventional processes are fairly moderate. Hydration of malic acid is exothermic. This heat is recovered and helps preheat the reaction. Steam is also used for concentration and purification of the malic acid stream. Steam usage is in the range of 2.5–5.0 kg steam per 1 kg malic acid produced. Electricity is used in the range of 0.5 kW·h per kg malic acid. Malic acid production generates low levels of solid, airborne, and liquid waste. Solid waste is primarily nontoxic malic acid salts resulting from regenerating carbon cells and ion-exchange resins. Airborne emissions are primarily particulates. A 1% malic acid solution is readily biodegradable, with a BOD (5) of 5300 mg/L.

Shipping and Storage. Malic acid is shipped in 50-lb, 100-lb, and 25-kg, multiwall paper bags or 100-lb (45.5 kg) fiber drums. A technical-grade, 50% solution may be shipped in tank cars or tank trucks. Malic acid can be stored in dry form without difficulty, although conditions of high humidity and elevated temperatures should be avoided to prevent caking.

Economic Aspects. Malic acid is manufactured in over 10 countries, with 1992 worldwide production estimated at approximately 33,000 t, distributed as follows: 44.4%, North America; 52.1%, Far East; and 3.5%, Africa.

The production is primarily used for food (26.6%) and beverages (54.7%); however, some industrial applications (18.7%) exist, eg, coatings, polymers, and resins. (Historical patterns of use in the United States have been stable and are as noted in parentheses.) Over the past few years, the list price of malic acid has been stable. In the United States, the current list price for malic acid is $1.79/kg, delivered and packaged in 50-lb (22.7-kg) bags (39).

Specifications and Analysis. R,S-Malic acid that is sold in the United States meets the specifications of the *Food Chemicals Codex* and *National Formulary*, which are listed in Table 5 (40,41). Malic acid is available in the following U.S. standard sieve sizes:

Granular	min	100 wt % through 2.00 mm (10 mesh) sieve
	max	10 wt % through 0.30 mm (50 mesh) sieve
Fine granular	min	99 wt % through 0.71 mm (25 mesh) sieve
	max	5 wt % through 0.15 mm (100 mesh) sieve
Powder	min	2 wt % through 0.18 mm (80 mesh) sieve
	max	5 wt % through 0.15 mm (100 mesh) sieve

Aqueous titration with 1N sodium hydroxide is the usual malic acid assay. Maleic and fumaric acid are determined by a polarographic method. Analytical methods have been described (40).

Health and Safety. The U.S. FDA has affirmed R,S- and S(−)-malic acid as substances that are generally recognized as safe (GRAS) as flavor enhancers, flavoring agents and adjuvants, and as pH control agents at levels ranging from 6.9% for hard candy to 0.7% for miscellaneous food uses (42). R,S- and

Table 5. Malic Acid Specifications

Parameter or substance	Food Chemicals Codex[a]	National Formulary[b]
assay	not less than 99.0% as $C_4H_6O_5$	99.0%–100.5%
arsenic	≤3 ppm	
fumaric acid	≤1.0%	≤1.0%
heavy metals (as Pb)	≤0.002%	≤0.002%
lead	≤10 ppm	
maleic acid	≤0.05%	≤0.05%
residue on ignition	≤0.1%	≤0.1%
water-insoluble matter	≤0.1%	≤0.1%
specific rotation		−0.10–0.10°
organic volatile impurities		meets requirements of NF method I

[a]Ref. 40.
[b]Ref. 41.

$S(-)$-malic acid may not be used in baby foods. Malic acid is also cleared to correct natural acid deficiencies in juice or wine (43).

Uses. R,S-Malic acid is utilized in a variety of food and beverage and some industrial applications because of its unique combination of properties. These include having unusual taste-blending characteristics, flavor-fixing qualities, the ability to retain sour taste longer, high water solubility, and chelating and buffering properties. Malic acid is also a reactive intermediate in chemical synthesis.

Beverages.   Malic acid is increasingly being employed in both liquid and powder drinks as a flavor enhancer, pH buffer, and to increase the effectiveness of antimicrobial preservatives. Having a greater acid taste than citric acid, malic acid is often substituted for citric acid and contributes to the beverage formulation by intensifying and imparting improved taste to fruit flavors. In low calorie drink formulations, malic acid suppresses the bitter aftertaste that saccharin can often impart. This is attributed to the flavor profile of malic acid, which provides a stronger, yet slower and longer flavor release in comparison to citric acid.

Candy.   Its low melting point and sugar inversion properties make malic acid a desirable acidulant, especially in hard candy products (44,45). Due to their insolubility, hard water salts can cause clouding of the finished product. However, because of the higher solubility of calcium malate [17482-42-7] relative to alternative acidulants, clarity of the finished product is enhanced. Additionally, in sugar confectionery products where acidulation may exceed 2.0%, malic acid can provide economic benefits.

Other Food Uses.   Jellies, jams, and preserves use malic acid to balance flavor and adjust pH for pectin set. Canned fruits and vegetables employ malic acid in combination with ascorbic acid to produce a synergistic effect that aids in the reduction of browning. Wine and cider producers use malic acid in malolactic fermentation to provide bouquet and for pH adjustment.

Miscellaneous.   Malic acid is used in pharmaceuticals (qv), cosmetics (qv), dentifrices (qv), metal cleaning, electroless plating (46), wash-and-wear textile finishing (47–49), for stabilization of heat-sensitive copying paper (50), as an inhibitor of gelation, livering, and agglomeration in cellulose nitrate liqueurs, and in many other applications.

Thiomalic Acid

Thiomalic acid [70-49-5] (mercaptosuccinic acid), $C_4H_6O_4S$, mol wt = 150.2, is a sulfur analogue of malic acid. The properties of the crystalline, solid thiomalic acids are given in Table 6. The racemic acid has the following acid dissociation constants at 25°C: pK_{a1}=3.30; pK_{a2}=4.94.

R,S-Thiomalic acid [644-87-1] can be prepared from bromosuccinic acid by reaction with K_2S (51,52). The two enantiomers can be obtained from the corresponding optically active potassium bromosuccinates (52,53). Salts or amides of maleic acid react with NaSH to give the thiomalic derivatives (54). In the presence of a ferric salt in aqueous ammoniacal solution, thiomalic acid is oxidized to the disulfide (55). Oxidation with diluted HNO_3 gives sulfosuccinic acid (56). The thiomalic acids form salts, chelates, and esters. R,S-Thiomalic acid gives a red color with $FeCl_3$ + NH_4OH (57). Reaction with aqueous AuCN gives aurothiomalic acid (58).

Thiomalic acid is a skin sensitizer (59) and an antidote in heavy-metal poisoning (60). Traditionally, it was a component of cold permanent hair-waving solutions (see HAIR PREPARATIONS) (61–62) and of rust-removing (63) and corrosion-inhibiting compositions (see CORROSION AND CORROSION CONTROL) (64). Sodium aurothiomalate [12244-57-4] (Myochrisin) and other gold thiomalate complexes have antiarthritic properties (65) (see GOLD AND GOLD COMPOUNDS). The well-known insecticide, malathion [121-75-5], is the thiomalate S-ethyl ester of O,O-dimethylphosphonodithioic acid (see INSECT CONTROL TECHNOLOGY).

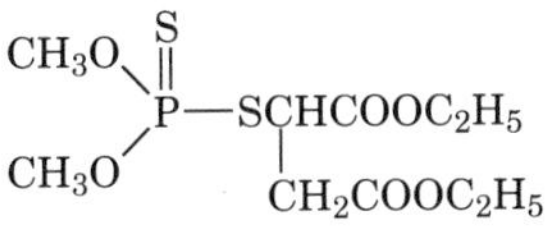

$$\begin{array}{c} CH_3O \\ \quad\quad\quad\;\;P-SCHCOOC_2H_5 \\ CH_3O \quad\quad\quad | \\ \quad\quad\quad\quad CH_2COOC_2H_5 \end{array}$$

Table 6. Properties of Thiomalic Acids

| Acid | CAS Registry Number | Mp, °C | Solubility | | $[\alpha]_D^{17a}$ |
			Water	Ethanol	
R,S-	[644-87-1]	151	very sol	very sol	
R	[20182-99-4]	154	sol	sol	+64.4°
S	[74708-34-2]	152–153	sol	slightly sol	−64.8°

^{a}5% acid in ethanol.

Tartaric Acid

Tartaric acid [526-83-0] (2,3-dihydroxybutanedioic acid, 2,3-dihydroxysuccinic acid), $C_4H_6O_6$, is a dihydroxy dicarboxylic acid with two chiral centers. It exists as the dextro- and levorotatory acid: the meso form (which is inactive owing to internal compensation), and the racemic mixture (which is commonly known as racemic acid). The commercial product in the United States is the natural, dextrorotatory form, $(R\text{-}R^*,R^*)$-tartaric acid (L(+)-tartaric acid) [87-69-4]. This enantiomer occurs in grapes as its acid potassium salt (cream of tartar). In the fermentation of wine (qv), this salt forms deposits in the vats; free crystallized tartaric acid was first obtained from such fermentation residues by Scheele in 1769.

In Europe, South Africa, and Japan, racemic (R^*,R^*)-tartaric acid [133-37-9] (DL-tartaric acid) is also produced commercially via maleic anhydride oxidation.

Physical Properties.   When crystallized from aqueous solutions above 5°C, natural $(R\text{-}R^*,R^*)$-tartaric acid is obtained in the anhydrous form. Below 5°C, tartaric acid forms a monohydrate which is unstable at room temperature. The optical rotation of an aqueous solution varies with concentration. It is stable in air and racemizes with great ease on heating. Some of the physical properties of $(R\text{-}R^*,R^*)$-tartaric acid are listed in Table 7.

The solubility of $(R\text{-}R^*,R^*)$-tartaric acid in water varies from 115 g/100 g H_2O at 0°C to 343 g/100 g H_2O at 100°C. One hundred grams of absolute ethanol dissolves 20.4 g of tartaric acid at 18°C, and 100 g of ethyl ether dissolves 0.3 g at 18°C. Densities (d_4^{15}) of $(R\text{-}R^*,R^*)$-tartaric acid solutions at 15°C range from 1.0045 for a 1 wt % solution to 1.296 for a 50 wt % solution (67).

Some physical properties of the four enantiomeric tartaric acids are compared in Table 8.

Chemical Properties.   The notation used by *Chemical Abstracts* to reflect the configuration of tartaric acid is as follows: $(R\text{-}R^*,R^*)$-tartaric acid [87-69-4] (**4**); $(S\text{-}R^*,R^*)$-tartaric acid [147-71-7] (**5**); and *meso*-tartaric acid [147-73-9] (**6**). Racemic acid is an equimolar mixture of the two optically active enantiomers and, hence, like the meso acid, is optically inactive.

Table 7. Physical Properties of (R-R*,R*)-Tartaric Acid

Property	Value
mol wt	150.086
appearance	colorless or translucent crystals
crystal system	monoclinic
mp, °C (anhydrous)	169–170
d^{20}, g/cm^3	1.76
$[\alpha]_D^{20}$ (for concentration, c, from 20–50%)	$15.050 - 0.1535c$
K_{a1} at 25°C	1.04×10^{-3}
K_{a2} at 25°C	4.55×10^{-5}
heat of combustion, kJ/mol[a]	1080[b]
heat of solution, kJ/mol[a]	-13.8[c]

[a]To convert kJ to kcal, divide by 4.184.
[b]Ref. 66.
[c]Ref. 3.

Table 8. Physical Properties of Tartaric Acids

Properties	Natural	Levorotatory	Racemic	Meso
mp, °C (anhydrous)	169–170	169–170	205–206	159–160
soly, g/100 g H_2O (in water, at 20°C)	139	139	20.6	125
soly of acid potassium salt, g/100 g H_2O (at 25°C)	0.84	0.84	0.72	16.7
soly of Ca salt, g/100 g	0.020^a	0.025^b	0.004^a	0.034^b
mol water in hydrate of calcium salt	4	4	8	3

^{a}At 25°C.
^{b}At 20°C.

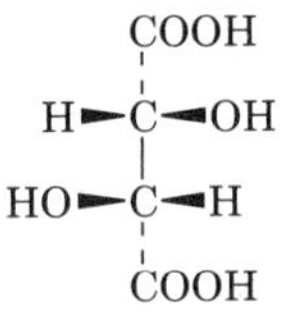

COOH   COOH   COOH

H►C◄OH   HO►C◄H   H►C◄OH

HO►C◄H   H►C◄OH   H►C◄OH

COOH   COOH   COOH

(4) 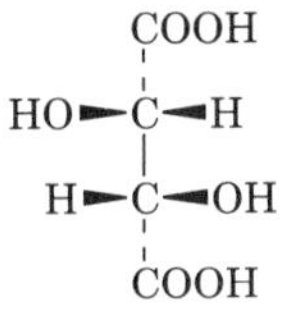   **(5)** 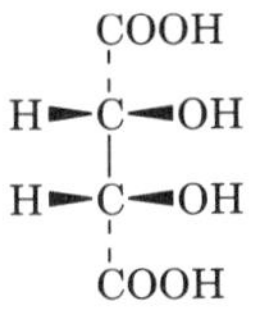   **(6)**

Reactions. When free $(R\text{-}R^*,R^*)$-tartaric acid (**4**) is heated above its melting point, amorphous anhydrides are formed which, on boiling with water, regenerate the acid. Further heating causes simultaneous formation of pyruvic acid, $CH_3COCOOH$; pyrotartaric acid, $HOOCCH_2CH(CH_3)COOH$; and, finally, a black, charred residue. In the presence of a ferrous salt and hydrogen peroxide, dihydroxymaleic acid [526-84-1] (**7**) is formed. Nitrating the acid yields a dinitro ester which, on hydrolysis, is converted to dihydroxytartaric acid [617-48-1] (**8**), which upon further oxidation yields tartronic acid [80-69-3] (**9**).

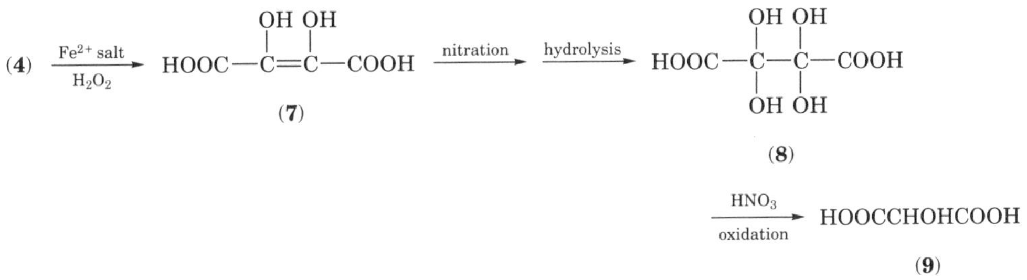

$$(4) \xrightarrow[H_2O_2]{Fe^{2+}\ salt} HOOC-\underset{OH}{C}=\underset{OH}{C}-COOH \xrightarrow{nitration} \xrightarrow{hydrolysis} HOOC-\underset{\underset{OH}{|}}{\overset{\overset{OH}{|}}{C}}-\underset{\underset{OH}{|}}{\overset{\overset{OH}{|}}{C}}-COOH$$

(7)   **(8)**

$$\xrightarrow[oxidation]{HNO_3} HOOCCHOHCOOH$$

(9)

Tartaric acid is reduced stepwise with concentrated hydriodic acid, first to $R(+)$-malic acid (**2**) and then to succinic acid [110-15-6].

$$(4) \xrightarrow{conc\ HI} (2) \xrightarrow{conc\ HI} HOOCCH_2CH_2COOH$$

In common with other hydroxy organic acids, tartaric acid complexes many metal ions. Formation constants for tartaric acid chelates with various metal ions are as follows: Ca, 2.9; Cu, 3.2; Mg, 1.4; and Zn, 2.7 (68). In aqueous solution, tartaric acid can be mildly corrosive toward carbon steels, but under normal conditions it is noncorrosive to stainless steels (Table 9) (27).

Occurrence. $(R-R^*,R^*)$-Tartaric acid occurs in the juice of the grape and in a few other fruits and plants. It is not as widely distributed as citric acid or $S(-)$-malic acid. The only commercial source is the residues from the wine industry. $(S-R^*,R^*)$-Tartaric acid has been found in the fruit and leaves of *Bauhinia reticulata*, a tree native to Mali (western Africa). Like the dextrorotatory acid, it forms anhydrous monoclinic crystals.

The racemic acid is not a primary product of plant processes but is formed readily from the dextrorotatory acid by heating alone or with strong alkali or strong acid. The methods by which such racemic compounds can be separated into the optically active modifications were devised by Pasteur and were applied first to the racemic acid. Racemic acid crystallizes as the dihydrate $(C_4H_6O_6)_2 \cdot 2H_2O$ in triclinic prisms. It becomes anhydrous on drying at 110°C and melts incongruently at 205°C. Calcium racemate, $(C_4H_4O_6Ca)_2$, is even less soluble in water than calcium tartrate [15808-04-5], thus a dilute racemic acid solution is precipitated by a saturated calcium sulfate solution, whereas active tartaric acid is not.

meso-Tartaric acid is not found in nature. It is obtained from the other isomers by prolonged boiling with caustic alkali. The free acid crystallizes as a monohydrate, $C_4H_6O_6 \cdot H_2O$, in monoclinic prisms. On drying at 110°C, it becomes anhydrous and melts at 159–160°C.

Synthesis. Racemic acid is obtained synthetically by treatment of maleic acid with hydrogen peroxide in the presence of a catalyst, eg, tungstic acid (69). Other synthetic routes that have been explored include the production of $(R-R^*,R^*)$-tartaric acid by bacterial fermentation of glucose (70) or 5-keto-D-gluconic acid (71), catalytic oxidation of 5-keto-D-gluconic acid with gaseous oxygen (72,73), and nitric acid oxidation of carbohydrates (qv), eg, glucose (74). Production of (R^*,R^*)-tartaric acid by catalytic chlorate oxidation of fumaric or maleic acid also has been described (75).

meso-Tartaric acid can be prepared by microbiological conversion of *trans*-epoxysuccinic acid [22734-83-4] (76). *cis*-Epoxysuccinic acid [16533-72-5] does not undergo this conversion. None of the foregoing processes have achieved commercial significance.

Manufacture. *(R-R*,R*)-Tartaric Acid and Its Salts.* The raw materials available for the manufacture of tartaric acid and tartrates are by-products of

Table 9. Corrosion by 25 wt % Solution of Tartaric Acid

Metal	Corrosion rate, mm/yr[a]	Temperature, °C
aluminum	<0.5	below 30
	>1.3	above 30
carbon steel	>1.3	[b]
stainless steel (304)	<0.05	below 200
stainless steel (316)	<0.05	below 200

[a]To convert mm/yr to mils penetration/yr (Mpy), divide by 2.54×10^{-2}.
[b]At all temperatures.

wine making. Crude tartars are recovered from the following sources: (*1*) The press cakes from grape juice, ie, unfermented (marcs) or partly fermented (pomace), are boiled with water, and alcohol, if present, is distilled. The hot mash is settled, decanted, and the clear liquor is cooled to crystallize. The recovered high test crude cream of tartar (vinaccia) has an 85–90% cream of tartar content. (*2*) Lees, which are the dried slimy sediments in the wine fermentation vats, consist of yeast cells, pectinous substances, and tartars. Their content of total tartaric acid equivalent ranges from 16–40%. (*3*) The crystalline crusts that form in the vats in the secondary fermentation period (argols) contain more than 40% tartaric acid; they are high in potassium hydrogen tartrate [*868-14-4*] and low in the calcium salt.

The chemical reactions involved in tartaric acid production are formation of calcium tartrate from crude potassium acid tartrate,

$$2 \ KHC_4H_4O_6 + Ca(OH)_2 + CaSO_4 \longrightarrow 2 \ CaC_4H_4O_6 + K_2SO_4 + 2 \ H_2O$$

formation of tartaric acid from calcium tartrate,

$$CaC_4H_4O_6 + H_2SO_4 \longrightarrow H_2C_4H_4O_6 + CaSO_4$$

formation of Rochelle salt [*304-59-6*] from argols,

$$2 \ KHC_4H_4O_6 + Na_2CO_3 \longrightarrow 2 \ KNaC_4H_4O_6 + CO_2 + H_2O$$

and formation of cream of tartar from tartaric acid and Rochelle salt (RS) liquors,

$$2 \ H_6C_4O_6 + 2 \ KNaC_4H_4O_6 + K_2SO_4 \longrightarrow 4 \ KHC_4H_4O_6 + Na_2SO_4$$

This process is summarized in Figure 4.

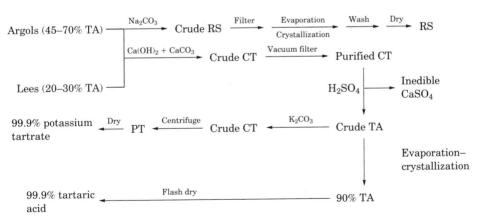

Fig. 4. Manufacturing process for (R-R^*, R^*)-tartaric acid (TA) and its salts, calcium tartrate (CT) and potassium tartrate (PT).

Production of the Rochelle salt involves roasting of the blended argols at 160–165°C, mixing with wash liquor from a previous charge, and treatment with sodium carbonate and afterward with potassium oxalate. Evaporation of the resulting filtrate, followed by centrifugation, washing, and drying, yields the finished product.

Cream of tartar [868-14-4] can be produced (1) directly from the argol–sodium carbonate cook; (2) by combining tartaric acid solutions (which are available from the manufacturer of tartaric acid) with Rochelle salt solution derived from suitable crude potassium bitartrate, eg, high grade argols or recovered cream of tartar; and (3) by saturating a water suspension of argols with sulfur dioxide. An accelerated process for removing potassium bitartrate from crude tartaric acid solutions has been described (77).

For the production of tartar emetic (antimony potassium tartrate [28300-74-5]), potassium bitartrate [868-14-4] and antimony oxide, Sb_2O_3, are added simultaneously to water in a stainless-steel reactor. The reaction mixture is diluted, filtered, and collected in jacketed granulators where crystallization takes place after cooling. Centrifuging, washing, and drying complete the process.

(R^*,R^*)-Tartaric (racemic) acid is obtained synthetically by epoxidation of maleic acid with hydrogen peroxide in the presence of a catalyst followed by hydrolysis of the resulting cis-epoxysuccinic acid (69,78–84). This commercial process is used in South Africa (83), and it involves the addition of 60% H_2O_2 to 40% aqueous maleic acid. The addition of a molybdenum catalyst permits the reaction to take place at 70°C in 20 h. The resulting cis-epoxysuccinic acid is hydrolyzed by boiling, and (R^*,R^*)-tartaric acid is isolated by cooling, centrifuging, washing, and drying. A 20% excess of maleic acid is used and the yield (based on H_2O_2) is 84%. Unreacted maleic acid is converted either to fumaric acid or malic acid. The tartaric acid in the mother liquor is recovered.

(R^*,R^*)-Tartaric acid also can be produced by racemization of $(R-R^*,R^*)$-tartaric acid in the presence of meso-tartaric acid (85). In this process, formation of meso-tartaric acid during racemization does not occur.

Economic Aspects.   The estimated total worldwide market for tartaric acid is 58,000 t and potassium bitartrate (acid basis) is 20,000 t. The majority of tartaric acid consumption, represented by beverage, food, and pharmaceutical applications, is shown in Table 10. Potassium bitartrate (cream of tartar) is primarily used in baking powders and mixes.

Table 10. 1991 Worldwide Tartaric Acid Market Segments[a]

Market	Percent
alcoholic beverages	30
emulsifiers	20
pharmaceuticals	15
foods	10
textiles	10
electrochemicals	10
others	5

[a]Does not include potassium bitartrate.

The 1991 U.S. market for tartaric acid and tartrate was estimated at 3600 t. There is no domestic U.S. producer; therefore, all product is imported from other countries. Imports into the United States for 1991 are summarized in Table 11.

In the United States, prices for tartaric acid have ranged from $1.10 to $6.61/kg since the early 1970s with little consistency. The 1993 list price of tartaric acid in the United States was $6.50/kg fob plant packaged in 100-lb (45.5-kg) bags (39). Much of the fluctuation is a result of the availability of raw materials for production.

Specifications and Analysis. $(R\text{-}R^*,R^*)$-Tartaric acid sold in the United States meets the specifications of the *Food Chemicals Codex* (40) and the *National Formulary* (41) (Table 12).

Tartaric acid is supplied as Fine-Granular and Powder in 45-kg bags. It should be stored in tightly closed containers. Test methods for tartaric acid and some tartrates have been described (40,87).

Health and Safety. The FDA affirmed $(R\text{-}R^*,R^*)$-tartaric acid as a generally-recognized-as-safe (GRAS) food substance (88).

Tartaric acid and tartrates are poorly absorbed from the intestine. Their metabolism is different from that of citric acid in that tartaric acid is only slightly oxidized. The acid that is absorbed is excreted unchanged in the urine. So far as

Table 11. 1991 Tartaric/Tartrate U.S. Imports From Other Countries[a]

Country	Tartaric acid imports, %	Potassium bitartrate imports, %
Italy	45.7	36.9
Spain	31.3	51.5
France	3.7	7.0
Argentina	5.9	
the Netherlands	4.3	
Chile	3.5	
other	5.6	4.6

[a]Ref. 86.

Table 12. Specifications for $(R\text{-}R^*,R^*)$-Tartaric Acid

Parameter	Food Chemicals Codex[a]	National Formulary[b]
specific rotation	$[\alpha]_D^{25} + 12.0° - 13.0°$[c]	between 12.0 and 13.0°
assay	not less than 99.7% of $C_4H_6O_6$ after drying	99.7% - 100.5%
arsenic (as As)	3 ppm max	
heavy metals (as Pb)	10 ppm max	≤0.001%
loss on drying	0.5% max	≤0.5%
oxalate	passes test	passes test
residue on ignition	0.05% max	≤0.1%
sulfate	passes test	passes test
organic volatile impurities		passes test

[a]Ref. 40.
[b]Ref. 41
[c]Determined in a solution containing 2 g in each 10 mL sample.

is known, all nutritional and physiological investigations have been made with the dextrorotatory enantiomer.

Uses. *Carbonated Beverages.* Tartaric acid has been used like citric acid as an acidulant in carbonated beverages (qv). However, it has almost been completely replaced in the marketplace by less expensive acidulants like phosphoric, citric, malic, and fumaric acids.

Wine Making. Wine making is one of the principal areas of tartaric acid use. There is a relationship between the size of the grape crop and its tartaric acid content when grapes are pressed. In poor harvest years, the tartaric acid content is low; in good harvest years, the tartaric acid content is high. Thus, in poor harvest years, tartaric acid often is added to correct acid deficiencies in wine.

Other Food. Tartaric acid is also used in the manufacture of gelatin (qv) desserts and in fruit jellies, especially in pectin jellies for candies where a low pH is necessary for proper setting. It is used as a starch modifier in starch jelly candies so that the product flows freely while being cast. It is used in hard candy because its melting point permits it to fuse into the "glass" and does not contribute to moisture.

Emulsifiers. A growing application for tartaric acid is in the production of DATEM esters (diacetyl tartaric esters of monoglycerides). These esters are used as dough conditioners in the baking industry. They allow the reduction or elimination of shortening, as well as adding dough strength to free-standing breads.

Pharmaceuticals. Tartaric acid is used in the manufacture of fine drug salts, as in effervescent salts. Tartar emetic [28300-74-5] is used in small doses as an expectorant in cough syrups. It can be used to treat infections caused by *Schistosoma japonium*.

Industrial Uses. Tartaric acid is used in photography, and its iron salts are used in blue copy paper. The diethyl and dibutyl esters are used in paints as lacquer solvents. In the textile industry, tartaric acid acts as a stabilizer in nylon dyeing and in cellusosic fiber bleaching with peroxide. It is used as a chelating agent for boron and other micronutrients in fertilizers. In the plastics industry, tartaric acid is used as a polymerization agent of methyl methacrylate, phenol–formaldehyde resins, PVC, and acrylonitrile. In metals, it is used as a complexing agent in metal cleaning for copper and alloys, aluminum, and ferrous metals. It is used in ceramics as a component in special clays. In the electronics industry, tartaric acid is used in the anodization of semiconductors of gallium arsenide.

Salts. Rochelle salt is used in the silvering of mirrors. Its properties of piezoelectricity make it valuable in electric oscillators. Medicinally, it is an ingredient of mild saline cathartic preparations, eg, compound effervescing powder. In food, it can be used as an emulsifying agent in the manufacture of process cheese.

Cream of tartar is used in baking powder and in prepared baking mixes (see BAKERY PROCESSES AND LEAVENING AGENTS). Its limited solubility at low temperatures inhibits the reaction with bicarbonate until baking temperature is reached, thus releasing a significant portion of the carbon dioxide at the optimum time.

BIBLIOGRAPHY

"Tartaric Acid" in *ECT* 1st ed., Vol. 13, pp. 645–656, by R. Pasternack, Chas. Pfizer & Co., Inc.; in *ECT* 2nd ed., Vol. 19, pp. 723–732, by D. F. Chichester, Chas. Pfizer & Co., Inc.; "Malic Acid" in *ECT* 2nd ed., Vol. 12, pp. 837–839, by W. E. Irwin, L. B. Lockwood, and M. F. Zienty, Miles Laboratories, Inc., Chemicals Division; "Hydroxy Dicarboxylic Acids" in *ECT* 3rd. ed., Vol. 13, pp. 103–121, by S. E. Berger, Allied Chemical Corp.

1. T. S. Patterson and C. Buchanan, *J. Chem. Soc.*, 3006 (1928).
2. A. E. Dunstan and F. B. Thole, *J. Chem. Soc.* **93**, 1815 (1908).
3. W. H. Gardner, *Food Acidulants*, Allied Chemical Corp., New York, 1966, p. 7.
4. Technical data, Haarmann & Reimer Corp., Elkhart, Ind., 1991.
5. E. Fischer, *Ber.* **29**(2), 1377 (1896).
6. M. G. J. W. Bremer, *Bull. Soc. Chim. Fr. Ser. 2* **25**, 6 (1876).
7. J. F. Stoddart in J. F. Stoddart, ed., *Comprehensive Organic Chemistry*, Vol. 1, Pergamon Press, Oxford, U.K., 1979, pp. 3–33.
8. W. D. Bancroft and H. L. Davis, *J. Phys. Chem.* **34**, 897 (1931).
9. J. A. Niewland, R. R. Vogt, and W. L. Foohey, *J. Am. Chem. Soc.* **52**, 1018 (1930).
10. U.S. Pat. 1,489,744 (Apr. 8, 1924), G. R. Downs and L. Weisberg (to Barrett Co.).
11. K. Freudenberg and A. Lux, *Ber.* **61B**, 1083 (1928).
12. T. Purdie and G. B. Neave, *J. Chem. Soc.* **97**, 1517 (1910).
13. T. Purdie and Pitkeathyly, *J. Chem. Soc.* **75**, 157 (1899).
14. L. H. Flett and W. H. Gardner, *Maleic Anhydride Derivatives*, John Wiley & Sons, Inc., New York, 1952, p. 64.
15. T. Curtius and C. von Hofe, *J. Prakt. Chem.* **95**(2), 210 (1917).
16. A. E. Arppe, *Ann. Chim.* **96**, 106 (1855).
17. R. Anschutz and Q. Wirtz, *Ann. Chem.* **239**, 137 (1887).
18. J. B. Tingle and S. J. Bates, *J. Am. Chem. Soc.* **32**, 1233 (1910).
19. H. von Pechmann, *Ann. Chem.* **264**, 272 (1891).
20. V. V. K. Sastry, *J. Indian Chem. Soc.* **19**, 403 (1942).
21. R. C. Shah and co-workers, *J. Indian Chem. Soc.* **14**, 717 (1937).
22. J. Smiles and E. W. McClellan, *J. Chem. Soc.* **119**, 1810 (1921).
23. H. J. Fenton and H. O. Jones, *J. Chem. Soc.* **77**, 77 (1900).
24. Fr. Pat. 849,852 (Dec. 4, 1939), H. Goldstein and A. Bonn.
25. F. H. Yorston, *Rec. Trav. Chim.* **46**, 711 (1927).
26. N. M. Shah, *J. Indian Chem. Soc.* **16**, 285 (1939).
27. P. A. Schweitzer, *Corrosion Resistance Tables*, 3rd ed., Marcel Dekker, Inc., New York, 1991.
28. F. Dwyer, *Chelating Agents and Metal Chelates*, Academic Press, Inc., New York, 1964.
29. T. J. Sausville, *Food Technol.* **19**, 67 (1965).
30. T. J. Sausville, *Glass Packer* **44**, 27 (1965).
31. L. H. Meyer, *Food Chemistry*, Reinhold Publishing Corp., New York, 1960, pp. 275–277.
32. I. Chibata, T. Tosa, and I. Takata, *Trends Biotechnol.* **1**, 9 (1983).
33. Jpn. Pat. 4360 (Dec. 16, 1950), K. Saito, Y. Ono, and Y. Mikawa.
34. U.S. Pat. 3,379,756 (Apr. 23, 1968), C. R. Ahlgren (to Allied Chemical Corp.).
35. U.S. Pat. 3,391,187 (July 2, 1968), M. A. Cullen, Jr. and M. R. Ingleman (to Allied Chemical Corp.).
36. U.S. Pat. 3,371,112 (Feb. 27, 1968), L. O. Winstrom and M. R. Ingleman (to Allied Chemical Corp.).
37. U.S. Pat. 3,983,170 (Sept. 28, 1976), S. Sumikawa and R. Maida (to International Organics, Inc. and Kyowa Hakko Kogyo Co., Ltd.).

38. U.S. Pat. 4,035,419 (July 12, 1977), S. Sumikawa, S. Sakaguchi, and T. Okiura.
39. *Chem. Mktg. Rep.*, 35 (Jan. 25, 1993).
40. *Food Chemicals Codex*, 3rd ed., 3rd Suppl., National Academy of Sciences, National Research Council, Washington, D.C., 1992.
41. *National Formulary XVII*, Suppl. 6, The United States Pharmacopeial Convention, Inc., Rockville, Md., Mar. 15, 1992.
42. *Code of Federal Regulations*, 21 CFR 184.1069, Office of the Federal Register, U.S. Government Printing Office, Washington, D.C., 1993.
43. Ref. 42, 27 CFR 24.246.
44. M. B. Sherman, *Manuf. Confect.* **45**(4), 39 (1965).
45. S. E. Berger and D. R. Carr, *Food Technol.* **20**, 1477–1478 (1966).
46. U.S. Pat. 2,935,425 (May 3, 1960), G. Gutzeit, P. Talmey, and W. G. Lee (to General American Transportation Corp.).
47. Brit. Pat. 1,375,537 (Nov. 27, 1974), H. M. Ferrarini (to Sun Chemical Corp.).
48. U.S. Pat. 3,788,804 (Jan. 29, 1974), R. J. Harper and co-workers (to U.S. Dept. of Agriculture).
49. A. G. Pierce, Jr., E. A. Boudreaux, and J. D. Reid, *Am. Dyestuff Rep.* **59**(5), 50 (1970).
50. U.S. Pat. 3,074,809 (Jan. 22, 1963), R. Owen (to Minnesota Mining and Manufacturing Co.).
51. L. Carius, *Ann. Chem.* **129**, 6 (1864).
52. B. Holmberg, *Ark. Kem. Min. Och. Geol.* **6**(1), 4 (1915); *Chem. Zentr.* **I**, 968 (1916).
53. P. A. Levene and L. A. Mikeska, *J. Biol Chem.* **60**, 687 (1924).
54. Brit. Pat. 670,702 (Apr. 23, 1952), R. L. Evans.
55. E. Biilmann, *Ann. Chem.* **348**, 132 (1906).
56. *Beil.* **3**, 439 (1921).
57. R. Andreasch, *Monatsh.* **49**, 131 (1918).
58. U.S. Pat. 2,370,593 (Feb. 27, 1945), N. R. Trenner and F. A. Bacher (to Merck & Co.).
59. J. G. Voss, *J. Invest. Dermatol.* **31**, 273 (1958).
60. F. Meidinger, *Arch. Intern. Pharmacodyn.* **76**, 351 (1948).
61. Ger. Pat. 1,035,856 (Aug. 7, 1958), J. H. Brant (to Gillette Co.).
62. A. Shansky, *Soap Cosmet. Chem. Spec.* **52**(9), 32 (1976).
63. U.S. Pat. 3,277,012 (Oct. 4, 1966), E. W. Krockow (to Dr. Spiess GmbH).
64. N. K. Patel and J. Franco, *Indian J. Technol.* **13**, 239 (1975).
65. G. Jasmin, *J. Pharmacol. Exptl. Therap.* **120**, 349 (1957).
66. M. S. Kharasch, *Bur. Std. J. Res.* **2**, 359 (1929).
67. H. W. Ockerman, *Source Book for Food Scientists*, The Avi Publishing Co., Inc., Westport, Conn., 1978, p. 276.
68. J. A. Dean, *Langes Handbook of Chemistry*, 12th ed., McGraw-Hill Book Co., Inc., New York, 1979.
69. J. M. Church and R. Blumberg, *Ind. Eng. Chem.* **43**, 1780 (1951).
70. U.S. Pat. 2,314,831 (Mar. 24, 1943), J. Kamlet (to Miles Laboratories).
71. U.S. Pat. 2,559,650 (July 10, 1951), L. B. Lockwood and G. E. N. Nelson (to the United States of America, as represented by the Secretary of Agriculture).
72. U.S. Pat. 2,197,021 (Apr. 16, 1940), R. Pasternack and E. V. Brown (to Chas. Pfizer & Co., Inc.).
73. U.S. Pat. 2,417,230 (Mar. 11, 1947), W. E. Barch (to Standard Brands).
74. U.S. Pat. 2,419,019 and 2,419,020 (Apr. 15, 1947), R. A. Hales (to Atlas Powder Co.).
75. U.S. Pat. 2,000,213 (May 7, 1935), G. Braun (to Standard Brands).
76. U.S. Pat. 2,947,665 (Aug. 2, 1960), J. W. Foster.
77. N. Ya. Novotel'nova, R. A. Yurchenko, and L. F. Petrova, *Khlebopek. Konditer. Promst.* (**1**), 30 (1978).
78. Brit. Pat. 1,442,748 (July 14, 1976), D. F. Lewis and A. B. Rodgriguez (to Imperial Chemical Industries Ltd.).

79. Jpn. Kokai 77 85,119 (July 15, 1977), N. Kazutani and co-workers (to Nippon Peroxide Co., Ltd.).

80. Jpn. Kokai 76,113,822 (Oct. 7, 1976), M. Kataoka, K. Hosoi, and M. Ono (to Toray Industries, Inc.).

81. Ger. Offen. 2,555,699 (July 1, 1976), K. Petritsch, P. Korl, and F. Pogoriach (to Oesterreichische Chemische Werke GmbH).

82. Ger. Offen. 2,543,333 (May 26, 1977), G. Prescher and G. Schreyer (to Deutsche Gold- und Silber-Scheideanstalt vorm. Roessler).

83. J. A. Bewsey, *Chem. Ind. (London)* **3**, 119 (1977).

84. Fr. Demande 2,285,370 (Apr. 16, 1976), (to Établissements Louis François).

85. Fr. Demande 2,303,778 (Sept. 8, 1976), M. Saotome and co-workers (to Nippon Per- oxide Co., Ltd.; Showa Chemical Co., Ltd.).

86. *Piers Import Data Base*, Journal of Commerce, New York, 1992.

87. W. Horwitz, ed., *Official Methods of Analysis of the Association of Official Analytical Chemists*, Washington, D.C., 1990.

88. Ref. 42, 21 CFR 184.1099.

GARY T. BLAIR
JEFFREY J. DEFRATIES
Haarmann & Reimer Corporation

HYGROMETRY. See DRYING.

HYPNOTICS, SEDATIVES, ANTICONVULSANTS, AND ANXIOLYTICS

The clinical effects observed with the variety of therapeutic agents that are used as hypnotics, sedatives, anxiolytics, and anticonvulsants reflect a spectrum of related activities at the cellular level characterized by a series of generalized responses within the central nervous system (CNS) that result in alterations in the dynamic state of brain function. Considerable advances have been made in the elucidation of the molecular events surrounding the activities of these agents and the γ-aminobutyric acid [56-12-2] (GABA)/benzodiazepine (BZ) receptor complex (1). Serotonin [50-67-9] (2), ion channel(s), and purinergic (3) recognition sites have been implicated in the mechanism of action of a variety of anxiolytics, hypnotics, sedatives, and anticonvulsants, but the molecular targets for many of these agents, both in terms of efficacy and side effect liabilities, remain unknown. Recombinant deoxyribonucleic acid (DNA) technology is being employed to determine the receptor subclasses and novel molecular targets within the CNS (4).

Hypnotic agents depress the CNS and are able to induce sleep when given at appropriate doses. Such agents typically act either by enhancing inhibitory neurotransmitter actions, eg, GABA, in the CNS or by inhibiting the actions of excitatory neurotransmitters, eg, glutamate (see NEUROREGULATORS). At lower doses, this class of compound can be sedating and can also have a calming or anxiolytic action. At higher doses, the traditional hypnotic agents, but not the BZs, can produce coma, a degree of anesthesia, and eventually death.

Anticonvulsants or antiepileptics are agents that prevent epileptic seizures or modulate the convulsant episodes elicited by seizure activity. Certain of these agents, eg, the BZs, are also hypnotics, anxiolytics, and sedatives, reinforcing the possibility of a common focus of action at the molecular level (1).

Anxiolytics are compounds that act primarily to relieve the symptoms of anxiety although such agents can also be used as anticonvulsants, sedatives, hypnotics, and anesthetic agents (see ANESTHETICS). The principal class of anxiolytics, the BZs, shows dependence liability (5) whereas newer agents such as buspirone [36505-84-7] and ritanserine [8705-43-2] produce antianxiety effects via central serotoninergic systems (6).

Hypnotics and Sedatives

Sleep disturbance or insomnia has become an increasingly prevalent condition in modern society. It is associated with a variety of mental states including chronic anxiety, depression, various dementias, stress, overwork, and aging. The condition is manifest by difficulty in falling asleep, a disturbed sleep cycle, awakening without any feeling of refreshment, and a general malaise. Whereas some chronic sleep disturbances may be attributable to viral infections, causes are generally asymptomatic and result in a poor quality of life. Superimposed on everyday stress factors that have the potential to interfere with sleep is the phenomenon of transcontinental and international jet travel that alters the circadian and ultradian mechanisms involved in sleep (7).

As a therapeutic class, hypnotics are nonselective CNS depressants that elicit drowsiness and a natural sleep state from which the individual can be aroused. The effects of hypnotics are generally dose-dependent.

Alcohols. Ethanol [64-17-5], C_2H_5OH (1) is a potent sleep inducing agent that in the form of wine (qv), liquor (see BEVERAGE SPIRITS, DISTILLED), and beer (qv) has been in recreational use since prehistory. Ethanol (qv) is a sedative, anxiolytic, and hypnotic agent. It can prove fatal in excess both acutely, owing to CNS depressive actions, and chronically as the result of progressive alcohol-induced tissue atrophy, most notably cirrhosis of the liver and neurological damage. Whereas ethanol is typically viewed as a CNS stimulant, this perceived effect results from a depression of tonic inhibitory neurotransmission processes. Chlorinated alcohols such as ethchlorvynol [113-18-8], C_7H_9ClO (2) and chloral hydrate [302-17-0], $C_2H_3Cl_3O_2$ (3) are also useful hypnotics. The latter compound also possesses anticonvulsant and muscle relaxant activities. Both ethanol derivatives are associated with nausea, vomiting, dizziness at high doses and a typical hangover effect reflecting a residual depression of the CNS. Chloral hydrate irritates the gastrointestinal tract and can cause ataxia, nightmares, and allergic reactions. Ethchlorvynol has many properties in common with chloral hydrate and in addition has hypotensive actions.

(1)   (2)   (3)

Barbiturates. The barbiturates represent a seminal class of hypnotic agents that have been largely replaced in clinical use by the benzodiazepines (BZs). The latter have a wider therapeutic index in terms of respiratory depression and residual depression and less abuse potential than the barbiturates. In addition, the BZs lack the spectrum of drug interactions that are seen with barbiturates. As a class, barbiturates have a broad spectrum of CNS depressant activity from mild sedation to general anesthesia. Representative compounds include amobarbital [57-43-2], $C_{11}H_{18}N_2O_3$ (4), pentobarbital [76-74-4], $C_{11}H_{18}N_2O_3$ (5), phenobarbital [50-06-6], $C_{12}H_{12}N_2O_3$ (6), and the 5-allyl-5-(1-methylbutyl) analogue, secobarbital [76-73-3], $C_{12}H_{17}N_2O_3$ (7). The barbiturates decrease sleep latency, decrease the number of awakenings and REM sleep period, and decrease body movement. With chronic use, the effects of the barbiturates can decrease by up to 50% within two weeks. The 5-phenyl substituted barbiturates, eg (6), are selective as anticonvulsant agents. Phenobarbital (6) is used in emergency rooms for the treatment of convulsions, whereas shorter-acting barbiturates are used as intravenous anesthetics.

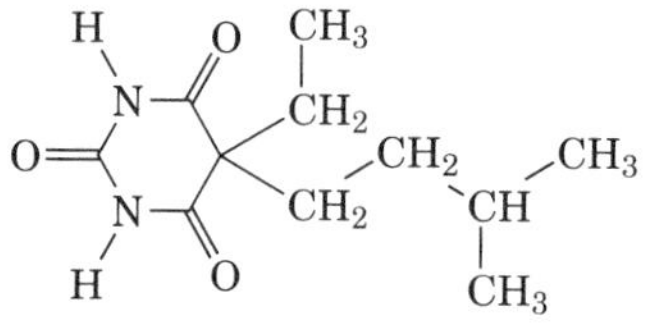

(4)

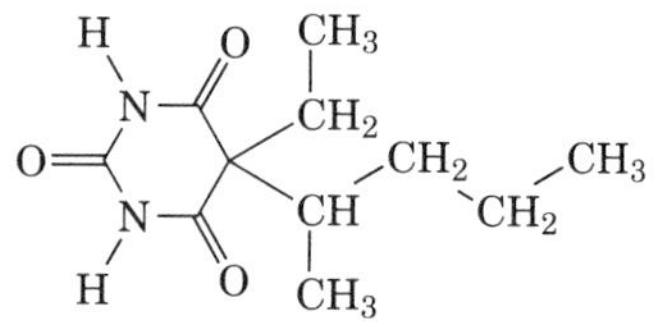

(5)

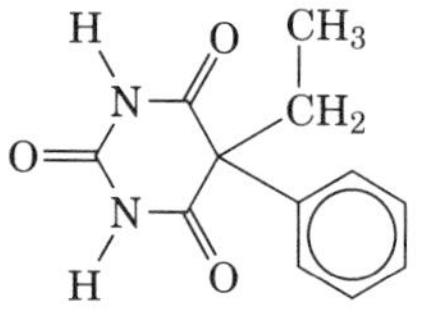

(6)

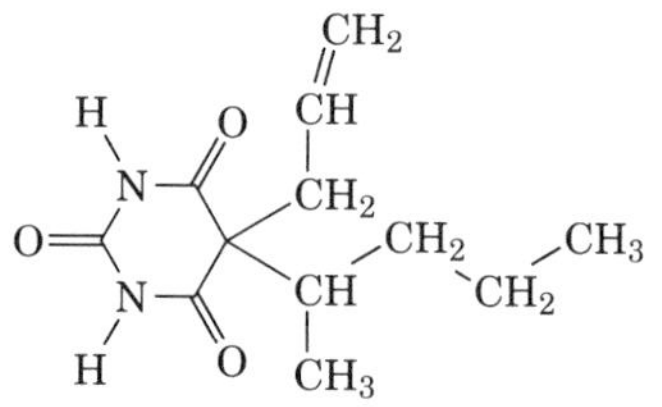

(7)

Mechanistically, the barbiturates interact with the $GABA_A$/BZ receptor complex to indirectly facilitate GABAergic neurotransmission (1). Hypnotic doses of barbiturates result in a hangover like that from usage of ethanol and the chlorinated alcohols. Barbiturate poisoning is a significant clinical problem resulting from deliberate or accidental overdosing.

Benzodiazepines. Many of the 1,4-BZs are routinely used as hypnotics. Examples include chlordiazepoxide [58-25-3], $C_{21}H_{14}Cl_2N_3O$ (**8**), clorazepate [57109-90-7], $C_{16}H_{13}ClN_2O_4$ (**9**), diazepam [439-14-5], $C_{16}H_{13}ClN_2O$ (**10**), flurazepam [17617-23-1], $C_{21}H_{23}ClFN_3O$ (**11**), lorazepam [846-49-1], $C_{15}H_{10}Cl_2N_2O$ (**12**), triazolam [28911-01-5], $C_{17}H_{12}Cl_2N_4$ (**13**), estazolam [29975-16-4], $C_{16}H_{11}ClN_4$ (**14**), quazepam [36735-22-5], $C_{17}H_{11}ClF_4N_2S$ (**15**), and temazepam [846-50-4], $C_{16}H_{13}ClN_2O_2$ (**16**), all shown in Figure 1. Unlike the barbiturates, the BZs are not general CNS depressants and are generally considered to be the safest available as hypnotics. As a class, BZs are also muscle relaxants and interact with alcohol (1). These last properties in addition to dependence liability (5) and amnesia limit their usefulness in the clinical setting.

Extensive research into the BZs stimulated by the identification of the $GABA_A$/BZ receptor complex (1) resulted in the identification of a number of non-BZ pharmacophores that interact with the central BZ/GABA receptor complex having a reduced incidence of the side effects seen with classical BZs. Zolpidem [82626-48-0], $C_{19}H_{21}N_3O$ (**17**) and zopiclone [43200-80-2], $C_{17}H_{17}ClN_6O_3$ (**18**) are sedative–hypnotics lacking anticonvulsant and anxiolytic activity that are thought to selectively interact with a BZ receptor subtype termed the $\omega 1$ or BZ1. Activation of the $\omega 2$ BZ or BZ2 receptor is responsible for the side effects of the BZs which include dependence liability, muscle relaxation, and alcohol and barbiturate potentiation. Rebound insomnia is a significant problem associated with the BZ hypnotics (8) that may be alleviated with the use of compounds like zolpidem and zopiclone.

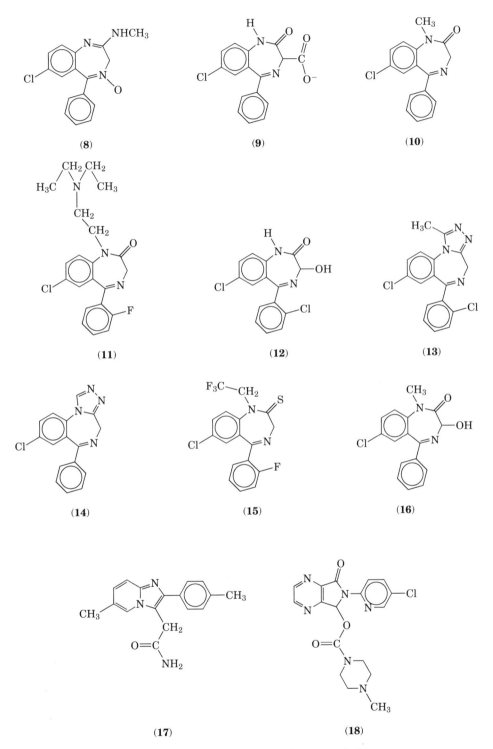

Fig. 1. Structures of 1,4-benzodiazepines and the newer non-BZ pharmacophores used as hypnotics. See text.

Miscellaneous Agents. Compounds having sedative–hypnotic properties include the phenothiazines (Fig. 2), methotrimeprazine [60-99-1], $C_{19}H_{24}N_2OS$ (**19**) and promethazine [60-87-7], $C_{17}H_{20}N_{20}S$ (**20**), methyprylon [125-64-4], $C_{10}H_{17}NO_2$ (**21**), glutethemide [77-21-4], $C_{13}H_{15}NO_2$ (**22**), methaqualone [72-44-6], $C_{16}H_{14}N_2O$ (**23**), meprobamate [57-53-4], $C_9H_{18}N_2O_4$ (**24**), carbromal [77-65-6], $C_7H_{13}BrN_2O_2$ (**25**), bromoisovalum [496-67-3], $C_6H_{11}BrN_2O_2$ (**26**), ethinamate [126-52-3], $C_9H_{13}NO_2$ (**27**), etomidate [33125-97-2], $C_{14}H_{16}N_2O_2$ (**28**), and paraldehyde [123-63-7], $C_6H_{12}O_3$ (**29**). Of these, glutethimide is addicting and shows little advantage in use as compared to the BZs or barbiturates whereas meprobamate is the most clinically useful.

A newer class of hypnotic at the preclinical stage as of this writing (ca 1993) are the neurosteroids, also known as the epalons and represented by (**30**) (9), $C_{21}H_{34}O_2$. These also interact with the GABA$_A$/BZ receptor complex, have shown interesting activity in preclinical models (10), and are undergoing clinical trials.

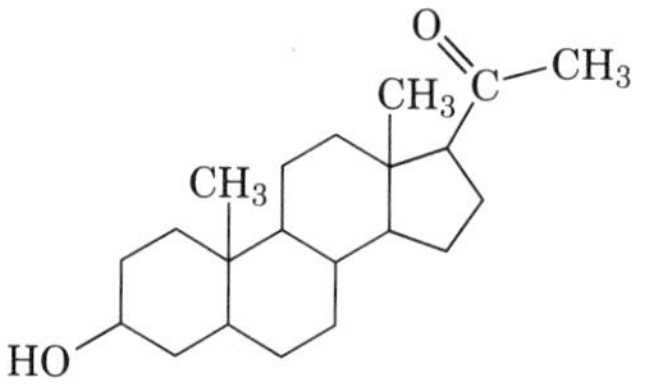

(**30**)

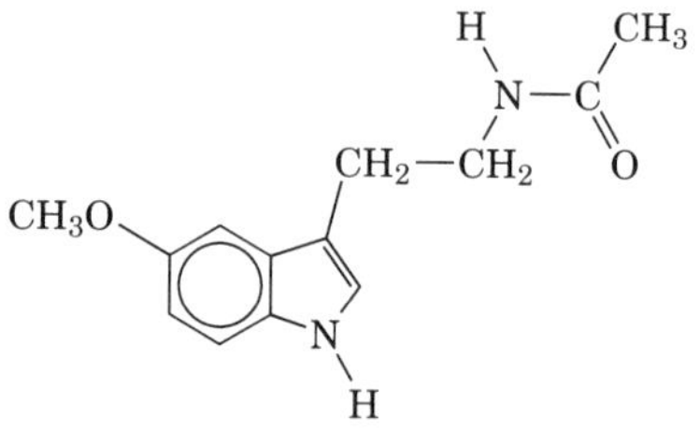

(**31**)

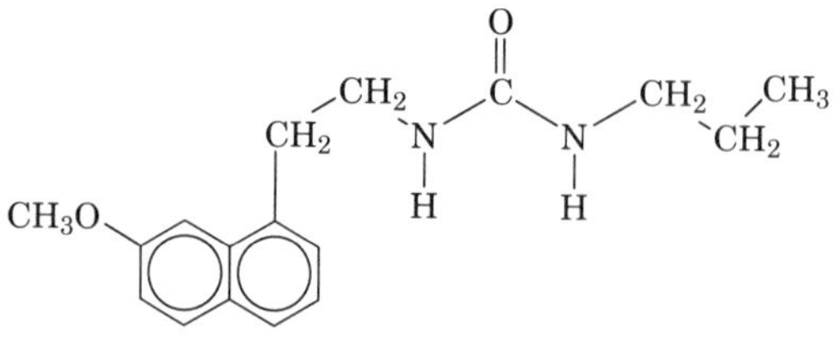

(**32**)

Melatonin [73-31-4], $C_{13}H_{16}N_2O_2$ (**31**) has marked effects on circadian rhythm (11). Novel ligands for melatonin receptors such as (**32**) (12), $C_{17}H_{16}N_2O_2$, have affinities in the range of 10^{-13} M, and have potential use as therapeutic agents in the treatment of the sleep disorders associated with jet lag. Such agents may also be useful in the treatment of seasonal affective disorder (SAD), the depression associated with the winter months. Histamine (see HISTAMINE AND HISTAMINE ANTAGONISTS), adenosine (see NUCLEIC ACIDS), and neuropeptides such as corticotropin-like intermediate lobe peptide (CLIP) and vasoactive intestinal polypeptide (VIP) have also been reported to have sedative–hypnotic activities (7).

BZ analogues that can act as antagonists or inverse agonists of the classical anxiolytic agonist, diazepam [439-14-5], have also been discovered as the result

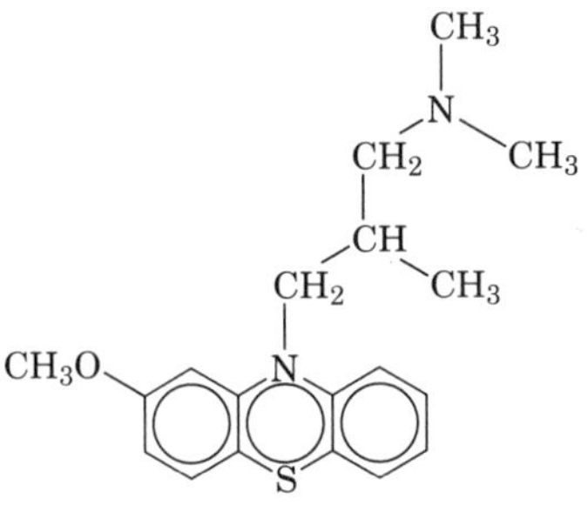

(**19**)

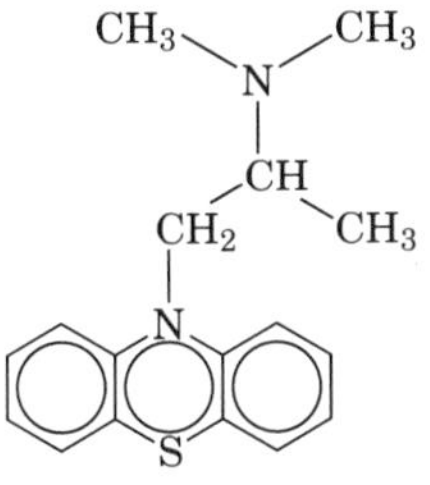

(**20**)

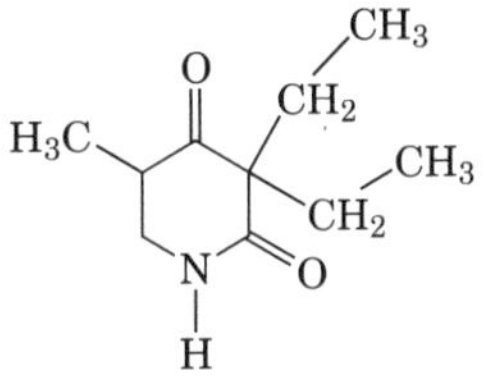

(**21**)

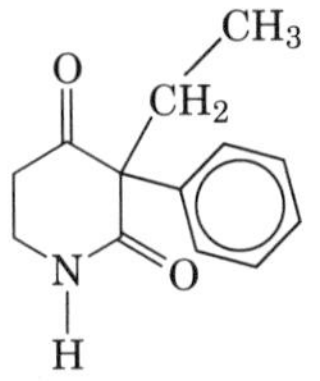

(**22**)

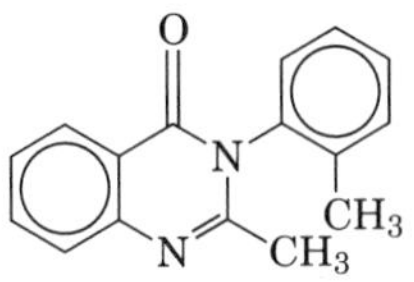

(**23**)

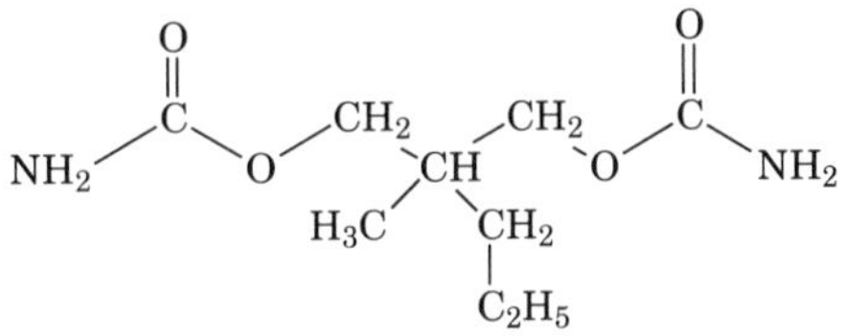

(**24**)

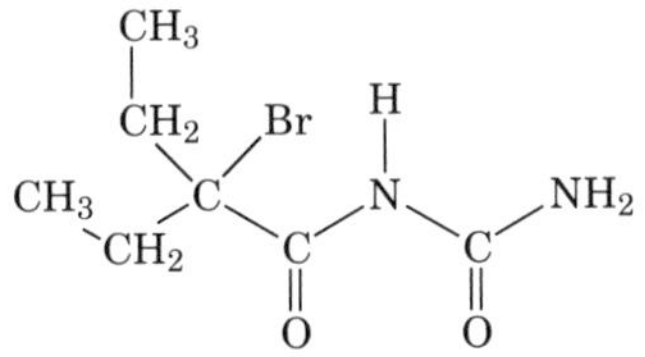

(**25**)

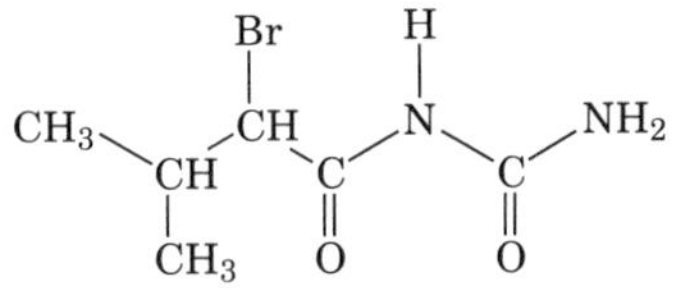

(**26**)

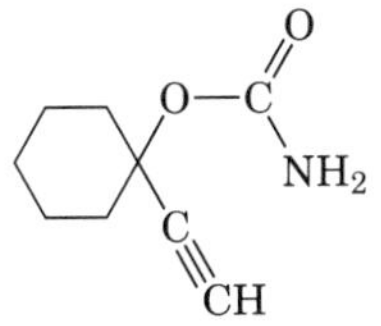

(**27**)

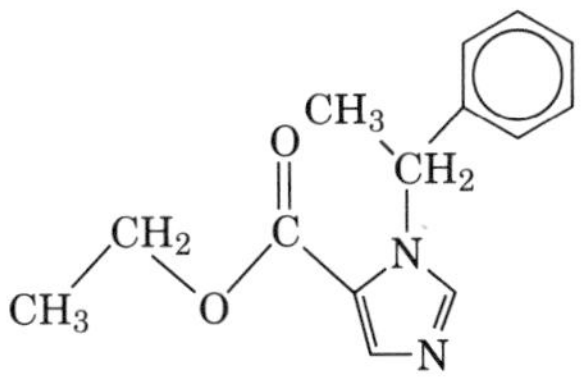

(**28**)

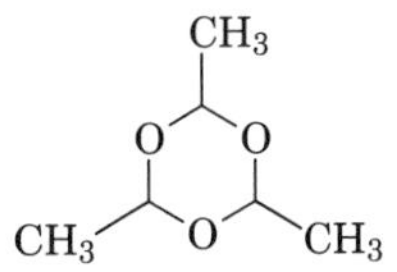

(**29**)

Fig. 2. Structures of phenothiazines having sedative–hypnotic properties. See text.

of the intensive effort following the discovery of the GABA/BZ receptor complex. Flumazenil [*78755-81-4*], $C_{15}H_{14}FN_3O_3$ (**33**), an imidazoBZ, is in clinical use as an analeptic. It is used in reversing the sedation associated with the BZs, specifically in terms of outpatient anesthesia (1).

(33)

Anticonvulsants

The neurological CNS disorder known as epilepsy encompasses a group of disorders associated with seizures and convulsions (13). Primary or idiopathic epilepsy refers to situations in which there is no apparent cause for seizures; secondary or symptomatic epilepsy can be traced to infections, adverse drug reactions, trauma, neoplasm, developmental abnormalities, or cerebrovascular disease. Convulsive seizures are associated with a generalized disturbance in cerebral function that may result from the abnormal discharge of a localized group of neurons in response to ill-defined endogenous or exogenous stimuli. Partial seizures may be ascribed to different lobes of the cerebral cortex.

Seizures are classified in terms of clinical manifestations into either partial or generalized according to guidelines established by the Commission on Classification and Terminology of the International League Against Epilepsy (14). The partial seizure category includes simple partial and complex partial seizures; the generalized seizure category includes absence seizures, atypical absence seizures, myoclonic, clonic, tonic, tonic–clonic (grand mal), and atonic seizures (13–15).

Simple partial seizures are characterized by convulsions confined to a specific muscle group, ie, sensory disturbances but no loss of consciousness. Complex partial seizures involve atypical electroencephalogram (EEG) activity, confused behavior, abnormalities in anterior temporal lobe function, and loss of consciousness. Absence or petit mal seizures are associated with a brief loss of consciousness associated with high voltage, synchronous, spike, and wave EEG patterns with clonic motor manifestations. Atypical absence seizures are slower in onset than absence seizures having a heterogenous EEG profile. Myoclonic seizures involve isolated clonic jerks associated with multiple EEG spiking, whereas clonic seizures involve a rhythmic contraction of all muscles, autonomic features, and loss of consciousness. Tonic seizures have similar manifestations to clonic seizures except that these are marked by opisthotonus, spasm of the back muscles, that causes the head and limbs to arch backward and the trunk forward, ie, the classical manifestation of an epileptic episode. Tonic–clonic or grand mal seizures involve major convulsive episodes with tonic, then clonic contractions of the muscles in the limbs, trunk, and head. Such attacks last 2–5 min. Atonic seizures are associated with a loss of postural tone.

The majority of patients experience only one type of seizure, although some individuals can have two or more seizure types (13). Considerable experience is required to determine the various seizure nuances in affected patients and assign medication accordingly. *Status epilepticus* describes the situation where seizures occur with no intervening periods of consciousness.

There are many classes of anticonvulsant agent in use, many associated with side effect liabilities of unknown etiology. Despite many years of clinical use, the mechanism of action of many anticonvulsant drugs, with the exception of the BZs, remains unclear and may reflect multiple effects on different systems, the summation of which results in the anticonvulsant activity. The pharmacophore structures involved are diverse and as of this writing there is little evidence for a common mechanism of action. Some consensus is evolving, however, in regard to effects on sodium and potassium channels (16) to reduce CNS excitation owing to convulsive episodes.

Barbiturates. Whereas the barbiturates, eg, phenobarbital (**6**), and bromides, were used as antiepileptic agents in the nineteenth century, the discovery and introduction of phenytoin [*57-41-0*], $C_{15}H_{12}N_2O_2$ (**34**), in the mid-1930s represented a significant therapeutic advance. Phenytoin is used in the treatment of all types of seizure except absence and is the drug of choice in *status epilepticus*. This drug is not very useful, however, in treating myoclonic or atonic seizures. Phenytoin exerts its anticonvulsant effects without causing a general depression of the CNS. Its mechanism of action has been linked to alterations in Na^+-K^+-adenosine triphosphatase (ATPase) activity and sodium channel conductance that result in a stabilization of excitation threshold (17,18). Overdosing with phenytoin generally leads to increased excitability of the CNS that can exacerbate epileptic symptoms and on a chronic basis leads to intellectual deterioration, memory impairment, and peripheral neuropathy and in some instances megablastic anemia (13). Phenytoin, like other anticonvulsants, is counterindicated in pregnancy owing to tetratogenic effects characterized as fetal anticonvulsant syndrome (19). Mephenytoin [*50-12-4*], $C_{12}H_{14}N_2O_2$ (**35**) and ethotoin [*86-35-1*], $C_{11}H_{12}N_2O_2$ (**36**) are other hydantoin anticonvulsant agents (see HYDANTOIN AND DERIVATIVES).

(**34**)   (**35**)   (**36**)

The barbiturate phenobarbital (**6**) is a long-acting anticonvulsant used against generalized and partial seizures but it is ineffective in absence seizures (13). The compound is thought to produce its effects via the GABA$_A$/BZ receptor complex resulting in an enhancement of inhibitory GABAergic transmission that results in an increased threshold to electrical and chemical convulsant stimuli. A concomitant depression of excitatory glutamatergic transmission is also produced by phenobarbital. Phenobarbital is sedating and, like phenytoin, has adverse effects on CNS function including ataxia and learning deficits (13). Unlike phenytoin, phenobarbital produces physical dependence and impotence. Methylphenobarbital [115-38-8], $C_7H_{15}N_2O_3$ (**37**) and metharbital [50-11-3], $C_9H_{14}N_2O_3$ (**38**) are *N*-methyl derivatives of phenobarbital that are also used as anticonvulsants.

(**37**)   (**38**)   (**39**)   (**40**)

Primidone [125-33-7], $C_{12}H_{14}N_2O_2$ (**39**) is an analogue of phenobarbital that is used for the treatment of generalized tonic–clonic seizures. It is metabolized in humans to phenobarbital (**6**) and phenylethylmalondiamide [7206-76-0], $C_{11}H_{14}N_2O_2$ (**40**) and these metabolites are probably responsible for its anticonvulsant actions. Primidone has many of the side effect liabilities seen with phenobarbital.

Succinimides. Ethosuximide [77-67-8], $C_7H_{11}NO_2$ (**41**) and the related succinimide, methsuximide [77-41-8], $C_{12}H_{13}NO_2$ (**42**) are used in absence seizure treatment. Like the other anticonvulsants discussed, the mechanism of action of the succinimides is unclear. Effects on T-type calcium channels and $Na^+–K^+$–ATPase activity have been reported (20). Ethosuximide has significant CNS and gastrointestinal (GI) side effect liabilities (13).

(**41**)   (**42**)

Benzodiazepines. Several BZs have anticonvulsant activity and are used for the treatment of epilepsy producing their anticonvulsant actions via interactions with the $GABA_A$/BZ receptor complex to enhance inhibitory GABAergic transmission (1). The anticonvulsant actions of the BZs tend to tolerate upon chronic usage in six months, and BZs also lead to withdrawal symptomatology. Other side effects include sedation, ataxia, and cognitive impairment.

Clorazepate (**9**) is a prodrug of the antiepileptic agent, desmethyldiazepam [1088-11-5], $C_{15}H_{11}ClN_2O$ (**43**) that can also be formed from chlordiazepoxide and diazepam. Clorazepate is used in the treatment of generalized or partial seizures. It appears to tolerate to a lesser extent than other BZs. Diazepam and lorazepam are used in the treatment of *status epilepticus*. Clonazepam [1622-61-3], $C_{15}H_{10}ClN_3O_3$ (**44**) is effective in the treatment of absence seizures.